DYNAMICS

For B.A./B.Sc. (Pass and Honours) Students of Mathematics

Dr. M.D. RAISINGHANIA
M.Sc., Ph.D.
Formerly Reader and Head
Department of Mathematics
S.D. (Post-graduate) College, Muzaffarnagar, U.P.

S. CHAND
PUBLISHING

S Chand And Company Limited
(ISO 9001 Certified Company)

S Chand And Company Limited

(ISO 9001 Certified Company)

Head Office: D-92, Sector–2, Noida – 201301, U.P. (India), Ph. 91-120-4682700

Registered Office: A-27, 2nd Floor, Mohan Co-operative Industrial Estate, New Delhi – 110 044, Phone: 011-49731800

www.**schandpublishing.com;** e-mail: **info@schandpublishing.com**

Marketing Offices:

Chennai	:	Ph: 23632120; chennai@schandpublishing.com
Guwahati	:	Ph: 2738811, 2735640; guwahati@schandpublishing.com
Hyderabad	:	Ph: 40186018; hyderabad@schandpublishing.com
Jalandhar	:	Ph: 4645630; jalandhar@schandpublishing.com
Kolkata	:	Ph: 23357458, 23353914; kolkata@schandpublishing.com
Lucknow	:	Ph: 4003633; lucknow@schandpublishing.com
Mumbai	:	Ph: 25000297; mumbai@schandpublishing.com
Patna	:	Ph: 2260011; patna@schandpublishing.com

First Edition 2006
Second Edition 2009; Third Edition 2013
Fourth Edition 2015
Reprints 2017, 2018, 2019; Reprint with updates 2020, 2021, 2022

Reprint 2024

ISBN: 978-81-219-2649-2 **Product Code:** H6DYN68MATH10ENAD15O

PRINTED IN INDIA

By Vikas Publishing House Private Limited, Plot 20/4, Site-IV, Industrial Area Sahibabad, Ghaziabad – 201 010 and Published by S Chand And Company Limited, A-27, 2nd Floor, Mohan Co-operative Industrial Estate, New Delhi – 110 044.

PREFACE TO THE FOURTH EDITION

Some important problems have been added and solved in the miscellaneous set of problem which was introduced in last edition of this book. Some articles have been re-written to bring more clarity in the discussion of theory. Referenced to the latest papers of various universities have been inserted at proper places.

I hope that these changes will make the book more useful.

Suggestions for further improvement of the book will be highly appreciated.

AUTHOR

PREFACE TO THE FIRST EDITION

This book on dynamics has been specially written to meet the requirements of under-graduate and Honours students of Mathematics, Physics and Engineering, and those preparing for I.A.S., I.F.S. and other competitive examinations. This book contains complete unified U.G.C. syllabus of mathematics recommended by the board constituted by the University Grants Commission of India.

The authors possess a very long and rich experience of teaching mathematics and have first hand experience of the problems and difficulties that students generally face.

The salient features of the book are :

* The matter has been presented in a simple and lucid language, so that students themselves shall be able to understand the solutions of the problems.
* Each chapter opens with necessary definitions and complete proofs of the standard theorems and results. These in turn are followed by solved examples which have been classified in various types and methods. This classification will help the students to revise the subject matter at the time of examination without losing any confidence.
* Care has been taken not to omit even a minor step so that the students can understand everything without the guidance of a teacher. Hence even an average student will be able to grasp the subject matter easily. Furthermore, the problems have been graded in every chapter to instill confidence in the students.
* Up-to-date yearwise reference of various examination papers have been given throughut the book.
* At the end of each chapter objective type questions have been given in keeping with the latest trend of question papers.
* This book serves as a suitable textbook, help book, reference book and question bank for all examinations.

In view of these special features, it is sincerely hoped that the book will surely serve its purpose.

We are grateful to Shri Ravindra Kumar Gupta, Managing Director, Shri Navin Joshi, Vice-President (Publishing) and Shri R.S. Saxena, Adviser (Publishing) for showing personal interest throughout the preparation of the book.

All valuable suggestions for further improvement of this book will be highly appreciated.

AUTHOR

Dedicated to the memory of
my parents and wife

CONTENTS

New-Unified U.G.C. Syllabus of Dynamics for B.A./B.Sc. (General and Honours) Part II Mathematics.

	Topics	*Chapter (s) covering the Topic*
(i)	Velocities and accelerations along radial and transverse directions, and along trangential and normal directions	5
(ii)	Simple harmonic motion. Elastic strings	3 and 4
(iii)	Motion on smooth plane curves	8 and 11
(iv)	Motion on rough plane curves	12
(v)	Motion in a resisting medium	12
(vi)	Motion of particles of varying mass	12
(vii)	Central orbits	9
(viii)	Kepler's laws of motion	10
(ix)	Motion of a particle in three dimensions. Acceleration in terms of different co-ordinate systems	13

Reference : S.L. Loney. An elementary treatise on the dynamics of a particle and of rigid bodies, Cambridge University Press, 1956.

1

Basic Concepts of Dynamics

1.1 Some basic definitions

1. Dynamics. It is that branch of mechanics which deals with the motion of bodies under the action of given forces.

2. Particle. A particle is a portion of matter which is indefinitely small in size *i.e.*, which for the purpose of our investigation is so small that the distance between its different parts may be neglected. It may be treated as a mathematical point endowed with mass.

3. Body. A portion of matter occupying finite space, *i.e.*, limited in every direction is called a body.

Rigid body : A rigid body is defined to be a collection of particles, the distance between any two of which remains invariant

4. Rest and Motion. A particle is said to be at rest if it does not change its position relative to its surroundings with time. If it changes its position with time, it is said to be in motion.

5. Path. The path of a particle is the curve drawn through the successive positions of the point during motion. If the path is a straight line, then the motion is called *rectilinear* and if the path is a curve, then the motion is said to be *curvilinear*.

6. Position. The position of a particle at a given time is determined by the point occupied by the particle on its path.

7. Displacement. The change of position of a particle in the given interval of time is known as its displacement in that interval. The displacement of a particle from P to Q will be denoted by the symbol $\overrightarrow{PQ}$. Note that displacement is a vector quality.

8. Velocity. The rate change of displacement with respect to time is known as velocity. It is a vector quantity because it has both magnitude and direction.

9. Speed. The mangnitude of velocity is called speed.

10. Average speed. Average speed of a particle in a given interval of time is given by the ratio of the total distance travelled by it to the total time taken.

Thus, average speed = (Total distance)/(Total time)

11. Uniform velocity. A particle is said to move with uniform velocity if it moves in a constant direction and covers equal distances in equal intervals of time, however small these intervals may be.

12. Acceleration and retardation. The rate of change of velocity with respect to time is called *acceleration*. It is a vector quantity as it has both magnitude and direction. If a particle moves such that its velocity decreases, then the acceleration is negative and is called *retardation*. Clearly, the velocity of a particle is uniform when its acceleration is zero.

13. Uniform acceleration. When velocity of a particle moving in a straight line changes by equal amounts in equal intervals of time, however small, the particle is said to move with uniform acceleration.

14. Expression for velocity at a point. Let a particle move along the line OX, where O is origin. Let P be the position of the particle at any time t such that $OP = s$. Then,

v = velocity at time $t = ds/dt$

The velocity at any time t is positive or negative according as the particle is moving towards right (*i.e.*, away from origin O) or towards left (*i.e.*, towards the origin O).

O ← s → P(t) ——— X

Expressions for acceleration at a point. Let a particle move along the line OX, where O is origin.

origin. Let P be the position of the particle at any time t and let v be its velocity at P. Then, if f is the acceleration of the particle at any time t at P, then

$$f = \frac{dv}{dt}, \quad f = v\frac{dv}{ds} \quad \text{and} \quad f = \frac{d^2 s}{dt^2}$$

Instantaneous rest. A particle is said to be at instantaneous rest at any time if and only if $v = 0$ but $f \neq 0$.

1.2 Motion in a straight line with constant acceleration

A particle moves along a straight line from a fixed point in it with initial velocity u and constant acceleration f, in its direction of motion. Let s be the distance of the particle from the origin at time t and v be the velocity attained at that time. Then

(i) $v = u + ft$ (ii) $s = ut + (1/2) ft^2$ (iii) $v^2 = u^2 + 2fs$.

[Note : Replace f be $-f$ if f is retardation in any problem].

(iv) Distance travelled in the nth second $= u + (1/2)(2n - 1) f$.

(v) The average velocity is the mean of the initial and the final velocities and is equal to the velocity at the middle of the interval *i.e.*,

$$\text{average velocity} = (u + v)/2 = u + f \cdot (t/2)$$

ILLUSTRATIVE SOLVED EXAMPLES

Ex. 1. *If a body starting from rest, moving with uniform acceleration describes 1000 cm in 10 seconds, then the acceleration with which it moves will be*

(a) 20 cm/sec², (b) 25 cm/sec², (c) 30 cm/sec², (d) 35 cm/sec². **[I.A.S. (Prel) 1993]**

Sol. Ans. (a) $s = ut + (1/2) f t^2 \Rightarrow 1000 = 0 + (1/2) \times f \times 10^2 \Rightarrow f = 20$

Ex. 2. *A train starts from a station with constant acceleration for 2 minutes and attains a constant speed. It then runs for 11 minutes at this speed and retards uniformly during the next 3 minutes and stops at the next station which is 9 km off. The maximum speed (in km/hour) attained by the train is (a) 30 (b) 35 (c) 40 (d) 45.* **[I.A.S. (Prel.) 1995]**

Sol. Ans. (c). Let the particle move from A to B with uniform acceleration f_1 for 2 minutes. Then, using $v = u + ft$ and $s = ut + (1/2) ft^2$, we get $V = f_1 \times 2$ and $s_1 = 0 + (1/2) f_1 \times 2^2$ so that $s_1 = V$. Motion from B to C is with uniform max. velocity V, so $s_2 = 11V$. Next, for motion from C to D with uniform retardation f_2 for 3 minutes, using $v = u - ft$ and $s = ut - (1/2) ft^2$, *we get* $0 = V - 3f_2$ and $s_3 = 3V - (1/2) \times f_2 \times 3^2$ so that $s_3 = (3/2) V$. Now, $9 = AD = s_1 + s_2 + s_3 = V + 11V + (3/2)V = (27/2)V$ so that

$$V = (2/3) \text{ km/minute} = (2/3) \times 60 \text{ km/hour} = 40 \text{ km/hour}.$$

Ex. 3. *If a particle starting with a velocity u and subject to a uniform acceleration 'a' travels for 'n' seconds, then the distance travelled by it in nth second is*

(a) u + (a/2) (2n – 1), (b) u + (a/2)(2 – n), (c) u + (n/2) (2a – 1), (d) u + (n/2) (2a + 1). **[I.A.S. (Prel.) 98]**

Sol. Ans. (a). Refer result (*iv*) of Art 1.2. Here $f = a$.

Ex. 4. *A bus starts from rest with an acceleration of 1 m/s². A man, who is 48 m behind the bus, starts running towards it with a uniform velocity of 10 m/s. He will be able to catch the bus in*

(a) 6 s (b) 7 s (c) 8 s (d) 9s. **[I.A.S. (Prel.) 2002]**

Sol. Ans. (c). Let M be the position of the man when the bus starts from B. Let the man be able to catch the bus at C after t sec. Then, for man, $MC = 10t$ and for bus, $BC = 0 + (1/2) \cdot 1 \cdot t^2$. Form figure $MC = MB + BC$ so that $10t = 48 + t^2/2$ or $t^2 - 20t + 96 = 0$ or $t = 8$ or 12. Required time is $t = 8$ when for the first time man catches bus at C. [Note that $t = 12$ corresponds to time when man and bus will be together at a later stage of further motion].

1.3 Vertical motion under gravity. When a body is let fall in vacuum towards the earth, it will move vertically downwards with an acceleration g (known as acceleration due to gravity). Its value

system is 32 ft/sec², in C.G.S. system is 981 cm/sec² and in M.K.S. system is 9.8 m/sec².

Downward motion. If a particle is projected vertically downwards from a point A with initial velocity u and h be the distance of the particle from A at time t and v be the velocity attained at that time, then

(*i*) $v = u + gt$ (*ii*) $h = ut + (1/2)\, gt^2$ (*iii*) $v^2 = u^2 + 2gh$

(*iv*) Distance travelled in the nth sec $= u + (g/2)\,(2n - 1)$.

In particular, if the body starts from rest or is simply let fall or dropped so that $u = 0$, then the above formulas become (*i*) $v = gt$, (*ii*) $h = (1/2)\, gt^2$, (*iii*) $v^2 = 2gh$ or $v = (2gh)^{1/2}$, (*iv*) distance travelled in the nth sec. $= (g/2)\,(2n - 1)$.

Upward motion. If a particle is projected vertically upwards from O with velocity u and h be the distance of the particle from O at time t and v be the velocity attained at that time, then

(*i*) $v = u - gt$ (*ii*) $h = ut - (1/2)\, gt^2$ (*iii*) $v^2 = u^2 - 2gh$

(*iv*) Greatest height attained during upward motion $= u^2/2g$

(*v*) Time to reach the greatest height $= u/g$

(*vi*) Time of flight (*i.e.*, the total time taken by the particle to reach the greatest height and then to return to the starting point again) $= 2u/g$.

(*vii*) Time of ascent = time of descent.

ILLUSTRATIVE SOLVED EXAMPLES

Ex. 1. *If a particle moves along a straight line according to the law* $s^2 = at^2 + 2bt + c$ *where s is the displacement at any instant, then how does the acceleration vary with displacement ?*

(*a*) *as* $1/s$ (*b*) *as* $1/s^2$ (*c*) *as* $1/s^3$ (*d*) *as* $1/s^4$ **[I.A.S. (Prel.) 2005]**

Sol. Ans. (*c*). We have $s = (at^2 + 2bt + c)^{1/2}$ ... (1)

From (1), $v = ds/dt = (at + b) / (at^2 + 2bt + c)^{1/2}$ ... (2)

$$\therefore \text{ acceleration} = f = \frac{dv}{dt} = \frac{a(at^2 + 2bt + c)^{1/2} - (at + b)^2\,(at^2 + 2bt + c)^{-1/2}}{at^2 + 2bt + c}$$

$$= \frac{a(at^2 + 2bt + c) - (at + b)^2}{(at^2 + 2bt + c)^{3/2}} = \frac{ac - b^2}{s^3}, \text{ using (1)}$$

Ex. 2. *Two particles of* m_1 *and* m_2 *gms are projected vertically upwards such that the velocity of projection of* m_1 *in double that of* m_2. *If the maximum height to which* m_1 *and* m_2 *rise be* h_1 *and* h_2 *respectively, then* (*a*) $h_1 = 2h_2$ (*b*) $2h_1 = h_2$ (*c*) $h_1 = 4h_2$ (*d*) $4h_1 = h_2$. **[I.A.S. (Prel.) 1996]**

Sol. Ans. (*c*). Let velocity of m_2 be u. Then by problem the velocity of m_1 is $2u$. Now, by problem $h_1 = (2u)^2/2g$ and $h_2 = u^2/2g$ so that $h_1 = 4\,(u^2/2g) = 4h_2$.

Ex. 3. *A particle is thrown vertically upwards with a velocity 30 m/sec. Find its greatest height.* **[Pune 2010]**

Sol. Greatest height $= \dfrac{u^2}{2g} = \dfrac{30 \times 30}{2 \times 9.8} = 45.91\text{m}$

Ex. 4. *If* $x = at^2 + bt + c$, *find velocity when* $t = 2$ **[Agra 2009]**

Sol. By definition, velocity $= dx/dt = 2at + b$

Hence, when $t = 2$, the required velocity $= 4a + b$

Ex. 5. *The velocity of a jet projected vertically upwards to reach a height of 4 km is* ($g = 9.8$ *m/sec*^{2}) (*a*) *28.0 km/s* (*b*) *2.8 km/s* (*c*) *0.28 km/s* (*d*) *0.028 km/s.*

[I.A.S. (Prel.) 2001]

Sol. Ans. (*c*). Max. height $H = u^2/2g \Rightarrow u^2 = 2gH = 2 \times 9.8 \times 4 \times 1000$ so that $u = 280$ m/sec $= (280/1000)$ km/sec. $= 0.28$ km/sec.

Ex. 6. *An object was thrown vertically downwards with the initial velocity* v_0. *If during the fifth second of its fall, it travels (3/2) times the distance it had travelled during the third second then the value of* v_0 *is given by* (*a*) *2g m/sec* (*b*) *(3g/2) m/sec* (*c*) *g m/sec* (*d*) *g/2 m/sec.*

[I.A.S. Prel. 2002]

Sol. Ans. (*b*). Since distance travelled in nth sec $= u + (g/2)(2n - 1)$, here $v_0 + (g/2)(2 \times 5 - 1) = (3/2)[v_0 + (g/2)(2 \times 3 - 1)]$ so $v_0 = (3/2)g$.

Ex. 7. *A stone falling vertically from rest travels half of its total path in the last second of its fall. The total time of its fall is (a) less than 2 seconds (b) equal to 2 seconds (c) greater than 2 seconds but less than 3 seconds (d) greater than 3 seconds.* **[I.A.S. Prel. 2003]**

Sol. Ans. (*c*). Let total time of fall be n seconds then by problem, $(g/2)(2n - 1) = (1/2) \times (1/2) gn^2$ so that $n^2 - 4n + 2 = 0$ giving $n = 1 + \sqrt{2}$ and $n = 1 - \sqrt{2}$. But $(1 - \sqrt{2})$ gives –ve value of time. So reject this value. Then the required time $= 1 + \sqrt{2}$ which is > 2 seconds but < 3 seconds.

1.4 Momentum of a body, force and Newton's laws of motion.

Momentum of a body. If m be the mass of the body moving with velocity v, then the momentum of body $= mv$. It is a vector quantity.

Force. Force is an action which changes or tends to change the state of rest or of uniform motion of a body in a straight line.

Impressed forces. The external forces acting upon a body are known as the impressed forces.

Internal forces. The mutual forces of action and reaction are known as internal forces.

Newtons laws of motion. They are stated as follows :

First law. *Every body (or particle) continues in its state of rest or of uniform motion in a straight line unless it is compelled by impressed forces to change that state.*

Second law. *The rate of change of momentum is directly proportional to the impressed force and takes place in the direction of the force.*

Third law. *To every action, there is an equal and opposite reaction.*

Same useful results and definitions

(*i*) **Inertia and law of inertia.** A body has no tendency of itself to change its state of rest or motion if it is kept free from the action of external forces. The inability of a body to change its position by itself is known as *inertia* and hence first law is also known as the *law of inertia*.

(*ii*) **Equation of motion or fundamental equation of dynamics** is given by $P = mf$, when P is the impressed force acting on a body (or particle) of mass m and f is its acceleration. Formula $P = mf$ is remembered as follows :

Force causing motion in absolute units = mass × acceleration

In vector notation $P = mf$ is written as $\mathbf{P} = m\mathbf{f}$.

(*iii*) Deduction of first law from second law. ***[Pune 2010]***

From Newton's law II, $$d(mv)/dt = kP, \quad \text{...(1)}$$
where m is the mass of the particle, v is the velocity, P is the external force and k is a constant. If there is no force acting on the body, then $P = 0$ and so (1) gives
$$d(mv)/dt = 0 \text{ so that } mv = c \text{ or } v = c/m = u, \text{ a constant.} \quad \text{...(2)}$$
where c is constant of integration. Now (2) shows that the velocity of the particle is constant and hence the particle continues to move in a straight line with constant speed. Again, if $u = o$, then (2) shows that $v = o$ *i.e.*, the particle continues to be at rest. This is nothing but a statement of the first law. The first law of motion can thus be deduced from the second law.

(*iv*) **The physical independence of forces.** Suppose a force $\mathbf{P}_1$ acting on a particle of mass m, produces an acceleration $\mathbf{f}_1$ in it. Similarly if a force $\mathbf{P}_2$ acting on the same particle, produces an acceleration $\mathbf{f}_2$ in it. Then,
$$m\mathbf{f}_1 = \mathbf{P}_1 \quad \text{and} \quad m\mathbf{f}_2 = \mathbf{P}_2$$
On adding these, $$m(\mathbf{f}_1 + \mathbf{f}_2) = \mathbf{P}_1 + \mathbf{P}_2,$$
showing that a force $\mathbf{P}_1 + \mathbf{P}_2$ acting on the particle produces an acceleration $\mathbf{f}_1 + \mathbf{f}_2$. Hence we see that the additional force $\mathbf{P}_2$ produces an additional acceleration $\mathbf{f}_2$, as it would do if acting alone on the particle. This result is known as the principle of physical independence of forces.

(*v*) **Unit of force**

Poundal. It is the absolute unit of force in the F.P.S. (Foot-Poundal-Second) system. Thus, a poundal is the force which produces an acceleration of 1 ft/sec^2 in a mass of one pound.

Dyne. It is the absolute unit of force in the C.G.S. (Centimeter-Gramme-Second) system. Thus, a dyne is the force which produces an acceleration 1 cm/sec^2 in a mass of one gramme.

Newton. It is the absolute units of force in the M.K.S. (Meter-Kilogram-Second) system. Thus, a Newton is the force which produces an acceleration of 1 m/sec^2 in a mass of one kilogram. A Newton in briefly written as N.

Note 1. One Newton = 10^5 dynes

Note 2. The above units of force are called absolute units because their values are the same everywhere and do not depends on the *g*, the acceleration due to gravity.

Gravitational units of force. Units of force which depend on the value of *g* are called gravitational units of force. In this system in F.P.S., it is called a pound weight or *lb*. wt, in C.G.S., it is called gram weight or gm. wt and in M.K.S., it is called kilogram weight or kg. wt.

Connection between absolute and gravitational units of force

$$1 \text{ lb. wt} = g \text{ poundals} = 32 \text{ poundals}$$
$$1 \text{ gm. wt} = g \text{ dynes} = 981 \text{ dynes}$$
$$1 \text{ kg. wt} = g \text{ Newtons} = 9.8 \text{ Newtons} = 9.8 \text{ N}.$$

Note. While using the formula $P = mf$, P is always measured in absolute units, *i.e.*, in poundals or dynes or Newtons.

ILLUSTRATIVE SOLVED EXAMPLES

Ex. 1. *Which one of the following statements is correct ? (a) Newton's three laws of motion are independent (b) first law of motion is contained in the second law as a special case (c) second law of motion can be deduced from the third law (d) third law of motion can be deduced from the second law.* **[I.A.S. (Prel.) 1997]**

Sol. Ans. (*b*) Refer result (*iii*) of Art 1.4.

Ex. 2. *Newton's second law of motion is given by (a) velocity = mass/force (b) acceleration = moving mass/force, (c) acceleration = moving force/mass (d) None of the above.* **[I.A.S. (Prel.) 1997]**

Sol. Ans. (*c*) Since $P = mf$ so $f = P/m$.

Ex. 3. *Newton's third law says (a) for every action there is an unequal and opposite reaction (b) to every action there is an equal and opposite reaction (c) to every action there is an equal and same reaction (d) None of the above.* **[I.A.S. (Prel.) 1998]**

Sol. Ans. (*b*). Refer Art 1.4, third law.

Ex. 4. *If a body of mass M kg at rest is acted upon by a constant force of W kg.wt, then in T seconds it moves through a distance of (a) gTW/2M meters (b) $gTW^2/2M$ meters (c) $g^2TW/2M$ meters (d) $gT^2W/2M$ metres.* **[I.A.S. (Prel.) 1996]**

Sol. Ans. (*d*). From $P = mf$, $Wg = Mf$ so that $f = Wg/M$. Using $s = ut + (1/2)ft^2$ we get,

$$s = 0 + (1/2)(Wg/M)T^2 = gT^2W/2M.$$

Ex. 5. *If a body of (1/2) kg is subjected to a force* $(\mathbf{i} + 2\mathbf{j} + \mathbf{k})$ *N and if the body is initially at rest, then its velocity after 4 second will be (a) $2\sqrt{6}$ m/sec (b) $4\sqrt{6}$ m/sec (c) $8\sqrt{6}$ m/sec (d) $16\sqrt{6}$ m/sec.* **[I.A.S. (Prel.) 1998]**

Sol. Ans. (*c*). Magnitude of force $= |\mathbf{i} + 2\mathbf{j} + \mathbf{k}| = (1^2 + 2^2 + 1^2)^{1/2} = \sqrt{6}$ Newton. Now $P = mf$ $\Rightarrow \sqrt{6} = (1/2)f \Rightarrow f = 2\sqrt{6}$ m/sec^2. Then by, $v = u + ft$, $v = 0 + (2\sqrt{6}) \times 4 = 8\sqrt{6}$ m/sec.

Ex. 6. *A body of mass 50 gm is acted upon by a constant force F = 100 dynes. The time required to move the body through a distance 25 cm from rest is (a) 10 s, (b) 5.5 s, (c) 5 sec, (d) 3.5 sec.* **[I.A.S. (Prel.) 2001]**

Sol. Ans. (*c*). $P = mf \Rightarrow 100 = 50f \Rightarrow f = 2$. Then

$$s = ut + (1/2)ft^2 \Rightarrow 25 = 0 + (1/2) \times 2 \times t^2 \Rightarrow t = 5 \text{ sec}.$$

Ex. 7. *The magnitude of a force which is acting on a body of mass 1 kg for 5 sec. to produce velocity of 1 m/sec in it, is (a) 1000 dynes, (b) 2000 dynes, (c) 10,000 dynes, (d) 20,000 dynes.*

[I.A.S. (Prel.) 1999]

Sol. Ans. (*d*). $v = u + ft \Rightarrow 100 = f \times 5 \Rightarrow f = 20.$

$\therefore \quad P = mf = 1000 \times 20 = 20{,}000$ dynes.

1.5 Reaction on the lift when a body of mass *m* is carried on a lift moving with an acceleration *f*

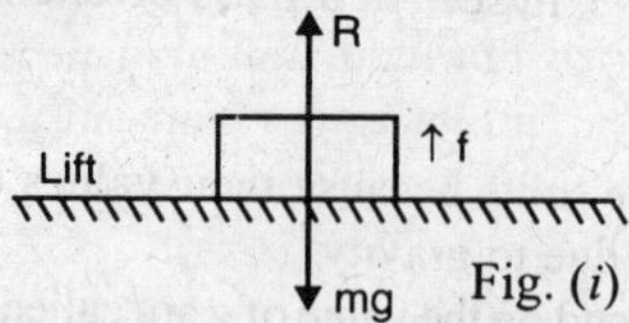

Fig. (*i*)

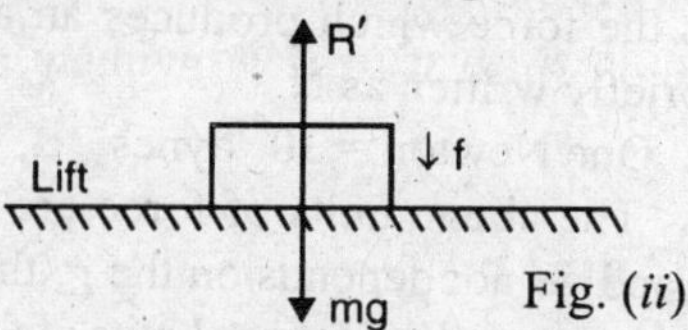

Fig. (*ii*)

Case I Lift moving up

R = reaction = $m(g + f)$

Case II. Lift moving down

R' = reaction = $m(g - f)$

Note 1. The reaction does not depend upon the velocity but it depends on the acceleration. When $f = 0$, $R = mg$ *i.e.*, when the lift is at rest or moves with constant velocity, the reaction is equal to the weight of the body.

Note 2. In case II, if $f = g$, then $R = 0$, *i.e.*, reaction vanishes and the contact becomes loose. Again, in case II, if $f > g$, R is negative. In this case the body is left behind and moves down freely under gravity.

Note 3. The pressure on the body is called the *apparent weight* of the body. Clearly, in case I, the apparent weight > actual weight whereas in case II, the apparent weight < actual weight.

ILLUSTRATIVE SOLVED EXAMPLES

Ex. 1. *A person weighing 80 kg is standing on a lift. The lift moves upwards with a uniform acceleration 4.9 m/s². The apparent weight (in kg) of the person is (a) 160 (b) 120 (c) 80 (d) 40.*

[I.A.S. (Prel.) 1994]

Sol. Ans. (*b*). $R = m(g + f) = 80(4.9 + 9.8) = 120\text{ g} = 120$ kg-wt.

Ex. 2. *If a body of mass m kg is carried by a lift moving with upward acceleration f, then the pressure on the plane of the lift is*

(*a*) $mf - mg$ (*b*) $mg - mf$ (*c*) $mg + mf$ (*d*) $(mg)(mf)$. **[I.A.S. (Prel.) 1999]**

Sol. Ans. (*c*). Refer case I of Art 1.5. Pressure = Reaction.

Ex. 3. *A person having a mass of 98 kg is descending in a lift with an acceleration of 2 m/s². The thrust of his feet on the lift while descending, is nearly equal to (take g = 9.8 m/s²) (a) 76.4 × 10⁶ dynes, (b) 96.0 × 10⁶ dynes, (c) 115.6 × 16⁶ dynes (d) 196 × 10⁶ dynes.* **[I.A.S. (Prel.) 2001]**

Sol. Ans. (*a*). By case II of Art. 1.5, Thrust = Reaction = $m(g - f) = 98(9.8 - 2) = 98 \times 7.8$ N $= 98 \times 7.8 \times 10^5$ dynes $= 76.4 \times 10^6$ dynes (nearly).

1.6 Motion of two bodies connected by a string

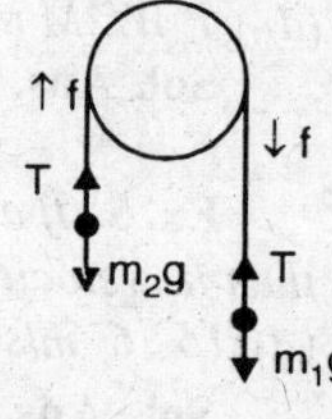

Case I. Two masses m_1 and m_2 $(m_1 > m_2)$ are suspended by a light inextensible and flexible string over a smooth, fixed, small, light pulley (or a peg). Then,

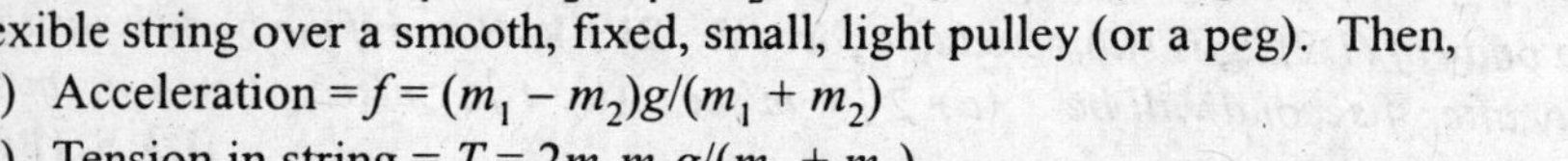

(*i*) Acceleration $= f = (m_1 - m_2)g/(m_1 + m_2)$

(*ii*) Tension in string $= T = 2m_1m_2g/(m_1 + m_2)$

(*iii*) Pressure on pulley $= 2T = 4m_1m_2g/(m_1 + m_2)$

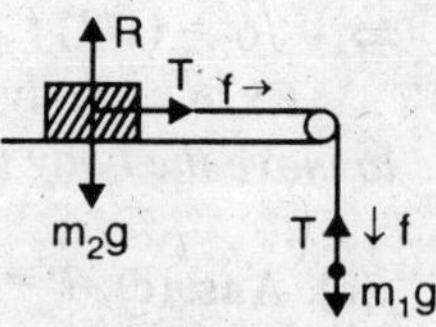

Case II. Two masses m_1, m_2 are connected by an inelastic string; m_2 is placed on a horizontal table and the string passes over a light smooth pulley at the edge of the table and m_1 is hanging freely. Then,

(*i*) Acceleration $= f = m_1g/(m_1 + m_2)$

(*ii*) Tension in string $= T = m_1m_2g/(m_1 + m_2)$

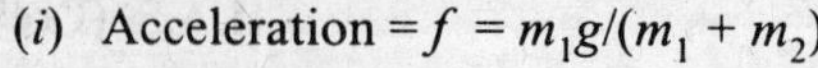

(*iii*) Pressure on pulley $= T\sqrt{2} = m_1m_2g\sqrt{2}/(m_1 + m_2)$

Example. *Two masses of 5 kg and 9 kg are fastened to ends of a cord passing over a frictionless pulley. The acceleration of resulting motion is*

(a) 2.8 m/sec^2 *(b)* 2.5 m/sec^2 *(c)* 5.6 m/sec^2 *(d)* 5.0 m/sec^2. **[I.A.S. (Prel.) 1993]**

Sol. Ans. (*a*). From case I of Art 1.6, $f = [(9-5) \times 9.8]/(9+5) = 2.8$ m/sec^2.

1.7 Composition and resolution of velocities

Resultant velocity and components of velocity. When a particle possesses a number of velocities, then the single velocity which is equivalent to the given velocities is called their *resultant*. The given velocities are known as the *components* of their resultant.

Resolved parts of a velocity. If a velocity be decomposed into two components at right angles to each other, the components are known as the *resolved* parts of the velocity.

Parallelogram law of velocities. If a moving point possesses simultaneously two velocities, which are represented in magnitude and direction by the sides of a parallelogram, drawn from a point, then the resultant velocity is represented in magnitude and direction by the diagonal of the parallelogram passing through that point.

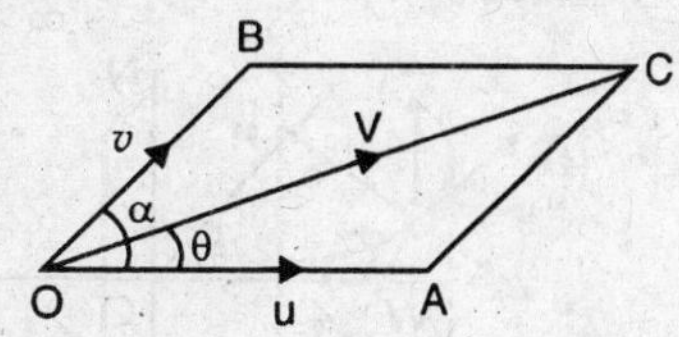

Let the two velocities u and v acting at an angle α be represented in magnitude, direction and sense by the two side OA and OB of the parallelogram $OACB$, then their resultant velocity V is represented by the diagonal OC. Let OC make an angle θ with OA. Then we have

$$V = (u^2 + v^2 + 2uv \cos \alpha)^{1/2}, \quad \tan \theta = (u \sin \alpha)/(u + v \cos \alpha) \qquad \text{...(1)}$$

Some particular cases of composition of velocities.

Case I. When u and v are perpendicular to each other, then $\alpha = 90°$

(1) $\Rightarrow$ $V = (u^2 + v^2)^{1/2}$ and $\tan \theta = u/v$.

Case II. When u and v are in the same direction, then $\alpha = 0$.

(1) $\Rightarrow$ $V = u + v$ and $\theta = 0$.

Case III. When u and v are in the opposite directions, then $\alpha = \pi$.

If $u > v$, (1) $\Rightarrow$ $V = u - v$ and V has the direction of u.

If $v > u$, (1) $\Rightarrow$ $V = v - u$ and V has the direction v.

Case IV. If $v = u$. Then (1) $\Rightarrow$ $V = 2u \cos(\alpha/2)$ and $\theta = \alpha/2$. Thus, the direction of resultant velocity bisects the angle between the two component velocities.

Components u and v of a velocity V in the directions making angles α and β with it on either side.

$u = (V \sin \beta)/\sin(\alpha + \beta)$, $v = (V \sin \alpha)/\sin(\alpha + \beta)$.

Resolved parts u and v of a velocity V in two directions at right angles to each other, when u makes an angle θ with V.

$$u = V \cos \theta \quad \text{and} \quad v = V \sin \theta$$

Note. The sum of the resolved parts of a number of velocities in a given direction is equal to the resolved part of their resultant in the same direction.

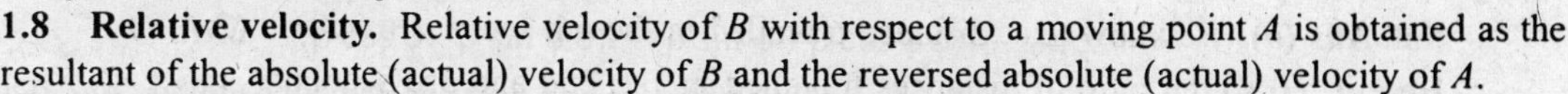

1.8 Relative velocity. Relative velocity of B with respect to a moving point A is obtained as the resultant of the absolute (actual) velocity of B and the reversed absolute (actual) velocity of A.

In vector notation, if **u** and **v** are the absolute velocity vectors of two points A and B respectively, then the relative velocity of B relative to A, is **v** – **u**.

Ex. 1. *The velocity of a boat relative to water is* $2\mathbf{i} + 2\mathbf{j}$ *and that of water relative to earth is* $\mathbf{i} - 2\mathbf{j}$*, where* **i**, **j** *represent velocities of one km an hour due to east and north respectively. The velocity of boat relative to earth is (a) 3 km an hour due east, (b)* $2 \times 2^{1/2} + 5^{1/2}$ *km an hour at 45° north of east, (c)* $(17)^{1/2}$ *km an hour at* $\tan^{-1} 4$ *north of east, (d)* $17^{1/2}$ *km an hour at* $\tan^{-1} 4$ *due north.*

[I.A.S. (Prel.) 2003]

Sol. Ans. (*a*). Let **u, v** and **w** be absolute velocity of boat, water and earth respectively. Then, given that

$$\mathbf{u}-\mathbf{v}=2\mathbf{i}+2\mathbf{j} \quad \text{and} \quad \mathbf{v}-\mathbf{w}=\mathbf{i}-2\mathbf{j}$$

Adding these, $(\mathbf{u}-\mathbf{v})+(\mathbf{v}-\mathbf{w})=2\mathbf{i}+2\mathbf{j}+\mathbf{i}-2\mathbf{j}$

$\therefore$ Velocity of boat relative to earth = $\mathbf{u}-\mathbf{w}=3\mathbf{i}$.

Ex. 2. *To a man walking at 4 km per hour along a road due west the wind appears to blow from the south while to a cyclist travelling in the same direction at 8 km per hour it appears to come from the south-west. What is the true direction and velocity of the wind?* **[I.F.S. 1998]**

Sol. Let v be the actual velocity of wind and let it blow in a direction making an angle θ south of west.

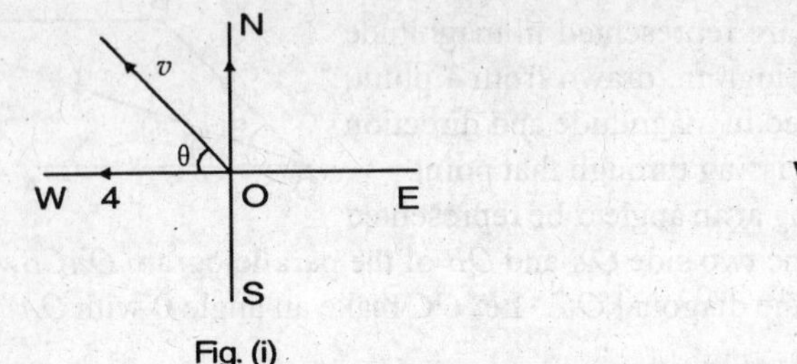

Fig. (i)

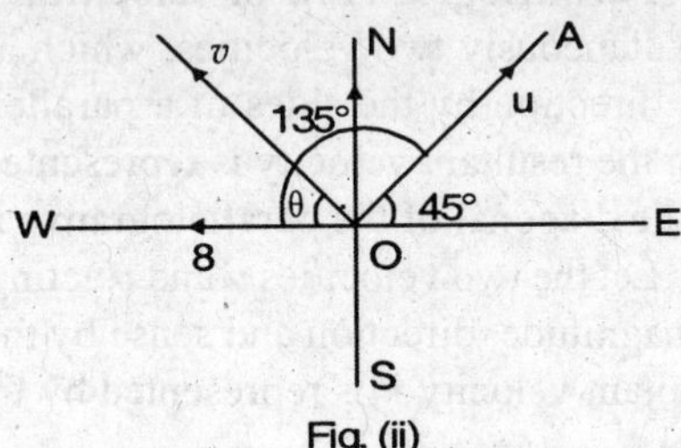

Fig. (ii)

Situation I. v is the resultant of the velocity (= 4 km/hr) of the person along OW and the apparent velocity of wind along OS (see figure (*i*)). Resolving velocities along OW, we have

$$v\cos\theta=4 \qquad \text{... (1)}$$

Situation II. In this case v is the resultant of the velocity (= 8 km/hr) of the person along OW and apparent velocity (u, say) of wind along OA making 45° with OE (see figure (*ii*)). Resolving velocities along OW and ON, we have

$$v\cos\theta=8+u\cos 135°=8-u/\sqrt{2} \quad \text{and} \quad v\sin\theta=u\sin 135°=u/\sqrt{2}$$

In order to eliminate u, adding the above equations, we get

$$v\cos\theta+v\sin\theta=8 \quad \text{or} \quad v\sin\theta=4, \text{ using (1)} \qquad \text{... (2)}$$

Squaring and adding (1) and (2), we get $\quad v^2=2\times 16 \Rightarrow v=4\sqrt{2}$ km/hr

Dividing (2) by (1), we get $\quad \tan\theta=1 \Rightarrow \theta=45°$

Hence wind blows with velocity $4\sqrt{2}$ km/hr from north-west.

MISCELLANEOUS PROBLEMS

Ex. 1. *The distance travelled by a particle, falling freely from rest in first, second and third seconds are in the ratio*

(a) 1 : 2 : 3 *(b) 1 : 3 : 5* *(c) 9 : 4 : 1* *(d) 1 : 4 : 9.* **[M.S. Univ. T.N. 2007]**

Sol. Ans (*b*). We know that, distance travelled in the nth sec = $u + (g/2) \times (2n-1)$

Here $u = 0$ so the ratio of distances travelled in 1st, 2nd and 3rd seconds

$$=\frac{1}{2}g\times 1^2:\frac{1}{2}g\times(2\times 2-1):\frac{1}{2}g\times(2\times 3-1)=1:3:5$$

Ex.2. *Two balls are projected simultaneously with the same velocity from the top of a tower, one vertically upwards and the other vertically downwards. If they reach the ground in time t_1 and t_2, then the height of the tower is $(1/2)\times g\,t_1\,t_2$* ***(Garhwal 2010)***

Sol. Let u be the common velocity of projection for both the balls and h the height of the tower. Then for the first ball, and the second ball, we have

$$h=-ut_1+(1/2)\times gt_1^2 \qquad \text{... (1)}$$

and $$h = ut_2 + (1/2) \times gt_2^2 \qquad \text{... (2)}$$

Multiplying (1) by t_2 and (2) by t_1 and then adding, we get

$$h(t_1 + t_2) = (1/2) \times g\, t_1\, t_2\, (t_1 + t_2) \quad \text{or} \quad h = (1/2) \times g\, t_1\, t_2$$

Ex. 3. *Two motor cars are moving along two roads perpendicular to each other, towards point of intersection of two roads. Their velocities at a particular time are v_1 and v_2 respectively while their distances from the crossing are s_1 and s_2 respectively. If the accelerations of the two cars be f_1 and f_2 respectively, then they shall avoid collision, if*

(*a*) $(s_1 f_2 - s_2 f_1)^2 \neq 2(v_2 f_1 - v_1 f_2)(v_2 s_1 - v_1 s_2)$ (*b*) $(s_1 v_2 - s_2 v_1)^2 \neq 2(f_1 s_2 - f_2 s_1)(v_2 f_1 - v_1 f_2)$

(*c*) $(v_2 f_1 - v_1 f_2) \neq 2(s_1 v_2 - s_2 v_1)(f_1 s_2 - f_2 s_1)$ (*d*) $(s_1 v_1 - s_2 v_2)^2 (f_1 s_2 - f_2 s_1)(v_2 f_1 - v_1 f_2) \neq 0$

[I.A.S. 2004]

Sol. Ans. (*a*) If possible, let the two cars collide after time t at the point of intersection. Then, using $s = ut + (1/2) \times ft^2$, for both the cars we have

$$s_1 = v_1 t + (1/2) \times f_1 t^2 \quad \text{or} \quad f_1 t^2 + 2v_1 t - 2s_1 = 0 \qquad \text{... (1)}$$

and $$s_2 = v_2 t + (1/2) \times f t^2 \quad \text{or} \quad f_2 t^2 + 2v_2 t - 2s_2 = 0 \qquad \text{... (2)}$$

Solving (1) and (2) for t^2 and t by cross-multiplication, we get

$$\frac{t^2}{4(-v_1 s_2 + v_2 s_1)} = \frac{t}{2(-s_1 f_2 + s_2 f_1)} = \frac{1}{2(f_1 v_2 - f_2 v_1)}$$

$$\Rightarrow \quad t^2 = (s_1 v_2 - s_2 v_1)/2(f_1 v_2 - f_2 v_1) = (s_2 f_1 - s_1 f_2)^2/(f_1 v_2 - f_2 v_1)^2$$

$$\Rightarrow \quad (s_1 f_2 - s_2 f_1)^2 = (v_2 f_1 - v_1 f_2)(v_2 s_1 - v_1 s_2) \qquad \text{... (3)}$$

Hence to avoid collision, the condition is given by

$$(s_1 f_2 - s_2 f_1)^2 \neq (v_2 f_1 - v_1 f_2)(v_2 s_1 - v_1 s_2)$$

Ex. 4. *If time t be regarded as a function of velocity v, prove that the rate of decrease of acceleration is given by $f^3 \times (d^2 t/dv^2)$, f being the acceleration.*

[Kanpur 2006, 07; Rohilkhand 2001]

Sol. Let $t = f(v)$, where f is an arbitrary function ... (1)

Differentiating (1) w.r.t. 't', we get $1 = \phi'(v) \times (dv/dt)$

$\therefore$ Acceleration $= f = dv/dt = 1/\phi'(v)$ ... (2)

Differentiating (2) w.r.t. 't', we get $$\frac{df}{dt} = -\frac{1}{\{\phi'(v)\}^2}\phi''(v)\frac{dv}{dt}$$

or $$\frac{df}{dt} = -\frac{\phi''(v)}{\{\phi'(v)\}^2} \times \frac{dv}{dt} = -f^3 \phi''(v), \text{ using (2)} \qquad \text{... (3)}$$

Now, (2) $\Rightarrow$ $dt/dv = \phi'(v)$ $\Rightarrow$ $d^2 t/dv^2 = \phi''(v)$ ... (4)

$\therefore$ From (3) and (4), $df/dt = -f^3 \times (d^2 t/dv^2)$,

which is negative. Hence the rate of decrease of acceleration is $f^3 \times (d^2 t/dv^2)$.

Ex. 5. *A point moves with uniform acceleation. If v_1, v_2, v_3 be average velocities in three successive intervals of time t_1, t_2, t_3, prove that $(v_1 - v_2)/(v_2 - v_3) = (t_1 + t_2)/(t_2 + t_3)$.*

[Kanpur 2019; Pune 2010; Rohilkhand 2003]

Sol. Let AB, BC, CD be the distances moved during successive interval t_1, t_2, t_3 respectively. Let u be the initial velocity of the particle at A and let f be its uniform acceleration. Using $v = u + ft$, we have

For motion from A to B velocity at $B = u + ft_1$

For motion from B to C velocity at $C = (u + ft_1) + ft_2 = u + f(t_1 + t_2)$

For motion from C to D velocity at $D = u + f(t_1 + t_2) + ft_3$

Since, average velocity during an interval = (1/2) × (initial velocity + final velocity), we get

$$v_1 = \{u + (u + ft_1)\}/2 = u + (1/2) \times ft_1$$

$$v_2 = \{(u + ft_1) + (u + ft_1 + ft_2)\}/2 = u + ft_1 + (1/2) \times ft_2$$

$$v_3 = \{(u + ft_1 + ft_2) + (u + ft_1 + ft_2 + ft_3)\}/2 = u + ft_1 + ft_2 + (1/2) \times ft_3$$

Then, $v_1 - v_2 = -(1/2) \times f(t_1 + t_2)$ and $v_2 - v_3 = -(1/2) \times f(t_2 + t_3)$

Dividing, we have $(v_1 - v_2)/(v_2 - v_3) = (t_1 + t_2)/(t_2 + t_3)$

Ex. 6. *A point moving in a straight line with uniform acceleration describes s_1 and s_2 in successive intervals t_1 and t_2 seconds. Prove that the acceleration is $2(s_2 t_1 - s_1 t_2)/t_1 t_2 (t_1 + t_2)$.* **[I.A.S. 2004, Kanpur 2007]**

Sol. Let the particle move along the line OX. Let $OA = s_1$ and $AB = s_2$. According to the given problem, the particle travels distances OA and AB in times t_1 and t_2 respectively. Thus, particle travels distance $s_1 + s_2$ in $(t_1 + t_2)$ seconds.

Let u be the velocity of the particle at O and let f be its uniform acceleration.

For motion from O to A: Using $s = ut + (1/2) \times ft^2$, we get

$$s_1 = ut_1 + (1/2) \times ft_1^2 \quad \text{so that} \quad s_1/t_1 = u + (1/2) \times ft_1 \qquad \text{... (1)}$$

For motion from O to B: Using $s = ut + (1/2) \times ft^2$, we get

$$s_1 + s_2 = u(t_1 + t_2) + (1/2) \times f(t_1 + t_2)^2 \Rightarrow (s_1 + s_2)/(t_1 + t_2) = u + (1/2) \times f(t_1 + t_2) \qquad \text{... (2)}$$

In order to eliminate u, subtracting (1) from (2), we get

$$\frac{s_1 + s_2}{t_1 + t_2} - \frac{s_1}{t_1} = \frac{1}{2} f t_2 \quad \text{or} \quad \frac{s_2 t_1 - s_1 t_2}{t_1(t_1 + t_2)} = \frac{1}{2} f t_2$$

Thus, $f = 2(s_2 t_1 - s_1 t_2)/t_1 t_2 (t_1 + t_2)$

Ex.7. *A particle moving with uniform acceleration in a straight line passes through the points P, Q, R. If $PQ = QR = b$ and if the time from P to Q is t_1, time from Q to R is t_2. Then, show that the acceleration is $2b(t_1 - t_2)/t_1 t_2 (t_1 + t_2)$.* ***[G.N.D.U. Amritsar 2010]***

Hint. Do as in Ex. 6 by taking P, Q, R in place of O, A, B and $s_1 = s_2 = b$.

Ex. 8. *A body travels a distance s in t seconds. It starts from rest and ends at rest. In this first part of the journey it moves with constant acceleration f and in the second part with constant retardation r. Show that $t = \{2s(1/f + 1/r)\}^{1/2}$.* ***[Delhi BA (Prog) III, 2011; I.F.S. 1997]***

Sol. Let OAB be the velocity time graph. Portions OA and AB correspond to motions with acceleration f and relardation r respectively. Let AM be perpendicular to OT.

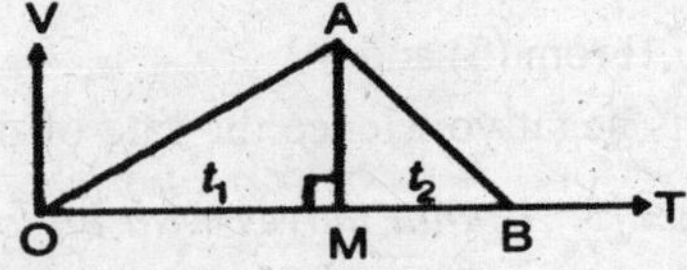

Let $OM = t_1, MB = t_2,$ $AM = v.$ Then $t = t_1 + t_2$ ... (1)

Now, $\Delta AOB = \dfrac{1}{2} \times OB \times AM \quad \Rightarrow \quad s = \dfrac{1}{2} tv \quad \Rightarrow \quad v = \dfrac{2s}{t}$ ... (2)

Again $\quad v/t_1 = f$ and $v/t_2 = r \quad \Rightarrow \quad t_1 = v/f$ and $t_2 = v/r$

$\therefore \quad t_1 + t_2 = v\left(\dfrac{1}{f} + \dfrac{1}{r}\right) \quad \Rightarrow \quad t = \dfrac{2s}{t}\left(\dfrac{1}{f} + \dfrac{1}{r}\right)$, using (1) and (2)

or $\quad t^2 = 2s(1/f + 1/r)$ so that $t = \{2s\,(1/f + 1/r)\}^{1/2}$

Ex. 9. *The velocity of a train increases at a constant rate f_1 from rest to v, then remains constant for an interval and finally decreases to zero at the constant rate f_2. If x is the total distance described, prove that the total time taken is $(x/v) + (v/2) \times (1/f_1 + 1/f_2)$.* **[I.F.S. 1995]**

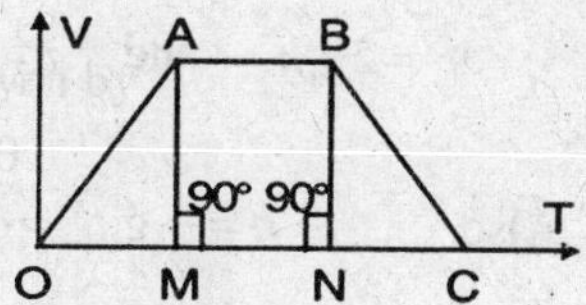

Sol. Let $OABC$, be the velocity-time graph, OA corresponding to motion during uniform acceleration, AB to motion during constant velocity and BC to motion during retardation. Let $AM \perp OT$ and $BN \perp OT$.

Let $\quad OM = t_1, MN = t_2, \quad NC = t_3$ and $AM = BN = v$.

$\therefore \quad$ Total time $= t_1 + t_2 + t_3 = t$., say ... (1)

Here $\quad f_1 = \text{acceleration} = v/t_1 = f_1 \quad \Rightarrow \quad t_1 = v/f_1$ ... (2)

$f_2 = \text{retardation} = v/t_3 = f_2 \quad \Rightarrow \quad t_3 = v/f_2$ ... (3)

Now, $\quad x = \text{total distance} = \Delta OAM + \text{rectangle } ABNM + \Delta BNC$

$\Rightarrow, \quad x = \dfrac{1}{2} v t_1 + v t_2 + \dfrac{1}{2} v t_3 \quad \Rightarrow \quad t_1 + 2t_2 + t_3 = \dfrac{2x}{v}$ ... (4)

Then, $\quad (4) \Rightarrow (t_1 + t_2 + t_3) + t_2 = 2x/v \quad \Rightarrow \quad t + t_2 = 2x/v$, using (1)

Thus, $\quad t_2 = 2x/v - t$ ... (5)

Substituting the values of t_1, t_3, t_2 given by (2), (3) and (5) in (1), we have

$v/f_1 + (2x/v - 1) + v/f_2 = t$ or $2t = 2x/v + v(1/f_1 + 1/f_2)$

Thus, $\quad$ total time $= t = x/v + (v/2) \times (1/f_1 + 1/f_2)$

Ex. 10. *A lift ascends with constant acceleration f, then with constant velocity and finally stops under constant retardation. If the total height ascended is h and total time occupied is t, show that the time during which the lift is ascending with constant velocity is $\{t^2 - (4h/f)\}^{1/2}$*

Sol. For figure, refer Ex. 9. Let $OABC$ be the velocity time graph, OA corresponding to motion during acceleration, AB to motion during constant velocity and BC to motion during retardation. Draw $AM \perp OT$ and $BN \perp OT$. Let $OM = t_1$, $MN = t_2$, $NC = t_3$ and $AM = BN = v$. Then, we have

$\therefore \quad$ Total time $= t_1 + t_2 + t_3 = t \quad$ (given) ... (1)

Acceleration $= f = v/t_1 \quad \Rightarrow \quad t_1 = v/f$ ... (2)

and $\quad$ Retardation $= f = v/t_3 \quad \Rightarrow \quad t_3 = v/f$ ... (3)

Here, $\quad h = \text{Total distance} = \Delta OAM + \text{rectangle } ABNM + \Delta BNC$

$\Rightarrow \quad h = (1/2) \times vt_1 + vt_2 + (1/2) \times vt_3 \quad \Rightarrow \quad v(t_1 + t_2 + t_3 + t_2) = 2h$ ... (4)

Since $\quad t_1 + t_2 + t_3 = t, \quad (4) \Rightarrow \quad v(t + t_2) = 2h$ ... (5)

Substituting the values of t_1 and t_3 from (2) and (3) in (6), we get

$$\frac{v}{f}+t_2+\frac{v}{f}=t \qquad \text{or} \qquad \frac{2v}{f}=t-t_2 \qquad \text{or} \qquad v=\frac{f(t-t_2)}{2} \quad \text{... (6)}$$

Using (6), $\quad$ (5) $\Rightarrow$ $\quad (1/2)\times f(t^2-t_2^2)=2h \quad$ or $\quad t^2-t_2^2=4h/f$

or $\quad t_2^2=t^2-4h/f \quad$ so that $\quad t_2=(t^2-4h/f)^{1/2}$

Ex. 11. *A particle takes T seconds less and acquires a velocity s m/sec more at one place than at another in falling from rest through the same distance. Show that the gravities* g_1 *and* g_2 *at the two places are related by* $g_1/g_2 = s^2/T^2$.

Sol. Let v_1 and v_2 respectively be the velocities acquired and t_1, t_2 respectively the times taken in falling a height h meters. Then, we have

$$v_1^2=2g_1h \quad \text{and} \quad v_2^2=2g_2h \quad \Rightarrow \quad v_1=\sqrt{2g_1h} \quad \text{and} \quad v_2=\sqrt{2g_2h} \quad \text{... (1)}$$

Also, $$h=\frac{1}{2}g_1t_1^2=\frac{1}{2}g_2t_2^2 \quad \Rightarrow \quad t_1=\sqrt{\frac{2h}{g_1}} \quad \text{and} \quad t_2=\sqrt{\frac{2h}{g_2}} \quad \text{... (2)}$$

Given $$s=v_1-v_2 \quad \Rightarrow \quad s=\sqrt{2g_1h}-\sqrt{2g_2h}, \text{ using (1)} \quad \text{... (3)}$$

Also, given $$T=t_2-t_1=\sqrt{\frac{2h}{g_2}}-\sqrt{\frac{2h}{g_1}}=\frac{\sqrt{2g_1h}-\sqrt{2g_2h}}{\sqrt{g_1g_2}}, \text{ using (2)}$$

or $\quad T=s/\sqrt{g_1g_2}$, using (3) $\quad$ or $\quad g_1g_2=s^2/T^2$

EXERCISES

1. If x, y, z be the distances described by a particle during the pth, qth and rth second respectively, prove that

(*i*) $(q-r)x+(r-p)y+(p-q)z=0$; $\qquad$ (*ii*) $p(y-z)+q(z-x)+r(x-y)=0$

2. If a point moving under uniform acceleration describes successive equal distances in times t_1, t_2, t_3 respectively, then prove that $1/t_1 - 1/t_2 + 1/t_3 = 3/(t_1+t_2+t_3)$.

3. O, A, B, C, D are five points in a straight line such that $OA = AB = BC = CD$. If a particle starts from rest with constant acceleration from O, prove that the times of describing AB, BC, CD are as $\sqrt{2}-1:\sqrt{3}-\sqrt{2}:\sqrt{4}-\sqrt{3}$.

4. If the coordinates of a point moving with constant acceleration be x_1, x_2, x_3 at the instants t_1, t_2, t_3 respectively, prove that the acceleration is

$$2[(x_2-x_3)\,t_1+(x_3-x_1)\,t_2+(x_1-x_2)\,t_3]/(t_2-t_3)\,(t_3-t_1)\,(t_1-t_2)$$

5. A particle starts from rest with acceleration f; at the end of time t it becomes $2f$, at the end of time $2t$ it becomes $3f$, and so on. Find the velocity at the end of time nt and the total distance described. **Ans.** Velocity = $(1/2)\times n(n+1)ft$; Distance = $(1/2)\times n\,(n+1)\,(2n+1)\,ft^2$

6. Two balls are projected simultaneously with the same velocity from the top of a tower, one vertically upwards and the other vertically downwards. If they reach the ground in times t_1, t_2 respectively, show that $\sqrt{t_1, t_2}$ is the time which each will take to reach the ground if simply let drop the top of the tower.

7. A lift performs first part of its ascent with uniform acceleration f and the remainder with uniform retardation $2f$. Prove that if h is the depth of shaft and t, the time of ascent, then $h = (ft^2)/3$.

8. For $1/p$ of the distane between two stations, a train is uniformly accelerated and for $1/q$ of the distance, it is uniformly retarded; it starts from rest at one station and comes to rest at another station. Prove that the ratio of its greatest velocity to its average velocity is $(1 + 1/p + 1/q) : 1$.

9. If $x = Ae^t + Be^{-t}$ is the displacement of a particle at any time t, then acceleration is
(*a*) x (*b*) $-x$ (*c*) x^2 (*d*) $-x^2$ **[M.S. Univ. T.N. 2007]**

Hint. $dx/dt = Ae^t - Be^{-t}$ and d^2x/dt^2 = acceleration $= Ae^t + Be^{-t} = x$. **Ans.** (*a*)

10. If the displacement of a particle moving in a straight line from the centre of force at time t is given by $x = a \sin t + b \cos t$. Find velocity and acceleration of the particle **[Meerut 2018]**

Sol. Given $$x = a \sin t + b \cos t \qquad \text{... (1)}$$

$$\text{velocity} = v = dx/dt = a \cos t - b \sin t \qquad \text{... (2)}$$

$$\text{Acceleration} = f = dv/dt = -(a \sin t + b \cos t), \text{ using (2)}$$

2

Work, Power, Energy and Impulse. System of Particles And Conservation Principles

2.1 Work. Definition. Work is said to be done by a force when its point of application undergoes a displacement.

2.2 Measurement of work done by a constant force

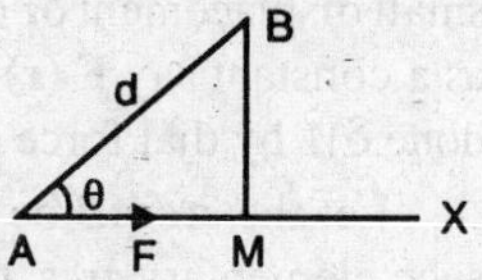

Let the point of application A be displaced to B such that AB $(= d$, say) makes an angle θ with the direction of constant force F in the direction AX. Draw BM perpendicular to AX, then the displacement in the direction of force is AM. Hence the work done W by F is given by

$$W = F.AM = AB \cos\theta = Fd\cos\theta \qquad ...(1)$$

i.e., $W = d.\ F\cos\theta$ = (Actual displacement) × (resolved part of force in the direction of actual displacement)

From (1), we have the following particular cases :

(*i*) When $\theta = 0$, *i.e.*, when the displacement is in the same direction as force, then $W = Fd$

(*ii*) When $\theta = \pi$, *i.e.*, when the displacement is opposite to the direction of the force, then $W = -Fd$

(*iii*) When $\theta = \pi/2$, *i.e.*, when the displacement is perpendicular to the direction of the force, then $W = 0$.

2.3 Measurement of work when the force is variable and displacement takes place along its line of action

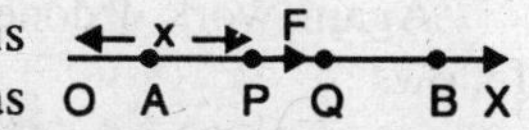

Let a variable force act along the line OX and suppose the particle is displaced along OX from A to B under the action of this force. With O as origin, let $OA = a$, $OB = b$, $OP = x$ and $OQ = x + \delta x$, where Q is a point in the neighborhood of P. Clearly, $PQ = \delta x$. Since PQ is small, the force F may be regarded as constant from P to Q.

Then, small amount of work δW, done by the force F as the particle moves from P to $Q = F.\ PQ$ *i.e.*, $\delta W = F\,\delta x$.

Hence the total amount of work done W by the variable force when the particle is displaced from A (where $x = a$) to B (where $x = b$) is given by

$$W = \int_a^b F dx$$

Example. *Prove that the work done in stretching an elastic string is equal to the product of the mean of the initial and final tensions and the extension.*

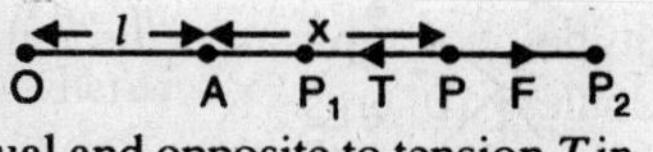

Let $OA = l$ be the natural length of the given elastic string whose modulus of elasticity is λ. Let the end O be fixed and the other end A be displaced to P, by applying a variable force F, the force being always equal and opposite to tension T in the string so that $F = \lambda(x/l)$ acting along OP, where $AP = x$. Let $AP_1 = x_1$ and $AP_2 = x_2$.

Hence work W done in stretching the string from P_1 to P_2 is given by

$$W = \int_{x_1}^{x_2} F dx = \int_{x_1}^{x_2} \frac{\lambda x}{l} dx = \frac{\lambda}{2l}\left[x^2\right]_{x_1}^{x_2} = \frac{\lambda}{2l}(x_2^2 - x_1^2)$$

$$= (\lambda/2l) \times (x_2 + x_1)(x_2 - x_1) = (x_2 - x_1) \times (1/2) \times (\lambda x_1/l + \lambda x_2/l)$$

$$= \{(T_1 + T_2)/2\} \times (x_2 - x_1),$$

where T_1 and T_2 are the initial and final tensions

Thus, work done = (Mean tension) × (extension)

2.4 Measurement of work when the force is variable and the point of application moves along a plane curve.

Let a variable force **F**, acting on a particle P, displace it from A to B along an arc AB. Let P and Q be neighboring points on this curve such that $\vec{OP} = \mathbf{r}$, $\vec{OQ} = \mathbf{r} + \delta\mathbf{r}$ so that $\vec{PQ} = \vec{OQ} - \vec{OP} = \mathbf{r} + \delta\mathbf{r} - \mathbf{r} = \delta\mathbf{r}$. During the small displacement $\delta\mathbf{r}$ of its point of application the force **F** may be regarded as a constant say **F** (**r**). By the rule of constant force (refer Art. 2.2), work done δW by that force during the displacement δr of its point of application is given by $\delta W = \mathbf{F}.\delta\mathbf{r}$.

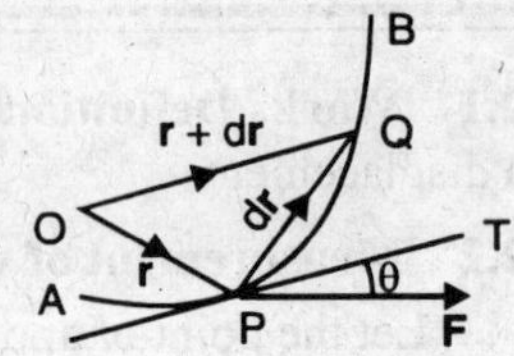

Let the arc AB be sub-divided into elementary arcs of the type PQ. Then, summing up for all such elementary arcs, the work done W by the force **F**, in the displacement from A to B along the arc AB, is given by

$$W = \int_A^B \mathbf{F}\cdot d\mathbf{r} = \int_A^B F_t\, ds \qquad ...(1)$$

where F_t is the resolved part of **F**, along the tangent PT at P(**r**) to the arc and s denotes the are length of the curve AB measured from some fixed point on the curve to any other point P whose position vector is **r**. The integration on the right hand side of (1) is performed along the arc AB.

2.5 Power of an agent. Definition. The rate of work done by the force is known as the power of the force or its agent. It is given by

$P = dW/dt = \mathbf{F}\cdot\mathbf{v}$, where **v** is the velocity of the particle (being acted by a force **F**)

If the motion takes place in a straight line so that the direction of **F** and **v** is the same, then.

$P = F$v, where v is the speed and F is magnitude of the force.

Again, work W done in an interval (t_1, t_2) can be expressed in terms of the power of the force, as follows

$$W = \int_{t_1}^{t_2} P dt$$

2.6 Units of Work. These are of two types :

***(i)* The absolute unit of work.** The absolute unit of work is the work done by an absolute unit of forces in moving its point of application through a distance of unit length in the direction of force.

In F.P.S. system, the absolute unit of work is the work done by a force of one poundal in moving its point of application through one foot in the direction of the force and is called a *foot-poundal.*

In C.G.S. system, the absolute unit of work is the work done by a force of one dyne in moving its point of application through one centimeter in the direction of force and is called an erg.

In M.K.S. system, the absolute unit of work is the work done by a force of one Newton in moving its point of application through one meter in the direction of force and is called a Joule. Clearly, One Joule = 10^7 ergs.

***(ii)* The gravitational (or practical) units of work.** The gravitational unit of work is the work done in lifting the weight of a unit mass through a height of unit length.

In F.P.S system, the gravitational unit of work is foot-pound. It is the work done in raising a mass of one pound vertically through a height one foot. Clearly, one foot-pound = g foot-poundals.

In C.G.S. system, the gravitational unit of work is gramme - centimeter. It is the work done in raising a mass of one gramme vertically through a height of one centimeter. Clearly one gramme centimeter = g ergs

In M.K.S. system, the gravitational unit of work is kilogram - meter. It is the work done in raising a mass of one kilogram vertically through height of one meter. Clearly, one kilogram-meter = $10^5 g$ ergs = $(g/100)$ Joules.

2.7 Units of Power

(*i*) In the F.P.S. system, the unit of power is Horse-Power (H.P.). An agent is said to be working at one Horse Power if it be doing 550. foot-pounds of work per second or 33000 foot-pounds of work per minute.

If F pound-weight is force applied and v ft/sec is the velocity of the particle, then H.P. = $(Fv)/550$

(*ii*) In M.K.S. system, the unit of power is Watt. It is the rate of doing one Joule of work per second.

1 Watt = 1 Jule per sec. = 10^7 ergs per second.

Also, one Horse Power = 746 watts (nearly)

If F kg-wt is force applied and v m/sec is the velocity of the particle, then

$$\text{H.P.} = (Fv)/75.$$

2.8 Kinetic energy (K.E). Definition. It is the capacity of the body to do work by virtue of its motion. It is measured by the amount of work done that the external forces will do to bring it to rest.

Measure of K.E. of a mass M, moving with velocity *u*. Let a particle of mass m move along a curve OB with velocity v at P at any time t. Let O be a fixed point and let are $OP = s$. We know that acceleration of the particle at P is $v(dv/ds)$. Then the force F, which acts opposite to the direction of motion, in order to bring it to rest, is given by $F = -mv(dv/ds)$. Let arc $OA = a$ and arc $OB = b$. Again, let velocity of particle at A and B be u and zero respectively.

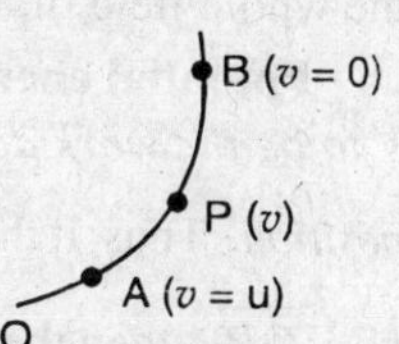

Now, by definition, K.E. at A is equal to the work done by the impressed force F from A to B. Thus, the required K.E.

$$= \int_a^b F ds = \int_{s=a}^{s=b} \left(-mv\frac{dv}{ds}\right) ds = -m\int_{v=u}^{v=0} v\,dv = \frac{1}{2}mu^2.$$

Units of K.E. Since the K.E. has been defined to be equal to work done in some way, the units for measuring K.E. are the same as those for work.

2.9 Principle of work-energy. *The gain in kinetic energy of a particle, in any motion, equals the work done by applied force (s).* **[Calcutta 1993, 96]**

Proof. Refer figure of Art. 2.4. Let a variable force **F**, acting on a particle P, displace it from A to B along the curve AB. Let m be the mass of the particle and **v** be velocity of the particle at P along tangent PT. Let v = | **v** | = magnitude of **v**. Then, acceleration of the particle at P in $d\mathbf{v}/dt$. Now, fundamental equation of dynamics gives $m\,(d\mathbf{v}/dt) = \mathbf{F}$ ...(1)

Let T be the K.E. of the particle at P. Then

$$T = (1/2)mv^2 = (1/2)\, m\mathbf{v}\cdot\mathbf{v} \quad \text{...(2)}$$

From (2), $\dfrac{dT}{dt} = \dfrac{d}{dt}\left(\dfrac{1}{2}m\mathbf{v}\cdot\mathbf{v}\right) = \left(\dfrac{m}{2}\right)\dfrac{d}{dt}(\mathbf{v}\cdot\mathbf{v}) = \left(\dfrac{m}{2}\right)\{(d\mathbf{v}/dt)\cdot\mathbf{v} + \mathbf{v}\cdot(d\mathbf{v}/dt)\}$

or $dT/dt = (m/2)\{2\mathbf{v}\cdot(d\mathbf{v}/dt)\} = \mathbf{v}\cdot\{m(d\mathbf{v}/dt)\}$

Thus, $dT/dt = \mathbf{v}\cdot\mathbf{F}$, using (1) ...(3)

If the particle is at A when $t = t_1$ and at B when $t = t_2$, then integrating over inteval (t_1, t_2), (3) gives

$$\int_{t_1}^{t_2} \frac{dT}{dt} dt = \int_{t_1}^{t_2} \mathbf{F}\cdot\mathbf{v}\,dt = \int_A^B \mathbf{F}\cdot d\mathbf{r}, \text{ as } \mathbf{v} = \frac{d\mathbf{r}}{dt},$$

where **r** is the position vector of P.

Thus, $$T_B - T_A = \int_A^B \mathbf{F}\cdot d\mathbf{r}, \qquad ...(4)$$

where T_A and T_B are kinetic energies at A and B respectively and R.H.S. represents work done by the force **F** in the displacement from A to B along the arc AB.

The relation (4) shows that the change in K.E. is equal to the work done by the forces in moving the particle from A to B. This relation is known as the *principle of work-energy.*

2.10 Energy. *Energy of a body is its capacity to do work*

It is of two kinds : Kinetic energy (K.E.) and potential energy (P.E.). We have already discussed K.E. in Art. 2.8. We now propose to discuss P.E.

2.11 Conservative and non-conservative forces. *A force is called conservative, if the work done by it in displacing its point of application from one given point to another depends upon these points only and not upon the path followed.*

We know that if a variable force **F** displaces its point of application from a point A to a point B, along a curve C then the work done by the force is given by $W = \int_A^B \mathbf{F}\cdot d\mathbf{r}$, where the integration is performed along the curve C. Then by definition, the force **F** is conservative if and only if the value of the above integral does not depend upon the curve C.

If **F** be uniquely defined at every point of a region of space, then the totality of all such vector forces **F** throughout the space is termed a *field of force.* If at every point of the space **F** is conservative, the whole field of force is said to be *conservative.*

Potential energy (P.E.). *In case of a conservative field of force, we define the potential energy V to be the work done by the force in moving the particle from its existing position to some standard position.* Thus, if the existing position be specified by the vector $\mathbf{r} = \overrightarrow{OP}$ and the standard position by $\mathbf{r}_o = \overrightarrow{OP_o}$, then the P.E. at P is given by $$V = \int_{\mathbf{r}}^{\mathbf{r}_o} \mathbf{F}\cdot d\mathbf{r}$$

Non-conservative forces. Forces which are not conservative are called non-conservative.

Two important properties of conservative forces. Each of these properties can also taken as an alternative definition of conservative forces.

Property I. *A force is called conservative, if the work done by it, in displacing its point of application along a closed curve, starting from a point on it, and arriving in the end, at the same point, is zero.*

Property II. *A necessary and sufficient condition that a field of force* $\mathbf{F} = X\mathbf{i} + Y\mathbf{j} + Z\mathbf{k}$ *be conservative is that there exists a differential function* $V(x, y, z)$ *such that* $X = -\partial V/\partial x$, $Y = -\partial V/\partial y$, $Z = -\partial V/dz$. *The function* $V(x, y, z)$ *is called the potential function of the conservative field.*

2.12 Examples of conservative and non conservative forces. A few examples of the conservative forces are force of gravity, tension, normal reaction, a constant force **F** and a central force **F** while a few examples of the non-conservative forces are force of friction and resistance of air.

Illustration 1. *The weight of particle is a conservative force:* Let C be any closed curve in a vertical plane, and let A, B be two pints on it. Let h be the vertical distance between them with B higher than A. Then, the work done from A to B (along C) is $-mgh$ while work done from B to A (along C) $= mgh$. Hence the total work done from A to B and B to A (along C) $= 0$. Therefore, weight of the particle is a conservative force.

Illustration 2. *Friction is a non-conservative force.* Let a particle of mass be kept at rest on a rough horizontal table. Let the frictional force be F. When the body is moved in a straight line from A to B, the work done by the force of friction F is $-F\cdot AB$, while if the body is moved back from B to A, the work done by the force of friction is $F\cdot AB$. Hence the total work done from A to B and B to A

along the straight line $AB = -F \cdot AB - F \cdot AB = -2F \cdot AB$, which is non zero. Therefore the frictional force is not conservative.

2.13 The Principle of conservation of Energy. *If a particle acted upon by a conservative system of forces moves along any path, the sum of its kinetic and potential energies remains constant.*

[Meerut 2005, 07; Delhi (H) 1998, 2001, 2004]

Proof. Suppose a conservative force **F** acts on a particle of mass m constrained to move along a curve C joining points A and B. Then work done in moving the particle from A to B is given by

A P F r O C B

$$W = \int_A^B \mathbf{F} \cdot d\mathbf{r} \qquad \text{...(1)}$$

Since F is conservative, a potential function V exists such that $\mathbf{F} = -\text{grad } V = -\{(\partial V/\partial x)\,\mathbf{i} + (\partial V/\partial y)\,\mathbf{j} + (\partial V/\partial z)\,\mathbf{k}\}$. Now, if $\mathbf{r} = x\mathbf{i} + y\mathbf{j} + z\mathbf{k}$, then $d\mathbf{r} = dx\,\mathbf{i} + dy\,\mathbf{j} + dz\,\mathbf{k}$ and hence

$$\mathbf{F} \cdot d\mathbf{r} = -\{(\partial V/\partial x)\,dx + (\partial V/\partial y)\,dy + (\partial V/\partial z)\,dz\} = -dV$$

Hence (1) gives $\qquad W = \int_A^B dV = V_A - V_B, \qquad$ ...(2)

where V_A and V_B denote the potential energies at A and B.

Let **v** be velocity of the particle at any time t at P. Let v, v_A, v_B be speeds of the particle at P, A and B respectively. Since $\mathbf{F} = m(d\mathbf{v}/dt)$ and $\mathbf{v} = d\mathbf{r}/dt$ (*i.e.*, $d\mathbf{r} = \mathbf{v}dt$), We have

$$\mathbf{F} \cdot d\mathbf{r} = m(d\mathbf{v}/dt) \cdot (\mathbf{v}\,dt) = m\mathbf{v} \cdot d\mathbf{v} = md\,(\mathbf{v}^2/2) = md\,(\mathbf{v}^2/2)$$

$\therefore$ By (1), $\quad W = \int_A^B m\,d(\mathbf{v}^2/2) = (m/2)[\mathbf{v}^2]_A^B = (1/2)m\mathbf{v}_B^2 - (1/2)\,m\mathbf{v}_A^2 \qquad$ ...(3)

From (2) and (3), $\quad V_A - V_B = (1/2)m\mathbf{v}_B^2 - (1/2)m\mathbf{v}_A^2$

or $\quad V_A + (1/2)\,m\mathbf{v}_A^2 = v_B + (1/2)\,m\mathbf{v}_B^2,$

showing that for a single particle in a conservative field of force the sum of the potential and kinetic energies is constant. This is the principle of conservation of energy established for a single particle

Clearly, for a system of particles moving in a conservative filed of force, the principle of conservation of energy would still hold, since it holds for each individual particle.

Example. *At a certain instant, a particle of mass m, moving freely in a vertical plane under gravity, is at a height h above the ground and has a speed v. Use the principle of energy to find its speed when it strikes the ground.* **[Delhi Math (H) 1995, 2001, 2004]**

Sol. Let a particle be at A, at a height h above the ground BC. Let v be its velocity at A and let it strike the ground at B with velocity v_1. Since the earth's gravitational field is conservative hence by the principle of conservation of energy,

A h B Ground C

$$\text{K.E. at } B + \text{P.E. at at } B = \text{K.E. at } A + \text{P.E. at } A$$

or $\quad (1/2)\,mv_1^2 + 0 = (1/2)\,mv^2 + mgh \Rightarrow v_1 = (v^2 + 2gh)^{1/2}$

2.14 Impulse of a force Definition. If a constant force **F** acts on a particle during the interval (t_0, t_1) then the vector $I = (t_1 - t_0)\,\mathbf{F}$ is called *the impulse of the force* **F** *during the interval* (t_0, t_1). Direction of **I** is same as that of **F**.

Similarly if a variable force **F** (t) acts on a particle during the interval (t_0, t_1) then the vector $I = \int_0^{t_1} \mathbf{F}\,dt$ is called the impulse of **F** (t) during the interval (t_0, t_1). Direction of **I** is that of the time average of **F** over the internal (t_0, t_1)

2.15 Impulse - momentum principle for a particle. *The change of linear momentum of a particle during a time interval is equal to the net impulse of the external forces during this interval.*

Proof. Let a force **F** act on a particle of mass m during the interval (t_0, t_1) and let **v** be velocity of the particle at any time t. Let $\mathbf{v}_0$ be the velocity of the particle at the beginning of the time interval

(t_0, t_1) and $\mathbf{v}_1$ be the velocity at the end of this time interval. Now, by the fundamental equation of dynamics, we have

$$m(d\mathbf{v}/dt) = \mathbf{F} \quad ...(1)$$

Let I be the impulse vector of the force during the time interval (t_0, t_1) Then,

$$I = \int_{t_0}^{t_1} \mathbf{F}\, dt = \int_{t_0}^{t_1} m \frac{d\mathbf{v}}{dt} dt = m\int_{v_0}^{v_1} d\mathbf{v} = m[\mathbf{v}]_{v_2}^{v_1}$$

$\therefore$ $\mathbf{I} = m\mathbf{v}_1 - m\mathbf{v}_0$ = Change in the momentum vector,

which is known as the *impulse-momentum principle.*

Note : The above principle is a re-statement of Newton's second law of motion in an integral form.

2.16 Units of impulse. Due to equivalence of impulse and the change in momentum, we adopt the same units for impulse as those used for momentum. Hence the absolute units of impulse are : in C.G.S. system, gm-cm/sec; in M.K.S. system, kg - m/sec and in F.P.S. system, pound-foot/sec.

2.17 Principle of linear momentum for a system of particles. *The rate of change of linear momentum of a system of particles equals the total impressed forces, that is, the vector sum of all external forces.* **[Delhi Math (H) 1994]**

Proof. Suppose the given system S consists of masses m_i, $i = 1, 2, ... n$ with respect to a fixed point O in space. Let $\mathbf{r}_i$ be the position vector and $\mathbf{v}_i(= d\mathbf{r}_i/dt)$ be the velocity of the particle. Then, by definition, the total linear momentum of the system is given by $\mathbf{M} = \sum_{i=1}^{n} m_i \mathbf{v}_i$ ...(1)

Two types of forces act on each particle. External forces, which are exerted by agents outside the system and internal forces, which are exerted by the other particles of the system. Thus, the ith particle is acted by a resultant external force ($\mathbf{F}_i$, say) and resultant internal force ($\mathbf{F}'_i$, say). Then the fundamental equation of dynamics for the ith particle is given by

$$\frac{d}{dt}(m_i \mathbf{v}_i) = \mathbf{F}_i + \mathbf{F}'_i \quad ...(2)$$

From (1), $$\frac{d\mathbf{M}}{dt} = \frac{d}{dt}\sum_{i=1}^{n} m_i \mathbf{v}_i = \sum_{i=1}^{n} \frac{d}{dt}(m_i \mathbf{v}_i)$$

Now, using (2), $$\frac{d\mathbf{M}}{dt} = \sum_{i=1}^{n} (\mathbf{F}_i + \mathbf{F}'_i) = \sum_{i=1}^{n} \mathbf{F}_i + \sum_{i=1}^{n} \mathbf{F}'_i \quad ...(3)$$

Let $\mathbf{F}_{ij}$ be the force on the ith particle due to the jth particle. Since action and reaction are equal and opposite, we have

$$\mathbf{F}_{ij} + \mathbf{F}_{ji} = \mathbf{0} \text{ for all } i \text{ and } j \quad \text{and} \quad \mathbf{F}_{ij} = \mathbf{0} \text{ when } i = j \quad ...(4)$$

Then, $$\mathbf{F}'_i = \sum_{j=1}^{n} \mathbf{F}_{ij} \text{ and so } \sum_{i=1}^{n} \mathbf{F}'_i = \sum_{i=1}^{n}\sum_{j=1}^{n} \mathbf{F}_{ij} = \sum_{\substack{(i,\,j)\\ i<j}} \left(\mathbf{F}_{ij} + \mathbf{F}_{ji}\right) = \mathbf{0}$$

Hence (3) becomes $$d\mathbf{M}/dt = \mathbf{F}, \quad ...(5)$$

where $\mathbf{F} = \Sigma \mathbf{F}_i$ = the total impressed (external) forces.

Relation (5) shows that the rate of change of linear momentum of the system is equal to the total impressed forces on the system.

Corollory I. Principle of linear momentum in scalar form.

The rate of change the resolved part of linear momentum of a system of particles, in any direction equals the sum of the resolved parts of the impressed forces in that direction.

Proof. Let **e** denote unit vector in a given direction is space and let **M** be the total linear momentum of the system of particles. Then the linear momentum of the system in the direction of **e** is given by **e.M**.

Now, $$d(\mathbf{e}\cdot\mathbf{M})/dt = \mathbf{e}\cdot(d\mathbf{M}/dt) + (d\mathbf{e}/dt)\cdot\mathbf{M} \qquad ...(6)$$

Since **e** is a constant vector, $$d\mathbf{e}/dt = \mathbf{0} \qquad ...(7)$$

Again, from (5), $$d\mathbf{M}/dt = \mathbf{F} = \Sigma\mathbf{F}_i \qquad ...(8)$$

Using (7) and (8), (6) gives $$d(\mathbf{e}\cdot\mathbf{M})/dt = \Sigma\mathbf{e}_i\mathbf{F}_i,$$

which proves the required result.

Note. As a particular case of the above corollary if motion takes place parallel to *xy*-plane, then we have

$$\frac{d}{dt}\sum_{i=1}^{n} m_i\left(\frac{dx_i}{dt}\right) = \sum_{i=1}^{n} X_i, \qquad \frac{d}{dt}\sum_{i=1}^{n} m_i\left(\frac{dy_i}{dt}\right) = \sum_{i=1}^{n} Y_i,$$

where m_i, (x_i, y_i) and (X_i, Y_i) are the mass, position and impressed forces of the *i*th particle.

Corollary II. Linear momentum and impulse. *The change in linear momentum of a system of particles over an interval equals the sum of the impulses of the impresses forces over the same interval.*

[Delhi Maths (H) 1992]

Proof. The motion of the system of particles is given by

$$d\mathbf{M}/dt = \mathbf{F}, \quad \text{where} \quad \mathbf{F} = \sum_{i=1}^{n} \mathbf{F}_i$$

In order to find the change in **M** over a time interval (t_0, t_1), we integrate the above equation over (t_0, t_1) and get

$$\int_{t_0}^{t_1} \frac{d\mathbf{M}}{dt}\,dt = \int_{t_0}^{t_1} \mathbf{F}\,dt \quad \text{or} \quad \mathbf{M}_1 - \mathbf{M}_0 = \sum_{i=1}^{n}\int_{t_0}^{t_1} F_i\,dt \qquad ...(9)$$

The R.H.S. of (9) represents the total impulse of the impressed forces over the interval (t_0, t_1). Thus (9) proves the result stated in corollary II.

Note. Continuous or discontinuous motion. In the above discussion we supposed that the forces are of finite magnitude and their action spread over an interval of time. Hence we supposed that the motion is *continuous*. Sometimes a very large force acting on a particle for a very short time produces a finite change in linear momentum instantaneously, without a change in position. Such a force is known as *impulsive force* and the motion which involves such changes are known as *discontinuous*.

An impulsive force (or an impulse). Definition. *It is a very large force, which acting for a very short time on a particle, instantaneously changes its velocity without changing its position.*

An impulsive force is measured by its impulse.

Corollary III. Principle of linear momentum (Discontinuous system/impulsive forces). *The change in the linear momentum of a system of particles, under impulsive forces, equals the vector sum of all the external impulses.*

Proof. Suppose the given system *S* consists of masses m_i, $i = 1, 2, ...n$. With respect to a fixed *O* in space, let $\mathbf{r}_i$ be the position vector of the *i*th particle and let its velocity change from $\mathbf{v}_i$ to $\mathbf{v}_i'$ under impulsive forces. Let $\mathbf{J}_i$ and $\mathbf{J}_i'$ be the external and internal impulse on this particle. Then we have

$$m_i\mathbf{v}_i' - m_i\mathbf{v}_i = \mathbf{J}_i + \mathbf{J}_i', \; i = 1, 2, ..., n \qquad ...(1)$$

Summing up for all the particles, (1) reduces to

$$\Sigma m_i\mathbf{v}_i' - \Sigma m_i\mathbf{v}_i = \Sigma\mathbf{J}_i + \Sigma\mathbf{J}_i' \qquad ...(2)$$

But action and reaction for any two particles are equal and opposite and so $\Sigma\mathbf{J}_i' = \mathbf{0}$. So (2) reduces to

$$\Sigma m_i\mathbf{v}_i' - \Sigma m_i\mathbf{v}_i = \Sigma\mathbf{J}_i \qquad ...(3)$$

If **M** denotes the linear momentum, (3) reduces to $\Delta\mathbf{M} = \Sigma\mathbf{J}_i$,

which proves the required result. Here $\Delta M = \Sigma m_i\,\mathbf{v}_i' - \Sigma\, m_i\mathbf{v}_i$

2.18 Principle of conservation of linear momentum. *If the sum of the resolved parts of the external forces, on a system of particles, resolved in a given direction be* zero, then *the resolved part of the total linear momentum* in *that direction remains constant.*

Proof. Let **M** be the total linear momentum and $\mathbf{F} = \Sigma \mathbf{F}_i$ be the vector sum of the impressed forces. Then for a given system of particles, we have

$$d\mathbf{M}/dt = \mathbf{F} \qquad \ldots(1)$$

Let **e** be a constant vector such that $\mathbf{e} \cdot \mathbf{F} = 0$

so that $\mathbf{e} \cdot \Sigma \mathbf{F}_i = 0$ or $\Sigma \mathbf{e} \cdot \mathbf{F}_i = 0$ ...(2)

Since **e** is a constant vector, $d\mathbf{e}/dt = \mathbf{0}$ ...(3)

From (1), (2) and (3), $d(\mathbf{e} \cdot \mathbf{M})/dt = (d\mathbf{e}/dt) \cdot \mathbf{M} + \mathbf{e} \cdot (d\mathbf{M}/dt) = \mathbf{e} \cdot \mathbf{F} = 0$

Hence $\mathbf{e} \cdot \mathbf{M} = C$ or $\Sigma \mathbf{e} \cdot (m_i \mathbf{v}_i) = C$, ...(4)

where $\mathbf{M} = \Sigma m_i \mathbf{v}_i$ and C is a constant. Relation (4) proves the required result.

2.19 Principle of conservation of linear momentum for discontinuous motion. *If the sum of the resolved parts of the external impulses on a system of particles, resolved in a given direction be zero, then the resolved part of the total linear momentum in that direction, is the same, before and after the action of the impulses.*

Proof. For a system of particles in discontinuous motion, we have

$$\Delta \mathbf{M} = \Sigma \mathbf{J}_i,$$

where $\Sigma \mathbf{M} = \mathbf{M}_1 - \mathbf{M}_0$ represents the change in linear momentum and $\Sigma \mathbf{J}_i$ is the sum of the external impulses. Let **e** be a constant vector such that $\Sigma \mathbf{e} \cdot \mathbf{J}_i = 0$. Then,

$$\Delta(\mathbf{e} \cdot \mathbf{M}) = \mathbf{e} \cdot \Delta \mathbf{M} = 0 \quad \text{so that} \quad \mathbf{e} \cdot \mathbf{M}_1 = \mathbf{e} \cdot \mathbf{M}_0,$$

which prove the required result.

Remark. From Art. 2.19 and 2.20, we find that whether change in motion is continuous (or discontinuous) if there are no external forces (or impulses), the linear momentum of the system is conserved. The presence or absence of internal forces in either case is immaterial, so far as the whole system is concerned.

2.20 Motion of a shot and gun. *The momentum generated in the shot is equal and opposite to that generated in the gun, if the latter is free to move.*

V ← M m → u

Proof. Let m be mass and u, the velocity of the shot and M the mass and V, the velocity of recoil of the gun. Then taking the direction of shot as positive, momentum of the gun and shot (before the shot is fired) $= m \times 0 + M \times 0 = 0$.

Momentum of the gun and shot (after the shot is fired) $= mu + M(-V)$

By the principle of conservation of linear momentum, we have

$$0 = mu - MV \quad \text{so that} \quad mu = MV$$

Ex. 1. *A bullet of mass 0.01 kg is fired from a rifle of mass 20 kg with speed of 100 m/sec. Velocity of recoil (in m/sec) is*

(a) 1 *(b) 0.05* *(c) 20* *(d) 0.01* **[IAS (Prel.) 1995]**

Sol. Ans. (*b*). Here $m = .01$, $M = 20$, $v = 100$ m/sec. Then, by result $mu = MV$, $V = (mu)/M = (.01 \times 100)/20 = .05$ m/sec.

Ex. 2. *If two bodies A and B at rest having masses in the ratio 2 : 1 are subjected to forces in the ratio 1 : 2 for equal interval of time, then (a) their moments will be equal, (b) their velocities will be equal, (c) their energies will be equal, (d) None of the above.* **[IAS (Prel.) 2000]**

Sol. Ans. (*d*) Let masses of A and B be $2m$ and m respectively and forces acting as them be F and $2F$ respectively. Let f_1 and f_2 be accelerations produced in A and B respectively. Then $F = 2mf_1$ and $2F$

$= mf_2 \Rightarrow f_1 = F/2m$ and $f_2 = 2F/m$

Let v_1 and v_2 be velocities acquired by A and B after moving from rest for time t. Then $v_1 = f_1 t = Ft/2m$, $v_2 = F_2 t = 2Ft/m$. So $v_1 \neq v_2$.

Now, momentum of $A = 2mv_1 = 2mFt/2m$ and momentum of $B = mv_2 = 2Fmt/m$. So they are not equal. Next kinetic energies of A and B are given by $(1/2) \times (2m) \times v_1^2$ and $(1/2) \times m v_2^2$ *i.e.*, $m(Ft/2m)^2$ and $(m/2) \times (4F^2t^2/m^2)$ which are not equal. So result (*d*) is correct.

2.21 Law of motion of mass centre. *The mass centre of system of particles, moves like a particle of mass equal to the total mass of the system, and acted on by a force which is equal to the vector sum of all the external forces acting on the system.* **[Delhi Maths (H) 2011]**

Proof. Consider a system of particles of masses m_i, $i = 1, 2, ..., n$. Referred to a fixed point O as origin, at any time t, let $\mathbf{r}_i$ be the position vector and $\mathbf{v}_i$ the velocity of the ith particle. Let $\mathbf{F}_i$ be external force on ith particle. Then, by the principle of linear momentum, we have

$$d\mathbf{M}/dt = \mathbf{F} \qquad ...(1)$$

where $\mathbf{M} = \Sigma m_i \mathbf{v}_i$ and $\mathbf{F} = \Sigma \mathbf{F}_i$ = total external force on the system.

Let m_c $(= \Sigma m_i)$ denote the total mass of the system and $\mathbf{r}_c$ denote the position of the mass centre of the system at any time t. Then, by definition

$$\mathbf{r}_c = (\Sigma m_i \mathbf{r}_i)/(\Sigma m_i) \quad \text{so that} \quad \Sigma m_i \mathbf{r}_i = m_c \mathbf{r}_c \qquad ...(2)$$

Differentiate both sides of (2) w.r.t. 't', we have

$$\Sigma m_i (d\mathbf{r}_i/dt) = m_c(d\mathbf{r}_c/dt) \quad \text{or} \quad \Sigma m_i \mathbf{v}_i = m_c \mathbf{v}_c$$

or

$$M = m_c \mathbf{v}_c \qquad ...(3)$$

Differentiative both sides of (3) w.r.t. 't' and using (1), we get

$$dM/dt = m_c(d\mathbf{v}_c/dt) \quad \text{or} \quad m_c(d\mathbf{v}_c/dt) = \mathbf{F}, \qquad ...(4)$$

which represents the equation of motion of a single particle of man m_c, located at the mass centre of the system, and moving with it.

Hence the law of motion of the mass centre is established.

Corollary. Uniform motion of mass centre. Let the external forces acting on a system of particles be such that $\mathbf{F} = \Sigma \mathbf{F}_i = \mathbf{0}$. Then from (4), we have

$$M(d\mathbf{v}_c/dt) = \mathbf{0} \text{ so that } \mathbf{v}_c = \mathbf{v}_o \text{ (a constant vector)}$$

Thus, when the vector sum of the external forces acting on a system of particles be zero, then the mass centre has uniform motion, that is, it moves with a constant velocity in a straight line.

2.22 Principle of angular momentum for a system of particles

The rate of change of the angular momentum of a system about a fixed point is equal to the total moment of the external forces about the same point. **[Mumbai 2010; Delhi Maths (H) 2009]**

Proof. Consider a system of particles of masses m_i, $i = 1, 2, ..., n$. Referred to a fixed point O as origin, at any time t, let $\mathbf{r}_i$ be the position vector and $\mathbf{v}_i$ the velocity of the ith particle P_i. Let $\mathbf{F}_i$ be external force on P_i. Then by definition, $m_i \mathbf{v}_i$ is the linear momentum of the particle and $\mathbf{r}_i \times (m_i \mathbf{v}_i)$ is the angular momentum of the ith particle about the origin O. Summing up for all particles, the vector

$$H(0) = \sum_{i=1}^{n} \mathbf{r}_i \times \mathbf{m}\mathbf{v}_i \qquad ...(1)$$

is known as the angular momentum of the system about O.

Let $\mathbf{F}_i$ be the external force and $\mathbf{F}_i'$ be the resultant internal force on the ith particle, due to all others. Then the equation of motion for the ith particle is

$$\frac{d}{dt}(m_i \mathbf{v}_i) = \mathbf{F}_i + \mathbf{F}_i \qquad ...(2)$$

Differentiating both sides of (1) w.r.t. 't', we get

$$\frac{d\mathbf{H}(0)}{dt} = \frac{d}{dt}\sum_{i=1}^{n}\mathbf{r}_i \times m\mathbf{v}_i = \sum_{i=1}^{n}\frac{d\mathbf{r}_i}{dt}\times m\mathbf{v}_i + \sum_{i=1}^{n}\mathbf{r}_i \times \frac{d}{dt}(m\mathbf{v}_i)$$

or $$\frac{d\mathbf{H}(0)}{dt} = \sum_{i=1}^{n}\mathbf{r}_i \times (\mathbf{F}_i + \mathbf{F}_i'),$$ using (2) and the fact that

$(d\mathbf{r}_i/dt)\times m\mathbf{v}_i = \mathbf{v}_i \times m\mathbf{v}_i = \mathbf{0}$ for each i

or $$\frac{d\mathbf{H}(0)}{dt} = \sum_{i=1}^{n}\mathbf{r}_i \times \mathbf{F}_i + \sum_{i=1}^{n}\mathbf{r}_i \times \mathbf{F}_i' \qquad \text{...(3)}$$

Let $\mathbf{P}_{ij}$ be the internal force on the ith particle due to the jth particle. Then, we have

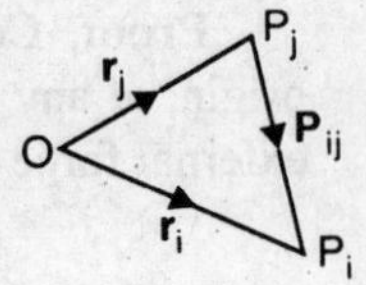

$$\mathbf{F}_i' = \sum_{j=1}^{n}\mathbf{P}_{ij} \quad (\text{with } \mathbf{P}_{jj} = \mathbf{O})$$

and so $$\mathbf{r}_i \times \mathbf{F}_i' = \sum_{j=1}^{n}\mathbf{r}_i \times \mathbf{P}_{ij} \qquad \text{...(4)}$$

Clearly R.H.S. of (4) contains $(n-1)$ terms of the form $\mathbf{r}_i \times \mathbf{P}_{ij}$. Similar number of terms will occur for every other particle of the system. So the expression $\sum_{i=1}^{n}\mathbf{r}_i \times \mathbf{F}_i'$ consists of $n(n-1)$ terms of the form $\mathbf{r}_i \times \boldsymbol{P}_{ij}$ or $n(n-1)/2$ terms of the form $(\mathbf{r}_i \times \mathbf{P}_{ij} + \mathbf{r}_i \times \mathbf{P}_{ji})$. Now, $\boldsymbol{P}_{ji} = -\boldsymbol{P}_{ij}$ and $(\mathbf{r}_i - \mathbf{r}_j)\times \mathbf{P}_{ij} = \mathbf{0}$, since $(\mathbf{r}_i - \mathbf{r}_j)$ and $\mathbf{P}_{ij}$ are collinear vectors

$$\therefore \quad \mathbf{r}_i \times \mathbf{P}_{ij} + \mathbf{r}_j \times \mathbf{P}_{ji} = (\mathbf{r}_i - \mathbf{r}_j)\times \mathbf{P}_{ij} = \mathbf{0},$$

which is true for each pair (i, j) of particles. Hence $\Sigma\mathbf{r}_i \times \mathbf{F}_i' = \mathbf{0}$ and so (3) reduces to

$$d\mathbf{H}(O)/dt = \mathbf{G}, \qquad \text{...(5)}$$

where $\mathbf{G} = \Sigma\mathbf{r}_i \times \mathbf{F}_i$ = total moment of all external forces about O.

Relation (5) establishes the principle of angular momentum.

Note. Another form of the principle of angular momentum. *The rate of change of the angular momentum of a system about a fixed line is equal to the total moment of the external forces about that line.*

Proof. Suppose that $\boldsymbol{e}$ is a unit vector along a fixed line passing through the fixed point O. Then, from relation (5), we have

$$d\mathbf{H}/dt = \boldsymbol{G} \Rightarrow (d\mathbf{H}/dt)\cdot\mathbf{e} = \mathbf{G}\cdot\mathbf{e}$$

$$\therefore \; (d\mathbf{H}/dt)\cdot\mathbf{e} + \mathbf{H}\cdot(d\mathbf{e}/dt) = \mathbf{G}\cdot\mathbf{e}, \text{ as } d\mathbf{e}/dt = \mathbf{0}, \; \mathbf{e} \text{ being a constant vector}$$

or $$\frac{d}{dt}(\mathbf{H}\cdot\mathbf{e}) = \sum_{i=1}^{n}(\mathbf{r}_i \times \mathbf{F}_i)\cdot\mathbf{e}, \text{ as } \mathbf{G} = \sum_{i=1}^{n}\mathbf{r}_i \times \mathbf{F}_i$$

This relation proves what we wished to prove.

2.23 Principle of angular momentum for motion relative to mass centre

The rate of change of angular momentum of a system of particles about the mass centre is equal to the total moment of the external forces about the mass centre. **[Delhi Maths (H) 1996, 2005, 08]**

Proof. Consider a system of particles of masses m_i, $i = 1, 2, ..., n$. Referred to a fixed point O as origin, at any time t, let $\mathbf{r}_i$ be the position vector and $\mathbf{v}_i$ the velocity of the ith particle P_i. Let $\mathbf{F}_i$ be external force on P_i. Then, by the principle of angular momentum of the system, we have

$$d\mathbf{H}(O)/dt = \mathbf{G}(O) \qquad \text{...(1)}$$

where $$\mathbf{H}(O) = \sum_{i=1}^{n}\mathbf{r}_i \times m_i\mathbf{v}_i \quad \text{and} \quad \mathbf{G}(O) = \sum_{i=1}^{n}\mathbf{r}_i \times \mathbf{F}_i, \qquad \text{...(2)}$$

$\mathbf{H}(O)$ being the angular momentum of the system about O.

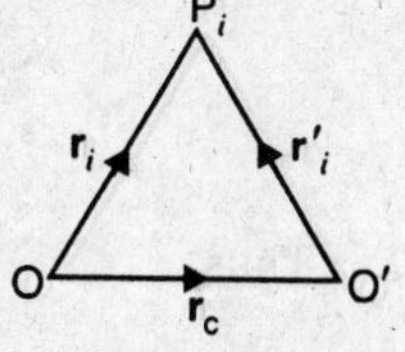

Let $\mathbf{r}_c$ be the position vector of the mass centre O' relative to O. Let $\mathbf{r}_i'$ be the position of vector of P_i relative to O' so that $\mathbf{r}_i = \mathbf{r}_c + \mathbf{r}_i'$ ...(3)

Let $\mathbf{a}_c$ be the acceleration and $\mathbf{v}_c$ the velocity of the mass centre. Relative to O', let $\mathbf{a}_i'$ be the acceleration and $\mathbf{v}_i'$ the velocity of P_i. By Newton's second law, the equation of motion of P_i relative to the mass centre O' is given by

$$m_i(\mathbf{a}_i' + \mathbf{a}_c) = \mathbf{F}_i + \mathbf{F}_i' \quad \text{or} \quad m_i\mathbf{a}_i' = \mathbf{F}_i + \mathbf{F}_i' - m_i\,\mathbf{a}_c, \qquad \text{...(4)}$$

where $\mathbf{F}_i'$ is the resultant internal force on P_i. Then the angular momentum of the system relative to mass centre O' is given by

$$\mathbf{H}(O') = \sum_{i=1}^{n} \mathbf{r}_i' \times m_i \mathbf{v}_i'$$

This gives
$$\frac{d\mathbf{H}(O')}{dt} = \sum_{i=1}^{n} \frac{d\mathbf{r}_i'}{dt} \times m_i\mathbf{v}_i' + \sum_{i=1}^{n} \mathbf{r}_i' \times m_i \frac{d\mathbf{v}_i'}{dt} \qquad \text{...(5)}$$

Now, $(d\mathbf{r}_i'/dt) \times m\mathbf{v}_i' = \mathbf{v}_i' \times m\mathbf{v}_i' = \mathbf{0}$ and $d\mathbf{v}_i'/dt = \mathbf{a}_i'$

$\therefore$ (5) gives
$$\frac{dH(O')}{dt} = \sum_{i=1}^{n} \mathbf{r}_i' \times m_i\mathbf{a}_i' = \sum_{i=1}^{n} \mathbf{r}_i' \times (\mathbf{F}_i + \mathbf{F}_i' - m_i\mathbf{a}_c), \text{ by (4)}$$

or
$$\frac{dH(O')}{dt} = \sum_{i=1}^{n} \mathbf{r}_i' \times \mathbf{F}_i + \sum_{i=1}^{n} \mathbf{r}_i' \times \mathbf{F}_i' - \left(\sum_{i=1}^{n} m_i r_i'\right) \times \mathbf{a}_c \qquad \text{...(6)}$$

Since O' is the mass centre, so $(\Sigma m_i \mathbf{r}_i')/\Sigma m_i = \mathbf{0}$.

Thus
$$\Sigma m_i \mathbf{r}_i' = \mathbf{0} \qquad \text{...(7)}$$

Let
$$\Sigma \mathbf{r}_i' \times \mathbf{F}_i = \mathbf{G}(O') \qquad \text{...(8)}$$

where $\mathbf{G}(O')$ is the total momentum of external forces about mass centre O'.

Let $\mathbf{P}_{ij}$ be the internal force on the ith particle due to the jth particle. Then

$$\mathbf{F}_i' = \sum_{j=1}^{n} \mathbf{P}_{ij} \qquad (\text{with } \mathbf{P}_{jj} = \mathbf{0})$$

and hence
$$\mathbf{r}_i' \times \mathbf{F}_i' = \sum_{j=1}^{n} \mathbf{r}_i' \times \mathbf{P}_{ij} \qquad \text{...(9)}$$

Clearly R.H.S. of (9) contain $(n-1)$ terms of the form $\mathbf{r}_i' \times \mathbf{P}_{ij}$. Similar number of terms will occur for every other particle of the system. Hence the expression $\sum_{i=1}^{n} \mathbf{r}_i' \times \mathbf{F}_i'$ consists of $n(n-1)$ terms of the form $\mathbf{r}_i' \times \mathbf{P}_{ij}$ or $n(n-1)/2$ terms of the form $(\mathbf{r}_i' \times \mathbf{P}_{ij} + \mathbf{r}_j \times \mathbf{P}_{ji})$. Now, $\mathbf{P}_{ij} = -\mathbf{P}_{ji}$ and $(\mathbf{r}_i' - \mathbf{r}_j') \times \mathbf{P}_{ij} = \mathbf{0}$, since $(\mathbf{r}_i' - \mathbf{r}_j')$ and $\mathbf{P}_{ij}$ are collinear vectors.

$$\therefore \qquad \mathbf{r}_i' \times \mathbf{P}_{ij} + \mathbf{r}_j' \times \mathbf{P}_{ji} = (\mathbf{r}_i' - \mathbf{r}_j') \times \mathbf{P}_{ij} = \mathbf{0}$$

which is true for each pair (i, j) of particles.

$$\therefore \qquad \sum_{i=1}^{n} \mathbf{r}_i' \times \mathbf{F}_i' = \mathbf{0} \qquad \text{...(10)}$$

Using (7), (8) and (10), (6) reduces to $\qquad d\mathbf{H}(O')/dt = \mathbf{G}(O')$.

The above vector equation establishes the principle of angular momentum for motion relative to the mass centre.

Corollary. Principle of angular momentum in scalar form. Let **e** be a constant unit vector. Then for the ith particle, $\mathbf{e} \cdot \mathbf{r}_i \times (\mathbf{m}_i\,\mathbf{v}_i)$ is the moment of linear momentum about a line through the fixed point O, in the direction **e** and $\mathbf{e} \cdot \mathbf{r}_i \times \mathbf{F}_i$ is the moment of the external force $\mathbf{F}_i$ about the same line.

Now,
$$\frac{d}{dt}\mathbf{e} \cdot \sum_{i=1}^{n} \mathbf{r}_i \times (m_i\mathbf{v}_i) = \frac{d}{dt}\mathbf{e} \cdot \mathbf{H}(O) = \mathbf{e} \cdot \frac{d}{dt}\mathbf{H}(O) \quad [\because \mathbf{e} \text{ being a constant vector} \Rightarrow d\mathbf{e}/dt = \mathbf{0}]$$

$$= \mathbf{e} \cdot \sum_{i=1}^{n} \mathbf{r}_i \times \mathbf{F}_i, \text{ by (1) and (2)}$$

Thus, $$\frac{d}{dt} \mathbf{e} \cdot \Sigma \mathbf{r}_i \times (m_i v_i) = \sum_{i=1}^{n} \mathbf{e} \cdot \mathbf{r}_r \times \mathbf{F}_i,$$

which represents the principle of angular momentum in scalar form.

Particular case of above corollary. For motion parallel to a fixed plane take OX and OY as rectangular axes in the plane of motion. Let the position, the velocity and the external force of the ith particle be given by (x_i, y_i), $(dx_i/dt, dy_i/dt)$ and (X_i, Y_i) respectively. Then the principle of angular momentum is represented by the scalar equation

$$\frac{d}{dt} \Sigma m_i \left(x_i \frac{dy_i}{dt} - y_i \frac{dx_i}{dt} \right) = \Sigma (x_i Y_i - y_i X_i)$$

MISCELLANEOUS PROBLEMS

Ex. 1. *A particle can descend along a straight smooth tube from point A to a point B where AB makes an angle 60° with horizontal. There is a smooth semicircular tube (whose diameter is AB) from A to B. If V_1, V_2 be the velocities of the particle having reached at B in two cases– (i) while descending along the straight path and (ii) while descending along the semicircular path, and if AB = l, then*

(a) $V_1^2 = \sqrt{3} gl = V_2^2$, *(b)* $V_1^2 = \sqrt{3} gl$, $V_2^2 = (\pi gl)/2$

(c) $V_1^2 = 2gl$, $V_2^2 = (\pi gl)/2$ *(d)* $V_1^2 = V_2^2 = gl$. **[I.A.S. (Prel.) 2004]**

Sol. Ans. (*a*) (*i*) Using $v^2 = u^2 + 2fs$ for motion from A to B, we have

$$V_1^2 = 0 + 2 \times g \cos 30° \times l = \sqrt{3} gl.$$

(*ii*) While moving from A to B along semicircular are ACB, by energy equation, we get

$$(1/2) \times m V_2^2 - (1/2) \times m \times 0^2 = mg \times AC$$

or

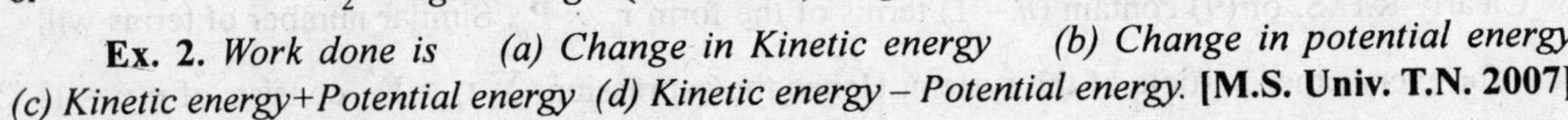

$$V_2^2 = 2gAC = 2g \times (AB \sin 60°) = 2gl \times (\sqrt{3}/2) = \sqrt{3} gl.$$

Ex. 2. *Work done is (a) Change in Kinetic energy (b) Change in potential energy (c) Kinetic energy+Potential energy (d) Kinetic energy – Potential energy.* **[M.S. Univ. T.N. 2007]**

Sol. Ans (*a*) Refer Art. 2.9

Ex. 3. *A body of mass 5 kg takes 2 seconds to fall from the top of a tower to the ground. Find the work done by gravity.*

Sol. Let h be the height of tower. Using $h = ut + (1/2) \times gt^2$, we get

$$h = (0 \times 2) + (1/2) \times 9.81 \times (2)^2 = 19.62 \text{ meters}$$

$\therefore$ Work done by gravity = mgh = 5 × g × 19.62 Joules = 5 × 19.62 kg-meters = 98.1 kg–meters

Ex. 4. *A car of mass 20 quintals and travelling at 15 m/sec is brought to rest in 25 meters by application of brakes. Find the work done by the force of resistance (assumed uniform) offered by the brakes.* ***[G.N.D.U. Amritsar 2010]***

Sol. Let f be the retardation produced by brakes. Using $v^2 = u^2 - 2fs$, we get

$(0)^2 = (15)^2 - (2 \times f \times 25)$ so that $f = (9/2)$ m/sec^2

F = Force of resistance = mf = 10 × 100 × (9/2) = 9000 newtons

$\therefore$ Required work done = 9000 × 25 Joules = (225000/9.8) kg-meters

= 22959 kg-meters (approximately)

Ex. 5. *A car of mass 750 kg is running up a hill of 1 in 30 at a steady speed of 36 km/hr, the friction is equal to the weight of 40 kg. Find the work done in 1 second.* **(I.A.S. 2004)**

Sol. Let α be the inclination of the hill to the horizontal plane.

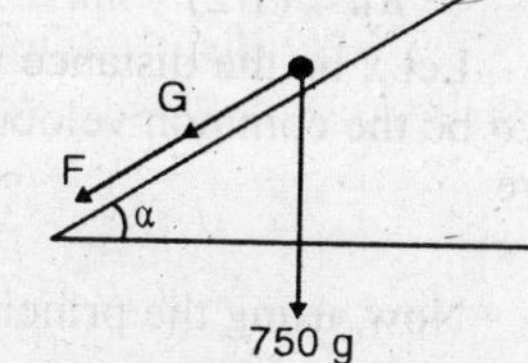

Then, by given probelm sin α = 1/30. Here, we have

R = Frictional resistance = 40g Newtons

and G = Component of weight down the slope

= mg sin α = 750 g × (1/30) = 259 Newtons

Since the train is moving at a steady rate of 36 km/hr, *i.e.*, (36 × 1000)/(60 × 60), *i.e.*, 10 m/sec, no force is being exerted to generate acceleration.

∴ Force exerted by car = Total resistance = (40 + 25)g Newtons

Hence, the required work done = 65 g × 10 Joules = 650 kg-meters

Ex. 6. *A body of mass $(m_1 + m_2)$ moving in a straight line is split into two parts of masses m_1 and m_2 by an internal explosion which generates kinetic energy E. If after the explosion, the two parts move in the same line as before, find their relative velocity.* **[I.A.S. 2005]**

Sol. Let u be the velocity of the body before explosion, v_1 and v_2 be the velocities of parts m_1 and m_2 after explosion. Using the principle of conservation of mementum, we get

$$m_1 v_1 + m_2 v_2 = (m_1 + m_2)u \quad \text{... (1)}$$

Since, kinetic energy after splitting = kinetic energy before splitting, we have

$$(1/2)\times m_1 v_1^2 + (1/2)\times m_2 v_2^2 = E + (1/2)\times(m_1+m_2)u^2$$

or

$$(m_1+m_2)\,(mv_1^2 + m_2 v_2^2) = 2E(m_1+m_2) + (m_1+m_2)^2 u^2 \quad \text{... (2)}$$

Squaring both sides of (1) gives $\qquad (m_1 v_1 + m_2 v_2)^2 = (m_1+m_2)^2 u^2 \quad$ ... (3)

Subtracting (3) from (1), $\qquad (m_1 + m_2)\ (m_1 v_1^2 + m_2 v_2^2) - (m_1 v_1 + m_2 v_2)^2 = 2E(m_1+m_2)$

or $m_1 m_2 (v_1 - v_2)^2 = 2E(m_1+m_2)$ $\qquad$ or $\qquad v_1 - v_2 = \{2E(m_1+m_2)/m_1 m_2\}^{1/2}$,

which is the required relative velocity.

Ex. 7. *A bullet of mass m moving with velocity u strikes a block of mass M which is free to move in the direction of motion of the bullet and is embedded in it. Show that a portion M/(M + m) of the kinetic energy is lost.* **[I.F.S. 1995]**

Sol. Let v be the common velocity of the block and the bullet when the bullet gets embedded in it. Then, using the principle of conservation of linear momentum, we have

$(m + M)\, v = mu$ $\qquad$ so that $\qquad v = (mu)/(m + M)$ ... (1)

∴ Loss of Kinetic energy

$$= \frac{1}{2}mu^2 - \frac{1}{2}(m+M)v^2 = \frac{1}{2}mu^2 - \frac{1}{2}(m+M)\times\frac{m^2u^2}{(m+M)^2}, \text{ using (1)}$$

$$= \frac{1}{2}mu^2\left(1 - \frac{m}{m+M}\right) = \frac{1}{2}\times\frac{m\,Mu^2}{m+M} = \frac{M}{m+M}\times\frac{1}{2}mu^2,$$

showing that a portion $M/(M + m)$ of the kinetic energy is lost.

Ex. 8. *If a shot of mass m gm striking fixed metal plate with velocity u penetrates it through a distance a cm, show that it will completely pierce through a plate, free to move, of mass M gm and the thickness b cm if b < (Ma)/(m + M), the resistane being supposed uniform.* **[I.F.S. 1998]**

Sol. Let R dynes be the resistance to penetration. Since the shot is reduced to rest, the principle of work and energy gives

$Ra = (1/2) \times mu^2$ so that $R = (mu^2)/2a$ ... (1)

Let x be the distance that a shot can penetrate into a plate of mass M which is free to move. Let v be the common velocity of the shot and the plate. Using the principle of work and energy, we have

$$Rx = (1/2) \times mn^2 - (1/2) \times (M \times m)\, v^2 \quad ...(2)$$

Now, using the principle of conservation of linear momentum, we have

$(m + M)v = mu$ so that $v = (mu)/(M + m)$ ... (3)

Using (1) and (3), (2) reduces to

$$\frac{mu^2}{2a} \times x = \frac{1}{2} mu^2 - \frac{1}{2} \times (M+m) \times \frac{m^2u^2}{(M+m)^2} \quad \text{or} \quad \frac{x}{a} = 1 - \frac{m}{m+M}$$

or $x/a = M/(m + M)$ so that $x = (Ma)/(m + M)$

The shot will pierce the second plate completely if

$b < x$ *i.e.*, $b < (Ma)/(m + M)$

Ex. 9. *A gun is mounted on a gun-carriage, movable on a smooth horizontal plane and the gun is elevated at the angle α to the horizon. A shot is fired and leaves the gun in a direction inclined at an angle θ to the horizon. If the mass of the gun and the carriage be n times that of shot, show that $\tan \theta = (1 + 1/n) \tan \alpha$.*

Sol. Let u be the velocity of the shot relative to the gun (along the barrel AB) and v, the velocity of recoil of the gun. Let the mass of the shot be m, then by problem, the mass of the gun and the carriage is nm.

Let V be the actual velocity of the shot which is inclined at an angle θ to the horizontal.

Here V is resultant of u and v. So, on resolving horizontally and vertically we have

$V \cos\theta = u \cos\alpha - v$ and $V \sin\theta = u \sin\alpha$ ... (1)

Dividing, these give $\tan\theta = (u \sin\alpha)/(u \cos\alpha - v)$ ... (2)

Using the principle of conservation of linear momentum in horizontal direction, we have

$mV \cos\theta = nmv$ or $m(u \cos\alpha - v) = nmv$, by (1)

or $mu \cos\alpha = (n+1)mv$ or $v = (u \cos\alpha)/(n + 1)$ ... (3)

Using (3), (2) reduces to

$$\tan\theta = \frac{u \sin\alpha}{u\cos\alpha - (u\cos\alpha)/(n+1)} = \frac{(n+1)\sin\alpha}{n\cos\alpha} = \left(1 + \frac{1}{n}\right)\sin\alpha$$

Ex. 10. *A train of mass M kg is ascending a smooth incline of 1 in n and when the velocity of train is v m/sec, its acceleration is f m/sec^2, prove that the effective power of the engine is Mv(nf + g)n watts.*

Sol. Let θ be the inclination of the plane OB with horizontal OA and let R be the normal reaction of the plane. Let the force exerted by the engine be P Newtons.

Then, equation of motion along the plane OB is given by

$Mf = P - Mg \sin\theta$ or $Mf = P - (mg)/n$, since $\sin\theta = 1/n$

Hence, $P = Mf + (Mg)/n = (M/n) \times (nf + g)$ newtons

$\therefore$ Power of the engine $= Pv$ watts $= (Mv/n) \times (nf + g)$ watts

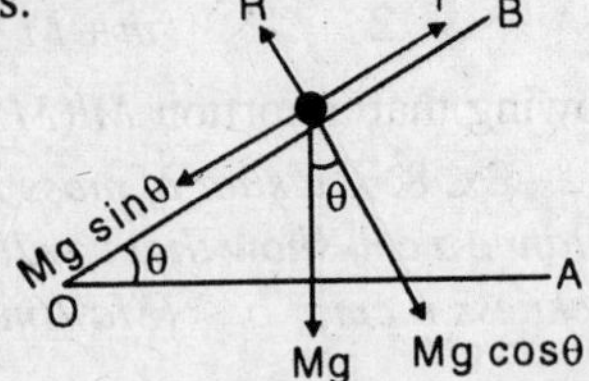

Ex. 11. *A shell lying a straight smooth horizontal tube, suddenly explodes and breaks into two portions of masses m_1 and m_2. If s is the distance apart in the tube of the masses after time t, show that the work done by the explosion is (1/2) × {(m_1 m_2)/(m_1 + m_2)} × (s^2/t^2).* **[I.A.S. 2008]**

Sol. Let v_1 and v_2 be the velocities of the masses m_1 and m_2 respectively after explosion. Using the principle of conservation of momentum, we have

$$-m_1 v_1 + m_2 v_2 = 0 \qquad \text{so that} \qquad v_2 = (m_1 v_1)/m_2 \qquad \text{... (1)}$$

∴ Relative speed of the two masses after explosion

$$= v_2 - (-v_1) = v_2 + v_1 = (m_1 v_1)/m_2 + v_1 = \{(m_1 + m_2)v_1\}/m_2, \text{ by (1)}$$

Given that $\qquad s$ = distance between the two masses after time t.

$$\therefore\ s = \text{relative speed} \times \text{time } t \quad \Rightarrow \quad s = \frac{m_1 + m_2}{m_2} v_1 \times t \;\Rightarrow\; v_1 = \frac{m_2 s}{t(m_1 + m_2)} \qquad \text{... (2)}$$

From (1) and (2),

$$v_2 = \frac{m_1}{m_2} \times \frac{m_2 s}{t(m_1 + m_1)} = \frac{m_1 s}{t(m_1 + m_2)} \qquad \text{... (3)}$$

Required work done by the explosion

$$= \text{Change in kinetic energy } = (1/2) \times m_1 v_1^2 + (1/2) \times m_2 v_2^2$$

$$= \frac{1}{2} m_1 \times \frac{m_2^2 s^2}{t^2 (m_1 + m_2)^2} + \frac{1}{2} m_2 \times \frac{m_1^2 s^2}{t^2 (m_1 + m_2)^2} = \frac{1}{2} \times \frac{m_1 m_2 s^2}{t^2 (m_1 + m_2)^2} \times (m_1 + m_2)$$

$$\Rightarrow = (1/2) \times \{(m_1 m_2)/(m_1 + m_2)\} \times (s^2/t^2).$$

Ex. 12. *A body falls from rest freely under gravity. If the speed is v when it has lost an amout P of gravitational potential energy, then what is the mass of the body ?*

(a) $P/2v^2$ *(b)* P/v^2 *(c)* P/v *(d)* $2P/v^2$. **[I.A.S. (Prel) 2008]**

Sol. Ans. (d) Let m be the mass of the body. Then by the principle of conservation of energy (see Art. 2.3), we have

Increase in Kinetic energy = loss in gravitational potential energy

$\Rightarrow (1/2) \times mv^2 - (1/2) \times m \times 0^2 = P$, *(given)* $\Rightarrow m = 2P/v^2$

EXERCISES

1. A shell of mass m is projected from a gun of mass M by an explosion which generates kinetic energy E. Prove that the initial velocity of the shell is $\{2EM/m\,(M + m)\}^{1/2}$, it being assumed that at the instant of explosion the gun is free to recoil.

2. A shot of mass m is fired from a gun of mass M with velocity u relative to the gun. Show that the actual velocities of the shot and the gun are $Mu/(M + m)$ and $mu/(M + m)$ respectively, and that their kinetic energies are inversely proportional to their masses.

3. A shot of mass m penetrates a thickness s into a fixed plate of mass M; prove that if M is free to move, the thickness penetrated is $(Ms)/(M + m)$.

4. A gun of mass M fires a shell of mass m horizontally and the energy of the explosion is such as would be sufficient to project the shot vertically to a height h. Show that the velocity of recoil of the gun is $\{2m^2gh/M(M + m)\}^{1/2}$.

5. A shell of mass M is moving with a velocity u in a line AB. An internal explosion, which generates an energy E breaks it into two masses m_1 and m_2, which move in AB. Show that their velocities are $u + (2m_2E/m_1M)^{1/2}$ and $u - (2m_1E/m_2M)^{1/2}$.

6. A mass m after falling freely through a distance 'a' begins to raise a mass M greater than itself and connected with it by means of an inextensible string passing over a fixed pulley. Show that M will have returned to its original position at the end of time $\{(2m)/(M - m)\} \times (2a/g)^{1/2}$.

7. An engine of H metric horse power is pulling up a train of mass M metric tons on an incline of 1 in m. If the force of friction be n kg. per metric ton, show that the maximum velocity of the train is $(75\ Hm)/M(1000 + mn)$ m/sec.

8. Define work, power and energy with units. **(Guwahati 2007)**

3

Rectilinear Motion with Variable Acceleration

3.1 Rectilinear motion. (Meerut 2004, 2007, 2008)

Definition. When a point (or particle) moves along a straight line, its motion is said to be a ***rectilinear motion***.

In the present chapter we shall discuss rectilinear motion under variable acceleration which may be function of time, velocity or distance.

Suppose a particle of mass m moves in a straight line such that it is at a distance s from the fixed point on the straight line at any time t and has velocity v at that instant, then the expressions for acceleration f are given by $f = dv/dt$ or $f = d^2s/dt^2$ or $f = v(dv/ds)$.

If P be the force acting upon the particle at that instant along the line of motion, then by Newton's second law of motion, of the equation of the motion of the particle is given by

I. $m(dv/dt) = P$ or **II.** $m(d^2s/dt^2) = P$ or **III.** $mv(dv/ds) = P$.

Clearly, if P is a function of time, equation I or II should be used

If P is a function of v, equation I or III should be used.

If P is a function of s, equation III should be used.

3.2 Motion under inverse square law

A particle moves in a straight line under an attraction towards a fixed point on the line, varying inversely as the square of the distance from the fixed point; to investigate the motion.

[Garhwal 2002, Kanpur 2010, 2016; Rohilkhand 1999]

Let a particle start from point A where $OA = a$, O being the fixed point (which is also centre of given force). Let P be the position of the particle at time t, where $OP = x$. Then acceleration at $P = \mu/x^2$ towards O where $\mu\ (> 0)$ is a constant. The equation of motion of the particle of mass m at P is given by

A′ O P A (x between O and P)

$$m(d^2x/dt^2) = -m(\mu/x^2) \quad \text{or} \quad d^2x/dt^2 = -\mu/x^2 \qquad ...(1)$$

[We have taken –ve sign on R.H.S. because d^2x/dt^2 is +ve in the direction of x increasing while in this problem μ/x^2 acts in the direction of x decreasing *i.e.*, along PO]

or $$v(dv/dx) = -\mu/x^2 \quad \text{or} \quad 2v\,dv = -(2\mu/x^2)dx$$

Integrating, $$v^2 = 2\mu/x + A, \text{ where } A \text{ is a constant.} \qquad ...(2)$$

Initially at A, when $x = a$, $v = 0$. So (2) given $A = -2\mu/a$.

$\therefore$ (2) becomes $$v^2 = 2\mu(1/x - 1/a) \qquad ...(3)$$

giving velocity at any distance from the centre of force O.

From (3), $$v = dx/dt = \left(\frac{2\mu}{a}\right)^{1/2}\left(\frac{a-x}{x}\right)^{1/2} \qquad ...(4)$$

[We have taken –ve sign on R.H.S. because as t increases, x decreases during motion of the particle from P to O]

Separating variables, (4) gives, $$dt = -\left(\frac{a}{2\mu}\right)^{1/2}\left(\frac{x}{a-x}\right)^{1/2} dx.$$

Putting $x = a\cos^2\theta$ so that $dx = -2a\cos\theta\sin\theta\,d\theta$, we get

$$dt = \left(\frac{a}{2\mu}\right)^{1/2}\left(\frac{a\cos^2\theta}{a - a\cos^2\theta}\right)^{1/2} .\,2a\sin\theta\cos\theta\,d\theta$$

or $\quad dt = a(a/2\mu)^{1/2}(1+\cos 2\theta)\,d\theta$

Integrating, $\quad t = a(a/2\mu)^{1/2}\{\theta + (1/2)\sin 2\theta\} + B$

or $\quad t = a\,(a/2\mu)^{1/2}(\theta + \sin\theta\cos\theta) + B$

or $\quad t = a(a/2\mu)^{1/2}\{\theta + \cos\theta\,(1-\cos^2\theta)^{1/2}\} + B$

Now, $x = a\cos^2\theta \Rightarrow \cos\theta = (x/a)^{1/2}$ and $\theta = \cos^{-1}(x/a)^{1/2}$

$\therefore \quad t = a\,(a/2\mu)^{1/2}\{\cos^{-1}(x/a)^{1/2} + (x/a)^{1/2}(1 - x/a)^{1/2}\} + B \qquad ...(5)$

Initially at A, when $t = 0, x = a$. So (5) gives $B = 0$

$\therefore$ By (5), $\quad t = a\,(a/2\mu)^{1/2}\{\cos^{-1}(x/a)^{1/2} + (x/a)^{1/2}(1 - x/a)^{1/2}\} \qquad ...(6)$

giving time taken by the particle, to move from A to P.

Putting $x = 0$ in (6), the time t' taken by the particle from A to O is

$$t' = a(a/2\mu)^{1/2}(\pi/2 + 0) = (\pi/2)\,(a^3/2\mu)^{1/2}$$

Putting $x = 0$ in (4), we find that the velocity at O is infinite and so the particle moves to the left of O towards A'. Since the particle is always attracted towards O, so while moving from O to A', the particle is acted upon by retardation which is inversely proportional to the square of the distance from O. Clearly, the particle will came to instantaneous rest at A' where $OA' = OA = a$ and then retrace its path and pass through O and will be at instantaneous rest at A. Thus the motion of the particle is oscillatory about O and time of one complete oscillation.

$$= 4 \times (\text{time from } A \text{ to } O) = 4t' = 2\pi(a^3/2\mu)^{1/2}, \text{ by (7)}.$$

3.3 Motion of a particle under the attraction of the earth

When a particle moves under the attraction of the earth, the acceleration on it, towards the centre of the earth will be as follows :

(*i*) When the particle moves outside the surface of the earth, the acceleration varies inversely as the square of the distance of the particle from the centre of the earth.

(*ii*) When the particles moves inside the earth, the acceleration varies as the distance of the particle from the centre of the earth.

(*iii*) the value of the acceleration at the surface of the earth is g. **(Meerut 2004)**

Type 1 : Solved examples based on Art. 3.2 and 3.3

Ex. 1. *A particle moves from rest in a straight line under an attractive force μ/(distance)^2. Show that if initial distance is 2a, the distance will be a after a time* $(1 + \pi/2)\,(a^3/\mu)^{1/2}$.

Sol. Proceed as in Art. 3. 2 upto equation (6) after replacing a by $2a$ for the present problem and get

$$t = 2a\,(a/\mu)^{1/2}\{\cos^{-1}(x/2a)^{1/2} + (x/2a)^{1/2}(1 - x/2a)^{1/2}\} \qquad ...(i)$$

giving time t from the initial position $x = 2a$ to any point at a distance x from the centre O. Let t_1 be the time from $x = 2a$ to $x = a$. Then putting $t = t_1$ and $x = a$ in (*i*), we have

$$t_1 = 2a\,(a/\mu)^{1/2}\{\cos^{-1}(1/2)^{1/2} + (1/2)^{1/2}(1 - 1/2)^{1/2}\} = 2a\,(a/\mu)^{1/2}(\pi/4 + 1/2)$$

Thus, the required time $= t_1 = (1 + \pi/2)\,(a^3/\mu)^{1/2}$.

Ex. 2. *Show that the time of descent to the centre of force, varying inversely as the square of the distance from the centre through first half of its initial distance is that through the last half as* $(\pi + 2) : (\pi - 2)$. **[Garhwal 2003; Kanpur 2008, Meerut 2005, 2017]**

Sol. Proceed as in Art 3.2 upto equation (6) and get

$$t = (a^3/2\mu)^{1/2}\{\cos^{-1}(x/a)^{1/2} + (x/a)^{1/2}(1 - x/a)^{1/2}\}, \qquad ...(i)$$

giving time taken to move from initial position A to any point P.

Let M be the middle point of OA and let t_1 be the time taken by the particle form A to M so that when $t = t_1, x = a/2$. Then (*i*) gives

$$t_1 = (a^3/2\mu)^{1/2}\{\cos^{-1}(1/\sqrt{2}) + (1/2)^{1/2}(1 - 1/2)^{1/2}\} = (a^3/2\mu)^{1/2}(\pi/4 + 1/2)$$

Again, let t_2 be the time from A to O so that when $t = t_2, x = 0$. Then (*i*) gives

$$t_2 = (a^3/2\mu)\,(\cos^{-1}0 + 0) = (a^3/2\mu)^{1/2}.\ (\pi/2)$$

Let t_3 be time from M to O. Then, we have

$$t_3 = t_2 - t_1 = (a^3/2\mu)^{1/2}(\pi/2) - (a^3/2\mu)^{1/2}(\pi/4 + 1/2) = (a^3/2\mu)(\pi/4 - 1/2)$$

$$\therefore \quad \frac{\text{time from } A \text{ to } M}{\text{time from } M \text{ to } O} = \frac{t_1}{t_3} = \frac{(a^3/2\mu)^{1/2}\{(\pi+2)/2\}}{(a^3/2\mu)^{1/2}\{(\pi-2)/2\}} = \frac{\pi+2}{\pi-2}.$$

Ex. 3. *Show that the time occupied by a body, under the acceleration μ/x^2 towards the origin, to fall from rest at distance a to distance x from the attracting centre can be put in the form.*

$$(a^3/2\mu)^{1/2}\ [\cos^{-1}(x/a)^{1/2} + \{(x/a)(1 - x/a)\}^{1/2}]$$

Prove also that the time occupied from $x = 3a/4$ to $a/4$ is one third of the whole of descent from a to O.

Sol. Try yourself as in Ex. 2.

Ex. 4. (***a***) *A particle falls towards the earth from infinity; show that its velocity on reaching the surface of the earth is the same as that which it would have acquired in falling with constant acceleration and through a distance equal to the earth's radius.*

(***b***) *A particle falls towards the earth from infinity. Find its velocity on reaching the earth.*

Sol. (***a***) Let O be the centre and a the radius of the earth. Let a vertical line OB through O meet the earth's surface at A. Let a particle fall from rest from infinity towards O along BO. Let P be its position at any time t such that $OP = x$. Then by Newton's law of gravitation, the acceleration of the particle at P is μ/x^2 towards O. Hence the equation of motion of the particle at P is

$$d^2x/dt^2 = -\mu/x^2 \qquad \text{...(1)}$$

Since (1) helds good at A where $x = a$ and $d^2x/dt^2 = -g$

$\therefore$ From $\quad -g = -\mu/a^2 \quad$ or $\quad \mu = ga^2 \qquad$ ...(2)

(1) becomes $\quad v(dv/dx) = -(ga^2)/x^2 \quad$ or $\quad 2vdv = -(2ga^2/x^2)dx$

Integrating, $\quad v^2 = (2ga^2)/x + A$, A is a constant. $\qquad$ ...(3)

Initially, when $x = \infty$, $v = 0$. So (3) gives $A = 0$.

$\therefore$ From (3), $\qquad v^2 = 2ga^2/x$

Putting $x = a$, the velocity V at surface of earth (*i.e.*, at A) is given by

$$V^2 = (2ga^2)/a \quad \text{or} \quad V = (2ga)^{1/2} \qquad \text{...(4)}$$

Let V_1 be the velocity acquired by the particle in falling a distance equal to the earth's radius a with constant acceleration g.

Then $\qquad V_1^2 = 0 + 2ga \quad$ or $\quad V_1 = (2ga)^{1/2} \qquad$ ...(5)

From (4) and (5), $V = V_1$, which is what we wished to prove.

(***b***) Do exactly as in part (*a*) upto equation (4).

Ex. 5. *A particle falls freely from infinity on the earth. Show that its velocity on the earth's surface is $(2ga)^{1/2}$, where a is the radius of the earth. What shall be its velocity on reaching the centre of the earth.* **[Agra 2014; Kanpur 2012]**

Sol. For the first part, refer Ex. 4 (*a*).

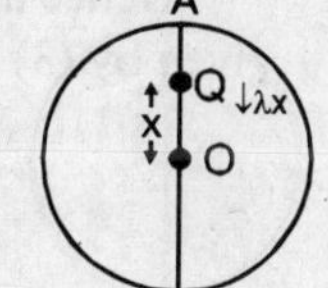

Solution of the second part. Let the particle be at Q at any time t such that $OQ = x$. Then we know that acceleration of the particle at Q is λx towards O. Hence the equation of motion at Q is $\qquad d^2x/dt^2 = -\lambda x \qquad$ ...(1)

Since (1) holds good at A where $x = a$ and $d^2x/dt^2 = -g$.

$\therefore$ From (1), $\quad -g = -\lambda a \quad$ so that $\quad \lambda = g/a \qquad$ ...(2)

$\therefore$ (1) gives $\quad v(dv/dx) = -(gx/a) \quad$ or $\quad 2vdv = -(2gx/a)dx$

Integrating, $\quad v^2 = -(g/a)x^2 + B$, B is a constant $\qquad$ ...(3)

Initially at A, $x = a$, $v = (2ga)^{1/2}$. Then (3) gives

$$2ga = -ga + B \text{ or } B = 3ga. \text{ So (3) becomes}$$

$$v^2 = 3ga - (g/a)x^2. \quad \text{...(4)}$$

Let $u = V$ when $x = 0$. Then (4) gives $\quad V^2 = 3ga.$

Thus, the velocity at centre $= V = (3ga)^{1/2}$

Ex. 6. (*a*) *Find the velocity of a particle with which if a particle is projected from the surface of the earth it may never return to it.*

(*b*) *Compute in kilometer per second the least velocity which will carry a body from the earth's surface to infinity.* **[Agra 2002]**

Sol. (*a*) Draw a figure as in Ex. 4(*a*) Here particle is projected from *A* along *OAB*. If *P* be its position at any time *t*, then do as in Ex. 4(*a*) upto equation (3).

$$\therefore \quad v^2 = (2ga^2)/x + A, A \text{ is a constant} \quad \text{...}(i)$$

Let $v = 0$ when $x = \infty$. Then (*i*) gives $A = 0$.

$$\therefore \text{ From } (i), \quad v^2 = (2ga^2)/x \quad \text{...}(ii)$$

Let $v = V$ when $x = a$. Then (*ii*) gives

$$V = \text{required velocity of projection} = (2ga)^{1/2} \quad \text{...}(iii)$$

(*b*) Take $a = 6400$ kilometers or 6400×1000 meters and $g = 9.8$ meters/sec^2. Then

$$V^2 = 2 \times 9.8 \times 6400 \times 1000 = 196 \times 64 \times 10000$$

Thus, $\quad V = 14 \times 8 \times 100$ meter/sec $= 11.2$ km/sec.

Remark. Escape velocity. Definition. *The velocity with which a particle is to be projected vertically upward from the centre of attraction so that it escapes the centre of attraction, is called the escape velocity.*

Results to be committed to memory. If R be radius of the earth, then escape velocity from the surface of earth is $(2gR)^{1/2}$. In particular, if $R = 6400$ km and $g = 9.8$ m/sec^2 then escape velocity = 11.2 km/sec. Thus, if a body is projected vertically upwards with a velocity of 11.2 km/sec or more the body will never come back to the earth's surface.

Ex. 7. *A particle is projected vertically upwards from the surface of earth with a velocity just sufficient to carry it to the infinity. Prove that the time it takes to reach a height h is (1/3) $(2a/g)^{1/2}$ $\{(1 + h/a)^{3/2} - 1\}$ where a is the radius of the earth.* **[Garhwal 2010, Meerut 2004, Purvanchal 2011]**

Sol. Refer figure of Ex. 4(*a*). Let O be the centre of the earth and A the point of projection on earth's surface. Let P be the position of the particle at any time t such that $OP = x$. Then the acceleration of the particle at P is μ/x^2 towards O. Hence the equation of motion of the particle at P is

$$d^2x/dt^2 = -\mu/x^2 \quad \text{...(1)}$$

At A, $\quad x = a$ and $d^2x/dx^2 = -g$. $\quad$ So (1) $\Rightarrow \mu = a^2 g$.

$\therefore$ (1) gives $\quad v(dv/dx) = -(a^2g)/x^2 \quad$ or $\quad 2v\,dv = -(2a^2g/x^2)dx$.

Integrating, $\quad v^2 = (2a^2g)/x + A$, A is a constant. $\quad$ ...(2)

Given that, when $x = \infty$, $v = 0$. So (2) gives $A = 0$

$\therefore$ (2) becomes $\quad (dx/dt)^2 = (2a^2g)/x \quad$ or $\quad dx/dt = (2a^2g/x)^{1/2}$

[+ve sign is taken because as t increases, x increases]

or $$dt = \{1/a\,(2g)^{1/2}\}x^{1/2}\,dx$$

Hence the required time t' from the surface of the earth to a height h *i.e.*, from $x = a$ to $x = a + h$ is given by (3) on integration as

$$t' = \frac{1}{a(2g)^{1/2}} \int_a^{a+h} x^{1/2}\,dx = \frac{1}{a(2g)^{1/2}} \left[\frac{2}{3}x^{3/2}\right]_a^{a+h}$$

or $$t' = \frac{2}{3a(2g)^{1/2}} \{(a+h)^{3/2} - a^{3/2}\} = \frac{1}{2}\left(\frac{2a}{g}\right)^{1/2} \left\{\left(1+\frac{h}{a}\right)^{3/2} - 1\right\}$$

Ex. 8. *If h be the height due to the velocity v at the earth's surface supposing its attraction* *constant and H the corresponding height when the variation of gravity is taken into account, prove that 1/h – 1/H = 1/r, where r is the radius of the earth.* **[Agra 2000; Kanpur 2007, Meerut 2013, 2019]**

Sol. Refer figure of Ex. 4(a). Here $OA = r$ (given). When variation of gravity in taken into account, let P be the position of the particle at any time t measured from the instant the particle is projected vertically upwards from the earth's surface with velocity v and let $OP = x$. Then acceleration of the particle at P is μ/x^2 towards O. Hence the equation of motion of the particle at P is

$$d^2x/dt^2 = -\mu/x^2 \qquad ...(1)$$

But at A, $x = r$ and $d^2x/dt^2 = -g$. So (1) gives $\mu = gr^2$

$\therefore$ From (1), $v(dv/dx) = -(gr^2/x^2)$ or $2vdv = -(2gr^2/x^2)dx$

Integrating, $v^2 = (2gr^2/x) + A$, A is a constant. ...(2)

Suppose the particle in this case rises upto the point B where $AB = H$. Then at B, $x = OB = OA + AB = r + H$ and $v = 0$. So (2) gives

$$0 = (2gr^2)/(r + H) + A \quad \text{or} \quad A = -(2gr^2)/(r + H).$$

$\therefore$ (2) gives, $v^2 = 2gr^2\{1/x - 1/(r + H)\}$ ...(3)

On the surface of the earth, *i.e.*, at A, $v = v$ (given) and $x = r$.

$\therefore$ (3) gives $v^2 = 2gr^2\ \{1/r - 1/(r + H)\}$ ...(4)

Next, suppose the acceleration due to gravity to be constant and equal to g. Then, by given problem, using $v^2 = u^2 - 2gh$, we get $0 = v^2 - 2gh$.

or $v^2 = 2gh$ ...(5)

From (4) and (5), $2gh = 2gr^2\{1/r - 1/(r + H)\}$

or $h = \dfrac{Hr}{H + r}$ or $\dfrac{1}{h} = \dfrac{H + r}{rH}$ or $\dfrac{1}{h} - \dfrac{1}{H} = \dfrac{1}{r}$

Ex. 9. *If the earth's attraction vary inversely as the square of the distance from its centre and g be its magnitude at the surface, the time of falling from a height h above the surface to the surface is*

$$\left(\frac{a+h}{2g}\right)^{1/2}\left\{\left(\frac{h}{a}\right)^{1/2} + \frac{a+h}{a}\sin^{-1}\left(\frac{h}{a+h}\right)^{1/2}\right\},$$ *where a is radius of the earth.*

[Lucknow 2001]

Sol. Refer figure of Ex. 4(a). Let O be the centre and a the radius of the earth. Let a vertical line OB through O meet the earth's surface at A, where $AB = h$. Let a particle fall from rest from B towards O along BO. Let P be its position at any time t such that $OP = x$. Then acceleration of the particle at P is μ/x^2. Hence the equation of motion of the particle at P is

$$d^2x/dt^2 = -\mu/x^2 \qquad ...(1)$$

At A, $x = a$ and $d^2x/dt^2 = -g$. So (1) gives $\mu = ga^2$.

$\therefore$ (1) becomes $v(dv/dx) = -(ga^2/x^2)$ or $2vdv = -(2ga^2/x^2)dx$.

Integrating, $v^2 = (2ga^2)/x + A$, A is a constant ...(2)

Initially at B, when $x = a + h = b$, (say), $v = 0$. So (2) gives

$$O = (2ga^2)/b + A \quad \text{or} \quad A = -(2ga^2)/b$$

$\therefore$ (2) gives $\left(\dfrac{dx}{dt}\right)^2 = 2ga^2\left(\dfrac{1}{x} - \dfrac{1}{b}\right)$ or $\dfrac{dx}{dt} = -a(2g)^{1/2}\left(\dfrac{1}{x} - \dfrac{1}{b}\right)^{1/2}$

[–ve sign is taken because as t increases, x decreases]

Separating variables, $$dt = -\frac{1}{a}\left(\frac{b}{2g}\right)^{1/2}\left(\frac{x}{b-x}\right)^{1/2} dx \qquad ...(3)$$

Let t_1 be time taken in moving from B to A. Then integrating (3) between the limits $x = a + h = b$ and $x = a$, we have

$$t_1 = \frac{1}{a}\left(\frac{b}{2g}\right)^{1/2}\int_b^a \left(\frac{x}{b-x}\right)^{1/2} dx \qquad ...(4)$$

Putting $x = b\cos^2\theta$ and $dx = -2b\cos\theta\sin\theta\, d\theta$, (4) gives

$$t_1 = \frac{1}{a}\left(\frac{b}{2g}\right)^{1/2} \int_0^{\cos^{-1}(a/b)^{1/2}} \left(\frac{b\cos^2\theta}{b\sin^2\theta}\right)^{1/2} . 2b\cos\theta\sin\theta\, d\theta$$

$$= \left(\frac{b}{a}\right)\left(\frac{b}{2g}\right)^{1/2} \int_0^{\cos^{-1}(a/b)^{1/2}} (1+\cos 2\theta)\, d\theta = \frac{b}{a}\left(\frac{b}{2g}\right)^{1/2}\left[\theta + \frac{1}{2}\sin 2\theta\right]_0^{\cos^{-1}(a/b)^{1/2}}$$

$$= \frac{b}{a}\left(\frac{b}{2g}\right)^{1/2} [\theta + \sin\theta\cos\theta]_0^{\cos^{-1}(a/b)^{1/2}} = \frac{b}{a}\left(\frac{b}{2g}\right)^{1/2}\left[\theta + \cos\theta\,(1-\cos^2\theta)^{1/2}\right]_0^{\cos^{-1}(a/b)^{1/2}}$$

$$= (b/a)\,(b/2g)^{1/2}\,\{\cos^{-1}(a/b)^{1/2} + (a/b)^{1/2}\,(1-a/b)^{1/2}\}$$

$$= (b/2g)^{1/2}\,\{(b/a)\cos^{-1}(a/b)^{1/2} + (b/a)^{1/2}\,(1-a/b)^{1/2}\}$$

$$= \left(\frac{a+h}{2g}\right)^{1/2}\left[\frac{a+h}{a}\cos^{-1}\left(\frac{a}{a+h}\right)^{1/2} + \left(\frac{a+h}{a}\right)^{1/2}\left\{1-\frac{a}{a+h}\right\}^{1/2}\right]$$

[Noting that $b = a + h$ as assumed]

$$= \left(\frac{a+h}{2g}\right)^{1/2}\left[\frac{a+h}{a}\sin^{-1}\left(1-\frac{a}{a+h}\right)^{1/2} + \left(\frac{a+h}{a}\right)^{1/2}\left(\frac{h}{a+h}\right)^{1/2}\right]$$

$$= \left(\frac{a+h}{2g}\right)^{1/2}\left[\frac{a+h}{a}\sin^{-1}\left(\frac{h}{a+h}\right)^{1/2} + \left(\frac{h}{a}\right)^{1/2}\right]$$

Ex. 10. *Assuming that a particle falling freely under gravity can penetrate the earth without meeting any resistance, show that a particle falling from rest at a distance b ($b > a$) from the centre of the earth would on reaching the centre acquire a velocity $\{ga\,(3b-2a)/b\}^{1/2}$ and the time to a travel from the surface to the centre of the earth is $(a/g)^{1/2}\sin^{-1}\{b/(3b-2a)\}^{1/2}$ where a is the radius of the earth.* **[Garhwal 93, Kanpur 95, Lucknow 98]**

Sol. Refer figure of Ex. 4(*a*). Let O be the centre and let a vertical line OB through O meet the earth's surface at A, where $OA = a$ and $OB = b$. Let be a particle fall from B towards O along BO. Let P be its position at any time t such that $OP = x$. Then acceleration of particle at P is μ/x^2. Hence the equation of motion of the particle is

$$d^2x/dt^2 = -\mu/x^2 \qquad ...(1)$$

At A, $x = a$ and $d^2x/dt^2 = -g$. So (1) gives $\mu = ga^2$

$\therefore$ (1) becomes $\quad v(dv/dx) = -(ga^2/x^2) \quad$ or $\quad 2v\,dv = -(2ga^2/x^2)dx$

Integrating, $\quad v^2 = (2ga^2/x) + A$, A is a constant ...(2)

Initially at B when $x = b$, $v = O$, So (2) gives

$0 = (2ga^2)/b + A \quad$ or $\quad A = -(2ga^2)/b$

$\therefore$ (2) gives $\quad v^2 = 2ga^2(1/x - 1/b)$ ...(3)

Let V be velocity of the particle at A so that when $x = a$, $v = V$. Then by (3)

$$V^2 = 2ga^2(1/a - 1/b) \qquad ...(4)$$

Motion of the particle inside the earth. Refer figure of Ex. 5. When the particle penetrates inside the earth the law of acceleration changes. Let Q be the position of the particle at any time t (while moving from A towards O). Then acceleration of the particle at Q is λx towards O. Hence its equation of motion is $\quad d^2x/dt^2 = -\lambda x$ ...(5)

At A, when $\quad x = a$, $d^2x/dt^2 = -g$. $\quad$ So (5) gives $\lambda = g/a$.

$\therefore$ (5) gives $\quad v(dv/dx) = -(gx/a) \quad$ or $\quad 2v\,dv = -(2gx/a)\,dx$.

Integrating $\quad v^2 = -(gx^2/a) + B$, B is a constant ...(6)

For motion from A to O, initially at A, when $x = a$, $v = V$. So (6) gives

$V^2 = -ga + B \quad$ or $\quad B = V^2 + ga$

$\therefore$ (6) gives $\quad v^2 = -(gx^2/a) + V^2 + ga$

or $\quad v^2 = -(gx^2/a) + 2ga^2\,(1/a - 1/b) + ga$, using (4)

or $\quad v^2 = ag(3b - 2a)/b - (gx^2/a)$ ...(7)

Let V_1 be velocity in reaching the centre O of the earth so that when $x = 0$, $v = 0$. Then (7) gives

$$V_1^2 = ag(3b-2a)/b \quad \text{or} \quad V_1 = \{ga(3b-2a)/b\}^{1/2}$$

Re-writing (7), $$\left(\frac{dx}{dt}\right)^2 = \frac{g}{a}\left\{\frac{a^2(3b-2a)}{b} - x^2\right\} = \frac{g}{a}(c^2 - x^2) \qquad ...(8)$$

where $$c^2 = (a^2/b)(3b-2a) \qquad ...(9)$$

From (8), $$dx/dt = -(g/a)^{1/2}(c^2 - x^2)^{1/2} \qquad ...(10)$$

[–ve sign is taken because as t increase, x decreases]

From (10), $$dt = -(a/g)^{1/2}\{dx/(c^2 - x^2)^{1/2}\} \qquad ...(11)$$

Let time taken from A to O be t'. So integrating (11) between $x = a$ to $x = 0$,

$$t' = -\left(\frac{a}{g}\right)^{1/2}\int_a^0 \frac{dx}{(c^2 - x^2)^{1/2}} = -\left(\frac{a}{g}\right)^{1/2}\left[\sin^{-1}\frac{x}{c}\right]_a^0 = \left(\frac{a}{g}\right)^{1/2}\sin^{-1}\frac{a}{c}$$

$$\therefore \quad \text{By (9)}, \; t' = \left(\frac{a}{g}\right)^{1/2}\sin^{-1}\frac{a}{\{(a^2/b)(3b-2a)\}^{1/2}} = \left(\frac{a}{g}\right)^{1/2}\sin^{-1}\left\{\frac{b}{(3b-2a)}\right\}^{1/2}$$

3.4 Rectilinear motion under miscellaneous laws of forces

Read article 3.1. We now present motion of a particle moving under a given law of force (or acceleration).

Ex. 1. *A particle moves under an acceleration varying as the distance and directed away from a fixed point. Discuss the motion.* **[Agra 2000]**

Or Discuss the motion $\ddot{x} = \mu x$ *and show that* $t = (1/\mu^{1/2})\log\{x + (x^2 - a^2)^{1/2}\}/a$ *where a is the distance, when the velocity is zero. Assume that* $\mu > 0$ **[Gorakhpur 1997]**

Sol. Let O be the fixed point on OX and A be the initial position of the particle so that $OA = a$. Let P be its position at any time t such that $OP = x$. Then the acceleration of the particle at P is μx along OP. Hence its equation of motion is

O ← a → A ← x → P X

$$d^2x/dt^2 = \mu x \quad \text{or} \quad v(dv/dx) = \mu x \quad \text{or} \quad 2v\,dv = 2\mu x\,dx.$$

Integrating, $$v^2 = \mu x^2 + A, \; A \text{ is a constant} \qquad ...(1)$$

Initially, when $x = a$, let $v = 0$. So (1) gives $A = -\mu a^2$.

$\therefore$ (1) gives $(dx/dt)^2 = \mu x^2 - \mu a^2$ or $dx/dt = \mu^{1/2}(x^2 - a^2)^{1/2}$

[+ve sign is taken because as t increases, x increases]

or $$dt = \frac{1}{\mu^{1/2}}\frac{dx}{(x^2 - a^2)^{1/2}} \quad \text{so that } t = \frac{1}{\mu^{1/2}}\log\frac{x + (x^2 - a^2)^{1/2}}{a} + B \qquad ...(2)$$

Initially, when $x = a$, $t = 0$. So (2) gives $B = 0$.

Then, (2) gives $$t = (1/\mu^{1/2})\log\{x + (x^2 - a^2)^{1/2}\}/a$$

Ex. 2. *If a particle is projected towards the centre of repulsion, varying as the distance from the centre, from a distance a from it with a velocity* $a\mu^{1/2}$*: prove that the particle will approach the centre but will never reach it.* **[Kumaun 1995]**

Sol. Refer figure of Ex. 1. Proceed as in Ex. 1. upto equation (1). Now proceed as follows. Initially at A, when $x = a$, $v = -a\mu^{1/2}$. Then (1) gives $A = 0$.

$\therefore$ (1) gives $(dx/dt)^2 = \mu x^2$ or $dx/dt = -\mu^{1/2}x$

[–ve sign is taken because as t increases, here x decreases]

Integrating it between the limits $x = a$ to $x = 0$, the time t_1 from A to 0 is given by

$$\int_0^{t_1} dt = -\frac{1}{\mu^{1/2}}\int_a^0 \frac{dx}{x} \quad \text{or } t_1 = -\frac{1}{\mu^{1/2}}[\log x]_a^0 = \frac{1}{\mu^{1/2}}(\log a - \log 0]$$

Then, $t_1 = \infty$, as $\log 0 = -\infty$. Thus, the particle will reach the centre O after infinite time *i.e.*, the particle will never reach the centre O.

Ex. 3(*a*). *A particle starts from rest at a distance a from the centre of force which attracts inversely as the distance. Prove that the time of arriving at the centre is a* $(\pi/2\mu)^{1/2}$.

[Garhwal 1995, I.F.S. 2003]

(b) A particle starts from rest at a distance 2c from the centre which attracts inversely as the distance. Find the time of arriving at the centre. **[Meerut 2017]**

Sol. (*a*) Refer figure of Ex.1. Let O he the centre of force and A be the initial position such that $OA = a$. Let P be its position at any time t such that $OP = x$. Then its equation of motion is

$d^2x/dt^2 = -\mu/x$ or $v(dv/dx) = -\mu/x$ or $2vdv = (-2\mu/x)\,dx$.

Integrating, $$v^2 = -2\mu \log x + A, \text{ where } A \text{ is a constant} \quad ...(1)$$

Initially at A, $x = a, v = 0$. So (1) gives $A = 2\mu \log a$.

$\therefore$ (1) gives $$(dx/dt)^2 = 2\mu(\log a - \log x) = 2\mu \log (a/x)$$

or $$dx/dt = -(2\mu)^{1/2}\{\log (a/x)\}^{1/2}$$

[–ve sign is taken because as t increase, here x decreases]

Separating variables, $$dt = -\frac{1}{(2\mu)^{1/2}} \frac{dx}{\{\log(a/x)\}^{1/2}}$$

Integrating between the limits $x = a$ to $x = 0$, the time t_1 from A to O is given by

$$\int_0^{t_1} dt = -\frac{1}{(2\mu)^{1/2}} \int_a^0 \frac{dx}{\{\log(a/x)\}^{1/2}} = -\frac{1}{(2\mu)^{1/2}} \int_0^{\infty} \frac{-2az^2\,dz}{z}$$

[Putting $\log (a/x) = z^2$ or $x = ae^{-z^2}$ and $dx = -2aze^{-z^2}dz$]

or $$t_1 = \frac{2a}{(2\mu)^{1/2}} \int_0^{\infty} e^{-z^2} dz = \frac{2a}{(2\mu)^{1/2}} \frac{\sqrt{\pi}}{2}, \text{ as } \int_0^{\infty} e^{-z^2} dz = \frac{\sqrt{\pi}}{2}$$

Thus, the required time $= t_1 = a(\pi/2\mu)^{1/2}$

(*b*) Do as in part (*a*). Here $a = 2c$. Hence, $t_1 = 2c(\pi/2\mu)^{1/2}$

Ex. 4 (*a*). *A particle moves in such a way that its acceleration varies inversely as the cube of the distance from a fixed point and is directed towards the fixed point; discuss the motion.*

(*b*) *A particle moves in a straight line from a distance a towards the centre of force, the force being μ/x^3 at a distance x from the centre. Show that it reaches the centre in time $a^2/\mu^{1/2}$.*

Sol. (*a*) Refer figure of Ex. 1. Let O be the centre of force and A be the initial position, where $OA = a$. Let P be its position at any time such that $OP = x$. Then its equation of motion is

$d^2x/dt^2 = -\mu/x^3$ or $v(dv/dx) = -\mu/x^3$ or $2vdv = -(2\mu/x^3)dx$

Integrating, $$v^2 = \mu/x^2 + A, \quad A \text{ is a constant.} \quad ...(1)$$

Initially, when $x = a$, let $v = 0$. So (1) gives $A = -\mu/a^2$.

$\therefore$ (1) gives $(dx/dt)^2 = \mu(1/x^2 - 1/a^2)$ or $dx/dt = -\mu^{1/2}(1/x^2 - 1/a^2)^{1/2}$

[–ve sign is taken because as t increases, here x decreases]

or $$dt = -\frac{a}{\mu^{1/2}} \frac{xdx}{(a^2 - x^2)^{1/2}} = \frac{a}{2\mu^{1/2}}(a^2 - x^2)^{-1/2}(-2x)\,dx$$

Integrating, $$t = (a/\mu^{1/2})(a^2 - x^2)^{1/2} + B, B \text{ is a constant} \quad ...(2)$$

Initially at A, when $x = a$, $t = 0$. So (2) gives $B = 0$

$\therefore$ (2) becomes $$t = (a/\mu^{1/2})(a^2 - x^2)^{1/2} \quad ...(3)$$

giving the position of the particle at any time t.

(*b*) Proceed as in part (*a*) upto equation (3). For the required time from A to O, put $x = 0$ in (3) and get $t = a^2/\mu^{1/2}$.

Ex. 4 (*c*). *A particle moves in a straight line towards a centre of force $\mu/(\text{distance})^3$ starting from rest at a distance a from the centre of force; show that the time of reaching a point distant b from the centre of force is $a\{(a^2 - b^2)/\mu\}^{1/2}$ and its velocity then is $\{\mu(a^2 - b^2)\}^{1/2}/ab$. Also show that the time to reach the centre is $a^2/\mu^{1/2}$.* **[Avadh 2011]**

Sol. Refer figure of Ex.1. Let the centre of force O be taken as origin. Let P be the position of the particle at any time t such that $OP = x$. Let the particle start from rest A where $OA = a$. Since the acceleration of particle at P is μ/x^3 towards O, we have

$$v(dv/dx) = -\mu/x^3 \quad \text{or} \quad 2v\,dv = -2\mu x^{-3}dx.$$

Integrating, $v^2 = \mu/x^2 + A$, A is a constant. ...(1)

Initially, when $x = a$ $v = 0$. So (1) gives $A = -\mu/a^2$

$\therefore$ (1) gives $(dx/dt)^2 = \mu(1/x^2 - 1/a^2)$ or $dx/dt = -\{\mu(a^2 - x^2)\}^{1/2}/ax$ ...(2)

Putting $x = b$ in (2), the velocity V at distance b from O is given by

$$V = \{\mu(a^2 - b^2)^{1/2}\}/ab \quad \text{(in magnitude)}$$

From (2) $\quad dt = -[ax/\{\mu(a^2 - x^2)^{1/2}\}]dx$ ...(3)

Integrating it between the limits $x = a$ to $x = b$, the time t_1 taken from $x = a$ to $x = b$ is given by

$$t_1 = -\frac{a}{\mu^{1/2}}\int_a^b \frac{x\,dx}{(a^2 - x^2)} = \frac{a}{2\mu^{1/2}}\int_a^b (a^2 - x^2)^{-1/2}(-2x)dx$$

$$= (a/2\mu^{1/2})[2(a^2 - x^2)^{1/2}]_a^b = a(a^2 - b^2)^{1/2}/\mu^{1/2}$$

Let t_2 be time taken in moving from A to O. Then integrating (3) between the limits $x = a$ to $x = 0$, the required time t_2 in given by

$$t_2 = -\frac{a}{\mu^{1/2}}\int_a^O \frac{x\,dx}{(a^2 - x^2)^{1/2}} = (a/2\mu^{1/2})\left[2(a^2 - x^2)^{1/2}\right]_a^0, \text{ as before}$$

or $\quad t_2 = (a/2\mu^{1/2}) \times (2a) = a^2/\mu^{1/2}$

Ex. 5 (*a*). *A particle whose mass is m is acted upon by a force $m\mu\,(x + a^4/x^3)$ towards the origin; if it starts from rest at a distance a show that it will arrive at the origin in time $\pi/(4\mu^{1/2})$.*

[Agra 1997; Rohilkhand 2001, 02, 11; Bundelkhand 94, Kolkata 93, Gorakhpur 96, 98; Garhwal 94, 96, Kumaun 97, 98: Kanpur 97; Lucknow 2010, Purvanchal 2010]

(*b*) *A particle of mass m is acted upon by a force $m(x + a^4/x^3)$ towards the origin. If it starts from rest at a distance a from the origin, show that the time taken by it to reach the origin is $\pi/4$.*

[I.A.S. 2001]

Sol. Refer figure of Ex. 1. Let centre of force O be taken as origin and let A be the initial position of the particle such that $OA = a$. Let P be its position at any time t such that $OP = x$. Then its equation of motion is

$$v(dv/dx) = -\mu(x + a^4/x^3) \quad \text{or} \quad 2v\,dv = -2\mu(x + a^4/x^3)\,dx$$

Integrating, $v^2 = \mu(-x^2 + a^4/x^2) + A$, A is a constant. ...(1)

Initially, when $x = a$, $v = 0$ so (1) gives $A = 0$.

$\therefore$ By (1), $(dx/dt)^2 = \mu(a^4 - x^4)/x^2$ or $dx/dt = -\mu^{1/2}(a^4 - x^4)^{1/2}/x$ ...(2)

[–ve sign is taken because as t increases, x decreases]

Integrating (2) between the limits $x = a$ to $x = 0$, the time t_1 from A to O is given by

$$\int_a^{t_1} dt = -\frac{1}{\mu^{1/2}}\int_a^0 \frac{x\,dx}{(a^4 - x^4)^{1/2}} = -\frac{1}{\mu^{1/2}}\int_{\pi/2}^0 \frac{(1/2)a^2\cos\theta\,d\theta}{a^2\cos\theta} = \frac{1}{2\mu^{1/2}}[\theta]_0^{\pi/2}$$

[Putting $x^2 = a^2\sin\theta$ and $2x\,dx = a^2\cos\theta\,d\theta$]

$\therefore$ $\quad t_1 = (1/2\mu^{1/2}) \times (\pi/2) = \pi/4\mu^{1/2}$ ...(3)

(*b*) Here $\mu = 1$. Hence required time $= \pi/4$, using (3)

Ex. 6. *A particle moves in a straight line, its acceleration directed towards a fixed point O in the line and is always equal to $\mu(a^5/x^2)^{1/3}$ when it is at a distance x from O. If it starts from rest at a distance a from O, show that it will arrive at O with a velocity $a(6\mu)^{1/2}$ after time $(8/15)(6/\mu)^{1/2}$.*

[Garhwal 2001]

Sol. Refer figure of Ex.1. Let centre of force O be taken as origin and let A be the initial position of the particle such that $OA = a$. Let P be its position at any time t such that $OP = x$. Then its equation of motion is

$$v(dv/dx) = -\mu a^{5/3}x^{-2/3} \quad \text{or} \quad 2v\,dv = -2\mu a^{5/3}x^{-2/3}\,dx.$$

Integrating, $v^2 = -6\mu a^{5/3}x^{1/3} + A$, A is a constant ...(1)

Initially at A, when $x = a$, $v = 0$ (given). So (1) $\Rightarrow A = 6\mu a^2$.

$\therefore$ (1) gives $(dx/dt)^2 = -6\mu a^{5/3}x^{1/3} + 6\mu a^2 = 6\mu\,a^{5/3}(a^{1/3} - x^{1/3})$ ...(2)

Suppose the particle arrives at O with velocity v_1. Then when $x = 0$, $v = v_1$ and so (2) gives

$v_1^2 = 6\mu a^2$ or $v_1 = a(6\mu)^{1/2}$.

From (2), $dx/dt = -(6\mu a^{5/3})^{1/2}(a^{1/3} - x^{1/3})^{1/2}$ ...(3)

[–ve sign is taken because as t increases, x decreases]

Integrating (3) between the limits $x = a$ to $x = 0$, the time t_1 taken from A to O is given by

$$\int_0^{t_1} dt = -\frac{1}{(6\mu\, a^{5/3})^{1/2}} \int_a^0 \frac{dx}{(a^{1/3} - x^{1/3})^{1/2}} = \frac{1}{(6\mu\, a^{5/3})^{1/2}} \int_0^{\pi/2} \frac{6a\cos^5\theta \sin\theta \, d\theta}{a^{1/6} . \sin\theta}$$

[Putting $x = a\cos^6\theta\, d\theta$ and $dx = -6a\cos^5\theta \sin\theta\, d\theta$]

$$\therefore \quad t_1 = \left(\frac{6}{\mu}\right)^{1/2} \int_0^{\pi/2} \cos^5\theta\, d\theta = \left(\frac{6}{\mu}\right)^{1/2} \times \frac{4.2}{5.3.1} = \frac{8}{15}\left(\frac{6}{\mu}\right)^{1/2}$$

Ex. 7. *A particle, moving in a straight line, is subject to a retardation of amount kv^n per unit mass, where v is the speed at time t. Show that, if $n < 1$, particle will come to rest at a distance $u^{2-n}/k(2-n)$ from the point of projection at time $u^{1-n}/k(1-n)$, where u is the initial speed. What happens when*

(i) $1 < n < 2$ (ii) $n > 2$? **[Delhi Maths (H) 2001, 2011]**

Sol. Let x be the distance described by the particle after a time t and v the velocity attained. Since its retardation is kv^n, its equation of motion is

$v(dv/dx) = -kv^2$ or $v^{1-n}dv = -k\,dx$ ...(1)

Integrating, $v^{2-n}/(2-n) = -kx + A$, A is a constants, $n \neq 2$ ...(2)

Initially, when $x = 0$, $v = u$ (given), so (2) $\Rightarrow A = u^{2-n}/(2-n)$

$\therefore$ (1) gives $x = (u^{2-n} - v^{2-n})/k(2-n)$, $n \neq 2$ ...(3)

Let the particle come to rest when $x = x_1$. So putting $x = x_1$ and $v = 0$ in (3),

the required distance $= x_1 = u^{2-n}/k(2-n)$

Since $v(dv/dx) = dv/dt$, (1) becomes

$dv/dt = -kv^n$ or $v^{-n}dv = -kdt$

Integrating, $v^{1-n}/(1-n) = -kt + B$, B is a constant, $n \neq 1$. ...(4)

Initially, when $t = 0$, $v = u$. So (4) $\Rightarrow B = u^{1-n}/(1-n)$

$\therefore$ (4) gives $t = (u^{1-n} - v^{1-n})/k(1-n)$, $n \neq 1$ ...(5)

Let the particle come to rest when $t = t_1$. So putting $t = t_1$ and $v = 0$ in (5), the required time $= t_1 = u^{1-n}/k(1-n)$, where $n \neq 1$.

Case I. Let $1 < n < 2$ and let $n = m + 1$ so that $0 < m < 1$.

Then (3) and (5) may be re-written as

$x = (u^{1-m} - v^{1-m})/k(1-m)$ ...(6)

and $t = -(u^{-m} - v^{-m})/\text{km}$ so that $v^m = 1/(u^{-m} + kt\,m)$ ...(7)

From (7), we find that $v \to 0$ as $t \to \infty$ and from (6) we find that $v \to 0$ as $x \to u^{1-m}/k(1-m)$.

Thus, when $1 < n < 2$, we see that the particle tends ultimately to rest at a finite distance $u^{1-m}/k(1-m)$ after an infinite time.

Case II. Let $n > 2$ and let $n = 2 + r$ when $r > 0$. Then,

by (3), $x = -(u^{-r} - v^{-r})/kr$ ...(8)

and by (5), $t = -(u^{-r-1} - v^{-r-1})/k(r+1)$ ...(9)

From (9), $kt(r+1) = -\dfrac{1}{u^{r+1}} + \dfrac{1}{v^{r+1}}$ or $v^{r+1} = \dfrac{1}{kt\,(r+1) + 1/u^{r+1}}$

$\therefore$ $v \to 0$ as $t \to \infty$ and from (8) as $v \to 0$, $x \to \infty$.

Thus in this case the particle will come to rest at an infinite distance and at an infinite period of time *i.e.,*, the particle will never come to rest in a finite time or after finite displacement.

Ex. 8. *A particle starts from rest and moves along a straight line with an acceleration f varying as t^n. If v be the velocity at a distance s from the starting point, shows that $(n+1)v^2 = (n+2)fs$.*

Sol. Here, f = acceleration $= d^2s/dt^2 = kt^n$ ...(1)

Integrating (1), $ds/dt = (kt^{n+1})/(n+1) + A$, A is a constant. ...(2)

Initially, when $t = 0$, ds/dt = velocity = 0 (given) so (2) $\Rightarrow A = 0$

$\therefore$ (2) given $\quad ds/dt = v = (kt^{n+1})/(n+1)$ ...(3)

Re-writing (3), $\quad ds = \{k/(n+1)\}\, t^{n+1}\, dt.$

Integrating, $\quad s = (kt^{n+2})/(n+1)(n+2)\, B$, B is constant ...(4)

Initially, when $t = 0$, $s = 0$. So (4) $\Rightarrow B = 0$.

$\therefore$ (4) gives $\quad s = (kt^{n+2})/(n+1)(n+2)$ ...(5)

Now, $\quad (n+1)v^2 = (n+1) \times \{k^2 t^{2n+2}/(n+1)^2\}$, using (3)

$\quad = kt^n \times \{kt^{n+2}/(n+1)\} = f(n+2)s$, by (1) and (5)

Thus, $\quad (n+1)\, v^2 = (n+2)\, fs$, as required.

Ex. 9. *A particle moves along the axis of x starting from rest at x = a. For an interval t_1 from the beginning of the motion the acceleration is $-\mu x$, for a subsequent time t_2 the acceleration is μx, and at the end of this interval the particle is at the origin, prove that $\tan(t_1\mu^{1/2}) \tanh(t_2\mu^{1/2}) = 1$.*

[Meerut 2005, 2014]

Sol. Let P be the position of the particle at any time t such that $OP = x$. Let A be its initial position and let $OA = a$. Let t_1 be time taken by the particle to move from A to B and t_2 be time from B to origin O such that $OB = b$.

Motion from A to B. Equation of motion is

$$v(dv/dx) = -\mu x \quad \text{or} \quad 2v\,dv = -2\mu x\,dx.$$

Integrating, $\quad v^2 = -\mu x^2 + A$, A is a constant ...(1)

Initially when $x = a$, $v = 0$ (given). So (1) $\Rightarrow A = \mu a^2$

$\therefore$(1) becomes $(dx/dt)^2 = \mu(a^2 - x^2)$ or $dx/dt = -\mu^{1/2}(a^2 - x^2)^{1/2}$ ...(2)

[$-$ve sign is taken because as t increases, x decreases]

Integrating (2) between $x = a$ to $x = b$, t_1 is given by

$$t_1 = -\frac{1}{\mu^{1/2}}\int_a^b \frac{dx}{(a^2 - x^2)^{1/2}} = \frac{1}{\mu^{1/2}}\left[\cos^{-1}\frac{x}{a}\right]_a^b = \frac{1}{\mu^{1/2}}\cos^{-1}\frac{b}{a} = \frac{1}{\mu^{1/2}}\tan^{-1}\frac{(a^2 - b^2)^{1/2}}{b}$$

$\therefore \quad \tan(t_1\mu^{1/2}) = (1/b)(a^2 - b^2)^{1/2}$ ...(3)

It V is the velocity at B, the by (2) $\quad V^2 = \mu(a^2 - b^2)$ ...(4)

For motion from B to O, Equation of motion is

$$v(dv/dx) = \mu x \quad \text{or} \quad 2v\,dv = 2\mu x\,dx$$

Integrating $\quad v^2 = \mu x^2 + B$, B is a constant. ...(5)

But at B, $x = b$ and $v = V = \mu^{1/2} \cdot (a^2 - b^2)^{1/2}$, so (5) gives

$$\mu(a^2 - b^2) = \mu b^2 + B \text{ so that } B = \mu(a^2 - 2b^2)$$

$\therefore$(4) becomes $(dx/dt)^2 = \mu x^2 + \mu(a^2 - 2b^2) = \mu(x^2 + a^2 - 2b^2)$

$$\text{or} \quad \frac{dx}{dt} = -\mu^{1/2}(x^2 + a^2 - 2b^2)^{1/2} \quad \text{or} \quad dt = -\frac{1}{\mu^{1/2}}\frac{dx}{\{x^2 + (a^2 - 2b^2)^{1/2}\}}.$$

Integrating between the limits $x = b$ to $x = 0$ for motion from B to O, the time t_2 from B to O is given by

$$t_2 = -\frac{1}{\mu^{1/2}}\int_b^0 \frac{dx}{\{x^2 + (a^2 - 2b^2)^{1/2}\}} = -\frac{1}{\mu^{1/2}}\left[\sinh^{-1}\frac{x}{(a^2 - 2b^2)^{1/2}}\right]_b^0$$

or $\quad t_2 = (1/\mu^{1/2})\sinh^{-1}\{b/(a^2 - 2b^2)^{1/2}\}$ or $\sinh(t_2\mu^{1/2}) = b/(a^2 - 2b^2)^{1/2}$

$\therefore \quad \cosh^2(t_2\mu^{1/2}) = 1 + \sinh^2(t_2\mu^{1/2}) = 1 + b^2/(a^2 - 2b^2) = (a^2 - b^2)/(a^2 - 2b^2)$

Thus, $\quad \sinh(t_2\mu^{1/2}) = b/(a^2 - 2b)^{1/2}$, $\cosh(t_2\mu^{1/2}) = (a^2 - b^2)^{1/2}/(a^2 - 2b^2)^{1/2}$

Dividing, $\quad \tanh(t_2\mu^{1/2}) = b/(a^2 - b^2)^{1/2}$ ...(6)

Multiplying (3) and (6), $\tan(t_1\mu^{1/2})\tanh(t_2\mu^{1/2}) = 1$

Ex. 10. **(*a*)** *A particle moves towards a centre of attraction starting from rest at a distance 'a' from the centre. If its velocity when at a distance 'x' from the centre varies as $\{a^2 - x^2)/x^2\}^{1/2}$, find the law of force.*

Sol. Since the particle moves towards the centre of attraction, hence according to problem, here

$$\frac{dx}{dt} = -\mu\left(\frac{a^2 - x^2}{x^2}\right)^{1/2} \quad \text{or} \quad \left(\frac{dx}{dt}\right)^2 = \mu^2\left(\frac{a^2}{x^2} - 1\right)$$

Differentiating both sides w.r.t. 't' we get

$$2\frac{dx}{dt}\frac{d^2x}{dt^2} = \mu^2\left(-\frac{2a^2}{x^3}\right)\frac{dx}{dt} \quad \text{or} \quad \frac{d^2x}{dt^2} = -\frac{a^2\mu^2}{x^3},$$

showing that the law of force is that of inverse cube of the distance and is directed towards the centre because this acceleration is negative.

Ex. 10. (*b*) *Assuming that at a distance x from a centre of force the speed v of a particle, moving in a straight line is given by the equation* $x = ae^{bv^2}$*, where a and b are constants. Find the law and nature of force.*

Sol. Here, $$x = a\,e^{bv^2}. \quad \text{...(1)}$$

Differentiating both sides of (1) w.r.t. 'x', we get

$$1 = ae^{bv^2}.\,2bvx(dv/dx) = 2bvx(dv/dx), \text{ using (1)}$$

or $v(dv/dx) = 1/(2bx)$ or $d^2x/dt^2 = 1/(2bx)$, $[\because v(dv/dx) = d^2x/dt^2]$

showing that the law of force is that of inverse distance and is directed away from the centre because this acceleration is positive.

Ex. 11. *A particle is attracted by a force to a fixed point varying inversely as (distance)^n. If the velocity acquired in falling from an infinite distance to a distance 'a' from the center be equal to the velocity acquired in falling from rest from distance 'a' to a distance 'a/4' prove that n = 3/2.*

Sol. Let the centre of attraction O be taken as origin. Let P be the position of the particle at any time such that $OP = x$. Then its equation of motion is

x
a/4
O B A P X
a

$$v(dv/dt) = -\mu/x^n \quad \text{or} \quad 2v\,dv = -2\mu x^{-n}dx$$

Integrating, $$v^2 = (2\mu)/(n-1)\,x^{n-1} + A, \; A \text{ is a constant} \quad \text{...(1)}$$

Let the particle fall from rest from infinity so that when $x = \infty$, $v = 0$. Then (1) gives $A = 0$. So (1) reduces to

$$v^2 = (2\mu)/(n-1)\,x^{n-1} \quad \text{...(2)}$$

Let V_1 be velocity of the particle at $x = a$. So by (2), we get

$$V_1^2 = (2\mu)/(n-1)\,a^{n-1} \quad \text{...(3)}$$

Next, consider motion of the particle starting from rest from $x = a$. Then when $x = a$, $v = 0$. So (1) gives $0 = (2\mu)/(n-1a^{n-1} + A$ so that $A = -(2\mu)/(n-1)a^{n-1}$

$\therefore$ (1) becomes, $$v^2 = \frac{2\mu}{n-1}\left(\frac{1}{x^{n-1}} - \frac{1}{a^{n-1}}\right) \quad \text{...(4)}$$

Let V_2 be the velocity in this case when $x = a/4$.

By (4), $$V_2^2 = \frac{2\mu}{n-1}\left\{\frac{1}{(a/4)^{n-1}} - \frac{1}{a^{n-1}}\right\} \quad \text{...(5)}$$

According to given problem, $V_1 = V_2$ or $V_1^2 = V_2^2$

$$\therefore \quad \frac{2\mu}{(n-1)\,a^{n-1}} = \frac{2\mu}{n-1}\left\{\frac{4^{n-1}}{a^{n-1}} - \frac{1}{a^{n-1}}\right\}$$

or $1 = 4^{n-1} - 1$ or $2^{2n} = 2^3$ or $2n = 3$ or $n = 3/2$

Ex. 12. *A particle moves in a straight line with an acceleration towards a fixed point in the straight line which is equal to ($\mu/x^2 - \lambda/x^3$) at a distance 'x' from the given point, the particle starts from rest at a distance 'a'. Show that it oscillates between this distance and the distance (λa)/($2\mu a - \lambda$) and the periodic time is ($2\pi\,\mu a^3$)/($2a\mu - \lambda$)$^{3/2}$.* **[Calcutta 2003]**

Sol. Refer figure of Ex. 9. Let O be the fixed point taken as origin and A the starting point such that $OA = a$. Let P be the position of the particle at any time t such that $OP = x$. Then the equation of motion of the particle is given by

$$v\frac{dv}{dx} = -\left(\frac{\mu}{x^2} - \frac{\lambda}{x^3}\right) \quad \text{or} \quad 2v\,dv = -\left(\frac{2\mu}{x^2} - \frac{2\lambda}{x^3}\right)dx$$

Integrating, $v^2 = 2\mu/x - \lambda/x^2 + A$, where A is a constant. ...(1)

Initially at A, when $x = a$, $v = 0$. Hence, from (1), we get

$0 = 2\mu/a - \lambda/a^2 + A$ or $A = -2\mu/a + \lambda/a^2$

$\therefore$ by (1), $v^2 = 2\mu(1/x - 1/a) - \lambda(1/x^2 - 1/a^2) = 2\mu(1/x - 1/a) - \lambda(1/x - 1/a)(1/x + 1/a)$

or $v^2 = (1/x - 1/a) \times \{2\mu - \lambda(1/x + 1/a)\} = \lambda(1/x - 1/a)(2\mu/\lambda - 1/x - 1/a)$

or $$v^2 = \lambda\left(\frac{1}{x} - \frac{1}{a}\right)\left(\frac{2a\mu - \lambda}{\lambda a} - \frac{1}{x}\right) = \lambda\left(\frac{1}{x} - \frac{1}{a}\right)\left(\frac{1}{b} - \frac{1}{x}\right) \quad ...(2)$$

where $1/b = (2a\mu - \lambda)/\lambda a$ *i.e.,* $b = \lambda a/(2a\mu - \lambda)$ ...(3)

Clearly, the particle comes to rest when $v = 0$. So, by (2)

$\lambda(1/x - 1/a)(1/b - 1/x) = 0$ giving $x = a$ and $x = b = \lambda a/(2a\mu - \lambda)$, by (3)

The above two values of x gives us the two positions of instantaneous rest. Thus the particle oscillates between $x = a$ and $x = \lambda a/(2\mu a - \lambda)$.

By (2), $$\left(\frac{dx}{dt}\right)^2 = \frac{\lambda}{ab}\frac{(a-x)(x-b)}{x^2} \quad \text{or} \quad \frac{dx}{dt} = -\left(\frac{\lambda}{ab}\right)^{1/2}\frac{\{(a-x)(x-b)\}^{1/2}}{x},$$

where –ve sign is taken because as t increases then x decreases]

or $dt = -(ab/\lambda)^{1/2} \times [x/\{(a-x)(x-b)\}^{1/2}]\,dx$ ...(4)

Let t_1 be the time taken by the particle in moving from one position of rest to another, *i.e.*, from $x = a$ to $x = b$. Then, integrating (4) between the limits $t = 0$ to $t = t_1$ and $x = a$ to $x = b$, we get

$$t_1 = -\left(\frac{ab}{\lambda}\right)^{1/2}\int_a^b \frac{x\,dx}{\{(a-x)(x-b)\}^{1/2}} = \left(\frac{ab}{\lambda}\right)^{1/2}\int_b^a \frac{x\,dx}{[-ab - \{x^2 - (a+b)x\}]^{1/2}}$$

$$= \left(\frac{ab}{\lambda}\right)^{1/2}\int_{x=b}^{a} \frac{x\,dx}{[(a-b)^2/4 - \{x - (a+b)/2\}^2]^{1/2}}$$

$$= \left(\frac{ab}{\lambda}\right)^{1/2}\int_{-(a-b)/2}^{(a-b)/2} \frac{\{(a+b)/2 + y\}\,dy}{[\{(a-b)/2\}^2 - y^2]^{1/2}}, \quad \text{putting } x - (a+b)/2 = y \text{ so that } dx = dy$$

$$= \left(\frac{ab}{\lambda}\right)^{1/2}\int_{-(a-b)/2}^{(a-b)/2} \frac{\{(a+b)/2\}\,dy}{[\{(a-b)/2\}^2 - y^2]^{1/2}} + \left(\frac{ab}{\lambda}\right)^{1/2}\int_{-(a-b)/2}^{(a+b)/2} \frac{y\,dy}{[\{(a-b)/2\}^2 - y^2]^{1/2}}$$

$$= \left(\frac{ab}{\lambda}\right)^{1/2} \times \frac{a+b}{2}\left[\sin^{-1}\frac{y}{(a-b)/2}\right]_{-(a-b)/2}^{(a-b)/2} + 0,$$

[the second integral vanishes because its integrand is an odd function.]

$$\therefore t_1 = \left(\frac{ab}{\lambda}\right)^{1/2} \times \frac{a+b}{2}\{\sin^{-1}1 - \sin^{-1}(-1)\} = \left(\frac{ab}{\lambda}\right)^{1/2} \times \frac{a+b}{2}(\pi/2 + \pi/2)$$

$\therefore$ The required periodic time of one complete oscillation

$= 2 \times$ (time taken to move from $x = a$ to $x = b$) $= 2t_1$

$$= 2 \times \frac{\pi(a+b)}{2} \times \left(\frac{ab}{\lambda}\right)^{1/2} = \pi\left(a + \frac{\lambda a}{2a\mu - \lambda}\right)\left(\frac{a}{\lambda} \times \frac{\lambda a}{2a\mu - \lambda}\right)^{1/2}, \text{ by (3)}$$

$$= \pi \times \frac{2a^2\mu}{2a\mu - \lambda} \times \frac{a}{(2a\mu - \lambda)^{1/2}} = \frac{2\pi\mu a^3}{(2a\mu - \lambda)^{3/2}}$$

Ex. 13. *A particle starts from rest at a distance b from a fixed point, under the action of a force through the fixed point, the law of which at a distance x in $\mu(1 - a/x)$ towards the point when $x > a$ but $\mu(a^2/x^2 - a/x)$ from the same point when $x < a$; prov that the particle will oscillate through a space $(b^2 - a^2)/b$.*

Sol. Let the fixed point be O and the particle start from B, such that $OB = b$ (given). The particle move towards the centre of force O. Let $OA = a$. Let P be the position of the particle at any time t such that $OP = x$.

O C A P(t) B
x

Consider the motion from B to A such that $x > a$. Since the law of foce when $x > a$ is given to be $\mu(1 - a/x)$ towards O, hence the equation of motion is given by

$$mv(dv/dx) = -\mu(1 - a/x) \quad \text{or} \quad 2v\,dv = -(2\mu/m)(1 - a/x)\,dx$$

Integrating, $\quad v^2 = -(2\mu/m) \times (x - a \log x) + A$, A is a constant. ...(1)

Initially at B, when $x = b$, $v = 0$. So (1) gives $A = (2\mu/m) \times (b - a \log b)$

So, from (1), $\quad v^2 = (2\mu/m) \times (b - a \log b - x + a \log x)$ ...(2)

Let v_1 be the velocity of the particle when it reaches A, where $x = a$. Then, from (2)

$$v_1^2 = (2\mu/m) \times (b - a \log b - a + a \log a) \quad ...(3)$$

When the particle crosses point A with velocity v_1 towards O, then it moves under the law of force $\mu(a^2/x^2 - a/x)$ at a distance x from O and directed away from O. Accordingly, the equation of motion (valid for $x < a$) is given by

$$mv(dv/dx) = \mu(a^2/x^2 - a/x) \quad \text{or} \quad 2v\,dv = (2\mu/m) \times (a^2/x^2 - a/x)\,dx$$

Integrating, $\quad v^2 = (2\mu/m) \times (-a^2/x - a \log x) + B$, B is a constant. ...(4)

But at the point A : When $x = a$, $v = v_1$. So, by (4)

$$v_1^2 = (2\mu/m) \times (-a - a \log a) + B \quad ...(5)$$

Subtracting (5) from (4), we have

$$v^2 - v_1^2 = (2\mu/m) \times (a + a \log a - a^2/x - a \log x) \quad ...(6)$$

If the particle comes to rest at the point C, where $x = c$, then putting $x = c$ and $v = 0$ in (6) we get

$$-v_1^2 = (2\mu/m) \times (a + a \log a - a^2/c - a \log c)$$

or $\quad -(2\mu/m) \times (b - a \log b - a + a \log a) = (2\mu/m) \times (a + a \log a - a^2/c - a \log c)$

or $\quad a \log b + a - b - a \log a = a + a \log a - a^2/c - a \log c$

or $\quad a^2/c + a \log c = b + a(2 \log a - \log b) \quad$ or $\quad a^2/c + a \log c = b + a \log (a^2/b)$

or $\quad \dfrac{a^2}{c} + a \log c = \dfrac{a^2}{(a^2/b)} + a \log (a^2/b) \Rightarrow c = a^2/b \Rightarrow OC = a^2/b$

But B and C are two positions of instantaneous rest of the particle. Hence the required distance through which the particle oscillates.

$$= BC = OB - OC = b - a^2/b = (b^2 - a^2)/b.$$

Ex. 14. *A particle of mass m moving in a straight line is acted upon by an attractive force which is expressed by the formula $m\mu a^2/x^2$ for values of $x \geq a$, and by the formula $m\mu x/a$ for $x \leq a$, where x is the distance from a fixed origin in the line. If the particle starts at a distance 2a from the origin, prove that it will reach the origin with velocity $(2\mu a)^{1/2}$. Prove further that the time taken to reach the origin is $(1 + 3\pi/4) \times (a/\mu)^{1/2}$.* **[Lucknow 1989]**

Sol. Let O be the fixed origin and the particle start from B, such that $OB = 2a$ (given). The particle moves towards the centre of force O. Let $OA = a$. Let P the position of the particle at any time t, such that $OP = x$.

O Q A P B
← x →

Motion from B to A. While moving from B to A, since the law of force for $x \geq a$ is given to be $m\mu a^2/x^2$ directed towards O, hence the equation of motion is given by

$$mv(dv/dx) = -\mu m a^2/x^2 \quad \text{or} \quad 2v\,dv = -\{(2\mu a^2)/x^2\}\,dx$$

Integrating, $\quad v^2 = (2\mu a^2)/x + A$, where A is a constant. ...(1)

Initially at B, when $x = 2a$, $v = 0$. So, by (1), $\quad A = -\mu a$.

$\therefore$ From (1), $\quad v^2 = (2\mu a^2)/x - \mu a = a\mu(2a - x)/x$ ...(2)

giving velocity of the particle beteween B and A. Let v_1 be the velocity of the particle when it reaches A, where $x = a$. Then, putting $x = a$ and $v = v_1$ in (2), we get

$$v_1^2 = \mu a \quad \text{or} \quad v_1 = -(\mu a)^{1/2} \quad ...(3)$$

its direction being towards the origin O.

Re-writing (2), $\quad (dx/dt)^2 = a\mu(2a - x)/x$ or $dx/dt = -\{a\mu(2a - x)/x\}^{1/2}$,

where the –ve sign is taken because as t increases then x decreases.

Separating variables, $\quad dt = -\dfrac{1}{(a\mu)^{1/2}}\left(\dfrac{x}{2a - x}\right)^{1/2} dx$ ...(4)

Let t_1 be the taken by the particle to move from B to A. So integrating (4) between the limits $t=0$ to $t=t_1$ and the corresponding limits $x=2a$ to $x=a$, we get

$$\int_0^{t_1} dt = -\frac{1}{(a\mu)^{1/2}}\int_{x=2a}^{a}\left(\frac{x}{2a-x}\right)^{1/2} dx \qquad \ldots(5)$$

Put $x=2a\cos^2\theta$ so that $dx=-4a\cos\theta\sin\theta\, d\theta$. Also, note that when $x=2a$ then $x=2a\cos^2\theta$ $\Rightarrow 2a=2a\cos^2\theta \Rightarrow \theta=0$ and when $x=a$ then $x=2a\cos^2\theta \Rightarrow a=2a\cos^2\theta \Rightarrow \cos\theta = 1/\sqrt{2} \Rightarrow$ $x=\pi/4$. Hence (5) reduces to

$$t_1 = -\frac{1}{(a\mu)^{1/2}}\int_0^{\pi/4}\frac{\cos\theta}{\sin\theta}\times(-4a\sin\theta\cos\theta)\,d\theta$$

$$= 2\left(\frac{a}{\mu}\right)^{1/2}\int_0^{\pi/4}(1+\cos 2\theta)\,d\theta = 2\left(\frac{a}{\mu}\right)^{1/2}\left[\theta+\frac{1}{2}\sin 2\theta\right]_0^{\pi/4}$$

$$\therefore \qquad t_1 = 2(a/\mu)^{1/2}(\pi/4+1/2) = (a/\mu)^{1/2}(\pi/2+1) \qquad \ldots(6)$$

Motion from *A* to *O*. When particle crosses OA with velocity v_1 (given by (3)), then particle moves towards O along AO. Let Q be its position at any time t such that $OQ=x$. While moving from A to O, since the law of force for $x \le a$ is given by $m\mu x/a$ directed towards O, hence the equation of the particle is

$$mv(dv/dx) = -m\mu x/a \qquad \text{or} \qquad 2v\,dv = -\{(2\mu x)/a\}\,dx$$

Integrating, $\qquad v^2 = -(\mu/a)x^2 + B$, B is a constant. $\qquad \ldots(7)$

But at A, when $x=a$, $v=v_1=-(\mu a)^{1/2}$. So (7) gives $B=2\mu a$.

$\therefore$ From (7), $\qquad v^2 = -(\mu/a)x^2 + 2\mu a = (\mu/a)\times(2a^2-x^2) \qquad \ldots(8)$

Eq. (8) gives the velocity of the particle between A and O. Let v_2 be the velocity of the particle at O. Then, putting $x=0$ and $v=v_2$ in (8), we have

$$v_2^2 = (\mu/a)\times(2a^2) \qquad \text{or} \qquad v_2^2 = 2a\mu,$$

showing that the particle reaches the origin O with velocity $(2a\mu)^{1/2}$.

Re-writing (8), $\left(\dfrac{dx}{dt}\right)^2 = \dfrac{\mu(2a^2-x^2)}{a}$ or $\dfrac{dx}{dt} = -\left(\dfrac{\mu}{a}\right)^{1/2}(2a^2-x^2)^{1/2}$,

where the $-$ve sign is taken because as t increases then x decreases.

Separating variables, $\qquad dt = -(a/\mu)^{1/2}\{dx/(2a^2-x^2)^{1/2}\} \qquad \ldots(9)$

Let t_2 be the time taken by the particle to move from A to O. So integrating (9) between the limits $t=0$ to $t=t_2$ and the corresponding limits $x=a$ to $x=0$, we have

$$\int_0^{t_2} dt_2 = -\left(\frac{a}{\mu}\right)^{1/2}\int_a^0 \frac{dx}{\{(a\sqrt{2})^2-x^2\}^{1/2}} \qquad \text{or} \qquad t_2 = \left(\frac{a}{\mu}\right)^{1/2}\left[\cos^{-1}\frac{x}{a\sqrt{2}}\right]_a^0$$

or $\quad t_2 = (a/\mu)^{1/2}\times(\pi/2-\pi/4) \quad$ or $\quad t_2 = (a/\mu)^{1/2}\times(\pi/4) \qquad \ldots(10)$

From (6) and (10), the required total time $(=t_1+t_2)$

$= (a/\mu)^{1/2}\times(\pi/2+1) + (a/\mu)^{1/2}\times(\pi/4) = (a/\mu)^{1/2}(3\pi/4+1)$

EXERCISE

1. A particle whose mass is 'm' is acted upon by a force $m\mu x^{-5/3}$ towards the centre. If it starts from rest at a distance 'a' from this centre, then show that it will arrive at the centre after time $2a^{4/3}/(3\mu)^{1/2}$

2. A particle P of mass m moves along a straight line through a point O and at any instant the distance OP is x. When $x > a$, the particle is attracted towards O by a force mk/x^2, and when $x < a$ the particle is repelled from O by a force mak/x^3. If the particle is released from rest at a distance $2a$ from O, show that it come to rest instantaneously, when $x = a/\sqrt{2}$ and find the time the particle takes to travel from $x=a$ to $x=a/\sqrt{2}$. ***[Delhi Maths (H) 2009; Mumbai 2010]*** **[Ans.** $(1/2)\times(a^3/k)^{1/2}$**]**

3. A particle of mass m moves on a straight line under the influence of a force directed towards the origin O on the line and proportional to the distance from O; the force at unit distance is of magnitude mk^2. The particle passes O with velocity u. If x is its coordinate at time t and v its velocity at that instant. Show that $v^2 + k^2x^2 = u^2$. **[Delhi B.Sc. III (Prog) 2009]**

OBJECTIVE PROBLEMS

Ex. 1. *Taking the radius of the earth to be 6.4 × 10⁸ cm and the value of 'g' to be 981 cm/sec², the escape velocity from the surface of the earth is*

(a) 11.2×10^5 cm/sec (b) 12.9×10^5 cm/sec (c) 8.1×10^5 cm/sec (d) 9.7×10^5 cm/sec

[I.A.S. (Prel.) 1993]

Sol. Ans. (*a*). Refer Ex. 6 (*b*) on page 3.4.

Ex. 2. *The escape velocity from the earth is about 11 km/sec. The escape velocity from the planet having twice the radius and the same mean density as the earth is about (a) 5.5 km/sec (b) 11 km/sec (c) 16.5 km/sec (d) 22 km/sec.* **[I.A.S. (Prel.) 1996]**

Sol. Ans. (*c*). Refer Ex. 6 (*b*) on page 3.4. Given $(2gR)^{1/2} = 11$.

Required escape vel. $= (2g \times 2R)^{1/2} = 2^{1/2}(2gR)^{1/2} = \sqrt{2} \times 11$ km/sec $= 16.5$ km/sec.

Ex. 3. *The escape velocity for a body projected vertically upwards is 11.2 km/sec. If the body is projected in a direction marking an angle of 60° with vertical, then the escape velocity will be*

(a) 11.2 km/sec (b) $5.6\sqrt{3}$ km/sec (c) 5.6 km/sec (d) None of the above.

[I.A.S. (Prel.) 1997]

Sol. Refer Ex. 6 (*b*) on page 3.4.

Ex. 4. *The escape velocity of a projectile from the earth is approximately (a) 101 km/sec (b) 7 km/sec (c) 11.2 km/sec (d) 112 km/sec.* **[I.A.S. 1999]**

Sol. Ans. (*c*). Refer Ex. 6 (*b*) on page 3.4.

Ex. 5. *A steam boat is moving with velocity v_1 when steam is shut off. If the retardation at any subsequent time is equal to the magnitude of the velocity at that time then the velocity in time t after the steam is shut off is*

(a) $v_1 e^t$ (b) $v_1 e^{-t}$ (c) $2v_1 e^t$ (d) $2v_1 e^{-t}$ **[I.A.S. (Prel.) 1999]**

Sol. Ans. (*b*). Given $dv/dt = -v$ or $(1/v)dv = -dt$

Integrating, $\log v - \log c = -t$ or $v = ce^{-t}$ ...(1)

Initially, when $t = 0$, $v = v_1$ so (1) gives $c = v_1$

Hence (1) reducers to $v = v_1 e^{-t}$.

Ex. 6. *If the acceleration due to gravity on the surface of the earth is g = 980 cm/sec² and radius of the earth R = 6400 km, then the escape vel. on the surface of the earth is*

(a) 112 km/sec, (b) $56\sqrt{2}$ km/sec., (c) 11.2 km/sec, (d) (11.2) $\sqrt{2}$ km/sec. **[I.A.S. (Prel.) 2001]**

Sol. Ans. (*c*). Refer Ex. 6(*b*) on page 3.4.

Ex. 7. *A particle of unit mass is travelling along the x-axis such that at t = 0, it is located at x = 0 and has speed v_o. If the particle is acted upon by a force which apposes the motion and has magnitude proportional to the square of the instantaneous speed, then the speed at time t is proportional to (a) 1 + kt, (b) $(1+t)^2$, (c) 1/t, (d) 1/(1 + kt), where k is a constant.*

[I.A.S. (Prel.) 2002]

Sol. Ans. (*d*). Let P be the position of the particle at any time t let v be its velocity at P. Then equation of the particle at P is given by

v_o v O P(t) X

$m(dv/dt) = -l\,v^2$, l being a constant.

or $dv/dt = -hv^2$, where $h = l/m$ = constant.

or $-v^{-2}\,dv = h\,dt$ so that $1/v = ht + c$ ...(1)

Initially at O, when $t = 0$, $v = v_0$. So (1) gives $1/v_0 = c$

$\therefore$ (1) becomes $1/v = ht + 1/v_0$ or $v = v_0/(1 + h v_o t)$

or $\qquad v = v_0/(1 + kt)$, where $k = hv_0$ = constant.

Since v_0 is a constant, v is proportional to $1/(1 + kt)$.

Ex. 8. *Fill in the blanks "………" so that the following statements are correct and complete.*

(*i*) *When a particle moves outside the surface of the earth, the acceleration varies inversely as the ………….. of the distance of the particle from the centre of the earth.*

(*ii*) *If a particle is projected towards the centre of repulsion varying as the distance from the centre, then its equation of motion is ………*

(*iii*) *When a particle moves inside the earth through a hole made in the earth, the acceleration varies directly as the ………….. of the particle from the centre of the earth.*

Ans. (*i*) square $\quad$ (*ii*) $d^2x/dt = \mu x$ $\quad$ (*iii*) distance

Ex. 9. True or false : Write 'T' for true and 'F' for false statement.

(*i*) *In a rectilinear motion if v is the velocity at time t at a distance x from a fixed point on the path, them* $d^2x/dt^2 = v(dv/dx)$.

(*ii*) *Assuming that at a distance x from the centre of force, the speed v of a particle, moving in a line is given by* $x = a\,e^{v^2}$, *a being a constant. Then acceleration varies directly as the distance of the particle from the centre of force.*

Ans. (*i*) T $\quad$ (*ii*) F

Ex. 10. *A particle is at rest at the origin. It moves along the x-axis with an accleration* $x - x^2$, *where x is the distance of the particle at time t. The particle next comes to rest after it has covered a distance*

(*a*) *1* $\quad$ (*b*) *1/2* $\quad$ (*c*) *3/2* $\quad$ (*d*) *2.* $\qquad$ **[I.A.S. (Prel.) 2004]**

Sol. Ans. (*c*). Given, acceleration $= v(dv/dx) = x - x^2$

or $\qquad 2v\,dv = (2x - 2x^2)dx \qquad$ so that $\qquad v^2 = x^2 - (2/3)x^3 + C \qquad$ ...(1)

Intially, when $x = 0, v = 0$. So (1) gives $C = 0$. Then (1) becomes

$v^2 = x^2 - (2/3)x^3$ and so the particle will come to rest when $0 = x^2 - (2/3)x^3$ or $x^2(1 - 2x/3) = 0$ giving $x = 0$ and $x = 3/2$. But $x = 0$ corresponds to the initial position. So the required value of distance $= x = 3/2$.

Ex. 11. *If the moon's radius is 1/4 of earth's radius, moon's mass is 1/81 of earth, and if* V_M, V_E *be respectively the escape velocities on the surface of moon and on the surface of earth, then*

(*a*) $V_E/V_M = 2.25$ (*b*) $V_E/V_M = 4.5$ $\quad$ (*c*) $V_E/V_M = 9$ $\quad$ (*d*) $V_E/V_M = 2/9$. $\quad$ **[I.A.S. (Prel.) 2004]**

Sol. Ans. (*c*). Escape velocity. Definition. We know that when a body is projected upwards then it comes down to the earth due to gravitational pull of the earth on it. Let us examine the possibility of projecting a body with velocity such that it will never come back to the earth. Clearly, it will be so if it can be given a velocity that will take it beyond the gravitational field of the earth. This velocity of the body is called the *escape velocity*.

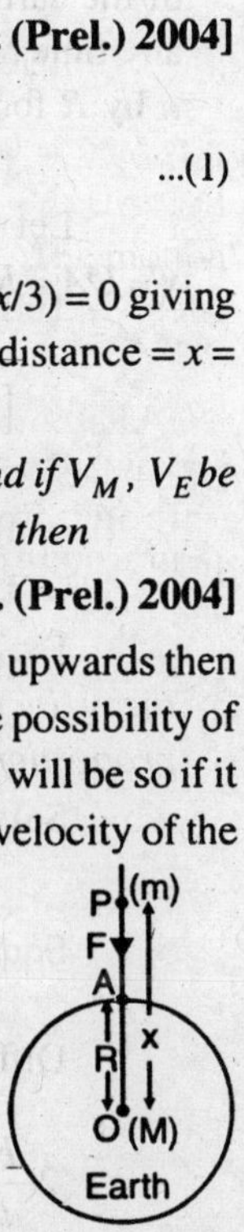

To find escape velocity on the surface of the earth

Let m be the mass of the body and M, that of the earth, the force F acting on the body is given by $F = (mMG)/x^2$, where G is the usual gravitational constant. Then, the motion of the body is given by

$$mv(dv/dx) = -F \quad \text{or} \quad mv(dv/dx) = -(mMG)/x^2$$

so that $\qquad 2v\,dv = -\{(2MG)/x^2\}dx \qquad$ ...(1)

Let R be the radius of the earth and V_E be the escape velocity from a point A on the surface of the earth. Then, by definition of escape velocity, when $x = R$, $v = V_E$ and when $x \to \infty$, $v \to 0$. Integrating (1) from $v = V_E$ to $v = 0$ and the corresponding limits $x = R$ to $x = \infty$, we get

$$\int_{V_E}^{0} 2v\,dv = -\int_{R}^{\infty} \frac{2MG}{x^2}\,dx \qquad \text{or} \qquad [v^2]_{V_E}^{0} = 2MG\left[\frac{1}{x}\right]_R^{\infty}$$

$$\text{or} \qquad V_E^2 = (2MG)/R \qquad \text{or} \qquad V_E = \{(2MG)/R\}^{1/2}$$

Next, consider moon in place of earth. Replacing M by M' (= mass of the moon, say), R by R' (= radius of the moon, say) and V_E by V_M in (2), we get

$$V_M = \{(2M'G)/R'\}^{1/2}$$

Dividing (2) by (3), $$\frac{V_E}{V_M} = \left(\frac{M}{M'} \times \frac{R'}{R}\right)^{1/2} = \left(81 \times \frac{1}{4}\right)^{1/2} = \frac{9}{2} = 4.5$$

[$\because$ given that $R' = R/4$ and $M' = M/81$]

Ex. 12. *A particle whose weight on the surface of the earth is W falls to the surface of the earth from a height equal to the diameter 2R of the earth. What is the work done by the earth's attraction ?*

(a) 2RW *(b) 2RW/3* *(c) 4 RW/3* *(d) 3RW/2* **[I.A.S. Prel 2007]**

Sol. Draw figure as in Ex. 4 (*a*) of page 3.3 by taking $OA = R$ and $AB = 2R$. Let O be the centre of the earth. Let a particle fall from B towards O along BO. Let P be the position of the particle at any time t such that $OP = x$. Then, as in Ex. 4 (*a*) page 3.3, proceed upto equation (3) by replacing a by R for the given problem. Thus, we get

$$2v\,dv = -(2gR^2/x^2)\,dx$$

Let the particle acquire a velocity V at A. Then, when $x = OB = 3R$, $v = 0$ and when $x = OA = R$, $v = V$. Integrating (1) between the corresponding limits, for motion from B to A, we have

$$[v^2]_0^V = \left[\frac{2gR^2}{x}\right]_{3R}^{R} \qquad \text{or} \qquad V^2 = 2gR^2\left(\frac{1}{R} - \frac{1}{3R}\right) = \frac{4Rg}{3}$$

$\therefore$ Required work done = change in K.E. in moving from B to A.

$= (1/2) \times m \times V^2 - (1/2) \times m \times o^2 = (1/2) \times (W/g) \times (4Rg/3) = 2RW/3$, as $W = mg$

Ex. 13. *A particle moves along a straight line and at a distance x from a fixed point on it, the velocity is* $\mu[(c-x)/x]^{1/2}$. *Prove that its acceleration is directed towards O and inversely proportional to the square of the distance.*

Sol. Here velocity $= dx/dt = \mu\{(c-x)/x\}^{1/2}$, (given) ...(1)

Squaring (1), $$\left(\frac{dx}{dt}\right)^2 = \mu\frac{c-x}{x} \qquad \text{or} \qquad \left(\frac{dx}{dt}\right)^2 = \mu\left(\frac{c}{x} - 1\right) \qquad ...(2)$$

Differentiating (2) w.r.t. 't' we have

$$2\frac{dx}{dt}\frac{d^2x}{dt^2} = -\frac{\mu c}{x^2}\frac{dx}{dt} \qquad \text{or} \qquad \frac{d^2x}{dt^2} = -\frac{\mu^2 c}{x^2},$$

showing that the acceleration d^2x/dt^2 is directed towards O and is inversely proportional to the square of the distance.

4

Simple Harmonic Motion

4.1 Simple Harmonic Motion. Definition. [Kanpur 2015; Bundelkhand 2006; Meerut 2016, 2019]

A particle is said to execute simple Harmonic Motion if it moves in a straight line such that its acceleration is always directed towards a fixed point in the line and is proportional to the distance of the particle from the fixed point.

Note. Simple harmonic motion is briefly written as S.H.M.

Expressions for acceleration, velocity and position of the particle at any instant and nature of motion of S.H.M. **[Avadh 2006; Lucknow 2009; Agra 2003; Rohilkhand 2011; Meerut 2010]**

Let O be the fixed point in the line $A'OA$ and let P denote the position of the particle at any time t such that $OP = x$. Let the particle start from rest from A where $OA = a$. Then acceleration at P is μx. Then, the equation of motion is

A′ P′ O ← x → P A

$$d^2x/dt^2 = -\mu x \qquad \text{or} \qquad v(dv/dx) = -\mu x, \qquad ...(1)$$

where we have taken a negative sign on the R.H.S. because d^2x/dt^2 is the acceleration in the direction of x increasing *i.e.*, in the direction OP while μx is the acceleration towards O, *i.e.*, in the direction PO.

Rewriting (1), $\qquad 2v dv = -2\mu x dx \quad$ so that $\quad v^2 = -\mu x^2 + A, \qquad ...(2)$

where A is a constant of integration and v is the velocity at P.

Now, initially, when $x = a$, $v = 0$. Then (2) gives $\qquad 0 = -\mu a + A$ or $A = \mu a^2$.

$\therefore$ (2) becomes $\qquad v^2 = \mu(a^2 - x^2), \qquad ...(3)$

which gives the velocity v at any displacement x.

From (3), $\qquad (dx/dt)^2 = \mu(a^2 - x^2) \quad$ or $\quad dx/dt = -\mu^{1/2}(a^2 - x^2)^{1/2} \qquad ...(4)$

taking –ve sign on R.H.S., since x decreases as t increases

Rewriting (4), $\quad -dx/(a^2 - x^2)^{1/2} = \mu^{1/2}\, dt \qquad$ so that $\qquad \cos^{-1}(x/a) = \mu^{1/2}\, t + B \qquad ...(5)$

where B is a constant of integration. Now, the following two cases arise :

Case I. If the time t is measured from the instant when the particle P is at A *i.e.*, if $x = a$ when $t = 0$, then from (5), $\cos^{-1} 1 = B$ so that $B = 0$. Hence (5) reduces to

$$\cos^{-1}(x/a) = t\,\mu^{12} \qquad \text{or} \qquad x = a\cos(t\,\mu^{1/2}) \qquad ...(6)$$

[Kanpur 2008; Guwahati 2007]

Case II. If the time t is measured from the instant when the particle is at O *i.e.*, if $x = 0$ when $t = 0$. Then for motion from 0 to A, (3), $\Rightarrow dx/dt = \mu^{1/2}\,(a^2 - x^2)^{1/2}$, taking the positive sign on R.H.S. as x increases with t. Integrating the diff. eqn. gives $\quad \sin^{-1}(x/a) = t\,\mu^{1/2} + C$.

Since $x = 0$, when $t = 0$ we get $C = 0$. Thus, $\quad \sin^{-1}(x/a) = t\,\mu^{1/2} \quad$ or $\quad x = a\sin(t\,\mu^{1/2}) \qquad ...(7)$

Equation (6) or (7) gives the position of the particle at any time t.

Nature of motion. Suppose a particle starts from rest at A where $OA = a$. Then the particle has maximum acceleration μa towards O (refer equation (1)). When the particle arrives at O, x is zero, and then by (4), the velocity $= -a\,\mu^{1/2}$. The particle thus passes through O and immediately the acceleration alters its direction and tends to reduce the velocity; also the velocity is zero at A' where $OA' = OA = a$. Hence the particle comes to an instantaneous rest at A' and then retraces its path, passes through O, and again is instantaneously at rest at A'. The whole motion of the particle is thus an oscillation from A to A' and back, continually repeated over and over again.

Some important definitions

(i) Amplitude **[Meerut 2012, 2015, 2017, 2019]**

In a S.H.M. the distance from the centre of force of the position of maximum displacement is known as the amplitude of S.H.M. Thus, the distance $AO = A'O = a$ is amplitude of S.H.M.

(ii) Maximum and minimum accelerations. Refer equation (1). The particle has maximum acceleration μa at the extreme position A or A' and has minimum acceleration zero at the centre of force (or mean position) O

(iii) Maximum and minimum velocities. Refer equation (3). The particle has maximum velocity $a\mu^{1/2}$ at O and minimum velocity zero at extreme positions A or A'.

(iv) Periodic motion and period of the motion **[Kanpur 2015; Meerut 2012, 2016]**

A particle is said to have a *periodic motion* when it moves in such a manner that after a certain fixed interval of time, it occupies the same position and moves, in the same direction with the same velocity. The fixed minimal interval of time for such a motion is called the *period of the motion.*

(v) *To show that S.H.M. is periodic with period $2\pi/\mu^{1/2}$, which is independent of the amplitude a.* **[Bundelkhand 2006; Rohilkhand 2010, 11; Agra 2003, 2006, 2007; Kanpur 2010, 16; Meerut 2016]**

In S.H.M., we have

$$x = a \sin t\mu^{1/2} = a \sin (t\mu^{1/2} + 2\pi) = a \sin (t + 2\pi/\mu^{1/2})\mu^{1/2} \quad \text{...(I)}$$

where a is the amplitude and x is measured from the mean position.

Differentiating both sides of (I) w.r.t., t, we have $\quad v = dx/dt = a\mu^{1/2} \cos t\mu^{1/2}$

or $\quad v = a\sqrt{\mu} \cos (t\sqrt{\mu} + 2\pi) = a \cos (t + 2\pi/\sqrt{\mu})\sqrt{\mu} \quad$ ...(II)

From (I) and (II), it follow that the values of x and v are repeated when t is increased by $2\pi/\sqrt{\mu}$, *i.e.*, after every interval of $2\pi/\sqrt{\mu}$, we have the same position and the same velocity in the same direction. Thus S.H.M. is periodic, the periodic T being $T = 2\pi/\sqrt{\mu}$ which is independent of the amplitude a.

(vi) Frequency. *The number of complete oscillations in one second is called frequency.* If n be the frequency and T, the time period, then

$$nT = 1 \quad \text{or} \quad n = 1/T = \sqrt{\mu}/2\pi$$ **[Avadh 2006]**

(vii) Phase and Epoch. Rewriting (1), $d^2x/dt^2 + \mu x = 0$, which is a linear differential eqn. with constant coefficients and its general solution is given by

$$x = a \cos (t\sqrt{\mu} + \varepsilon) \quad \text{...(I)}$$

The constant ε is called the starting phase or the epoch of the motion and the quantity $t\sqrt{\mu} + \varepsilon$ is called the argument of the motion. **[Rohilkhand 2006]**

The phase at any time t of a S.H.M. is the time that has elapsed since the particle was at its maximum distance in the positive direction. Form (I), x is maximum when $\cos (t\sqrt{\mu} + \varepsilon)$ is maximum *i.e.*, when $\cos (t\sqrt{\mu} + \varepsilon) = 1$.

Clearly x is maximum at time t_0 where $\cos (t_0\sqrt{\mu} + \varepsilon) = 1$.

or $\quad t_0\sqrt{\mu} + \varepsilon = 0 \quad$ or $\quad t_0 = -\varepsilon/\sqrt{\mu}. \quad$ ...(II)

$\therefore \quad$ Phase at time $t = t - t_0 = t + \varepsilon/\sqrt{\mu}$, using (II)

4.2 Geometrical representation of S.H.M. *If a particle describes a circle with constant angular velocity, the foot of the perpendicular from the particle on a diameter moves with S.H.M.* **[Kanpur 2011]**

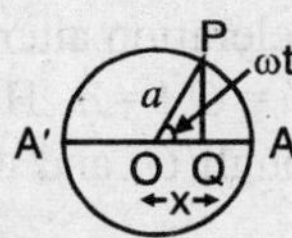

Proof. Let a particle move with uniform angular velocity ω round the circumference of a circle of radius a. Let AOA' be a fixed diameter of the circle. Let the particle start from A and reach P after time t. Then $\angle AOP = \omega t$. Let Q be the foot of the perpendicular drawn from P on AA'. If $OQ = x$. Then, from ΔOPQ we have

$$x = a \cos \omega t \quad ...(1)$$

From (1), $\quad v = dx/dt = -a\omega \sin \omega t \quad ...(2)$

From (2), $\quad d^2x/dt^2 = -a\omega^2 \cos \omega t = -\omega^2 x$, by (1) $\quad ...(3)$

(2) and (3) give the velocity and acceleration of Q at any time t.

From (3), it follows that Q executes a S.H.M. with centre at the origin O. From (1), it follows that the amplitude of this S.H.M. is a because the maximum value of x is a. Again the periodic time T of Q is the time taken by P is moving once along the whole circumference of the circle *i.e.*, the time taken by P to turn through an angle 2π with the uniform rate of ω.

$\therefore$ Periodic time of $Q = 2\pi/\omega$.

4.3 Compounding of two simple harmonic motions of the same period and in the same straight line

(M.S. Univ. T.N. 2007; Patna 2003)

Let $2\pi/n$ be the period of the given simple harmonic motions. Then the most general displacements of given motions can be written as a cos $(nt + \varepsilon)$ and $a' \cos (nt + \varepsilon')$, so that

$$x = a \cos (nt + \varepsilon) + a' \cos (nt + \varepsilon')$$

or $\quad x = a (\cos nt \cos \varepsilon - \sin nt \sin \varepsilon) + a'(\cos nt \cos \varepsilon' - \sin nt \sin \varepsilon')$

or $\quad x = (a \cos \varepsilon + a' \cos \varepsilon') \cos nt - (a \sin \varepsilon + a' \sin \varepsilon') \sin nt$

or $\quad x = A \cos E \cos nt - A \sin E \sin t \quad$ or $\quad x = A \cos (nt + E) \quad ...(1)$

where $\quad a \cos \varepsilon + a' \cos \varepsilon' = A \cos E,\ a \sin \varepsilon + a' \sin \varepsilon' = A \sin E \quad ...(2)$

Squaring and adding the relations of (2), we get

$$(a \cos \varepsilon + a' \cos \varepsilon')^2 + (a \sin \varepsilon + a' \sin \varepsilon')^2 = A^2$$

or $\quad A = \{a^2 + a'^2 + 2aa' \cos (\varepsilon - \varepsilon')\}^{1/2} \quad ...(3)$

On dividing, the relations (2) give

$$\tan E = (a \sin \varepsilon + a' \sin \varepsilon')/(a \cos \varepsilon + a' \cos \varepsilon') \quad ...(4)$$

(1) Shows that the resulting motion of two simple harmonic motions is also a S.H.M. of the same period whose amplitude A and epoch E are given by (3) and (4) respectively.

If we draw OP $(= a)$ at an angle $\in$ to a fixed line, and OQ $(= a')$ at an angle $\in'$ and complete the parallelogram $OPRQ$ then by equations (2) we see that OR represents amplitude A and that it is inclined at an angle E to the fixed line. The line representing the resultant of the two given motions is therefore the geometrical resultant of the lines representing the two component motions.

4.4 Some useful results to be committed to memory for direct applications

(*i*) Referred to the centre of force as origin, the equation of S.H.M. is

$$\ddot{x} = -\mu x, \text{ where } \ddot{x} = d^2x/dt^2 \text{ and } \mu\ (> 0) \text{ is a constant.}$$

(*ii*) Acceleration is maximum at extreme positions *i.e.*, when $x = a$ or $x = -a$ and maximum acceleration (in magnitude) is μa. **[Rohilkhand 2008]**

(*iii*) Velocity v at a distance x is given by $v^2 = \mu(a^2 - x^2)$. Velocity is maximum at the mean position and maximum velocity (in magnitude) $= a\sqrt{\mu}$.

(*iv*) $x = a \sin t\sqrt{\mu}$, if t is measured from the mean position.

(*v*) $x = a \cos t\sqrt{\mu}$, if t is measured from the extreme position.

(*vi*) Time period $= T = 2\pi/\sqrt{\mu}$. (*vii*) Frequency $= n = 1/T = \sqrt{\mu}/2\pi$.

(*viii*) Amplitude $= OA = O'A' = a$. (*ix*) Time from A to O = Time from O to A.

(*x*) Velocity $= dx/dt$; acceleration $= d^2x/dt^2 = dv/dt = v\ (dv/dx)$.

(*xi*) $\mu\ (> 0)$ is called the intensity of the given attractive force.

TYPE 1. SOLVED EXAMPLES BASED ON ART 4.1 TO 4.3

Ex. 1. *The maximum velocity of a body moving with S.H.M. is 2 m/sec and its period is 1/5 sec. What is its amplitude ?* **[Meerut 2018]**

Sol. Let the required amplitude be a meters. Then, here

$$\text{Maximum velocity} = a\,\mu^{1/2} = 2, \text{(given)} \quad ...(1)$$

Again, periodic time $= T = 2\pi/\mu^{1/2} = 1/5$ sec. (given) ...(2)

Multiplying (1) and (2), $2\pi a = 2/5$ or $a = 1/5\,\pi$ meters.

Ex. 2. (*a*) *The speed v of a point P which moves in a straight line is given by the relation $v^2 = a - bx^2$, where x is the distance of the point P from a fixed point on the path, a and b being constants. Show that the motion of P is simple harmonic and determine its amplitude and period.* **[Kanpur 2015]**

Sol. Given $v^2 = a - bx^2$ ...(1)

Diff. (1) w.r.t. 'x', $2v(dv/dx) = -2bx$ or $d^2x/dt^2 = -bx$, ...(2)

which is of the standard from $d^2x/dt^2 = -\mu x$ and hence (2) represents S.H.M. whose period $= 2\pi/\sqrt{\mu}$ $= 2\pi/\sqrt{b}$.

Now to find the amplitude we must find the distance from the centre of a point where the velocity is zero. So putting $v = 0$ in (*ii*), we get $0 = a - bx^2$ so that $x = \pm(a/b)^{1/2}$. Since here the centre is at origin hence the amplitude $= (a/b)^{1/2}$.

Ex. 2. (*b*) *A particle moves in a straight line and its velocity at a distance x from the origin is $k\,(a^2 - x^2)^{1/2}$, where a and k are constants. Prove that the motion is simple harmonic and find the amplitude and periodic time of the motion* **[Agra 2002; Meerut 2014]**

Hint. Try like Ex. 2. (*a*). **Ans.** Period $= 2\pi/k$, amplitude $= a$.

Ex. 3. (*a*) *The speed v of a particle moving along the axis of x is given by the relation $v^2 = n^2\,(8ax - x^2 - 12\,a^2)$. Show that the motion is simple harmonic with centre at $x = 4a$ and amplitude $= 2a$. Find also the time from*

(*i*) *$x = 4a$ to $x = 6a$* **[Delhi B.A. (Prog) III 2009; 2016]**

(*ii*) *$x = 5a$ to $x = 6a$.*

Sol. Given $v^2 = n^2\,(8ax - x^2 - 12a^2)$ ...(1)

Diff. (1) w.r.t. 'x', $2v\,(dv/dx) = n^2\,(8a - 2x)$ or $d^2x/dt^2 = n^2(4a - x)$

or $d^2x/dt^2 = -\,n^2\,(x - 4a)$ ...(2)

Let $x - 4a = X$ so that $d^2x/dt^2 = d^2X/dt^2$ ...(3)

Then (2) reduces to $d^2X/dt^2 = -\,n^2X$, ...(4)

which is the equation of a S.H.M. with centre at $X = 0$ *i.e.*, $x - 4a = 0$, *i.e.*, $x = 4a$ (Note that the centre is the point where the acceleration is zero).

Now to find the amplitude we must find the distance from the centre of a point where the velocity is zero. So putting $v = 0$ in (1), we get

$n^2\,(8ax - x^2 - 12a^2) = 0$ or $x^2 - 8ax + 12a^2 = 0$ giving $x = 6a, 2a$.

$\therefore$ The required amplitude $= 6a - 4a = 2a$.

(*i*) The required time taken in moving from $x = 4a$ to $x = 60$ is the time taken in moving from the centre of oscillation to the extreme end of the particle can reach *i.e.*, (1/4)th of periodic time of S.H.M. $= (1/4)\,(2\pi/n) = \pi/2n$.

(*ii*) The distances of the points $x = 5a$ and $x = 6a$ from the centre ($x = 4a$) are a and $2a$ respectively. If the time t be measured from the centre, the position at any time t is given by formula $x = a \sin t\,\mu^{1/2}$, so that $x = 2a \sin nt$.

Let $x = a$ when $t = t_1$, so $\sin nt_1 = 1/2$ or $nt_1 = \pi/6$,

and $x = 2a$ when $t = t_2$ so $\sin nt_2 = 1$ or $nt_2 = \pi/2$,

$\therefore$ The required time $= t_2 - t_1 = (1/n)\,(\pi/2 - \pi/6) = \pi/3n$.

Ex. 3(b). *The speed v of the point P which moves in a line is given by the relation $v^2 = a + 2bx - cx^2$, where x is the distance of the point P from a fixed point on the path, and a, b, c are constants. Show that the motion is simple harmonic if c is positive, determine the period and the amplitude of the motion.*

Sol. Try like Ex. 3(*a*). Period $= 2\pi/c^{1/2}$ amplitude $= (b^2 + ac)^{1/2}/c$.

Ex. 4. *Show that if the displacement of a particle in a straight line is expressed by the equation $x = a \cos nt + b \sin nt$, it describes a S.H.M. whose amplitude is $(a^2 + b^2)^{1/2}$ and period is $2\pi/n$.*

[Bundelkhand 2009; Meerut 2005, 07; Delhi BA (Prog.) III, 2012]

Sol. Given $x = a \cos nt + b \sin nt$ ...(1)

Diff. (1) w.r.t., $dx/dt = -an \sin nt + bn \cos nt$ ...(2)

Diff. (2) w.r.t., $d^2x/dt^2 = -an^2 \cos nt - bn^2 \sin nt = -n^2 (a \cos nt + b \sin nt)$

or $d^2x/dt^2 = -n^2 x$, by (1) ...(3)

Since (3) is of standard from $d^2x/dt^2 = -\mu x$, it follows that (3) represents a S.H.M. for which $\mu = n^2$. Its period is $= 2\pi/\mu^{1/2} = 2\pi/n$.

Also amplitude of the motion is the value of x for which $dx/dt = 0$.

$\therefore$ From (2), $-an \sin nt + bn \cos nt = 0$ or $\tan nt = b/a$

so that $\sin nt = b/(a^2 + b^2)^{1/2}$ and $\cos nt = a/(a^2 + b^2)^{1/2}$ ...(4)

Substituting these values of sin nt and cos nt in (1), we get the

$$\text{amplitude} = a\{a/(a^2 + b^2)^{1/2}\} + b\{b/(a^2 + b^2)^{1/2}\} = (a^2 + b^2)^{1/2}.$$

Ex. 5(*a*). *A point moving in a straight line with S.H.M. has velocities v_1 and v_2 when its distances from the centre are x_1 and x_2. Show that the period of motion is $2\pi \{(x_1^2 - x_2^2)/(v_2^2 - v_1^2)\}^{1/2}$.*

[Kanpur 2018; Avadh 2011; Bundelkhand 2011; Guwahati 2007; Agra 2006; Meerut 2016]

Sol. Let the equation of S.H.M., with centre O as origin be $d^2x/dt^2 = -\mu x$. If a be the amplitude of the motion, then the velocity v of the particle when it is at a distance x from the mean position is given by

$$v^2 = \mu(a^2 - x^2) \quad ...(1)$$

Here the particle has a velocity v_1 at a distance x_1 from the centre and v_2 at a distance x_2 from the centre. So by (1)

$$v_1^2 = \mu(a^2 - x_1^2) \quad \text{and} \quad v_2^2 = \mu(a^2 - x_2^2)$$

$$\therefore \quad v_1^2 - v_2^2 = \mu(x_1^2 - x_2^2) \quad \text{or} \quad \mu = (v_2^2 - v_1^2)/(x_1^2 - x_2^2)$$

$$\therefore \quad T = \text{Periodic time} = 2\pi/\mu^{1/2} = 2\pi\{(x_1^2 - x_2^2)/(v_2^2 - v_1^2)\}^{1/2}.$$

Ex. 5(*b*). *In a S.H.M. if the velocities at distances b and c from the centre of force be respectively u and v. Then prove that the frequency n of oscillation is given by $4\pi^2 n^2 (b^2 - c^2) = v^2 - u^2$.*

Sol. Proceed as in Ex. 5(*a*) with $x_1 = b, x_2 = c, v_1 = u, v_2 = v$.

Then $T = 2\pi\{(b^2 - c^2)/(v^2 - u^2)\}^{1/2}$. But $n = 1/T$. So we get

$$1/n = 2\pi\{(b^2 - c^2)/(v^2 - u^2)\}^{1/2} \text{ or } 4\pi^2 n^2 (b^2 - c^2) = v^2 - u^2.$$

Ex. 6. *Show that the particle executing S.H.M. requires one sixth of its period to move from the position of maximum displacement to one in which the displacement is half the amplitude.*

[Lucknow 2006; Rohilkhand 2006; Kanpur 2012]

Sol. Let the equation of S.H.M. with centre O as origin be

$$d^2x/dt^2 = -\mu x$$

Then time period $T = 2\pi/\mu^{1/2}$ ...(1)

←— a —→ ←a/2→
A′ O M A

Let A be the position of the maximum displacement and M be the mid point of OA. If $OA = a$, then $AM = a/2$.

Let t be the time taken by the particle to move from A to M, then using $x = a \cos (t\, \mu^{1/2})$, we have

$a/2 = a \cos (t\, \mu^{1/2})$ or $t\, \mu^{1/2} = \pi/3$ or $t = \pi/3\, \mu^{1/2}$ ...(2)

From (1) and (2), $t/T = 1/6$ or $t = T/6$, as required.

Ex. 7(*a*). *A particle is moving with S.H.M. and while making an excursion from one position of rest to the other, its distances from the middle point of its path at three consecutive seconds are observed to be x_1, x_2, x_3, prove that the time of a complete oscillation is $2\pi/\cos^{-1}\{(x_1 + x_3)/2x_2\}$.*

[Meerut 2016, 2018; Avadh 2008; Bundelkhand 2006; Kanpur 2014, 2015]

Sol. With the middle point of the path as origin, let the equation of the S.H.M. be $d^2x/dt^2 = -\mu x$. Let a be the amplitude of the motion. If the time t be measured from the position of instantaneous rest $x = a$, we have

$$x = a \cos (t\, \mu^{1/2}), \quad ...(1)$$

where x is the distance of the particle from the centre at time t. Let the particle be at distances x_1, x_2 and x_3 from the middle point (*i.e.*, centre) at times t, $t+1$ and $t+2$ seconds respectively. Then, by (1), we get

$$x_1 = a \cos (t\, \mu^{1/2}), \quad ...(2)$$
$$x_2 = a \cos (t+1)\mu^{1/2} \quad ...(3)$$
and
$$x_3 = a \cos (t+2)\mu^{1/2} \quad ...(4)$$

$\therefore$ $x_1 + x_3 = a \{\cos (t+2)\mu^{1/2} + \cos t\, \mu^{1/2}\} = 2a \cos (t+1)\mu^{1/2} \cos \mu^{1/2}$

or $x_1 + x_3 = 2x_2 \cos \mu^{1/2}$ or $\mu^{1/2} = \cos^{-1} \{(x_1 + x_3)/2x_2\}$

$\therefore$ Periodic time$= 2\pi/\mu^{1/2} = 2\pi/\cos^{-1} \{(x_1 + x_3)/2x_2\}$

Ex. 7(*b*). *At the end of three successive seconds the distances of a point moving with S.H.M. from the mean position measured in the same direction are 1, 5 and 5. Show that the period of a complete oscillation is $2\pi/\theta$ where $\cos \theta = 3/5$.* **[Delhi B.Sc. III (Prog) 2009]**

(*c*) *At the end of three successive seconds, the distances of a point moving with S.H.M. from its mean position measured in the same direction are 1, 3 and 4. Show that the period of complete oscillation is $2\pi/\cos^{-1}(5/6)$.*

Sol. Try as in Ex. 7(*a*).

Ex. 8. *A particle moves with S.H.M. in a straight line. In the first second after starting from rest it travels a distance equal to 'b' and in the next second it travels a distance 'c' in the same direction. Prove that the amplitude of the motion is $2b^2/(3b-c)$.* **[Bangalore 92]**

A′ O C B A

Sol. With O as origin, let the equation of the S.H.M. be $d^2x/dt^2 = -\mu x$. Let a be the amplitude of motion so that $OA = OA' = a$. If the time t be measured from the position A of instantaneous rest, we have

$$x = a \cos t\, \mu^{1/2}, \quad ...(1)$$

where x is the distance of the particle from the centre at time t. By problem, when $t = 1$, $x = OB = a - b$, where $AB = b$ and when $t = 2$, $x = OC = a - (b + c)$.

$\therefore$ By (1), $a - b = a \cos \mu^{1/2}$ ...(2)

and $a - (b+c) = a \cos 2\mu^{1/2}$ or $a - b - c = a(2\cos^2 \mu^{1/2} - 1)$

or $a - b - c = 2a\{(a-b)/a\}^2 - a$, using (2)

or $2a - b - c = 2(a-b)^2/a$ or $a(2a - b - c) = 2(a^2 - 2ab + b^2)$

or $a(3b - c) = 2b^2$ or $a = 2b^2/(3b-c)$.

Ex. 9. *A particle starts from rest under an acceleration k^2x directed towards a fixed point and after time t another particle starts from the same position under the same acceleration. Show that the particles will collide at time $\pi/k + t/2$ after the start of the first particle provided $t < 2\pi/k$.*

Sol. With fixed point as origin, the equation of motion of the particle is $d^2x/dt^2 = -k^2x$, which represents a S.H.M. Hence its period $= 2\pi/k$. The given condition $t < 2\pi/k$ implies that the second particle starts before the first particle has made one complete oscillation. Let the particles meet after time t' of the start of the second particle. Then, using formula $x = a \cos t\, \mu^{1/2}$ for displacement after time t, we must have

$$a \cos \{k(t + t')\} = a \cos kt' \quad \text{or} \quad \cos \{k(t+t')\} = \cos kt'$$

or $k(t + t') = 2\pi - kt'$ or $t' = \pi/2 - t/2$.

$\therefore$ The required time$= t + t' = t + \pi/k - t/2 = \pi/k + t/2$

Ex. 10. *A point P moves in a straight line with S.H.M., the centre of motion being O and extreme position being A. A point is taken in OA such that $2OQ^2 = OA^2$. Show that time from A to Q is the same as that from Q to O.*

A′ O Q A

Sol. Let the equation of S.H.M. be $d^2x/dt^2 = -\mu x$. Let a be the amplitude of the motion. If t is measured from A, then we have

$$x = a \cos t\, \mu^{1/2}, \quad ...(1)$$

where x is the distance of the particle from the centre at time t. Let $OQ = x$. Then, since $2OQ^2 = OA^2$, we have $2x^2 = a^2$ so that $x = a/\sqrt{2}$. Let t_1 be time taken from A to Q so that when $t = t_1$, $x = OQ = a/\sqrt{2}$. So (1) gives

$$a/\sqrt{2} = a \cos t_1\mu^{1/2} \quad \text{or} \quad t_1\mu^{1/2} = \pi/4 \quad \text{or} \quad t_1 = \pi/4\mu^{1/2} \qquad ...(2)$$

Now, t_2 = time from A to O = (1/4) periodic time = (1/4) $(2\pi/\mu^{1/2}) = \pi/2\mu^{1/2}$

$\therefore \quad t_3$ = time Q to $O = t_2 - t_1 = \pi/2\mu^{1/2} - \pi/4\mu^{1/2} = \pi/4\mu^{1/2}$ ...(3)

From (2) and (3), $t_1 = t_3$ so that the time taken from A to Q is the same as that from Q to O.

Ex. 11. *A particle is moving with S.H.M. of amplitude a and periodic time T. Prove that* $\int_0^T v^2 dt = (2\pi^2 a^2)/T$. **[Delhi B.A. (Prog.) III, 2015, 16, 17; Agra 2003]**

Sol. Let the equation of S.H.M. be $d^2x/dt^2 = -\mu x$. Then

$$v^2 = \mu(a^2 - x^2), \qquad x = a \cos t\mu^{1/2} \qquad ...(1)$$

and $\quad T = 2\pi/\mu^{1/2} \quad$ so that $\quad \mu^{1/2} = 2\pi/T$ ...(2)

From (1), $\quad v^2 = \mu(a^2 - a^2 \cos^2 t\mu^{1/2}) = \mu a^2 \sin^2 t\mu^{1/2}$

or $\quad v^2 = (4\pi^2/T^2)\, a^2 \sin^2 (2\pi t/T)$, by (2).

$$\therefore \quad \int_0^T v^2 dt = \int_0^T \frac{4\pi^2 a^2}{T^2} \sin^2 \frac{2\pi t}{T}\, dt = \frac{2\pi^2 a^2}{T^2} \int_0^T \left(1 - \cos \frac{4\pi t}{T}\right) dt$$

$$= (2\pi^2 a^2/T^2)\,[t - (T/4\pi) \sin (4\pi t/T)]_0^T = 2\pi^2 a^2/T$$

Ex. 12. *Show that in a S.H.M. of amplitude a and period T the velocity v at a distance x from the centre is given by* $v^2T^2 = 4\pi^2 (a^2 - x^2)$. *Find the new amplitude if the velocity were doubled when the particle is at a distance a/2 from the centre; the period remaining unaltered.*

Sol. With centre as origin, let the equation of S.H.M. be $d^2x/dt^2 = -\mu x$. Let a be the amplitude and v be velocity at a distance x from the centre. Then

$$v^2 = \mu(a^2 - x^2) \qquad ...(1)$$

and $\quad$ the time period $= T = 2\pi/\sqrt{\mu}$ or $\sqrt{\mu} = 2\pi/T$. ...(2)

Using (2), (1) gives $\quad v^2 = (4\pi^2/T^2)(a^2 - x^2) \quad$ or $\quad v^2T^2 = 4\pi^2(a^2 - x^2)$ ...(3)

Let $v = v_1$ when $x = a/2$. Then (3) gives

$$v_1^2T^2 = 4\pi^2(a^2 - a^2/4) \quad \text{or} \quad v_1^2T^2 = 3\pi^2a^2 \qquad ...(4)$$

By problem, when new amplitude is a_1 (say), then when $x = a/2$, $v = 2v_1$. Since period T remains unchanged, so from (3), we have

$$4v_1^2T^2 = 4\pi^2(a_1^2 - a^2/4) \quad \text{or} \quad 4 \times 3\pi a^2 = 4\pi^2(a_1^2 - a^2/4), \text{ by (4)}$$

or $\quad a_1^2 - a^2/4 = 3a^2 \quad$ or $\quad a_1^2 = 3a^2 + a^2/4 \quad$ or $\quad a_1 = (a/2)\,(13)^{1/2}$

Ex. 13. *A body moving in a straight line OAB with S.H.M. has zero velocity when at the points A and B whose distances from O are a and b respectively and has velocity v when half way between them. Show that the complete period is* $\pi(b - a)/v$.

Sol. Since velocities at A and B are zero, so A and B are extreme positions.

$\therefore \quad$ amplitude $= (AB)/2 = (OB - OA)/2 = (b - a)/2$

O A C B

Let C be the mid-point of AB. Since velocity at C is v, so v is the maximum velocity as it is the velocity at the mean position.

$\therefore \quad v = (\text{amplitude}) \times \mu^{1/2} = \{(b - a)/2\}\mu^{1/2} \quad$ or $\quad \mu^{1/2} = 2v/(b - a)$

$\therefore$ Periodic time $= T = 2\pi/\mu^{1/2} = \pi(b - a)/v$.

Ex. 14. *A particle is performing a S.H.M. of period T about a centre O and it passes through a point P where OP = b with velocity v in the direction OP. Prove that the time which elapses before it returns to P is* $(T/\pi) \tan^{-1} (vT/2\pi b)$. **[Kumaun 2001, 02; Purvanchal 2008; Rohilkhand 2000;] Meerut 2008, 2012; Purvanchal 2007; Nagpur 2010; Delhi B.A. (Prog.) III, 2014, 15, 17]**

Sol. With centre O as origin, let the equation of S.H.M. be $d^2x/dt^2 = -\mu x$. Let a be the amplitude. If the time be measured from extreme position A and v be velocity at any distance x from O, then we have

← b →
A′ O P A

Periodic time $= T = 2\pi/\mu^{1/2} \quad$ or $\quad \mu^{1/2} = 2\pi/T$ ...(1)

Let time from A to P be t_1. Then using $x = a \cos t\mu^{1/2}$ at P, where $x = b$ and $t = t_1$, we get

$$b = a \cos t_1\mu^{1/2} \qquad ...(2)$$

Let $v = v$ when $x = b$ at P. Then using $v^2 = \mu(a^2 - x^2)$, we get

$$v^2 = \mu(a^2 - b^2) = \mu(a^2 - a^2 \cos^2 t_1\mu^{1/2}) = \mu a^2 \sin^2 t_1\mu^{1/2}$$

$\therefore$ $$v = \mu^{1/2}\, a \sin t_1\mu^{1/2} \qquad ...(3)$$

Dividing (3) by (2), $$v/b = \mu^{1/2} \tan t_1\mu^{1/2} = (2\pi/T) \tan (2\pi t_1/T), \text{ by (1)}$$

$\therefore$ $\tan (2\pi t_1/T) = vT/2\pi b$ or $t_1 = (T/2\pi) \tan^{-1} (vT/2\pi b)$ ...(4)

$$\text{Required time} = \text{time from } P \text{ to } A + \text{time from } A \text{ to } P = t_1 + t_1 = 2t_1 = (T/\pi) \tan^{-1} (vT/2\pi b).$$

Ex. 15. *A point executes S.H.M. such that in two of its positions velocities are u, v and the two corresponding accelerations are α, β; show that the distance between the two positions is* $(v^2 - u^2)/(\alpha + \beta)$ *and the amplitude of the motion is* $\{(v^2 - u^2)(\alpha^2 v^2 - \beta^2 u^2)\}^{1/2}/(\beta^2 - \alpha^2)$. *Also find the time period.* **[Kanpur 2011, Meerut 2009, 11]**

Sol. Let μ be the intensity and a the amplitude of S.H.M. Let x_1 and x_2 be the distances from the centre of the two positions where u and v are the velocities and α and β are the accelerations respectively. But, the velocity v of the particle at a distance x from the centre is given by $v^2 = \mu(a^2 - x^2)$ and the accelerations at distance x from the centre is μx. Hence we have

$$u^2 = \mu (a^2 - x_1^2) \quad ...(i) \qquad v^2 = \mu(a^2 - x_2^2) \quad ...(ii)$$

$$\alpha = \mu x_1 \quad ...(iii) \qquad \beta = \mu x_2 \quad (iv)$$

Subtracting (*i*) from (*ii*), $$v^2 - u^2 = \mu (x_1^2 - x_2^2) \qquad ...(v)$$

Adding (*iii*) and (*iv*), $$\alpha + \beta = \mu (x_1 + x_2) \qquad ...(vi)$$

Dividing (*v*) and (*vi*), $$(v^2 - u^2)/(\alpha + \beta) = x_1 - x_2$$

$\therefore$ Required distance $= x_1 - x_2 = (v^2 - u^2)/(\alpha + \beta)$ ...(*vii*)

Now, subtracting (*iii*) from (*iv*), we have

$$\beta - \alpha = \mu(x_2 - x_1) = \mu(u^2 - v^2)/(\alpha + \beta), \text{ using (vii)}$$

$\therefore$ $$\mu = (\beta^2 - \alpha^2)/(u^2 - v^2) \qquad ...(viii)$$

Hence, $$\text{time period} = T = 2\pi/\mu^{1/2} = 2\pi\{(u^2 - v^2)/(\beta^2 - \alpha^2)\}^{1/2}$$

Now, from (*i*), $$u^2/\mu = a^2 - x_1^2 \quad \text{or} \quad a^2 = u^2/\mu + x_1^2$$

$\therefore$ $$a^2 = u^2/\mu + \alpha^2/\mu^2, \text{ since by (iii), } x_1 = \alpha/\mu$$

or $$a^2 = u^2 \frac{u^2 - v^2}{\beta^2 - \alpha^2} + \alpha^2 \left(\frac{u^2 - v^2}{\beta^2 - \alpha^2}\right)^2 = \frac{u^2 - v^2}{\beta^2 - \alpha^2}\left\{u^2 + \frac{\alpha^2 (u^2 - v^2)}{\beta^2 - \alpha^2}\right\}, \text{ by (viii)}$$

or $$a^2 = \{(u^2 - v^2)(\beta^2 u^2 - \alpha^2 v^2)\}/(\beta^2 - \alpha^2)^2$$

$\therefore$ $$\text{amplitude} = a = \{(u^2 - v^2)(\beta^2 u^2 - \alpha^2 v^2)\}^{1/2}/(\beta^2 - \alpha^2)$$

Ex. 16. (a) *If in a S.H.M. u, v, w be the velocities at distance a, b, c from a fixed point on the straight line which is not the centre of force, show that the period T is given by the equation*

$$\frac{4\pi^2}{T^2}(a - b)(b - c)(c - a) = \begin{vmatrix} u^2 & v^2 & w^2 \\ a & b & c \\ 1 & 1 & 1 \end{vmatrix}$$

[Rohilkhand 2006, 08, 11, Avadh 2008; Purvanchal 2011; Agra 2007; Kanpur 2009, 2014, 2017; Lucknow 2001]

Sol. Let O be the centre of force and O' the fixed point from which the distances of the points A, B and C are given to be a, b, c respectively. Let $OO' = d$. Here $O'A = a$, $O'B = b$, $OC' = c$ so that $OA = d + a$, $OB = d + b$ and $OC = d + c$. Let μ be the intensity of the S.H.M. whose amplitude is k. Then the velocity V at a distance x is given by $V^2 = \mu(k^2 - x^2)$. Using this formula for point A where $x = OA = d + a$ and $V = u$, we get

d
O O′ A B C

$$u^2 = \mu\{k^2 - (d + a)^2\} \quad \text{or} \quad u^2/\mu = k^2 - (d^2 + 2da + a^2)$$

or $$(u^2/\mu + a^2) + 2ad + d^2 - k^2 = 0 \qquad ...(1)$$

Similarly for B, $$(v^2/\mu + b^2) + 2bd + d^2 - k^2 = 0 \qquad ...(2)$$

and for C, $$(w^2/\mu + b^2) + 2cd + d^2 - k^2 = 0 \qquad ...(3)$$

Eliminating $2d$ and $(d^2 - k^2)$ from (1), (2) and (3), we get

$$\begin{vmatrix} u^2/\mu + a^2 & a & 1 \\ v^2/\mu + b^2 & b & 1 \\ w^2/\mu + c^2 & c & 1 \end{vmatrix} = 0$$

or $$\begin{vmatrix} u^2/\mu & a & 1 \\ v^2/\mu & b & 1 \\ w^2/\mu & c & 1 \end{vmatrix} + \begin{vmatrix} a^2 & a & 1 \\ b^2 & b & 1 \\ c^2 & c & 1 \end{vmatrix} = 0$$

or $$\begin{vmatrix} a^2 & a & 1 \\ b^2 & b & 1 \\ c^2 & c & 1 \end{vmatrix} = -\frac{1}{\mu}\begin{vmatrix} u^2 & a & 1 \\ v^2 & b & 1 \\ w^2 & c & 1 \end{vmatrix}$$

Operating $R_2 - R_1$ and $R_3 - R_1$ on L.H.S. and interchanging rows and column on R.H.S., we get

$$\begin{vmatrix} a^2 & a & 1 \\ b^2 - a^2 & b - a & 0 \\ c^2 - a^2 & c - a & 0 \end{vmatrix} = -\frac{1}{\mu}\begin{vmatrix} u^2 & v^2 & w^2 \\ a & b & c \\ 1 & 1 & 1 \end{vmatrix}$$

or $$-(b-a)(c-a)\begin{vmatrix} b+a & 1 \\ c+a & 1 \end{vmatrix} = -\frac{1}{\mu}\begin{vmatrix} u^2 & v^2 & w^2 \\ a & b & c \\ 1 & 1 & 1 \end{vmatrix}$$

or $$\mu(a-b)(b-c)(c-a) = \begin{vmatrix} u^2 & v^2 & w^2 \\ a & b & c \\ 1 & 1 & 1 \end{vmatrix} \qquad \ldots(4)$$

But $T = 2\pi/\mu^{1/2}$ so that $\mu = 4\pi^2/T^2$. So (4) gives

$$\frac{4\pi^2}{T^2}(a-b)(b-c)(c-a) = \begin{vmatrix} u^2 & v^2 & w^2 \\ a & b & c \\ 1 & 1 & 1 \end{vmatrix} \qquad \ldots(5)$$

Ex. 16. (*b*) *In a S.H.M., the velocities at distances a, b, c from a fixed point on the straight line of motion (not coinciding with the centre of foce) are u, v, w respectively. Prove that the time period T is given by*

$$4\pi^2(b-c)(c-a)(a-b) = T^2\Sigma(b-c)u^2.$$

Sol. Proceed exactly as in Ex. 16(*a*) upto equation (5). Expanding the determinant on the R.H.S. of (5), we get

$$(4\pi^2/T^2)(b-c)(c-a)(a-b) = (b-c)u^2 + (c-a)v^2 + (a-b)w^2$$

or $$4\pi^2(b-c)(c-a)(a-b) = T^2\Sigma(b-c)u^2, \text{ as required.}$$

Ex. 17. *A particle rests in equilibrium under the attraction of two centres of force which attract directly as the distance, their intensities being μ and μ′; the particle is displaced slightly towards one of them, show that the time of a small oscillation is $2\pi/(\mu + \mu')^{1/2}$.*

[Avadh 2007; Lucknow 2007, 09; Agra 2003; Kanpur 2000; Nagpur 2010]

O E← x →P O′

Sol. Let O and O' be the two given centres of forces, their intensities being μ and μ' respectively. Let a particle of mass m be in equilibrium at E under the attraction of the given forces. Let $OE = a$ and $O'E = a'$. Then the forces acting on the particle at E are $m\mu a$ towards O and $m\mu' a'$ towards O'. Since the particle is in equilibrium at E, we have $m\mu a = m\mu' a'$ or $\mu a = \mu' a'$ ...(1)

Let the particle be slightly displaced towards O' and let it go. Let P be its position at any time t such that $EP = x$. Then the forces acting on the particle at P are $m\mu(a + x)$ towards O and $m\mu'(a' - x)$ towards O'. Then the equation of motion of the particle is

$m(d^2x/dt^2) = m\mu'(a' - x) - m\mu(a + x)$ or $d^2x/dt^2 = \mu' a' - \mu a - (\mu' + \mu)x$

Thus, $d^2x/dt^2 = -(\mu + \mu')x$, as by (1), $\mu a = \mu' a'$

This represents S.H.M. of period $2\pi/(\mu + \mu')^{1/2}$

Ex. 18. *A particle of mass m is attached to a light wire which is stretched tightly between two fixed points with a tension T. If a, b be the distances of the particle from the two ends, prove that the period of small transverse oscillation of mass m is* $2\pi\{mab/T(a+b)\}^{1/2}$. **(Avadh 2009)**

Prove also that for a wire of given length the period is longest when the particle is attached to the middle point.

Sol. Let A and B be two fixed points of the given wire and let a particle of m be attached at the point O of the wire where $OA = a$ and $OB = b$. Let the particle be given a small transverse displacement (*i.e.*, a displacement at right angles to AB) and let go. Let P be the position of the particle at any time t such that $OP = x$. Since the displacement is small, it follows that the tension T in the string in the displaced position is same as in the original position.

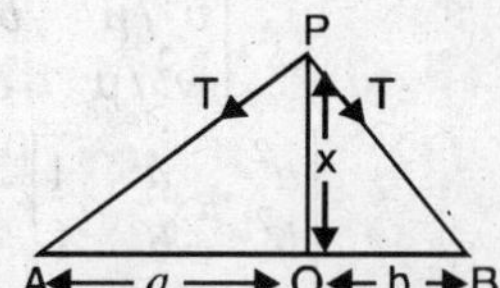

From figure, $AP = (a^2 + x^2)^{1/2}$ and $BP = (b^2 + x^2)^{1/2}$

Resultant force on the particle at P along PO

$= T\cos\angle APO + T\cos\angle BPO = T.\{x/(a^2+x^2)^{1/2}\} + T.\{x/(b^2+x^2)^{1/2}\}$

$= T\,(x/a)\,(1 + x^2/a^2)^{-1/2} + T(x/b)\,(1 + x^2/b^2)^{-1/2}$

$= T\,[(x/a)\,\{1 - (1/2)\,(x^2/a^2) + \ldots\} + (x/b)\,\{1 - (1/2)\,(x^2/b^2) + \ldots\}]$

$= T(x/a + x/b)$, neglecting higher powers of x/a and x/b as x is small

Now, the equation of motion of the particle is

$m(d^2x/dt^2) = -\{T(a+b)x\}/ab$ or $d^2x/dt^2 = -\{T(a+b)/mab\}x$,

which is the standard form of a S.H.M. with centre at the origin.

Required periodic time $= 2\pi/\{T(a+b)/mab\}^{1/2} = 2\pi\{mab/T(a+b)\}^{1/2}$ ...(1)

2nd part. If the length of the wire is c so that $a + b = c$, then from (1) we see that periodic time is maximum if ab is maximum. We know that when $a + b = c =$ constant, then ab is maximum if $a = b = c/2$. Hence the periodic time is longest when the particle is attached to the middle point.

Ex. 19. *A body is attached to one end of an inelastic string, and the other end moves in a vertical line wih S.H.M. of amplitude a, making n oscillations per second. Show that the string will not remain tight during the motion unless* $n^2 < g/(4\pi^2 a)$.

Sol. Let the string remain tight during the motion so that the body of mass m (say) also moves in an identical S.H.M. Suppose the body moves in S.H.M. between A and A' and O is the centre of motion. Then, $OA = OA' = a$. Let P be the position of the body at any time such that $OP = x$. Let T be tension in the string. Then the equation of motion of the body is

$m(d^2x/dt^2) = T - mg$ or $T = mg + m(d^2x/dt^2)$ ...(1)

showing that T is least when d^2x/dt^2 is least. Since the least value of d^2x/dt^2 is $-\mu a$, it follows that the least value of $T = mg - m\mu a$.

Obviously, the string will remain tight if this least tension is positive, *i.e.*,

if $T > 0$, *i.e.*, $mg - m\mu a > 0$

or $g - \mu a > 0$ or $\mu < g/a$ ...(2)

Since the body is making n oscillations per second,

$\therefore$ period $= 1/n = 2\pi/\sqrt{\mu}$ or $\mu = 4\pi^2 n^2$ ...(3)

From (2) and (3), $4\pi^2 n^2 < g/a$ or $n^2 < g/(4\pi^2 a)$.

Ex. 20. **(*a*)** *A particle of mass m is placed on a horizontal board, which is made to execute vertical simple harmonic oscillations of period T and amplitude a. Show that the particle does not lose contact with the board at any time if* $a < gT^2/4\pi^2$. **(Delhi B.A. (Prog.) III, 2014]**

(b) *A horizontal shelf is moved up and down with S.H.M. of period 1/2 sec. What is the amplitude admissible in order that a weight placed on the shelf may not be jerked off.* **(Guwahati 2007)**

(c) *A horizontal shelf executes S.H.M. of period one second in a vertical line, and its amplitude is 0.25 m. Show that a book weighing 2 kg and resting on the shelf will not leave it. Find the greatest and least pressure exerted by the book on the shelf.*

A R P O A' mg x

Sol. (*a*) Let a body of mass m be in contact with the board so that the body also moves in an identical S.H.M. of the board. Let the board move in S.H.M. between A and A' and O is the centre of motion. Then amplitude $= OA = OA' = a$. Let P be the position of the body at any time t such that $OP = x$. Let R be the reaction of the board. Then, equation of motion of the body is

$$m(d^2x/dt^2) = R - mg \quad \text{or} \quad R = mg + m(d^2x/dt^2) \qquad \ldots(1)$$

showing that R is least when d^2x/dt^2 is least. Since the least value of d^2x/dt^2 is $-\mu a$, it follows that the least value of $R = mg - m\mu a$.

Obviously the body will not be jerked off if this least value of R remains non-negative, *i.e.*, if

$$R \geq 0 \quad \text{or} \quad mg - m\mu a \geq 0 \quad \text{or} \quad \mu \leq g/a \qquad \ldots(2)$$

Now, $\quad T = \text{ time period} = 2\pi/\sqrt{\mu} \Rightarrow \mu = 4\pi^2/T^2 \qquad \ldots(3)$

$\therefore$ (2) and (3) give $\;4\pi^2/T^2 \leq g/a \quad \text{or} \quad a \leq gT^2/4\pi^2 \qquad \ldots(4)$

(*b*) Do as in part (*a*) upto equation (4). Here, given that $T = 1/2$ so that (4) reduces to $a \leq g/16\pi^2$. So the greatest value of the amplitude $= a = g/16\pi^2$.

(*c*) Proceed as in part (*a*) upto equation (4). Here $T = 1$ and $a = 0.25$ m $= (1/4)$ m. Then (4) reduces to $1/4 \leq g/4\pi^2$ or $\pi^2 \leq g$, which is true and hence the look will not leave the shelf.

Since maximum and minimum value of d^2x/dt^2 are μa and $-\mu a$ respectively.

So, by (1), $\quad R_{max} = m(g + \mu a) \quad \text{and} \quad R_{min} = \mu(g - \mu a) \qquad \ldots(5)$

But $\; a = (1/4)$ m and $T = 1 \quad$ or $\quad 2\pi/\sqrt{\mu} = 1$ so that $\mu = 4\pi^2$

$\therefore$ (5) $\Rightarrow \quad R_{max} = 2(g + \pi^2) \quad$ and $\quad R_{min} = 2(g - \pi^2)$, in absolute units.

EXERCISE

1. A particle executes S.H.M. If when at distances of 3 cm and 4 cm from the mean position its velocities are 8 cm/sec and 6 cm/sec, find its period and maximum acceleration.

[Ans. π sec; 20 cm/sec^2]

2. A body moves in a straight line with S.H.M. If T is the time period, v the velocity and f the acceleration in any position, show that the expression $f^2T^2 + 4\pi^2v^2$ is of constant value.

3. A particle executing simple harmonic motion has magnitude a, show that the distance of the point from the centre at which the velocity is half the maximum is $a\sqrt{3}/2$.

[Kanpur 2019; Rohilkhand 2008; G.N.D.U. Amritsar 2010]

4. A particle is executing S.H.M. of period T about centre O and it passes through a point P with velocity v in the direction OP. If the particle returns to P in time t, show that the distance OP is given by

$$OP = (2\pi/vT)\cot(vt/T).$$

5. In a S.H.M. of period $2\pi/w$, if the initial displacement be x_0 and initial velocity u_0, prove that

(*i*) amplitude $= (x_0^2 + u_0^2/w^2)^{1/2}$

(*ii*) position at time $t = (x_0^2 + u_0^2/w^2)^{1/2} \cos\{wt - \tan^{-1}(u_0/wx_0)\}$

(*iii*) time to the position of rest $= (1/w)\tan^{-1}(u_0/wx_0)$ **(Meerut 2007)**

6. A particle of mass m is constrained to execute a S.H.M. under a force towards O, of magnitude $m\omega^2x$, x being the particle's displacement from O. When passing through O its velocity is V and when its velocity has become $V/2$ in the same direction, an impulse I is applied to the particle in the direction of its motion. Assuming the same law of force, find the time and total distance travelled from O to first positon of instaneous rest.

Ans. Distance $= x = -(I\sqrt{3}/2m\omega)\cos\omega t + (V/\omega + I/2m\omega)\sin\omega t$ and time t is given by $\tan\omega t = -(2mV + I)/I\sqrt{3}$

7. If a be the amplitude and T, the period of a particle executing S.H.M. in a straight line through the centre of force, show that the time taken by the particle to travel a distance x from the centre of force is $(T/2\pi)\sin^{-1}(x/a)$ and that the velocity in that position is $2\pi(a^2 - x^2)^{1/2}/T$.

4.5 Elastic strings and spiral springs. Hooke's law (Agra 2003; Meerut 2005)

When an elastic string (or a spiral spring) is fixed at one end and pulled at the other end, then the length of the string (or spring) increases and this increase in length is known as its extension. The force which tends to restore an elastic string (or spring) to its natural shape is called tension in the string (or spring) and is given by an experimental law known as Hooke's law, which states : "The tension of an elastic string (or spring) is proportional to the extension of the string (or spring) beyond its natural length".

If x is the stretched length of an elastic string (or a spiral spring) of natural length l, then by Hooke's law the tension in the string (or spring) is given by $T = \lambda(x - l)/l$, where λ is called the modulus of elasticity, and it depends on the material of the string (or spring). Note carefully that the direction of the tension is always opposite to the extension.

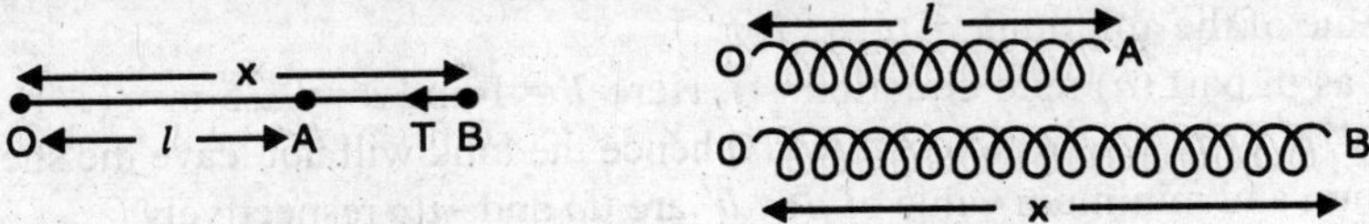

(*i*) Elastic string (*ii*) Spriral spring

Theorem. *Work done against the tension in stretching a light elastic string, is equal to the product of its extension and the mean of its final and initial tensions.*

Proof. Refer solved example on page 2.1, Chapter 2.

4.6 Type 2 : Problems based on particle attached to a horizontal elastic string

Ex. 1. *One end of an elastic string whose modulus of elasticity is λ and whose natural length is a is tied to a fixed point on a smooth horizontal table, and the other end is tied to a mass m lying on the table. The particle is pulled to a distance where the extension of string becomes b and then let go; describe the character of motion and find the period of one complete oscillation.*

(Avadh 2009; Kanpur 2003; Kumaun 94; Rohilkhand 97)

Sol. Let a string OA of natural length l lie on a small horizontal table. Let O be the fixed end and a particle of mass m be attached at A. The mass m is pulled upto B, where $AB = b$ and then let go. Let P be the position of the particle at any time t, where $AP = x$. Then, by Hooke's law, the tension T in the string at P $= (\lambda/a)x$, acting towards O. Hence the equation of motion of the particle at P is.

$$m(d^2x/dt^2) = -(\lambda/a)x \quad \text{or} \quad d^2x/dt^2 = -(\lambda/am)x, \qquad \ldots(1)$$

showing that the motion of the particle is simple harmonic with centre at the origin A. The equation (1) holds good as long as the string is stretched. Clearly the string becomes slack just as the particle reaches A. Therefore equation (1) holds good for motion from B to A.

Re-writing (1), $v(dv/dx) = -(\lambda/am)x$ or $2v\,dv = -(2\lambda/am)x\,dx$

Integrating it, $v^2 = -(\lambda/am)x^2 + C$, where C is a constant. ...(2)

Initially at B, when $x = b$, $v = 0$. So (2) gives $C = (\lambda/am)b^2$.

So (2) becomes $v^2 = (\lambda/am)(b^2 - x^2)$, ...(3)

giving velocity at point between B and A. Let V be velocity at A. So at A, when $x = 0$, $v = V$. Then (3) gives

$$V = (\lambda/am)^{1/2}b, \text{ in the direction } AO. \qquad \ldots(4)$$

Discussion of motion. The motion from B to A is simple harmonic. When the particle reaches A, the string becomes slack and there is no more S.H.M. of the particle. But due to velocity V at A given by (4) in the direction AO, the particle continues to move to the left of A. Since no force acts on the particle during motion from A to A' where $OA' = OA = a$, the particle moves with uniform velocity V (by Newton's first law of motion) from A upto A'. The string will be again tight when the particle reaches A'. Beyond A' the particle moves under retardation due to tension in the string which would try to pull it towards O. Hence the particle after some time would again come to an instantaneous rest at B', where $A'B' = AB = b$. Clearly the time from A' to B' will be the same as the time from B to A. At B' the velocity will be zero but attraction will be $\lambda b/am$ towards O, so the particle would begin to move towards O. Then particle will retrace it path and it will again come to instantaneous rest at B. Thus the particle will continue to oscillate between B and B'.

Now, t_1 = Time from B to $A = (1/4) \times$ [time period of S.H.M. given by (1)]

or $$t_1 = (1/4) \times \{2\pi/(\lambda/am)^{1/2}\} = (\pi/2)\,(am/\lambda)^{1/2} \qquad \ldots(5)$$

Also, t_2 = time from A to O with uniform velocity V

$$= \frac{OA}{V} = \frac{a}{(\lambda/am)^{1/2}\,b} = \left(\frac{am}{\lambda}\right)^{1/2} \cdot \frac{a}{b}, \text{ by (4)} \qquad \ldots(6)$$

Hence time taken by the particle in moving from B to O

$$= t_1 + t_2 = (\pi/2)\,(am/\lambda)^{1/2} + (am/\lambda)^{1/2}\,(a/b), \text{ by (5) and (6)} \qquad \ldots(7)$$

$\therefore$ Required time of one complete oscillation

$$= 4 \times (\text{time from } B \text{ to } O) = 4(t_1 + t_2)$$
$$= 4(\pi/2 + a/b)\,(am/\lambda)^{1/2} = 2(\pi + 2a/b)\,(am/\lambda)^{1/2}, \text{ by (7)}$$

Ex. 2. *One end of an elastic string (modulus of elasticity λ) whose natural length is a, is fixed to a point on a smooth horizontal table and the other is tied to a particle of mass m, which is lying on the table. The particle is pulled to a distance from the point of attachment of the string equal to twice its natural length and then let go. Find the time of complete oscillation.*

Sol. Do like Ex. 1. Here $b = a$. **[Ans.** $2(\pi + 2)\,(am/\lambda)^{1/2}$**]**

Ex. 3. *A and B are two points in a horizontal plane distant l apart and are joined by a light elastic string of natural length l_o and modulus λ, where $l_o < l$. The string has a particle of mass m attached to its mid-point. This particle is drawn aside a small distance and released. Show that the ensuring motion of the particle is approximately simple harmonic with period* $\pi\{mll_o/\lambda(l - l_o)\}^{1/2}$.

[Delhi B.Sc. (Prog) III, 2010, 11; Delhi Maths (H) 2006]

Sol. Here a light elastic string is stretched tightly between the fixed points A and B. Let a particle of mass m be attached at the middle-point O of AB so that $AO = BO = l/2$. Let the particle be displaced slightly perpendicular to AB and then let go. Let P be the position of the particle at any time t, where $OP = x$. Let T be tension in the string PA and PB. Let $\angle APO = \angle BPO = \theta$. From ΔAOP and ΔBOP, $AP = BP = (x^2 + l^2/4)^{1/2}$ and so

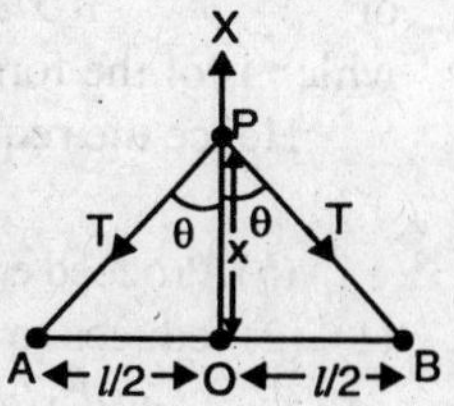

$$\cos\theta = OP/AP = x/(x^2 + l^2/4)^{1/2}.$$

$$\therefore \cos\theta = \frac{x}{(l/2)(1 + 4x^2/l^2)^{1/2}} = \frac{x}{(l/2)}\left(1 + \frac{4x^2}{l^2}\right)^{-1/2}$$
$$= (2x/l)\,(1 - 2x^2/l^2 + \ldots) = 2x/l$$

[Neglecting higher powers of x/l, being small]

Extension in string AP

$$= AP - (l_0/2) = (l^2/4 + x^2)^{1/2} - (l_0/2) = (l/2)(1 + 4x^2/l^2)^{1/2} - (l_0/2)$$
$$= (l/2)\,(1 + 2x^2/l^2 + \ldots) - (l_0/2) = (l - l_0)/2$$

[Neglecting higher powers of x/l, being small]

$$\therefore \quad T = \lambda\frac{\text{extension}}{\text{natural length}} = \frac{\lambda(l - l_0)}{l_0}, \text{ upto first order of approximation}$$

Resultant force acting on the particle along $PO = T\cos\theta + T\cos\theta = 2T\cos\theta = 2\{\lambda(l - l_0)/l_0\}\,(2x/l) = 4\lambda(l - l_0)x/ll_0$ ...(1)

The equation of motion of the particle is

$$m\frac{d^2x}{dt^2} = -2T\cos\theta \quad \text{or} \quad \frac{d^2x}{dt^2} = -\frac{4\lambda(l - l_0)}{mll_0}x, \text{ by (1)}$$

This is the standard equation of a S.H.M. with centre at origin.

$\therefore$ Required time period $= 2\pi/\{4\lambda(l-l_0)/mll_0\}^{1/2} = \pi\{mll_0/\lambda(l-l_0)\}^{1/2}$

Ex. 4(*a*) *A particle of mass 'm' executes S.H.M. in the line joining the points A and B on a smooth table and is connected with these points by elastic strings whose tensions in equilibrium are each T, show that the time of an oscillation is $2\pi[mll'/T(l+l')]^{1/2}$, where l, l' are the extensions of the string beyond their natural lengths.* **(Purvanchal 2007; Garhwal 2007; Rohilkhand 2010)**

(*b*) *A particle is performing S.H.M. in the line joining two points A and B on a smooth plane and is connected with these points by elastic strings of natural lengths a and a'; the moduli of elasticity being λ and λ' respectively. Show that the periodic time is given by $2\pi\{m/(\lambda/a+\lambda'/a')\}^{1/2}$.*

Sol. (*a*) Let the given particle of mass m be in equilibrium position at O being pulled by two horizontal strings OA and OB whose other ends are connected to two fixed points A and B. Let a, a' be the natural lengths, l, l' be extensions beyond the natural lengths and λ, λ' be the modulus of elasticity of the string OA and OB respectively. Then, in the position of equilibrium (refer figure (*i*)), we have

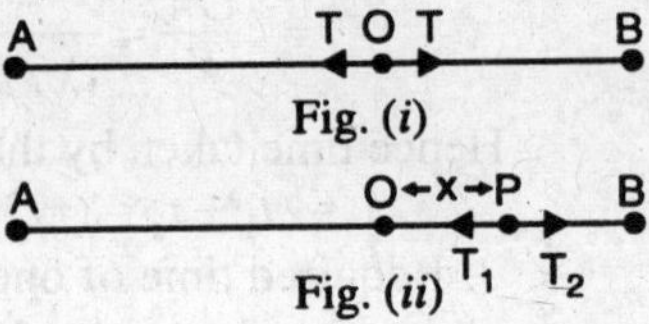

$$\text{Tension in the string } OA = \text{Tension in the string } OB$$

or $$\lambda(l/a) = \lambda'(l'/a') = T \text{ (given)} \quad \ldots(1)$$

Now, suppose the particle is slightly displaced to the position P and then let go (see figure (*ii*)). Let P be the position of the particle at any time t, where $OP = x$. Then, we have

$$T_1 = \text{tension in the string } AP = \lambda(l+x)/a, \text{ acting along } PA$$

and $$T_2 = \text{tension in the string } PB = \lambda'(l'-x)/a', \text{ acting along } PB.$$

Then, the equation of motion of the particle at P is

$$m(d^2x/dt^2) = T_2 - T_1 = \lambda'(l'-x)/a' - \lambda(l+x)/a$$

or $$m(d^2x/dt^2) = \lambda'(l'/a') - \lambda(l/a) - (\lambda/a + \lambda'/a')x \quad \ldots(2)$$

From (1), $$\lambda/a = T/l \quad \text{and} \quad \lambda'/a' = T/l' \quad \ldots(3)$$

Using (1) and (3), (2) reduces to

$$m(d^2x/dt^2) = T - T - (T/l + T/l')x = -T(l+l')x/ll'$$

or $$d^2x/dt^2 = -\{T(l+l')/mll'\}x,$$

which is of the form of the standard equation of S.H.M. $d^2x/dt^2 = -\mu x$.

Hence the required time of one complete oscillation

$$= T = 2\pi/\mu^{1/2} = 2\pi\{mll'/T(l+l')\}^{1/2}$$

(*b*) Proceed exactly as in part (*a*) upto relation (2). Now complete the solution as follows. Using (1), (2) reduces to

$$m(d^2x/dt^2) = -(\lambda/a + \lambda'/a')x \quad \text{or} \quad d^2x/dt^2 = -\{(\lambda/a + \lambda'/a')/m\}x$$

which is of the form of the standard equation of S.H.M. $d^2x/dt^2 = -\mu x$.

$\therefore$ $$\text{Required time period} = 2\pi/\mu^{1/2} = 2\pi\{m/(\lambda/a + \lambda'/a')\}^{1/2}$$

Ex. 4. (*c*) *Two light elastic string are fastened to a particle of mass m and their ends to fixed points so that the strings are taut. The modulus of each is λ, the tension T and lengths a and b. Show that the period of an oscillation along the line of the strings is $2\pi[mab/\{(T+\lambda)(a+b)\}]^{1/2}$.* **(Agra 1993)**

Sol. Proceed as in Ex. 4(*a*) and refer figures of that exercise. Here $\lambda = \lambda'$. Let a', b' be the natural lengths of OA and OB. Then in figure (*i*), $OA(=a)$ and $OB(=b)$ are extended lengths. Then, as before, equation (1) of Ex. 4(*a*) becomes

$$\lambda(a-a')/a' = \lambda(b-b')/b' = T$$

Also, $T_1 = \lambda(a+x-a')/a'$ and $T_2 = \lambda(b-x-b')/b'$ $\quad \ldots(1)$

Equation of motion of the particle at P is

$$m(d^2x/dt^2) = T_2 - T_1 = \lambda\{(b-x-b')/b' - (a+x-a'/a'\}$$

or $$m(d^2x/dt^2) = -\lambda(1/b' + 1/a')x + \lambda(b-b')/b' - \lambda(a-a')/a'$$

or $$m(d^2x/dt^2) = -\lambda(1/a' + 1/b')x, \text{ using (1)} \quad \ldots(2)$$

Now, (1) $\Rightarrow$ $\lambda a - \lambda a' = a'T$ and $\lambda b - \lambda b' = b'T$

$\therefore$ $a'(T+\lambda) = \lambda a$ and $b'(T+\lambda) = \lambda b$. So (2) becomes

$$m(d^2x/dt^2) = -\lambda\{(T+\lambda)/\lambda a + (T+\lambda)/\lambda b\}x = -\{(T+\lambda)(a+b)/ab\}x$$

or $$d^2x/dt^2 = -\{(T+\lambda)(a+b)/abm\}x, \qquad \ldots(3)$$

showing that the particle executes S.H.M. with centre at the origin O. Since the particle is slightly displaced from its equilibrium position O, it follows that during the entire motion of the particle both the strings remain tant and the entire motion is given by (2). Hence the required period T' of one complete oscillation is given by

$$T = \frac{2\pi}{\sqrt{\mu}} = \frac{2\pi}{\{(T+\lambda)(a+b)/mab\}^{1/2}} = 2\pi\left\{\frac{mab}{(T+\lambda)(a+b)}\right\}^{1/2}$$

Ex. 5. *An elastic string of natural length (a + b) where a > b and modulus of elasticity λ has a particle of mass m attached to it at a distance a from one end, which is fixed to a point A of a smooth horizontal plane. The other end of the string is fixed to a point B so that the string is just unstretched. If the particle be held at B and then released, show that it will oscillate to and fro through a distance* $b(\sqrt{a}+\sqrt{b})/\sqrt{a}$ *in a periodic time* $\pi(\sqrt{a}+\sqrt{b})(m/\lambda)^{1/2}$ **(I.A.S. 2003)**

Sol. Let O be the point where the particle of mass m is attached such that $OA = a$ and $OB = b$, where $a > b$. When the particle is held at B, the portion AO of the string is stretched whereas the portion OB is slack and hence when the particle is released from B, it moves towards O starting from rest at B.

Fig. (*i*): A —— a —— O ←x→ P —— B (T at P; O to B is b)

Fig. (*ii*): A —— E —— P′ ←y→ O —— B (T′ at P′; A to O is a, O to B is b)

Let P be the position of the particle between O and B at any time t such that $OP = x$ [refer figure (*i*)]. Then the tension in the string $AP = T = \lambda(x/a)$ acts towards O and the tension in the string PB is zero because it is slack. Hence the equation of motion of the particle at P is

$$m(d^2x/dt^2) = -T = -\lambda(x/a) \quad \text{or} \quad d^2x/dt^2 = -(\lambda/am)x \qquad \ldots(1)$$

which represents a S.H.M. with centre at O and amplitude $OB(=b)$.

Let t_1 be the time taken by the particle from B to O. Then

$$t_1 = (1/4) \times \text{time period of the S.H.M. given by (1)}$$

or $$t_1 = (1/4) \times \{2\pi/(\lambda/am)^{1/2}\} = (\pi/2)(am/\lambda)^{1/2} \qquad \ldots(2)$$

Re-writing (1), $v(dv/dx) = -(\lambda/am)x$ or $2v\,dv = -(2\lambda/am)x\,dx$

Integrating, $v^2 = -(\lambda/am)x^2 + C$, where C is a constant $\qquad \ldots(3)$

But Initially at B, $x = b$, $v = 0$. So (3) gives $C = \lambda b^2/am$.

Then (3) becomes $v^2 = (\lambda/am)(b^2 - x^2) \qquad \ldots(4)$

Let V be the velocity of the pariticle at O, where $x = 0$.

Then, by (4) $V^2 = (\lambda/am)b^2$ or $V = b(\lambda/am)^{1/2} \qquad \ldots(5)$

Starting from B, when the particle reaches O with velocity V given by (5), then the particle crosses O and moves towards the left of O. Let P' be the position of the particle at any time t, since it starts from O to the left of it and let $OP' = y$ [refer figure (*ii*)]. Then the tension in the string OP' is $T' = \lambda(y/b)$ acting towards O and the tension in the string $P'A = 0$ because it is stack. The equation of motion of the particle at P' is

$$m(d^2y/dt^2) = -T' = -\lambda(y/b) \quad \text{or} \quad d^2y/dt^2 = -(\lambda/bm)y \qquad \ldots(6)$$

Re-writing (6), $v(dv/dy) = -(\lambda/bm)y$ or $2v\,dv = -(2\lambda/bm)y\,dy$

Integrating, $v^2 = -(\lambda/bm)y^2 + D$, where D is a constant $\qquad \ldots(7)$

But at O, $y = 0$, $v^2 = V^2 = \lambda b^2/am$. So (7) gives $D = \lambda b^2/am$

$\therefore$ (7) becomes $v^2 = \lambda b^2/am - \lambda y^2/bm = (\lambda/am)(b^2/a - y^2) \qquad \ldots(8)$

While moving along OA, the particle is subjected to retardation given by (6). Suppose the particle comes to instantaneous rest at the point E between O and A such that $OE = c$. Then, at E, $y = c$ and $v = 0$.

So (8) gives $\quad O = (\lambda/am)(b^2/a - c^2) \quad$ or $\quad c = b(b/a)^{1/2}$

From E the particle retraces its path and comes to an instaneous rest at B. Thus the particle oscillates to and fro through a distance BE given by

$$BE = BO + OE = b + c = b + b(b/a)^{1/2} = b(\sqrt{a} + \sqrt{b})/\sqrt{a}.$$

Equation (6) represents a S.H.M. with centre at O. Let t_2 be the time taken by the particle from O to E, Then

$$t_2 = (1/4) \times \text{time period of the S.H.M. given by (6)}$$

or $$t_2 = (1/4) \times \{2\pi/(\lambda/bm)^{1/2}\} = (\pi/2)(bm/\lambda)^{1/2} \quad \ldots(9)$$

Hence, the required time period for making a complete oscillation between B and E

$$= 2(\text{time taken from } B \text{ to } E) = 2(t_1 + t_2)$$

$$= 2[(\pi/2)(am/\lambda)^{1/2} + (\pi/2)(bm/\lambda)^{1/2}] = \pi(m/\lambda)^{1/2}(\sqrt{a} + \sqrt{b})$$

Ex. 6. *Two particles of masses m_1 and m_2 are tied to the ends of an elastic string of natural length a and modulus of elasticity λ. They are placed on a smooth table so that the string is just taut and m_2 is projected with any velocity directly away from m_1. Prove that the string will become slack after the lapse of time $\pi\{am_1m_2/\lambda(m_1 + m_2)\}^{1/2}$.*

Ex. 7. (*a*) *An elastic string of natural length 2l can just support a certain weight when it is stretched till its whole length is 3l. One end of the string is now attached to a point on a smooth horizontal table and the same weight is attached to the other end. Prove that if the weight is pulled to any distance and then let go, the string will become slack again after a time $(\pi/2)(l/g)^{1/2}$*

[Mumbai 2010]

Sol. Refer figure (*i*). Let m be the mass which, hanging vertically, stretches the string from natural length OA (= $2l$) to the length OE (= $3l$), so that extension = $AE = 3l - 2l = l$. If λ be the modulus of elasticity, then by Hooke's law.

Tension in string at equilibrium position $E = T = \lambda(l/2l) = \lambda/2$. For equilibrium position E, $T = mg$ or $\lambda/2 = mg$ or $\lambda = 2mg$. Now, refer figure (*ii*)

Let the given string OA of natural length $2l$ be placed on a horizontal table with one end O being fixed. Let mass m be attached at A. Let the mass be pulled to any point B and then let go. Let P be the position of the particle at any time t where $AP = x$. Tension in the string $OP = T' = \lambda(x/2l)$. Then equation of motion of the particle at P is

$$m(d^2x/dt^2) = -T' = -\lambda(x/2l) = -(2mg)(x/2l), \text{ as } \lambda = 2mg$$

or $$d^2x/dt^2 = -(g/l)x, \text{ which represents a S.H.M.}$$

Fig. (*i*)

Fig. (*ii*)

As soon as the string passes A, it becomes slack.

Hence, the required time = time for motion from B to A

$= (1/4) \times (\text{time period}) = (1/4) \times (2\pi/\mu^{1/2}) = (1/4) \times (2\pi/(g/l)^{1/2}) = (\pi/2)(l/g)^{1/2}$

Ex. 7(*b*) *An elastic string of natural length l just supports a certain weight when it is stretched till its whole length is 3l/2. One end of the string is now attached to a smooth table and the same weight is attached to the other end. Prove that if the weight is pulled to any distance b and let go, the string will become slack again after a time $(\pi/2)(l/g)^{1/2}$*

Sol. Try yourself as in Ex. 7(*a*).

Ex. 8. *A particle is attached to the middle point of a uniform elastic string which is stretched between two points A and B on a smooth horizontal table. If this particle is pulled to some point C between A and B and then liberated, show that the motion is simple hormonic and find the period of the motion. What is the amplitude ?*

Hint : Let a = natural length and na be stretched length of string AB. Let m be mass of the particle and λ be the modulus of elasticity of the string. Let O be the middle point of AB and let the particle be displaced to some point C in AB and then released. [**Ans.** time period $= \pi(am/\lambda)^{1/2}$; amplitude = OC]

Ex. 9. *A light elastic string of elasticity modulus λ is stretched to double its length and is tied to two fixed points distant 2a apart. A particle of mass m, tied to its middle point, is displaced in the line of the* *string through a distance equal to half its distance from the fixed and released. Prove that the time of a complete oscillation is $\pi(am/\lambda)^{1/2}$ and maximum velocity is $(a\lambda/m)^{1/2}$.* **[Bhopal 2004]**

4.7 Particle suspended by an elastic string. (Vertical elastic string)

A light elastic string of natural length 'a' and modulus of elasticity λ is suspended by one end, to the other end is, tied particle of mass m. The particle is slightly pulled down and released, find the motion. **[Guwahati 2007]**

Sol. Let one end of an elastic string of natural length a and modulus of elasticity λ be attached to the fixed point O and with the other end A, a mass m be attached (refer figure (*i*)).

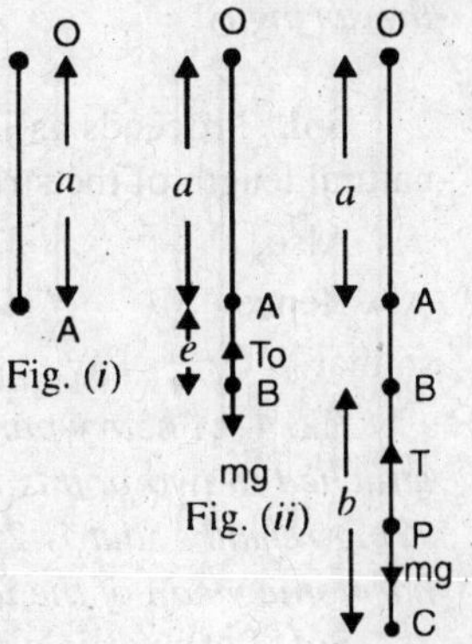

Due to weight mg of the particle the string OA is stretched and if B is the position of equilibrium of the particle such that $AB = e$, then tension T_0 in the string will balance the weight of the particle. Refer figure (*ii*). Thus, at B, we get

$$mg = T_0 \quad \text{or} \quad mg = \lambda(e/a) \qquad \ldots(1)$$

Fig. (*i*) Fig. (*ii*) Fig. (*iii*)

Let the particle be now pulled down a further distance $BC(= b$, say) and released. Let P be the position of the particle at any subsequent time t. Let $BP = x$ and let T be tension in the string. Then equation of motion of the particle is

$$m(d^2x/dt^2) = mg - T = mg - \lambda(e + x)/a = mg - \lambda(e/a) - \lambda(x/a)$$

or $$m(d^2x/dt^2) = -\lambda(x/a), \text{ using (1)}$$

or $$d^2x/dt^2 = -(\lambda/am)x \qquad \ldots(2)$$

which is of standard form $d^2x/dt^2 = -\mu x$ of S.H.M., where $\mu = \lambda/am$.

Here centre of oscillation is B, from which x is measured and amplitude $= BC = b$. The periodic time T of S.H.M. represented by (2) is given by

$$T = 2\pi/\mu^{1/2} = 2\pi/(\lambda/am)^{1/2} = 2\pi(am/\lambda)^{1/2} = 2\pi(e/g)^{1/2}, \text{ by (1)} \qquad \ldots(3)$$

Remark 1. The particle will execute S.H.M. so long as it is below A *i.e.*, the string is tight. If BC is greater than BA, the particle rises above the point A and thus, the string becomes slack. The particle moves under the force of gravity only. Thus, there are two situations.

Situation I. When $BC \le BA$. Then the particle will not rise above A and it will come to instantaneous rest before or just reaching A. The entire motion will be S.H.M. with centre at B, amplitude BC and period T given by (3).

Situation II. When $BC > AB$. Then the particle will rise above A (and so the string will become slack) and hence the motion will be simple harmonic only upto A and part of the motion above A will be free motion under gravity.

Remark 2. While dealing with spiral spring note that the law of compression is the same as that of extension. Hence the tension operates even when the particle rises above A. Hence the equation (2) holds throughout the motion and motion is S.H.M. of period given by (3).

TYPE 3 : SOLVED EXAMPLES ON VERTICAL ELASTIC STRINGS

Ex. 1. *A light elastic string of natural length l is hung by one end and to the other end are tied successively particles of masses m_1 and m_2. If t_1 and t_2 be the periods and c_1, c_2 the statical extensions corresponding to these weights, prove that $g(t_1^2 - t_2^2) = 4\pi^2(c_1 - c_2)$.* **[Bundelkhand 2010]**

Sol. Proceed as in Art. 4.7 upto equation (3) by taking $e(= AB)$ as statical extension corresponding to mass m. Then, time period T is given by

$$T = 2\pi(e/g)^{1/2} \qquad \ldots(i)$$

Here when $m = m_1$, $e = c_1$, $T = t_1$ and when $m = m_2$, $e = c_2$, $T = t_2$

So by (*i*), $t_1 = 2\pi(c_1/g)^{2}$ and $t_2 = 2\pi(c_2/g)^{1/2}$

Thus, $t_1^2 - t_2^2 = 4\pi^2(c_1/g - c_2/g)$ or $g(t_1^2 - t_2^2) = 4\pi^2(c_1 - c_2)$

Ex. 2. *An elastic string without weight of which the unstretched length is l and modulus of elasticity is ng, is suspended by one end and a mass m is attached to the other end. Show that the time of small vertical oscillation is $2\pi(ml/ng)$* **[Bundelkhand 92; Rohilkhand 2007]**

Sol. Proceed as in Art. 4.7 upto equation (3) by assuming that BC is very small so that $BC < BA$ and hence the entire motion is S.H.M. whose time period is given by $T = 2\pi(am/\lambda)^{1/2}$ ...(*i*)

Here $a = l$ and $\lambda = ng$. So (*i*) gives $T = 2\pi\,(ml/ng)^{1/2}$

Ex. 3. *A mass m hangs from a fixed point by a light spring and is given a small vertical displacement. Show that the motion is simple harmonic. If l is the length of the spring when the system is in equilibrium and n the number of oscillations per second, show that the natural length of the spring is*

$$l - (g/4\pi^2 n^2).$$ **[Rohilkhand 2002; Avadh 1994]**

Sol. Proceeds as in Art. 4.7 upto equation (3) noting that here $AB = e = l - a$, where a is the natural length of the string.

Also, $\quad 1/n = T = 2\pi\{(l-a)/g\}^{1/2}$, using (3)

Hence $\quad 1/n^2 = (4\pi^2/g)\,(l-a) \quad$ or $\quad g/4\pi^2 n^2 = l - a$

so that $\quad a = l - (g/4\pi^2 n^2)$, as required

Ex. 4. *A heavy particle is attached to one point of a uniform string. The ends of the string are attached to two points in a vertical line. Show that the period of a vertical oscillation in which the string remains taut is $2\pi(mh/2\lambda)^{1/2}$, where λ is the coefficient of elasticity of the string and h the harmonic mean of the unstreched lengths of the two parts of the string.*

Sol. Let a particle of mass m be attached to a point O of a string whose ends have been attached to fixed points A_1 and A_2 in a vertical line. The string is taut and the particle is in equilibrium at O. Let $OA_1 = a_1'$, $OA_2 = a_2'$ and let a_1, a_2 be the natural lengths of the stretched portions OA_1 and OA_2 of the string. Let T_1 and T_2 be tensions in portions OA_1 and OA_2 as shown in figure (*i*). Then, at O in equilibrium position, we have

$$\lambda(a_1' - a_1)/a_1 = mg + \lambda(a_2' - a_2)/a_2 \qquad \text{...(1)}$$

Fig. (*i*) Fig. (*ii*)

Let the particle be given a small displacement towards A_2 and then let go. Let P be its position at any time t such that $OP = x$. Let tensions in the strings A_1P and A_2P be T_1' and T_2' respectively. Then equation of motion of the particle at P is

$$m(d^2x/dt^2) = mg + T_2' - T_1'$$

or $\quad m(d^2x/dt^2) = mg + \lambda(a_2' - x - a_2)/a_2 - \lambda(a_1' + x - a_1)/a_1$

or $\quad m(d^2x/dt^2) = mg + \lambda(a_2' - a_2)/a_2 - \lambda(a_1' - a_1)/a_1 - \lambda(1/a_2 + 1/a_1)x$

or $\quad m(d^2x/dt^2) = -\lambda(a_1 + a_2)x/a_1 a_2$, using (1) ...(2)

Since h is harmonic mean of a_1 and a_2, we have

$$h = 2a_1a_2/(a_1 + a_2) \quad \text{so that} \quad (a_1 + a_2)/a_1 a_1 = 2/h \qquad \text{...(3)}$$

Using (3), (2) reduces to $\quad d^2x/dt^2 = -(2\lambda/hm)x$,

showing that the motion is simple harmonic of period

$$T = 2\pi/\sqrt{\mu} = 2\pi/(2\lambda/hm)^{1/2} = 2\pi(mh/2\lambda)^{1/2}$$

Ex. 5(*a*). *A light elastic string whose natural length has one end fixed to a point O, and to the other end is attached a weight which in equilibrium would produce an extension e. Show that if the weight be let fall from rest at O, it will come to stay instantaneously at a point distant $(2ae + e^2)^{1/2}$ below the position of equilibrium.* **[Kumaun 2003]**

(*b*) *A light elastic string of natural length l has one extremity fixed at a point O and the other attached to a stone, the weight of which in equilibrium would extend to a length l_1. Show that if the stone is dropped from rest at O, it will come to instantaneous rest at a depth $(l_1^2 - l^2)^{1/2}$ below the equilibrium position.* **[Lucknow 2011; Meerut 1999]**

(*c*) *A light elastic string of natural length a has one extremity fixed at a point O and the other attached to a body of mass m. The equilibrium length of the string with the body attached is a sec θ. Show that if the body be dropped from rest at O it will come to instantaneous rest at a depth a tan θ below the position of equilibrium.*

Sol. (*a*) For figure refer Art. 4.7. OA is the natural length of a string whose one end is fixed at O. See figure (*i*). Let B be the position of equilibrium of a stone of mass attached to the other end of the string and $AB = e$. Then at B, the tension T_0 of the string will balance the weight so that (refer figure (*ii*))

$$mg = T_0 \quad \text{or} \quad mg = \lambda(e/a) \qquad \ldots(1)$$

If the weight is dropped from rest at O, then motion from O to A will be due to gravity only as there will be no tension, the string being slack. If V be the velocity acquired by the weight at A, then by "$v^2 = u^2 + 2gh$", we have $V^2 = 0 + 2ga$ so that $V = (2ga)^{1/2}$ in vertically downward direction.

When the stone falls below A, the tension of the string comes into play. Since the force of tension acting vertically upwards remain less than the weight of the stone acting vertically downwards throughout the motion from A to B, hence the velocity goes on increasing. But when the stone begins to fall below B, its velocity goes on decreasing because now the force of tension exceeds the weight of the stone. Suppose the stone comes to an instantaneous rest at C, where $BC = b$.

Let P be position of the particle at any time t such that $BP = x$. Refer figure (*iii*). Let T be the tension in the string. Then equation of motion is

$$m(d^2x/dt^2) = mg - T = mg - \lambda(e + x)/a = mg - \lambda(e/a) - \lambda(x/a)$$

or
$$d^2x/dt^2 = -(g/e)x, \text{ using (1)} \qquad \ldots(2)$$

The equation (2) holds good so long as the string is stretched *i.e.*, for the motion of the stone between A and C. Re-writing (2), we have

$$v(dv/dx) = -(g/e)x \quad \text{or} \quad 2v\,dv = -(2g/e)x\,dx$$

Integrating, $\quad v^2 = -(g/e)x^2 + C$, C being a constant $\qquad \ldots(3)$

At A, when $x = BA = -e$, $v = V = (2ga)^{1/2}$. So (3) gives

$$2ga = -(g/e)e^2 + C \quad \text{or} \quad C = g(2a + e). \text{ So (3) becomes}$$

$$v^2 = g(2a + e) - (g/e)x^2 \qquad \ldots(4)$$

At C, when $x = b$, $v = 0$. Then (4) gives

$$0 = g(2a + e) - (g/e)b^2 \quad \text{or} \quad BC = b = (2ae + e^2)^{1/2} \qquad \ldots(5)$$

(*b*) Proceed as in part (*a*) upto (5). Here $a = l$ and $e = l_1 - l$. Hence from (5), the required distance $= BC = \{2l(l_1 - l) + (l_1 - l)^2\}^{1/2} = (l_1^2 - l^2)^{1/2}$

(*c*) Proceed as in part (*a*) upto (5). Here $e = a \sec\theta - a = a(\sec\theta - 1)$. Hence required distance $= BC = \{2a^2(\sec\theta - 1) + a^2(\sec\theta - 1)^2\}^{1/2}$

$$= [a^2(\sec^2\theta - 1)\}^{1/2} = a\tan\theta.$$

Ex. 6. *A heavy particle attached to a fixed point by an elastic string hangs freely, stretching the string by a quantity e. It is drawn down by an additional distance f and then let go; determine the height to which it will arise if* $f^2 - e^2 = 4ae$, *a being the unstretched length of the string.*

Sol. Do exactly as in Art. 4.7. upto equation (2). Then proceed as follow. Note that here $BC = f$ and $AB = e$ (given). Since we are given $f^2 - e^2 = 4ae$ = positive quantity, it follows that $f > e$ *i.e.*, $BC > BA$. Hence when the particle, while moving in S.H.M., reaches the point A, its velocity is not zero. But at A the string becomes slack and hence in its upward motion beyond A, the particle will move freely under gravity. Now, from (1), $\lambda/am = g/e$. Hence (2) reduces to

$$v(dv/dx) = -(g/e)x \quad \text{or} \quad 2v\,dv = -(2g/e)x\,dx$$

Integrating, $\quad v^2 = -(g/e)x^2 + C$, C being a constant $\qquad \ldots(3)$

At C, when $x = BC = f$, $v = 0$. So (3) gives $C = (g/e)f^2$

Then (3) becomes $\quad v^2 = (g/e)(f^2 - x^2) \qquad \ldots(4)$

giving velocity at point from C to A. Let V be velocity of the particle at A. Then, at $A : x = -e$, $v^2 = V^2$. So (4) gives, $V^2 = (g/e)(f^2 - e^2)$. Then, (4) reduces to

$$V^2 = (g/e)(f^2 - e^2) = (g/e)(4ae) = 4ag, \text{ as } f^2 - e^2 = 4ae \text{ (given)}$$

Now, consider motion of the particle from A with velocity V moving freely under gravity in upward direction. Let H be the maximum height attained by it.

Then, using $v^2 = u^2 - 2gh$, we get $0 = V^2 - 2gH$ or $2gH = 4ag$ or $H = 2a$

Hence the total height to which the particle rises above C

$$= CB + BA + H = f + e + 2a.$$

Ex. 7. (a) *A heavy particle of mass m is attached to one end of an elastic string of natural length a, whose modulus of elasticity is equal to the weight of the particle and the other end is fixed at O. The particle is let fall from O. Show that a part of motion is simple harmonic and that the greatest depth of the particle below O is* $(2 + \sqrt{3})\,a$. *Show that this depth is attained in time* $\{\sqrt{2} + \pi - \cos^{-1}(1/\sqrt{3})\}\,(a/g)^{1/2}$. **[Gorakhpur 96, 98; Lucknow 2001]**

(b) *One end of an elastic string is fastened at O; to the other end is fastened a particle heavy enough to stretch the string to double its natural length a. Show that if the particle is dropped from O, it will descend a distance* $(2 + \sqrt{3})a$ *before coming to rest.*

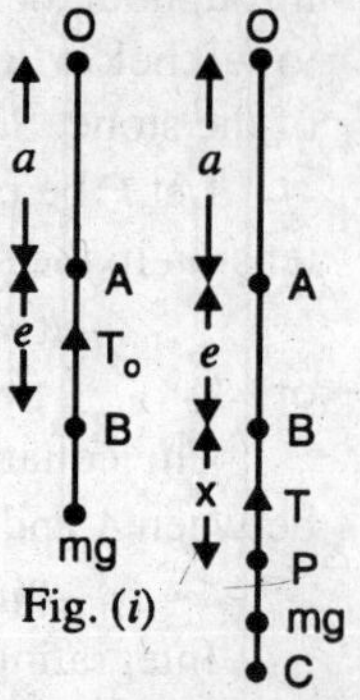

Sol. (a) Let $OA = a$ be the natural length of the string whose one end O is fixed. Let B be the position of equilibrium of a particle of mass m attached to the other end A of the string and let $AB = e$ = statical extension as shown in Fig. (*i*). Then, at B, the weight mg of the particle = the tension T_o in the string.

$\therefore$ $mg = \lambda(e/a) = (mg)\,(e/a)$, as $\lambda = mg$ (given) ...(1)

Thus, here $e = a$ ...(2)

Fig. (*i*)

Fig. (*ii*)

When the particle is let fall from rest at O, then motion from O to A is under gravity only, string being slack. If V is the velocity of the particle when it reaches A. then by using formula $v^2 = u^2 + 2gh$, we get

$V^2 = 0 + 2ga$ so that $V = (2ga)^{1/2}$ in downward direction ...(3)

When the particle falls below A, string is taut and tension of string begins to operate. While moving from A to B the velocity of the particle goes on increasing because the tension in the string is less than the weight of the particle. But while moving below B, the velocity of the particle goes on decreasing because now the tension in string is greater than the weight of the particle. Suppose the particle comes to an instantaneous rest at C, where $BC = b$ (say)

During the motion of the particle below B, let P be the position of the particle after any time t such that $BP = x$. Refer figure (*ii*). At P, the forces acting on the particle are its weight mg acting vertically downwards and tension $T = \lambda(e + x)/a$, acting vertically upwards. Hence, the equation of motion is

$$m(d^2x/dt^2) = mg - T = mg - \lambda(e + x)/a = mg - \lambda(e/a) - \lambda(x/a)$$

Using (1), the above equation becomes $d^2x/dt^2 = -\lambda(x/am)$

or $d^2x/dt^2 = -(mg)\,(x/am) = -(g/a)x$, as $\lambda = mg$ ...(4)

showing that the motion of the particle is simple harmonic with centre at B and amplitude BC and helds good from B to C.

Re-writing (4), $v(dv/dx) = (g/a)x$ or $2v\,dv = -(2g/a)x\,dx$

Integrating, $v^2 = -(g/a)x^2 + C$, C being a constant ...(5)

At A, when $x = -e = -a$, $v = V = (2ga)^{1/2}$. Hence (5) reduces to

$$2ga = -(g/a)a^2 + C \quad \text{so that} \quad C = 3ga.$$

Then, (5) becomes $v^2 = 3ga - (g/a)x^2 = (g/a)\,(3a^2 - x^2)$, ...(6)

which gives velocity of the particle at any point between A and C. At C, when $x = BC = b$, $v = 0$. So (6) reduces to $0 = (g/a)\,(3a^2 - b^2)$ or $b = a\sqrt{3}$.

$\therefore$ Total distance fallen below O $= OC = OA + AB + BC = a + e + b$

$$= a + a + a\sqrt{3} = a(2 + \sqrt{3}). \quad ...(7)$$

Here $BC = b = a\sqrt{3}$ and $AB = e = a$. So $BC > AB$. Hence the particle will rise above A, and the motion will be simple harmonic upto A and above A the particle will move freely under gravity. Thus, the part of the motion from C to A and back from A to C is simple harmonic. The rest of the motion from A to O and back from O to A is under gravity only.

To find time from *O* to *C*. Let t_1 be time taken by the particle in moving from O to A. Using $h = ut + (1/2)gt^2$, we get $a = 0 + (1/2)g\,t_1^2$ so that $t_1 = (2a/g)^{1/2}$.

Let t_2 be time taken by the particle from A to C.

From (6), $\qquad v = dx/dt = (g/a)^{1/2}(3a^2 - x^2)^{1/2}$,

where positive sign is taken because x increases as t incrérases.

Separating variables, $\qquad dt = (a/g)^{1/2}\dfrac{dx}{(3a^2-x^2)^{1/2}} \qquad \ldots(8)$

Integrating (8) between the limits $t = 0$ to $t = t_2$ and $x = -BA = -e = -a$ to $x = BC = b = a\sqrt{3}$, we get

$$\int_0^{t_2} dt = \left(\frac{a}{g}\right)^{1/2}\int_{-a}^{a\sqrt{3}}\frac{dx}{\{(a\sqrt{3})^2-x^2\}^{1/2}} = \left(\frac{a}{g}\right)^{1/2}\left[\sin^{-1}\frac{x}{a\sqrt{3}}\right]_{-a}^{a\sqrt{3}}$$

or $\quad t_2 = (a/g)^{1/2}\{\sin^{-1}1 - \sin^{-1}(-1/\sqrt{3})\} = (a/g)^{1/2}\{\pi/2 + \sin^{-1}(1/\sqrt{3})\}$

$\qquad = (a/g)^{1/2}\{\pi/2 + \pi/2 - \cos^{-1}(1/\sqrt{3})\} = (a/g)^{1/2}\{\pi - \cos^{-1}(1/\sqrt{3})\}$

$\therefore$ Required time $= t_1 + t_2 = (a/g)^{1/2}\sqrt{2} + (a/g)^{1/2}\{\pi - \cos^{-1}(1/\sqrt{3})\}$

$\qquad = (a/g)^{1/2}\{\sqrt{2} + \pi - \cos^{-1}(1/\sqrt{3})\}$

(b) Proceed as in part (*a*). Here note that $e = AB = OB - OA = 2a - a = a$. Also, at A, $mg = \lambda(a/a)$ so that $\lambda = mg$. Now do exactly as in part (*a*) upto equation (7) only for the present problem.

Ex. 8. *One end of a light elastic string of natural length a and modulus of elasticity 2mg is attached to a fixed point O and other end to a mass m. The particle initially held at rest at A, is let fall. Show that greatest extension of the string is a(1 + $\sqrt{5}$)/2 during the motion and show that the particle will reach back O again after a time ($\pi + 2 - \tan^{-1} 2$) $(2a/g)^{1/2}$* **[Gorakhpur 97; I.A.S. 96]**

Hint. Proceed exactly as in Ex. 7(*a*). Here $\lambda = 2mg$

$\therefore$ As before, at B $\qquad mg = \lambda(e/a) = (2mg)(e/a) \qquad \ldots(1)$

Thus, we have $\qquad e = a/2 \qquad \ldots(2)$

As before, $\qquad V = \text{velocity at } B = (2ga)^{1/2} \qquad \ldots(3)$

As before, the equation of motion of the particle at P is

$$m(d^2x/dt^2) = mg - T = mg - \lambda(e+x)/a = mg - \lambda(e/a) - \lambda(x/a)$$

Using (1), the above equation becomes $\qquad d^2x/dt^2 = -\lambda(x/am)$

or $\qquad dx^2/dt^2 = -(2mg)(x/am) = -(2g/a)x$, as $\lambda = 2mg \qquad \ldots(4)$

Re-writing (4), $\quad v(dv/dx) = -(2g/a)x \quad$ or $\quad 2v\,dv = -(4g/a)x\,dx$

Integrating, $\qquad v^2 = -(2g/a)x^2 + C$, C being a constant $\qquad \ldots(5)$

At A, where $x = -e = -a/2$, $v = V = (2ga)^{1/2}$. Hence (5) reduces to

$\qquad 2ga = -(2g/a)(a^2/4) + C \qquad$ or $\qquad C = 5ag/4$

Then (5) becomes $\qquad v^2 = (5ag/4) - (2g/a)x^2 = (2g/a)(5a^2/4 - x^2) \qquad \ldots(6)$

At C, when $x = BC = b$, $v = 0$. So (6) reduces to

$\qquad 0 = (2g/a)(5a^2/4 - b^2) \qquad$ so that $\qquad b = (a/2)\sqrt{5}$

Hence, the required greatest extension of the string

$\qquad = AC = AB + BC = e + b = a/2 + (a/2)\sqrt{5} = a(1 + \sqrt{5})/2$

To find time from *O* to *C*. As in Ex. 7(*a*), $t_1 = (2a/g)^{1/2}$

From (6), $\qquad dt = \left(\dfrac{a}{2g}\right)^{1/2}\dfrac{dx}{\{(a\sqrt{5}/2)^2 - x^2\}^{1/2}} \qquad \ldots(7)$

If t_2 is time from A to C, then from (7), as before

$$\int_0^{t_2} dt = \left(\frac{a}{2g}\right)^{1/2}\int_{-a/2}^{(a/2)\sqrt{5}}\frac{dx}{\{(a\sqrt{5}/2)^2-x^2\}^{1/2}} = \left(\frac{a}{2g}\right)^{1/2}\left[\sin^{-1}\frac{x}{(a/2)\sqrt{5}}\right]_{-a/2}^{(a/2)/\sqrt{5}}$$

or $t_2 = (a/2g)^{1/2}\{\sin^{-1} 1 - \sin^{-1}(-1/\sqrt{5})\} = (a/2g)^{1/2}\{\pi/2 + \sin^{-1}(1/\sqrt{5})\}$

$= (a/2g)^{1/2}\{\pi/2 + \tan^{-1}(1/2)\} = (a/2g)^{1/2}(\pi/2 + \cot^{-1} 2)$

$= (a/2g)^{1/2}(\pi/2 + \pi/2 - \tan^{-1} 2) = (a/2g)^{1/2}(\pi - \tan^{-1} 2)$

$\therefore$ The required total time to return back to $O = 2(t_1 + t_2)$

$- 2\{(2a/g)^{1/2} + (a/2g)^{1/2}(\pi - \tan^{-1} 2)\} = (2a/g)^{1/2}(2 + \pi - \tan^{-1} 2)$

Ex. 9. *A heavy particle is attached to one end of an elastic string, the other end of which is fixed at O. The modulus of elasticity of the string is equal to the weight of the particle. The string is drawn vertically down till it is four times its natural length and then let go*

(a) *Show that the particle will return to this point in time $(a/g)^{1/2}(2\sqrt{3} + 4\pi/3)$, where a is the natural length of the string.* **[Meerut 2004; Kanpur 2000, 06; Agra 91; Avadh 95; Bundelkhand 94; Gorakhpur 95; Lucknow 97]**

(b) *Find height above O through which the particle rises.* **[Lucknow 1996]**

Sol. For figures, refer Ex. 7(*a*). Let $OA = a$ be the natural length of the string whose one end O is fixed. Let B be the position of equilibrium of a particle of mass m attached to the other end A of the string and let $AB = e$ as shown in figure (*i*) of Ex. 7(*a*). Then at B,

the weight of the particle = the tension T_o in the string.

$\therefore \quad mg = \lambda(e/a) = (mg)(e/a)$ as $\lambda = mg$ (given) ...(1)

Thus, here $e = a$ ...(2)

Now the particle is pulled down to a point C such that OC $= 4a$ (given) and then let go. The particle will start to move towards B from rest from C. Let P be its position after time t, where $BP = x$. Refer figure (*ii*) of Ex. 7(*a*). At P the forces acting on the particle are its weight mg acting vertically downwards and tension $T = \lambda(x + e)/a$ acting vertically upwards. Then, the equation of motion of the particle at P is

$m(d^2x/dt^2) = mg - T = mg - \lambda(e + x)/a = mg - \lambda(e/a) - \lambda(x/a)$

or $m(d^2x/dt^2) = -\lambda(x/a)$ *i.e.*, $d^2x(dt^2 = -(\lambda/am)x$, using (1)

or $d^2x/dt^2 = -(mg/am)x = -(g/a)x$, as $\lambda = mg$ ...(3)

or $v(dv/dx) = -(g/a)x$ or $2v\,dv = -(2g/a)x\,dx$

Integrating, $v^2 = -(g/a)x^2 + C$, C being a constant ...(4)

At the point C, when $x = BC = OC - OB = 4a - (a + e) = 4a - (a + a)$ *i.e.*, $x = 2a$, $v = 0$. Hence (4) reduces to $0 = -(g/a)(2a)^2 + C$ so that $C = 4ag$.

From (4), $v^2 = 4ag - (g/a)x^2$ or $(dx/dt)^2 = (g/a)(4a^2 - x^2)$ ...(5)

When the particle reaches A, let the velocity of the particle be V. Then, putting $x = -a$ and $v = V$ in (5), we get

$V^2 = (g/a)(4a^2 - a^2)$ or $V = (3ag)^{1/2}$ ...(6)

From (5), $$\frac{dx}{dt} = -\left(\frac{g}{a}\right)^{1/2}(4a^2 - x^2) \text{ or } dt = -\left(\frac{a}{g}\right)^{1/2}\frac{dx}{(4a^2 - x^2)^{1/2}}$$ ...(7)

where we have taken negative sign on R.H.S. due to the fact that in moving from C towards B, x decreases as t increases.

Let t_1 be the time taken from C to A. Then integrating (7) between $t = 0$ to $t = t_1$ and corresponding limits $x = 2a$ to $x = -a$, we get

$$\int_0^{t_1} dt = -\left(\frac{a}{g}\right)^{1/2}\int_{2a}^{-a}\frac{dx}{(4a^2 - x^2)^{1/2}} = \left(\frac{a}{g}\right)^{1/2}\left[\cos^{-1}\frac{x}{2a}\right]_{2a}^{-a}$$

or $t_1 = (a/g)^{1/2}\{\cos^{-1}(-1/2) - \cos^{-1} 1\} = (a/g)^{1/2}\{\pi - \cos^{-1}(1/2)\}$

Thus, $t_1 = (a/g)^{1/2}(\pi - \pi/3) = (a/g)^{1/2}(2\pi/3)$

Thus particle has velocity V in upward direction and it moves above A. But the string becomes slack in upward motion from A so the S.H.M. ceases at A and the particle moves vertically upwards freely under gravity till its velocity V is destroyed. Let t_2 be the time taken by the particle from A till its velocity V becomes zero. Then, using formula $v = u - gt$, we get

$0 = V - gt_2$ or $t_2 = V/g = (3ag)^{1/2}/g = (3a/g)^{1/2}$.

Conditions being the same, the particle will take time t_2 in falling freely back to A. Again from A to C the time taken by the particle will be t_1 (which was taken by it to move from C to A). So the required time

$$= 2(t_1 + t_2) = 2\{(a/g)^{1/2}\,(2\pi/3) + (3a/g)^{1/2}\} = (a/g)^{1/2}\,(4\pi/3 + 2\sqrt{3})$$

(*b*) Proceed as in part (*a*) upto equation (6) only. The particle rises above O and moves freely under gravity till its velocity V is destroyed. Let H be the maximum height attained. Then, using formula $v^2 = u^2 - 2gh$, we get $0 = V^2 - 2gh$ so that $H = V^2/2g = (3ag)/2g = 3a/2$

Ex. 10. *An elastic string has natural length l_o and is held vertically with the upper end fixed. A heavy particle attached to its lower end produces a static deflection d of the string. If the particle is pulled down further distance c below equilibrium and released, show that in the case when $d(d + 4l_o) > c^2 > d^2$, the time taken for the particle to return to its original position of release is $2(d/g)^{1/2}(\pi - \theta + \tan\theta)$, where $\cos\theta = d/c$, and θ is acute.* **[Delhi B.Sc. 1996, 1998, 2003]**

Sol. We proceed as in Ex. 9(*a*). Let $OA = l_o$ be natural length of the string whose one end O is fixed. Let B be the position of equilibrium of a particle of mass m attached to the other end A of the string such that $AB = d$ (given) as shown in figure (*i*). Then at B, the weight mg of the particle = the tension T_o in the string

i.e., $mg = \lambda(d/l_o)$ so that $\lambda = mgl_o/d$ ...(1)

Now the particle is pulled down to a point C such that $BC = c$ (given) and then let go. The particle will start to move towards B from rest from C. Let P be its position after time t, where $BP = x$ as shown in figure (*ii*). At P the forces acting on the particle are its weight mg acting vertically downwards and tension $T = \lambda(x + d)/l_o$, acting vertically upwards. Then, the equation of motion of the particle is

Fig. (*i*)

Fig. (*ii*)

$$m(d^2x/dt^2) = mg - T = mg - \lambda(x + d)/l_o = mg - \lambda(d/l_o) - \lambda(x/l_o)$$

or $\quad m(d^2x/dt^2) = -\lambda(x/l_o) = -(mgl_o/d)\,(x/l_o)$, by (1)

Thus, $\quad d^2x/dt^2 = -(g/d)x$ or $[D^2 + (g/d)]x = O$, where $D \equiv d/dt$ ...(2)

which holds only so long as the string is taut. The general solution of (1) is

$$x = A\cos t(g/d)^{1/2} + B\sin t(g/d)^{1/2} \quad ...(3)$$

Diff. (3) w.r.t. 't', we have

$$dx/dt = -A(g/d)^{1/2}\sin t(g/d)^{1/2} + B(g/d)^{1/2}\cos t\,(g/d)^{1/2} \quad ...(4)$$

But at C, when $t = 0$, $x = c$ and dx/dt = velocity = 0. Hence (3) and (4) give $c = A$ and $0 = B(g/d)^{1/2}$ and so $A = c$ and $B = 0$.

Then (3) becomes $\quad x = c\cos t\,(g/d)^{1/2}$, ...(5)

and (4) becomes $\quad dx/dt = -c(g/d)^{1/2}\sin t\,(g/d)^{1/2}$ ...(6)

Now (5) holds only as long as the string is taut. But when $x = -d$ at A, the string has attained its natural length. If t_1 be the time taken by the particle to move from C to A, then putting $t = t_1$ and $x = -d$ in (5), we get

$-d = C\cos t_1(g/d)^{1/2}$ or $\cos t_1(g/d)^{1/2} = -d/c = -\cos\theta$, as $\cos\theta = d/c$ (given)

or $\quad \cos t_1(g/d)^{1/2} = \cos(\pi - \theta)$ so that $t_1(g/d)^{1/2} = \pi - \theta$

Thus, $\quad t_1 = (d/g)^{1/2}(\pi - \theta)$. ...(7)

At this instant, the velocity of the string at point A

$= -c(g/d)^{1/2}\sin t_1(g/d)^{1/2} = -c(g/d)^{1/2}\sin(\pi - \theta) = -c(g/d)^{1/2}\sin\theta$, by (7)

Thus at time $t = t_1$, the particle is rising against gravity with an upward velocity of $= c(g/d)^{1/2}\sin\theta$. Let t_2 be the further time taken for the particle to attain its greatest height H, say. Using $v^2 = u^2 - 2gh$, we get

$0 = V^2 - 2gh$ or $H = V^2/2g = (1/2g)\,c^2(g/d)\sin^2\theta = (1/2)\,(c^2/d)\,(1 - \cos^2\theta)$

or $\quad H = (c^2/2d)\,(1 - d^2/c^2)$, as $\cos\theta = d/c$ (given) ...(8)

If the string does not tighten in this phase of the motion, then we must have $H < 2l_o$, *i.e.*,

$$(c^2/2d)(1 - d^2/c^2) < 2l_o \text{ giving } d(d + 4l_o) > c^2 \quad \text{...(9)}$$

Due to given condition (2), the conditoin (9) holds. Hence in the subsequent upward flight of the particle, the string does not tighten. Let t_2 be the time taken by the particle to move from A with velocity V till its velocity is destroyed at its greatest height. Then, by $v = u - gt$, we get $0 = V - gt_2$ or $t_2 = V/g$

Thus, $t_2 = (1/g)\, c(g/d)^{1/2} \sin\theta = (c/g)(g/d)^{1/2} \cos\theta \tan\theta$

or $t_2 = (c/g)(g/d)^{1/2}(d/c)\tan\theta = (d/g)^{1/2}\tan\theta$, as $\cos\theta = d/c$ ...(8)

Hence the total time from C to the highest point attained

$$= t_1 + t_2 = (d/g)^{1/2}(\pi - \theta + \tan\theta), \text{ by (7) and (10).}$$

Hence the required total time taken by the particle to return to its original position

$$= 2(t_1 + t_2) = 2(d/g)^{1/2}(\pi - \theta + \tan\theta)$$

Ex. 11. *A particle of mass m is attached to one end of an elastic string of natural length a and modulus of elasticity 2mg, whose other end is fixed at O. The particle is let fall from A, when A is vertically above O and OA = a. Show that its velocity will be zero art B, where OB = 3a. Calculate also the time from A to B.* **[Meerut 2014]**

Sol. Let $OC = a$ be the natural length of the string whose one end is fixed at O. Let λ be the modulus of elasticity of string and a mass m be attached to the other end C of the string. Here, given that

$$\lambda = 2mg \quad \text{...(1)}$$

Let D be the position of equilibrium of the particle such that $CD = e$ as shown in figure (*i*). Then at D, the weight of the particle = the tension T_o in the string.

$\therefore \quad mg = \lambda(e/a) = 2mg(e/a)$, by (1) ...(2)

Thus, here $\quad e = a/2$ ...(3)

Fig. (*i*)

Fig. (*ii*)

When the particle is let fall from A (where A is vertically above A) such that $OA = a$ as shown in figure (*ii*)), the motion from A to C will be under gravity only, the string being loose from A to C. Let V be the velocity of the particle when it reaches the point C, then $V^2 = 0 + 2g.2a$.

Thus, $\quad V = 2(ga)^{1/2}$ in the downward direction ...(4)

When the particle falls below C, string is taut and tension of the string begins to operate. While moving from C to D the velocity of the particle goes on increasing as the tension in the string is less than the weight of the particle. But while moving below C, the velocity of the particle goes on decreasing as now the tension in the string is greater than the weight of the particle. Suppose the partcle comes to an instantaneous rest at B, where $DB = b$ (say).

During the motion of the particle below D, let P be the position of the particle after time t such that $DP = x$. At P, the forces acting on the particle are its weight mg acting vertically downwards and tension $T = \lambda(e + x)/a$, acting vertically upwards. Then, the equation of motion of the particle is

$$m(d^2x/dt^2) = mg - T = mg - \lambda(e + x)/a = mg - \lambda(e/a) - \lambda(x/a)$$

or $\quad m(d^2x/dt^2) = -(2mg)(x/a)$, using (1) and (2)

or $\quad d^2x/dt^2 = -(2g/a)x$ ...(5)

which represents a S.H.M. and holds good for the motion from C to B.

Re-writing (5), $\quad v(dv/dv) = -(2g/a)x \quad$ or $\quad 2v\,dv = -(4g/a)x\,dx$

Integrating, $\quad v^2 = -(2g/a)x^2 + C$, C being a constant ...(6)

But at C, when $x = -DC = -e = -a/2$, $v^2 = V^2 = 4ag$. So (6) reduces to

$$4ag = -(2g/a)(a^2/4) + C \text{ so that } C = 9ag/2. \text{ So (6) becomes}$$

$$v^2 = 9ag/2 - (2g/a)x^2 \quad \text{or} \quad (dx/dt)^2 = (2g/a)(9a^2/4 - x^2) \quad \text{...(7)}$$

But at B, when $x = DB = b$, $v = 0$. So (7) gives

$$0 = (2g/a)(9a^2/4 - b^2) \quad \text{so that} \quad b = 3a/2 \quad \text{...(8)}$$

$\therefore OB = OC + CD + DB = a + e + b = a + a/2 + 3a/2$, by (3) and (8)

Thus, $OB = 3a$, which is what we wished to prove.

To find time from A to B. Let t_1 be time from A to C, then from $v = u + gt$,

$$V = 0 + gt_1 \quad \text{so that} \quad t_1 = V/g = 2(a/g)^{1/2}, \text{ by (1)} \qquad \text{...(9)}$$

From (7), $\dfrac{dx}{dt} = \left(\dfrac{2g}{a}\right)^{1/2}\left(\dfrac{9a^2}{4} - x^2\right)^{1/2}$ or $dt = \left(\dfrac{a}{2g}\right)^{1/2}\dfrac{dx}{(9a^2/4 - x^2)^{1/2}}$ ...(10)

where positive sign is taken because x increases as t increases.

Integrating (10) from C to B, the time t_2 from C to B is given by

$$t_2 = \left(\frac{a}{2g}\right)^{1/2}\int_{-a/2}^{3a/2}\frac{dx}{\{(3a/2)^2 - x^2\}^{1/2}} = \left(\frac{a}{2g}\right)^{1/2}\left(\sin^{-1}\frac{x}{3a/2}\right)_{-a/2}^{3a/2}$$

$$t_2 = (a/2g)^{1/2}\{\sin^{-1} 1 - \sin^{-1}(-1/3)\} = (a/2g)^{1/2}(\pi/2 + \sin^{-1}(1/3)) \qquad \text{...(11)}$$

$\therefore$ Required time from A to B

= time from A to C + time from C to B = $t_1 + t_2 = 2(a/g)^{1/2} + (a/2g)^{1/2}\{\pi/2 + \sin^{-1}(1/3)\}$

$(1/2)(a/2g)^{1/2}\{4\sqrt{2} + \pi + 2\sin^{-1}(1/3)\}$

Ex. 12. *A heavy particle of mass m is attached to one end of an elastic string of natural length a, whose other end is fixed at O. The particle is let fall from rest at O. Show that the part of the motion is simple harmonic and if the greatest depth of the particle below O is a $\cot^2\theta/2$, the modulus of elasticity of the string is $(1/2)\, mg \tan^2\theta$ and that the particle attains this depth in time $(2a/g)^{1/2}$ $\{1 + (\pi - \theta)\cot\theta\}$, where θ is a positive angle.*

Sol. We shall proceed as in Ex. 7(*a*). For figures refer Ex. 7(*a*). Let $OA = a$ be the natural length of the string whose one end O is fixed. Let B be the position of equilibrium of a particle of mass m attached to the other end A of the string and let $AB = e$ as shown in fig. (*i*) of Ex. 7(*a*). Then, at B

the weight mg of the particle = the tension T_o in the string.

$\therefore \quad mg = \lambda(e/a)$, λ being the modulus of elasticity of the string. ...(1)

When the particle is let fall from O, then the motion from O to A is under gravity only, string being slack. If V is the velocity of the particle when it reaches A, then by using the formula $v^2 = u^2 + 2gh$, we get

$$V^2 = 0 + 2ga \text{ so that } V = (2ga)^{1/2} \text{ in downward direction.} \qquad \text{...(2)}$$

When the particle falls below A, the string is taut and tension of string begins to operate. While moving from A to B the velocity of the particle goes on increasing because the tension in the string is less than the weight of the particle. But while moving below B, the velocity of the particle goes on decreasing because now the tension in string is greater than the weight of the particle. Suppose the particle comes to an instantaneous rest at C.

During the motion of the particle below B, let P be the position of the particle after any time t such that $BP = x$ as shown in figure (*ii*) of Ex. 7(*a*). At P, the forces acting on the particle are its weight mg acting vertically downwards and tension $T = \lambda(x + e)/a$, acting vertically upwards. Hence, the equation of motion of the particle at P is

$$m(dx^2/dt^2) = mg - T = mg - \lambda(e + x)/a = mg - \lambda(e/a) - \lambda(x/a)$$

or $$m(d^2x/dt^2) = -(\lambda/a)x = -(mg/e)x, \text{ using (1)}$$

Thus, $$d^2x/dt^2 = -(g/e)x \qquad \text{...(3)}$$

which holds so long as the string is taut. (3) represents a S.H.M. with centre at B and amplitude BC. Thus the motion of the particle below A is S.H.M.

Re-writing (3), $v(dv/dx) = -(g/e)x$ or $2v\,dv = -(2g/e)x\,dx$

Integrating, $v^2 = -(g/e)x^2 + C$, C being a constant ...(4)

But at A, when $x = -BA = -e$, $v = V = (2ga)^{1/2}$, by (2). So (4) gives

$$2ga = -(g/e)e^2 + C \quad \text{so that} \quad C = 2ga + ge.$$

Then (4) becomes $v^2 = 2ga + ge - (g/e)x^2$, ...(5)

which gives the velocity of the particle at any point between A and C. Given that $OC = a \cot^2 \theta/2$. Hence at C, when $x = BC = OC - OB = OC - (OA + AB) = a \cot^2 \theta/2 - (a + e)$, $v = 0$. Hence (5) reduces to

$$0 = 2ga + ge - (g/e)\{(a \cot^2 \theta/2 - a) - e\}^2$$

or $$0 = 2ga + ge - (g/e)[(a \cot^2 \theta/2 - a)^2 + e^2 - 2ea(\cot^2 \theta/2 - 1)\}$$

or $$0 = -(g/e)(a \cot^2 \theta/2 - a)^2 + 2ga \cot^2 \theta/2$$

or $$0 = -(\lambda/ma)(a \cot^2 \theta/2 - a)^2 + 2ga \cot^2 \theta/2, \text{ as by (1) } g/e = \lambda/ma$$

$$\therefore \quad \lambda = \frac{2mg\, a^2 \cot^2 \theta/2}{a^2(\cot^2 \theta/2 - 1)^2} = \frac{2mg \cot^2 \theta/2}{(\cos^2 \theta/2 - \sin^2 \theta/2)^2} \times \sin^4 \theta/2$$

or $$\lambda = \frac{1}{2}\frac{mg(2\sin\theta/2\cos\theta/2)^2}{\cos^2\theta} = \frac{1}{2}mg\frac{\sin^2\theta}{\cos^2\theta} = \frac{1}{2}mg\tan^2\theta \qquad \ldots(6)$$

From (1), $\lambda = (mga)/e$. Hence (6) reduces to

$$(mga)/e = (1/2)\, mg \tan^2 \theta \quad \text{so that} \quad e = 2a \cot^2 \theta \qquad \ldots(7)$$

From (5), $$(dx/dt)^2 = (g/e)(2ea + e^2 - x^2)$$

or $$(dx/dt^2) = (g/2a)\tan^2\theta\,(4a^2 \cot^2 \theta + 4a^2 \cot^4 \theta - x^2), \text{ by (7)}$$

or $$(dx/dt)^2 = (g/2a)\tan^2\theta\,[4a^2 \cot^2 \theta\,(1 + \cot^2 \theta) - x^2\}$$

or $$(dx/dt)^2 = (g/2a)\tan^2\theta\,(4a^2 \cot^2 \theta \operatorname{cosec}^2 \theta - x^2)$$

or $$dx/dt = (g/2a)\tan\theta\,\{(2a \cot\theta \operatorname{cosec}\theta)^2 - x^2\}^{1/2},$$

where positive sign is taken because x increases as t increases.

or $$dt = \left(\frac{2a}{g}\right)^{1/2} \cot\theta \frac{dx}{\{(2a\cot\theta\operatorname{cosec}\theta)^2 - x^2\}^{1/2}} \qquad \ldots(8)$$

Now, $BC = OC - OB = OC - (OA + AB) = OC - (a + e)$

$= a \cot^2 \theta/2 - a - 2a \cot^2 \theta$, using (7)

$= a(\cos^2 \theta/2 - \sin^2 \theta/2)/\sin^2 \theta/2 - 2a \cos^2 \theta/\sin^2 \theta$

$= a \cos\theta/\sin^2 \theta/2 - 2a \cos^2 \theta/\sin^2 \theta = (a \cos\theta/\sin^2\theta)(4 \cos^2 \theta/2 - 2\cos\theta)$

$= (a \cos\theta/\sin^2\theta)\{2(1 + \cos\theta) - 2\cos\theta\} = 2a \cot\theta \operatorname{cosec}\theta$

Let t_1 be the time taken by the particle from A to C. Then integrating (8) between the limits $t = 0$ to $t = t_1$ and the corresponding limits $x = -e = -2a \cot^2 \theta$ to $x = BC = 2a \cot\theta \operatorname{cosec}\theta$, we get

$$\int_0^{t_1} dx = \left(\frac{2a}{g}\right)^{1/2} \cot\theta \int_{-2a\cot^2\theta}^{2a\cot\theta\operatorname{cosec}\theta} \frac{dx}{\{(2a\cot\theta\operatorname{cosec}\theta)^2 - x^2\}^{1/2}}$$

or $$t_1 = (2a/g)^{1/2}\cot\theta\left[\sin^{-1}(x/2a\cot\theta\operatorname{cosec}\theta)\right]_{-2a\cot^2\theta}^{2a\cot\theta\operatorname{cosec}\theta}$$

$= (2a/g)^{1/2} \cot\theta\,[\sin^{-1} 1 - \sin^{-1}(-\cot\theta/\operatorname{cosec}\theta)]$

$= (2a/g)^{1/2} \cot\theta\,[\pi/2 + \sin^{-1}(\cos\theta)] = (2a/g)^{1/2} \cot\theta\,\{\pi/2 + \sin^{-1}\sin(\pi/2 - \theta)\}$

$= (2a/g)^{1/2} \cot\theta\,(\pi/2 + \pi/2 - \theta) = (2a/g)^{1/2}(\pi - \theta)\cot\theta$

Again if t_2 be the time from O to A while moving under gravity, then, by $h = ut + (1/2)gt^2$, we get $\quad a = 0 + (1/2)gt_2^2 \quad$ or $\quad t_2 = (2a/g)^{1/2}$

The required time $= t_1 + t_2$

$$= (2a/g)^{1/2}(\pi - \theta)\cot\theta + (2a/g)^{1/2} = (2a/g)^{1/2}[1 + (\pi - \theta)\cot\theta]$$

Ex. 13. *Two bodies of masses M and M′, are attached to the lower end of an elastic string whose upper end is fixed and hang at rest, M′ falls off; show that the distance of M from the upper end of the string at time t is $a + b + c \cos (g/b)^{1/2}\, t$, where a is unstretched length of the string. b and c the distances by which it would be stretched when supporting M and M′ respectively.*

[Avadh 2007, 09; Purvanchal 2006; Lucknow 2008]

Sol. Let $OA = a$ be the unstretched length (*i.e.*, natural length) of the string whose one end O is fixed. Let B be the position of equilibrium of a particle of mass M attached to the other end A of the string where $AB = b$ (given) as shown in figure (*i*). Then, at B, the weight Mg of the particle = the tension T_o in the string.

$\therefore \qquad Mg = \lambda(b/a)$ ...(1)

Similarly, $\qquad M'g = \lambda(c/a)$ ...(2)

Adding (1) and (2), $\qquad (M+M')g = \lambda\{(b+c)/a\}$,

showing that the string will be stretched by the distance $(b+c)$ when supporting both the masses M and M' at the lower end. If OC be the stretched length of the string when both the masses M and M' are attached to its lower end, then from figure (*ii*), we have

$$AC = b + c \text{ and hence } BC = AC - AB = (b+c) - b = c$$

Fig. (*i*)

Fig. (*ii*)

When the mass M' falls off at C, then the mass M will move towards B starting with zero velocity at C. Let P be the position of the particle of M at a subsequent time t, where $BP = x$. At P the forces acting on the particle of mass M are its weight Mg acting vertically downwards and tenstion $T = \lambda\,(b+x)/a$, acting vertically upwards. Hence the equation of motion of the particle of mass M at P is

$$M(d^2x/dt^2) = Mg - T = Mg - \lambda(b+x)/a = Mg - \lambda(b/a) - \lambda(x/a)$$

or $\qquad M(d^2x/dt^2) = -\lambda(x/a) = -(Mg\,a/b)\,(x/a)$, using (1)

or $\qquad d^2x/dt^2 = -(g/b)x \quad$ or $\quad \{D^2 + (g/b)\}x = 0$, where $D \equiv d/dt$

Its general solution is $\qquad x = A\cos(g/b)^{1/2}\,t + B\sin(g/b)^{1/2}\,t$ ...(3)

Differentiating (3) w.r.t. 't', we have

$$dx/dt = -A(g/b)^{1/2}\sin(g/b)^{1/2}\,t + B(g/b)^{1/2}\cos(g/b)^{1/2}\,t \qquad ...(4)$$

But at C, when $t = 0$, $x = c$ and dx/dt = velocity = 0. So (3) and (4) give $c = A$ and $0 = B(g/b)^{1/2}$. Thus $A = c$, $B = 0$.

$\therefore$ (3) gives $\qquad x = c\cos(g/b)^{1/2}\,t$ ...(5)

Hence the required distance of M from $O = OP = OA + AB + BP$

$$= a + b + x = a + b + c\cos(g/b)^{1/2}\,t, \text{ by (5)}$$

Ex. 14. *A smooth pulley is suspended from a fixed point by a spring of natural length l and modulus of elasticity ng. If masses m_1 and m_2 hang at the ends of a light inextensible string passing round the pulley, show that the pulley executes S.H.M. about the centre whose depth below the point of suspension is $l(1 + 2M/n)$, where M is the harmonic mean between m_1 and m_2.*

Sol. Let O be the fixed point from which the pulley is suspended by a spring OA of natural length l. Let B be the position of equilibrium of the pulley when masses m_1 and m_2 hang at the ends of the given string passing round the pulley as shown in figure (*i*). Let $m_2 > m_1$, T' the tension in the string and f the common acceleration of the particles m_1, m_2. Then the equations of motion of masses m_1 and m_2 respectively are given by

$m_2g - T' = m_2f$ and $T' - m_1g = m_1f$

Multiplying these equations by m_1 and m_2 respectively, we get

$m_1m_2g - m_1T' = m_1m_2f \qquad$ and $\qquad m_2T' - m_1m_2g = m_1m_2f$

Thus $\quad m_1m_2g - m_1T' = m_2T' - m_1m_2g$

or $\qquad T' = (2m_1m_2g)/(m_1 + m_2)$

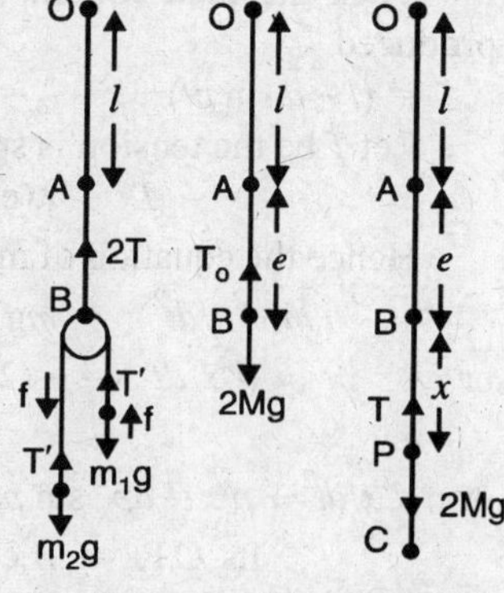

Fig. (*i*) Fig. (*ii*) Fig. (*iii*)

But M is the harmonic mean of m_1 and m_2, so

$$M = (2m_1m_2)/(m_1 + m_2) \text{ and hence } T' = Mg$$

$\therefore$ Pressure on the pulley due to masses m_1, $m_2 = 2T' = 2Mg$, showing that the pulley, which itself is light, behaves like a particle of mass $2M$.

Hence the given problem reduces to the vertical motion of a mass $2M$ attached to the end A of the spring OA whose other end is fixed at O as shown in figure (*ii*). At the equilibrium position B, we have $AB = e$

the weight $2Mg$ of the particle of mass $2M$ = the tension T_o in the spring.

$\therefore$ $2Mg = T_o = \lambda(e/l) = ng(e/l)$, as $\lambda = ng$ (given) ...(1)

Thus, $e = (2Ml)/n$ ...(2)

Now, let the particle of mass $2M$ be slightly pulled down and then let go. Let P be the position of the particle at any time such that $BP = x$. Then, at P, the forces acting on the particle (of mass $2M$) are its weight $2Mg$ acting vertically downwards and tension $T = \lambda(e + x)/l$, acting vertically downwards. Hence the equation of motion of the particle of mass $2M$ at P is

$$2M(d^2x/dt^2) = 2Mg - T = 2Mg - \lambda(e+x)/l = 2Mg - \lambda(e/l) - \lambda(x/l)$$

or $$2M(d^2x/dt^2) = -\lambda(x/l), \text{ using (1)}$$

or $$d^2x/dt^2 = -(ng/2Ml)x, \text{ as } \lambda = ng \qquad \text{...(3)}$$

(3) represents a S.H.M. about the centre B. Hence the pulley executes S.H.M. about centre B whose depth below the point of suspension O is given by

$$OB = OA + AB = l + e = l + (2Ml)/n = l(1 + 2M/n), \text{ by (2)}$$

Ex. 15. *A heavy particle of mass m hangs at the lower end of a light elastic spring of natural length l_o and modulus 2mg. At zero time the system is in equilibrium and the upper end of the spring is made to execute vertical oscillations so that its downward displacement at time t later is a sin pt, where $p^2 = 2g/l_o$. Find the displacement of the particle at this instant.*

Sol. Figure (*i*) shows the given elastic spring OA of natural length l_o. Figure (*ii*) shows the system in equilibrium at time $t = 0$. Let OB be the spring in the equilibrium position. Then the weight mg of the particle will balance tension T_o in the spring. If l is the extended length of the spring, then with $\lambda = 2mg$, we have

$$mg = \lambda(l - l_o)/l_o = 2mg(l - l_o)/l_o$$

and hence $l = 3l_o/2$ ...(1)

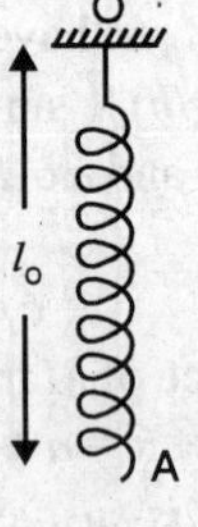

Fig. (*i*)

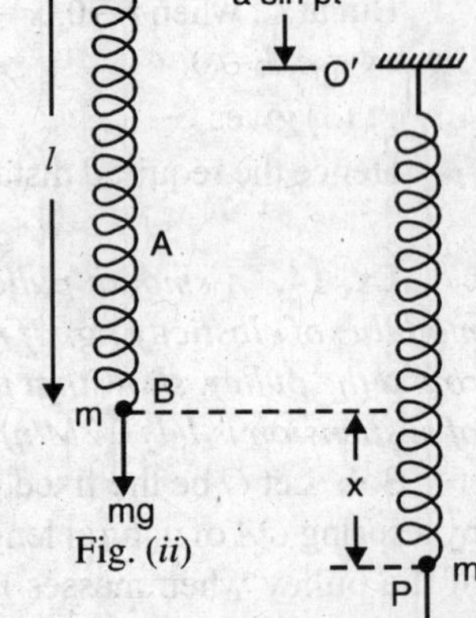

Fig. (*ii*) Fig. (*iii*)

Now, the upper end is made to execute vertical oscillations, and its downward acceleration at the instant t is $OO' = a \sin pt$. In figure (*iii*), the situation at time t is depicted, the particle having been displaced a distance x.

Then the total stretch of the spring, *i.e.*, extension produced

$= (l - a \sin pt) + x - l_o = 3l_o/2 - a \sin pt + x - l_o$, by (1)

Let T be the tension is spring at P, then

$$T = \lambda(\text{extension})/l_o = (2mg/l_o)\,(l_o/2 - a \sin pt + x)$$

Hence the equation of motion of mass m at P is

$$m(d^2x/dt^2) = mg - T = mg - (2mg/l_o)\,(l_o/2 - a \sin pt + x)$$

or $$d^2x/dt^2 = -(2g/l_o)\,(x - a \sin pt) = -p^2(x - a \sin pt)$$

[$\because$ given that $p^2 = 2g/l_o$]

or $d^2x/dt^2 + p^2x = ap^2 \sin pt$ or $(D^2 + p^2)x = ap^2 \sin pt$, $D = d/dt$ ...(2)

Its C.F. $= A \cos pt + B \sin pt$

$$\text{P.I.} = \frac{1}{D^2 + p^2} ap^2 \sin pt = ap^2 \frac{1}{D^2 + p^2} \sin pt = ap^2\left(-\frac{t}{2p} \cos pt\right)$$

Hence the complete solution of (2) is given by

$$x = A \cos pt + B \sin pt - (1/2)\, apt \cos pt \qquad \text{...(4)}$$

Initially at $t = 0$, $x = 0$. So (4) gives $A = 0$. Thus, we have

$$x = B \sin pt - (1/2)apt \cos pt \qquad \text{...(5)}$$

Differentiating (5) w.r.t. 't' we get

$$dx/dt \;=\; Bp\cos pt - (1/2)ap\cos pt + (1/2)ap^2 t\sin pt \qquad ...(6)$$

Initially at $t = 0$, $dx/dt = 0$. So (6) gives $0 = Bp - (1/2)ap$ so that $B = a/2$.

$\therefore\; x = (a/2)\sin pt - (1/2)\,apt\cos pt = (a/2)(\sin pt - pt\cos pt)$, by (4)

$$= \frac{1}{2}\frac{ap^2}{p^2}(\sin pt - pt\cos pt) = \frac{a}{2p^2}\left(\frac{2g}{l_o}\right)(\sin pt - pt\cos pt),\ \text{as } p^2 = \frac{2g}{l_0}$$

$$\therefore \qquad x \;=\; (ga/p^2 l_o)(\sin pt - pt\cos pt),$$

which is the required displacement at the instant t.

Ex. 16. (*a*) *A particle of mass m is attached to the mid-point of an unstretched elastic string of natural length a and modulus mg. The string, with mass attached, is then stretched between two points in the same vertical line, distant 2a apart. Find the position of equilibrium of the particle.*

If the particle is now slightly displaced from its equilibrium position in either a vertical or a horizontal direction, show that in each case the ensuing motion is simple harmonic, but that the period of a horizontal oscillation is $(15/7)^{1/2}$ times the period of the vertical oscillation.

[Delhi Maths (H) 1990]

(*b*) *A particle of mass m is attached to the mid-point of an unstreched elastic string of natural length a and modulus mg. The string with the mass attached is then stetched between two points on the same vertical line, distant 2a apart. Find the position of equilibrium. If the particle is now slightly displaced from its equilibrium position in a vertical direction, show that the ensuing motion is simple harmonic with time period $\pi(a/g)^{1/2}$.* **[Delhi Maths (H) 2004]**

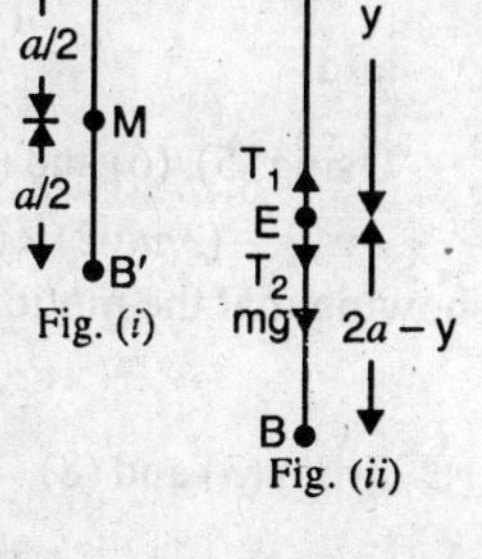

Fig. (*i*) Fig. (*ii*)

Sol. (*a*) Figure (*i*) shows the given string $A'B'$ of length a in its unstrecthed position. Due to mass m attached at its middle point M, the string $A'B'$ will behave as made up of two string $A'M$ and $B'M$, each of natural length $a/2$. Let the string $A'B'$ be strecthed between two points A and B in the same vertical line such that $AB = 2a$ as shown in fig. (*ii*). Clearly, due to the mass attached the original upper part $A'M$ of the string will be extended more in downward direction and let in the equilibrium position AE be its new position. As a result, the original lower part $B'M$ of the string will take the new position BE. Thus, E is the desired equilibrium position of the particle of mass m.

Let $AE = y$. Then $BE = 2a - y$. Given that λ = modulus of elasticity = mg. At the equilibrium position E, the forces acting on the particle of mass m are (*i*) its weight mg acting vertically downwards (*ii*) tension T_1 in the string AE (*iii*) tension T_2 in the string BE. Using Hooke's law and noting that natural length of each of portions AE and BE is $a/2$, we have

$$T_1 = \lambda\frac{y - a/2}{a/2} = \frac{mg(2y-a)}{a},\quad T_2 = \lambda\frac{(2a-y) - a/2}{a/2} = \frac{mg(3a-2y)}{a} \qquad ...(1)$$

For static equilibrium at E, $\qquad T_1 = mg + T_2$

$\therefore \qquad (1/a)\times mg\,(2y - a) \;=\; mg + (1/a)\times mg\,(3a - 2y)$, using (1)

This given $y = 5a/2$, which is the desired position of equilibrium.

From figure (*ii*), $AE = y = 5a/4$ and $BE = 2a - y = 2a - (5a/4) = 3a/4$.

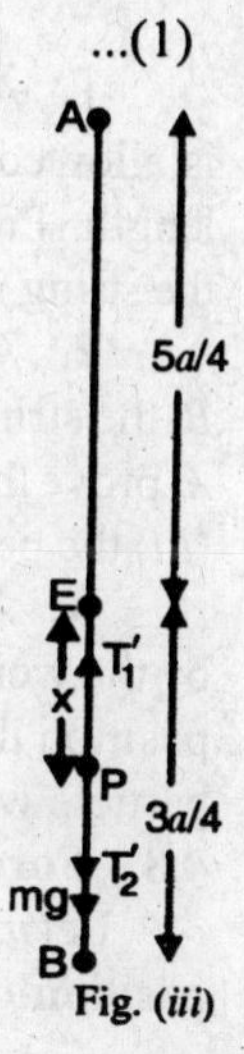

Fig. (*iii*)

Case I. When the particle is slightly displaced from its equilibrium position *E* in vertically downward direction *EB* and let go. Then the particle will begin to move in the vertically upward direction as shown in fig. (*iii*). Let P be the position of the particle at any time t such that $EP = x$. When the particle is at P, the forces acting on it are (*i*) its weight mg acting vertically downwards (*ii*) tension T_1' in the string AP (*iii*) tension T_2' in the string BP. By Hooke's law, we have

$$T_1' = \lambda\frac{(5a/4+x) - a/2}{a/2} = \frac{mg(3a+4x)}{2a};\quad T_2' = \lambda\frac{(3a/4-x) - a/2}{a/2} = \frac{mg(a-4x)}{2a} \qquad ...(2)$$

Now, the equation of motion of the particel at P is given by

$$m\ddot{x} = mg + T_2' - T_1'$$

or $$m\ddot{x} = mg + (mg/2a)(a - 4x) - (mg/2a) \times (3a + 4x)$$

or $$\ddot{x} = g(1 + 1/2 - 2x/a - 3/2 - 2x/a) \quad \text{or} \quad \ddot{x} = -(4g/a)x,$$

showing that the motion is simple harmonic of time period T_1 given by

$$T_1 = 2\pi/(4g/a)^{1/2} = \pi(a/g)^{1/2} \qquad ...(3)$$

Case II. When the particle is slightly displaced from its equilibrium position E in the horizontal direction EX and let go. Then the particle will begin to move towards E along PE as shown in figure (*iv*). Let P be the position of the particle at any time t such that $EP = x$. When the particle is at P, the forces acting on the particle are (*i*) its weight mg acting vertically downwards (*ii*) tension T_1'' in string PA (*iii*) tension T_2'' in string PB. Let $\angle APE = \alpha$ and $\angle BPE = \beta$. So resultant force acting on the particle at P along $EP = -T_1'' \cos\alpha - T_2'' \cos\beta$ and hence the equation of motion of the particle of mass m at P along EP is

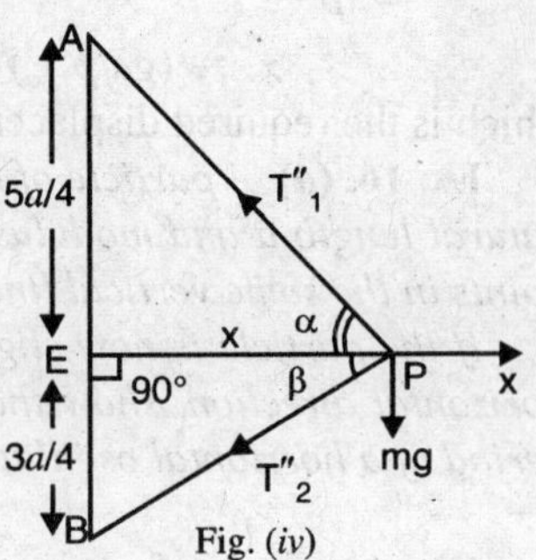

Fig. (*iv*)

$$m\ddot{x} = -T_1'' \cos\alpha - T_2'' \cos\beta \qquad ...(4)$$

From ΔAPE, $AP = \{(5a/4)^2 + x^2\}^{1/2} = (5a/4)(1 + 16x^2/25a^2)^{1/2}$

$\therefore \quad AP = (5a/4)\{1 + (1/2) \times (16x^2/25a^2) + ...\} = 5a/4,$

upto the first order of approximation, since x is small.

Similarly, we have $BP = 3a/4$, nearly. From figure, we have

$$\cos\alpha = x/AP = x/(5a/4) = 4x/5a. \text{ Similarly, } \cos\beta = 4x/3a \qquad ...(5)$$

Also, $$T_1'' = \frac{\lambda(AP - a/2)}{a/2} = \frac{mg(5a/4 - a/2)}{a/2} = \frac{3mg}{2} \qquad ...(6)$$

and $$T_2'' = \frac{\lambda(BP - a/2)}{a/2} = \frac{mg(3a/4 - a/2)}{a/2} = \frac{mg}{2} \qquad ...(7)$$

Using (5), (6) and (7), (4) reduces to

$$m\ddot{x} = -(3mg/2) \times (4x/5a) - (mg/2) \times (4x/3a) \quad \text{or} \quad \ddot{x} = -(28g/15a)x,$$

showing that the motion is simple harmonic of time period T_2, given by

$$T_2 = 2\pi/(28g/15a)^{1/2} = \pi(15a/7g)^{1/2} \qquad ...(8)$$

From (3) and (8), $$\frac{T_2}{T_1} = \pi\left(\frac{15a}{7g}\right)^{1/2} \times \frac{1}{\pi}\left(\frac{g}{a}\right)^{1/2} = \left(\frac{15}{7}\right)^{1/2}$$

Thus, $T_2 = (15/7)^{1/2} T_1$, which is what we wished to prove.

(*b*) Proceed exactly as in part (*a*) upto equation (3) only.

EXERCISE

1. A heavy particle is attached to an inextensible string to a fixed point from which the particle is allowed to fall freely; when the particle is in its lowest position the string is of twice its natural length. Prove that the modulus is four times the weight of the particle and show the time during which the string is extended beyond its natural length is $(1/2) \times (a/g)^{1/2} (\pi - \sin^{-1} 1/3)$.

2. A light elastic string AB of length l is fixed at A and is such that if a weight w be attached to B, the string will be stretched to a length $2l$. If a weight $w/4$ be attached to B and let fall from the level of A, prove that (*i*) the amplitude of the S.H.M. that ensure is $3l/4$ (*ii*) the distance through which it falls is $2l$ (*iii*) the period of oscillation is $(l/4g)^{1/2} (4\sqrt{2} + \pi + 2 \sin^{-1} 1/3)$. **[Meerut 2013, 2019]**

3. An elastic string supporting a heavy particle hangs in equilibrium, the elongation produced by the weight of the particle being equal to e. The particle is now pulled down below the equilibrium position through a distance 'd' greater than 'e' and then released. Show that the height to which the particle will rise after the string ceases to be taut is $(d^2 - e^2)/2e$.

4.8 Forced vibrations of harmonic oscillator

Let m be the mass of a particle, executing S.H.M., along OX with mean position at O. Let P be the position of the particle at any time t such that $OP = x$. Let $-mp^2x$ be the controlling force. In addition to the controlling force, let a disturbing (or driving) force mF also acts on the particle. Then the equation of motion of the particle at point P is given by

$-mp^2x$ mF O ← x → P X

$$m(d^2x/dt^2) = -mp^2x + mF \qquad \text{or} \qquad (D^2+p^2)x = F,\ D \equiv d/dt \qquad \ldots(1)$$

Now, the C.F. of (1) $= A\cos pt + B\sin pt$, ...(2)

where A and B are arbitrary constants.

While finding P.I. of (1), we consider the following cases :

Case I. Where $F = K$ = a constant. Then, for (1)

$$\text{P.I.} = \frac{1}{D^2+p^2}K e^{o\cdot t} = K\frac{1}{o^2+p^2}K e^{o\cdot t} = \frac{K}{p^2}$$

Hence the general solution of (1) is given by

$$x = A\cos pt + B\sin pt + K/p^2, \qquad \ldots(3)$$

which represents oscillatory motion of the same period as before, but with its mean position at $x = K/p^2$. It follows that the effect of a constant disturbing force on a harmonic oscillator is just to displace its mean position.

Case II. Let $F = K\cos wt$, $(w \neq p)$, which is periodic function with its period different from the original period of the oscillator.

$$\text{Then, P.I. of (1)} = \frac{1}{D^2+p^2}K\cos wt = \frac{K}{p^2-w^2}\cos wt$$

Hence the general solution of (1) is given by

$$x = A\cos pt + B\sin pt + \{K/(p^2-w^2)\}\cos wt,\ w \neq p \qquad \ldots(4)$$

Case III. Let $F = K\cos pt$ which is periodic function with the same period as the original period of oscillation.

$$\text{Then, P.I. of (1)} = \frac{1}{D^2+p^2}K\cos pt = K\left(\frac{t}{2p}\sin pt\right).$$

Hence the general solution of (1) is given by

$$x = A\cos pt + B\sin pt + (K/2p)\,t\sin pt \qquad \ldots(6)$$

From (6) we find that the amplitude of the motion increases continuously with time. The motion is oscillatory, which is obtained by superposition of an undisturbed S.H.M. and a second S.H.M. of the same period as disturbing force.

Example. *A particle of mass m oscillates in a line with natural period $2\pi/n$. If an applied force F cos pt now acts in the line so that the particle is instantaneously at rest at zero time at a distance d from the centre of oscillation, prove that the displacement of the particle at a subsequent time t is* $d\cos nt + \{F/m(n^2-p^2)\}(\cos pt - \cos nt)$ **[Delhi Maths (H) 1999, 2006]**

Sol. Let m be mass of a particle, executing S.H.M., along OX with mean position O. Since time period of S.H.M. is $2\pi/n$, $-mn^2x$ will be controlling force on the particle. Since an applied force $F\cos pt$ now acts along OX, the equation of motion of the particle at point P is given by

$-mn^2x$ $F\cos pt$ O ← x → P X

$$m(d^2x/dt^2) = -m\,n^2x + F\cos pt \quad \text{or} \quad (D^2+n^2)x = (F/m)\cos pt, \qquad \ldots(1)$$

where $D \equiv d/dt$. The complementary function (C.F.) of (1) is given by

$$\text{C.F.} = A\cos nt + B\sin nt,\ A, B \text{ being arbitrary constants} \qquad \ldots(2)$$

$$\text{Also, P.I. of (1)} = \frac{1}{D^2+n^2}\frac{F}{m}\cos pt = \frac{F}{m}\frac{1}{n^2-p^2}\cos pt$$

Hence the general solution of (1) is given by

$$x = A\cos nt + B\sin nt + \{F/m(n^2-p^2)\}\cos pt \qquad \ldots(3)$$

Differentiating both sides of (3) w.r.t. 't', we get

$$dx/dt = -An\sin nt + Bn\cos nt - \{Fp/m(n^2-p^2)\}\sin pt \qquad \ldots(4)$$

We are given that, initially when $t = 0$, $x = d$, $dx/dt = 0$.

Hence (3) gives $d = A + F/m(n^2 - p^2)$ or $A = d - \{F/m(n^2 - p^2)\}$ ...(5)

and (4) gives $0 = Bn$ so that $B = 0$. ...(6)

Substituting the values of A and B given by (5) and (6) in (3), we get

$$x = \left\{d - \frac{F}{m(n^2 - p^2)}\right\}\cos nt + \frac{F}{m(n^2 - p^2)}\cos pt$$

or $$x = d\cos nt + \{F/m(n^2 - p^2)\}(\cos pt - \cos nt),$$

which is the required displacement of the particle at a subsequent time t.

4.9 Damped harmonic Oscillation (or Resisted S.H.M.) Definition.

If a particle moves in a straight line under the action of a controlling force towards a fixed point in a line, proportional to its distance from the fixed point and the motion is resisted by a force (known as *damping force*) proportional to the velocity, then particle is said to *execute damped harmonic oscillation.*

Ex. 1. *Discuss the motion of a simple harmonic oscillation for which the controlling force is mw^2x and the damping force is $-2mk\dot{x}$, where m is the mass of the particle, w and k are constants.*

Sol. Suppose a particle of mass m is moving in a straight line OX under a controlling force mw^2x towards a fixed point O of the line OX, where x is distance of the particle from O at time t and in addition suppose a damping force $2mk\dot{x}$ acts in the direction opposting the velocity $\dot{x}$, where $k > 0$. Then the resultant force on m in the sense of x increasing is $-(mw^2x + 2mk\dot{x})$. Hence the equation of motion of the particle at P is

$- mw^2x$ ←
O ← x → P ——X
$- 2mk\dot{x}$ ←

$$m(d^2x/dt^2) = -mw^2x - 2mk(dx/dt), \text{ as } \dot{x} = dx/dt$$

or $$(D^2 + 2kD + w^2)x = 0, D \equiv d/dt \qquad ...(1)$$

The auxillary equation of (1) is $D^2 + 2kD + w^2 = 0$, giving

$$D = \{-2k \pm (4k^2 - 4w^2)^{1/2}\}/2 \quad \text{or} \quad D = -k \pm (k^2 - w^2)^{1/2} \qquad ...(2)$$

We now discuss three cases : (*i*) $k < w$ (*ii*) $k > w$ (*iii*) $k = w$.

Case I. Let $K < w$. Let $k^2 - w^2 = -\alpha^2$, where α is real. Then by (2), the roots of the auxiliary equation of (1) are $-k \pm i\alpha$. Hence solution of (1) is

$x = e^{-kt}(C_1 \cos \alpha t + C_2 \sin \alpha t)$, C_1, C_2 being arbitrary constants. ...(3)

For physical reasons, we assume C_1, C_2 to be real constants.

From (3), $dx/dt = -ke^{-kt}(C_1 \cos \alpha t + C_2 \sin \alpha t) + e^{-kt}(-C_1\alpha \sin \alpha t + C_2\alpha \cos \alpha t)$ ...(4)

Suppose initially, when $t = 0$, $x = a$ and $dx/dt = 0$. Then (3) and (4) give

$$a = C_1 \text{ and } 0 = -kC_1 + C_2\alpha \Rightarrow C_1 = a, C_2 = ka/\alpha.$$

∴ (3) becomes $$x = e^{-kt}\{a \cos \alpha t + (ka/\alpha)\sin \alpha t\}$$

or $$x = be^{-kt}\cos(\alpha t + \varepsilon), \qquad ...(5)$$

where b and ε are constants which can be expressed in terms of a, α and k.

Equation (5) does not represent a S.H.M. But the motion is periodic with period $2\pi/\alpha$. The factor e^{-kt}, $k > 0$ indicates a general decay of the oscillations. In fact, $x \to 0$ as $t \to \infty$. So be^{-kt} is damped amplitude as it decays with time. This type of motion is called *damped oscillatory motion.* This case corresponds to small damping (or light damping or under damping), because after a long time the oscillation as a whole ultimately dies out.

Since $k < w$, so $k^2 - w^2 = -\alpha^2 \Rightarrow \alpha^2 = w^2 - k^2 \Rightarrow \alpha^2 < w^2 \Rightarrow \alpha < w \Rightarrow 2\pi/\alpha > 2\pi/w$, showing that the period of the damped harmonic oscillation slightly exceeds that of the natural oscillation (*i.e.,* oscillation due to S.H.M. only)

(5) shows that the motion is oscillatory, with an amplitude, that is continuously decreasing (decaying).

We have already assumed that the particle initially started from rest at an extreme position $x = a$. We now propose to find the subsequent extreme positions. For maxima or minima of x, $dx/dt = 0$, and so by (5)

$-kbe^{-kt}\cos(\alpha t+\varepsilon)-k\alpha e^{-kt}\sin(\alpha t+\varepsilon)=0$ so that $\tan(\alpha t+\varepsilon)=-k/\alpha$, showing that the instant t, at which x goes to an extreme position are given by

$$\alpha t_n+\varepsilon=n\pi-\tan^{-1}(k/\alpha),\ n \text{ being any positive integer,} \qquad \ldots(6)$$

showing that the successive times of reaching an extreme position differ by π/α. If $x=x_n$, be the extreme position, at the instant t_n, then

$$x_n=be^{-kt_n}\cos(n\pi-\tan^{-1}k/\alpha)=(-1)^n\,be^{-kt_n}\cos(\tan^{-1}k/\alpha) \qquad \ldots(7)$$

(7) shows that the successive extreme positions are alternately on the right or left of $x=0$. Again by (7), we find that

$$|x_{n+1}/x_n|=e^{-k(t_{n+1}-t_n)}=e^{-k\pi/\alpha},$$

showing that the ratios of successive swings to opposite sides, form a geometric progression with common ratio $e^{-k\pi/\alpha}$.

Case II. Let $k>w$. Let $k^2-w^2=\beta^2$, where β is real. Then by (2), the roots of the auxiliary equation of (1) are $-k\pm\beta$. Hence solution of (1) is

$$x=C_3e^{-(k-\beta)t}+C_4e^{-(k+\beta)t},\ C_3, C_4 \text{ being arbitrary constants.} \qquad \ldots(8)$$

Since $\quad \beta^2=k^2-w^2<k^2$, so $\quad \beta<k \quad$ or $\quad (k-\beta)>0$.

Since $k-\beta>0$ and $k+\beta>0$, both the exponentials on R.H.S. of (8) decay with time. Thus $x\to 0$ as $t\to\infty$.

From (8), $\qquad dx/dt=-C_3(k-\beta)e^{-(k-\beta)t}-C_4(k+\beta)e^{-(k+\beta)t}$

$\therefore \qquad dx/dt=0 \Rightarrow e^{-kt}[-C_3(k-\beta)e^{\beta t}-C_4(k+\beta)e^{-\beta t}\}=0$

or $\qquad C_3(\beta-k)e^{\beta t}=C_4(k+\beta)e^{-\beta t} \quad$ or $\quad e^{2\beta t}=C_3(\beta-k)/C_4(\beta+k)$,

which has at most one solution. Thus, the velocity vanishes, once only, if at all. The motion is non-oscillatory or dead beat. This case corresponds to *large damping* (*or over damping*).

Case III. Let $k=w$. In this case by (2), the roots of the auxiliary equation of (1) are each equal to $-k$. Hence the general solution of (1) is given by

$$x=(C_5t+C_6)e^{-kt},\ C_5, C_6 \text{ being arbitrary constants.} \qquad \ldots(9)$$

Now, $\qquad \lim\limits_{t\to\infty} t\,e^{-kt}=\lim\limits_{t\to\infty}\dfrac{t}{e^{kt}}=\lim\limits_{t\to\infty}\dfrac{1}{k\,e^{kt}}=0$, as $k>0$. Also, $\lim\limits_{t\to\infty} e^{-kt}=0$.

Hence, (9) shows that $x\to 0$ as $t\to\infty$.

From (9), $\qquad dx/dt=C_5e^{-kt}-k(C_5t+C_6)e^{-kt}$

$\therefore \quad dx/dt = 0$ only when $\quad [C_5-(C_5t+C_6)k]e^{-kt}=0$

so that $\qquad t=(C_5-kC_6)/C_5k$,

showing that the velocity vanishes once. Here the motion is *non-oscillatory* and is known as *critically damped.*

Ex. 2. *Discuss the effect of a damping force $-2\mu\dot{x}$ acting on a particle executing a S.H.M. given by $\ddot{x}=-p^2x$, when $\mu<p$. Also calculate the logarithmic decrement.*

Sol. Proceed as in Ex. 1 noting that in our present problem, we have $k=\mu$ and $w=p$. Since $\mu<p$ so $k<w$, hence you have to discuss case I of Ex. 1 only and get $|x_{n+1}/x_n|=e^{-\mu\pi/\alpha}$. $\qquad \ldots(i)$

The logarithmic decrement of the damped oscillation is defined as the logarithmic of the reciprocal of the ratio $|x_{n+1}/x_n|$ to the base 10. Hence, from (*i*), the required logarithmic decrement

$$=\log_{10}|x_n/x_{n+1}|=\log_{10}e^{\mu\pi/\alpha}=(\mu\pi/\alpha)\log_{10}e$$

Ex. 3. *A particle of unit mass moves in a straight line and is attracted to an origin O in the line by a force w^2x, where x is the displacement from O. There is also a damping force $2k\dot{x}$, where $\dot{x}$ is the speed. Find the condition that the motion should be periodic, and show that if this condition is satisfied, the successive maximum displacements from the origin form a geometric progression with common ratio $e^{-k\pi/\alpha}$ where $\alpha=(w^2-k^2)^{1/2}$.*

Sol. Refer case I of Ex. 1 for the solution.

Ex. 4. (Forced vibrations of a damped harmonic oscillator) *A particle of mass m is moving in a straight line under a controlling force $-mw^2x$, a damping force $-2mk\dot{x}$, and a periodic force or* *disturbing force mF cos pt. Investigate its motion where k < w and show that the harmonic motion ultimately dies out.* **[Delhi Maths (H) 1995, 2008]**

Sol. Let P be the position of the particle at any time t such that $OP = x$, where O is fixed point on OX. Then the equation of motion of the particle is

$$m(d^2x/dt^2) = -mw^2x - 2mk(dx/dt) + mF \cos pt$$

or $\quad (D^2 + 2kD + w^2)x = F \cos pt$, where $D \equiv d/dt$ ...(1)

$-m\omega^2 x \leftarrow$, $-2mk\dot{x} \leftarrow$, O, x, P, X, mF cos pt

Now, Auxiliary equation of (1) is $D^2 + 2kD + w^2 = 0$, giving

$$D = \{-2k \pm (4k^2 - 4w^2)^{1/2}\}/2 = -k \pm (k^2 - w^2)^{1/2} \quad ...(2)$$

Given that $k < w$. Let $\quad k^2 - w^2 = -\alpha^2$, where α is real. ...(3)

$\therefore$ C.F. of (1) is $= ae^{-kt} \cos(\alpha t + \varepsilon)$, where a and ε are arbitrary constants.

P.I. of (1)

$$= \frac{1}{D^2 + 2kD + w^2} F \cos pt = F \frac{1}{-p^2 + 2kD + w^2} \cos pt = F \frac{1}{(\omega^2 - p^2) + 2kD} \cos pt$$

$$= F \frac{(w^2 - p^2) - 2kD}{(w^2 - p^2)^2 - 4k^2D^2} \cos pt = F \frac{(w^2 - p^2)\cos pt + 2kp \sin pt}{(w^2 - p^2)^2 + 4k^2p^2} \quad ...(4)$$

Let $\quad F(w^2 - p^2)/\{(w^2 - p^2)^2 + 4k^2p^2\} = b \cos \varepsilon'$ ...(5)

and $\quad 2kpF/\{(w^2 - p^2)^2 + 4k^2p^2\} = -b \sin \varepsilon'$ ...(6)

Squaring and adding (5) and (6), $\quad b^2 = F^2/\{(w^2 - p^2)^2 + 4k^2p^2\}$ ...(7)

Dividing (6) by (5), $\quad \tan \varepsilon' = -(2kp)/(w^2 - p^2)$ ...(8)

Using (5) and (6), (4) reduces to

P.I. of (1) $= b \cos pt \cos \varepsilon' - b \sin pt \sin \varepsilon' = b \cos(pt + \varepsilon')$.

Hence the general situation of (1) is x = C.F. + P.I., *i.e.*,

$$x = a e^{-kt} \cos(\alpha t + \varepsilon) + b \cos(pt + \varepsilon'), \quad ...(9)$$

where b and ε' are given by the relations (7) and (8).

Equation (9) shows that the motion is made up of two periodic motions of periods $2\pi/\alpha$ and $2\pi/p$ respectively. But the amplitude $a e^{-kt}$ of the former tends to zero as $t \to \infty$. Thus the free harmonic motion ultimately dies out. It is consequently called the *transient*. The second part $b \cos(pt + \varepsilon')$ of solution does not vanish permanently at any time, however large ; it is called the *steady state*. Thus, ultimately the particle performs a forced oscillation of amplitude b and period $2\pi/p$. Hence the steady solution of the forced damped harmonic oscillator is given by $x = b \cos(pt + \varepsilon')$.

Ex. 5. *Discuss the motion of a simple oscillator under light damping for which controlling force is $-mn^2x$, damping force is $-2m\mu\dot{x}$ and disturbing force is mk cos pt where m is the mass of a particle and n, μ, k are constants. Here $\dot{x} = dx/dt$* **[Delhi Maths (H) 2003]**

Sol. Do like Ex. 4 by taking $w = n$, $k = \mu$ and $F = k$.

OBJECTIVE PROBLEMS

Ex. 1. *A point moves with S.H.M. whose period is 4 sec, if it starts from rest at a distance 4 m from the centre of its path then the time it takes before it has described 2 meters is*

(a) 1/3 sec (b) 2/3 sec (c) 3/4 sec (d) 4/5 sec. **[I.A.S. Prel. 1993]**

Sol. (*b*). Since the particle starts from extreme position, we have

$$x = a \cos t \mu^{1/2} \quad ...(1)$$

Given that $T = 2\pi/\mu^{1/2} = 4$ so that $\mu^{1/2} = \pi/2$. Also given that $a = 4$ m and $x = 2$ m. So (1) reduces to

$2 = 4\cos(t\pi/2) \Rightarrow \cos(t\pi/2) = 1/2 \Rightarrow t\pi/2 = \pi/3 \Rightarrow t = 2/3$

Ex. 2. *A particle executing a S.H.M. has a acceleration 8 cm/sec^2 when it is at a distance 2 cm for the centre. The time period will be*

(a) 2/π sec (b) 1/π sec (c) π/2 ssec (d) π sec.

Sol. Ans. (*d*) Acceleration $= \mu x = 8 \Rightarrow \mu \times 2 \Rightarrow \mu = 4$

So period $= 2\pi/\mu^{1/2} = 2\pi/2 = \pi$ sec

Ex. 3. *A particle moving in S.H.M. has a speed of 8 cm/sec and 6 cm/sec at distances 3 cm and 4 cm respectively from the centre of the path. Period of its motion will be (a) π/2 sec (b) 2π sec (c) π sec (d) π/3 sec.*

Sol. Ans. (*c*) Using $v^2 = \mu(a^2 - x^2)$, here $64 = \mu(a^2 - 9)$ and $36 = \mu(a^2 - 16)$. Subtracting, $64 - 36 = -9\mu + 16\mu \Rightarrow \mu = 4$. So period $= 2\pi/\mu^{1/2} = \pi$ sec.

Ex. 4. *The amplitude of a particle executing S.H.M. is 20 cm and time period is 1 sec. The maximum velocity is (a) 10π cm/sec (b) 20π cm/sec (c) 30π cm/sec (d) 40π cm/sec.*

Sol. Ans. (*d*) Here T = time period $= 2\pi/\mu^{1/2} = 1 \Rightarrow \mu^{1/2} = 2\pi$. Also amplitude $= a = 20$ cm. Maximum velocity $= a\mu^{1/2} = 20 \times 2\pi = 40\pi$ cm/sec

Ex. 5. *If the equation of motion of a particle executing a S.H.M. is $d^2x/dt^2 + 16x = 0$, then the frequency will be (a) π/2 (b) π/8 (c) 2/π (d) 8/π*

Sol. Ans. (*c*) Here $d^2x/dt^2 = -16x$. Comparing with standard equation of S.H.M. $d^2x/dt^2 = -\mu x$, here $\mu = 16$. Frequency $= 1/T = \mu^{1/2}/2\pi = 4/2\pi = 2/\pi$

Ex. 6. *A particle is executing a S.H.M. such that its period of oscillation is π sec. If its maximum acceleration is 8 cm/sec^2, then its amplitude will be (a) 2 cm (b) 4 cm (c) 8 cm (d) 16 cm.*

Sol. Ans. (*a*). Here time period $= \pi = 2\pi/\mu^{1/2} \Rightarrow \mu = 4$. Now maximum acceleration $= \mu a \Rightarrow 8 = 4 \times a \Rightarrow a = 2$.

Ex. 7. *A particle starting from rest moving with S.H.M. has a period of 18 sec., travels 25 cms in 3 sec. Find the amplitude and maximum velocity.* **[Bangalore 2004]**

Sol. T = time period $= 2\pi/\mu^{1/2} \Rightarrow 18 = 2\pi/\mu^{1/2} \Rightarrow \mu^{1/2} = \pi/9$. ....(1)

Now, $x = a\cos t\,\mu^{1/2} \Rightarrow 25 = a\cos(3 \times \pi/9) \Rightarrow a$ = amplitude = 50

Also, maximum velocity $= a\,\mu^{1/2} = 50 \times \pi/9 = 50\,\pi/9$ cm/sec

Ex. 8. *A particle moving with S.H.M. has speeds 8 and 6 at distacnes 3 and 4 respectively from the mean position. Show that the period is π and maximum acceleration is 20.* **[Bangalore 2004]**

Hint. Procced as in Ex. 9, maximum acceleration is μa.

Ex. 9. *Complete the following statements :*

(i) In a S.H.M., the intensity of force is μ, then time taken to make one complete oscillation is .... **[Kanpur 2006, Meerut 2004]**

(ii) In a S.H.M. the velocity is maximum at the ...... **[Meerut 2004, 05]**

(iii) The maximum velocity of a body moving with S.H.M. is 2 ft/sec and its period is 1/5 sec. The amplitide is ...... **[Meerut 2005]**

(iv) The relation between time period and frequency in S.H.M. is .... **[Meerut 2005]**

Ans. *(i)* $2\pi/\sqrt{\mu}$ *(ii)* mean position *(iii)* $(1/50\pi^2)$ feat *(iv)* $nT = 1$

Ex. 10. *In an elastic string, Hooke's law is*

(*a*) $T = \lambda(a/l)$ (*b*) $T = \lambda(l/a)$ (*c*) $T = \lambda l$ (*d*) $T = \lambda(dl/a)$,

where l is the natural ength of the string and a is the extension in the string beyond its natural length. **(Garhwal 2004)**

Sol. Ans. (*a*). Refer Art. 4.5 page 4.12

Ex. 11. *If T is the time period of a simple harmonic motion of amplitude a and intensity of force* μ, *then* (*a*) $T = \pi/2\sqrt{\mu}$ (*b*) $T = 2\pi/\sqrt{\mu}$ (*c*) $T = a\sqrt{\mu}$ (*d*) $T = 2\pi\sqrt{\mu}$

Sol. Ans. (*b*). Refer page 4.2. **(Bundelkhand 2007; Rohilkhand 2010)**

Ex. 12. *If in a simple harmonic motion of amplitude a and intensity of force* μ, *the displacement of the particle from the centre of force at time t is x, then its equation of motion is*

(*a*) $dx/dt = \mu x$ (*b*) $dx/dt = -\mu x$ (*c*) $d^2x/dt^2 = \mu x$ (*d*) $d^2x/dt^2 = -\mu x$

Sol. Ans. (*d*). Refer Art. 4.1, page 4.1. **(Avadh 2006; Rohilkhand 2007)**

Ex. 13. *A particle moving in a straight line OAB with S.H.M. has zero velocity at the points A and B whose distances from O are a and b respectively, and has velocity v when half way between them. If the time period of the motion is T, then*

(*a*) $T = \{\pi(b-a)\}/v$ (*b*) $T = (\pi v)/(b-a)$ (*c*) $T = (\pi b)/(v-a)$ (*d*) $T = \{\pi(a-b)\}/v$

Sol. Ans. (*a*). Refer solved Ex. 13, page 4.7 **(Garhwal 2002)**

Ex. 14. *The speed v of a particle moving along the axis of x is given by the relation* $v^2 = n^2(8bx - x^2 - 12b^2)$. *The motion is simple harmonic with its centre at*

(*a*) $x = 4b$ (*b*) $x = 0$ (*c*) $x = -4b$ (*d*) $x = 2b$ **(Avadh 2006; Rohilkhand 2007)**

Sol. Ans. (*a*). Refer solved Ex. 3(*a*), page 4.4

Ex. 15. *What is the value of the period of S.H.M.?*

(*a*) $T = \pi/\sqrt{\mu}$ (*b*) $T = \pi/2\sqrt{\mu}$ (*c*) $T = \mu/\sqrt{\pi}$ (*d*) $T = 2\pi/\sqrt{\mu}$ **(Agra 2012)**

Sol. Ans. (*d*). Refer page 4.2.

MISCELLANEOUS PROBLEMS

Ex. 1. *A particle starts from rest under an acceleration* k^2x *directed towards a fixed point and after time* t_1, *another particle starts from the same position under the same acceleration. Show that the particles will collide at time* $(\pi/k) + (t_1/2)$ *after the start of the first particle provided* $t_1 < 2\pi/k$. **(Lucknow 2019)**

Sol. For the first particle, the equation of motion of the particle moving with simple harmonic motion is given by

$$d^2x/dt^2 = -k^2x \quad \text{...(1)}$$

and hence from Art. 4.1, it follows that the time period T of the motion is given by

$$T = 2\pi/k \quad \text{...(2)}$$

The condition $t_1 < 2\pi/k$ indicates that the second particle starts before the first particle has made one complete revolution. If the first particle starts from $x = a$ at $t = 0$, then the solution of (1) is given by (refer Art. 4.1)

$$x = a\cos(kt) \quad \text{...(3)}$$

Similarly, for the second particle, we have

$$x = a\cos\{k(t - t_1)\} \quad \text{...(4)}$$

Hence, if the two particle meet each other after time t, then we have

$$x = a \cos \{k(t - t_1)\} = a \cos (kt)$$

and hence either $\quad k(t - t_1) = kt \quad$ or $\quad 2\pi - k(t - t_1) = kt$

Now, $k(t - t_1) = kt \Rightarrow t_1 = 0$, which is absurd as $t_1 \neq 0$.

Therefore, we must consider the second option, namely,

$2\pi - k(t - t_1) = kt \quad$ giving $\quad 2kt = 2\pi + kt_1 \quad$ so that $\quad t = (\pi/k) \neq (t_1/2)$.

Ex. 2. *A body moving with simple harmonic motion has an amplitude a and time period T. Show that the velocity v at a distance x from the mean position is given by* $v^2T^2 = 4\pi^2(a^2 - x^2)$.

If the velocity is trebled when the distance from the mean position in 2a/3, the period being unaltered, find the new amplitude. **(Nagpur 2018)**

Sol. Let the body be moving from A to A' in simple harmonic motion such that O is its mean position and $OA = OA' = a$. While moving along OA, let P be its position at any time t and let $OP = x$. Let the motion of the body be given by $d^2x/dt^2 = -\mu x$. Then, by Art. 4.1, the velocity v at P is given by

$$v^2 = \mu(a^2 - x^2) \quad \text{...(1)}$$

and time period T is given by

$$T = 2\pi/\sqrt{\mu} \quad \text{so that} \quad \mu = 4\pi^2/T^2 \quad \text{...(2)}$$

Substituting the value of μ given by (2) in (1), we have

$$v^2T^2 = 4\pi^2 (a^2 - x^2) \quad \text{...(3)}$$

Second part : Let $v = V$ when $x = 2a/3$. Then, (3) yields

$$V^2T^2 = 4\pi^2(a^2 - 4a^2/9) \quad \text{giving} \quad V^2T^2 = (20/9) \times \pi^2a^2 \quad \text{...(4)}$$

In the subsequent motion, let b be the new amplitude, so that by hypothesis at this very point, the velocity in $3V$ while the time period T remains the same. Therefore, (3) yields

$$9V^2T^2 = 4\pi^2(b^2 - 4a^2/a) \quad \text{giving} \quad 9 \times (20/9) \times \pi^2a^2 = 4\pi^2(b^2 - 4a^2/9), \text{ using (4)}$$

or $\quad b^2 - (4a^2/9) = 5a^2 \quad$ or $\quad b^2 = (49a^2)/9 \quad$ or $\quad b = 7a/3$,

showing that the required new amplitude $= b = 7a/3$.

Ex. 3. *A particle P of mass m moves in a straight line OCP being attracted by a force μ m PC, always directed towards C while C moves along OC with a constant acceleration f. If initially C was at rest at the origin O, and P was at a distance c from O moving velocity V, prove that the distance of P from O at any time t is given by*

$$(f/\mu + c) \cos (t\sqrt{\mu}) + (V/\sqrt{\mu}) \sin (t\sqrt{\mu}) - (f/\mu) + (1/2) \times ft^2$$ **(Agra 2018)**

5

Plane Kinematics

5.1 Carterian rectangular components of velocity and acceleration of a particle moving along a plane curve.

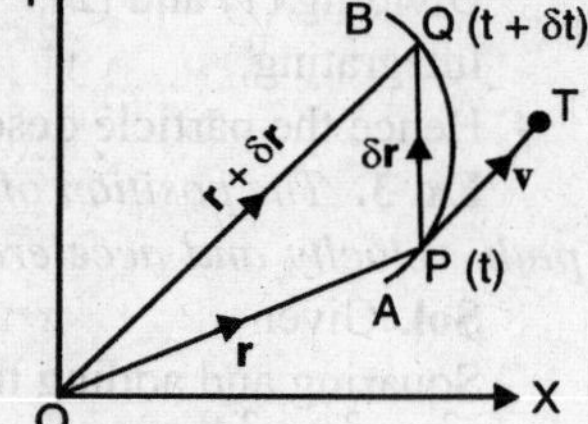

Let *OXY* be a rectangular frame in the plane and let **i**, **j** be unit vectors along *OX*, *OY* respectively. Consider the motion of a particle along the curve *AB* in *xy*-plane. Let *P* and *Q* be the positions of the particle at the instants t and $t + \delta t$ respectively. Let the position vectors of *P* and *Q* be **r** and **r** + *d***r** respectively. Then, we have

$$\mathbf{r} = x\,\mathbf{i} + y\,\mathbf{j}, \qquad \ldots(1)$$

where (x, y) are the carterian coordinates of the point *P*.

Now, by definition, the velocity **v** of the particle at time *t* is

$$\mathbf{v} = \lim_{\delta t \to 0} \frac{\delta \mathbf{r}}{\delta t} = \frac{d\mathbf{r}}{dt} = \frac{d}{dt}(x\mathbf{i} + y\mathbf{j}) = \dot{x}\,\mathbf{i} + \dot{y}\,\mathbf{j}, \qquad \ldots(2)$$

where we have denoted differential coefficient with regard to time by dot, thus $\dot{x}$ means dx/dt and $\dot{y}$ means dy/dt.

Also, note that
$$\mathbf{v} = \frac{d\mathbf{r}}{ds}\frac{ds}{dt} = \hat{\mathbf{t}}\,\frac{ds}{dt}, \qquad \ldots(3)$$

where $\hat{\mathbf{t}} = d\mathbf{r}/ds$ = unit vector along tangent *PT* to curve *AB*.

From (3), we see that the velocity at *P* is along tangent to the curve *P*.

Again, by definition, the acceleration ***a*** of the particle at time *t* is

$$\boldsymbol{a} = \lim_{\delta t \to 0} \frac{\delta \mathbf{v}}{\delta t} = \frac{d\mathbf{v}}{dt} = \frac{d}{dt}(\dot{x}\,\mathbf{i} + \dot{y}\,\mathbf{j}) = \ddot{x}\,\mathbf{i} + \ddot{y}\,\mathbf{j} \qquad \ldots(4)$$

where $\ddot{x}$ means d^2x/dt^2 and $\ddot{y}$ means d^2y/dt^2.

From (2) and (4), we not that

$dx/dt = \dot{x}$ = component of the velocity of *P* along *OX*

$dy/dt = \dot{y}$ = component of the velocity of *P* along *OY*.

$d^2x/dt^2 = \ddot{x}$ = component of the acceleration of *P* along *OX*.

$d^2y/dt^2 = \ddot{y}$ = component of the acceleration of *P* along *OY*.

Also, the magnitude of the velocity = $|\mathbf{v}| = v = (\dot{x}^2 + \dot{y}^2)^{1/2}$

and the magnitude of acceleration = $|\boldsymbol{a}| = (\ddot{x}^2 + \ddot{y}^2)^{1/2}$

Note $\ddot{x}$ can be re-written as follows

$$\ddot{x} = \frac{d^2x}{dt^2} = \frac{d}{dt}\left(\frac{dx}{dt}\right) = \frac{d\dot{x}}{dt} = \frac{d\dot{x}}{dx}\frac{dx}{dt} = \dot{x}\frac{d\dot{x}}{dx}$$

Similarly, $\ddot{y} = d^2y/dt^2 = d\dot{y}/dt = \dot{y}\;(d\dot{y}/dy)$.

TYPE 1 : SOLVED EXAMPLES BASED ON ART 5.1

Ex. 1. *A point moves in a plane, its velocities parallel to the axes of x and y being $u + ay$ and $v + bx$ respectively. Show that it moves in a conic section.* **[I.F.S. 1994]**

Sol. Given $dx/dt = u + ay$ and $dy/dt = v + bx$.

Now, $\dfrac{dx}{dt}\cdot\dfrac{dt}{dy} = \dfrac{u+ay}{v+bx}$ or $2(v+bx)\,dx - 2(u+ay)dy = 0$

Integrating, $2vx + bx^2 - 2uy - ay^2 + c = 0$, which is a conic section (Recall that a general equation of second degree in x and y represents a conic section)

Ex. 2. *A particle is moving with a constant velocity parallel to the axis of y and a velocity proportional to y parallel to the x-axis. Prove that it will describes a parabola.*

Sol. Given $dy/dt = k$, k being some constant. ...(1)

and $dx/dt \propto y$ so that $dx/dt = ly$, ...(2)

where l is the constant of proportionality.

Dividing (1) and (2), $dy/dx = k/ly$ or $2ydy = (2k/l)dx$

Integrating, $y^2 = ax + c$, where $a = 2k/l$ = constant.

Hence the particle describes a parabola $y^2 = ax + c$.

Ex. 3. *The position of a moving point at time t is given by $x = a\cos t$ and $y = a \sin t$. Find its path, velocity and acceleration.* **[Rohilkhand 2008; Agra 2003; Meerut 2006]**

Sol. Given $x = a\cos t$, $y = a\sin t$ ...(1)

Squaring and adding the above relations, we get

$x^2 + y^2 = a^2$, which is the required path of the particle.

Differentiating w.r.t. 't' the relations (1), we get

$$\dot{x} = -a\sin t \quad\text{and}\quad \dot{y} = a\cos t \qquad ...(2)$$

So the required velocity $= (\dot{x}^2 + \dot{y}^2)^{1/2} = a$, making an angle $\tan^{-1}(\dot{y}/\dot{x})$ *i.e.*,

$\tan^{-1}(-\cot t)$ or $\tan^{-1}\tan(\pi/2 + t)$ or $(\pi/2 + t)$ with x-axis.

Again, differentiating w.r.t. 't' the relations (2), we get

$$\ddot{x} = -a\cos t \quad\text{and}\quad \ddot{y} = -a\sin t.$$

Hence the required acceleration $= (\ddot{x}^2 + \ddot{y}^2)^{1/2} = a$, making an angle $\tan^{-1}(\ddot{y}/\ddot{x})$ or $\tan^{-1}\tan t$ or t with x-axis.

Ex. 4. *A particle moves in the curve $y = a\log\sec(x/a)$ in such a way that the tangent to the curve rotates uniformly; prove that the resultant acceleration of the particle varies as the square of the radius of curvature.* **[Lucknow 2001, 08; Rohilkhand 2008; Agra 2003, 08]**

Sol. The equation of the given path is $y = a\log\sec(x/a)$. ...(1)

Since the tangent to the curve rotates uniformly, so

$$d\psi/dt = \text{constant} = c, \text{ say.} \qquad ...(2)$$

Diff. (1), $\dfrac{dy}{dx} = \dfrac{a}{\sec(x/a)}\sec\dfrac{x}{a}\tan\dfrac{x}{a}\cdot\dfrac{1}{a}$ so that $\dfrac{dy}{dx} = \tan\dfrac{x}{a}$ ...(3)

Diff. (3), $d^2y/dx^2 = (1/a)\sec^2(x/a)$. ...(4)

$$\therefore\ \rho = \text{the radius of curvature} = \frac{\{1 + (dy/dx)^2\}^{3/2}}{d^2y/dx^2} = \frac{\{1 + \tan^2(x/a)\}^{3/2}}{(1/a)\sec^2(x/a)}$$

Thus, $\rho = \dfrac{a\sec^3(x/a)}{\sec^2(x/a)} = a\sec(x/a)$. ...(5)

Since $dy/dx = \tan\psi$, (3) reduces to $\tan\psi = \tan(x/a)$

or $\psi = x/a$ so that $x = a\psi$ ...(6)

Diff. (6) w.r.t. 't' $dx/dt = a(d\psi/dt) = ac$, by (2) ...(7)

Diff. (7), $\ddot{x} = d^2x/dt^2 = 0$, since ac is a constant. ...(8)

Now, $\dfrac{dy}{dt} = \dfrac{dy}{dx}\cdot\dfrac{dx}{dt} = \tan\left(\dfrac{x}{a}\right).ac$, using (3) and (7)

$\therefore\ \ddot{y} = d^2y/dt^2 = ac\sec^2(x/a).(1/a)(dx/dt) = ac^2\sec^2(x/a)$, by (7) ...(9)

Now, resultant velocity $= (\ddot{x}^2 + \ddot{y}^2)^{1/2} = ac^2\sec^2(x/a)$, by (8) and (9)

$= (c^2/a)\{a\sec(x/a)\}^2 = (c^2/a)\rho^2$, by (5)

Hence, resultant acceleration varies as square of radius of curvature.

Ex. 5. *A particle is acted only a force parallel to the axis of y whose acceleration is λy and is initially projected with a velocity $a\sqrt{\lambda}$ parallel to the axis of x at a point where y = a. Prove that it will describe the catenary*

$$y = a \cosh (x/a).$$ **[Meerut 1998]**

Sol. Here, given that $\qquad \ddot{y} = d^2y/dt^2 = \lambda y \qquad$ ...(1)

Since there is no force parallel to the x-axis, it follows that

$$\ddot{x} = d^2x/dt^2 = 0 \qquad ...(2)$$

Integrating (2), $dx/dt = c$, where c is constant of integration.

But, given that initially, $dx/dt = a\sqrt{\lambda}$. Hence $c = a\sqrt{\lambda}$.

Thus, $\qquad dx/dt = a\sqrt{\lambda} \qquad$ ...(3)

Re-writing (1), $\quad \dot{y}\,(d\dot{y}/dy) = \lambda y \quad$ or $\quad 2\dot{y}d\dot{y} = 2\lambda y dy.$

Integrating, $\quad \dot{y}^2 = \lambda y^2 + c'$, c' is constant of integration $\qquad$ ...(4)

Since initially at $y = a$, there is no velocity parallel to y-axis, we have $\dot{y} = 0$ when $y = a$. So (4) gives $c' = -\lambda a^2$.

$\therefore$ From (4), $\quad (dy/dt)^2 = \lambda(y^2 - a^2) \quad$ or $\quad dy/dt = \sqrt{\lambda}\,(y^2 - a^2)^{1/2} \qquad$ ...(5)

where the positive sign is put on R.H.S. because the particle is moving in the direction of y increasing.

Dividing (5) by (3), $\quad \dfrac{dy}{dx} = \dfrac{(y^2 - a^2)^{1/2}}{a} \quad$ or $\quad \dfrac{dy}{(y^2 - a^2)^{1/2}} = \dfrac{dx}{a}.$

Integrating, $\quad \cosh^{-1}(y/a) = (x/a) + c''$, where c'' is a constant. $\qquad$ ...(6)

Let us take $x = 0$ when $y = a$. Then (6) gives $c'' = 0$. So (6) gives

$\cosh^{-1}(y/a) = x/a \quad$ or $\quad y = a \cosh (x/a)$, which is a catenary.

Ex. 6. *A particle is acted on by a force parallel to the axis of y whose acceleration (always towards the x-axis) is μy^{-2} and when y = a, it is projected parallel to the x-axis with velocity $(\mu/a)^{1/2}$. Prove that it will describe a cycloid.*

Sol. Given that $\qquad \ddot{y} = d^2y/dt^2 = -\mu y^{-2} \qquad$ ... (1)

where the negative sign is put on the R.H.S. because the acceleration is always towards the x-axis (*i.e.*, in the direction of y decreasing).

Since there is no force parallel to x-axis, it follows that

$$\ddot{x} = d^2x/dt^2 = 0 \qquad ...(2)$$

Integrating (2), $dx/dt = c$, where c is constant of integration.

But, given that initially, $dx/dt = (\mu/a)^{1/2}$. Hence $c = (\mu/a)^{1/2}$

Thus, $\qquad dx/dt = (\mu/a)^{1/2} \qquad$ ...(3)

Re-writing (1), $\dot{y}\,(d\dot{y}/dy) = -\mu/y^2 \quad$ or $\quad 2\dot{y}\,d\dot{y} = -2\mu\, y^{-2}\, dy$

Integrating, $\quad \dot{y}^2 = (2\mu/y) + c'$, c' being constant of integration $\qquad$ ...(4)

Since initially at $y = a$, there is no velocity parallel to the y-axis, we have $\dot{y} = 0$ when $y = a$. So (4) gives $c' = -(2\mu/a)$. Hence (4) gives

$$\left(\frac{dy}{dt}\right)^2 = \frac{2\mu}{y} - \frac{2\mu}{a} = \frac{2\mu}{a}\left(\frac{a-y}{y}\right) \quad \text{or} \quad \frac{dy}{dt} = -\left(\frac{2\mu}{a}\right)^{1/2}\left(\frac{a-y}{y}\right)^{1/2} \qquad ...(5)$$

where the negative sign is put on R.H.S. because the particle is moving in the direction of y decreasing.

Dividing (5) by (3), $\quad \dfrac{dy}{dx} = -\left(\dfrac{a-y}{y}\right)^{1/2} \quad$ or $\quad dx = -\left(\dfrac{y}{a-y}\right)^{1/2} dy \qquad$ ...(6)

Let $\quad y = a\cos^2\theta \quad$ so that $\quad dy = -2a\cos\theta\sin\theta\, d\theta \qquad$ ...(7)

Using (7), (6) gives $\quad dx = 2a\cos^2\theta\, d\theta = a(1 + \cos 2\theta)\, d\theta.$

Integrating $\; x = a[\theta + (1/2)\sin 2\theta] + c'' = (a/2)(2\theta + \sin 2\theta) + c'' \qquad$ ...(8)

Initially, when $y = a$, let us take $x = 0$. Now, when $y = a$ then $y = a\cos^2\theta \Rightarrow a = a\cos^2\theta \Rightarrow \cos\theta = 1 \Rightarrow \theta = 0$. So putting $x = 0$ and $\theta = 0$ in (8), we get $c'' = 0$. Hence (8) reduces to $\qquad x = (a/2)(2\theta + \sin 2\theta)$ ...(9)

Rewriting (7), $\qquad y = (a/2)(1 + \cos 2\theta)$ ...(10)

(9) and (10) give parametric equation of a cycloid. Hence the required path of the particle is a cycloid.

EXERCISE

1. The co-ordinates of a moving point at time t are given by $x = at^2, y = 2at$. Find the magnitude and direction of its resultant velocity and acceleration.

[**Ans.** Vel. $= 2a(t^2 + 1)^{1/2}$ at $\tan^{-1}(1/t)$, acceleration $= 2a$ at $\pi/2$ with x-axis.]

2. The coordinates of a moving point at time t are given by $x = a(2t + \sin 2t), y = a(1 - \cos 2t)$, prove that its acceleration is constant and find the direction of motion at time t.

[Meerut 2015, 2016; Agra 2008; Rohilkhand 2003]

[**Ans.** Acceleration $= 4a$, at an angle t with x-axis.]

3. Prove that the parabola $y^2 = 4ax$ can be described under a constant force parallel to axis of y and a force proportional to y parallel to the axis of x.

[**Hint.** Take $\ddot{y} = \lambda,\ \ddot{x} = \mu y, \lambda, \mu$ being constant and $x = y = 0$ at $t = 0$]

5.2 Angular velocity and angular acceleration

[Rohilkhand 2003, 06]

Angular-velocity. Definition. Let P be a moving point in a plane in which O is a fixed point and OX is a fixed line through O. Then the angular velocity of P about O (or of the line OP) in the plane XOP is the rate of change of the angle XOP. Let $P(r, \theta)$ and $Q(r + \delta r, \theta + \delta\theta)$ be the positions of the moving point at the instants t and $t + \delta t$ respectively. Then the angle turned by the particle in time $\delta t =$ $\delta\theta$.

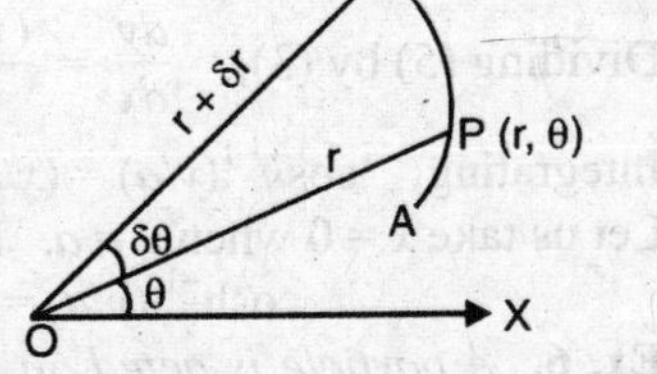

$\therefore$ Average rate of change of the angle of P about $O = \delta\theta/\delta t$.

So the angular velocity of the point P about $O = \displaystyle\lim_{\delta t \to 0} \frac{\delta\theta}{\delta t} = \frac{d\theta}{dt} = \dot{\theta}$,

where the dot placed over θ denotes differentiaion w.r.t. 't'

Since the angular velocity has magnitude and direction, it is a vector quantity represented by $\boldsymbol{\omega}$. Its magnitude is $d\theta/dt$ or ω and its direction is perpendicular to the plane XOP. The unit of angular velocity is radian/sec.

Angular acceleration. Definition. The rate of change of the angular velocity is known as angular acceleration.

$$\text{Thus, angular acceleration of } P = \frac{d}{dt}\left(\frac{d\theta}{dt}\right) = \frac{d^2\theta}{dt^2} = \ddot{\theta},$$

where two dots placed over θ denotes differentiation of θ twice w.r.t. 't'.

Angular acceleration is also a vector quantity and its unit is radian/sec^2.

5.3 Relation between angular velocity and linear velocity

[Agra 2005, 06, 09, 11; Meerut 2005; Purvanchal 2011; Kanpur 2018]

Let v be the velocity of a point P moving in a plane curve AB at any time t and let its polar coordinates be (r, θ). Now $v = (ds/dt)$ is along the tangent PT to the path at P.

Let the radius vector OP make an angle $\phi (= \angle OPM)$ with the tangent PT. Let $OM (= p)$ be the perpendicular distance O fom the tangent MT at P. Then, we know that

$$p = r \sin\phi \quad \text{and} \quad r(d\theta/ds) = \sin\phi. \text{ From these,}$$

$$d\theta/ds = (\sin\phi)/r = p/r^2. \qquad \text{...(1)}$$

$$\therefore \quad \text{Angular velocity of } P = \frac{d\theta}{dt} = \frac{d\theta}{ds}\cdot\frac{ds}{dt} = \frac{pv}{r^2}, \text{ as } \frac{ds}{dt} = v \qquad \text{...(2)}$$

Since $p/r = \sin\phi$, (2) may also be written as

The angular velocity of P about O.

$$= \frac{v \sin \varphi}{r} = \frac{\text{The resolved part of the vel. of } P \perp \text{ to } OP}{OP}$$

Note. If A and B are both in motion, then the angular velocity of B relative to A

$$= \frac{\text{the resolved part of the vel. of } B \text{ relative to A} \perp \text{ to } OB}{AB}$$

5.4 Angular velocity and angular acceleration of direction of motion

We know that the direction of motion (or the direction of velocity) is tangential. Hence the angular velocity of direction of motion is the angular velocity of tangent AP. Let P and Q be the positions of the moving point at the instant t and $t + \delta t$ respectively. Let the tangents AP and BQ at P and Q to the path CD of the particle make angles ψ and $\psi + \delta\psi$ with the x-axis (or the initial line).

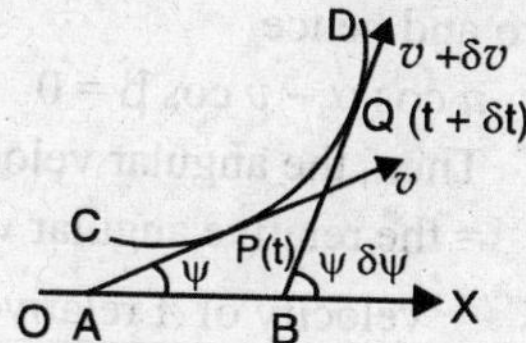

Then the angular velocity of direction of motion $= d\psi/dt$

and the angular acceleration of direction of motion $= d^2\psi/dt^2$.

Note. Since $ds/dt = v$ = velocity of P at time t and $\rho = ds/d\psi$

$\therefore$ The angular-velocity of direction of motion $= \dfrac{d\psi}{dt} = \dfrac{d\psi}{ds}\dfrac{ds}{dt} = \dfrac{v}{\rho}$.

TYPE 2 : SOLVED EXAMPLES BASED ON ART. 5.2, 5.3 AND 5.4

Ex. 1. *A point moves along a circle with constant speed. Prove that its angular velocity about any point of the circle is half of that about the centre.*

Sol. Let O be any point on the circle. Let O be the pole and the diameter OA be the initial line. Let the moving point be at P any time t such that $\angle POA = \theta$. Then, by geometry, we know that $\angle PCA = 2\theta$. Now, by definition,

Angular velocity of P about $O = d\theta/dt$ ...(1)

and angular velocity of P about $C = d(2\theta)/dt = 2\,(d\theta/dt)$ ...(2)

From (1) and (2), angular velocity of P about $O = (1/2) \times$ angular velocity of P about C. This proves the required result.

Ex. 2. *A particle describes a parabola with uniform speed; show that its angular-velocity about the focus S at any point P, varies inversely as $(SP)^{3/2}$.* **[Meerut 2013]**

Sol. The pedal equation of a parabola referred to its focus S as pole can be taken as

$$p^2 = ar. \qquad \text{...(1)}$$

Let v be given the constant speed of the particle at point $P(r, \theta)$. Then the angular velocity $d\theta/dt$ of the particle about S (*i.e.*, about pole) is given by

$$\frac{d\theta}{dt} = \frac{vp}{r^2} = \frac{v(ar)^{1/2}}{r^2} = \frac{va^{1/2}}{r^{3/2}} = \frac{va^{1/2}}{(SP)^{3/2}}, \text{ using (1)}$$

showing that angular velocity of P varies inversely as $(SP)^{3/2}$.

Ex. 3. *Prove that the angular velocity of a prejectile about the focus of its path varies inversely as its distance from the focus.* **[Meerut 2014; Agra 2010]**

Sol. Since the path of a projectile is a parabola (whose vertex is A, focus is S and the direction is BC), the pedal equation of this parabolic path referred to its focus S as pole can be taken as $p^2 = ar$.

Let v be the velocity of the projectile at $P(r, \theta)$. Then by definition of parabola, $MP = PS = r$, where PM is perpendicular to the directrix BC.

We know that the velocity of the projectile at any point of its path is equal to the velocity acquired in falling freely from the directrix to the point. Hence

$$v = (2g.\, MP)^{1/2} = (2gr)^{1/2}.$$

Now, the angular velocity of P about S (*i.e.*, about pole) is given by

$$\frac{d\theta}{dt} = \frac{vp}{r^2} = \frac{(2gr)^{1/2}.(ar)^{1/2}}{r^2} = \frac{(2ag)^{1/2}}{r} = \frac{(2ag)^{1/2}}{SP},$$

showing that angular velocity of P varies inversely as SP.

Ex. 4. *The line joining two points A, B is of constant length a and the velocities of A, B are in directions which make angles α and β respectively with AB. Prove that the angular velocity of A is {u sin (α − β)}/a cos β, where u is the velocity of A.* **[Agra 1994]**

Sol. Let v be the velocity of B. Since the distance AB always remains constant (= a given), it follows that the velocity of A relative to B in the direction AB must be zero and hence

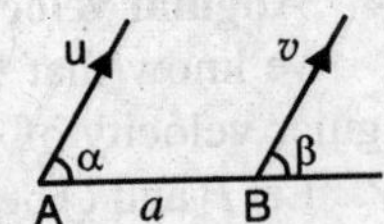

$u \cos \alpha - v \cos \beta = 0$ so that $v = (u \cos \alpha)/\cos \beta$ ...(1)

Then, the angular velocity of AB

= the relative angular velocity of A with respect to B.

$$= \frac{\text{velocity of } A \text{ relative to } B \text{ in a direction perpendicular to } AB}{AB}$$

$$= \frac{u \sin \alpha - v \sin \beta}{a} = \frac{u \sin \alpha - (u \cos \alpha / \cos \beta) \sin \beta}{a}, \text{ using (1)}$$

$$= \frac{u\,(\sin \alpha \cos \beta - \cos \alpha \sin \beta)}{a \cos \beta} = \frac{u \sin (\alpha - \beta)}{a \cos \beta}$$

Ex. 5. *Prove that the angular acceleration of the direction of motion of a point moving in a plane is* $(v/\rho) \times (dv/ds) - (v^2/\rho^2) \times (d\rho/ds)$.

[Avadh 2009; Lucknow 2006; Kanpur 2003, Agra 2005, 09, 10]

Sol. If the tangent at any point of the path makes an angle ψ with the x-axis, then ψ gives the direction of motion. The angular acceleration of the direction of motion is given by $d^2\psi/dt^2$.

We know that $d\psi/dt = v/\rho$. ...(1)

$$\therefore \quad \frac{d^2\psi}{dt^2} = \frac{d}{dt}\left(\frac{d\psi}{dt}\right) = \frac{d}{dt}\left(\frac{v}{\rho}\right) = \frac{d}{ds}\left(\frac{v}{\rho}\right)\frac{ds}{dt} = \left(\frac{1}{\rho}\frac{dv}{ds} - \frac{v}{\rho^2}\frac{d\rho}{ds}\right) v, \text{ as } \frac{ds}{dt} = v$$

or $d^2\psi/dt^2 = (v/\rho) \times (dv/ds) - (v^2/\rho^2) \times (d\rho/ds)$, as desired.

Ex. 6. *A rod moves with its ends sliding on rectangular axes OX and OY. If (x, y) are the coordinates at any instant of a point P on the rod and if the angular velocity ω of the rod is constant, show that the component accelerations of P along the axes are* $-x\omega^2$ *and* $-y\omega^2$, *and that the resultant acceleration is* $OP.\ \omega^2$ *towards O.* **[Kumaun 1994]**

Sol. Let a rod AB of length $(a + b)$ slide on OX and OY and let P be the fixed point on AB such that $BP = a$ and $AP = b$. At any time, let $\angle BAO = \theta$ and (x, y) be coordinates of P. Let $\angle POX = \alpha$. Given that

$$\dot{\theta} = d\theta/dt = \omega = \text{constant}. \quad ...(1)$$

From triangle BNP and PMA, we find

$$x = a \cos \theta \text{ and } y = b \sin \theta. \quad ...(2)$$

Differentiating (2) w.r.t. 't' and using (1), we have

$$\dot{x} = -a \sin \theta.\ \dot{\theta} = -a\omega \sin \theta, \quad \dot{y} = b \cos \theta.\ \dot{\theta} = b\omega \cos \theta \quad ...(3)$$

Again, diff. (3) w.r.t. 't' and using (1) and (2) we have

$$\ddot{x} = -a\omega \cos \theta\ \dot{\theta} = -\omega^2 x, \quad \ddot{y} = -b\omega \sin \theta\ \dot{\theta} = -\omega^2 y \quad ...(4)$$

$\therefore$ Resultant acceleration $= (\ddot{x}^2 + \ddot{y}^2)^{1/2} = (\omega^4 x^2 + \omega^4 y^2)^{1/2} = \omega^2 (x^2 + y^2)^{1/2}$

$= \omega^2.OP$, since $OP = (x^2 + y^2)^{1/2}$

This resultant makes an angle $\tan^{-1}(\ddot{y}/\ddot{x}) = \tan^{-1}(y/x) = \tan^{-1}(\tan \alpha) = \alpha = \angle POX$ with x-axis, showing that the resultant acceleration acts along PO.

Ex. 7. *A wheel rolls along a straight road with constant speed v. Show that the actual velocity of P is v. (AP/CP), where A is the point of contact of the wheel with the road and C is the centre of the wheel. Also find its direction. Find also the angular velocity of P relative to A.*

Sol. At any instant, let P be any point on the rim, where $\angle PCB = \theta$ and B is the highest point on the rim. Centre moves in a straight line parallel to the ground with velocity v. Then, the velocity of P is the resultant of the velocity of P relative to C and the actual velocity of C.

When the wheel has turned through an angle θ, the centre has moved through a distance x so that $x = a\theta$, where a is the radius of the wheel. Hence $\qquad v = \dot{x} = a\dot{\theta}$.

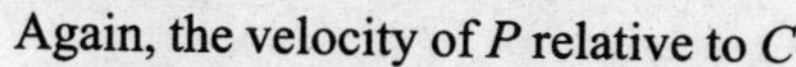

Again, the velocity of P relative to C

= velocity of P in a circle of radius $a = a\dot{\theta}$ along tangent PT

Thus point P possesses two velocities (*i*) v parallel to the road *i.e.*, parallel to PE and (*ii*) v along the tangent PT at P to the wheel.

By geometry, $\angle EPT = \angle PCB = \theta$.

$\therefore\quad V$ = The actual velocity of P = resultant of v along PE and v along PT

or $\quad V = (v^2 + v^2 + 2v^2 \cos\theta)^{1/2} = \{2v^2(1+\cos\theta)\}^{1/2} = \{2v^2 . 2\cos^2(\theta/2)\}^{1/2}$

or $$V = 2v\cos\theta/2 = v \cdot \frac{2a\cos\theta/2}{a} = v \cdot \frac{AP}{CP}, \qquad ...(1)$$

[$\because$ from ΔAPB, $AP/AB = AP/2a = \cos\theta/2 \Rightarrow AP = 2a\cos\theta/2$]

The direction of velocity V of P bisects the angle between the velocities at P and hence it makes angle $\theta/2$ with the horizontal. Since the line BP makes an angle $\pi/2 - \theta/2$ with the vertical, hence it makes an angle $\theta/2$ with the horizontal.

Hence the direction of V is along BPD. Also PD is perpendicular to AP.

From (1), the actual velocity of $A = 2v\cos\pi/2 = 0$ as for A, $\theta = \pi$

Now, the angular velocity of P relative to A

$$= \frac{\text{velocity of } P \text{ relative to } A \text{ in a direction } \perp AP}{AP} \qquad ...(2)$$

Now actual velocity of A is zero. Therefore, the velocity of P relative to A is equal to the actual velocity of $P = v.(AP/CP)$ whose direction is $\perp AP$.

Then from (2), the angular velocity of P relative to A

$$= \frac{v.(AP/CP)}{AP} = \frac{v}{CP} = \frac{v}{a}.$$

Ex. 8. *A wheel of radius b rolls without slipping along a straight road. If the centre of the wheel has a uniform velocity v, find at any instant the velocity and acceleration of the two points of the rim which are at a height h above the road. Examine in particular the cases h = 0, h = 2b.*

[Delhi B.Sc. (Prog) III 2010; Delhi Maths (H) 2007, 10]

Sol. At any time time t, let $P(x, y)$ be any point on the rim, where $\angle PCA = \theta$ and A is the lowest point of the rim. Initially at $t = 0$, let P be at the origin O. Since there is no slipping, we have

$$OA = \text{arc } PA = b\theta.$$

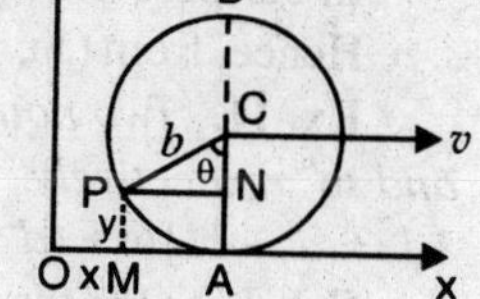

Let $PM \perp OA$ and $PN \perp CA$. Then,

$$x = OM = OA - MA = b\theta - PN = b\theta - b\sin\theta \qquad ...(1)$$

and $$y = PM = NA = CA - CN = b - b\cos\theta \qquad ...(2)$$

The path of P is given by (1) and (2) which are parametric equations of a cycloid.

Since the centre of the wheel has uniform velocity v, we have

$$\therefore \quad v = \frac{d}{dt}(OA) = \frac{d}{dt}(b\theta) = b\dot{\theta} \text{ so that } \dot{\theta} = \frac{v}{b} \qquad ...(3)$$

From (1) $\dot{x} = b\dot{\theta} - b\cos\theta\,\dot{\theta} = v(1-\cos\theta)$, using (3)
From (2), $\dot{y} = b\sin\theta\,\dot{\theta} = v\sin\theta$, using (3) ...(4)

From (3) and (4) we have

$$\ddot{x} = (v\sin\theta)\,\dot{\theta} = (v^2/b)\sin\theta,\ \ddot{y} = (v\cos\theta)\,\dot{\theta} = (v^2/b)\cos\theta. \quad ...(5)$$

To find the expressions for the velocity and acceleration of two points of the rim which are at a height h above the road, we find θ corresponding to $y = h$. So, by (2)

$$h = b(1-\cos\theta) \quad \text{so that} \quad \cos\theta = (b-h)/b. \quad ...(6)$$

$$\therefore \quad \sin\theta = \pm(1-\cos^2\theta)^{1/2} = \pm\left\{1-\left(\frac{b-h}{b}\right)^2\right\}^{1/2} = \pm\frac{(2bh-h^2)^{1/2}}{b} \quad ...(7)$$

Using (6) and (7), (4) reduces to

$$\dot{x} = v\left(1-\frac{b-h}{b}\right) = \frac{vh}{b} \text{ and } \dot{y} = \pm\frac{v}{b}(2bh-h^2)^{1/2} \quad ...(8)$$

giving the rquired components of velocity at height h above the road.

Using (6) and (7), (5) reduces to

$$\ddot{x} = \pm(v^2/b^2)(2bh-h^2)^{1/2} \text{ and } \ddot{y} = (v^2/b^2)(b-h), \quad ...(9)$$

giving the required components of acceleration at height h above the road.

Particular case I. Let $h = 0$. Then from (8), $\dot{x} = \dot{y} = 0$. Hence in this case the point P (called instantaneous centre of rotation) has no velocity. Thus it always remains at rest.

Particular case II. When $h = 2b = AB$. Then from (8), $\dot{x} = 2v$, $\dot{y} = 0$. Then the point P will be at the heighest point B of the rim and it possesses velocity $= (\dot{x}^2+\dot{y}^2)^{1/2} = 2v$, which is double the velocity of the centre of the wheel.

Also we observe that in both cases the acceleration is always v^2/b towards the centre of the wheel.

Ex. 9. *A particle describes a curve with constant velocity and its angular velocity about a given fixed point O varies inversely as its distance from O; show that the curve is an equiangular spiral.* **[I.A.S. 2002, Lucknow 1994]**

Sol. Taking O is pole, let the position of the particle at any time t be $P(r, \theta)$. Let v be the given constant velocity of the particle. Then, by given problem,

$$d\theta/dt \propto 1/r \text{ so that } d\theta/dt = k/r, \text{ where } k \text{ is a constant.}$$

or $\quad vp/r^2 = k/r \quad$ so that $\quad p = (k/v)r \quad$ (where k/v = constant),

which is pedal equation of an equiangular spiral.

Note. $p = ar$ is pedal equation of an equiangular spiral, a being constant.

Ex. 10. *If the velocity of a point moving in a plane curve varies as the radius of curvature, show that the direction of motion revolves with constant angular velocity.* **[Purvanchal 95]**

Sol. From Art. 5.4, we know that

the angular velocity of the direction of motion $= d\psi/dt = v/\rho$. ...(1)

In our problem $\quad v \propto \rho \quad$ so that $\quad v/\rho = k$, k being constant.

Hence, from (1), $\quad d\psi/dt = k =$ constant.

Ex. 11. *Two points are describing concentric circles of radii a and b with angular velocity ω and ω′ respectively. Prove that angular velocity of line joining them, when its length is d, is* $\{d^2+a^2-b^2\}\omega + (d^2-a^2+b^2)\,\omega'\}/2d^2$.

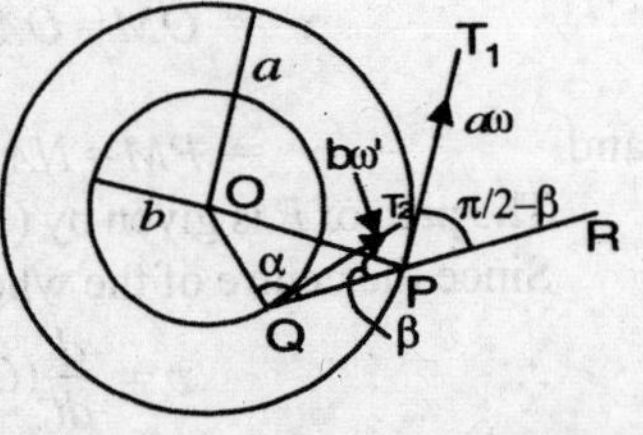

Sol. Let O be the common centre of two concentric circles, with $b < a$. Let P, Q be the positions of the particles at any time t. Then velocities of the particles at P and Q are $a\omega$ and $b\omega'$ along tangents PT_1 and QT_2 respectively. Let $\angle OQP = \alpha$ and $\angle OPQ = \beta$. From figure $\angle T_1PR = \pi - (\pi/2) - \beta = \pi/2 - \beta$ and $\angle T_2QP = \alpha - \pi/2$. Hence, we have

Component of velocity of $P \perp PQ = a\omega \sin(\pi/2 - \beta) = a\omega \cos \beta$
and component of velocity of $Q \perp QP = b\omega' \sin(\alpha - \pi/2) = -b\omega' \cos \alpha$

$\therefore$ Angular velocity of P about Q

$$= \frac{a\omega \cos \beta - (-b\omega' \cos \alpha)}{PQ} = \frac{a\omega \cos \beta + b\omega' \cos \alpha}{d}, \text{ as } PQ = d \text{ (given)}$$

From ΔOPQ, $\cos \beta = \dfrac{d^2 + a^2 - b^2}{2da}$ and $\cos \alpha = \dfrac{d^2 + b^2 - a^2}{2db}$

$\therefore$ The required angular velocity of the line PQ

$$= \frac{a\omega}{d} \cdot \frac{d^2 + a^2 - b^2}{2da} + \frac{b\omega'}{d} \cdot \frac{d^2 + b^2 - a^2}{2db} = \frac{\omega(d^2 + a^2 - b^2) + \omega'(d^2 + b^2 - a^2)}{2d^2}$$

EXERCISE

1. If P is any point on the cirucmference of a circle, centre C, which rolls with angular velocity ω on a fixed circle of centre O, and radius a, prove that the angular velocity of P about O is $\omega - (a\omega/OP) \cos \theta$, where $\theta = \angle COP$.

2. Two points are moving with uniform velocities u, v in perpendicular lines OX and OY, the motions being towards O. If initially, their distances from the origin are a and b respectively, find the angular velocity of the line joining them at the end of t sec., and show that it is geatest when $t = (au + bv)/(u^2 + v^2)$. **[Ans.** $\omega = (av - bu)/\{(a - ut)^2 + (b - vt)^2\}$**]**

3. A particle falls down a straight line $x = a$, starting from the axis of x. If the distance from the axis of x be $(1/2) ft^2$ at time t, find the angular velocity and acceleration of the line joning the particle to the origin. How far has the particle dropped when the angular acceleration becomes zero?

[Ans. Angular velocity $= (4aft)/(4a^2 + f^2t^4)$, angular acceleration $= 4af(4a^2 - 3f^2t^4)/(4a^2 + f^2t^4)^2$ and required distance $= a/\sqrt{3}$ **)]**

4. A point P is moving along a fixed straight line AB with uniform velocity v. Show that its angular velocity about O is inversely proportional to OP^2. **[Bundelkhand 2007]**

5.5 Rate of change of a unit vector in a plane. Let **i** and **j** be unit vectors along OX and OY repsectively. Let P be the position a particle at any time t the such that $OP = 1$ and $\angle POX = \theta$.

Let $\mathbf{e}_r$ and $\mathbf{e}_\theta$ be unit vectors along and perpendicular to OP which are inclined at angles θ and $\pi/2 + \theta$ with OX respectively. Then, we know that

$$\mathbf{e}_r = (\cos \theta)\, \mathbf{i} + (\sin \theta) \mathbf{j} \qquad \text{...(1)}$$

and $\mathbf{e}_\theta = \{\cos(\pi/2 + \theta)\}\mathbf{i} + \{\sin(\pi/2 + \theta)\}\mathbf{j} = -(\sin \theta)\, \mathbf{i} + (\cos \theta) \mathbf{j}$ ...(2)

Here the vectors $\mathbf{e}_r$ and $\mathbf{e}_\theta$ are are functions of θ, where θ is function of t. Remember that **i** and **j** are constant vectors and hence.

$$d\mathbf{i}/dt = \mathbf{0} \quad \text{and} \quad d\mathbf{j}/dt = \mathbf{0} \qquad \text{...(3)}$$

Differentiation (1) and (2) w.r.t. 't' we have

$$d\mathbf{e}_r/dt = (-\sin \theta)\, \dot{\theta}\, \mathbf{i} + (\cos \theta)\, \dot{\theta}\, \mathbf{j} = \mathbf{e}_\theta\, \dot{\theta}, \text{ by (2)} \qquad \text{...(4)}$$

and
$$d\mathbf{e}_\theta/dt = -(\cos \theta)\, \dot{\theta}\, \mathbf{i} - (\sin \theta)\, \dot{\theta}\, \mathbf{j} = -\mathbf{e}_r\, \dot{\theta}, \text{ by (1)} \qquad \text{...(5)}$$

Note that $\mathbf{e}_r$ and $\mathbf{e}_\theta$ are unit vectors in the directions of r increasing and θ increasing respectively.

Note. We shall use the following results directly

$$d\mathbf{e}_r/dt = \mathbf{e}_\theta\, \dot{\theta} \quad \text{and} \quad d\mathbf{e}_\theta/dt = -\mathbf{e}_r\, \dot{\theta}. \qquad \text{...(6)}$$

Corollary. Let **t** and **n** be the unit vectors along the tangent PM and normal PN respectively at any point $P(s, \psi)$ at any time t of a plane curve AB. Let $\angle PTX = \psi$. Then proceeding as above, we have

$$d\mathbf{t}/dt = \mathbf{n}\, \dot{\psi} \quad \text{and} \quad d\mathbf{n}/dt = -\mathbf{t}\, \dot{\psi}$$

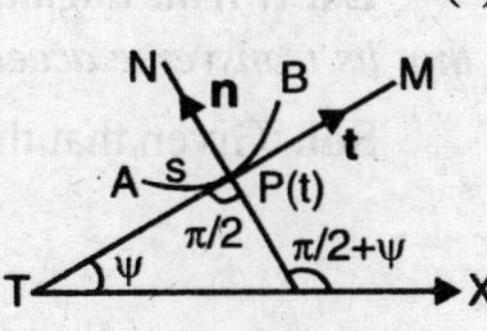

Note that **t** is in the direction of s increasing and **n** is in the direction in which ψ increases *i.e.*, in the direction of inwards drawn normal.

Note. Here s is the arcual distance AP of P measured from a fixed point A on the path AB.

5.6 Radial and transverse components of velocity and acceleration

[M.S. Univ. TN 2007; Patna 2003; Bhopal 2004; Kanpur 2015, 17; Kerala 2001, 02; Rohilkhand 2002, 03; Agra 2009, 11, 12; Garhwal 2010; Pune 2010; Meerut 2016, 17, 18; Purvanchal 2006, 08, 11; Guwahati 2007; Gujarat univ. 2004; Avadh 2006; Bundelkhand 2010; Lucknow 2009, 10]

Let $P(r, \theta)$ be the position of a particle moving on curve $r = f(\theta)$ at any time t. Let O be the pole and OX the initial line. Let $\mathbf{e_r}$ and $\mathbf{e_\theta}$ be unit vectors along OP and perpendicular to it respectively in the sense of r and θ increasing respectively. Then, we know that (refer Art. 5.5)

$$d\mathbf{e_r}/dt = \mathbf{e_\theta}\dot{\theta} \quad \text{and} \quad d\mathbf{e_\theta}/dt = -\mathbf{e_r}\dot{\theta} \quad \text{...(1)}$$

Now, the position vector $\mathbf{r}$ of P is given by

$$\mathbf{r} = \overrightarrow{OP} = OP\,\mathbf{e_r} = r\,\mathbf{e_r} \quad \text{...(2)}$$

If $\mathbf{v}$ is the velocity vector of the particle P, then

$$\mathbf{v} = d\mathbf{r}/dt = d(r\,\mathbf{e_r})/dt = \dot{r}\,\mathbf{e_r} + r\,(d\mathbf{e_r}/dt)$$

or $$\mathbf{v} = \dot{r}\,\mathbf{e_r} + r(\mathbf{e_\theta}\dot{\theta}) = \dot{r}\,\mathbf{e_r} + r\dot{\theta}\,\mathbf{e_\theta}, \text{ by (1)} \quad \text{...(3)}$$

Hence, the radial and transverse components (v_r, v_θ) of velocity of a particle $P(r, \theta)$, at any time t, are given by

$v_r = \dot{r} = dr/dt$ = radial component of velocity at $P(r, \theta)$ in the direction of $\mathbf{e}_r$ *i.e.*, in the direction of r increasing

and $v_\theta = r\dot{\theta} = r(d\theta/dt)$ = transverse component of velocity at $P(r, \theta)$ in the direction of $\mathbf{e}_\theta$ *i.e.*, in the direction of θ increasing.

Resultant velocity $= (v_r^2 + v_\theta^2)^{1/2} = \{\dot{r}^2 + (r\dot{\theta})^2\}^{1/2}$

Determination of radial and transverse acceleration

If $\mathbf{f}$ is the acceleration vector of the particle P, then

$$f = \frac{d\mathbf{v}}{dt} = \frac{d}{dt}(\dot{r}\,\mathbf{e_r}) + \frac{d}{dt}(r\dot{\theta}\,\mathbf{e_\theta}), \text{ by (3)}$$

$$= \ddot{r}\,\mathbf{e_r} + \dot{r}\,(d\mathbf{e_r}/dt) + (\dot{r}\,\dot{\theta} + r\,\ddot{\theta})\,\mathbf{e_\theta} + r\dot{\theta}\,(d\mathbf{e_\theta}/dt)$$

$$= \ddot{r}\,\mathbf{e_r} + \dot{r}\,\mathbf{e_\theta}\,\dot{\theta} + (\dot{r}\dot{\theta} + r\ddot{\theta})\mathbf{e_\theta} + r\dot{\theta}\,(-\mathbf{e_r}\,\dot{\theta}) = (\ddot{r} - r\dot{\theta}^2)\,\mathbf{e_r} + (2\dot{r}\,\dot{\theta} + r\,\ddot{\theta})\,\mathbf{e_\theta}$$

Hence, the radial and transverse components (f_r, f_θ) of acceleration of a particle at $P(r, \theta)$, at any time t, are given by

$$f_r = \ddot{r} - r\dot{\theta}^2 = d^2r/dt^2 - r(d\theta/dt)^2$$ = radial component of acceleration at $P(r, \theta)$ in the direction of r increasing

$$f_\theta = 2\dot{r}\,\dot{\theta} + r\,\ddot{\theta} = \frac{1}{r}\frac{d}{dt}(r^2\,\dot{\theta})$$ = transverse component of acceleration at $P(r, \theta)$ in the direction of θ increasing.

TYPE 3 : SOLVED EXAMPLES BASED ON ART 5.6

In what follows, we shall use the following important results:

(*i*) Radial velocity $= dr/dt$ (*ii*) Transverse vel. $= r(d\theta/dt)$ (*iii*) Radial acceleration $= (d^2r/dt^2) - r(d\theta/dt)^2$ (*iv*) Transverse acceleration $= 2\dfrac{dr}{dt}\dfrac{d\theta}{dt} + r\dfrac{d^2\theta}{dt^2} = \dfrac{1}{r}\dfrac{d}{dt}(r^2\dot{\theta})$

(*v*) Angular velocity $= d\theta/dt = vp/r^2$, (*vi*) $p = r\sin\phi$.

Ex. 1. *If the angular velocity of a point moving in a plane curve be constant about a fixed origin, show that its transverse acceleration varies as its radial velocity.* *(Kanpur 2009, 11, 13, 14; Agra 2012, 13)*

Sol. Given that the angular velocity $= \dot{\theta} = \omega$ (constant), say ...(1)

From (2), $\ddot{\theta} = d\omega/dt = 0$ ...(2)

$\therefore$ the transverse acceleration of the point $= 2\dot{r}\,\dot{\theta} + r\,\ddot{\theta} = 2\dot{r}\omega$, by (1) and (2)

So, the transverse acceleration = (2ω). (radial velocity)

⇒ The transverse acceleration varies as radial velocity, as 2ω is a constant.

Ex. 2. *A point P describes a curve $r = ae^{\theta}$ with a constant angular velocity about its pole. Find the radial and transverse accelerations and resultant acceleration of the point P.*

[Agra 2014; Mumbai 2010; Nagpur 2010; Kanpur 2009]

Sol. Given path of the particle is $r = ae^{\theta}$ ...(1)

Also, the angular velocity of $P = \dot{\theta} = \omega$ (const.), say ...(2)

From (1), $\dot{r} = ae^{\theta}\,\dot{\theta} = r\omega$, by (1) and (2) ...(3)

From (3), $\ddot{r} = \omega\dot{r} = r\omega^2$, by (3) ...(4)

From (2) $\ddot{\theta} = d^2\theta/dt^2 = d\omega/dt = 0$ ...(5)

$\therefore$ f_r = Radial acceleration $= \ddot{r} - r\dot{\theta}^2 = r\omega^2 - r\omega^2 = 0$, by (4) and (5) and f_θ = Transverse acceleration $= 2\dot{r}\,\dot{\theta} + r\,\ddot{\theta} = 2r\omega^2$, by (2) and (3)

$\therefore$ Resultant acceleration $= (f_r^2 + f_\theta^2)^{1/2} = f_\theta = 2\,r\omega^2$.

Ex. 3. *A point describes a circle of radius a with a uniform speed v; show that the radial and transverse acceleration are $-(v^2/a)\cos\theta$ and $-(v^2/a)\sin\theta$, if a diameter is taken as initial line and one end of this diameter as pole.* **[Rohilkhand 2001, Lucknow 2009]**

Sol. Let horizontal *OA* be taken as initial line, *O* being at pole. Let *C* be centre of the circle. Then the polar equation of the circle is

$$r = 2a\cos\theta \quad ...(1)$$

Let $P(r, \theta)$ be the position of the point at any time t and v be its uniform velocity along tangent PT at P. Let PN be ⊥ to OP. Then, by geomtery, $\angle NPT = \theta$.

Radial velocity $= \dot{r} = -v\sin$ ...(2)

and transverse velocity $= r\dot{\theta} = v\cos\theta$ ...(3)

From (2), $\ddot{r} = -v\cos\theta\,\dot{\theta} = -v\cos\theta\{v\cos\theta)/r\}$, by (3)

or $\ddot{r} = -(v^2\cos^2\theta)/r$ ...(4)

Hence, the radial acceleration at P

$$= \ddot{r} - r\,\dot{\theta}^2 = -\frac{v^2\cos^2\theta}{r} - r\frac{(v\cos\theta)^2}{r}, \text{ using (3) and (4)}$$

$$= -\frac{2v^2\cos^2\theta}{r} = -\frac{2v^2\cos^2\theta}{2\,a\cos\theta} = -\frac{v^2\cos\theta}{a}, \text{ since by (1), } r = 2a\cos\theta$$

Again, using (3), the transverse acceleration at P

$$= \frac{1}{r}\frac{d}{dt}(r^2\,\dot{\theta}) = \frac{1}{r}\frac{d}{dt}\left\{\frac{r^2(v\cos\theta)}{r}\right\} = \frac{v}{r}\frac{d}{dt}(r\cos\theta) = \frac{v}{r}(\cos\theta\,\dot{r} - r\sin\theta\,\dot{\theta})$$

$= (v/r)\,[\cos\theta\,(-v\sin\theta) - r\sin\theta\,\{(v\cos\theta)/r]$ by (2) and (3)

$$= -\frac{2v^2\sin\theta\cos\theta}{r} = -\frac{2v^2\sin\theta\cos\theta}{2a\cos\theta} = -\frac{v^2\sin\theta}{a}, \text{ by (1)}$$

Ex. 4. **(*a*)** *If the curve is equiangular spiral $r = ae^{\theta\cot\alpha}$ and if the radius vector to the particle has constant angular velocity, show that the resultant acceleration of the particle makes an angle 2α with the radius vector and is of magnitude v^2/r, where v is the speed of the particle.*

[Agra 2011; Delhi B.A. (Prog) III 2010, 2014; Mumbai 2010]

(*b*) *A point P describes an equiangular spiral $r = ae^{\theta\cot\alpha}$ with constant angular velocity about the pole O. Find its acceleration and show that its direction makes the same angle with the tangent at P as the radius vector OP makes with the tangent.*

Sol. (*a*) Let the given particle be at $P(r, \theta)$ at any t on the given curve AB given by $r = a\, e^{\theta \cot \alpha}$ ...(1)

Also, angular velocity $= \dot{\theta}$ = constant $= \omega$, say ...(2)

From (1), $\dot{r} = a \cot \alpha\, e^{\theta \cot \alpha}\, \dot{\theta} = r\omega \cot \alpha$ ...(3)

The radial and transverse components of velocity are given by

$v_r = \dot{r} = r\omega \cot \alpha$ and $v_\theta = r\,\dot{\theta} = r\omega$, by (2) and (3)

Let v be the magnitude of the resultant velocity. Then

$$v^2 = v_r^2 + v_\theta^2 = r^2\omega^2 \cot^2\alpha + r^2\omega^2 = r^2\omega^2 \operatorname{cosec}^2 \alpha \quad \text{...(4)}$$

$\therefore$ From (3), $\ddot{r} = \dot{r}\, \omega \cot \alpha = r\omega^2 \cot^2\alpha$, using (3)

Now, radial acceleration at P is given by

$$f_r = \ddot{r} - r\dot{\theta}^2 = r\omega^2 \cot^2\alpha - r\omega^2 = r\omega^2 (\cot^2\alpha - 1) \quad \text{...(5)}$$

and transverse acceleration is given by

$$f_\theta = 2\dot{r}\,\dot{\theta} + r\,\ddot{\theta} = 2\,(r\omega \cot \alpha)\,\omega + o, \text{ by (1) and (2)} \quad \text{...(6)}$$

Let f be the magnitude of the resultant acceleration of the particle and β, the angle it makes with the radius vector. Then

$$f^2 = f_r^2 + f_\theta^2 = r^2\,\omega^4\,(\cot^2\alpha - 1)^2 + 4r^2\omega^4 \cot^2\alpha, \text{ using (5) and (6)}$$

$$= r^2\omega^4\,(\cot^4\alpha + 1 - 2\cot^2\alpha + 4\cot^2 a) = r^2\omega^4 (\cot^2\alpha + 1)^2 = r^2\omega^4 \operatorname{cosec}^4 \alpha$$

Hence, $f = r\omega^2 \operatorname{cosec}^2 \alpha = v^2/r$, using (4)

Let f act along PD. Let PQ be OP produced and let PN be $\perp PQ$. Then $\angle DPQ = \beta$.

$$\therefore \quad \tan \beta = \frac{f_\theta}{f_r} = \frac{2r\omega^2 \cot \alpha}{r\omega^2\,(\cot^2 \alpha - 1)} = \frac{2 \tan \alpha}{1 - \tan^2 \alpha} = \tan 2\alpha, \text{ by (5) and (6)}$$

Hence $\beta = 2\alpha$, showing that f makes an angle 2α with PQ.

(*b*) Proceed as in part (*a*) to find β and f. Let PT be tangent at P to the curve (1). Then, we know that v acts along PT.

Let ϕ be angle which PT makes with PQ (*i.e.*, radius vector OP produced, then $\phi = \angle TPQ$ and is given by

$$\tan \phi = \frac{v_\theta}{v_r} = \frac{r\omega}{r\omega \cot \alpha} = \tan \alpha \quad \text{so that} \quad \phi = \alpha$$

Hence the angle which the direction of acceleration makes with PT

$= \angle DPT = \angle DPQ - \angle TPQ = \beta - \phi = 2\alpha - \alpha = \alpha$, as required.

Ex. 5. (*a*) *The velocities of a particle along and perpendicular to the radius vector from a fixed origin are λr^2 and $\mu\theta^2$, where λ and μ are constants, find the polar equation of the path of the particle and also its radial and transverse accelrations in terms of r and θ only.*

[Meerut 2016, 2018; Avadh 2008; Agra 2001, 08; Kanpur 2010; Kerala 2001]

(*b*) *The velocities of a particle along and perpendicular to the radius vector are λr and $\mu\theta$; find the path and show that the acceleration along and perpendicular to the radius vector are $\lambda^2 r - \mu^2\theta^2/2$ and $\mu\theta\,(\lambda + \mu/r)$.* **[Lucknow 2010; Rohilkhand 2006, 09; Guwahati 2007; Gujrat Univ. 2004]**

(*c*) *The velocities of a particle along and perpendicular to the radius vector from a fixed origin are λr and $\lambda a\theta$. Find the path and show that the acceleration along and perpendicular to the radius vector are $\lambda^2 (r - a^2\theta^2/r)$ and $\lambda^2\, a\,\theta\,(1 + a/r)$.* **[Lucknow 2001]**

Sol. (*a*) Given that radial velocity $= \dot{r} = \lambda r^2$ ...(1)

and transverse velocity $= r\dot{\theta} = \mu\theta^2$ ...(2)

Dividing (1) by (2), $\dfrac{dr/dt}{r\,(d\theta/dt)} = \dfrac{\lambda r^2}{\mu\theta^2}$ or $\mu r^{-3}\, dr = \lambda\theta^{-2}\, d\theta$.

Integrating, $-(\mu/2r^2) = -(\lambda/\theta) + c$, where c is constant of integration.

Hence the equation of the path is $c + \mu/2r^2 = \lambda/\theta$.

Now, radial acceleration $(= \ddot{r} - r\dot{\theta}^2)$

$$= \frac{d}{dt}\left(\frac{dr}{dt}\right) - r\,\dot{\theta}^2 = \frac{d}{dt}(\lambda r^2) - r\left(\frac{\mu\theta^2}{r}\right)^2, \text{ using (1) and (2)}$$

$= 2\lambda r\dot{r} - (\mu^2\theta^2/r) = 2\lambda r.\lambda r^2 - (\mu^2\theta^2/r) = 2\lambda^2 r^3 - (1/r)\,\mu^2\theta^2$, by (1)

and tansverse acceleration

$$= \frac{1}{r}\frac{d}{dt}\left(r^2\frac{d\theta}{dt}\right) = \frac{1}{r}\frac{d}{dt}\left\{r^2\left(\frac{\mu\theta^2}{r}\right)\right\} = \frac{\mu}{r}\frac{d}{dt}(r\theta^2), \text{ by (2)}$$

$= (\mu/r)\ (\dot{r}\,\theta^2 + 2r\theta\,\dot{\theta}) = (\mu/r)\ [\lambda r^2\cdot\theta^2 + 2r\theta\cdot(\mu\theta^2/r)]$, by (1) and (2)

$= (\mu/r)\ (\lambda r^2\theta^2) + (\mu/r)\ (2\mu\theta^3) = \lambda\mu r\theta^2 + 2(1/r)\mu^2\theta^3$.

(*b*) Proceed as in part (*a*). The path is $\theta = ce^{-\mu/\lambda r}$

(*c*) Proceed as in part (*a*). The path is $\theta = ce^{-a/r}$.

Ex. 6. (*a*) *A particle moves along a circle $r = 2a\cos\theta$ in such a way that its acceleration towards the origin is always zero. Show that the transverse acceleration varies as the fifth power of* cosec θ. **[Meerut 2016; Avadh 2007; Kanpur 2006, 2014]**

(*b*) *A particle moves along a circle $r = 2a\cos\theta$ in such a way that its acceleration towards the origin in always zero. Prove that $d^2\theta/dt^2 = -2\cot\theta\,(d\theta/dt)^2$.* **[Agra 2011; Nagpur 2010; Rohilkhand 2003]**

(*c*) *A particle moves along a circle $r = 2a\cos\theta$ in such a way that acceleration towards the origin is zero. Find the angular acceleration.*

Sol. The given curve is $r = 2a\cos\theta$. ...(1)

Given that acceleration towards the origin (*i.e.*, pole) is always zero. Hence the radial acceleration must be zero. Hence, we have

$$\ddot{r} - r\,\dot{\theta}^2 = 0 \qquad ...(2)$$

From (1), $\dot{r} = -2a\sin\theta.\dot{\theta}$ ...(3)

From (3), $\ddot{r} = -2a\,(\sin\theta.\ddot{\theta} + \cos\theta.\dot{\theta})$ ...(4)

Substituting the values of r and $\ddot{r}$ given by (1) and (4) in (1), we get

$$-2a\,(\sin\theta.\ddot{\theta} + \cos\theta.\dot{\theta}) - (2a\cos\theta).\dot{\theta}^2 = 0 \quad \text{or} \quad \sin\theta.\ddot{\theta} + (2\cos\theta)\,\dot{\theta}^2 = 0$$

or $$\frac{\ddot{\theta}}{\dot{\theta}} = -2\cot\theta\frac{d\theta}{dt} \qquad \text{or} \qquad \frac{\ddot{\theta}}{\dot{\theta}}\,dt = -2\cot\theta\,d\theta.$$

Integrating, $\log\dot{\theta} = -2\log\sin\theta + \log c = \log(c/\sin^2\theta)$

or $$\dot{\theta} = c\,\text{cosec}^2\,\theta \qquad ...(5)$$

Now, the transverse acceleration

$$= \frac{1}{r}\frac{d}{dt}(r^2\dot{\theta}) = \frac{1}{2a\cos\theta}\frac{d}{dt}(4a^2\cos^2\theta\cdot c\,\text{cosec}^2\,\theta\,\dot{\theta}), \text{ by (1) and (5)}$$

$$= 2ac\sec\theta\,\frac{d}{dt}(\cot^2\theta) = 2ac\sec\theta.(-2\cot\theta\,\text{cosec}^2\,\theta\,\dot{\theta})$$

$= -4ac\,\text{cosec}^3\theta.\,(c\,\text{cosec}^2\theta)$, using (5)

Since $(-4ac)$ is constant, so the transverse acceleration varies as $\text{cosec}^5\theta$.

(*b*) Proceed as in part (*a*) upto relation (5).

From (5), $\ddot{\theta} = -2c\,\text{cosec}\,\theta\,(\text{cosec}\,\theta\cot\theta)\,\dot{\theta} = -2(c\,\text{cosec}^2.\theta)^2\cot\theta$

or $d^2\theta/dt^2 = -2(d\theta/dt)^2\cot\theta$, by (5)

(*c*) Refer part (*b*). We are to find $d^2\theta/dt^2$ as before.

Ex. 7. *A small bead slides with constant speed v on a smooth wire in the shape of the cardiod $r = a(1 + \cos\theta)$. Show that the angular velocity is $(v/2a)\sec(\theta/2)$ and that the radial component of acceleration is constant.* **(Purvanchal 2007; Meerut 2017)**

Sol. Path of the particle is $r = a(1 + \cos\theta)$ or $r = 2a\cos^2(\theta/2)$. ...(1)

From (1), $\log r = \log(2a) + \log\cos^2(\theta/2) = \log(2a) + 2\log\cos(\theta/2)$

Diff.w.r.t. θ, $\dfrac{1}{r}\dfrac{dr}{d\theta} = 2\dfrac{(-1/2)\sin(\theta/2)}{\cos(\theta/2)}$ or $\cot\phi = -\tan\dfrac{\theta}{2} = \cot\left(\dfrac{\pi}{2} + \dfrac{\theta}{2}\right)$

$\therefore\ \theta = \pi/2 + \theta/2$ and so $p = r\sin\phi = r\sin(\pi/2 + \theta/2) = r\cos(\theta/2)$ ...(2)

Now, the angular velocity of the bead

$$= \dot\theta = \frac{vp}{r^2} = \frac{vr\cos(\theta/2)}{r^2} = \frac{v\cos(\theta/2)}{2a\cos^2(\theta/2)} = \frac{v}{2a}\sec(\theta/2), \quad ...(3)$$

[Using (1) and (2)]

By (1), $\dot r = -a\sin\theta.\ \dot\theta = -2a\sin(\theta/2)\cos(\theta/2)\,(v/2a)\sec(\theta/2)$, by (3) ...(4)

or $\qquad \dot r = -v\sin(\theta/2)$.

By (4), $\ddot r = -v\cos(\theta/2).\,(1/2)\,\dot\theta = -(v/2)\cos(\theta/2a)\cdot(v/2a)\sec(\theta/2)$

or $\qquad \ddot r = -(v^2/4a)$ ...(5)

Hence the radial acceleration $= \ddot r - r\dot\theta^2$

$= -(v^2/4a) - 2a\cos^2(\theta/2).\,(v/2a)^2\sec^2(\theta/2)$, using (1), (3) and (5)

$= -(v^2/4a) - v^2/2a = -(3v^2/4a)$, which is constant.

Ex. 8. *A point moves on a parabola $2a = r(1 + \cos\theta)$ in such a manner that the componenet of velocity at right angles to the radius vector from the focus is constant. Show that the acceleration of the point is constant in magnitude.* **[Rohilkhand 1999; Garhwal 94, Lucknow 91]**

Sol. Given $2a = r\cos(1 + \cos\theta) = 2r\cos^2(\theta/2)$ or $r = a\sec^2(\theta/2)$, ...(1)

whose focus is at pole. Hence, by problem, transverse velocity is constant.

Thus, here $r\dot\theta$ = const. $= k$, say $\quad$ or $\quad \dot\theta = k/r$. ...(2)

From (1), $\dot r = a.\,2\sec(\theta/2).\sec(\theta/2)\tan(\theta/2)\cdot(1/2)\,\dot\theta$

$\qquad = a\sec^2(\theta/2)\tan(\theta/2)\,\dot\theta = r\tan(\theta/2)\cdot(k/r)$, by (1) and (2)

Thus $\qquad \dot r = k\tan(\theta/2)$ ...(3)

From (3), $\ddot r = k\sec^2(\theta/2)\cdot(1/2)\,\dot\theta = (k/2)\sec^2(\theta/2)\cdot(k/r)$, by (2)

or $\qquad \ddot r = (k/2)\cdot(r/a)\cdot(k/r) = k^2/2a$, using (1)

$\therefore\ f_r$ = Radial acceleration $= \ddot r - r\dot\theta^2 = (k^2/2a) - r(k/r)^2 = k^2(r - 2a)/2ar$

and $\ f_\theta$ = transverse acceleration of the point

$$= \frac{1}{r}\frac{d}{dt}(r^2\dot\theta) = \frac{1}{r}\frac{d}{dt}\left(r^2\cdot\frac{k}{r}\right) = \frac{k}{r}\cdot\dot r = \frac{k}{r}\cdot k\tan(\theta/2), \text{ by (3)}$$

$= (k^2/r)\tan(\theta/2) = (k^2/r)(\sec^2\theta/2 - 1)^{1/2} = (k^2/r)(r/a - 1)^{1/2}$ by (1)

Hence the resultant acceleration f of the point is given by

$$f^2 = f_r^2 + f_\theta^2 = \frac{k^4(r-2a)^2}{4a^2r^2} + \frac{k^4(r-a)}{r^2a} = \frac{k^4(r-2a)^2 + 4k^4a(r-a)}{4a^2r^2}$$

$$= \frac{k^4(r^2 + 4a^2 - 4ar + 4ar - 4a^2)}{4a^2r^2} = \frac{k^4r^2}{4a^2r^2} = \frac{k^4}{4a^2}$$

Thus, $\qquad f = k^2/2a$, which is constant.

Ex. 9. (*a*) *Show that the path of a point P which possesses two constant velocities u and v, the first of which is in a fixed direction and the other is perpendicular to the radius OP drawn from a fixed point O, is a conic section whose focus is O and the eccentricity is u/v.* **[I.A.S. 2008]**

(*b*) *A point P moves so that it always has a velocity compounded of V_1 parallel to a fixed line and V_2 in the transverse direction perpendicular to OP, where O is a fixed point. prove that its orbit is a conic having O as focus and discuss the cases : $V_1 > V_2$, $V_1 = V_2$, $V_1 < V_2$.* **[Delhi 2004, 2011]**

(*c*) *A point which is rowed with const. velocity u starts from a point A on the bank of a river which flows with const. velocity v, and it points always towards a point B on the other bank exactly opposite to A : find the equation of the path of the boat. If v = u, show that the path is a parabola whose focus is B.*

(*d*) *A vessel streams at a constant speed u along a straight line while another vessel streaming at a constant speed v, always moves at right angles to the radius vector to the former. Show that the path of either vessel relative to the other is a conic section of eccentricity u/v.* **[Gorakhpur 98]**

Sol. (*a*) Let O be taken as pole and let the fixed direction be that of the initial line OX so that the line of action of u is along PR, where PR is parallel to OX. Here $P(r, \theta)$ is the position of the point at any time t.

The resolved parts of u along and perpendicular to radius vector OP are $u \cos \theta$ and $u \sin \theta$ respectively.

Hence the radial velocity of $P = dr/dt = u \cos \theta$ ...(1)

and the transversal velocity of $P = r(d\theta/dt) = v - u \sin \theta$ ...(2)

Dividing (1) by (2), we have

$$\frac{dr}{r d\theta} = \frac{u \cos \theta}{v - u \sin \theta} \quad \text{or} \quad \frac{dr}{r} = \frac{u \cos \theta}{v - u \sin v} d\theta.$$

Integrating, $\log r = -\log (v - u \sin \theta) + \log c$, c is constant of integration

or $\quad r = c/(v - u \sin v) \quad$ or $\quad c/r = v - u \sin \theta$

$$\text{or} \quad \frac{(c/v)}{r} = 1 - \frac{u}{v} \sin \theta \quad \text{or} \quad \frac{(c/v)}{r} = 1 + \frac{u}{v} \cos \left(\frac{\pi}{2} + \theta\right), \qquad ...(3)$$

which is of the form $l/r = 1 + e \cos (\theta + \alpha)$. Hence (3) represents a conic whose focus is at O and whose eccentricity is u/v.

(*b*) Proceed as in part (*a*). Here $u = V_1$ and $v = V_2$. Then path is

$$\frac{(c/V_2)}{r} = 1 + \frac{V_1}{V_2} \cos \left(\frac{\pi}{2} + \theta\right), \qquad ...(i)$$

which is conic section with O as pole and eccentricity $= e = V_1/V_2$.

Particular cases : $V_1 > V_2 \Rightarrow V_1/V_2 > 1 \Rightarrow e > 1 \Rightarrow$ (*i*) is hyperbola

$V_1 = V_2 \Rightarrow V_1/V_2 = 1 \Rightarrow e = 1 \Rightarrow$ (*i*) is parabola

$V_1 < V_2 \Rightarrow V_1/V_2 < 1 \Rightarrow e < 1 \Rightarrow$ (*i*) is ellipse

(*c*) Let B be taken as pole and the bank BX as initial line. Let $P(r, \theta)$ be the position of the boat at any time t. Then the boat at P will possess two velocities, u along PB and v parallel to BX (in figure v is along PR where PR is parallel to BX). The resolved parts of v along and perpendicular to radius vector BP are $v \cos \theta$ and $v \sin \theta$ as shown in the figure.

Now, the radial velcity of $P = dr/dt = v \cos \theta - u$ ...(1)

and the transverse velocity of $P = r(d\theta/dt) = -v \sin \theta$ ...(2)

Dividing (1) by (2), we have

$$\frac{dr}{r d\theta} = \frac{v \cos \theta - u}{-v \sin \theta} \quad \text{or} \quad \frac{dr}{r} = \left(\frac{u}{v} \operatorname{cosec} \theta - \cot \theta\right) d\theta$$

Integrating $\log r = (u/v) \log \tan (\theta/2) - \log \sin \theta + \log c$, where c is constant of integration.

or $\quad \log\{(r/c) \sin \theta\} = \log (\tan \theta/2)^{u/v} \quad$ or $\quad (r/c) \sin \theta = (\tan \theta/2)^{u/v}$, ...(3)

which is the required equation of the path of the boat.

When $v = u$, (3) reduces to $\quad r \sin \theta = c \tan(\theta/2)$

$$\text{or} \quad \frac{c}{r} = \frac{\sin (\theta/2)}{\tan (\theta/2)} = \frac{2 \sin (\theta/2) \cos (\theta/2)}{\sin (\theta/2)/\cos (\theta/2)} = 2 \cos^2(\theta/2) = 1 + \cos \theta,$$

i.e., $\quad c/r = 1 + \cos \theta$, which is a parabola whose focus is at B.

(*d*) Refer figures of part (*a*) Suppose the first vessel move along PR with velocity u and the second vessel move along PN with velocity v. We now obtain the path of the first vessel with respect to the second. Then, the first vessel will possess two velocities u along PR and v along PN. Now, proceed as in part (a) to get the required result.

Ex. 10. *The acceleration of a moving point in a plane curve is resolved into two components, one parallel to the initial line and the other along the radius vector, prove that these components are given by*

$$-\frac{1}{r\sin\theta}\frac{d}{dt}(r^2\dot{\theta}) \quad \text{and} \quad \frac{\cot\theta}{r}\frac{d}{dt}(r^2\dot{\theta})+\ddot{r}-r\dot{\theta}^2$$

Sol. Let $P(r, \theta)$ be the position of a moving point at any time t. Let f_1 and f_2 be the components of the acceleration of the point P parallel to initial line OX and along the radius vector OP respectively.

Now, resolving the accelerations of P in the radial direction OP and transverse direction PN, we obtain.

$$f_2+f_1\cos\theta = \text{Radial acceleration} = \ddot{r}-r\dot{\theta}^2 \qquad \text{...(1)}$$

$$-f_1\sin\theta = \text{Transverse accleration} = \frac{1}{r}\frac{d}{dt}(r^2\dot{\theta}) \qquad \text{...(2)}$$

From (2),
$$f_1 = -\frac{1}{r\sin\theta}\frac{d}{dt}(r^2\dot{\theta}) \qquad \text{...(3)}$$

From (1),
$$f_2 = -f_1\cos\theta+\ddot{r}-r\dot{\theta}^2$$

or
$$f_2 = \frac{\cot\theta}{r}\frac{d}{dt}(r^2\dot{\theta})+\ddot{r}-r\dot{\theta}^2, \text{ using (3)} \qquad \text{...(4)}$$

(3) and (4) given the required components of acceleration.

Ex. 11. *Starting from* $x = r\cos\theta$, $y = r\sin\theta$, *calculate* $\ddot{x}$ *and* $\ddot{y}$. *Hence, by resolving the acceleration vector along and perpendicular to the radius vector, determine the radial and transverse components of acceleration.*

Sol. Let (x, y) and (r, θ) be the carterian and polar coordinates of the moving point P at any time t. Then

$$x = r\cos\theta \quad \text{and} \quad y = r\sin\theta \qquad \text{...(1)}$$

Differentiating (1) twice w.r.t 't', we get

$$\dot{x} = \dot{r}\cos\theta - r\sin\theta\,\dot{\theta}, \quad \dot{y} = \dot{r}\sin\theta + r\cos\theta\,\dot{\theta} \qquad \text{...(2)}$$

$$(2) \Rightarrow \quad \ddot{x} = \ddot{r}\cos\theta - \dot{r}\sin\theta\,\dot{\theta} - (\dot{r}\sin\theta\,\dot{\theta} + r\cos\theta\,\dot{\theta}^2 + r\sin\theta\,\ddot{\theta})$$

$$\ddot{y} = \ddot{r}\sin\theta + \dot{r}\cos\theta\,\dot{\theta} + \dot{r}\cos\theta\,\dot{\theta} - r\sin\theta\,\dot{\theta}^2 + r\cos\theta\,\ddot{\theta}$$

Thus,
$$\ddot{x} = (\ddot{r}-r\dot{\theta}^2)\cos\theta - (2\dot{r}\dot{\theta}+r\ddot{\theta})\sin\theta \qquad \text{...(3)}$$

and
$$\ddot{y} = (\ddot{r}-r\dot{\theta}^2)\sin\theta + (2\dot{r}\dot{\theta}+r\ddot{\theta})\cos\theta \qquad \text{...(4)}$$

Now resolving the acccelerations $\ddot{x}$ and $\ddot{y}$ of P in the radial direction OP and transverse direction PN, we have

Radial acceleration of $P = \ddot{x}\cos\theta + \ddot{y}\cos(\pi/2-\theta)$

$$= \ddot{x}\cos\theta + \ddot{y}\sin\theta = \ddot{r}-r\dot{\theta}^2, \text{ using (3) and (4)}$$

and transverse acceleration of $P = \ddot{y}\sin(\pi/2-\theta) - \ddot{x}\sin\theta$

$$= \ddot{y}\cos\theta - \ddot{x}\sin\theta = 2\dot{r}\dot{\theta}+r\ddot{\theta}, \text{ using (3) and (4)}$$

Ex. 12. *A insect crawls at a constant rate u along the spoke of cart wheel of radius a, the cart is moving with velocity v. Find the acceleration along and perpendicular to the spoke.* **[Kumaun 2003; Kanpur 2007, 2019]**

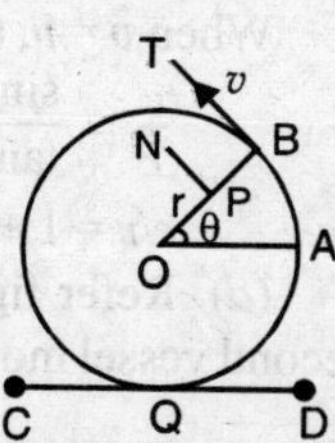

Sol. Suppose that initially the position of the insect be at centre O and the spoke be OA (in horizontal position parallel to the ground CQD).

Let OB be the position of the spoke after time t and let P be the position of the insect at that instant. Then $OP = ut$.

With O as pole and OA as initial line, let coordinates of P be (r, θ) so that

$$\angle POA = \theta \qquad \text{and} \qquad r = OP = ut. \qquad ...(1)$$

Since the cart moves with velocity v, the velocity of the point B of the wheel relative to O is also v and is along the tangent BT to the wheel at B.

Then the angular velocity of B relative to O

$$= \frac{d\theta}{dt} = \frac{\text{The velocity of } B \text{ relative to } O \text{ in a direction} \perp OB}{OB} = \frac{v}{OB} = \frac{v}{a} \qquad ...(2)$$

Hence, the acceleration of the insect along the spoke

= the radial acceleration of P

$$= \frac{d^2 r}{dt^2} - r\left(\frac{d\theta}{dt}\right)^2 = \frac{d^2(ut)}{dt^2} - ut\left(\frac{v}{a}\right)^2, \text{ using (1) and (2)}$$

$$= 0 - ut\,(v/a)^2 = -utv^2/a^2$$

and the acceleration of the insect $\perp$ to the spoke

= the transverse acceleration of P

$$= \frac{1}{r}\frac{d}{dt}\left(r^2\frac{d\theta}{dt}\right) = \frac{1}{ut}\frac{d}{dt}\left(u^2t^2 \cdot \frac{v}{a}\right) = \frac{1}{ut}\cdot\frac{2u^2vt}{a} = \frac{2uv}{a}, \text{ by (1) and (2)}$$

Ex. 13. (*a*) *A straight smooth tube revolves with angular velocity* ω *in a horizontal plane about one extrimity which is fixed. If at zero time a particle inside it be at a distance a from a fixed end and moving with velocity V along the tube, show that its distance at time t is a cosh wt + (V/ω) sinh wt.*

(*b*) *A smooth tube rotates in a horizontal plane about a point in itself with uniform angular velocity ω. At t = 0. The particle inside the tube is at rest relative to tube and at a distance 'a' from the point about which the tube rotates. Show that at time t the distance of the point of rotation is a cosh ωt, i.e.,* $a(e^{\omega t} + e^{-\omega t})/2$. *Find the force the tube is exerting on the particle.*

(*c*) *A particles moves in a smooth straight horizontal tube which is made to rotate with constant angular velocity ω about a vertical axis which intersects the tube. Prove that the distance of the particle from the axis is given by* $r = Ae^{\omega t} + Be^{-\omega t}$, *where A and B are constants depending on the initial position of the particle. If when t = 0 the particle is at a distance r = b from the axis, what velocity must it have along the tube in order that after a very long interval of time it may be very close to axis ?*

Sol. (*a*) Let OX be the initial position of the tube and A that of the particle inside it. After time t, let OX' be the new position of the tube and P that of the particle such that $OP = r$ and $\angle POX = \theta$.

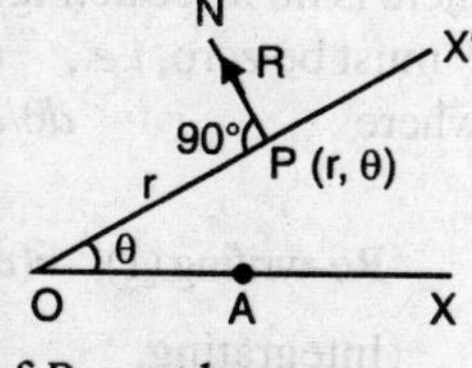

Since the tube is smooth, the only force acting on the particle at P in the horizontal plane of motion is the normal reaction R between the tube and the particle acting in the transverse sense PN at P. Thus, there is in no force acting on the particle at P along the radius vector OP and hence radial acceleration of P must be zero :

$$\therefore \qquad (d^2r/dt^2) - r\,(d\theta/dt)^2 = 0 \quad \text{or} \quad (d^2r/dt^2) - \omega^2 r = 0, \qquad ...(1)$$

where $\quad d\theta/dt$ = angular velocity of $P = \omega$ = constant (given) ...(2)

Re-writing (1), $\quad (D^2 - \omega^2)\,r = 0$, where $D \equiv d/dt$

Its general solution is $\qquad r = A\cosh \omega t + B\sinh \omega t$ ...(3)

From (3), $\qquad dr/dt = Aw\sinh \omega t + Bw\cosh \omega t$ ...(4)

Initially, $t = 0$, $r = OA = a$ and $dr/dt = V$, so (3) and (4) give

$a = A$ and $V = B\omega$ so that $A = a$, $B = V/\omega$. So from (3), we get

$r = a\cosh \omega t + (V/\omega)\sinh \omega t$, which is the desired distance.

(*b*) Proceed as in part (*a*) upto equation (4). In this part, $V = 0$ and so as before $A = a$. and $B = 0$. So from (3) $\qquad r = a\cosh \omega t = (a/2) \times (e^{\omega t} + e^{-\omega t})$. ...(5)

In order to find the force R exerted by the tube on the particle, we require the acceleration in the transverse sense at P. *i.e.*, transverse component f_θ of the acceleration of P.

Now, from (2), $\ddot{\theta} = O$. From (5), $dr/dt = a\omega \sinh \omega t$ ...(6)

$\therefore \quad f_\theta = 2\dot{r}\,\dot{\theta} + r\,\ddot{\theta} = 2(a\omega \sinh \omega t)\,\omega = 2a\,\omega^2 \sinh \omega t$, by (2) and (6)

It m be the mass of the particle, then

$R = mf_\theta = m\,(2a\omega^2 \sinh \omega t) = 2am\,\omega^2 \sinh \omega t$

(c) Refer figure of part (*a*). Here we have b in place of a. Proceed as in part (*a*) upto equation (2). Then, for present problem, we proceed as follows :

Re-writing (1), $\quad (D^2 - \omega^2)\,r = 0$, where $D \equiv d/dt$

whose solution is $\quad r = A\,e^{\omega t} + B\,e^{-\omega t}$ ...(3)

Initially, given that $r = b$ when $t = 0$. So (3) gives $A + B = b$ ...(4)

Re-writing (3), $\quad re^{-\omega t} = A + B\,e^{-2\omega t}$ ...(5)

Taking limit as $t \to \infty$ on both sides of (4) and using the given fact that $r \to 0$ as $t \to \infty$, we have $0 = A + 0$ so that $A = 0$. Then (4) gives $B = b$. Hence (3) reduces to

$$r = b\,e^{-\omega t} \quad \text{so that} \quad \dot{r} = -b\omega e^{-\omega t} \qquad ...(6)$$

At $t = 0$, from (6) $\dot{r} = -b\omega$ which is the required velocity.

Here negative sign means that the particle should have velocity in the directio of r decreasing, *i.e.*, along PO.

Ex. 14. (*a*) *A ring which can slide on a thin long smooth rod rests at a distance d from one end O. The rod is then set revolving uniformly about O in a horizontal plane, show that in space the ring describes the curve* $r = d \cosh \theta$.

(*b*) *A small ring is at rest on a smooth straight horizontal rod of length a at a distance b from one end O of the rod. The rod is then suddenly set rotating in a horizontal plane about the end O with constant angular velocity* ω. *Prove that the ring will leave the rod with velocity* $\omega(2a^2 - b^2)^{1/2}$ *after a time* $(1/\omega)\cosh^{-1}(a/b)$. **[Meerut 1995]**

Sol. (*a*) Refer figure of part (*a*) of Ex. 13. Let OX be the initial position of the rod and A that of the ring. After time t, let OX' be the new position of the rod and P that of the ring such that $OP = r$ and $\angle POX = \theta$.

Since the rod is smooth, the only force acting on the ring at P on the horizontal plane of motion is the normal reaction R between the rod and the ring acting in a direction perpendicular to OP. Thus, there is no force acting on the particle at P along the radius vector OP and hence radial acceleration of P must be zero, *i.e.*, $\quad (d^2r/dt^2) - r(d\theta/dt)^2 = 0 \quad$ or $\quad d^2r/dt^2 = r\omega^2$ ...(1)

where $\quad d\theta/dt =$ angular velocity of $P = \omega =$ constant ...(2)

[The rod revolves uniformly $\Rightarrow d\theta/dt$ is constant]

Re-writing (1), $\quad \dot{r}(d\dot{r}/dr) = r\omega^2 \quad$ or $\quad 2\dot{r}d\dot{r} = 2r\omega^2\,dr$ ...(3)

Integrating, $\quad \dot{r}^2 = r^2\omega^2 + c$, c being constant of integration ...(4)

Initially, $\dot{r} = 0$ when $r = d$. So $0 = d^2\omega^2 + c \quad$ or $\quad c = -d^2\omega^2$

$\therefore$ From (4), $\quad \dot{r}^2 = \omega^2(r^2 - d^2) \quad$ or $\quad dr/dt = \omega(r^2 - d^2)^{1/2}$

or $$\frac{dr}{(r^2 - d^2)^{1/2}} = \omega\,dt = d\theta, \text{ using (2)}$$

Integrating, $\quad \cosh^{-1}(r/d) = \theta + c'$, when c' is constant of integration.

Initially, when $\theta = 0$, $r = d$. So $\cosh^{-1}(1) = 0 + c' \quad$ or $\quad c' = 0$

Hence $\quad \cosh^{-1}(r/d) = \theta \quad$ so that $\quad r = d\cosh\theta$,

which is the equation of the curve described by the ring in space.

(*b*) Proceed as in part (*a*) (noting that here $d = b$ and $OX = OX' = a$) upto equation (4). For the remaining part of solution, proceed as follows.

Initially, $\dot{r} = 0$ when $r = b \quad$ so $\quad 0 = b^2\omega^2 + c \quad$ or $\quad c = -a^2\omega^2$.

$\therefore$ From (4), $\dot{r}^2 = \omega^2(r^2 - b^2)$ ...(5)

The velocity of the ring at P

$= \{\dot{r}^2 + (r\dot{\theta})^2\}^{1/2} = \{\omega^2(r^2 - b^2) + r^2\omega^2\}^{1/2}$, using (2) and (5), ...(6)

When the particle leaves the rod at the end X', $r = OX'$ = length of the rod = a. Hence velocity of the ring at the time of leaving the rod can be obtained by replacing r by a in (6) and is given by $\{\omega^2(a^2 - b^2) + a^2\omega^2\}^{1/2}$ *i.e.*, $\omega(2a^2 - b^2)^{1/2}$.

From (5), $dr/dt = \omega(r^2 - b^2)^{1/2}$ or $dr/(r^2 - b^2)^{1/2} = \omega dt$

where we have taken positive sign because r increases as t increases.

Integrating $\cosh^{-1}(r/b) = \omega t + c$, where c is an arbitrary constant.

Initially, $r = b$ when $t = 0$, so $\cosh^{-1}(1) = 0 + c$ or $c = 0$. ...(7)

Thus, $\cosh^{-1}(r/b) = \omega t$ or $t = (1/\omega)\cosh^{-1}(r/b)$

Now, when the ring leaves the rod at X', $r = a$ and so that corresponding time, from, (7), is $t = (1/\omega)\cosh^{-1}(a/b)$.

Ex. 15. *If a rod which always passes through the origin rotates with uniform angular velocity* ω, *while one end describes the curve* $r = a + b\,e^{\theta}$, *show that the radial acceleration of any point of the rod is the same at every instant, and the radial velocity is the same at every point at a given instant.*

Sol. Let AB be the rod which always passes through the origin O and whose end A describes the curve

$$r = a + b\,e^{\theta} \qquad ...(1)$$

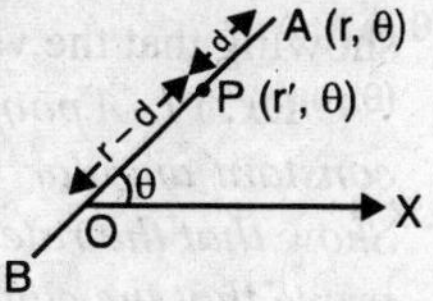

At any time t, let AB be the position of the rod. Let P be any point of the rod, such that $AP = d$. Let polar coordinates of A and P be (r, θ) and (r', θ) respectively, where $r' = r - d$.

Given, angular velocity of the rod $= d\theta/dt = \omega$ = constant, say. ...(2)

Then, radial acceleration f_r of P is given by

$$f_r = \frac{d^2r'}{dt^2} - r'\left(\frac{d\theta}{dt}\right)^2 = \frac{d^2(r-d)}{dt^2} - (r-d)\omega^2 = \frac{d^2r}{dt^2} - (r-d)\omega^2 \qquad ...(3)$$

Now, by (1), $\dot{r} = b\,e^{\theta}\,\dot{\theta} = b\omega\,e^{\theta}$, by (2) ...(4)

From (4), $\ddot{r} = b\omega\,e^{\theta}\,\dot{\theta} = b\omega^2 e^{\theta} = \omega^2(r - a)$, by (1) ...(5)

Using (5), (3) gives $f_r = \omega^2(r - a) - \omega^2(r - d) = \omega^2(d - a)$,

which is constant being independent of t. Hence the radial acceleration of any point of the rod is same at every instant.

Again, the radial velocity of $P = \dfrac{dr'}{dt} = \dfrac{d(r-d)}{dt} = \dfrac{dr}{dt}$ ...(6)

From (1), $dr/dt = b\,e^{\theta}\,\dot{\theta} = w(r - a)$, by (1) and (2)

$\therefore$ By (6), the radial velocity of $P = w(r - a)$, which is independent of d. So radial velocity of every point of the rod is the same at a given instant.

Ex. 16. *If the radial and transverse velocities of a particle are always proportional to each other. Then,*

(i) Show that the path is an equiangular spiral. **[Pune 2010; Kanpur 2002, Meerut 2005]**

(ii) If in addition the radial and transverse accelerations are always proportional to each other, show that the velcoity of the particle varies as some power of the radius vector.

[Purvanchal 2009; Agra 2002]

Sol. (*i*) Given radial velocity = k (the transverse velocity)

so that $dr/dt = kr(d\theta/dt)$, k is some constant ...(1)

From (1), $(1/r)dr = k d\theta$ so that $\log r - \log c = k\theta$

or $\log(r/c) = k\theta$ or $r = ce^{k\theta}$, c being constant of integration.

This is the equation of the path which is an equiangular spiral.

(*ii*) Here, in addition to (1), we have

radial acceleration = λ (transverse accelerations), λ being a constant.

i.e.,
$$\frac{d^2r}{dt^2}-r\left(\frac{d\theta}{dt}\right)^2=\lambda\frac{1}{r}\frac{d}{dt}\left(r^2\frac{d\theta}{dt}\right)$$

or
$$\frac{d^2r}{dt^2}-r\left(\frac{1}{kr}\frac{dr}{dt}\right)^2=\frac{\lambda}{r}\frac{d}{dt}\left(r^2\cdot\frac{1}{kr}\frac{dr}{dt}\right),\text{ as }\frac{d\theta}{dt}=\frac{1}{k}\frac{dr}{dt}\text{ by (1)}$$

or
$$\frac{d^2r}{dt^2}-\frac{1}{k^2r}\left(\frac{dr}{dt}\right)^2=\frac{\lambda}{kr}\frac{d}{dt}\left(r\frac{dr}{dt}\right)=\frac{\lambda}{kr}\left(r\frac{d^2r}{dt^2}+\frac{dr}{dt}\cdot\frac{dr}{dt}\right)$$

or
$$(1-\lambda/k)\,\ddot{r}=(1/k^2+\lambda/k)\,(1/r)\,\dot{r}^2\quad\text{or}\quad\ddot{r}=(n/r)\,\dot{r}^2$$

[Taking $(1/k^2+\lambda/k)/(1-\lambda/k)=n=$ constant, say]

or
$$\dot{r}\frac{d\dot{r}}{dr}=\frac{n}{r}\dot{r}^2\quad\text{or}\quad\frac{d\dot{r}}{\dot{r}}=\frac{n}{r}dr$$

Integrating, $\log\dot{r}=n\log r+\log c$, when c is constant of integration

or
$$\log\dot{r}=\log c+\log r^n\quad\text{or}\quad\dot{r}=cr^n\qquad\text{...(2)}$$

Using (2), (1) gives
$$cr^n=kr\dot{\theta}\quad\text{or}\quad r\dot{\theta}=(c/k)r^n\qquad\text{...(3)}$$

∴ The resultant velocity of the particle
$$=\{\dot{r}^2+(r\dot{\theta})^2\}^{1/2}=\{c^2r^{2n}+(c^2/k^2)\,r^{2n}\}^{1/2}=r^n\,c\,(1+1/k^2)^{1/2},$$
showing that the velocity varies as some power of radius vector.

Ex. 17. *A point starts from the origin in the direction of the initial line with velocity f/ω and with constant angular velocity ω about the origin and with constant negative radial acceleration $-f$. Show that the rate of growth of the radius velocity is never positive, but tends to the limit zero and prove that the equation of the path is $\omega^2r=f(1-e^{-\theta})$.*

Sol. Given that angular velocity $=d\theta/dt=\omega=$ const. ...(1)

and radial acceleration $=\ddot{r}-r\dot{\theta}^2=-f$ ...(2)

From (1) and (2), $\dot{r}(d\dot{r}/dr)=r\omega^2-f$ or $2\dot{r}d\dot{r}=(2r\omega^2-2f)dr$

Integrating, $\dot{r}^2=r^2\omega^2-2fr+c$, where c is a constant.

Initially, $r=0$ and $\dot{r}=f/\omega$ (given) so $c=f^2/\omega^2$.

∴
$$(dr/dt)^2=r^2\omega^2-2fr+f^2/\omega^2=(f/\omega-r\omega)^2$$

or $dr/dt=f/\omega-r\omega$, where the positive sign is taken because the particle moves in the direction of r increasing

or
$$\frac{dr}{d\theta}\frac{d\theta}{dt}=\frac{f}{\omega}-r\omega\quad\text{or}\quad\frac{dr}{d\theta}\cdot\omega=\frac{f}{\omega}-r\omega,\text{ by (1)}$$

or
$$\frac{dr}{d\theta}=\frac{f}{\omega^2}-r\quad\text{or}\quad\frac{dr}{(f/\omega^2)-r}=d\theta$$

Integrating, $-\log\{(f/\omega^2)-r\}+\log A=\theta$, when A is any constant.

Initially, $r=0$ and $\theta=0$ so $A=f/\omega^2$ and hence we get
$$\log(f/\omega^2)-\log\{(f-\omega^2r)/\omega^2\}=\theta\quad\text{or}\quad\log\{f/(f-\omega^2r)\}=\theta$$

or
$$f/(f-\omega^2r)=e^{\theta}\quad\text{or}\quad fe^{-\theta}=f-\omega^2r\quad\text{or}\quad\omega^2r=f(1-e^{-\theta})\qquad\text{...(3)}$$

From (1) and (2),
$$\ddot{r}=r\omega^2-f=f(1-e^{-\theta})-f,\text{ by (3)}\qquad\text{...(4)}$$

∴ The rate of growth of radial velocity.
$$=\frac{d}{dt}\left(\frac{dr}{dt}\right)=\frac{d^2r}{dt^2}=-fe^{-\theta}=-f/e^{\theta},$$
which is never positive for all values of θ and tends to zero when $\theta\to\infty$.

EXERCISE

1. A particle is constrained to move along the equiangular spiral $r=ae^{b\theta}$ so that the radius vector moves with constant angular velocity ω. Determine the radial and transeversal components of velocity and acceleration. **(Kanpur 2008, 19)** [**Ans.** $\omega ab\,e^{b\omega t}$, $a\omega\,e^{b\omega t}$, $(b^2-1)\,a\omega^2\,e^{b\omega t}$, $2ab\omega^2\,e^{b\omega t}$]

2. A point moves in a parabola, in such a manner that the component of velocity at right angles to the radius vector from the focus is constant. Show that the acceleration of the point is constant in magnitude.

3. If the angular velocity ω of a particle about the origin is consant and the rate of change of acceleration is directed wholly along the radius vector, prove that $d^2r/dt^2 = (\omega^2 r)/3$.

4. A particle moves in a circular path of radius a so that its angular velocity about a fixed point in the circumference is constant and equal to ω. Show that the resultant acceleration of the particle at every point is of constant magnitude $4a\omega^2$.

5.7 Tangential and normal components of velocity and acceleration.

[Kanpur 2018, 19; Mumbai 2010; Calcutta 2003, Agra 2009; Bundelkhand 2011; Delhi B. Sc. (Prog.) III 2008, 11; Kanpur 2015, 16; Meerut 2006; Bangaluru 2004]

Let *P* be the position of a particle moving on curve *AB* (with intrisic equation $s = f(\psi)$ at any time such that *s* is the arcual distance of *P* from *A* and ψ is the angle which the tangent *PT* makes with *x*-axis. Then (*s*, ψ) are the intrinsic coordinates of *P*.

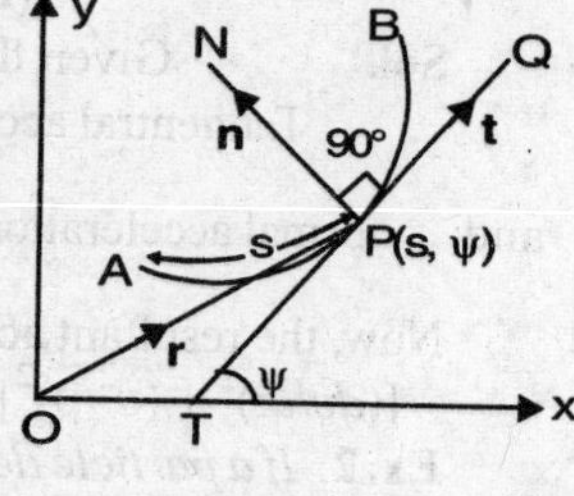

Let **t** denote the unit vector along *PQ* (*i.e.*, along the tangent at *P* in the direction of *s* increasing) and **n** the unit vector along the normal at *P* in the direction of ψ increasing (*i.e.*, in the direction of inwards drawn normal *PN*)

With the origin *O* as pole, let the position vector of *P* be **r**. Then, from vector calculus, we know that $\mathbf{t} = d\mathbf{r}/ds$ ...(1)

Let **v** be the velocity vector of the particle at *P*. Then

$$\mathbf{v} = \frac{d\mathbf{r}}{dt} = \frac{d\mathbf{r}}{ds}\frac{ds}{dt} = \frac{ds}{dt}\mathbf{t} = \frac{ds}{dt}\mathbf{t} + 0\mathbf{n}, \qquad ...(2)$$

showing that tangential and normal components of velocity are *ds*/*dt* and 0 respectively. Hence the magnitude *v* of the resultant velocity, *i.e.*, | **v** | at *P* is given by $v = ds/dt$ along the tangent at *P* to the path of the particle.

Tangential and normal components of acceleration

Let **f** be the acceleration of the particle at *P*. Then, by (2), we have

$$\mathbf{f} = \frac{d\mathbf{v}}{dt} = \frac{d}{dt}(v\mathbf{t}), \text{ where } v = |\mathbf{v}| = \frac{ds}{dt}$$

$$= \frac{dv}{dt}\mathbf{t} + v\frac{d\mathbf{t}}{dt} = \frac{dv}{dt}\mathbf{t} + v\frac{d\psi}{dt}\mathbf{n}, \qquad [\because \text{By corollary of Art. 5.5, } d\mathbf{t}/dt = (d\psi/dt)\,\mathbf{n}]$$

$$\therefore \quad \mathbf{f} = \frac{dv}{dt}\mathbf{t} + v\frac{d\psi}{ds}\frac{ds}{dt}\mathbf{n} = \frac{dv}{dt}\mathbf{t} + \frac{v^2}{\rho}\mathbf{n} \qquad ...(3)$$

[∵ ρ = radius of curvature at $P = ds/d\psi$]

Relation (3) expresses **f** as a linear combination of the vectors **t** and **n**. Hence the coefficients of **t** and **n** in the above linear combination are respectively the tangential and normal acceleration of the particle at *P*. Thus,

the tangential acceleration at $P = dv/dt$, which is positive in the direction of *s* increasing

and the normal acceleration at $P = v^2/\rho$,

which is the positive in the direction of inwards drawn normal.

Remark. The tangential acceleration can also expressed in two ways. These forms will be used directly in problems.

From I. Tangential acceleration $= \dfrac{dv}{dt} = \dfrac{d}{dt}\left(\dfrac{ds}{dt}\right) = \dfrac{d^2s}{dt^2} = \ddot{s}$

From II. Tangential acceleration $= \dfrac{dv}{dt} = \dfrac{dv}{ds}\dfrac{ds}{dt} = v\dfrac{dv}{ds}$

Resultant acceleration $= \left\{(d^2s/dt^2)^2 + (v^2/\rho)^2\right\}^{1/2} = \left\{(dv/dt)^2 + (v^2/\rho)^2\right\}^{1/2}$

An important particular case. *If a particle describes a circle of radius a with a uniform speed v, show that its acceleration at any point of the path is v^2/a and is directed towards the centre of the circle.*

[Agra 2002; Remember this result for direct applications for a circle]

Proof : Since v is constant, tangential acceleration $= dv/dt = 0$.

Again, the normal acceleration $= v^2/\rho = v^2/a$

[$\because$ For circle, ρ = radius of curvature = radius = a]

$\therefore$ Resultant acceleration $= \{(dv/dt)^2 + (v^2/\rho)^2\}^{1/2} = v^2/a$,

showing that the resultant acceleration reduces to only normal acceleration. But the normal acceleration at a point is directed towards the inward drawn normal at that point. Hence the accleration at any point of circular path is v^2/a which in directed toward the centre of the circle.

TYPE 4 : SOLVED EXAMPLES BASED ON ART. 5.1

Ex. 1. *Prove that the acceleration of a point moving in a curve with uniform speed is $\rho(d\psi/dt)^2$.*

[(Bundelkhand 2009; Lucknow 2008, 10; Rohilkhand 2008; Kanpur 2003]

Sol. Given that speed $= v = ds/dt =$ constant $= k$, say ...(1)

$\therefore$ Tangentral acceleration $= dv/dt = 0$, by (1) ...(2)

and normal acceleration $= \dfrac{v^2}{\rho} = \rho\left(\dfrac{v}{\rho}\right)^2 = \rho\left(\dfrac{ds/dt}{ds/d\psi}\right)^2 = \rho\left(\dfrac{d\psi}{dt}\right)^2$ ...(3)

Now, the resultant acceleration

$= [(dv/dt)^2 + (v^2/\rho)^2]^2 = v^2/\rho = \rho(d\psi/dt)^2$, by (2) and (3)

Ex. 2. *If a particle describes the cycloid $s = 4a \sin\psi$ with uniform speed v. Find its acceleration at any point.* **[Agra 98, Meerut 96, 99]**

Sol. The given curve is $s = 4a \sin\psi$ ...(1)

and the given speed $= v = ds/dt =$ constant. ...(2)

From (2), the tangential acceleration $= dv/dt = 0$ ...(3)

Diff. (1) w.r.t. 'ψ', $ds/d\psi = 4a\cos\psi = 4a(1 - \sin^2\psi)^{1/2}$

or $\rho = 4a(1 - s^2/16a^2)^{1/2}$ or $\rho = (16a^2 - s^2)^{1/2}$ ...(4)

Hence,the normal acceleration $= v^2/\rho = v^2/(16a^2 - s^2)^{1/2}$ ...(5)

Now, the resultant acceleration at any point

$= \{(dv/dt)^2 + (v^2/\rho)^2\}^{1/2} = v^2/\rho = v^2/(16\,a^2 - s^2)^{1/2}$, by (3) and (5).

Ex. 3. *A particle describes a plane curve with a constant speed and its acceleration is constant in magnitude. Prove that the path is a circle.*

Sol. Let v be the constant speed of the particle.

Then, its tangential acceleration $= dv/dt = 0$ Also,its normal acceleration $= v^2/\rho$

$\therefore$ Resultant acceleration $= \{(dv/dt)^2 + (v^2/\rho)^2\}^{1/2} = v^2/\rho$

Given that $v^2/\rho =$ constant $= k$, say.

Hence $\rho = v^2/k =$ constant, as v is constant,

showing that the path of the particle is such that its radius of curvature is constant and hence the path of the particle must be a circle. (Remember that for a circle, radius of curvature = radius of the circle)

Ex. 4. *If the tangential and normal accelerations of a particle describing a plane curve be constant throughout, prove that the radius of curvature at any point is given by $\rho = (at + b)^2$.*

[Avadh 2008; Bundelkhand 2008; Meerut 2010]

Sol. Given that $dv/dt =$ constant $= \lambda$, say ...(1)

and $v^2/\rho =$ constant $= \mu$, say ....(2)

Integrating (1), $v = \lambda t + c$, where c is constant of integration ...(3)

From (2) and (3) $(\lambda t + c)^2/\rho = \mu$ or $\rho = \{(\lambda t + c)/\mu^{1/2}\}^2$

or $\rho = (a\,t + b)^2$, where $a = \lambda/\mu^{1/2}$ and $b = c/\mu^{1/2}$

Ex. 5. (*a*) *Apoint moves along the arc of a cycloid in such a manner that the tangent at it rotates with a constant angular velocity. Show that the acceleration of the moving point is constant.* **[Agra 93]**

(*b*) *The rate of change of direction of velocity of a particle moving in a cycloid is constant. Prove that acceleration must be constant in magnitude.*

Sol. (*a*) The intrinsic equation of a cycloid is $s = 4\sin\psi$ ...(1)

Since the tangent rotates with constant angular velocity,so

$$d\psi/dt = \text{constant} = \omega, \text{ say} \qquad ...(2)$$

Diff. (1) w.r.t. '*t*' $ds/dt = 4a\cos\psi.(d\psi/dt) = 4a\omega\cos\psi$, by (2)

Thus, $v = ds/dt = 4a\omega\cos\psi$ ...(3)

From (3), $dv/dt = -4a\omega\sin\psi.(d\psi/dt) = -4a\omega^2\sin\psi$, by (2) ...(4)

Again, by (1), $ds/d\psi = 4a\cos\psi$ so that $\rho = 4a\cos\psi$ ...(5)

$\therefore\ v^2/\rho = (4a\omega\cos\psi)^2/(4a\cos\psi) = 4a\omega^2\cos\psi$, by (3) and (5) ...(6)

and so resultant acceleration of the moving particle.

$= \{(dv/dt)^2 + (v^2/\rho)^2\}^{1/2} = (16a^2\omega^4\sin^2\psi + 16a^2\omega^4\cos^2\psi)^{1/2} = 4a\omega^2$, showing that the acceleration is constant in magnitude.

(*b*) Since the direction of velocity is always along the tangent, it follows from the given problem that the angular velocity of the tangent is constant *i.e.*, $d\psi/dt$ = constant as in part (*a*). Now proceed as in part (*a*).

Ex. 6. (*a*) *A point moves in a plane curve, so that its tangential and normal acceleration are equal and the angular velocity of the tangent is constant. Find the curve.* **[Mumbai 2010; Avadh 2011, Agra 2012; Lucknow 2010; Kanpur 2013, 17; Meerut 2012, Bhopal 2004]**

(*b*) *If tangential and normal acceleration components be equal, prove that the velocity varies as* e^{ψ}. **[Kanpur 2006, 15]**

(*c*) *A particle moves in a plane in such a manner that its tangential and normal accelerations are always equal and its velocity varies as exp* ($\tan^{-1} s/c$), *s being the length of the arc measured from a fixed point on the curve. Find the path.* **[Meerut 2011; Agra 2008]**

Sol. (*a*) Tangential accel = normal accel $\Rightarrow v(dv/ds) = v^2/\rho$ ...(1)

and angular velocity of tangent is constant $\Rightarrow d\psi/dt$ = const. = ω, say ...(2)

From (1), $dv/ds = v(d\psi/ds)$ or $(1/v)dv = d\psi$

Integrating, $\log v - \log c = \psi$ or $v = c\,e^{\psi}$ ...(3)

From (3), $\dfrac{ds}{dt} = c\,e^{\psi}$ or $\dfrac{ds}{d\psi}\cdot\dfrac{d\psi}{dt} = c\,e^{\psi}$ or $\dfrac{ds}{d\psi}\cdot\omega = c\,e^{\psi}$, by (2)

or $ds = (c/\omega)e^{\psi}\,d\psi$ so that $s = A\,e^{\psi} + \beta$, ...(4)

where $A\ (= c/\omega)$ and B are arbitrary constants. (4) is the required equation.

(*b*) Proceed as in part (*a*) upto equation (3). From (3), we see that the velocity varies as e^{ψ}.

(*c*) Tangential accel. = normal accel. $\Rightarrow v(dv/ds) = v^2/\rho$ ...(1)

and also given that velocity $= v = k\ e^{\tan^{-1}(s/c)}$, ...(2)

where k is some constant. Since $\rho = ds/d\psi$, (1) reduces to

$dv/ds = v(d\psi/ds)$ or $(1/v)/dv = d\psi$

Integrating, $\log v = \psi + A$, where A is constant of integration.

or $\log\{k\,e^{\tan^{-1}(s/c)}\} = \psi + A$, using (2)

or $\log k + \tan^{-1}(s/c) = \psi + A$ or $\tan^{-1}(s/c) = \psi + B$, ...(3)

where $B\ (= A - \log k)$ is an arbitrary constant. To find B, we assume that $s = 0$ when $\psi = 0$. Then (3) gives $B = 0$.

So by (3), $\tan^{-1}(s/c) = \psi$ or $s = c\tan\psi$ is the required path.

Ex. 7. *A particle moves in a catenary $s = c\tan\psi$, the direction of its acceleration at any point makes equal angles with the tangent and the normal to the path at that point. If the speed at the vertex (where $\psi = 0$) be u, show that the velcoity and acceleration at any point ψ are*

$$ue^{\psi} \text{ and } (\sqrt{2}/c)\,u^2e^{2\psi}\cos^2\psi.$$

Sol. Since the direction of the resultant acceleration makes equal angles with the tangent and normal, hence the tangential and normal accelerations must be equal.

Thus, $v(dv/ds) = v^2/\rho = v^2(d\psi/ds)$ or $(1/v)dv = d\psi$ ...(1)

Integrating, $\log v - \log c = \psi$ or $v = c\,e^{\psi}$ ...(2)

Given that $v = u$, $\psi = 0$ so (2) gives $c = u$ and hence $v = u\,e^{\psi}$.

Given $s = c \tan \psi$ so $\rho = ds/d\psi = c \sec^2 \psi$ ...(3)

Now, the resultant acceleration of the particle

$$= \{(v dv/ds)^2 + (v^2/\rho)^2\}^{1/2} = \{(v^2/\rho)^2 + (v^2/\rho)^2\}^{1/2}, \text{ by (1)}$$

$$= \frac{\sqrt{2}\,v^2}{\rho} = \frac{\sqrt{2}\,u^2\,e^{2\psi}}{c \sec^2 \psi} = \frac{\sqrt{2}}{c}\,u^2\,e^{2\psi} \cos^2 \psi, \text{ by (2) and (3).}$$

Ex. 8. *A particle describes a curve (for which s and ψ varnish simultaneously) with uniform speed v. If the acceleration at any point s be $v^2c/(s^2 + c^2)$, find the intrinsic equation of the curve.*

[Avadh 2011; Bundelkhand 2010; Kanpur 2009, 2013; Meerut 2009, 2017]

Sol. Given that $ds/dt = v =$ constant, so $dv/dt = 0$

∴ Acceleration at any point 's'

$$= \{(dv/dt)^2 + (v^2/\rho)^2\}^{1/2} = \{0 + (v^2/\rho)^2\}^{1/2} = v^2/\rho$$

But it is given that the acceleration at point 's' is $v^2c/(s^2 + c^2)$

$$\therefore \quad \frac{v^2 c}{s^2 + c^2} = \frac{v^2}{\rho} = \frac{v^2 d\psi}{ds} \quad \text{or} \quad d\psi = \frac{c\,ds}{s^2 + c^2}, \quad \text{as } \rho = \frac{ds}{d\psi}$$

Integrating, $\psi + A = \tan^{-1}(s/c)$, where A is constant of integration

When $\psi = 0$, $s = 0$ (given) so $A = \tan^{-1} 0 = 0$.

∴ $\psi = \tan^{-1}(s/c)$ or $s = c \tan \psi$, which is the required curve.

Ex. 9. *A particle is moving in a parabola with uniform angular velocity about the focus, prove that its normal acceleration at any point is proportional to the radius of curvature of the its path at that point.*

[Guwahati 2007 ; Agra 2001, 2010; Kanpur 2000, Lucknow 2009; Rohilkhand 2007; Meerut 2000, 2006]

Sol. The pedal equation of the parabola referred to the focus as pole is given by

$$p^2 = ar \quad ...(1)$$

Given that angular velocity about focus (*i.e.*, about pole) is constant, so

$$d\theta/dt = vp/r^2 = \text{constant} = \omega, \text{ say} \quad ...(2)$$

From (1), $2p(dp/dr) = a$ or $dr/dp = 2p/a$ ...(3)

∴ Radius of curvature $= \rho = r(dr/dp) = r(2p/a) = (2pr)/a$ ...(4)

Now, (Normal acceleration)/ (radius of curvature ρ)

$$= \frac{v^2/\rho}{\rho} = \frac{v^2}{\rho^2} = \frac{(r^2\omega/p)^2}{(2pr/a)^2}, \text{ using (2) and (4)}$$

$$= (r^2\omega^2 a^2)/4p^4 = (r^2\omega^2 a^2)/4\,(ar)^2, \text{ by (1)}$$

∴ Normal acceleration $= (\omega^2/4)\,\rho$, where $\omega^2/4$ is a constant.

Hence normal acceleration is proportional to ρ.

Ex. 10. *A curve is described by a particle having a constant acceleration in a direction inclined at a constant angle to the tangent. Show that the curve is an equiangular spiral.*

[Agra 2002; Purvanchal 2009]

Sol. Let f be the constant acceleration of the particle and α the constant angle which the direction of f makes with the tangent PT at any point $P(s, \psi)$ of its path at any time t.

Resolving that acceleration f along the tangent TP and normal PN at P,

$f \cos \alpha$ = tangential acceleration $= v(dv/ds)$ ...(1)

and $f \sin \alpha$ = normal acceleration $= v^2/\rho$ ...(2)

Dividing (2) by (1), $$\tan \alpha = \frac{v^2/\rho}{v(dv/ds)} = \frac{v\,ds}{\rho\,dv}$$

or $$\tan \alpha = v\left(\frac{d\psi}{ds}\right)\left(\frac{ds}{dv}\right), \text{ as } \rho = \frac{ds}{d\psi}$$

or $$(1/v)dv = \cot \alpha \, d\psi \text{ so that } v = c\, e^{\psi \cot \alpha}, \qquad \text{...(3)}$$

where c is constant of integration. Re-writing (2), we have

$$\rho = \frac{v^2}{f \sin \alpha} \quad \text{or} \quad \frac{ds}{d\psi} = \frac{c^2 e^{2\psi \cot \alpha}}{f \sin \alpha}, \text{ using (3)}$$

or $$ds = A\, e^{2\psi \cot \alpha}\, d\psi, \text{ where } A = c^2/(f \sin \alpha) = \text{constant}$$

Integrating, $$s = (A/2 \cot \alpha)\, e^{2\psi \cot \alpha} + B = De^{2\psi \cot \alpha} + B,$$

where $D\,(= A/2 \cot \alpha)$ and B are artibitrary constants. This intrinsic equation of an equiangular spiral. Thus, required curve is an equiangular spiral.

Ex. 11. *A particle describes a cylcoid with uniform speed. Prove that the normal acceleration at any point varies inversely as the square root of the distance from the base of the cycloid.*

[*Rohilkhand 2001*]

Sol. In the adjoining figure we have shown an inverted cycloid whose intrinsic equation is

$$s = 4a \sin \psi \qquad \text{...(1)}$$

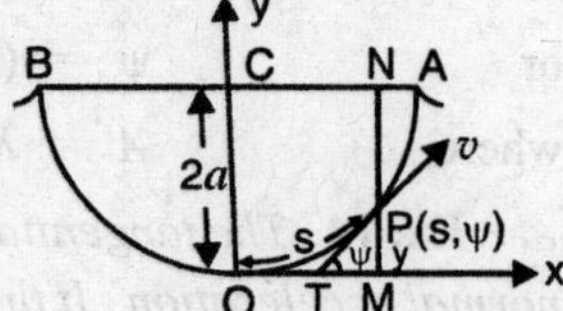

The point O is the vertex of the cycloid and the A and B are the cusps. AB is called the base line. Here $OC = 2a$. Let $P(s, \psi)$ be the position of the particle at any time t such that $MP = y = y$-coordinate of P. Then

$$s^2 = 8ay \qquad \text{...(2)}$$

Also, given that, the speed $= v = ds/dt =$ constant.

From (1), $$\rho = ds/d\psi = 4a \cos \psi = 4a\,(1 - \sin^2 \psi)^{1/2} = 4a(1 - s^2/16a^2)^{1/2}$$

$\therefore$ Normal acceleration $$= v^2/\rho = v^2/(16a^2 - s^2)^{1/2} \qquad \text{...(3)}$$

Let $PM \perp OX$ and $PN \perp AB$. Then, from figure, we have

$$PN = MN - PM = 2a - y = 2a - (s^2/8a) = (16a^2 - s^2)/8a, \text{ by (2)}$$

Hence, $16a^2 - s^2 = 8a \times PN$ or $(16a^2 - s^2)^{1/2} = (8a\, PN)^{1/2}$

$\therefore$ From (3), normal acceleration $= v^2/(8a\, PN)^{1/2} = (v^4/8a)^{1/2} \cdot (1/\sqrt{PN})$, showing that normal acceleration varies inversely as the square root the distance PN from the base.

Ex. 12. *A point moves in a plane curve so that its tangential aceleration is constant, and the magnitudes of the tangential velocity and normal acceleration are in a constant ratio ; find the intrisic equation of the curve.* **[Meerut 2015; Rohilkhand 2000; Kanpur 2012, 17]**

Sol. Here tangential acceleration $= v\,(dv/ds) =$ constant $= \lambda$, say ...(1)

and $v/(v^2/\rho) =$ constant $= \mu$, say so that $\rho = \mu v$ ...(2)

From (1), $$2v dv = 2\lambda ds \quad \text{so that} \quad v^2 = 2\lambda s + c$$

$\therefore$ $$v = (2\lambda s + c)^{1/2}, \text{ c being arbitrary constant.} \qquad \text{...(3)}$$

From (2) and (3), $$ds/d\psi = \mu\,(2\lambda s + c)^{1/2} \quad \text{or} \quad (2\lambda s + c)^{-1/2} ds = \mu\, d\psi$$

Integrating, $$(2\lambda s + c)^{1/2}/\lambda = \mu\psi + A, \text{ when } A \text{ is a constant}$$

or $$(2\lambda s + c) = (\lambda\mu\psi + A\lambda)^2 = \lambda^2\mu^2\psi^2 + 2\,A\,\lambda\mu\psi + A^2\lambda^2$$

or $$s = (\lambda^2\mu^2\psi^2 + 2A\,\lambda\mu\psi + A^2\lambda^2 - c)/2\lambda$$

or $$s = B\psi^2 + D\psi + E, \text{ when } B = \lambda\mu^2/2, D = A\,\mu, E = (A^2\lambda^2 - c)/2\lambda$$

This is the required intrinsic equation of the curve.

Ex. 13. *A particle describes a plane curve. If the tangent and normal accelerations are each constant throughout the motion prove that the angle ψ, through which the direction of motion turns in time t is given by ψ = A log (1 + Bt).*

[Avadh 2006; Purvanchal 2006, 11; Lucknow 2007, 11; Delhi B.A. (Prog.) III, 2012]

Sol. Here, tangential acceleration $= dv/dt =$ constant λ, say ...(1)

and normal acceleration $= v^2/\rho =$ constant $= \mu$, say ...(2)

Integrating (1), $v = \lambda t + c$, where c is constant of integration ...(3)

Rewriting (2), $v \cdot \dfrac{v}{\rho} = \mu$ or $(\lambda t + c) \cdot \dfrac{ds/dt}{ds/d\psi} = \mu$, by (1) and (2)

or $(\lambda t + c)\,(d\psi/dt) = \mu$ or $d\psi = \{\mu/(\lambda t + c)\}\,dt$

Intregrating, $\psi = (\mu/\lambda) \log (\lambda t + c) + D$, where D is a constant ...(4)

Let $\psi = 0$ when $t = 0$. Then $0 = (\mu/\lambda) \log c + D$ or $D = -(\mu/\lambda) \log c$

$\therefore$ From (4), $\psi = (\lambda/\mu) \{\log (\lambda t + c) - \log c\} = (\lambda\mu) \log \{(\lambda t + c)/c\}$

or $\psi = (\lambda/\mu) \log \{1 + (\lambda/c)t\} = \mathrm{A} \log (1 + Bt)$,

where, $A = \lambda/\mu$ and $B = \lambda/c$ are some new constants.

Ex. 14. *The tangential acceleration of a particle moving along a circle of radius a is k times the normal acceleration. If the speed at a certain time is u, prove tht it will return to the same point after a time (a/ku) (1 − $e^{-2\pi k}$).* **[Kanpur 2018; Calcutta 2003, Purvanchal 2007]**

Sol. Since tangential aceleration $= k$ (normal acceleration),

$\therefore \quad v\dfrac{dv}{ds} = \dfrac{kv^2}{\rho}$ or $v\dfrac{dv}{ds} = \dfrac{kv^2}{(ds/d\psi)}$ or $\dfrac{1}{v}dv = k d\psi$

Integrating, $\log v - \log c = k\psi$ or $v/c = e^{k\psi}$ or $v = ce^{k\psi}$ ...(1)

Again, for a circle of radius a, $s = a\psi$ or $\psi = s/a$ ...(2)

Using (2), (1) reduces to $v = c\,e^{(k/a)s}$ ...(3)

Let $v = 0$ when $s = 0$. So by (3), $u = c$ and hence (3) becomes

$v = u\,e^{(k/a)s}$ or $ds/dt = u\,e^{(k/a)s}$ or $dt = (1/u)\,e^{-(k/a)s} ds$

Integrating, $t = A + (1/u)(-a/k)\,e^{-(k/a)s}$ where A is a constant.

Initially, when $t = 0, s = 0$. So $0 = A - (a/ku)$ or $A = a/ku$

$\therefore \quad t = (a/ku) - (a/ku)\,e^{-(k/a)s} = (a/ku)\,\{(1 - e^{-(k/a)s}\}$ ...(4)

Let T be the time when the particle will return to the point from where it started (*i.e.*, $s = 0$). Then, when $t = T$, $s = 2\pi a$, *i.e.* the circumference of the cirlce. So from (4), we get $T = (a/ku)\,(1 - e^{-2\pi k})$, which is the required time.

Ex. 15. *A particle, projected with velocity u, acted on by a force which produces a constant acceleration f in the plane of the motion. Obtain the intrinsic equation of curve described, and show that the particle will be moving in opposite direction to that of the projection at time (2u/f cos α) ($e^{\pi \cot \alpha}$ − 1).*

Sol. For figure refer solved Ex. 10. Let f be the constant acceleration of the particle and α the constant angle which the direction of f makes with the tangent PT at any point $P(s, \psi)$ of its path at any time t. Resolving f along PT and normal PN at P, we have

$$f\cos\alpha = \text{tangential acceleration} = v(dv/ds) \quad ...(1)$$

and
$$f\sin\alpha = \text{normal acceleration} = v^2/\rho = v^2(d\psi/ds) \quad ...(2)$$

Dividing (1) by (2), $\cot\alpha = (1/v)(dv/d\psi)$ or $(1/v)dv = \cot\alpha\, d\psi$

Integrating, $\log v - \log c = \cot\alpha.\psi$ or $v = ce^{\psi\cot\alpha}$ ...(3)

Initially, when $v = u$, let $\psi = 0$. So by (3), $u = c$.

$\therefore$ From (3), $v = u\,e^{\psi\cot\alpha}$ ...(4)

Now, from (2), $ds = \dfrac{v^2}{f\sin\alpha}d\psi = \dfrac{u^2e^{2\psi\cot\alpha}}{f\sin\alpha}d\psi$, by (4)

Integrating, $s = (u^2/f\sin\alpha)\times(1/2\cot\alpha)\,e^{2\psi\cot\alpha} + A$ ...(5)

Let $s = 0$ when $\psi = 0$. So $A = -(u^2/2f\cos\alpha)$ and (5) gives

$$s = \frac{u^2}{2f\cos\alpha}e^{2\psi\cot\alpha} - \frac{u^2}{2f\cos\alpha} = \frac{u^2}{2f\cos\alpha}(e^{2\psi\cot\alpha} - 1), \quad ...(6)$$

which is the required intrinsic equation of the path of the particle.

Re-writing (1), we have $dv/dt = f\cos\alpha$ or $dv = (f\cos\alpha)\,dt$.

Integrating, $v = (f\cos\alpha)\,t + B$, where B is a constant. ...(7)

Initially, when $t = 0$, $v = u$, So $u = 0 + B$ or $B = u$ ...(8)

$\therefore$ From (7), $v = ft\cos\alpha + u$

From (4) and (8) $ft\cos\alpha + u = ue^{\psi\cot\alpha}$ or $t = (u/f\cos\alpha)(e^{\psi\cot\alpha} - 1)$

Note that the particle will be moving in a direction opposite to that of the projection (where $\psi = 0$), when we have $\psi = \pi$ and hence putting $\psi = \pi$ in the above relation, the required time T is given by $T = (u/f\cos\alpha)(e^{\pi\cot\alpha} - 1)$

EXERCISE

1. One point describes the diameter AB of a circle with constant velocity and another the semi circumference AB from rest with constant tangential acceleration. They start together from A and arrive together at B. Show that the velocities at B are $\pi : 1$.

2. The direction of the acceleration of a particle moving in a cycloid makes with the normal an angle equal to that which the tangent to the cycloid at the point makes with tangent at the vertex and is in the same sense. Prove that the tangent at the point turns uniformly, and the magnitude of the acceleration is constant.

MISCELLANEOUS PROBLEMS

Ex. 1. *A particle is moving in a circle of radius a with velocity v. The normal acceleration is (a) d^2s/dt^2 (b) $v(dv/ds)$ (c) dv/dt (d) v^2/a.* **[Kanpur 2001]**

Sol. Ans. (*d*) Refer page 5.22.

Ex. 2. *The coordinates of a moving point at time t are given by $x = at^2$, $y = 2at$. At time t, the magnitude of its resultant acceleration is (a) a (b) 2a (c) 3a (d) None of the above.* **[Ans. (*d*)]**

Ex. 3. *A point P describes, with constant angular velocity ω about O, the equiangular spiral*

$r = ae^{\theta}$, O being the pole of the spiral. The transverse acceleration of P is (a) 0 (b) ωr^2 (c) $2\omega^2 r$ (d) None of these. **[Ans. (c)]**

Ex. 4. *A particle moves along the curve $r = e^{\theta}$, prove that radial velocity is equal to the transverse velocity.* **(Bangalore 2004)**

Hint : Use results given on page 5.10.

5. (*i*) If v is the linear velocity, then angular velocity ω =

(*a*) vr (*b*) v/r (*c*) vp/r^2 (*d*) r/v **[M.S. Univ. T.N. 2007]**

(*ii*) Relation between angular velocity ω and linear velocity v is

(*a*) $\omega = v/r$ (*b*) $v = \omega/r$ (*c*) $\omega = r/v$ (*d*) None of these **[Agra 2006]**

(*iii*) Transverse component of acceleration is

(*a*) $\frac{d}{dt}(r^2\theta)$ (*b*) $\frac{d}{dr}(r^2\dot{\theta})$ (*c*) $\frac{1}{r}\frac{d}{dt}(r^2\dot{\theta})$ (*d*) None of these **[Agra 2006]**

(*iv*) If v be velocity of a point P moving in a plane curve and (r, θ) its coordinates referred to a fixed point O, then angular velocity at P is

(*a*) vp/r^2 (b) vp (c) v/p (d) none of these **[Garhwal 2010]**

Ans. (*i*) (*c*) (*ii*) (*d*) (*iii*) (*c*) (*iv*) (d)

6. Show that the resultant velocity of the particle at any point of the curve is along the tangent at that point **[Agra 2005]**

7. A body rotates with uniform angular acceleration α. If ω is the angular velocity when the body has turned through an angle θ from rest, show that $\omega^2 = 2\alpha\theta$. **[Agra 2003]**

[Sol. Given that $\ddot{\theta} = \alpha$ and $\dot{\theta} = \omega$ ... (1)

Now, $\ddot{\theta} = \alpha \Rightarrow \dot{\theta}(d\dot{\theta}/d\theta) = \alpha \Rightarrow 2\dot{\theta}\, d\dot{\theta} = 2\alpha\, d\theta$ ... (2)

Integrating (2), $(\dot{\theta})^2 = 2\alpha\theta + c$, c being an arbitrary constant ... (3)

Initially, when $t = 0$, let $\theta = 0$, $\dot{\theta} = 0$, (as body has turned through an angle θ from rest).

Then (3), gives $c = 0$. Hence, from (3), $(\dot{\theta})^2 = 2\alpha\theta$ or $\omega^2 = 2\alpha\theta$, using (1)

8. *For a moving particle, its coordinates at time t seconds are $x = 3t - 4t^2$, $y = 4t - 8t^2$. Find its path and velocity at t seconds.* **[Guwahati 2007]**

Sol. Given $4t^2 - 3t + x = 0$ and $8t^2 - 4t + y = 0$ ... (1)

Solving (1) for t^2 and t, we get $t^2 = (4x - 3y)/8$ and $t = (2x - y)/2$ ...(2)

Eliminating t from the equations of (2), $(4x - 3y)/8 = (2x - y)^2/4$

or $2(2x - y)^2 = 4x - 3y$ or $8x^2 - 16xy + 2y^2 - 4x + 3y = 0$,

which is the equation of path of the given moving particle.

From (1), $dx/dt = 3 - 8t$ and $dy/dt = 4 - 16t$

$\therefore$ Required velocity = $\{(dx/dt)^2 + (dy/dt)^2\}^{1/2} = (25 - 176t + 320)^{1/2}$.

9. *Prove that for a particle moving along a curve, the tangential and normal components of velicity are $(\dot{s}, 0)$ and of acceleration are $(dv/dt, v^2/\rho)$, where v is speed and ρ is the ratios of* *curvature of the curve at the point. Here* $\dot{s} = ds/dt$ **(Mumbai 2010; Delhi B.Sc. (Prog.) II 2008)**

Hint. Refer Art. 5.7 page 5.21

10. *Transversal component of acceleration is*

(a) $\frac{d}{dt}(r^2\dot{\theta})$ (b) $\frac{1}{r}\frac{d}{dt}(r^2\dot{\theta})$ (c) $\frac{d}{dt}(r^2\theta)$ (d) *None of these* **[Agra 2007]**

Sol. Ans. (b) Refer Art. 5.6

11. *Relation between angular velocity and linear velocity is*

(a) $\dot{\theta} = \frac{r^2}{vp}$ (b) $v = \frac{\dot{\theta}p}{r^2}$ (c) $\dot{\theta} = \frac{vp}{r^2}$ (d) $v = \frac{p}{\dot{\theta}r^2}$ **[Agra 2012]**

Sol. Ans. (c) Refer Art. 5.3

Ex. 12. *When* $x = a(2t + \sin 2t)$, $y = a(1 - \cos 2t)$, *then the acceleration of the particle is*

(a) *constant and equal to 4a* (b) *variable and equal to t*

(c) *constant and equal to a* (d) *none of these* **(Agra 2007)**

Sol. Ans. (*a*). Here $dx/dt = a(2 + 2\cos 2t)$ **and** $dy/dt = 2a \sin 2t$

$\Rightarrow$ $d^2x/dt^2 = -4a \sin 2t$ and $d^2y/dt^2 = 4a \cos 2t$

Hence, the required acceleration = $\{(d^2x/dt^2)^2 + (d^2y/dt)^2\}^{1/2} = (16a^2\sin^2 2t + 16a^2\cos^2 2t)^{1/2} = 4a$.

Ex. 13. *If a particle moves along a circle of radius r with centre at pole, then the radial acceleration is given by*

(a) $r^2\dot{\theta}$ (b) $\ddot{r} - r\dot{\theta}^2$ (c) $\ddot{r} + r\dot{\theta}^2$ (d) $r\dot{\theta}^2$ **(Garhwal 2003; Rohilkhand 2006)**

Sol. Ans. (*b*). Refer Art. 5.6 on page 5.10

Ex. 14. *Radial acceleration at a point is*

(a) $\ddot{r} - r\dot{\theta}^2$ (b) $\ddot{r} - r\ddot{\theta}$ (c) $\dot{r} - r\dot{\theta}$ (d) $\ddot{r} - r\dot{\theta}$

(Agra 2006, 08; Bundelkhand 2008; Garhwal 2004;)

Sol. Ans. (*a*). Refer Art. 5.6 on page 5.10

Ex. 15. *If (s, ψ) be the intrisic coordinates of the position of a particle moving along a plane curve at time t and v its velocity at time t, then the tangential component of its acceleration a time t is:*

(a) ds/dt (b) d^2s/dt^2 (c) $v(dv/dt)$ (d) dv/ds **(Bundelkhand 2007)**

Sol. Ans. (*b*). Refer Form I on page 5.21

Ex. 16. *If (s, ψ) be the intrisic coordinates of the position of a particle moving along a plane curve at time t and v be its velocity at that instant, then the normal component of its acceleration at time t is:*

(a) $v(d\psi/dt)$ (b) $v(dv/ds)$ (c) $d^2\psi/dt^2$ (d) $(d^2s/dt^2)\,(d\psi/dt)$ **(Rohilkhand 2010)**

Sol. Ans. (*a*). Refer page 5.21. Normal component of acceleration

$$= \frac{v^2}{\rho} = v \times v \times \frac{1}{\rho} = v \times \frac{ds}{dt} \times \frac{d\psi}{ds} = v\frac{d\psi}{dt}, \quad \text{as} \quad \rho = \frac{ds}{d\psi}$$

Ex. 17. *If (x, y) be the position of a particle moving in a plane curve at any time, then the magnitude of its acceleration at time t is:*

(a) $d^2x/dt^2 + d^2y/dt^2$ (b) $\{(dx/dt)^2 + (dy/dt)^2\}^{1/2}$

(c) $\{(d^2x/dt^2)^2 + (d^2y/dt^2)^2\}^{1/2}$ (d) $dx/dt + dy/dt$ **(Bundelkhand 2009)**

Sol. Ans. (c). Refer Art. 5.1, page 5.1

Ex. 18. *The magnitude of the resultant acceleration of a particle moving in a plane curve at time t is:*

(*a*) $\{(d^2s/dt^2)^2 + (v^2/\rho)^2\}^{1/2}$ (*b*) $(d^2s/dt^2) + (v^2/\rho)$

(*c*) $(d^2s/dt^2)^2 + (v^2/\rho)^2$ (*d*) $\{(ds/dt)^2 + (v^2/\rho)^2\}^{1/2}$ **(Bundelkhand 2008)**

Sol. Ans. (*a*). Refer Art. 5.21.

Ex. 19. *The angular velocity of a particle moving in a parabolic path about the focus of its path varies as follows as its distance from the focus:*

(*a*) *Inversely proportional* (*b*) *Directly proportional*

(*c*) *inversely proportional to the square*

(*d*) *Directly proportional to the square* **(Garhwal 2002; Rohilkand 2011)**

Sol. Ans. (*a*). Refer solved Ex. 3 on page 5.5.

Ex. 20. *If R.A. denotes radial acceleration, then*

(*a*) $R.A. = \dot{r}$ (*b*) $R.A. = r\dot{\theta}$ (*c*) $R.A. = \ddot{r} - r\dot{\theta}^2$ (*d*) $R.A. = 2\dot{r}\dot{\theta} - r\ddot{\theta}^2$ **(Agra 2011)**

Sol. Ans. (*c*) Refer Art. 5.6 on page 5.10.

Ex. 21. *Angula velocity of a particle is given as*

(*a*) $\dot{\theta} = vp$ (*b*) $\dot{\theta} = (vp)r$ (*c*) $\dot{\theta} = (vp)/r^2$ (*d*) $\dot{\theta} = (vp^2)/r$ **(Agra 2011)**

Sol. Ans. (*c*). Refer Art. 5.3, page 5.4.

Ex. 22. *What is the value of tangential acceleration?*

(*a*) dt/dv (*b*) v^2/t (*c*) dv/dt (*d*) $1/v^2$ **(Agra 2012)**

Sol. Ans. (*c*). Refer Art. 5.7, page 5.21

6

Projectiles

6.1 Motion of a particle in a vertical plane. If a ball is thrown from a point, into the air, inclined at an acute angle to the horizontal, then it describes a curved path and comes back to the ground. A body so projected is called the *projectile,* path described by the particle is called its *trajectory. Velocity of projection* is the velocity with which the particle is projected. *Angle of projection* is the angle which the direction of projection makes with the horizontal. *Range* is the distance measured between the point of projection and the point where the projectile hits a given plane through the point of projection. If the given plane is horizontal, the range is called *horizontal* range. *Time of flight* is the time taken to complete a particular range, *i.e.,* the time between the instant of projection and the instant when it hits a fixed plane through the point of projection.

In order to get the trajectory in simplified form, we assume that (*i*) the body projected is so small that it can be treated as particle (*ii*) the gravitational force on the particle remains constant during its time of flight. (*iii*) the projectile moves in vacuum, *i.e.,* air resistance is neglected.

6.2 The motion of a projectile and its trajectory. *A particle of mass m is projected from a fixed point into vacuum with velocity u, in a direction making an angle* $\alpha (\neq \pi/2)$ *with the horizontal. To find its motion and to show the its path is a parabola.* **[Pune 2010; M.S. Univ. T.N. 2007, Bhopal 2004, Bangalore 2004; Kerala 2001]**

Let a particle of mass m be projected from a point O with velocity u in a direction making an angle α with the horizontal. Let O be taken as origin and the horizontal and vertical lines through O in the plane of motion of the projectile be taken as the axes of x and y respectively. Let $P(x, y)$ be the position of the projectile at any time t.

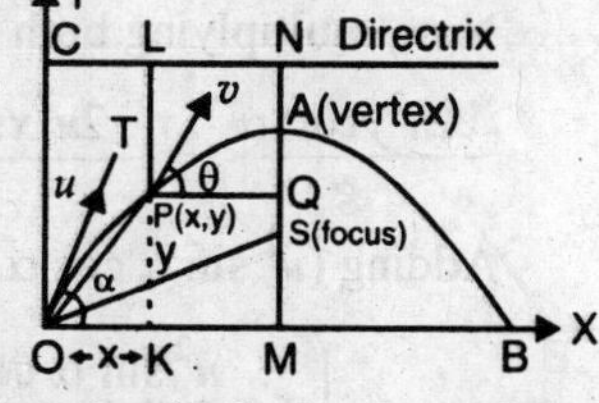

Resolved parts of acceleration of the projectile at P are $\ddot{x}$ and $\ddot{y}$ along OX and OY respectively. Since the only force acting on the particle at P is its weight mg, acting vertically downwards, hence the equation of motion in the horizontal and vertical directions respectively are

$$m\ddot{x} = O \qquad \text{and} \qquad m\ddot{y} = -mg$$

i.e.,
$$\ddot{x} = d^2x/dt^2 = 0 \qquad ...(1)$$
and
$$\ddot{y} = d^2y/dt^2 = -g \qquad ...(2)$$

Integrating (1), $\dot{x} = dx/dt = A$, where A is constant of integration. ...(3)

But initially, at O, when $t = 0$, $\dot{x}$ = horizontal component of velocity = $u \cos \alpha$. Hence (3) gives $A = u \cos \alpha$ and so (3) becomes

$$\dot{x} = dx/dt = u \cos \alpha \qquad ...(4)$$

Integrating (2) w.r.t. 't', $\quad \dot{y} = dy/dt = -gt + B$ ...(5)

As before, at O, $t = 0$, $\dot{y}$ = vertical components of velocity = $u \sin \alpha$. Hence (5) gives $B = u \sin \alpha$ and so (5) becomes

$$\dot{y} = dy/dt = u \sin \alpha - gt \qquad ...(6)$$

From (4), we find that the horizontal component of velocity will remain constant and equal to u cos a throughout the motion.

Equation (4) and (6) give the components of velocity in the horizontal and vertical directions at any time t.

Integrating (4) and (6) w.r.t. 't', we get

$x = u \cos \alpha . t + C$ and $y = u \sin \alpha . t - (1/2) g t^2 + D$ ...(7)

Initially, at O, $t = 0$, $x = 0$ and $y = 0$. So (7) gives $C = 0$, $D = 0$.

Thus $x = u \cos \alpha . t$ ...(8)

and $y = u \sin \alpha . t - (1/2) g t^2$ ...(9)

Equation (8) and (9) give the position of the particle at any time t, i.e., the coordinates (x, y) of P in terms of t.

The equation of path described by the projectile is obtained by eliminating the variable t between (8) and (9). From (8), $t = x/(u \cos \alpha)$

Substituting this value of t in (9), we get

$$y = u \sin \alpha \left(\frac{x}{u \cos \alpha} \right) - \frac{g}{2} \left(\frac{x}{u \cos \alpha} \right)^2 \quad \text{or} \quad y = x \tan \alpha - \frac{g x^2}{2u^2 \cos^2 x} \qquad \text{...(10)}$$

which is the required equation of the trajectory. The equation (10) is a second degree in x and y in which the second degree terms are in perfect square and hence it represents a parabola.

Remark. Equation of trajectory in a particular case. Suppose a particle is projected from O horizontally with velocity u. Take O as origin and horizontal and downward vertical lines OX and OY as coordinate axes. If $P(x, y)$ be the position of the particle at any time t. Then considering motion from O to P in horizontal and vertical directions, we get $x = ut$

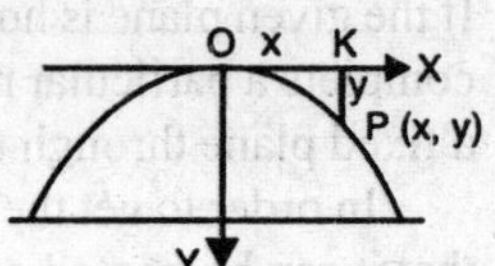

and $y = (1/2) g t^2$ giving $y = (g x^2)/2u^2$

which is the equation of the trajectory **Commit this result to memory.**

6.3 To find the latus rectum, the vertex, the focus and the height of the directrix of the trajectory.

[Bangalore 2004; Bhopal 2004; G.N.D.U. Amritsar 2010]]

Refer Art. 6.2 and its figure. The equation of the trajectory is

$$y = x \tan x - (g/2u^2 \cos^2 \alpha) x^2 \qquad \text{...(1)}$$

Now, multiplying both sides by $-(2u^2 \cos^2 \alpha/g)$, we get

$$-\frac{2u^2 y \cos^2 \alpha}{g} = -\frac{2u^2 x \sin \alpha \cos \alpha}{g} + x^2 \quad \text{or} \quad x^2 - \frac{2u^2 x \sin \alpha \cos \alpha}{g} = -\frac{2u^2 y \cos^2 \alpha}{g}$$

Adding $(u^2 \sin \alpha \cos \alpha/g)^2$ on both sides, we obtain

$$\left(x - \frac{u^2 \sin \alpha \cos \alpha}{g} \right)^2 = -\frac{2u^2 \cos^2 \alpha}{g} \left(y - \frac{u^2 \sin^2 \alpha}{2g} \right), \qquad \text{...(2)}$$

which is a parabola of the form $(x - h)^2 = -4a(y - k)$

$$\text{Its vertex } A \text{ is } \left(\frac{u^2 \sin \alpha \cos \alpha}{g}, \frac{u^2 \sin^2 \alpha}{2g} \right), \text{ i.e. } \left(\frac{u^2 \sin 2\alpha}{2g}, \frac{u^2 \sin^2 \alpha}{2g} \right) \qquad \text{...(3)}$$

$$\text{and latus rectum} = \frac{2u^2 \cos^2 \alpha}{g} = \frac{2}{g} (\text{Horizontal component of velocity})^2 \qquad \text{...(4)}$$

To find the height of the directrix LN

Height of the directrix above $OX = MN = MA + AN$

= Ordinate of A + (1/4) (latus rectum)

$= (u^2 \sin^2 \alpha)/2g + (1/4) \times \{(2u^2 \cos^2 \alpha)/g\} = u^2/2g$ ...(5)

Hence equation of directrix is $y = u^2/2g$ ...(6)

Remark. From (6), it follows that all the trajectories which have the same point of projection and the same initial velocity have a common directrix.

To find the coordinates of focus S

Abscissa of focus S $= OM =$ abscissa of vertex $A = (u^2 \sin \alpha \cos \alpha)/g$

Ordinate of S $= MS = MA - SA = MA -$ (latus rectum)/4

$=$ Ordinate of vertex $A - (1/4) \times \{(2u^2 \cos^2 \alpha)/g\}$

$= (u^2 \sin^2 \alpha)/2g - (u^2 \cos^2 \alpha)/2g = -(u^2/2g)(\cos^2 \alpha - \sin^2 \alpha) = -(u^2/2g) \cos 2\alpha$

Hence the coordinates of focus *S* are

$$\left(\frac{u^2 \sin \alpha \cos \alpha}{g}, -\frac{u^2 \cos 2\alpha}{2g}\right), \text{ i.e., } \left(\frac{u^2 \sin 2\alpha}{2g}, \frac{u^2 \cos 2\alpha}{2g}\right) \qquad ...(7)$$

We now dicuss the position of *S*. From figure, we see that *S* lies above, on or blow the *x*-axis according as the ordinate of $S >, =, \text{ or } < O$

i.e., according as $-(u^2 \cos 2\alpha)/2g >, =, \text{ or } < 0$

i.e., according as $\cos 2\alpha <, =, \text{ or } > 0$

i.e., according as $2\alpha >, =, < \pi/2$

i.e., according as $\alpha >, =, < \pi/4$

6.4 A particle is projected with velocity u making an angle α with horizontal

To find (*i*) Time of flight (*ii*) the horizontal range (*iii*) the maximum horizontal range (*iv*) the direction of projection for a given horizontal range (v) the greatest height attained.

[Guwahati 2007; M.S. Univ. T.N. 2007; Bangalore 1997; 2004]

(*i*) Refer Art. 6.2 and its figure. We know that the height of the particle at *P* at any time *t* is given by

$$y = u \sin \alpha . t - (1/2) g t^2 \qquad ...(1)$$

Now time of flight, *T* (say), is the time which the particle takes in reaching the horizontal plane through the point of projection. Putting $y = 0$ in (1),

$0 = u \sin \alpha . t - (1/2) g t^2$ so that $t = 0$ or $t = (2u \sin \alpha)/g$

But $t = 0$ corresponds to the time when the particle is at the point of projection *O*. Hence, the time of flight *T* is given by

$$T = (2u \sin \alpha)/g = (2/g) \times (\text{vertical comp. of velocity}) \qquad ...(2)$$

(*ii*) Let $R(= OB)$ be the horizontal range. Then *R* is the horizontal distance described by the particle in time *T* (*i.e.*, time of flight) with a constant horizontal velocity $u \cos \alpha$.

$$\therefore \quad R = (u \cos \alpha) . \frac{2u \sin \alpha}{g} = \frac{2u^2 \sin \alpha \cos \alpha}{g} = \frac{u^2 \sin 2\alpha}{g} \qquad ...(3)$$

The above formula for range can be re-written as

Range = (2/g) (Horizontal comp. of vel.) (vertical comp. of vel.)

(*iii*) The horizontal range *R* is maximum when $\sin 2\alpha$ is maximum *i.e.*, when $\sin 2\alpha = 1$ or $2\alpha = \pi/2$ or $\alpha = \pi/4$.

Hence the horizontal range is maximum when the angle of projection is π/4 and then maximum horizontal range = u^2/g.

(*iv*) We have, $R = \dfrac{u^2 \sin 2\alpha}{g} = \dfrac{u^2 \sin(\pi - 2\alpha)}{g} = \dfrac{u^2 \sin\{2(\pi/2 - \alpha)\}}{g}$,

showing that the horizontal range *R* remains the same when the angle of projection is α or $\pi/2 - \alpha$. Hence to get a given horizontal range for a given velocity of projection, there are two possible directions of projection. The inclination, say α and β of these two directions of projection to the horizontal are complementary angles. Thus,

$$\alpha + \beta = \pi/2 = \pi/4 + \pi/4 \quad \text{so that} \quad \beta - \pi/4 = \pi/4 - \alpha,$$

showing that the possible directions of projection for a given range are equally inclined to the direction of projection for the maximum range. **[(M.S. Univ. T.N. 2007]**

(*v*) From figure, the greatest height attained

$$= MA = \text{ordinate of } P = (u^2 \sin^2 \alpha)/2g = (1/2g) \times (\text{vertical comp. of vel.})^2$$

6.5 Velocity and direction of motion of a projectile after a given time *t*

Refer figure of Art. 6.2. Let *u* be the velocity of projection and α the angle of projection. Let particle be at $P(x, y)$ after time *t*. Let *v* be velocity at *P* and let θ be the angle which the direction of velocity at *P* makes with horizontal. Then, we have

$$v \cos \theta = u \cos \alpha \qquad ...(1)$$

and

$$v \sin \theta = u \sin \alpha - gt \qquad ...(2)$$

Squaring and adding (1) and (2), we have

$$v^2 = u^2 \cos^2 \alpha + (u \sin \alpha - gt)^2 = u^2 - 2u\,gt \sin \alpha + g^2t^2$$

Then $\quad v = (u^2 - 2u\,gt \sin \alpha + g^2t^2)^{1/2},$

which gives magnitude of velocity at any time t.

Dividing (2) by (1), $\quad \tan \theta = \dfrac{u \sin \alpha - gt}{u \cos \alpha}$ or $\theta = \tan^{-1} \dfrac{u \sin \alpha - gt}{u \cos \alpha},$

which gives the direction of motion after any time t.

Velocity and direction of motion of a projectile when it is at height y above the horizontal passing through the point of projection.

Refer figure of Art. 6.2. We have

$$v \cos \theta = u \cos \alpha \quad \ldots(3)$$

and $\quad v^2 \sin^2 \theta = u^2 \sin^2 \alpha - 2gy \quad \ldots(4)$

Squaring (3), $\quad v^2 \cos^2 \theta = u^2 \cos^2 \alpha \quad \ldots(5)$

Adding (4) and (5), $\quad v^2 = u^2 - 2gy \quad$ or $\quad v = (u^2 - 2gy)^{1/2},$

which gives magnitude of velocity at any height y above OM.

From (4), $\quad v \sin \theta = (u^2 \sin^2 \alpha - 2gy)^{1/2} \quad \ldots(6)$

Dividing (6) by (3), $\quad \tan \theta = (u^2 \sin^2 \alpha - 2gy)^{1/2}/(u \cos \alpha) \quad \ldots(7)$

which gives the direction of motion after reaching a height y.

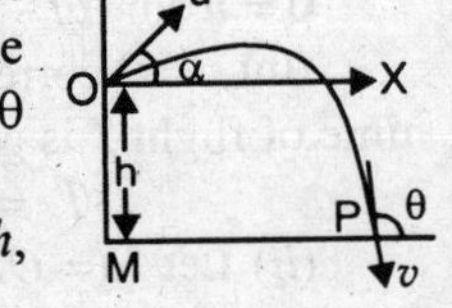

Remark 1. Suppose a particle is projected from the top of a hill at height h. Let u be the velocity of projection and α the angle of projection. Let v be the velocity at P after falling a depth h and θ be its direction. Then, as before $v \cos \theta = u \cos \alpha$ and $v^2 \sin^2 \theta = u^2 \sin^2 \alpha + 2gh$

These give $v^2 \cos^2 \theta + v^2 \sin^2 \theta = u^2 \cos^2 \alpha + u^2 \sin^2 \alpha + 2gh$ or $v^2 = u^2 + 2gh,$ which gives the magnitude of velocity at a depth h.

Remark 2. All results of this article should be committed to memory.

6.6 Velocity at any point of the trajectory. *To show that the velocity of a projectile at any point of its path is that due to a fall from the directrix to that point.*

Refer Art. 6.2 and its figure. Take the point of projection O as origin, the horizontal and vertical lines OX and OY in the plane of projection as the x-axis and y-axis. Let $P(x, y)$ be the position of the particle at any time t. Let, v the velocity at P, make an angle θ with a horizontal line PQ as shown in the figure.

Then $\quad v \cos \theta = u \cos \alpha \quad$ and $\quad v \sin \theta = u \sin \alpha - gt \quad \ldots(1)$

and $\quad y = u \sin \theta . t - (1/2)gt^2 \quad \ldots(2)$

Squaring and adding the relations of (1), we have

$$\begin{aligned} v^2 &= u^2 \cos^2 \alpha + (u \sin \alpha - gt)^2 = u^2 - 2u\,gt \sin \alpha + g^2t^2 \\ &= u^2 - 2g\{u \sin \alpha . t - (1/2)gt^2\} = u^2 - 2gy, \text{ by (2)} \\ &= 2g(u^2/2g - y) = 2g\,(\text{Height of directrix} - \text{Ordinate of } P) \end{aligned}$$

$\therefore \; v^2 = 2g(LK - PK) = 2g\,LP = 2g \times$ depth of P below directrix $\quad \ldots(3)$

Let V be the velocity of the particle at P which is falling freely from L. Then,

$$V^2 = 0^2 + 2g.LP = v^2, \text{ using (3)}$$

Thus $v = V$, showing that the velocity v at any point P of the path is the same as would be acquired in falling from a vertical distance from the directrix to that point.

6.7 Some geometrical properties of parabola for direct applications

(*i*) The distance of any point on the parabola from its focus is always equal to its distance from the directrix.

(*ii*) The tangents at the extremities of any focal chord of a parabola intersect at right angles on the directrix.

(*iii*) The tangent at any point bisects the angle between the perpendicular from the point on the directrix and the focal distance of the point.

(*iv*) The line joining the point of intersection of the tangents at the extremities of any chord of a parabola to the middle point of the chord is parallel to the axis of the parabola.

TYPE 1 : SOLVED EXAMPLES BASED ON ART. 6.1 TO 6.7

Ex. 1. (*a*) *A particle is projected with a velocity u so that the range on a horizontal plane is twice the greatest height attained. Find the range.*

Sol. Let α be the angle of projection. Given that,

Range on horizontal plane $= 2 \times$ greatest height attained

$\therefore \quad (2u^2 \sin\alpha \cos\alpha)/g = 2 \times \{(u^2 \sin^2\alpha)/g\}$ so that $\tan\alpha = 2$

Hence the required horizontal range

$$= \frac{u^2 \sin 2\alpha}{g} = \frac{u^2}{g} \cdot \frac{2\tan\alpha}{1+\tan^2\alpha} = \frac{u^2}{g} \cdot \frac{2\times 2}{1+4} = \frac{4u^2}{5g}$$

Ex. 1. (*b*) *A body is projected at such an angle that the horizontal range is four times the greatest height. Find angle of projection.* **[I.A.S. (Prel.) 2000]**

Sol. Proceed as in part (*a*). **Ans.** $\alpha = 45°$

Ex. 2. (*a*) *If the maximum horizontal range for a particle is R, show that the greatest heigh attained is R/4.* **[Delhi B.Sc. II (Prog) 2009]**

Sol. Let u be the velocity of projection. Then, given that

$$R = \text{maximum horizontal range} = u^2/g \qquad \ldots(1)$$

Also, we know that for maximum range, $\alpha = \pi/4$. $\quad\ldots(2)$

Hence the corresponding greatest height attained

$$= \frac{u^2 \sin^2\alpha}{2g} = \frac{R\sin^2(\pi/4)}{2} = \frac{R\times(1/2)}{2} = \frac{R}{4}, \text{ by (1) and (2)}$$

Ex. 2. (*b*) *A man throws a ball h meters vertically upwards. Show that the greatest horizontal distance he can throw it is 2h.* **[Delhi B.Sc. (II) Prog. 2009]**

Sol. Let u be the velocity of projection. Then $h = u^2/2g$ or $u^2 = 2gh$.

Hence the greatest horizontal distance $= R = u^2/g = (2gh)/g = 2h$.

Ex. 3. *If t be the time in which a particle reaches a point P in its path and t′ the time from P till it reaches the horizontal plane through the point of projection, show that height of P above the horizontal plane is (1/2)gt t′ and maximum height attained is (g/8) (t + t′)* **[I.A.S. 1995]**

Sol. Let B be the point of projection, u the velocity of projection and α the angle of projection. Let OC be the horizontal range. Suppose time taken by the particle in moving from B to P is t and the time from P to C is t'. Then $t + t'$ is the time of flight and so

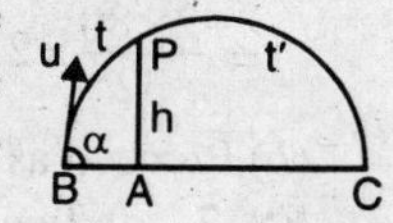

$t + t' = (2u\sin\alpha)/g$ so that $u\sin\alpha = (g/2)(t+t')$ $\quad\ldots(1)$

Now, $\quad h = u\sin\alpha . t - (1/2)gt^2 = (g/2)(t+t')t - (1/2)gt^2$, using (1)

Thus, height of P above the horizontal plane $= PA = h = (1/2)gtt'$.

Maximum height $= (u\sin\alpha)^2/2g = (1/2g) \times (g/2)^2 \times (t+t')^2$, by (1)

$= (g/8)(t+t')^2$

Ex. 4. *Obtain the equation of the path of a projectile and show that it may be written in the form $yR/(xR - x^2) = \tan\alpha$, where R is the horizontal range and α the angle of projection.*

Sol. Refer Art. 6.2. Equation of trajectory is

$$y = x\tan\alpha - (g/2u^2\cos^2\alpha)x^2 \qquad \ldots(1)$$

But $\quad R = (2u^2 \sin\alpha\cos\alpha)/g$ so that $u^2 = gR/(2\sin\alpha\cos\alpha)$ $\quad\ldots(2)$

Substituting this value of u^2 in (1), we get

$$y = x\tan\alpha - \frac{gx^2}{2\cos^2\alpha} \times \frac{2\sin\alpha\cos\alpha}{Rg} = x\tan\alpha - \frac{x^2\tan\alpha}{R} = \frac{\tan\alpha}{R}(Rx - x^2)$$

or $\quad yR/(xR - x^2) = \tan\alpha$, which is the required path

Ex. 5. (*a*) *A particle is projected so as to have a range R on the horizontal plane through the point of projection. If α_1, α_2 are the possible angles of projection and t_1, t_2 the corresponding times of flight, show that $(t_1^2 - t_2^2)/(t_1^2 + t_2^2) = \sin(\alpha_1 - \alpha_2)/\sin(\alpha_1 + \alpha_2)$.* **[I.F.S. 1995]**

Sol. Let u be the velocity of projection. For a range R, the two possible angles of projections are α_1 and α_2, when $\alpha_1 + \alpha_2 = \pi/2$.

Now, $t_1 = (2u \sin \alpha_1)/g$ and $t_2 = (2u \sin \alpha_2)/g$

Dividing, $t_1/t_2 = (\sin \alpha_1)/(\sin \alpha_2)$ or $t_1^2/t_2^2 = (\sin^2 \alpha_1)/\sin^2 \alpha_2)$

Applying componendo and dividendo, we have

$$\frac{t_1^2 - t_2^2}{t_1^2 + t_2^2} = \frac{\sin^2 \alpha_1 - \sin^2 \alpha_2}{\sin^2 \alpha_1 + \sin^2 \alpha_2} = \frac{\sin(\alpha_1+\alpha_2)\sin(\alpha_1-\alpha_2)}{\sin^2 \alpha + \sin^2(\pi/2 - \alpha_1)}, \text{ as } \alpha_1 + \alpha_2 = \frac{\pi}{2}$$

$$= \frac{\sin(\pi/2)\sin(\alpha_1-\alpha_2)}{\sin^2 \alpha_1 + \cos^2 \alpha_1} = \frac{\sin(\alpha_1-\alpha_2)}{\sin(\pi/2)} = \frac{\sin(\alpha_1-\alpha_2)}{\sin(\alpha_1+\alpha_2)}$$

Ex. 5. (*b*) *If the maximum horizontal range of a given gun is R, find the firing angle which should be used to hit a target located at a distance R/2 on the same level.* **[Delhi B.A. (Prog.) III, 2013]**

Sol. Try yourself. Ans. 15° and 75°.

Ex. 6. (*a*) *A shell bursts on contact with the ground and pieces from it fly in all directions with all velocities upto 80 units. Show that a man 100 units away is in danger for a time* $5\sqrt{2}/2$ *units if g = 32 units.* **[I.A.S. 1997]**

(*b*) *A shell explodes on contact with the ground and its fragments are observed to fly in all directions with speed 49 m/sec. Find the difference between the two times at which fragments may fall at a point 24g meters away.*

Sol. (*a*) Let α be the angle of projection to get the range = 100 units

Then, $100 = \dfrac{u^2 \sin 2\alpha}{g}$ or $100 = \dfrac{(80)^2 \sin 2\alpha}{32}$ or $\sin 2\alpha = \dfrac{1}{2}$

Hence $2\alpha = 30°$ or $150°$ so that $\alpha = 15°$ or $75°$

Thus for the same range 100 units, there will be two directions of projection, namely, 15° and 75°. Let T_1 and T_2 be the times taken by two pieces to reach the man. Then $T_1 = (2u \sin 15°)/g$ and $T_2 = (2u \sin 75°)/g$

The required time during which the man will remain in danger

$= T_2 - T_1 = (2u/g)(\sin 75° - \sin 15°) = (2u/g)(2 \sin 45° \sin 30°)$

$$= \frac{2\times 80}{32} \times 2 \times \frac{1}{\sqrt{2}} \times \frac{1}{2} = \frac{5}{\sqrt{2}} = \frac{5\sqrt{2}}{2} \text{ units}$$

(*b*) Proceed as in part (*a*). **Ans.** 2 seconds.

Ex. 7. (a) *Two bodies are projected from the same point with the same velocity but in different directions. If the range in each case be R and the times of flight be t and t′, prove that R = (g/2)tt′.*

(b) *If h and h′ be the greatest heights in the two paths of a particle with a given velocity for a given range R, prove that* $R = 4\sqrt{hh'}$.

Sol. (*a*) Let u be the velocity of projection of both the bodies. Then for getting the same range R, the angles of projections of the two bodies are α and $90° - \alpha$. Hence, we have

(*a*) $t = (2u \sin \alpha)/g$ and $t' = \{2u \sin(\pi/2 - \alpha)\}/g = (2u \cos \alpha)/g$

Now, $$\frac{1}{2}gtt' = \frac{1}{2}g \times \frac{2u \sin \alpha}{g} \times \frac{2u \cos \alpha}{g} = \frac{2u^2 \sin \alpha \cos \alpha}{g} = R$$

(*b*) $h = (u^2 \sin^2 \alpha)/(2g$ and $h' = \{u^2 \sin^2(\pi/2 - \alpha)\}/2g = (u^2 \cos^2 \alpha)/2g$

Now, $$hh' = \frac{u^2 \sin^2 \alpha}{2g} \times \frac{u^2 \cos^2 \alpha}{2g} = \frac{1}{16}\left(\frac{2u^2 \sin \alpha \cos \alpha}{g}\right)^2 = \frac{R^2}{16} \Rightarrow R = 4\sqrt{hh'}$$

Ex. 8. *If the velocity at the maximum height of a projectile is half of its initial velocity u, then its horizontal range will be*

(a) $3u^2/2g$ *(b)* $2g/3u^2$ *(c)* $u^2\sqrt{3}/2g$ *(d)* $g/u^2\sqrt{3}$. **[I.A.S. (Prel.) 2000]**

Sol. Ans. (*c*) Since the vertical component of velocity of a projectile vanishes at the maximum height, hence velocity of particle at maximum height = horizontal component of velocity = $u \cos \alpha$; where α in the angle of projection. By given condition of the problem, we have

$u \cos \alpha = (1/2)u$ so that $\cos \alpha = 1/2$ or $\alpha = 60°$

$\therefore$ Horizontal range $= \dfrac{2u^2 \sin \alpha \cos \alpha}{g} = \dfrac{2u^2 \sin 60° \cos 60°}{g} = \dfrac{u^2\sqrt{3}}{2g}$

Ex. 9. (*a*) *From a point on the ground at a distance x from the foot of a vertical wall, a ball is thrown at an angle of 45° which just clears the top of the wall and afterwards strikes the ground at a distance y on the other side. Find the height of the wall.*

(*b*) *A particle is projected from a point on the ground so as to pass over a vertical wall of height h at a horizontal distance 'a' from the point of projection. Show that it strikes the ground at a distance* $(2U^2h)/ga$ *beyond the wall, U being the horizontal component of velocity.*

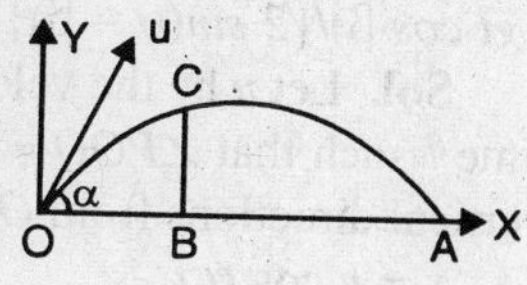

Sol. (*a*) Let the point of projection O be taken as the origin and the horizontal and vertical lines through O as axes of x and y. Let u be the velocity of projection. Here the angle of projection $= \alpha = 45°$. The equation of trajectory OCA is given by

$$y = x \tan 45° - \frac{gx^2}{2u^2 \cos^2 45°} \quad \text{or} \quad y = x - \frac{gx^2}{u^2} \qquad \ldots(1)$$

Let CB be wall and B be its foot. Given that $OB = x$ and $BA = y$. Let $BC = h$. Then, clearly $C(x, h)$ and $A(x + y, 0)$ both lie on (1)

$\therefore$ $h = x - (gx^2)/u^2$ ...(2)

and $0 = (x + y) - \{g(x + y)^2/u\}$ so that $u^2 = g(x + y)$ ...(3)

From (2) and (3), $h = x - \dfrac{gx^2}{g(x+y)} = x - \dfrac{x^2}{x+y} = \dfrac{xy}{x+y}$

(*b*) Try yourself.

Ex. 10. *A particle is projected in a direction making on angle θ with the horizon. If it passes through the points* (x_1, y_1) *and* (x_2, y_2) *referred to horizontal and vertical axes through the point of projection, then prove that* $\tan \theta = (x_2^2 y_1 - x_1^2 y_2)/x_1x_2(x_2 - x_1)$ *and the constant horizontal component of velocity is* $\{gx_1x_2(x_2 - x_1)/2(x_2y_1 - x_1y_2)\}^{1/2}$.

Sol. Let u be the velocity of projection. Then equation of trajectory is

$$y = x \tan \theta - (gx^2)/2u^2 \cos^2 \theta \qquad \ldots(1)$$

Since (1) passes through (x_1, y_1) and (x_2, y_2), we have

$$y_1 = x_1 \tan \theta - (gx_1^2)/2u^2 \cos^2 \theta \qquad \ldots(2)$$

and

$$y_2 = x_2 \tan \theta - (gx_2^2)/2u^2 \cos^2 \theta \qquad \ldots(3)$$

Multiplying (2) by x_2^2 and (3) by x_1^2 and then subtracting, we get

$y_1x_2^2 - y_2x_1^2 = (x_1x_2^2 - x_2x_1^2)\tan \theta$ or $\tan \theta = (x_2^2 y_1 - x_1^2 y_2)/x_1x_2(x_2 - x_1)$

Next, multiplying (2) by x_2 and (3) by x_1 and then subtracting, we get

$$y_1x_2 - y_2x_1 = \frac{g(x_2^2x_1 - x_1^2x_2)}{2u^2 \cos^2 \theta} \quad \text{or} \quad u \cos \theta = \left\{\frac{gx_1x_2(x_2 - x_1)}{2(x_2y_1 - x_1y_2)}\right\}^{1/2}$$

Ex. 11(*a*). *A particle aimed at a mark which is in a horizontal plane through the point of projection, falls 'a' metres short of it when the elevation is α and goes 'b' meters too far when the elevation is β. Show that, if the velocity of projection be the same in all cases, the proper elevation is* $(1/2) \sin^{-1}$ $\{(a \sin 2\beta + b \sin 2\alpha)/(a + b)\}$ **[Bangalore 1996]**

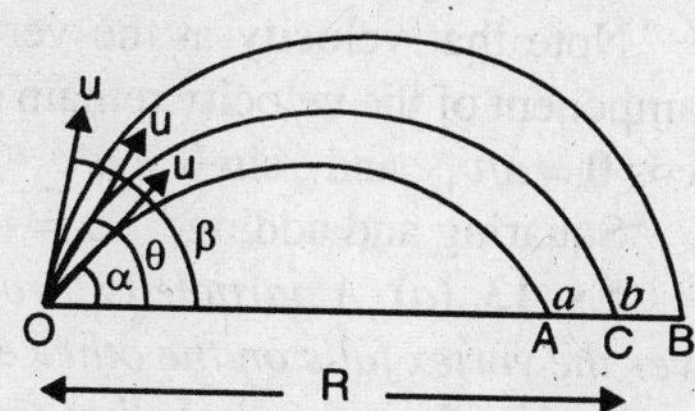

Sol. Let O be the point of projection of the projectile which is aimed at C such that $OC = R$ (say). Let θ be the required elevation and u, the velocity of projection which is the same in all cases.

When the angle of elevation is α, the particle falls at A and when the angle of elevation is β, it falls at B. Then, $OA = R - a$ and $OB = R + b$. Hence, we have

$R - a = (u^2 \sin 2\theta)/2g$ and $R - b = (u^2 \sin 2\beta)/2g$

Multiplying the above equations by b and a respectively and adding

$$R(a + b) = (u^2/g)(b \sin 2\alpha + a \sin 2\beta) \quad \text{...(1)}$$

Also, $R = (u^2 \sin 2\theta)/g$ so that $u^2/g = R/\sin 2\theta$ ...(2)

From (1) and (2), $R(a + b) = R(b \sin 2\alpha + a \sin 2\beta)/\sin 2\theta$

or $$\sin 2\theta = \frac{a \sin 2\beta + b \sin 2\alpha}{a+b} \quad \text{or} \quad \theta = \frac{1}{2}\sin^{-1}\frac{a \sin 2\beta + b \sin 2\alpha}{a+b}$$

Ex. 11(b). *A particle is projected from O at an elevation α and after t seconds to have an elevation β as seen from the point of projection. Prove that the initial velocity was (gt cos β)/{2 sin(α – β)}.*

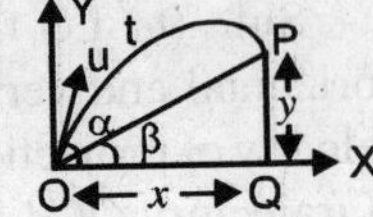

Sol. Let u be the velocity of the projection. Let the particle be at $P(x, y)$ after time t such that $\angle POQ = \beta$. Considering motion of the particle in horizontal and vertical directions from O to P, we have

$x = u \cos \alpha . t$ and $y = u \sin \alpha . t - (1/2)gt^2$

Now, $$\frac{\sin \beta}{\cos \beta} = \tan \beta = \frac{y}{x} = \frac{t\{u \sin \alpha - (1/2)gt\}}{u \cos \alpha \cdot t} = \frac{2u \sin \alpha - gt}{2u \cos \alpha}$$

or $$\cos \beta(2u \sin \alpha - gt) = 2u \cos \alpha \sin \beta$$

or $$2u(\sin \alpha \cos \beta - \cos \alpha \sin \beta) = gt \cos \beta$$

or $2u \sin(\alpha - \beta) = gt \cos \beta$ or $u = (gt \cos \beta)/2 \sin(\alpha - \beta)$

Ex. 12. (*a*) *If at any instant the velocity of a particle be u, and its direction of motion makes an angle θ to the horizontal, then show that it will be moving at right angles to the direction after time (u/g) cosec θ.*

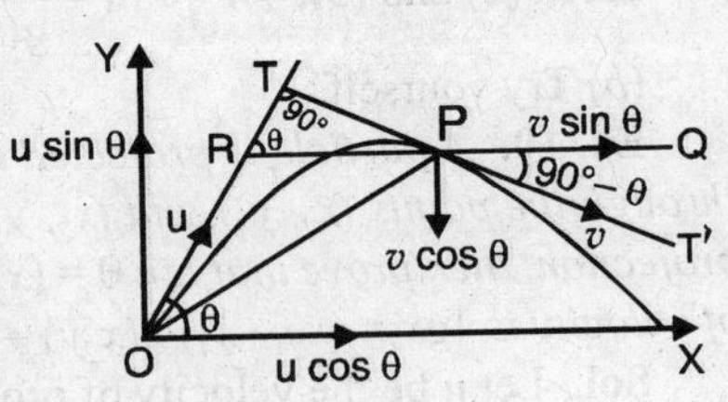

Sol. Let at O the velocity of the particle be u, making an angle θ with the horizontal. Let the particle be at P after time t such that the direction of velocity v at P along PT' is at right angle to the direction of velocity at O of the same particle. Let PQ be horizontal line through P. Let QP produced meet OT at R. Then, $\angle TRP = \theta$, $\angle PTR = 90°$ and so $\angle TPR = 90° - \theta$ and hence $\angle QPT' = \angle TPR = 90° - \theta$.

Since horizontal component of velocity remains constant throughout the motion,

$u \cos \theta = v \cos (90° - \theta)$ so that $v = (u \cos \theta)/\sin \theta$ ...(1)

Consider vertical motion from O to P, we have

$-v \cos \theta = u \sin \theta - gt$ or $gt = u \sin \theta + v \cos \theta$

or $gt = u \sin \theta + \{(u \cos \theta)/\sin \theta\}\cos \theta = (u/\sin \theta)(\sin^2 \theta + \cos^2 \theta)$

so that $gt = u/\sin \theta$ or $t = (u/g) \operatorname{cosec} \theta$

Ex. 12. (*b*) *v_1 and v_2 be the velocities at the ends of a focal chord of a projectile's path and u the velocity at the vertex of the path, then show that*

$$1/v_1^2 + 1/v_2^2 = 1/u^2.$$ **[Delhi B.A. (Prog) III, 2011, 2015, 2017]**

Sol. Draw a figure as in Ex. 12. (*a*). For the present problem, here $u = v_1$, $v = v_2$. Let OP be the given focal chord. Let the direction OT of motion of the projectile at O make an angle θ with the horizontal. Since the tangents at the ends of a focal chord of trajectory (which is a parabola) intersect each other at right angles, hence $\angle OTT' = 90°$ as shown in the figure. Here PT' is direction of motion at P. From figure $\angle TRP = \theta$, $\angle RTP = 90°$ and hence $\angle QPT' = \angle TPR = 90° - \theta$.

Note that velocity at the vertex of the path is in horizontal direction only. Since horizontal component of the velocity remain constant throughout the motion, $v_1 \cos \theta = v_2 \cos(90° - \theta) = u$ so $\cos \theta = u/v_1$ and $\sin \theta = u/v_2$

Squaring and adding, $1 = u^2/v_1^2 + u^2/v_2^2$ or $1/v_1^2 + 1/v_2^2 = 1/u^2$

Ex. 13. (*a*) *A particle is thrown over a triangle from one end of a horizontal base and grazing over the vertex falls on the other end of the base. If A, B be the base angles of the triangle and α the angle of projection, prove that tan α = tan A + tan B.* **[Pune 2010]**

(*b*) *A particle is thrown over an isosceles right angled ΔABC, right angled at C, from one end A of the horizontal base and grazing the vertex C falls at B. Show that the angle of projection is* $\tan^{-1} 2$.

Sol. (*a*) Take the horizontal base AB as x-axis and upward drawn vertical line through A as y-axis. Let $CD \perp AB$ and let $CD = p$, $AB = c$. Let u be the velocity of projection and α be the angle of projection. From ΔACD and BCD, $AD = p \cot A$ and $BD = p \cot B$.

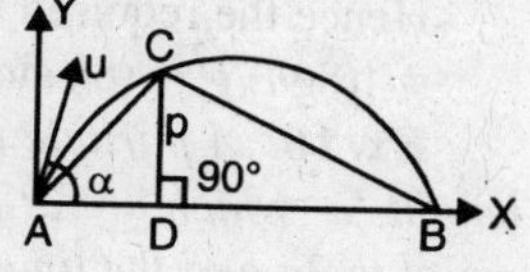

$$\therefore \quad AB = c = AD + BD = p(\cot A + \cot B) \quad \ldots(1)$$

Equation of the trajectory is $\quad y = x \tan\alpha - (gx^2)/(2u^2 \cos^2\alpha) \quad \ldots(2)$

Since point $C(p \cot A, p)$ lies on (1), we have

$$p = p \cot A \tan\alpha - \frac{gp^2 \cot^2 A}{2u^2 \cos^2\alpha} \quad \text{or} \quad 1 = \cot A \tan\alpha - \frac{gp \cot^2 A}{2u^2 \cos^2\alpha} \quad \ldots(3)$$

Again, $p(\cot A + \cot B) = AB = \text{range} = (2u^2 \sin\alpha \cos\alpha)/g \quad \ldots(4)$

so that $\quad pg/2u^2 = (\sin\alpha\cos\alpha)/(\cot A + \cot B) \quad \ldots(4')$

Substituting the value of $pg/2u^2$ from (4′) in (3), we get

$$1 = \cot A \tan\alpha - \frac{\cot^2 A}{\cos^2\alpha} \cdot \frac{\sin\alpha\cos\alpha}{\cot A + \cot B} = \cot A \tan\alpha - \frac{\cot^2 A \tan\alpha}{\cot A + \cot B}$$

$$\text{or} \quad 1 = \tan\alpha\left(\cot A - \frac{\cot^2 A}{\cot A + \cot B}\right) \quad \text{or} \quad 1 = \frac{\tan\alpha \cot A \cot B}{\cot A + \cot B}$$

or $\quad \tan\alpha = (\cot A + \cot B)/(\cot A \cot B) \quad$ or $\quad \tan\alpha = \tan A + \tan B \quad \ldots(5)$

(*b*) **Hint :** Proceed as in part (*a*). Here $A = B = 45°$. So from result (5),

$$\tan\alpha = \tan 45° + \tan 45° = 2 \quad \text{or} \quad \alpha = \tan^{-1} 2.$$

Ex. 14. (*a*) *A shot is fired from a gun on the toip of a cliff 90 m high, with a velocity of 80 m/sec at an elevation of 30°. Find the horizontal distance from the vertical line through gun and point, where the shot strikes the ground.*

(*b*) *A cricket ball thrown by a man from a height of 1.8 m at an angle of 30° with the horizontal with a speed of 18 m/sec is caught by another field man at a height of 60 cm from the ground. How far were the two men apart ?*

Sol. (*a*) Let O be the top of the cliff. Take O as origin and the horizontal and vertical lines through O as axes. Let B be the foot of the cliff and let A be point where shot strikes the ground such that $AB = a$. Given $OB = 90$ m. Considering the vertical motion from O to A,

$$y = (u \sin\alpha)t - (1/2)gt^2 \quad \text{or} \quad -90 = 80(\sin 30°)t - (1/2)(9.8)t^2$$

or $\quad 49t^2 - 400t - 900 = 0 \quad$ giving $\quad t = 10, -90/49.$

Rejecting negative value of t as t cannot be negative, $t = 10$ sec.

Considering the Horizontal motion from O to A, we have

$$a = BA = (u\cos\alpha)t = (80 \cos 30°) \times 10 = 80 \times (\sqrt{3}/2) \times 10 = 400\sqrt{3} \text{ m}$$

(*b*) Try yourself. Ans. 30.55 minutes (nearly)

Ex. 15. *The maximum height of a projectile is h and the angle of projection is α. Find out the difference of time when it is at a high of h* $\sin^2\alpha$.

Sol. Let O be the point of projection. Let velocity of projection be u. Let P and Q be at the same height $h\sin^2\alpha$ so that $PA = QB = h\sin^2\alpha$. Let t be the time taken by the projectile to reach a height $h\sin^2\alpha$. Then,

$$h\sin^2\alpha = u\sin\alpha . t - (1/2)gt^2 \quad \text{or} \quad gt^2 - 2u\sin\alpha . t + 2h\sin^2\alpha = 0 \quad \ldots(1)$$

Again, $h = \text{maximum height} = (u^2 \sin^2\alpha)/2g \Rightarrow u\sin\alpha = (2gh)^{1/2} \quad \ldots(2)$

Using (2), (1) becomes $\quad gt^2 - 2(2gh)^{1/2} t + 2h\sin^2\alpha = 0 \quad \ldots(3)$

Let times taken by the projectile in moving from O to P and O to Q be t_1 and t_2 respectively. Then, t_1, t_2 are roots of (3).

$\therefore \quad t_1 + t_2 = 2(2gh)^{1/2}/g \quad$ and $\quad t_1 t_2 = (2h \sin^2 \alpha)/g$

Hence the required time $= t_1 - t_2 = \{(t_1 + t_2)^2 - 4t_1 t_2\}^{1/2}$

$= \{(8gh)/g^2 - (8h \sin^2 \alpha)/g\}^{1/2} = 2(2h/g)^{1/2} (1 - \sin^2 \alpha)^{1/2} = 2(2h/g)^{1/2} \cos \alpha$

Ex. 16. *A particle is projected with a velocity $2(ag)^{1/2}$ so that it just clears two walls of equal height 'a' which are at a distance '2a' from each other. Show that the latus rectum of the path is equal to 2a and the time between the walls is $2(a/g)^{1/2}$.*

Sol. Refer figure of Ex. 15. Let the angle of projection be α. Here velocity of projection $= u = 2(ag)^{1/2}$. Let PA and QB be two walls of height a such that $AB = 2a$. Let O be the point of projection.

Now, the height of the directrix above $P = 2a - a = a$.

Since the distances a point on the parabola from the directrix and focus are same, hence if S be the focus of the trajectory (which is a parabola), then $PS = a$ and S must lie at middle point of PQ. Thus PQ = latus rectum $= 2a$.

Also, $\quad$ latus rectum $= (2u^2 \cos^2 \alpha)/g = 2a \quad$ so $\quad u \cos \alpha = (ag)^{1/2}$

Consider motion from P to Q in horizontal direction. Then

$$2a = PQ = (u \cos \alpha)t = (ag)^{1/2} t \quad \text{so that} \quad t = 2(a/g)^{1/2}$$

Ex. 17. *A body is projected at an angle α to the horizontal so as just to clear two walls of equal height 'a' at a distance '2a' from each other. Show that the range is equal to 2a $\cot(\alpha/2)$.* **[Guwahati 2007]**

Sol. Try yourself.

Ex. 18. *Prove that if a particle is projected from O at an elevation α, and if after time t, the particle is at P, then $\tan \beta = (1/2)(\tan \alpha + \tan \theta)$, where β and θ are respectively the inclination to the horizontal of OP and of the direction of motion of the particle when at P.* **[Avadh 1990]**

Sol. Let u be the velocity of projection at O and let v be the velocity of the projectile at $P(x, y)$ after time t.

Then, $v \cos \theta = u \cos \alpha$ and $v \sin \theta = u \sin \alpha - gt$. On dividing, these give

$$\tan \theta = \frac{u \sin \alpha - gt}{u \cos \alpha} = \tan \alpha - \frac{gt}{u \cos \alpha} \qquad \ldots(1)$$

Also, we have $\quad x = (u \cos \alpha).t$

and $\quad y = (u \sin \alpha)t - (1/2)gt^2$

$\therefore$ From figure, $\tan \beta = \dfrac{y}{x} = \dfrac{u \sin \alpha t - (1/2)gt^2}{(u \cos \alpha)t} = \tan \alpha - \dfrac{gt}{2u \cos \alpha}$

or $$2 \tan \beta = 2 \tan \alpha - (gt)/u \cos \alpha \qquad \ldots(2)$$

Subtracting (1) from (2), $2 \tan \beta - \tan \theta = \tan \alpha$ or $\tan \beta = (1/2)(\tan \alpha + \tan \theta)$.

Ex. 19. *Two particles are projected from the same point in the same vertical plane with equal velocities. If t, t' be the times taken to reach the other common point of their paths and T, T' times to the highest points, show that (tT + t'T') is independent of the direction of projection.*

Sol. Try yourself.

Ex. 20. *Two particles are simultaneously projected in the same vertical plane from the same point with velocities u and v at angles α and β to the horizon. Show that (i) the line joining them moves parallel to itself.*

(ii) the time that elapses before their velocities are parallel is

$$\{uv \sin(\alpha - \beta)\}/g(v \cos \beta - u \cos \alpha)$$

(iii) the time that elapses before their transits through the other common point is $2uv \sin(\alpha - \beta)/g(u \cos \alpha + v \cos \beta)$ **[I.A.S. 1999]**

Sol. (i) Let the common point of projection O be origin. Take horizontal and upward vertical lines through O as coordinate axes. Let P and Q be the respective positions of two particles at any time t. If (x_1, y_1) and (x_2, y_2) be the coordinate of P and Q respectively. Then, we have

$$x_1 = (u \cos \alpha)t, \qquad y_1 = (u \sin \alpha)t - (1/2)gt^2 \qquad \ldots(1)$$

and $$x_2 = (v \cos \beta)t, \qquad y_2 = (v \sin \beta)t - (1/2)gt^2 \qquad \ldots(2)$$

$$\therefore \text{ Slope of } PQ = \frac{y_2 - y_1}{x_2 - x_1} = \frac{(v \sin \beta - u \sin \alpha)t}{(v \cos \beta - u \cos \alpha)t} = \frac{v \sin \beta - u \sin \alpha}{v \cos \beta - u \cos \alpha}$$

This is independent of t and hence the line PQ has a constant direction and hence the line PQ moves parallel to itself.

(ii) Let θ_1 and θ_2 be the respective directions of motion of the two particles at time t. Then, we know that

$\tan \theta_1 = (u \sin \alpha - gt)/u \cos \alpha$ and $\tan \theta_2 = (v \sin \beta - gt)/v \cos \beta$

These directions will be parallel, if $\tan \theta_1 = \tan \theta_2$,

i.e., if $(u \sin \alpha - gt)/u \cos \alpha = (v \sin \beta - gt)/v \cos \beta$

i.e., if $uv \sin \alpha \cos \beta - gvt \cos \beta = uv \cos \alpha \sin \beta - ugt \cos \alpha$

i.e., if $gt(v \cos \beta - u \cos \alpha) = uv(\sin \alpha \cos \beta - \cos \alpha \sin \beta)$

i.e., if $t = \{uv \sin (\alpha - \beta)\}/g(v \cos \beta - u \cos \alpha)$

(iii) Let t_1 and t_2 be the times to reach the common point (h, k). Then

$$h = (u \cos \alpha)t_1 = (v \cos \beta)t_2 \qquad \ldots(1)$$

and $$k = (u \sin \alpha)t_1 - (1/2)gt_1^2 = (v \sin \beta)t_2 - (1/2)gt_2^2 \qquad \ldots(2)$$

From (2), $$u \sin \alpha . t_1 - v \sin \beta . t_2 = (1/2)g\,(t_2^2 - t_1^2)$$

or $$(1/2)\, g(t_2 + t_1)\,(t_2 - t_1) = u \sin \alpha . t_1 - v \sin \beta . t_2$$

$$\therefore \qquad t_2 - t_1 = \frac{2}{g}\,\frac{u \sin \alpha . t_1 - v \sin \beta . t_2}{t_2 + t_1} \qquad \ldots(3)$$

Now, by (1), $t_1/(v \cos \beta) = t_2/(u \cos \alpha) = l$, say

so that $t_1 = lv \cos \beta$ and $t_2 = lu \cos \alpha$

Substituting these values of t_1 and t_2 in R.H.S. of (3), we get

$$t_2 - t_1 = \frac{2}{g}\,\frac{u \sin \alpha . lv \cos \beta - v \sin \beta . lu \cos \alpha}{lv \cos \beta + lu \cos \alpha} = \frac{2uv \sin(\alpha - \beta)}{g(u \cos \alpha + v \cos \beta)}$$

Ex. 21. *A ball is projected so as just to clear two walls the first of height a at a distance b from the point of projection and the second of height b at a distance a from the point of projection. Show that the range on the horizontal plane is $(a^2 + ab + b^2)/(a + b)$, and that the angle of projection exceeds $\tan^{-1} 3$.* **[Bangalore 1994, 95]**

Sol. Take the point of projection O as origin. Let horizontal and upward vertical lines through O be taken as coordinate axes. Then equation of the path of the ball is $y = x \tan \alpha - (gx^2)/(2u^2 \cos^2 \alpha)$

i.e., $$2yu^2 \cos^2 \alpha = 2xu^2 \sin \alpha \cos \alpha - gx^2 \qquad \ldots(1)$$

Since the ball is just to clear two walls, hence the uppermost points (b, a) and (a, b) of these walls must lie on (1) and hence

$$2au^2 \cos^2 \alpha = 2b\,u^2 \sin \alpha \cos \alpha - gb^2 \qquad \ldots(2)$$

and $$2b\,u^2 \cos^2 \alpha = 2a\,u^2 \sin \alpha \cos \alpha - ga^2 \qquad \ldots(3)$$

Multiplying (2) by b, (3) by a and then subtracting, we get

$$0 = 2(b^2 - a^2)u^2 \sin \alpha \cos \alpha - g(b^3 - a^3) \qquad \ldots(4)$$

$$\therefore \quad R = \text{horizontal range} = \frac{2u^2 \sin \alpha \cos \alpha}{g} = \frac{b^3 - a^3}{b^2 - a^2}, \text{ using (4)}$$

$$= \{(b - a)\,(b^2 + ab + a^2)\}/(b - a)\,(b + a) = (b^2 + ab + a^2)/(b + a)$$

Again, multiplying (2) by a^2, (3) by b^2 and then subtracting, we get

$$2(a^3 - b^3)u^2 \cos^2 \alpha = 2ab(a - b) \sin \alpha \cos \alpha$$

$$\therefore \quad \tan \alpha = \frac{a^3 - b^3}{ab(a - b)} = \frac{a^2 + ab + b^2}{ab} = \frac{a}{b} + \frac{b}{a} + 1 = \left\{\left(\frac{a}{b}\right)^{1/2} - \left(\frac{b}{a}\right)^{1/2}\right\}^2 + 3$$

so that $\tan \alpha > 3$ and hence $\alpha > \tan^{-1} 3$.

Ex. 22. *A gun is fixed from the sea level out to sea. It is then mounted on a battery h meters higher up and fired at the same elevation α. Show that the range is increased by* $(1/2)\{(1 + 2gh/u^2 \sin^2 \alpha)^{1/2} - 1\}$ *of itself, u being the velocity of projection.* **[I.A.S. 1994]**

Sol. Let O' be a point of the sea level. Let R_1 be its range when gun is fired from it. ThenOA

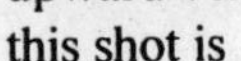

$= R_1 = (2u^2 \sin \alpha \cos \alpha)/g$ ...(1)

Let O be a point at a height h above the sea level. Let R_2 be range $(= OB)$ on the sea level when the shot is fired from O. Referred to the horizontal and upward vertical lines OX and OY as coordinate axes, the equation of the path of this shot is

$$y = x \tan \alpha - (gx^2)/2u^2 \cos^2 \alpha \quad ...(2)$$

Since $B(R_2, -h)$ lies on (2), we have

$$-h = R_2 \tan \alpha - (gR_2^2)/2u^2 \cos^2 \alpha$$

or

$$gR_2^2 - (2u^2 \sin \alpha \cos \alpha).R_2 - 2u^2 h \cos^2 \alpha = 0$$

$$\therefore \quad R_2 = \{2u^2 \sin \alpha \cos \alpha \pm (4u^4 \sin^2 \alpha \cos^2 \alpha + 8u^2 gh \cos^2 \alpha)^{1/2}\}/2g$$

$$= \{2u^2 \sin \alpha \cos \alpha + 2u^2 \sin \alpha \cos \alpha\,(1 + 2gh/u^2 \sin^2 \alpha)^{1/2}\}/2g$$

[Since R_2 is +ve, we reject –ve sign]

$$= \{R_1 g + R_1 g\,(1 + 2gh/u^2 \sin^2 \alpha)^{1/2}\}/2g, \text{ by (1)}$$

Thus,

$$R_2 = (R_1/2)\,\{1 + (1 + 2gh/u^2 \sin^2 \alpha)^{1/2}\}$$

On substracting R_1 from both sides, we get

$$\therefore \quad R_2 - R_1 = (R_1/2)\,\{1 + (1 + 2gh/u^2 \sin^2 \alpha)^{1/2}\} - R_1$$

Hence the fraction by which R_1 increases is

$$(R_2 - R_1)/R_1 = (1/2)\,\{(1 + 2gh/u^2 \sin^2 \alpha)^{1/2} - 1\}.$$

Ex. 23. *Shots fired simultaneously from the top and bottom of a vertical cliff with elevations α and β respectively strike an object simultaneously, at the same point. Show that if a is the horizontal distance of the object from the cliff the height of the cliff is a(tan β – tan α).*

Sol. Try yourself.

Ex. 24. *From a tower an object was observed on the ground at a depression φ below the bottom. A gun was fired at an elevation α, but the shot missing the object struck the ground at a point whose depression was ψ. Prove that the correct elevation θ of the gun is given by*

$$[\cos \theta \sin(\theta + \phi)]/\cos \alpha \sin(\alpha + \psi) = (\cos^2 \theta \sin \psi)/\cos^2 \psi \sin \phi$$

Sol. Try yourself.

Ex. 25. (*a*) *A stone is projected with velocity u from a height h to hit a point in the level at a horizontal distance R from the point of projection. Show that the angle of projection is given by*

$$R^2 \tan^2 \alpha - (2u^2/g) R \tan \alpha + R^2 - (2hu^2/g) = 0.$$

Hence deduce that the maximum range on the level for this velocity is $\{(u^4/g^2) + (2hu^2/g)\}^{1/2}$ *and that if R' is this maximum range and α the angle of projection to give the maximum range, then* $\tan \alpha = u^2/gR'$ *and* $\tan 2\alpha = R'/h$.

(*b*) *A shot putter at the top A of a cliff of height h above sea level throws a shot at a speed V and at an angle of elevation α with the horizontal. The shot describes a parabolic trajectory under gravity, in the absence of air resistance, and enters the sea at B. If the horizontal distance R of B from foot of the cliff is to be maximum, show that* $R_{max} = (V^2/g)\,\{1 + (2gh)/V^2\}^{1/2}$

[Delhi Maths (H) 1997]

(*c*) *A particle is projected under gravity with velocity* $(2ga)^{1/2}$ *from a point at a distance h above a level plane. Show that the angle of projection α for the maximum range on the plane is given by* $\tan^2 \alpha = a/(a + h)$, *and that maximum range is* $2\{a(a + h)\}^{1/2}$.

(*d*) *A gun is mounted on a hill of height h above a level plane, show that, if the resistance of air is neglected, the greatest horizontal range for given muzzle velocity u is obtained by firing at an angle of elevation α such that* $\operatorname{cosec}^2 \alpha = 2\,(1 + gh/u^2)$. **[Delhi B.A. III (Prog) 2009]**

Sol. (*a*) Let the point of projection be A at height h above the ground. Let O be foot of the perpendicular drawn on the ground OB. Take A as origin and horizontal and upward vertical lines as coordinate axes. Then equation of path is

$$y = x \tan \alpha - (gx^2)/2u^2 \cos^2 \alpha \qquad ...(1)$$

Suppose the stone hits the ground at B whose coordinates are $(R, -h)$. Since it lies on (1), we have

$$-h = R \tan \alpha - (gR^2/2u^2)(1 + \tan^2 \alpha)$$

or $\quad R^2 \tan^2 \alpha - (2u^2/g) R \tan \alpha + R^2 - (2u^2 h/g) = 0, \quad ...(2)$

which gives the required angles of projection. Since roots of (1) must be real, so its discriminant ≥ 0, *i.e.*,

$$\frac{4u^4R^2}{g^2} - 4R^2\left(R^2 - \frac{2u^2h}{g}\right) \geq 0 \qquad \text{or} \qquad \frac{u^4}{g^2} - R^2 - \frac{2u^2h}{g} \geq 0$$

$$\text{or} \qquad \frac{u^4}{g^2} + \frac{2u^2h}{g} \geq R^2 \qquad \text{or} \qquad R \leq \left(\frac{u^4}{g^2} + \frac{2u^2h}{g}\right)^{1/2} \qquad ...(3)$$

If R' be the maximum range, then R' is given by

$$R' = R_{\max} = \left(\frac{u^4}{g^2} + \frac{2u^2h}{g}\right)^{1/2} = \frac{u^2}{g}\left(1 + \frac{2gh}{u^2}\right)^{1/2} \qquad ...(4)$$

Now if R' is the max. range and α the corresponding angle of projection, then from (2), we have

$$R'^2 \tan^2 \alpha - (2u^2/g) R' \tan \alpha + R'^2 - (2u^2h)/g = 0$$

or $\quad R'^2 \tan^2 \alpha - (2u^2/g)R' \tan \alpha + (u^4/g^2) + (2u^2h/g) - (2u^2h/g) = 0$, by (4)

or $\quad g^2R'^2 \tan^2 \alpha - 2u^2 gR' \tan \alpha + u^4 = 0 \quad$ or $\quad (u^2 - gR' \tan \alpha)^2 = 0$

so that $\quad u^2 - gR' \tan \alpha = 0 \quad$ giving $\quad \tan \alpha = u^2/gR' \qquad ...(5)$

Now, $$\tan 2\alpha = \frac{2 \tan \alpha}{1 - \tan^2 \alpha} = \frac{2(u^2/gR')}{1 - (u^2/gR')^2} = \frac{2u^2 gR''}{g^2R'^2 - u^4}$$

$$= \frac{2u^2 gR'}{g^2(u^4/g^2 + 2u^2h/g) - u^4}, \text{ using (4)}$$

Thus, $\quad \tan 2\alpha = (2u^2 gR')/(2u^2hg) = R'/h$

(*b*) Proceed exactly as in part (*a*) upto equation (4). Here $u = V$

(*c*) Proceed as in part (*a*). Here $u = (2ag)^{1/2}$. Use (4) and (5).

(*d*) Proceed upto equation (5). From (5), $\quad \cot \alpha = gR'/u^2 \qquad ...(6)$

$\therefore \quad \operatorname{cosec}^2 \alpha = 1 + \cot^2 \alpha = 1 + (g^2 R'^2)/u^4$, using (6)

$\quad = 1 + (g^2/u^4) \times (u^4/g^2)(1 + 2gh/u^2)$, using (4)

Thus, $\quad \operatorname{cosec}^2 \alpha = 2 + (2gh)/u^2 = 2(1 + gh/u^2)$

Ex. 26. *The radii of the front and hind wheels of a carriage are 'a' and 'b' and 'c' is the distance between the axle-trees. A particle of mud driven from the highest point of the hind wheel is observed to alight on the highest point of the front wheel. Show that the velocity of carriage is*

$$[g[(c + b - a)(c + a - b)]/4(b - a)]^{1/2}$$ **[Avadh 1993]**

Sol. Let u be the velocity of carriage. Then the velocity of the highest point O of the hind wheel is $2u$ horizontally. It follows that the actual velocity of the mud particle while driven from O is $2u$ horizontally. But the carriage is also moving horizontally with velocity u. So the horizontal velocity of the mud particle relative to carriage $= 2u - u = u$, acting horizontally.

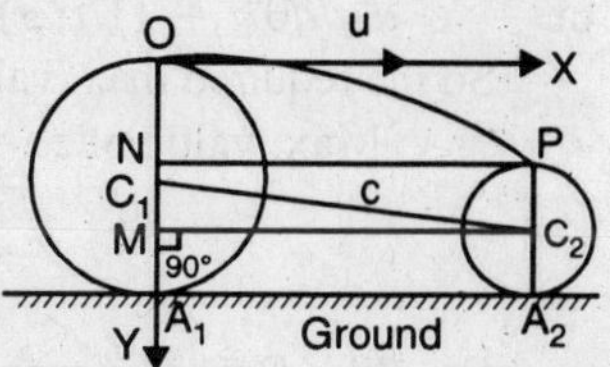

Let P be the highest point of the front wheel. Let C_1 and C_2 be centres of hind and front wheels respectively. Let $C_2M \perp OA_1$ and $PN \perp OA_1$. Then, from figure $ON = OA_1 - A_1N = OA_1 - PA_2 = 2b - 2a = 2(b - a)$, since $C_1A_1 = C_1O = b$ and $C_2A_2 = C_2P = a$. Also $C_1M = C_1A_1 - MA_1 = C_1A_1 - C_2A_2 = b - a$.

So, from right angled ΔC_1MC_2, $C_2M^2 = C_1C_2^2 - C_1M^2 = c^2 - (b-a)^2$,
Again, $PN^2 = C_2M^2 = c^2 - (b-a)^2$ so $PN = \{c^2 - (b-a)^2\}^{1/2}$
since, $C_1 C_2$ = distance between the axle trees = c (given).

Take O as origin and horizontal and downward vertical lines through O as coordinate axes. Then, equation of the path of the mud particle is $y = gx^2/2u^2$. The coordinates of the mud particle at P are (NP, ON) which lies on its path.

$\therefore \quad ON = (g \times NP^2)/2u^2 \quad$ or $\quad u^2 = (g \times NP^2)/2ON$

or
$$u^2 = \frac{g\{c^2 - (b-a)^2\}}{4(b-a)} = \frac{g(c+b-a)(c-b+a)}{4(b-a)}$$

[Putting values of NP and ON]

or $\quad u = [g\{(c+b-a)(c+a-b)\}/4(a-b)]^{1/2}$

Ex. 27. *Mud is thrown off from the tyre of a wheel (radius a) of a car traveling at a speed v, where $v^2 > ga$. Neglecting the resistance of the air, show that no mud can rise higher than a height $a + (v^2/2g) + (ga^2/2v^2)$ above the ground.* **[Delhi B.Sc. (Prog) III 2011; Delhi B.Sc. (Hons) 2005]**

Sol. Let C be the centre of the wheel moving on the ground EK. Suppose at any time t mud be at P. Let mud move with velocity v along the tangent PT to the wheel at P making an angle θ with the horizontal PX. Take P as origin and horizontal and upward vertical lines as coordinate axes.

Then the mud can rise only upto point H, where HM is the maximum height attained by the mud in its trajectory.

Thus, $\quad HM = (v^2 \sin^2\theta)/2g \quad \ldots(1)$

Draw PN perpendicular to horizontal line CN. Then $\angle CPN = \theta$ and so from ΔPCN, $PN = PC \cos\theta = a\cos\theta$.

Draw HK perpendicular to ground EK and CD perpendicular to HK. Let $HK = z$. Then z is height to which the mud can rise above the ground.

From figure, $\quad z = HK = HM + MD + DK = HM + PN + CE$

Thus, $\quad z = (v^2 \sin^2\theta)/2g + a\cos\theta + a \quad \ldots(2)$

From (1), $\quad dz/d\theta = (v^2 \sin\theta\cos\theta)/g - a\sin\theta \quad \ldots(3)$

For max. and min. of z, $\quad dz/d\theta = 0$, *i.e.*,

$(v^2 \sin\theta\cos\theta)/g - a\sin\theta = 0 \quad$ or $\quad \sin\theta\,\{(v^2/g)\cos\theta - a\} = 0 \quad \ldots(4)$

Since $\theta \neq 0$, so $\sin\theta \neq 0$. Then (4) $\Rightarrow \cos\theta = (ga)/v^2$

Given that $v^2 > ga$ so that $(ga)/v^2 < 1$ and hence $\cos\theta < 1$

Hence, $\quad \cos\theta = (ga)/v^2$ gives a possible value of θ for max. or min.

Re-writing (3), $\quad dz/d\theta = (v^2/2g)\sin 2\theta - a\sin\theta$

$\therefore \quad d^2z/d\theta^2 = (v^2/g)\cos 2\theta - a\cos\theta = (v^2/g)(2\cos^2\theta - 1) - a\cos\theta$

When $\cos\theta = ga/v^2$, the value of $d^2z/d\theta^2$ is given by

$$d^2z/d\theta^2 = \frac{v^2}{g}\left(\frac{2g^2a^2}{v^4} - 1\right) - \frac{ga^2}{v^2} = \frac{ga^2}{v^2} - \frac{v^2}{g} = \frac{1}{v^2 g}(g^2a^2 - v^4)$$

or $\quad d^2z/d\theta^2 = (1/v^2 g)(ga + v^2)(ga - v^2) < 0$, as $v^2 > ga$

So the required max. value of z is obtained by putting $\cos\theta = (ga)/v^2$ in (2)

$\therefore \quad$ Max. value of $z = (v^2/2g)(1 - \cos^2\theta) + a\cos\theta + a$

$$= \frac{v^2}{2g}\left(1 - \frac{g^2a^2}{v^4}\right) + a \times \frac{ga}{v^2} + a = a + \frac{v^2}{2g} + \frac{ga^2}{2v^2}$$

Ex. 28. *Drops of water are thrown tangentially off the rim of a wheel of radius a which is rotating in a vertical plane about its axis, which is fixed and horizontal, with angular velocity $\omega > (g/a)^{1/2}$. Show that if a horizontal ceiling, at height $h > a$ above the axis of the wheel, is not to be spattered with water, $a\omega < \{gh + g(h^2 - a^2)^{1/2}\}^{1/2}$ the effects of air resistance being neglected*

[Delhi Maths (H) 2000, 09]

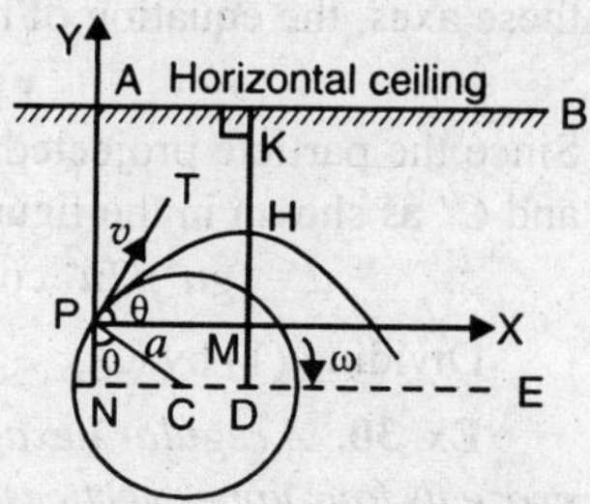

Sol. Let C be the centre of the wheel moving with angular velocity ω as shown in the figure. Suppose at any time a drop of water be at P. Let the drop move with velocity v along the tangent PT to the wheel at P making an angle θ with the horizontal. Take P as origin and horizontal and upward vertical lines as coordinate axes. Then the drop can rise only upto point H where HM is the maximum height attained by the drop in its trajectory. Thus, $HM = (v^2 \sin^2 \theta)/2g$ ...(1)

Let ND be the horizontal line passing through the axis of the wheel. Then, clearly CD passes through C. Let $HK \perp AB$, where AB is the horizontal ceiling such that $KD = h > a$ (given). Let $PN \perp ND$. Then, from ΔPNC, $PN = a \cos \theta$ and hence $MD = PN = a \cos \theta$. Let $HD = z$. Then z is height to which the drop can rise above the axis of the wheel. From figure, we have

$$z = HD = HM + MD = (v^2 \sin^2 \theta)/2g + a \cos \theta \qquad \text{...(2)}$$

$$\therefore \quad dz/d\theta = (v^2/g) \sin \theta \cos \theta - a \sin \theta = \sin \theta\{(v^2/g) \cos \theta - a\} \qquad \text{...(3)}$$

For Max. and Min. of z, $dz/d\theta = 0$ so that $\sin \theta = 0$ or $\cos \theta = (ag)/v^2$

But $\theta \neq 0$ and hence $\sin \theta \neq 0$ and for max. on min. of z, we have

$$\cos \theta = (ag)/v^2 \qquad \text{...(4)}$$

For circular motion about centre, the angular velocity ω is given by $\omega = v/a$. Given that $\omega > (g/a)^{1/2}$. Hence, we have

$$v/a > (g/a)^{1/2} \quad \text{or} \quad v^2/a^2 > g/a \quad \text{or} \quad ag < v^2 \qquad \text{...(5)}$$

showing that (4) gives possible values of θ for max. or min.

Re-writing (3), $\quad dz/d\theta = (v^2/2g) \sin 2\theta - a \sin \theta$

$$\therefore \quad d^2z/d\theta^2 = (v^2/g)(\cos 2\theta) - a \cos \theta = (v^2/g)(2\cos^2 \theta - 1) - a \cos \theta$$

$$= \frac{v^2}{g}\left(\frac{2a^2g^2}{v^4} - 1\right) - a\left(\frac{ag}{v^2}\right) = \frac{ga^2}{v^2} - \frac{v^2}{g} = \frac{1}{gv^2}(g^2a^2 - v^4)$$

$$= (1/gv^2)(ga + v^2)(ga - v^2) < 0, \text{ as } ga < v^2 \text{ by (5)}$$

Hence the maximum value of z is given by putting $\cos \theta = (ga/)v^2$ in (2).

$$\therefore \quad \text{Max. value of } z = (v^2/2g)(1 - \cos^2 \theta) + a \cos \theta$$

$$= \left(\frac{v^2}{2g}\right)\left(1 - \frac{g^2a^2}{v^4}\right) + a\left(\frac{ga}{v}\right) = \frac{v^2}{2g} + \frac{ga^2}{v^2} \qquad \text{...(6)}$$

From figure, it is clear that the horizontal ceiling AB at a height $h(> a)$ will not be spattered with water, if

Max. value of $z < h \quad$ or $\quad (v^2/2g) + (ga^2/v^2) < h$, by (6)

or $\quad (a^2\omega^2/2g) + (ga^2/2a^2\omega^2) < h$, since $v = a\omega$

or $\quad a^2\omega^4 + g^2 < 2\omega^2 gh \quad$ or $\quad a^4\omega^4 - 2a^2\omega^2 gh < -a^2g^2$

or $\quad a^2\omega^2 - 2a^2\omega^2 gh + g^2h^2 < g^2h^2 - a^2g^2 \quad$ or $\quad (a^2\omega^2 - gh)^2 < g^2(h^2 - a^2)$

or $\quad a^2\omega^2 - gh < g(h^2 - a^2)^{1/2} \quad$ or $\quad a\omega < \{gh + g(h^2 - a^2)^{1/2}\}^{1/2}$

Ex. 29. *A particle is to be projected so as just to pass through three equal rings, of diameter d, placed in parallel vertical planes at distance a apart, with their highest points in a horizontal straight line at a hight h above the point of projection. Prove that the elevation must be* $\tan^{-1}(2\sqrt{hd}/a)$.

Sol. Let AA', BB' and CC' be three equal rings, of diameter d so that $AA' = BB' = CC' = d$. These rings are parallel such that $A'B' = B'C' = a$. Here B is the highest point of the middle ring and hence the particle must be moving horizontally at D. Given that $BM = h$.

Let u be the velocity of projection and α be its inclination to the horizontal. Then, $h = (u^2 \sin^2 \alpha)/2g \quad$ so that $u^2 \sin^2 \alpha = 2gh$ ...(1)

At B, the velocity of the particle is $u \cos \alpha$ in the horizontal direction. Take B as origin and horizontal and downward vertical lines BX' and BY' through B as coordinate axes. Then referred to these axes, the equation of the trajectory is

$$y = (gx^2)/2u^2 \cos^2 \alpha \qquad \text{...(2)}$$

Since the particle projected just passes through the given rings, hence (2) must pass through A', B and C' as shown in the figure. Hence the point $C'(a, d)$ lies on (2).

$$\therefore \quad d = (ga^2)/2u^2 \cos^2 \alpha \quad \text{or} \quad u^2 \cos^2 \alpha = (ga^2)/2d \qquad \text{...(3)}$$

Dividing (1) by (2), $\tan^2 \alpha = (4hd)/a$ or $\alpha = \tan^{-1} (2\sqrt{hd} / a)$

Ex. 30. *A regular hexagon stands on one side on the ground and a particle is projected so as to graze its four upper vertices. Show that the velocity of the particle on reaching the ground to its least velocity as $\sqrt{31}$ to $\sqrt{3}$.*

Sol. Let P be the point of projection. Let the particle strike the ground at Q after grazing the four upper vertices E, F, A and B of the hexagon $ABCDEF$. Let velocity at the highest H of the tajectory be u. Then u is the minimum velocity of the particle in its entire motion from P to Q. Let V be the velocity of the particle at Q. Then, we are required to find $V : u$.

By energy equation, for motion for H to Q, we have

$$(1/2)mV^2 - (1/2)mu^2 = mg.HL$$

or
$$V^2 = u^2 + 2g.HL$$

or
$$V^2/u^2 = 1 + (gh.HL)/u^2 \qquad \text{...(1)}$$

With H as origin and horizontal and downward vertical lines as coordinate axes, the equation of the path is

$$y = (gx^2)/2u^2 \quad \text{or} \quad 2u^2y = gx^2 \qquad \text{...(2)}$$

Let $2a$ be the length of each side of the hexagon. Let $HG \perp AF$ and let $HG = b$. Then coordinates of A are (a, b). Again $OB = OK + KB = GA + AB \cos 60° = a + a = 2a$. Also, $HO = HG + GO = b + AK = b + AB \sin 60° = b + a\sqrt{3}$. So the coordinates of B are $(2a, b + a\sqrt{3})$.

Since $A(a, b)$ and $B(2a, b + a\sqrt{3})$ lie on (2), we get

$$2u^2b = a^2g \qquad \text{...(3)}$$

and $\quad 2u^2(b + a\sqrt{3}) = 4a^2g \quad$ or $\quad u^2(b + a\sqrt{3}) = 2a^2g \qquad$ ...(4)

Dividing (4) by (3), $(b + a\sqrt{3})/b = 4$ giving $b = a/\sqrt{3}$ ...(5)

Using this value of b in (3), we get $u^2 = (\sqrt{3}/2)\, ag$ ...(6)

From, figure, $HL = HG + GO + OL = b + 2GO = b + 2AK$

$$= b + 2 \times (2a \sin 60°) = b + 2a\sqrt{3} = (a/\sqrt{3}) + 2a\sqrt{3} = (7/\sqrt{3})ag.$$

Using (6) and the above value of HL in (1), we get

$$\frac{V^2}{u^2} = 1 + \frac{2g(7/\sqrt{3})ag}{(\sqrt{3}/2)ag} = 1 + \frac{28}{3} \Rightarrow \frac{V}{u} = \frac{\sqrt{31}}{\sqrt{3}}$$

Ex. 31. *An aeroplane is flying with constant velocity v and at a constant height h. Show that, if a gun is fired point blank at the aeroplane after it has passed directly over the gun and when it's angle of elevation as seen from the gun is α, the shell will hit the aeroplane provided $2(V \cos a - v)\, v \tan^2 \alpha = gh$ where V is the initial velocity of the shot.*

Sol. Let A_0 be the position of the aeroplane when the shot was fired from O. Gun is fired point blank means the initial velocity V of the shot has its direction along OA_0, *i.e.*, the angle of projection for shot is α.

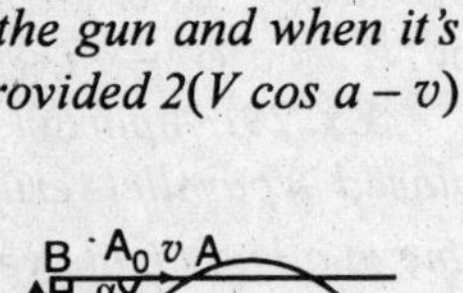

Let the aeroplane be flying at a height h along the line BA_0 where $\angle A_0BO = 90°$. Initially when shot was at O, then aeroplane was at A_0. Suppose the shot hits the aeroplane at A after time t.

The horizontal distance moved by aeroplane in time $t = A_0A = vt$

Now, from ΔA_0BO, $BA_0 = h \cot \alpha$ so $BA = BA_0 + A_0A = h \cot \alpha + vt$ ...(1)

Considering motion of shot from O to P in horizontal and vertical directions,

$$BA = V \cos \alpha . t \quad \text{or} \quad h \cot \alpha + vt = V \cos \alpha . t, \text{ using (1)}$$

so that $$t = (h \cot \alpha)/(V \cos \alpha - v) \qquad ...(2)$$

and $$h = V \sin \alpha . t - \frac{gt^2}{2} = \frac{Vh \sin \alpha \cot \alpha}{V \cos \alpha - v} - \frac{gh^2 \cot^2 \alpha}{2(V \cos \alpha - v)^2}$$

or $$2(V \cos \alpha - v)^2 = 2V \cos \alpha (V \cos \alpha - v) - gh \cot^2 \alpha$$

or $$gh \cot^2 \alpha = 2(V \cos \alpha - v) \{V \cos \alpha - (V \cos \alpha - v)\}$$

or $$gh \cot^2 \alpha = 2(V \cos \alpha - v)v \quad \text{or} \quad 2(V \cos \alpha - v)\, v \tan^2 \alpha = gh$$

Ex. 32. (*a*) *A stone in thrown in such a manner that it would just hit a bird at the top of a tree and afterwards reach a height double that of the tree. If at the moment of throwing the stone the bird flies away horizontally, show that not withstanding this, the stone will hit the bird if its horizontal velocity to that of the bird be as* $(\sqrt{2} + 1) : 2$.

Ex. 32. (*b*) *A shot is fired with velocity V at an elevation α so as to hit a bird sitting at the top of a pole of height h. However the bird immediately starts flying horizontally away from the gun with velocity v. Show that it will escape being hit if* $(2V \cos \alpha - v)(V^2 \sin^2 \alpha - 2gh)^{1/2} = vV \sin \alpha$

Sol. Try yourself.

Ex. 32. (*c*) *A shell fired with vlocity V at an elevation θ hits an airship at a height h from the ground which is moving, horizontally away from the gun with velocity v. Show that if* $(2V \cos \theta - v)(v^2 \sin^2 \theta - 2gh)^{1/2} = vV \sin \theta$ *the shell might have also hit the ship if the latter had remained stationary in the position it occupied when the gun was actually fired.* **[I.F.S. 2003]**

Sol. Try yourself.

Ex. 33. (*a*) *Prove that if particles be projected from the same point in the same plane so as to describe equal parabolas, the vertices of their paths lie on a parabola. Find the vertex and latus rectum of this parabola.*

Sol. Take the common point of projection O as the origin and the horizontal and vertical lines OX and OY in the vertical plane of projection as the co-ordinate axes. Let u be the velocity of projection and α the angle of projection of one of the trajectories. Then if (x_1, y_1) be the coordinates of the vertex then

$$x_1 = (u^2 \sin \alpha \cos \alpha)/g \qquad ...(1)$$

and $$y_1 = (u^2 \sin^2 \alpha)/2g \qquad ...(2)$$

Since the parabolas described for different value of u and α are equal, hence their latusrecta will be equal.

$$\therefore \quad (2u^2 \cos^2 \alpha)/g = \text{constant} \quad \text{or} \quad u \cos \alpha = \text{constant} = K \text{ (say)} \qquad ...(3)$$

To find locus of (x, y), we shall eleminate parameters u and α with help of three relations of (1), (2) and (3).

Using (3), (1) gives $x_1 = (K/g)(u \sin \alpha)$ or $(u \sin \alpha) = gx_1/K$

Putting this value in (2), $$y_1 = \left(\frac{gx_1}{K}\right)^2 \times \frac{1}{2g} \quad \text{or} \quad x_1^2 = \frac{2K^2}{g} y_1$$

$\therefore$ the locus of (x_1, y_1) is $$x^2 = (2K^2/g)y,$$

which is a parabola whose vertex is (0, 0) and latus rectum is $2K^2/g$.

Ex. 33. (*b*) *A number of particles are projected simultaneously from the point with equal velocities in the same vertical plane in different directions. Find the locus of the foci of their paths.*

[I.F.S. 2000]

Sol. Take the common point O as the origin and the horizontal and vertical lines OX and OY in the vertical plane of projection as the coordinate axes. Let α the angle of projection of one of the trajectories. Let u be the velocity of projection of all particles. Then, if (x_1, y_1) be the coordinates of the focus, then

$$x_1 = (u^2/2g) \sin 2\alpha \quad \text{and} \quad y_1 = -(u^2/2g) \cos 2\alpha$$

Squaring and adding these, $x_1^2 + y_1^2 = (u^2/2g)^2$

Hence the required locus of foci is circle $x^2 + y^2 = (u^2/2g)^2$

Ex. 34. (*a*) *Particles are projected from the same point in a vertical plane with velocity* $(2gk)^{1/2}$. *Prove that locus of the vertices of their paths is the ellipse* $x^2 + 4y(y - k) = 0$. **[Delhi B.A. (Prog) III, 2013]**

Sol. Try yourself.

Ex. 34. (*b*) *Particles are projected from the same point in a vertical plane with velocity which vary as* $1/(\sin \alpha)^{1/2}$, α *being the angle of projection. Find the locus of the vertices of the parabolas described.*

Sol. Try yourself. **Ans.** $x^2/(k^2/g^2) + y^2/(k^4/4g^2) = 1$, which is an ellipse, k being a constant.

Ex. 35. (*a*) *A gun is fired from a point O to hit a car, which is moving horizontally with uniform speed U away from O. At the instant when the gun is fired, the distance of the car from O is b. Prove that if the car is to be hit, the horizontal component u of the velocity of projection of the shell, and the vertical component v must satisfy the relation* $v(u - U) = (1/2)bg$.

Sol. Let the car be at P when the gun is fired from O. Then $OP = b$ (given). Suppose the shot strikes the car at A after time t. Then, PA = distance moved by the car in time $t = Ut$. So $OA = OP + PA = b + Ut$.

Let V be the velocity of projection of the shell. Then, u and v are given as its components along OX and OY respectively. We know that

Horizontal range = $(2/g) \times$ (Horizontal comp. of vel.) $\times$ (vertical comp. of vel.)

$\therefore \quad b + Ut = OA$ = horizontal range $= (2/g) \times uv$

or $\quad b + U(2v/g) = (2uv)/g$, since t = time of flight $= (2v)/g$

or $\quad bg = 2uv - 2Uv \quad$ or $\quad v(u - U) = (1/2)bg$

Ex. 35. (*b*) *A shot fired with velocity V at an elevation* α *strikes a point P on a horizontal plane through the point of projection. If the point P is receding from the gun with velocity U, show that the elevation must be changed to* θ, *where* $\sin 2\theta = \sin 2\alpha + (2U/V) \sin \theta$.

Sol. Draw figure as in Ex. 35 (*a*). When the point P is at rest, then $OP = (V^2 \sin 2\alpha)/g$. Now P moves along PA with velocity U, let shot strikes it at A after the elevation of the gun is changed from α to θ. So $OA = (V^2 \sin 2\theta)/g$ and the corresponding time of flight $= T = (2V \sin \theta)/g$. Cleraly the point P must move from P to A in time T so that $PA = UT$.

i.e., $\quad PA = (2UV \sin \theta)/g$, on putting value of T.

From figure, $OA = OP + PA$ so $\quad \dfrac{V^2 \sin 2\theta}{g} = \dfrac{V^2 \sin 2\alpha}{g} + \dfrac{2UV \sin \theta}{g}$

giving $\quad \sin 2\theta = \sin 2\alpha + (2U/V) \sin \theta$

EXERCISE

1. If r be the horizontal range of a projectile and h its greatest height, prove that the initial speed is $\{2g(h + r^2/16h)\}^{1/2}$. **[Delhi 2008]**

2. Prove that if the time of flight of a bullet over a horizontal range R is T seconds, the inclination of the velocity of projection is $\tan^{-1}(gT^2/2R)$. **[Delhi B.A. (Prog) III, 2011, 2013]**

3. A particle is projected upward in a direction inclined at 60° to the horizontal. Show that the velocity of the particle when at its greatest height is half its initial velocity.

4. A particle just clears a wall of height b at a distance a and strikes the ground at a distance c from the point of projection. Prove that the angle of projection is $\tan^{-1}\{bc/a(c - a)\}$ and the velocity of projection V is given by $2V^2/g = \{a^2(c - a)^2 + b^2c^2\}/ab(c - a)$. **[Delhi B.A. (Prog) III, 2014]**

5. A particle is projected from a point at a height $3h$ above a horizontal plane, the direction of projection making an angle α with the horizon. Show that if the greatest height above the point of projection is h, then the horizontal distance travelled before striking the plane is $6h \cot \alpha$.

6. Show that the locus of the foci of all trajectories which pass through two gives points is a hyperbola.

6.8 To determine the possible directions in which a particle be projected to pass through a given point **(I.A.S. 2004)**

(*i*) Let u be the velocity of projection and α the angle of projection of the projectile projected from O. Take O as origin and horizontal and vertical lines through O as coordinate axes. Let coordinates of the given point P be (a, b). The equation of the trajectory is given by

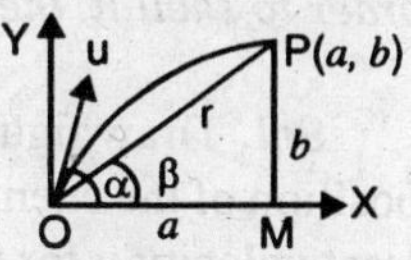

$$y = x \tan\alpha - (gx^2)/2u^2 \cos^2\alpha$$

or $$y = x \tan\alpha - (gx^2/2u^2)(1 + \tan^2\alpha) \quad ...(1)$$

Since P(a, b) lies on (1), we have

$$b = a \tan\alpha - (ga^2/2u^2)(1 + \tan^2\alpha) = 0$$

or $$(2u^2 b)/ga^2 = (2u^2/ga)\tan\alpha - (1 + \tan^2\alpha) = 0$$

or $$\tan^2\alpha - (2u^2/ga)\tan\alpha + 1 + (2u^2 b)/ga^2 = 0, \quad ...(2)$$

which is a quadratic equation in $\tan\alpha$ giving us two values of $\tan\alpha$ and consequently two values of α, corresponding to each of which we get a direction of projection provided the roots of (2) are real and distinct.

(*ii*) **To find the least velocity of projection to hit a given print.** In order to be able to hit a given point P from a given point O, the roots of (2) must be real for which its discriminant ≥ 0, that is,

$$4u^2/g^2a^2 - 4(1 + 2u^2 b/ga^2) \geq 0 \quad \text{or} \quad u^4 - g^2a^2(1 + 2u^2 b/ga^2) \geq 0$$

or $$u^4 - 2u^2 gb \geq g^2a^2 \quad \text{or} \quad u^4 - 2u^2 gb + g^2b^2 \geq g^2a^2 + g^2b^2$$

or $$(u^2 - gb)^2 > g^2(a^2 + b^2) \quad \text{or} \quad u^2 > gb + g(a^2 + b^2)^{1/2}$$

or $$u^2 \geq g\{b + (a^2 + b^2)^{1/2}\} \quad \text{or} \quad u^2 > g(b + r), \quad ...(3)$$

where $r = OP = (a^2 + b^2)^{1/2}$ from ΔOPM.

From (3), the least value of the velocity of projection from O in order that it may pass through P is given by

$$u = \{g(b + r)\}^{1/2} \text{ or } u = [g\{b + (a^2 + b^2)^{1/2}\}]^{1/2} \text{ or } u = \{g(PM + OP)\}^{1/2} \quad ...(4)$$

If OP makes an angle β with the horizontal, then from ΔOPM, $b = PM = r \sin\beta$. Then, (4) reduces to

Least velocity of projected to hit P = $\{gr(1 + \sin\beta)\}^{1/2}$ ...(5)

Note. If the point P to be hit lies below the point of projection O, then replacing b by $-b$ (or β by $-\beta$) in the above results (4) and (5), we get

$$\text{the least velocity} = \{g(r - b)\}^{1/2} = \{gr(1 - \sin\beta)\}^{1/2} \quad ...(6)$$

(*iii*) **Time required to hit a given point.** Let a particle be projected from a given point O with a given velocity u at a given angle of projection to pass through a given point $P(a, b)$.

Let t be time taken by the particle to move from O to P. Then considering the horizontal and vertical motions of the particle from O to P, we have

$$a = u \cos\alpha . t \quad ...(7)$$

and $$b = u \sin\alpha . t - (1/2)gt^2 \quad \text{or} \quad b + (1/2)gt^2 = u \sin\alpha . t \quad ...(8)$$

Squaring and adding (7) and (8), we get

$$a^2 + b^2 + gbt^2 + (1/4)g^2t^4 = u^2t^2$$

or $$g^2t^4 + 4(gb - u^2) + 4(a^2 + b^2) = 0$$

or $$t^4 + (4/g^2)(gb - u^2) + (4/g^2)(a^2 + b^2) = 0, \quad ...(9)$$

which is quadratic in t^2 and thus gives two possible values of t^2 and consequently two possible values of t hit the given point P. If corresponding to the two directions of projections to hit P the possible values of t be t_1 and t_2, then the roots of (9) will be t_1^2 and t_2^2. Hence from the theory of quadratic equation,

$$t_1^2 + t_2^2 = -(4/g^2)(gb - u^2) \equiv (4/g^2)(u^2 - gb) \quad ...(10)$$

and $$t_1^2 t_2^2 = (4/g^2)(a^2 + b^2) = 4r^2/g^2, \text{ where } OP = r \quad ...(11)$$

From (11), $$t_1 t_2 = 2r/g = (2.OP)/g \quad ...(12)$$

TYPE 2 : SOLVED EXAMPLES BASED ON ART. 6.8

Ex. 1. *The angular elevation of an enemy's position on a hill h meters high is β. Show that in order to shell it, the initial velocity of the projectile must not be less than* $\{gh(1 + \operatorname{cosec} \beta)\}^{1/2}$.

[Purvanchal 1997]

Sol. Draw figure as in Art. 6.8 by taking $b = h$. Here PM is a hill of height h meters and P is the position of the enemy. Let O be the point from which the enemy's position is to be shelled. Then, the least velocity u for the required purpose is given by

$$u = \{g(PM + OP)\}^{1/2} = \{g(h + h \operatorname{cosec} \beta)\}^{1/2} = \{gh(h + \operatorname{cosec} \beta)\}^{1/2}$$

[∵ $PM = h$ and from ΔOPM, $OP = PM \operatorname{cosec} \beta = h \operatorname{cosec} \beta$]

Ex. 2. *Show that the product of the two times of flight from P to Q with a given velocity of projection is (2PQ)/g.* **[Avadh 1993]**

Sol. Refer part (*iii*) of Art. 6.8. In figure take O and P as P and Q respectively for the present problem. Then, the result follows from relation (12).

Ex. 3. *If* t_1 *and* t_2 *be the times of flight from A to B and β the inclination of AB to the horizontal, prove that* $t_1^2 + 2t_1t_2 \sin \beta + t_2^2$ *is indepenent of β.*

Sol. Draw figures as in Art. 6.8. Take A and B in place of O and P respectively. Then, as in part (*iii*), do upto relation (11). Then

$$t_1^2 + t_2^2 = (4/g^2)(u^2 - gb) \quad \text{and} \quad t_1^2 t_2^2 = (4/g^2)(a^2 + b^2) \qquad \dots(1)$$

From figure, $\sin \beta = b/(a^2 + b^2)^{1/2}$ ...(2)

$$\therefore \quad t_1^2 + 2t_1t_2 \sin \beta + t_2^2 = (t_1^2 + t_2^2) + 2(t_1t_2) \sin \beta$$

$$= \frac{4(u^2 - gb)}{g^2} + \frac{4(a^2 + b^2)^{1/2}}{g} \times \frac{b}{(a^2 + b^2)^{1/2}}, \text{ by (1) and (2)}$$

$$= (4u^2/g^2) - (4b/g) + (4b/g) = 4u^2/g^2,$$

which is independent of a and b and hence independent of β.

Ex. 4. *Two points P and Q are at a distance a apart, their heights above the ground being* h_1 *and* h_2. *Prove that the least velocity with which a particle can be thrown from the ground level, so as to pass through both the points, is* $\{g(a + h_1 + h_2)\}^{1/2}$

6.9 Range and time of flight up an inclined plane

A particle is projected with velocity u making an angle α with the horizon from a point O on an inclined plane inclined at an angle β to the horizon, in the vertical plane through the line of greatest slope. If the particle strikes the inclined plane, to determine the range and the time of flight.

[Delhi Maths (Hon.) 1996; Bangalore 1992, 93]

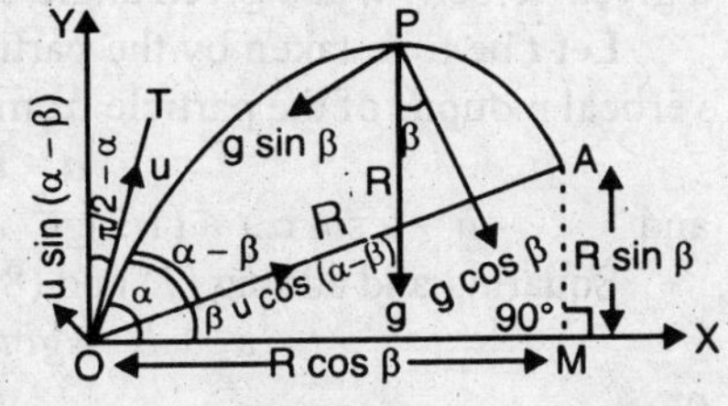

Suppose the particle strike the inclined plane at A. Let $OA = R$. Let T be the time of flight from O to A. As shown in the figure, the components of initial velocity of the particle along and perpendicular to the inclined plane are $u \cos(\alpha - \beta)$ and $u \sin(\alpha - \beta)$ respectively. Again, the component of g along the inlcined is $g \sin \beta$ (down the plane) and the component of g perpendicular to the inclined plane is $g \sin \beta$ (along the downward normal to the plane OA) Let time taken from O to A be T.

While moving from O to A the displacement of the particle perpendicular to OA is zero. So considering motion of the particle from O to A perpendicular to OA and using the formula "$s = ut + (1/2)ft^2$", we have

$$0 = u \sin(\alpha - \beta).T - (1/2)g \cos \beta.T^2 \quad \text{or} \quad T\{g \cos \beta.T - 2u \sin(\alpha - \beta)\} = 0$$

Since $T = 0$ gives time from O to O, hence time from O to A is given by

$$\therefore \quad \textbf{T = time of flight up the inclined plane} = \frac{2u \sin(\alpha - \beta)}{g \cos \beta} \qquad \dots(1)$$

Again, considering the motion of the particle from O to A up the inclined plane OA and using the formula "$s = ut + (1/2)ft^2$", we get

$$R = u\cos(\alpha-\beta).T-(1/2)\,g\sin\beta\,T^2 = T\{u\cos(\alpha-\beta)-(1/2)\,g\sin\beta.T\}$$

$$= \frac{2u\sin(\alpha-\beta)}{g\cos\beta}\left\{u\cos(\alpha-\beta)-\frac{g\sin\beta}{2}\times\frac{2u\sin(\alpha-\beta)}{g\cos\beta}\right\}, \text{ using (1)}$$

$$= \frac{2u^2\sin(\alpha-\beta)}{g\cos^2\beta}\{\cos\beta\cos(\alpha-\beta)-\sin\beta\sin(\alpha-\beta)\}$$

$$\therefore\ \mathbf{R = \text{range up an incline plane} = OA = \frac{2u^2\sin(\alpha-\beta)\cos\alpha}{g\cos^2\beta}} \qquad \text{...(2)}$$

Maximum range up an inclined plane. Re-writing (2), we get

$$R = \frac{u^2}{g\cos^2\beta}\{2\sin(\alpha-\beta)\cos\alpha\} = \frac{u^2}{g\cos^2\beta}\{\sin(2\alpha-\beta)-\sin\beta\} \qquad \text{...(3)}$$

From (3) we see that for given fixed values of u and β, R is maximum where $\sin(2\alpha-\beta)$ is maximum *i.e.*, when $2\alpha-\beta=\pi/2$

Hence for maximum range up an inclined plane, $\alpha = \pi/4+\beta/2$ ...(4)

$$\textbf{Maximum range up an inclined plane} = \frac{u^2(1-\sin\beta)}{g(1-\sin^2\beta)} = \frac{u^2}{g(1+\sin\beta)} \qquad \text{...(5)}$$

Remark. Re-writing (4), $\alpha-\beta=\pi/2-\alpha$, showing that the angle between the direction of projection OT and the inclined plane is the same as the angle between the direction of projection and vertical line OY. *Thus in the case of maximum range on the inclined plane the direction of projection bisects the angle between the vertical line and the inclined plane.*

Again, we know that vertical line OY at O is perpendicular to the directrix of the parabolic path of the particle and the direction of projection at O is along the tangent OT to the parabolic path at O. Since in the case of a parabola the tangent at any point bisects the angle between the focal distance of that point and the perpendicular from that point to the directrix, hence the line of greatest slope through O must pass through the focus.

It follows that when the range on an inclined plane is maximum, the focus of the parabolic path lies on the inclined plane.

6.10 Range and time of flight down an inclined plane. Let O be the point of projection, u the velocity of projection and α the angle of projection with horizon. Let the inclined plane make an angle β with the horizon. Proceed as in Art. 6.9 to show that

$$\therefore\ \mathbf{T = \text{time of flight down the inclined plane} = \frac{2u\sin(\alpha+\beta)}{g\cos\beta}} \qquad \text{...(1)}$$

$$\therefore\ \mathbf{R = \text{range down an inclined plane} = \frac{2u^2\sin(\alpha+\beta)\cos\alpha}{g\cos^2\beta}} \qquad \text{...(2)}$$

From (2), $\quad R = (u^2/g\cos^2\beta)\{\sin(2\alpha+\beta)+\sin\beta\}$ ...(3)

For given u and β, R is maximum when $\quad \sin(2\alpha+\beta)=1$

or $\quad 2\alpha+\beta = \pi/2 \quad$ so that $\quad \boldsymbol{\alpha = \pi/4-\beta/2},$ ...(4)

which gives the angle of projection for maximum range.

$\therefore$ **Maximum range down an inclined plane = $u^2/g(1-\sin\beta)$** ...(5)

Remark. On comparising the results (1), (2) (4) and (5) of this article with the corresponding results (1), (2), (4) and (5) of the previous article, we note that if we replace β by $-\beta$ in the results of motion up an inclined plane, we get the corresponding results for motion down the inclined plane.

TYPE 3 : SOLVED EXAMPLES BASED ON ART. 6.9 AND 6.10

Ex. 1. *For a given velocity of projection the maximum range down an inclined plane is three times the range up the inclined plane. Show that the inclination of the plane to the horizontal is 30°.*

(Purvanchal 1997)

Sol. Let u be the velocity of projection and β the required inclination of the plane to the horizontal. By the problem, we have

Max. range down the plane $= 3 \times$ max. range up the plane

$$\therefore \quad u^2/g(1 - \sin\beta) = 3 \times \{u^2/g(1 + \sin\beta)\}$$

or $\quad 1 + \sin\beta = 3(1 - \sin\beta)$ or $\sin\beta = 1/2$ so that $\beta = 30°$

Ex. 2. *Show that if a gun be situated on an inclined plane, the maximum range in a direction at right angles to the line of greatest slope is a harmonic mean between the maximum ranges up and down the plane respectively.*

Sol. Let O be the point of projection and u the velocity of projection. Let β be the inclination of the plane to the horizontal. Then, we have

$$\text{Max. range up the plane} = u^2/g(1 + \sin\beta) = R_1, \text{ say} \qquad \ldots(1)$$

and $$\text{max. range down the plane} = u^2/g(1 - \sin\beta) = R_2 \text{ say} \qquad \ldots(2)$$

Since the line of greatest slope through O is the line lying in the inclined plane and at right angles to the line in which the inclined plane meets the horizontal plane, hence the direction through O at right angles to the line of greatest slope is a horizontal direction. If R_3 is maximum range in the horizontal direction, then we have

$R_3 =$ max. range in horizontal direction with vel. of projection $u = u^2/g$ $\qquad \ldots(3)$

Now, $\dfrac{1}{R_1} + \dfrac{1}{R_2} = \dfrac{g(1+\sin\beta)}{u^2} + \dfrac{g(1-\sin\beta)}{u^2} = \dfrac{2g}{u^2} = \dfrac{2}{R_3}$, by (1), (2) and (3)

Thus, the harmonic mean of R_1 and R_2 is R_3 as required.

Ex. 3. (*a*) *Show that the greatest range up an inclined plane through the point of projection is equal to the distance through which a particle could fall freely during the corresponding time of flight.*

(*b*) *Show that the time of a particle describing an arc of its path cut off by a focal chord is equal to the time of falling vertically from rest through a height equal to the chord.*

(*c*) *If R is the maximum range up an inclined plane and T the corresponding time of flight, prove that* $R = (1/2)gT^2$.

Sol. (*a*) Let u be the velocity of projection and β be the inclination of the plane to the horizontal. We know that max. range R up the inclined plane is

$$R = u^2/g(1 + \sin\beta) \qquad \ldots(1)$$

and for the max. range the angle of projection α is given by $\alpha = \pi/4 + \beta/2$. Hence the corresponding time of flight to cover range R is given by

$$T = \frac{2u\sin(\alpha-\beta)}{g\cos\beta} = \frac{2u\sin(\pi/4+\beta/2-\beta)}{g\cos\beta} = \frac{2u\sin(\pi/4-\beta/2)}{g\cos\beta}$$

$\therefore$ The distance moved (under gravity) by a particle in this time T

$$= \frac{1}{2}gT^2 = \frac{g}{2}\cdot\frac{4u^2\sin^2(\pi/4-\beta/2)}{g\cos^2\beta} = \frac{u^2\{1-\cos(\pi/2-\beta)\}}{g(1-\sin^2\beta)} = \frac{u^2}{g(1+\sin\beta)}$$

$= R =$ max. range up the plane, by (1)

(***b***) Refer remark of Art. 6.9. We know that when the range on an inclined plane is maximum, the focus of the parabolic path lies on the inclined plane *i.e.*, max. range is the length of the focal chord through the point of projection. Let the given focal chord make an angle β with the horizontal. Now do exactly as in part (*a*).

(*c*) Proceed as in part (*a*).

Ex. 4. *If from a point on the side of a hill two bodies are projected in the vertical plane through the line of greatest slope with the same velocity but in directions at right angles to each other, show that the difference of their ranges is independent of their angles of projection.*

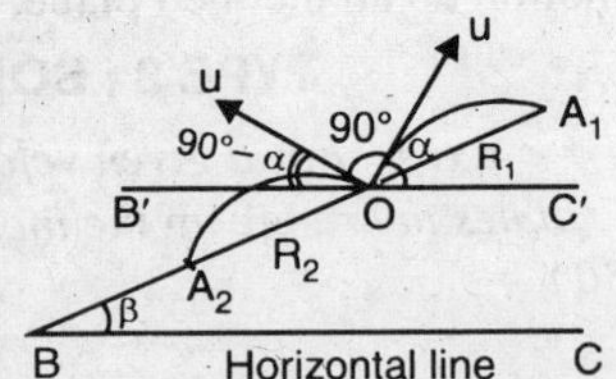

Sol. Let O be the point of projection and u the velocity of projection. Let β be the inclination of the plane to the horizontal BC. Let $B'C'$ be horizontal line through O. Let first particle be projected from O making an angle α with $B'C'$. If $R_1 (= OA_1)$ is the range up the hill, then

$$R_1 = (2u^2/g \cos^2 \beta) \cos \alpha \sin (\alpha - \beta) \quad \ldots(1)$$

Now the second particle projected from O with velocity u makes 90° with the direction of the first particle. So, the direction of the second particle will be inclined at $\pi/2 - \alpha$ with $B'C'$ as shown in the figure. If $R_2 (= OA_2)$ be the range down the hill for the second particle, then

$$R_2 = (2u^2/g \cos^2 \beta) \cos (\pi/2 - \alpha) \sin (\pi/2 - \alpha + \beta)$$

[Replace α by $\pi/2 - \alpha$ in result (2) of Art. 6.10]

or $$R_2 = (2u^2/g \cos^2 \beta) \sin \alpha . \cos (\alpha - \beta) \quad \ldots(2)$$

From (1) and (2), we get

$$R_2 - R_1 = (2u^2/g \cos^2 \beta) \{\sin \alpha \cos (\alpha - \beta) - \cos \alpha \sin (\alpha - \beta)\}$$

$$= (2u^2/g \cos^2 \beta) \sin \{\alpha - (\alpha - \beta)\} = (2u^2 \sin \beta)/g \cos^2 \beta,$$

which is independent of the angle of projection α.

Ex. 5. *A stone is thrown at an angle α with the horizontal from a point in a plane whose inclination to the horziontal is β, the trajectory lying in the vertical plane containing the line of greatest slope. Show that if γ be the elevation of that point of the path which is most distant from the inclined plane, then $2 \tan \gamma = \tan \alpha + \tan \beta$.* **[I.A.S. 1996]**

Sol. Let O be the point of projection, u the velocity of projection and α the angle of projection. OA is the given inclined plane such that $\angle AOX = \beta$. Take O as origin and the horizontal and vertical lines through O as coordinate axes. Let (h, k) be the point of the trajectory which is most distant from OA. Then the tangent PT at P to the trajectory will be parallel to OA. Given that $\angle POX = \gamma$. Now, the equation of the trajectory is

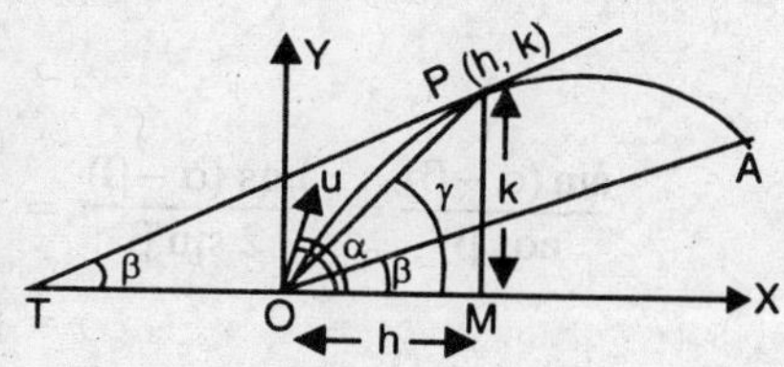

$$y = x \tan \alpha - (gx^2)/2u^2 \cos^2 \alpha \quad \ldots(1)$$

From (1), $$dy/dx = \tan \alpha - (gx)/u^2 \cos^2 \alpha, \quad \ldots(2)$$

giving slope of tangent at any point (x, y) of the trajectory.

Since slope of PT at (h, k) is $\tan \alpha$, (2) gives

$$\tan \beta = \tan \alpha - (gh)/u^2 \cos^2 \alpha \quad \ldots(3)$$

Since $P(h, k)$ lies on (1), $$k = h \tan \alpha - (gh^2)/2u^2 \cos^2 \alpha$$

or $$k/h = \tan \alpha - (gh)/2u^2 \cos^2 \alpha \quad \ldots(4)$$

From ΔOPM, $$k/h = \tan \gamma \quad \ldots(5)$$

Re-writing (3), $$(gh)/u^2 \cos^2 \alpha = \tan \alpha - \tan \beta \quad \ldots(6)$$

Using (5) and (6) in (4), $$\tan \gamma = \tan \alpha - (1/2)(\tan \alpha - \tan \beta)$$

or $2 \tan \gamma = 2 \tan \alpha - \tan \alpha + \tan \beta$ or $2 \tan \gamma = \tan \alpha + \tan \beta$

Ex. 6. (*a*) *A particle is projected at an angle α with the horiziontal from the foot of the plane, whose inclination to the horizontal is β. Show that it will strike the plane at right angles if $\cot \beta = 2 \tan (\alpha - \beta)$.* **[Bangalore 1997, 94, I.A.S. 1989, Chennai 98]**

(*b*) *A shot is fired at an angle α to the horizontal up an hill of inclination β to the horizontal. Show that it strikes the hill (i) normally if $\tan \alpha = 2 \tan \beta + \cot \beta$ (ii) horizontall if $\tan \alpha = 2 \tan \beta$.*

Sol. (*a*) Refer figure of Art. 6.9 and do upto relation (1) as in Art. 6.9.

Thus, $$\text{time of flight} = T = (2u/g \cos \beta) \sin (\alpha - \beta) \quad \ldots(1)$$

Since the particle strikes the plane OA at right angles at A, hence the direction of velocity of the particle at A is perpendicular to OA and so the component of velocity of the particle at A along OA is zero. So considering the motion of the particle from O to A along OA and using the formula "$v = u + ft$",

$$0 = u \cos (\alpha - \beta) - g \sin \beta . T \quad \text{or} \quad T = (u/g \sin \beta) \cos (\alpha - \beta) \quad \ldots(2)$$

From (1) and (2), $$(2u/g \cos \beta) \sin (\alpha - \beta) = (u/g \sin \beta) \cos (\alpha - \beta)$$

or $$2\tan(\alpha-\beta)=\cot\beta \qquad \ldots(3)$$

(*b*) Try yourself.

Ex. 7. *A particle is projected from O at an elevation α. Prove that there are two positions A on its path at which the direction of velocity is perpendicular to OA. Prove that for real positions to exist, α is not less that* $\tan^{-1}(1/3)$; *also if in the two positions OA makes angles* θ_1, θ_2 *with the horizontal, prove that* $\theta_1+\theta_2=\alpha$.

Sol. Try yourself.

Ex. 8. *A particle is projected with a velocity u from a point on an inclined plane whose inclination to the horizon is β, and strikes it at right angles. Show that*

(*i*) *the time of flight is* $(2u)/g(1+3\sin^2\beta)^{1/2}$. **[Bangalore 1997]**

(*ii*) *the range on the inclined plane is* $(2u^2\sin\beta)/g(1+3\sin^2\beta)$.

(*iii*) *the vertical height of the point struct, above the point of projection is* $(2u^2\sin^2\beta)/g(1+3\sin^2\beta)$.

Sol. Refer figure of Art. 6.9. Now proceed as in part (*a*) of Ex. 6 to show that if particle strikes *OA* at right angles then

$$2\tan(\alpha-\beta)=\cot\beta$$

$$\Rightarrow \frac{\sin(\alpha-\beta)}{\cos\beta}=\frac{\cos(\alpha-\beta)}{2\sin\beta}=\frac{\{\sin^2(\alpha-\beta)+\cos^2(\alpha-\beta)\}^{1/2}}{(\cos^2\beta+4\sin^2\beta)^{1/2}}$$

$$\left[\because \text{If } \frac{a}{b}=\frac{c}{d}, \text{ then each fraction}=\frac{(a^2+b^2)^{1/2}}{(c^2+d^2)^{1/2}}\right]$$

$$\therefore \frac{\sin(\alpha-\beta)}{\cos\beta}=\frac{\cos(\alpha-\beta)}{2\sin\beta}=\frac{1}{(1-\sin^2\beta+4\sin^2\beta)^{1/2}}$$

$$=1/(1+3\sin^2\beta)^{1/2} \qquad \ldots(1)$$

(*i*) $$T=\text{Time of flight}=\frac{2u\sin(\alpha-\beta)}{g\cos\beta}=\frac{2u}{g\cos\beta}\times\frac{\cos\beta}{(1+3\sin^2\beta)^{1/2}}$$

$$=(2u)/g(1+3\sin^2\beta)^{1/2} \qquad \text{[using (1)]}$$

(*ii*) R = range up the inclined plane = OA

$$=\frac{2u^2\sin(\alpha-\beta)\cos\alpha}{g\cos^2\beta}=\frac{2u^2\sin(\alpha-\beta)}{g\cos^2\beta}\cos\{(\alpha-\beta)+\beta\}$$

$$=\frac{2u^2\sin(\alpha-\beta)}{g\cos^2\beta}\{\cos(\alpha-\beta)\cos\beta-\sin(\alpha-\beta)\sin\beta\}$$

$$=\frac{2u^2}{g\cos^2\beta}\cdot\frac{\cos\beta}{(1+3\sin^2\beta)^{1/2}}\left[\frac{2\sin\beta}{(1+3\sin^2\beta)^{1/2}}\cos\beta-\frac{\cos\beta}{(1+3\sin^2\beta)^{1/2}}\sin\beta\right]$$

[using (1)]

$$\therefore \quad R=(2u^2\sin\beta)/g(1+3\sin^2\beta) \qquad \ldots(2)$$

(*iii*) Required vertical distance = AM (see figure)

$$=R\sin\beta=(2u^2\sin^2\beta)/g(1+3\sin^2\beta), \text{ using (2)}$$

Ex. 9. *Prove that the minimum velocity required to project a particle from a height h to fall at a distance d from the point of projection is* $\{g(d-h)\}^{1/2}$.

u
O
d
h
β
A
B
X

Sol. Let *O* be the point of projection and *A* the point where the particle strikes the inclined plane *OA*. Given $OB=h$, $OA=d$. Let $\angle OAB=\beta$.

For minimum initial velocity *u* of the particle, *OA* must be the maximum range down the plane.

$\therefore\ d=OA=$ max. range down the plane $=u^2/g(1-\sin\beta)$ ...(1)

Also from ΔOAB, $\sin\beta=h/d$ ...(2)

From (1) and (2), $\quad d = \dfrac{u^2}{g(1-h/d)}$ or $u = \{g(d-h)\}^{1/2}$

Ex. 10. (*a*) *A fort and a ship are both armed with guns give their projectiles a muzzle velocity $(2gk)^{1/2}$ and guns in the fort are at height h above the ship. If d_1 and d_2 are the greatest horizontal ranges at which the fort and ship, respectively, can engage, prove that $d_1/d_2 = \{(k+h)/(k-h)\}^{1/2}$*

(*b*) *A fort is on the edge of a cliff of height h. Show that there is an annular region of area $8\pi hk$ in which the fort is out of the range of the ship, but the ship is not out of the range of the fort, where $(2gk)^{1/2}$ is the velocity of the shells used by both.*

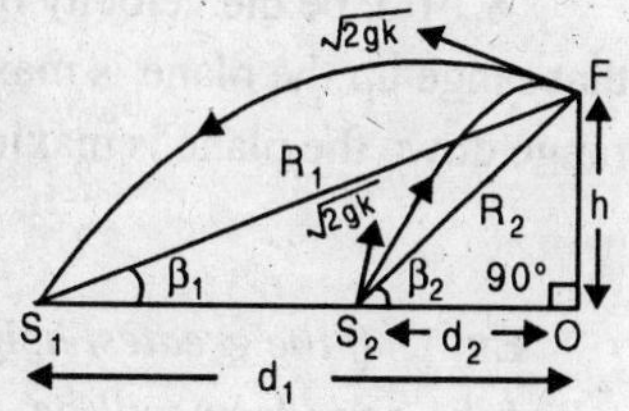

Sol. (*a*) Let F be the fort, where $OF = h$. Let S_1 be the farthest position of the ship where it can be hit from the fort. Let $OS_1 = d_1$, $FS_1 = R_1$ and $\angle FS_1O = \beta_1$. From ΔFS_1O it is clear that for fixed value of h, d_1 is maximum when R_1 is maximum. Now, we have

$$R_1 = \text{max. range down the plane } FS_1 = u^2/g(1-\sin\beta_1)$$

$\therefore \quad R_1 g(1-\sin\beta_1) = u^2 \quad$ or $\quad R_1 g(1-h/R_1) = 2gk$

or $\quad R_1 - h = 2k \quad$ or $\quad R_1 = h + 2k$

From ΔFS_1O, $\quad d_1^2 = OS_1^2 = R_1^2 - h^2 = (h+2k)^2 - h^2 = 4k(k+h)$

Thus $\quad d_1 = 2\{k(k+h)\}^{1/2}$ ... (1)

Now, let S_2 be the farthest position of the ship from where the fort could be hit. Let $OS_2 = d_2$, $FS_2 = R_2$ and $\angle FS_2O = \beta_2$. Then as before d_2 is maximum where R_2 is maximum. As, before, we have

$$R_2 = \text{max. range up the plane} = \frac{u^2}{g(1+\sin\beta_2)} = \frac{2gk}{g(1+h/R_2)}$$

or $\quad R_2 + h = 2k \quad$ so that $\quad R_2 = 2k - h$

$\therefore$ from ΔOS_2F, $\quad d_2^2 = OS_2^2 = R_2^2 - h^2 = (2k-h)^2 - h^2 = 4k(k-h)$

Thus, $\quad d_2 = 2\{k(k-h)\}^{1/2}$ ...(2)

From (1) and (2), $\quad d_1/d_2 = \{(k+h)/(k-h)\}^{1/2}$

(*b*) Refer figure of part (*a*). Doing as before, get

$$d_1^2 = 4k(k+h) \quad \text{and} \quad d_2^2 = 4k(k-h) \qquad \text{...(3)}$$

From figure, note that if the ship is anywhere between S_1 and S_2 then the fort will be out of the range of ship but the ship is not out of range of the fort. If the line OS_1 revolves about O, then there would be an annular region bounded by the concentric circles with centre at O and radii d_1 and d_2 in which the fort is out of range of the ship while the ship is not out of the range of the fort.

$\therefore$ The required annular region

$= \pi(d_1^2 - d_2^2) = \pi\{4k^2 + 4kh - (4k^2 - 4kh)\} = 8\pi kh$, by (3)

Ex. 11. *A point is projected with a velocity 25 ms^{-1} so as to strike at right angle a plane through the point of projection inclined at 45° to the horizontal. Prove that the range on the given plane in $25\sqrt{2}$ meters.* **[Bangalore 2004]**

Sol. Here $u = 25$ ms^{-1}, $g = 10$ ms^{-2} and $\beta = 45°$. Refer part (*ii*) of Ex. 8. Then required range

$$= \frac{2u^2}{g} \times \frac{\sin\beta}{1+3\sin^2\beta} = \frac{2\times(25)^2}{10} \times \frac{(1/\sqrt{2})}{1+(3/2)} = 25\sqrt{2} \text{ meters}$$

EXERCISES

1. A particle is projected with velocity u from a point on a plane inclined at an angle β to the horizontal. If r and r' are its maximum ranges up and down the plane, prove that $1/r + 1/r'$ is independent of the inclination of the plane.

2. Show that for a given velocity of projection the maximum range down a plane of inclination α is greater than up the plane in the ratio $(1 + \sin \alpha) : (1 - \sin \alpha)$

3. The line joining C to D is inclined at an angle α to the horizontal. Show that the least velocity required to shoot from C to D is $\tan(\pi/4 + \pi/2)$ times the least velocity required to short from D to C.

4. If u be the velocity of projection and v_1 the velocity of striking the plane when projected so that range up the plane is maximum and v_2 the velocity of striking the plane when projected so that range down the plane is maximum, prove that $u^2 = v_1 v_2$.

MISCELLANEOUS PROBLEMS

Ex. 1. *If the greatest height to which a man can throw a stone is h, then the greatest distance to which he can throw will be*

(a) h/2 *(b) h* *(c) 2h* *(d) 4h.* **[I.A.S. Prel. 1998]**

Sol. Ans. (c). Refer Ex. 2(*b*) on page 6.6.

Ex. 2. *A bullet is shot with a velocity 200 ms^{-1}. Find the angle at which it is shot, if the greatest height attained by it is 1500 meters.* **[Bangalore 2004]**

Sol. Here $u = 200$ ms^{-1}, $g = 10$ ms^{-2}. We want α such that

$$1500 = (u^2 \sin^2 \alpha/2g) = (200 \times 200 \times \sin^2 \alpha)/(2 \times 10), \text{ giving}$$

$$\sin^2 \alpha = 3/4 \quad \text{or} \quad \sin \alpha = \sqrt{3}/2 \quad \text{or} \quad \alpha = 60°. \textbf{ Ans.}$$

Ex. 3. *A particle is projected with a velocity 80 ms^{-1} at an angle 60° from the foot of a plane at angle 30°. Find the range on the inclined plane.* **[Bangalore 2004]**

Sol. Here $u = 80$ ms^{-1}, $g = 10$ ms^{-2}, $\alpha = 60°$ and $\beta = 30°$.

$$\therefore \text{ Required range} = \frac{2u^2 \sin(\alpha - \beta) \cos \alpha}{g \cos^2 \beta} = \frac{2 \times (80)^2 \sin 30° \cos 60°}{10 \times \cos^2 30°}$$
$$= 2 \times 80 \times 80 \times (1/4) \times (4/3) \times (1/10) = (1280/3)\text{ m}$$

Ex. 4. *A particle is projected with velocity v at an angle (<45°) to the horizontal and reaches a point on the horizontal distance R from the point of projection. What is the greatest height (h) attained during the path of projectile ?*

(a) $h = (v^2/4g) \times \{1 - \sqrt{1 - (g^2R^2/v^4)}\}$ *(b)* $h = (v^2/4g) \times \{1 + \sqrt{1 - (g^2R^2/v^4)}\}$

(c) $h = (v^2/2g) \times \{1 - \sqrt{1 - (g^2R^2/v^4)}\}$ *(d)* $h = (v^2/2g) \times \{1 + \sqrt{1 - (g^2R^2/v^4)}\}$

Sol. Ans. (*a*). Here $R = (v^2 \sin 2\alpha)/g$ so that $\sin 2\alpha = gR/v^2$ ... (1)

$$\therefore \; h = \frac{v^2 \sin^2 \alpha}{2g} = \frac{v^2}{4g}(1 - \cos 2\alpha) = \frac{v^2}{4g}\{1 - \sqrt{1 - \sin^2 2\alpha}\} = \frac{v^2}{4g}\{1 - \sqrt{1 - (g^2R^2/v^4)}\}$$

Ex. 5. *A particle is projected with initial velocity V from a point O at an angle of elevation α. Show that the range on a plane through O inclined at β to horizontal* $(\beta < \alpha < \pi/2)$ *is* $V^2\{\sin(2\alpha - \beta) - \sin\beta\}/g\cos^2\beta$ *and hence deduce that for a given velocity of projection V, the maximum range is* $V^2/g(1 + \sin\beta) = R$, *(say).*

If we now vary β, show that the locus of the point having polar co-ordinates (R, β) is parabola having O as focus, semi-latus vectum V^2/g and axis $\beta = \pi/2$. **[Delhi Maths (Hons.) 2007]**

Sol. For the first part, refer Art. 6.9 upto equation (5).

Second part Here $$V^2/g(1+\sin\beta) = R, \qquad \text{... } (i)$$

where *V, g* are constant quantities and β, *R* are variable quantities. Re-writing (*i*), we have

$$\frac{V^2}{gR} = 1+\sin\beta \qquad \text{or} \qquad \frac{(V^2/g)}{R} = 1+\cos(\pi/2-\beta) \qquad \text{... } (ii)$$

Comparing (*ii*) with $l/r = 1+e\cos(\pi/2-\theta)$, we find that locus of (*R*, β) is a parabola having *O* as focus, semi-latus rectum V^2/g and axis $\beta = \pi/2$.

Ex. 6. *A particle passes through the point (x, y). If R is the horizontal range, prove that the angle of projection is* $\tan^{-1}\left(\frac{y}{x}\frac{R}{R-x}\right)$. **[Guwahati 2007]**

Sol. Proceed as in solved Ex. 4, page 6.5 and show that

$$\frac{yR}{xR-x^2} = \tan\alpha \qquad \text{giving} \qquad \alpha = \tan^{-1}\left(\frac{y}{x}\frac{R}{R-x}\right)$$

Ex. 7. *If a particle with intial velocity u is projected upward in a direction inclined at an angle 60° to the horizontal, then what is the velocity when it is at the greatest height?*

(*a*) *u* (*b*) 2*u* (*c*) *u*/2 (*d*) $u\sqrt{3}/2$

Sol. Ans. (c) Let *v* be the velocity at the greatest height. Since horizontal component of the velocity of projectile remains unaltered throughout the motion, we have

$$v = u\cos 60° \Rightarrow v = u/2$$

Ex. 8. *An aeroplane is flying with uniform speed v, in an arc of a vertical circle of the radius 'a', where centre is at height h vertically above a point O of the ground. If a bomb is dropped from the aeroplane, when at a height Y and strikes the ground at O, show that Y satisfies the equation* $kY^2 + (a^2 - 2hk)Y + k(h^2 - a^2) = 0$ *where* $k = h + (a^2 g/2v^2)$

[Delhi Maths. (H) 2008, 10]

Sol. Let P be the point in the lower half of the circle, from which the bomb is dropped. Let C be the center of circle of radius *a* such that OC = h, where O is on the ground. Let Px and Py be taken as coordinate axes with P as origin as shown in the figure. Let ∠*PCM* = θ. Then, velocity of the bomb will be *v* making an angle θ with Px. Note that at the time of dropping the bomb, the velocity of aeroplane and the bomb will be the same in magnitude and direction.

Equation of the trajectory is given by

$$y = x\tan\theta + (gx^2/2v^2\cos^2\theta) \qquad \text{... (1)}$$

By hypothesis the bomb strikes the ground at O whose coordinates are (PM, Y), i.e., (*a* sin θ, Y), where PM is perpendicular to OC. Hence, we must have

$$Y = (a\sin\theta)\tan\theta + \frac{ga^2\sin^2\theta}{2v^2\cos^2\theta} = a\frac{\sin^2\theta}{\cos\theta} + \frac{ga^2}{2v^2}\tan^2\theta$$

or $\qquad Y = \{a \cos \theta + (ga^2 / 2v^2)\} \times (\sec^2 \theta - 1)$ ... (2)

From ΔPCM, $\qquad CM = a\cos\theta \quad$ or $\quad h - Y = a\cos\theta$ ... (3)

Thus, $\qquad \sec\theta = a/(h-Y)$. Also given $\qquad k - h = a^2 g / 2v^2$ ... (4)

Using (3) and (4), (2) reduces to

$$Y = (h - Y + k - h)\ \{a^2/(h-Y)^2 - 1\} \quad \text{or} \quad Y = (k - Y)\ \{a^2/(h-Y)^2 - 1\}$$

or $\qquad Y\ (h-Y)^2 = (k-Y)\ \{a^2 - (h-Y)^2\} \quad$ or $\quad k(h-Y)^2 - a^2(k-Y) = 0$

or $\qquad kY^2 + (a^2 - 2hk)\ Y + k\ (h^2 - a^2) = 0$

Ex. 9. *Find the greatest distance that a stone can be thrown inside a horizontal tunnel 10 ft high with a velocity of projection 80 ft/sec. Find also the corresponding time of flight.*

Sol. Let the velocity of projection be u inclined at an angle α to the horizontal.

Then, $\qquad R$ = horizontal range = $R = (2u^2 \sin\alpha \cos\alpha)/g$ ...(1)

Also, $\qquad H$ = maximum height attained = $(u^2 \sin^2 \alpha)/2g$ ... (2)

From (1), it follows that R increases with α. However, the trajectory is to remain within the given tunnel whose height in 10 ft. Hence,

$H \le 10 \quad$ or $\quad (u^2 \sin^2 \alpha)/2g \le 10 \quad$ or $\quad \sin^2 \alpha \le 1/10$, using (2)

Hence, maximum value of $\alpha = \sin^{-1}(1/\sqrt{10})$ ...(2)

$\therefore$ from (1), the required greatest range $= \dfrac{(80)^2}{32} \times 2 \times \dfrac{1}{\sqrt{10}} \times \dfrac{3}{\sqrt{10}} = 120\ ft$

[Note that $\sin\alpha = 1/\sqrt{10} \Rightarrow \cos\alpha = (1 - \sin^2\alpha)^{1/2} = 3/\sqrt{10}$]

Also, the corresponding time of flight $= \dfrac{2u \sin\alpha}{g} = \dfrac{2 \times 80}{32} \times \dfrac{1}{\sqrt{10}} = \dfrac{\sqrt{10}}{2} = 1.58$ sec.

Ex. 10. *A particle just clears a wall of height b at a distance a and strikes the ground at a distance c from the point of pojection. Prove that the angle of projection is* $\tan^{-1}\{bc/a(c-a)\}$ *and the velocity of projection V is given by* $2V^2/g = \{a^2(c-a)^2 + b^2c^2\}/ab(c-a)$.

Sol. Equation of trajectory is given by $\qquad y = x\tan\alpha - (gx^2)/(2V^2\cos^2\alpha)$ ...(1)

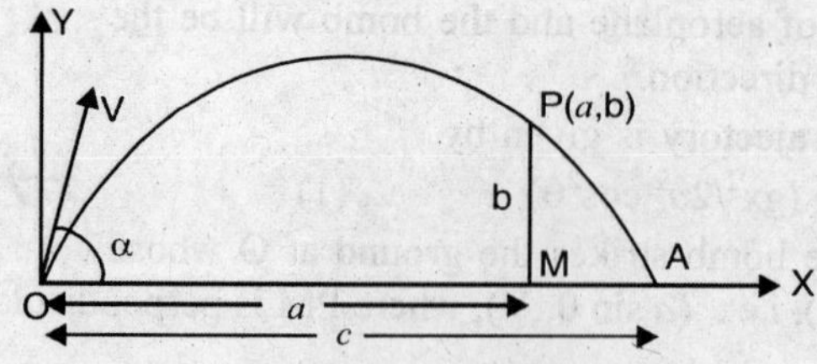

Let PM be a wall of height b such that PM = b and OM = a. Then, P (a, b) lies on (1) and hence

$$b = a\tan\alpha - (ga^2)/(2V^2\cos^2\alpha) \quad ...(2)$$

Now, c = horizontal range = $(V^2 \sin 2\alpha)/g \Rightarrow g/V^2 = (\sin 2\alpha)/c$ ... (3)

Using (3), (2) yields $$b = a\tan\alpha - \frac{a^2}{2\cos^2\alpha} \times \frac{\sin 2\alpha}{c} = a\tan\alpha - \frac{a^2}{2\cos^2\alpha} \times \frac{2\sin\alpha\cos\alpha}{c}$$

or $$b = a\tan\alpha - \frac{a^2}{c}\tan\alpha = a\tan\alpha\left(1 - \frac{a}{c}\right) \quad \text{giving} \quad \tan\alpha = \frac{bc}{a(c-a)} \quad ...(4)$$

From (4), α = the required angle of projection = $\tan^{-1}\{bc/a(c-a)\}$

Now, (3) $$\Rightarrow \sin 2\alpha = \frac{gc}{V^2} \quad \text{or} \quad \frac{2\tan\alpha}{1+\tan^2\alpha} = \frac{gc}{V^2} \quad \text{or} \quad \frac{\dfrac{2bc}{a(c-a)}}{1 + \dfrac{b^2c^2}{a^2(c-a)^2}} = \frac{gc}{V^2}$$

or $$\frac{2abc(c-a)}{a^2(c-a)^2 + b^2c^2} = \frac{gc}{V^2} \quad \text{or} \quad \frac{2V^2}{g} = \frac{a^2(c-a)^2 + b^2c^2}{ab(c-a)}$$

7

Direct and Oblique Impact (Collision of elastic bodies)

7.1 Direct and Oblique Impact. Two bodies are said to *impinge directly* when the direction of motion of each is along the common normal at the point at which they touch. They are said to *impinge obliquely* when the direction of motion of either, or both, is not along the common normal, at the point of contact. The direction of the common normal is called the *line of action.* In the case of two spheres the common normal is the line joining their centres.

7.2 Three Fundamental principles of impact. **(M.S. Univ. T.N. 2007)**

I. Newton's experimental law or Newton's law of restitution

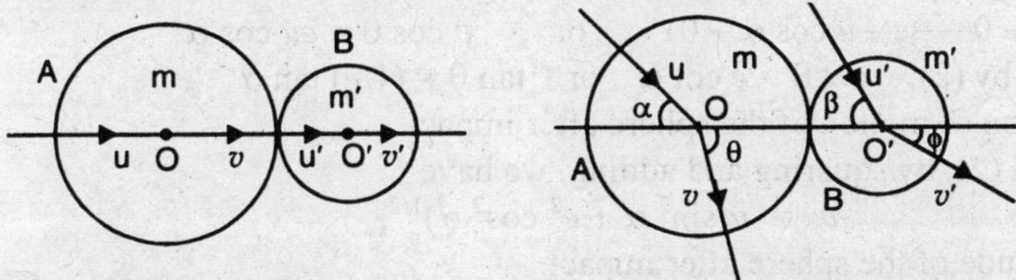

Fig. (*i*) Direct impact **Fig. (*ii*) Oblique impact**

***(i)* For direct impact.** When two bodies impringe directly [see figure (*i*)], their relative velocity after impact is in a constant ratio to their relative velocity before impact, and is in the opposite direction.

Thus, if the velcities of two impinging bodies be u, u' before impact and v, v' after impact, then

$$(v - v')/(u - u') = -e \quad \text{or} \quad v - v' = -e(u - u'),$$

where e is the constant quantity which depends on the substances of which the bodies are made and is independent of their sizes. e is called the *modulus or coefficient of elasticity (or restitution or resilence) of the bodies.* e lies between 0 and 1. When $e = 0$, the body is said to be *inelastic* whereas when $e = 1$, the body is said to be *perfectly elastic.*

***(ii)* For Oblique Impact.** When two bodies impinge obliquely, their relative velocity resolved along their common normal after impact is in a constant ratio to their relative velocity before impact resolved in the same direction and in the opposite direction. Thus, if two bodies of masses m and m' be moving with velocities u, u' in directions, inclined at angles α, and β respectively to their common normal and v, v' be their velocities in directions θ, ϕ respectively after impact [See figure (*ii*)], then

$$v \cos\theta - v' \cos\phi = -e(u \cos\alpha - u' \cos\beta)$$

II. Motion of two smooth bodies perpendicular to the line of impact.

When two smooth bodies impinge, there is no tangential action between them. As there is no impulsive force along the common tangent, there is no change in momentum in that direction. Hence there is no change of velocity for either body along the tangent, in other words, the velocity of either body in a direction perpendicular to the common normal is unaffected by impact. Thus the resolved part of the velocity of each body along the tangent at the point of contact, before impact is equal to its resolved part after the impact. In case of oblique impact (see figure (*ii*)), we have

$$u \sin\alpha = v \sin\theta \quad \text{and} \quad u' \sin\beta = v' \sin\phi$$

III. Principle of conservation of momentum. The algebraic sum of the final momenta of two impinging bodies is equal to the algebraic sum of their initial momenta, both being measured along the common normal. Thus, for case

(*i*) of direct impact, $mv + m'v' = mu + m'u'$

(*ii*) of oblique impact, $mv \cos\theta + m'v' \cos\phi = mu \cos\alpha + m'u' \cos\beta$

7.3 Impact on a fixed plane. *A smooth sphere, or particle, whose mass is m and whose coefficient of restitution is e, impinge obliquely on a fixed plane; find the change in its motion.*

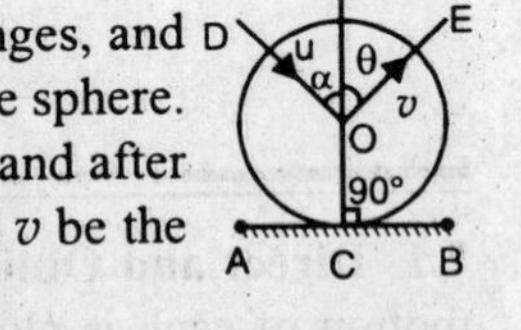

Let AB be the smooth fixed plane, C the point at which the sphere impinges, and CN the normal to the plane at C so that CN passes through the centre, O, of the sphere. Let DO and OE be the directions of motion of the centre of the sphere before and after impact, and let the angles NOD and NOE be α and θ respectively. Let u and v be the velocities of the sphere before and after impact as shown in the figure.

Since the plane is smooth, there is no force parallel to the plane. Hence the resolved parts of velocities of the sphere in a direction parallel to AB remain unaltered before and after impact and so we have

$$v \sin\theta = u \sin\alpha \quad \ldots(1)$$

By Newton's experimental law, for the sphere and the plane (which is fixed before and after impact), we have along common normal CO

$$v\cos\theta - 0 = -e(-u\cos\alpha - 0) \quad \text{or} \quad v\cos\theta = eu\cos\alpha \quad \ldots(2)$$

Dividing (1) by (2), $\cot\theta = e\cot\alpha$ or $\tan\theta = (1/e)\tan\alpha$ ...(3)

giving the direction of motion of the sphere after impact.

From (1) and (2), by squaring and adding, we have

$$v = u(\sin^2\alpha + e^2\cos^2\alpha)^{1/2}, \quad \ldots(4)$$

giving the magnitude of the sphere after impact.

Remark (*i*). *The angles* α *and* θ *are known as the angles of incidence and reflexion respectively.*

(*ii*) If the impact is direct, we have $\alpha = 0$. Thus, from (3) and (4), we have $\theta = 0$ and $v = eu$. Hence, if *a sphere impinges directly on a smooth plane, it rebounds with velocity eu.*

(*iii*) If the sphere is perfectly elastic, *i.e.*, $e = 1$. Then, from (3) and (4), we have $\theta = \alpha$ and $v = u$. *Hence, if a perfectly elastic sphere impinges on a fixed smooth plane, its velocity is unaltered in magnitude by impact and angle of reflexion is equal to the angle of incidence.*

(*iv*) If the sphere is perfectly inelastic, *i.e.*, $e = 0$, then from (3) and (4), we have $\theta = \pi/2$ and $v = u \sin\alpha$. *Hence, if a sphere impinges on a perfectly inelastic fixed plane, it does not rebound but simply slides along the plane, its velocity parallel to the plane remaining unaltered.*

(*v*) Loss in K.E. of the sphere

$= (1/2)mu^2 - (1/2)mv^2 = (1/2)mu^2 - (1/2)mu^2(\sin^2\alpha + e^2\cos^2\alpha)$, by (4)

$= (1/2)mu^2(1 - \sin^2\alpha - e^2\cos^2\alpha) = (1/2)mu^2(1 - e^2)\cos^2\alpha$

(*vi*) The impulse of the force of impact on the plane is equal and opposite to the impulse of the force of impact on the sphere, and is therefore measured by the change of momentum of the sphere perpendicular to the plane.

$\therefore$ the impulse of the blow $= m(v\cos\theta) - m(-u\cos\alpha)$

$= m(v\cos\theta + u\cos\alpha) = m(eu\cos\alpha + u\cos\alpha) = m(1+e)\,u\cos\alpha$, by (1)

Note. Some useful results of motion under gravity (see chapter 1)

I. Velocity attained due to a fall from a height $h = (2gh)^{1/2}$

II. When a particle is projected vertically upwards with velocity u, then

Greatest height attained by the particle $= u^2/2g$

Time to reach the greatest height $= u/g$; Total time of flight $= 2u/g$

Time of ascent = Time of descent $= u/g$

EXAMPLES BASED ON ART. 7.3

Ex. 1. (a) *A heavy elastic ball drops from the ceiling of a room, and after rebounding twice from the floor, reaches a height equal to half that of ceiling, show that the coefficient of restitution is* $1/2^{1/4}$.

Sol. Let h be the height of ceiling and e be the coefficient of restitution.

Then, u = Velocity on reaching the floor due to a fall from height $h = (2gh)^{1/2}$

$\therefore$ Velocity after first rebound = eu

Then, Velocity on reaching the floor before second impact = eu

$\therefore$ Velcity after second rebound = $e \cdot eu = e^2u$

Given that the maximum height attained after second impact = $h/2$

$\therefore$ $(e^2u)^2/2g = h/2$ or $e^4 \cdot 2gh = gh$ or $e^4 = 1/2$ or $e = 1/2^{1/4}$

Ex. 1. (b) *An elastic ball hits a horizontal floor vertically with a speed of u and is allowed to strike the floor another two times after which it rebounds from the floor with a velocity 27u/64. Neglecting the resistance of air, the coefficient of elasticity between the ball and the floor is*

(a) 4/5 *(b) 3/4* *(c) 2/3* *(d) 9/26.* **(I.A.S. Prel. 2003)**

Sol. Ans. (b) Let e be the required coefficient of elasticity.

Velocity after first rebound = eu, velocity after second rebound = e^2u

and velocity after third rebound = e^3u

Thus by given data, $e^3u = 27u/64 \Rightarrow e = 3/4$

Ex. 2. *At what angle must a body whose elasticity is 1/3 be incident on a perfectly hard plane so that the angle between the directions before and after impact be at right angles.*

Sol. Refer figure of Art. 7.3. Let α be the angle of incidence and θ be the angle of reflection. Then given that $\theta + \alpha = 90°$. Hence $\theta = 90° - \alpha$. So

$\tan\theta = (1/e)\tan\alpha$ or $\tan(90° - \alpha) = 3\tan\alpha$, as $e = 1/3$ (given)

or $\cot\alpha = 3\tan\alpha$ or $\tan^2\alpha = 1/3$ or $\tan\alpha = 1/\sqrt{3}$ or $\alpha = 30°$

Ex. 3. *A particle falls from a height h in time t upon a fixed horizontal plane. It rebounds and reaches the maximum height h' in time t', show that*

(i) $t' = et$ *and* $h' = e^2h$

(ii) the whole distance (up and down) described by the particle before it has finished rebounding is $(1 + e^2)h/(1 - e^2)$ *(Guwahati 2007)*

(iii) and that the time that elapses is $\{(1 + e)/(1 - e)\}^{1/2} \times (2h/g)^{1/2}$.

Sol. *(i)* Let the particle acquire a velocity v after falling through a height h in time t. Then, we have

using $v = u + gt$, we get $v = 0 + gt$ or $t = v/g$ ...(1)

and using $v^2 = u^2 + 2gh$, we get $v^2 = 0 + 2gh$ or $h = v^2/2g$ ...(2)

Since velocity of rebound = $e \times$ velocity before rebound,

so the particle rebounds with velocity ev. Since the max. height attained is h' in time t' (given), so

$t' = (ev)/g = (etg)/g = et$, using (1)

and $h' = (ev)^2/2g = (e^2 \cdot 2gh)/2g = e^2h$, using (2)

(ii) The height to which the particle rises after the first rebound is e^2h. Similarly, the heights to which the particle rises after the second, third etc. rebounds are e^4h, e^6h and so on. Distances covered between the first and second rebounds, second and third rebounds, etc. are $2(e^2h)$, $2(e^4h)$ and so on.

$\therefore$ Total distance covered before the particle has finished rebounding

$= h + 2(e^2h) + 2(e^4h) + 2(e^6h) + \ldots$ ad. int.

$= h + \dfrac{2e^2h}{1-e^2}$, since sum of an infinite G.P. $(a + ar + ar^2 + \ldots) = \dfrac{a}{1-r}$

$= \{(1 + e^2)/(1 - e^2)\}h$

(iii) The time of flight (up and down) after the first impact is $2(et)$. Similarly, the times of flight after the second, third etc. impacts are $2(e^2t)$, $2(e^3t)$ and so on.

$\therefore$ Total time taken before the particle has finished rebounding

$= t + 2(et) + 2(e^2t) + 2(e^3t) + \ldots$ ad. int.

$$= t + \frac{2et}{1-e} = \frac{1+e}{1-e}t = \frac{1+e}{1-e}\left(\frac{2h}{g}\right)^{1/2}, \text{ as } t = \left(\frac{2h}{g}\right)^{1/2}, \text{ by (1) and (2)}$$

Ex. 4. *From a point on a smooth horizontal plane, a ball is projected with velocity u at an angle α to the horizon. Show that it will keep rebounding from the plane for a time (2u sin α)/g(1 – e) and will have a range (u² sin 2α)/g(1 – e), e being the coefficient of elasticity.*

Sol. Let the ball be projected from O and let it strike the plane at $A_1, A_2, A_3, \ldots$ at the time of first, second, third etc. impacts between the ball and the plane.

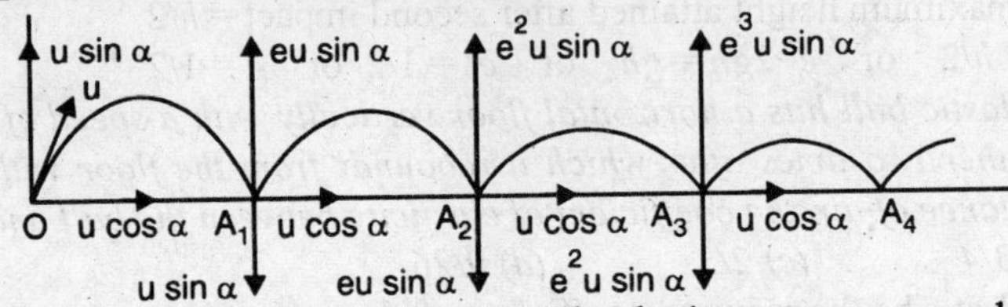

The vertical and horizontal components of initial velocity are $u \cos \alpha$ and $u \sin \alpha$ respectively. The ball describes trajectory from O to A_1 and after striking at A_1, the ball again describes another trajectory from A_1 to A_2 and so on.

Every time the particle rebounds after striking the plane, the vertical comónent of velocity becomes e times of that before striking. Hence the initial vertical velocities after the first, second, ... rebounds are $eu \sin \alpha$, $e^2u \sin \alpha$, $e^3u \sin \alpha$, ... But time of flight = $(2/g)$ (vertical component of velocity)

$\therefore$ Total time that elapsed before the ball ceases to rebound

$$T = \frac{2u \sin \alpha}{g} + \frac{2eu \sin \alpha}{g} + \frac{2e^2u \sin \alpha}{g} + \ldots \text{ ad. inf.}$$

or $$T = \frac{2u \sin \alpha}{g}(1 + e + e^2 + \ldots \text{ ad. inf}) = \frac{2u \sin \alpha}{g(1-e)}$$

During this time the horizontal velocity, being unaltered by the impacts, is always $u \cos \alpha$. So the required horizontal distance described

$$= (u \cos \alpha) \times T = u \cos \alpha \cdot \frac{2u \sin \alpha}{g(1-e)} = \frac{u^2 \sin 2\alpha}{g(1-e)}$$

Ex. 5. (*a*) *A ball is projected with a velocity u at an elevation α from a point distant d from a smooth vertical wall in a plane perpendicular to it. After rebounding from the wall, it returns to the point of projection. Prove that u² sin 2α = gd(1 + 1/e).*

(*b*) *A ball is projected in a plane perpendicular to a smooth vertical wall so that after rebounding, it passes through the point of projection. If the point is at a maximum distance d from the wall, prove that u, the velocity of projection is given by u² = gd(1 + 1/e) where e is the coefficient of restitution.*

(*c*) *A imperfectly elastic ball is projected with velocity (gh)^{1/2} at angle α with horizon, so that it strikes a vertical wall distant d from the point of projection, and returns to the point of projection. Show that the coefficient of restitution between the ball and the wall is d/(h sin 2α – d).*

Sol. (*a*) Let O be the point of projection and AB, the wall. The vertical and horizontal components of initial velocity u at O are $u \cos \alpha$ and $u \sin \alpha$. Let t_1 be the time taken by the ball to reach the wall and t_2 the time taken by it to come back to the point of projection O. After striking wall at A, horizontal component of velocity become $eu \cos \alpha$.

Since the horizontal component of velocity $u \cos \alpha$ from O to A and $eu \cos \alpha$ from A to O remain unaltered, so

$$t_1 = d/(u \cos \alpha) \qquad \text{and} \qquad t_2 = d/(eu \cos \alpha) \qquad \ldots(1)$$

Clearly total distance travelled during $(t_1 + t_2)$ is zero. Since the vertical component of velocity $u \sin \alpha$ is not altered by impact at A. So considering motion from O to A and from A to O back in vertical direction and using $h = ut - (1/2)gt^2$, we get $0 = u \sin\alpha \cdot (t_1 + t_2) - (1/2)g(t_1 + t_2)^2$ or $t_1 + t_2 = (2u/g)\sin\alpha$

or $$\frac{d}{u\cos\alpha} + \frac{d}{eu\cos\alpha} = \frac{2u\sin\alpha}{g} \quad \text{or} \quad u^2 \sin 2\alpha = gd\left(1 + \frac{1}{e}\right) \qquad \text{...(2)}$$

(*b*) Since *d* is maximum horizontal range (given). So here $\alpha = 45°$. So proceed as in part (*a*) and put $\alpha = 45°$ in (2). Then, $u^2 = gd(1 + 1/e)$.

(*c*) Here $u = (gh)^{1/2}$. So (2) reduces to

$$gh\sin 2\alpha = gd\left(1 + \frac{1}{e}\right) \quad \text{or} \quad 1 + \frac{1}{e} = \frac{h\sin 2\alpha}{d} \quad \text{or} \quad e = \frac{d}{h\sin 2\alpha - d}$$

Ex. 6. *A particle is projected with velocity u from a point on the ground so as to strike a smooth vertical wall and after impact to stirke the ground at a point mid way between the point of projection and the wall. Show that the angle of projection is (1/2) $\sin^{-1}$ {(1 + 2e)ag/2eu²} where a is the distance of the wall from the point of projection and e is the coefficient of restitution between the particle and the wall.*

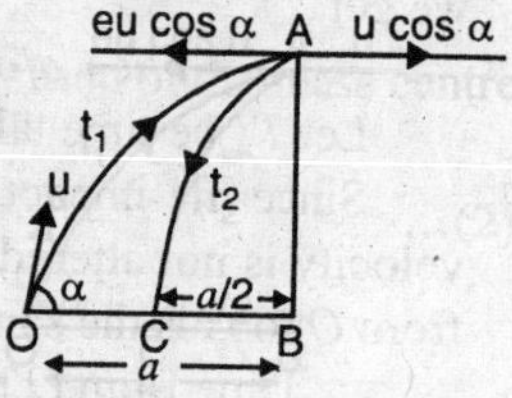

Sol. Let *O* be the point of projection and *AB*, the wall. Given $OB = a$. Let *C* be the middle point of *OB* so that $BC = a/2$. Let α be the angle of projection. Horizontal component of velocity for the first trajectory *OA* is $u\cos\alpha$ which remains unaltered.

$\therefore \quad t_1$ = time from *O* to $A = a/u\cos\alpha$ ...(1)

After striking the wall at *C*, the horizontal component of velocity of the particle becomes $eu\cos\alpha$. Given that the particle strikes the ground at *C*. So

t_2 = time from *A* to $C = (a/2)/(eu\cos\alpha) = a/(2eu\cos\alpha)$ ...(2)

The vertical component of velocity is not altered by the impact at *A*. Hence, in time $t_1 + t_2$ the particle starting with vertical velocity $u\sin\alpha$ from *O* returns to *C*. Since *O* and *C* are on the same horizontal level, vertical distance travelled by the particle is zero. So by using $h = ut - (1/2)gt^2$, we get

$0 = u\sin\alpha \cdot (t_1 + t_2) - (1/2)g(t_1 + t_2)^2$ or $t_1 + t_2 = (2u/g)\cdot\sin\alpha$

or $a/(u\cos\alpha) + a/(2eu\cos\alpha) = (2u\sin\alpha)/g$, by (1) and (2)

or $ag(2e + 1) = 4u^2 e\sin\alpha\cos\alpha$ or $\sin 2\alpha = (1 + 2e)ag/2eu^2$

or $2\alpha = \sin^{-1}[(1 + 2e)ag/2eu^2\}$ or $\alpha = (1/2)\sin^{-1}\{(1 + 2e)ag/2eu^2\}$

Ex. 7. *A ball is projected from a point in one of two smooth parallel vertical walls against the other in a plane perpendicular to both after being reflected at each wall impinge again on the second at a point in the same horizontal plane as it started from. Show that be² = a(1 + e + e²), where e is the coefficient of elasticity, b the free range on a horizontal plane and a the distance between the walls.*

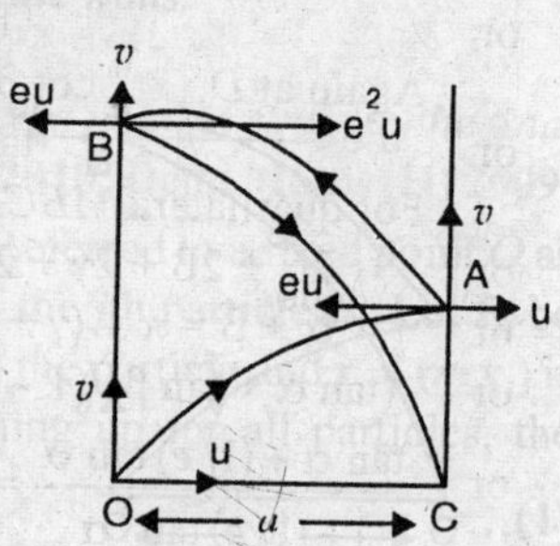

Sol. Let *OB* and *CA* be two parallel walls, where *O* and *C* are on the same horizontal plane such that $OC = a$. Let the ball be projected from *O* and let horizontal and vertical components of the velocity of ball be *u* and *v* respectively. It then strikes the second wall at *A*, then first wall at *B* and strikes again the second wall at *C*.

The ball strikes at *A* with a horizontal velocity *u*, which after impact becomes *eu*. With this horizontal velocity *eu*, the ball strikes at *B*, which after impact becomes *e*(*eu*) *i.e.*, e^2u. Horizontal distance moved during motion from *O* to *A*, *A* to *B* and *B* to *C* is the same, namely *a*. If *T* is the total time taken in moving from *O* to *C*, we get

$T = a/u + a/eu + a/e^2u = (a/e^2u)(e^2 + e + 1)$ ...(1)

Throughout the motion from *O* to *C*, the vertical velocity *v* remains unaltered due to impact with walls at *B* at *C*. Also the vertical distance travelled in time *T* is zero. So considering motion from *O* to *C* in vertical direction and using $h = ut - (1/2)gt^2$, we get

$0 = v\cdot T - (1/2)gT^2$ or $2v = gT$

or $2v = g(a/e^2u)(e^2 + e + 1)$ or $(2uv/g)e^2 = a(e^2 + e + 1)$, by (1) ...(2)

Given that b = free range on horizontal plane $= (2uv)/g$ ...(3)

From (2) and (3), $be^2 = a(e^2 + e + 1)$.

Ex. 8. *A particle is projected from a given point at the foot of one of the two smooth vertical walls, so as after three impacts, it may return to the point of projection. If the last impact be direct, prove that $e^3 + e^2 + e = 1$.* **(Purvanchal 1994, 96, 98)**

Sol. Let OB and CD be parallel walls, where O and D are on the same horizontal plane such that $OD = a$. Let the ball be projected from O and let horizontal and vertical components of the velocity of the ball be u and v respectively. It then strikes the second wall at A, then first wall at B, then again the second wall at C and finally strikes the first wall at O.

After striking at A, B and C the horizontal components of the velocities of the ball become eu, e^2u and e^3u respectively. Horizontal distance moved during motion from O to A, A to B and B to C is the same, namely, a. If T is the total time taken in moving from O to C, we get

$$T = a/u + a/eu + a/e^2u = (a/e^2u)\,(e^2 + e + 1) \qquad \ldots(1)$$

Let T' be time taken from C to O. Then $T' = a/e^3u$ ...(2)

Since the impact at C is direct *i.e.*, at the right angles to the wall and vertical component of velocity is not altered by impacts on walls at A and B, therefore the vertical height moved in rising from O to C is the same as in falling from C to O.

$\therefore$ Time from O to C = Time from C to O or $T = T'$, ...(3)

From (1), (2) and (3), $(a/e^2u)\,(e^2 + e + 1) = a/e^3u$ or $e^3 + e^2 + e = 1$

Ex. 9. *A imperfectly elastic ball, reflected at the circumference of a circle, describes the sides of an inscribed quadrilateral in order. Prove that two of the angles of the quadrilateral are right angles. Also find the initial direction of projection.*

Sol. Let $ABCD$ be quadrilateral inscribed in a circle with centre O. Let the ball be projected from A at an angle α with radius OA. Let β, γ and δ be the angles of reflection at the points B, C and D respectively. The angles of incidence at B, C and D are clearly α, β and γ respectively.

Refer result (3) of Art. 7.3. Accordingly, we have

cotangent of angle of reflection = $e \times$ cotangent of angle of incidence

$\therefore$ At B, $\cot\beta = e\cot\alpha$ or $\tan\beta = (1/e)\tan\alpha$ ...(1)

Similarly, at C, $\cot\gamma = e\cot\beta$ or $\tan\gamma = (1/e)\tan\beta$

or $\tan\gamma = (1/e)^2\tan\alpha$, using (1) ...(2)

Again at D, $\cot\delta = e\cot\gamma$ or $\tan\delta = (1/e)\tan\gamma$

or $\tan\delta = (1/e)^3\tan\alpha$, by (2) ...(3)

For quarilateral $ABCD$, sum of the angles $A, B, C, D = 2\pi$

or $2\alpha + 2\beta + 2\gamma + 2\delta = 2\pi$ or $\alpha + \beta + \gamma + \delta = \pi$ ...(4)

or $\alpha + \beta = \pi - (\gamma + \delta)$ or $\tan(\alpha + \beta) = -\tan(\gamma + \delta)$

or $(\tan\alpha + \tan\beta)/(1 - \tan\alpha\tan\beta) = -(\tan\gamma + \tan\delta)/(1 - \tan\gamma\tan\delta)$

or $$\frac{\tan\alpha + (1/e)\tan\alpha}{1-(1/e)\tan^2\alpha} = -\frac{(1/e^2)\tan\alpha + (1/e^3)\tan\alpha}{1-(1/e^5)\tan^2\alpha}, \text{ by (1), (2) and (3)}$$

or $$\frac{e+1}{e-\tan^2\alpha} = -\frac{e^2(e+1)}{e^5-\tan^2\alpha} \quad\text{or}\quad e^5 - \tan^2\alpha = -e^2(e - \tan^2\alpha)$$

or $(e^2 + 1)\tan^2\alpha = e^5 + e^3$ or $\tan^2\alpha = e^3$, ...(5)

giving the initial direction of projection.

Now, $$\tan(\beta+\gamma) = \frac{\tan\beta + \tan\gamma}{1-\tan\beta\tan\gamma} = \frac{(1/e)\tan\alpha + (1/e^2)\tan\alpha}{1-(1/e^3)\tan^2\alpha}, \text{ by (1) and (2)}$$

or $\tan(\beta + \gamma) = \{(1/e + 1/e^2)\tan\alpha\}/(1 - 1)$, using (5)

Hence, $\beta + \gamma = \pi/2$ or $\angle BCD = \pi/2$ ...(6)

Then from (4), $\alpha + \delta = \pi - (\beta + \gamma) = \pi - \pi/2 = \pi/2 \Rightarrow \angle BAD = \pi/2$

Thus two angles $\angle BAD$ and $\angle BCD$ are right angles.

Ex. 10. *A smooth circular table is surrounded by a smooth rim whose interior surface is vertical. Show that ball whose coefficient of restitution is e, projected along table from a point in the rim making an angle α with radius through point will return to the point of projection after two impacts if* $\tan^2 \alpha = e^3/(1 + e + e^2)$. *Prove also that when the ball returns to the point of projection its velocity is to the original velocity as* $e^{3/2} : 1$. **[Gorakhpur 96]**

Sol. Let the ball be projected from A with a velocity u making angle α with radius OA through O, O being the centre of the circular table. Suppose the ball strikes the table at B and C and then returns to A. Let the velocities after impact at B and C be v and w respectively. Let the angles of reflection at B and C be β and γ respectively. Then, by geometry, the angles of incidence at B and C are α and β respectively. Refer result (3) of Art. 7.3. So, we have

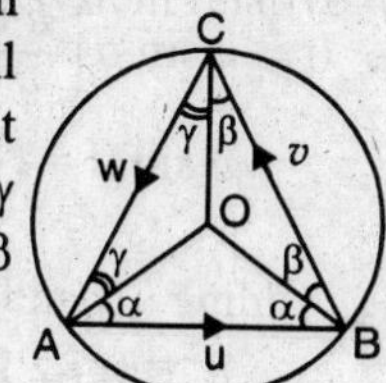

At B, $\cot \beta = e \cot \alpha$ or $\tan \beta = (1/e) \tan \alpha$ ...(1)

At C, $\cot \gamma = e \cot \beta = e^2 \cot \alpha$ or $\tan \gamma = (1/e^2) \tan \alpha$ ...(2)

In ΔABC, $\angle A + \angle B + \angle C = \pi$ or $2\alpha + 2\beta + 2\gamma = \pi$

or $\beta + \gamma = \pi/2 - \alpha$ or $\tan (\beta + \gamma) = \tan (\pi/2 - \alpha) = \cot \alpha$

or $\dfrac{\tan \beta + \tan \gamma}{1 - \tan \beta \tan \gamma} = \cot \alpha$ or $\dfrac{(1/e) \tan \alpha + (1/e^2) \tan \alpha}{1 - (1/e) \tan \alpha \cdot (1/e^2) \tan \alpha} = \dfrac{1}{\tan \alpha}$

or $\left(\dfrac{1}{e} + \dfrac{1}{e^2}\right) \tan^2 \alpha = 1 - \dfrac{1}{e^3} \tan^2 \alpha$ or $\left(\dfrac{1}{e} + \dfrac{1}{e^2} + \dfrac{1}{e^3}\right) \tan^2 \alpha$

or $\{(1 + e + e^2)/e^3\} \tan^2 \alpha = 1$ or $\tan^2 \alpha = e^3/(1 + e + e^2)$ ...(3)

The rim being smooth, there is no force along tangents at B and C. Hence the tangential components of velocities at B and C is not altered and hence

$v \sin \beta = u \sin \alpha$ and $w \sin \gamma = v \sin \beta$ so $w \sin \gamma = u \sin \alpha$

$\therefore$ $w/u = \sin \alpha \operatorname{cosec} \gamma = \sin \alpha \, (1 + \cot^2 \gamma)^{1/2}$

or $w/u = \sin \alpha \, (1 + e^4 \cot^2 \alpha)$, using (2) ...(4)

From (3), $\dfrac{\sin^2 \alpha}{\cos^2 \alpha} = \dfrac{e^3}{1+e+e^2} \Rightarrow \dfrac{\sin^2 \alpha}{e^3} = \dfrac{\cos^2 \alpha}{1+e+e^2} = \dfrac{\sin^2 \alpha + \cos^2 \alpha}{1+e+e^2+e^3}$ so that $\sin^2 \alpha = e^3/(1 + e + e^2 + e^3)$. Hence (4) reduces to

$$\frac{w}{u} = \frac{e^{3/2}}{(1+e+e^2+e^3)^{1/2}} \cdot \left\{1 + \frac{e^4(1+e+e^2)}{e^3}\right\}^{1/2} = e^{3/2}, \text{ by (3)}$$

$\therefore$ $\dfrac{\text{velocity of return to } A}{\text{original velocity}} = \dfrac{w}{u} = \dfrac{e^{3/2}}{1}$, as required.

Ex. 11. *A perfectly elastic ball is projected from the foot of a plane inclined at an angle θ to the horizon. If after striking the plane at a distance l from the point of projection, it rebounds and returns to its former path, show that the velocity of projection is* $\{gl(1 + 3 \sin^2 \theta)/2 \sin \theta\}^{1/2}$.

Sol. Let the ball be projected from O and strikes the plane at A where $OA = l$ (given). Let the direction of projection make an angle α with the inclined plane.

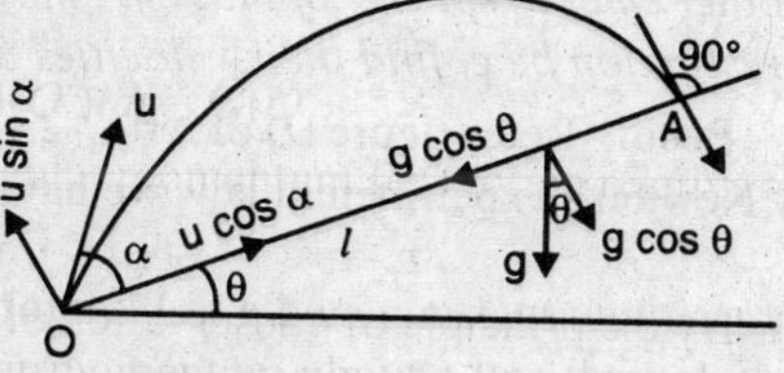

The ball can retrace its path only if the direction of motion at the time of impact at A is perpendicular to the plane. Hence the component of velocity along the plane vanishes at A. Hence for motion along the plane from O to A, using $v = u - ft$, we have

$0 = u \cos \alpha - g \sin \theta \cdot t$ or $t = (u \cos \alpha)/g \sin \theta$ ...(1)

During this time distance moved along the plane is zero.

So for motion along the plane from O to A, using $s = ut - (1/2)ft^2$, we get

$$l = u\cos\alpha\left(\frac{u\cos\alpha}{g\sin\theta}\right) - \frac{1}{2}g\sin\theta\left(\frac{u\cos\alpha}{g\sin\theta}\right)^2 = \frac{u^2\cos^2\alpha}{2g\sin\theta}$$

or $$u^2\cos^2\alpha = 2gl\sin\theta \qquad \ldots(2)$$

During the time t given by (1), the distance moved perpendicular to the inclined plane with retardation $g\cos\theta$ is zero. So, using $s = ut - (1/2)ft^2$, for motion from O to A in direction perpendicular to the plane, we get

$$0 = u\sin\alpha\left(\frac{u\cos\alpha}{g\sin\theta}\right) - \frac{1}{2}g\cos\theta\left(\frac{u\cos\alpha}{g\sin\theta}\right)^2 \quad \text{or } u\sin\alpha = \frac{1}{2}g\cos\theta\cdot\frac{u\cos\theta}{g\sin\theta}$$

or $$u\sin\alpha = (1/2)u\cos\alpha\cot\theta = (1/2)\,(2gl\sin\theta)^{1/2}\cot\theta, \text{ by (2)}$$

or $$u^2\sin^2\alpha = (gl\cos^2\theta)/2\sin\theta \qquad \ldots(3)$$

Adding (2) and (3), $$u^2 = gl\{2\sin\theta + (\cos^2\theta)/2\sin\theta\}$$

or $$u^2 = (gl/2\sin\theta)\,(4\sin^2\theta + \cos^2\theta) = (gl/2\sin\theta)\,(1 + 3\sin^2\theta)$$

or $$u = \{gl(1 + 3\sin^2\theta)/2\sin\theta\}^{1/2}, \text{ as required}$$

EXERCISE

1. A ball is projected in a horizontal plane and makes one rebound, show that if the second range is equal to the greatest height, the angle of projection is $\tan^{-1}(4e)$, e being the coefficient of elasticity.

2. A particle is projected at an angle α to the horizontal to strike a smooth vertical wall and after rebounding it passes through the point of projection. If ϕ be the angle of inclination to the horizontal at which the particle rebounds from the wall, e the coefficient of restitution, show that $e(e + 1)\tan\phi = (1 - e)\tan\alpha$

3. A ball is projected from a point in a horizontal plane; after one rebound from the plane it strikes directly against a vertical wall; after two more rebound it returns to the point of projection. Prove that $e = 2(1 - e^2)$

4. A square table $ABCD$ whose side is 'a' has raised edges. A particle of elasticity e is projected from a point P in AB and hits the sides BC, CD, DA in Q, R, S, prove that PQ, RS are parallel. If α be the angle QPB and $BP = x$, pove that if the particle returns to P, then $x(1 - e) = a(1 - e\cot\alpha)$.

5. From one corner A of a rectangular billiard table $ABCD$ a ball is projected in a direction making an angle α with the side AB; it strikes first in the side BC, then DC, then BC again and then returns to A. Prove that

$$AB/AD = (e^2\cot\alpha)/(1 + e^2)$$

6. A particle of elasticity e is dropped from a vertical height 'a' upon the highest point of a plane which is of length b and is inclined at an angle α to the horizon and descends to the bottom in three jumps, prove that

$$b = 4ae\,(1 + e)\,(1 + e^2)\,(1 + e + e^2)\sin\alpha$$

7.4 Direct impact of two spheres. *A smooth sphere, of mass m, impinges directly with velocity u on another smooth sphere, of mass m′, moving in the same direction with velocity u′. If the coefficient of restitution be e, find their velocities after impact.* **(M.S. Univ. T.N. 2007)**

Proof. Refer figure (*i*) of Art. 7.2. Let v and v' be the velocities of the two spheres after impact. By Newton's experimental law, we have

$$v - v' = -e(u - u') \qquad \ldots(1)$$

By the principle of conservation of linear momentum, we get

$$mv + m'v' = mu + m'u' \qquad \ldots(2)$$

Multiplying (1) by m' and adding to (2), we get

$$(m + m')v = (m - em')u + m'(1 + e)u', \text{ giving}$$

$$v = \{(m - em')u + m'(1 + e)u'\}/(m + m') \qquad \ldots(3)$$

Again, multiplying (1) by m and subtracting from (2), we get

$$(m+m')v' = m(1+e)u+(m'-em)u', \text{ giving}$$

$$v' = \{m(1+e)u+(m'-em)u'\}/(m+m') \qquad ...(4)$$

(3) and (4) give the velocities after impact.

Cor. 1. *If two equal and perfectly elastic spheres moving with velocities u and u′ impinge directly, show that they interchange their velocities after impact.* **(M.S. Univ. T.N. 2007)**

Proceed as above by taking $m' = m$ and $e = 1$ in (1) and (2). Solve them to get $v = u'$ and $v' = u$. Hence proved.

Cor. 2. Impulse of the blow on the sphere of mass m

= Change in its momentum = $m(v-u)$

$$= m\left[\frac{(m-em')u+m'(1+e)u'}{m+m'}-u\right] = \frac{m}{m+m'}(m'u'+em'u'-m'u-em'u)$$

$$= \frac{mm'}{m+m'}\{u'(1+e)-u(1+e)\} = \frac{mm'}{m+m'}(1+e)(u'-u)$$

The impulse of the blow on the other ball is equal and opposite to this.

7.5 Loss of kinetic energy by direct impact.

Two spheres of given masses moving with given velocities impinge directly. Show that there is always loss of kinetic energy unless the elasticity is perfect. **(Kerala 2001)**

Proof. Refer figure of Art. 7.2. Let m, m' be the masses of the spheres and u, u' their velocities. When they impinge, let v, v' be their velocities after impact. Let e be the coefficient of restitution. Then, we have

By the principle of conservation of linear momentum, we get

$$mv + m'v' = mu + m'u' \qquad ...(1)$$

By Newton's experimental law, we get

$$v - v' = -e(u-u') \qquad ...(2)$$

To the square of (1) add the square of (2) multiplied by mm′; then we get

$$(mv + m'v')^2 + mm'(v-v')^2 = (mu + m'u')^2 + mm'e^2(u-u')^2$$

or $m^2v^2 + m'^2v'^2 + mm'v'^2 + mm'v^2 = (mu + m'u')^2 + mm'e^2(u-u')^2 + mm'(u-u') - mm'(u-u')^2$

or $m(m+m')v^2 + m'(m+m')v'^2 = (mu+m'u')^2 + mm'(u-u')^2 - mm'(1-e^2)(u-u')^2$

or $(m+m')(mv^2 + m'v'^2) = m^2u^2 + m'^2u'^2 + mm'u^2 + mm'u'^2 - mm'(1-e^2)(u-u')^2$

or $(m+m')(mv^2 + m'v'^2) = m(m+m')u^2 + m'(m+m')u'^2 - mm'(1-e^2)(u-u')^2$

or $mv^2 + m'v'^2 = mu^2 + m'u^2 - \{mm'/(m+m')\}(1-e^2)(u-u')^2$

or $\left(\frac{1}{2}mu^2+\frac{1}{2}m'u'^2\right)-\left(\frac{1}{2}mv^2+\frac{1}{2}m'v'^2\right) = \frac{mm'}{2(m+m')}(1-e^2)(u-u')^2$

or [Total K.E. before impact] – [Total K.E. after impact]

$$= \{mm'/2(m+m')\}(1-e^2)(u-u')^2,$$

showing that the loss of K.E. due to impact

$$= \{mm'/2(m+m')\}(1-e^2)(u-u')^2 \qquad ...(3)$$

and this loss does not vanish unless $e = 1$, *i.e.*, unless the balls are perfectly elastic.

Corollary. *If the inelastic spheres have direct impact then the kinetic energy lost by the impact is that of a body whose mass is half the harmonic mean between those spheres, and whose velocity equals their relative velocity before impact.*

Proof. Putting $e = 0$ in result (3) of above article,

the loss in K.E. = (1/2) $\{mm'/(m+m')\}(u-u')^2$, where $u-u'$ is relative velocity of spheres and $mm/(m+m')$ is the half of the harmonic mean of m and m'. Hence the result.

EXAMPLES BASED ON ART. 7.4 AND 7.5

Ex. 1. *Two elastic spheres impinge directly with equal and opposite velocities. Find the ratio of their masses so that one of them may be reduced to rest by the impact, the coefficient of elasticity being e.*

Sol. Let m, m' be the masses of two spheres A and B. Let A and B be moving with equal velocity u in opposite directions as shown in the figure. Let the sphere A be reduced to rest after impact and let v be the velocity of sphere B after impact. Then, by Newton's experiment law

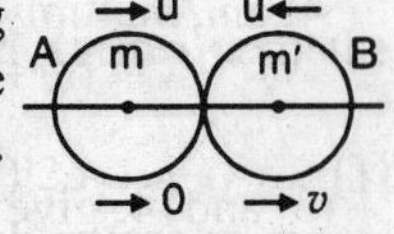

$$v - 0 = -e(-u - u) \quad \text{or} \quad v = 2eu \qquad \ldots(1)$$

By law of conservation of linear momentum, we have

$$mu - m'u = m'v \quad \text{or} \quad mu = m'(u + v) = m'(u + 2eu), \text{ by (1)}$$

or $\quad m/m' = 1 + 2e \quad$ so that $\quad m : m' = (1 + 2e) : 1$

Ex. 2. *A ball impinges directly upon another ball at rest and is itself reduced to rest by impact. If half of the initial K.E. is destroyed in the collision, find the coefficient of restitution.*

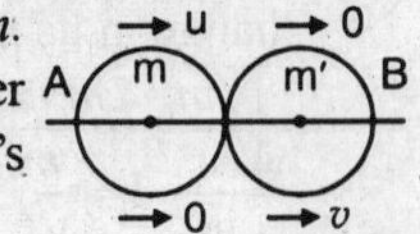

Sol. Let u be the velocity of the ball A before impact and v, that of the ball B after impact. Let m, m' be their masses and let e be the coefficient of restitution. By Newton's experimental law,

$$v - 0 = -e(0 - u) \quad \text{or} \quad v = eu \qquad \ldots(1)$$

By law of conservation of momentum $m'v = mu$ $\qquad \ldots(2)$

Given, $\quad$ loss in K.E. = half of the initial K.E.

$\therefore \quad$ Initial K.E. – Final K.E. = (1/2) × initial K.E.

or $\quad$ (1/2) × Initial K.E. = Final K.E. $\quad$ or $\quad (1/2) \times (1/2)mu^2 = (1/2)m'v^2 \qquad \ldots(3)$

Multiplying the corresponding sides of (1) and (2), $m'v^2 = meu^2 \qquad \ldots(4)$

Using (4), (3) becomes $\quad (1/4)mu^2 = (1/2)meu^2 \quad$ or $\quad e = 1/2$

Ex. 3. *Three perfectly elastic balls of masses m_1, m_2, m_3 are placed in a straight line. The first impinges directly on the second with a velocity u and then the second impinges the third. Show that the third ball will move after impact with velocity u if $(m_1 + m_2)(m_2 + m_3) = 4m_1m_2$.*

Sol. Here $e = 1$. Consider direct impact between the first and second balls. Let v, v' be the velocities after impact. Then, by Newton's experimental law,

$$v' - v = -(0 - u) \quad \text{or} \quad v' - v = u \qquad \ldots(1)$$

By law of conservation of momentum, $\quad m_1v + m_2v' = m_1u \qquad \ldots(2)$

Multiplying (1) by m_1 and then adding to (2), we get

$$(m_1 + m_2)v' = 2m_1u \quad \text{or} \quad v' = (2m_1u)/(m_1 + m_2) \qquad \ldots(3)$$

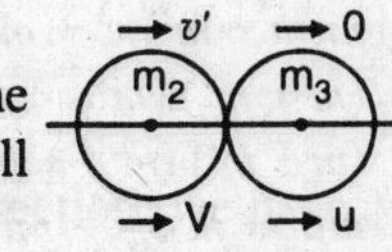

Now the second sphere impinges on the third ball with velocity v'. Let V be the velocity of the second ball after impact. Also note that the velocity of the third ball after impact is given to be u. So by Newton's experimental law,

$$u - V = -(0 - v') \quad \text{or} \quad u - V = v' \qquad \ldots(4)$$

By law of conservation of momentum, $\quad m_2V + m_3u = m_2v' \qquad \ldots(5)$

Multiplying (4) be m_2 and then adding to (5), we get

$$(m_2 + m_3)u = 2m_2v' \quad \text{or} \quad (m_2 + m_3)u = 2m_2 \times \frac{2m_1u}{m_1 + m_2}, \text{ by (3)}$$

or $$(m_1 + m_2)(m_2 + m_3) = 4m_1m_2$$

Ex. 4. *An imperfectly elastic sphere of mass m moving with velocity u impinges on another sphere of mass m' at rest. The second sphere afterwards strikes a vertical plane at right angles to its path. Show that there will be no further impact of the spheres if $m(1 + e + ee') < em'$, where e and e' are the coefficients of restitution between the spheres and between the sphere and the plane respectively.*

Ex. 5. *Three equal spheres are in a straight line on a table and one of them moves towards the other two which are at rest and not in contact; if e = 1/2, find how many impacts will take place and show that the ultimate speeds of the spheres are in the ratios 13 : 15 : 36.*

Sol. Let the spheres be A, B, C, each of mass m and let A move with velocity u to strike B which is at rest. Let v, v' be the velocities of A and B after impact. Then by Newton's experimental law and the principle of conservation of momentum, we have

$v - v' = -(1/2)(u - 0)$ or $v - v' = -(1/2)u$ ...(1)

and $mv + mv' = mu$ or $v + v' = u$ ...(2)

From (1) and (2), $v = u/4$, $v' = 3u/4$. Thus $v' > v$. ...(3)

Since B moves faster than A after impact, so A is left behind and B strikes C (which is at rest) with velocity v'. Let V, V' be the velocities of B and C after impact. Then, as usual,

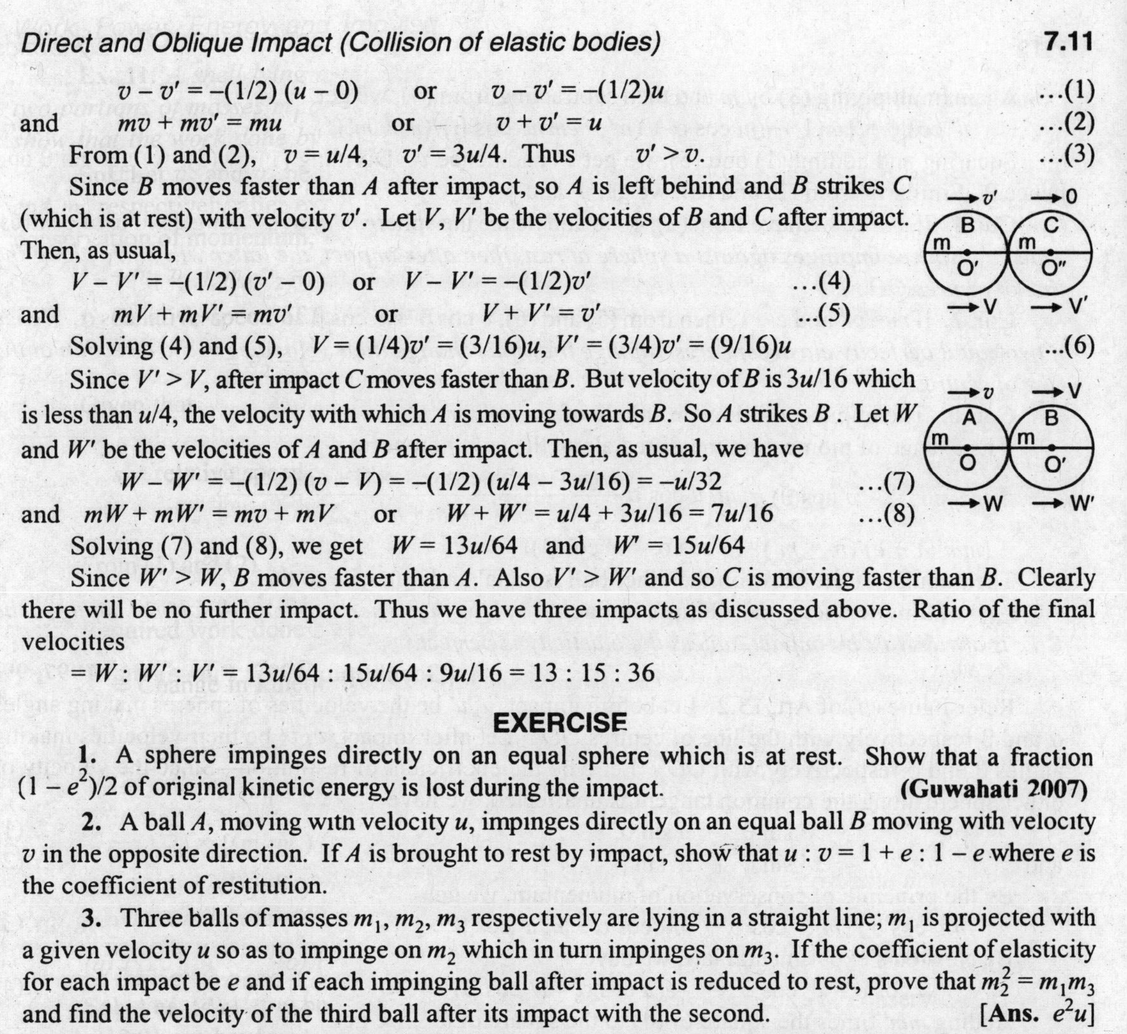

$V - V' = -(1/2)(v' - 0)$ or $V - V' = -(1/2)v'$ ...(4)

and $mV + mV' = mv'$ or $V + V' = v'$ ...(5)

Solving (4) and (5), $V = (1/4)v' = (3/16)u$, $V' = (3/4)v' = (9/16)u$ ...(6)

Since $V' > V$, after impact C moves faster than B. But velocity of B is $3u/16$ which is less than $u/4$, the velocity with which A is moving towards B. So A strikes B. Let W and W' be the velocities of A and B after impact. Then, as usual, we have

$W - W' = -(1/2)(v - V) = -(1/2)(u/4 - 3u/16) = -u/32$ ...(7)

and $mW + mW' = mv + mV$ or $W + W' = u/4 + 3u/16 = 7u/16$ ...(8)

Solving (7) and (8), we get $W = 13u/64$ and $W' = 15u/64$

Since $W' > W$, B moves faster than A. Also $V' > W'$ and so C is moving faster than B. Clearly there will be no further impact. Thus we have three impacts as discussed above. Ratio of the final velocities

$= W : W' : V' = 13u/64 : 15u/64 : 9u/16 = 13 : 15 : 36$

EXERCISE

1. A sphere impinges directly on an equal sphere which is at rest. Show that a fraction $(1 - e^2)/2$ of original kinetic energy is lost during the impact. **(Guwahati 2007)**

2. A ball A, moving with velocity u, impinges directly on an equal ball B moving with velocity v in the opposite direction. If A is brought to rest by impact, show that $u : v = 1 + e : 1 - e$ where e is the coefficient of restitution.

3. Three balls of masses m_1, m_2, m_3 respectively are lying in a straight line; m_1 is projected with a given velocity u so as to impinge on m_2 which in turn impinges on m_3. If the coefficient of elasticity for each impact be e and if each impinging ball after impact is reduced to rest, prove that $m_2^2 = m_1 m_3$ and find the velocity of the third ball after its impact with the second. **[Ans. e^2u]**

4. The masses of three spheres A, B, C are $7m$, $7m$, m; their coefficient of restitution is unity, their centres are in straight line and C lies between A and B. Initially A and B are at rest and C is given a velocity along the line of centres towards A. Show that it strikes A twice and B once and the final velocities of A, B, C are proportional to 21 : 21 : 1.

9.6 Oblique impact of two spheres. *A smooth sphere, of mass m, impinges with velocity u obliquely on a smooth sphere of mass m', moving with velocity u'. If the directions of motion before impact make angles α and β respectively with the line joining the centres of the spheres, and if the coefficient of restitution be e, find the velocities and directions of motion after impact.*

Refer figure (*ii*) of Art. 7.2. Let the velocities of the spheres after impact be v and v' in directions inclined at angles θ and φ respectively to the line of centres OO'. Since the spheres are smooth, there is no force perpendicular to the common normal OO', and therefore the velocities along the common tangent are unaltered. Hence, $v \sin\theta = u \sin\alpha$ ...(1)

and $v' \sin\phi = u' \sin\beta$ ...(2)

By Newton's experimental law, we have

$v\cos\theta - v'\cos\phi = -e(u\cos\alpha - u'\cos\beta)$ ...(3)

By the principle of conservation of momentum, we have

$mv\cos\theta + m'v'\cos\phi = mu\cos\alpha + m'u'\cos\beta$ ...(4)

Multiplying (3) by m' and then adding to (4), we get

$v\cos\theta = \{(m - em')u\cos\alpha + m'(1 + e)u'\cos\beta\}/(m + m')$ ...(5)

Again, multiplying (3) by m and then subtracting from (4), we get

$$v' \cos \phi = \{m(1+e)u \cos \alpha + (m' - em)u' \cos \beta\}/(m+m') \quad \ldots(6)$$

Squaring and adding (1) and (5), we get v^2 and hence v. Dividing (1) by (5), we get $\tan \theta$ and hence θ. Similarly from (2) and (6), we get v' and ϕ.

Cor. 1. If $u' = 0$, we have from (2), $\phi = 0$ and hence the sphere m' moves along the line of centres. *Thus, if a sphere impinges against a sphere at rest, then after impact, the latter will move along the line of centres OO′.*

Cor. 2. If $m = m'$ and $e = 1$, then from (5) and (6), $v \cos \theta = u' \cos \beta$ and $v' \cos \phi = u \cos \alpha$. *Hence if two equal perfectly elastic spheres impinge they interchange their velocities in the direction of the line of centres.*

Cor. 3. The impulse of blow on sphere of mass m

= its change of momentum measured along the common normal

$$= m(u \cos \alpha - v \cos \theta) = m\left\{u \cos \alpha - \frac{(m - em')u \cos \alpha + m'(1+e)u' \cos \beta}{m+m'}\right\}$$

$$= \{mm'(1+e)/(m+m')\}\,(u \cos \alpha - u' \cos \beta)$$

The impulse of the blow on the other ball is equal and opposite to this.

7.7 Loss of kinetic energy by oblique impact. *To show that when two smooth spheres impinge, the K.E. is always lost by impact, unless the elasticity is perfect.*

[Gorakhpur 1997; Purvanchal 1997, 99]

Refer figure (*ii*) of Art. 13.2. Let before impact, u, u' be the velocities of spheres making angles α and β respectively with the line of centres OO'. Let after impact, v, v' be their velocities making angles θ and ϕ respectively with OO'. Let e be the coefficient of restitution. Since the velocity of either sphere along the common tangent is unaffected, we have

$$v \sin \theta = u \sin \alpha \quad \ldots(1)$$

and

$$v' \sin \phi = u' \sin \beta \quad \ldots(2)$$

By the principle of conservation of momentum, we get

$$mv \cos \theta + m'v' \cos \phi = mu \cos \alpha + m'u' \cos \beta \quad \ldots(3)$$

By Newton's experimental law, we have

$$v \cos \theta - v' \cos \phi = -e(u \cos \alpha - u' \cos \beta) \quad \ldots(4)$$

Adding mm' times the square of (4) to the square of (3), we get

$$(mv \cos \theta + m'v' \cos \phi)^2 + mm(v \cos \theta - v \cos \phi)^2 = (mu \cos \alpha + m'u' \cos \beta)^2 + mm'e^2(u \cos \alpha - u' \cos \beta)^2$$

or

$$(m^2 + mm')v^2 \cos^2 \theta + (m'^2 + mm')v'^2 \cos^2 \phi = (mu \cos \alpha + m'u' \cos \beta)^2 + mm'(u \cos \alpha - u' \cos \beta)^2 + mm'e^2(u \cos \alpha - u' \cos \beta)^2 - mm'(u \cos \alpha - u' \cos \beta)^2$$

or

$$(m+m')(mv^2 \cos^2 \theta + m'v'^2 \cos^2 \phi) = (m^2 + mm')u^2 \cos^2 \alpha + (m'^2 + mm)u'^2 \cos^2 \beta - mm'(1-e^2)(u \cos \alpha - u' \cos \beta)^2$$

or

$$(m+m')(mv^2 \cos^2 \theta + m'v'^2 \cos^2 \phi) = (m+m')(mu^2 \cos^2 \alpha + m'u'^2 \cos^2 \beta) - mm'(1-e^2)(u \cos \alpha - u' \cos \beta)^2$$

or

$$(1/2)mv^2 \cos^2 \theta + (1/2)m'v'^2 \cos^2 \phi = (1/2)mu^2 \cos \alpha + (1/2)m'u'^2 \cos^2 \beta - (1/2) \times \{mm'/(m+m')\}(1-e^2)(u \cos \alpha - u' \cos \beta)^2 \quad \ldots(5)$$

From (1),

$$(1/2)mv^2 \sin^2 \theta = (1/2)mu^2 \sin^2 \alpha \quad \ldots(6)$$

From (2),

$$(1/2)m'v'^2 \sin^2 \phi = (1/2)m'u'^2 \sin^2 \beta \quad \ldots(7)$$

Adding (5), (6) and (7), we have

$$(1/2)mv^2 + (1/2)m'v'^2 = (1/2)mu^2 + (1/2)m'u'^2 - (1/2) \times \{mm'/(m+m')\}(1-e^2)(u \cos \alpha - u' \cos \beta)^2$$

or

$$\{(1/2)mu^2 + (1/2)m'u'^2\} - \{(1/2)mv^2 + (1/2)m'v'^2\} = (1/2) \times \{mm'/(m+m')\}(1-e^2)(u \cos \alpha - u' \cos \beta)^2,$$

$\therefore$ K.E. before impact – K.E. after impact

$= \text{the loss of K.E.} = (1/2) \times \{mm'/(m + m')\} (1 - e^2) (u \cos \alpha - u \cos \beta)^2,$ ...(8)

which does not vanish unless $e = 1$, that is, unless the balls are perfectly elastic.

Remark. Re-writing (8), the amount of K.E. lost by impact

$= \{mm'/2(m + m')\} (1 - e^2)V^2$, where $V = u \cos \alpha - u' \cos \beta$ is the relative velocity of the two spheres before impact in the direction of the common normal.

SOLVED EXAMPLES BASED ON ART. 7.6 AND ART. 7.7

Ex. 1. *The velocity of a ball A, after its oblique impact on a ball B, has the same magnitude and direction as the velocity of B before impact. Show that the ratio of their masses is equal to the coefficient of restitution.*

Sol. Let m, m' be the masses of the balls A and B. Refer figure (*ii*) of Art. 7.2. Let u, u' be their velocities before impact making angles α and β respectively with the line of centres OO'. If v, v' be their velocities after impact making angle θ and ϕ with OO', then in the present problem, we have

$$v = u' \quad \text{and} \quad \theta = \beta \qquad ...(1)$$

By Newton's experimental law, we have

$$v \cos \theta - v' \cos \phi = -e(u \cos \alpha - u' \cos \beta)$$

or
$$u' \cos \beta - v' \cos \phi = -e(u \cos \alpha - u' \cos \beta), \text{ by (1)} \qquad ...(2)$$

By the principle of conservation of momentum, we have

$$mv \cos \theta + m'v' \cos \phi = mu \cos \alpha + m'u' \cos \beta$$

or
$$mu' \cos \beta + m'v' \cos \phi = mu \cos \alpha + m'u' \cos \beta, \text{ by (1)}$$

or
$$m'(u' \cos \beta - v' \cos \phi) = -m(u \cos \alpha - u' \cos \beta) \qquad ...(3)$$

Dividing (3) by (2), $m' = m/e$ or $m : m' = e : 1$

Ex. 2. (*a*) *A ball impinges on another at rest. Prove that if the coefficient of restitution be equal to the ratio of their masses, the balls will leave in directions at right angles to each other. What will happen in the case of direct collision ?*

Sol. Let the masses of the balls be em and m where e is the coefficient of restitution. Let u be velocity of the first ball before impact, making an angle α with the line of centres OO'. The ball of mass m is initially at rest and as such, will move along the common normal OO' after impact with velocity v', say. Let v be the velocity of the first sphere after impact, makng an angle θ with the OO'. Then, by principle of conservation of momentum, we get

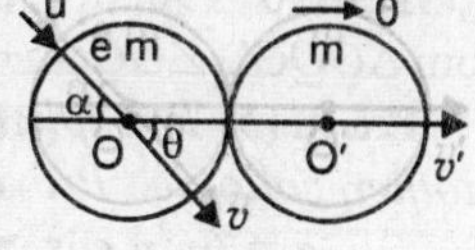

$$em \cdot v \cos \theta + mv' = em \cdot u \cos \alpha \quad \text{or} \quad ev \cos \theta + v' = eu \cos \alpha \qquad ...(1)$$

By Newton's experimental law, we get

$$v \cos \theta - v' = -e(u \cos \alpha - 0) \quad \text{or} \quad v \cos \theta - v' = -eu \cos \alpha \qquad ...(2)$$

Adding (1) and (2), $(1 + e)v \cos \theta = 0$ so that $\cos \theta = 0$, as $e \neq -1$. Hence, $\theta = 90°$, *i.e.*, the balls will leave in directions at right angles to each other.

In the case of direct impact, (1) and (2) become

$$ev + v' = eu \qquad ...(3)$$

and
$$v - v' = -eu \qquad ...(4)$$

Adding (3), and (4), $(1 + e)v = 0$ so $v = 0$ *i.e.*, first ball comes to rest. Putting $v = 0$ in (4), $v' = eu$, showing that the second ball moves with the e times original velocity of the first ball.

Ex. 2. (*b*) *A sphere collides obliquely with another sphere of equal mass which is initially at rest, both spheres being smooth and perfectly elastic. Show that their paths after collision are at right angles.*

Hint. Take $e = 1$ and proceed as in Ex. 2. (*a*).

Ex. 2. (*c*) *A ball of mass M strikes another ball of mass m at rest. If they separate in mutually perpendicular directions, prove that the coefficient of elasticity is equal to the ratio of their masses.*

Hint. Refer figure of Ex. 2. (*a*). Take M in place of em and take $\theta = 90°$ for this problem. Proceed as before to get $e = M/m$

Ex. 3. (*a*) *Two equal, smooth and perfectly elastic spheres moving at right angles to one another impinge obliquely. Show that after impact, they will still move at right angles to each other.*

[Gorakhpur 92; Purvanchal 96, 99]

Sol. Let m be the mass of each sphere and u, u' their velocities before impact making angles α and β with line of centres OO'. Here the spheres A and B are moving in directions at right angles to each other. So, from $\Delta OO'P$, by geometry, $\beta = 90^\circ + \alpha$. Let v, v' be the velocities of spheres after impact making angles θ and ϕ with OO'.

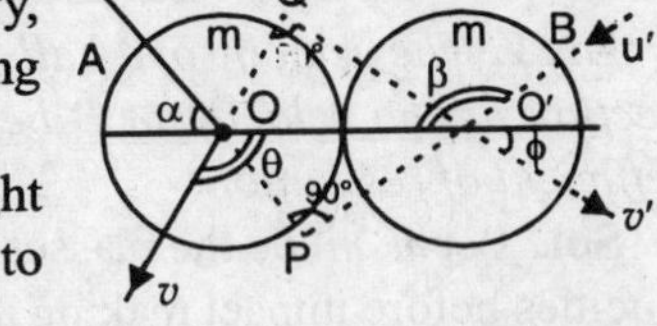

The spheres being smooth, there is no force between them at right angles to the common normal OO'. So the velocities at right angles to OO' remain unaltered due to impact. Hence, we have

$$v \sin\theta = u \sin\alpha \qquad \text{...(1)}$$

and $\quad v' \sin\phi = u' \sin\beta \quad$ or $\quad v' \sin\phi = u' \cos\alpha$, as $\beta = 90^\circ + \alpha$ ...(2)

By principle of conservation of momentum, we get

$$mv\cos\theta + m'v'\cos\phi = mu\cos\alpha + mu'\cos\beta$$

or $\quad v\cos\theta + v'\cos\phi = u\cos\alpha - u'\sin\alpha$, as $\beta = 90^\circ + \alpha$ ...(3)

By Newton's experimental law, we have

$$v\cos\theta - v'\cos\phi = -(u\cos\alpha - u'\cos\beta)$$

or $\quad v\cos\theta - v'\cos\phi = -(u\cos\alpha + u'\sin\alpha)$, as $\beta = 90^\circ + \alpha$ ...(4)

Subtracting (4) from (3), $2v'\cos\phi = 2u\cos\alpha$ or $v'\cos\phi = u\cos\alpha$ ...(5)

Adding (3) and (4), $\quad 2v\cos\theta = -2u'\sin\alpha$ or $v\cos\theta = -u'\sin\alpha$ ...(6)

Dividing (1) by (6), $\quad \tan\theta = -u/u'$ ...(7)

Dividing (2) by (5), $\quad \tan\phi = u'/u$ ...(8)

From (7) and (8), $\quad \tan\theta\tan\phi = -1$ or $\tan\theta = -\cot\phi = \tan(90^\circ + \phi)$,

Thus, $\theta = 90^\circ + \phi$, showing that the given sphere still move at right angles even after impact as from $\Delta OQO'$, $\angle OQO' + \phi = \theta = 90^\circ + \phi \Rightarrow \angle OQO' = 90^\circ$

Ex. 3. (*b*) *Two spheres of masses m and m', moving with velocities u and u' at right angles to one another, colloide. Prove that if their directions after impact are also at right angles, then*

$$e = (m^2 u\cos\alpha + m'^2 u'\sin\alpha)/mm'(u\cos\alpha + u'\sin\alpha),$$

where the direction of u makes angle α *with line of centres.* **[Purvanchal 1992]**

Sol. Refer figure of Ex. 3. (*a*). Let the direction of motion of the second sphere before impact make an angle β with the line of centres OO'. Let v, v' be their velocities after impact making angles θ and ϕ with OO'. Since their directions of motion before and after impact are at right angles to each other, so

$$\beta = 90^\circ + \alpha \quad \text{and} \quad \theta = 90^\circ + \phi, \text{ by geometry} \qquad \text{...(1)}$$

The spheres being smooth, there is no force between them at right angles to the common normal OO'. Hence the velocities at right angles to OO' remain unaltered due to impact. Thus, we have

$u\sin\alpha = v\sin\theta \quad$ or $\quad u\sin\alpha = v\cos\phi$, by (1) ...(2)

and $\quad u'\sin\beta = v'\sin\phi \quad$ or $\quad u'\cos\alpha = v'\sin\phi$, by (1) ...(3)

By the principle of conservation of momentum, we get

$$mv\cos\theta + m'v'\cos\phi = mu\cos\alpha + m'u'\cos\beta$$

or $\quad -mv\sin\phi + m'v'\cos\phi = mu\cos\alpha - m'u'\sin\alpha$, by (1) ...(4)

By Newton's experimental law, we have

$$v\cos\theta - v'\cos\phi = -e(u\cos\alpha - u'\cos\beta)$$

or $\quad -v\sin\phi - v'\cos\phi = -e(u\cos\alpha + u'\sin\alpha)$, by (1) ...(5)

or $$e = \frac{v\sin\phi + v'\cos\phi}{u\cos\alpha + u'\sin\alpha} = \frac{\dfrac{u\sin\alpha}{\cos\phi}\cdot\sin\phi + \dfrac{u'\cos\alpha}{\sin\phi}\cdot\cos\phi}{u\cos\alpha + u'\sin\alpha}$$

[On putting values of v and v' given by (2) and (3)]

or $\quad e = (u \sin \alpha \tan \phi + u' \cos \alpha \cot \phi)/(u \cos \alpha + u' \sin \alpha)$ ...(6)

From (4), $\quad m(u \cos \alpha + v \sin \phi) = m'(v' \cos \phi + u' \sin \alpha)$

or $m\left(u \cos \alpha + \dfrac{u \sin \alpha}{\cos \phi} \cdot \sin \phi\right) = m'\left(\dfrac{u' \cos \alpha}{\sin \phi} \cdot \cos \phi + u' \sin \alpha\right)$, by (2) and (3)

or $\quad mu \cos (\alpha - \phi)/\cos \phi = m'u' \cos (\alpha - \phi)/\sin \phi$

giving $\quad \tan \phi = m'u'/mu \quad$ and $\quad \cot \phi = mu/m'u'$ ...(7)

Substituting these values of $\tan \phi$ and $\cot \phi$ in (6), we get

$$e = \frac{(m'u'/mu)u \sin \alpha + (mu/m'u')u' \cos \alpha}{u \cos \alpha + u' \sin \alpha} = \frac{m^2 u \cos \alpha + m'^2 u' \sin \alpha}{mm'(u \cos \alpha + u' \sin \alpha)}$$

Ex. 4. *Two equal smooth spheres of radius r, move with same speed in opposite directions in parallel lines which are at a distance c apart; prove that direction of motion of each deviates on account of impact, through a right angle if $(1 + e)c^2 = 4er^2$, e being the coefficient of restitution.*

[Purvanchal 98]

Sol. Before impact, let u be the velocity of each sphere making an angle α with line OO' joining their centre O, O' but their velocities are given to be in opposite directions as shown in the figure. Due to symmetry their velocities after impact will also be equal in magnitude v (say). Since they deviate through a right angle, so after impact their directions will make angles $90° - \alpha$ with OO' as shown in figure. By Newton's experimental law, we have

$$-v \cos (90° - \alpha) - \{v \cos (90° - \alpha)\} = -e\{u \cos \alpha - (-u \cos \alpha)\}$$

or $\quad 2v \sin \alpha = 2eu \cos \alpha \quad$ or $\quad v \sin \alpha = eu \cos \alpha$ ...(1)

Since the spheres are smooth, their velocities in a direction perpendicular to their common normal OO' remain unaltered. Hence, we have

$u \sin \alpha = v \sin (90° - \alpha) \quad$ or $\quad v \cos \alpha = u \sin \alpha$ ...(2)

Dividing (1) by (2), $\quad \tan \alpha = e \cot \alpha \quad$ or $\quad \tan^2 \alpha = e$ ...(3)

Let $O'P \perp OP$. Given $O'P = c$. Also $OO' = 2r$. So from $\Delta OO'P$, $OP^2 = (2r)^2 - c^2$ so that $OP = (4r^2 - c^2)^{1/2}$. From $\Delta OO'P$,

$\tan \alpha = O'P/OP = c/(4r^2 - c^2)^{1/2}$ and hence (3) reduces to

$c^2/(4r^2 - c^2) = e \quad$ or $\quad c^2 = (4r^2 - c^2)e \quad$ or $\quad 4er^2 = c^2(1 + e)$

Ex. 5. (*a*) *Two equal balls of radius a are in contact and are struck simultaneously by a ball of radius c moving in the direction of their common tangent; if all the balls be of the same material the coefficient of elasticity being e, prove that the impinging ball will be reduced to rest if*

$2e = c^2(a + c)2/a^3(2a + c)$

[Purvanchal 1993, 99]

Sol. Let two equal spheres with centre A and B be in contact with each other at D. Let a third ball with centre C be moving with velocity u, say along the common tangent CD of the spheres with centres A and B. Given that spheres with centres A and B are at rest before impact and the sphere with centre C is reduced to rest after impact.

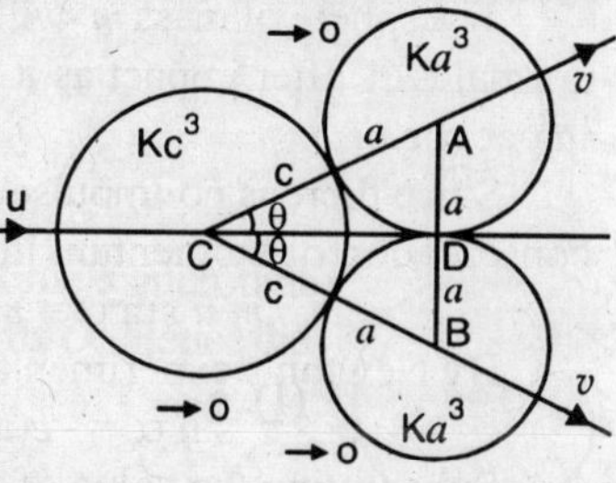

Let CA and CB be inclined at an angle θ to the common tangent CD. By symmetry, velocities of spheres with centres A and B will be equal after impact, say each equal to v and will move along CA and CB as shown in figure.

Since the balls are of the same material, their masses are proportional to their cubes. So the mass of the ball of radius a will be $k \cdot (4/3)\pi a^3$ or Ka^3, where $K = (4/3)\pi k$ is a constant. Similarly, the mass of the ball of radius c will be Kc^3.

By law of conservation of momentum, the total momentum of all the three spheres along CD remains constant

$\therefore \quad Ka^3 \cdot v \cos\theta + Ka^3 \cdot v \cos\theta = Kc^3 \cdot u \quad$ or $\quad c^3 u = 2a^3 v \cos\theta \qquad \ldots(1)$

Using Newton's experimental law for the spheres with centres B and C,

$v - 0 = -e(0 - u\cos\theta) \qquad$ or $\qquad v = eu\cos\theta \qquad \ldots(2)$

Multiplying (1) and (2), $c^3 uv = 2a^3 euv \cos^2\theta$ or $\quad c^3 = 2a^3 e \cos^2\theta \qquad \ldots(3)$

From ΔACD, $\qquad \sin\theta = a/(a+c)$

$$\therefore \quad \cos^2\theta = 1 - \sin^2\theta = 1 - \frac{a^2}{(a+c)^2} = \frac{c^2 + 2ac}{(a+c)^2} = \frac{c(2a+c)}{(a+c)^2} \qquad \ldots(4)$$

Using (4), (3) gives $\quad c^3 = \dfrac{2a^3 ec(2a+c)}{(a+c)^2} \quad$ or $\quad 2e = \dfrac{c^2(a+c)^2}{a^3(2a+c)}$

Ex. 5. (*b*) *Two equal spheres of mass m are in contact on a smooth horizontal table. A third ball of mass m′ impinges symmetrically on them and is reduced to rest. Prove that e = 2m′/3m and find the loss of K.E. due to impact.*

Sol. Draw figure as in Ex. 5(*a*). Hence $c = a$. Hence ΔABC is an equilateral triangle and so $2\theta = 60°$ or $\theta = 30°$. Here take $Kc^3 = m'$ and $Ka^3 = m$.

By law of conservation of momentum, the total momentum of all the three spheres along CD remains constant.

$\therefore \quad mv\cos 30° + mv\cos 30° + m' \cdot 0 = m' \cdot u + m \cdot 0 + m \cdot 0$

or $\quad 2mv(\sqrt{3}/2) = m'u \qquad$ or $\qquad \sqrt{3}mv = m'u \qquad \ldots(1)$

By Newton's experimental law, for the collision of spheres with centres A and C,

$v - 0 = -e(0 - u\cos 30°) \qquad$ or $\qquad v = (\sqrt{3}/2)eu \qquad \ldots(2)$

From (1) and (2), $\quad \sqrt{3}m \cdot (\sqrt{3}/2)eu = m'u \quad$ or $\quad e = 2m'/3m \qquad \ldots(3)$

Now K.E. before impact $= (1/2)m'u^2$

and $\qquad$ K.E. after impact $= 2 \cdot (1/2)mv^2 = m(\sqrt{3}/2)^2 e^2 u^2$, by (1)

$= (3/4)me^2u^2 = (3/4) \cdot (2m'/3e)e^2u^2$, by (3)

$= (1/2)m'eu^2$

Loss of K.E. $= (1/2)m'u^2 - (1/2)m'eu^2 = (1/2)m'(1-e)u^2$

Ex. 6. *A smooth spherical ball of mass m is tied to a fixed point by a light inextensible string and another spherical ball of mass m′ impinges directly on it with velocity u in a direction making an acute angle α with the string. Prove that the velocity with which m begins to move is*

$$\{m'(1+e)u \sin\alpha\}/(m + m'\sin^2\alpha)$$

Sol. Let O and O' be centres of spheres of masses m and m' respectively. Due to the string, the sphere of mass m can move only perpendicularly to the string (*i.e.*, the horizontal direction) after impact with velocity v', say.

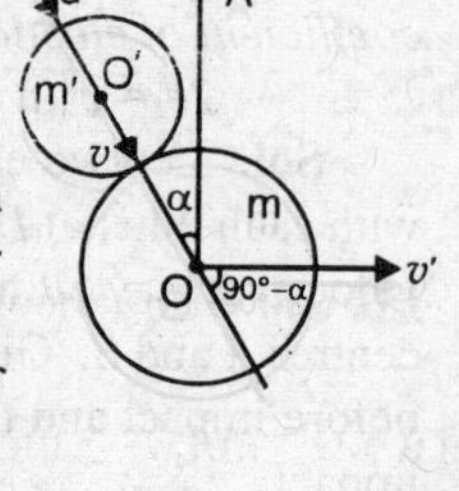

The sphere of mass m' will keep on moving in the same direction of common normal OO' after impact as it was moving before impact. Let its velocity after impact be v.

Since there is no impulse perpendicular to the string, hence by principle of conservation of momentum in the horizontal direction, we have

$m'u\sin\alpha = mv' + m'v\sin\alpha \qquad \ldots(1)$

By Newton's experimental law along OO',

$v - v'\sin\alpha = -e(u - 0) \qquad$ or $\qquad v = v'\sin\alpha - eu \qquad \ldots(2)$

Substituting the value of v from (2) in (1), we have

$m'u\sin\alpha = mv' + m'(v'\sin\alpha - eu)\sin\alpha$

or $\quad m'(1+e)u\sin\alpha = v'(m + m'\sin^2\alpha)$

or $\quad v' = $ required velocity $= \{m'(1+e)u\sin\alpha\}/(m + m'\sin^2\alpha)$

Ex. 7. (*a*) *A smooth ball impinges an another equal ball at rest in a direction making an angle α with the line of centres. If the coefficient of elasticity of two balls is 1/3, prove that the angle through which the direction of the impinging ball is deviated is $\tan^{-1}\{2\tan\alpha/(1+3\tan^2\alpha)\}$.* **[Purvanchal 97]**

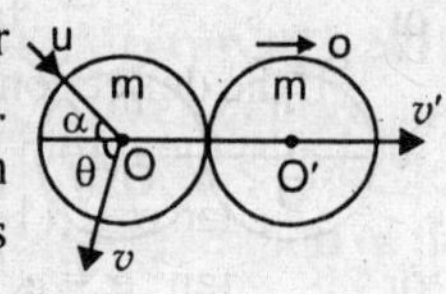

Sol. Let u, v be the velocities of the impinging sphere before and after impact respectively. Let direction of v make an θ with the line of centres OO'. The second sphere being at rest before impact will move along the common normal OO' after impact because the force on it during the period of impact is only along the common normal. Let v' be its velocity after impact.

Since there is no force in the direction perpendicular to the common normal, hence the velocities at right angles to the common normal for the impinging sphere will remain unaltered due to impact.

Thus, $\quad u \sin \alpha = v \sin \theta$ ...(1)

Let m be mass of each sphere. So by the principle of conservation of momentum,

$mv' + m(-v \cos \theta) = m\cdot 0 + mu \cos \alpha$ or $v' - v \cos \theta = u \cos \alpha$ ...(2)

Given $e = 1/3$. By Newton's experimental law, we get

$v' - (-v \cos \theta) = -(1/3)(0 - u \cos \alpha)$ or $v' + v \cos \theta = (1/3) u \cos \alpha$ ...(3)

Subtracting (2) from (3), $2v \cos \theta = -(2/3) u \cos \alpha$ or $v \cos \theta = -(1/3) u \cos \alpha$

or $(u \sin \alpha / \sin \theta) \cos \theta = -(1/3) u \cos \alpha$, as by (1), $v = u \sin \alpha / \sin \theta$

or $\quad 3 \sin \alpha \cot \theta = -\cos \alpha \quad$ or $\quad \tan \theta = -3 \tan \alpha$ ...(4)

Now, $\quad \tan(\theta + \alpha) = \dfrac{\tan \theta + \tan \alpha}{1 - \tan \theta \tan \alpha} = \dfrac{-3 \tan \alpha + \tan \alpha}{1 - (-3 \tan \alpha) \tan \alpha}$, by (4)

or $\quad \tan(\theta + \alpha) = -\dfrac{2 \tan \alpha}{1 + 3 \tan^2 \alpha} \quad$ or $\quad \tan(\theta + \alpha) = \dfrac{2 \tan \alpha}{1 + 3 \tan^2 \alpha}$, numerically

(negative sign shows that $\theta + \alpha$ is an obtuse angle)

Thus, the required angle $= \theta + \alpha = \tan^{-1} \{2 \tan \alpha / (1 + 3 \tan^2 \alpha)\}$

Ex. 7. (*b*) *A smooth sphere of mass m, travelling with velocity u, impinges obliquely on a smooth sphere of mass M at rest, its original line of motion making an angle α with the line of centres, at the moment of impact. Show that the sphere of mass m will be deflected through a right angle if $\tan^2 \alpha = (eM - m)/(M + m)$.*

Hint. Proceed as in Ex. 7(*a*).

Ex. 8. *A smooth ball impinges on another smooth equal ball at rest in a direction that makes an angle α with the line of centres at the moment of impact. Prove that if D be the angle through which the direction of motion of the impinging ball is deviated, then $\tan D = \{(1 + e) \tan \alpha\}/(1 - e + 2 \tan^2 \alpha)\}$.*

If the direction of the impinging ball bisects the angle between the future directions of motion of the balls, then prove that $\tan \alpha = e^{1/2}$.

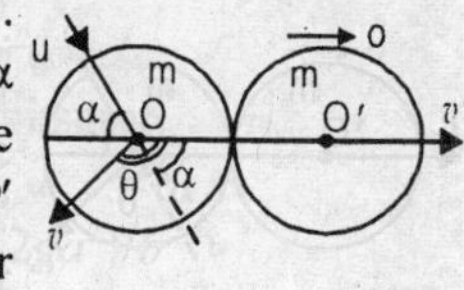

Sol. Let m be mass of each ball and let OO' be their line of centres O, O'. Let u be the velocity before impact of the impinging ball, making an angle α will OO'. Let v be its velocity after impact making an angle θ with OO'. The second ball with centre O' is initially at rest and as such, will move along OO' after impact with velocity v', say. Since the velocity in a direction perpendicular to OO' will remain unaltered for sphere of mass m, we have

$$v \sin \theta = u \sin \alpha \quad \text{...(1)}$$

By principle of conservation of momentum, we get

$mv \cos \theta + mv' = mu \cos \alpha + m\cdot 0 \quad$ or $\quad v \cos \theta + v' = u \cos \alpha$ ...(2)

By Newton's experimental law, we have

$v \cos \theta - v' = -e(u \cos \alpha - 0) \quad$ or $\quad v \cos \theta - v' = -eu \cos \alpha$ ...(3)

Adding (2) and (3), $2v \cos \theta = (1 - e) u \cos \alpha$ or $v \cos \theta = \{(1 - e)/2\} u \cos \alpha$ ...(4)

Dividing (1) by (4), $\quad \tan \theta = (2 \tan \alpha)/(1 - e)$ ...(5)

Since $\quad D = \theta - \alpha \quad$ so $\quad \tan D = \tan(\theta - \alpha)$

or $\quad \tan D = \dfrac{\tan \theta - \tan \alpha}{1 + \tan \theta \tan \alpha} = \dfrac{\{(2 \tan \alpha)/(1 - e)\} - \tan \alpha}{1 + \{(2 \tan \alpha)/(1 - e)\} \tan \alpha}$

or $\tan D = \{(1-e)\tan\alpha\}/(1-e+2\tan^2\alpha)$ ...(6)

If the deflection of the impinging ball bisects the angle between the future directions of motion of the balls, then from figure $D=\alpha$ and (6) reduces to

$$\tan\alpha = \{(1+e)\tan\alpha\}/(1-e+2\tan^2\alpha) \quad \text{or} \quad 1-e+2\tan^2\alpha = 1+e$$

or $\tan^2\alpha = e$ or $\tan\alpha = e^{1/2}$, as required.

EXERCISE

1. Two equal smooth elastic spheres moving in opposite parallel directions with equal velocities impinge on one another; if inclination of their direction of motion to line of centres be $\tan^{-1}\sqrt{e}$ where e is the coefficient of elasticity, show that their directions of motion will be turned through a right angle.

2. A smooth ball impinges on another smooth equal ball at rest in a direction that bisects the angle between the subsequent directions of the two balls. If the direction of motion of the impinging ball before impact makes an angle α with the line of centres, prove that $\tan\alpha = \sqrt{3}$.

3. Two equal smooth spheres moving along parallel lines in opposite directions with velocities u and v, collide with the line of centres at an angle α with their direction of motion. If after impact their lines of motion are at right angles to one another, show that $(u-v)^2/(u+v)^2 = \sin^2\alpha + e^2\cos^2\alpha$.

4. Two equal balls are lying in contact on a smooth table and a third equal ball moving along their common tangent, strikes them simultaneously. Prove that $(3/5)(1-e^2)$ of its K.E. is lost by impact, e being the coefficient of restitution.

5. Two balls whose masses are m and m' are moving in the directions, making angles α, α' with the line of centres, collide. If their directions of motion after the impact be perpendicular to their direction before impact; prove that $e = (m\sin^2\alpha' + m'\sin^2\alpha)/(m\cos^2\alpha' + m'\cos^2\alpha)$

MISCELLANEOUS PROBLEMS

1. A particle is projected from a point on an inclined plane and at the rth impact, it strikes the plane perpendicularly and at the nth impact is at the point of projection. Show that

$$e^n - 2e^r + 1 = 0$$ **(M.S. Univ. T.N. 2007)**

8

Constrained Motion in Vertical and Horizontal Circles

8.1. Constrained motion. **[Agra 2010; Meerut 2011, 18, 19]**

When a particle in motion under given forces is compelled to remain in a particular path or in contact with a given surface or curve, its motion is known as a *constrained motion*.

Some examples of constrained motion are (*i*) a particle moving down a rough or smooth inclined plane (*ii*) a particle sliding down a curve under gravity. (*iii*) a particle tied to one end of a string and made to move in a particular way.

In the present chapter we propose to discuss constrained motion in vertical and horizontal circles.

PART I : CONSTRAINED MOTION IN A VERTICLE CIRCLE

8.2 Motion on the outside of a smooth vertical circle

(*i*) *A particle slides down the outside of a smooth vertical circle starting from rest at the highest point, to discuss the motion.* **[Meerut 2014, 2019]**

(*ii*) *Determine the path the particle will follow after it leaves the contact with the circle.* **[Agra 2009; Garhwal 2002; Kumaun 2002]**

Let A be the highest point of the circle with centre O and radius a and AOB be its vertical diameter. Let P be the position of the particle at any time t, having velocity v along tangent PT where $\angle AOP = \theta$. Let $PN \perp AB$ and

$$AN = h = OA - ON = a - a\cos\theta = a(1 - \cos\theta). \quad \ldots(1)$$

We now apply the principle of conservation of energy to the motion from A to P. As R is always perpendicular to the direction of motion, we have

$$(1/2)mv^2 - (1/2)m.0^2 = mg.AN \quad \text{or} \quad v^2 = 2ga(1 - \cos\theta) \quad \ldots(2)$$

The external forces acting on the particle are the normal reaction R of the circle along OP and weight mg of the particle acting vertically downwards. We know that the resolved parts of the acceleration along the tangent PT and along the normal PO at P are d^2s/dt^2 (or dv/dt or $v\,dv/ds$) and v^2/a. Hence the equation of motion in the inward normal sense is

$$mv^2/a = mg\cos\theta - R \quad \ldots(3)$$

From (3), $R = mg\cos\theta - (m/a)v^2 = mg\cos\theta - (m/a)\times 2ga(1 - \cos\theta)$, by (2)

or $$R = mg(3\cos\theta - 2) \quad \ldots(4)$$

When the particle leaves the circle, then $R = 0$ and so by (4)

$$3\cos\theta - 2 = 0 \text{ giving } \cos\theta = 2/3 \quad \text{or} \quad \theta = \cos^{-1}(2/3) \quad \ldots(5)$$

At the time of leaving the circle the depth of P below A

$$= AN = h = a(1 - \cos\theta) = a(1 - 2/3) = a/3, \quad \ldots(6)$$

showing that the particle leaves the circle when it has descended through a height equal to the one-third of the radius.

Putting $\cos\theta = 2/3$ in (1), the velocity of the particle at the time of leaving the circle is given by

$$v^2 = 2ga(1 - 2/3) = (2/3)ga \quad \text{or} \quad v = (2ga/3)^{1/2} \quad \ldots(7)$$

(*ii*) **Equation of the trajectory described by particle after leaving the circular path at *P*.** Take the point P where the particle leaves the circular path as origin, the horizontal line as the x-axis and downward vertical line as y-axis as shown in the figure. Then, equation of the required parabolic path is given by

$$y = x\tan\theta + (gx^2)/2v^2\cos^2\theta, \quad \ldots(8)$$

where θ is given by (5) so $\cos\theta = 2/3$ and $\tan\theta = \sqrt{5}/2$ and v is given by (7) *i.e.*, $v = (2ga/3)^{1/3}$. So (8) gives

$$y = x\frac{\sqrt{5}}{2} + \frac{gx^2}{2\times(2ga/3)\times(4/9)} \qquad \text{or} \qquad y = \frac{\sqrt{5}}{2}x + \frac{27x^2}{16a} \qquad \dots(9)$$

Note. All results of projectile should be remembered for this chapter.

TYPE 1 : SOLVED EXAMPLES BASED ON ART. 8.2

Ex. 1. *If a particle slides down the outside of a smooth vertical circle, starting from rest at the highest point; show that it will leave the circle after describing vertically a distance equal to one third of the radius of the circle* **[Gorakhpur 1995; Rohilkhand 1995]**

Sol. Refer Art. 8.2. Do upto equation (6) for this problem.

Ex. 2. *A heavy particle is allowed to slide down a smooth vertical circle of radius 27a from rest at the highest point. Show that on leaving the circle it moves in a parabola of latus rectum 16a.*

[Agra 2002, 08; Lucknow 1997, Purvanchal 1997]

Sol. Proceed exactly as in Art. 8.2 upto equation (7) by replacing a by 27 for this problem. Then (5) and (7) give $\cos\theta = 2/3$ and $v^2 = (2/3)\times g\times(27a) = 18ag$. Now, the latus rectum of the parabola subsequently described

$$= \frac{2v^2\cos^2\theta}{g} = \frac{2\times 18ag\times(4/9)}{g} = 16a.$$

Ex. 3. *A particle is placed at the highest point of a smooth vertical circle of radius a and is allowed to slide down starting with a negligible velocity. Prove that it will leave the circle after describing vertically a distance equal to one third of the radius. Find the position of the directrix and the focus of the parabola subsequently described and show that its latus rectum is (16/27)a.*

[Agra 2005; Gorakhpur 1994]

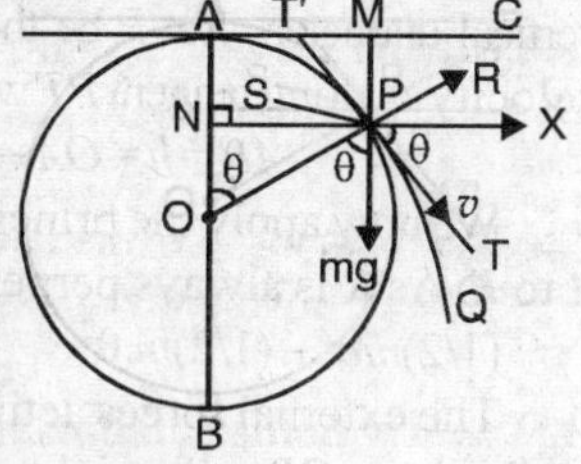

Sol. Proceed exactly as in Art. 8.2 upto equation (7). Then the latus rectum of the parabola subsequently described

$$= \frac{2v^2\cos^2\theta}{g} = \frac{2}{g}\cdot\frac{2ag}{3}\cdot\frac{4}{9} = \frac{16a}{27}, \text{ using (5) and (7)}$$

To find the position of the directrix. We know that in a parabolic path (PQ, say) of a projectile the velocity at any point of its path is equal to that due to fall from the directrix to that point. Hence, if H be the height of the directrix above P, then the velocity acquired in falling a distance H under gravity $= (2gH)^{1/2}$. Hence, using result (7) of Art. 8.2., must have

$$(2ag/3)^{1/2} = (2gH)^{1/3} \text{ or } H = a/3 = h = AN \text{ by (6),}$$

showing that the directrix is a horizontal line AM through A.

To find the position of focus *S*. Let $PM \perp AM$ and let $T'T$ be tangent at P to the parabolic path PQ at P. If S be the focus of the parabola subsequently described, then using the geometrical properties of a parabola, we must have $SP = PM = a/3$ and $\angle SPT' = \angle MPT'$. This will determine the position of focus S of the parabola.

Ex. 4. *A particle slides down the arc of a smooth vertical circle of radius a, being slightly displaced from rest at the highest point. Find where it will leave the circle and prove that it will strike a horizontal plane through the lowest point of the circle at a distance* $5(\sqrt{5}+4\sqrt{2})a/27$.

[Agra 1995, 99; I.A.S. 1995]

Sol. First of all do as in Art. 8.2 upto equation (8). Let the particle strike the horizontal plane through the lowest point B of the circle at the point E. Let $PD \perp BE$. Then $PD = NB = NO + OB = a\cos\theta + a = a\times(2/3) + a = 5a/3$. Let $DE = b$. Then coordinates of E referred to PX and PY as coordinate axes are $(b, 5a/3)$. Since the point E lies on the parabolic path (8), we have

$$5a/3 = (b\sqrt{5})/2 + 27b^2/16a \qquad \text{or} \qquad 81b^2 + 24\sqrt{5}ab - 80a^2 = 0$$

$$\therefore \quad b = \frac{-24\sqrt{5}a \pm \{(-24\sqrt{5}a)^2 + 4\times 81\times 81a^2\}^{1/2}}{2\times 81} = \frac{-4\sqrt{5}a \pm 20\sqrt{2}a}{27}$$

Rejecting negative value of b, $\qquad b = (20\sqrt{2}a - 4\sqrt{5}a)/27$

$\therefore$ distance of E from the lowest point B of the circle

$$= BE = BD + DE = PN + b = a \sin\theta + b, \text{ where } \cos\theta = 2/3$$
$$= (\sqrt{5}/3)a + (20\sqrt{2} - 4\sqrt{5})a/27 = 5(\sqrt{5} + 4\sqrt{2})a/27$$

Ex. 5. (*a*) *A particle slides down along the outside of the arc of smooth vertical circle of radius a. If it starts with velocity u at the highest point, find the angle θ which the radius vector to the particle makes with the vertical when the particle leaves the circle.*

(*b*) *A particle is projected from the highest point of a vertical circle with tangential velocity* $\{(1/2)\, ga(3\sqrt{2} - 4)\}^{1/2}$*; show that it will leave the circle when the radius to the circle makes an angle of 45° with the upward vertical, a being the radius of the circle.*

(*c*) *A particle is projected with speed* $(ag/2)^{1/2}$ *from the highest point of the outside of a fixed smooth sphere of radius a. Show that it will leave the sphere at the point whose vertical distance below the point of projection is a/6. Also find the magnitude of the velocity of the particle when it strikes the plane upon which the sphere stands.* **[Meerut 2018]**

Sol. (*a*) Proceed as in Art. 8.2 upto equation (3). Here velocity at A is u and so equation (2) for this problem becomes

$$(1/2)mv^2 - (1/2)mu^2 = mg\, AN \quad \text{or} \quad v^2 = u^2 + 2ga(1 - \cos\theta) \qquad \ldots(2)$$

Also, equation (3) as before is
$$mv^2/a = mg\cos\theta - R \qquad \ldots(3)$$

Now proceed as follows. From (2) and (3), we get

$$R = mg\cos\theta - mv^2/a = mg\cos\theta - (m/a)\{u^2 + 2ga(1 - \cos\theta)\}$$

or
$$R = (m/a)(3ga\cos\theta - u^2 - 2ga) \qquad \ldots(4)$$

When the particle leaves the circle, then $R = 0$ and so by (4), we get

$$3ga\cos\theta - u^2 - 2ga = 0 \quad \text{or} \quad \cos\theta = (u^2 + 2ga)/3ga \qquad \ldots(5)$$

Thus,
$$\theta = \cos^{-1}\{(u^2 + 2ga)/3ga\} \qquad \ldots(6)$$

(*b*) Hint. Do as in part (*a*). In equation (6), take $u = \{(1/2)\, ga(3\sqrt{2} - 4)\}^{1/2}$ so that $\theta = \cos^{-1}(1/\sqrt{2})$ or $\theta = 45°$.

(*c*) First of all do as in part (*a*) upto equation (5). Here $u = (ag/2)^{1/2}$. Then (5) reduces to $\cos\theta = \{(1/2)ag + 2ag\}/3ga$ or $\cos\theta = 5/6$.

$\therefore$ Vertical distance of P below the point of projection A

$$= AN = a(1 - \cos\theta) = a(1 - 5/6) = a/6.$$

Now the particle after leaving contact with the sphere describes parabolic path PE. Let the particle strike the plane, upon which the sphere stands, at E.

Then, vertical distance fallen by the particle in moving from P to E

$$= PD = NB = AB - AN = 2a - a/6 = 11a/6 \qquad \ldots(i)$$

Putting $\cos\theta = 5/6$ and $u = (ag/2)^{1/2}$ in equation (2), the velocity v at P at the time of leaving the sphere is given by

$$v^2 = ag/2 + 2ag(1 - 5/6) \quad \text{or} \quad v^2 = 5ag/6 \qquad \ldots(ii)$$

Let V be the velocity of the particle at E. Then using the principle of conservation of energy to the motion from P to E, we get

$$(1/2)mV^2 - (1/2)mv^2 = mg.NB \quad \text{or} \quad V^2 = v^2 + 2g.NB$$

or
$$V^2 = (5ag/6) + 2g \times (11a/6), \text{ by } (i) \text{ and } (ii)$$

or
$$V^2 = (9/4) \times (2ag) \quad \text{so that} \quad V = (3/2) \times (2ag)^{1/2}$$

Ex. 6. (*a*) *A particle is placed on the outside of a smooth vertical circle. If the particle starts from a point whose angular distance is α from the highest point of the circle, show that it will fly off the curve when the corresponding radius makes an angle* $\cos^{-1}\{(2/3)\cos\alpha]$ *with the vertical and would subsequently describe a parabola of latus rectum* $(16a\cos^2\alpha)/27$.

[Meerut 2005, 08, 12, 2019]

(*b*) *A particle moves down along the outside of a smooth vertical circle of radius a. If it starts from rest at a depth a/2 below the highest point, prove that it leaves the circle at a distance a/3 above the centre.* **[Agra 2010]**

(*c*) *A particle of mass m slides down on the outside of a smooth vertical circle of radius 'a' units starting from rest from a point at which the radius makes an angle* α *with the vertical. Discuss the motion.* **[Bangalore 2004]**

Sol. (*a*) Let A be the highest point of the circle with centre O and radius a and AOB be its vertical diameter. Let the particle starts from rest from Q when $\angle AOQ = \alpha$. Let P be the position of the particle at any time t, having velocity v along tangent PT where $\angle AOP = \theta$. Let $QN \perp AB$ and $PM \perp AB$. Then, from figure, we find

$$MN = ON - OM = a\cos\alpha - a\cos\theta = a(\cos\alpha - \cos\theta) \qquad \ldots(1)$$

Applying the principle of conservation of energy to the motion of the particle from Q to P, we get

$$(1/2)mv^2 - (1/2)m.0^2 = mg \times NM \quad \text{or} \quad v^2 = 2ga(\cos\alpha - \cos\theta), \text{ by (1)} \qquad \ldots(2)$$

The external forces acting on the particle P are the normal reaction R of the circle along OP and weight mg of the particle acting vertically downwards. So the equation of the particle along inward normal is

$$mv^2/a = mg\cos\theta - R \quad \text{or} \quad R = mg\cos\theta - mv^2/a$$

$$\therefore R = mg\cos\theta - 2mg(\cos\alpha - \cos\theta) \quad \text{or} \quad R = mg\,(3\cos\theta - 2\cos\alpha) \qquad \ldots(3)$$

When the particle leaves the circle, then $R = 0$ and so by (3)

$$3\cos\theta - 2\cos\alpha = 0 \text{ or } \cos\theta = (2/3)\cos\alpha \text{ or } \theta = \cos^{-1}\{(2/3)\cos\alpha)\}$$

(***b***) Try yourself. Given that $AN = a/3$. We require OM.

(***c***) Proceed as in part (*a*).

Ex. 7. *A particle moves under gravity in a vertical circle sliding down the convex side of the smooth circular arc. If the initial velocity is that due to fall to the starting point from a height h above the centre, show that it will fly off circle when at a height 2h/3 above the centre.*

[Rohilkhand 2009: Lucknow 1998]

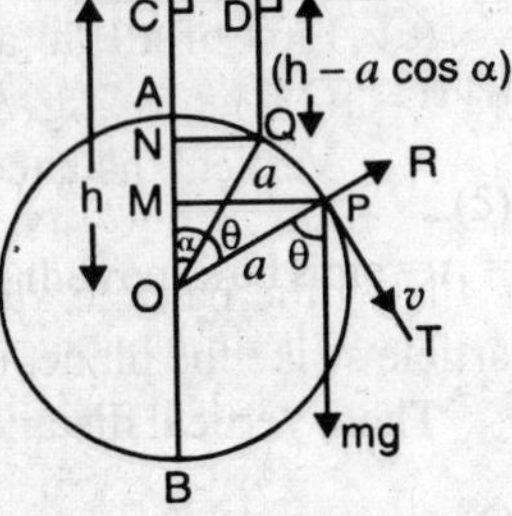

Sol. Let A be the highest point of the circle with centre O and radius a and AOB be its vertical diameter. Let the particle starts from Q where $\angle AOQ = \alpha$. Let CD be a horizontal line such that $OC = h$, $QN \perp AB$ and $QD \perp CD$. Then, from figure, we see that $QD = ND = OC - ON = h - a\cos\alpha$. So, by given problem, the initial velocity of the particle at

$$Q = \{2g(h - a\cos\alpha)\}^{1/2} = u, \text{ say} \qquad \ldots(1)$$

Let P be the position of the particle at any time t, having velocity v along tangent PT where $\angle AOP = \theta$. Let $PM \perp AB$. Then, from figure, we see that

$$NM = ON - OM = a\cos\alpha - a\cos\theta = a(\cos\alpha - \cos\theta) \qquad \ldots(2)$$

By principle of conservation of energy to the motion from Q to P, we have

$$(1/2)mv^2 - (1/2)mu^2 = mg \times NM \quad \text{or} \quad v^2 = u^2 + 2g \times NM$$

or $\quad v^2 = 2g(h - a\cos\alpha) + 2ga(\cos\alpha - \cos\theta)$, using (1) and (2)

or $$v^2 = 2g(h - a\cos\theta) \qquad \ldots(3)$$

The external forces acting on the particle P are the normal reaction R of the circle along OP and weight mg of the particle acting vertically downwards. Hence the equation of motion along inward normal is

$$mv^2/a = mg\cos\theta - R \quad \text{or} \quad R = mg\cos\theta - mv^2/a$$

or $\quad R = mg\cos\theta - (2mg/a)(h - a\cos\theta) = (mg/a)(3a\cos\theta - 2h)$

When the particle leaves the circle, $R = 0$ and so

$$3a\cos\theta - 2h = 0 \quad \text{giving} \quad \cos\theta = 2h/3a \qquad \ldots(4)$$

$\therefore$ Required height above the centre $= OM = a\cos\theta = 2h/3$, by (4).

Ex. 8. (*a*) *A roadway of a bridge over a canal is in the form of a circular arc of radius 20 meters. Find the greatest speed in km.p.h. with which a motor-cyclist can cross the bridge without leaving the ground at the highest point.*

(*b*) *On a canal of breadth 20 meters, there is a bridge in the form of a circular arc, the middle point of which is at a height 2.5 meters from either end. Find the greatest speed with which a motor-cyclist can pass the bridge safely if the height of the combined body, i.e., of man and the motor cycle be 3/4 meter above the point of contact of the wheel with the ground.*

Sol. (*a*) Let m be mass of the motorcycle and its rider. Let PQ be canal in form of circular arc of radius 20 meters and centre O. Let v be the velocity at A along tangent PT. The external forces acting on the motorcycle and its rider are the normal reaction R of the circle along OA and weight mg acting vertically downwards. So the equation of motion along inward normal is

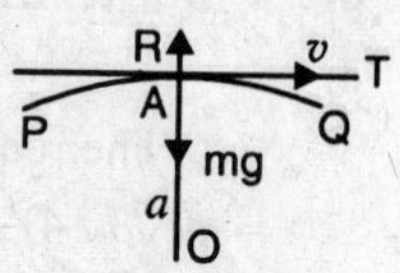

$$mv^2/20 = mg - R \quad \text{or} \quad R = mg - mv^2/20$$

The motor cyclist will not leave the ground at the highest point if $R \geq 0$ *i.e.*, $mg - mv^2/20 \geq 0$ or $v^2 \leq 20 \times 9.8$ or $v^2 \leq 196$

$\therefore$ greatest velocity $=$ 14 m/sec $= (14 \times 60 \times 60)/1000 = 50.4$ km.p.h.

(***b***) Try yourself. **Ans.** 14.7 m/sec (approx.)

Ex. 9. *A body is projected along the arc of a smooth circle of radius a from the highest point with velocity* $(1/2) \times (ag)^{1/2}$; *find where it leave the circle and prove that it will strike a horizontal plane through the centre of the circle at a distance* $(9\sqrt{39} + 7\sqrt{7})a/64$, *from the centre.*

Sol. Let A be the highest point of the circle with centre O and radius a and AOB be its vertical diameter. Let P be the position of the body at any time t, having velocity v along PT where $\angle AOP = \theta$. Let $PN \perp AB$. Then, from figure,

$$AN = OA - ON = a - a\cos\theta = a(1 - \cos\theta)$$

Initially body starts from A with velocity $(1/2) \times (ag)^{1/2}$. Applying the principle of conservation of energy to the motion from A to P, we get

$$(1/2)mv^2 - (1/2)m \times (1/4)ag = mg \times AN$$

or $$v^2 = (1/4)ag + 2ga(1 - \cos\theta)$$

or $$v^2 = (9/4)ag - 2ag\cos\theta = ag(9/4 - 2\cos\theta) \quad \ldots(1)$$

External forces acting on the body are the normal reaction R along OP and weight mg of the body acting vertically downwards. Then the equation of motion of the body along inward normal PO is given by

$$mv^2/a = mg\cos\theta - R \quad \text{so that} \quad R = mg\cos\theta - mv^2/a$$

or $$R = mg\cos\theta - (m/a)ag\,(9/4 - 2\cos\theta), \text{ using (1)}$$

or $$R = 3mg\cos\theta - (9/4)mg = 3mg(\cos\theta - 3/4) \quad \ldots(2)$$

The body leaves the circle if $R = 0$ so that

$$3mg(\cos\theta - 3/4) = 0 \quad \text{giving} \quad \cos\theta = 3/4 \quad \ldots(3)$$

So from (1), the velocity of the body along PT at the time of leaving the circle is given by

$$v^2 = ag\{9/4 - 2 \times (3/4)\} = (3/4)ag \quad \ldots(4)$$

Let PD be the parabolic path described by the body soon after leaving the circle at P with velocity v given by (4) at an angle θ given (3). Let OD be a horizontal line passing through the centre of the circle. Let PC be parallel to AB and $DC \perp PC$. Then $PC = ON = a\cos\theta = a \times (3/4) = 3a/4$, by (3)

Take P as the origin and horizontal line PX and vertical line (in downwards direction) PY as coordinate axes. Then if $CD = b$, then coordinates of D are (CD, PC) or $(b, 3a/4)$. Equation of the parabolic path of the body is

$$y = x\tan\theta + (gx^2)/2v^2\cos^2\theta \quad \ldots(5)$$

From (3), $\cos\theta = 3/4$ so $\tan\theta = \sqrt{7}/3 \quad \ldots(6)$

Using (4) and (6), (5) reduces to

$$y = \frac{\sqrt{7}x}{3} + \frac{gx^2}{2 \times (3/4)ag \times (9/16)} \quad \text{or} \quad y = \frac{\sqrt{7}x}{3} + \frac{32x^2}{27a} \quad \ldots(7)$$

Since $D(b, 3a/4)$ lies on (7), we get

$$3a/4 = (\sqrt{7}/3)b + (32b^2)/27a \quad \text{or} \quad 128b^2 + 36\sqrt{7}ab - 81a^2 = 0$$

$$\therefore \quad b = [-36\sqrt{7}a \pm \{36 \times 36 \times 7a^2 + 4 \times 128 \times 81a^2\}^{1/2}]/(2 \times 128)$$

or $b = (1/256)\{-36\sqrt{7} + 36\sqrt{39}a\}$, reject –ve sign as b cannot be negative

or $$b = 9(\sqrt{39} - \sqrt{7})a/64$$

$\therefore$ The required distance $OD = OC + CD = a \sin\theta + b$

$= a(\sqrt{7}/4) + 9(\sqrt{39} - \sqrt{7})a/64$, as $\cos\theta = 3/4 \Rightarrow \sin\theta = \sqrt{7}/4$

$= (9\sqrt{39} + 7\sqrt{7})a/64$

Ex. 10. *A heavy particle slides under gravity down the inside of a smooth vertical tube held in a vertical plane. It starts from the highest point with velocity $(2ag)^{1/2}$ where a is the radius of the circle. Prove that when in the subsequent motion the vertical component of the acceleration is maximum, the pressure on the curve is equal to twice the weight of the particle.*

[Bundelkhand 1991; Garhwal 1990; Gorakhpur 1994; Meerut 1992]

Sol. Let A be the highest point of the tube with centre O and radius a and AOB be its vertical diameter. Let P be the position of the particle at any time t, having velocity v along tangent PT where $\angle POA = \theta$. Let $PN \perp AB$. Then $AN = OA - ON = a - a\cos\theta = a(1 - \cos\theta)$. Initially the particle starts from A with velocity $(2ag)^{1/2}$. Applying the principle of conservation of energy to the motion from A to P,

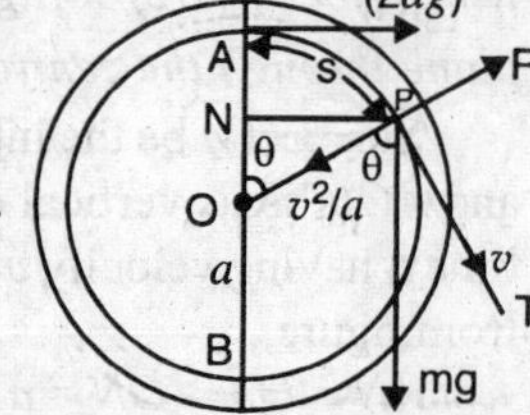

$$(1/2)mv^2 - (1/2)m \times (2ag) = mg \times AN$$

or $$v^2 = 2ag + 2ga(1 - \cos\theta)$$

or $$v^2 = 4ag - 2ag\cos\theta \quad \text{or} \quad v^2/a = 2g(2 - \cos\theta) \quad \ldots(1)$$

Now $\ddot{s}$ and v^2/a are the components of accleration at P along tangent PT and inward normal PO respectively. Hence, if f be the vertical component of acceleration at P, then

$$f = \ddot{s}\cos(90^\circ - \theta) + (v^2/a)\sin\theta = \ddot{s}\sin\theta + (v^2/a)\cos\theta \quad \ldots(2)$$

Again, the external forces acting on the particle are normal reaction R along OP and weight mg of the body acting vertically downwards. Then equation of motion of the particle along PT and PO are given by

$$m\ddot{s} = mg\sin\theta \quad \text{or} \quad \ddot{s} = g\sin\theta \quad \ldots(3)$$

and $$mv^2/a = mg\cos\theta - R \quad \text{or} \quad R = mg\cos\theta - mv^2/a \quad \ldots(4)$$

From (3) and (4), $R = mg\cos\theta - 2mg(2 - \cos\theta) = mg(3\cos\theta - 4)$ ...(5)

From (1), (2) and (3), we have

$$f = g\sin^2\theta + 2g\cos\theta(2 - \cos\theta) = g(4\cos\theta - 2\cos^2\theta + \sin^2\theta) \quad \ldots(6)$$

From (6), $$df/d\theta = g(-4\sin\theta + 4\cos\theta\sin\theta + 2\sin\theta\cos\theta)$$

or $$df/d\theta = 2g\sin\theta(3\cos\theta - 2) \quad \ldots(7)$$

For maximum or minimum of f, we have $df/d\theta = 0$

or $\sin\theta(3\cos\theta - 2) = 0$ so that either $\sin\theta = 0$ or $\cos\theta = 2/3$

But $\sin\theta = 0$ gives $\theta = 0$ which corresonds to initial position A.

So we consider $\cos\theta = 2/3$ only for max. or min. of f.

From (7), $$d^2f/d\theta^2 = 2g\cos\theta(3\cos\theta - 2) + 2g\sin\theta(-3\sin\theta)$$

or $d^2f/d\theta^2 = 6g\cos^2\theta - 4g\cos\theta - 6g(1 - \cos^2\theta) = 12g\cos^2\theta - 4g\cos\theta - 6g$

When $\cos\theta = 2/3$, $\quad d^2f/d\theta^2 = 12g \times (4/9) - 4g \times (2/3) - 6g = -(10/3)g = -$ve,

showing that f is maximum when $\cos\theta = 2/3$ and hence the corresponding value of R can be obtained from (5) and is given by

$R = mg(2 - 4) = -2mg =$ twice the weight of the particle, –ve sign indicates that the reaction has changed direction at that point.

8.3 Motion on the inside of a smooth vertical circle. *A particle is projected from the lowest point with velocity u and moves along the inside of the arc of a smooth vertical circle, to discuss the subsequent motion.* **(Agra 2000, 05, 09; Rohilkhand 2000, 01, 02)**

Consider the motion of a heavy particle of mass m placed on the inside of a smooth vertical circle (with centre O and radius a) and projected with velocity u from its lowest point A. Let P be the position of the particle at any time t, such that $\angle POA = \theta$. Let v be velocity of the particle along tangent PT. Let $PN \perp AO$.

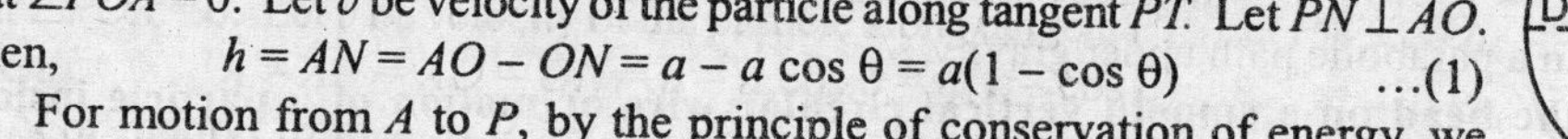

Then, $\quad h = AN = AO - ON = a - a\cos\theta = a(1-\cos\theta)$ ...(1)

For motion from A to P, by the principle of conservation of energy, we have

$$(1/2)mv^2 - (1/2)mu^2 = (-mg)h \quad \text{or} \quad v^2 = u^2 - 2ga(1-\cos\theta) \qquad ...(2)$$

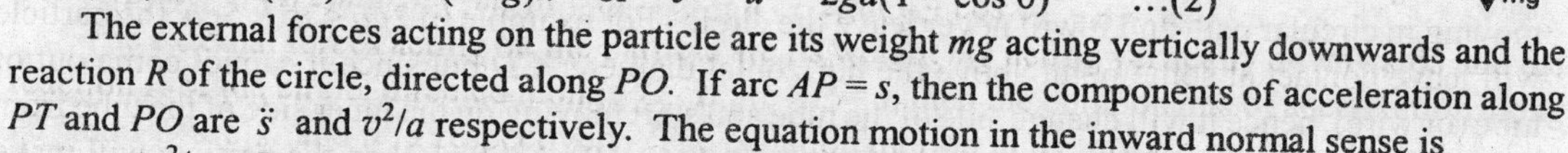

The external forces acting on the particle are its weight mg acting vertically downwards and the reaction R of the circle, directed along PO. If arc $AP = s$, then the components of acceleration along PT and PO are $\ddot{s}$ and v^2/a respectively. The equation motion in the inward normal sense is

$$mv^2/a = R - mg\cos\theta \quad \text{or} \quad R = mg\cos\theta + (m/a)v^2 \qquad ...(3)'$$

or $\quad R = mg\cos\theta + (m/a)\{u^2 - 2ga(1-\cos\theta)\}$, by (2)

or $\quad R = (m/a)(u^2 - 2ga + 3ga\cos\theta)$ ...(3)

From (1), $\quad h = a(1-\cos\theta) \quad$ or $\quad \cos\theta = (a-h)/a$ ...(4)

Using (4), (2) becomes $\quad v^2 = u^2 - 2gh$ ...(5)

Using (4), (3) becomes $\quad R = (m/a)\{u^2 - 2ga + 3g(a-h)\}$

or $\quad R = (m/a)(u^2 + ga - 3gh)$ ...(6)

Velocity v of the particle at point P is given by (2) or (5) and reaction R is given by (3) or (6). From these expressions of v and R, we find that v and R both decrease as h increases.

At A, $h = 0$, $\quad$ so $\quad v^2 = u^2 \quad$ and $\quad R = (m/a)(u^2 + ga)$ ...(7)

At B, $h = 2a$, $\quad$ so $\quad v^2 = u^2 - 4ga \quad$ and $\quad R = (m/a)(u^2 - 5ga)$ ...(8)

Let $v = 0$ when $h = h_1$ and $R = 0$ when $h = h_2$.

$\therefore \quad$ From (5), $0 = u^2 - 2gh_1$, $\quad$ so that $\quad h_1 = u^2/2g$ ...(9)

and from (6), $\quad 0 = (m/a)(u^2 + ga - 3gh_2) \quad$ or $\quad h_2 = (u^2 + ga)/3g$ ...(10)

We now discuss the following situations :

Case I. Neither velocity nor reaction vanish before reaching the highest point B. If $u^2 \geq 5ga$, then from (8), we see that v and R do not vanish till the particle reaches the highest point. *Hence the particle will* make complete revolutions along the circle if the velocity *of projection at the lowest point is at least* $(5ga)^{1/2}$

Case II A. A suppose velocity vanishes before reaction has vanished.

Then $\quad h_1 < h_2 \quad$ or $\quad u^2/2g < (u^2 + ga)/3g$, by (9) and (10)

or $\quad 3u^2 < 2u^2 + 2ga \quad$ or $\quad u^2 < 2ga$ ...(11)

In this case, the particle comes to a momentary rest, and then coming back retraces its path to the lowest point, moves to the other side and comes back again. The motion is oscillatory.

Case II B. Suppose velocity and reaction vanish simultaneously.

Then $\quad h_1 = h_2 \quad$ or $\quad u^2/2g = (u^2 + g)/3a \quad$ or $\quad u^2 = 2ga$ ...(12)

When $u^2 = 2ga$, from (9) and (10), $\quad h_1 = h_2 = a$,

showing that the particle will be momentarily at rest, at point C (which is an end of the horizontal diameter CD) and then moving back retraces its path, reaching the lowest point with velocity u and then rising upto D on the other side, and then comes back. The motion is oscillatory.

Case III. Suppose reaction vanishes before velocity has vanished.

Then $\quad h_1 > h_2 \quad$ or $\quad u^2 > 2ga$, as in Case I ...(13)

Assume that $u^2 < 5ga$. Thus, if $2ga < u^2 < 5ga$, *i.e.*, $(2ga)^{1/2} < u < (5ga)^{1/2}$, then the particle will go above the horizontal diameter CD but will not reach the highest point B. Hence the particle, before reaching the highest point, leaves the circle at some point above the level of the horizontal diameter. It then describes parabolic path thereafter.

Some important results to be committed to memory

1. $\cos\theta = ON/OP = (a-h)/a$, where $AN = h$, $OP = a$, $h = a(1-\cos\theta)$
2. $v^2 = u^2 - 2gh$ or $v^2 = u^2 - 2ga + 2ga\cos\theta$
3. $R = (m/a)(u^2 - 3gh + ga)$ or $R = (m/a)(u^2 - 2ga + 3ga\cos\theta)$
4. If $u^2 \geq 5ga$, then particle makes complete revolutions
5. If $u^2 \leq 2ga$, then particle oscillates about the lowest point
6. If $(2ga)^{1/2} < u < (5ga)^{1/2}$, then the particle leaves the circle before reaching the highest point and then moves in a parabolic path under gravity.

8.4 Motion of the bead on a smooth vertical circular wire or motion of a particle inside a vertical circular tube. Refer Art. 8.3. In the case of motion of a bead on a smooth vertical wire or that of a particle moving inside a smooth vertical circular tube, the bead or the particle keeps on moving in circular path and it will never leave the circle. For example, if $2ga < u^2 < 5ga$, the particle will leave contact at some point above the horizontal diameter CD but since it cannot come out of tube (or wire), hence it will change side of the tube and start moving on the inner side of the tube from C to B (in place of outer side during motion from A to C). If $u^2 \geq 5ga$, then the particle will make complete revolutions by moving partly on the outer side and partly on the inner side of the tube. Thus,

(*i*) If $u^2 \geq 5ga$, the bead makes complete revolutions

(*ii*) If $u^2 < 5ga$, then bead oscillates about the lowest point

8.5 Motion of a particle attached to the end of a string. *A heavy particle is tied to one end of a light inextensible string whose other end is attached to a fixed point. It is projected horizontally with a given velocity u from its vertical position; to discuss the subsequent motion.*

[Kanpur 2018]

Proceed exactly as in Art. 8.3. The motion is exactly the same as that of a particle moving inside a smooth vertical circle. In entire discussion of Art. 8.3, replace R by T and the radius of the circle by the length of the string. Thus, as before if a particle is projected from its lowest position, hanging at rest, with horizontal velocity u, it will

(*i*) make complete revolutions if $u \geq (5ga)^{1/2}$

(*ii*) oscillate about the lowest point if $u \leq (2ga)^{1/2}$

(*iii*) leave the circular path and describe a parabola if $(2ga)^{1/2} < u < (5ga)^{1/2}$. In this case string becomes slack somewhere between C and B.

TYPE 2 : SOLVED EXAMPLES BASED ON ART. 8.3, 8.4 AND 8.5

Ex. 1. *Discuss motion of a particle which is projected from the lowest point of a smooth vertical circle with velocity u along the inside of this circle, where* $2ga < u^2 < 5ga$ *where a is radius of the circle.*

Sol. Proceed as in Art. 8.3 and discuss its case III for the present problem.

Ex. 2. *A light string is attached to a fixed point O and carries at its free end a particle of mass m. The particle is describing complete revolution about O under gravity and the string is just taut when the particle is vertically above O. Find the tension in the string when in a horizontal position.*

Sol. First proceed as in Art. 8.3 upto equation (3) by replacing R by tension T in the string OA of length a and get

$$T = (m/a)(u^2 - 2ga + 3ga\cos\theta) \qquad \ldots(3)$$

By the given problem the string is just taut at B, *i.e.*, $T = 0$ when $\theta = \pi$. Then (3) gives

$$0 = (m/a)(u^2 - 2ga - 3ga) \quad \text{or} \quad u^2 = 5ga$$

Tension T' in the string when in horizontal position can be obtained by putting $T = T'$, $\theta = 90°$ and $u^2 = 5ga$ in (3)

$$\therefore \quad T' = (m/a)(5ga - 2ga + 3ga\cos 90°) \quad \text{or} \quad T' = 3mg.$$

Ex. 3. (*a*) *Find the least velocity that should be given to the body of a pendulum of length a in its lowest position so that it should move in a complete vertical circle.* **[Gorakhpur 1998]**

(*b*) *A particle is projected from the lowest point with velocity u and moves along the inside of a smooth vertical circle of radius a. Find the maximum velocity when the particle moves round the circle and the minimum pressure at the lowest point.*

Sol. **(*a*)** Refer Art. 8.5 and discuss its case (*i*) as in Art. 8.3.

(*b*) Refer Art. 8.3.

Ex. 4. *A heavy particle of mass m is tied to one end of an string of length a, whose other end is tied to a fixed point and the string hangs vertically. It is then projected from its vertical position of equilibrium with a horizontal velocity u so that it just makes a complete revolution. Show that* $u^2 = 5ga$ *and that the tension in the string at its lowest position is 6 mg.* **[Lucknow 1994, 96]**

Sol. First do as in Art. 8.3 upto case I by replacing R by T and word 'radius a' by "length a of the string". For tension at the lowest point, replace θ by 0 and u^2 by $5ga$ is (3) and get

$$T = (m/a)\,(5ga - 2ga + 3ga \cos 0^\circ) = 6mg$$

Ex. 5. (*a*) *A heavy particle hanging vertically from a point by a light inextensible string of length l is started so as to make a complete revolution in a vertical plane. Prove that the sum of the tensions at the ends of any diameter is constant.* **[I.A.S. 1998, Kumaun 1996]**

(*b*) *A heavy particle hangs vertically from a fixed point by a light inextensible string of length a. It is projected with velocity* $(5ag)^{1/2}$. *Show that the sum of tensions at the ends of any diameter is constant.*

(*c*) *A heavy particle is constrained to move in a vertical circle of radius a. Show that the sum of the reactions at the ends of a diameter is constant.*

Sol. **(*a*)** First do as in Art. 8.3 upto equation (3) by replacing R by T and 'radius a' by "length a of the string. Thus tension T in the string at a point P(refer figure of Art. 8.3) where $\angle AOP = \theta$ is given by

$$T = (m/a)\,(u^2 - 2ga + 3ga \cos \theta) \qquad \ldots(3)$$

In the figure of Art. 8.3, draw POP' as diameter. Then, tension T' at the other end P' of diameter POP' can be obtained by replacing θ by $\theta + \pi$ in (3).

$$\therefore \quad T' = (m/a)\,\{u^2 - 2ga + 3ga \cos(\theta + \pi)\} = (m/a)\,(u^2 - 2g - 3ga \cos \theta) \qquad \ldots(4)$$

Adding (3) and (4), $\qquad T + T' = (m/a)\,(2u^2 - 4g)$,

which is constant, because it is independent of θ.

(*b*) Since $u = (5ag)^{1/2}$, the particle makes complete revolutions. Rest of the solution is the same as in part (*a*).

(*c*) Replace T by R in solution of part (*a*).

Ex. 6. *A heavy particle of weight W, attached to a fixed point by a light intextensible string, describes a circle in a vertical plane. The tension in the string has the values mW and nW respectively when the particle is at the highest and lowest points in the path. Show that n = m + 6.*

[Kanpur 2019; Luckow 2000; Meerut 2009]

Sol. First do as in Art. 8.3 by replacing R by T, "radius a" by "string of length a" and mass m by W/g upto equation (3). Refer figure of Art. 8.3.

$$\therefore \qquad T = (W/ga)\,(u^2 - 2ga + 3ga \cos \theta) \qquad \ldots(1)$$

Given that $T = mW$ at B where $\theta = \pi$ and $T = nW$ at A where $\theta = 0$. Hence from (1), we obtain

$$mW = (W/ag)\,(u^2 - 2ga + 3ag \cos \pi) \quad \text{or} \quad m = (1/ag)\,(u^2 - 5ga) \qquad \ldots(2)$$

and $$nW = (W/ag)\,(u^2 - 2ga + 3ag \cos 0) \quad \text{or} \quad n = (1/ag)\,(u^2 + ag) \qquad \ldots(3)$$

Subtracting (2) from (3), $\qquad n - m = 6 \quad$ or $\quad n = m + 6$

Ex. 7. *A heavy particle hangs from a fixed point O, by a string of length a. It is projected horizontally with velocity* $u^2 = (2 + \sqrt{3})ag$; *show that the string becomes slack when it has described an angle* $\cos^{-1}\left(-1/\sqrt{3}\right)$. **[Gorhwal 2001; Agra 2000, 11; Kanpur 2007, 08]**

Sol. First do as in Art. 8.3 upto equation (3) by replacing R by T and "radius a" by "length a of the string". Then, we have

$$T = (m/a)(u^2 - 2ga + 3ga\cos\theta) \quad \text{...(3)}$$

Here $u^2 = (2 + \sqrt{3})ag$. Hence (3) reduces to

$T = (m/a)(2ag + \sqrt{3}\,ag - 2ag + 3ga\cos\theta) = (m/a)(\sqrt{3}\,ag + 3ag\cos\theta)$

String becomes slack when $T = 0$ and so above equation gives

$$0 = (m/a)(\sqrt{3}ag)(1 + \sqrt{3}\cos\theta) \quad \text{or} \quad \theta = \cos^{-1}(-1/\sqrt{3})$$

Ex. 8. (*a*) *A tumbler containing some water is swung in a vertical circle of radius a by means of a string attached to it. Show that in order to prevent water from spilling outside, velocity at the lowest point must be at least* $(5ga)^{1/2}$.

(*b*) *A glass containing some water is suspended by a string of length a and made to turn along a vertical circle of that radius. Show that in order to prevent water from spilling outside, velocity at highest point must be at least* $(ga)^{1/2}$

Sol. (*a*) Do as in Arg. 8.3 upto equation (3) by replacing R by T and "radius a" by "string of length a". Then, we get

$$T = (m/a)(u^2 - 2ga + 3ga\cos\theta) \quad \text{...(3)}$$

Let $T = T'$ when $\theta = \pi$ *i.e.*, at highest point B. Then

$$T' = (m/a)(u^2 - 2g + 3ga\cos\pi) = (m/a)(u^2 - 5ga)$$

In order that the water may not spill out, $T' \geq 0$

i.e., $(m/a)(u^2 - 5ga) \geq 0$ or $u \geq (5ga)^{1/2}$

$\therefore$ The required least velocity $= (5ga)^{1/2}$

(*b*) First do as in part (*a*) to show $u^2 \geq 5ga$ ...(*i*)

As in Art. 8.3, velocity at any point P is given by

$$v^2 = u^2 - 2ga(1 - \cos\theta)$$

Let velocity at the highest point B to V. So $v = V$ when $\theta = \pi$

$\therefore$ $V^2 = u^2 - 2ga(1 - \cos\pi) = u^2 = 4ga$ or $u^2 = V^2 + 4ga$ ...(*ii*)

From (*i*) and (*ii*), $V^2 + 4ga \geq 5ga$ or $V^2 \geq ga$ or $V \geq (ga)^{1/2}$,

showing the least value of the velocity at the highest point is $(ga)^{1/2}$.

Ex. 9. *A particle moving on the inside of a vertical circle just reaches the horizontal diameter. Show that the reaction at any point varies as the depth below the horizontal diameter.*

Sol. Do as in Art. 8.3 upto equation (6). Since the particle just reaches the end C of horizontal diameter, so $v = 0$ when $h = a$ and so (5) gives

$0 = u^2 - 2ga$ so that $u^2 = 2ga$. Putting $u^2 = 2ga$ in (6), we have

$$R = (m/a)(2ga + ga - 3gh) \quad \text{or} \quad R = (3mg/a)(a - h)$$

or $\quad R = (3mg/a).ON$ or R varies as ON,

showing that reaction at P varies as the depth ON below CD.

Ex. 10. *A simple pendulum is started to make complete revolutions. Show that if the tension in the string when the bob is in the lowest position is k times the tension when the bob is in the highest position, the velocities in these positions being* u_1 *and* u_2 *respectively, prove* $u_1^2/u_2^2 = (5k + 1)(k + 5)$.

[Meerut 2004]

Sol. First do as in Art. 8.3 upto equation (3) by replacing R by T and "radius a" by "string of length a" upto equation (3). Here $u = u_1$ (given).

$$\therefore \quad v^2 = u_1^2 - 2ga(1 - \cos\theta) \quad \text{...(2)}$$

and $$T = (m/a)(u_1^2 - 2ga + 3ga\cos\theta) \quad \text{...(3)}$$

Here $v = u_2$ when $\theta = \pi$ *i.e.*, highest point B.

$\therefore$ From (2), $u_2^2 = u_1^2 - 2ga(1 - \cos\pi)$ or $u_2^2 = u_1^2 - 4ga$ ...(4)

Let $T = T'$ at B where $\theta = \pi$. Then, by problem, $T = kT'$ at A where $\theta = 0$. Hence, from (3), we have

$$T' = (m/a)(u_1^2 - 2ga + 3ga\cos\pi) = (m/a)(u_1^2 - 5ga) \quad \text{...(5)}$$

and $$kT' = (m/a)(u_1^2 - 2ga + 3ga\cos 0) = (m/a)(u_1^2 + ga) \quad \text{...(6)}$$

Dividing (6) by (5), $k = (u_1^2 + ga)/(u_1^2 - 5ga)$ or $u_1^2(k - 1) = ga(5k + 1)$

or $\quad u_1^2(k-1) = (1/4)(u_1^2 - u_2^2)(5k+1)$, since by (4), $ga = (u_1^2 - u_2^2)/4$

or $\quad 4u_1^2(k-1) = u_1^2(5k+1) - u_2^2(5k+1)$ or $u_1^2/u_2^2 = (5k+1)/(k+5)$.

Ex. 11. *A particle of mass m gm is suspended from a fixed point by a string a cm long. It is projected horizontally with a velocity $2(ag)^{1/2}$ cm/sec. Find the height of the particle above the point of suspension when the string becomes slack. Find also the tension in the string when the particle is at a depth a/2 below the point of suspension.* **[I.F.S. 1997]**

Sol. Do as in Art. 8.3 upto equation (6) by replacing R by T, "radius a" by "string of length a" and u by $2(ga)^{1/2}$. Then we have

$$v^2 = u^2 - 2gh \quad \text{or} \quad v^2 = 4ga - 2gh \qquad \ldots(5)$$

and $$T = (m/a)(u^2 + ga - 3gh) \quad \text{or} \quad T = (m/a)(5ga - 3gh) \qquad \ldots(6)$$

The string becomes slack, when $T = 0$. So, by (6), $5ga - 3gh = 0$ or $h = 5a/3$. Also the height of point of suspension above the lowest point $= a$.

$\therefore$ The height of the particle above the point of suspension when the string becomes slack $= (5a/3) - a = 2a/3$.

When the particle is at a depth $a/2$ below the point of suspension *i.e.*, $ON = a/2$, then $h = AN = a - a/2 = a/2$. Then the corresponding value of T is given (6) by putting $h = a/2$.

$$\therefore \quad T = (m/a)(5ga - 3ga/2) = (7mg)/a$$

Ex. 12. *A heavy particle of mass m makes complete revolutions in a smooth circular tube fixed in a vertical plane. Its greatest speed is n times its least. Prove that the pressure in the tube when the particle is moving vertically is $\{2(n^2+1)mg\}/(n^2-1)$ and that its speed then is $\{(n^2+1)/2\}^{1/2}$ times its speed at the highest point.*

Sol. Do as in Art. 8.3 upto equation (3). Thus, get

$$v^2 = u^2 - 2ga + 2ga\cos\theta \qquad \ldots(2)$$

and $$R = (m/a)(u^2 - 2ga + 3ga\cos\theta) \qquad \ldots(3)$$

Let v_1 and v_2 be the greatest and least speeds. Then, by (2)

$$v = v_1,\ \theta = 0 \quad \text{so} \quad v_1^2 = u^2 \qquad \ldots(4)$$

and $$v = v_2,\ \theta = \pi \quad \text{so} \quad v_2^2 = u^2 - 4ag \qquad \ldots(5)$$

Given $\quad v_1 = nv_2 \quad i.e.\ v_1^2 = n^2 v_2^2 \quad \text{so} \quad u^2 = n^2(u^2 - 4ag)$

Thus, $$u^2 = (4agn^2)/(n^2-1) \qquad \ldots(6)$$

When the particle is moving vertically at C, where $\theta = \pi/2$, then corresponding value of R, from (3), is given by

$$R = \frac{m}{a}(u^2 - 2ga) = \frac{m}{a}\left(\frac{4agn^2}{n^2-1} - 2ga\right) = \frac{2(n^2+1)mg}{n^2-1}$$

Let V_1 be speed at C, so from (2), with $\theta = \pi/2$

$$V_1^2 = u^2 - 2ga = \frac{4ag\,n^2}{n^2-1} - 2ga \quad \text{or} \quad V_1 = \left\{\frac{2ag(n^2+1)}{n^2-1}\right\}^{1/2} \qquad \ldots(7)$$

Let V_2 be speed at B, so from form (2), with $\theta = \pi$,

$$\therefore \quad V_2^2 = u^2 - 4ga = \frac{4ag\,n^2}{n^2-1} - 4ga \quad \text{or} \quad V_2 = \left(\frac{4ag}{n^2-1}\right)^{1/2} \qquad \ldots(8)$$

Thus, $V_1/V_2 = \{(n^2+1)/2\}^{1/2}$ or $V_1 = \{(n^2+1)/2\}^{1/2} V_2$, as required.

Ex. 13. *Find the greatest angle through which a person can oscillate on a swing, ropes of which can support twice the person's weight.*

If the ropes are strong enough and he swings through 180° and if v is his speed at any point, prove that the tension in the rope at that point is $(3mv^2)/2a$ where m is the mass of the person and a is the length of the rope.

Sol. Do as in Art. 8.3 upto equation (3) by replacing R by T and "radius a" by "string of length a" and get

$$v^2 = u^2 - 3ga + 2ga\cos\theta \qquad \ldots(1)$$

and $T = (m/a)(u^2 - 2ga + 3ga \cos \theta)$ ...(2)

At the lowest point $\theta = 0$ and tension is maximum and equal to $2mg$ (given). So putting $T = 2mg$ and $\theta = 0$ in (2), we get

$$2mg = (m/a)(u^2 - 2ga + 3ga \cos 0) \quad \text{or} \quad u^2 = ga \quad ...(3)$$

From (1), $v^2 = ga - 2ga + 2ga \cos \theta = 2ga \cos \theta - ga$ ...(4)

Putting $v = 0$ in (4), $0 = ga(2 \cos \theta - 1)$ or $\theta = 60°$,

showing that the person can swing through an angle 60° from the vertical AB on one side of the lowest point. Similarly on the other side of the vertical AB he can swing through an angle 60°. Hence the greatest angle through which the man can oscillate = 60° + 60° = 120°.

Second part. If the rope is strong enough and the person can swing through an angle of 180° *i.e.*, through an angle of 90° on each side of the vertical AB, then $v = 0$ when $\theta = 90°$ and so by (1),

$$0 = u^2 - 2ga + 2ga \cos 90° \quad \text{or} \quad u^2 = 2ga \quad ...(5)$$

Putting $u^2 = 2ga$ in (1), $v^2 = 2ga \cos \theta$ or $\cos \theta = v^2/2ga$ ...(6)

$\therefore$ from (2), $T = (m/a)\{2ga - 2ga + 3ga \times (v^2/2ga)\}$, by (5) and (6)

Thus, $T = (3mv^2)/2a$, as required.

Ex. 14. (*a*) *A particle makes complete revolutions in a vertical circle. If ω_1, ω_2 be the greatest and least angular velocities and R_1, R_2 the greatest and least reactions, prove that when the particle projected from the lowest point of the circle makes an angle θ at the centre, its angular velocity is $\{\omega_1^2 \cos^2(\theta/2) + \omega_2^2 \sin^2(\theta/2)\}^{1/2}$ and that the reaction is $R_1 \cos^2(\theta/2) + R_2 \sin^2(\theta/2)$.*

[Agra 1994; Rohilkhand 1995]

(*b*) *A simple pendulum is started to make complete revolutions. If ω_1 and ω_2 be its greatest and least angular velocities and T_1 and T_2 the greatest and the least tensions, prove that when the pendulum makes an angle θ with the vertical, the angular velocity is $\{\omega_1^2 \cos^2(\theta/2) + \omega_2^2 \sin^2(\theta/2)\}^{1/2}$ and the tension is $T_1 \cos^2(\theta/2) + T_2 \sin^2(\theta/2)$.*

Sol. (*a*) First do as in Art. 8.3 upto equation (3). Thus

$$v^2 = u^2 - 2ga + 2ga \cos \theta \quad ...(2)$$

and $$R = mg \cos \theta + (m/a)v^2 \quad ...(3)$$

Let ω (= $d\theta/dt$) be the angular velocity of the particle at any t. Then $v = a\omega$ and so (2) reduces to

$$a^2\omega^2 = u^2 - 2ga + 2ga \cos \theta \quad ...(4)$$

From (4), we see that $\omega = \omega_1$ when $\theta = 0$ and $\omega = \omega_2$ when $\theta = \pi$

$$\therefore \quad a^2\omega_1^2 = u^2 - 2ga + 2ga \cos 0 \quad \text{or} \quad a^2\omega_1^2 = u^2 \quad ...(5)$$

and $$a^2\omega_2^2 = u^2 - 2ga + 2ga \cos \pi \quad \text{or} \quad a^2\omega_2^2 = u^2 - 4ga \quad ...(6)$$

Subtraction (6) from (5), $a^2(\omega_1^2 - \omega_2^2) = 4ga$ ...(7)

Subtraction (4) from (5), $a^2(\omega_1^2 - \omega^2) = 2ga(1 - \cos \theta) = 4ga \sin^2(\theta/2)$ ...(8)

Dividing (8) by (7), $(\omega_1^2 - \omega^2)/(\omega_1^2 - \omega_2^2) = \sin^2(\theta/2)$

or $\omega_1^2 - \omega^2 = (\omega_1^2 - \omega_2^2) \sin^2(\theta/2)$ or $\omega^2 = \omega_1^2 \cos^2(\theta/2) + \omega_2^2 \sin^2(\theta/2)$

$$\therefore \quad \omega = \{\omega_1^2 \cos^2(\theta/2) + \omega_2^2 \sin^2(\theta/2)\}^{1/2} \quad ...(9)$$

Since $v = a\omega$, (3) may be re-written as

$$R = mg \cos \theta + (m/a)a^2\omega^2 \quad \text{or} \quad R = ma\omega^2 + mg \cos \theta \quad ...(10)$$

Now, when $\theta = 0$, $R = R_1$, $\omega = \omega_1$ and when $\theta = \pi$, $R = R_2$, $\omega = \omega_2$

$$\therefore \quad R_1 = ma\omega_1^2 + mg \quad \text{and} \quad R_2 = ma\omega_2^2 - mg \quad ...(11)$$

Re-writing (10) and putting value of w from (9) we have

$$R = ma\{\omega_1^2 \cos^2(\theta/2) + \omega_2^2 \sin^2(\theta/2)\} + mg\{\cos^2(\theta/2) - \sin^2(\theta/2)\}$$

$$= (ma\omega_1^2 + mg) \cos^2(\theta/2) + (mg\omega_2^2 - mg) \sin^2(\theta/2)$$

$$= R_1 \cos^2(\theta/2) + R_2 \sin^2(\theta/2), \text{ using (11)}$$

(*b*) Proceed as in part (*a*) by replacing R by T.

Ex. 15. *A heavy particle is free to move in a vertical circle of radius a and is projected with velocity u from the lowest point A of the circle. If v be its velocity when it has risen to a vertical h,* *prove that $v^2 = u^2 - 2gh$. If it just reaches B, show that $u = (g/a)^{1/2} AB$. Also show that tension at A is to tension at B as $(2a^2 + 2AB^2); (2a^2 - AB^2)$.*

Sol. Do as in Art. 8.3 upto (6) by replacing R by T. P by B, B by E and "radius a" by "string of length a". So equations (5) and (6) for the given problem are (Refer the adjoining figure for this problem)

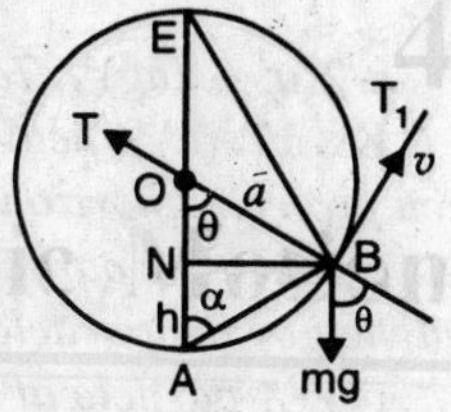

$$v^2 = u^2 - 2gh \qquad \text{...(5)}$$

and
$$T = (m/a)(u^2 + ga - 3gh) \qquad \text{...(6)}$$

Let $\angle BAE = \alpha$. From right angled ΔABE, $\cos\alpha = AB/AE = AB/2a$. Here $h = AN$. Since the particle just reaches B, so $v = 0$ at B.

$\therefore$ From (5), $\quad 0 = u^2 - 2gh \quad$ or $\quad u^2 = 2gh$

$\therefore \quad u^2 = 2g.AN = 2g \times AB\cos\alpha = 2g\, AB \times (AB/2a)$

or $\quad u^2 = (g/a) \times AB^2 \quad$ so that $\quad u = (g/a)^{1/2} \times AB \qquad$ ...(7)

Let T and T' be the values of tensions at B and A respectively. Tension T' at A can be obtained by putting $h = 0$ in (6). Thus $T' = (m/a)(u^2 + ga) \qquad$ ...(8)

From (6) and (8), $$\frac{T'}{T} = \frac{u^2 + ga}{u^2 + ga - 3gh} = \frac{(g/a)AB^2 + ga}{(g/a)\times AB^2 + ga - 3g\,AB\cos\alpha},$$

[Using (7) and noting that $AN = h = AB\cos\alpha$]

or $$\frac{T'}{T} = \frac{AB^2 + a^2}{AB^2 + a^2 - 3a.AB \times (AB/2a)} = \frac{2AB^2 + 2a^2}{2AB^2 + 2a^2 - 3AB^2} = \frac{2a^2 + 2AB^2}{2a^2 - AB^2}$$

Ex. 16. *A particle is projected along the inside of a smooth vertical circle of radius a, the velocity at the lowest point being u. Show that if $2ga < u^2 < 5ag$, the particle will leave the circle before arriving at the highest point and will describe a parabola whose latus rectum is $2(u^2 - 2ag)^3/27a^3g^3$.* **[Delhi B.A. /B.Sc. (Hons) 2003; Agra 2011]**

Sol. Let P be the position of the particle at any time t, such that $\angle AOP = \theta$. Let v be its velocity at P along tangent PT. R is normal reaction acting along PO and mg is its weight acting vertically downwards.

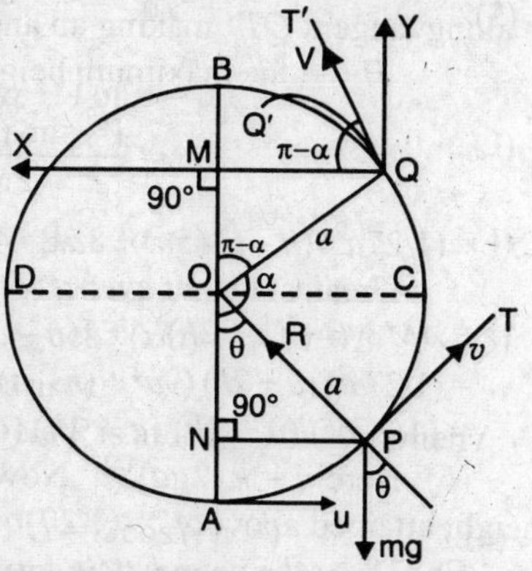

For motion from A to P, by the principle of conservation of energy,

$$(1/2)mv^2 - (1/2)mu^2 = (-mg) \times AN$$

$$v^2 = u^2 - 2g(OA - ON)$$

or $$v^2 = u^2 - 2g(a - a\cos\theta)$$

Thus, $$v^2 = u^2 - 2ga + 2ga\cos\theta \qquad \text{...(1)}$$

Equation of motion in the inward drawn normal PO is

$mv^2/a = R - mg\cos\theta \quad$ or $\quad R = mg\cos\theta + (m/a)v^2$

or $\quad R = mg\cos\theta + (m/a)(u^2 - 2ga + 2ga\cos\theta)$, using (1)

Thus, $\quad R = (m/a)(u^2 - 2ga + 3ga\cos\theta) \qquad$ ...(2)

Let the particle leaves the circle at Q where $\theta = \alpha$ and $R = 0$. So, by (2)

$0 = (m/a)(u^2 - 2ga + 3ga\cos\alpha)\quad$ or $\quad \cos\alpha = -(u^2 - 2ga)/3ga \qquad$ ...(3)

Since $2ag < u^2 < 5ag$ (given), so $\cos\alpha$ is negative and $|\cos\alpha| < 1$, showing that α is real and $\pi/2 < \alpha < \pi$. Hence the particle rises above the horizontal diameter CD and leaves the circular path at Q before arriving at the highest point B. Let V be velocity at Q. Putting $\theta = \alpha$ and $v = V$ in (1), we get

$V^2 = u^2 - 2ga + 2ga\cos\alpha = u^2 - 2ga - \{2ga(u^2 - 2ga)\}/3ag$, by (3)

or $\quad V^2 = (u^2 - 2ga)(1 - 2/3) \quad$ or $\quad V^2 = (u^2 - 2ag)/3 \qquad$ ...(4)

By geometry $\angle QOB = \pi - \alpha$. Let QT' be tangent at Q to the circular path. Then after leaving the circular path, initial velocity of the particle for parabolic path will be V along QT'. By geometry, $\angle T'QX = \angle QOB = \pi - \alpha$, where QX is horizontal line through Q. Let QY be vertical line through Q.

The latus rectum of the parabolic path

$$= \frac{2}{g}\{V\cos(\pi-\alpha)\}^2 = \frac{2}{g}V^2\cos^2\alpha = \frac{2}{g}\cdot\frac{u^2-2ag}{3}\cdot\left(\frac{u^2-2ga}{3ga}\right)^2, \text{ by (3) and (4)}$$

$= 2(u^2 - 2ag)^3/27a^2g^3$.

Ex. 17. (*a*) *A heavy particle hangs by a inextensible string of length a from a fixed point and is then projected horizontally with a velocity* $(2gh)^{1/2}$. *If* $5a/2 > h > a$, *prove that the circular motion ceases when the particle has reached the height* $(a + 2h)/3$. *Prove also that the greatest height ever reached by the particle above the point of projection is* $\{(4a - h)(a + 2h)^2\}/27a^2$.

(*b*) *A particle at the end of a string of length a, the upper end of which is fixed is projected horizontally with a velocity which is k times that due to a fall a. If the particle leaves the circular path, show that it does so at a height* $(1 + 2k^2)/3$ *above the lowest point. Also find the greatest height that the particle attains during the entire motion.* **[Lucknow 1991]**

Sol. (*a*) Refer figure of Ex. 16. Do as in that exercise upto equation (3) by replacing R by T. Here $u = (2gh)^{1/2}$. Hence (3) becomes

$$\cos\alpha = -(2gh - 2ga)/3ga \quad \text{or} \quad \cos\alpha = -(2h - 2a)3a \qquad \ldots(4)$$

Since $5a/2 > h > a$ (given), so $\cos\alpha$ is negative and $|\cos\alpha| < 1$, showing that α is real and $\pi/2 < \alpha < \pi$. Hence the particle rises above the horizontal diameter CD and leaves the circular path at Q before arriving at the highest point B. Let V be velocity at Q. Putting $\theta = \alpha$, $v = V$ and $u = (2gh)^{1/2}$ in (1),

$$V^2 = 2gh - 2ga + 2ga\cos\alpha = 2g(h-a) - 2ga.(2/3a)(h-a)$$

or $$V^2 = 2g(h-a)(1 - 2/3) \quad \text{or} \quad V^2 = 2g(h-a)/3 \qquad \ldots(5)$$

Height of Q above $A = AM = AO + OM = a + a\cos(\pi - \alpha) = a - a\cos\alpha$

$$\therefore \quad AM = a + (1/3)(2h - 2a) = (a + 2h)/3 \qquad \ldots(6)$$

When the circular motion ceases at Q, the particle begins to describe parabolic path with velocity V along tangent QT' making an angle $(\pi - \alpha)$ with the horizontal line QX.

$\therefore \quad H$ = the maximum height reached by the particle above Q

$$= \frac{V^2\sin^2(\pi-\alpha)}{2g} = \frac{V^2(1-\cos^2\alpha)}{2g} = \frac{2g(h-a)}{6g}\left\{1 - \frac{(2h-2a)^2}{9a^2}\right\}, \text{ by (4), (5)}$$

$$= (1/27a^2)(h-a)(5a^2 + 8ah - 4h^2) = \{(h-a)(a+2h)(5a-2h)\}/27a^2 \qquad \ldots(7)$$

$\therefore$ Greatest height ever reached by the particle above A

$= AM + H = (a+2h)/3 + \{(h-a)(a+2h)(5a-2h)\}/27a^2$, by (6) and (7)

$= (1/27a^2)(a+2h)\{9a^2 + (h-a)(5a-2h)\} = (1/27a^2)(a+2h)(4a^2 + 7ah - 2h^2\}$

$= (1/27a^2)(a + 2h)(a + 2h)(4a - h) = (1/27a^2)(4a - h)(a + 2h)^2$

(*b*) Here $u = k.(2ga)^{1/2}$. Now proceed exactly as in part (*a*) with this value of u and get max. height attained above $A = (1/27)a(1 + 2k^2)^2(4 - k^2)$.

Ex. 18. *A heavy particle hanging vertically from a fixed point by a light inextensible cord of length a is struck by a horizontal blow which imparts it a velocity* $2(ga)^{1/2}$, *prove that the cord becomes slack when the particle has risen to a height 2a/3 above the fixed point. Also find the height of the highest point of the parabola subsequently described.* **[Rohilkhand 1997]**

Sol. Proceed as in Ex. 16 upto equatioin (2) by replacing u by $2(ga)^{1/2}$ and R by T. Then, as before, we have

$$v^2 = u^2 - 2ga + 2ga\cos\theta = 4ga - 2ga + 2ga\cos\theta \qquad \ldots(1)$$

and $$T = (m/a)(u^2 - 2ga + 3ga\cos\theta) = (m/a)(4ga - 2ga + 3ga\cos\theta) \qquad \ldots(2)$$

Let the cord becomes slack at the point Q, where $\theta = \alpha$ and $T = 0$. Hence from (2), α is given by

$$0 = (m/a)(2ga + 3ga\cos\alpha) \quad \text{or} \quad \cos\alpha = -2/3 \qquad \ldots(3)$$

Let $QM \perp AB$. Here $\angle QOM = \pi - \alpha$. So, we have

$$OM = \text{height of } Q \text{ above } O = a\cos(\pi - \alpha) = -a\cos\alpha = 2a/3$$

Let V be velocity at Q. Putting $\theta = \alpha$ and $v = V$ in (1), we get

$$V^2 = 2ga + 2a\cos\alpha = 2ga + 2a(-2/3) = 2ga/3 \qquad ...(4)$$

When the string becomes slack at Q, the circular motion ceases at Q and the particle begins to describe parabolic path with velocity V along tangent QT' making an angle $(\pi - \alpha)$ with the horizontal line.

$\therefore$ H = the maximum height reached by the particle above Q

$$= \frac{V^2 \sin^2(\pi - \alpha)}{2g} = \frac{V^2(1-\cos^2\alpha)}{2g} = \frac{(2ga)/3}{2g}\left(1 - \frac{4}{9}\right) = \frac{5a}{27}, \text{ by (3) and (4)}$$

$\therefore$ Height of the highest point of the parabolic path above O

$$= OM + H = 2a/3 + 5a/27 = 23a/27.$$

Ex. 19. (*a*) *A particle inside and at the lowest point of a fixed smooth hollow sphere of radius a is projected horizontally with velocity* $(7ag/2)^{1/2}$. *Show that it will leave the sphere at a height* $3a/2$ *above the lowest point and its subsequent path meets the sphere again at the point of projection.*

[Delhi B.A. III Prog. 2010; Bangalore 2004]

(*b*) *A particle tied to a string of length a is projected from its lowest point, so that after leaving the circular path it describes a free path passing through the lowest point. Prove that the velocity of projection is* $(7ag/2)^{1/2}$.

(*c*) *A particle is projected along the inside of a smooth vertical circle of radius a from the lowest point. Show that the velocity of projection required in order that after leaving the circle, the particle may pass through the centre is* $(ag/2)^{1/2}(\sqrt{3}+1)$.

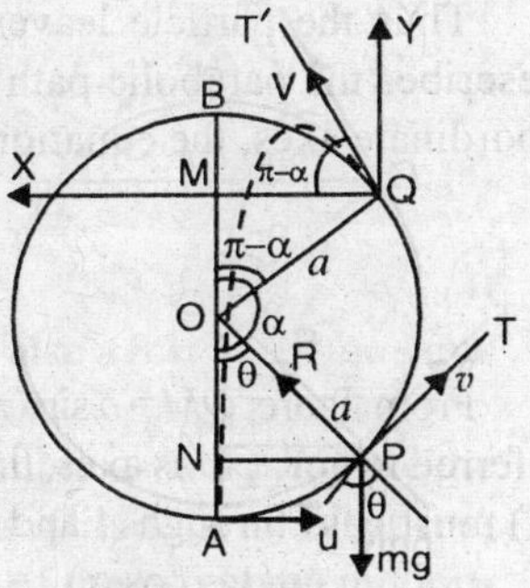

Sol. (*a*) Let the figure represent a section of the sphere by a vertical diametral plane through the direction of projection. Let P be the position of the particle at any time t, such that $\angle AOP = \theta$. Let u and v denote velocities of the particle at A and P respectively. For motion from A to P, by the principle of energy, we have

$$(1/2)mv^2 - (1/2)mu^2 = (-mg)AN;$$

where $\quad AN = OA - ON = a - a\cos\theta$

or $\quad v^2 = u^2 - 2g\,AN$ or $v^2 = u^2 - 2ga(1-\cos\theta) \qquad ...(1)$

External forces acting on the particle are its weight mg acting vertically downwards and the reaction of the circle, directed along PO. So equation of motion of the along PO is given by

$$mv^2/a = R - mg\cos\theta \quad \text{or} \quad R = mg\cos\theta + (m/a)v^2$$

or $\quad R = mg\cos\theta + (m(a)(u^2 - 2ga + 2ga\cos\theta)$, by (1)

or $\quad R = (m(a)(u^2 - 2ga + 3ga\cos\theta) \qquad ...(2)$

Let the particle leave the sphere at Q where $\theta = \alpha$ and $R = 0$. Also, for the present problem, $u = (7ag/2)^{1/2}$ (given). So (2) gives

$$0 = (m/a)(7ag/2 - 2ga + 3ga\cos\alpha) = 3mg(\cos\alpha + 1/2)$$

so that $\quad \cos\alpha = -1/2 \quad$ or $\quad \alpha = 120° \qquad ...(3)$

$\therefore$ Height of the particle at Q above the lowest point A

$= AM = AO + OM = a + a\cos(\pi - \alpha)$, where $\angle QOM = \pi - \alpha$

$= a - a\cos\alpha = a + a(-1/2) = 3a/2$, by (3)

Let V be the velocity of the particle at Q along tangent QT'. So putting $v = V$, $\theta = \alpha = 120°$ and $u = (7ag/2)^{1/2}$ in (1), we have

$$V^2 = 7ag/2 - 2ga + 2ga\cos 120° = (3/2)ag - ag = ag/2 \qquad ...(4)$$

The particle now moves freely under gravity with velocity V making an angle $(\pi - \alpha)$ *i.e.*, 60° with horizontal line QX and describes trajectory QA. With Q as origin and horizontal and vertical lines QX and QY as coordinate axes, the equation of the trajectory is given by

$$y = x\tan(\pi - \alpha) - \frac{gx^2}{2V^2\cos^2(\pi-\alpha)} \quad \text{or} \quad y = -x\tan\alpha - \frac{gx^2}{2V^2\cos^2\alpha}$$

or $\quad y = -x \tan 120^\circ - \dfrac{gx^2}{2(ag/2)\cos^2 120^\circ}$ $\quad$ or $\quad y = \sqrt{3}x - \dfrac{4x^2}{a}$ ...(5)

From figure, $QM = a \sin(\pi - \alpha) = a \sin 60^\circ = a\sqrt{3}/2$. Also $AM = 3a/2$.

Hence referred to QX and QY as axes, the coordinates of A are $(a\sqrt{3}/2, -3a/2)$. Putting $x = a\sqrt{3}/2$ and $y = -3a/2$ in (5), we get

$$-\frac{3a}{2} = \sqrt{3}\left(\frac{a\sqrt{3}}{2}\right) - \frac{4}{a}\left(\frac{a\sqrt{3}}{2}\right)^2 \quad \text{or} \quad -\frac{3a}{2} = \frac{3a}{2} - 3a \quad \text{or} \quad -\frac{3a}{2} = -\frac{3a}{2},$$

showing that coordinates of A satisfy (5). So the particle after leaving the sphere at Q describes a trajectory passing through point of projection A.

(*b*) Refer figure of part (*a*). Do as in part (*a*) upto equation (2) by replacing R by T. Thus, we have, as before

$$v^2 = u^2 - 2ga + 2ga \cos\theta \quad \text{...(1)}$$

and $\quad T = (m/a)(u^2 - 2ga + 3ga \cos\theta)$ ...(2)

Let the particle leave the circle at Q where $\theta = \alpha$ and $R = 0$

$\therefore\ 0 = (m/a)(u^2 - 2ga + 3ga \cos\alpha)$ $\quad$ or $\quad \cos\alpha = -(u^2 - 2ag)/3ag$ ...(3)

Let V be the velocity of the particle at Q along tangent QT'. So putting $v = V$ and $\theta = \alpha$ and $R = 0$ in (2), we get

$mV^2/a = -mg \cos\alpha$ $\quad$ or $\quad V^2 = -ag \cos\alpha$ ...(4)

Thus the particle leaves the circle at Q with velocity V at an angle $\pi - \alpha$ with horizon and describes the parabolic path QA. With Q as origin and horizontal and vertical lines QX and QY as coordinate axes, the equation of the parabola path QA is given by

$$y = x\tan(\pi - \alpha) - \frac{gx^2}{2V^2\cos^2(\pi - \alpha)} \quad \text{or} \quad y = -x\tan\alpha - \frac{gx^2}{2(-ag\cos\alpha)\cos^2\alpha},$$

or $\quad y = -x\tan\alpha + x^2/(2a\cos^3\alpha)$ ...(5)

From figure, $QM = a\sin(\pi - \alpha) = a\sin\alpha$ and $AM = AO + OM = a + a\cos(\pi - \alpha) = a(1 - \cos\alpha)$. Hence referred to QX, QY as axes, the coordinates of A are $\{a\sin\alpha, -a(1 - \cos\alpha)\}$. By problem the parabola (5) must pass through A and so coordinates of A must satisfy (5). Thus, we must have

$$-a(1 - \cos\alpha) = -a\sin\alpha\tan\alpha + (a^2\sin^2\alpha)/(2a\cos^3\alpha)$$

or $\quad \dfrac{\sin^2\alpha}{2\cos^3\alpha} = \dfrac{\sin^2\alpha}{\cos\alpha} + \cos\alpha - 1 = \dfrac{\sin^2\alpha + \cos^2\alpha - \cos\alpha}{\cos\alpha} = \dfrac{1 - \cos\alpha}{\cos\alpha}$

or $\quad \sin^2\alpha = 2\cos^2\alpha(1 - \cos\alpha)$ $\quad$ or $\quad 1 - \cos^2\alpha = 2\cos^2\alpha(1 - \cos\alpha)$

or $\quad (1 - \cos\alpha)(1 + \cos\alpha) = 2\cos^2\alpha(1 - \cos\alpha)$

or $\quad 1 + \cos\alpha = 2\cos^2\alpha$, since $(1 - \cos\alpha) \neq 0$

or $\quad 2\cos^2\alpha - \cos\alpha - 1 = 0$ $\quad$ or $\quad (2\cos\alpha + 1)(\cos\alpha - 1) = 0$

or $\quad 2\cos\alpha + 1 = 0$ $\quad$ or $\quad -2(u^2 - 2ag)/3ag + 1 = 0$, by (3)

or $\quad -2u^2 + 4ag + 3ag = 0$ $\quad$ or $\quad u = (7ag/2)^{1/2}$

(*c*) This is similar to part (*b*). You have to replace T by R and do exactly upto equation (5) as in part (*b*). In figure modify parabolic path as QO in place of QA as in figure of part (*b*) because in this problem, parabolic both must pass through centre O of the sphere. From figure, $QM = a\sin(\pi - \alpha) = a\sin\alpha$ and $OM = a\cos(\pi - \alpha) = -a\cos\alpha$. Hence referred to QX, QY as coordinate axes, the coordinates of O are $(a\sin\alpha, a\cos\alpha)$. By problem (5) must pass through O and so coordinate of O must satisfy (5). Thus, we get

$$a\cos\alpha = -a\sin\alpha\tan\alpha + (a^2\sin^2\alpha)/(2a\cos^3\alpha)$$

or $\quad \dfrac{\sin^2\alpha}{2\cos^3\alpha} = \dfrac{\sin^2\alpha}{\cos\alpha} + \cos\alpha = \dfrac{\sin^2\alpha + \cos^2\alpha}{\cos\alpha} = \dfrac{1}{\cos\alpha}$

or $\quad \sin^2\alpha = 2\cos^2\alpha$ $\quad$ or $\quad 1 - \cos^2\alpha = 2\cos^2\alpha$ $\quad$ or $\quad \cos\alpha = -(1/\sqrt{3})$, ...(6)

where we have taken –ve sign for cos α as $\pi/2 < \alpha < \pi$.

From (3) and (6), $-(1/\sqrt{3}) = -(u^2 - 2ga)/3ag$ or $u^2 = (2+\sqrt{3})ag$

or $$u^2 = ag(4+2\sqrt{3})/2 = (ag/2)(1+\sqrt{3})^2$$

Thus, $$u = (ag/2)^{1/2}(1+\sqrt{3}), \text{ as desired.}$$

Ex. 20. (*a*) *A particle is free to move on a smooth vertical circular wire of radius a. It is projected from the lowest point with velocity just sufficient to carry it to the highest point. Show that the reaction between the particle and wire is zero after a time* $(a/g)^{1/2} \log(\sqrt{5}+\sqrt{6})$. **[Purvanchal 2006 Bundelkhand 2006; Garhwal 2003, 04; Kumaun 2001, 03; Rohilkhand 2011]**

(*b*) *A heavy bead slides on smooth circular wire of radius a. It is projected from the lowest point with a velocity just sufficient to carry it to the highest point, prove that the radius through the lead in time t will turn through an angle* $2\tan^{-1}\{\sinh t(g/a)^{1/2}\}$ *and that the bead will take an infinite time to reach the highest point.* **[Rohilkhand 2003; Kanpur 2012]**

Sol. (*a*) Let *APB* be a vertical circular wire with *A* as its lowest point and *O* as centre. Let a particle of mass *m* be projected from *A* with velocity *u*. Let *P* be the position of the particle at any time *t*, where $\angle AOP = \theta$. Let v be its velocity at *P* along tangent *PT*. For motion from *A* to *P*, by the principle of conservation of energy,

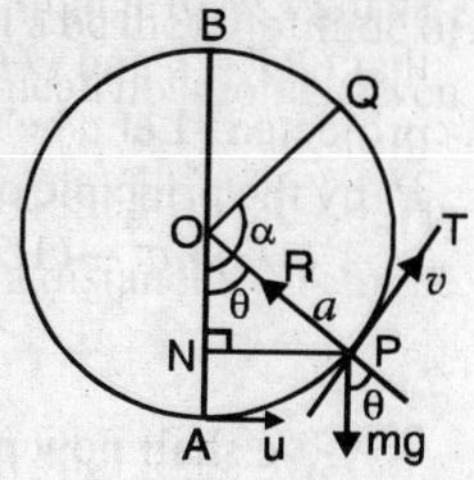

$$(1/2)mv^2 - (1/2)mu^2 = (-mg).AN,$$

where $AN = OA - ON = a - a\cos\theta$

or $v^2 = u^2 - 2ga(1-\cos\theta)$ or $v^2 = u^2 - 2ga + 2ga\cos\theta$ ...(1)

By problem, when $\theta = \pi$ at *B*, $v = 0$. So (1) gives

$$0 = u^2 - 2ga - 2ga \quad \text{or} \quad u^2 = 4ga$$

$\therefore$ (1) becomes $v^2 = 2ga + 2ga\cos\theta = 2ga(1+\cos\theta) = 4ga\cos^2(\theta/2)$ ...(2)

Equation of motion along inward drawn normal *PO* is

$mv^2/a = R - mg\cos\theta$ or $R = mg\cos\theta + (m/a)\times 2ga(1+\cos\theta)$, by (2)

or $$R = mg(3\cos\theta + 2),$$ ...(3)

Let reaction $R = 0$ at *Q* where $\theta = \alpha$. Then, from (3), we get

$$0 = mg(3\cos\alpha + 2) \quad \text{or} \quad \cos\alpha = -2/3$$ ...(4)

Let arc $AP = s$. Then $s = a\theta$ and $v = ds/dt = a\,d\theta/dt$

$\therefore$ (2) gives $a^2(d\theta/dt)^2 = 4ga\cos^2(\theta/2)$ or $d\theta/dt = 2(g/a)^{1/2}\cos\theta/2$, ...(5)

taking +ve sign before the radical sign as θ increases when *t* increases

or $$dt = (1/2)(a/g)^{1/2}\sec\theta/2\,d\theta$$ ...(6)

Let *T* be the time taken by the particle to move from *A* to *Q*. Then integrating (6) from $\theta = 0$ to $\theta = \alpha$ and $t = 0$ to $t = T$, we get

$$\int_0^T dt = \frac{1}{2}\left(\frac{a}{g}\right)^{1/2}\int_0^\alpha \sec\frac{\theta}{2}d\theta = \left(\frac{a}{g}\right)^{1/2}\left[\log\left(\sec\frac{\theta}{2}+\tan\frac{\theta}{2}\right)\right]_0^\alpha$$

or $$T = (a/g)^{1/2}\{\log(\sec\alpha/2 + \tan\alpha/2)\}$$ ...(7)

From (4), $2\cos^2\alpha/2 - 1 = -2/3$ or $\cos^2\alpha/2 = 1/6$ or $\cos\alpha/2 = 1/\sqrt{6}$

$\therefore$ $\sec\alpha/2 = \sqrt{6}$ and $\tan\alpha/2 = (\sec^2\alpha/2 - 1)^{1/2} = (6-1)^{1/2} = \sqrt{5}$

$\therefore$ (7) becomes T = the required time $= (a/g)^{1/2}\log(\sqrt{6}+\sqrt{5})$

(*b*) Proceed exactly as in part (*a*) upto equation (6). Then, the time *t* from *A* to *P* is given by

$$t = \frac{1}{2}\left(\frac{a}{g}\right)^{1/2}\int_0^\theta \sec\frac{\theta}{2}d\theta = \left(\frac{a}{g}\right)^{1/2}\left[\log(\sec\frac{\theta}{2}+\tan\frac{\theta}{2})\right]_0^\theta$$

$= (a/g)^{1/2}\log(\sec\theta/2 + \tan\theta/2) = (a/g)^{1/2}\log\{(1+\tan^2\theta/2)^{1/2} + \tan\theta/2\}$ ... (8)

$\therefore$ $t = (a/g)^{1/2}\sinh^{-1}(\tan\theta/2)$, as $\sinh^{-1}z = \log\{z + (1+z^2)^{1/2}\}$

or $t(g/a)^{1/2} = \sinh^{-1}(\tan\theta/2)$ or $\tan\theta/2 = \sinh\{t(g/a)^{1/2}\}$

or $\qquad \theta/2 = \tan^{-1}[\sinh\{t(g/a)^{1/2}\}]$ or $\theta = 2\tan^{-1}[\sinh\{t(g/a)^{1/2}\}]$

Let T be the time taken by the bead to move from A to B. Then, integrating (6) from $\theta = 0$ to $\theta = \pi$, we get

$$\int_0^T dt = \frac{1}{2}\left(\frac{a}{g}\right)^{1/2}\int_0^{\pi}\sec\frac{\theta}{2}\,d\theta = \left(\frac{a}{g}\right)^{1/2}\left[\log(\sec\frac{\theta}{2}+\tan\frac{\theta}{2})\right]_0^{\pi}$$

or $$T = (a/g)^{1/2}\lim_{\theta\to\pi}\log(\sec\theta/2+\tan\theta/2) = \infty$$

showing that bead take infinite time to reach the highest point B.

Ex. 21. *A particle is hanging from a fixed point O by means of a string of length a. There is a small nail at O′ in the same horizontal line with O at a distance b(< a) from O. Find the minimum velocity with which the particle should be projected from the lowest point in order that it may make a complete revolution round the nail without the string becoming slack.*

Sol. Let a particle of mass m hang from a fixed point O by mens of a string OA of length a. Let OC be horizontal line through O and O' be such that $OO' = b$ and $O'C = a - b$. Let u be the velocity with which the particle is projected. Let v be its velocity at P, where $\angle AOP = \theta$. For motion from A to P, by the principle of conservation of energy, we get

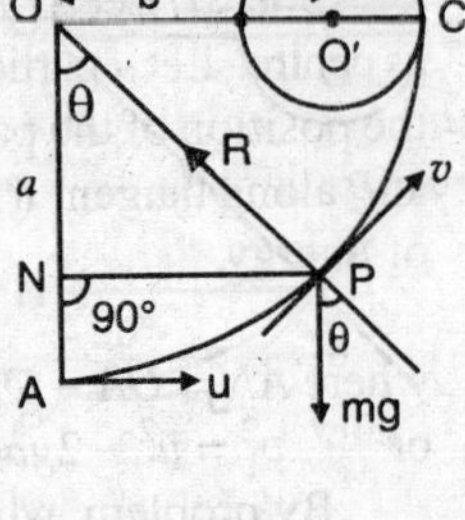

$$(1/2)mv^2 - (1/2)mu^2 = (-mg).AN, \text{ where}$$
$$AN = OA - ON = a - a\cos\theta$$
$$\therefore \qquad v^2 = u^2 - 2ga(1-\cos\theta) \qquad \ldots(1)$$

We shall now prove an important result and use it in this problem. For the particle to make complete revolution of circle of radius a, we must have

$$u^2 \geq 5ga \qquad \ldots(2)$$

Let velocity at C be V. Putting $\theta = \pi/2$ in (1), we get

$$V^2 = u^2 - 2ga \quad \text{or} \quad u^2 = V^2 + 2ga \qquad \ldots(3)$$

$\therefore$ (2) reduces to $\qquad V^2 + 2ga \geq 5ga \quad$ or $\quad V^2 \geq 3ga,$

showing that condition for making complete revolution round a circle of radius a is $V^2 \geq 3ga$ where V is velocity at the end of horizontal diameter.

Now, when string starting from A reaches at C, the particle moves in circle of radius $a - b$. Using the above result just proved, in order that it may make complete revolution round the nail, must have

$$V^2 \geq 3g(a-b) \qquad \ldots(4)$$

From (3) and (4), $\quad u^2 - 2ga \geq 3g(a-b) \quad$ or $\quad u^2 \geq g(5a-3b),$

showing that required minimum velocity of projection of the particle at the lowest point is $\{g(5a-3b)\}^{1/2}$.

Ex. 22. (a) *A particle is projected with velocity $u > (2ag)^{1/2}$ from the lowest point of a smooth vertical circle of radius a. If leaves the circular path at P and afterwords moving on a parabolic path in vacuum again meets the circle in Q. Prove that arc QB = 3 arc PB, were B is the highest point of the circle.* **[Agra 1990; Kanpur 2003]**

(b) *A particle is projected from the lowest point of a vertical circle with velocity u so as to move along the inside of the circle. If the particle leaves the circle, find where the particle will meet the circle again.*

[Garhwal 1996; Rohilkhand 1999]

Sol. (a) Consider the motion of a particle of mass m placed on the inside of a smooth vertical circle (with centre O and radius a) and projected with velocity u from its lowest point A. Let AB be the vertical diameter of the circle so that B is the highest point of the circle. Let P' be the position of the particle at any time t such that $\angle P'OA = \theta$. Let v, be the velocity of the particle at P' along the tangent $P'T'$. Let $P'N$ be perpendicular to AB. Then, $AN = OA - ON = a - a\cos\theta = a(1-\cos\theta)$. For motion from A to P, by the principle of conservation of energy, we have

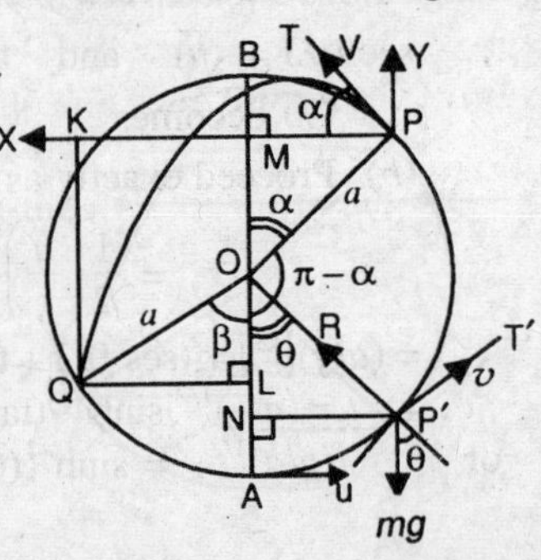

$(1/2)\, mv^2 - (1/2)mu^2 = (-mg) \times AN$ or $v^2 = u^2 - 2ga\,(1 - \cos\theta)$ ...(1)

The external forces acting on the particle are its weight mg acting vertically downwards and the reaction R of the circle, directed along $P'O$. Then, the equation of motion of the particle along the inward drawn normal $P'O$ is given by

$$m(v^2/a) = R - mg\cos\theta \quad \text{or} \quad R = mg\cos\theta - (m/a)v^2$$

or $R = mg\cos\theta + (m/a)(u^2 - 2ga + 2ga\cos\theta) = (m/a)(u^2 - 2ga + 3ga\cos\theta)$ ...(3)

Let the particle leave the circle at P where $\angle POB = \alpha$. Thus, $R = 0$ when $\theta = \pi - \alpha$. So (2) reduces to

$$0 = (m/a)\,\{u^2 - 2ga + 3ga\cos(\pi - \alpha)\} \text{ so that } u^2 = 2ga + 3ga\cos\alpha \quad ...(3)$$

Let V be the velocity of the particle at P. Then $v = V$ when $\theta = \pi - \alpha$ and so (1) reduces to

$$V^2 = u^2 - 2ga + 2\,ga\cos(\pi - \alpha) = u^2 - 2ga - 2ga\cos\alpha.$$

or $\quad V^2 = (2ga + 3ga\cos\alpha) - 2ga - 2ga\cos\alpha$, using (3)

Thus, $\quad V^2 = ga\cos\alpha$ ...(4)

Let PT be the tangent at P to the circular path. Then after leaving the circular path the initial velocity of the particle for parabolic path will be V along PT. By geometry, $\angle TPX = \angle POB = \alpha$, where PX in horizontal line through P. Let PY be vertical line through P. With P as origin and PX, PY as coordinate axes, the equation of the parabolic path of the particle is given by

$$y = x\tan\alpha - \frac{gx^2}{2V^2\cos^2\alpha} \quad \text{or} \quad y = x\tan\alpha - \frac{x^2}{2a\cos^3\alpha} \text{ by (4)} \quad ...(5)$$

Let the path of the particle (5) meet the circle in Q where $\angle QOA = \beta$. Let QL be perpendicular to AB and QK be perpendicular to PX. Then from figure, we get

$$PK = PM + MK = PM + QL = a\sin\alpha + a\sin\beta$$

and $\quad QK = ML = OM + OL = a\cos\alpha + a\cos\beta$

Referred to PX and PY as coordinate axes, the coordinates of Q are $(PK, -QK)$, *i.e.*, $(a(\sin\alpha + \sin\beta), -a(\cos\alpha + \cos\beta))$.

Since Q lies on (5), we must have

$$-a(\cos\alpha + \cos\beta) = a(\sin\alpha + \sin\beta)\tan\alpha - (a^2/2a\cos^3\alpha) \times (\sin\alpha + \sin\beta)^2$$

or $\quad (1/2) \times (\sin\alpha + \sin\beta)^2 = \cos^3\alpha\,\{\tan\alpha\,(\sin\alpha + \sin\beta) + \cos\alpha + \cos\beta\}$

$\quad = \cos^3\alpha\,\{(\sin\alpha/\cos\alpha) \times (\sin\alpha + \sin\beta) + \cos\alpha + \cos\beta\}$

$\quad = \cos^2\alpha\,\{\sin\alpha\,(\sin\alpha + \sin\beta) + \cos\alpha\,(\cos\alpha + \cos\beta)\}$

or $\quad (1/2) \times (\sin\alpha + \sin\beta)^2 = \cos^2\alpha\,(1 + \cos\alpha\cos\beta + \sin\alpha\sin\beta)$

$\therefore \quad (\sin\alpha + \sin\beta)^2 = 2\cos^2\alpha\,\{1 + \cos(\beta - \alpha)\}$

or $\quad [2\sin\{(\beta + \alpha)/2\}\cos\{(\beta - \alpha)/2\}]^2 = 2\cos^2\alpha \times 2\cos^2\{(\beta - \alpha)/2\}$

so that $\quad \sin^2\{(\beta + \alpha)/2\} = \cos^2\alpha$, as $\cos^2\{(\beta - \alpha)/2\} \neq 0$

Hence $\quad \sin\{(\beta + \alpha)/2\} = \cos\alpha$ or $\sin\{(\beta + \alpha)/2\} = \sin(\pi/2 - \alpha)$

so that $\quad (\beta + \alpha)/2 = \pi/2 - \alpha$ or $\beta + \alpha = \pi - 2\alpha$ or $\pi - \beta = 3\alpha$

Thus, $\quad \angle QOB = 3 \times \angle POB$ so that arc $QB = 3 \times$ arc PB.

Note : Since $u > (2ag)^{1/2}$, it follows that the particle will rise above the horizontal diameter passing through O and later on leave the circular path at P and meet it again at Q.

(*b*) Proceed as part (*a*) to find Q such that arc $QB = 3 \times$ arc PB.

EXERCISE

1. A particle of mass m oscillates through the angle 2α inside a fixed smooth vertical circle of radius a. If the pressure at any point be equal to $mg\,(1 + 3v^2/2ga)$ where v is velocity of particle at that point, then show that $\theta = \pi/2$.

2. A stone of weight W is tied to one end of a string and is describing a circle in a vertical plane, the other end of the string being fixed. If the maximum speed of the stone is twice the minimum speed, prove that when the string is horizontal, its tension is $10\,W/3$. Also find the tension when the string makes an angle of 45° with the downward vertical. **[Ans.** $(10/3 + 3/\sqrt{2})\,W$ **]**

3. A simple pendulum of length l is set swinging in a vertical plane so as to make complete revolutions. In the subsequent motion, ω is the angular velocity of the particle at any point of its path. If maximum angular velocity is greater than ω by x and minimum angular velocity is less by the same amount prove that $l\omega x = g$.

4. A particle is projeted along the inner surface of a smooth vertical circle of radius a, its velocity at the lowest point being $(ag/5)\sqrt{95}$; show that ti will leave the circle at an angular distance $\cos^{-1}(3/5)$ from the highest point and its velocity then is $(1/5) \times (15ag)^{1/2}$.

5. A particle attached to a fixed *peg O* by a string of length l, is lifted up with the string horizontal and then let go. Prove that when the string makes an angle θ with the horizontal, the resultant acceleration is $g(1 + 3\sin^2\theta)^{1/2}$. **[I.A.S. 2007; Agra 1994]**

6. A particle attached to a fixed *peg O* by a string of length l, is let fall from a point in the horizontal line through O at a distance $l\cos\theta$ from O; show that its velocity when it is vertically below O is $\{2gl(1-\sin^3\theta)\}^{1/2}$.

7. A particle is projected from the lowest point inside a smooth circle of radius a with a velocity due to a height h above the centre. Find the point where it leaves the circle and show that it afterwards pass through the

(*a*) centre if $h = a\sqrt{3}/2$ (*b*) the lowest point if $h = 3a/4$.

8. A particle is projected along the inside of a smooth fixed sphere, from its lowest point, with a velocity equal to due to falling freely down the vertical diameter of the sphere. Show that the particle will leave the sphere and afterwards pass vertically over the point of projection at a distance equal to (25/32) of the diameter. **[Agra 1993]**

9. Find velocity with which a particle must be projected along the interior of a smooth vertical hoop of raius a from the lowest point in order that it may leave hoop at an angular distance of 30° from the vertical. Show that it will strike hoop again at the extremity of the horizontal diameter. **[Avadh 1999]**

10. Show that a particle projected with velocity $(2ag)^{1/2}$ from the lowest point of a vertical circle of radius a and moving inside it will just reach the end of the horizontal diameter; while if projected with velocity $(5ag)^{1/2}$ it will reach the highest point. Prove that the reaction at any point in the first case is proportioal to the depth below the horizontal diameter and in the second case to the depth below the highest point.

11. In an oscillating pendulum the tension of the string when the bob is in its lowest position is n times the weight of the bob. Prove that the angle of swing on each side of the vertical is $\cot^{-1}\{(3-n)/2\}$. **[Agra 2012]**

12. A particle slides outside a smooth fixed smooth vertical circle starting from rest at the top, and another is projected from the lowest point, inside the circle with just sufficient velocity to carry it to the top, pove that both leave circle at the same point and proceed to describe the parts of the same parabola.

13. A particle moves on the inner side of smooth circular hoop of radius a whose plane is vertical. It is projected from the bottom with a velocity insufficient to carry it round the hoop so that it comes off at an angular distance α from the highest point. Show that it crosses the vertical diameter of the hoop at a distance $(a/2)\sec^3\alpha(1-\cos\alpha)^2(1+2\cos\alpha)$ below the top of the hoop.

14. A heavy particle is attached to a fixed point by a fine string of length a; the particle is projected horizontally from the lowest point with velocity $\{ag(2+3\sqrt{3}/2)\}^{1/2}$. Prove that the string would first become slack when inclined to the upward vertical at an angle of 30°, will become tight again, when horizontal, and slack again when inclined to the upward vertical at an angle $\cos^{-1}(3\sqrt{3}/g)$.

8.6 Simple pendulum [Avadh 1986; Gorakhpur 1991; Purvanchal 1990]

Definition. If a heavy particle attached to one end of a light string, the other end of the string being fixed, oscillates in a vertical circle about is lowest position, then the system is known as a *simple pendulum.*

Discussion of motion of a simple pendulum. Let O be the fixed point, l the length of the string, A the lowest position of the particle of mass m. Let P be the position of the particle at any time t, when $\angle AOP = \theta$. Let arc $AP = s$ so $s = l\theta$ and $d^2s/dt^2 = l(d^2\theta/dt^2)$. The forces acting on the particle are its weight mg acting vertically downwards through P, and T the tension of the string acting along PO. Resolved part of acceleration of particle along tangent PT' at P is d^2s/dt^2 *i.e.,* $l(d^2\theta/dt^2)$. Hence the equation of motion in the direction of PT' is

$$ml(d^2\theta/dt^2) = -mg \sin\theta \quad \text{or} \quad d^2\theta/dt^2 = -(g/l)\sin\theta \quad ...(1)$$

If θ be small, then $\sin\theta = \theta$ and so (1) becomes

$$d^2\theta/dt^2 = -(g/l)\theta, \quad ...(2)$$

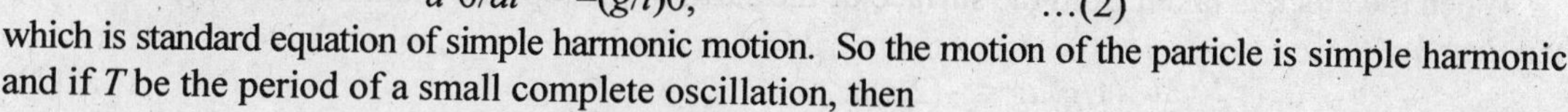

which is standard equation of simple harmonic motion. So the motion of the particle is simple harmonic and if T be the period of a small complete oscillation, then

$$T = 2\pi(l/g)^{1/2} \quad ...(3)$$

Remark 1. The length l of the string is called the length of the simple pendulum. The time period T depends upon l and g but is independent of θ, provided it is small.

Remark 2. Let B and B' be the extremities of the path of the simple pendulum.

Then, time from B to B' and back to $B = T = 2\pi(l/g)^{1/2}$

$\therefore$ Time from B to $B' = T/2 = \pi(l/g)^{1/2}$

Since the pendulum of a clock beats at each extremity, so time between two consecutive beats $= \pi(l/g)^{1/2}$

Second's pendulum. Definition. A second's pendulum is that which oscillates from rest to rest in one second *i.e.,* half of its time period is one second. In other words, a simple pendulum is called a second's pendulum if the time interval between two consecutive beats is one second and such a clock is said to be correct. Thus, for a second's pendulum, we have

$$\pi(l/g)^{1/2} = 1 \quad \text{so that} \quad l = g/\pi^2$$

In C.G.S. system, $g = 981$ cm/sec^2, so $l = 981/9.87 = 9.4$ cm

In M.K.S. system, $g = 9.8$ m/sec^2, so $l = 9.8/9.87 = 0.992$ meters

8.7 Formulae for gain or loss of beats (time) by a clock **[Delhi 2008; M.S. Univ. T.N. 2007]**

Let n be the number of beats in a given time interval T. Since $T = n \times$ (time of one beat), we have

$$T = n \times \pi(l/g)^{1/2} \quad \text{or} \quad n = (T/\pi) \times (g/l)^{1/2} \quad ...(1)$$

From (1), $\log n = \log(T/\pi) + (1/2)(\log g - \log l)$ ...(2)

Taking differentials of both sides of (2) and noting that t/π is constant,

we get $$\frac{dn}{n} = \frac{1}{2}\left(\frac{dg}{g} - \frac{dl}{l}\right) \quad \text{or} \quad dn = \frac{n}{2}\left(\frac{dg}{g} - \frac{dl}{l}\right) \quad ...(3)$$

where dn, dg and dl represent small variations in n, g and l respectively. We now discuss two particular cases.

Case (*i*) When only l changes and g remains constant *i.e.,* $dg = 0$. [Delhi B.A. III (Prog.) 2008]

Then, (3) reduces to $$dn = -(n/2l)dl, \quad ...(4)$$

showing that dn is $-$ve if dl is +ve. Hence if the length of the pendulum is increased and g remains constant, then there will be a decrease in the number of beats in any given time interval and so the clock goes slow.

Similarly, (4) shows that dn is +ve if dl is $-$ve. Hence if the length of the pendulum is decreased and g remains constant, then there will be an increase in the number of beats in any given time interval and so the clock goes fast.

Case (*ii*) When only g changes and l remains constant *i.e.,* $dl = 0$. [Delhi B.A. III (Prog) 2008]

Then, (3) reduces to $$dn = (n/2g)dg, \quad ...(5)$$

showing that dn is +ve if dg is +ve. Hence if g is increased and l remains constant, then there will be an increase in the number of beats in any given time interval and so that clock goes fast.

Similarly, (5) shows that dn is $-$ve if dg is $-$ve. Hence if g is decreased and l remains constant, then there will be a decrease in the number of beats in any given time interval and so that clock goes slow.

8.8 Formulae for gain or loss of beats (time) by clock when it is taken to a point above or below the surface of the earth.

Situation (*i*) When the clock is carried to a mountain at a height *h* above the surface of the earth. By Newton's gravitational law, we know that outside the surface of the earth g varies inversely as the square of the distance from the centre of the earth. Thus, $g = \mu/r^2$, where r is the distance of the point outside the surface of the earth from the centre of the earth and μ is a constant.

Now, $\quad g = \mu/r^2 \Rightarrow \log g = \log \mu - 2 \log r \Rightarrow \dfrac{dg}{g} = -\dfrac{2dr}{r}$

When the clock is taken from the surface of the earth to the top of a mountain of height h (say), then $dr = h$.

$\therefore \quad (1/g)dg = -(2/r)h \qquad \ldots(1)$

$\therefore \quad dn = (n/2g)dg \Rightarrow dn = -(n/r)h$, using (1) $\qquad \ldots(2)$

From (2), it follows that if a clock is carried to a mountain, it loses number of beats in any time interval and so the clock becomes slow.

Situation (*ii*) When the clock is carried to a mine at a depth *d* below the surface of the earth. We know that inside the surface of the earth, g varies directly as the distance from the centre of the earth. Thus, $g = \mu r$, where r is the distance of the point from the centre of earth and μ is a constant.

Now, $\quad g = \mu r \Rightarrow \log g = \log \mu + \log r \Rightarrow (1/g)dg = (1/r)dr$

If d be the depth below the surface of the earth, then $dr = -d$ and so

$(1/g)dg = -(d/r) \qquad \ldots(3)$

$\therefore \quad dn = (n/2g)dg \Rightarrow dn = -(n/2r)d$, by (3) $\qquad \ldots(4)$

From (4), it follows that if a clock is carried to a mine, it loses number of beats in any time interval and so the clock becomes slow.

TYPE 3 : SOLVE EXAMPLES BASED ON ART. 8.6, 8.7 AND 8.8

Ex. 1. *Find how many seconds a clock would lose per day if the length of its pendulum is increased in the ratio 900 : 901.*

Sol. Let l be the original length and dl be increase in it. Then, given

$l/(l + dl) = 900/901 \quad$ or $\quad (l + dl)/l = (900 + 1)/900$

or $\quad 1 + (1/l)dl = 1 + 1/900 \quad$ so that $\quad (1/l)dl = 1/900$

Let n = number of beats (or seconds) per day = $24 \times 60 \times 60 = 86400$

Let dn denote the change in the number of beats. Then

$dn = -(n/2)(1/l)dl = -(86400/2) \times (1/900) = -48$

Hence the number of seconds lost by the clock per day = 48

Ex. 2. *If a pendulum of length l makes n complete oscillation in a given time, show that if*

(i) g is changed to g + g′, the number of oscillations gained is ng′/2g

(ii) l is changed to l + l′, the number of oscillations lost is nl′/2l

Sol. (*i*) When g is changed to $g + g'$, then $dg = g'$ and l is constant.

$\therefore \quad dn = (n/2g).dg = (n/2g)g'$, [Refer result (5) of Art. 8.7]

showing that the number of oscillations gained = $ng'/2g$.

(*ii*) When l is changed to $l + l'$, the $dl = l'$ and g is constant.

$\therefore \quad dn = -(n/2l)dl = -(n/2l)l'$, [Refer result (4) of Art. 8.7]

showing that the number of oscillations lost = $nl'/2l$.

Ex. 3. (*a*) *A pendulum whose length is l, makes m complete oscillations in 24 hours. When its length is slightly altered, it makes m + n oscillations in 24 hours. Show that the diminution in length is (2nl)/m nearly.* **[G.N.D.U. 1995]**

Sol. Let the length be increased by l'. Let T be the given interval of time, *i.e.*, 24 hours, then

$T = m \times 2\pi(l/g)^{1/2} \quad$ or $\quad m = (T/2\pi) \times (g/l)^{1/2} \qquad \ldots(1)$

When l changes to $l + l'$, m changes to $m + n$, so (1) becomes

$$m+n=\frac{T}{2\pi}\left(\frac{g}{l+l'}\right)^{1/2}=\frac{T}{2\pi}\left\{\frac{g}{l(1+l'/l)}\right\}^{1/2}=\frac{T}{2\pi}\left(\frac{g}{l}\right)^{1/2}\left(1+\frac{l'}{l}\right)^{-1/2}$$

or $\quad m+n = m\{1-(1/2)(l'/l)+....\}$, nearly, [using (1)]

or $\quad n = -(ml')/2l \quad$ or $\quad l'=-(2nl)/m$,

showing that the diminution in the length of the pendulum = $2nl/m$

Ex. 3. (*b*) *A simple pendulum has a period T. When the string is lengthened by small fraction 1/n of its length, the period is T', show that*

$$1/n = 2(T'-T)/T \text{ approximately.}$$

Sol. Given $\quad T=2\pi(l/g)^{1/2}$ ...(1)

When l becomes $l + l/n$, T becomes T' so, by (1)

$$T'=2\pi\left\{\frac{l+l/n}{g}\right\}^{1/2}=2\pi\left(\frac{l}{g}\right)^{1/2}\left(1+\frac{1}{n}\right)^{1/2}=T\left(1+\frac{1}{2n}\right) \text{ approximately}$$

or $\quad T'-T = T/2n \quad$ or $\quad 1/n=2(T'-T)/T$ approximaetly.

Ex. 4. *A clock loses five seconds a day, find the alternation required in length of its pendulum in order that it may keep correct time.* **[Gorakhpur 1996, 98; Puryanchal 1997]**

Sol. Here $\quad n$ = number of seconds per day = $24 \times 60 \times 60 = 86400$

and $\quad dn$ = number of seconds lost = -5 (given)

Now, $\quad 1 = \pi(l/g)^{1/2} \quad$ so that $\quad l=g/\pi^2$ ...(1)

But $\quad dn=-\dfrac{n}{2}\dfrac{dl}{l} \quad$ so $\quad -5=-\dfrac{86400}{2}\cdot\dfrac{dl}{g/\pi^2}$, by (1)

$$\therefore \quad dl=\frac{10g}{86400}=\frac{10\times 981}{86400\times 9.87}=0.0115 \text{ cm}$$

Hence the length of pendulum must be decreased by 0.0115 cm.

Ex. 5. *Show that in an incorrect second's pendulum of a clock which loses x seconds in a day must be shortended by x/432 percent of its original length in order to keep a correct time.*

Sol. Here $\quad n = 24 \times 60 \times 60 = 86400 \quad$ and $\quad dn=-x$

$$\therefore \quad dn=-\frac{n}{2}\frac{dl}{l}\Rightarrow -x=-\frac{86400}{2}\frac{dl}{l}\Rightarrow dl=\frac{xl}{43200},$$

showing that the length of the pendulum must be shortended by $(x/32000)l$ *i.e.* $x/320$ percent of its length.

Ex. 6. *A second's pendulum which gains 10 seconds per day at one place, loses 10 seconds per day at another. Compare the values of gravity at two places.* **[Gorakhpur 1995]**

Sol. Let g be the gravity when the pendulum beats seconds. At first place where the pendulum gains 10 seconds per day, let the incorrect value of g be $g+g_1$. Thus, $dg=g_1$, $dn=10$ and $n = 24 \times 60 \times 60 = 86400$. Then,

$$dn=\frac{n}{2}\frac{dg}{g}\Rightarrow 10=\frac{86400}{2}\frac{g_1}{g}\Rightarrow\frac{g_1}{g}=\frac{1}{4320}$$

or $\quad 1+g_1/g=1+1/4320 \quad$ or $\quad (g+g_1)/g=4321/4320$ ...(1)

At the second place where the pendulum loses 10 seconds per day, let the incorrect value of g be $g+g_2$ so that $dg=g_2$, $n=86400$ and $n=-10$. Then

$$\frac{dn}{n}=\frac{n}{2}\frac{dg}{g}\Rightarrow -10=\frac{86400}{2}\frac{g_2}{g}\Rightarrow\frac{g_2}{g}=-\frac{1}{4320}$$

or $\quad 1+g_2/g=1-(1/4320) \quad$ or $\quad (g+g_2)/g=4319/4320$ ...(2)

Dividing (1) by (2), $\quad (g+g_1)/(g+g_2) = 4321/4319$, as required.

Ex. 7. *A clock is taken from one place on the earth's surface to another. If the value of g is increased by one percent, find what increase must be made in the length of the pendulum to keep the time of the clock correct.*

Sol. Given dg = 1% of $g=g/100 \Rightarrow (1/g)\,dg=1/100$

We have, $dn=(n/2)\{(1/g)\,dg-(1/l)\,dl\}$, see result (3) of Art. 8.7

Since the clock keeps correct time at the second place, so $dn = 0$

$$0 = (n/2)\{1/100 - (1/l)\,dl\} \Rightarrow dl = l/100,$$

showing that the length of the pendulum must be increased by one percent.

Ex. 8. (*a*) *Assuming that acceleration due to gravity varies inversely as the square of its distance from the centre of the earth, find approximately the height of a mountain at the top of which a pendulum, which beats seconds at the sea level, loses 20 seconds a day. The radius of the earth be taken as* 64×10^5 *meters.*

(*b*) *A clock which keeps correct time at the surface of the earth loses 10 seconds a day when taken down a mine, find its depth. The radius of the earth being taken as* 64×10^5 *meters.*

Sol. In what follows we shall use results (2) and (4) of Art. 8.8

(*a*) Here $dn = -20$, $r = 64 \times 10^5$ meters, $n = 24 \times 60 \times 60 = 86400$

$$\therefore \quad dn = -\frac{nh}{r} \Rightarrow -20 = -\frac{86400 \times h}{64 \times 10^5} \Rightarrow h = 1481 \text{ meters}$$

(*b*) Here $dn = -10$, $r = 64 \times 10$ meters, $n = 24 \times 60 \times 60 = 86400$

$$\therefore \quad dn = -\frac{nd}{2r} \Rightarrow -10 = -\frac{86400 \times d}{2 \times 64 \times 10^5} \Rightarrow d = 1481 \text{ meters}$$

Ex. 9. *A pendulum which beats seconds at the surface of the earth is carried to the top of a mountain 5 km high, how many seconds will it lose or gain per day ? What correction in its present length be made so that it may beat seconds at the top of the mountain ?*

Sol. Here $n = 24 \times 60 \times 60 = 86400$, $h = 5$ km, $r = 6400$ km.

Using result (2) of Art. 8.8, $dn = -\dfrac{nh}{r} \Rightarrow dn = -\dfrac{86400 \times 5}{6400} = -67.5$ seconds,

showing that the clock will lose 67.5 seconds per day on the top of mountain.

We have $dn = (n/2)\{(1/g)dg - (1/l)dl\}$, by result (3) of Art. 8.7

But $dn = 0$ because the pendulum is to keep correct time

$$\therefore \quad 0 = (n/2)\{(1/g)dg - (1/l)dl\} = 0 \quad \text{or} \quad (1/l)dl = (1/g)dg$$

or $\quad (1/l)dl = -2h/r$, as $(1/g)dg = -2h/r$, by result (1) of Art. 8.8

$$\therefore \quad dl = -(2hl)/r = -(2 \times 5 \times l)/6400 = -(1/640)l,$$

showing that to keep correct time at the top of the mountain, the present length should be decreased by (1/640) of itself.

Ex. 10. *A second's pendulum was too long on a given day by a quantity* α*, it was then over corrected so as to become too short by* α *during the next day. Prove that if* l *is the correct length, then the number of minutes gained in two days was nearly* $(1080\alpha^2)/l^2$. **[Kanpur 1998]**

Sol. Let l be the correct length of the pendulum, then

$$1 = \pi(l/g)^{1/2} \quad \ldots(1)$$

Let n be the number of beats made by the pendulum on the first day when its length is $l + \alpha$ in place of l, then

$$86400 = n\pi\left(\frac{l+\alpha}{g}\right)^{1/2} = n\pi\left(\frac{l}{g}\right)^{1/2}\left(1+\frac{\alpha}{l}\right)^{1/2} = n\left(1+\frac{\alpha}{l}\right)^{1/2}, \text{ by (1)}$$

$$\therefore \quad \frac{n}{86400} = \left(1+\frac{\alpha}{l}\right)^{2} = 1 - \frac{1}{2}\frac{\alpha}{l} + \frac{1.3}{2.4}\frac{\alpha^2}{l^2} + \ldots, \text{ nearly}$$

$\therefore$ Gain of beets in the first day

$$= n - 86400 = 86400\left(1 - \frac{1}{2}\frac{\alpha}{l} + \frac{3}{8}\frac{\alpha^2}{l^2}\right) - 86400 = 86400\left(-\frac{1}{2}\frac{\alpha}{l} + \frac{3}{8}\frac{\alpha^2}{l^2}\right) \quad \ldots(2)$$

Let n' be the number of beats made by the pendulum on the next day when its length is $l - \alpha$ in place of l, then

$$86400 = n'\pi\left(\frac{l-\alpha}{g}\right)^{1/2} = n\pi\left(\frac{l}{g}\right)^{1/2}\left(1-\frac{\alpha}{l}\right)^{1/2} = n'\left(1-\frac{\alpha}{l}\right)^{1/2} \text{ by (1)}$$

$$\therefore \quad \frac{n'}{86400} = \left(1 - \frac{\alpha}{l}\right)^{-1/2} = 1 + \frac{1}{2}\cdot\frac{\alpha}{l} + \frac{1\cdot 3}{2\cdot 4}\cdot\frac{\alpha^2}{l^2} + ..., \text{ nearly}$$

$\therefore$ gain of beats on the second day

$$= n' - 86400 = 86400\left(1 + \frac{1}{2}\frac{\alpha}{l} + \frac{3}{8}\frac{\alpha^2}{l^2}\right) - 86400 = 86400\left(-\frac{1}{2}\frac{\alpha}{l} - \frac{3}{8}\frac{\alpha^2}{l^2}\right) \qquad ...(3)$$

Adding (2) and (3), the pendulum gains
$= 2 \times 86400 \times (3/8) \times (\alpha^2/l^2)$ seconds nearly $= 1080(\alpha^2/l^2)$ minutes nearly.

Ex. 11. (*a*) *A pendulum of length l hangs against a wall inclined at an angle α to the horizontal. Show that the time of complete oscillation is* $2\pi(l/g \sin \alpha)^{1/2}$ **[Gorakhpur 1992, 1994]**

(*b*) *A pendulum of length l has one end of a string fastened to a nail on a smooth plane inclined to horizon at an angle α. With the string and the weight on the plane, its time of oscillation is t seconds. If the pendulum of length l′ oscillates in one second when suspended vertically, prove that* $\alpha = \sin^{-1}(l/l't^2)$.

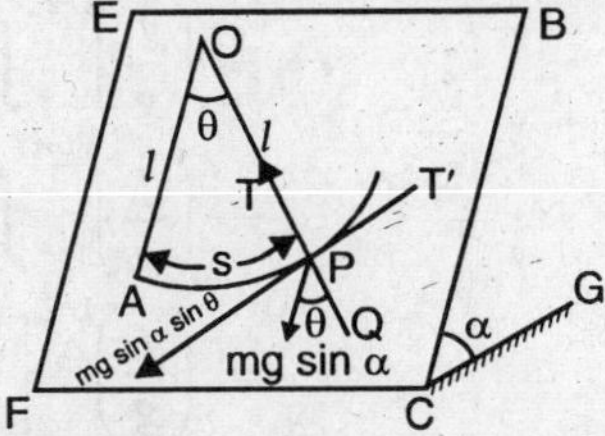

Sol. (*a*) Let *CBEF* be the wall inclined at an angle α to horizontal *CG*. Let *CB* be the line of greatest slope of the wall. Let *OA* be the initial position of the pendulum where *O* is the point of suspension. Then *OA* will be line of greatest slope, where *OA* is parallel to *CB*. Let *m* be mass of the bob of the pendulum. Let *P* be its position of any time *t*, where $\angle POA = \theta$. Let arc $AP = s$ so that $s = l\theta$ and $d^2s/dt^2 = l(d^2\theta/dt^2)$.

The forces acting on the bob are its weight *mg* sin α along the line of greatest slope and *T*, the tension of the string acting along *PO*. Resolved part of acceleration of the particle along tangent *PT′* is d^2s/dt^2 *i.e.,* $l(d^2\theta/dt^2)$. Hence the equation of motion in the direction of *PT′* is

$$ml(d^2\theta/dt^2) = -mg \sin \alpha \sin \theta \quad \text{or} \quad d^2\theta/dt^2 = -\{(g/l) \sin \alpha\}\, \theta,$$

[Assume that sin θ is small so that sin θ = θ]

which is standard equation of S.H.M. and so the required time of complete oscillation

$$= 2\pi/(g \sin \alpha/l)^{1/2} = 2\pi\,(l/g \sin \alpha)^{1/2}$$

(*b*) Proceed as in part (*a*). Let *O* be the *peg*. Then

$$t = 2\pi(l/g \sin \alpha)^{1/2} \quad \text{or} \quad t^2 = (4\pi^2 l)/g \sin \alpha \qquad ...(1)$$

In second case, when $\theta = 90°$ and $l = l'$, $t = 1$

$$\therefore \quad 1 = 2\pi(l'/g)^{1/2} \quad \text{or} \quad 1 = (4\pi^2 l')/g \qquad ...(2)$$

Dividing (1) by (2), $\quad t^2 = (l/l')\sin \alpha \quad$ or $\quad \sin \alpha = l/l't^2 \quad$ or $\quad \alpha = \sin^{-1}(l/l't^2)$

Ex. 12. (*a*) *If a simple pendulum oscillates through an angle α on either side of the mean position, show that the period of an oscillation is* (*i*) $2\pi(l/g)^{1/2}\,(1 + 4 \sin^2 \alpha/2)$ *when α is so small that* $\sin^2 \alpha/2$ *and higher powers of sin α /2 may be neglected.* (*ii*) $2\pi(l/g)^{1/2}\,(1 + \alpha^2/16)$ *if higher powers of α are neglected.* **[Delhi Maths (H) 2006)]**

(*b*) *In a simple pendulum, show that the period T is given by* $T = 2\pi(l/g)^{1/2}\,\{1 + k^2/4 + 9k^4/16 + ...)$, *where* $k = \sin \alpha/2$ *and α is the amplitude.*

Sol. (*a*) First proceed as in Art. 8.6 upto equation (1). Thus

$$\frac{d^2\theta}{dt^2} = -\frac{g \sin \theta}{l} \quad \text{or} \quad \frac{\dot\theta\, d\dot\theta}{d\theta} = -\frac{g \sin \theta}{l} \quad \text{or} \quad 2\dot\theta\, d\dot\theta = \frac{-2g \sin \theta\, d\theta}{l}$$

Integrating, $\quad \dot\theta^2 = (2g/l) \cos \theta + C \qquad ...(1)$

If the pendulum oscillates through an angle α so that at extreme position *B*, when θ = α, v = velocity at $B = 0$ or $l\dot\theta = 0$ or $\dot\theta = 0$. Then (1) gives

$$0 = (2g/l) \cos \alpha + C \text{ or } C = -(2g/l) \cos \alpha. \text{ So (1) reduces to}$$

$$\left(\frac{d\theta}{dt}\right)^2 = \left(\frac{2g}{l}\right)(\cos \theta - \cos \alpha) = \frac{4g}{l}\left(\sin^2 \frac{\alpha}{2} - \sin^2 \frac{\theta}{2}\right)$$

$$\therefore \quad dt = \frac{1}{2}\left(\frac{l}{g}\right)^{1/2} \frac{d\theta}{(\sin^2\alpha/2 - \sin^2\theta/2)^{1/2}} \quad \ldots(2)$$

Let t' be the time taken from the lowest position A to the extreme position B so that $t = 0$ when $\theta = 0$ and $t = t'$ when $\theta = \alpha$. Then integrating (2), we get

$$\int_0^{t'} dt = \frac{1}{2}\left(\frac{l}{g}\right)^{1/2} \int_0^{\alpha} \frac{d\theta}{(\sin^2\alpha/2 - \sin^2\theta/2)^{1/2}} \quad \ldots(3)$$

Putting $\sin\theta/2 = \sin\alpha/2 \sin\phi$ and $(1/2)\cos\theta/2\, d\theta = \sin\alpha/2 \cos\phi\, d\phi$, (3) gives

$$t' = \frac{1}{2}\left(\frac{l}{g}\right)^{1/2} \int_0^{\pi/2} \frac{2\sin\alpha/2\cos\phi\, d\phi}{\cos\theta/2\,(\sin^2\alpha/2 - \sin^2\alpha/2\sin^2\phi)^{1/2}}$$

$$= \left(\frac{l}{g}\right)^{1/2} \int_0^{\pi/2} \frac{d\phi}{\cos\theta/2} = \left(\frac{l}{g}\right)^{1/2} \int_0^{\pi/2} \frac{d\phi}{(1-\sin^2\theta/2)^{1/2}}$$

$$= \left(\frac{l}{g}\right)^{1/2} \int_0^{\pi/2} (1-\sin^2\frac{\alpha}{2}\sin^2\phi)^{-1/2} d\phi$$

$$= \left(\frac{l}{g}\right)^{1/2} \int_0^{\pi/2} \left[1+\frac{1}{2}\sin^2\frac{\alpha}{2}\sin^2\phi + \frac{1.3}{2.4}\sin^4\frac{\alpha}{2}\sin^4\phi + \ldots\right] d\phi$$

$$= \left(\frac{l}{g}\right)^{1/2} \left[\frac{\pi}{2} + \frac{1}{2}\sin^2\frac{\alpha}{2}\cdot\frac{1}{2}\cdot\frac{\pi}{2} + \frac{1.3}{2.4}\cdot\sin^4\frac{\alpha}{2}\cdot\frac{3.1}{4.2}\cdot\frac{\pi}{2} + \ldots\right]$$

[Using Walli's formulas]

$$= (\pi/2)(l/g)^{1/2}\{1 + (1/4)\sin^2\alpha/2 + (9/64)\sin^4\alpha/2 + \ldots\} \quad \ldots(4)$$

$= (\pi/2)(l/g)^{1/2}\{1 + (1/4)\sin^2\alpha/2\}$, neglecting powers of $\sin\alpha/2$ higher than 2

If T be the time of complete oscillation, *i.e.*, period, then

$$T = 4t' = 2\pi(l/g)^{1/2}\{1 + (1/4)\sin^2\alpha/2\}$$

or $\quad T = 2\pi(l/g)(1 + \alpha^2/16)$, if higher powers of α be neglected

(*b*) Take $\sin\alpha/2 = k$ in (4). Then period is $T = 4t'$, when t' is given by (4).

EXERCISE

1. If a clock be taken to the top of a mountain or a depth inside a mine, in either case, show that the clock will go slow.

2. A pendulum is carried to the top of a mountain half a mile high. How many seconds will it lose per day ? By how much its present length be shortened so that it may beat seconds at the top of the mountain. **[Purvanchal 96, 99]**

OBJECTIVE PROBLEMS

Ex. 1. *A particle with mass 'm' is tied to one end of light inextensible string of length 'l' is displaced from its vertical position of equilibrium with a velocity 'u', then (a) $T/m = (u^2 + lg - 3gh)/l$ (b) the particle will oscillate if $u^2 > 5lg$ (c) the particle will leave circular path if $5lg > u^2 > lg$ (d) the particle will make revolutions if $u^2 = 2gl$ (given T = tension at a height h ascended above the lowest paint).* **[I.A.S. (Prel.) 1993]**

Sol. Ans. (*a*) Refer Result (6) of Art. 8.3. Here $R = T$, $a = l$.

Ex. 2. *A particle projected from the lowest point with velocity u along the inside of the arc of the smooth vertical circle of radius r. It will oscillate about the lowest point if*

(a) $u^2 > 2gr$ (b) $u^2 < 2gr$ (c) $u^2 > 5gr$ (d) $u^2 = 5gr$ **[I.A.S. (Prel.) 1994]**

Sol. Ans. (*b*). Refer Case II A of Art. 8.3, page 8.7.

Ex. 3. *If a particle starts from rest at the highest point of a smooth vertical circular wire of radius 'a'. Then (1) the particle leaves the wire at a depth a/3 from the highest point. (2) The tangent at the point on the circle, where the particle leaves the wire, makes with the horizontal angle $\cos^{-1}(2/3)$. (3) The subsequent path has latus rectum (16/27)a. (4) The velocity of the particle at any height h from the lowest point is $(gh)^{1/2}$.*

Select the correct answer using the codes given below (a) 1, 2, 3 and 4 (b) 1, 2 and 3 (c) 1 and 3 (d) 1 and 4. **[I.A.S. Prel. 1995]**

Sol. Ans. (*b*). Refer Art. 8.2, page 8.1 and Ex. 3 on page 8.2.

Ex. 4. *A heavy particle of mass m is projected along the inside of a smooth vertical circle of radius a from its lowest point with velocity u. The least value of u for describing the complete circle is*

(a) $(ag)^{1/2}$ *(b)*$(2ag)^{1/2}$ *(c)* $(4ag)^{1/2}$ *(d)* $(5ag)^{1/2}$. **[Kanpur 2001, 2003]** **Ans. (*d*)**

Ex. 5. *A heavy particle of mass m is projected along the inside of a smooth vertical circle of radius a from its lowest point with velocity u. If after rising above the horizontal diameter of the circle the particle is to leave the circle, then we must have*

(a) $u < (2ag)^{1/2}$ *(b)* $u = (2ag)^{1/2}$ *(c)* $u > (5ag)^{1/2}$ *(d)* $(2ag)^{1/2} < u < (5ag)^{1/2}$. **Ans. (*d*)**

Ex. 6. *The greatest angle through which a person can oscillate on a swing the ropes of which can support twice the person's weight at rest is*

(a) 180° *(b)* 120° *(c)* 90° *(d)* 60°. **Ans. (*b*)**

Ex. 7. *How many seconds a clock would lose per day if the length of its pendulum were increased in the ratio 900 : 901 ? (a) 50 seconds (b) 48 seconds (c) 46 seconds (d) 44 seconds.*

Ans. (*d*)

Ex. 8. *A pendulum beats seconds accurately at a place whre g is 981 cm/sec². If taken to a place where g is 982 cm/sec², the seconds gained by the pendulum per day is (a) 44 (b) 42 (c) 40 (d) none of the above.* **Ans. (*a*)**

Ex. 9. Fill in the blanks "......" so that the following statements are complete and correct

(*i*) *If a particle is compelled to move along a given curve on a surface, then its motion is called a ......* **[Meerut 2004]**

(*ii*) *A seconds pendulum is that which oscillates from rest to rest in ......*

Ans. (*i*) Constrained motion (*ii*) One second

Ex. 10. Write 'T' for true and 'F' for false statement :

(*i*) *A heavy particle of mass m is projected along the inside of a smooth vertical circle of radius a from its lowest point with velocity* $(6ag)^{1/2}$. *The particle will leave the circle at some point of its path.*

(*ii*) *A heavy particle of mass m hanging vertically from a point by a light inextensible string of length l is started as so as to make a complete revolution in a vertical plane, then the sum of tensions at the ends of any diameter is equal to 2mg.* **Ans.** (*ii*) F (*ii*) F.

PART II : CONSTRAINED MOTION IN A HORIZONTAL CIRCLE

8.9 Motion in a Horizontal Circle with Uniform Speed. Consider the motion of a particle with uniform speed v along a cicle whose centre is O and radius a. Let OA be a fixed radius of the cirlce and let $P(a, \theta)$ be the position of the particle at any time t (taking O as pole and OA as initial line). Then for P, we have

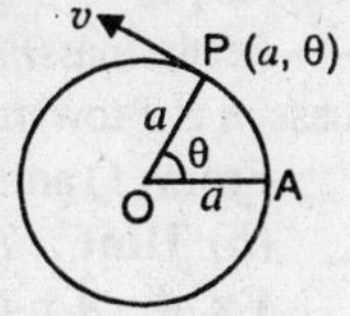

$$r = a \quad \text{and} \quad v = a\,\dot{\theta} \quad \text{or} \quad \dot{\theta} = v/a = \text{constant } \omega, \text{ say} \qquad \ldots (1)$$

$\therefore$ Radial acceleration of $P = \ddot{r} - r\dot{\theta}^2 = 0 - a\,(v/a)^2 = -(v^2/a)$, by (1) ...(2)

and transverse acceleration of $P = 2\dot{r}\dot{\theta} + r\ddot{\theta} = 0$, by (1)

Thus if a particle moves along a circle of radius a with uniform speed v, then it has acceleration v^2/a *or* $a\omega^2$ *directed along radius towards the centre.*

Hence, the external force F acts along PO and the equation of motion is $m(v^2/a) = F$. *Therefore, a particle can describe a circle of radius a, with uniform speed v, if a force* $F = mv^2/a$ *is applied on it, along the radius from it to the centre.*

Remark. The force $F\ (= mv^2/a)$ is called the *centripetal force.* If ω be the angular velocity of P, then $v = a\omega$ and so $F = ma\omega^2$. Let n be the angular frequence *i.e.* number of revolutions per second, then $\omega = 2n\pi$ and so $F = 4\pi^2 n^2\, a\, m$.

Thus, $F = mv^2/a = ma\omega^2 = 4\pi^2 n^2\ am,$ ...(3)

which is provided by the force of constraint, tension of spring (or string) acting on the end to which it is tied the particle or the reaction of a smooth wire along which the particle is moving.

SOLVED EXAMPLES

Ex. 1. *It a particle describes a circle of radius 'a' uniformly making one revolution in 'n' seconds, prove that its acceleration is directed towards the centre and is $(4\pi^2 a)/n^2$.*

[G.N.D.U. Amritsar 2010]

Sol. Proceed as in Art 8.9. Then acceleration $= a\omega^2$ ...(1)

Here number of revolution in 1 sec $= 1/n \Rightarrow \omega = (1/n).\ 2\pi$ ...(2)

Using (2), (1) given acceleration $= a\,(2\pi/n)^2 = (4\pi^2 a)/n^2$.

Ex. 2. (*a*) *A particle of mass 5gm describes a circle in a horizontal plane round a point with which it is connected by a string of 16 cm. If in one second the string makes 6 revolutions, find acceleration and tension in string.*

(*b*) *A particle of mass 50 gm in attached to a fixed point on a smooth floor by means of a string 70 cm long. If the particle is moving with a uniform speed of 12 r. p.m., find the tension in the string.*

Sol. (*a*) Refer figure of Art 8.9. Here particle of mass 5 gm is at *P* and *OP* is string, where *O* is a fixed points. Given $a = OP = 16$ cm and $\omega = 6$ r.p.s. $= 6 \times 2\pi$ radians /sec $= 12\pi$ radians/sec.

$\therefore$ Normal acceleration $= a\omega^2 = 16 \times 144\pi^2 = 2304\ \pi^2$ cm/sec^2.

If *T* be the tension in the string, then

$$T = mr\omega^2 = 5 \times 16 \times 144\pi^2 = 11520\ \pi^2 \text{ dynes}$$
$$= 11520 \times (22/7)^2 \times (1/980) \text{ gm. wt.} = 5.46 \text{ gm.wt.}$$

(*b*) **Hint :** $\omega = 12$ r.p.m $= (12 \times 2\pi)/60$ radians/sec, $T = mr\omega^2 = 5.64$ gm.wt.

Ex. 3. (*a*) *A particle of mass m is fastened to one end of a string passing through a hole in a smooth table and supports at its other end a particle of mass M. Find the number of revolutions which the particle on the table must make in a circle of radius a in order to keep the hanging mass at rest.*

(*b*) *A particle of mass m on a smooth table is fastened to one end a fine string which passes through a small hole in the table and supports at its other end a particle of mass 4 m. It the first particle be held at a distance a from the hole, find the velocity with which mass m must be projected so as to describe a circle of radius a.*

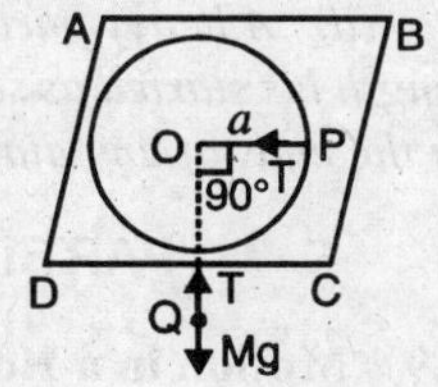

Sol. (*a*). Let *POQ* be the string which passes through the hole *O* of table *ABCD*. Masses *m* and *M* are attached at ends *P* and *Q* respectively. Since the table is smooth, the tension in the two portions *PO* and *QO* is the same, say *T*. As *M* is at rest, so $T = Mg$ ...(1)

Let the mass *m* make *n* revolutin per second. Then, $w = 2\pi n$. Since the only force on mass *m* is *T* towards *O*, so we have $T = ma\omega^2 = 4\ m\ a\ \pi^2\ n^2$ ...(2)

From (1) and (2), $4\ ma\pi^2 n^2 = Mg \Rightarrow n = (1/2\pi) \times (Mg/ma)^{1/2}$

(*b*) **Hint :** $T = mv^2/a = 4mg \Rightarrow v = 2(ag)^{1/2}$

Ex. 4. *A particle attached to a fixed point by a string one meter long describes a cirle in a horizontal plane through the point. The string can bear a tension equal to twenty times the weight of the particle. Show that the greatest possible number of revolutions per second is just over two.*

Sol. Let mass of the particle be *m* kg. So maximum tension in string $= 20 \times mg$. Here $a =$ radius of circular path = one meter. Let *n* be the greatest possible number of revolutions per second, then $\omega = 2\pi n$ radians/sec. Given condition is $m \times 1 \times (2\pi n^2) \leq 20\ gm \Rightarrow n^2 < 5\ g/\pi^2$

$\therefore \quad n^2 \leq 5 \times 9.8 \times (7/22)^2$ or $n \leq 49/22$ or $n \leq 2.2$,

showing that greatest number of revolutions made per second is just over 2.

Ex. 5. *A particle of mass m moves on a smooth table being attached to a fixed point on the table by an elastic string of modulus λ and the natural length l. Find the velocity of the particle if it is describing a cirle of radius a.*

Sol. Let T be the tension in the string and v be the velocity of particle when it is describing a circle of radius a (see Fig. of Art. 8.9). Then $T = mv^2/a$. Again by Hooke's law, $T = \lambda(a - l)/l$. Hence, we have

$$\frac{mv^2}{a} = \lambda\,\frac{a-l}{a} \Rightarrow v = \frac{\lambda\, a\,(a-l)}{ma} \Rightarrow v = \left\{\frac{\lambda a(a-l)}{ma}\right\}^{1/2}$$

8.10 Conical pendulum. Definition. **[G.N.D.U. Amritsar 2010; M.S. Unive. T.N. 2007]**

When a particle is tied to one end of a rod or a string, the other end of which is tied to a fixed point O, and the particle describes a circle in a horizontal plane, so that the string or the rod describes a cone whose axis is the vertical line through O, then the system is called a conical pendulum.

Discussion of motion of a conical pendulum.

Let OP be the string, the end O being fixed and a particle of mass m attached to the end P. Let T be the tension in the string and θ its inclination to the vertical. Let the particle describe a horizontal circle of radius a (and centre at N) with uniform angular velocity ω.

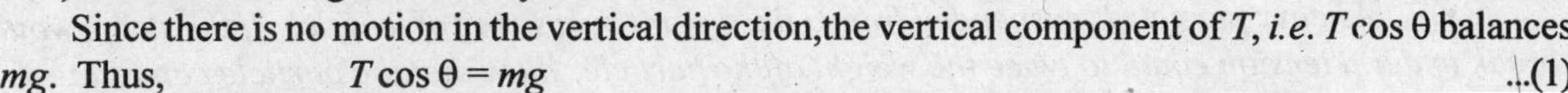

Since there is no motion in the vertical direction, the vertical component of T, *i.e.* $T\cos\theta$ balances mg. Thus, $\qquad T\cos\theta = mg$ ...(1)

The horizontal component of T, *i.e.*, $T\sin\theta$ supplies the necessory central force $ma\omega^2$ to make the particle move in a circle

$\therefore \qquad T\sin\theta = ma\omega^2 = m\,(l\sin\theta)\omega^2$, as $a = l\sin\theta$ ...(1)′

or $\qquad \sin\theta\,(T - ml\omega^2) = 0$

Now $\sin\theta \neq 0$, for if $\sin\theta = 0$ then $\theta = 0$ and the string hangs vertically. The system is no longer a conical pendulum. So the above equation reduces to

$$T = ml\omega^2 \qquad \text{...(2)}$$

Using (2), (1) reduces to $\qquad \cos\theta = g/\omega^2 l$ ...(3)

For real value of θ, $\qquad \cos\theta \le 1 \quad$ or $\quad g/l\omega^2 \le 1$.

When $g = l\omega^2$, then $\cos\theta = 1 \;\Rightarrow\; \theta = 0$ *.i.e.* the string will be hanging vertically and there will be no conical pendulum.

Thus, the particle moves in a conical pendulum only if

$$g/l\omega^2 \le 1 \quad \text{or} \quad \omega^2 \ge g/l \quad \text{or} \quad \omega \ge (g/l)^{1/2} \qquad \text{...(4)}$$

Now, h = the depth of the particle below $O = ON = l\cos\theta$. So (3) gives

$$h/l = g/\omega^2 l \quad \text{or} \quad \omega^2 = g/h, \quad \text{or} \quad \omega = (g/h)^{1/2} \qquad \text{...(5)}$$

showing that the angular velocity is independent of the length of the string and varies inversely as the square root of the depth of the particle below O.

If T' is time of one revolution of the conical pendulum then

$$T' = 2\pi/\omega = 2\pi\,(h/g)^{1/2} \qquad \text{...(6)}$$

If θ is very small, then $h = l$ approx. and $\qquad T' = 2\pi\,(l/g)^{1/2}$

If n be the number of oscillations per second, then

$$\omega = 2\pi n \quad \text{and so} \quad T = 4\,m\pi^2 n^2\, l \text{ absolute units.} \qquad \text{...(7)}$$

Cerollary. If the particle in the above case be constrained to move on a smooth horizontal table, then (1) has to be modified. If R be the normal reaction of the table on the particle, then R will act along PQ (opposite the direction of weight mg of the particle. Draw a fresh figure to indicate R along PQ).

Since then in no vertical motion of the particle, we have

$$T\cos\theta + R = mg \quad \text{or} \quad R = mg - T\cos\theta = mg - 4m\pi^2 n^2 l\cos\theta \text{ by (7)}$$

If the particle is to remain in contact with the table, then

$$R \ge O \quad \text{and so} \quad g \ge 4\,\pi^2 n^2 l\cos\theta \qquad \text{...(8)}$$

TYPE 1 : SOLVED EXAMPLES ON ART 8.10

Ex. 1. *A mass of 1 kg is suspended by a string of length 1 meter from a fixed point. The mass revolves in a horizontal circle with centre vertically below the fixed point. Find the greatest number of* *compete revolutions per second when the greatest tension the string can bear in 20 π^2 kg wt [$g = 9.8$ m/sec^2]*

Sol. Let n be the number of revolutions per seconds. Then $\omega = 2\pi n$ radians/sec. Here $T = ml\omega^2 = 1 \times 1 \times 4\pi^2 n^2$ Newton. Given that the max. tension of $T = 20\pi^2$ kg. $wt = 20\pi^2 \times 9.8$ Newton. Clearly, $4n^2\pi^2 \le 20\pi^2 \times 9.8 \Rightarrow n^2 \le 49 \Rightarrow n \le 7$, showing that the max. number of complete revolutions is 7.

Ex. 2. *A heavy particle is suspended from a fixed point O by a string of length l. If string would break under a tension equal to twice the weight of the particle, find the greatest angular velocity at which the string and particle can rotate as a conical pendulum without the string breaking.*

[Delhi Maths (H) 2011]

Sol. Proceed as in Art 8.10 upto equation (2). Then do as follows :

Since maximum tension of string is $2mg$ (given), for a possible motion as conical pendulum, we must have $\qquad ml\omega^2 \le 2mg \quad$ or $\quad \omega \le (2g/l)^{1/2}$ ...(*i*)

showing that the maximum value of ω is $(2g/l)^{1/2}$

Ex. 3. *A heavy particle is suspended from a fixed point by a string of length l. If the string would break under a tension equal to twice the weight of the particle, show that the particle can rotate as a conical pendulum, without the string breaking, if* $g/l < \omega^2 < 2g/l$

Sol. First do as in Art 8.10 upto equation (4). Thus, $\omega < (g/l)^{1/2}$ or $(g/l)^{1/2} < \omega$. Also, doing as in Ex. 2 upto relation (*i*), we have $\omega \le (2g/l)^{1/2}$ in order that the string may not break. So the required value of ω is given by

$$(g/l)^{1/2} < \omega \quad \text{and} \quad \omega < (2g/l)^{1/2} \quad \textit{i.e.} \quad (g/l)^{1/2} < \omega < (2g/l)^{1/2}.$$

Ex. 4. (*a*) *If the velocity of the bob of a conical pendulum is v and the length l, prove that the inclination of the string to the vertical is given by*

$$gl \sin^2\theta = v^2 \cos\theta.$$

(*b*) *A particle is tied to the end of a string of length 12 cm and revolves as a conical pendulum. If the string is always inclined to the vertical at an angle of 30°, find the velocity of the particle.*

Sol. Do as in Art 8.10 upto equation (1)′ and get

$$T\cos\theta = mg \qquad \text{...(i)}$$

and $\quad T\sin\theta = m a\omega^2 = mv^2/a = (mv^2)/l \sin\theta$, as $a = l\sin\theta$ ...(*ii*)

Dividing (*ii*) by (*i*), $\quad (\sin\theta)/\cos\theta = (gl\sin\theta)/mv^2$

or $\quad gl\sin^2\theta = v^2\cos\theta$ ...(*iii*)

(*b*) Proceed as in part (*a*) upto (*iii*). Here $\alpha = 30°$, $l = 12$.

Then (*iii*) $\Rightarrow$ $980 \times 12 \times \sin^2 30° = v^2 \cos 30° \quad \Rightarrow \quad v = 58.24$ cm/sec.

Ex. 5. (*a*) *A particle of mass m is attached by a light rod of length l to a fixed point at a height l/2 above a smooth horizontal table on which the particle is describing a circle with velocity v. Show that there will be pressure on the table if $v^2 < (3/2)\,gl$ and find this pressure.*

(*b*) *A particle moving in a conical pendulum is also constrained to move on a smooth horizontal table at a depth h below the fixed end of the string. Show the particle will not leave contact with the table if the number of revolutions per second is less than $(1/2\pi) \times (g/h)^{1/2}$.*

(*c*) *A particle of mass m attached to one end of a sting of length l whose other end is fixed to a point at a height h above a smooth horizontal table. The particle is made to revolve in a circle on the table so as to make n revolutions per second. Find the greatest value of n is order that the particle may remain in contact with the table.*

Sol. (*a*) Let OP be the light rod, the end O being fixed and a particle of mass m atached to the end P. Suppose that the mass m lies on a horizontal table at depth h below O and $OP = l$. Let T be tension in the rod and θ its inclination to the vertical. Let R be normal reaction (pressure) of the table on the particle. Let the particle describe a horizontal circle of radius a with uniform angular velocity ω and making n revolutions per second so that $\omega = 2\pi n$.

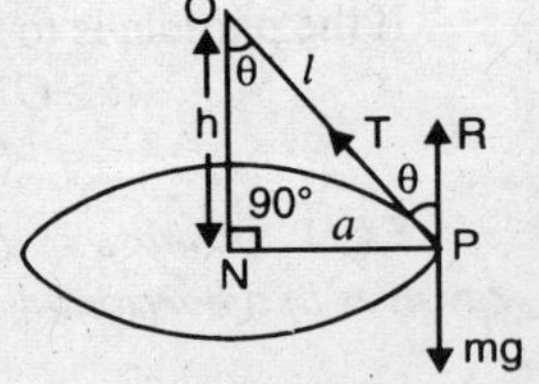

Since there in *no* motion of mass m in vertical direction, so

$T \cos\theta + R = mg$ or $R = mg - T\cos\theta$ ...(1)

For circular horizontal motion, we must have

$T \sin\theta = ma\omega^2 = mv^2/a$, as $\omega = v/a$ ...(2)

In the present problem $h = l/2$. From $\Delta\, PON$, $\cos\theta = h/l = 1/2$ and so $\theta = 60°$. Then $a = l \tan 60°$ $= (l/2)\sqrt{3}$. Then (2) reduces to

$T(\sqrt{3}/2) = mv^2 \times (2/l\sqrt{3})$ or $T = 4\,mv^2/3l$.

$\therefore$ (1) gives $R = mg - (4\,mv^2/3l)\cos 60° = (mg/3l)(3gl - 2v^2)$.

There will be pressure on the table if

$R > O \Rightarrow 3\,gl - 2v^2 > 0 \Rightarrow 2v^2 < 3gl \Rightarrow v^2 < (3gl)/2$.

(*b*) First of all do upto equation (2) as in part (*a*). Then proceed as folows :

Since $\omega = 2\pi n$ and $a = h\tan\theta$, (2) reduces to

$T\sin\theta = m\,(h\tan\theta) \times (2\pi n)^2$ or $T\cos\theta = 4\,m\,\pi^2 n^2 h$, as $\theta \neq 0$

$\therefore$ (1) becomes $R = mg - 4\,m\pi^2 n^2 h$.

The particle will not leave contact with the table, when $R \geq 0$

i.e. $mg - 4\,m\,\pi^2 n^2 h \geq 0$ or $n^2 \leq g/4\pi^2 h$ or $n \leq (g/h)^{1/2} \times (1/2\pi)$

This shows that n should not be less than $(1/2\pi) \times (g/h)^{1/2}$.

(*c*) Do as in part (*b*), then the greatest value of n in order that the particle remain in contact with table is $(1/2\pi) \times (g/h)^{1/2}$.

Ex. 6. *A horizontal bar AB of length 'a' is made to rotate with a constant angular velocity* ω *about a vertical axis through the end B. If a particle is attached to A by a string of length l, and the string makes an angle* θ *with the vertical when the motion is steady, prove that*

$\tan\theta = (\omega^2/g)(a + l\sin\theta)$ *or* $a\cot\theta + l\cos\theta = g/\omega^2$

Sol. Here AP is the string, the end A being attached to the given bar AB. Let T be the tension in the string and θ its inclination to the vertical. Then the particle describes a horizontal circle of radius PN with uniform angular velocity ω. Since there is no motion of mass m is vertical direction, so

$T\cos\theta = mg$ ...(1)

Now, form figure, $PN = PQ + QN = l\sin\theta + a$.

For horizontal circular motion of mass m, we have

$T\sin\theta = m\,(PN)\,\omega^2 = m\,(l\sin\theta + a)\,\omega^2$ ...(2)

Dividing (2) by (1), $\tan\theta = (\omega^2/g)(a + l\sin\theta)$ ...(3)

Re-writing (3), $\cot\theta\,(a + l\sin\theta) = g/\omega^2$ or $a\cot\theta + l\cos\theta = g/\omega^2$.

Ex. 7. *A concial pendulum consists of a light elastic string of natural length l and modulus* λ *attached to a bob of mass m. Show that when the pendulum rotates with an angular velocity* ω, *the inclination to the vertical is given by* $\lambda l\omega^2\cos\theta = g(\lambda - m\omega^2 l)$.

Sol. Refer figure of Art 8.10. In that figure take $l + x$ in place of l. Here x is the extension in the string when it is describing a conical pendulum.

By Hooke's law, $T = \lambda\,x/l$ so that $x = lT/\lambda$ ...(1)

Proceed as in Art. 8.10 upto equiton (2) with $l + x$ in place of l. Then we have

$T\cos\theta = mg$ or $T = (mg)/\cos\theta$ ...(2)

and $T = m\,(l + x)\,\omega^2$ ...(3)

From (1) and (2), $x = (lmg)/\lambda\cos\theta$ ...(4)

Using (2) and (4), (3) reduces to

$$\frac{mg}{\cos\theta} = m\left(l + \frac{lmg}{\lambda\cos\theta}\right)\omega^2 \quad \text{or} \quad g\lambda = \omega^2\,(l\lambda\cos\theta + lmg)$$

or $g\lambda = l\,\lambda\,\omega^2\cos\theta + lmg\omega^2$ or $\lambda l\omega^2\cos\theta = g\,(\lambda - m\omega^2\,l)$

Ex. 8. (***a***) *An elastic string has a natural length l and is streched by an amount 'a' when a mass m is suspended from it. Show that the length of the string will be equal to* $gl/(g - 4\pi^2\,na)$ *when, with* *weight attached, it makes n revolutions per second as a conical pendulum.*

(*b*) *An elastic string of naturnal length l is extended by an amount x when it supports a mass m at rest, and is extended by an amount y when it is rotating as a conical pendulum carrying a particle of the same mass. Show that* $gy = w^2 x(l + y)$. **[Delhi B.A. (Prog.) III, 2011)**

Sol. (*a*). Let $O'L$ be elastic string of natural length l supporting a mass m at rest. Due to mass m the string is extended by LM (= a, given), so by Hooke's law, tension T' in string is given by

$$T' = \lambda a/l, \text{ where } \lambda \text{ is modulas of elasticity}$$

Since mass m is at rest so $T' = mg$.

Thus, $T' = \lambda a/l = mg \Rightarrow \lambda = (mgl)/a$ ...(1)

Now the same string takes shape OP (Refer Fig. of Art 8.10) and mass m at P revolves in a conical pendulum. Let extension be equal to y. Then length of the string OP becomes $l + y$ during making horizontal circular motion.

$$\therefore \quad T = \frac{\lambda y}{l} = \frac{mgl}{a}\cdot\frac{y}{l} = \frac{mgy}{a}, \text{ by (1)} \qquad ...(2)$$

As in Art 8.10, $T = m(l+x)\omega^2 = m(l+x)(2\pi n)^2$ ...(3)

From (2) and (3), $4\pi^2 n^2 m(l+y) = (mgy)/a$

or $gy - 4n^2\pi^2 ay = 4\pi^2 n^2 al$ or $y = (4n^2\pi^2 al)/(g - 4n^2\pi^2 a)$

$$\therefore \quad \text{Length of string} = l + y = l + \frac{4n^2\pi^2 al}{g - 4\pi^2 n^2 a} = \frac{gl}{g - 4\pi^2 n^2 a}.$$

(*b*) Proced as in part (*a*). Here we have x in place of a. So as before $\lambda = (mgl)/x$. When elasic string is rotating as a conical pendulum, extension is y.

$$\therefore \quad T = \frac{\lambda y}{l} = \frac{mgl}{x}\cdot\frac{y}{l} = \frac{mgy}{x} \qquad ...(i)$$

Also, $T = m(l+y)\omega^2$ so $(mgy)/x = m(l+y)\omega^2$

Thus, $gy = \omega^2 x(l+y)$, as required.

Ex. 9. *A string of length l has its ends attached to two points A and B in the same vertical line at a distance 'a' apart. If a smooth bead P on the string rotates uniformly about B so that BP is always horizontal, prove that the angular velocity is* $l\{(2g)/a(l^2 - a^2)\}^{1/2}$. *Show also that P is nearer to B than A by* a^2/l.

Sol. Here APB is string such that $AP + PB = l$. Let $PB = r$. Then $AP = l - r$. From ΔABP, $a^2 + r^2 = (l - r^2)$

or $a^2 = l^2 - 2lr$ or $r = (l^2 - a^2)/2l$ ...(1)

Since the bead is smooth, so the tension T is the same in portions PA and PB. Since there is no vertical motion of bead, so

$T\cos\theta = mg$ or $T = (mg)/\cos\theta$ ...(2)

For the bead making circular motion of radius r, we have

$T + T\sin\theta = mr\omega^2$ or $T(1 + \sin\theta) = mr\omega^2$. ...(3)

From (2) and (3), $\dfrac{mg}{\cos\theta}(1 + \sin\theta) = mr\omega^2$ or $\dfrac{g(l-r)}{a}\left(1 + \dfrac{r}{l-r}\right) = r\omega^2$

[$\because$ from ΔBAP, $\sin\theta = r/(l - r)$ and $\cos\theta = a/(l - r)$]

$\omega^2 = (2l^2 g)/a(l^2 - a^2)$ or $\omega = l\{(2g)/a(l^2 - a^2)\}^{1/2}$

P is nearer to B than A by $PA - PB = (l - r) - r = l - 2\{(l^2 - a^2)/2l\}$, by (1)

$= a^2/l$, which is what we wished to prove.

Ex. 10. *One end of a string 2l long is fastened to a point A on a fixed smooth vertical rod; the other to a small ring P of mass m which slides on the rod; another mass m′ is fastened to the middle point Q of the string and revolves with velocity v in a horizontal circle so that the angle AQP is a right angle. Prove that* $v^2 = lg(2m + m')/m'\sqrt{2}$.

Sol. Here AQP is string of length $2l$ such that $AQ = QP = l$, (given). Since $\angle AQP = 90°$, so $\angle QAP = \angle QPA = 45° = \theta$, say. $QN = AQ \sin\theta = l \sin 45° = l/\sqrt{2}$. Let T and T' be the tensions in the strings AQ and PQ respectively.

The ring of mass m at P is at rest under action of three forces, namely R, mg and T' as shown in figure. Resolving along PA, for the equilibrium of P, we have

$$T' \cos 45° = mg \quad \text{or} \quad T' = mg\sqrt{2} \qquad \ldots(1)$$

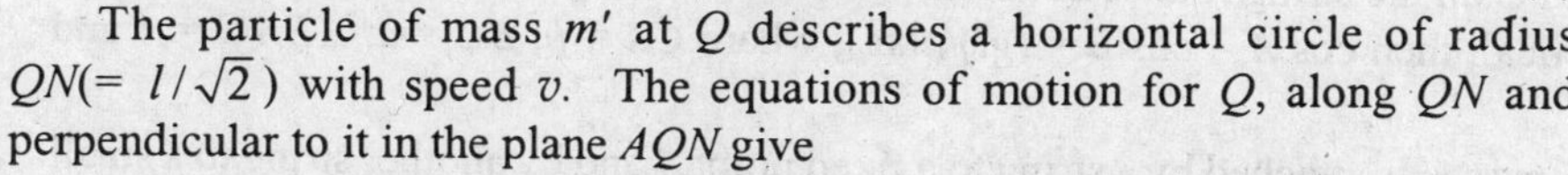

The particle of mass m' at Q describes a horizontal circle of radius $QN (= l/\sqrt{2})$ with speed v. The equations of motion for Q, along QN and perpendicular to it in the plane AQN give

$$0 = m'g + T' \cos 45° - T \cos 45° \quad \text{or} \quad T - T' = m'g\sqrt{2} \qquad \ldots(2)$$

and $\quad m'v^2/QN = T \sin 45° + T' \sin 45° \quad$ or $\quad m'v^2\sqrt{2}/l = (T + T')/\sqrt{2}$

i.e., $$T + T' = (2m'v^2)/l \qquad \ldots(3)$$

Subtracting (2) from (3), $\quad 2T' = (2m'v^2)/l - m'g\sqrt{2}$

or $\quad 2mg\sqrt{2} = (2m'v^2)/l - m'g\sqrt{2}$, using (1)

or $\quad (2m + m')gl\sqrt{2} = 2m'v^2 \quad$ or $\quad v^2 = lg(2m + m')/m'\sqrt{2}$

Ex. 11. (*a*) *A particle of mass m′ is attached to the middle of a light string of length 2l one end of which is fastened to a fixed point and the other end to a smooth ring of mass m which slides on a smooth vertical rod. If the particle describes a horizontal circle with uniform angular velocity ω about the rod, prove that the inclination of both portions of the string to the vertical is*

$$\cos^{-1}\{(m' + 2m)g/m'l\omega^2\}.$$

(*b*) *If m = m′ and particle moves with uniform angular velocity about the rod, show that the inclination of both portions of the string to the vertical is* $\cos^{-1}(3g/l\omega^2)$.

Sol. (*a*) Proceed as in Ex. 10. Here $\angle AQP \neq 90°$. So $\theta \neq 45°$. For equilibrium of P, as before

$$T' \cos\theta = mg \qquad \ldots(1)$$

The particle Q describes a horizontal circle of radius $QN (= l \sin\theta)$ with angular velocity ω. The equations of motion for Q, along QN and perpendicular to it in the plane AQN give

$$0 = m'g + T' \cos\theta - T \cos\theta \quad \text{or} \quad T - T' = (m'g)/\cos\theta \qquad \ldots(2)$$

and $\quad m'.QN.\omega^2 = T \sin\theta + T' \sin\theta \quad$ or $\quad m'l \sin\theta.\omega^2 = (T + T') \sin\theta$

i.e., $$T + T' = m'l\omega^2 \qquad \ldots(3)$$

Subtracting (2) from (3), $\quad 2T' = m'l\omega^2 - (m'g)/\cos\theta$

or $\quad 2.\dfrac{mg}{\cos\theta} = m'l\omega^2 - \dfrac{m'g}{\cos\theta} \quad$ or $\quad m'l\omega^2 = \dfrac{(m' + 2m)g}{\cos\theta}$, by (1)

or $\quad \cos\theta = (m' + 2m)g/m'l\omega^2 \quad$ or $\quad \theta = \cos^{-1}\{(m' + 2m)g/m'l\omega^2\}$

(*b*) Do exactly in the same way as part (*a*) with $m = m'$.

Then $\quad \theta = \cos^{-1}\{(3mg)/ml\omega^2\} \quad$ or $\quad \theta = \cos^{-1}(3g/l\omega^2)$.

EXERCISE

1. In a conical pendulum, l is the length of the string, m the mass of the particle, ω the angular velocity, v the linear velocity and n the number of revolutions per second made by revolving mass. Show that

(*i*) the string makes an angle $\cos^{-1}(g/l\omega^2)$ with the vertical

(*ii*) ω must be greater than $(g/l)^{1/2}$

(*iii*) $h = g/(4\pi^2 n^2)$, where h is the depth of the particle below the fixed point. Also time period of revolution $= 2\pi(h/g)^{1/2}$

(*iv*) $\tan\theta = v^2/ga$, where a is the radius of the horizontal circle described by the particle.

2. A particle of mass m is attached to elastic string of natural length 'a' and modulus of elasticity equal to the weight of the particle. Show that the least value of the angular velocity with which it can move in a conical pendulum is $(g/2a)^{1/2}$

3. A particle is tied by two equal strings each of length l to two fixed points A and B in the same vertical line, about which the particle revolves with uniform angular velocity ω. If the strings are both both taut and inclined at angle θ to the vertical, prove that the ratio of their tensions is

$$(l\omega^2 \cos\theta + g) : (l\omega^2 \cos\theta - g)$$

4. A string of length l has its ends fastened to two points A and B in the same vertical line, A being above B. A bead C on the string rotates about AB with uniform angular velocity ω. Show that if the string is not vertical, then $\cos A - \cos B = (gl)/ab\omega^2$, where $CA = b$, $BC = a$, $\angle BAC = A$ and $\angle ABC = B$.

5. A particle of mass m is attached by a string to a fixed point A and by another string to a small smooth ring of mass m' which can slide along a smooth vertical rod passing through A, both strings being tight and inextensible. If the lengths of the vertical projections of the strings are x and y when the whole system is revolving with angular velocity ω about the rod, prove that

$$m\omega^2/g = (m + m')/x + (m'/y).$$

8.11 Relative rest inside a rotating sphere

A hollow smooth sphere is rotating with uniform angular velocity ω about a vertical diameter, to find the position of relative rest of a particle placed in it.

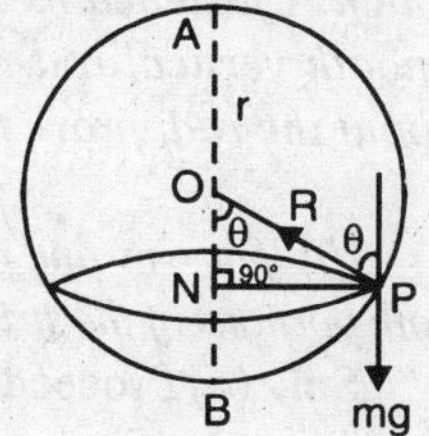

Let a hollow smooth sphere with centre O and radius r be rotating with constant angular velocity ω about its vertical diameter AB. Let P be the position of relative rest of a particle of m such that $\angle BOP = \theta$, then P describes a circle with centre N and radius $NP(= r\sin\theta)$. The forces acting on the particle are (*i*) Normal reaction R on the particle, acting along PO (*ii*) Weight mg of the particle, acting vertically downwards. Since there is no motion of the particle in vertical direction so $R\cos\theta = mg$ ...(1)

Considering the motion of the particle along PN, we get

$R\sin\theta = m.r\sin\theta.\omega^2$ or $(R - mr\omega^2)\sin\theta = 0$ ...(2)

Now, (2) $\Rightarrow$ either $\sin\theta = 0$ or $R - mr\omega^2 = 0$

If $\sin\theta = 0$, then $\theta = 0$ and hence the position of relative rest is at the lowest point B of the sphere and in this case from (1), $R = mg$.

If $\sin\theta \neq 0$, then $R = mr\omega^2$ or $(mg)/\cos\theta = mr\omega^2$, by (1)

Thus, $\cos\theta = g/r\omega^2$ or $\theta = \cos^{-1}(g/r\omega^2)$ ...(3)

Since $\cos\theta \leq 1$, so $g/r\omega^2 \leq 1$ or $\omega^2 \geq g/r$ or $\omega \geq (g/r)^{1/2}$ ...(4

Hence the particle is at relative rest at the lowest point or at a point the radius through which is inclined at an angle $\cos^{-1}(g/r\omega^2)$ for which $\omega > (g/r)^{1/2}$ [$\because$ $\omega = (g/r)^{1/2} \Rightarrow \cos\theta = 1 \Rightarrow \theta = 0$ which corresponds to the lowest point B]

TYPE 2 : SOLVED EXAMPLES

Ex. 1. *A particle moving on the inside of a smooth hollow sphere of radius r, describes a horizontal circle at a distance r/2 below the centre of the sphere. Show that its speed is $(1/2) \times (6gr)^{1/2}$.*

Sol. Refer to figure of Art. 8.11. Here, given that $OP = r$, $ON = r/2$. Hence from ΔOPN, $\cos\theta = ON/OP = 1/2$ so that $\theta = 60°$. As in Art. 8.11 proceed upto (1), which now becomes

$R\cos 60° = mg$ or $R/2 = mg$ or $R = 2mg$ ...(*i*)

Considering the motion of the particle along PN, we get

$R\sin 60° = mv^2/(r\sin 60°)$ or $R = (4mv^2)/3r$ ...(*ii*)

From (*i*) and (*ii*), $(4mv^2)/3r = 2mg$ or $v = (1/2) \times (6gr)^{1/2}$

Ex. 2. *A smooth hollow sphere of radius 3 cm rotates with uniform angular velocity about the vertical diameter. A heavy particle placed inside the sphere is observed to be at relative rest at a distance 3.5 cm below the centre of the sphere. Find the speed with which the particle moves.*

Hint. Do as in Ex. 1. Here $r = 3$ cm, Take $g = 980$, velocity $= (1/2) \times (6gr)^{1/2} = (1/2) \times (6 \times 980 \times 3)^{1/2} = 21\sqrt{10}$ cm/sec

MORE PROBLEMS ON HORIZONTAL CIRCULAR MOTION

Ex. 1. *A body in moving in a horizontal circle making n revolution per second on the inner surface of a smooth cone of semi-vertical angle* α*, with its axis vertical and vertex downwards. Show that*

(*i*) *the height of the particle above the vertex is* $(g \cot^2 \alpha)/4\pi^2 n^2$

(*ii*) *the distance of the body from the axis of the cone is* $(g \cot \alpha)/4\pi^2 n^2$

Sol. Let O be the vertex of the cone, OA the axis, P the position of the particle at any time t, $PN \perp OA$ and PM the normal to the surface of cone at P meeting OA in M. Since $\angle MPO = 90°$ hence $\angle MPN = \angle PON = \alpha$. Let m be the mass of the particle and R the normal reaction of the surface on it along PM. Since there is no motion of the particle in vertical direction, so

$$R \sin \alpha = mg \qquad \ldots(1)$$

Now, P describes a circular motion with centre N and radius NP. Considering the motion of the particle along PN, we get

$$R \cos \alpha = m(PN)\omega^2 = m \times PN \times (2\pi n)^2, \text{ as } \omega = 2\pi n \qquad \ldots(2)$$

Dividing (2) by (1), $\cot \alpha = (4\pi^2 n^2 \times PN)/g \Rightarrow PN = (g \cot \alpha)/4\pi^2 n^2 \qquad \ldots(3)$

This gives the distance of the body from the axis of the cone.

From ΔOPN, $ON/PN = \cot \alpha$ so that $ON = PN \cot \alpha$

or $\quad ON = (g \cot^2 \alpha)/4\pi^2 n^2$, using (3)

This gives the height of the body above the vertex.

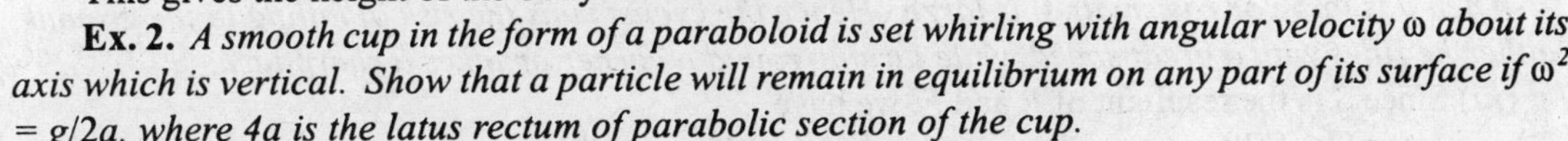

Ex. 2. *A smooth cup in the form of a paraboloid is set whirling with angular velocity* ω *about its axis which is vertical. Show that a particle will remain in equilibrium on any part of its surface if* $\omega^2 = g/2a$*, where 4a is the latus rectum of parabolic section of the cup.*

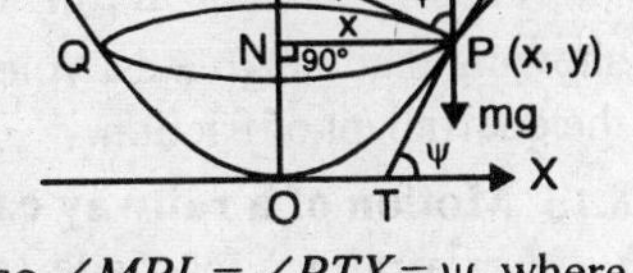

Sol. Let POQ represent the vertical cross section of the paraboloid. Let its equation be

$$x^2 = 4ay \qquad \ldots(1)$$

Clearly the given cup is generated by revolving (1) about y-axis. Let O be the vertex of (1), OY the axis, P the position of the particle of mass m at any time t, $PN \perp OY$ and PT and PM are tangent and normal at P. Let $\angle PTX = \psi$ and (x, y) be coordinates of P. Since $\angle MPT = 90°$, so $\angle MPL = \angle PTX = \psi$, where PL is parallel to OY. Let R be normal reaction on the particle.

Since there is no vertical motion of the particle, so

$$R \cos \psi = mg \qquad \ldots(2)$$

Now P describes a circle with centre N and radius x with constant angular velocity ω. Considering the motion of the particle along PN, we have

$$R \sin \psi = m\omega^2 x \qquad \ldots(3)$$

Dividing (3) by (2), $\quad \tan \psi = (\omega^2 x)/g \qquad \ldots(4)$

From (1), $\quad \tan \psi = dy/dx = x/2a \qquad \ldots(5)$

From (4) and (5), $\quad \omega^2 x/g = x/2a$ or $\omega^2 = g/2a$ or $\omega = (g/2a)^{1/2}$

which is independent of the position of P. Hence the particle will remain in equilibrium at any point of the given surface.

8.12 Motion of a bicyclist on a curved path (or nearly circular path)

To show that (*i*) *a cyclist moving in a circular path must bend towards the centre of the circle.* (*ii*) *If* θ *be the inclination of the rider and cycle to the vertical, r the radius of the circle and v the magnitude of the velocity, show that* $\tan \theta = v^2/rg$ (*iii*) *The greatest velocity with which a circle of r can be described is* $(gr\mu)^{1/2}$ *where* μ *is the coefficient of friction.*

Proof. (*i*) Suppose, if possible, the cyclist does not bend. Let rider and cycle be represented by the vertical line AGB, where G is the centre of gravity of the mass m of rider and cycle together. Then mass m is being acted by two forces : (*i*) R, the resultant reaction of the ground acting at B (*ii*) weight

mg acting vertically downwards through G. The resultant of these two forces must be mv^2/r acting horizontally through G towards the centre of the circular path, where v is the velocity and r the radius of the path.

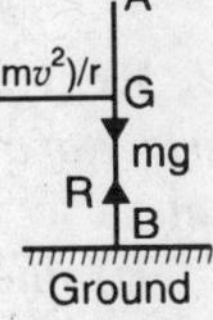

Since mv^2/r and mg both pass through G, it follows that R must also pass through G, *i.e.*, R acts vertically along BA. Now, the resultant of mg and R must act vertically and so there is no horizontal force to produce the centripetal force mv^2/r for circular motion. Hence the cyclist has to bend, so that the reaction, which has to act along BG, may become inclined to the vertical and have a component in a horizontal direction to supply the force mv^2/r.

(*ii*) Let AB represent a section of the cycle and the rider, through the centre of gravity G of the two together. The external forces acting on the system (machine) are the weight mg acting vertically downwards through G and the total reaction S of the ground which has components R and F along the vertical and horizontal directions.

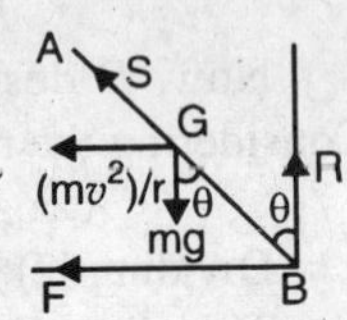

If the rider and cycle of mass m be moving with uniform speed v and r be the radius of the circular path, then there must be a force mv^2/r acting towards the centre of motion. In this case this necessary force which has to act at G is the resultant of S and mg. Since mg acts at G, so S must also pass through G. Since mv^2/r is the resultant of S and mg, we have

$$S \cos \theta = mg \quad \ldots(1)$$

and

$$S \sin \theta = mv^2/r \quad \ldots(2)$$

Dividing (2) by (1), $\tan \theta = v^2/rg$ ...(3)

When v is large or r is small, θ is large. Hence the cyclist runs the risk of falling to the ground if either he rides with a great speed while turning round a corner or takes a sharp turn.

(*iii*) Since S is the resultant of R and F, we have

$\tan \theta = F/R$ and so $v^2/rg = F/R$, using (3) ...(4)

Now, $F \le \mu R \Rightarrow F/R \le \mu \Rightarrow v^2/rg \le \mu$, by (4)

Thus, $v^2 \le \mu gr$ or $v \le (\mu gr)^{1/2}$, ...(5)

showing that the greatest velocity with which a circle of radius r can be described is $(\mu gr)^{1/2}$, μ being the coefficient of friction.

8.13 Motion of a railway carriage or a motor car on a curved level track

Let a railway carriage (or a motor car) be moving in a horizontal circular track of radius r with constant speed v. Let $ABCD$ be the section of the carriage by a plane through G, its centre of gravity and perpendicular to its track PGQ. A and B represent the points where the wheels touch the rails.

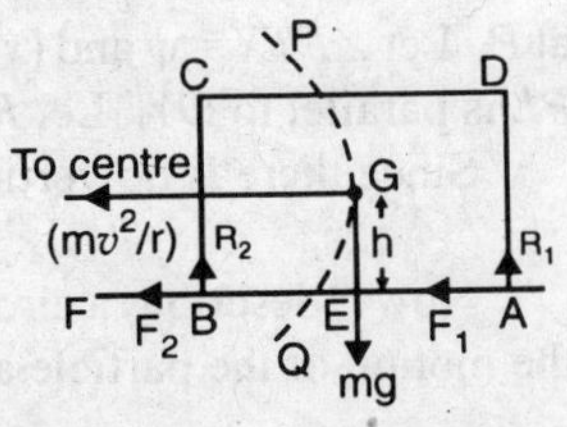

The external forces acting on the carriage are normal reactions R_1 and R_2 along AD and BC in vertical directions, weight mg acting vertically downwards through G, and the lateral thrusts (or friction forces in the case of car) F_1 and F_2 along AB and BF, both in horizontal direction.

Since G describes a horizontal circle of radius r with speed v, the centripetal force mv^2/r acting at G in horizontal direction is the resultant effect of the above five forces, namely R_1, R_2 mg, F_1 and F_2.

Resoving in the horizontal and vertical directions, we get

$$R_1 + R_2 = mg \quad \text{(statical equation)} \quad \ldots(1)$$

and

$$F_1 + F_2 = mv^2/r \quad \text{(dynamical equation)} \quad \ldots(2)$$

Let $AB = 2a$ and h be the height of G above AB. Taking moments about A and B, we get

$$(mv^2/r) \times h = mga - R_2 \times 2a \quad \text{and} \quad (mv^2/r) \times h = R_1 \times 2a - mga$$

Thus, $R_1 = (m/2) \times (g + v^2h/ar)$, $R_2 = (m/2) \times (g - v^2h/ar)$ ...(3)

Equation (2) and (3) determine the unknown quantities F_1, F_2, R_1 and R_2 in terms of known quantities v, h, r, a and m.

Corollary 1. $R_1 : R_2 = (m/2) \times (g + v^2h/ar) : (m/2) \times (g - v^2h/ar)$

or $R_1 : R_2 = (gar + v^2h) : (gar - v^2h)$

which is the ratio in which the weight of the carriage is divided between the inner and outer rails.

Corollary 2. Upsetting of the carriage. From (3), we see that R_1 is always positive and hence the outer wheel never leaves the contact of rails. Again, from (3), we observe that R_2 decreases as v increases and R_2 becomes negative if $(m/2) \times (g - v^2h/ar) < 0$ or $v > (gra/h)^{1/2}$ ...(4)

Thus when $v = (gra/h)^{1/2}$, the inner wheels no longer remains in contact with the track, the carriage begins to tilt about the outer, wheels. From the expression of v we see that v is large when a is large and h is small so that the danger of upsetting can be minimised by keeping the centre of gravity of the carriage as low as possible and the distance between the wheels as large as possible. These being fixed for any particular carriage, the danger is further minimised by keeping v small and r large *i.e.*, by reducing the velocity of the carriage and avoiding sharp turns.

Once the track is laid down and the carriage designed, the only way to avoid upsetting is to keep the speed $< (gra/h)^{1/2}$.

For upsetting of carriage, $v^2 > (gra)/h$, by (4)

or $gra < v^2h$ or $r < v^2h/ag$, ...(5)

showing that for not upsetting the carriage, $r \geq v^2h/ag$ ...(6)

(6) shows that the radius of the smallest circle which the centre of gravity can describe without the inner wheels leaving the rails is v^2h/ag.

8.14 Motion of a railway carriage (or a motor car) on a curved banked up track. We know that that when the rails are at level, the necessary centripetal force mv^2/r towards the centre of the track is supplied by lateral thrusts (or friction in case of car). Since actions and reactions are equal and opposite, the wheels also exert an equal pressure on the rails which results in the wear and tear of wheels and the rails. In order to avoid these undesirable effects, the plane of the track is banked up, *i.e.*, it is made inclined to the horizontal by raising the outer rails. Such arrangement will not produce lateral thrust.

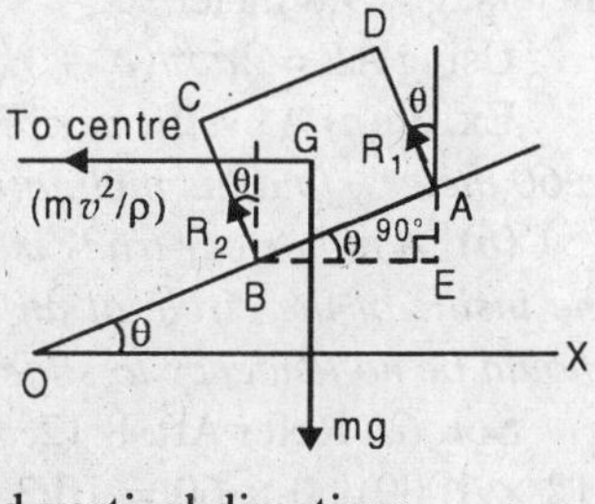

Let a railway carriage (or a motor car) be moving with uniform speed v in circular path of radius r. Let $ABCD$ be the section of the carriage by a plane through G, its centre of gravity and perpendicular to its track. Let the floor of the carriage be inclined at θ to the horizontal. Let R_1, R_2 be the normal reactions perpendicular to the floor. Let m be mass of the carriage.

In order that there may be no lateral thrust, the necessary centripetal force mv^2/r towards the centre of the horizontal path must be resultant of three external forces R_1, R_2 and mg. Resolving in the horizontal and vertical directions,

$R_1 \cos\theta + R_2 \cos\theta = mg$ (statical equation) ...(1)

and $R_1 \cos\theta + R_2 \sin\theta = mv^2/r$ (dynamical equation) ...(2)

Dividing (2) by (1), $\tan\theta = v^2/rg$, ...(3)

which gives the required inclination θ of the track.

Now, there being no rotation of the carriage in this plane, we have $0 = R_1a - R_2a$, where $AB = 2a$. Thus $R_1 = R_2$ and so (1) $\Rightarrow$ $R_1 = R_2 = (1/2)mg \sec\theta$.

With r and θ fixed, in order that there may be no lateral thrust, the speed must be $v = (gr \tan\theta)^{1/2}$.

Corollary. The height of the outer rail A above the inner rail B is AE, where BE is horizontal. Clearly $\angle ABE = \theta$. Now, from (3), $\tan\theta = v^2/rg$ so that $\sin\theta = v^2/(v^4 + r^2g^2)^{1/2}$. Let AB $(= 2a)$ be the breadth between the rails.

Hence if there is no lateral thrust on the rails, then outer rail should be raised to height $= AE = AB \sin\theta = 2av^2/(v^4 + r^2g^2)^{1/2}$.

8.15 Lateral thrust of the rails on the wheels of a carriage on a banked-up track. While banking up the railway track, the angle θ between the floor of the carriage and the horizontal is adjusted corresponding to a velocity v with which the trains are expected to travel in circular path of radius r.

Then, we know that $\tan\theta = v^2/rg$ ...(1)

If, however, the actual velocity is V, different from v, lateral thrusts appear again. Draw figure as in Art. 8.14. In the same figure, show yourself two additional forces as follows. Lateral thrusts F_1 and F_2 acting at A and B along AB and BO respectively. Let $ABCD$ be the section of the carriage by a plane through G, its centre of gravity and perpendicular to its track.

Since G describes a horizontal circle of radius r with speed V, the centripetal force acting at G is the resultant of all the forces F_1, F_2, R_1, R_2 and mg acting on the carriage. Resolving the forces along AB, we get

$(mV^2/r)\cos\theta = F_1 + F_2 + mg\sin\theta$ or $F_1 + F_2 = (mV^2/r)\cos\theta - mg\sin\theta$

or $F_1 + F_2 = m\cos\theta(V^2/r - g\tan\theta) = m\cos\theta(V^2/r - v^2/r)$, using (1)

Thus, F = total lateral thrust $= (m/r)\cos\theta(V^2 - v^2)$ ...(2)

If $V > v$, then F is positive and acts along AB and is caused by the pressure of outer rail at A while if $V < v$, then F is negative and acts along BA and is caused by the inner rail at B.

Now, $\tan\theta = v^2/rg \Rightarrow \cos\theta = (rg)/(v^4 + r^2g^2)^{1/2}$ ...(3)

From (2) and (3), $F = \dfrac{m}{r}\times\dfrac{rg}{(v^4+v^2g^2)^{1/2}}(V^2-v^2) = \dfrac{mg(V^2-v^2)}{(v^4+r^2g^2)^{1/2}}$ ...(4)

TYPE 3 : SOLVED EXAMPLES

Ex. 1. *An automobile travels round a curve of radius r. If h is the height of the centre of gravity above the ground and 2b the width between the wheels, show that it will overturn if the speed exceeds $(grb/h)^{1/2}$. (Assuming that no side-slipping takes place).* **[Delhi B.A. (Prog.) II, 2012]**

Sol. Refer Art. 8.13. Take $a = b$. Do upto equation (4).

Ex. 2. *A train is moving with a speed of 48 km per hour on a circular track of radius 820 meters; if the distance between the rails be 1.68 meters, find how much outer rail is to be raised over the inner if there is no side thrust on the rails.*

Hint. Refer corollary of Art. 8.14. Here $v = 48$ km/hour $= 40/3$ m/sec, $r = 820$ meters, $2a = 1.68$ meters, $g = 9.8$ meter/sec^2.

Using $AE = 2av^2/(v^4 + r^2g^2)^{1/2}$, finally get $AE = 0.37$ cm.

Ex. 3. (*a*) *A cyclist is riding a circular track at the rate of 12 km/hour. If the radius of the track is 60 meters, find his inclination to the vertical.*

(*b*) *A motor car track is in the form of a circle of 250 meters radius and is sloping downwards the inside of the circle at an angle $\tan^{-1}(1/5)$. At what speed must a car run along it so that there should be no tendency to slide up ?*

Sol. (*a*) Refer Art. 8.12. Then inclination θ is given by $\tan\theta = v^2/rg$, where $v = 12$ km/hour $= (12 \times 1000)/60 \times 60 = 10/3$ m/sec. $r = 60$ meters, $g = 9.8$ m/sec^2. Then $\tan\theta = 0.0189$ or $\theta = \tan^{-1}(0.0189)$

(*b*) **Hint.** Use Art. 8.12. Here $\tan\theta = 1/5$, $r = 250$. Hence $v^2/rg = \tan\theta$ gives $v^2 = rg\tan\theta = 250 \times 9.8 \times (1/5) = 490 \Rightarrow v = 7\sqrt{10}$ m/sec.

Ex. 4. (*a*) *A cyclist rides with speed v on a circular level track of radius r. Find the coefficient of friction between the tyre and the ground.*

(*b*) *A cyclist riding at 7 m/sec has to turn a corner. What is the least radius of the circle he can describe if the coefficient of friction between the tyre and the road is 1/3.*

Ans. (*a*) Refer Art. 8.12. Let m be the mass of the rider and cycle.

Since, centripetal force = force of friction, we have

$$mv^2/r = \mu mg \Rightarrow \mu = v^2/rg$$

(*b*) Let m be the mass of the rider and cycle. Give $v = 7$ m/sec and $\mu = 1/3$. Since $mv^2/r = \mu mg$, we have

$$(m \times 49)/r = (1/3) \times 9.8 \times m \Rightarrow r = 15 \text{ meters}$$

Ex. 5. *An engine of weight W is at rest on a railway curve of radius r, which is banked up to suit a speed v; show that there is an internal lateral thrust on the rail of magnitude $Wv^2/(v^4 + r^2g^2)^{1/2}$.*

Hint. Proceed as in Art. 8.15. Here $mg = W$, $v = 0$ and $V = v$. So from relation (4) of Art. 8.15, we get required lateral thrust F.

Ex. 6. *A car takes a banked corner of a racing track at a speed V, the laternal gradient α being designed to reduce the tendency to side slip to zero, for a lower speed U. Show that the coefficient of friction necessary to prevent side slip for the greater speed V must be at least* $\{(V^2 - U^2) \sin\alpha \cos\alpha\}/(V^2 \sin^2\alpha + U^2 \cos^2\alpha)$.

Sol. Let $ABCD$ be the section of the carriage by a plane through G, its centre of gravity and perpendicular to its track. Let $AB = 2a$ and m be mass of the carriage. Let V be the speed and r the radius of corner. The forces acting on the carriage are : (*i*) its weight mg acting vertically downwards through G (*ii*) the normal reaction R_1 and R_2 perpendicular to AB (*iii*) F_1 and F_2 are frictional forces acting at A and B along AB.

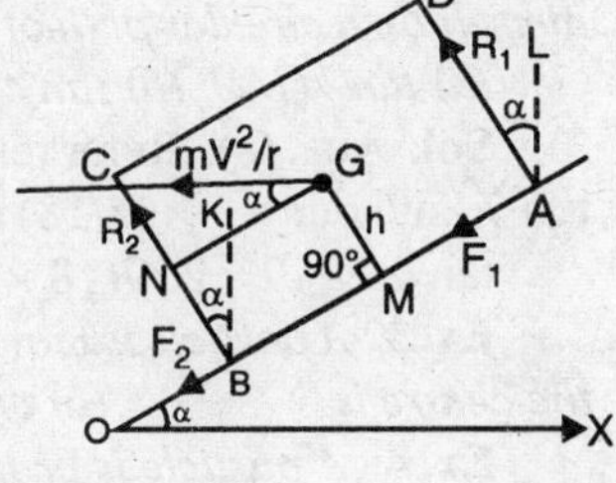

The centripetal force mV^2/r acting at G is the resultant of five forces R_1, R_2, F_1, F_2 and mg. Resolving forces along and perpendicular to AB, we get

$$(mV^2/r)\cos\alpha = F_1 + F_2 + mg \sin\alpha \quad ...(1)$$

and
$$(mV^2/r)\sin\alpha = R_1 + R_2 - mg\cos\alpha \quad ...(2)$$

Let $GM \perp AB$ and $GM = h$. As there is no rotation about G

$$\therefore \quad -R_1 a + R_2 a + F_1 h + F_2 h = 0 \quad ...(3)$$

From (1), $F_1 + F_2 = (mV^2/r)\cos\alpha - mg\sin\alpha = mg\cos\alpha\,(V^2/gr - \tan\alpha)$ ...(4)

and from (2), $R_1 + R_2 = mg\cos\alpha + (mV^2/r)\sin\alpha = mg\cos\alpha\{1 + (V^2/gr)\tan\alpha\}$ ...(5)

Now the tendency to side slip is least, when $F_1 + F_2 = 0$

i.e., when $V = U$ and $\tan\alpha = U^2/gr$ or $gr = U^2\cot\alpha$ ...(6)

Substituting the value of gr given by (6) in (4) and (5), we get

$$F_1 + F_2 = mg\cos\alpha\left(\frac{V^2}{U^2}\tan\alpha - \tan\alpha\right) = mg\sin\alpha\left(\frac{V^2}{U^2} - 1\right)$$

and
$$R_1 + R_2 = mg\cos\alpha\{1 + (V^2/U^2)\tan^2\alpha\}$$

$$\therefore \quad \frac{F_1 + F_2}{R_1 + R_2} = \tan\alpha\frac{(V^2/U^2) - 1}{1 + (V^2/U^2)\tan^2\alpha} = \frac{(V^2 - U^2)\sin\alpha\cos\alpha}{U^2\cos^2\alpha + V^2\sin^2\alpha} \quad ...(7)$$

There will be no slipping in the present case too with speed $V(> U)$ provided we have

$$F_1 \le \mu R_1 \text{ and } F_2 \le \mu R_2 \quad i.e., \quad F_1 + F_2 \le \mu(R_1 + R_2)$$

or $\mu \ge (F_1 + F_2)/(R_1 + R_2)$ or $\mu \ge \{(V^2 - U^2)\sin\alpha\cos\alpha\}/(V^2\sin^2\alpha + U^2\cos^2\alpha)$

So least coeff. of friction is $\{(V^2 - U^2)\sin\alpha\cos\alpha\}/(V^2\sin^2\alpha + U^2\cos^2\alpha)$.

EXERCISE

1. A curved railway track of radius r, is banked up at angle α, so that on a train moving with speed U, there is no lateral thrust. Show that for a train of mass m moving at speed $V(\ne U)$, there is a lateral thrust given by

$$S = (m/r)\cos\alpha\,(V^2 - U^2).$$

2. A curve on a railway line is banked up so that the lateral thrust on the inner rail due to a train moving with speed V_1 is equal to thrust on the outer rail, if the train is moving with speed $V_2(> V_1)$. Show that there will be no lateral thrust on either rail, where train is moving with speed v, given by $v^2 = (V_1^2 + V_2^2)/2$.

3. If the maximum and minimum speeds of trains on a track of radius r are v_1 and v_2 respectively, show that the outer lateral thrust of the fastest train is equal to the inner lateral thrust of the slowest one if the outer rail is banked up at an angle $\tan^{-1}\{(v_1^2 + v^2)/2gr\}$.

4. A train of weight mg is at rest on a railway curve of radius a, which is banked up to suit a speed u; show that there is an inward lateral thrust on rails of magnitude $mgu^2/(u^4 + a^2g^2)^{1/2}$. If now the engine moves with velocity v, prove that this lateral thrust is $mg(v^2 - u^2)/(u^4 + a^2g^2)^{1/2}$

MISCELLANEOUS PROBLEMS

Ex. 1. *If a particle of mass 4 gm moves in a horizontal circle under the action of a force of magnitude 400 dynes towards the centre of the circle with a speed 20 cm/sec, then the radius of the circle is* (a) 2 cm, (b) 4 cm, (c) 8 cm, (d) 10 cm.

Sol. Ans. (b) Refer Art. 8.9. Here $F = 400$ dynes, $m = 4$ gm and $v = 20$ cm/sec. If a is the radius of the circle, then
$$F = mv^2/a \Rightarrow 400 = (4 \times 20 \times 20)/a \Rightarrow a = 4 \text{ cm}.$$

Ex. 2. *A cyclist moving at uniform speed of 7 km per hour inclines by cycle to the vertical so as to keep himself on a circular path of radius 10 m. The angle of inclination to the vertical is ($g = 9.8$ m/sec²).*
(a) $\tan^{-1}(1/2)$, (b) $\tan^{-1}(25/162)$, (c) $\tan^{-1}(25/648)$, (d) $\tan^{-1}(25/324)$. **[I.A.S. Prel. 2000]**

Sol. Ans. (c). Refer result (3) of Art. 8.12. The required angle θ is given by $\tan\theta = v^2/rg$. Here $v = 7$ km/hour $= 7 \times (5/18)$ m/sec, $r = 10$ m and $g = 9.8$ m/sec². So we have
$$\tan\theta = (49 \times 25)/(18 \times 18 \times 10 \times 9.8)) = 25/648 \Rightarrow \theta = \tan^{-1}(25/648).$$

Ex. 3. *A particle is moving in a circle of radius r with uniform speed v. The acceleration towards the centre is* (a) v/r (b) v^2/r (c) vr (d) v/r^2.

Ex. 4. *A particle is projected from the lowest point with a velocity u and moves along the inside of the arc of smooth circle. In order that the particle may not leave contact with the arc of the circle.*

(a) $u^2 > 3gr$ (b) $u^2 > 5gr$ (c) $u^2 < 5gr$ (d) $u^2 \le 5gr$

Sol. Ans. (b) Refer Art. 8.3.

Ex. 5. *A smooth heavy bead moves along a wire which is bent in a circle of radius a in a vertical plane. The bead starts from rest from position where the radius makes an angle 60° with the upward vertical. What is the velocity of the bead when it reaches the lowest point (the wire is fixed in space).*

(a) $\sqrt{3ga}$ (b) $\sqrt{2ga}$ (c) $2\sqrt{ga}$ (d) $\sqrt{5ga}$

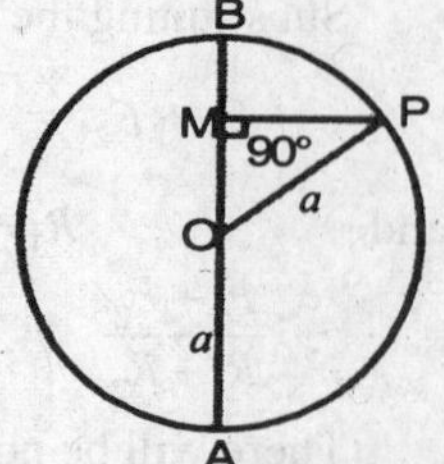

Sol. Ans. (a) Let the figure represent the given circular wire fixed in vertical plane. Let O be its centre and let AB be its vertical diameter. Let P be the initial position of the bead. Given that $\angle POB = 60°$. From figure, $AM = AO + OM = a + a\cos 60° = 3a/2$. Let V be the velocity of the bead on reaching the lowest point A of the wire. Then, by the principle of work energy (See Art. 2.9), we have

The gain in kinetic energy of the bead = workdone by external force mg
$$\Rightarrow \quad (1/2) \times m \times V^2 - (1/2) \times m \times (0)^2 = mg \times (PM) \quad \Rightarrow \quad V^2/2 = g \times (3a/2)$$
$$\Rightarrow \quad V^2 = 3ga \quad \Rightarrow \quad V = \sqrt{3ga}$$

Ex. 6. *A particle attached to a fixed peg O by a string of length l, is lifted up with the string horizontal and then let go. Prove that when the string makes an angle θ with the horizontal, the resultant acceleration is $g\,(1 + 3\sin^2\theta)^{1/2}$.*

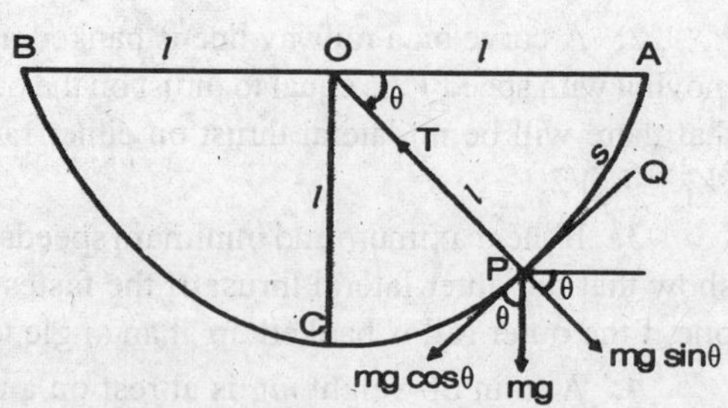

Sol. Let a particle of mass m be attached to a string of length l whose other end is attached to a fixed peg O. According to the given problem, initially the string is held in the horizontal position OA, where $OA = l$. Then, the particle starts from rest at A and moves in a circle whose centre is O and radius l. Let P be the position of the particle at any time t such that $\angle AOP = \theta$ and arc $AP = s$. The forces acting on the particle at P are : (*i*) weight mg acting vertically downwards (*ii*) the tension T in the string along OP.

Resolving forces along the tangent QP at P, we have

$m(d^2s/dt^2) = mg\cos\theta$ or $d^2s/dt^2 = g\cos\theta$ ... (1)

Since $s = l\theta$ we have $d^2s/dt^2 = l(d^2\theta/dt^2)$ ... (2)

From (1) and (2), we have $l(d^2\theta/dt^2) = g\cos\theta$ ... (3)

Multiplying both sides of (3) by $2l(d\theta/dt)$ and then integrating, we have

$$l^2(d\theta/dt)^2 = 2gl\sin\theta + c,\ c \text{ being an arbitrary constant}$$

or $v^2 = 2gl\sin\theta + c,$ since $v = \text{velocity of the particle} = l(d\theta/d\theta)$ ... (4)

But initially at the point A, when $\theta = 0$, $v = 0$. So (4) $\Rightarrow$ $c = a$

Hence from (4), we get $v^2 = 2gl\sin\theta$ $\Rightarrow$ $v^2/l = 2g\sin\theta$ ... (5)

The required resultant acceleration of the particle at P

$= \{(\text{Tangential acceleration})^2 + (\text{Normal acceleration})^2\}^{1/2}$

$= \{(d^2s/dt^2)^2 + (v^2/\rho)^2\}^2 = \{(d^2s/dt^2)^2 + (v^2/l)^2\}^2$, as $\rho = l$, for a circle

$= \{(g\cos\theta)^2 + (2g\sin\theta)^2\}^{1/2} = g(1-\sin^2\theta + 4\sin^2\theta)^{1/2} = g(1+3\sin^2\theta)^{1/2}$, using (1) and (5)

Ex. 7. *A pendulum of length 'l' makes 'n' complete oscillations in a given time, show that*

(i) If g is changed to $g + g'$, *the number of oscillations gained is* $(ng')/2g$

(ii) If length l is changed to $l + l'$, *the number of oscillations lost is* $(nl')/2l$

[Delhi B.A. (Prog) III 2008]

Ex. 8. *A particle is allowed to slide down a smooth circle of radius 27 a from rest at the highest point. Show that in leaving the circle it moves in a parabola of latusrectum 16a.*

Hint. proceed as in Ex. 3, page 8.2 **[Agra 2008]**

Ex. 9. *A heavy particle is tied to one end of a light in extensible string whose other end is attached to a fixed point. The particle completes the circle if the velocity of projection at the lowest point is at least as:*

(a) $\sqrt{6ag}$ *(b)* $\sqrt{3ag}$ *(c)* $\sqrt{5ag}$ *(d)* $\sqrt{2ag}$ **(Garhwal 2002)**

Sol. Ans. (*c*). Refer Art. 8.5, page 8.8.

Ex. 10. *In order to rise above the horizontal diameter of a vertical circle, the velocity of projection u of a heavy particle tied to a light inextensible string of length a will be related to the length of the string as follows:*

(a) $u^2 \geq 2ag$ *(b)* $u^2 > 2ag$ *(c)* $u^2 < 2ag$ *(d)* $u^2 \leq 2ag$ **(Garhwal 2003)**

Sol. Ans. (*b*). Refer Art. 8.3. page 8.7.

Ex. 11. *In order to oscillate, the velocity of projection u of a heavy particle tied to a light inextensible string of length a is:*

(a) $u^2 \leq 2ag$ *(b)* $u^2 > 2ag$ *(c)* $u^2 > 4ag$ *(d)* $u^2 = 2ag$ **(Garhwal 2004)**

Sol. Ans. (*a*). Refer Art. 8.5. page 8.8.

Ex. 12. *A particle slides down the outside of a smooth vertical circle of radius a, starting from rest at the highest point. It will leave the circle after descending vertically a distance equal to* *(a) a/3* *(b) a/4* *(c) a/2* *(d) 2a/3* **(Agra 2010)**

Sol. Ans. (*a*). Refer Art. 8.2, page 8.1

Ex. 13. *If u is initial velocity at lowest point of a circle, then velocity required to reach the highest point of the circle is*

(*a*) $u = \sqrt{2ag}$ (*b*) $u = \sqrt{3ag}$ (*c*) $u = \sqrt{4ag}$ (*d*) $u = \sqrt{5ag}$ **(Agra 2011)**

Sol. Ans. (*d*). Refer Art. 8.3, page 8.7.

9

Central Orbits

9.1 Central force and Central orbits. Definitions. **[Bundelkhand 2008; Kumaun 2002; Kanpur 2015; Agra 2007; Delhi 2007; Pune 2010; Meerut 2005, 07; Ranchi 2010]**

A force whose line of action always passes through a fixed point, is called a *central force.* The fixed point is known as the *centre of force.* The path of a particle, moving under a central force is called a *central* orbit.

9.2 Theorem. *A central orbit is always a plane curve*

[Agra 2007; Meerut 2005, 2017, 2018; Bundelkhand 2008]

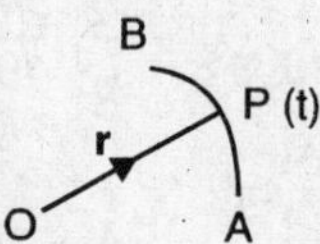

Proof. Take the centre of force O as the origin of vectors. Let $\mathbf{r}$ be the position vector of the particle P moving in a central orbit AB at any instant t. Since the particle moves under the action of a central force with centre at O, hence the only force acting on the particle at P is along OP or PO. Hence acceleration vector $d^2\mathbf{r}/dt^2$ of P must be parallel to $\mathbf{r}$ and so

$$(d^2\mathbf{r}/dt^2)\times\mathbf{r} = \mathbf{0} \quad \text{or} \quad (d^2\mathbf{r}/dt^2)\times\mathbf{r} + (d\mathbf{r}/dt)\times(d\mathbf{r}/dt) = \mathbf{0}$$

[Note that $(d\mathbf{r}/dt)\times(d\mathbf{r}/dt) = \mathbf{0}$]

or $$\frac{d}{dt}\left(\frac{d\mathbf{r}}{dt}\times\mathbf{r}\right) = \mathbf{0} \Rightarrow \frac{d\mathbf{r}}{dt}\times\mathbf{r} = \text{a constant vector} = \mathbf{h}, \text{ say} \qquad \ldots(1)$$

Taking dot product of both sides of (1) with $\mathbf{r}$, we have

$$\mathbf{r}\cdot\{(d\mathbf{r}/dt)\times\mathbf{r}\} = \mathbf{r}\cdot\mathbf{h} \quad \text{or} \quad \mathbf{r}\cdot\mathbf{h} = 0, \qquad \ldots(2)$$

using the fact that scalar triple product involving two equal vectors vanishes.

Now, (2) shows that $\mathbf{r}$ is always perpendicular to a constant vector $\mathbf{h}$. Hence OP always lies in a plane through O. Thus the path of P, a central orbit, is a plane curve.

9.3A Differential equation of a central orbit. (Polar form) **[Kanpur 2002, 2017; Bangalore 2004; Garhwal 2004; Mumbai 2010; Bhopal 2004 Kumaun 2001; Agra 2001, 2007, 2008; Rohilkhand 2002; Meerut 2004, 14; Delhi B.A. (Prog.) III 2010, 2017]**

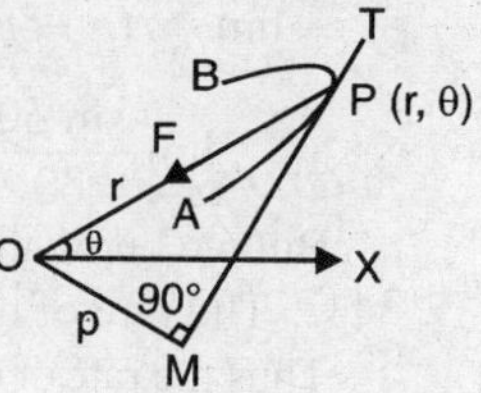

Suppose a particle moves in a plane curve AB under an acceleration F which is always directed towards a fixed point O in the plane. Take the centre of force O as the pole and a fixed line OX in the plane of the orbit as initial line. Let $P(r, \theta)$ be the position of the particle at any time t.

Since the acceleration F is always directed towards O, hence the particle at P has only radial acceleration (towards O) of magnitude F and the transverse component of the acceleration of the particle is zero. Hence the equations of motion in the radial and transverse directions are given by

$$\ddot{r} - r\dot{\theta}^2 = -F \qquad \ldots(1)$$

and $$\frac{1}{r}\frac{d}{dt}(r^2\dot{\theta}) = 0 \qquad \ldots(2)$$

Integrating (2), $$r^2\dot{\theta} = \text{constant} = h, \text{ say} \qquad \ldots(3)$$

Let $$r = 1/u \qquad \ldots(4)$$

Using (4), (3) reduces to $$d\theta/dt = h/r^2 = hu^2 \qquad \ldots(5)$$

From (4), $$\frac{dr}{dt} = -\frac{1}{u^2}\frac{du}{dt} = -\frac{1}{u^2}\frac{du}{d\theta}\frac{d\theta}{dt} = -h\frac{du}{d\theta}, \text{ by (5)} \qquad \ldots(6)$$

Using (4) (5) and (6), (1) can be re-written as

$$-h^2u^2(d^2u/d\theta^2) - (1/u)\,(u^2h)^2 = -F \quad \text{or} \quad h^2u^2(d^2u/d\theta^2) + h^2u^3 = F$$

or $\qquad d^2u/d\theta^2 + u = F/h^2u^2$ ...(7)

which is the required differential equation of a central orbit in polar form referred to the centre of force as the pole.

9.3B Differential equation of a central orbit (Pedal form) **[Mumbai 2010; Ranchi 2010 Meerut 2004,05 ; Purvanchel 2006, 09; M.S. Univ. T.N. 2007; Kanpur 2001, 03, 06; Agra 2007, 08; Avadh 2006; Bundelkhand 2011; Kumaun 2002]**

Refer figure of Art. 9.3A Proceed as in that article upto relation (7). Now proceed as follows. Let PT be tangent at P and let OM be perpendicular on PT so that $OM = p$. Then by calculus, we have

$$1/p^2 = 1/r^2 + (1/r^4)(dr/d\theta)^2 \qquad ...(8)$$

From (4), $u = 1/r$ so that $du/d\theta = -(1/r^2)(dr/d\theta)$ ...(9)

Using (4) and (9), (8) becomes $1/p^2 = u^2 + (du/d\theta)^2$ ...(10)

Differentiating (10) with respect to 'θ' we have

$$-\frac{2}{p^3}\frac{dp}{d\theta} = 2u\frac{du}{d\theta} + 2\frac{du}{d\theta}\frac{d^2u}{d\theta^2} \quad \text{or} \quad -\frac{1}{p^3}\frac{dp}{d\theta} = \frac{du}{d\theta}\left(\frac{d^2u}{d\theta} + u\right)$$

or $$-\frac{1}{p^3}\frac{dp}{d\theta} = -\frac{1}{r^2}\frac{dr}{d\theta}\left(\frac{F}{h^2u^2}\right), \text{ using (7) and (9)}$$

or $(1/p^3) \times (dp/d\theta) = (F/h^2) \times (dr/d\theta)$ or $(h^2/p^3) \times (dp/dr) = F,$ ...(11)

which is the equation of the central orbit in the pedal form.

9.4 Rate of description of the sectorial area. Areal velocity. *In every central orbit, the sectorial area traced out by the radius vector to the centre of force increases uniformly per unit of time.* **[Purvanchal 2002; Agra 2000, 01; Rohilkhand 2000; Nagpur 2010; I.F.S. 1998; Meerut 2003]**

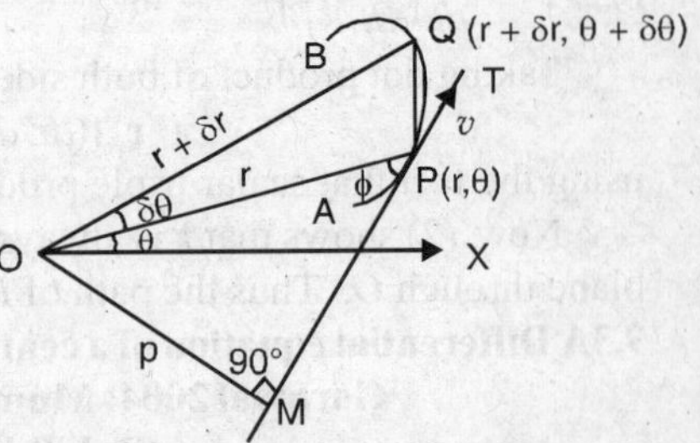

Let AB be the central orbit. Take the centre of force O as the pole and a fixed line OX in the plane of the orbit as initial line. Let $P(r, \theta)$ and $Q(r + \delta r, \theta + \delta\theta)$ be the positions of a moving point on AB at times t and $t + \delta t$ respectively. Sectorial area swept out by OP in time δt is ΔOPQ.

Hence the rate of description of the sectorial area

$$= \lim_{\delta t \to 0} \frac{\text{Sectorial area } \Delta OPQ}{\delta t} = \lim_{\delta t \to 0} \frac{(1/2) \times r(r + \delta r) \sin \delta\theta}{\delta t}$$

$$= \lim_{\delta t \to 0} \frac{1}{2} r(r + \delta r) \frac{\sin \delta\theta}{\delta\theta} \times \frac{\delta\theta}{\delta t} = \frac{1}{2} r^2 \frac{d\theta}{dt},$$

as $$\lim_{\delta t \to 0} \frac{\sin \delta\theta}{\delta\theta} = \lim_{\delta\theta \to 0} \frac{\sin \delta\theta}{\delta\theta} = 1$$

But we know that $r^2(d\theta/dt) = \text{constant} = h$ ...(1)

$\therefore$ The rate of description of sectorial area $= h/2 =$ constant.

Thus the rate of description of the sectorial area is constant and is equal to $h/2$.

Areal velocity. The rate of description of the sectorial area is known as areal velocity of the particle about the fixed point O. Using the result just proved, we find that areal velocity about the origin of a particle, moving under an attractive central force, remains constant for all time.

9.5 To show that the linear velocity varies inversely as the perpendicular from the centre upon the tangent to the path i.e., to show that $v = h/p$. **[Lucknow 2011; Mumbai 2010]**

Refer figure of Art. 9.4 Let the velocity v of the particle at P be along tangent PT and $OM = p =$ perpendicular from O on tangent PT.

Then $p = r \sin \phi$, $\quad r(d\theta/ds) = \sin \phi \quad$ and $\quad ds/dt = v$ ...(1)

Now, for a central orbit, we have

$$h = r^2 \frac{d\theta}{dt} = r\left(r\frac{d\theta}{ds}\right)\frac{ds}{dt} = r \sin \phi \times v = pv, \text{ using (1)}$$

Thus, $v = h/p$ so that $v \propto 1/p$, ...(2)
showing that the linear velocity v at P varies inversely as the perpendicular p from the centre of the orbit upon the tangent to the path.

Corollary 1. *To derive an expression for velocity* **[Mumbai 2010]**

We know that $v = h/p$ and $1/p^2 = u^2 + (du/d\theta)^2$

Hence $$v^2 = h^2\{u^2 + (du/d\theta)^2\} \qquad ...(3)$$

Corollary 2. *To show that the angular momentum about the origin (centre) of a particle, moving under an attractive central force, remains constant for all time.* **[Pune 2010]**

Proof. We know that the angular momentum of a particle at any instant t about a point O is the moment of its linear momentum at that instant about O. Now linear momentum of the particle of mass m moving in central orbit with velocity v along tangent PT is mv.

∴ Angular momentum of the particle at P(or at an instant t)
$= (mv) \times OM = mvp = m(vp) = mh$, as $pv = h$ = constant, as m and h are both constants

9.6 Energy equation

[Delhi B.A. (Prog.) III, 2012 Delhi B.Sc 2001]

Suppose a particle of mass m is moving under an attractive central force F per unit mass. Referred to the centre of force as pole, let $P(r, \theta)$ be the position of the particle at any time t. Let T and V be respectively the kinetic and potential energies of the particle per unit mass. Then by the principle of conservation of energy,
$$T + V = E \quad \text{(a constant)} \qquad ...(1)$$

Now, $T = (1/2)v^2 = (h^2/2)\{u^2 + (du/d\theta)^2\}$, by Cor. I of Art. 9.5 ...(2)

Let the attractive central force F be a function of r alone so that the system is conservative.

Then, $dV/dr = F$ or $V = \int_{r_0}^{r} F\,dr$, where r_0 is some constant. ...(3)

Using (2), (1) becomes $(h^2/2)\{u^2 + (du/d\theta)^2\} = E - V$

or $$(du/d\theta)^2 + u^2 = 2(E - V)/h^2 \qquad ...(4)$$

This is the energy equation for the orbit of a particle moving under an attractive central force, $F = dV/dr$. *E and h* are constants which can be determined with help of initial conditions.

9.7 Different forms of velocity at a point in central orbit

Form I. Velocity in a circle. The phrase "Velocit in a circle" at any point of a central orbit means the velocity necessary to describe a circle, passing through that point and with centre at the centre of force, while moving under the action of prescribed central force F per unit mass.

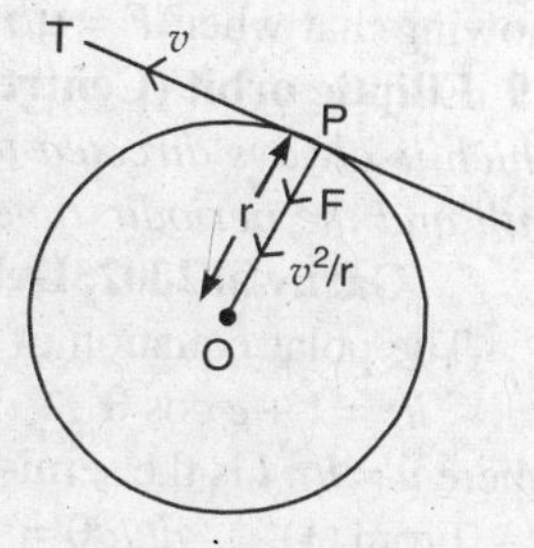

Let the centre of force O be taken as pole. Let F be central acceleration, directed towards O, at any point P of the orbit where OP = r. Let v be velocity in a circle at P. Then, v is the velocity at the point P of a particle which moves, under the same acceleration F, in a circle with centre at O. Since for a circle with centre at the pole O, the radius vector. OP is also normal to the circle at P, hence it follows that

$F = v^2/r$ or $v^2 = r$P, noting that for a circle, $\rho = r$

Thus, while moving under a central attractive acceleration F, the velocity V in a circle at a distance a from the centre of force is given by
$$V^2 = a\,[F]_{r=a} \qquad ...(1)$$

Form II. "Velocity from infinity" means the velocity that would be acquired by the particle in falling with given acceleration F from infinity to the point under consideration. Thus, if V be the velocity acquired in falling from rest at infinity to a point distant a from the centre of force O, then
$$V^2 = -2\int_{\infty}^{a} F\,dr \qquad ...(2)$$

Form III. "Velocity of fall to the point of projection" mean the velocity V that would be acquired by the particle in falling from the centre of replusion under a repulsive force $(-F)$ (per unit mass) to a point at a distance a so that
$$V^2 = 2\int_{0}^{a} (-F)\,dr \qquad ...(3)$$

Remark. Remember results (1), (2) and (3) for direct applications in problems of this chapter.

9.8 Escape velocity of a particle under a central force

Let v_0 be the speed at $r = a$ of a particle subject to a central force $F(r)$ per unit mass and let v be the velocity at a distance r from the centre of attraction. To show by energy methods or otherwise that

$$v^2 = v_0^2 - V^2 + 2\int_r^{\infty} F(r)\, dr \text{ , where } V^2 = 2\int_a^{\infty} F(r)$$

V is known as the escape velocity.

Taking $F(r) = \mu/r^2$, to show that the escape velocity is $(2\mu/a)^{1/2}$.

Proof. Let *a* particle of unit mass be moving in central orbit *AB*. Let $P(r, \theta)$ be its position at any time *t*. Then equation motion of the particle along tangent *PT* is given by

$$v\frac{dv}{ds} = -F(r)\cos\phi = -F(r)\frac{dr}{ds}, \text{ as } \cos\phi = \frac{dr}{ds}$$

or $\quad 2v\, dv = -2F(r)\, dr \;\Rightarrow\; \int_{v_0}^{v} 2v\, dv = -2\int_a^r F(r)\, dr$

or $\quad v^2 = v_0^2 - 2\int_a^r F(r)\, dr = v_0^2 - 2\left[\int_a^{\infty} F(r)\, dr - \int_r^{\infty} F(r)\, dr\right]$

Thus, $\quad v^2 = v_0^2 - V^2 + 2\int_r^{\infty} F(r)\, dr, \quad \text{...(1)}$

where $\quad V^2 = 2\int_a^{\infty} F(r)\, dr, \quad \text{...(2)}$

From (1), we find that as $r \to \infty$, $\quad v^2 = v_0^2 - V^2$,

showing that v is real only when $v_0 \geq V$. This velocity V is known as the *escape velocity* because if $v_0 \geq V$, the particle can escape to infinity.

Particular Case. When $F = \mu/r^2$, (2) reduces to

$$V^2 = 2\int_a^{\infty} \frac{\mu}{r^2}\, dr = -2\mu[1/r]_a^{\infty} = \frac{2\mu}{a},$$

showing that when $F = \mu/r^2$, the escape velocity is $(2\mu/a)^{1/2}$.

9.9 Elliptic orbit (Centre of force being the focus).

A particle moves in an ellipse under a force which is always directed towards its focus, to find the law of force, the velocity at any point of its path and the periodic time. **[Delhi Maths (H) 2006; Agra 2010, 14; Kanpur 2002, 03 Garhwal 2002; Delhi B.A. III Prog 2010; Pune 2010; Purvanchal 2003; Rohilkhand 2002]**

The polar equation of the ellipse referred to focus as pole is

$l/r = 1 + e\cos\theta \quad$ or $\quad u = (1 + e\cos\theta)/l, \quad \text{...(1)}$

where $u = l/r$, l is the semi-latus rectum and $e < 1$.

From (1), $\quad du/d\theta = -(e/l)\sin\theta \quad$ and $\quad d^2u/d\theta^2 = -(e/l)\cos\theta \quad \text{...(2)}$

We know that the differential equation of central orbit is

$$\frac{d^2u}{d\theta} + u = \frac{F}{h^2u^2} \quad \text{or} \quad -\frac{e\cos\theta}{l} + \frac{1 + e\cos\theta}{l} = \frac{F}{h^2u^2}, \text{ by (1) and (2)}$$

or $\quad \dfrac{1}{l} = \dfrac{F}{h^2u^2} \quad$ or $\quad F = \dfrac{h^2u^2}{l} = \dfrac{\mu}{r^2}$, where $\mu = \dfrac{h^2}{l} \quad \text{...(3)}$

(3) shows that the acceleration F varies inversely as the square of the distance of the particle from the focus.

Also, from $\mu = h^2/l$, we have $\quad h = (\mu l)^{1/2} \quad \text{...(4)}$

Velocity v of the particle at any point in central orbit is given by

$$v^2 = h^2\{u^2 + (du/d\theta)^2\}$$

$$\therefore \quad v^2 = \mu l\left\{\left(\frac{1 + e\cos\theta}{l}\right)^2 + \left(-\frac{e\sin\theta}{l}\right)^2\right\}, \text{ by (1), (2) and (4)}$$

$= (\mu/l)(1 + 2e\cos\theta + e^2\cos^2\theta + e^2\sin^2\theta) = (\mu/l)(1 + 2e\cos\theta + e^2)$
$= (\mu/l)\{2(1 + e\cos\theta) + (e^2 - 1)\}$

or $\quad v^2 = \mu\{2(1 + e\cos\theta)/l - (1 - e^2)/l\}$ ...(5)

If $2a$, $2b$ be axes of the ellipse, then we have

$$l = \frac{b^2}{a} = \frac{a^2(1-e^2)}{a} = a(1-e^2) \quad \text{so that} \quad \frac{1-e^2}{l} = \frac{1}{a} \qquad \text{...(6)}$$

Using (6), (3) reduces to $\quad h^2 = \mu a(1 - e^2)$ ...(6)′

Using (6), (5) becomes $\quad v^2 = \mu(2/r - 1/a)$ ...(7)

From (7), note that $\quad v^2 < 2\mu/r.$

Finally, if T be the periodic time, *i.e.*, the time taken by the particle in describing the whole arc of the ellipse, then $(1/2)h \times T = \pi ab$, where πab is the area of the ellipse and the rate of description of the arc is $h/2$ (Refer Art. 9.4)

Thus, $\quad (h/2) \times T = \pi ab \quad$ or $\quad T = (2\pi ab)/h = (2\pi ab)/(\mu l)^{1/2}$

$$\text{or} \quad T = \frac{2\pi ab}{\{\mu(b^2/a)\}^{1/2}} = \frac{2\pi a^{3/2}}{\mu^{1/2}}, \text{ as } l = \frac{b^2}{a}, \qquad \text{... (8)}$$

showing that period for one complete revolution is proportional to $a^{3/2}$, a being semi-major axis.

Note 1. For hyperbolic orbit with focus at the centre of force, proceed as above.

Note that $e > l$ and $l = \dfrac{b^2}{a} = \dfrac{a^2(e^2 - l)}{a} = a(e^2 - 1)$ so $\dfrac{e^2 - 1}{l} = \dfrac{1}{a}$. **[Agra 2010]**

As before, $\quad F = \mu/r^2 \quad$ and $\quad h = (\mu l)^{1/2}.$

Then (7) is modified as $\quad v^2 = \mu(2/r + 1/a).$

From the above relation, note that $\quad v^2 > 2\mu/r$

Note 2. For parabolic orbit with focus as the centre of force, proceed as above. Here $e = 1$. $F = \mu/r^2$ and (7) modifies as $v^2 = 2\mu/r$.

9.10 Working rule to get the law of force when the central orbit is known

Case I. If the polar equation of the orbit is given, then put $u = 1/r$ and find $d^2u/d\theta^2$ from the given equation. Put values of u and $d^2u/d\theta^2$ in the differential equation of the orbit in polar form, namely,

$$d^2u/d\theta^2 + u = F/h^2u^2 \quad \text{so that} \quad F = h^2u^2(u + d^2u/d\theta^2), \qquad \text{...(1)}$$

where F is the central acceleration assumed to be attractive. If the value of F obtained from (1) is positive, the force is attractive and if the value of F is negative, the force is repulsive.

Case II. If the pedal equation of the orbit is given, then find dp/dr from the given equation and put in the differential equation of the orbit in pedeal form, namely, $F = (h^2/p^3)(dp/dr)$ ...(2)

Then we get the value and nature of force as in case I.

SOLVED EXAMPLES BASED ON ART 9.1 TO ART 9.10

Ex. 1. (*a*) *A particle describes the equiangular spiral $r = ae^{\theta \cot \alpha}$ under a force F to the pole. Find the law of force.* **[Agra 2011; Lucknow 2003; Kanpur 2014, 15; Rohilkhand 2003, Meerut 2014, 2017; Bundelkhand 2006; Purvanchal 2002, 03, 06]**

(*b*) *A particle describes the angular spiral $r = a\, e^{m\theta}$ (a and m are constants) under a force F to the pole. Prove that F varies inversely as the cube of distance.* **[Agra 2011]**

Sol. (*a*) Given $\quad r = a\,e^{\theta\cos\theta} \quad$ or $\quad 1/u = a\,e^{\theta\cot\alpha}$, as $r = 1/u$ ...(1)

Thus, $\quad u = (1/a)\,e^{-\theta\cot\alpha}$ ...(2)

From (2), $\quad du/d\theta = -(1/a)\cot\alpha\, e^{-\theta\cot\alpha} = -u\cot\alpha$, by (2) ...(3)

From (3), $\quad d^2u/d\theta^2 = -\cot\alpha\cdot(du/d\theta) = u\cot^2\alpha$, by (3) ...(4)

The diff. eqn. of the central orbit is $\quad d^2u/d\theta^2 + u = F/h^2u^2$

or $\quad u\cot^2\alpha + u = F/h^2u^2 \quad$ or $\quad F = h^2u^3(1 + \cot^2\alpha)$, using (4)

or $\quad F = (h^2\operatorname{cosec}^2\alpha)/r^3$, as $u = 1/r$ ...(5)

From (5), we see that F is positive. Hence there is attractive force which varies inversely as the cube of the distance from the pole.

Ex. 2. (*a*) *A particle describes the curve $r^n = a^n \cos n\theta$ under a force to the pole. Find the law of force.*

[Avadh 2006; Meerut 2019; Rohilkhand 2002; Purvanchel 2008, 2010]

Sol. Given $\quad r^n = a^n \cos n\theta \quad$ or $\quad 1/a^n = u^n \cos n\theta$, as $r = 1/u$ ...(1)

Diff. (1) w.r.t. 'θ' $\quad 0 = u^n(-n \sin n\theta) + nu^{n-1}(du/d\theta) \cos n\theta$

so that $\quad du/d\theta = u \tan n\theta$ ...(2)

From (2), $\quad d^2u/d\theta^2 = u.n \sec^2 n\theta + (du/d\theta).\tan n\theta$

or $\quad d^2u/d\theta^2 = un \sec^2 n\theta + u \tan^2 n\theta$, by (2) ...(3)

The diff. eqn. of the central orbit is $\quad d^2u/d\theta + u = F/h^2u^2$

or $\quad un \sec^2 n\theta + u \tan^2 n\theta + u = F/h^2u^2$, using (3)

or $\quad u(n+1) \sec^2 n\theta = F/h^2u^2 \quad$ or $\quad F = (h^2/r^3).(n+1).(a^n/r^n)^2$, by (1)

or $\quad F = \{h^2(n+1)a^{2n}\}/r^{2n+3}$, showing that there is an attractive force which varies inversely as the $(2n+3)$th power of the distance from the pole.

Ex. 2. (*b*) *A particle describe the curve $r^2 = a^2 \cos 2\theta$ under a force directed to the pole; show that the force is proportional to $1/r^7$.* **[Kanpur 2010; Agra 2002, 03, 12; Bangalore 2004]**

Sol. Try as in Ex. 2(*a*) by taking $n = 2$.

Ex. 2. (*c*) *Find the law of force towards the pole under which the curve $r^3 = a^3 \cos 3\theta$ can be described.* **[Purvanchal 2002]**

Sol. Try as in Ex. 2(*a*) by taking $n = 3$.

Ex. 3. (*a*) *A particle describes a circle, pole on its circumference, under a force F to the pole. Find the law of force.*

(*b*) *A particle describes the curve $r = 2a \cos\theta$ under the force F to the pole. Find the law of force.* **[Bundelkhand 2008; Garhwal 2010]**

Sol. (*a*) Let a be the radius of the given circle. Taking pole on the circumference of the circle and diameter through the pole as the initial line, the equation of the circle is given by

$$r = 2a \cos\theta \quad \text{or} \quad 1/2a = u \cos\theta, \text{ where } r = 1/u \qquad ...(1)$$

Diff. (1) w.r.t. 'θ', $\quad 0 = (du/d\theta).\cos\theta + u.(-\sin\theta)$

Thus, $\quad du/d\theta = u \tan\theta$ ...(2)

From (2), $\quad d^2u/d\theta^2 = u \sec^2\theta + (du/d\theta) \tan\theta = u \sec^2\theta + u \tan^2\theta$, by (2)

Putting the value of $d^2u/d\theta^2$ in equation of central orbit, namely,

$F/h^2u^2 = u + d^2u/d\theta^2$, we get $\quad F/h^2u^2 = u + u \sec^2\theta + u \tan^2\theta$

or $\quad F = h^2u^2(u^2 \sec^2\theta + u^2 \sec^2\theta) = 2h^2u^3 \sec^2\theta = 2h^2u^3 (2au)^2$, by (1)

or $\quad F = 8a^2h^2u^5 \quad$ or $\quad F = (8a^2h^2)/r^5$, showing that there is an attractive force which varies inversely as the fifth power of the distance from the pole.

(*b*) Given curve is $r = 2a \cos\theta$. Now, do as in part (*a*).

Ex. 4. *A particle describes the curve $r^2 = a^2 \sin 2\theta$ under a force F to the pole. Find the law of force.*

Sol. Given $\quad r^2 = a^2 \sin 2\theta \quad$ or $\quad 1/a^2 = u^2 \sin 2\theta$, as $r = 1/u$ ...(1)

Diff. (1) w.r. 'θ' $\quad 0 = 2u(du/d\theta) \sin 2\theta + 2u^2 \cos 2\theta$

Thus, $\quad du/d\theta = -u \cot 2\theta$ ...(2)

From (2), $\quad d^2u/d\theta^2 = -(du/d\theta) \cot 2\theta + 2u \operatorname{cosec}^2 2\theta$

or $\quad d^2u/d\theta^2 = u \cot^2 2\theta + 2u \operatorname{cosec}^2 2\theta$, by (2) ...(3)

The diff. eqn. of the central orbit is

$$d^2u/d\theta^2 + u = F/h^2u^2$$

or $\quad u \cot^2 2\theta + 2u \operatorname{cosec}^2 2\theta + u = F/h^2u^2$, using (3)

or $\quad F = h^2u^3(\cot^2 2\theta + 2 \operatorname{cosec}^2 2\theta + 1) = 3h^2u^3 \operatorname{cosec}^2 2\theta$

or $\quad F = 3h^2u^3(a^2u^2)^2 \quad$ or $\quad F = (3h^2a^4)/r^7$, using (1)

Hence, there is an attractive force which varies inversely as the seventh power of the distance from the pole.

Ex. 5. *A particle describes the curve $r^n \cos n\theta = a^n$ under a force F to the pole. Find the law of force.* **[Avadh 2008; Kanpur 2010, 11]**

Sol. Given $r^n \cos n\theta = a^n$ or $a^n u^n = \cos n\theta$, as $r = 1/u$ ...(1)

Taking logrithms of both sides of (1), $n \log a + n \log u = \log \cos n\theta$

Diff. it w.r.t. 'θ', $(n/u)(du/d\theta) = -n \tan n\theta$ or $du/d\theta = -u \tan n\theta$ ...(2)

Diff. (2) w.r.t. 'θ', $d^2u/d\theta^2 = -(du/d\theta)\tan\theta - u.n \sec^2 n\theta$

or $d^2u/d\theta^2 = u \tan^2 n\theta - un \sec^2 n\theta$, using (2) ...(3)

The diff. eqn. of the central orbit is $d^2u/d\theta^2 + u = F/h^2u^2$

or $u \tan^2 n\theta - un \sec^2 n\theta + u = F/h^2u^2$, using (3)

or $F/h^2u^2 = u \sec^2 n\theta - un \sec^2 n\theta = u(1-n)\sec^2 n\theta = u(1-n)(1/a^n u^n)^2$, by (1)

or $F = h^2u^3(1-n)/a^{2n}u^{2n} = \{h^2(1-n)/a^{2n}\}r^{2n-3}$,

showing that there is attractive force which varies as r^{2n-3}.

Ex. 6. *A particle describes the cardiod $r = a(1 + \cos\theta)$ under a central force to the pole. Find the law of force.* **[Agra 2002; Gorakhpur 96; Rohilkhand 2001]**

Sol. Given $r = a(1 + \cos\theta)$ or $1 = 2au \cos^2 \theta/2$, as $r = 1/u$ ...(1)

Taking logarithm of both sides of (1), $0 = \log 2a + \log u + 2 \log \cos \theta/2$

Diff. it w.r.t. 'θ', $0 = (1/u)(du/d\theta) - \tan\theta/2$ or $du/d\theta = u \tan\theta/2$ ...(2)

From (2), $d^2u/d\theta^2 = (du/d\theta)\tan\theta/2 + (u/2)\sec^2\theta/2$

or $d^2u/d\theta^2 = u\tan^2\theta/2 + (u/2)\sec^2\theta/2$, using (2)

Putting this in equation of central orbit $d^2u/d\theta^2 + u = F/h^2u^2$, we get

$F = h^2u^2\{u\tan^2\theta/2 + (u/2)\sec^2\theta/2 + u\} = h^2u^3(1 + 1/2)\sec^2\theta/2$

or $F = (3/2)h^2u^3 \times (2au)$, using (1).

Thus $F = (3h^2a)/r^4$, showing that the force varies inversely as the fourth power of the distance from the pole.

Ex. 7. (a) *A particle describes the curve $au = \tanh(\theta/\sqrt{2})$ under a force F to the pole. Find the law of force.* **(Kumaun 2003; Purvanchel 2007)**

Sol. Given $au = \tanh(\theta/\sqrt{2})$, where $r = 1/u$ ...(1)

Diff. (1) w.r.t. 'θ', $a(du/d\theta) = (1/\sqrt{2})\,\text{sech}^2(\theta/\sqrt{2})$ ...(2)

From (2), $a(d^2u/d\theta^2) = -\text{sech}^2(\theta/\sqrt{2})\tanh(\theta/\sqrt{2})$ ...(3)

The diff. eqn. of the central orbit is $d^2u/d\theta^2 + u = F/h^2u^2$

or $-(1/a)\,\text{sech}^2(\theta/\sqrt{2})\tanh(\theta/\sqrt{2}) + (1/a)\tanh(\theta/\sqrt{2}) = F/h^2u^2$, by (1) & (3)

or $(1/a)\tanh(\theta/\sqrt{2})(1 - \text{sech}^2\,\theta/\sqrt{2}) = F/h^2u^2$

or $F = h^2u^2(1/a)\tanh^3(\theta/\sqrt{2}) = (h^2u^2/a)\,a^3u^3$, by (1)

or $F = (h^2a^2)/r^5$, showing that the force varies inversely as the fifth power of the distance from the pole.

Ex. 7. (b) *A particle describes the curve $r = a(1 + \cosh\theta)/(\cosh\theta - 2)$ under a force F to the pole. Show that the low of force is $F \propto 1/r^4$.* **[I.A.S. 2003]**

Or *A particle describes the curve $au = (\cosh\theta - 2)/(\cosh\theta + 1)$ under a force F to the pole. Find the law of force.*

Sol. Let $r = 1/u$.

Then the given curve can be re-written as

$1/u = a(1 + \cosh\theta)/(\cosh\theta - 2)$ or $au = (\cosh\theta - 2)/(\cosh\theta + 1)$ ...(1)

Re-writing (1), $au = \dfrac{(\cosh\theta + 1) - 3}{\cosh\theta + 1} = 1 - \dfrac{3}{\cosh\theta + 1}$ ...(2)

Differentiating both sides of (2) w.r.t. 'θ', we get

$(a/3)(du/d\theta) = \sinh\theta/(\cosh\theta + 1)^2$ ...(3)

Differentiating both sides of (3) w.r.t. 'θ', we get

$$\frac{a}{3} \times \frac{d^2u}{d\theta^2} = \frac{(\cosh\theta + 1)^2 \cosh\theta - \sinh\theta \times 2(\cosh\theta + 1)\sinh\theta}{(\cosh\theta + 1)^4}$$

$$= \frac{(\cosh\theta+1)\cosh\theta - 2\sinh^2\theta}{(\cosh\theta+1)^3} = \frac{\cosh^2\theta+\cosh\theta-2(\cosh^2\theta-1)}{(\cosh\theta+1)^3}$$

$$= \frac{2+\cosh\theta-\cosh^2\theta}{(\cosh\theta+1)^3} = \frac{(\cosh\theta+1)(2-\cosh\theta)}{(\cosh\theta+1)^3}$$

$\therefore \quad d^2u/d\theta^2 = 3(2 - \cosh\theta)/a(\cosh\theta + 1)^2$ ...(4)

The differential equation of the central orbit is $F/h^2u^2 = u + d^2u/d\theta^2$

$$\therefore \quad \frac{F}{h^2u^2} = \frac{\cosh\theta-2}{a(\cosh\theta+1)} + \frac{3(2-\cosh\theta)}{a(\cosh\theta+1)^2}, \text{ by (1) and (4)}$$

or $$\frac{aF}{h^2u^2} = \frac{(\cosh\theta-2)(\cosh\theta+1)+3(2-\cosh\theta)}{(\cosh\theta+1)^2} = \frac{\cosh^2\theta-4\cosh\theta+4}{(\cosh\theta+1)^2}$$

or $\quad aF/h^2u^2 = (\cosh\theta - 2)^2/(\cosh\theta + 1)^2 = a^2u^2$, using (1)

Thus, $\quad F = ah^2u^4 = ah^2/r^4 \quad$ or $\quad F \propto 1/r^4$.

Ex. 8. *A particle describes the curve $p^2 = ar$ under a central force to the pole. Find the law of force.* **[Garhwal 2004; Kanpur 2007, 10; Kolkata 1995; Meerut 2003]**

Sol. Given $\quad p^2 = ar \quad$ or $\quad 1/p^2 = (1/a)(1/r)$ ...(1)

Diff. (1) w.r.t. 'r' $\quad -(2/p^3)(dp/dr) = -(1/a)(1/r^2)$

or $\quad (h^2/p^3)(dp/dr) = (h^2/2a)(1/r^2)$ ...(2)

Diff. eqn. of central orbit in pedal form is $\quad F = (h^2/p^3)(dp/dr)$

or $\quad F = (h^2/2a)(1/r^2)$, using (1)

Hence the attractive force varies inversely as the square of the distance from the pole.

Ex. 9. *A particle describes the curve $pa^n = r^{n+1}$ under a central force to the pole. Find the law of force.* **(Rohilkhand 2003; Agra 2003)**

Sol. Given $\quad pa^n = r^{n+1} \quad$ or $\quad 1/p^2 = a^{2n}\, r^{-2n-2}$

Diff. w.r.t. 'r', $\quad -(2/p^3)(dp/dr) = -a^{2n}(2n+2)r^{-2n-3}$

or $\quad (h^2/p^3)(dp/dr) = \{h^2a^{2n}(n+1)\}/r^{2n+3}$ ...(1)

Then pedal equation $F = (h^2/p^3)(dp/dr)$ of central orbit gives

$F = \{h^2a^{2n}(n+1)\}/r^{2n+3}$, showing that there is attractive force which varies inversely as r^{2n+3}.

Ex. 10. *If the central force varies as the cube of the distance from a fixed point, find the orbit.* **[Gorakhpur 93; Meerut 97]**

Sol. Let the fixed point be pole. Then $F = \mu r^3$, where μ is a constant.

The diff. eqn. of a central orbit in pedal form is

$(h^2/p^3)(dp/dr) = F \quad$ or $\quad (h^2/p^3)(dp/dr) = \mu r^3$

or $\quad 4h^2p^{-3}dp = 4\mu r^3\, dr$

Integrating, $\quad -(2h^2)/p^2 = \mu r^4 - C$, where C is a constant

or $\quad (2h^2)/p^2 = C - \mu r^4$, which is equation of the orbit in pedal form.

Ex. 11. *If the central force varies as the distance from a fixed point, then show that the orbit is a conic, centre being the pole.* **[Delhi Maths. (H) 2000; Avadh 94; Kumaun 91]**

Sol. Let the fixed point be pole. Then $F = \mu r$, where μ is a constant.

The diff. eqn. of a central orbit in pedal form is

$(h^2/p^3)(dp/dr) = F \quad$ or $\quad (h^2/p^3)(dp/dr) = \mu r$

or $\quad 2h^2p^{-3}dp = 2\mu r\, dr$

Integrating, $\quad h^2/p^2 = C - \mu r^2$, where C is a constant ...(1)

Now, (1) is of the form $\quad (a^2b^2)/p^2 = a^2 + b^2 - r^2$, ...(2)

where $\quad h^2 = \mu a^2b^2 \quad$ and $\quad C = \mu(a^2 + b^2)$ ...(3)

We know that (2) is the pedal equation of an ellipse with centre as pole and a, b are given by (3).

Ex. 12. *In an orbit described under a force to a centre, the velocity at any point is inversely proportional to the distance of the point from the centre of force. Show that the path in an equiangular spiral.*

Sol. Given $v \propto 1/r$ *i.e.,* $v = k/r$, ...(1)
where v is the velocity of the particle at point P at distance r from the centre of force (*i.e.,* pole) and k is a constant.

For central orbit, $v = h/p$, ...(2)
where p is the length of the perpendicular from pole to the tangent at point P of the orbit. From (1) and (2), $k/r = h/p$ or $p = ar$, where $a = h/k$ = constant

$p = ar$ is the pedal equation of equiangular spiral. Hence the result.

Ex. 13. *A particle describes an ellipse under a force μ/r^2 and has a velocity V at a distnce R from the centre of force. Show that the time period is $(2\pi/\mu^{1/2})\,(2/R - V^2/\mu)^{-3/2}$.*

[Gorakhpur 99, 96; Purvanchal 92]

Sol. Refer Art. 9.9. Then the velocity of the particle, describing an elliptic orbit, at a distance r from the centre of force is given by

$$v^2 = \mu(2/r - 1/a) \qquad \text{...(1)}$$

Here $v = V$ where $r = R$. Then (1) gives $V^2 = \mu(2/R - 1/a)$

or $V^2/\mu = 2/R - 1/a$ or $1/a = 2/R - V^2/\mu$ ...(2)

The required time period T is given by

$$T = (2\pi/\mu^{1/2})a^{3/2} = (2\pi/\mu^{1/2})\,(2/R - V^2/\mu)^{-3/2}, \text{ by (2)}$$

Ex. 14. *A particle of mass m describes elliptical orbit of semi-major axis a under a force $m\mu/r^2$ directed to a focus. Prove that the time average of reciprocal distance is $(1/T)\int(1/r)dt = 1/a$ and deduce that the time average of the square of the speed is $(1/T)\int v^2 dt = \mu/a$. The integrals are evaluated for a complete revolution.* **[Delhi Maths (H) 2002]**

Sol. The equation of the ellipse with focus at pole is

$$l/r = 1 + e\cos\theta, \text{ where } e < 1 \qquad \text{...(1)}$$

and l = semi-latus rectum of ellipse $= b^2/a = a(1 - e^2)$ ...(2)

Now, $$\int \frac{dt}{r} = \int r \frac{1}{r^2(d\theta/dt)}\,d\theta = \frac{1}{h}\int r\,d\theta, \text{ as } r^2\frac{d\theta}{dt} = h = \text{a constant}$$

Since the integeral is evaluated for a complete revolution, so

$$\int \frac{dt}{r} = \frac{1}{h}\int_0^{2\pi} \frac{l}{1 + e\cos\theta}\,d\theta = \frac{2l}{h}\int_0^{\pi} \frac{d\theta}{1 + e\cos\theta}, \text{ as } \cos(2\pi - \theta) = \cos\theta$$

$$= \frac{2l}{h}\int_0^{\pi} \frac{d\theta}{1 + e(1 - \tan^2\theta/2)/(1 + \tan^2\theta/2)}$$

$$= \frac{2l}{h}\int_0^{\pi} \frac{\sec^2\theta/2\,d\theta}{1 + \tan^2\theta/2 + e(1 - \tan^2\theta/2)} = \frac{4l}{h}\int_0^{\infty} \frac{du}{(1+e) + (1-e)u^2}$$

[Putting $\tan\theta/2 = u$ and $(\sec^2\theta/2)\,d\theta = 2du$]

$$= \frac{4l}{h}\frac{1}{(1+e)^{1/2}(1-e)^{1/2}}\left[\tan^{-1}\frac{(1-e)^{1/2}u}{(1+e)^{1/2}}\right]_0^{\infty} = \frac{4l}{h(1-e^2)^{1/2}}\frac{\pi}{2} \qquad \text{...(3)}$$

Also, $T = (2\pi\,ab)/h$ = time period for an ellipse ...(4)

$$\therefore \quad \frac{1}{T}\int \frac{dt}{r} = \frac{h}{2\pi ab}\cdot\frac{2\pi l}{h(1-e^2)^{1/2}} = \frac{a(1-e^2)}{ab(1-e^2)^{1/2}} = \frac{1}{a}, \text{ by (3) \& (4)} \qquad \text{...(5)}$$

[$\because b^2 = a^2(1 - e^2) \Rightarrow b = a(1 - e^2)^{1/2}$]

For ellipse orbit, $v^2 = \mu(2/r - (1/a))$, by Art. 9.9

$$\therefore \quad \frac{1}{T}\int v^2 dt = \frac{1}{T}\int \mu\left(\frac{2}{r} - \frac{1}{a}\right)dt = 2\mu\left(\frac{1}{T}\int\frac{dt}{r}\right) - \frac{\mu}{a}\left(\frac{1}{T}\int dt\right)$$

$= 2\mu \times (1/a) - (\mu/a) \times (1/T) \times T = (2\mu/a) - (\mu/a) = \mu/a$,

using (5) and fact that integral is evaluated for a complete revolution.

Ex. 15. *(i) If $F = \mu/r^5 = \mu u^5$, find the speed v with which the particle can describe the circle $r = a$. (ii) If the particle moves under this attraction $F = \mu u^5$ with the same areal constant as in the circular path, and $dr/dt = -(3v/4\sqrt{2})$ when $r = 2a$, $\theta = 0$, find the equation of the spiral path of the particle and show that as $\theta \to \infty$, the path is arymptotic to the circle $r = a$.*

Sol. *(i)* If v be the velocity of the particle in circular motion of radius a, then equation of motion along the inward normal is

$$v^2/a = F \quad \text{or} \quad v^2/a = \mu/a^5, \quad \text{as} \quad F = \mu/r^5 = \mu/a^5 \text{ where } r = a \qquad \ldots(1)$$

Then, $$v = \mu^{1/2}/a^2 \qquad \ldots(2)$$

(ii) For the central orbit, $h = pv \Rightarrow h = a\mu^{1/2}/a^2 = \mu^{1/2}/a$

The diff. eqn. of the central orbit is $\quad d^2u/d\theta^2 + u = F/h^2u^2$

or $$d^2u/d\theta^2 + u = (\mu u^5)/(\mu a^{-2}u^2), \text{ using (1)}$$

or $$d^2u/d\theta^2 + u = a^2u^3 \qquad \ldots(3)$$

Multiplying both sides of (3) by $2(du/d\theta)$ and integrating, we get

$$(du/d\theta)^2 + u^2 = (1/2)a^2u^4 + C, \text{ where } C \text{ is a constant} \qquad \ldots(4)$$

Now, $$\frac{du}{d\theta} = \frac{du}{dt}\frac{dt}{d\theta} = \frac{d}{dt}\left(\frac{1}{r}\right)\frac{dt}{d\theta} = -\frac{1}{r^2}\frac{dr}{dt}\frac{1}{\dot\theta} = -\frac{\dot r}{h}, \text{ as } h = r^2\dot\theta \qquad \ldots(5)$$

Given that when $r = 2a$ *i.e.*, $u = 1/2a$, and $\theta = 0$ then $\dot r = -3v/4\sqrt{2}$

So (5) becomes $$\frac{du}{d\theta} = -\frac{1}{h}\left(-\frac{3v}{4\sqrt{2}}\right) = \frac{3}{4h\sqrt{2}} \times \frac{\mu^{1/2}}{a^2}, \text{ by (1)}$$

or $$du/d\theta = (3ha)/(4ha^2\sqrt{2}), \text{ using (2)}$$

Thus, when $u = 1/2a$, $du/d\theta = 3/4a\sqrt{2} > 0$. Hence (4) gives

$$9/32a^2 + 1/4a^2 = 1/32a^2 + C \quad \text{so that} \quad C = 1/2a^2$$

$\therefore$ (4) becomes $(du/d\theta)^2 + u^2 = (a^2u^4)/2 + 1/2a^2 = (a^4u^4 + 1)/2a^2$

or $$\left(\frac{du}{d\theta}\right)^2 = \frac{a^4u^4 + 1}{2a^2} - u^2 = \frac{(a^2u^2 - 1)^2}{2a^2} \quad \text{or} \quad \frac{du}{d\theta} = -\frac{a^2u^2 - 1}{a\sqrt{2}}, \qquad \ldots(6)$$

where we have taken negative sign in order that the initial condition namely $u = 1/2a$, $du/d\theta = 3/4a\sqrt{2} > 0$ may be satisfied.

From (6), $$d\theta = -\frac{a\sqrt{2}\,du}{a^2u^2 - 1} = \frac{a}{\sqrt{2}}\left[\frac{1}{1+au} + \frac{1}{1-au}\right]$$

Integrating, $$\theta + A = (a/\sqrt{2})(1/a)\{\log(1 + au) - \log(1 - au)\}$$

or $$\theta + A = (1/\sqrt{2})\log\{(1 + au)/(1 - au)\} \qquad \ldots(7)$$

Initially, when $\theta = 0$, $u = 1/2a$. Then (7) reduces to

$A = (1/\sqrt{2})\log\{(1 + 1/2)/(1 - 1/2)\} \quad$ or $\quad A = (1/\sqrt{2})\log 3$

(7) gives $$\theta = \frac{1}{\sqrt{2}}\log\frac{1+au}{3(1-au)} = \frac{1}{\sqrt{2}}\log\frac{r+a}{3(r-a)},$$

or $$(r + a)/3(r - a) = e^{\theta\sqrt{2}} \quad \text{or} \quad (r - a)/(r + a) = (1/3)e^{-\theta\sqrt{2}} \qquad \ldots(8)$$

which is the required spiral path of the particle.

Now as $\theta \to \infty$, (8) shows that $(r - a) \to 0$ and hence the path is asymptotic to the circle $r = a$.

Ex. 16. *The velocity at any point of a central orbit is (1/n)th of what it would be for a circular orbit at the same distance. Show that central force varies as $1/r^{2n^2+1}$ and that the equation of the orbit is $r^{n^2-1} = a^{n^2-1}\cos(n^2 - 1)\theta$.* **[Agra 2006; Meerut 1998]**

Sol. Under the same central force F, let v and V be the velocities at a distance r from the centre of force in central orbit and circular orbit respectively.

Then, we know that $\quad V^2/r = F \quad$ and $\quad v^2 = h^2\{(du/d\theta)^2 + u^2\} \qquad \ldots(1)$

Also given that $\quad v = V/n \Rightarrow v^2 = V^2/n^2 = (Fr)/n^2 = F/n^2u \qquad \ldots(2)$

From (1) and (2), $$F/u = n^2h^2\{(du/d\theta)^2 + u^2\} \quad \ldots(3)$$

Differentiative both sides of (3) w.r.t. 'θ', we get

$$\frac{1}{u}\frac{dF}{d\theta} - \frac{F}{u^2}\frac{du}{d\theta} = h^2n^2\left[2\frac{du}{d\theta}\frac{d^2u}{d\theta^2} + 2u\frac{du}{d\theta}\right]$$

or $$\frac{1}{u}\frac{dF}{du}\frac{du}{d\theta} - \frac{F}{u^2}\frac{du}{d\theta} = 2h^2n^2\frac{du}{d\theta}\left(\frac{d^2u}{d\theta^2} + u\right)$$

or $$\frac{dF}{du} = \frac{F}{u} + 2h^2n^2u\left(\frac{d^2u}{d\theta^2} + u\right) = \frac{F}{u} + 2h^2n^2u\left(\frac{F}{h^2u^2}\right)$$

[Since for a central orbit, $d^2u/d\theta^2 + u = F/h^2u^2$]

or $$\frac{dF}{du} = \frac{(2n^2+1)F}{u} \qquad \text{or} \qquad \frac{dF}{F} = (2n^2+1)\frac{du}{u}$$

Integrating, $\log F - \log C = (2n^2 + 1) \log u$, C is a constant

or $$F = Cu^{2n^2+1} = C/r^{2n^2+1} \Rightarrow F \propto 1/r^{2n^2+1}$$

To find orbit. Putting $F = Cu^{2n^2+1}$ in (3), we get

$$Cu^{2n^2} = n^2h^2\left[\left(\frac{du}{d\theta}\right)^2 + u^2\right] \quad \text{or} \quad \left(\frac{du}{d\theta}\right)^2 = u^2\left[\frac{Cu^{2n^2-2}}{h^2n^2} - 1\right] \quad \ldots(4)$$

Assume that $C = h^2n^2a^{2n^2-2}$. Then (4) becomes

$$\left(\frac{du}{d\theta}\right)^2 = u^2(a^{2n^2-2}u^{2n^2-2} - 1) \quad \text{or} \quad \frac{du}{u\left\{\left(a^{n^2-1}u^{n^2-1}\right)^2 - 1\right\}^{1/2}} = d\theta \quad \ldots(5)$$

Put $a^{n^2-1}u^{n^2-1} = \sec z$ so that $(n^2 - 1)\, a^{n^2-1}u^{n^2-2}\,du = \sec z \tan dz$.

Thus (5) gives $$\frac{\sec z \tan z\, dz}{u(n^2-1)a^{n^2-1}u^{n^2-2}(\sec^2 z - 1)^{1/2}} = d\theta$$

or $$\frac{\sec z \tan z\, dz}{\sec z \tan z} = (n^2 - 1)d\theta \qquad \text{or} \qquad dz = (n^2 - 1)d\theta$$

Integrating, $z = (n^2 - 1)\theta + A$, where A is constant ...(6)

Let z and θ be so chosen that $A = 0$. Then (6) becomes

$$z = (n^2 - 1)\theta \quad \text{or} \quad \sec z = \sec(n^2 - 1)\theta \quad \text{or} \quad a^{n^2-1}u^{n^2-1} = \sec(n^2-1)\theta$$

or $$a^{n^2-1} = r^{n^2-1}\sec(n^2-1)\theta \quad \text{or} \quad r^{n^2-1} = a^{n^2-1}\cos(n^2-1)\theta$$

Ex. 17. *Show that only law for a central attraction, for which the velocity in a circle at any distance is equal to the velocity in falling from infinity to the distance, is that of inverse square cube.*

Sol. If v_1 be the velocity for circular orbit of radius r, F the central attraction, then we know that

$$v_1^2 = Fr \quad \ldots(1)$$

Again, if v_2 be the velocity acquired in falling from infinity to the distance r from centre of force (pole), then

$$v_2^2 = -2\int_{\infty}^{r} F\, dr \quad \ldots(2)$$

But by the given problem, $v_1 = v_2$ or $$Fr = -2\int_{\infty}^{r} F\, dr \quad \ldots(3)$$

Diff. (3) w.r.t. 'r', $F + r(dF/dr) = -2F$ or $(1/F)dF = -(3/r)dr$

Integrating, $\log F - \log \mu = -3 \log r$, μ being a constant

or $\log (F/\mu) = \log r^{-3} = \log (1/r^3)$ or $F = \mu/r^3$

EXERCISE

1. (*a*) A particle describes the curve $r^n = A \cos n\theta + B \sin n\theta$ under a force F to the pole. Find the law of force. **[Kanpur 2016, 2019]**

(*b*) A particle describes the curve $r = a \sin n\theta$, under a force F to the pole. Find the law of force. **[Meerut 2015; Bundelkhand 2009, 10; Rohilkhand 2006; Kanpur 2017; Bhopal 2004]**

(*c*) Find the law of force F towards the pole under which the curve $r = a \cosh n\theta$ is described.

[Ans. (*a*) $F \alpha\, 1/r^{2n-3}$ (*b*) $F \propto 2n^2a^2r^{-5} - (n^2-1)r^{-3}$ (*c*) $F \alpha\, 1/r^3$**]**

2. Find the law of force F towards the pole under which following curves are describes : (*i*) $r^2 = 2ap$ (*ii*) $b^2/p^2 = (2a/r) - 1$ (*iii*) $(a^2b^2)/p^2 = a^2 + b^2 - r^2$ **[Rohilkhand 2008]**

[Ans. (*i*) $F \alpha\, 1/r^5$ (*ii*) $F \propto 1/r^2$ (*iii*) $F \alpha\, r$**]**

3. If particle describes the curve $au = (\cosh\theta - 2)/(\cosh\theta + 1)$ under a force F to the pole. Find the law of force. **[Ans.** $F \alpha\, 1/r^5$**]**

4. A particle describes a curve whose equation is $r^2 = ap$, under a force to the pole. Find the law of force. **[Agra 2007]** **[Ans.** $F \alpha\, 1/r^5$**]**

5. A particle (of unit mass) describes an equiangular spiral of an angle α and a force which is always in the direction perpendicular to straight line joining particle to pole of the spiral, show that the force is $\mu r^{2\sec^2\alpha - 3}$ and that the rate of description of sectorial area about the pole is $(1/2)\,(\mu \sin\alpha\cos\alpha)^{1/2}\ r^{\sec^2\alpha}$ **[Kanpur 2003]**

6. (*i*) A particle is describing an ellipse of eccentricity e about a centre of force at a focus. Prove with usual notations, $v^2 = \mu(2/r - 1/a)$ and $h^2 = \mu a(1 - e^2)$.

(*ii*) When the particle is at one end of a minor axis, its velocity is doubled. Prove that the new path is a hyperbola of eccentricity $(9 - 8e^2)^{1/2}$.

9.11 Apse, apse line, apsidal distance and apsidal angle **[Agra 2014; Kanpur 2014]**

Apse. An apse is a point in a central orbit at which the radius vector drawn from the centre of force to the moving particle has a maximum or minimum value. **[Ranchi 2010; Meerut 2005]**

Apse line. The line through the centre of force and an apse is called an apse line.

Apsidal distance. The length of radius vector to an apse is called an apsidal distance. **[Lucknow 2010]**

Apsidal angle. The angle between the radii vectors to two consecutive apses is called an apisdal angle.

Perihelion and aphelion points. The point of nearest approach to the centre of force is known as *perihelion* and the point of farthest approach from centre, if finite is known as *aphclion.*

Theorem I. *At an apse the radius vector is perpendicular to the tangent i.e., at an apse the particle moves at right angles to the radius vector.* **[Agra 2014]**

Proof. By definition of an apse, r is maximum or minimum and hence $u = 1/r$ is minimum or maximum at an apse and so $du/d\theta = 0$.

If p be the perpendicular from the centre of force upon the tangent to the path at any point whose distance is r from pole, then

$$1/p^2 = u^2 + (du/d\theta)^2 \qquad \ldots(1)$$

At an apse $du/d\theta = 0$,so (1) reduces to $1/p^2 = u^2 = 1/r^2$ so that $p = r$, showing that perpendicular distance p in the case of the apse is equal to the radius vector. So at an apse particle is moving at right angles to the radius vector.

Theorem II. *When the central acceleration is a single valued function of the distance (i.e., when the acceleration is a function of the distance only and is always the same at the same distance), every apse line divides the orbit into two equal and similar portions and thus there can only be two apsidal distances.*

Proof. Let the central acceleration F be a single valued function of r. Equation of central orbit is

$$(d^2u/d\theta^2) + u = F/h^2u^2 \quad \text{or} \quad h^2(d^2u/d\theta^2 + u) = F/u^2 \qquad \ldots(1)$$

Multiplying both sides of (1) by $2(du/d\theta)$ and then integrating, we get

$$v^2 = h^2\{(du/d\theta)^2 + u^2\} = 2\int \frac{F}{u^2}\,du + c \text{ where } c \text{ is an arbitrary constant}$$

or $$v^2 = c - 2\int F\,dr, \text{ as } \frac{1}{u} = r \Rightarrow -\frac{du}{u^2} = dr \qquad ...(2)$$

Let O be the centre of force and ABC be a part of the central orbit having three consecutive cusps A, B and C. Let V be the velocity of the particle at B. Then, if the velocity of the particle were reversed at B, it would describe the path BPA. For, as the acceleration $F(r)$ depends on the distance from O only, the velocity given by (2) would depend only the distance from O and not on the direction of motion.

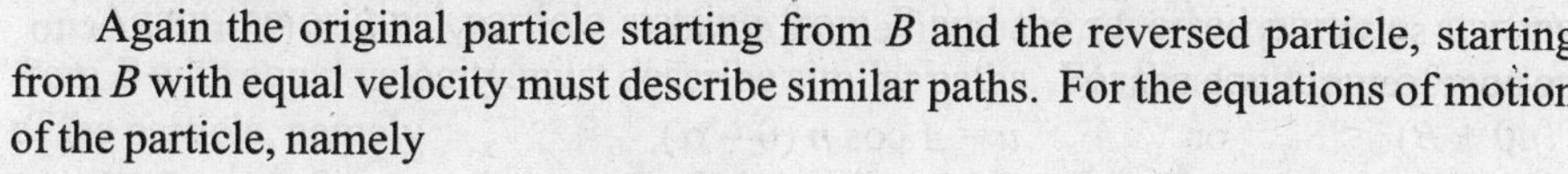

Again the original particle starting from B and the reversed particle, starting from B with equal velocity must describe similar paths. For the equations of motion of the particle, namely

$$\ddot{r} - r\dot{\theta}^2 = -F \qquad \text{and} \qquad r^2\dot{\theta} = \text{constant} = h,$$

which do not depend on the direction of motion, show that the values of r and θ at any time for the first particle (*i.e.*, OP' and $\angle BOP'$) are equal to the same quantities at the same time t for the second particle (*i.e.*, OP and $\angle BOP$).

It follows that the curves $BP'C$ and BPA are exactly the same and any one can be obtained by rotating the other about the line OB. Since A and C are the points where the radius vector is perpendicular to the tangent, hence $OA = OC$. Similarly, if D were the next apse after C, we would have got $OB = OD$ and so on. Thus we find that there are only two different apsidal distances.

Theorem III. *When the central acceleration varies as some integral power of the distance, say μu^n, then to show that there are at most two apsidal distances.* **[Meerut 2017]**

Proof. Let the central acceleration $= F = \mu u^n$, where n is an integer. The differential equation of the path is

$$h^2(u + d^2u/d\theta^2) = F/u^2 = \mu u^{n-2}, \text{ as } F = \mu u^n$$

Multiplying both sides by $2(du/d\theta)$ and integrating, we get

$$h^2\{u^2 + (du/d\theta)^2\} = (2\mu\, u^{n-1})/(n-1) + (2C\mu)/(n-1) \qquad ...(1)$$

The particle is at an apse when $du/d\theta = 0$. Then (1) gives

$$h^2u^2 = \frac{2\mu\, u^{n-1}}{n-1} + C \qquad \text{or} \qquad u^{n-1} - \frac{n-1}{2}\frac{h^2}{\mu}u^2 + C = 0 \qquad ...(2)$$

Whatever be the values of n and C, (2) cannot have more than two changes of sign, and hence, by Descartes' rule, (2) cannot have more than two positive roots. If follows that there are at most two values of u(or r) *i.e.*, at most two apsidal distances.

An expression for apidal angle. From energy equation, we have (Refer equation (4) of Art. 9.6)

$$(du/d\theta)^2 + u^2 = (2/h^2)(E - V) \qquad ...(1)$$

At an apse, $du/d\theta = 0$. So by (1), the apsidal distances are given by

$$u = (2/h^2)(E - V), \text{ where } u = 1/r \qquad ...(2)$$

Let u_1, u_2 be consecutive roots of equation (2), where $u_1 < u_2$.

When u is increasing, we have $du/d\theta = \{f(u)\}^{1/2}$,

where $$f(u) = (2/h^2)(E - V) - u^2$$

Hence the apsidal angle, between radii $1/u_1$ and $1/u_2$ is

$$\theta = \int_{u_1}^{u_2} \frac{1}{\{f(u)\}^{1/2}}\,du$$

With help of known values of $f(u)$, u_1 and u_2, the above expression can be evaluated to give the required apsidal angle θ.

9.12 A particle moves with a central acceleration μ/distance)³, to find the path and to distinguish the cases

Here $F = \mu/r^3 = \mu u^3$. Putting this in the differential equation $d^2u/d\theta^2 + u = F/h^2u^2$ of a central orbit, we obtain

$$d^2u/d\theta^2 + u = \mu u/h^2 \qquad \text{or} \qquad d^2u/d\theta^2 - (\mu/h^2 - 1)u = 0 \qquad ...(1)$$

Three cases arise :

Case I. Let $h^2 < \mu$ so that $(\mu/h^2 - 1)$ is positive and equal to n^2, say. Then (1) reduces to $d^2u/d\theta^2 - n^2u = 0$ whose general solution is

$$u = Ae^{n\theta} + Be^{-n\theta} \quad \text{or} \quad u = L\cosh n\theta + M\sinh n\theta, \quad \text{...(2)}$$

where A, B, L, M are arbitrary constants.

This is a spiral curve with an infinite number of convolutions about the pole. In the particular case when A or B vanishes, it is an equiangular spiral.

Case II. Let $h^2 > \mu$ so that $(\mu/h^2 - 1)$ is negative and equal to $-n^2$, say. Then (1) reduces to $d^2u/d\theta^2 + n^2u = 0$ whose general solution is

$$u = A\cos(n\theta + B) \quad \text{or} \quad u = A\cos n(\theta - \alpha),$$

where A and α are arbitrary constants. The apse is given by $du/d\theta = 0$ *i.e.*, $-An\sin n(\theta - \alpha) = 0$ which gives $\theta = \alpha$. So apse is given by $u = A$.

Case III. Let $h^2 = \mu$. Then (1) reduce to $d^2u/d\theta^2 = 0$ whose solution is

$$u = A\theta + B = A(\theta - \alpha),\ A, a \text{ being arbitrary constants.}$$

This represents a reciprocal spiral in general. In the particular case when $A = 0$, it represents a circle.

9.13 Given the law of force, to find the orbit. Working rule

Step I. For a given central force F, we use the differential equation

$$h^2(d^2u/d\theta^2 + u) = F/u^2 \quad \text{or} \quad (h^2/p^3)(dp/dr) = F,$$

whichever is convenient. In case the force is repulsive, F must be taken with negative sign.

Step II. At an apse, $dr/d\theta = 0$, $du/d\theta = 0$ and $p = r$.

Step III. While finding orbit, we use the following results directly. Remember them with usual notations :

(*i*) For a central orbit $r^2\dot{\theta} = h = pv$

(*ii*) $p = r\sin\phi$, $1/p^2 = u^2 + (du/d\theta)^2$ (*iii*) Velocity in a circle of radius $a = V = (Fa)^{1/2}$

(*iv*) Velocity V at a point at a distance a from pole from infinity is given by

$$V^2 = -2\int_{\infty}^{a} F\,dr$$

(*v*) Velocity V of fall to the point of projection from pole to a point at a distance a from pole is given by $\quad V^2 = 2\int_0^a F\,dr$

9.14 Solved examples based on Art. 9.13

TYPE I. Ex. 1. (*a*) *A particle is projected from an apse at a distance a under the law of force μ/r^5. If the velocity of projection be $(\mu/2a^4)^{1/2}$, show that the equation of the orbit is a circle.*

Sol. Here, the central force per unit mass $= F = \mu/r^5 = \mu u^5$, if $u = 1/r$. **[Agra 2011]**

Hence the differential equation of the central orbit becomes

$$h^2(d^2u/d\theta^2 + u) = F/u^2 \quad \text{or} \quad h^2(d^2u/d\theta^2 + u) = \mu u^3$$

Multiplying both sides by $2(du/d\theta)$ and then integrating, we get

$$v^2 = h^2\{(du/d\theta)^2 + u^2\} = (\mu/2)u^4 + A,\ A \text{ is a constant} \quad \text{...(1)}$$

Initially at an apse; $r = a$, $u = 1/a$, $du/d\theta = 0$, $v = (\mu/2a^4)^{1/2}$

$\therefore$ (1) $\Rightarrow \mu/2a^4 = h^2/a^2 = \mu/2a^4 + A \Rightarrow A = 0,\ h^2 = \mu/2a^2$

Then (1) gives $\quad (\mu/2a^2)\{(du/d\theta)^2 + u^2\} = (\mu/2)u^4$

or $\quad (du/d\theta)^2 + u^2 = a^2u^4 \quad$ or $\quad (du/d\theta)^2 = u^2(a^2u^2 - 1) \quad$ ...(2)

But $u = 1/r$ so $du/d\theta = -(1/r^2)(dr/d\theta)$. Then (2) gives

$$(1/r^4)(dr/d\theta)^2 = (1/r^2)(a^2/r^2 - 1) \quad \text{or} \quad (dr/d\theta)^2 = (a^2 - r^2) \quad \text{...(3)}$$

or $\quad dr/d\theta = -(a^2 - r^2)^{1/2}$, taking –ve sign*

$$\therefore \quad d\theta = -\frac{dr}{(a^2 - r^2)^{1/2}} \quad \text{so that} \quad \theta + B = \cos^{-1}\frac{r}{a} \quad \text{...(4)}$$

If θ be measured from the initial radius vector, then $\theta = 0$ when $r = a$. So (4) gives $B = 0$. Hence

*If we take +ve sign on the R.H.S. of (3), we get the same result. This fact will be used in many problems of this chapter.

(4) becomes

$$\theta = \cos^{-1}(r/a) \quad \text{or} \quad r = a\cos\theta,$$

which is equation of a circle in polar form.

Ex. 1. (*b*) *A particle is projected from an apse at a distance a under the law of force* μ/r^5. *If the velocity of projection be* $(\mu/a^4)^{1/2}$, *show that the equation of the orbit is the circle* $r = a$, *i.e.*, $x^2 + y^2 = a^2$.

Sol. Proceed exactly as in Ex. 1(*a*).

Ex. 1. (*c*) *A particle moves under a force* $m\mu\{3au^4 - 2(a^2 - b^2)u^5\}$. $a > b$ *and is projected from an apse at a distance* $(a + b)$ *with velocity* $\mu^{1/2}/(a + b)$. *Show that the equation of its path is* $r = a + b\cos\theta$. **[Agra 2008; Rohilkhand 2003; Garhwal 2002; Lucknow 2001, 06, 09, 11; Purvanchal 2008]**

Sol. Here central force per unit mass $= F = \mu\{3au^4 - 2(a^2 - b^2)u^5\}$

Hence the differential equation of the central orbit becomes

$$h^2(d^2u/d\theta^2 + u) = F/u^2 = \mu\{3au^2 - 2(a^2 - b^2)u^3\}$$

Multiplying both sides by $2(du/d\theta)$ and then integrating, we get

$$v^2 = h^2\{(du/d\theta)^2 + u^2\} = \mu(2au^3 - (a^2 - b^2)u^4\} + A \qquad \ldots(1)$$

Initially at an apse, $u = 1/r = 1/(a + b)$, $du/d\theta = 0$, $v = \mu^{1/2}/(a + b)$. So

$$(1) \Rightarrow \frac{\mu}{(a+b)^2} = \frac{h^2}{(a+b)^2} = \mu\left\{\frac{2a}{(a+b)^3} - \frac{(a^2-b^2)}{(a+b)^4}\right\} + A \Rightarrow h^2 = \mu,\ A = 0$$

$\therefore$ (1) gives $\qquad \mu\{(du/d\theta)^2 + u^2\} = \mu\{2au^3 - (a^2 - b^2)u^4\}$

or $\qquad (du/d\theta)^2 = -u^2 + 2au^3 - (a^2 - b^2)u^4 \qquad \ldots(2)$

But $u = 1/r$ so that $du/d\theta = -(1/r^2)(dr/d\theta)$. Then (2) gives

$$\frac{1}{r^4}\left(\frac{dr}{d\theta}\right)^2 = -\frac{1}{r^2} + \frac{2a}{r^3} - \frac{a^2 - b^2}{r^4} = \frac{1}{r^4}\{-r^2 + 2ar - (a^2 - b^2)\}$$

or $\qquad (dr/d\theta)^2 = b^2 - (r^2 - 2ar + a^2) = b^2 - (r - a)^2$

or $\qquad \dfrac{dr}{d\theta} = -\{b^2 - (r - a)^2\}^{1/2} \quad$ or $\quad d\theta = -\dfrac{dr}{\{b^2 - (r-a)^2\}^{1/2}}$

Integrating, $\qquad \theta + B = \cos^{-1}\{(r - a)/b\} \qquad \ldots(3)$

Initially at an apse when $r = a + b$ (given), let $\theta = 0$. So (3) gives

$0 + B = \cos^{-1} 1$ or $B = 0$. Thus (3) becomes

$\theta = \cos^{-1}\{(r - a)/b\}$ or $(r - a)/b = \cos\theta$ or $r = a + b\cos\theta$

Ex. 1. (*d*) *A particle is moving with central acceleration* $\mu(r^5 - c^4 r)$ *being projected from an apse at a distance c with velocity* $c^3(2\mu/3)^{1/2}$, *show that its path is the curve* $x^4 + y^4 = c^4$. **[Bundelkhand 2007; Garhwal 2003; Purvanchel 2010; Kanpur 2019; Agra 2012; Meerut 2013; Rohilkhand 2019]**

Sol. Here central force per unit mass $= F = \mu(u^{-5} - c^4u^{-1})$, where $r = 1/u$. Hence the differential equation of the central orbit becomes

$$h^2(d^2u/d\theta^2 + u) = F/u^2 = \mu(u^{-7} - c^4u^{-3}).$$

Multiplying both sides by $2(du/d\theta)$ and then integrating, we get

$$v^2 = h^2\{(du/d\theta)^2 + u^2\} = \mu\{-(1/3u^6) + (c^4/u^2)\} + A \qquad \ldots(1)$$

Initially at an apse, $u = 1/r = 1/c$, $du/d\theta = 0$, $v = c^3(2\mu/3)^{1/2}$

$$\therefore (1) \Rightarrow \frac{2\mu c^6}{3} = \frac{h^2}{c^2} = \mu\left(-\frac{c^6}{3} + c^6\right) + A \Rightarrow h^2 = (2/3)\mu c^8,\ A = 0$$

$\therefore$ (1) gives $\qquad (2/3)\mu c^8\{(du/d\theta)^2 + u^2\} = \mu\{-(1/3u^6) + (c^4/u^2)\}$

or $\qquad c^8\left(\dfrac{du}{d\theta}\right)^2 = -\dfrac{1}{2u^6} + \dfrac{3c^4}{2u^2} - c^8u^2 = \dfrac{1}{u^6}\left(-\dfrac{1}{2} + \dfrac{3c^4u^4}{2} - c^8u^8\right)$

$$= \frac{1}{u^6}\left[-\frac{1}{2} - \left(c^8u^8 - \frac{3}{2}c^4u^4 + \frac{9}{16}\right) + \frac{9}{16}\right] = \frac{1}{u^6}\left[\left(\frac{1}{4}\right)^2 - \left(c^4u^4 - \frac{3}{4}\right)^2\right]$$

or $c^4u^3\dfrac{du}{d\theta}=\left\{\left(\dfrac{1}{4}\right)^2-\left(c^4u^4-\dfrac{3}{4}\right)^2\right\}^{1/2}$ or $4d\theta=\dfrac{4c^4u^3\,du}{\{(1/4)^2-(c^4u^4-3/4)^2\}^{1/2}}$

Putting $c^4u^4-3/4=z$ so that $4c^4u^3\,du=dz$, we get

$$4d\theta=\frac{dz}{\{(1/4)^2-z^2\}^{1/2}} \quad \text{so that} \quad 4\theta+B=\sin^{-1}4z \qquad \text{...(2)}$$

or $\quad 4\theta+B=\sin^{-1}(4c^4u^4-3)$, as $z=c^4u^4-3/4$ ...(3)

Initially at an apse, when $u=1/c$, let $\theta=0$.

So (3) gives $B=\sin^{-1}1=\pi/2$. Then (3) becomes

$4\theta+\pi/2=\sin^{-1}(4c^4u^4-3)$ or $\sin(\pi/2+4\theta)=4c^4u^4-3$

or $\cos 4\theta=4c^4u^4-3$ or $4c^4u^4=3+\cos 4\theta$

or $4c^4/r^4=3+2\cos^2 2\theta-1=2(1+\cos^2 2\theta)$

or $4c^4/r^4=2\{(\sin^2\theta+\cos^2\theta)^2+(\cos^2\theta-\sin^2\theta)^2\}$

or $c^4=r^4(\cos^4\theta+\sin^4\theta)=(r\cos\theta)^4+(r\sin\theta)^4$

or $c^4=x^4+y^4$ as $x=r\cos\theta,\ y=r\sin\theta$

Ex. 1. (e) *A particle moves with a central acceleration $\mu(r+a^4/r^3)$ being projected from an apse at a distance 'a' with a velocity $2a\mu^{1/2}$. Prove that it describes the curve $r^2(2+\cos\theta\sqrt{3})=3a^2$.* **[Kanpur 2018]**

Sol. Here central acceleration per unit mass $=F=\mu(u^{-1}+a^4u^3)$, if $u=1/r$

Hence the differential equation of the central orbit becomes

$$h^2(d^2u/d\theta^2+u)=F/u^2=\mu(u^{-3}+a^4u)$$

Multiplying both sides by $2(du/d\theta)$ and integrating, we get

$$v^2=h^2\{(du/d\theta)^2+u^2\}=\mu(-1/u^2+a^4u^2)+A \qquad \text{...(1)}$$

Initially at an apse, $u=1/r=1/a$, $du/d\theta=0$, $v=2(\mu a)^{1/2}$

$\therefore$ (1) $\Rightarrow 4\mu a^2=h^2/a^2=\mu(-a^2+a^2)+A \quad\Rightarrow\quad h^2=4\mu a^4,\ A=4\mu a^2$

Then (1) becomes $\quad 4\mu a^4\{(du/d\theta)^2+u^2\}=\mu(-1/u^2+a^4u^2)+4\mu a^2$

or $\quad 4a^4(du/d\theta)^2=-4a^4u^2-1/u^2+a^4u^2+4a^2=(-1-3a^4u^4+4a^2u^2)/u^2$

or $\quad 4a^4u^2(du/d\theta)^2=-(a^2u^2\sqrt{3}-2/\sqrt{3})^2+4/3-1=1/3-(a^2u^2\sqrt{3}-2/\sqrt{3})^2$

or $\quad 2a^2u(du/d\theta)=-\{(1/3)-(a^2u^2\sqrt{3}-2/\sqrt{3})^2\}^{1/2}$

or $$\sqrt{3}\,d\theta=\frac{-2\sqrt{3}a^2u\,du}{\{(1/3)-(a^2u^2\sqrt{3}-2/\sqrt{3})^2\}^{1/3}} \qquad \text{...(2)}$$

Putting $a^2u^2\sqrt{3}-2/\sqrt{3}=z$ so that $(2a^2u\sqrt{3})\,du=dz$, (2) gives

$$\sqrt{3}\,d\theta=\frac{-dz}{(1/3-z^2)^{1/2}} \quad \text{so that} \quad \theta\sqrt{3}+B=\cos^{-1}(z\sqrt{3}) \qquad \text{...(3)}$$

Initially at an apse, $u=1/a$ and let $\theta=0$.

But $u=1/a\Rightarrow z=\sqrt{3}-(2/\sqrt{3})=1/\sqrt{3}$. So (3) gives $B=\cos^{-1}1=0$

Then (3) becomes $\quad \cos^{-1}(z\sqrt{3})=\theta\sqrt{3}$ or $z\sqrt{3}=\cos(\theta\sqrt{3})$

or $\quad \sqrt{3}(a^2u^2\sqrt{3}-2/\sqrt{3})=\cos(\theta\sqrt{3})$ or $3a^2u^2-2=\cos(\theta\sqrt{3})$

or $\quad 3a^2u^2=2+\cos(\theta\sqrt{3})$ or $3a^2=r^2(2+\cos\theta\sqrt{3})$

Ex. 1. (f) *A particle, subject to a central force per unit mass equal to $\mu\{2(a^2+b^2)u^5-3a^2b^2u^7\}$ is projected at a distance a with velocity $\mu^{1/2}/a$ in a direction at right angles to the initial distance, show that the path is the curve $r^2=a^2\cos^2\theta+b^2\sin^2\theta$.* **[Agra 2003]**

Sol. Here central acceleration per unit mass $=F=\mu\{2(a^2+b^2)u^5-3a^2b^2u^7\}$. Hence the differential equation of the central orbit becomes

$$h^2(d^2u/d\theta^2 + u) = F/u^2 = \mu\{2(a^2+b^2)u^3 - 3a^2b^2u^5\}$$

Multiplying both sides by 2(du/dθ) and integrating, we get

$$v^2 = h^2\{(du/d\theta)^2 + u^2\} = \mu\{(a^2+b^2)u^4 - a^2b^2u^6\} + C \qquad \ldots(1)$$

Initially at an apse, $u = 1/r = 1/a$, $du/d\theta = 0$ and $v = \mu^{1/2}/a$.

$$\therefore (1) \Rightarrow \frac{\mu}{a^2} = \frac{h^2}{a^2} = \mu\left(\frac{a^2+b^2}{a^4} - \frac{a^2b^2}{a^6}\right) + c \Rightarrow h^2 = \mu \text{ and } C = 0$$

Then (1) becomes $\mu\{(du/d\theta)^2 + u^2\} = \mu\{(a^2+b^2)u^4 - a^2b^2u^6\}$

or $\quad (du/d\theta)^2 = (a^2+b^2)u^4 - a^2b^2u^6 - u^2 = u^2(1 - a^2u^2)(b^2u^2 - 1)$

or $\quad du/d\theta = u\{(1 - a^2u^2)(b^2u^2 - 1)\}^{1/2}$, taking +ve sign ...(2)

Here $\quad u = 1/r \quad$ so that $\quad du/d\theta = -(1/r^2)(dr/d\theta)$

$\therefore$ (2) becomes $\quad -(1/r^2)(dr/d\theta) = (1/r)(1/r^2)\{(r^2 - a^2)(b^2 - r^2)\}^{1/2}$

$$\text{or} \quad d\theta = -\frac{r\,dr}{\{(r^2-a^2)(b^2-r^2)\}^{1/2}} \quad \text{or } \theta + B = -\int \frac{r\,dr}{\{(r^2-a^2)(b^2-r^2)\}^{1/2}} \qquad \ldots(3)$$

Putting $\quad r^2 = a^2\cos^2\phi + b^2\sin^2\phi$, we get $\quad 2r\,dr = (b^2 - a^2)\sin 2\phi\, d\phi$,
$r^2 - a^2 = (b^2 - a^2)\sin^2\phi$ and $b^2 - r^2 = (b^2 - a^2)\cos^2\phi$. So (3) reduces to

$$\theta + B = -\int d\phi \qquad \text{or} \qquad \theta + B = -\phi \qquad \ldots(4)$$

Initially at an apse when $r = a$, $a^2 = a^2\cos^2\phi + b^2\sin^2\phi$ so that $(a^2 - b^2)\sin^2\phi = 0$ giving $\phi = 0$ as $a \neq b$. Thus, at an apse, when $r = a$, $\phi = 0$ and let $\theta = 0$. So (4) gives $B = 0$.

Then (4) becomes $\theta = -\phi$ and so $r^2 = a^2\cos^2\phi + b^2\sin^2\phi$ reduces to
$r^2 = a^2\cos^2(-\theta) + b^2\sin^2(-\theta) \qquad$ or $\qquad r^2 = a^2\cos^2\theta + b^2\sin^2\theta$

Ex. 1. (g) *If $F = \mu(u^2 - au^3)$, where $a > 0$, be the central force per unit mass and a particle is projected from an apse at a distance 'a' from the centre of force with a velocity $(\mu c/a^2)^{1/2}$, where $a > c$, prove that the other apsidal distance of the orbit is $a(a + c)/(a - c)$ and find the apsidal angle.*

[Delhi Maths (H) 2004, 10]

Sol. Here central force per unit mass $= F = \mu(u^2 - au^3)$

Hence the differential equation of the central orbit becomes

$$h^2(d^2u/d\theta^2 + u) = F/u^2 = \mu(1 - au)$$

Multiplying both sides by $2(du/d\theta)$ and integrating, we get

$$v^2 = h^2\{(du/d\theta)^2 + u^2\} = \mu(2u - au^2) + A \qquad \ldots(1)$$

Initially at an apse $u = 1/r = 1/a$, $du/d\theta = 0$ and $v = (\mu c/a^2)^{1/2}$

$\therefore$ (1) $\Rightarrow \quad \mu c/a^2 = h^2/a^2 = \mu/a + A \Rightarrow h^2 = \mu c$, $A = \mu c/a^2 - \mu/a$

$$\text{Then (1) becomes} \quad \mu c\left\{\left(\frac{du}{d\theta}\right)^2 + u^2\right\} = \mu(2u - au^2) + \frac{\mu c}{a^2} - \frac{\mu}{a}$$

or $\quad (du/d\theta)^2 + u^2 = 2u/c - au^2/c + 1/a^2 - 1/ac$

$$\text{or} \quad \left(\frac{du}{d\theta}\right)^2 = -\left(\frac{a+c}{c}u^2 - \frac{2u}{c} + \frac{a-c}{a^2c}\right) = -\frac{a+c}{c}\left\{u^2 - \frac{2u}{a+c} + \frac{a-c}{a^2(a+c)}\right\}$$

$$\text{or} \quad \left(\frac{du}{d\theta}\right)^2 = -\frac{a+c}{c}\left(u - \frac{1}{a}\right)\left\{u - \frac{a-c}{a(a+c)}\right\} \qquad \ldots(2)$$

Let $b = a(a + c)/(a - c)$. Since $a > c$, hence $b > 0$. Then (2) becomes

$$(du/d\theta)^2 = -(a/c + 1)(u - 1/a)(u - 1/b) \qquad \ldots(3)$$

At an apse $du/d\theta = 0$. So apsidal distances are given by

$$-(a/c + 1)(u - 1/a)(u - 1/b) = 0 \quad \text{or} \quad u = 1/a, 1/b \quad \text{or} \quad r = a, b.$$

Now $r = a$ is given apsidal distance. So the required other apsidal distance is given by $r = b = (a - c)/a(a + c)$. Clearly $b > a$ or $1/b < 1/a$. Let u be increasing so that $1/b < u < 1/a$. Then, (3) gives

$$(du/d\theta)^2 = (a/c + 1)(1/a - u)(u - 1/b), \text{ as } 1/b < u < 1/a$$

$$\therefore \quad du/d\theta = (a/c + 1)^{1/2}\{(1/a - u)(u - 1/b)\}^{1/2}$$

Hence the required apsidal angle is given by

$$\theta = \left(\frac{c}{a+c}\right)^{1/2} \int_{1/b}^{1/a} \frac{du}{\{(1/a-u)(u-1/b)\}^{1/2}} \qquad \text{...(4)}$$

Put $1/a - u = z^2$ so that $du = -2z\,dz$. Then, we have

$$u - 1/b = 1/a - z^2 - 1/b = (1/a - 1/b) - z^2 = t^2 - z^2,$$

where $$t^2 = (1/a - 1/b) > 0 \text{ as } 1/a > 1/b$$

When $u = 1/b$, $z^2 = 1/a - 1/b = t^2$ or $z = t$ and when $u = 1/a$, $z = 0$

$\therefore$ (4) reduces to $$\theta = \left(\frac{c}{a+c}\right)^{1/2} \int_t^0 \frac{(-2dz)}{(t^2-z^2)^{1/2}}$$

or $$\theta = \left(\frac{c}{a+c}\right)^{1/2} (-2)\left[\sin^{-1}\frac{z}{t}\right]_t^0 = \pi\left(\frac{c}{a+c}\right)^{1/2}, \text{ as } \sin^{-1} 1 = \frac{\pi}{2}.$$

Ex. 1. (*h*) *Find the law of force to the pole when the path is the cardioid $r = a(1 - \cos\theta)$, and prove that if F be force at the apse and v the velocity there, then $3v^2 = 4aF$.*

[Kolkata 1993; Lucknow 1994; I.A.S. 2001]

Sol. Given $r = a(1 - \cos\theta)$ or $r = 2a \sin^2 \theta/2$ ...(1)

Taking logarithms of both sides, $\log r = \log 2a + 2 \log \sin \theta/2$

Diff. it w.r.t. 'θ', $(1/r)\,(dr/d\theta) = \cot(\theta/2)$

so that $\cot\phi = \cot\theta/2$ giving $\phi = \theta/2$ ...(2)

Now, $p = r \sin\phi$ gives $p = r \sin\theta/2 = r(r/2a)^{1/2}$, by (1)

Thus, pedal equation of (1) is $2ap^2 = r^3$ or $1/p^2 = 2a/r^3$ ...(3)

Diff. (3) w.r.t. 'r' $$-\frac{2}{p^3}\frac{dp}{dr} = -\frac{6a}{r^4} \quad\text{or}\quad \frac{h^2}{p^3}\frac{dp}{dr} = \frac{3ah^2}{r^4} \qquad \text{...(4)}$$

Now, from the pedal equation of a central orbit, we get

$F = (h^2/p^3)\,(dp/dr) = (3ah^2)/r^4$ so that $F \propto 1/r^4$, *i.e.*, the attractive force varies inversely as the fourth power of the distance from the pole.

Second part. At an apse, $dr/d\theta = 0$ or $a \sin\theta = 0$, since $r = a(1 - \cos\theta)$. Now, $\sin\theta = 0 \Rightarrow \theta = 0$ or π. Since $\theta = 0$ gives $r = 0$ so that $\theta = 0$ corresponds to cusp O of the cardioid (1). Hence $\theta = \pi$ is the apse where $r = a(1 - \cos\pi) = 2a$ and $p = 2a$.

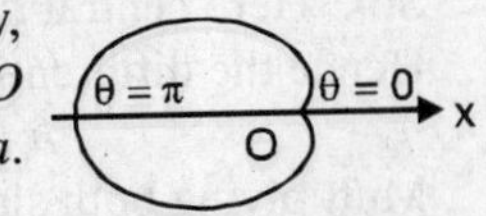

But $h = pv \Rightarrow h = 2av$. Hence force at the apse where $r = 2a$, is given by

$$F = \frac{3ah^2}{r^4} = \frac{3a \times (2av)^2}{(2a)^4} = \frac{3v^2}{4a} \quad \text{so that} \quad 3v^2 = 4aF.$$

Ex. 1. (*i*) *A particle moves under a central force $m\lambda(3a^3u^4 + 8au^2)$. It is projected from an apse at a distance a from the centre of force with velocity $(10\lambda)^{1/2}$. Show that the second apsidal distance is half the first and that the equation to the path is $2r = a\{1 + \text{sech}\,(\theta/\sqrt{5})\}$.* **[Kanpur 2002]**

Ex. 1. (*j*) *A Particle moves under a central force $m\lambda(3a^3u^4 + 8au^2)$. It is projected from an apse at a distance a from the centre of force with velocity $\sqrt{10\lambda}$. Show that the second apsidal distance is half the first and that the equation to the path is $2r = a\{1 + \text{sech}\,(\theta/\sqrt{5})\}$.*

Ex. 1. (*k*) *A particle of mass m moves under a central attractive force $m\mu(5/r^3 + 8c^2/r^5)$, and is projected from an apse at a distance a with velocity $(3/c)\sqrt{\mu}$, prove that orbit is $r = c\cos(2\theta/3)$.*

[Garhwal 2010; Calcutta 2003]

Ex. 1. (*l*) *A particle moves with central acceleration $\mu u^2 + \lambda u^3$ and the velocity of projection at a distance R is V. Show that the particle will ultimately go off to infinitey if $V^2 > 2\mu/R + \lambda/R^2$.*

[I.F.S. 2001]

Sol. Differnetial equation of the path is

$$h^2(d^2u/d\theta^2 + u) = F/u^2 \quad\text{or}\quad h^2(d^2u/d\theta^2 + u) = (\mu u^2 + \lambda u^3)/u^2$$

Multiplying both sides by $2(du/d\theta)$ and integrating, we get

$$v^2 = h^2\{(du/d\theta)^2 + u^2\} = 2\,\mu u + \lambda u^2 + A \qquad \text{...(1)}$$

Initially, when $u = 1/R$, $v = V$ (given). So (1) $\Rightarrow A = V^2 - 2\mu/R - \lambda/R^2$ ...(2)

$\therefore$ By (1), $h^2(du/d\theta)^2 = 2\mu u + \lambda u^2 - u^2h^2 + A$

or $$h^2\left(\frac{du}{d\theta}\right)^2 = u^2(\lambda - h^2) + 2\mu u + A = (\lambda - h^2)\left(u^2 + 2u\frac{\mu}{\lambda - h^2} + \frac{A}{\lambda - h^2}\right)$$

or $$\frac{du}{[\{u + \mu/(\lambda - h^2)\}^2 + A/(\lambda - h^2) - \mu^2/(\lambda - h^2)^2]^{1/2}} = \frac{(\lambda - h^2)^{1/2}}{h}d\theta$$

Integrating and using formula $\int \frac{dx}{(x^2 + a^2)^{1/2}} = \log\{x + (x^2 + a^2)^{1/2}\}$, we get

$$\log[u + \mu/(\lambda - h^2) + \sqrt{\{u + \mu/(\lambda - h^2)\}^2 + A/(\lambda - h^2) - \mu^2/(\lambda - h^2)^2}\,]$$
$$= (1/h)(\lambda - h^2)^{1/2}\,\theta + B,$$ where B is an arbitrary constant.

The particle will ultimately go to infinity if a value of θ can be found for which $r \to \infty$ *i.e.*, $u = 0$. Putting $u = 0$ in the above equation, we have

$$\log[\mu/(\lambda - h^2) + \sqrt{A}/(\lambda - h^2)^{1/2}\} = (\theta/h)(\lambda - h^2)^{1/2} + B$$

For a real value of θ to exist, A must be positive.

So by (2), $A = V^2 - 2\mu/R - \lambda/R^2 > 0$ giving $V^2 > 2\mu/R + \lambda/R^2$

TYPE II. Ex. 2. (*a*) *A particle moves with a central acceleration which varies inversely as the cube of the distance. If it be projected from an apse at a distance a from the origin with a velocity which is $\sqrt{2}$ times the velocity for a circle of radius a show that the equation to its path is $r\cos(\theta/\sqrt{2}) = a$.* **[Rohilkhand 2010; Kanpur 2006, 07; Meerut 2000; Purvanchal 2008]**

Sol. Here central acceleration $= F = \mu/r^3 = \mu u^3$, μ is constant ...(1)

If V is the velocity for a circle of radius a, then

$$V^2/a = [F]_{r=a} = \mu/a^3 \quad \text{so that} \quad V = (\mu/a^2)^{1/2}$$

$\therefore$ the velocity of projection $= v_1$ (say) $= \sqrt{2}V = (2\mu/a^2)^{1/2}$ ...(2)

The differential equation of the central orbit becomes

$$h^2(d^2u/d\theta^2 + u) = F/u^2 \quad \text{or} \quad h^2(d^2u/d\theta^2 + u) = \mu u$$

Multiplying both sides by $2(du/d\theta)$ and integrating, we get

$v^2 = h^2\{(du/d\theta)^2 + u^2\} = \mu u^2 + A$, A is a constant. ...(3)

Initially at an apse, $r = a$, $u = 1/a$, $du/d\theta = 0$, $v = v_1 = (2\mu/a^2)^{1/2}$

$\therefore$ (1) $\Rightarrow$ $2\mu/a^2 = h^2/a^2 = \mu/a^2 + A \Rightarrow h^2 = 2\mu$, $A = \mu/a^2$.

Then (1) gives $2\mu\{(du/d\theta)^2 + u^2\} = \mu u^2 + \mu/a^2$

or $2(du/d\theta)^2 = 1/a^2 + u^2 - 2u^2 = (1 - a^2u^2)/a^2$

or $-a\sqrt{2}\,(du/d\theta) = (1 - a^2u^2)^{1/2}$ or $(1/\sqrt{2})\,d\theta = -\{a/(1 - a^2u^2)^{1/2}\}\,du$

Integrating, $\theta/\sqrt{2} + B = \cos^{-1}(au)$, where B is a constant. ...(4)

Initially at an apse, when $u = 1/a$, let $\theta = 0$. Then (4) gives $B = 0$.

Hence (4) $\Rightarrow \cos(\theta/\sqrt{2}) = au = a/r$ or $a = r\cos(\theta/\sqrt{2})$

Ex. 2. (*b*) *If the law of force be $\mu(u^4 - 10au^5/9)$ and the particle be projected from an apse at a distance 5a with a velocity equal to $(5/7)^{1/2}$ of that in a circle at the same distance. Show that the orbit is the limacon $r = a(3 + 2\cos\theta)$.* **[Garhwal 2001, 03; Agra 2010]**

Sol. Here central acceleration $= F = \mu(u^4 - 10au^5/9)$ ...(1)

If V is velocity for a circle of radius $5a$, then

$$V^2/5a = [F]_{r=5a} = [F]_{u=1/5a} = \mu\left\{\frac{1}{(5a)^4} - \frac{10a}{9\times(5a)^5}\right\} = \frac{7\mu}{9\times(5a)^4}$$

$$\therefore \quad V = [7\mu/\{9\times(5a)^3\}]^{1/2}$$

If v_1 is the velocity of projection of the particle, then

$$v_1 = \left(\frac{5}{7}\right)^{1/2} V = \left(\frac{5}{7}\right)^{1/2} \times \left\{\frac{7\mu}{9\times(5a)^3}\right\}^{1/2} = \left(\frac{\mu}{225a^3}\right)^{1/2} \quad ...(3)$$

The differential equation of the central orbit becomes

$h^2(d^2u/d\theta^2 + u) = F/u^2$ or $h^2(d^2u/d\theta^2 + u) = \mu(u^2 - 10au^3/9)$

Multiplying both sides by $2(du/d\theta)$ and integrating, we get

$$v^2 = h^2\{(du/d\theta)^2 + u^2\} = \mu(2u^3/3 - 5au^4/9) + A \quad \ldots(4)$$

Initially at an apse, $u = 1/r = 1/5a$, $du/d\theta = 0$, $v^2 = v_1^2 = \mu/225a^3$

$$\therefore (4) \Rightarrow \frac{\mu}{225a^3} = \frac{h^2}{25a^2} = \mu\left\{\frac{2}{3}\left(\frac{1}{5a}\right)^3 - \frac{5a}{9}\left(\frac{1}{5a}\right)^4\right\} + A$$

Solving these for h^2 and μ, we get $A = 0$, $h^2 = \mu/9a$

Then (4) gives $(\mu/9a)\{(du/d\theta)^2 + u^2\} = \mu(2u^3/3 - 5au^4/9)$

or $(du/d\theta)^2 = 6au^3 - 5a^2u^4 - u^2$ ...(5)

Putting $u = 1/r$ and $du/d\theta = -(1/r^2)(dr/d\theta)$, (5) gives

$$\frac{1}{r^4}\left(\frac{dr}{d\theta}\right)^2 = \frac{6a}{r^3} - \frac{5a^2}{r^4} - \frac{1}{r^2} \quad \text{or} \quad \left(\frac{dr}{d\theta}\right)^2 = 6ar - 5a^2 - r^2$$

$$\text{or} \quad \left(\frac{dr}{d\theta}\right)^2 = 4a^2 - (r-3a)^2 \quad \text{or} \quad d\theta = -\frac{dr}{\{(2a)^2 - (r-3a)^2\}^{1/2}}$$

Integrating, $\theta + B = \cos^{-1}\{(r - 3a)/2a\}$, B is a constant ...(6)

Initially at an apse, when $r = 5a$, let $\theta = 0$. Then (6) $\Rightarrow B = 0$.

$\therefore$ (6) becomes $(r - 3a)/2a = \cos\theta$ or $r = a(3 + 2\cos\theta)$

Ex. 2. (c) *A particle moves in a curve under a central acceleration so that its velocity at any point is equal to that in a circle at the same distance and under the same attraction. Show that the law of force is that of inverse cube, and the path is an equiangular spiral.* **[Purvanchel 2007]**

Sol. Let the central acceleration per unit mass $= F = u^2\phi'(u)$ ...(1)

The differential equation of the central orbit is

$$h^2\{d^2u/d\theta^2 + u\} = F/u^2 = \phi'(u)$$

Multiplying both sides by $2(du/d\theta)$ and integrating, we get

$$v^2 = h^2\{(du/d\theta)^2 + u^2\} = 2\phi(u) + A, \ A \text{ is a constant} \quad \ldots(2)$$

By given problem, the velocity at any point is equal to that in a circle at the same distance. So velocity v is given by

$v^2/r = F$ or $v^2 = rF = (1/u)\,u^2\phi'(u)$ or $v^2 = u\phi'(u)$ ...(3)

Using (3), (2) gives $u\phi'(u) = 2\phi(u) + A$

$$\text{or} \quad \frac{\phi'(u)}{u^2} - \frac{2\phi(u)}{u^3} = \frac{A}{u^3} \quad \text{or} \quad \frac{d}{du}\left\{\frac{1}{u^2}\phi(u)\right\} = \frac{A}{u^3}$$

Integrating, $(1/u^2)\,\phi(u) = -(A/2u^2) + B$, where B is a constant

or $\phi(u) = -A/2 + Bu^2$ ...(4)

Diff. (4) w.r.t. 'u', $\phi'(u) = 3Bu$

$\therefore$ (1) $\Rightarrow F = u^2\phi'(u) = u^2 \times 2Bu = (2B)/r^3$, showing that the acceleration varies inversely as the cube of the distance from the centre of force.

To determine equation of the path

The differential equation of a central orbit in pedal form is

$$\frac{h^2}{p^3}\frac{dp}{dr} = F \quad \text{or} \quad \frac{h^2}{p^3}\frac{db}{dr} = \frac{2B}{r^3}, \quad \text{as } F = \frac{2B}{r^3}$$

or $-(2h^2/p^3)dp = -(4B/r^3)dr$

Integrating, $h^2/p^2 = 2B/r^2 + C$, C is a constant ...(5)

Assume that $p \to \infty$ as $r \to \infty$. Then (5) gives $C = 0$.

Now (5) becomes $h^2/p^2 = 2B/r^2$ or $p^2 = (h^2/2B)r^2$ or $p^2 = a^2r^2$, where $a^2 = h^2/2B$. Thus, $p = ar$, which is the pedal equation of an equiangular spiral, is the required path.

Ex. 2. (*d*) *A particle P, of mass m, is moving in a circle, radius a and centre O, under a force* $\mu m(r + 2a^3/r^2)$ *directed towards O. If P is acted on by an impulse tangential to the path and of magnitude* $(3\mu\, m^2a^2)^{1/2}$*, show that the velocity of P is immediately doubled, and that the greatest and least distances from O in the ensuing motion are a and 3a.* **[Delhi Maths (H) 2002, 07]**

Sol. Here, central force per unit mass $= F = \mu(r + 2a^3/r^2) = \mu(1/u + 2a^3u^2)$

Let V be the velocity of the particle P in a circle of radius a under the same acceleration. Then, we know that

$$V^2/a = [F]_{r=a} = \mu(a + 2a) = 3a\mu \qquad \text{or} \qquad V = (3\mu a^2)^{1/2} \qquad \ldots(1)$$

Let V_1 be the velocity when P is acted on by the given impulse. Since V_1 is the velocity of the particle just after the action of imulse, hence we have

$$mV_1 - m\cdot 0 = (3\mu m^2a^2)^{1/2} \qquad \text{so that} \qquad V_1 = (3\mu a^2)^{1/2}, \qquad \ldots(2)$$

which is of the same magnitude as given by (1). With help of (1) and (2), the velocity v_1 after impact is given by

$$v_1 = V + V_1 = (3a^2\mu)^{1/2} + (3a^2\mu)^{1/2} \qquad \text{or} \qquad v_1 = 2(3a^2\mu)^{1/2} \qquad \ldots(3)$$

The differential equation of the central orbit is

$$h^2(d^2u/d\theta^2 + u) = F/u^2 = \mu(1/u^3 + 2a^3)$$

Multiplying both sides by $2(du/d\theta)$ and integrating, we get

$$v^2 = h^2\{(du/d\theta)^2 + u^2\} = \mu(-1/u^2 + 4a^3u) + A \qquad \ldots(4)$$

Since the velocity is acting tangentially to the circle at P, the direction of motion is at right angles to the initial distance $OP(= a)$. Hence P is an apse, where $du/d\theta = 0$.

Then, initially at apse P : $u = 1/r = 1/a$, $du/d\theta = 0$, $v = v_1 = 2(3a^2\mu)^{1/2}$.

$\therefore$ (4) $\Rightarrow$ $12a^2\mu = h^2/a^2 = \mu(-1/a^2 + 4a^2) + A \Rightarrow h = 12a^4\mu,\ A = 9a^2\mu$

Then (4) gives $\quad 12a^4\mu\{(du/d\theta)^2 + u^2\} = \mu(-1/u^2 + 4a^3u) + A$

or $\quad (du/d\theta)^2 + u^2 = -(1/12a^4u^2) + (u/3a) + (3/4a^2) \qquad \ldots(5)$

The required greatest and least distance from O in the ensuring motion are apsidal distances which are given by $du/d\theta = 0$.

Then (5) reduces to

$$u^2 = -(1/12a^4u^2) + (u/3a) + (3/4a^2) \quad \text{or} \quad 12a^4u^4 - 4a^3u^3 - 9a^2u^2 + 1 = 0$$

or $\quad (au - 1)(3au - 1)(2au + 1)^2 = 0$, on factorization of L.H.S.

Thus, $\quad u = 1/a,\ 1/3a \qquad$ (since $u \neq -1/2a$, being negative)

Hence the required maximum and minimum distances are given

$$u = 1/r = 1/a \quad \text{and} \quad u = 1/r = 1/3a \quad i.e., \quad r = a \quad \text{and} \quad r = 3a.$$

Ex. 2. (*e*) *A particle moves with central acceleration* $\mu(r + 2a^3/r^2)$ *being projected from an apse at a distance a with twice the velocity for a circle at that distance; find the apsidal distance and show that the equation to the path is* $\theta/2 = \tan^{-1}(t\sqrt{3}) - (1/\sqrt{5})\tan^{-1}\{t(5/3)^{1/2}\}$*, where* $t^2 = (r - a)/(3a - r)$.

Sol. Left as an exercise.

TYPE III. Ex. 3. (*a*) *A particle is projected from an apse at a distance a with the velocity from infinity, the acceleration being* μu^7*; show that the equation to its path is* $r^2 = a^2\cos 2\theta$.

[Kanpur 2000; Lucknow 2008; Mumbai 2010; Meerut 2001; Purvanchal 2002, 03]

Sol. Here, the central acceleration $= F = \mu u^7 = \mu/r^7 \qquad \ldots(1)$

If V is the velocity at a distance a acquired in falling from rest from infinity under the same acceleration, then we know that

$$V^2 = -2\int_{\infty}^{a} F\,dr = -2\int_{\infty}^{a}\frac{\mu}{r^7}\,dr = \frac{\mu}{3}\left[\frac{1}{r^6}\right]_{\infty}^{a} = \frac{\mu}{3a^6} \qquad \ldots(2)$$

The differential equation of the path is

$$h^2(d^2u/d\theta^2 + u) = F/u^2 \qquad \text{or} \qquad h^2(d^2u/d\theta^2 + u) = \mu u^5$$

Multiplying both sides by $2(du/d\theta)$ and integrating, we get

$$v^2 = h^2\{(du/d\theta)^2 + u^2\} = (1/3)\mu u^6 + A, \ A \text{ is a constant} \qquad \ldots(3)$$

Initially at an apse; $u = 1/r = 1/a$, $du/d\theta = 0$, $v = V = (\mu/3a^6)^{1/2}$

$\therefore$ (3) $\Rightarrow$ $\mu/3a^6 = h^2/a^2 = \mu/3a^6 + A$ giving $h^2 = \mu/3a^4, A = 0$

$\therefore$ (3) gives $(\mu/3a^4)\{(du/d\theta)^2 + u^2\} = \mu u^6/3$

or $(du/d\theta)^2 + u^2 = a^4u^6$ or $(du/d\theta)^2 = a^4u^6 - u^2$ ...(4)

Now, $u = 1/r$ so that $du/d\theta = -(1/r^2)(dr/d\theta)$. So (4) gives

$$\frac{1}{r^4}\left(\frac{dr}{d\theta}\right)^2 = \frac{a^4}{r^6} - \frac{1}{r^2} = \frac{a^4 - r^4}{r^6} \quad \text{or} \quad d\theta = -\frac{r\,dr}{(a^4 - r^4)^{1/2}} \quad ...(5)$$

Putting $r^2 = t$ and $2r\,dr = dt$, (5) reduces to

$$-\frac{dt}{(a^4 - t^2)} = 2d\theta \quad \text{so that} \quad \cos^{-1}\left(\frac{t}{a^2}\right) = 2\theta + B$$

or $\cos^{-1}(r^2/a^2) = 2\theta + B$, as $t = r^2$ ...(6)

Initially when $r = a$, let $\theta = 0$. Then (6) $\Rightarrow B = 0$

Hence (6) gives $\cos^{-1}(r^2/a^2) = 2\theta$ or $r^2 = a^2\cos 2\theta$

Ex. 3. (b) *A particle is projected from an apse at a distance a with the velocity from infinity under the action of a central acceleration μ/r^{2n+3}. Prove that the equation of the path is $r^n = a^n \cos n\theta$.* **[Purvanchal 2009; Agra 2002, 11; Kolkata 94; Meerut 2002]**

Sol. Here, the central acceleration $= F = \mu/r^{2n+3} = \mu u^{2n+3}$ ...(1)

If V is the velocity of the particle at a distance a acquired in falling from infinity under the same acceleration, then, we know that

$$V^2 = -2\int_\infty^a F\,dr = -2\int_\infty^a \frac{\mu\,dr}{r^{2n+3}} = -2\mu\left[\frac{r^{-2n-2}}{-2n-2}\right]_\infty^a = \frac{\mu}{(n+1)a^{2n+2}}$$

The differential equation of the path is

$h^2(d^2u/d\theta^2 + u) = F/u^2 = (1/u^2) \times \mu u^{2n+3} = \mu u^{2n+1}$

Multiplying both sides by $2(du/d\theta)$ and integrating, we get

$v^2 = h^2\{du/d\theta)^2 + u^2\} = \{\mu/(n+1)\}u^{2n+2} + A$ ...(2)

Initially at apse : $u = 1/r = 1/a$, $du/d\theta = 0$, $v^2 = V^2 = \mu/(n+1)a^{2n+2}$

$$\therefore (2) \Rightarrow \quad \frac{\mu}{(n+1)a^{2n+2}} = \frac{h^2}{a^2} = \frac{\mu}{n+1}\frac{1}{a^{2n+2}} + A \Rightarrow h^2 = \frac{\mu}{(n+1)a^{2n}}, A = 0$$

Then (2) gives
$$\frac{\mu}{(n+1)a^{2n}}\left\{\left(\frac{du}{d\theta}\right)^2 + u^2\right\} = \frac{\mu}{(n+1)}u^{2n+2}$$

or $(du/d\theta)^2 = a^{2n}u^{2n+2} - u^2$ ...(3)

But $u = 1/r$ so that $du/d\theta = -(1/r^2)(dr/d\theta)$. So (3) becomes

$$\frac{1}{r^4}\left(\frac{dr}{d\theta}\right)^2 = \frac{a^{2n}}{r^{2n+2}} - \frac{1}{r^2} = \frac{a^{2n} - r^{2n}}{r^{2n+2}} \quad \text{or} \quad d\theta = -\frac{r^{n-1}dr}{(a^{2n} - r^{2n})^{1/2}} \quad ...(4)$$

Putting $r^n = t$ and $nr^{n-1}\,dr = dt$, (4) reduces to

$$d\theta = -\frac{dt}{n(a^{2n} - t^2)^{1/2}} \quad \text{or} \quad n\,d\theta = -\frac{dt}{\{(a^n)^2 - t^2\}^{1/2}}$$

Integrating, $n\theta + B = \cos^{-1}(t/a^n)$ or $n\theta + B = \cos^{-1}(r^n/a^n)$ ...(5)

Initially, when $r = a$, let $\theta = 0$. Then (5) gives $B = 0$.

Now, (5) becomes $n\theta = \cos^{-1}(r^n/a^n)$ or $r^n = a^n\cos n\theta$.

Ex. 3. (c) *A particle is projected at right angles at a distance a under a central force μ/r^n per unit mass with velocity from infinity. Find the path of the orbit.* **[Delhi Maths (H) 1998]**

Sol. Here, central acceleration per unit mass $= \mu/r^n = \mu u^n$ ...(1)

If V is the velocity of the particle at a distance a acquired in falling from infinity under the same acceleration, then we know that

$$V^2 = -2\int_\infty^a F\,dr = -2\int_\infty^a \frac{\mu\,dr}{r^n} = -2\mu\left[\frac{r^{-n+1}}{-n+1}\right]_\infty^a = \frac{2\mu}{(n-1)a^{n-1}}, n \neq 1 \quad ...(2)$$

The differential equation of the path is

$$h^2(d^2u/d\theta^2 + u) = F/u^2 = \mu u^{n-2}, \text{ by (1)}$$

Multiplying both sides by $2(du/d\theta)$ and integrating, we get

$$v^2 = h^2\{(du/d\theta)^2 + u^2\} = \{2\mu/(n-1)\}\, u^{n-1} + A \quad \ldots(3)$$

Initially at an apse : $u = 1/r = 1/a$, $du/d\theta = 0$, $v^2 = V^2 = 2\mu/(n-1)a^{n-1}$.

$$\therefore \quad (3) \Rightarrow \frac{2\mu}{(n-1)a^{n-1}} = \frac{h^2}{a^2} = \frac{2\mu}{(n-1)a^{n-1}} + A \text{ giving } h^2 = \frac{2\mu}{(n-1)a^{n-3}},\, A = 0$$

Then, (3) gives
$$\frac{2\mu}{(n-1)a^{n-3}}\left\{\left(\frac{du}{d\theta}\right)^2 + u^2\right\} = \frac{2\mu u^{n-1}}{n-1}, \quad n \neq 1$$

or $\quad (du/d\theta)^2 + u^2 = a^{n-3}\, u^{n-1} \quad$ or $\quad (du/d\theta)^2 = u^2(a^{n-3}\, u^{n-3} - 1)$

or $$\frac{du}{d\theta} = \pm u\{(au)^{n-3} - 1\}^{1/2} \quad \text{or} \quad \frac{du}{u\{(au)^{n-3} - 1\}^{1/2}} = \pm u\, d\theta \quad \ldots(4)$$

Putting $\quad (au)^{n-3} = \sec^2 t \quad$ *i.e.*, $\quad au = \sec^{2/(n-3)} t \quad \ldots(5)$

Then $\quad a\, du = \{2/(n-3)\} \sec^{2/(n-3)} t \tan t \, dt \quad \ldots(6)$

$\therefore$ (4) gives
$$\frac{2}{n-3}\int \frac{\sec^{2/(n-3)} t \tan t \, dt}{\sec^{n(n-3)} t \tan t} = \pm\theta + B$$

or $\quad \{2/(n-3)\}t = \pm\theta + B$, B is a constant. $\quad \ldots(7)$

Initially, when $u = 1/a$, let $\theta = 0$. Then (5) $\Rightarrow \sec^2 t = 1 \Rightarrow t = 0$.

Putting $t = 0$ and $\theta = 0$ in (7), we get $B = 0$. Then (7) gives

$t = \pm\{(n-3)\theta/2\}$ $\quad$ so that $\quad \sec\{(n-3)\theta/2\} = \sec t$

or $\quad \sec^{2/(n-3)}\{(n-3)\theta/2\} = \sec^{2/(n-3)} t = au = a/r$

or $\quad r = a \cos^{2/(n-3)}\{(n-3)\theta/2\} \quad$ or $\quad r^{(n-3)/2} = a^{(n-3)/2} \cos\{(n-3)\theta/2\}$,

is the desired orbit, where $n \neq 1$ and $n \neq 3$.

Ex. 3. (*d*) *Show that only law for a central attraction for which the velocity in a circle at a distance is equal to the velocity acquired in falling from infinity to the distance is that of inverse cube.*

Sol. Hint. Take $\quad F = \phi'(r)$.

Ex. 3. (*e*) *A particle moving with a central acceleration $\mu/(distance)^3$ is projected from an apse at a distance a with velocity V; show that the path is*

$$r \cosh\left\{\frac{(\mu - a^2V^2)^{1/2}}{aV}\theta\right\} = a \quad \text{or} \quad r \cos\left\{\frac{(a^2V^2 - \mu)^{1/2}}{aV}\theta\right\} = a$$

according as V is < or > the velocity from infinity. **Bundelkhand 2008|**

Sol. Here, central acceleration $= F = \mu/r^3 = \mu u^3$, where $u = 1/r \quad \ldots(1)$

Hence the differential equation of the central orbit becomes

$$h^2(d^2u/d\theta^2 + u) = F/u^2 = \mu u, \text{ by (1)}$$

Multiplying both sides by $2(du/d\theta)$ and then integrating, we get

$$v^2 = h^2\{(du/d\theta)^2 + u^2\} = \mu u^2 + A, \; A \text{ is a constant.} \quad \ldots(2)$$

Initially at an apse : $u = 1/r = 1/a$, $du/d\theta = 0$ and $v = V$.

$\therefore$ (2) $\Rightarrow V^2 = h^2/a^2 = \mu/a^2 + A$ so that $h^2 = a^2V^2$, $A = (V^2a^2 - \mu)/a^2$

Then (2) gives $\quad a^2V^2\{(du/d\theta)^2 + u^2\} = \mu u^2 + (V^2a^2 - \mu)/a^2$

or $\quad a^2V^2(du/d\theta)^2 = -a^2V^2u^2 + \mu u^2 + (V^2a^2 - \mu)/a^2 = (a^2V^2 - \mu)(-u^2 + 1/a^2)$

or $\quad a^4V^2(du/d\theta)^2 = (a^2V^2 - \mu)(1 - a^2u^2) \quad \ldots(3)$

Let V_1 be the velocity acquired by the particle in falling from infinity to the distance a. Then, we have

$$V_1^2 = -2\int_\infty^a F\, dr = -2\int_\infty^a \frac{\mu}{r^3} dr = -2\left[-\frac{\mu}{2r^2}\right]_\infty^a = \frac{\mu}{a^2} \quad \ldots(4)$$

Case I. Let $V < V_1$, *i.e.*, V is < the velocity from infinity

Then $\quad V^2 < V_1^2 \Rightarrow V^2 < \mu/a^2 \Rightarrow \mu - a^2V^2 > 0 \quad \ldots(5)$

Re-writing (5), $a^4V^2(du/d\theta)^2 = (\mu - a^2V^2)(a^2u^2 - 1)$

so that $\dfrac{(\mu - a^2V^2)^{1/2}}{aV} d\theta = \dfrac{a\,du}{(a^2u^2-1)^{1/2}}$

Integrating, $(\mu - a^2V^2)^{1/2}(\theta/aV) = \cosh^{-1} au + B$, B is a constant ...(6)

Initially, when $u = 1/a$, let $\theta = 0$. Then (6) gives $B = 0$, as $\cosh^{-1} 1 = 0$

∴ (6) gives $\cosh\{(\mu - a^2V^2)^{1/2}(\theta/aV)\} = au = a/r$

or $r\cosh\{(\mu - a^2V^2)^{1/2}(\theta/aV)\} = a$ is the required path.

Case II. Let $V > V_1$, *i.e.*, V is > the velocity from infinity

Then $V^2 > V_1^2 \Rightarrow V^2 > \mu/a^2 \Rightarrow a^2V^2 - \mu > 0$ ...(7)

Then, from (3), $\dfrac{(a^2V^2-\mu)^{1/2}}{aV} d\theta = -\dfrac{a\,du}{(1-a^2u^2)^{1/2}}$

Integrating, $(a^2V^2 - \mu)^{1/2}(\theta/aV) = \cos^{-1}(au) + C$, C is a constant ...(8)

Initially, when $u = 1/a$, let $\theta = 0$. Then (8) gives $C = 0$

∴ (8) gives $\cos\{(a^2V^2 - \mu)^{1/2}(\theta/aV)\} = au = a/r$

or $r\cos\{(a^2V^2 - \mu)^{1/2}(\theta/aV)\} = a$ is the required path.

TYPE IV. Ex. 4. (*a*) *A particle moving under a constant force from a centre is projected at a distance a in a direction perpendicular to the radius vector with velocity acquired in falling to the point of projection from the centre, show that its path is* $(a/r)^3 = \cos^2(3\theta/2)$.

Also show that the particle will ultimately move in a straight line through the origin in the same way as if its path had always been this line.

If the velocity of projection be double that in the previous case show that the path is

$$\theta/2 = \tan^{-1}\{(r-a)/a\}^{1/2} - (1/\sqrt{3})\tan^{-1}\{(r-a)/3a\}^{1/2}$$

Sol. Here, repulsive central acceleration $= F = -f$, if f is a +ve const.

Let V be the velocity of the particle acquired in falling from the centre of repulsion under a force $(-F)$ to a distance a. Then, we have

$$V^2 = 2\int_0^a (-F)\,dr = 2\int_0^a (f)\,dr = 2af \quad \text{or} \quad V = (2af)^{1/2} \qquad ...(1)$$

The differnetial equation of the central orbit is

$$h^2(d^2u/d\theta^2 + u) = F/u^2 = -f/u^2$$

Multiplying both sides by $2(du/d\theta)$ and then integrating, we get

$v^2 = h^2\{(du/d\theta)^2 + u^2\} = 2f/u + A$, A is a constant. ...(2)

Initially the particle is projected in a direction perpendicular to the radius vector. Hence the point of projection is an apse. So initially at an apse, when $u = 1/r = 1/a$, $du/d\theta = 0$, $v = V = (2af)^{1/2}$

∴ (2) ⇒ $2af = h^2/a^2 = 2af + A$ so that $h^2 = 2fa^3$, $A = 0$.

Then (2) gives $2fa^3\{(du/d\theta)^2 + u^2\} = 2f/u$

or $a^3\left(\dfrac{du}{d\theta}\right)^2 = -a^3u^2 + \dfrac{1}{u} = \dfrac{1-a^3u^3}{u}$ or $d\theta = -\dfrac{a^{3/2}u^{1/2}\,du}{(1-a^3u^3)^{1/2}}$

Putting $a^{3/2}u^{3/2} = t$ and $(3/2)a^{3/2}u^{1/2}\,du = dt$, we get

$(3/2)d\theta = -\{1/(1-t^2)^{1/2}\}\,dt$ so that $(3/2)\theta = \cos^{-1} t + B$

or $(3/2)\theta = \cos^{-1}(a^{2/3}u^{3/2}) + B$, where B is a constant. ...(3)

Initially when $u = 1/a$, let $\theta = 0$. Then (3) gives $B = 0$

∴ (3) gives $\cos(3\theta/2) = a^{3/2}u^{3/2} = (a/r)^{3/2}$ or $(a/r)^3 = \cos^2(3\theta/2)$ ...(4)

Second Part. From (4), we find that as $r \to \infty$, $\cos(3\theta/2) \to 0$, *i.e.*, $3\theta/2 \to \pi/2$ or $\theta \to \pi/3$. It follows that the particle ultimately moves in a straight line through the centre of force, inclined at an angle $\pi/3$, in the same way as if its path had always been this line $\theta = \pi/3$.

Third Part. Proceed upto equation (2) as before. To find the value of A, use new conditions : $u = 1/a$, $du/d\theta = 0$, $v = 2V = 2(2af)^{1/2}$.

Now complete the solution yourself.

Ex. 4. (*b*) *A particle is acted on by a central repulsive force which varies as the nth power of this distance. If the velocity at any point be equal to that which would be acquired in falling from the centre to the point, show that the equation to the path is of the form $r^{(n+3)/2} \cos\{(n+3)\theta/2\}$ = constant.*

Sol. Try yourself as in Ex. 4. (*a*).

TYPE V. Ex. 5. (*a*) *In a central orbit the force is $\mu u^3(3 + 2a^2u^2)$; if the particle be projected at distance a with a velocity $(5\mu/a^2)^{1/2}$ in a direction making an angle $\tan^{-1}(1/2)$ with the radius vector, show that the equation to the path is $r = a \tan\theta$.* **[Kanpur 2011, 2016]**

Sol. Here, the central acceleration $F = \mu u^3(3 + 2a^2u^2)$ ...(1)

The differential equation of the central orbit is

$$h^2(d^2u/d\theta^2 + u) = F/u^2 = \mu(3u + 2a^2u^3), \text{ by (1)}$$

Multiplying both sides by $2(du/d\theta)$ and then integrating, we get

$$v^2 = h^2\{(du/d\theta)^2 + u^2\} = \mu(3u^2 + a^2u^4) + A, \ A \text{ is a constant} \quad ...(2)$$

According to the given problem, initially $\phi = \tan^{-1}(1/2)$ or $\tan\phi = 1/2$ so that $\sin\phi = 1/\sqrt{5}$. Also initially, $r = a$. Hence $p = r\sin\phi$ gives

$$p = a/\sqrt{5} \quad \text{or} \quad 1/p^2 = 5/a^2 \quad \text{or} \quad (du/d\theta)^2 + u^2 = 5/a^2$$

Thus, when $u = 1/r = 1/a$, $(du/d\theta)^2 + u^2 = 5/a^2$, $v = (5\mu/a^2)^{1/2}$. So (2) gives

$$5\mu/a^2 = h^2 \times (5/a^2) = \mu(3/a^2 + 1/a^2) + A \Rightarrow h^2 = \mu, \ A = \mu/a^2$$

$\therefore$ (2) gives $\quad \mu\{(du/d\theta)^2 + u^2\} = \mu(3u^2 + a^2u^4) + \mu/a^2$

$$\text{or} \quad \left(\frac{du}{d\theta}\right)^2 = 2u^2 + a^2u^4 + \frac{1}{a^2} = \frac{a^4u^4 + 2a^2u^2 + 1}{a^2} = \frac{(a^2u^2+1)^2}{a^2}$$

$$\text{or} \quad \frac{du}{d\theta} = -\frac{a^2u^2+1}{a} \quad \text{or} \quad -\frac{a\,du}{a^2u^2+1} = d\theta$$

Integrating, $\quad \cot^{-1}(au) = \theta + B \quad$ or $\quad \cot^{-1}(a/r) = \theta + B$ ...(3)

Initially, when $r = a$, let $\theta = \pi/4$. Then (3) gives $B = 0$.

$\therefore$ (3) gives $\quad a/r = \cot\theta \quad$ or $\quad r/a = \tan\theta \quad$ or $\quad r = a\tan\theta$.

Ex. 5. (*b*) *A particle subject to a force producing an acceleration $\mu(r + 2a)/r^5$ towards the origin is projected from the point $(a, 0)$ with a velocity equal to the velocity from infinity at an angle $\cot^{-1} 2$ with the initial line; show that the equation to the path is $r = a(1 + 2\cos\theta)$.* **(Avadh 2008)**

Sol. Here, central acceleration $= F = \mu(1/r^4 + 2a/r^5) = \mu(u^4 + 2au^5)$.

If V is the velocity of the particle acquired in falling from infinity under F to the point of projection situated at a distance a from the centre, then

$$V^2 = -2\int_\infty^a F\,dr = -2\int_\infty^a \mu\left(\frac{1}{r^4} + \frac{2a}{r^5}\right)dr = -2\mu\left[-\frac{1}{3r^3} - \frac{2a}{4r^4}\right]_\infty^a$$

or $\quad V^2 = 2\mu(1/3a^3 + 1/2a^3) = 5\mu/3a^3 \quad$ or $\quad V = (5\mu/3a^3)^{1/2}$ ...(1)

The differential equation of the central orbit is

$$h^2(d^2u/d\theta^2 + u) = F/u^2 = (\mu/u^2)(u^4 + 2au^5) = \mu(u^2 + 2au^3)$$

Multiplying both sides by $2(du/d\theta)$ and then integrating, we get

$$v^2 = h^2\{(du/d\theta)^2 + u^2\} = \mu(2u^3/3 + au^4) + A, \ A \text{ is a constant.} \quad ...(2)$$

According the given problem, initially $\phi = \cot^{-1} 2$ or $\cot\phi = 2$ so that so that $\sin\phi = 1/\sqrt{5}$. Also, initially $r = a$. Hence $p = r\sin\phi$ gives

$$p = a/\sqrt{5} \quad \text{or} \quad 1/p^2 = 5/a^2 \quad \text{or} \quad (du/d\theta)^2 + u^2 = 5/a^2.$$

Thus, when $u = 1/r = 1/a$, $(du/d\theta)^2 + u^2 = 5/a^2$, $v = V = (5\mu/3a^3)^{1/2}$

$\therefore$ (2) gives $\quad 5\mu/3a^3 = 5h^2/a^2 = \mu(2/3a^3 + a/a^4) + A \Rightarrow h^2 = \mu/3a, \ A = 0$

Then (2) gives $\quad (\mu/3a)\{(du/d\theta)^2 + u^2\} = \mu(2u^3/3 + au^4)$

or $\quad (du/d\theta)^2 = 2au^3 + 3a^2u^4 - u^2$ ...(3)

Putting $u = 1/r$ and $du/d\theta = -(1/r^2)(dr/d\theta)$, (3) reduces to

$$\frac{1}{r^4}\left(\frac{dr}{d\theta}\right)^2 = \frac{2a}{r^3}+\frac{3a^2}{r^4}-\frac{1}{r^2} = \frac{2ar+3a^2-r^2}{r^4} = \frac{(2a)^2-(r-a)^2}{r^4}$$

or $\dfrac{dr}{d\theta} = \{(2a)^2-(r-a)^2\}^{1/2}$ or $d\theta = \dfrac{dr}{\{(2a)^2-(r-a)^2\}^{1/2}}$

Integrating, $\theta + B = \sin^{-1}\{(r-a)/2a\}$ ...(4)

Initially, when $r = a$, let $\theta = 0$.

Then (4) gives $B = 0$. So, by (4)

$\theta = \sin^{-1}\{(r-a)/2a\}$ or $(r-a)/2a = \sin\theta$ or $r = a(1 + 2\sin\theta)$

Ex. 5. (*c*) *A particle of mass m moves under a central force* $m\mu/(\text{distance})^3$ *and is projected at a distance a from the centre of force with velocity which at an angle α to the radius vector would be acquired by a fall from rest at infinity to the point of projection, prove that the orbit is equiangular spiral.*

Sol. Here, central acceleration per unit mass $= F = \mu/r^3 = \mu u^3$

If V is the velocity of the particle at a distance a in falling from rest from infinity under the same acceleration, then

$$V^2 = -2\int_{\infty}^{a} F\,dr = -2\int_{\infty}^{a}\frac{\mu}{r^3}dr = -2\mu\left[-\frac{1}{2r^2}\right]_{\infty}^{a} = \frac{\mu}{a^2} \quad\text{or}\quad V = \frac{\mu^{1/2}}{a}$$

The differential equation of the central orbit is

$$h^2(d^2u/d\theta^2 + u) = F/u^2 = \mu u^3/u^2 = \mu u$$

Multiplying both sides by $2(du/d\theta)$ and then integrating, we get

$$v^2 = h^2\{(du/d\theta)^2 + u^2\} = \mu u^2 + A, \text{ when } A \text{ is a constant.} \quad ...(1)$$

According to the given problem, initially $\phi = \alpha$ and $r = a$ and so $p = r\sin\phi \Rightarrow p = a\sin\alpha \Rightarrow 1/p^2 = 1/(a^2\sin^2\alpha) \Rightarrow (du/d\theta)^2 + u^2 = 1/a^2\sin^2\alpha$.

Thus, when $u = 1/r = 1/a$, $(du/d\theta)^2 + u^2 = 1/(a^2\sin^2\alpha)$, $v = \mu^{1/2}/a$.

$\therefore$ (1) $\Rightarrow$ $\mu/a^2 = h^2/(a^2\sin^2\alpha) = \mu/a^2 + A$ so that $h^2 = \mu\sin^2\alpha$, $A = 0$

Then (1) gives $\mu\sin^2\alpha\{(du/d\theta)^2 + u^2\} = \mu u^2$

or $(du/d\theta)^2 = u^2\operatorname{cosec}^2\alpha - u^2 = u^2(\operatorname{cosec}^2\alpha - 1) = u^2\cot^2\alpha$ ...(2)

Putting $u = 1/r$ and $du/d\theta = -(1/r^2)(dr/d\theta)$, (2) becomes

$(1/r^4)(dr/d\theta)^2 = (1/r^2)\cot^2\alpha$ or $(1/r)\,dr = (\cot\alpha)\,d\theta$

Integrating, $\log r = \theta\cot\alpha + B$, B is a constant. ...(3)

Initially, when $r = a$, let $\theta = 0$. Then (3) $\Rightarrow B = \log a$

So (3) becomes $\log r = \theta\cot\alpha + \log a$ or $\log(r/a) = \theta\cot\alpha$

Thus, $r = ae^{\theta\cot\alpha}$, which is an equiangular spiral.

Ex. 5. (*d*) *A particle acted by a central attractive force* μu^3 *is projected with velocity* $\mu^{1/2}/a$ *at an angle π/4 with its initial distance a from the centre of force. Show that its orbit is the equiangular spiral* $r = ae^{-\theta}$.

Hint. Do as in Ex. 5. (*c*). Here r decreases while θ increases so take –ve sign while finding square root of $(dr/d\theta)^2$.

Ex. 5. (*e*) *A particle moves with a central acceleration* $\mu(u^5 - a^2u^7/8)$*; it is projected at a distance a with a velocity* $(25/7)^{1/2}$ *times the velocity for a circle at that distance and at an inclination* $\tan^{-1}(4/3)$ *to the radius vector, show that its path is the curve* $4r^2 - a^2 = 3a^2/(1-\theta)^2$.

Sol. Here, central acceleration $= F = \mu(u^5 - a^2u^7/8) = \mu(1/r^5 - a^2/8r^7)$

Let V be velocity for a circle at a distance a under the same acceleration F.

Then, $\dfrac{V^2}{a} = [F]_{r=a} = \mu\left(\dfrac{1}{r^5} - \dfrac{a^2}{8r^7}\right)_{r=a} = \dfrac{7\mu}{8a^5} \Rightarrow V = (7\mu/8a^5)^{1/2}$

$\therefore$ Velocity of projection $= (25/7)^{1/2}\times V = (25/7)^{1/2}\times(7\mu/8a^5)^{1/2} = (25\mu/8a^4)^{1/2}$

The differential equation of the central orbit is

$$h^2(d^2u/d\theta^2 + u) = F/u^2 = (\mu/u^2)(u^5 - a^2u^7/8) = \mu(u^3 - a^2u^5/8)$$

Multiplying both sides by $2(du/d\theta)$ and then integrating, we get

$$v^2 = h^2\{(du/d\theta)^2 + u^2\} = \mu(u^4/2 - a^2u^6/24) + A, \text{ A is constant.} \quad ...(1)$$

According to the given problem, initially $\phi = \tan^{-1}(4/3)$ or $\tan\phi = 4/3$ so that $\sin\phi = 4/5$. Also, initially, $r = a$. Hence $p = r\sin\phi$ gives

$$p = 4a/5 \quad \text{or} \quad 1/p^2 = 25/16a^2 \quad \text{or} \quad (du/d\theta)^2 + u^2 = 25/16a^2$$

Thus, when $u = 1/r = 1/a$, $(du/d\theta)^2 + u^2 = 25/16a^2$, $v = (25\mu/8a^4)^{1/2}$

$$\therefore (1) \Rightarrow \frac{25\mu}{8a^4} = \frac{25h^2}{16a^2} = \mu\left(\frac{1}{2a^4} - \frac{a^2}{24a^6}\right) + A \Rightarrow h^2 = \frac{2\mu}{a^2},\ A = \frac{8\mu}{3a^4}$$

Then (1) gives $$\frac{2\mu}{a^2}\left\{\left(\frac{du}{d\theta}\right)^2 + u^2\right\} = \mu\left(\frac{u^4}{2} - \frac{a^2u^6}{24}\right) + \frac{8\mu}{3a^4}$$

or $$\left(\frac{du}{d\theta}\right)^2 = \frac{a^2u^4}{4} - \frac{a^4u^6}{48} - \frac{4}{3a^2} - u^2$$

or $$(du/d\theta)^2 = (64 - 48a^2u^2 + 12a^4u^4 - a^6u^6)/48a^2 = (4 - a^2u^2)^3/48a^2$$

Putting $u = 1/r$ so that $du/d\theta = -(1/r^2)(dr/d\theta)$, we get

$$\frac{1}{r^4}\left(\frac{dr}{d\theta}\right)^2 = \frac{1}{48a^2}\left(4 - \frac{a^2}{r^2}\right)^3 = \frac{1}{48a^2r^6}(4r^2 - a^2)^3$$

or $$\left(\frac{dr}{d\theta}\right)^2 = \frac{(4r^2 - a^2)^3}{48a^2r^2} \quad \text{or} \quad \frac{dr}{d\theta} = \frac{(4r^2 - a^2)^{3/2}}{4ar\sqrt{3}}$$

or $$d\theta = \frac{4\sqrt{3}\,ar\,dr}{(4r^2 - a^2)^{3/2}} \quad \text{or} \quad d\theta = \left(\frac{a\sqrt{3}}{2}\right)(4r^2 - a^2)^{-3/2}(8r)\,dr$$

Integrating, $$\theta + B = (a\sqrt{3}/2)(-1/12)(4r^2 - a^2)^{-1/2} \quad ...(2)$$

Initially, when $r = a$, let $\theta = 0$. Then (2) gives $B = -1$.

Then (2) gives $$\theta - 1 = -\frac{a\sqrt{3}}{(4r^2 - a^2)^{1/2}} \quad \text{or} \quad 4r^2 - a^2 = \frac{3a^2}{(1-\theta)^2}$$

Ex. 5. (*f*) *A particle moves with a central acceleration $\mu/(\text{distance})^5$ and projected from the apse at a distance a with a velocity equal to n times that which would be acquired in falling from infinity; show that the other apsidal distance is $a/(n^2 - 1)^{1/2}$.* **[Purvanchel 2011; I.F.S. 2003]**

If n = 1 and particle be projected in any direction show that the path is a circle passing through the centre of force. **[Rohilkhand 2011]**

Sol. Try yourself.

Ex. 5. (*g*) *If the acceleration at a distance r is μ/r^5 and the particle is projected at a distance a from the centre of force with velocity $(\mu/2a^4)^{1/2}$, prove that the orbit is a circle through O of diameter a cosec α, where α is the inclination of the direction of projection to the radius vector.*

Sol. Try yourself.

TYPE VI. Ex. 6. (*a*) *A particle subject to the central acceleration $(\mu/r^3) + f$ is projected from an apse at a distance 'a' with the velocity $\mu^{1/2}/a$; prove that at any subsequent time t, $r = a - (1/2)ft^2$.*

Sol. Here, central acceleration $= F = \mu/r^3 + f = \mu u^3 + f$, if $u = 1/r$.

The differential equation of the central orbit is

$$h^2(d^2u/d\theta^2 + u) = F/u^2 = (\mu u^3 + f)/u^2 = \mu u + f/u^2$$

Multiplying both sides by $2(du/d\theta)$ and then integrating, we get

$$v^2 = h^2\{(du/d\theta)^2 + u^2\} = \mu u^2 - (2f/u) + A, \text{ A is a constant.} \quad ...(1)$$

Initially at an apse : $u = 1/r = 1/a$, $du/d\theta = 0$, $v = \mu^{1/2}/a$.

$$\therefore (1) \Rightarrow \mu/a^2 = h^2/a^2 = \mu/a^2 - 2fa + A \Rightarrow h^2 = \mu,\ A = 2fa \quad ...(2)$$

Then (1) gives $$\mu\{(du/d\theta)^2 + u^2\} = \mu u^2 - (2f/u) + 2fa$$

or $$\mu(du/d\theta)^2 = 2fa - (2f/u) \quad ...(3)$$

Putting $u = 1/r$ and $du/d\theta = -(1/r^2)(dr/d\theta)$, (2) gives

$$(\mu/r^4)(dr/d\theta)^2 = 2fa - 2fr = 2f(a-r)$$

or $$(dr/d\theta)^2 = (2fr^4/\mu)(a-r) \quad \text{or} \quad dr/d\theta = (2f/\mu)^{1/2} r^2 (a-r)^{1/2} \qquad \ldots(4)$$

$\therefore$ Now, $$h = r^2 \frac{d\theta}{dt} = r^2 \frac{d\theta}{dr}\frac{dr}{dt}$$

or $$\mu^{1/2} = r^2(\mu/2f)^{1/2} \frac{1}{r^2(a-r)^{1/2}} \frac{dr}{dt}, \text{ by (2) and (4)}$$

or $$dt = (1/\sqrt{2f})(a-r)^{-1/2} dr$$

Integrating, $$t = -(1/\sqrt{2f})\, 2(a-r)^{1/2} + B, \; B \text{ is constant.} \qquad \ldots(5)$$

Initially, when $t = 0$, $r = a$. So (5) gives $B = 0$. Then (5) becomes

$t = -(1/\sqrt{2f}) \times 2(a-r)^{1/2}$ or $t^2 = (2/f)(a-r)$ or $r = a - (1/2)ft^2$

Ex. 6. (*b*) *A particle moves under repulsive force mμ/(distance)^3 and is projected from an apse at a distance a with a velocity V, show that the equation to the path is r cos pθ = a, and that the angle θ described in time t is (1/p) tan^{-1} (pVt/a), where $p^2 = (\mu + a^2V^2)/(a^2V^2)$.*

[Rohilkhand 2001, 02; Agra 1992; Meerut 2003]

Sol. Here central acceleration $= F = -\mu/r^3 = -\mu u^3$, where $u = 1/r$.

The differential equation of the central orbit is

$$h^2(d^2u/d\theta^2 + u) = F/u^2 = -\mu u.$$

Multiplying both sides by $2(du/d\theta)$ and then integrating, we get

$$v^2 = h^2\{(du/d\theta)^2 + u^2\} = -\mu u^2 + A, \; A \text{ is a constant.} \qquad \ldots(1)$$

Initially at an apse : $u = 1/r = 1/a$, $du/d\theta = 0$, $v = V$. So (1) gives

$$V^2 = h^2/a^2 = -\mu/a^2 + A \quad \text{so that} \quad h^2 = a^2V^2, \; A = V^2 + \mu/a^2 \qquad \ldots(2)$$

$\therefore$ (1) becomes $a^2V^2\{(du/d\theta)^2 + u^2\} = -\mu u^2 + V^2 + \mu/a^2$

or $$a^2V^2(du/d\theta)^2 = -(a^2V^2 + \mu)u^2 + (a^2V^2 + \mu)/a^2$$

or $$a^2\left(\frac{du}{d\theta}\right)^2 = \frac{(a^2V^2 + \mu)(1 - a^2u^2)}{a^2V^2} = p^2(1 - a^2u^2), \text{ where } p^2 = \frac{\mu + a^2V^2}{a^2V^2}$$

or $$a\frac{du}{d\theta} = -p(1 - a^2u^2)^{1/2} \quad \text{or} \quad p\, d\theta = -\frac{a\, du}{(1 - a^2u^2)^{1/2}}$$

Integrating, $$p\theta + B = \cos^{-1}(au) \quad \text{or} \quad p\theta + B = \cos^{-1}(a/r) \qquad \ldots(3)$$

Initially, when $r = a$, let $\theta = 0$. Then (3) gives $B = 0$

$\therefore$ (2) becomes $$\cos p\theta = a/r \quad \text{or} \quad r\cos p\theta = a \qquad \ldots(4)$$

Second Part. In a central orbit, we know that

$$h = r^2(d\theta/dt) \quad \text{or} \quad aV = a^2 \sec^2 p\theta\, (d\theta/dt), \text{ by (2) and (4)}$$

or $$dt = (a/V) \times \sec^2 p\theta\, d\theta$$

Integrating, $$t + C = (a/pV)\tan p\theta, \; C \text{ is a constant.} \qquad \ldots(5)$$

Initially, when $t = 0$, let $\theta = 0$. Then (5) gives $C = 0$

So (5) becomes $$\tan p\theta = pVt/a \quad \text{or} \quad \theta = (1/p)\tan^{-1}(pVt/a)$$

Ex. 6. (*c*) *A particle of mass m moves under a central attractive force mμ(5/r^3 + 8c^2/r^5) and is prjected from an apse at a distance c with velocity (3μ$^{1/2}$)/c; prove that the orbit is r = c cos (2θ/3) and that it will arrive at the origin after a time πc^2/(8μ$^{1/2}$).* **[Garhwal 2001; Kolkata 2003]**

Sol. Here central acceleration $= F = \mu(5u^3 + 8c^2u^5)$, where $u = 1/r$

Hence differential equation of the central orbit is

$$h^2(d^2u/d\theta^2 + u) = F/u^2 = \mu(5u + 8c^2u^3)$$

Multiplying both sides by $2(du/d\theta)$ and then integrating, we get

$$v^2 = h^2\{(du/d\theta)^2 + u^2\} = \mu(5u^2 + 4c^2u^4) + A \qquad \ldots(1)$$

Initially at an apse, $u = 1/r = 1/a$, $du/d\theta = 0$, $v = (3\mu^{1/2})/c$

$$\therefore (1) \Rightarrow \quad 9\mu/c^2 = h^2/c^2 = \mu(5/c^2 + 4/c^2) + A \Rightarrow h^2 = 9\mu, \; A = 0 \qquad \ldots(2)$$

Then, (1) gives $$9\mu\{(du/d\theta)^2 + u^2\} = \mu(5u^2 + 4c^2u^4)$$

or $$9(du/d\theta)^2 = 4c^2u^4 - 4u^2 \qquad \ldots(3)$$

Putting $u = 1/r$ and $du/d\theta = -(1/r^2)(dr/d\theta)$, (3) reduces to
$(9/r^4)(dr/d\theta)^2 = 4(c^2/r^4 - 1/r^2)$ or $9(dr/d\theta)^2 = 4(c^2 - r^2)$

or $$\frac{dr}{d\theta} = -\frac{2(c^2-r^2)^{1/2}}{3} \quad \text{or} \quad \frac{2}{3}d\theta = -\frac{dr}{(c^2-r^2)^{1/2}}$$

Integrating, $2\theta/3 + B = \cos^{-1}(r/c)$, B is a constant.
Initially, when $r = c$, let $\theta = 0$. Then $B = \cos^{-1} 1 = 0$.
Then (3) gives $\cos(2\theta/3) = r/c$ or $r = c\cos(2\theta/3)$ ...(4)

Second Part. For central orbit, $h = r^2(d\theta/dt)$
or $3\mu^{1/2} = c^2\cos^2(2\theta/3)\cdot(d\theta/dt)$, by (2) and (4)
or $dt = (c^2/3\mu^{1/2})\cos^2(2\theta/3)\, d\theta$ ...(5)

Puting $r = 0$ in (4), we get $\cos(2\theta/3) = 0$ or $2\theta/3 = \pi/2$ or $\theta = 3\pi/4$. Thus, at point of projection, we have $t = 0$, $\theta = 0$ (by our assumption) and at centre of force O; where $r = 0$, $\theta = 3\pi/4$ and let $t = t_1$. So t_1 is the required time from the point of projection upto centre of force. So integrating (5) we have

$$\int_0^{t_1} dt = \int_0^{3\pi/4} \frac{c^2}{2\mu^{1/2}}\cos^2\frac{2\theta}{3}\, d\theta \quad \text{...(6)}$$

Put $2\theta/3 = \phi$ so that $(2/3)d\theta = d\phi$. When $\theta = 0$, $\phi = 0$ and when $\theta = 3\pi/4$, $\phi = \pi/2$. So (6) reduces to

$$t_1 = \int_0^{\pi/2} \frac{c^2}{3\mu^{1/2}}\cos^2\phi \times \left(\frac{3}{2}\right) d\phi = \frac{c^2}{2\mu^{1/2}}\int_0^{\pi/2}\cos^2\phi\, d\phi = \frac{c^2}{2\mu^{1/2}}\cdot\frac{1}{2}\cdot\frac{\pi}{2}$$

Thus, $t_1 = (\pi c^2/8\mu^{1/2})$, which is the required time.

Ex. 6. (*d*) *A particle, acted on by a repulsive central force $\mu r/(r^2 - 9c^2)^2$; is projected from an apse at a distance c with velocity $(\mu/8c^2)^{1/2}$. Show that it will describe a three cusped hyocycloid and that the time to the cusp is $(4/3)\pi c^2(2/\mu)^{1/2}$.* **[I.A.S. 1982]**

Sol. Here central acceleration per unit mass $= F = -\mu r/(r^2 - 9c^2)^2$, where negative sign is taken because the force is repulsive.

Hence the differential equation of central orbit in polar form is

$$\frac{h^2}{p^3}\frac{dp}{dr} = F = -\frac{\mu r}{(r^2-9c^2)^{1/2}} \quad \text{or} \quad -\frac{2h^2 dp}{p^3} = \frac{2\mu r\, dr}{(r^2-9c^2)^2}$$

Integrating, $$v^2 = \frac{h^2}{p^2} = -\frac{\mu}{(r^2-9c^2)} + A, \text{ where } A \text{ is constant.} \quad \text{...(1)}$$

At an apse, $p = r = c$, $v = (\mu/8c^2)^{1/2}$. So (1) becomes

$$\frac{\mu}{8c^2} = \frac{h^2}{c^2} = -\frac{\mu}{c^2-9c^2} + A \quad \text{so that} \quad h^2 = \frac{\mu}{8} \text{ and } A = 0. \quad \text{...(2)}$$

Then (1) gives $\mu/8p^2 = -\mu/(r^2 - 9c^2)$ or $8p^2 = 9c^2 - r^2$,
which is pedal equation of a three-cusped hypocycloid. ...(3)

Second part. For a central orbit, $v = ds/dt = h/p$
Hence $h\,dt = p\,ds = p(ds/dr)dr = (p/\cos\phi)dr$, as $dr/ds = \cos\phi$

or $$h\,dt = \frac{p\,dr}{(1-\sin^2\phi)^{1/2}} = \frac{p\,dr}{\{1-(p/r)^2\}^{1/2}}, \text{ as } p = r\sin\phi$$

or $$h\,dt = \frac{pr\,dr}{(r^2-p^2)^{1/2}} = \frac{p(-8p)\,dp}{(9c^2-8p^2-p^2)^{1/2}}, \quad \text{... (4)}$$

as by (3), $r^2 = 9c^2 - 8p^2$ and $r\,dr = -8p\,dp$

Now we are to find time t_1 from the point of projection (*i.e.*, apse where $p = c$) to the cusp (where $p = 0$). So integrating (4), we get

$$h\int_0^{t_1} dt = -\frac{8}{3}\int_c^0 \frac{p^2\,dp}{(c^2-p^2)^{1/2}} \quad \text{or} \quad t_1 = \frac{8}{3h}\int_0^c \frac{p^2\,dp}{(c^2-p^2)^{1/2}}$$

or $$t_1 = \frac{8}{3h}\int_0^{\pi/2} \frac{c^2 \sin^2\theta \cdot c\cos\theta\, d\theta}{c\cos\theta}, \text{ putting } p = c\sin\theta,\ dp = c\cos\theta\, d\theta$$

or $$t_1 = \frac{8c^2}{3h}\int_0^{\pi/2} \sin^2\theta\, d\theta = \frac{8c^2}{3h}\cdot\frac{1}{2}\cdot\frac{\pi}{2} = \frac{2\pi c^2}{3h} = \frac{2\pi c^2}{3(\mu^{1/2}/2\sqrt{2})}, \text{ by (2)}.$$

Thus, $t_1 = (4\pi c^2/3) \times (2/\mu)^{1/2}$, which is the required time.

Ex. 6. (*e*) *A particle moves with a central acceleration $\mu(3u^3 + a^2u^5)$ being projected from a distance a at an angle 45° with a velocity equal to that in a circle at the same distance. Prove that the time to the centre of force is $(a^2/\sqrt{2}\mu)\ (2 - \pi/2)$.*

Ex. 6. (*f*) *A particle is projected with velocity $(2\mu/3c^3)^{1/2}$ from a point P in a field of attractive force μ/r^4 to a point O distant c from P, where r denotes the distance from O. If the direction of projection makes an angle 45° with PO, prove that the orbit is a cardiod and the particle will arrive at O after a time $(3\pi/4 - 2)\ (3c^5/\mu)^{1/2}$.*

Ex. 6. (*g*) *A particle under a central acceleration μ/r^7 is projected from a point at a distance a with a velocity from infinity at an angle α with the initial line. Prove that the time of arriving at the centre of force is $(1/4)\ (a/V)\sec^2(\alpha/2)$, where V is the velocity of projection.*

Ex. 6. (*h*) *Show that a particle can describe a rectangular hyperbola under a force from a fixed centre varying as the distance and show that the time the radius vector to the particle from the centre takes in sweeping out an angle θ from vertex is given by $\tan\theta = \tanh(t\mu^{1/2})$, μ being acceleration at unit distance.*

Ex. 6. (*i*) *A particle under a central acceleration μr^2 is projected at a distance a with a velocity from infinity. Prove that the orbit is an equi-angular spiral and the time to reach a distance r from the centre is given by $(1/2) \times (r^2 - a^2) \times (1/\mu^{1/2})\sec\beta$, where β is the angle in the spiral between the tangent and radius vector.*

Ex. 6. (*j*) *Show that if the law of force is $\mu(r - a)^{-2}$ and the particle is projected with a velocity from infinity from a point at a distance c, (2a > c > a), at an angle $2\cos^{-1}(a/c)^{1/2}$, it will describe the orbit $\theta/2 = \tanh^{-1}\{(r - a)/a\}^{1/2} - \tan^{-1}\{(r - a)/a\}^{1/2}$.*

Sol. Try the above exercises yourself.

MISCELLANEOUS PROBLEMS ON CENTRAL ORBIT

1. A particle of unit mass is describing an elliptic or orbit of axes $2a$ and $2b$ under an attraction ur it the centre C when it is at a distance r from C, prove that its speed is gien by the $v^2 = \mu(a^2 + b^2 - r^2)$ and that the moment of the velocity about C is $\sqrt{\mu}\, ab$. If in the arbit a= 2b and the intensity of attraction is suddenly increased in the ratio1 : 4, when the particle is at a distance 3b/2 from C, prove that the axes of new orbit are : $(\sqrt{79} \pm \sqrt{15})b1/4$. **[Delhi Maths (H) 2005]**

2. A particle is attached to a fixed point on a smooth horizontal plane by an elastic string of natural length a. Initially the particle is at rest on the plane with the string just taut and it is projected horizontally in a direction perpendicular to the string with a kinetic energy equal to the potential energy of the string when its extension is $3a\sqrt{2}$. Prove that the second apsidal distance is equal to $3a$.

3. A particle is attracted towards a fixed point called centre by a force μ/r^2 per unit mass, μ being a constant and r the distance from the centre. It is projected from a position P with a velocity of magnitude v_0 making an angle α with OP. Assuming that $OP < 2\mu/v_0^2$, show that the orbit is an ellipse. Determine (in terms of μ, v_0, α and the distance OP) the eccentricity of the orbit and the inclination of the major axis to OP. **[Delhi Maths (H) 1996]**

9.15 Stability of Circular orbits.

With a central attractive force ($F > 0$), we can get a circular orbit of any radius, by projecting the particle at right angles to the radius vector with that velocity which makes $h^2 = (1/u^3)F$. But the question arises : Are these circular orbits stable or unstable ? In other words, if slightly disturbed, will the remaining orbit lie close to the original orbit, or will it deviate from it. In what follows we shall assume that that the position and velocity of the particle have been disturbed and discuss the resulting motion under the original central force.

9.16 Condition for stable circular orbit. *A particle describes a path which is nearly a circle about the centre of force (= μu^n) at its centre; find the condition that this may be a stable motion.*

[Gorakhpur 1998; Delhi Math (H) 2004; 08

The differential equation of the central orbit is

$$d^2u/d\theta^2 + u = F/h^2u^2 = \mu u^{n-2}/h^2, \text{ as } F = \mu u^n \text{ (given)} \qquad \ldots(1)$$

If the path is a circle of radius $1/c$, then $u = c$ so that $d^2u/d\theta^2 = 0$.

So (1) becomes $c = \mu c^{n-2}/h^2$ or $h^2 = \mu c^{n-3}$ ...(2)

Using (2), (1) becomes $u + d^2u/d\theta^2 = (1/c^{n-3})u^{n-2}$ ...(3)

Suppose the particle is to be slightly displaced from the circular path in such a way that h remains unaltered (for example, suppose it is given a small additional velocity in a direction away from the centre of force by means of a blow, the perpendicular velocity being unaltered)

Putting $u = c + x$, where x is small in (3), we get

$$c + x + \frac{d^2x}{d\theta^2} = \frac{1}{c^{n-3}}(c+x)^{n-2} = \frac{c^{n-2}}{c^{n-3}}\left(1 + \frac{x}{c}\right)^{n-2}$$

or $c + x + (d^2x/d\theta^2) = c\{1 + (n-2) \times (x/c) + \ldots\} = c + (n-2)x$, ...(4)

where we have neglected squares and higher powers of x, *i.e.*, , assuming that x is always small. Then (4) reduces

$$d^2x/d\theta^2 = -(3-n)x \qquad \ldots(5)$$

Three cases arises :

Case I. When $n = 3$. Solution of (5) is $x = A\theta + B$,
so that x continually increases as θ increases; hence x is not always small and the orbit does not continue to be nearly circular. Thus circular orbit is unstable.

Case II. When $n > 3$, solution of (5) is $x = Ae^{\theta(n-3)^{1/2}} + Be^{-\theta(n-3)^{1/3}}$
so that x continually increases as θ increases; hence as in case I the orbit does not continue to be nearly circular. Thus circular orbit is unstable.

Case III. When $n < 3$, solution of (5) is $x = A\cos\{\theta(3-n)^{1/2} + B\}$, ...(6)
where A and B are constants. When $n < 3$, (5) is standard equation for simple harmonic motion. Hence (6) gives a stable orbit, *i.e.*, small oscillations about the circular orbit occur.

When $n < 3$, the approximation to the path is $u = c + x$

i.e., $u = c + A\cos\{\theta(3-n)^{1/2} + B\}$ ...(7)

The apsidal distances are given by $du/d\theta = 0$, *i.e.*, by
$\sin\{\theta(3-n)^{1/2} + B\} = 0$ giving $\theta(3-n)^{1/2} + B = r\pi$, $r = 0, 1, 2, \ldots$ ...(8)

Let $\theta_0, \theta_1, \theta_2, \ldots$ be the values of θ corresponding to $r = 0, 1, 2, \ldots$ respectively. Then (8) gives

$$\theta_0(3-n)^{1/2} + B = 0,\ \theta_1(3-n)^{1/2} + B = \pi,\ \theta_2(3-n)^{1/2} + B = 2\pi, \ldots$$

$$\therefore \qquad \theta_1 - \theta_0 = \theta_2 - \theta_1 = \theta_3 - \theta_2 = \ldots = \pi/(3-n)^{1/2},$$

showing that the apsidal angle of the path $= \pi/(3-n)^{1/2}$ ... (9)

Since the maximum and minimum values of $\cos\{\theta(3-n)^{1/2} + B\}$ are +1 and −1 respectively, hence (7) shows that the maximum and minimum values of u are $c + A$ and $c - A$ respectively. Hence the motion is included between these values and apses are $c + A$ and $c - A$.

Note. The circular orbits under the attractive force μu^n are stable if, and only if,

$$n < 3. \qquad \ldots(10)$$

Thus the law of direct distance ($n = -1$) and the law of the inverse square ($n = 2$) give stable circular orbits; the law of the inverse cube ($n = 3$) gives unstable circular orbits.

9.17 General case of condition for stability of circular orbit under central force. *A particle describes a path which is nearly a circle about the centre of force F(u) at its centre; to find the condition that this may be a stable motion.* **[Delhi Math (H) 1997, 1999, 2001, 2002]**

The differential equation of the central orbit is

$$d^2u/d\theta^2 + u = F(u)/h^2u^2 \quad ...(1)$$

If the path is a circle of radius $1/c$, then $u = c$ and $d^2u/d\theta^2 = 0$

So (1) becomes $\quad c = F(c)/h^2c^2 \quad$ or $\quad h^2 = (1/c^3)\,F(c) \quad ...(2)$

Using (2), (1) becomes $\quad u + d^2u/d\theta^2 = (c^3/u^2)\,\{F(u)/F(c)\} \quad ...(3)$

Suppose the particle is to be slightly displaced from the circular path in such a way that h remain unchanged.

Putting $u = c + x$, where x is small in (3), we get

$$c+x+\frac{d^2x}{d\theta^2}=\frac{c^3}{F(c)}\times\frac{F(c+x)}{(c+x)^2}=\frac{C}{F(c)}F(c+x)\left(1+\frac{x}{c}\right)^{-2} \quad ...(4)$$

Using Taylor's expansion and Binomial theorem, we get

$$F(c+x) = F(c) + x\,F'(c) \quad \text{and} \quad (1 + x/c)^{-2} = 1 - 2x/c, \quad ...(5)$$

neglecting squares and higher powers of x, *i.e.*, assuming that x is always small. Using (5), (4) reduces to

$$c+x+\frac{d^2x}{d\theta^2}=\frac{c}{F(c)}\{F(c)+x\,F'(c)\}\left(1-\frac{2x}{c}\right)$$

or $$c+x+\frac{d^2x}{d\theta^2}=\frac{c}{F(c)}\left\{F(c)+xF'(c)-\frac{2x\,F(c)}{c}\right\},\ \text{neglecting } x^2$$

or $$c+x+\frac{d^2x}{d\theta^2}=c+c\frac{F'(c)}{F(c)}x-2x \quad \text{or} \quad \frac{d^2x}{d\theta^2}=-\left\{3-\frac{cF'(c)}{F(c)}\right\}x$$

or $$d^2x/d\theta^2 = -\mu x, \quad ...(6)$$

where $$\mu = 3 - cF'(c)/F(c) \quad ...(7)$$

Three cases arise :

Case I. When $\mu = 0$ *i.e.*, $cF'(c)/F(c) = 3$, solution of (6) is $x = A\theta + B$ so that x continually increases as θ increases. Hence the circular orbit is unstable.

Case II. When $\mu < 0$. Let $\mu = -\lambda^2$, $\lambda > 0$. Then solution of (6) is $x = Ae^{\lambda\theta} + Be^{-\lambda\theta}$ so that x continually increasea as θ increases. Hence the circular orbit is unstable.

Cae III. When $\mu > 0$, (6) represents a simple harmonic solution and solution of (6) is

$$x = A\cos(\theta\sqrt{\mu} + B) \quad ...(8)$$

(8) gives a stable orbit *i.e.*, small oscillations about the circular orbit occur.

Hence the circular orbit of radius $1/c$ is stable if and only if

$$\mu > 0, \quad i.e., \quad cF'(c)/F(c) < 3 \quad ...(9)$$

Remark 1. Proceeding as in previous article, we have

$$\text{Apsidal angle} = \pi/\{3 - cF'(c)/F(c)\}^{1/2} \quad ...(10)$$

Remark 2. If $f(r)$ is the central force in place of $F(u)$, where $r = 1/u$. Then

$$F(u) = f(1/r) \quad \text{so that} \quad F'(u) = f'(1/u) \times (-1/u^2)$$

Then the stability condition (9) can be re-written as

$$\frac{cf'(1/c)\cdot(-1/c^2)}{f(1/c)}<3 \quad \text{or} \quad f'\left(\frac{1}{c}\right)+3c\,f\left(\frac{1}{c}\right)>0 \quad ...(11)$$

Let $1/c = a$ = radius of the circular orbit. Then from (11), we see that a circular orbit of radius a, with central force $f(r)$ is stable if and only if

$$f'(a) + (3/a)\,f(a) > 0$$

Remember this result for direct application.

SOLVED EXAMPLES BASED ON ART. 9.16 AND 9.17

Ex. 1. *A particle of mass m moves in a central field of attractive force of which the intensity is* $mkr^{-2}e^{-r^2}$ *where k is a constant. Prove that a circular orbit of radius r is stable if and only if* $r^2 < 1/2$. **[Delhi Maths (H), 2003, 2005]**

Sol. Here $f(r) = kr^{-2}\,e^{-r^2}$ is the central force per unit mass. Since the force is attractive type, so $k > 0$.

Now, $\quad f'(r) = -2kr^{-3}\,e^{-r^2} + kr^{-2}e^{-r^2}(-2r) = -2kr^{-3}\,e^{-r^2}(1 + r^2)$

We know that a circular orbit of radius r, with central force $f(r)$ is stable if and only if (See Remark 2 in Art. 9.17)

$$f'(r) + (3/r) f(r) > 0$$

or $\quad -2kr^{-3}(1+r^2)+3kr^{-3}e^{-r^2} > 0$, putting values of $f(r)$ and $f'(r)$

or $\quad kr^{-3}e^{-r^2}(1-2r^2)>0 \quad$ or $\quad 1-2r^2>0$, as $kr^{-3}e^{-r^2}$ is a +ve quantity

or $\quad 1-2r^2>0 \quad$ or $\quad r^2<1/2$, as required

Ex. 2. *A particle moves under the influence of a centre which attracts it with a force $(b/r^2 + c/r^4)$ per unit mass b and c being positive constants and r the distance from the centre. The particle moves in a circular orbit of radius a. Prove that the motion is stable if and only if, $a^2b > c$.*

[Delhi B.A. (Prog) III 2007, 2017; Delhi B.Sc. (Hons.) 2001, 06, 09]

Sol. Here $f(r) = (b/r^2 + c/r^4)$ is the central force per unit mass. Then,

$$f'(r) = -2b/r^3 - 4c/r^5$$

We know that a circular orbit of radius a, with central force $f(r)$ is stable if and only if (see remark 2 in Art. 9.17).

$$f'(a) + (3/a) f(a) > 0 \qquad \text{...(1)}$$

Here, $\quad f(a) = b/a^2 + c/a^4 \quad$ and $\quad f'(a) = -2b/a^3 - 4c/a^5$.

$\therefore$ (1) becomes $\quad -2b/a^3 - 4c/a^5 + (3/a)\times(b/a^2 + c/a^4) > 0$

or $\quad b/a^3 - c/a^4 > 0 \quad$ or $\quad a^2b - c > 0 \quad$ or $\quad a^2b > c$

Ex. 3. *Find the velocity necessary for the description of a circular orbit of radius a, under a central force $(2\mu a^2u^5 - \mu u^3)$. Show that the orbit is unstable.*

Sol. Here $\quad f(r) = \mu(2a^2/r^5 - 1/r^3)$, where $r = 1/u$ ...(1)

The required velocity is given by

$$v^2/a = f(a) \quad \text{or} \quad v^2 = a\mu(2/a^3 - 1/a^3) = \mu/a^2$$

From (1), $\quad f'(r) = \mu(-10a^2/r^6 + 3/r^4)$ ...(2)

$$\therefore \quad f'(a)+\frac{3f(a)}{a}=\mu\left(-\frac{10}{a^4}+\frac{3}{a^4}\right)+\frac{3\mu}{a}\left(\frac{2}{a^3}-\frac{1}{a^3}\right)=-\frac{4\mu}{a^4}<0,$$

as $\mu > 0$. So circular orbit of radius a is unstable, by refer remark 2 of Art. 9.17.

Ex. 4. *A particle describes a circle under the force μ/r^2 per unit mass towards the centre. Show that the motion is stable and the aspidal angle of the path is π.*

Sol. Refer Art. 9.16. Here central force $= \mu/r^2 = \mu u^2$ and hence $n = 2$ for the present problem. Since $n < 3$, the circular orbit is stable by result (10). Again, by result (9), the required apsidal angle $= \pi/(3-2)^{1/2} = \pi$.

Ex. 5. *If the nearly circular orbit of a particle be $p^2(a^{m-2} - r^{m-2}) = b^m$, show that the apsidal angle is $\pi/m^{1/2}$ nearly.*

Sol. Given orbit is

$$b^m/p^2 = a^{m-2} - r^{m-2} \qquad \text{...(1)}$$

Differentiating both sides (1) w.r.t. 'r', we get

$$-\frac{2b^m}{p^3}\frac{dp}{dr}=-(m-2)r^{m-3} \quad \text{or} \quad \frac{1}{p^3}\frac{dp}{dr}=\left(\frac{m-2}{2b^m}\right)r^{r-3} \qquad \text{...(2)}$$

We know that equation of the central orbit is pedal form is

$$F=\frac{h^2}{p^3}\frac{dp}{dr} \quad \text{or} \quad F=\frac{h^2(m-2)}{2b^m}r^{m-3}=\mu u^{3-m}\text{, by (2)} \qquad \text{...(3)}$$

where $\quad \mu = h^2(m-2)/2b^m$ ...(4)

(3) gives central force for the present problem. Refer Art. 9.16. Here $n = 3 - m$. So by result (9), the required apsidal angle $= \pi/\{3-(3-m)\}^{1/2} = \pi/m^{1/2}$.

Ex. 6. *Obtain the condition for stability of a circular orbit under a central field of attractive force. Hence if the central attractive force is $(\mu/r^2)e^{-\lambda r}$, where μ and λ are positive constants, prove that a circular orbit of radius a is stable provided $a\lambda < 1$. Also show that if $a\lambda$ is small, the apse-line is a nearly circular orbit of mean radius a will advance in each revolution through an angle $\pi a\lambda$ approximately.*

[Delhi B.Sc. (Hons.) 2004]

Sol. First part : Refer Art. 9.17. Thus a circular orbit of radius a, with central force $f(r)$ is stable if

$$f'(a) + (3/a)\,f(a) > 0. \qquad \ldots(1)$$

Second part : Here central attractive force $= f(r) = (\mu/r^2)e^{-\lambda r}$. $\ldots(2)$

From (2), $f'(r) = -(2\mu/r^3)e^{-\lambda r} - (\lambda\mu/r^2)e^{-\lambda r} = -(\mu/r^3) \times (2 + \lambda r)e^{-\lambda r}$ $\ldots(3)$

$\therefore$ From (2) and (3), $f(a) = (\mu/a^2)e^{-\lambda a}, f'(a) = -(\mu/a^3) \times (2 + \lambda a)e^{-\lambda a}$

$\therefore$ (1) becomes $-(\mu/a^3) \times (2 + \lambda a)e^{-\lambda a} + (3/a) \times (\mu/a^2)e^{-\lambda a} > 0$

or $(\mu/a^3) \times (1 - a\lambda)e^{-\lambda a} > 0$ or $1 - a\lambda > 0$ or $a\lambda < 1$,

which is the condition for circular orbit of radius a to be stable.

Third part : At an apse $r = a$, we know that the velocity v in circle of radius a is given by

$v^2/a = f(a)$ or $v^2 = a \times (\mu/a^2)e^{-\lambda a} = (\mu/a)e^{-\lambda a}$, by (2) $\ldots(4)$

Now, at the apse, $p = a$. Also, for central orbit $h = vp$. Then, we get

$h^2 = v^2p^2$ or $h^2 = (\mu/a)e^{-\lambda a}a^2 = a\mu\, e^{-\lambda a}$, by (4) $\ldots(5)$

The differential equation of the central orbit is

$$d^2u/d\theta^2 + u = F/h^2u^2 \qquad \ldots(6)$$

Let $u = x + 1/a$, $\ldots(7)$

where x is assumed to be a small displacement such that squares of x and higher powers of x may be neglected.

Since $u = 1/r$, so $F = \mu u^2\, e^{-\lambda/u}$, by (2) $\ldots(8)$

Using (5), (7) and (8), (6) reduces to

$$\frac{d^2x}{d\theta^2} + x + \frac{1}{a} = \frac{\mu u^2 e^{-\lambda u^{-1}}}{a\mu e^{-\lambda a} u^2} = \frac{1}{a} e^{\lambda a} e^{-\lambda u^{-1}} \qquad \ldots(9)$$

From (7), $u = (1 + ax)/a$ or $u^{-1} = a(1 + ax)^{-1}$

$\therefore$ $e^{-\lambda u^{-1}} = e^{-\lambda a(1 + ax)^{-1}} = e^{-\lambda a(1 - ax)}$

[neglecting x^2 and higher powers of x, x being small]

or $e^{-\lambda u^{-1}} = e^{-\lambda a}\, e^{\lambda a^2 x} = e^{-\lambda a}(1 + \lambda a^2 x)$, $\ldots(10)$

[neglecting x^2 and higher powers of x, x being small]

Using (10), (9) reduces to

$$d^2x/d\theta^2 + x + 1/a = (1/a) \times e^{\lambda a} \times e^{-\lambda a}\,(1 + \lambda a^2 x)$$

or $$d^2x/d\theta^2 + (1 - \lambda a)x = 0 \qquad \ldots(11)$$

Since $\lambda a < 1$, so $(1 - \lambda a)$ is a positive constant and hence (11) represents a simple harmonic motion. General solution of (11) is given by

$x = A\cos\{\theta(1 - \lambda a)^{1/2} + B\}$, where A and B are constants. $\ldots(12)$

Assume that the apse-line advances by an angle α in one revolution. Then, we must have

$[x]_{\theta = 0} = [x]_{\theta = 2\pi + \alpha}$

$\therefore$ $A\cos B = A\cos\{(1 - \lambda a)^{1/2}\,(2\pi + \alpha) + B\}$, by (12) $\ldots(13)$

But $(1 - \lambda a)^{1/2} = 1 - (1/2)\lambda a$, neglecting $(\lambda a)^2$ and higher powers of λa, as λa is given to be small

$\therefore$ (13) gives $\cos B = \cos\{(1 - \lambda a/2)\,(2\pi + \alpha) + B\}$

or $\cos B = \cos\{2\pi + \alpha(1 - \lambda a/2) + B - \pi a\lambda\}$

or $B = \alpha(1 - \lambda a/2) + B - \pi a\lambda$ or $\alpha(1 - \lambda a/2) = \pi a\lambda$

or $\alpha = \pi a\lambda(1 - \lambda a/2)^{-1} = \pi a\lambda(1 + \lambda a/2)$, neglecting $(\alpha\lambda)^2$ and higher powers of small quantity $a\lambda$

or $\alpha = \pi a\lambda$, approximately on neglecting $(\alpha\lambda)^2$, $\alpha\lambda$ being small.

EXERCISE

1. A particle moves in a nearly circular orbit of radius a under the central force $\mu r^3(3+2r^2)$, show that the apsidal angle is $\pi\{(3+2a^2)/(18+16a^2)\}^{1/2}$.

Hint. Use remark 1 of Art. 9.17.

2. A particle is moving in a circular orbit of radius a under a force $\mu u^3(2a^2u^2-1)$ towards the centre. Show that the orbit is unstable and if slightly disturbed, it will take the form $r = a \tanh\theta$ or $r = a \coth\theta$.

3. One end of an elastic string of unstretched length a is tied to a point on the top of a smooth table, and a point attached to the other end can move freely on the table. If the path be nearly circular of radius b, show that its apsidal angle is approximately $\pi[(b-a)/(4b-3a)\}^{1/2}$.

MISCELLANEOUS PROBLEMS

Ex. 1. *The equation of the path of a particle moving in a central orbit is*

(a) $F = h^2u^2\{u + (du/d\theta)^2\}$ *(b)* $F = h^2u^2\{u^2 + d^2u/d\theta^2)$

(c) $F = h^2u^2\{u + d^2u/d\theta^2\}$ *(d) None of above.*

Sol. Ans. (c) Refer equation (7) of Art. 9.3 A.

Ex. 2. *The pedal equation of the path of a central orbit is*

(a) $F = (h^2/p^3)\,(dp/dr)$ *(b)* $F = (p^2/h^2)\,(dr/dp)$

(c) $F = (h^2/p^3)\,(dr/dp)$ *(d) None of the above.* **[Agra 2010, 12]**

Sol. Ans. (a) Refer equation (11) of Art. 9.3B.

Ex. 3. *The periodic time of a planet moving under inverse square law of acceleration is (a)* $\pi a(a/\mu)^{1/2}$ *(b)* $2\pi a/\mu^{1/2}$ *(c)* $2\pi a(a/\mu)^{1/2}$ *(d)* $\pi(a/\mu)^{1/2}$.

Sol. Ans. (c) Refer Art. 9.9.

Ex. 4. *A particle moves in a plane with an acceleration which is always directed towards a fixed point O in the plane. It is found that the path is given by* $r = a(1 + 2\sin\theta)$. *The apsidal distances are equal to (a) 2a and a/2 (b) 3a and a (c) 4a and 2a (d) 5a and 3a.*

Sol. Ans. (b) $r = a(1+2\sin\theta) \Rightarrow dr/d\theta = 2a\cos\theta$. For apsidal distances, $dr/d\theta = 0$ so $\cos\theta = 0$ giving $\theta = \pi/2$ or $3\pi/2$. When $\theta = \pi/2$, $r = a(1+2) = 3a$; when $\theta = 3\pi/2$, $r = a(1-2) = -a$. So apsidal distances are a, $3a$.

Ex. 5. *What is the angle between radius vector and tangent at an apse ?* **[Meerut 2003]**

Sol. Ans. 90°. Refer theorem I on page 9.12.

Ex. 6. *Fill in the blanks "..............." so that the following statements are correct.*

(i) If a particle moves in an ellipse under a force which is always directed towards the focus, then the periodic time T is .... **[Kanpur 2002, 03]**

(ii) The differential equation of central orbit in polar form is .... **[Meerut 2005]**

(iii) The rate of description of sectorial area by radius vector to the centre of force is .... **[Meerut 2005]**

Ans. *(i)* $\{2\pi a^{3/2}\}/\mu^{1/2}$ *(ii)* $d^2u/d\theta^2 + u = F/h^2u^2$ *(iii)* constant

Ex. 7. *The polar equation of the central orbit is*

(a) $\dfrac{h^2}{p^3}\dfrac{dp}{dr} = P$ *(b)* $h^2u^2\left(\dfrac{d^2u}{d\theta^2} + u\right) = P$ *(c)* $hu^2 = \dfrac{d\theta}{dt}$ *(iv)* None of these

Sol. Ans. (b). Refer Art. 9.3A. Here $F = P$ **[Kanpur 2003]**

Ex. 8. *The differential equation* $(h^2/p^3) \times (dp/dr) = P$ *of the central orbit is*

(a) polar form (b) pedal form (c) curvilinear form (d) cartesian form. **[Kanpur 2002]**

Sol. Ans. (b) Refer Art. 9.3 B. Here $F = P$.

Ex. 9. *The differential equation* $h^2u^2(d^2u/d\theta^2+u)=P$ *of central orbit is in*
(a) cartesian form *(b) polar form* *(c) pedal form* *(d) curvilinear form* **[Kanpur 2007]**

Sol. Ans. (*b*) Refer Art. 9.3 A. Here $F = P$.

Ex. 10. *The rate of description of sectorial area is defined as*
(a) radial velocity *(b) transversal velocity* *(c) areal velocity* *(d) angular velocity.* **[Kanpur 2001]**

Sol. Ans. (*c*) Refer Art. 9.4

Ex. 11. *An apse is point in a central orbit at which the radius vector drawn from the centre of force to the orbit has a*

(a) constant velocity *(b) zero velocity*
(c) maximum or minimum velocity *(d) None of these* **[Kanpur 2002]**

Sol. Ans. (*c*) Refer Art. 9.11.

Ex. 12. *If the central force is* μ/r^5, *prove that the velocities at the two apsidal distances satisfy the relation* $v_1^2+v_2^2=2h^4/\mu$

Sol. Here central acceleration $= F = \mu/r^5 = \mu u^5$, where $u = 1/r$

The differential equation of the central orbit becomes

$$h^2(d^2u/d\theta^2+u)=F/u^2 \qquad \text{or} \qquad h^2(d^2u/d\theta^2+u)=\mu u^3$$

Multiplying both sides by $2(du/d\theta)$ and integrating, we get

$$v^2=h^2\{(du/d\theta)^2+u^2\}=(\mu u^4)/2+A, \ A \text{ being an arbitrary constant} \qquad \text{... (1)}$$

At an apse, $du/d\theta=0$, So (1) reduces to

$$h^2u^2=(\mu u^4)/2+A \qquad \text{or} \qquad \mu(u^2)^2-2h^2u^2+2A=0, \qquad \text{... (2)}$$

which is a quadratic is u^2. Let its roots be u_1^2 and u_2^2. Thus,

$$u_1^2+u_2^2=2h^2/\mu \qquad \text{and} \qquad u_1^2u_2^2=2A/\mu \qquad \text{... (3)}$$

From (1), $$v^2=(\mu u^4)/2+A \qquad \text{... (4)}$$

At $u = u_1$, let $v = v_1$ and at $u = u_2$, let $v = v_2$. Then, (4) gives

$$v_1^2=(\mu v_1^4)/2+A \qquad \text{and} \qquad v_2^2=(\mu v_2^4)/2+A$$

$$\therefore\ v_1^2+v_2^2=(\mu/2)\times(u_1^4+u_2^2)+2A=(\mu/2)\times(u_1^4+u_2^4)+(\mu u_1^2u_2^2)/2, \text{ using (3)}$$

$$=(\mu/2)\times(u_1^4+u_2^4+2u_1^2u_2^2)=(\mu/2)\times(u_1^2+u_2^2)^2=(\mu/2)\times(2h^2/\mu)^2=(2h^4)/\mu, \text{ using (2)}$$

Ex. 13. *Find the law of force in orbit* $r = a\sec^2(\theta/2)$, *the pole being the centre of attraction.* **[Agra 2006]**

Sol. Let $r = 1/u$. Then $r = a\sec^2(\theta/2) \Rightarrow 1/u = a\sec^2(\theta/2) \Rightarrow u = (1+\cos\theta)/2a$

$$\therefore \qquad du/d\theta=-(1/2a)\times\sin\theta \qquad \text{and} \qquad d^2u/d\theta^2=-(1/2a)\times\cos\theta$$

Putting the above values in equation of central orbit $d^2u/d\theta^2+u=F/h^2u^2$, we get

$$-(1/2a)\times\cos\theta+(1/2a)\times(1+\cos\theta)=F/h^2u^2 \qquad \text{or} \qquad 1/2a = Fr^2/h^2$$

$\Rightarrow F=(h^2/2a)\times(1/r^2) \Rightarrow$ Force varies inversely as square of this distance from the pole.

Ex. 14. *A particle is attracted towards a fixed centre by a force μ/r^2 per unit mass, μ being a constant and r the distance from the centre. It is projected from a position P with the velocity of magnitude v_0, making an angle α with OP. Assuming that $OP < 2\mu/v_0^2$, show that the orbit is an ellipse, determine (in terms of μ, v_0, α and the distance OP) the eccentricity of the orbit and the inclination of the major axis to OP.*

[Delhi Maths (H) 1996, 2007]

Sol. Let O be taken as pole. Let $P(a, 0)$ be the initial position of the particle at $t = 0$. Let $Q(r, \theta)$ be the position of the particle at any time t. Let PT be tangent to the path of the particle and let OM be perpendicular to PT. Let $OM = p_0$. Given that $\angle OPM = \alpha$. Then from figure, we have $OM = p_0 = a \sin \alpha$. Also, we know that $pv = h$ (a constant) so that $p_0 v_0 = h$.

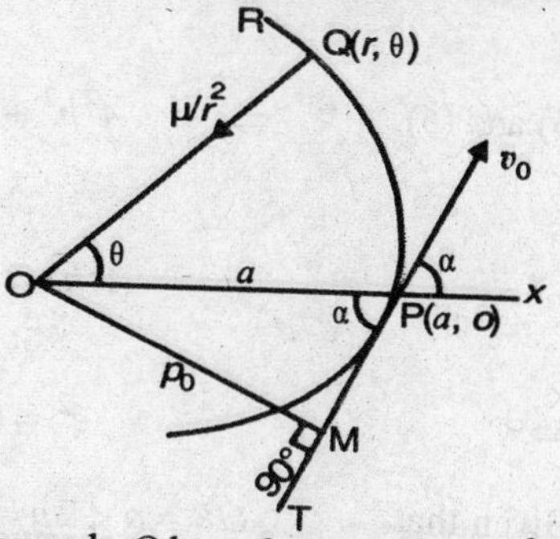

Thus, $\qquad h = av_0 \sin\alpha \qquad$ ... (1)

Also, the particle of unit mass at Q is being attracted towards O by a force $= F = \mu/r^2 = \mu u^2$, where $u = 1/r$. The differential equation $d^2u/d\theta^2 + u = F/h^2u^2$ of the central orbit reduces to

$$\frac{d^2u}{d\theta} + u = \frac{\mu}{h^2} \quad \text{or} \quad \frac{d^2}{d\theta^2}\left(u - \frac{\mu}{h^2}\right) + \left(u - \frac{\mu}{h^2}\right) = 0, \quad \text{as } \frac{\mu}{h^2} \text{ is a constant}$$

or $\qquad (D^2 + 1)(u - \mu/h^2) = 0, \qquad$ where $\qquad D \equiv d/d\theta,$

which general solution is $\qquad u - \mu/h^2 = A\cos(\theta - \theta_0)$

or $\qquad u = \mu/h^2 + A\cos(\theta - \theta_0)$, A and θ_0 being arbitrary constants $\qquad$ ... (2)

Re-writing (2), we have $\qquad \dfrac{h^2/\mu}{r} = 1 + \dfrac{Ah^2}{\mu}\cos(\theta - \theta_0) \qquad$ ... (3)

Comparing (3) with standard equation $l/r = 1 + e\cos(\theta - \theta_0)$ of a conic, we see that (3) represents a conic with semi-latus rectum $= l = h^2/\mu$ and eccentricity $= e = Ah^2/\mu$, having axis of the conic inclined at an angle θ_0 to the initial line OX.

We now proceed to show that (3) is an ellipse and for that purpose we must show that

$$e = Ah^2/\mu < 1.$$

The energy equation of the central orbit is given by

$$v^2 = -2\int F dr + B, \quad B \text{ being an arbitrary constant}$$

or $\qquad v^2 = -2\mu\displaystyle\int \frac{1}{r^2} dr + B \quad$ or $\quad v^2 = \dfrac{2\mu}{r} + B, \quad$ as $\quad F = \dfrac{\mu}{r^2} \qquad$ ... (4)

Initially at P, when $r = a$, $v = v_0$, so $\qquad (4) \Rightarrow \qquad B = v_0^2 - (2\mu/a)$

$\therefore$ (4) reduces to $\qquad v = 2\mu(1/r - 1/a) + v_0^2$

or $\qquad v^2 = 2\mu\left\{\dfrac{\mu}{h^2} + A\cos(\theta - \theta_0)\right\} - \dfrac{2\mu}{a} + v_0^2$, using (2) $\qquad$ ... (5)

Again, we have $$v^2 = h^2\{(du/d\theta)^2 + u^2\}$$

or $$v^2 = h^2\left[\{-A\sin(\theta-\theta_0)\}^2 + \{\mu/h^2 + A\cos(\theta-\theta_0)\}^2\right], \text{ using (2)}$$

or $$v^2 = h^2\{A^2 + \mu^2/h^4 + (2A\mu/h^2)\times\cos(\theta-\theta_0)\} \qquad \text{... (6)}$$

(5) and (6) $$\Rightarrow \quad A^2h^2 + \frac{\mu^2}{h^2} + 2A\mu\cos(\theta-\theta_0) = 2\mu\left\{\frac{\mu}{h^2} + A\cos(\theta-\theta_0)\right\} - \frac{2\mu}{a} + v_0^2$$

or $$A^2h^2 = \frac{\mu^2}{h^2} - \frac{2\mu}{a} + v_0^2 = \frac{\mu^2}{h^2}\left\{1 - \frac{h^2}{\mu^2}\left(\frac{2\mu}{a} - v_0^2\right)\right\}$$

$$\Rightarrow \quad e^2 = (Ah^2/\mu)^2 = 1 - (h^2/\mu^2)\times(2\mu/a - v_a^2) \qquad \text{... (7)}$$

Given that $OP = a < 2\mu/v_0^2 \quad \Rightarrow \quad 2\mu/v_0^2 > a \quad \Rightarrow \quad 2\mu/a - v_0^2 > 0$

Hence (7) $\Rightarrow \quad e^2 < 1 \quad \Rightarrow \quad e < 1 \quad \Rightarrow$ (3) is an ellipse

From (7), $$e = \{1 - (h^2/\mu^2)\times(2\mu/a - v_0^2)\}^{1/2}$$

or $$e = \left\{1 - \frac{a^2v_0^2\sin^2\alpha}{\mu^2}\left(\frac{2\mu}{a} - v_0^2\right)\right\}^{1/2}, \text{ using (1)} \qquad \text{... (8)}$$

giving the required eccentricity of the ellipse in terms of μ, v_0, α and OP $(= a)$

We now proceed to find the incination of the major axis to OP. (the initial line). Since the radius vector is perpendicular to the tangent at the extremity, it follows that an extremity of the major axis is an apse.

Therefore $\quad du/d\theta = 0 \quad$ or $\quad -A\sin(\theta-\theta_0) = 0$, using (2)

Hence $\theta = \theta_0$, showing that the major axis is inclined at an angle θ_0 to OP. Since $P(a, 0)$ lies on (3), we have

$$\frac{h^2/\mu}{a} = 1 + \frac{Ah^2}{\mu}\cos(-\theta_0) = 1 + e\cos\theta_0, \quad \text{as} \quad e = \frac{Ah^2}{\mu}$$

$$\therefore \quad \cos\theta_0 = \frac{1}{e}\left(\frac{h^2}{\mu a} - 1\right) = \frac{1}{e}\left(\frac{a^2v_0^2\sin^2\alpha}{\mu a} - 1\right), \text{ using (1)}$$

or $\quad \cos\theta_0 = (v_0^2\, a\sin^2\alpha - \mu)/\mu e \quad$ or $\quad \theta_0 = \cos^{-1}\{(v_0^2 a\sin^2\alpha - \mu)/\mu e\}$

which is the required inclination of the major axis to OP. Here e is given (8)

Ex 15. Find the law of force and velocity at any point when the orbit is the circle $r^2 = 2ap$ under a force to the pole on the circumference. **[Agra 2007]**

Ex. 16. *The mass of sun is derived by considering the earth to travel in a circular orbit of radius r and period T about the sun. Assuming Universal gravitational constant to be G, the calculated mass of the sun is* kr^3/GT^2. *What is the value of k?*

(a) $4\pi^2/3$ (b) π^2 (c) $4\pi^2$ (d) 4π **[I.A.S. (Prel.) 2008]**

Sol. Ans. (c) Let masses of sun and earth be M and m respectively. The mutual force of attraction $= (mMG)/r^2$

Hence, Force per unit mass on earth $= (MG)/r^2$

But attractive force per unit mass in central orbit $= \mu/r^2$

Hence, $\mu/r^2 = (MG)/r^2$ so that $\mu = MG$

We know that (see page 9.5), $T = (2\pi r^{3/2})/\sqrt{\mu}$

Therefore, $T = 2\pi \times \dfrac{r^{3/2}}{\sqrt{MG}}$ so that $M = \dfrac{4\pi^2 r^3}{GT^2} = \dfrac{kr^3}{GT^2}$ (given)

Thus, we have $k = 4\pi^2$

Ex. 17. *If the central force is a single-valued function of the distance, then the number of apsidal distances will be:*

(a) 2 (b) 1 (c) 3 (d) 0 **(Garhwal 2003)**

Sol. Ans. (*a*). Refer theorem II, page 9.12.

Ex. 18. *While describing the central orbit the following part of the acceleration is taken as zero: (a) Radial (b) Transverse (c) Tangential (d) Normal* **(Garhwal 2002, Rohilkhand 2011)**

Sol. Ans. (*b*). Refer Art. 9.3A, page 9.1.

Ex. 19. *The pedal equation of the central orbit is:*

(a) $(h/p) \times (dp/dr) = F$ (b) $(p^2/h^3) \times (dp/dr) = F$
(c) $(p/h) \times (dp/dr) = F$ (d) $(h^2/p^3) \times (dp/dr) = F$

(Bundelkhand 2006; Garhwal 2004; Rohilkhand 2008)

Sol. Ans. (*d*). Refer Art. 9.3B, page 9.2

Ex. 20. *In a central orbit at an apse*

(a) $p = 2r$ (b) $p = r$ (c) $p = 1/r$ (d) $p = r^2$ **(Agra 2010, Bundelkhand 2008)**

Sol. Ans. (*b*). Refer theorem I, page 9.12.

Ex. 21. *Referred to the centre of force as pole, the differential equation of a central orbit is*

(a) $d^2u/d\theta^2 + u^2 = (P/h^2u^2)$ (b) $du/d\theta + u = (P/hu)$
(c) $d^2u/d\theta^2 + u = (P/h^2u^2)$ (d) $d^2u/d\theta^2 + u = (P/h^2u)$

(Agra 2011; Bundelkhand 2007, 09)

Sol. Ans. (*c*). Refer Art. 9.3A, page 9.1. Note that here F = P.

Ex. 22. *In a central orbit if v is the linear velocity at any point of the path, then*

(a) $v^2 = h^2(u^2 + d^2u/d\theta^2)$ (b) $v^2 = h^2\{u^2 + (du/d\theta)^2\}$
(c) $v^2 = h^2(u^2 + du/d\theta)$ (d) $v^2 = h^2(u + du/d\theta)$ **(Bundelkhand 2008)**

Sol. Ans. (*b*). Refer Corollary 1 on page 9.3.

10

The Inverse Square Law (Planetary Motion)

10.1 Newton's law of gravitation. **[Meerut 2018]**

Every particle of the universe attracts every other particle with a force proportional to the product of their masses and inversely proportional to the square of their distance apart.

Thus for two particles of masses m_1, m_2 distant r apart, the force of attraction F between them is given by $F = (\gamma\, m_1 m_2)/r^2$ where γ is called the gravitational constant or the universal constant (sometimes denoted by G).

Remark. The motion of all planets in the solar system is governed by this law. In the present chapter we shall consider motion of all planets about the sun or the motion of the moon (the satellite of earth) about the earth in central orbits when the force of attraction varies inversely as the square of the distance from the centre of force.

10.2 Motion under the inverse square law. *To show that the path of particle which is moving so that its acceleration is always directed to a fixed point and is equal to $\mu/(distance)^2$ is a conic section and to distinguish between the three cases that arise.* **[Agra 2009; Calcutta 2002]**

Since the acceleration is always directed to a fixed point, hence the path is central orbit under given central acceleration $= F = \mu/r^2$.

The differential equation of the path in pedal form is

$$\frac{h^2}{p^3}\frac{dp}{dr} = F = \frac{\mu}{r^2} \qquad \text{or} \qquad -\frac{2h^2}{p^3}dp = -\frac{2\mu}{r^2}dr$$

Integrating, $\qquad v^2 = h^2/p^2 = 2\mu/r + A$, where A is a constant, $\qquad$ ...(1)

(using the result $v = h/p$ for a central orbit.)

We know that the pedal equations referred to focus as pole of ellipse, parabola and hyperbola (that branch which is nearer to the focus taken as pole), are

$$b^2/p^2 = 2a/r - 1, \qquad p^2 = ar \qquad \text{and} \qquad b^2/p^2 = 2a/r + 1, \qquad \text{...(2)}$$

respectively, where $2a, 2b$ are the lengths of major or minor axes in the case of ellipse, $4a$ is the length of latus rectum in the case of parabola and $2a, 2b$ are the lengths of transverse and conjugate axes in the case of hyperbola.

Since (1) can take any of three forms given by (2), three cases arise :

Case I. Elliptic path. Compare (1) with $b^2/p^2 = 2a/r - 1$, then

$$\frac{h^2}{b^2} = \frac{2\mu}{2a} = \frac{A}{-1} \qquad \text{so that} \qquad h^2 = \frac{\mu b^2}{a} \qquad \text{and} \qquad A = -\frac{\mu}{a}$$

Substituting these values in (1), for elliptic path, we have

$$v^2 = 2\mu/r - \mu/a \qquad \text{or} \qquad v^2 = \mu(2/r - 1/a) \qquad \text{...(3)}$$

$\therefore$ From (3), for elliptic path, we have $\qquad v^2 < 2\mu/r \qquad$ ...(4)

Case II. Parabolic path. Compare (1) with $p^2 = ar$, *i.e.*, $1/p^2 = (1/a)/r$. Then

$$\frac{h^2}{1} = \frac{2\mu}{1/a} \quad \text{and} \quad A = 0 \quad \text{so that} \quad h^2 = 2a\mu \quad \text{and} \quad A = 0$$

Substituting these values (1), for parabolic path, we have

$$v^2 = 2\mu/r \qquad \text{...(5)}$$

Case III. Hyperbolic path. Compare (1) with $b^2/p^2 = 2a/r + 1$. Then

$$\frac{h^2}{a^2} = \frac{2\mu}{2a} = \frac{A}{1} \quad \text{so that} \quad h^2 = \frac{\mu b^2}{a} \quad \text{and} \quad A = \frac{\mu}{a}.$$

Substituting these values in (1), for hyperbolic path, we have

$$v^2 = 2\mu/r + \mu/r \quad \text{or} \quad v^2 = \mu(2/r + 1/a) \quad \ldots(6)$$

$\therefore$ From (6), for hyperbolic path, we have $\quad v^2 > 2\mu/r \quad \ldots(7)$

From the above three cases we find that (1) always represents a conic section whose focus is at the centre of force. Again the path is ellipse, parabola or hyperbola according as constant A is negative, zero or positive.

Note carefully that in each of the above three cases the magnitude of the velocity at any point is independent of the direction of the velocity at the point.

We have proved that

(*i*) if $v^2 = \mu(2/r - 1/a)$, then the path is elliptic
(*ii*) if $v^2 = 2\mu/r$, then the path is parabolic
and (*iii*) if $v^2 = \mu(2/r + 1/a)$, then the path is hyperbolic

We have also proved that

(*i*) if $v^2 < 2\mu/r$, the path is elliptic
(*ii*) if $v^2 = 2\mu/r$, the path is parabolic
and (*iii*) if $v^2 > 2\mu/r$, the path is hyperbolic

If l is semi-latus rectum in all the three cases. Then

(*i*) $h^2 = \mu(b^2/a) = \mu l$, for the elliptic path
(*ii*) $h^2 = 2a\mu = \mu l$, for the parabolic path
(*iii*) $h^2 = \mu(b^2/a) = \mu l$, for the hyperbolic path

Thus, in all three cases, $\quad h^2 = \mu l \quad$ or $\quad h = (\mu l)^{1/2}$.

10.3 Kepler's laws of planetary motion

[Agra 2006, 12; Bhopal 2004; Kanpur 2019; Meerut 2016, 19; Mumbai 2010]

The astronomer Kepler discovered three laws connecting the motions of the various planets about the sun. They are :

1. Each planet describes an ellipse having the sun in one of its foci.
2. The radius vector drawn from the sun to the planet sweeps out equal areas in equal times.
3. The squares of the periodic times of the various planets are proportional to the cubes of the major axes of their orbits.

[Bangalore 2004]

10.4 Deductions from Kepler' laws

Deduction 1. From the second law the rate of description of secotiral area is constant which is true only in central orbits. Hence the acceleration of each planet, and therefore the force on it, is directed towards the sun.

Deduction 2. From a result of central orbits, we know that if the orbit is an ellipse under an acceleration towards one of its foci, then the law of acceleration is that of the inverse square of the distance from the focus (which is taken at centre of force). So from the first law it follows that acceleration of each planet varies inversely as the square of its distance from the sun.

Deduction 3. From a result of central orbits, we know that periodic time T under inverse square law for an elliptic orbit is given by

$$T = (2\pi a^{3/2})/\mu^{1/2} \quad \text{or} \quad T^2 = (\pi^2/2\mu)\cdot(2a)^3, \text{ where } 2a = \text{major axis}$$

i.e., squares of the periodic time varies as the cube of the major axis.

But according to the third Kepler's law, $T^2 \propto (2a)^3$. Hence it follows that the absolute acceleration (*i.e.,* the acceleration at unit distance from the sun) is the same for all planets.

10.5 To establish e quivalence between Kepler's laws for planetary motion and Newton's law of gravitation.

[Gujarat 2004]

(*i*) Deductions of three laws of Kepler from Newton's law of gravitation. Let us treat the sun as fixed particle in space and take its centre at the pole. From Newton's law of gravitation, the force F, per unit mass of the planet is given by $F = \mu/r^2 = \mu u^2$, where $u = 1/r$ and μ is a positive constant.

The differential equation of the central orbit is

$$\frac{d^2u}{d\theta^2} + u = \frac{F}{h^2u^2} = \frac{\mu u^2}{h^2u^2} \quad \text{or} \quad \frac{d^2}{d\theta^2}\left(u - \frac{\mu}{h^2}\right) + \left(u - \frac{\mu}{h^2}\right) = 0,$$

whose general solution is given by

$$u - \mu/h^2 = A\cos(\theta - \alpha),\ A,\ \alpha \text{ being constants.}$$

or
$$\frac{h^2/\mu}{r} = 1 + \frac{Ah^2}{\mu}\cos(\theta - \alpha), \qquad \ldots(1)$$

which represents polar equation of a conic with the sun at one focus, having l = semi-latus rectum = h^2/μ and e = eccentricity = Ah^2/μ. Since the only closed conics are ellipses, it follows that the planet must describe an ellipse with the sun at one of focus. Thus, the first law of Kepler has been deduced.

We know that when a particle moves in a central orbit, the rate of change of sectorial area traced out by the radius vector joining the particle to a fixed point is constant ($=h/2$). Thus, the second law of Kepler has been deduced.

From a result of central orbits, we know that periodic time T, under inverse square law for an elliptic orbit is given by

$$T = (2\pi a^{3/2})/\mu^{1/2} \quad \text{or} \quad T^2 = (\pi^2/2\mu) \times (2a)^3, \text{ where } 2a = \text{major axis,}$$

showing that the squares of the periodic times are proportional to the cubes of the major axis of the elliptic orbit. Thus, third law of Kepler has been deduced.

***(ii)* Deductions of Newton's law of gravitation from three laws of Kepler.** Kepler's second law tells us that the areal velocity is constant *i.e.*, $h/2$ is constant. Since $r^2\dot{\theta} = h$, we have $r^2\dot{\theta}$ = constant and so

$$\text{the transverse component of acceleration} = \frac{1}{r}\frac{d}{dt}(r^2\dot{\theta}) = 0,$$

showing the planet has only radial acceleration towards the sun. Hence the force must be directed towards the sun.

From the Kepler's first law, the path of the planet is elliptic and so its equation with pole at sun as focus can be taken as

$$l/r = 1 + e\cos\theta \quad \text{or} \quad u = (1/l)(1 + e\cos\theta),\ e < 1,\ u = 1/r \qquad \ldots(2)$$

From (2), $\quad du/d\theta = -(e/l)\sin\theta, \quad d^2u/d\theta^2 = -(e/l)\cos\theta \qquad \ldots(3)$

The force F per unit mass for the central orbits is given by

$$F = h^2u^2\left(\frac{d^2u}{d\theta^2} + u\right) = h^2u^2\left(-\frac{e\cos\theta}{l} + \frac{1}{l} + \frac{e}{l}\cos\theta\right), \text{ by (2) and (3)}$$

or
$$F = \frac{h^2u^2}{l} = \frac{h^2/l}{r^2}, \qquad \ldots(4)$$

Thus for each planet the force varies inversely as the square of the distance. But it remains to prove that the force is of the form

$$mF = m\mu/r^2, \qquad \ldots(5)$$

where m is the mass of the planet and μ a constant, the same for all the planets. To prove this, we shall now use Kepler's third law. We know that for an elliptic orbit, described under a central force directed to a focus, the periodic time T is given by $\quad T = (2\pi ab)/h$

and hence
$$\frac{T^2}{(2a)^3} = \frac{(4\pi^2a^2b^2)/h^2}{8a^3} = \frac{\pi^2}{2h^2} \times \frac{b^2}{a} = \frac{\pi^2 l}{2h^2}, \qquad \ldots(6)$$

where l = semi-latus rectum = b^2/a. But by Kepler's third law $T^2/(2a)^3$ is constant, the same for all planets. Hence, from (6) we conclude that h^2/l is the same for all planets, and so, by (4), $F = \mu/r^2$, where constant μ is the same for all planets. Hence (5) is true, and Newton's law of gravitation is thus deduced as a consequence of Kepler's laws.

10.6 Necessary modification of Kepler's third law

(Gorakhpur 95, 97)

Let P and S be the masses of the planet and sun respectively and r the distance (variable) between them. Then the force of mutual attraction between them is

$$F = (\gamma SP)/r^2, \gamma \text{ being an absolute constant.} \qquad \ldots(1)$$

Now the planet's acceleration is F/P *i.e.*, $\gamma S/r^2$ towards the sun and the sun's acceleration is F/S *i.e.*, $\gamma P/r^2$ towards the planet. Hence the acceleration of the planet relative to the sun

$$= \frac{\gamma S}{r^2} + \frac{\gamma P}{r^2} = \frac{\gamma(S+P)}{r^2} = \frac{\mu}{r^2}, \text{ where } \mu = \gamma(S+P) \qquad \ldots(2)$$

If T be the time period of the relative orbit, then

$$T = (2\pi a^{3/2})/\mu^{1/2} = (2\pi a^{3/2})/\{\gamma(S+P)\}^{1/2}, \text{ by (2)}$$

and so
$$T^2/(2a)^3 = \pi^2/2\gamma(S+P) \qquad \ldots(3)$$

The expression on the R.H.S. of (3) depends on the mass P of the planet. Accordingly the third law of Kepler needs a modification to that extent.

10.7 Some useful results of ellipse to be remembered

Let S, H be two foci of ellipse whose semi-major and semi-minor axes are a and b. Let e be its eccentricity and l be its semi-latus rectum. Let C be centre. SN, HM be perpendiculars on tangent $T'T$. Let PG be normal at P. Let AA' and BB' be major and minor axes. Then

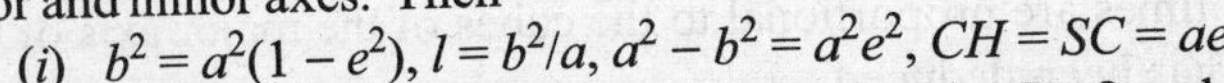

(*i*) $b^2 = a^2(1 - e^2)$, $l = b^2/a$, $a^2 - b^2 = a^2e^2$, $CH = SC = ae$

(*ii*) The sum of the focal distances of any point $P = 2a$ = length of its major axis so that $SP + HP = 2a$. Hence if $SP = r$, then $HP = 2a - r$.

(*iii*) SB = distance of extremity $B\,(0, b)$ of minor axis from the focus $S(ae, 0) = a$ $[\therefore SB^2 = CB^2 + CS^2 = b^2 + a^2e^2 = a^2(1-e^2) + a^2e^2 = a^2]$

(*iv*) $SN \times HM = b^2$. Also M and N lie on the auxiliary circle $x^2 + y^2 = a^2$ of ellipse $x^2/a^2 + y^2/b^2 = 1$. Hence $CN = CM = a$

(*v*) The tangent and normal at any point are each equally inclined to focal radii of that point so that $\angle MPH = \angle TPS$ and $\angle SPG = \angle HPG$

(*vi*) $AS = CA - CS = a - ae$, $\qquad A'S = CA' + CS = a + ae$

TYPE 1 : SOLVED EXAMPLES

Ex. 1. *If v_1 and v_2 are the linear velocities of a planet when it is respectively nearest and farthest from the sun, prove that $(1 - e)v_1 = (1 + e)v_2$.* **[Meerut 2017; Avadh 2007, 11; Kanpur 2013, 17; Agra 2011; Guwahati 2007; Calcutta 2002; Purvanchal 2002]**

Sol. Refer figure of Art. 10.7. We know that the path of the planet is an ellipse with the sun at its focus and the velocity v of the planet at any distance r from the focus S (the sun) is given by

$$v^2 = \mu(2/r - 1/a), \text{ a being semi-major axis} \qquad \ldots(1)$$

According to the given problem v_1 and v_2 are the velocities of the planet at A and A' which are nearest and farthest from the sun. Then $SA = a(1 - e)$ and $SA' = a(1 + e)$ by result (*vi*) of Art. 10.7. Putting these values in (1), we get

$$v_1^2 = \mu\left(\frac{2}{SA} - \frac{1}{a}\right) = \mu\left\{\frac{1}{a(1-e)} - \frac{1}{a}\right\} = \mu\frac{1+e}{a(1-e)} \qquad \ldots(2)$$

and
$$v_2^2 = \mu\left(\frac{2}{SA'} - \frac{1}{a}\right) = \mu\left\{\frac{1}{a(1+e)} - \frac{1}{a}\right\} = \mu\frac{1-e}{a(1+e)} \qquad \ldots(3)$$

Dividing (2) by (3), $\quad v_1^2/v_2^2 = (1 + e)^2/(1 - e)^2$

or $\quad v_1/v_2 = (1 + e)/(1 - e) \quad$ or $\quad (1 - e)v_1 = (1 + e)v_2$

Ex. 2. *Show that an unresisted particle falling to the earth's surface from a great distance would acquire a velocity $(2ga)^{1/2}$, where a is the radius of earth.*

Sol. For any particle on the surface of the earth, we know that the acceleration due to gravity, g, is given by

$\mu/a^2 = g$ and hence $\mu = a^2 g$ ...(1)

Since the particle is falling unresisted to the earth's surface from a great distance, hence its path must be a parabola about earth's centre as focus.

But the velocity for a parabolic path $v = (2\mu/r)^{1/2}$ ...(2)

At the surface earth, $r = a$ and $\mu = a^2 g$, by (1)

So (2) reduces to $v = (2a^2 g/a)^{1/2} = (2ga)^{1/2}$

Ex. 3. *A particle is projected from the earth's surface with velocity v. Show that if the dimunition of gravity is taken into account, but the resistance of the air is neglected, the path is an ellipse, of major axis $2ga^2/(2ga - v^2)$, where a is the earth's radius.*

Sol. If $2a_1$ is the major axis of the ellipse described by the particle, then its velocity V at a distance r from the centre of the earth is given by $V^2 = \mu(2/r - 1/a_1)$. For the given problem, $V = v$ and $r = a$. Hence, we get

$$v^2 = \mu(2/a - 1/a_1) \quad \text{...(1)}$$

Also for any particle on the surface of the earth, we have

$g = \mu/a^2$ or $\mu = a^2 g$ ...(2)

From (1) and (2), $v^2 = a^2 g(2/a - 1/a_1) = 2ag - (a^2 g)/a_1$

or $a_1(2ag - v^2) = a^2 g$ or $2a_1 = 2a^2 g/(2ag - v^2)$

Ex. 4. *A particle describes an ellipse as a central orbit about the focus. Prove that the velocity at the end of the minor axis is the geometric mean between the velocities at the ends of any diameter.*

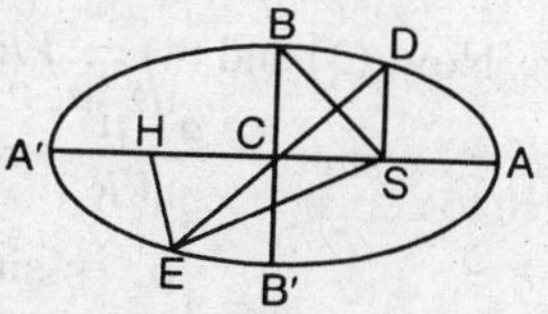

Sol. Let BB' be the minor axis of ellipse with S and H as foci and C as centre. Let DE be any diameter. Then $SB = a$ (Refer result (*iii*) of Art. 10.7). If V be the velocity of the particle at B, then we have $V^2 = \mu(2/r - 1/a) = \mu(2/a - 1/a)$, since $r = SB = a$ for B

Thus, $V^2 = \mu/a$ ...(1)

Join SD, SE and EH. Then, in triangles CDS and CHE, we have $CS = CH$, $CD = CE$ and $\angle SCD = \angle HCE$. Hence ΔCDS and ΔCHE are congruent and hence $SD = HE$, by geometry.

Now, $SD + SE = HE + SE$, as $SD = HE$

$= 2a$, being the sum of the focal distances of E

Thus, $SD + SE = 2a$ ...(2)

Let V_1 and V_2 be the velocities of the particle at D and E respectively. Then

$V_1^2 = \mu(2/SD - 1/a)$ and $V_2^2 = \mu(2/SE - 1/a)$

$\therefore$ $V_1^2 V_2^2 = \mu^2(2/SD - 1/a)(2/SE - 1/a)$

$$= \mu^2\left[\frac{4}{SD \times SE} - \frac{2}{a}\left(\frac{1}{SD} + \frac{1}{SE}\right) + \frac{1}{a^2}\right] = \mu^2\left[\frac{4}{SD \times SE} - \frac{2}{a}\frac{SE + SD}{SD \times SE} + \frac{1}{a^2}\right]$$

$$= \mu^2\left[\frac{4}{SD \times SE} - \frac{2}{a}\left(\frac{2a}{SD \cdot SE}\right) + \frac{1}{a^2}\right], \text{ using (2)}$$

$= \mu^2/a^2 = V^4$, using (1)

Thus, $V^2 = V_1 V_2$, *i.e.*, V is the geometric mean of V_1 and V_2.

Ex. 5. *A particle describes an ellipse under a force to the focus S. When the particle is at one extremity of the minor axis, its kinetic energy is doubled without any change in the direction of motion. Prove that the particle proceeds to describe a parabola.*

Sol. Refer figure of Ex. 4. If v and V be the velocities at a distance r from the focus S in an elliptic and parabolic path respectively, then we have

$$v^2 = \mu(2/r - 1/a) \quad \text{...(1)}$$

and

$$V^2 = 2\mu/r \quad \text{...(2)}$$

Let v_1 be the velocity of the particle (moving in an elliptic path) at B (where $r = SB = a$, by result (*iii*) of Art. 10.7), then

$v_1^2 = \mu(2/a - 1/a)$ or $v_1^2 = \mu/a$ ...(3)

Let v_2 be the velocity of the particle when its kinetic energy is doubled at B. If m be mass of the particle, then we have

$$(1/2)mv_2^2 = 2 \times \{(1/2)mv_1^2\} \Rightarrow v_2^2 = 2v_1^2 = 2\mu/a = 2\mu/SB \quad \text{...(4)}$$

From (2), the velocity V at B for parabolic path is given by

$$V^2 = 2\mu/SB \quad \text{...(5)}$$

From (4) and (5), $v_2 = V$, showing that the subsequent path of the particle at B must be a parabola.

Ex. 6. *A particle moves with a central acceleration $\mu/(distance)^2$; it is projected with velocity V at a distance R. Show that its path is rectangular hyperbola if the angle of projection is $\sin^{-1} [\mu/VR\{V^2 - (2\mu/R)\}^{1/2}]$* **[Kolkata 96; Gorakhpur 95, 98; Purvanchal 97]**

Sol. Here the particle describes a hyperbola under the acceleration $\mu/(\text{distance})^2$. We know that the velocity at a distance r from the centre of force is $v^2 = \mu(2/r + 1/a)$. For this problem, $r = R$, $v = V$. So we get

$$V^2 = \mu(2/R + 1/a) \qquad \text{giving} \qquad V^2 - (2\mu/R) + \mu/a \quad \text{...(1)}$$

Let α be the required angle of projection, at the point of projection. Thus, initially we have $r = R$ and $\phi = \alpha$. Then

$$h = vp \Rightarrow h = vr \sin\phi \Rightarrow h = VR \sin\alpha \quad \text{...(2)}$$

Also, $$h = (\mu l)^{1/2} = \{\mu(b^2/a)\}^{1/2} = (\mu a)^{1/2} \quad \text{...(3)}$$

[$\because$ for a rectangular hyperbola, $b = a$]

Now, (3) and (4) $\Rightarrow$ $VR \sin\alpha = (\mu a)^{1/2}$. Hence, we have

$$\therefore \quad \sin\alpha = \frac{a^{1/2}\mu^{1/2}}{VR} = \frac{\mu a^{1/2}}{VR(\mu)^{1/2}} = \frac{\mu}{VR(\mu/a)^{1/2}} = \frac{\mu}{VR\{V^2 - (2\mu/R)\}^{1/2}} \text{ by (1)}$$

$$\therefore \quad \alpha = \sin^{-1} [\mu/VR\{V^2 - (2\mu/R)\}^{1/2}], \text{ as required.}$$

Ex. 7. *Prove that the time taken by the earth to travel over half its orbit, remote from the sun, separated by the minor axis is two days more than half the year. The eccentricity of the orbit is 1/60.* **[Kanpur 2003; I.A.S. 1989]**

Sol. We know that the path of the earth round the sun is an ellipse with sun at focus S. Let AA' and BB' be major and minor axes of ellipse and C its centre. In figure, the remote half from the sun separated by the minor axis is $BA'B'$. Since the elliptic path is described in one year, so

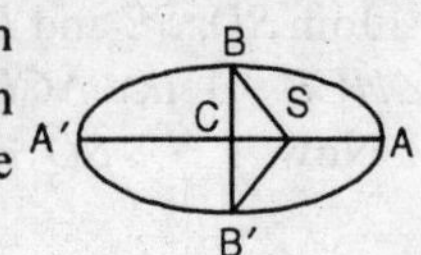

$$\text{The rate of description of area} = \frac{\text{area of ellipse}}{\text{one year}} = \frac{\pi ab}{\text{one year}} \quad \text{...(1)}$$

Let T be the time taken in describing the arc $BA'B'$ *i.e.* the sectorial area $SBA'B'S$. Then, we have

$$\text{The rate of description of area } SBA'B'S = (\text{area } SBA'B'S)/T \quad \text{...(2)}$$

In central orbit the rate of description of the secotiral area is constant. So,

$$(\text{area } SBA'B'S)/T = (\pi ab)/\text{one year, by (1) and (2)}$$

$$\therefore \quad T/(\text{one year}) = (\text{sectorial area } SBA'B'S)/\pi ab \quad \text{...(3)}$$

Now, sectorial area $SBA'B'S$

= half the area of ellipse $BA'B'$ + area of $\Delta BSB'$

$= (\pi ab)/2 + (1/2) \times (ae) \times b = (\pi ab)/2 + abe = ab(\pi/2 + e)$

$$\therefore \quad \frac{\text{sectorial area } SBA'B'S}{\pi ab} = \frac{ab(\pi/2 + e)}{\pi ab} = \frac{1}{2} + \frac{e}{\pi}$$

$\therefore$ (3) gives $\quad T/(\text{one year}) = 1/2 + e/\pi$

or $\quad T = (1/2) \text{ year} + (e/\pi) \text{ year, where } e = 1/60 \text{ (given)}$

$$= \frac{1}{2} \text{ year} + \left(\frac{1}{60} \times \frac{7}{22}\right) \times 365 \text{ days, as 1 year} = 365 \text{ days}$$

$$= \text{half year} + 2 \text{ days (nearly)}$$

Ex. 8. *Two particles of masses m_1 and m_2 moving in coplanar parabolas round the sun, collide at right angles and coalesce when their common distance from the sun is R. Show that the subsequent path of the combined particle is an ellipse of major axis $(m_1 + m_2)^2 R/2m_1m_2$.* **[I.A.S. 1998]**

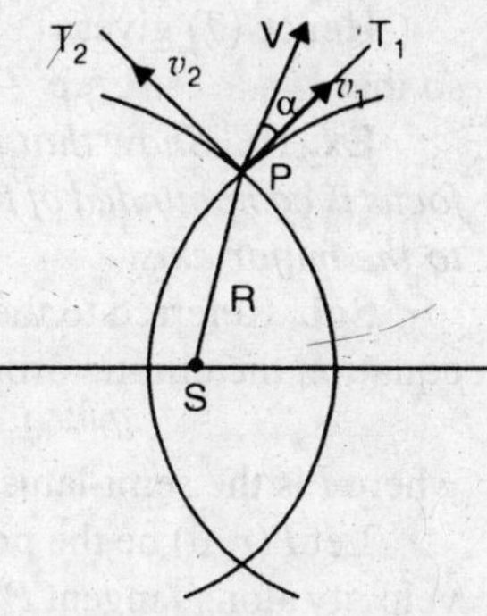

Sol. Let v_1 and v_2 be the velocities of the particles of masses m_1 and m_2 respectively before collision at the common point P whose distance from the sun S is R, *i.e.*, $SP = R$. Since both the particles are describing parabolas with S as focus, so at P, we have

$$v_1^2 = 2\mu/R = v_2^2 \qquad \ldots(1)$$

Let V be the velocity of the common body, of mass $(m_1 + m_2)$ into which the moving particle coalesce after collision, in the direction making an angle α (say) with that of v_1.

Given that directions of v_1 and v_2 are at right angles, *i.e.* $\angle T_1PT_2 = 90°$, where PT_1 and PT_2 are tangents to two parabolas at P. Then by the principle of conservation of momentum along PT_1 and PT_2, we have

$$(m_1 + m_2)V \cos \alpha = m_1 v_1 + m_2 \cdot 0 \quad \text{and} \quad (m_1 + m_2)V \sin \alpha = m_1 \cdot 0 + m_2 v_2$$

Squaring and adding these, $\quad (m_1 + m_2)^2 V^2 = m_1^2 v_1^2 + m_2^2 v_2^2$

or $\quad (m_1 + m_2)^2 V^2 = (m_1^2 + m_2^2) v_1^2$, using (1)

or $\quad V^2 = (m_1^2 + m_2^2) v_1^2/(m_1 + m_2)^2 \qquad \ldots(2)$

Since $(m_1 + m_2)^2 < m_1^2 + m_2^2$, (2) $\Rightarrow V^2 < v_1^2 \Rightarrow V^2 < 2\mu/R$, by (1)

Since $V^2 < 2\mu/R$, the path traced out by the combined body after collision must be an ellipse and for this elliptic path at P, we have

$$V^2 = \mu(2/R - 1/a), \text{ a being length of semi major axis}$$

or $$\frac{(m_1^2 + m_2^2)v_1^2}{(m_1 + m_2)^2} = \mu\left(\frac{2}{R} - \frac{1}{a}\right), \text{ by (2)}$$

or $$\frac{(m_1^2 + m_2^2)}{(m_1 + m_2)^2}\frac{2\mu}{R} = \mu\left(\frac{2}{R} - \frac{1}{a}\right), \text{ by (1)}$$

or $$\frac{1}{a} = \frac{2}{R}\left[1 - \frac{m_1^2 + m_2^2}{(m_1 + m_2)^2}\right] = \frac{2}{R} \times \frac{2m_1m_2}{(m_1 + m_2)^2}$$

$\therefore \quad 2a =$ length or major axis $= (m_1 + m_2)^2 R/2m_1m_2$.

Ex. 9. *A comet describing a parabola about the sun, when nearest to it suddenly breaks up, without gain or loss of kinetic energy into two equal portions one of which describes a circle; prove that the other will describe a hyperbola of eccentricity 2.* **[I.A.S. 2001]**

Sol. Let $4a$ be the latus rectum of the parabolic path of the comet about the sun S as focus. The comet is nearest to the sun when it is at vertex A so that $SA = a$. If V is the velocity the comet at A, then $\quad V^2 = 2\mu/SA \quad$ or $\quad V^2 = 2\mu/a \qquad \ldots(1)$

Let $2m$ be the mass of the comet which breaks into two equal parts m and m at A and let their velocities at A be v_1 and v_2. Since there is no loss or gain of kinetic energy, we have

$$(1/2) \times (2m)V^2 = (1/2)mv_1^2 + (1/2)mv_2^2 \text{ or } v_1^2 + v_2^2 = 4\mu/a, \text{ by (1)} \qquad \ldots(2)$$

If the portion of the comet with velocity v_1 at A describes a circle about S,

then $\quad v_1^2/a = \mu/a^2 \quad$ so that $\quad v_1^2 = \mu/a \qquad \ldots(3)$

Using (3), (2) gives $\quad \mu/a + v_2^2 = 4\mu/a \quad$ or $\quad v_2^2 = 3\mu/a, \qquad \ldots(4)$

showing that the portion of the comet with velocity v_2 at A satisfies condition $v_2^2 > 2\mu/a$ and hence this portion with velocity v_2 will describe a hyperbola with $2a_1$ and $2b_1$ as lengths of transverse and conjugate axes respectively (say). Now, in hyperbolic orbit, velocity v_2 at A is given by

$$v_2^2 = \mu(2/SA + 1/a_1) = \mu(2/a + 1/a_1), \text{ as } SA = a \qquad \ldots(5)$$

From (4) and (5), $\quad 3\mu/a = \mu(2/a + 1/a_1) \quad$ so that $\quad a_1 = a \qquad \ldots(6)$

For a hyperbolic orbit, we have $\quad vp = h = (\mu l)^{1/2} \qquad \ldots(7)$

At the point A, $p = a = a_1$, $v = v_2 = (3\mu/a)^{1/2} = (3\mu/a_1)^{1/2}$, $l = b_1^2/a_1 = a_1^2(e^2 - 1)/a_1$

Hence (7) gives $\quad (3\mu/a_1)^{1/2} a_1 = \{\mu a_1(e^2 - 1)\}^{1/2}$

so that $\quad 3 = e^2 - 1 \quad$ or $\quad e^2 = 4 \quad$ or $\quad e = 2.$

Ex. 10. *Show that the velocity of a particle moving in an ellipse about a centre of force in the focus is compounded of two constant velocities v/h perpendicular to the radius and μe/h perpendicular to the major axis.* **[Lucknow 97; Purvanchal 91]**

Sol. Referred to the focus S (*i.e.*, the centre of force) as pole, let the equation the elliptic orbit be

$$l/r = 1 + e\cos\theta, \qquad \ldots(1)$$

where l is the semi-latus rectum of the ellipse.

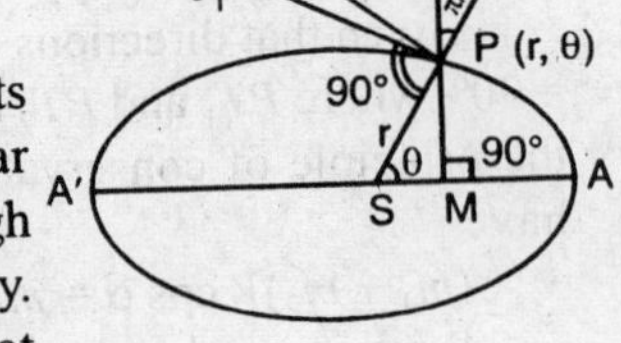

Let $P(r, \theta)$ be the position of the particle at any time t and let v be its velocity along tangent PT at P. Let PE be SP produced, PC perpendicular to SP and MD be perpendicular line on the major axis AA' (passing through P). Let the components of v along PC and PD be v_1 and v_2 respectively. Thus, v is the resultant of v_1 and v_2. Resolving the velocities v_1 and v_2 at P along and perpendicular to the radius vector SP, we have

$$\text{the radial velocity} = dr/dt = v_2\cos(\pi/2 - \theta) = v_2\sin\theta \qquad \ldots(2)$$

$$\text{and the transverse velocity} = r(d\theta/dt) = v_1 + v_2\sin(\pi/2 - \theta) = v_1 + v_2\cos\theta \qquad \ldots(3)$$

Differentiating (1) w.r.t. 't', we have

$$-\frac{l}{r^2}\frac{dr}{dt} = -e\sin\theta\frac{d\theta}{dt} = -e\sin\theta\left(\frac{h}{r^2}\right) \quad \text{or} \quad \frac{dr}{dt} = \frac{eh}{l}\sin\theta \qquad \ldots(4)$$

$$[\because \text{ in a central orbit, } r^2(d\theta/dt) = h]$$

From (2) and (4), $\quad v_2\sin\theta = (eh/l)\sin\theta \quad$ or $\quad v_2 = eh/l$

$$\text{or} \qquad v_2 = \frac{eh}{(h^2/\mu)} = \frac{e\mu}{h} = \text{constant, as } h^2 = \mu l \qquad \ldots(5)$$

For central orbit, $\quad r^2(d\theta/dt) = h \quad$ so that $\quad r(d\theta/dt) = h/r \qquad \ldots(6)$

Using (5) and (6), (3) reduces to

$$\frac{h}{r} = v_1 + \frac{e\mu}{h}\cdot\cos\theta = v_1 + \frac{\mu}{h}\left(\frac{l}{r} - 1\right), \text{ using (1)}$$

$$\therefore \qquad v_1 = h/r - \mu l/hr + \mu/h = h/r - h^2/hr + \mu/h, \text{ as } h^2 = \mu l$$

$$\text{or} \qquad v_1 = \mu/h = \text{constant} \qquad \ldots(7)$$

From (5) and (7), the required results follow.

Ex. 11. *A planet is describing an ellipse about the sun as focus, show that its velocity away from the sun is greatest when the radius vector to the planet is at right angles to the major axis of the path and that it then is 2π aeT/(1 – e²)^{1/2}, whose 2a is the major axis, e the eccentricity and T the periodic time.* **[Purvanchal 2003]**

Sol. Referred to the sun as focus as pole, let the equation of the elliptic orbit be

$$l/r = 1 + e\cos\theta \quad \text{or} \quad lu = 1 + e\cos\theta \qquad \ldots(1)$$

where $u = 1/r$ and l is the semi-latus rectum of the ellipse. Then the velocity of planet away from sun (pole) is clearly dr/dt. We wish to find the maximum value of dr/dt.

For a central orbit, $\quad h = r^2\dot{\theta} \quad$ so that $\quad \dot{\theta} = hu^2, \quad u = 1/r \qquad \ldots(2)$

Now, $\quad r = 1/u \Rightarrow dr/d\theta = -(1/u^2)(du/d\theta) \qquad \ldots(3)$

Then, $$\frac{dr}{dt} = \frac{dr}{d\theta}\cdot\frac{d\theta}{dt} = -\frac{1}{u^2}\frac{du}{d\theta}(hu^2), \text{ by (2) and (3)}$$

Thus, $$dr/d\theta = -h(du/d\theta). \qquad \ldots(4)$$

For maximum value of dr/dt, we have

$$\frac{d}{d\theta}\left(\frac{dr}{dt}\right) = 0 \quad \text{or} \quad \frac{d}{d\theta}\left(-h\frac{du}{d\theta}\right) = 0 \quad \text{or} \quad \frac{d^2u}{d\theta^2} = 0 \qquad \ldots(5)$$

From (1), $\quad du/d\theta = -(e/l)\sin\theta \quad$ and $\quad d^2u/d\theta^2 = -(e/l)\cos\theta \qquad \ldots(6)$

(5) and (6) $\Rightarrow \quad -(e/l)\cos\theta = 0 \Rightarrow \cos\theta = 0 \Rightarrow \theta = \pi/2,$

showing that the velocity away from the sun, *i.e.*, dr/dt is maximum when $\theta = \pi/2$, *i.e.*, when the radius vector to the point is at right angles to the major axis.

We now find the maximum value of the required velocity. When $\theta = \pi/2$, from (6), for maximum value of dr/dt, we have

$$du/d\theta = -(e/l)\sin(\pi/2) = -e/l \qquad \text{...(7)}$$

From (2) and (7), $$\left(\frac{dr}{dt}\right)_{\max} = \frac{he}{l} = (\mu l)^{1/2} \times \frac{e}{l} = (\mu/l)^{1/2} \times e \qquad \text{...(8)}$$

But $l = b^2/a = a(1-e^2)$ and $T = (2\pi a^{3/2})/\mu^{1/2}$

Hence $(l)^{1/2} = \{a(1-e)^2\}^{1/2}$ and $(\mu)^{1/2} = (2\pi a^{3/2})/T$

Substituting these values in (8), we get

$$\left(\frac{dr}{dt}\right)_{\max} = \frac{2\pi a^{3/2} e}{T\{a(1-e^2)\}^{1/2}} = \frac{2\pi a e}{T(1-e^2)^{1/2}}$$

Ex. 12. *A planet of mass M and periodic time T, when at its greatest distance from the sun; comes into collision with a meteor of mass m moving in the same orbit in the opposite direction with velocity v, if m/M be small, show that the major axis of the planet's path is reduced by* $(4m/M) \times (vT/\pi) \times \{(1-e)/(1+e)\}^{1/2}$. **[Kanpur 2000; Purvanchal 95, 97]**

Sol. The planet is at greatest distance from the sun S when at vertex A' where $SA' = A'C + CS = a + ae$. Before coming into collision at A', velocities of both the planet and the meteor will be v (since the orbit for both is same, so they will possess same velocity at A').

Now, $$v^2 = \mu\left[\frac{2}{SA'} - \frac{1}{a}\right] = \mu\left[\frac{2}{a(1+e)} - \frac{1}{a}\right] = \frac{\mu}{a}\left[\frac{1-e}{1+e}\right] \qquad \text{...(1)}$$

Let V be the common velocity after the impact. Then by the principle of conservation of momentum, we have

$(M+m)V = Mv - mv$ or $V = (M-m)v/M+m)$

or $V = (1 - m/M)(1 + m/M)^{-1} v = (1 - m/M)(1 - m/M)v$, as m/M is small

or $V = (1 - 2m/M)v$, since m/M is small ...(2)

From (2), $V < v$ so that $V^2 < v^2$ or $V^2 < \mu(2/SA' - 1/a)$, using (1)

Thus $V^2 < 2\mu/SA'$ and hence after collision the combined body will move in new elliptic path of major axis $2a'$ (say). So for this new orbit at A', we have

$$V^2 = \mu\left(\frac{2}{a+ae} - \frac{1}{a'}\right) \quad \text{or} \quad \left(1 - \frac{2m}{M}\right)^2 v^2 = \mu\left(\frac{2}{a+ae} - \frac{1}{a'}\right), \text{ by (2)}$$

or $$\left(1 - \frac{4m}{M}\right)v^2 = \mu\left(\frac{2}{a+ae} - \frac{1}{a'}\right), \text{ since } \frac{m}{M} \text{ is small}$$

or $$\left(1 - \frac{4m}{M}\right)\frac{\mu}{a}\left(\frac{1-e}{1+e}\right) = \mu\left(\frac{2}{a+ae} - \frac{1}{a'}\right), \text{ using (1)}$$

or $$\frac{1}{a'} = \frac{2}{a+ae} - \left(1 - \frac{4m}{M}\right) \times \frac{1}{a} \times \left(\frac{1-e}{1+e}\right) = \frac{1}{a}\left[\frac{2}{1+e} - \left(1 - \frac{4m}{M}\right)\left(\frac{1-e}{1+e}\right)\right]$$

or $$\frac{1}{a'} = \frac{1}{a}\left[\frac{2}{1+e} - \frac{1-e}{1+e} + \frac{4m}{M}\left(\frac{1-e}{1+e}\right)\right] = \frac{1}{a}\left[1 + \frac{4m}{M}\left(\frac{1-e}{1+e}\right)\right]$$

or $$a' = a\left[1 + \frac{4m}{M}\left(\frac{1-e}{1+e}\right)\right]^{-1} = a\left[1 - \frac{4m}{M}\left(\frac{1-e}{1+e}\right)\right], \text{ as } m/M \text{ is small}$$

or $$2a' - 2a = -[4(m/M) \times \{(1-e)/(1+e)\} \times 2a]$$

Hence the reduction in the major axis of the new ellipse

$= 4(m/M) \times \{(1-e)/(1+e)\} \times 2a$ ...(3)

We know that the periodic time of the planet $= T = (2\pi a^{3/2})/\mu^{1/2}$ ...(4)

$$\therefore \quad vT = \left(\frac{\mu}{a}\right)^{1/2} \times \left(\frac{1-e}{1+e}\right)^{1/2} \times \frac{2\pi a^{3/2}}{\mu^{1/2}}, \text{ using (1) and (4)}$$

so that $\qquad 2a = (vT/\pi) \times \{(1+e)/(1-e)\}^{1/2}$ ...(5)

Putting this value of $2a$ in (3), reduction in major axis

$$= \frac{4m}{M} \times \frac{1-e}{1+e} \times \frac{vT}{\pi} \times \left(\frac{1+e}{1-e}\right)^{1/2} = \frac{4mvT}{M\pi}\left(\frac{1-e}{1+e}\right)^{1/2}$$

Ex. 13. *Two inelastic particles of masses 3m and m are describing the same ellipse of eccentricity e in opposite directions under a force to the focus. They collide and coalesce at an extremity of minor axis. Prove that the eccentricity of the new orbit is $(1/4) \times (9 + 7e^2)^{1/2}$.*

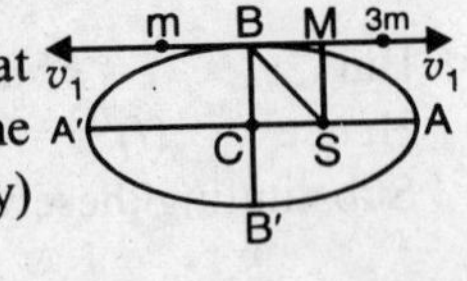

Sol. Suppose the given particles collide at extremity B of minor axis so that $SB = a$ (refer result (*iii*) of Art. 10.7), S being focus of the elliptic path. Since the orbit for the particles is the same, so they will possess the same velocity v_1, (say) given by

$$v_1^2 = \mu(2/SB - 1/a) = \mu(2/a - 1/a) = \mu/a \qquad ...(1)$$

Suppose that after impact when they coalesce and form one particle of mass $4m$, the velocity be V. Then by principle of conservation of momentum,

$$4mV = 3mv_1 - mv_1 \quad \text{so that} \quad V^2 = v_1^2/4 = \mu/4a, \text{ by (1)} \qquad ...(2)$$

Now, $V^2 = \mu/4a \Rightarrow V^2 < 2\mu/a$ and hence after collision the combined particle will move in new elliptic path of major axis and minor axes $2a'$ and $2b'$ respectively (say). Hence for the new orbit at B, we have

$$V^2 = \mu(2/SB - 1/a') \quad \text{or} \quad \mu/4a = 2(2/a - 1/a'), \text{ by (2)}$$

or $\qquad 1/a' = 2/a - 1/4a = 7/4a \qquad$ so that $\qquad a' = 4a/7$ ...(3)

Let SM be perpendicular to the tangent to ellipse at B. Then, clearly, $p = SM = b$. For central orbits, we have

$$vp = h = (\mu l)^{1/2}, \qquad ...(4)$$

Using (4) at B, $\qquad V \times SM = \{\mu \times (b'^2/a')\}^{1/2}$, as $l = b'^2/a'$

or $\qquad V^2b^2 = \mu a'^2(1-e'^2)/a'$, where $SM = b$, $b'^2 = a'^2(1-e'^2)$

or $\qquad (\mu/4a) \times a^2(1-e^2) = \mu a'(1-e'^2) = (4a\mu/7) \times (1-e'^2)$, by (2) and (3)

[Here e and e' are eccentricities of old and new orbits]

or $\quad 1 - e^2 = (16/7)(1-e'^2)$ or $e'^2 = 1 - (7/16) \times (1-e^2) = (1/16) \times (9 + 7e^2)$

Thus, $\quad e'$ = eccentricity of new orbit $= (1/4) \times (9 + 7e^2)^{1/2}$.

Ex. 14. *A body is describing an ellipse of eccentricity e under the action of a force tending to a focus and when at the nearer apse the centre of force is transferred to the other focus. Prove that the new eccentricity of the new orbit is $e(3+e)/(1-e)$.* **[Avadh 2008; I.A.S. 2009]**

Sol. For figure refer Ex. 4. A is the nearer apse with centre of force at S. Then $SA = a - ae = a(1-e)$. Then velocity v at A is given by

$$v^2 = \mu\left(\frac{2}{SA} - \frac{1}{a}\right) = \mu\left(\frac{1}{a-ae} - \frac{1}{a}\right) = \frac{\mu}{a}\frac{1+e}{1-e} \qquad ...(1)$$

Now the centre of force is shifted to H the other focus and with this point as centre of force, the velocity at A remaining unchanged. Then the body will describe another elliptic orbit whose major axis is a' and eccentricity is e'(say). Now velocity v at A for this new orbit is given by

$$v^2 = \mu\left[\frac{2}{1+AH} - \frac{1}{a'}\right] = \mu\left[\frac{2}{a+ae} - \frac{1}{a'}\right], \text{ as } AH = AC + CH = a + ae$$

or $$\frac{\mu}{a} \times \frac{1+e}{1-e} = \mu\left[\frac{2}{a(1+e)} - \frac{1}{a'}\right], \text{ using (1)} \qquad ...(2)$$

But for the new orbit, H is the focus and A is the nearer apse, so we have $AH = a' - a'e'$. But for the old orbit $AH = a + ae$. Thus, we have

$$a + ae = a' - a'e' \qquad \text{so that} \qquad a' = a(1+e)/(1-e').$$

Substituting this value of a' in (2), we obtain

$$\frac{1+e}{a(1-e)} = \frac{2}{a(1+e)} - \frac{1-e'}{a(1+e)} \qquad \text{or} \qquad \frac{(1+e)^2}{1-e} = 2 - (1-e')$$

or $$e' = \frac{(1+e)^2}{1-e} - 1 = \frac{e^2+3e}{1-e} = \frac{e(3+e)}{1-e}$$

Ex. 15. *If the velocity of a body in an elliptic orbit of major axis 2a, is the same at a certain point P, whether the orbit is being described in a periodic time T about one focus S or in periodic time T′ about the other focus S′, prove that SP = (2aT′)/(T + T′) and S′P = (2aT)/(T + T′)*

Sol. Since the sum of the focal distances of any point P of the ellipse is $2a$, so if $SP = r$, then $S'P = 2a - r$. Taking S as the focus, then we have

$$v^2 = \mu(2/r - 1/a) \quad \text{and} \quad T = (2\pi\, a^{3/2})/\mu^{1/2} \qquad \ldots(1)$$

Again, taking S' as focus, then as $S'P = 2a - r$, we have

$$v^2 = \mu'\{2/(2a-r) - 1/a\} \quad \text{and} \quad T' = (2\pi\, a^{3/2})/\mu'^{3/2} \qquad \ldots(2)$$

From (1) and (2), $\mu(2/r - 1/a) = \mu'\{2/(2a-r) - 1/a\}$ ...(3)

Again, from (1) and (2), $\mu = 4\pi^2 a^3/T^2$ and $\mu' = 4\pi^2 a^3/T'^2$ ...(4)

Putting values of μ and μ' from (4) in (3), we get

$$\frac{4\pi^2 a^3}{T^2}\left(\frac{2}{r} - \frac{1}{a}\right) = \frac{4\pi a^3}{T'^2}\left(\frac{2}{2a-r} - \frac{1}{a}\right) \quad \text{or} \quad \frac{2a-r}{rT^2} = \frac{r}{(2a-r)T'^2}$$

or $$T'^2(2a-r)^2 = T^2r^2 \quad \text{or} \quad T'(2a-r) = Tr \quad \text{or} \quad 2aT' = r(T+T')$$

$$\therefore \quad SP = r = (2aT')/(T+T') \qquad \ldots(5)$$

Hence $$S'P = 2a - r = 2a - \frac{2aT'}{T+T'} = \frac{2aT}{T+T'}, \text{ by (5)}$$

Ex. 16. *If a planet were suddenly stopped in its orbit supposed circular, show that it would fall into the sun in time which is $\sqrt{2}/8$ times the period of the planet's revolution.*

[Rohilkhand 2008; I.F.S. 2002; Purvanchal 2002]

Sol. Suppose a planet in its circular orbit of radius a and centre S (the sun) be stopped at position P. Then it moves from rest from P towards sun along straight line PS. Let at any time t it be at P' where $SP' = r$. Then its acceleration $= \mu/r^2$ along $P'S$ (along which r is decreases as t increases). Hence equation of motion of the planet is given by

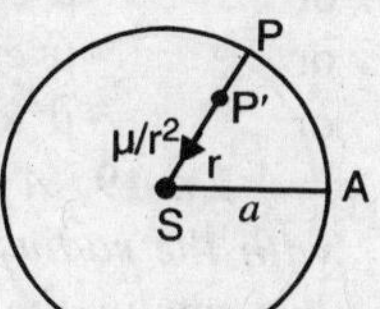

$$v(dv/dr) = -\mu/r^2 \quad \text{or} \quad 2v\,dv = -(2\mu/r^2)\,dr$$

Integrating, $v^2 = 2\mu/r + A$, A is a constant. ...(1)

Initially at P, when $r = a$, $v = 0$. So (1) gives $A = -\mu/a$

Then (1) becomes $v^2 = 2\mu(1/r - 1/a)$ or $(dr/dt)^2 = (2\mu/a)\{(a-r)/a\}$

or $$\frac{dr}{dt} = -\left(\frac{2\mu}{a}\right)^{1/2}\left(\frac{a-r}{r}\right)^{1/2} \quad \text{or} \quad dt = -\left(\frac{a}{2\mu}\right)^{1/2}\left(\frac{r}{a-r}\right)^{1/2} dr \qquad \ldots(2)$$

[–ve sign is taken because r decreases as t increases]

Let T' be the time taken by the planet to move from P to S. Then, since $t = 0$, when $r = a$ and $t = T'$ when $r = 0$, integration of (2) gives

$$\int_0^{T'} dt = -\left(\frac{a}{2\mu}\right)^{1/2}\int_a^0\left(\frac{r}{a-r}\right)^{1/2} dr = \left(\frac{a}{2\mu}\right)^{1/2}\int_0^{\pi/2}\left(\frac{a\cos^2\theta}{a\sin^2\theta}\right)^{1/2}\cdot 2a\cos\theta\sin\theta\, d\theta$$

[putting $r = a\cos^2\theta$ so that $dr = -2a\cos\theta\sin\theta\, d\theta$]

or $$T' = 2a\left(\frac{a}{2\mu}\right)^{1/2}\int_0^{\pi/2}\cos^2\theta\, d\theta = 2a\left(\frac{a}{2\mu}\right)^{1/2}\cdot\frac{1}{2}\cdot\frac{\pi}{2} = \frac{\pi a^{3/2}}{2\sqrt{2}\mu^{1/2}}$$

We know that the period of the planet's revolution $= T = (2\pi a^{3/2})/\sqrt{\mu}$

$$\therefore \quad T'/T = 1/4\sqrt{2} = \sqrt{2}/8 \quad \text{or} \quad T' = (\sqrt{2}/8)T,$$

showing that the time taken from P to S is $(\sqrt{2}/8)$ times the period of revolution.

Ex. 17. *If ω be the angular velocity of a planet at the nearer end of the major axis, prove that its period is given by $(2\pi/\omega) \times \{(1+e)/(1-e)^3\}^{1/2}$.* **[Meerut 2015; Mumbai 2010]**

Sol. For figure refer Art. 10.7. At the nearer end A, $r = SA = a - ae = a(1-e)$. For central orbit, we have $r^2\dot{\theta} = h = (\mu l)^{1/2}$ in which for A, $r = a(1-e)$, $\dot{\theta} = \omega$

$\therefore$ $a^2(1-e)^2\omega = \{\mu a(1-e^2)\}^{1/2}$, as $l = b^2/a = a^2(1-e^2)/a = a(1-e^2)$

Hence $$\mu^{1/2} = \frac{a^2(1-e)^2\omega}{\{a(1-e)(1+e)\}^{1/2}} = a^{3/2}\frac{(1-e)^{3/2}}{(1+e)^{1/2}}\omega \quad \ldots(1)$$

We know that periodic time T for elliptic orbit is given by

$$T = \frac{2\pi a^{3/2}}{\mu^{1/2}} = \frac{2\pi}{\omega}\left\{\frac{1+e}{(1-e)^3}\right\}^{1/2}, \text{ using (1)}$$

Ex. 18. *Prove that if when the particle is at a distance r from the focus its velocity is v in a direction making an angle ϕ with the radius vector, then $e^2\mu^2 = (v^2r-\mu)^2\sin^2\phi + \mu^2\cos^2\phi$.*

Sol. Since $h = pv$ and $p = r\sin\phi$, we have

$$h = rv\sin\phi \quad \ldots(1)$$

Again, $h = (\mu l)^{1/2}$, where $l = b^2/a = a^2(1-e^2)/a = a(1-e^2)$

$\therefore$ $h^2 = \mu a(1-e^2)$ or $r^2v^2\sin^2\phi = \mu a(1-e^2)$, by (1)

Thus, $$1/a = \mu(1-e^2)/(v^2r^2\sin^2\phi) \quad \ldots(2)$$

Now, $$v^2 = \mu\left(\frac{2}{r}-\frac{1}{a}\right) = \mu\left[\frac{2}{r}-\frac{\mu(1-e^2)}{v^2r^2\sin^2\phi}\right], \text{ by (2)}$$

or $$v^2 = \frac{2\mu}{r}-\frac{\mu^2(1-e^2)}{v^2r^2\sin^2\phi} \quad \text{or} \quad \left(v^2-\frac{2\mu}{r}\right)v^2r^2\sin^2\phi = -\mu^2(1-e^2)$$

or $$(v^2r-2\mu)\,v^2r\sin^2\phi = \mu^2e^2-\mu^2$$

or $$\mu^2e^2 = (v^2r-2\mu)\,v^2r\sin^2\phi+\mu^2 = (v^4r^2-2\mu v^2r)\sin^2\phi+\mu^2$$

or $$\mu^2e^2 = (v^4r^2-2\mu v^2r+\mu^2)\sin^2\phi-\mu^2\sin^2\phi+\mu^2$$

or $$\mu^2e^2 = (v^2r-\mu)^2\sin^2\phi+\mu^2\cos^2\phi$$

Ex. 19. *A particle is projected from a point P with velocity V in a direction making an angle α with the radius vector. If V were V', the orbit would be a parabola. Prove that if the orbit is hyperbolic, the angle between the asymptotes is $2\tan^{-1}\{(2V/V')(V^2/V'^2-1)^{1/2}\sin\alpha\}$.*

Sol. Given that, for parabolic orbit, $$V'^2 = 2\mu/r \quad \ldots(1)$$

and for hyperbolic orbit, $$V^2 = \mu(2/r+1/a) \quad \ldots(2)$$

Using (1), (2) gives $$V^2 = V'^2+\mu/a = V'^2+(rV'^2)/2a$$

$[\because (1) \Rightarrow \mu = rV'^2/2]$

Thus, $$V^2 = V'^2(1+r/2a) \quad \ldots(3)$$

Let β be angle between the asymptotes of the hyperbolic orbit, then from coordinate geometry, we have $\beta = \tan^{-1}(b/a)$.

Now, $$\frac{b}{a} = \left(\frac{b^2}{a^2}\right)^{1/2} = \left(\frac{l}{a}\right)^{1/2} = \left(\frac{\mu l}{\mu a}\right)^{1/2}, \text{ as } l = \frac{b^2}{a}$$

$$= \left(\frac{h^2}{\mu a}\right)^{1/2} = \left(\frac{v^2p^2}{\mu a}\right)^{1/2}, \text{ as } h^2 = vp \text{ for a central orbit}$$

$$= \left\{\frac{1}{\mu a}\times\mu\left(\frac{2}{r}+\frac{1}{a}\right)\times r^2\sin^2\alpha\right\}^{1/2}, \text{ as } p = r\sin\alpha,\ v = V \text{ which is given by (2)}$$

$$= \left\{2\left(1+\frac{r}{2a}\right)\times 2\times\frac{r}{2a}\times\sin^2\alpha\right\}^{1/2} = \left\{2\times\frac{V^2}{V'^2}\times 2\left(\frac{V^2}{V'^2}-1\right)\sin^2\alpha\right\}^{1/2}, \text{ by (3)}$$

$\therefore$ $\beta = \tan^{-1}(b/a) = \tan^{-1}\{2(V/V')(V^2/V'^2-1)^{1/2}\sin\alpha\}$

Ex. 20. *A comet is moving in a parabola about the sun at focus when at the end of its latus rectum its velocity suddenly becomes altered in the ratio n : 1; where n < 1. Show that the comet will describe an ellipse whose eccentricity is $(1-2n^2+2n^4)^{1/2}$ and whose major axis is $l/(1-n^2)$, where*

2l was the latus rectum of the parabolic orbit. **[Calcutta 2002]**

Sol. Let LL' be the latus rectum of the parabolic orbit about the sun S as focus. Let v be velocity of the comet at L, where $SL = l$.

$\therefore \quad v^2 = 2\mu/SL \quad$ or $\quad v^2 = 2\mu/l$ ...(1)

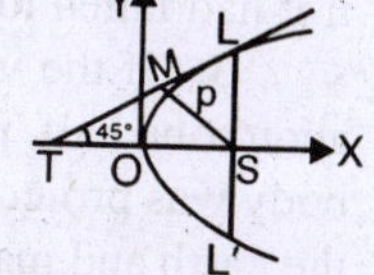

Let V be the altered velocity at L so that

$V/v = n/1 \quad$ so that $\quad V = nv$ ...(2)

Since $\quad n < 1$, (2) $\Rightarrow V < v \Rightarrow V^2 < v^2 \Rightarrow V^2 < 2\mu/l$ ...(3)

From (3), it follows that the subsequent path of the comet will be elliptic. Let e be its eccentricity and $2a'$ and $2b'$ major and minor axes. Then at L, for this new elliptic path, we have

$V^2 = \mu(2/SL - 1/a') \quad$ or $\quad n^2v^2 = \mu(2/l - 1/a')$, by (2) ...(4)

Dividing (4) by (1), $\quad n^2 = (l/2) \times (2/l - 1/a') \quad$ or $\quad n^2 = 1 - l/2a'$

or $\quad l/2a' = 1 - n^2 \quad$ so that $\quad 2a' = l/(1 - n^2)$ ...(5)

With vertex O as origin and OX and a line OY perpendicular to it as coordinate axes, let equation of the parabola be $y^2 = 4ax$ so that $dy/dx = 2a/y$. Then $\tan\psi = 2a/y$. At $L(a, 2a)$, $\tan\psi = 1$ so that $\psi = 45°$. Thus $\angle LTS = 45°$ and hence $\angle TLS = 45°$. Let SM be perpendicular to the tangent LT. Then, from ΔSLM, we have $SM = p = SL \sin 45° = l/\sqrt{2}$. Let l' be the latus rectum of the elliptic orbit. Then $l' = b'^2/a' = a'^2(1 - e'^2)/a' \quad$ or $\quad l' = a'(1 - e')$

Using formula $pV = (\mu l')^{1/2}$ for elliptic orbit, we get

$SM^2 \times V^2 = \mu l'^2 \quad$ or $\quad (l^2/2) \times (n^2v^2) = \mu a'(1 - e'^2)$, by (2)

or $\quad (l^2/2) \times n^2 \times (2\mu/l) = \mu a'(1 - e'^2)$, by (1)

or $\quad n^2 l = l(1 - e'^2)/2(1 - n^2)$, using (5)

or $\quad 2n^2(1 - n^2) = 1 - e^2 \quad$ so that $\quad e = (1 - 2n^2 + 2n^4)^{1/2}$

Ex. 21. *A particle is describing a parabolic orbit (latus rectum 4a) about a centre of force (μ) in the focus and on its arriving at a distance r from the focus moving towards the vertex the centre of the force ceases to act for a certain time T. When the force begins again to operate, prove that the new orbit will be an ellipse, parabola or hyperbola, according as*

$$T <=> 2r\{(r - a)/2\mu\}^{1/2}$$

Sol. Let the particle by moving with velocity v along tangent PT and when at P the force ceases to act for time T and the particle arrives at Q

such that $\quad PQ = vT \quad$ where $\quad v^2 = 2\mu/r \quad$ and $\quad r = SP$ ...(1)

Now, the pedal equation of parabola is $\quad p^2 = ar$ ...(2)

Then, $\quad p = r\sin\phi \Rightarrow \sin\phi = p/r = (ar)^{1/2}/r = (a/r)^{1/2}$, by (2) ...(3)

Also the particle has velocity v at Q. Hence the particle will now describe an ellipse, parabola or hyperbola according as

$v^2 <=> 2\mu/SQ \quad$ or $\quad 2\mu/r <=> 2\mu/SQ$, by (1)

or $\quad SQ <=> r \quad$ or $\quad SQ^2 <=> r^2$ ...(4)

From triangle PSQ by cosine formula, we have

$$\begin{aligned} SQ^2 &= SP^2 + PQ^2 - 2SP \cdot PQ\cos\phi \\ &= r^2 + v^2T^2 - 2r \cdot vT(1 - \sin^2\phi)^{1/2} \\ &= r^2 + v^2T^2 - 2rvT(1 - a/r)^{1/2}, \text{ by (3)} \end{aligned}$$

Thus, $SQ^2 = r^2 + v^2T^2 - 2vT(r^2 - ar)^{1/2}$ ...(5)

Hence from (4) and (5), we see that the particle will describe an ellipse, parabola or hyperbola according as

$r^2 + v^2T^2 - 2vT(r^2 - ar)^{1/2} <=> r^2 \quad$ or $\quad v^2T^2 <=> 2vT(r^2 - ar)^{1/2}$

or $\quad vT <=> 2(r^2 - ar)^{1/2} \quad$ or $\quad T(2\mu/r)^{1/2} <=> 2(r^2 - ar)^{1/2}$, by (1)

or $\quad T <=> 2r\{(r - a)/2\mu\}^{1/2}$

EXERCISE

1. By employing Kepler's law of planetary motion prove that a planet is acted upon by a central force directed towards the sun and that the force follows inverse square law. **[Bangalore 2004]**

Hint. Refer deductions 1 and 2 of Art 10.4.

2. Show that the velocity of a planet at any point of its orbit is the same as it would have been if it had fallen to the point from rest at a distance from the sun equal to the length of the major axis.

3. If the velocity of the earth at point of its orbit, assumed to be circular, were increased by about one half, prove that it would describe a parabola about the sun as focus. Show also that, if a body was projected from the earth with a velocity exceeding 7 miles per second, it will not return to the earth and may even leave the solar system. **[Gorakhpur 1998]**

4. A particle describes an ellipse about a centre of force at the focus; show that at any point of its path, the angular velocity about the other focus varies inversely as the square of the normal at that point.

5. A particle is describing a parabola under a force to the focus. It meets and coalesces with another particle of n times mass which was at rest before the impact. Show that the composite body will describe an ellipse whose eccentricity is given by $1 - e^2 = \{4n(n+1)/(n+1)^4\} \cos^2 (\theta/2)$, where θ is measured from the apse in the parabola.

10.8 To find the time of description of a given arc of a parabola starting from the vertex.

[Lucknow 1996]

The polar equation of a parabola of latus rectum $4a$ referred to its focus S as pole and the axis as initial line is given by

$$2a/r = 1 + \cos\theta = 2\cos^2\theta/2 \quad \text{or} \quad r = a\sec^2\theta/2 \qquad \ldots(1)$$

For a central orbit, $h = r^2(d\theta/dt)$

$$\therefore \quad dt = \frac{r^2}{h}d\theta = \frac{a^2\sec^4\theta/2}{(\mu l)^{1/2}}d\theta, \text{ by (1) and the result } h^2 = \mu l$$

or $$dt = \{a^2/(2a\mu)^{1/2}\}\sec^4(\theta/2)\,d\theta, \text{ as } l = 2a \text{ for a parabola} \qquad \ldots(2)$$

Let t be the time taken from the vertex A (where $\theta = 0$) to any point P (where $\theta = \theta$). Then, from (2), we get

$$\int_0 dt = (a^3/2\mu)^{1/2}\int_0^\theta \sec^4\frac{1}{2}\theta\,d\theta$$

or $$t = (a^3/2\mu)^{1/2}\int_0^\theta\left(1+\tan^2\frac{1}{2}\theta\right)\sec^2\frac{1}{2}\theta\,d\theta$$

or $$t = (a^3/2\mu)^{1/2}\,[2\tan\theta/2 + (2/3)\tan^3\theta/2]_0^\theta$$

$$\therefore \quad \text{Required time} = t = (2a^3/\mu)^{1/2}\{\tan\theta/2 + (1/3)\tan^3\theta/2\} \qquad \ldots(3)$$

P (r,θ)
r
θ
S A

10.9 To find the time of description of a given arc of an elliptic orbit starting from the nearer end of the major axis

[Gorakhpur 1998]

The polar equation of an ellipse of semi-latus rectum l referred to its focus S as pole and SA as initial line is

$$l/r = 1 + e\cos\theta, \text{ where } e < 1 \qquad \ldots(1)$$

For a central orbit, $h = r^2(d\theta/dt)$

$$\therefore \quad dt = \frac{r^2}{h}d\theta = \frac{l^2}{h}\times\frac{d\theta}{(1+e\cos\theta)^2} \qquad \ldots(2)$$

P(r, θ)
r
θ
A′ C S A X

Let t be the time taken from the vertex A (where $\theta = 0$) to any point P (where $\theta = 0$). Then from (2)

$$\int_0 dt = \frac{l^2}{h}\int_0^\theta\frac{d\theta}{(1+e\cos\theta)^2} \qquad \ldots(3)$$

Now, $$\frac{d}{d\theta}\frac{\sin\theta}{1+e\cos\theta} = \frac{\cos\theta(1+e\cos\theta) - \sin\theta(-e\sin\theta)}{(1+e\cos\theta)^2}$$

$$= \frac{e+\cos\theta}{(1+e\cos\theta)^2} = \frac{e^2+e\cos\theta}{e(1+e\cos\theta)^2} = \frac{(1+e\cos\theta)-(1-e^2)}{e(1+e\cos\theta)^2}$$

Thus, $$\frac{d}{d\theta}\frac{\sin\theta}{1+e\cos\theta}=\frac{1}{e(1+e\cos\theta)}-\frac{1-e^2}{e(1+e\cos\theta)^2}$$

or $$\frac{1-e^2}{e(1+e\cos\theta)^2}=\frac{1}{e(1+e\cos\theta)}-\frac{d}{d\theta}\frac{\sin\theta}{1+e\cos\theta} \quad ...(4)$$

Integrating both sides of (4), we have

$$\frac{1-e^2}{e}\int\frac{d\theta}{(1+e\cos\theta)^2}=\frac{1}{e}\int\frac{d\theta}{1+e\cos\theta}-\frac{\sin\theta}{1+e\cos\theta} \quad ...(5)$$

Now, $\int\frac{d\theta}{1+e\cos\theta}$

$$=\int\frac{d\theta}{1+e\{(1-\tan^2\theta/2)/(1+\tan^2\theta/2)\}}=\int\frac{\sec^2\theta/2\,d\theta}{(1+e)+(1-e)\tan^2\theta/2}$$

$$=\int\frac{2dz}{(1+e)+(1-e)z^2},\ \text{putting } \tan\theta/2=z \text{ and } \sec^2\theta/2\,d\theta=2dz$$

$$=\frac{2}{(1+e)^{1/2}(1-e)^{1/2}}\tan^{-1}\frac{(1-e)^{1/2}z}{(1+e)^{1/2}}=\frac{2}{(1-e^2)^{1/2}}\tan^{-1}\left\{\left(\frac{1-e}{1+e}\right)^{1/2}\tan\frac{\theta}{2}\right\}$$

Using the above result in (5), we get

$$\frac{1-e^2}{e}\int\frac{d\theta}{(1+e\cos\theta)^2}=\frac{2}{e(1-e^2)^{1/2}}\tan^{-1}\left\{\left(\frac{1-e}{1+e}\right)^{1/2}\tan\frac{\theta}{2}\right\}-\frac{\sin\theta}{1+e\cos\theta}$$

or $$\int\frac{d\theta}{(1+e\cos\theta)^2}=\frac{2}{(1-e^2)^{3/2}}\tan^{-1}\left\{\left(\frac{1-e}{1+e}\right)^{1/2}\tan\frac{\theta}{2}\right\}-\frac{e}{1-e^2}\frac{\sin\theta}{1+e\cos\theta}$$

Using the above result in (3), the required time t is given by

$$t=\frac{l^2}{h}\left[\frac{2}{(1-e^2)^{3/2}}\tan^{-1}\left\{\left(\frac{1-e}{1+e}\right)^{1/2}\tan\frac{\theta}{2}\right\}-\frac{e}{1-e^2}\frac{\sin\theta}{1+e\cos\theta}\right]_0^\theta$$

$$t=\frac{l^2}{(\mu l)^{1/2}}\frac{1}{(1-e^2)^{3/2}}\left[2\tan^{-1}\left\{\left(\frac{1-e}{1+e}\right)^{1/2}\tan\frac{\theta}{2}\right\}-e(1-e^2)^{1/2}\frac{\sin\theta}{1+e\cos\theta}\right]$$

But $l^2/(\mu l)^{1/2}=l^{3/2}/\mu^{1/2}$ and $l=b^2/a=a^2(1-e^2)/a=a(1-e^2)$

$\therefore$ $l^2/(\mu l)^{1/2}=a^{3/2}(1-e^2)^{3/2}/\mu^{1/2}$ and so required time t is given by

$$t=\frac{a^{3/2}}{\mu^{1/2}}\left[2\tan^{-1}\left\{\left(\frac{1-e}{1+e}\right)^{1/2}\tan\frac{\theta}{2}\right\}-e(1-e^2)^{1/2}\frac{\sin\theta}{1+e\cos\theta}\right] \quad ...(6)$$

10.10 To find the time of description of a given arc of a hyperbolic orbit starting from the vertex.

The polar equation of a hyperbola of semi-latus rectum l referred to its focus S as pole and SA as initial line (A being the vertex) is (draw figure yourself)

$$l/r=1+e\cos\theta,\ \text{where } e>1 \quad ...(1)$$

For a central orbit, $h=r^2\,(d\theta/dt)$

$$\therefore\quad dt=\frac{r^2}{h}d\theta=\frac{l^2}{h}\frac{d\theta}{(1+e\cos\theta)^2}=\frac{l^2}{(\mu l)^{1/2}}\frac{d\theta}{(1+e\cos\theta)^2}$$

Integrating, the required time t from the vertex A (where $\theta=0$) to any point $P(r,\theta)$ is given by

$$t=\frac{l^{3/2}}{\mu^{1/2}}\int_0^\theta\frac{d\theta}{(1+e\cos\theta)^2} \quad ...(2)$$

Now, $$\frac{d}{d\theta}\frac{\sin\theta}{1+e\cos\theta}=\frac{\cos\theta(1+e\cos\theta)-\sin\theta(-e\sin\theta)}{(1+e\cos\theta)^2}$$

$$= \frac{e+\cos\theta}{(1+e\cos\theta)^2} = \frac{e^2+e\cos\theta}{e(1+e\cos\theta)^2} = \frac{(1+e\cos\theta)+(e^2-1)}{e(1+e\cos\theta)^2}$$

Thus, $$\frac{d}{d\theta}\frac{\sin\theta}{1+e\cos\theta} = \frac{1}{e(1+e\cos\theta)} + \frac{e^2-1}{e(1+e\cos\theta)^2}$$

or $$\frac{e^2-1}{e(1+e\cos\theta)^2} = \frac{d}{d\theta}\frac{\sin\theta}{1+e\cos\theta} - \frac{1}{e(1+e\cos\theta)} \quad \ldots(3)$$

Integrating both sides of (3), we have

$$\frac{e^2-1}{2}\int\frac{d\theta}{(1+e\cos\theta)^2} = \frac{\sin\theta}{1+e\cos\theta} - \frac{1}{e}\int\frac{d\theta}{1+e\cos\theta} \quad \ldots(4)$$

Now, $$\int\frac{d\theta}{1+e\cos\theta}$$

$$= \int\frac{d\theta}{1+e\{(1-\tan^2\theta/2)/(1+\tan^2\theta/2)\}} = \int\frac{\sec^2\theta/2\,d\theta}{(e+1)-(e-1)\tan^2\theta/2}$$

$$= \int\frac{2dz}{(e+1)-(e-1)z^2}, \text{ putting } \tan\theta/2 = z \text{ and } \sec^2\theta/2\,d\theta = 2dz$$

$$= \frac{2}{(e-1)^{1/2}}\cdot\frac{1}{2(e+1)^{1/2}}\log\frac{(e+1)^{1/2}+(e-1)^{1/2}z}{(e+1)^{1/2}-(e-1)^{1/2}z}$$

$$= \frac{1}{(e^2-1)^{1/2}}\log\frac{(e+1)^{1/2}+(e-1)^{1/2}\tan\theta/2}{(e+1)^{1/2}-(e-1)^{1/2}\tan\theta/2} \quad \ldots(5)$$

Using (5), (4) reduces to

$$\frac{e^2-1}{e}\int\frac{d\theta}{(1+e\cos\theta)^2} = \frac{\sin\theta}{1+e\cos\theta} - \frac{1}{e(e^2-1)^{1/2}}\log\frac{(e+1)^{1/2}+(e-1)^{1/2}\tan\theta/2}{(e+1)^{1/2}-(e-1)^{1/2}\tan\theta/2}$$

or $$\int\frac{d\theta}{(1+e\cos\theta)^2} = \frac{e}{e^2-1}\cdot\frac{\sin\theta}{1+e\cos\theta} - \frac{1}{(e^2-1)^{3/2}}\log\frac{(e+1)^{1/2}+(e-1)^{1/2}\tan\theta/2}{(e+1)^{1/2}-(e-1)^{1/2}\tan\theta/2}$$

Hence (2) reduces to

$$t = \frac{l^{3/2}}{\mu^{1/2}}\left[\frac{e}{e^2-1}\frac{\sin\theta}{1+e\cos\theta} - \frac{1}{(e^2-1)^{3/2}}\log\frac{(e+1)^{1/2}+(e-1)^{1/2}\tan\theta/2}{(e+1)^{1/2}-(e-1)^{1/2}\tan\theta/2}\right]_0^\theta$$

But $$l = b^2/a = \{a^2(e^2-1)\}/a = a(e^2-1)$$

$$\therefore t = \frac{a^{3/2}(e^2-1)^{3/2}}{\mu^{1/2}}\left[\frac{e}{e^2-1}\frac{\sin\theta}{1+e\cos\theta} - \frac{1}{(e^2-1)^{3/2}}\log\frac{(e+1)^{1/2}+(e-1)^{1/2}\tan\theta/2}{(e+1)^{1/2}-(e-1)^{1/2}\tan\theta/2}\right]$$

or $$t = \frac{a^{3/2}}{\mu^{1/2}}\left[e(e^2-1)^{1/2}\frac{\sin\theta}{1+e\cos\theta} - \log\frac{(e+1)^{1/2}+(e-1)^{1/2}\tan\theta/2}{(e+1)^{1/2}-(e-1)^{1/2}\tan\theta/2}\right]$$

TYPE 2 : SOLVED EXAMPLES BASED ON ART. 10.8, 10.9 & 10.10

Ex. 1. *Prove that in a paraboic orbit the time taken to move from the vertex to a point distant r from the focus is* $(1/3\sqrt{\mu}^{\,1/2})(r+l)(2r-l)^{1/2}$, *where 2l is the latus-rectum.*

Sol. Proceed as in 10.8 upto relation (3) for the present problem. We shall express time t given by relation (3) in terms of r.

From (1), $\sec^2\theta/2 = 2r/l$ so that $1+\tan^2\theta/2 = 2r/l$ or $\tan^2\theta/2 = 2r/l - 1$

Thus, $\tan^2\theta/2 = (2r-l)/l$ and $\tan\theta/2 = \{(2r-l)/l\}^{1/2}$ ...(i)

Again, $l = 2a$ so that $a = l/2$. Then relation (3) of Art. 10.8 becomes

$$t = (l^3/4\mu)^{1/2}\times(1/3)\times\tan(\theta/2)\times(3+\tan^2\theta/2)$$

$$= \frac{1}{6}\frac{l\sqrt{l}}{\sqrt{\mu}}\left(\frac{2r-l}{l}\right)^{1/2}\left\{3+\frac{2r-l}{l}\right\} = \frac{1}{3\sqrt{\mu}}(r+l)(2r-l)^{1/2}, \text{ using (i)}$$

Ex. 2. *Show that if the time of describing an arc bounded by a focal chord of a parabolic orbit under Newtorian law be T, then T varies as (focal chord)$^{3/2}$.*

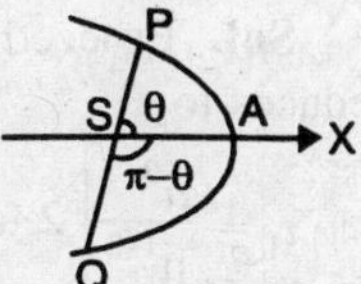

Sol. Proceed exactly as in Art. 10.8. Then for time t taken from Q to P (where PQ is a focal chord) is obtained by taking limits for θ from $\theta = -(\pi - \theta)$ to $\theta = \theta$ in place of $\theta = 0$ to $\theta = \theta$. Hence the required time is given by

$$t = (a^3/2\mu)^{1/2}\,[2\tan\theta/2 + (2/3)\tan^3\theta/2]^{\theta}_{-(\pi-\theta)}$$

$$= \left(\frac{2a^3}{\mu}\right)^{1/2}\left[\tan\frac{\theta}{2}+\frac{1}{3}\tan^3\frac{\theta}{2} - \tan\left\{-\left(\frac{\pi}{2}-\frac{\theta}{2}\right)\right\} - \tan^3\left\{-\left(\frac{\pi}{2}-\frac{\theta}{2}\right)\right\}\right]$$

$$= (2a^3/\mu)^{1/2}\,\{\tan\theta/2 + \cot\theta/2 + (1/3)(\tan^3\theta/2 + \cot^3\theta/2)\}$$

$$= (2a^3/\mu)^{1/2} \times (1/3) \times \{\tan^3\theta/2 + \cot^3\theta/2 + 3(\tan\theta/2 + \cot\theta/2)\}$$

$$= (2a^3/\mu)^{1/2} \times (1/3) \times (\tan\theta/2 + \cot\theta/2)^3, \text{ as } a^3 + b^3 + 3ab\,(a+b) = (a+b)^3$$

$$= \frac{1}{3}\left(\frac{2a^3}{\mu}\right)^{1/2}\left(\frac{\sin\theta/2}{\cos\theta/2}+\frac{\cos\theta/2}{\sin\theta/2}\right)^3 = \frac{1}{3}\left(\frac{2a^3}{\mu}\right)^{1/2}\frac{8}{\sin^3\theta} \qquad \ldots(4)$$

Using $r = 2a/(1+\cos\theta)$ for P and Q, we have

$$SP = \frac{2a}{1+\cos\theta}, \qquad SQ = \frac{2a}{1+\cos\{-(\pi-\theta)\}} = \frac{2a}{1-\cos\theta}$$

Let length of focal chord PQ be b, *i.e.*, $PQ = b$,

Then, $$b = SP + SQ = 2a\left[\frac{1}{1+\cos\theta}+\frac{1}{1-\cos\theta}\right] = \frac{4a}{\sin^2\theta} \qquad \ldots(5)$$

From (5), $\sin\theta = (4a/b)^{1/2}$ so that $\sin^3\theta = (4a)^{3/2}/b^{3/2}$

$\therefore$ from (4), $$t = \frac{8}{3}\left(\frac{2a^3}{\mu}\right)^{1/2}\cdot\frac{b^{3/2}}{(4a)^{3/2}} = \frac{1}{3}\left(\frac{2}{\mu}\right)^{1/2} b^{3/2}$$

Thus, t varies as $(PQ)^{3/2}$ *i.e.* t varies as (focal chord)$^{3/2}$

Ex. 3. *Prove that time taken to describe two portions into which an ellipse is divided by the latus rectum through the centre of force are in the ratio*

$$\{\cos^{-1} e - e(1-e^2)^{1/2}\} : \{\pi - \cos^{-1} e + e(1-e^2)^{1/2}\}$$

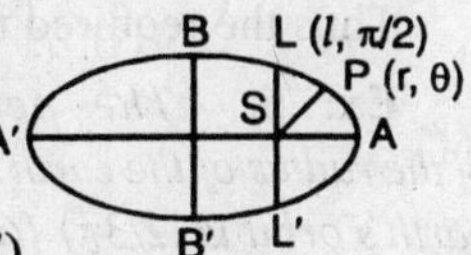

Sol. Proceed as in Art. 10.9 upto equation (6), which gives time t of describing as arc AP. Let t_1 be time of describing arc AL where LL' is latus rectum of the ellipse. Then putting $\theta = \pi/2$ in eqn. (6) of Art. 10.9, we get

$$t_1 = (a^{3/2}/\sqrt{\mu})\,[\,2\tan^{-1}\{(1-e)/(1+e)\}^{1/2} - e(1-e^2)^{1/2}] \qquad \ldots(7)$$

Taking $e = \cos\alpha$, we have

$$2\tan^{-1}\left(\frac{1-e}{1+e}\right)^{1/2} = 2\tan^{-1}\left(\frac{1-\cos\alpha}{1+\cos\alpha}\right)^{1/2} = 2\tan^{-1}\tan\frac{\alpha}{2} = 2\cdot\frac{\alpha}{2} = \alpha = \cos^{-1} e$$

$\therefore$ (7) reduces to $$t_1 = (a^{3/2}/\sqrt{\mu})\,[\cos^{-1} e - e(1-e^2)^{1/2}\} \qquad \ldots(8)$$

Clearly, T_1 = time of describing are $L'AL = 2t_1$

Now, T = time of describing the whole elliptic path $= (2\pi a^{3/2})/\sqrt{\mu}$

$$\therefore \quad T_2 = \text{time of describing the arc } LA'L = T - 2t_1$$
$$= (2\pi a^{3/2}/\mu^{1/2}) - 2(a^{3/2}/\mu^{1/2})\,\{\cos^{-1} e - e(1-e^2)^{1/2}\}$$
$$= (2a^{3/2}/\mu^{1/2})\,\{\pi - \cos^{-1} e + e(1-e^2)^{1/2}\} \qquad \ldots(9)$$

$\therefore$ Required ratio $= T_1 : T_2 = 2t_1 : T_2$

$= (2a^{3/2}/\mu^{1/2})\,\{\cos^{-1} e - e(1-e^2)^{1/2}\} : (2a^{3/2}/\mu^{1/2})\,\{\pi - \cos^{-1} e + e(1-e^2)^{1/2}\}$
$= \{\cos^{-1} e - e(1-e^2)^{1/2}\} : \{\pi - \cos^{-1} e + e(1-e^2)^{1/2}\}$

Ex. 4. *If the period of a planet be 365 days and eccentricity 1/60, then show that the times of describing two halves of the orbit bounded by the latus rectum through the centre of force are (365/2) {1 ± 1/15π} nearly.*

Sol. Proceed as in Ex. 3 upto equation (7). Now we proceed as follows. Since $e = 1/60$, (7) reduces to

$$t_1 = \frac{a^{3/2}}{\mu^{1/2}}\left[2\tan^{-1}\left\{\left(1-\frac{1}{60}\right)^{1/2}\left(1+\frac{1}{60}\right)^{-1/2}\right\} - \frac{1}{60}\left\{1-\left(\frac{1}{60}\right)^2\right\}^{1/2}\right]$$

$$= \frac{a^{3/2}}{\mu^{1/2}}\left[2\tan^{-1}\left\{\left(1-\frac{1}{2}\cdot\frac{1}{60}\right)\left(1-\frac{1}{2}\cdot\frac{1}{60}\right)\right\} - \frac{1}{60}\left\{1-\frac{1}{2}\cdot\left(\frac{1}{60}\right)^2\right\}\right]$$

[neglecting second and higher powers of 1/60]

$\therefore\ t_1 = (a^{3/2}/\mu^{1/2})\ \{2\tan^{-1}(1 - 1/60) - 1/60\}$, nearly, as before ...(8)

Since $\tan^{-1} 1 = \pi/4$ so $\tan^{-1}(1 - 1/60) = \pi/4 - \alpha$, where α is small

$\therefore\ 1 - 1/60 = \tan(\pi/4 - \alpha) = (1 - \tan\alpha)/(1 + \tan\alpha)$
$= (1 - \tan\alpha)(1 + \tan\alpha)^{-1} = (1 - \tan\alpha)(1 - \tan\alpha)$

[Neglect second and higher powers of $\tan\alpha$ because α is small]

$= 1 - 2\tan\alpha$, nearly, as before

Thus, $1 - 1/60 = 1 - 2\tan\alpha$ so that $\tan\alpha = 1/120$ ...(9)

Since α is very small so we can take $\tan\alpha = \alpha$, nearly

Thus, $\tan^{-1}(1 - 1/60) = \pi/4 - \alpha = \pi/4 - \tan\alpha = \pi/4 - 1/120$ ...(10)

Given that periodic time $= T = (2\pi a^{3/2})/\mu^{1/2} = 365$

Hence $a^{3/2}/\mu^{1/2} = 365/2\pi$ ...(11)

Using (10) and (11), (8) reduces to

$$t_1 = \frac{365}{2\pi}\left[2\left(\frac{\pi}{4}-\frac{1}{120}\right)-\frac{1}{60}\right] = \frac{365}{4}\left[1-\frac{1}{15\pi}\right]$$

$\therefore\ T_1$ = time of describing arc $L'AL = 2t_1 = (365/2)(1 - 1/15\pi)$

Let T_2 be time to describe arc $LA'L$. Then, we have

$T_2 = 365 - T_1 = 365 - (365/2)(1 - 1/15\pi) = (365/2)(1 + 1/15\pi)$

Thus the required times are $(365/2)(1 \pm 1/15\pi)$ nearly.

Ex. 5. *The perihelion distance of a comet describing a parabolic orbit is 1/n of the radius of the earth's path supposed circular, show that the time that the comet will remain within the earth's orbit is (2/3π) {(n + 2)/n} {(n – 1)/2n}^{1/2} of a year.*

Ex. 6. *Assuming the eccentricity e of a planet's orbit is a small fraction show that the ratio of the time taken by the planet to travel over the halves of its orbit separated by the minor axis is nearly 1 + (4e/π).* **[Delhi B.Sc. (Prog) III 2011; I.F.S. 2003]**

10.11 Tangential disturbing force. Disturbed elliptic motion.

[Gorakhpur 1997; 99; Purvanchal 1990, 99]

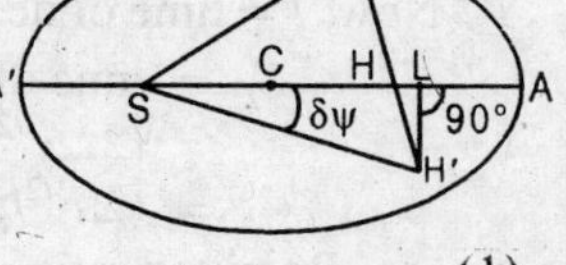

Let APA' be the path of a particle moving about a centre of force at S, and let H be the other focus. When the particle has arrived at P let its velocity v along the tangent is changed to $v + \delta v$ along the tangent because of a certain disturbing force along the tangent. If $2a$ and $2a'$ be the major axes before and after the disturbing force, then

$$v^2 = \mu(2/SP - 1/a) \qquad ...(1)$$

and $$(v+\delta v)^2 = \mu(2/SP - 1/a') \quad \ldots(2)$$

Differentiating (1) and noting that SP is constant, we get

$2v\,\delta v = (1/a^2)\,\delta a$ so that $\delta a = (2a^2 v\,\delta v)/\mu$, ...(3)

Relation (3) gives the increase in semi-major axis, *i.e.*, , δa.

Since the direction of motion at P remains unchanged, it follows that the perpendicular from S on the tangent at P, *i.e.*, p remains unchanged. Hence the relation $vp = h$ gives $p\,\delta v = \delta h$ so that

$$(h/v)\,\delta v = \delta h \quad \ldots(4)$$

For central orbits, $h^2 = \mu l$, ...(5)

where l = semi-latus rectum $= b^2/a = a^2(1-e^2)/a = a(1-e^2)$

Then, (5) becomes $h^2 = \mu a(1-e^2)$ ...(6)

Differentiating (6), $2h\,\delta h = \mu\{(1-e^2)\,\delta a - 2ae\,\delta e\}$

or $2ae\mu\,\delta e = \mu(1-e^2)\,\delta a - 2h\,\delta h$ ...(7)

Putting for δa and δh from (3) and (4) in (7), we get

$$\begin{aligned} 2\mu ae\,\delta e &= \mu(1-e)^2\cdot(2a^2 v\,\delta v/\mu)\cdot 2h\cdot(h\,\delta v/v) \\ &= 2a^2\cdot(1-e^2)v\,\delta v - \{2\mu a(1-e^2)\,\delta v\}/v, \text{ by (6)} \\ &= 2a(1-e^2)\,\{(av^2-\mu)/v\}\,\delta v \end{aligned}$$

or $$\delta e = \frac{1-e^2}{e}\cdot\frac{av^2-\mu}{\mu}\cdot\frac{\delta v}{v}, \quad \ldots(8)$$

which gives us the change in eccentricity.

Since the direction of motion remains unchanged, hence the other focus H' will lie on PH after disturbance. From figure, we find that

$$HH' = H'P - HP = (H'P + SP) - (HP + SP)$$

or $$HH' = 2a' - 2a = 2(a'-a) = 2\delta a \quad \ldots(9)$$

Let $\delta\psi$ be the angle HSH' through which the major axis turns. From figure, if $\angle H'HL = \alpha$, then

$$\begin{aligned} \delta\psi &= \tan\delta\psi = \frac{H'L}{SL} = \frac{HH'\sin\alpha}{2ae + HH'\cos\alpha}, \text{ as } SH = 2ae \\ &= (2\cdot\delta a\cdot\sin\alpha)/2ae, \text{ using (9) and the fact that } HH' \text{ is small} \\ &= (2va^2\,\delta v\sin\alpha)/\mu ae, \text{ using (3)} \end{aligned}$$

Thus, $$\delta\psi = (2va\cdot\delta v\sin\alpha)/\mu e \quad \ldots(10)$$

Again, we know that periodic time T is given by

$T = 2\pi a^{3/2}/\mu^{1/2}$ so that $\log T = \log(2\pi/\mu^{1/2}) + (3/2)\log a$

Differentiating it, $$\frac{\delta T}{T} = \frac{3}{2a}\delta a = \frac{3va\,\delta v}{3}, \text{ by (3)} \quad \ldots(11)$$

The results (3), (4), (8), (10) and (11) give the various changes due to disturbance in the elliptic path.

MISCELLANEOUS PROBLEMS

Ex. 1. *The square of the periodic times of the various planets are proportional to p power of the major axes of their orbits, then p is* **[Kanpur 2002]**

(*a*) 2 (*b*) 3 (c) 4 (*d*) 5

Sol. Ans. (*b*) Refer Art 10.3

Ex. 2. *According to Kepler's third law of motion, which identity holds good*

(a) $T^2 \propto a^3$ (b) $T^2 \propto 1/a^3$ (c) $a^2 \propto T^3$ (d) $a^2 \propto 1/T^3$ **[Agra 2007]**

Sol. Ans. (a). Refer Art. 10.3

11

Constrained motion in a plane

11.1 Introduction. When a particle in motion under given forces, does not follow its free path (trajectory) under these forces but is compelled to remain in contact with *a* a given surface or curve, then the particle is said to be in motion under constain. The constrain may be one-sided, as for example when a heavy particle slides on the inside of a circular wire and is free to break contact with the surface on the inside of the circle but cannot go outside. There will be then be a normal pressure inwards exerted by the curve on the particle so long as contact persists, and the pressure will vanish at the point where the particle leaves the curve. On the other hand if the constraint is two-sided as when a particle moves in a fine tube or a bead moves along a wire, then the normal reaction may vanish and change sign but the particle never leaves the prescribed path.

11.2 Motion of a heavy particle on a smooth curve in a vertical plane. Let a particle of mass m be constrained to move along the given curve AB. Let A be a fixed point on the curve and P be the position of the particle at any time t such that arc $AP = s$ and $\angle PCX = \psi$, where PT is tangent to curve at P. The components of acceleration of the particle along PT and PN are $\ddot{s}$ and v^2/ρ, where v is the velocity at P along PT and ρ, the radius of curvature at P. The external forces acting on the particle are its weight mg acting vertically downwards and normal reaction R at P along PN. Then the equations of motion in tangential and inward drawn normal sense are given by

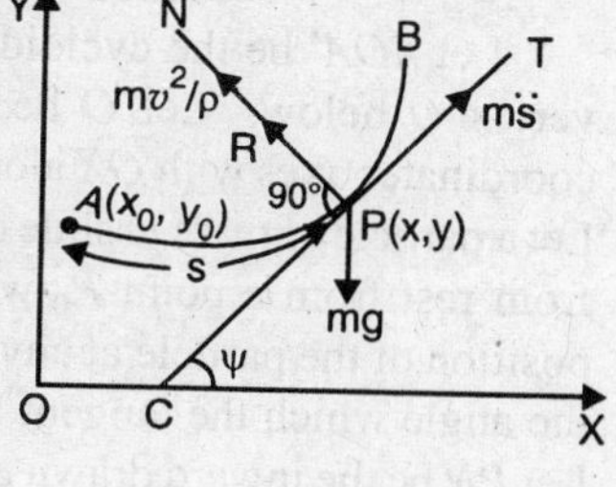

$$m\ddot{s} = -mg \sin \psi \qquad \ldots(1)$$

and

$$m(v^2/\rho) = R - mg \cos \psi \qquad \ldots(2)$$

Since $\sin \psi = dy/ds$ and $\ddot{s} = v(dv/ds)$ (1) may be re-written as

$$v(dv/ds) = -g(dy/ds) \qquad \text{or} \qquad 2v\,dv = -2g\,dy$$

Integrating, $v^2 = -2gy + C$, C being the constant of integration. ...(3)

If coordinates of A be (x_0, y_0) and u be the velocity of particle at A, then

$$u^2 = -2gy_0 + C \qquad \text{or} \qquad C = u^2 + 2gy_0$$

$\therefore$ From (3), $v^2 = u^2 - 2g(y - y_0)$ or $v^2 = u^2 - 2gh$, ...(4)

where $h = y - y_0$ = height of P above A.

Putting the value of v given by (4) in (2), we get R. Thus, velocity v and R can be found at any point P.

Note. When the external force system is conservative, then the equation (4) can also be easily obtained by using the equation of energy. Thus for motion for A to P, by the equation of energy, we have

$$(1/2)mv^2 - (1/2)mu^2 = (-mg)\cdot h \quad \text{or} \quad v^2 = u^2 - 2gh, \text{ which is the same as (4)}$$

Note 2. Equation of motion for general case. Let the particle be acted on by external force F and let its components along the tangent and normal be F_t and F_n respectively. Also, due to constraint, let R be the normal reaction. The equations of motion along the tangent and normal are

$$mv(dv/ds) = F_t \qquad \text{and} \qquad m(v^2/\rho) = F_n + R$$

When F and the curve is known, F_t, F_n and ρ can be obtained. Then v and R can be obtained by above equations. When the external force is conservative and the curve is smooth, we can use the equation of energy to find the velocity v at any point on the given curve.

11.3 Cycloid and related results for direct application. A cycloid is a curve which is traced out by a point on the circumference of a circle of radius a as the circle rolls along a fixed straight line. An arch AOA' is shown in the figure. O is origin and it is called vertex. The points A and A' are cusps and the line AA' joining them is called the base of the cycloid. OY is the axis of the cycloid. Let $P(x, y)$ be any point on the cycloid and PN be perpendicular to OX. Then $ON = x$ and $NP = y$. Let TT' be tangent at P, where $PT'X = \psi$ and arc $OP = s$. Then, remember the following facts :

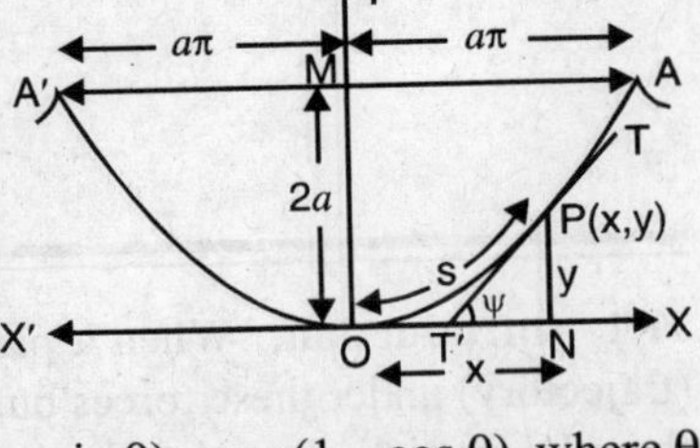

(i) *Parametric equations of the cycloid* are given by $x = a(\theta + \sin\theta)$, $y = a(1 - \cos\theta)$, where θ is parameter. Also note that $\theta = 2\psi$. **(Bundelkhand 2008, Kanpur 2010; 2014)**

(ii) *Intrinsic equation of the cycloid is* $s = 4a \sin\psi$

(iii) Arc $OA = 4a$ and height of the cycloid $= OM = 2a$

(iv) At O, $\psi = 0$, $s = 0$ while at A, $\psi = \pi/2$, $s = 4a$. Also, $MA = MA' = a\pi$

(v) Relation between s and y is $s^2 = 8ay$.

11.4 Motion on a smooth cycloid in a vertical plane. *A particle slides down a smooth cycloid whose axis is vertical and vertex downwards. Discuss the motion.* **[Bundelkhand 2007, 08, Meerut 2013, 2016; Agra 2014, 17; Kanpur 2010, 2019; Rohilkhand 2003, 06]**

Let AOA' be the cycloid with its base AA' horizontal and vertex O below. Let O be taken as origin and OX, OY be coordinates axes with OX along the tangent at O, to the cycloid. Let a particle of mass m slide down the arc of the cycloid starting from rest from a point P_0 where arc $OP_0 = s_0$. Let P be the position of the particle at any time t, where arc $OP = s$ and ψ be the angle which the tangent PT to the cycloid makes with OX. Let PN be the inward drawn normal at OP.

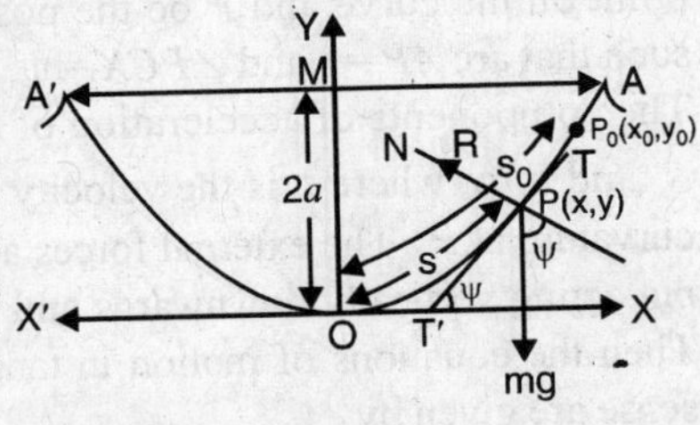

The external forces acting on the particle at P are its weight mg acting vertically downwards and the normal reaction R directed along PN. Then the equations of motion along PT and PN are

$$m(d^2s/dt^2) = -mg \sin\psi \quad \ldots(1)$$

and

$$m(v^2/\rho) = R - mg\cos\psi \quad \ldots(2)$$

Also the equation of cycloid is $s = 4a \sin\psi$ ...(3)

Using (3), (1) reduces to $d^2s/dt^2 = -(g/4a)s$, ...(4)

which is standard equation of a simple harmonic motion with centre at O. Hence the particle will oscillate in S.H.M. about the vertex O and time of period T of this S.H.M. is given by

$$T = 2\pi/(g/4a)^{1/2} \quad \text{or} \quad T = 4\pi(a/g)^{1/2}, \quad \ldots(5)$$

which is independent of the amplitude *i.e.,* the initial displacement s_0. It follows that from whatever point the particle may be allowed to slide down the arc of a smooth cycloid, the time period remain the same. Such a motion is known as *isochronous motion.*

Re-writing (4) $\{D^2 + (g/4a)\}\, s = 0$, where $D \equiv d/dt$ ...(6)

which is linear differential equation with constant coefficients. Its auxiliary equation is $D^2 + (g/4a) = 0$ so that $D = \pm i(g/4a)$ and its general solution is

$$s = A \cos\{t(g/4a)^{1/2}\} + B \sin\{t(g/4a)^{1/2}\}, \quad \ldots(7)$$

where A and B are arbitrary constants.

Initially at P_0, when $t = 0$, $s = s_0$, so (7) gives $A = s_0$ ...(8)

Differentiating (7) w.r.t. 't', we get

$$ds/dt = v = -A(g/4a)^{1/2} \sin\{t(g/4a)^{1/2}\} + B(g/4a)^{1/2} \cos\{t(g/4a)^{1/2}\} \quad \ldots(9)$$

Initially at P_0, when $t = 0$, $ds/dt = v = 0$. So (9) gives $B = 0$. ...(10)

Using (8) and (10), (7) gives $s = s_0 \cos\{t(g/4a)^{1/2}\}$, ...(11)

which gives, the position of the particle at any instant.

Using (8) and (10), (9) gives $v = -s_0(g/4a)^{1/2} \sin \{t(g/4a)^{1/2}\}$

which gives velocity at any time t at P.

Since $ds^2/dt^2 = v\, dv/ds$, (4) may be re-written as

$$v(dv/ds) = -(g/4a)s \qquad \text{or} \qquad 2v\, dv = -(g/4a) \times 2s\, ds$$

Integrating, $v^2 = C - (g/4a)s^2$, C being an arbitrary constant. ...(12)

Initially at P_0, when $s = s_0$, $v = 0$, so (12) gives $C = (g/4a)s_0^2$

$\therefore$ (12) reduces to $v^2 = (g/4a)(s_0^2 - s^2)$,

which gives the velocity at any position P. ...(13)

From (3), $\rho = ds/d\psi = 4a \cos \psi$ ...(14)

If $s_0 = 4a \sin \alpha$, (13) gives $v^2 = 4ga(\sin^2 \alpha - \sin^2 \psi)$ ...(15)

Substituting the values of ρ and v^2 from (14) and (15) in (2), we get

$R = mg \cos \psi + m \cdot 4ga(\sin^2 \alpha - \sin^2 \psi) \cdot (1/4a \cos \psi)$

$R = mg(\cos^2 \psi + \sin^2 \alpha - \sin^2 \psi)/\cos \psi = mg(\cos 2\psi + \sin^2 \alpha) \sec \psi$, ...(16)

which gives the reaction R at any point P.

Equation (11), (15) and (16) determine the motion completely.

11.5 Cycloidal pendulum. If a particle is constrained to move on a cycloid in a vertical plane, with its vertex below, and the tangent at the vertex horizontal, then the arrangement constitutes a *cycloidal pendulum*.

In Art. 11.4 we have shown that the motion of the particle is simple harmonic motion of period $T = 4\pi(a/g)^{1/2}$, which is independent of the amplitude. The oscillations of a cycloidal pendulum are called *isochronous* because the period is the same for large as well as for small oscillations.

TYPE 1 : SOLVED EXAMPLES

Ex. 1. *Show that the motion of a particle on a smooth cycloid, wnose axis is vertical and vertex downwards is S.H.M* **[Kanpur 2011]**

Sol. Refer Art. 11.4.

Ex. 2. *Describe the motion of a particle describing a cycloidal pendulum.*

Sol. Ref. Art. 11.4 and 11.5. **[Gorakhpur 1997]**

Ex. 3. *Prove that a cycloidal pendulum is perfectly isochronous.*

Sol. Refer Art. 11.5. **[Rohilkhand 1994]**

Ex. 4. *A particle falls down a smooth cycloid with its vertex at the lowest point, prove that the motion of the particle is S.H.M. with period $T = 4\pi(a/g)^{1/2}$, where $x = a(\theta + \sin \theta)$ and $y = a(1 - \cos \theta)$ is equation of the cycloid.*

Sol. Refer Art. 11.4 upto equation (5). **[Bangalore 1994]**

Ex. 5. (a) *A particle slides down a smooth cycloid whose axis is vertical and vertex downwards, starting from rest at the cusp. Find the velocity of the particle and reaction on it at any point of the cycloid.* **[Avadh 1998]**

Sol. Let AOA' be the cycloid with its base AA' horizontal and vertex O below. Let a particle start from the cusp A from rest and let $P(x, y)$ be its position at any time t, where OX, OY are coordinate axes with OX as tangent at O, to the cycloid. Let arc $OP = s$ and ψ be the angle made by tangent PT with OX. Let PN be inward drawn normal at P.

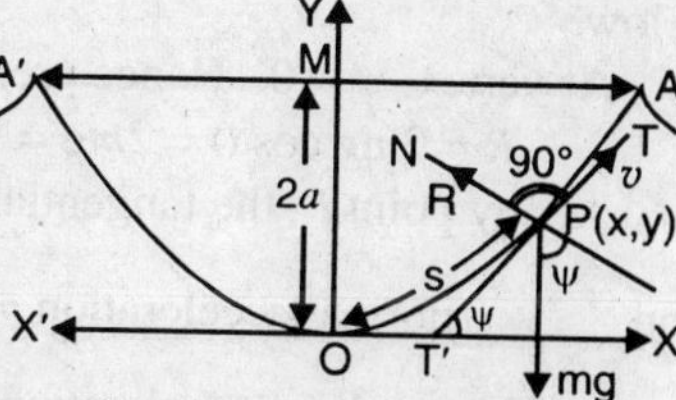

The external forces acting on the particle at P are its weight mg acting vertically downwards and the normal reaction R directed along inward drawn normal PN. Then equations of motion along PT and PN are

$$mv(dv/ds) = -mg \sin \psi \qquad ...(1)$$

and

$$m(v^2/\rho) = R - mg \cos \psi \qquad ...(2)$$

Also, the equation of cycloid is $s = 4a \sin \psi$ ...(3)

From (1) and (3), $v(dv/ds)=-g(s/4a)$ or $2v\,dv=-(g/4a)\cdot 2v\,dv$

Integrating, $v^2 = C-(g/4a)s^2$, C being an arbitrary constant. ...(4)

But initially at the cusp A, when $s=4a$, $v=0$. So (4) gives $C=4ag$.

$\therefore$ (4) reduces to $v^2=4ag-(g/4a)s^2=4ag-4ag\sin^2\psi$, by (3)

or $\qquad v^2=4ag(1-\sin^2\psi)$ or $v^2=4ag\cos^2\psi$ ...(5)

From (3), $\qquad \rho=ds/d\psi=4a\cos\psi$ ...(6)

Substituting the values of v^2 and ρ from (5) and (6) in (2), we get

$$R=\frac{mv^2}{\rho}+mg\cos\psi=m\frac{4ag\cos^2\psi}{4a\cos\psi}+mg\cos\psi=2mg\cos\psi \qquad ...(7)$$

Relations (5) and (7) give velocity and reaction at any point P.

Ex. 5. (b) *A particle oscillates from cusp to cusp of a smooth cycloid whose axis is vertical and vertex lowest. Show that the velocity v at any point P is equal to the resolved part of the velocity V at the vertex along the tangent at P i.e., $v=V\cos\psi$.*

Sol. Proceed as in Ex. 5(*a*) upto Eq. (5). Here you need not write equation (2). Thus, we have

$$v^2=4ag\cos^2\psi \qquad ...(5)$$

Given that $v=V$ when $\psi=0$ at vertex 0. So (5) gives $V^2=4ag$ ...(6)

From (5) and (6), $\quad v^2=V^2\cos^2\psi$ or $v=V\cos\psi$

Ex. 5. (c) *Show that for a particle, sliding down the arc and starting from the cusp of a smooth cycloid whose vertex is lowest, the vertical velocity is maximum when it has described the half the vertical height.*

Sol. Proceed as in Ex. 5(*a*) upto equation (5). Here you need not write equation (2). From (5), we have

$$V=2(ag)^{1/2}\cos\psi \qquad ...(6)$$

Let V be the vertical component of the velocity v at P. Then, from figure of Ex. 5(*a*), we have $V=v\cos(90°-\psi)=2(ag)^{1/2}\cos\psi\sin\psi=(ag)^{1/2}\sin 2\psi$, which is maximum when $\sin 2\psi=1$ or $2\psi=90°$ or $\psi=45°$. When $\psi=45°$, from (3) of Ex. 5(*a*), $s=4a\sin 45°=2\sqrt{2}a$

Putting $s=2\sqrt{2}a$ in the relation $s^2=8ab$, we have $8a^2=8ay$ so that $y=a$. Hence at the point where the vertical velocity V is maximum, we have $y=a$.

$\therefore$ The required vertical depth upto this point

= (the y-coordinate of A) $-a=2a-a=a=(1/2)\times 2a=(1/2)\times OM$

= half of the vertical height of the cycloid.

Ex. 5. (d) *A heavy particle slides down a smooth cycloid starting from rest at the cusp, the axis being vertical and vertex downwards, prove that the magnitude of the acceleration is equal to g at every point of the path and the pressure when the particle arrives at the vertex is equal to twice the weight of the particle.* **[Kanpur 2012; Meerut 2010; Agra 2012, 14; Purvanchal 2007]**

Sol. First of all proceed upto equation (7) as in Ex. 5(*a*). Now do the remaining part as given below.

At vertex, $\psi=0$. Hence pressure R from (7) is given by

$R=2mg\cos 0=2mg$ = twice the weight of the particle.

At any point P, the tangential acceleration $=v(dv/ds)=-g\sin\psi$, by (1)

and $\qquad$ normal acceleration $=\dfrac{v^2}{\rho}=\dfrac{4ag\cos^2\psi}{4a\cos\psi}=g\cos\psi \quad [\because$ from (3), $\rho=ds/d\psi=4a\cos\psi]$

$\therefore$ The resultant acceleration at any point P

$=\{(\text{tangential acceleration})^2+(\text{normal acceleration})^2\}^{1/2}=(g^2\sin^2\psi+g^2\cos^2\psi)^{1/2}=g.$

Ex. 6. *A particle oscillates in a cycloid under gravity, the amplitude of the motion being b, and period being T. Show that its velocity at any time t measured from a position of rest is $(2\pi b/T)\sin(2\pi t/T)$.* **[I.A.S. 2003]**

Sol. Proceed as in Art. 11.4 upto equation (11)′ after replacing s_0 by b for this problem. Negative sign in (11)′ indicates that as time increases s decreases. Hence magnitude of velocity at P is given by (11)′ as

$$v = (b/2) \times (g/a)^{1/2} \sin \{(t/2) \times (g/2)^{1/2}\} \qquad \ldots(i)$$

From (5), $(g/a)^{1/2} = 4\pi/T$. Hence (i) may be rewritten as

$$v = (b/2) \times (4\pi/T) \sin \{(t/2) \times (4\pi/T) = (2\pi b/T) \sin (2\pi t/T)$$

For the solution of this problem you need not write equation (2).

Ex. 7. *A particle is projected with velocity V from the cusp of a smooth inverted cycloid down the arc, show that the time of reaching the vertex is* $2(a/g)^{1/2} \tan^{-1} \{(4ag)^{1/2}/V\}$. **[Purvanchal 2011; Meerut 2015; Garhwal 2002; Kumaun 2002; Rohilkhand 2003]**

Sol. Proceed as in Art. 11.4 upto equation (7) noting that starting point P_0 is now at cusp A. For the solution of this problem you need not write equation (2) and (5). We now proceed with equation (7) of Art. 11.4 as follows.

Initially at A when $t = 0$, $s = 4a$, so (7) gives $\qquad A = 4a. \qquad \ldots(8)$

Differentiating (7) w.r.t. 't', we get

$$ds/dt = v = -A(g/4a)^{1/2} \sin \{t(g/4a)^{1/2}\} + B(g/4a)^{1/2} \cos \{t(g/4a)^{1/2}\} \qquad \ldots(9)$$

Initially at A when $t = 0$, $v = -V$, so (9) gives

$$-V = B(g/4a)^{1/2} \quad \text{or} \quad B = -V(4a/g)^{1/2} \qquad \ldots(10)$$

Using (8) and (10), (7) reduces to

$$s = 4a \cos \{t(g/4a)^{1/2}\} - V(4a/g)^{1/2} \sin \{t(g/4a)^{1/2}\} \qquad \ldots(11)$$

Let t' be the time taken by the particle in moving from A to the vertex O. Then, putting $s = 0$ and $t = t'$ in (11), we get

$$0 = 4a \cos \{t'(g/4a)^{1/2}\} - V(4a/g)^{1/2} \sin \{t'(g/4a)^{1/2}\}$$

or $\tan \{t'(g/4a)^{1/2}\} = (4ag)^{1/2}/V$ or $t'(g/4a)^{1/2} = \tan^{-1} \{(4ag)^{1/2}/V\}$

or $t' = 2(a/g)^{1/2} \tan^{-1} \{(4ag)^{1/2}/V\}$, which gives the required time.

Ex. 8. *A particle starts from rest at the cusp of a smooth cycloid whose axis is vertical and vertex downwards. Prove that when it has fallen through half the distance measured along the arc to the vertex, two-thirds of the time of descent will have elapsed.*

[Kanpur 2014, 2019; Lucknow 2001, 08; Rohilkhand 2007]

Sol. Proceed as in Art. 11.4 upto equation (7) noting that the starting point P_0 is now at cusp A. For the solution of this problem, you need not write equation (2). We now proved with equation (7) of Art. 11.4 as follows.

Initially at A where $t = 0$, $s = 4a$, so (7) gives $\qquad A = 4a \qquad \ldots(8)$

Differentiating (7) w.r.t. 't', we get

$$ds/dt = v = -A(g/4a)^{1/2} \sin \{t(g/4a)^{1/2}\} + B(g/4a)^{1/2} \cos \{t(g/4a)^{1/2}\} \qquad \ldots(9)$$

Initially at A when $t = 0$, $v = 0$, so (9) gives, $\qquad B = 0 \qquad \ldots(10)$

Using (8) and (10), (7) gives $\qquad s = 4a \cos \{t(g/4a)^{1/2}\} \qquad \ldots(11)$

Let t_1 be the time taken to move from the cusp A to half the distance along the arc to the vertex *i.e.*, to $s = 2a$. Then putting $t = t_1$ and $s = 2a$, (11) gives

$$2a = 4a \cos \{t_1(g/4a)^{1/2}\} \Rightarrow t_1(g/4a)^{1/2} = \pi/3 \text{ or } t_1 = (2/3)\pi\,(a/g)^{1/2} \qquad \ldots(12)$$

Now, if t_2 is the time from the cusp A to the vertex O then, clearly,

$$t_2 = T/4 = (1/4)\cdot 4\pi(a/g)^{1/2} = \pi(a/g)^{1/2}, \text{ by (5)} \qquad \ldots(13)$$

From (12) and (13), $t_1 = (2/3)t_2$, as desired.

Ex. 9. *A particle slides down the arc of a smooth cycloid whose axis is vertical and vertex lowest, starting at rest from the cusp. Prove that the time occupied in falling down the first half of the vertical height is equal to the time of falling down the second half.* **[Rohilkhand 2000; Kanpur 2001; Meerut 2007, 2017; Bundelkhand 2011; Purvanchal 2006]**

Sol. Proceed as in Art. 11.4 upto equation (7) noting that the starting point P_0 is now at cusp A. For the solution of this problem, do not write equation (2). With equation (7), now proceed as follows.

Initially at A when $t = 0, s = 4a$, so (7) gives $\qquad A = 4a.$ ...(8)

Differentiating (7) w.r.t. 't', we get

$$ds/dt = v = -A(g/4a)^{1/2} \sin \{t(g/4a)^{1/2}\} + B(g/4a)^{1/2} \cos \{t(g/4a)^{1/2}\} \qquad ...(9)$$

Initially at A, when $t = 0$, $v = 0$, so (9) gives $\qquad B = 0$ ...(10)

Using (8) and (10), (7) gives $\qquad s = 4a \cos \{t(g/4a)^{1/2}\}$ ...(11)

Since $s^2 = 8ay$, we have $\qquad s = 2(2ay)^{1/2}$ ...(12)

Using (12), (11) becomes $\quad 2(2ay)^{1/2} = 4a \cos \{t(g/4a)^{1/2}\}$

or $\cos \{t(g/4a)^{1/2}\} = (y/2a)^{1/2}$ or $t = (4a/g)^{1/2} \cos^{-1} \{(y/2a)^{1/2}\}$ ...(13)

The vertical height of the cycloid is $2a$. At the point, where the particle has fallen down the first half of the vertical height of the cycloid, $y = a$. Let t_1 be the time taken by the particle in falling down the first half of the vertical height, then t_1 can be obtained by putting $t = t_1$ and $y = a$ in (13). Thus,

$$t_1 = (4a/g)^{1/2} \cos^{-1} (1/\sqrt{2}) = 2(a/g)^{1/2} (\pi/4) = (1/2)\pi (1/g)^{1/2} \qquad ...(14)$$

Now, if t_2 is the time from cusp A to the vertex, then

$$t_2 = T/4 = (1/4)\cdot 4\pi(a/g)^{1/2} = \pi(a/g)^{1/2}, \text{ by (5)} \qquad ...(15)$$

From (14) and (15), $t_1 = (1/2)t_2$, showing that the time occupied in falling down the first half of the vertical height is equal to the time of falling down the second half.

Ex. 10. (*a*) *If a particle starts from rest at a given point of cycloid with its axis vertical and vertex downwards, prove that it falls 1/n of the vertical distance to the lowest point in time* $2(a/g)^{1/2} \sin^{-1} (1/\sqrt{n})$*, where a is the radius of the generating circle.*

[Agra 2000, 02, 05; Kanpur 1999; Purvanchal 1996]

Sol. Proceed as in Art. 11.4 upto equation (11). For the solution of this problem, do not write equations (2) and (5). With equaton (11), now proceed as follows. Re-writing (11), we get

$$\cos \{t(g/4a)^{1/2}\} = s/s_0 \quad \text{or} \quad t = 2(a/g)^{1/2} \cos^{-1} (s/s_0) \qquad ...(12)$$

Since $\quad s^2 = 8ay \quad$ so $\quad s_0^2 = 8ay_0$. Thus $\quad s/s_0 = (y/y_0)^{1/2}$

Then, (12) reduces to $\quad t = 2(a/g)^{1/2} \cos^{-1} (y/y_0)$ ...(13)

Let the particle take time t' in falling a distance $(1/n)y_0$ so that when $t = t'$, $y = y_0 - y_0/n = (1 - 1/n)y_0$ and so (13) gives

$$t' = 2(a/g)^{1/2} \cos^{-1}(1 - 1/n)^{1/2} = 2(a/g)^{1/2} \sin^{-1} \{1 - (1 - 1/n)\}^{1/2},$$

[using formula, $\cos^{-1} x = \sin^{-1} (1 - x^2)^{1/2}$]

or $\quad t' = 2(a/g)^{1/2} \sin^{-1} (1/\sqrt{n})$, which is the required time.

Ex. 10. (*b*) *A particle slides down the arc of a smooth cycloid whose axis is vertical and vertex lowest, starting from rest at a given point of the cycloid. Prove that the time occupied in falling down the first half of the vertical height to the lowest point is equal to the time of falling the second half.*

Sol. Proceed as in Ex. 10(*a*) by replacing n by 2. **[Meerut 2007]**

Ex. 11. *A particle starts from rest at any point P_0 in the arc of a smooth cycloid $s = 4a \sin \psi$ whose axis is vertical and vertex O downwards; prove that the time of descent to the vertex is* $\pi(a/g)^{1/2}$.

Show that if the particle is projected from P_0 downwards along the curve with velocity equal to that with which it reaches O when starting from rest at P_0, it will now reach O in half the time taken in the preceding case. **[Meerut 2006; Rohilkhand 1997]**

Sol. Proceed as in Art. 11.4 upto equation (11)$'$. For the solution of this problem, do not write equation (2). Let t_1 be time taken to move from P_0 to O,

$$t_1 = T/4 = (1/4)\cdot 4\pi(a/g)^{1/2} = \pi(a/g)^{1/2}, \text{ as required.} \qquad ...(12)$$

Let $v = v_1$ be velocity at O. So $v = v_1$ when $t = t_1$. Then, from (11)$'$,

$$v_1 = -s_0(g/4a)^{1/2} \sin \{t_1(g/4a)\}^{1/2} = -s_0(g/4a)^{1/2} \sin (\pi/2), \text{ using (12)}$$

Thus, $\qquad v_1 = -(s_0/2)(g/a)^{1/2}$ ...(13)

Second Part. Now suppose that the particle starts from P_0 with velocity v_1 given by (13). Then, when $t = 0$, $v = v_1$ at P_0 and so relation (9) now gives

$$-(s_0/2)(g/a)^{1/2} = (B/2) \times (g/a)^{1/2} \quad \text{or} \quad B = -s_0$$

With $A = s_0$ as before and $B = -s_0$, (7) gives

$$s = s_0 \cos\{t(g/4a)^{1/2}\} - s_0 \sin\{t(g/4a)^{1/2}\} \qquad \ldots(14)$$

Let t_2 be the time taken by the particle to reach vertex O in this case. Then $s = 0$ when $t = t_2$ and hence (14) gives

$$0 = s_0 \cos(t_2(g/4a)^{1/2}\} - s_0 \sin\{t_2(g/4a)^{1/2}\} \quad \text{or} \quad \tan\{t_2(g/4a)^{1/2}\} = 1$$

or $\quad (t_2/2) \times (g/a)^{1/2} = \pi/4 \quad$ or $\quad t_2 = (\pi/2) \times (a/g)^{1/2} = (1/2)t_1$., using (12)

Ex. 12. *Two particles are let drop from the cusp of a cycloid down the curve at an interval of time t; prove that they will meet in time* $2\pi(a/g)^{1/2} + t/2$.

[Garhwal 2003; Rohilkhand 2001, 03; Agra 2007, 09; Kanpur 2015]

Sol. Refer figure of Ex. 5(*a*). Let AOA' be the cycloid with the base AA' horizontal and vertex O below. Let a particle start from the cusp A from rest and P be its position at any time t. Take OX, OY as coordinate axes with OX as tangent at O to the cycloid. Let arc $OP = s$ and ψ be the angle made by tangent PT with OX. The external forces acting on the particle at P are its weight mg acting vertically downwards and the normal reaction R directed along the inward drawn normal PN. Then the equation of motion along tangent PT is

$$m(d^2s/dt^2) = -mg \sin\psi \quad \text{or} \quad d^2s/dt^2 + (g/4a)\, s = 0, \text{ as } s = 4a \sin\psi$$

Re-writing it, $\quad \{D^2 + (g/4a)\}s = 0$, when $D = d/dt$,

which is linear differential equation with constant coefficient. The roots of its auxiliary equation are $\pm i(g/4a)$ and hence its general solution is given by

$$s = A \cos\{t(g/4a)^{1/2}\} + B \sin\{t(g/4a)^{1/2}\}, \qquad \ldots(1)$$

where A and B are arbitrary constants.

Initially at cusp A, when $t = 0$, $s = 4a$, so (1) gives $\quad A = 4a \qquad \ldots(2)$

Differentiating (7) w.r.t. 't', we get

$$ds/dt = v = -A(g/4a)^{1/2} \sin\{t(g/4a)^{1/2}\} + B(g/4a)^{1/2} \cos\{t(g/4a)^{1/2}\} \qquad \ldots(3)$$

Initially at cusp A, when $t = 0$, $v = 0$, so (3) gives $\quad B = 0 \qquad \ldots(4)$

Hence the position of the first particle at any time t is

$$s_1 = 4a \cos\{t(g/4a)^{1/2}\}, \text{ using (2) and (4)} \qquad \ldots(5)$$

Similarly for the second particle, measuring time from the instant it is dropped from cusp, we have

$$s_2 = 4a \cos\{t(g/4a)^{1/2}\} \qquad \ldots(6)$$

Let t' be the time, when they meet, measured from the instant, the first, particle is let drop and let s' be arcual distance of the point where they meet. Then for the first particle, we have

$$s' = 4a \cos\{t'(g/4a)^{1/2}\} \qquad \ldots(7)$$

By problem, for the second particle, this instant is given by $t' - t$ from the moment it was dropped. So, we have

$$\therefore \quad s' = 4a \cos\{(t' - t)\,(g/4a)^{1/2}\} \qquad \ldots(8)$$

From (7) and (8) $\quad \cos\{t'(g/4a)^{1/2}\} = \cos\{(t' - t)\,(g/4a)^{1/2}\}$

$\Rightarrow \quad (g/4a)^{1/2}(t' - t) = 2\pi - (g/4a)^{1/2}t' \quad$ or $\quad 2t' - t = 2\pi(4a/g)^{1/2}$

$\therefore \quad 2t' = t + 4\pi(a/g)^{1/2} \quad$ so that $\quad t' = 2\pi(a/g)^{1/2} + t/2$

Ex. 13. *A particle is placed very near the vertex of a smooth cycloid whose axis is vertical and vertex upwards and is allowed to run down the curve. Prove that*

(*i*) *The particle will leave the curve when it is moving in a direction making an angle 45° downwards with the horizontal.*

(*ii*) *The particle will leave the curve when it has fallen through half the vertical height of the cycloid.* **[Bundlekhand 2010]**

(*iii*) *The latus rectum of the parabola subsequently described is equal to the height the cycloid.*

(*iv*) *It falls upon the base of cycloid at a distance* $(\sqrt{3} + \pi/2)$ *a from the centre of the base, a being radius of the generating circle.*

Sol. Let a particle of mass m, starting from O, slide down the arc of a smooth cycloid whose axis

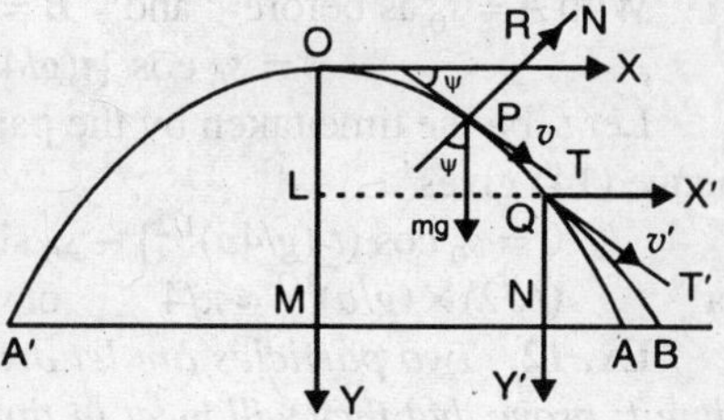

axis OM is vertical and vertex O is upwards. Let P be the position of the particle at any time t such arc $OP = s$ and the tangent PT at P to the cycloid makes an angle ψ with the tangent OX at the vertex. The external forces acting on the particle at P are its weight mg acting vertically downwards and the reaction R acting along the outwards drawn normal PN at P. Then the equations of the motion along the tangent and normal are

$$mv(dv/ds) = mg \sin \psi \quad \ldots(1)$$

and $$m(v^2/\rho) = mg \cos \psi - R \quad \ldots(2)$$

Also, for the cycloid $$s = 4a \sin \psi \quad \ldots(3)$$

From (1) and (3), $v(dv/ds) = g(s/4a)$ or $2v\, dv = (g/4a) \cdot 2s\, ds$

Integrating, $v^2 = (g/4a)v^2 + C$, C being an arbitrary constant ...(4)

Initially at O, $s = 0$ and $v = 0$. So (4) gives $C = 0$. Then, (4) gives

$$v^2 = (g/4a)s^2 = (g/4a) \times (4a \sin \psi)^2 \quad \text{or} \quad v^2 = 4ag \sin^2 \psi \quad \ldots(5)$$

From (3), $$\rho = ds/d\psi = 4a \cos \psi \quad \ldots(6)$$

Using (5) and (6), (2) gives $R = mg \cos \psi - (m \cdot 4ag \sin^2 \psi)/(4a \cos \psi)$

or $$R = (mg/\cos \psi)(\cos^2 \psi - \sin^2 \psi) \quad \ldots(7)$$

Part (*i*). When the particle leaves the cycloid at Q, we have $R = 0$.

$\therefore$ from (7) $mg(\cos^2 \psi - \cos^2 \psi) \sec \psi = 0$ or $\cos^2 \psi - \sin^2 \psi = 0$

or $\tan^2 \psi = 1 \Rightarrow \tan \psi = 1 \Rightarrow \psi = 45°$,

showing the particle will leave the cycloid when it is moving in a direction making an angle 45° downwards with the horizontal.

Part (*ii*). For a cycloid, we have $$s^2 = 8ay \quad \ldots(8)$$

From (3) and (8), $$y = s^2/8a = (4a \sin \psi)^2/8a = 2a \sin^2 \psi \quad \ldots(9)$$

Putting $\psi = 45°$ in (9), OL = depth of Q below OX = y-coordinate of $Q = 2a \sin^2 45° = 2a(1/\sqrt{2})^2 = a = (1/2) \times 2a = OM/2$, showing that the particle leaves cycloid when it has fallen through half the vertical height OM of cycloid.

Part (*iii*). Let v' be the velocity at Q. Then $v = v'$ when $\psi = 45°$ and so from (5), we get $v'^2 = 4ag \sin^2(45°) = 2ag$. Thus, the particle leaves the cycloid at Q with velocity $v'(= (2ag)^{1/2})$ in a direction making an angle 45° downwards with the horizontal. Beyond Q the particle describes a parabolic path.

$$\therefore \quad \text{latus rectum of this parabola} = \frac{2(v' \cos 45°)^2}{g} = \frac{2 \times 2ag \times (1/2)}{g}$$

$$= 2a = \text{height of the cycloid}$$

Part (*iv*). Taking Q as origin and horizontal and vertical (in downwards sense) lines QX' and QY' as coordinate axes, the equation of the parabolic path is

$$y = x \tan 45° + \frac{gx^2}{2v'^2 \cos^2 45°} = x + \frac{gx^2}{2 \cdot 2ag \cdot (1/2)} = x + \frac{x^2}{2a}$$

or $$2ay = 2ax + x^2 \quad \text{or} \quad x^2 + 2ax - 2ay = 0 \quad \ldots(10)$$

Suppose after leaving the cycloid at Q the particle strikes the base AA' at B. Now $LM = a$ and let $NB = x'$. Then coordinates of B referred to QX' and QY' as axes can be taken as (x', a). Since B lies on (10), we have

$$x'^2 + 2ax' - 2a^2 = 0 \quad \text{or} \quad x' = \{-2a \pm (4a^2 + 8a^2)^{1/2}\}/2 = -a \pm a\sqrt{3}$$

$\therefore$ $x' = a\sqrt{3} - a$, negative value of x' being inadmissible.

Thus, $x' = a(\sqrt{3} - 1)$ or $NB = a(\sqrt{3} - 1)$ ...(11)

The parametric equations of the cycloid with O as origin are given by

$$x = a(\theta + \sin \theta),\ y = a(1 - \cos \theta), \quad \ldots(12)$$

where θ is parameter and $$\theta = 2\psi \quad \ldots(13)$$

At Q, $\psi = \pi/4$ so $\qquad \theta = 2 \times (\pi/4) = \pi/2$

$\therefore$ from (12), x-coordinate of $Q = LQ = a(\pi/2 + \sin \pi/2) = a\,(\pi/2 + 1)$ ...(14)

Hence horizontal distance of B from the centre M of base BA

$= MN + NB = LQ + NB = a(1 + \pi/2) + a(\sqrt{3} - 1)$, by (11) and (14)

$= a\pi/2 + a\sqrt{3} = (\sqrt{3} + \pi/2)a$, as required.

Ex. 14. *A particle is moving on a smooth curve under gravity and its velocity varies as the arcual distance from the highest point. Prove that the curve is a cycloid.*

[Bundelkhand 1992; Purvanchal 1990; Rohilkhand 2007]

Sol. Let O be the highest point of the given curve AA' and P the position of the particle at any time t, such that arc $OP = s$ and the tangent PT at P to the curve makes an angle ψ with the tangent OX at point O of the curve.

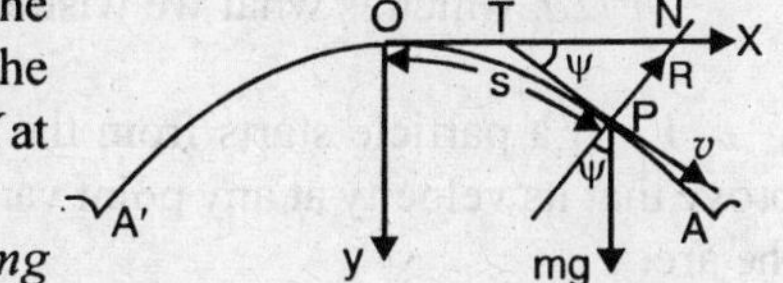

The external forces acting on the particle are its weight mg acting vertically downwards and the reaction R acting along the outwards drawn normal PN at P. Then, the equation of motion in the tangential direction is

$$m(d^2s/dt^2) = mg \sin \psi \qquad \text{or} \qquad d^2s/dt^2 = g \sin \psi \qquad ...(1)$$

By given problem, velocity at $P = ds/dt = ks$ ...(2)

Differentiating (2) w.r.t. 't', $\quad d^2s/dt^2 = k(ds/dt) = k \times (ks)$, by (2)

$\therefore$ from (1), $\quad k^2 s = g \sin \psi \qquad$ or $\qquad s = 4a \sin \psi$, where $4a = g/k^2$

This is intrisic equation of a cycloid.

Ex. 15. *A cycloid is placed with its axis vertical and vertex downwards and a heavy particle is projected from the cusp up the concave side of the curve with a velocity $(2gh)^{1/2}$. Prove that the latus rectum of the parabola described after leaving the arc is $h^2/2a$, where a is the radius of the generating circle.* **[Rohilkhand 1998]**

Sol. Let a particle of mass m be projected with velocity $(2gh)^{1/2}$ from the cusp A up the concave side of the cycloid. Let P be the position of the particle at any time t such that arc $OP = s$ and ψ be the angle at which tangent PT at P to the cycloid is inclined to OX which is tangent at vertex of the cycloid. Let v be the velocity of the particle at P along PT. The external forces acting on the particle at P are its weight mg acting vertically downwards and reaction R, directed along inward drawn normal PN. Then the equations of motion along the tangent and normal are

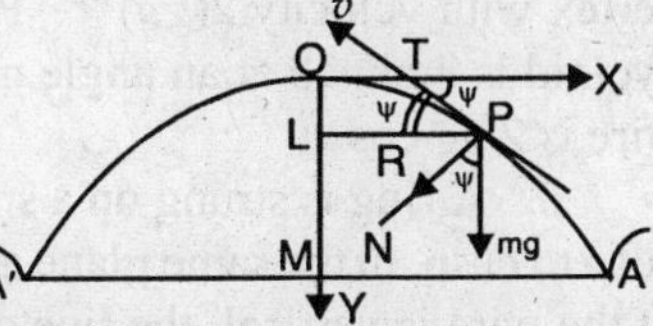

$$mv(dv/ds) = mg \sin \psi \qquad \text{or} \qquad v(dv/ds) = g \sin \psi \qquad ...(1)$$

and $\quad m(v^2/\rho) = R + mg \cos \psi \qquad$ or $\qquad R = m(v^2/\rho) - mg \cos \psi$ ...(2)

From a cycloid, we have $\qquad s = 4a \sin \psi$ ...(3)

Using (3), (1) gives $\quad v(dv/ds) = (g/4a)g \quad$ or $\quad 2v\,dv = (g/4a)\cdot 2s\,ds$

Integrating, $v^2 = (g/4a)\cdot s^2 + C$, C being an arbitrary constant. ...(4)

Initially at cusp A, $\quad s = 4a$ and $v = -(2gh)^{1/2}$. Then (4) gives

$$2gh = (g/4a) \times (16a^2) + C \quad \text{or} \quad C = 2gh - 4ga. \text{ Thus, (4) becomes}$$

$$v^2 = (g/4a)s^2 + 2gh - 4ga = (g/4a) \times (4a \sin \psi)^2 + 2gh - 4ga, \text{ by (3)}$$

or $\quad v^2 = 4ag \sin^2 \psi + 2gh - 4ag \qquad$ or $\qquad v^2 = 2gh - 4ag \cos^2 \psi$ ...(5)

From (3), $\qquad \rho = ds/d\psi = 4a \cos \psi$ ...(6)

Substituting the values of v^2 and ρ from (5) and (6) in (2), we get

$$R = \frac{m(2gh - 4ag\cos^2\psi)}{4a\cos\psi} - mg\cos\psi = \frac{mg(h - 2a\cos^2\psi - 2a\cos^2\psi)}{2a\cos\psi}$$

or $\qquad R = (mg/2a)\,(h - 4a\cos^2 \psi) \sec \psi$ ...(7)

The particle leaves the cycloid when $R = 0$. Hence the corresponding value of ψ can be obtained by (7) with $R = 0$.

$\therefore \quad 0 = (mg/2a)\,(h - 4a\cos^2\psi)\sec\psi \qquad \text{or} \qquad \cos^2\psi = h/4a$ ...(8)

The corresponding value of velocity at P at the time of leaving the cycloid is given by using (8) and so it is given by (5) as

$$v^2 = 2gh - 4ag \times (h/4a) = gh \qquad ...(9)$$

Thus, the particle leaves the cycloid at P with velocity v given by (9) inclined at angle ψ to the horizontal where ψ is given by (8). Beyond this point the particle describes a parabolic path whose latus rectum

$= (2/g)$ (square of the horizontal component of velocity at $P)^2$

$= (2/g) \times (v\cos\psi)^2 = (2/g) \times (ag) \times (h/4a)$, by (8) and (9)

$= h^2/2a$, which is what we wished to prove.

EXERCISE

1. If a particle starts from the vertex of a cycloid whose axis is vertical and vertex upward, prove that its velocity at any point varies as the distance of that point from the vertex measured along the arc.

2. Two particles connected by a fine string are constrained to move in a cycloidal tube in a vertical plane, the axis of the cycloid being vertical and vertex upwards; prove that the tension of the string is constant throughout the motion. **(I.A.S. 2005)**

3. If a particle slides down from rest at the vertex of cycloid, whose axis is vertical and vertex upwards, show that the velocity at any point is due to fall from the tangent at the vertex. **[Kanpur 2001]**

4. If a cycloid in fixed in a vertical plane with its vertex O as the highest point, show that a particle on the convex side projected horizontally from O with velocity u is at the arcual distance s from O at the instant t gives by $s = (u/p)\sinh pt$ where $p^2 = 4a/g$.

5. A bead of mass m is threaded on a smooth wire in the form of the cycloid $s = 4a\sin\psi$. The wire is fixed in a vertical plane with its vertex as the lowest point. The bead is projected from the vertex with velocity $2(ga)^{1/2}$. Prove that when the bead reaches a point at which the tangent to the cycloid is inclined at an angle ψ to the horizontal, its speed is $2(ga)^{1/2}\cos\psi$ and the reaction of the wire is $2mg\cos\psi$.

6. A ring is strung on a smooth closed wire which is in the shape of two equal cycloids joined cusp to cusp, in the same plane and symmetrically situated with respect to the line of cusps. The plane of the wire is vertical, the line of cusps horizontal, and the radius of the generating circle is a. The ring starts from the highest point with velocity v. Prove that the times from the upper vertex to the cusp, and from the cusp to the lower vertex are respectively

$$2(a/g)^{1/2}\sinh^{-1}\{(1/v) \times (4ag)^{1/2}\} \text{ and } 2(a/g)^{1/2}\sin^{-1}\{(4ag)/(v^2 + 8ag)\}$$

11.6 Motion under constraint in various curves

Read carefully Art. 11.1 and Art. 11.2. The method of solving can be understood by the following solved examples.

TYPE 2 : SOLVED EXAMPLES

Ex. 1. (a) *A bead of mass m slides on a smooth wire in the form of a parabola with axis vertical and vertex downward. If the bead starts from rest at an end of the latus rectum (of length 4p), find the speed with which it passes through the vertex. Find also the reaction of the wire on the bead at this point.* **[Delhi B.A. (Prog) II 2007, 2009; Delhi Maths (Hons.) 2007, 09]**

(b) *A smooth wire bent in the form of a parabola with its axis vertical and vertex downwards. A particle of mass m oscillates on the wire coming to rest at the extremities of the latus rectum. Show that the reaction of the wire on the particle when passing through the vertex is 2mg.* **[Delhi Maths. (Hons.) 1998]**

Sol. (a) Let the equation of the wire be

$$x^2 = 4py, \qquad ...(1)$$

whose latus rectum $= LL' = 4p$. Then its focus S will be at $(0, p)$. Let a particle fall from rest at the end L of

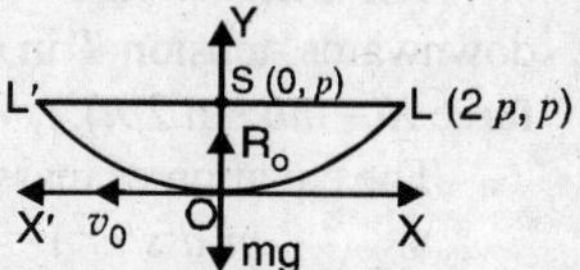

the latus rectum LL' to the vertex O. Let velocity of the bead at O be v_0. Then for motion from L to O, using the equation of energy,

$$(1/2)mv_0^2 - (1/2) \times m \times 0^2 = mg.SO \quad \text{or} \quad v_0^2 = 2gp, \quad \ldots(1)$$

which gives the required velocity with which bead crosses O.

The external forces acting on the bead at O are its weight mg acting vertically downwards and the inward normal reaction R_0 directed along OY. Let ρ_0 be the radius of curvature at O. Then the net inward force towards the centre of curvature is $R_0 - mg$ and the acceleration along OS is v_0^2/ρ.

$$\therefore \quad R_0 - mg = mv_0^2/\rho_0 \quad \text{or} \quad R_0 = mg + 2mgp/\rho_0, \text{ by (1)} \quad \ldots(2)$$

From (1), $\quad dy/dx = x/2p \quad$ and $\quad d^2y/dx^2 = 1/2p$

$\therefore$ At $O(0, 0)$, $\quad (dy/dx)_0 = 0 \quad$ and $\quad (d^2y/dx^2)_0 = 1/2p \quad \ldots(3)$

$$\therefore \quad \rho_0 = \{1 + (dy/dx)_0^2\}^{3/2}/(d^2y/dx^2)_0 = 2p, \text{ by (3)}$$

$\therefore$ from (2), $\quad R_0 = mg + mg = 2mg$, as required.

(*b*) Proceed as in part (*a*).

Ex. 2. *A horizontal smooth fixed plane wire has the shape of the catenary y = c cosh (x/c) and carries a small ring which is attracted towards the directrix with a force proportional to the distance y from the directrix. Show that the motion of the ring is simple harmonic and that the normal reaction varies inversely as y^2.*

Sol. Given $\quad y = c \cosh (x/c) \quad \ldots(1)$

For the catenary (1) with O as origin, OX is directrix and OY is its axis. $A(0, c)$ is its vertex. Let the ring of mass m be at $P(x, y)$ at any time t, where arc $AP = s$. Let the tangent PT at P be inclined at angle ψ with OX. Then for catenary (1), we have

$$y = c \sec \psi, \quad s = c \tan \psi \quad \ldots(2)$$

Let $PM \perp OX$. Then $PM = y$. Since the ring is attracted towards the directrix (*i.e.,* OX) with a force proportional to the distance of P from directrix, it follows that a force $k{\cdot}PM$ *i.e.,* ky will act on the ring directed along PM. Let R be normal reaction of the wire along the ring at P directed along inward drawn normal PN. Then, the equation of motion of along PT is given by

$$m(d^2s/dt^2) = -ky \sin \psi \quad \text{or} \quad d^2s/dt^2 = -(k/m) \times (c \sec \psi) \sin \psi,$$

by (2)

$$\text{or} \quad d^2s/dt^2 = -(k/m)s, \text{ by (2)} \quad \ldots(3)$$

Since $k > 0$, (3) represnets a S.H.M. with period $2\pi(m/k)^{1/2}$

Rewriting (3), $\quad v(dv/ds) = -(k/m)s \quad$ or $\quad 2v\, dv = -(k/m) \times 2s\, ds$

Integrating, $\quad v^2 = A - (k/m)s^2$, A being arbitrary constant. $\quad \ldots(4)$

Now, the equation of motion along normal PN is given by

$$mv^2/\rho = R - ky \cos \psi \quad \text{or} \quad R = mv^2/\rho + ky \cos \psi \quad \ldots(5)$$

From (2), $\quad \rho = ds/d\psi = c \sec^2 \psi \quad \ldots(6)$

Using (2), (4) and (6), (5) becomes

$$\begin{aligned} R &= (m/c) \cos^2 \psi \{A - (k/m)s^2\} + k(c \sec \psi) \cos \psi \\ &= (Am/c) \cos^2 \psi - (k/c) \cos^2 \psi \times (c \tan \psi)^2 + kc, \text{ by (2)} \\ &= (Am/c) \cos^2 \psi - kc \sin^2 \psi + kc = (Am/c) \cos^2 \psi + kc \cos^2 \psi \\ &= (Am/c + kc) \cos^2 \psi = (Am/c + kc) \times (c^2/y^2), \text{ as } y = c \sec \psi \end{aligned}$$

Since $(Am/c + kc)c^2$ is a constant, so R varies as $1/y^2$, as required

Ex. 3. *A simple pendulum of mass m and length a is hanging in equilibrium. At t = 0, a small horizontal disturbing force X comes into operation and continues to act, varying with time according to the formula X = mb sin 2pt where p^2 = g/a. Find a formula giving the position of the pendulum at any time.* **[Delhi B.A. (Prog.) III, 2013; Delhi B.Sc. (Hons.), 2000]**

Sol. Let P be the position of the simple pendulum at any time t, where $\angle AOP = \theta$ and arc $AP = s = a\theta$, OA being the initial vertical position of equilibrium.

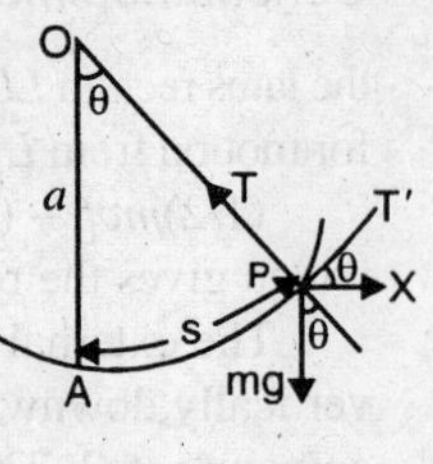

. The external forces acting on mass at P are its weight mg acting vertically downwards, tension T in the string along PO and the small horizontal disturbing force $X(= mb \sin 2pt)$.

The equation of mass m at P along tangent PT' is

$$m(d^2s/dt^2) = -mg \sin \theta + X \cos \theta = -mg\theta + mb \sin 2pt$$

$$[\because \text{ For small oscillations, } \sin \theta = \theta \text{ and } \cos \theta = 1]$$

or $\quad d^2s/dt^2 = -(g/a)s + b \sin 2pt, \quad$ as $\quad s = a\theta \Rightarrow \theta = s/a$

or $\quad d^2s/dt^2 = -p^2s + b \sin 2pt$, as $p^2 = g/a$

or $\quad (D^2 + p^2)s = b \sin 2pt$, where $D \equiv d/dt$ ...(1)

The roots of the auxilliary equation $D^2 + p^2 = 0$ of (1) are $\pm ip$.

$\therefore$ Complementary function = C.F. of (1) = $A \cos pt + B \sin pt$,

where A and B are arbitrary constants.

Also, particular integral of (1) = $\dfrac{1}{D^2 + p^2} b \sin 2pt = b \dfrac{1}{-(2p)^2 + p^2} \sin 2pt$

$$= -(b/3p^2) \sin 2pt$$

Hence the general solution of (1) is y = C.F. + P.I.

or $\quad s = A \cos pt + B \sin pt - (b/3p^2) \sin 2pt$ ...(2)

Initially at A, $t = 0, s = 0$ so (2) gives $\quad A = 0$

Putting $A = 0$ in (2) and then differentiating w.r.t. 't', we get

$$ds/dt = v = Bp \cos pt - (2bp/3p^2) \cos 2pt \quad ...(3)$$

Initially at A, $t = 0$, $v = 0$, so (3) gives $\quad B = 2b/3p^2$

Putting $A = 0$, $B = 2b/3p^2$ in (2), the required position is given by

$$s = (2b/3p^2) \sin pt - (b/3p^2) \sin 2pt = (2b/3p^2) \sin pt \,(1 - \cos pt)$$

Ex. 4. *A particle slides down a catenary, whose plane is vertical and vertex upwards, the velocity at any point being due to the fall from the directrix, prove that pressure at any point varies inversely as the distance of that point from the directrix.* **[Lucknow 1997, 2001]**

Sol. Let the intrisic equation of the catenary be

$$s = c \tan \psi, \quad ...(1)$$

where arc $AP = s$, P being the position of the particle on (1) at any time t. ψ is the angle between tangent PT at P and directrix OX of (1) where $OA = c$. OY is the axis of (1). Let (x, y) be coordinates of P. Then, we have $\quad y = c \sec \psi \quad$ ...(2)

The external forces acting on the particle are its weight mg acting vertically downwards and reaction R, directed along the outward drawn normal PN. The equation of motion of the particle along inward drawn normal at P is

$$mv^2/\rho = mg \cos \psi - R \quad \text{or} \quad R = mg \cos \psi - mv^2/\rho \quad ...(3)$$

Here $PM = y$ and by problem velocity at P being due to fall from the directrix, hence velocity v at P is given by $\quad v^2 = 2gy = 2gc \sec \psi$, by (2).

From (2), $\quad \rho = ds/d\psi = c \cdot \sec^2 \psi$. Also, $v^2 = 2gc \sec \psi$.

$\therefore$ (3) gives $\quad R = mg \cos \psi - (m \cdot 2gc \sec \psi)/(c \sec^2 \psi)$

or $\quad R = mg \cos \psi - 2mg \cos \psi = -mg \cos \psi = -mg(c/y)$, by (2)

Thus, $R = (-mgc)/y$, where $(-mgc)$ is a constant. This shows that pressure R at any point varies inversely as distance y of P from the directrix.

Ex. 5. *A particle falls from rest at the vertex of an inverted catenary; prove that the particle will leave the curve when the path described is $\sqrt{3}$ times the vertical distance through which the particle has fallen.*

Sol. Refer figure of Ex. 4. Proceed as in Ex. 4 upto equation (3). Then for the present problem, proceed as follows.

Let v be velocity at P along PT. Let $PN \perp OY$. Then $AN = ON - OA = y - c$. For motion from

A to P, by equation of energy, we have

$$(1/2)mv^2 - (1/2)m \cdot 0^2 = mg.AN \quad \text{or} \quad v^2 = 2g(y-c) \qquad \ldots(4)$$

From (1), $\quad \psi = ds/d\psi = c \sec^2 \psi \qquad \ldots(5)$

Using (4) and (5), (3) gives $\quad R = mg \cos\psi - \{2mg(y-c)\}/c \sec^2 \psi$

The particle will leave the curve when $R = 0$, *i.e.*, when

$$mg \cos\psi - \{2mg(y-c)/c\} \cos^2\psi = 0 \quad \text{or} \quad c \sec\psi - 2(y-c) = 0$$

or $\quad c \sec\psi - 2(c\sec\psi - c) = 0 \quad$ or $\quad \sec\psi = 2 \qquad \ldots(6)$

From (1), arc $AP = s = c\tan\psi = c(\sec^2\psi - 1)^{1/2} = c\sqrt{3}$, by (6)

and from (2), $\quad y = c\sec\psi = 2c \quad$ so that $\quad AN = y - c = 2c - c = c$

$\therefore \quad$ (arc AP)/$AN = (c\sqrt{3})/c = \sqrt{3} \Rightarrow$ arc $AP = \sqrt{3}\, AN$,

showing that when the particle leaves the curve the path described AP is $\sqrt{3}$ times the vertical height AN through which the particle has fallen.

Ex. 6. *A small bead, of mass m, moves on a smooth circular wire of radius a being acted by a central attraction mμ/(distance)^2 to a point within the circle situated at a distance b from the centre. Show that, in order that the bead may move completely round the circle, its velocity at the point of the wire nearest the centre of force must not be less than* $\{4\mu b)/(a^2 - b^2)\}^{1/2}$. **[Bhopal 2004; Lucknow 99]**

Sol. Let C be the centre of the circular wire and AB be its diameter. Let the centre of force be O where $CO = b$ (given). Let P be the position of the bead at any time t and let v be its velocity at P along tangent PT. Let $OP = r$. Then, by given problem force acting on P is $m\mu/r^2$ directed along PO. Let v_1 be velocity of the particle at A which is nearest to centre of force O. For motion from A to P, using the equation of energy, we have

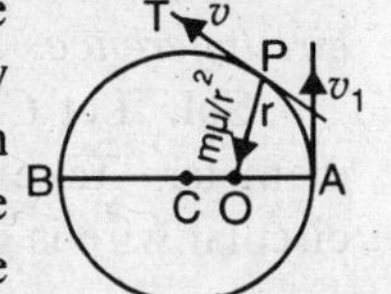

$$\left(\frac{1}{2}\right)mv^2 - \left(\frac{1}{2}\right)mv_1^2 = -\int_{a-b}^{r} \frac{m\mu}{r^2}\, dr = m\mu\left(\frac{1}{r} - \frac{1}{a-b}\right) \qquad \ldots(1)$$

Let velocity of the particle at B at v_2. Then $v = v_2$ when $r = a + b$

$$\therefore \text{ From (1),} \quad \left(\frac{1}{2}\right)mv_2^2 - \left(\frac{1}{2}\right)mv_1^2 = m\mu\left(\frac{1}{a+b} - \frac{1}{a-b}\right)$$

or $\quad (1/2)(v_2^2 - v_1^2) = -(2m\mu)/(a^2 - b^2) \quad$ or $\quad v_2^2 = v_1^2 - (4\mu b)/(a^2 - b^2)$

In order that the particle may move completely round the circle, its velocity v_2 at B must non-negative.

$\therefore \quad v_2 \geq 0 \quad$ or $\quad v_1^2 - (4\mu b)/(a^2 - b^2) \geq 0 \quad$ or $\quad v_1 \geq \{4\mu b/(a^2 - b^2)\}^{1/2}$,

showing that the necessary velocity at A must not be less than $\{(4\mu b)/(a^2 - b^2)\}^{1/2}$.

Ex. 7. *A particle of mass m moves in a smooth circular tube of radius a, under the action of a force, equal to mμ × distance, to a point inside the tube at a distance c from its centre; if the particle be placed very near by at its greatest distance from the centre of force, show that it will describe the quadrant ending at its least distance in time* $(a/\mu c)^{1/2} \log(1 + \sqrt{2})$ **[I.A.S. 1981]**

Sol. Let C be the centre of the circular tube of radius a and AD be its diameter. Let B be the middle point of arc AD. Let the centre of force be O where $OC = c$ (given). Let P be the position of the particle at any time t and v be its velocity at P along tangent PT. Let $OP = r$, then by given problem, force acting on P is $m\mu r$ directed along PO. Let $\angle ACP = \theta$ and arc $AP = s$. Then $s = a\theta$ and $v = ds/dt = a(d\theta/dt)$. The particle starts from rest from point A, where A is at the greatest distance from the centre of force O. Given $OC = c$ and so $OA = OC + CA = c + a$. For motion from A to P, using the equation of energy, we have

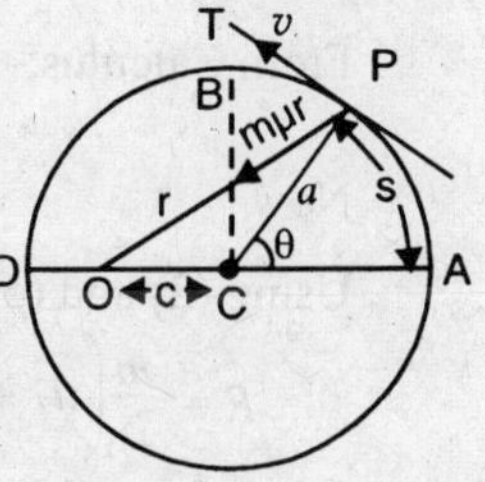

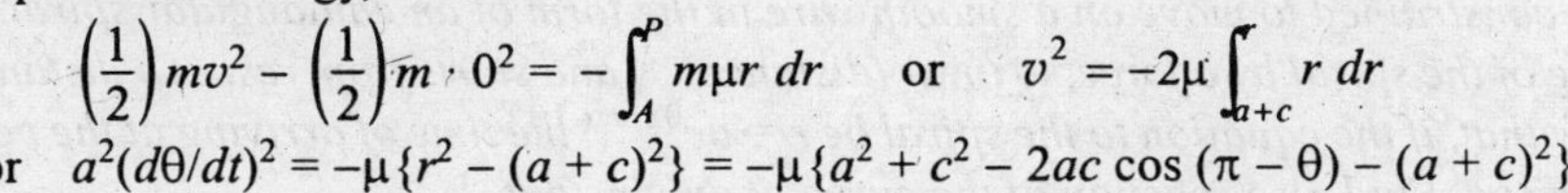

$$\left(\frac{1}{2}\right)mv^2 - \left(\frac{1}{2}\right)m \cdot 0^2 = -\int_A^P m\mu r\, dr \quad \text{or} \quad v^2 = -2\mu\int_{a+c} r\, dr$$

or $\quad a^2(d\theta/dt)^2 = -\mu\{r^2 - (a+c)^2\} = -\mu\{a^2 + c^2 - 2ac\cos(\pi - \theta) - (a+c)^2\}$

or $\quad a^2(d\theta/dt)^2 = 2\mu ac(1-\cos\theta) = 4\mu ac\sin^2(\theta/2)$

or $\quad d\theta/dt = 2(\mu c/a)^{1/2}\sin(\theta/2)$

taking positive sign before the radical sign as θ increases as t increases.

or $\quad 2(\mu c/a)^{1/2}dt = \operatorname{cosec}(\theta/2)\,d\theta \qquad \ldots(1)$

We are to find the time (t', say) of describing the quadrant BD ending at the least distance $OD(=a-c)$ from the centre of force, *i.e.*, ending at B. Clearly $\theta=\pi/2$ at B and $\theta=\pi$ at D. Hence integrating (1) and using the corresponding limits, we have

$$2\left(\frac{\mu c}{a}\right)^{1/2}\int_0^{t_1} dt = \int_{\pi/2}^{\pi}\operatorname{cosec}(\theta/2)\,d\theta$$

$$2(\mu c/a)^{1/2}t_1 = [2\log\tan\theta/4]_{\pi/2}^{\pi} = 2\log\tan\pi/4 - 2\log\tan\pi/8$$

$$= 2\log\cot\pi/8 = \log\cot^2\frac{\pi}{8} = \log\frac{2\cos^2\pi/8}{2\sin^2\pi/8} = \log\frac{1+\cos\pi/4}{1-\cos\pi/4}$$

or $\quad t_1 = \dfrac{1}{2}\left(\dfrac{a}{\mu c}\right)^{1/2}\log\dfrac{\sqrt{2}+1}{\sqrt{2}-1} = \dfrac{1}{2}\left(\dfrac{a}{\mu c}\right)^{1/2}\log(\sqrt{2}+1)^2 = \left(\dfrac{a}{\mu c}\right)^{1/2}\log(\sqrt{2}+1)$

Ex. 8. *A small bead is projected with any velocity along a smooth circular wire under the action of a force varying inversely as the fifth power of the distance from a centre of force situated on the circumference. Prove that the pressure on the wire is constant.*

Sol. Let C be the centre of the circular wire of radius a and OA be its diameter. Take O as pole and OA as initial line. Then polar equation of the circular wire is given by

$$r = 2a\cos\theta \qquad \ldots(1)$$

Here O is the centre of force. Let P be the position of the particle at any time t and let v be its velocity at P along tangent $T'T$. Let the particle be initially at A and let u be its velocity of projection. Here $OP = r$ and $\angle POA = \theta$. Then by the given problem, force acting on P is $m\mu/r^5$ directed along PO. Let R be pressure on wire acting along PC. For the motion from A to P, using the equation of energy, we have

$$\left(\frac{1}{2}\right)mv^2 - \left(\frac{1}{2}\right)mu^2 = -\int_A^P \frac{m\mu}{r^5}\,dr \quad \text{or} \quad v^2 = u^2 - 2\mu\int_{2a}^{r} r^{-5}\,dr$$

or $\quad v^2 = u^2 + \dfrac{\mu}{2}\left[\dfrac{1}{r^4}\right]_{2a}^{r} \quad$ or $\quad v^2 = u^2 + \dfrac{\mu}{2}\left(\dfrac{1}{r^4} - \dfrac{1}{16a^4}\right)$

or $\quad v^2 = u^2 + \mu/(2r^4) - \mu/(32a^4) \qquad \ldots(2)$

Equation of motion of the bead along normal is given by

$mv^2/a = R + (m\mu/r^5)\cos(\phi - 90^\circ)$, as $\quad \angle OPC = \angle OPT' - \angle CPT' = \phi - 90^\circ \qquad \ldots(3)$

or $\quad R = (mv^2/a) - (m\mu/r^5)\sin\phi \qquad \ldots(4)$

From (1), $\quad dr/d\theta = -2a\sin\theta$

From Calculus, $\quad \dfrac{ds}{d\theta} = \left\{r^2 + \left(\dfrac{dr}{d\theta}\right)^2\right\}^{1/2} = (4a^2\cos^2\theta + 4a^2\sin^2\theta)^{1/2} = 2a$

Now, $\quad \sin\phi = r\dfrac{d\theta}{ds} = r\times(1/2a) = r/2a \qquad \ldots(5)$

Using (2) and (5), (3) reduces to

$$R = \frac{m}{a}\left(u^2 + \frac{\mu}{2r^4} - \frac{\mu}{32a^4}\right) - \frac{m\mu}{r^5}\times\frac{r}{2a} = \frac{mu^2}{a} - \frac{m\mu}{32a^5},$$

which is independent of r and θ and hence pressure is constant.

Ex. 9. *A bead is constrained to move on a smooth wire in the form of an equiangular spiral. It is attracted to the pole of the spiral by a force, = (mμ) (distance)$^{-2}$, and starts from rest at a distance b from the pole. Show that, if the equation to the spiral be $r = ae^{\theta\cot\alpha}$ the time of arriving at the pole is $(\pi/2)\times(b^2/2\mu)^{1/2}\sec\alpha$. Find also reaction of the curve at any instant.*

Sol. Let O be the pole, OX be initial line and equation of the given curve be

$$r = ae^{\theta \cot \alpha} \qquad \ldots(1)$$

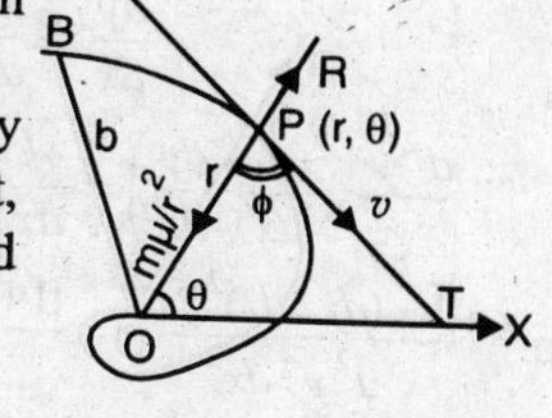

Let $P(r, \theta)$ be the position of the bead at any time t and v be the velocity at P along tangent PT. Let $\angle OPT = \phi$. Let the bead be initially at B at rest, where $OB = b$. By the given problem, force acting on P is $m\mu/r^2$ directed along PO. For the motion from B to P, the equation of energy gives

$$\left(\frac{1}{2}\right)mv^2 - \left(\frac{1}{2}\right)m.0^2 = -\int_b^r \frac{m\mu}{r^2}\,dr \quad \text{or} \quad v^2 = 2\mu\left(\frac{1}{r} - \frac{1}{b}\right) \qquad \ldots(2)$$

From (1), $\log r = \log a + \theta \cot \alpha$. Diff. it w.r.t. '$\theta$', we get

$$(1/r)\,(dr/d\theta) = \cot \alpha \Rightarrow \cot \phi = \cot \alpha \Rightarrow \phi = \alpha \qquad \ldots(3)$$

Now, from Calculus, $dr/ds = \cos \phi \Rightarrow dr/ds = \cos \alpha$, by (3)

$\therefore \quad ds = \sec \alpha\, dr \Rightarrow ds/dt = \sec \alpha\,(dr/dt) \Rightarrow v = \sec \alpha\,(dr/dt)$

$\therefore \quad (dr/dt)^2 = v^2 \cos^2 \alpha \quad$ or $\quad (dr/dt)^2 = 2\mu \cos^2 \alpha\, \{(b - r)/br\}$, by (2)

or $\quad dr/dt = -(2\mu)^{1/2} \cos \alpha\, \{(b - r)/br\}^{1/2}$, taking –ve sign as r decreases when t increases

or $$(2\mu)^{1/2} \cos \alpha\, dt = -\{br/(b - r)\}^{1/2}\, dr \qquad \ldots(4)$$

Let t' be time taken by the bead in moving from B to O. Hence integrating both sides of (4) and using the corresponding limits, we get

$$(2\mu)^{1/2} \cos \alpha \int_0^{t'} dt = -\int_b^0 \left(\frac{br}{b - r}\right)^{1/2} dr = \int_0^b \left(\frac{br}{b - r}\right)^{1/2} dr \qquad \ldots(5)$$

or $$(2\mu)^{1/2}\, t' \cos \alpha = \int_0^{\pi/2} \left(\frac{b^2 \sin^2 \theta}{b \cos^2 \theta}\right)^{1/2} .\, 2b \sin \theta \cos \theta\, d\theta, \text{ on putting}$$

$$r = b \sin^2 \theta \text{ and } dr = 2b \sin \theta \cos \theta\, d\theta$$

or $$(2\mu)^{1/2}\, t' \cos \alpha = 2b^{3/2} \int_0^{\pi/2} \sin^2 \theta\, d\theta = 2b^{3/2} . \frac{\pi}{4}$$

$\therefore \quad t' = (\pi/2) \times (b^3/2\mu)^{1/2} \sec \alpha$, which is the required time.

Equation of motion along normal is $\quad mv^2/\rho = (m\mu/r^2) \sin \phi - R \qquad \ldots(6)$

Now, $p = r \sin \phi \Rightarrow p = r \sin \alpha$. Then $\rho = r(dr/dp) = r \operatorname{cosec} \alpha \qquad \ldots(7)$

Using (2), (3) and (7), (6) reduces

$R = (m\mu/r^2) \sin \alpha - (m/r) \sin \alpha \cdot 2\mu\,(1/r - 1/b) = (m\mu/r) \sin \alpha\,(1/r - 2/r + 2/b)$

or $\quad R = (m\mu/r) \times (2/b - 2/r) \sin \alpha$, which is the required reaction.

Ex. 10. *A particle moves on the outside of an ellipse of eccentricity e whose major axis is vertical, starting from rest at the highest point. Show that it will leave the curve at a point where eccentric angle is ϕ where $e^2 \cos^3 \phi = 3 \cos \phi - 2$.* **[Bundelkhand 1990; Kumaun 1991]**

Sol. Let O be the centre and AA', BB' be major and minor axes of ellipse

$$x = a \cos \phi, y = b \sin \phi \qquad \ldots(1)$$

Let a particle start from the highest point A and let $P\,(a \cos \phi, b \sin \phi)$ be its position at any time. Let v be its velocity along tangent PT which is inclined at an angle ψ with OX. Let $PN \perp AA'$ For motion from A to P, by the equation of energy, we have

$$(1/2)mv^2 - (1/2)m\,.0^2 = mg \cdot AN,$$

where $\quad AN = OA - ON = a - a \cos \phi = a(1 - \cos \phi)$.

$\therefore$ $$v^2 = 2ga(1 - \cos \phi) \qquad \ldots(2)$$

Let $PM \perp OY$. Then, by geometry $\angle TPM = \angle TT'O = \pi - \psi$

The external forces acting on the particle are its weight mg acting vertically downwards and reaction R acting along the outward drawn normal. Then equation of motion along the inward drawn normal is

$mv^2/\rho = mg \sin(\pi - \psi) - R \quad$ or $\quad R = mg \sin \psi - m(v^2/\rho) \qquad \ldots(3)$

From (1), $\quad dx/d\phi = -a \sin \phi$, and $\quad dy/d\phi = b \cos \phi$

$$\therefore \quad \frac{dy}{dx} = \frac{dy/d\phi}{dx/d\phi} = \frac{b\cos\phi}{-a\sin\phi} = -\frac{b}{a}\cot\phi \qquad ...(4)$$

$$\frac{d^2y}{dx^2} = \frac{d}{dx}\left(\frac{dy}{dx}\right) = \frac{d}{dx}\left(-\frac{b}{a}\cot\phi\right) = \frac{d}{d\phi}\left(-\frac{b}{a}\cot\phi\right)\frac{d\phi}{dx} = \frac{(b/a)\,\text{cosec}^2\phi}{(-a\sin\phi)}$$

$$\rho = \frac{\{1+(dy/dx)^2\}^{3/2}}{d^2y/dx^2} = \frac{\{1+(b^2\cos^2\phi)/a^2\sin^2\phi\}^{3/2}}{(-b/a^2\sin^3\phi)} = -\frac{(a^2\sin^2\phi+b^2\cos^2\phi)^{3/2}}{ab}$$

or $\rho = (a^2 \sin^2 \phi + b^2 \cos^2 \phi)^{3/2}/ab$, numerically. ...(5)

From (4), $\cot \psi = -(a \sin \phi)/(b \cos \phi)$

$\therefore$ $\text{cosec}^2 \psi = 1 + \cot^2 \psi = (b^2 \cos^2 \phi + a^2 \sin^2 \phi)/b^2 \cos^2 \phi$

Since $\pi/2 < \psi < \pi$, $\sin \psi$ is positive, and hence

$\sin \psi = (b \cos \phi)/(a^2 \sin^2 \phi + b^2 \cos^2 \phi)^{1/2}$ ...(6)

The particle will leave the ellipse if $R = 0$. Then (3) gives

$mg \sin \psi = (mv^2)/\rho$ or $g\rho \sin \psi = v^2$

$$\text{or} \quad \frac{g(a^2\sin^2\phi+b^2\cos^2\phi)^{3/2}}{ab} \times \frac{b\cos\phi}{(a^2\sin^2\phi+b^2\cos^2\phi)^{1/2}} = 2ga(1-\cos\phi)$$

[Using (2), (5) and (6)]

or $\cos \phi\, (a^2 \sin^2 \phi + b^2 \cos^2 \phi) = 2a^2(1 - \cos \phi)$

or $\cos \phi\, \{a^2 \sin^2 \phi + a^2(1 - e^2) \cos^2 \phi\} = 2a^2(1 - \cos \phi)$

or $\cos \phi\, (1 - e^2 \cos^2 \phi) = 2(1 - \cos \phi)$ or $e \cos^3 \phi = 3 \cos \phi - 2$

Ex. 11. *From the lowest point of a smooth hollow elliptic cylinder of axes 2a, 2b with minor axis vertical, particle is projected in a vertical plane perpendicular to the axis of the cylinder; show that it will leave the cylinder if the velocity of projection lies between* $(2gb)^{1/2}$ *and* $\{g(a^2 + 4b^2)/b\}^{1/2}$

Sol. The figure shows a section of the elliptic cylinder by a vertical plane passing through the major and minor axes of the generating ellipse $ABA'B'$ and containing the direction of projection. Let O be centre the elliptic section.

$x = a \cos \phi, \quad y = b \sin \phi$ ...(1)

Let a particle start from the lowest point B' and let $P(a \cos \phi, b \sin \phi)$ be its position at any time t. Let v be its velocity along tangent PT which is inclined at an angle ψ with OX. Let $PN \perp BB'$. For motion from B' to P, by the equation of energy,

$(1/2) \times mv^2 - (1/2) \times mu^2 = (-mg) \times B'N$ or $v^2 = u^2 - 2g(y + b)$ ...(2)

The external forces acting on the particle are its weight mg acting vertically downwards and reaction R acting along inward drawn normal. Then equation of motion along the inward drawn normal is given by

$mv^2/\rho = R + mg \cos (\psi - \pi)$ or $R = mv^2/\rho + mg \cos \psi$ ...(3)

Proceed as in Ex. 10, to prove that $dy/dx = -(b/a) \cot \phi$ ...(4)

and $\rho = (a^2 \sin^2 \phi + b^2 \cos^2 \phi)^{3/2}/ab$ ...(5)

From (4), $\tan \psi = -(b \cos \phi)/(a \sin \phi)$

$\therefore$ $\sec^2 \psi = 1 + \tan^2 \psi = (a^2 \sin^2 \phi + b^2 \cos^2 \phi)/a^2 \sin^2 \phi$

Since $\pi/2 < \psi < \pi$, $\cos \psi$ is negative, so

$\cos \psi = -(a \sin \phi)/(a^2 \sin^2 \phi + b^2 \cos^2 \phi)^{1/2}$ ...(6)

The particle will leave the ellipse if $R = 0$. Then (3) gives

$0 = mv^2/\rho + mg \cos \psi$ or $v^2 + \rho g \cos \psi = 0$

$$\text{or} \quad u^2 - 2g(y+b) - g \times \frac{(a^2\sin^2\phi+b^2\cos^2\phi)^{3/2}}{ab} \times \frac{a\sin\phi}{(a^2\sin^2\phi+b^2\cos^2\phi)^{1/2}} = 0$$

or $u^2 = 2g(b \sin \phi + b) + (g/b) \sin \phi\, (a^2 \sin^2 \phi + b^2 \cos^2 \phi)$ ...(7)

If u_A be the velocity of projection in order that the particle may leave at A (where $\phi = 0$) and u_B in order that it may leave at B (where $\phi = \pi/2$), (7) gives

$$u_A^2 = 2gb \quad \text{and} \quad u_B^2 = 4gb + ga^2/b = g(a^2 + 4b^2)/b$$

Hence for the particle to leave somewhere in the arc AB, we have

$$u_A < u < u_B \quad \text{or} \quad (2gb)^{1/2} < u < \{g(a^2 + 4b^2)/b\}^{1/2}$$

Ex. 12. *A particle is constrained to move under gravity on a parabola whose axis is vertical and vertex upwards. Show that the pressure on the curve in any position of the parabola is numerically $m(u_0^2 - gl)/\rho$, where u_0 is the velocity at the vertex, $2l$ the latus rectum, and ρ the radius of curvature at the point. Discuss the case $u_0^2 = gl$.*

Sol. Let $P(x, y)$ be the position of the particle at any time t, referred to the vertex O as origin and OX, OY as x-axis and y-axis respectively. Let v be velocity a P along the tangent PT which is inclined at an angle ψ with OX. Initially the particle is projected from O with velocity u_0. Let $PN \perp OX$. For motion from O to P, the equation of energy gives

$$(1/2)mv^2 - (1/2)mu_0^2 = mg \cdot ON \quad \text{or} \quad v^2 = u_0^2 + 2gx \qquad \ldots(1)$$

The external forces on the particle at P are its weight mg acting vertically downwards and reaction R. Then, the equation of motion along the inward drawn normal is given by

$$mv^2/\rho = mg \sin(\pi - \psi) - R \quad \text{or} \quad R = mg \sin\psi - mv^2/\rho \qquad \ldots(2)$$

Equation of parabola is

$$y^2 = 2lx \qquad \ldots(3)$$

From (3), $dy/dx = l/y$ and $d^2y/dx^2 = -l^2/y^3$. So, numerically,

$$\therefore \ \rho = \frac{\{1 + (dy/dx)^2\}^{3/2}}{d^2y/dx^2} = \frac{(1 + l^2/y^2)^{3/2}}{l^2/y^3} = \frac{(2lx + l^2)^{3/2}}{l^2} = \frac{(2x + l)^{3/2}}{l^{1/2}}$$

Also, $\tan\psi = \dfrac{dy}{dx} = \dfrac{l}{y} \Rightarrow \sin\psi = \dfrac{l}{(l^2 + y^2)^{1/2}} = \dfrac{l}{(l^2 + 2lx)^{1/2}} = \dfrac{l^{1/2}}{(l + 2x)^{1/2}}$

$$\therefore \ \rho \sin\psi = \{(2x + l)^{3/2}/l^{1/2}\} \times \{l^{1/2}/(2x + l)^{1/2}\} = 2x + l \qquad \ldots(4)$$

From (2), $R = (m/\rho)\,\{g\rho \sin\psi - v^2\} = (m/\rho)\,\{g(2x + l) - u_0^2 - 2gx\}$, [Using (1) and (4)]

or $$R = (m/\rho)\,(gl - u_0^2), \qquad \ldots(5)$$

showing that the pressure at any point is $(m/\rho) \times (gl - u_0^2)$ numerically.

If $u_0^2 < gl$, R is +ve and the particle remains in contact with the given parabola throughout the motion. But if $u_0^2 > gl$, the pressure is –ve and this motion is not possible *i.e.*, it will not describe the given parabola. However, if $u_0^2 > gl$ and the particle be projected from the vertex along the inner side of the parabola, it will describe the given parabola.

When $u_0^2 = gl$ and $\rho \neq 0$, then $R = 0$ and hence the particle moves in the form of a parabola without constraint just like a projectile.

Ex. 13. *A smooth parabolic curve is fixed with its plane vertical and axis horizontal, show that a heavy particle placed on the arc at a height above the axis equal to the latus rectum, will run off the curve on reaching at the end of the latus rectum and will then describe a parabola having the same latus rectum as the given arc.*

Sol. Let the equation given parabola be

$$y^2 = 4ax \qquad \ldots(1)$$

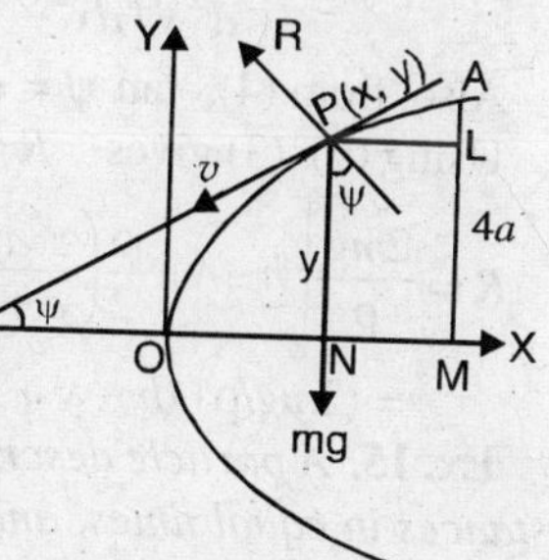

Let $PN \perp OX$, $AM \perp OX$ and $PL \perp AM$. Let A be a point on (1), where $AM = 4a$. Then initially a particle starts from A from rest. Let $P(x, y)$ be its position at any time t and let v be its velocity at P along tangent PT which is inclined at an angle of ψ with OX. From figure $AL = AM - LM = AM - PN = 4a - y$. For motion from A to P, the equation of energy gives

$$(1/2)mv^2 - (1/2)m \cdot 0^2 = mg \times AL = mg \cdot (4a - y)$$

or $$v^2 = 2g(4a - y) \qquad \ldots(2)$$

The external forces acting on the particle are its weight mg acting vertically downwards and reaction R along outward drawn normal. Then equation motion along inward drawn normal at P is given as

$$mv^2/\rho = mg \cos \psi - R$$

or $R = mg \cos \psi - 2(m/\rho)g\,(4a - y)$, by (2) ...(3)

From (1), $dy/dx = 2a/y$ and $d^2y/dx^2 = -4a^2/y^3$

$$\therefore\ \rho = \frac{\{1+(dy/dx)^2\}^{3/2}}{d^2y/dx^2} = \frac{(1+4a^2/y^2)^{3/2}}{-(4a^2/y^3)} = \frac{(4a^2+y^2)^{3/2}}{4a^2}, \text{ numerically} \quad ...(4)$$

Again, $\tan \psi = dy/dx = 2a/y$ so $\cos \psi = y/(4a^2 + y^2)^{1/2}$ ...(5)

The particle will leave the parabola when $R = 0$. Then (3) gives

$0 = (mg/\rho)\,\{\rho \cos \psi - 2(4a - y)\}$ or $\rho \cos \psi = 2(4a - y)$

or $\{(4a^2 + y^2)^{3/2}/4a^2\} \times \{y/(4a^2 + y^2)^{1/2}\} = 2(4a - y)$, by (4) and (5)

or $y(4a^2 + y^2) = 8a^2(4a - y)$ or $y^3 + 12a^2y - 32a^3 = 0$,

which is satisfied by $y = 2a$. Hence the particle will leave the given parabola at P if $y = 2a$. Beyond P, it will describe a parabolic path.

Putting $y = 2a$ in (2), the velocity v_1 at the time of leaving the given parabola is given by $v_1^2 = 2g(4a - 2a) = 2ga$. Let $\psi = \psi_1$, when $y = 2a$. Then (5) gives $\tan \psi_1 = 1$ so that $\psi_1 = \pi/4$.

Now, the latus rectum of the free parabolic path

$= (2/g) \times (\text{Horizontal component of velocity})^2 = (2/g) \times (v_1 \cos \psi_1)^2$

$= (2/g) \times v_1^2 \cos^2 \psi_1 = (2/g)\cdot 2ga \cdot \cos^2 (\pi/4) = 4a$,

showing that the latus rectum of the free parabola path is same as that of the given parabolic curve.

Ex. 14. *A smooth parabolic tube is placed, vertex downwards, in a vertical plane; a particle slides down the tube from rest under the influence of gravity. Prove that in any position the reaction of the tube is $2w(h + a)/\rho$, where w is the weight of the particle, ρ the radius of curvature, $4a$ the latus rectum, and h the original vertical height of the particle above the vertex.* **[I.A.S. 2008]**

Sol. Let the equation of the parabolic tube be

$$x^2 = 4ay \quad ...(1)$$

Let $BM \perp OX$ and $PN \perp BM$. Let B be a point on (1), where $BM = h$. Then initially a particle starts from B from rest. Let $P(x, y)$ be its position at any time t and let v be its velocity at P along tangent PT which is inclined at an angle ψ with OX. For motion from B to P, the equation of energy reduces to

$(1/2)\times mv^2 - (1/2)\times m \times o^2 = mg\cdot BN$ or $v^2 = 2g(h - y)$ ...(2)

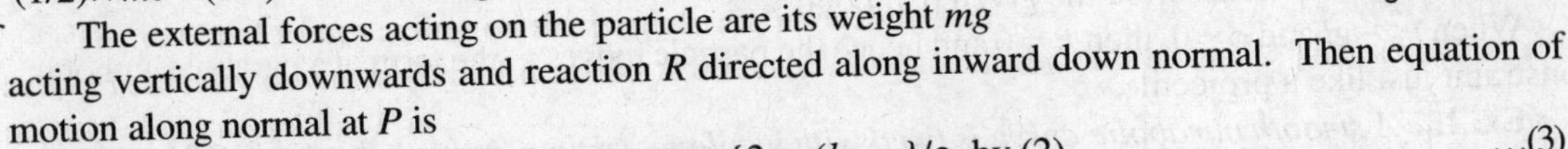

The external forces acting on the particle are its weight mg acting vertically downwards and reaction R directed along inward down normal. Then equation of motion along normal at P is

$mv^2/\rho = R - mg \cos \psi$ or $R = mg \cos \psi + \{2mg(h - y\}/\rho$, by (2) ...(3)

From (1), $dy/dx = x/2a = 2(ay)^{1/2}/2a = (y/a)^{1/2}$ and $d^2y/dx^2 = 1/2a$ ...(4)

$$\therefore\ \rho = \frac{\{1+(dy/dx)^2\}^{3/2}}{d^2y/dx^2} = \frac{(1+y/a)^{3/2}}{1/2a} = 2a\left(1+\frac{y}{a}\right)^{3/2} \quad ...(5)$$

Also, from (4), $\tan \psi = dy/dx = y^{1/2}/a^{1/2}$ and so $\cos \psi = a^{1/2}/(a + y)^{1/2}$ ...(6)

Using (6), (3) gives $R = mg\{a/(a + y)\}^{1/2} + \{2mg(h - y)\}/\rho$

$$\text{or } R = \frac{2mg}{\rho}\left[h - y + \frac{\rho}{2}\left(\frac{a}{a+y}\right)^{1/2}\right] = \frac{2mg}{\rho}\left[h - y + \frac{2a(a+y)^{3/2}}{2a^{3/2}} \times \left(\frac{a}{a+y}\right)^{1/2}\right], \text{ using (5)}$$

$= (2mg/\rho)\,(h - y + a + y) = (2w/\rho)\,(h + a)$, where $w = mg$

Ex. 15. *A particle describes a smooth curve under the action of gravity, describing equal vertical distances in equal times, and starting in a vertical direction. Show that the curve is a semi-cubical parabola the tangent at the cusp of which is vertical.*

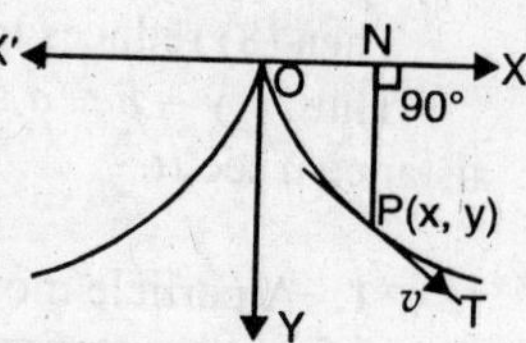

Sol. Let the particle be initially at O. Take O as origin and horizontal line OX and vertical downwards line OY as coordinate axes. Let $P(x, y)$ be the position of the particle at any time t and let v be its velocity at P along tangent PT. Let dx/dt and dy/dt be horizontal and vertical components of v. Then,

$$v^2 = (dx/dt)^2 + (dy/dt)^2 \quad \text{...(1)}$$

Since the particle describes equal vertical distances in equal times, hence its vertical component of velocity dy/dt remains constant throughout the motion,

Hence, $\quad dy/dt = \text{const.} = k$, say $\quad$ ...(2)

Let $PN \perp OX$. Then $PN = y$. For motion from O to P, the equation of energy reduces to

$(1/2)mv^2 - (1/2)mk^2 = mg \times PN \quad$ or $\quad v^2 - k^2 = 2gy$

or $\quad (dx/dt)^2 + (dy/dt^2) - (dy/dt)^2 = 2gy \quad$ or $\quad dx/dt = (2gy)^{1/2}$, using (1) and (2)

or $\quad \dfrac{dx}{dy} \cdot \dfrac{dy}{dt} = (2gy)^{1/2} \quad$ or $\quad k\dfrac{dx}{dy} = (2gy)^{1/2}$, using (2)

or $\quad dx = \{(1/k) \times (2g)^{1/2}\}\, y^{1/2}\, dy$

Integrating, $\quad x = (1/k) \times (2g)^{1/2} \times y^{3/2} \times (2/3) + C$

Initially, let when $x = 0, y = 0$. So $C = 0$

$\therefore\ x = (2/3k) \times (2g)^{1/2} \times y^{3/2}$ or $ax^2 = y^3$, where $a\ (= 9k^2/8g)$ is a constant. This equation represents a semi-cubical parabola whose cusp is at the origin and tangent there is y-axis, which is a vertical line.

Ex. 16. *A particle slides down the smooth curve $y = a \sinh (x/a)$. The axis of x being horizontal and the axis of y downwards, starting from rest at the point where the tangent is inclined at α to the horizon, show that it will leave the curve when it has fallen through a vertical distance $a \sec \alpha$.*

Sol. Let OAP be the given curve $\quad y = a \sinh (x/a) \quad$ ...(1)

Let A be a point on (1) such that $AM = h$, where $AM \perp OX$. Let the tangent AT' at A be inclined at an angle α to OX. Let $P(x, y)$ be the position of the particle at any time t and let v be the velocity of the particle at P along tangent PT which is inclined at an angle ψ with OX. Initially the particle is at rest at A. Let $PN \perp OX$ and $AL \perp PN$. Then $LP = NP - NL = NP - MA = y - h$ For motion from A to P, the equation of energy reduces to

$(1/2)mv^2 - (1/2)m \cdot 0^2 = mg \times LP \quad$ or $\quad v^2 = 2g(y - h) \quad$ ...(2)

The external forces acting on the particle at P are its weight mg acting vertically downwards and reaction R acting along the outward normal at P. Then equation of motion along normal is

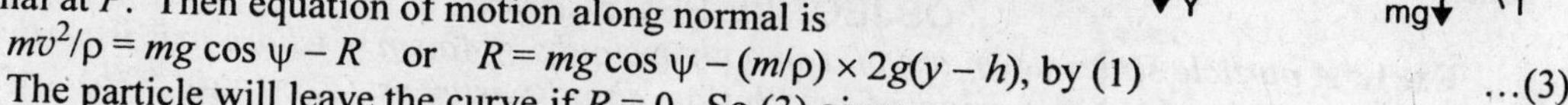

$mv^2/\rho = mg \cos \psi - R \quad$ or $\quad R = mg \cos \psi - (m/\rho) \times 2g(y - h)$, by (1) $\quad$ ...(3)

The particle will leave the curve if $R = 0$. So (3) gives

$0 = mg \cos \psi - (2mg/\rho) \times (h - y) = 0 \quad$ or $\quad \rho \cos \psi = 2(h - y) \quad$ ...(4)

From (1), $\quad dy/dx = \cosh (x/a) \quad$ and $\quad d^2y/dx^2 = (1/a) \sinh (x/a) \quad$ ...(5)

$$\therefore \quad \rho = \frac{\{1 + (dy/dx)^2\}^{3/2}}{d^2y/dx^2} = \frac{(1 + \cosh^2 x/a)^{3/2}}{(1/a)\sinh (x/a)} \quad \text{...(6)}$$

Also, by (5), $\tan \psi = dy/dx = \cosh x/a$ so $\cos \psi = 1/(1 + \cosh^2 x/a)^{1/2} \quad$ ...(7)

$$\text{From (1), (6) and (7), } \rho \cos \psi = \frac{1 + \cosh^2 x/a}{(1/a)\sinh (x/a)} = \frac{2 + \sinh^2 x/a}{(1/a)\sinh x/a} = \frac{2 + y^2/a^2}{y/a^2}$$

Hence (4) reduces to $\quad (2a^2 + y^2)/y = 2(h - y) \quad$ or $\quad y^2 - 2hy = 2a^2$

or $\quad y^2 - 2hy + h^2 = h^2 + 2a^2 \quad$ or $\quad (y - h)^2 = h^2 + 2a^2 \quad$ ...(8)

Now, $\tan \psi = \cosh x/a = (1 + \sinh^2 x/a)^{1/2} = (1 + y^2/a^2)^{1/2} = (y^2 + a^2)^{1/2}/a$

At A, $\psi = \alpha$ and $y = h$. Hence the above relation becomes

$\tan \alpha = (h^2 + a^2)^{1/2}/a \quad$ or $\quad h^2 = a^2 \tan^2 \alpha - a^2.$

Then (8) reduces to $(y-h)^2 = a^2 \tan^2 \alpha - a^2 + 2a^2 = a^2 \sec^2 \alpha$

Thus, $y - h = a \sec \alpha$, showing that the particle will leave the curve after falling the vertical distance $a \sec \alpha$.

EXERCISE

1. A particle moves in a smooth tube in the form of a catenary being attracted to the directrix with a force proportional to the distance from the directrix. Prove that the period of oscillation is independent of the amplitude.

2. A particle slides on curve $x = 2\{a(y-a)\}^{1/2}$ with a velocity due to a fall from the x-axis which is horizontal, the y-axis being downwards. Prove that pressure on the curve at any time is zero so that the curve is a free path under gravity.

3. A parabola of latus rectum $4a$, is placed with axis horizontal and plane vertical and a heavy smooth particle is projected from the vertex with a velocity due to a height h and moves on the concave side of the arc. Prove that the vertical height attained before leaving the arc is two-thirds of the greatest height and if θ be the angle described about the focus before leaving the curve, then $h = a(\tan^2 \theta/2 + 3 \tan \theta/2)$

4. From the lowest point of a smooth hollow cylinder whose cross-section is one-half of the lemnicate $r^2 = a^2 \cos 2\theta$, with axis vertical and node downwards, a particle is projected with velocity V along the inner surface in the plane of the cross-section; show that it will make a complete revolution if $3V^2 > 7ag$.

5. A particle is projected from the vertex of a smooth parabolic tube of latus rectum $4a$ along the tube and is acted upon by a repulsive force mgr/c from the focus; if the velocity of projection is that which would be acquired in moving from focus to the vertex, prove that the time of describing an angle θ about the focus is $2(c/g)^{1/2} \log \tan \{(\pi + \theta)/4\}$.

6. A small bead moves on a thin elliptic wire under a force to focus equal to $\mu/r^2 + \lambda/r^3$. It is projected from a point on the wire distant R from the focus with a velocity which would cause it to describe the ellipse freely under force μ/r^2; show that the reaction of the wire is $(\lambda/\rho) \times (1/r^2 - 1/ar + 1/R^2)$, where ρ is the radius of curvature.

7. A heavy ring of mass m is free to move on a smooth fixed parabolic wire of latus rectum $4a$ whose axis is vertical and vertex upwards and is attached to one end of an elastic string of natural length a and modulus of elasticity $2mg$, whose other end is fixed at the focus. The ring is projected from the vertex of the parabola with velocity $2(ga)^{1/2}$. Prove that when focal distance is r, its velocity is $\{(2gr/a) \times (3a - r)\}^{1/2}$ and the pressure on the wire is $\{mg(3r - 4a)\}/(ar)^{1/2}$

8. An elastic string of modulus λ is attached at an end to a focus of a smooth wire in the shape of an ellipse of latus rectum $2l$ and major axis $2a$. The other end of the string is attached to a small ring of unit mass which slides on the wire which is fixed with its plane horizontal. If the ring be slightly displaced from its position of unstable equilibrium at the end of the major axis of the ellipse, show that its angular velocity about the focus when the string becomes slack is $\{\lambda l(a-l)/a^3\}^{1/2}$.

OBJECTIVE PROBLEMS

Ex. 1. *A particle slides down arc of a smooth cycloid $s = 4a \sin \psi$ whose axis is vertical and vertex downwards, then its period of oscillation about vertex is (a) $\pi(a/g)^{1/2}$ (b) $2\pi(a/g)^{1/2}$ (c) $4\pi(a/g)^{1/2}$ (d) None of these.* **Ans. (c)**

Ex. 2. *A particle of mass m slides down a smooth cycloid whose axis is vertical and vertex downwards. Then the reaction at the lowest point is (a) mg (b) 2mg (c) 4mg (d) None of these.* **Ans. (*b*)**

Ex. 3. Fill in the blanks ".........." so that the following statements are complete and correct.

(*i*) *In the case of the motion of a heavy particle on the arc of a smooth cycloid the time of descent to the vertex is .... of the initial displacement of the particle.*

(*ii*) *When a particle moving on a curve leaves it at any point, the normal reaction between the particle and curve at that point is ....*

Ans. (*i*) independent (*ii*) zero.

Ex. 4. Write 'T' for true and 'F' for false statement

(*i*) *The cycloidal pendulum is perfectly isochronous.*

(*ii*) *A heavy particle oscillates from cusp to cusp of a smooth cycloid whose axis is vertical and vertex lowest. The velocity v at any point P is equal to the resolved part of the velocity V at the vertex along the tangent at P.*

Ans. (*i*) T (*ii*) T

MISCELLANEOUS PROBLEMS

1. *A particle starts from rest from any point P in the arc of a smooth cycloid* $s = 4a \sin \psi$, *whose axis is vertical and vertex A downward. Prove that the time of descending the vertex is* $\pi(a/g)^{1/2}$.

[Meerut 2006]

Sol. Refer figure of Art 11.4. Proceed as in Art. 11.4 upto equation (10). Let t' be the time taken by the particle in moving from its initial position P_0 to the vertex O. Then, putting $s = 0$ and $t = t'$ in (10) we obtain

$$0 = s_0 \cos\{t'(g/4a)^{1/2}\} \quad \Rightarrow \quad t'(g/4a)^{1/2} = \pi/2 \quad \Rightarrow \quad t' = \pi(a/g)^{1/2}.$$

2. *Two particles connected by a fine string are constrained to move in a cycloidal tube in a vertical plane, the axis of the cycloid being vertical vertex upwards, prove that that the tension of the string is constant throughout the motion.*

[I.A.S. 2005]

Sol. Let the equation of the cycloid be

$$s = 4a \sin \psi, \qquad \text{... (1)}$$

where s is measured from the vertex. Here the axis OY of the cycloid is vertical and vertex O is upwards. Let two particles P_1 (of mass m_1) and P_2 (of mass m_2) be connected by a fine string P_1P_2. At any time t, let arc $OP_1 = s$ and arc $OP_2 = s + l$, where l is the length of the string. Let tangents at P_1 and P_2 be inclined at angles ψ_1 and ψ_2 with x-axis. Then, using (1), we have

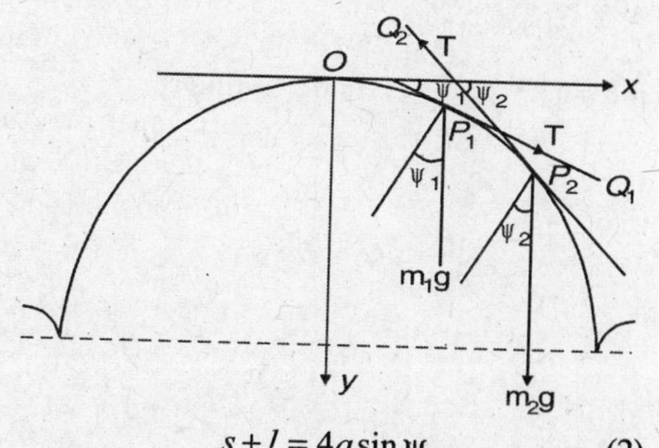

$$s = 4a \sin \psi_1 \qquad \text{and} \qquad s + l = 4a \sin \psi_2 \qquad \text{... (2)}$$

The external forces acting on the particle P_1 are its weight $m_1 g$ acting vertically downwards, tension T along tangent P_1Q_1 to arc P_1P_2 and reaction acting along outwards drawn normal at P_1. Then, equation of motion of mass m_1 at P_1 along tangent is given by

$$m_1(d^2s/dt^2) = T + m_1 g \sin \psi_1 \qquad \text{or} \qquad d^2s/dt^2 = T/m_1 + g \times (s/4a), \text{ by (2)} \quad \text{... (3)}$$

Proceeding as before and replacing s by $s + l$ and m_1 by m_2 the equation of motion for m_2 along tangent is

$$m_2 \frac{d^2(s+l)}{dt^2} = m_2 g \sin \psi_2 - T \qquad \text{or} \qquad m_2 \frac{d^2 s}{dt^2} = m_2 g \times \frac{s+l}{4a} - T, \text{ by (2)}$$

or

$$d^2s/dt^2 = -(T/m_2) + (g/4a) \times (s + l) \qquad \text{... (4)}$$

Subtracting (4) from (3), we get $\quad 0 = T(1/m_1 + 1/m_2) - (gl/4a)$

$$\Rightarrow \qquad T = (gl/4a) \times \{(m_1 m_2)/(m_1 + m_2)\} = \text{constant}$$

3. *If a particle falls down a smooth cycloid under its own weight starting from cusp, the acceleration of particle at every point is*

(*a*) $g \sin \psi$ (*b*) $g \cos \psi$ (*c*) g (*d*) $g/3$ **[Agra 2011, 14]**

Sol. Ans. (*c*). Refer solved Ex. 5 (*d*), page 11.4.

4. Find the periodic time in cycloid as motion. **[Agra 2012]**

Hint. Refer Art 11.4, page 11.2.

Ans. Periodic time = $T = 4\pi\,(a/g)^{1/2}$

12

Motion in a resisting medium
Motion on rough curves
Motion when the mass moving varies

12.1 Type 1 : Solved examples based on motion in resisting medium in a horizontal line.

Ex. 1. *A particle, moving in a straight line, is subjected to a resistance kv^3, where v is the velocity. Show that if v is the velocity at time t when the distance is s, $v = u/(1 + kus)$ and $t = (s/u) + (1/2)ks^2$, where u is the initial velocity.*

[Kumaun 2001; Kanpur 2007]

Sol. Let the point of projection O be taken as origin. Let v be the velocity at any time t at a distance s from O. Here resistance (per unit mass) is kv^3 and equation of motion of the particle is

$$1.v(dv/ds) = -1 \cdot kv^3 \quad \text{or} \quad (1/v^2)dv = -k\, ds$$

Integrating, $-(1/v) = -ks + C$, C being arbitrary constant. ...(1)

Initially at O, when $s = 0$, $v = u$ so (1) gives $\quad C = -(1/u)$

$\therefore$ (1) gives $\quad -(1/v) = -ks - (1/u) \quad$ or $\quad v = u/(1 + kus)$ ...(2)

Re-writing (2), $\quad ds/dt = u/(1 + kus) \quad$ or $\quad dt = (ks + 1/u)\, ds$

Integrating, $\quad t = (1/2)ks^2 + s/u + A$, where A is a constant. ...(3)

Initially at O, when $s = 0$, $t = 0$. So (3) gives $\quad A = 0$

Then (3) reduces to $\quad t = (s/u) + (1/2)ks^2$

Ex. 2. *A particle is projected with velocity V along a smooth horizontal plane in a medium whose resistance per unit mass is μ times the cube of the velocity. Show that the distance it has described in time t is $(1/\mu V)\,\{(1 + 2\mu V^2 t)^{1/2} - 1\}$ and that its velocity then is $V/(1 + 2\mu V^2 t)^{1/2}$.*

[Purvanchal 2008, 10; Nagpur 2010]

Sol. Proceed as in Ex. 1. Here the equation of motion is

$$m(dv/dt) = -m\mu v^2 \quad \text{or} \quad (1/v^3)dv = -\mu t,\ m \text{ being mass of the particle.}$$

Integrating, $\quad -(1/2v^2) = -\mu t + C$, where C is a constant. ...(1)

Initially, when $t = 0$, $v = V$ so (1) gives $\quad A = -(1/2V^2)$

So (1) becomes $\ -(1/2v^2) = -\mu t - (1/2V^2) \quad$ or $\quad v = V/(1 + 2\mu V^2 t)$, ...(2)

giving the velocity of the particle at any time t.

Re-writing (2), $\quad ds/dt = V/(1 + 2\mu V^2 t)^{1/2} \quad$ or $\quad ds = V(1 + 2\mu V^2 t)^{-1/2}\, dt$

Integrating, $\quad s = (1/\mu V)\,(1 + 2\mu V^2 t)^{1/2} + B$, where B is a constant.

Initially, when $t = 0$, $s = 0$. So we have $\quad B = -(1/\mu V)$

Thus, $\quad s = (1/\mu V)\,(1 + 2\mu V^2 t)^{1/2} - (1/\mu V) = (1/\mu V)\,\{(1 + 2\mu V^2 t)^{1/2} - 1\}$,

giving the distance described in time t.

Ex. 3. *A particle is projected with velocity u along a smooth horioznta1 plane in a medium whose resistance per unit mass is $k \times$ (velocity), show that the velocity after a time t and the distnace s in that time are given by $v = u\, e^{-kt}$ and $s = (u/k)\,(1 - e^{-kt})$.*

Sol. Proceed exactly as in Ex. 2.

12.2 Motion in resisting medium in a vertical line

Terminal or limiting velocity. Definition.

[Rohilkhand 2009; Kanpur 2017, 18; Meerut 2017, 18]

When a particle falls under gravity in a medium, whose resistance varies as the nth power of the velocity, the equation of motion is

$$dv/dt = g - kv^n, \qquad \text{...(1)}$$

where v is the velocity of the particle at time t. Let $dv/dt = 0$ when $v = V$. Then, from (1), $0 = g - kV^n$ or $V = (g/k)^{1/n}$. This will be the maximum possible velocity and it is known as *limiting or terminal velocity.*

If a particle were initially projected downwards with this velocity V, the initial acceleration is zero. The velocity does not change and the force of gravity is always balanced by the resistance. Hence particle descends with uniform velocity V. If the particle be projected with a velocity less than V, gravity overcomes resistance and the velocity in the course of motion increases upto V. On the other hand, if the velocity of projection exceeds V, resistance overcomes gravity and the velocity is decreased. In the case, when a particle is projected upwards, both gravity and resistance combine to reduce the initial velocity and bring it to rest; afterwards the particle descends in the manner described above.

12.3 Motion in a medium whose resistance varies as the velocity

Case I. Descent. *A particle is falling under gravity in a medium whose resistance varies as the velocity; to find the motion.* **[Avadh 2006; Bundelkhand 2009; Kanpur 2011, 2016; Agra 2006]**

Let a particle of mass m fall from rest under gravity from the fixed point O. Let P be the position of the particle at any time t such that $OP = x$. Let v be its velocity at P. Here kv is the force of resistance per unit mass so that $F (= mkv)$ is resistance of the medium on the particle acting in the vertical upward direction. The weight mg of the particle acts in vertically downward direction. Then the equation of motion of the particle at any time t is

$$m\ddot{x} = mg - mkv \quad \text{or} \quad \ddot{x} = g(1 - kv/g) \qquad \text{...(1)}$$

Let V be the terminal velocity so that when $v = V$, $\ddot{x} = 0$.

Then from (1), $\quad 0 = g(1 - kV/g) \quad$ or $\quad k = g/V \qquad$ ...(2)

Using (2), (1) becomes $\quad \ddot{x} = g(1 - v/V) \quad$ or $\quad \ddot{x} = (g/V)(V - v) \qquad$ ...(3)

To find relation between v and x. Since $\ddot{x} = v(dv/dx)$, (3) reduces to

$$v(dv/dx) = (g/V)(V - v) \quad \text{or} \quad \{v/(V - v)\}\, dv = (g/V)\, dx$$

or

$$\left(\frac{V}{V - v} - 1\right) dv = \frac{g}{V} dx$$

Integrating, $\quad -V \log(V - v) - v = gx/V + A$, where A is a constant. $\qquad$ ...(4)

Initially at O, when $x = 0, v = 0$. So (4) gives $\quad A = -V \log V$

Then (4) becomes $\quad -V \log (V - v) - v = gx/V - V \log V$

or

$$\frac{gx}{V} = V \log \frac{V}{V - v} - v \quad \text{or} \quad x = \frac{V^2}{g} \log \frac{V}{V - v} - \frac{Vv}{g} \qquad \text{...(5)}$$

To find relation between v and t. Since $\ddot{x} = dv/dt$, (3) gives

$$\frac{dv}{dt} = \frac{v}{V}(V - v) \quad \text{or} \quad V \frac{dv}{V - v} = gt$$

Integrating, $\quad -V \log (V - v) = gt + B$, where B is a constant. $\qquad$ ...(6)

Initially at O, when $t = 0, v = 0$. So (6) gives $\quad B = -V \log V$

Then (6) becomes $\quad -V \log (V - v) = gt - V \log V$

or $\quad V[\log (V - v) - \log V] = -gt \quad$ or $\quad V - v = Ve^{-gt/V}$ or $v = V(1 - e^{-gt/V}) \qquad$ ...(7)

To find relation between x and t. Since $v = dx/dt$, (7) gives

$$dx/dt = V - Ve^{-gt/V} \quad \text{or} \quad dx = (V - Ve^{-gt/V})\, dt$$

Integrating, $\quad x = Vt + (V^2/g)\, e^{-gt/V} + C$, where C is constant. $\qquad$ ...(8)

Initially at O, when $t = 0, x = 0$. So (8) gives $\quad C = -(V^2/g)$

Then (8) gives $\quad x = Vt - (V^2/g)(1 - e^{-gt/V}) \qquad$ ...(9)

Case II. Ascent. *A particle is projected vertically upwards with a velocity U in a medium whose resistance varies as the velocity; to find the motion.* **[Avadh 2008; Purvanchal 2007; Garhwal 2010]**

Let a particle of mass m be projected vertically upwards with velocity U from the fixed point O'.

Let Q be the position of the particle at any time t such that $O'Q = y$. Let v be its velocity at Q. Here kv is the force of resistance per unit mass so that $F(= mkv)$ is resistance of the medium on the particle acting in vertically downward direction. The weight mg of particle acts in vertically downward direction. Then the equation of motion of the particle at any time t is

$$m\ddot{y} = -mg - mkv \quad \text{or} \quad \ddot{y} = -g(1 + kv/g) \qquad \ldots(10)$$

O, R, Q (t), y, mg, F, O'

Let V be the terminal velocity of the particle during its downward motion *i.e.*, the velocity when the acceleration of the particle during its downward motion is zero. Then $0 = mg - mkV$ or $k = g/V$. So (10) becomes

$$\ddot{y} = -g(1 + v/V) \quad \text{or} \quad \ddot{y} = -(g/V)(V + v) \qquad \ldots(11)$$

To find relation between v and y. As $\ddot{y} = v(dv/dy)$, (11) gives

$$v\frac{dv}{dy} = -\frac{g(V+v)}{V} \quad \text{or} \quad -\frac{g\,dy}{V} = \frac{v\,dv}{V+v} = \left(1 - \frac{V}{V+v}\right)dv$$

Integrating, $v - V\log(V + v) = -gy/V + D$, where D is a constant. ...(12)

Initially at O', when $y = 0$, $v = U$. So (12) gives $D = U - V\log(V + U)$

Then (12) becomes $v - V\log(V + v) = -gy/V + U - V\log(V + U)$

$$\text{or} \quad \frac{gy}{V} = V\log\frac{V+v}{V+U} + U - v \quad \text{or} \quad y = \frac{V^2}{g}\log\frac{V+v}{V+U} + \frac{V}{g}(U - v) \qquad \ldots(13)$$

Let O be the highest point reached by the particle and let $O'O = h$. Then, when $y = h$, $v = 0$. So (13) gives

$$h = (V^2/g)\log\{V/(V+U)\} + (VU/g) \qquad \ldots(14)$$

To find relation between v and t. **[Kanpur 2012, 13]**

$$\text{Since } \ddot{y} = \frac{dv}{dt}, \text{ (11) gives} \quad \frac{dv}{dt} = -\frac{g(V+v)}{V} \quad \text{or} \quad \frac{dv}{V+v} = -\frac{g}{V}dt$$

Integrating, $\log(V + v) = -(g/V)t + E$, where E is a constant. ...(15)

Initially at O', when $t = 0$, $v = U$. So (15) gives $E = \log(V + U)$

Then (15) gives $\log(V + v) = -(g/V)t + \log(V + U)$

$$\text{or} \quad \log\frac{V+v}{V+U} = -\frac{gt}{V} \quad \text{or} \quad V + v = (V + U)e^{-gt/V}$$

$$\text{or} \quad v = (V + U)e^{-gt/V} - V \qquad \ldots(16)$$

To find relation between y and t. As $v = dy/dt$, (16) gives

$$dy/dt = (V + U)e^{-gt/V} - V \quad \text{or} \quad dy = \{(V + U)e^{-gt/V} - V\}\,dt$$

Integrating, $y = -(V/g)(V + U)e^{-gt/V} - Vt + F$, F is a const. ...(17)

Initially at O', when $t = 0$, $y = 0$. So (17) gives $F = (V/g)(V + U)$

Then (17) gives $y = -(V/g)(V + U)e^{-gt/V} - Vt + (V/g)(V + U)$ ...(18)

TYPE 2 : SOLVED EXAMPLES BASED ON ART. 12.3

Ex. 1. *A particle of mass m is projected vertically downwards under gravity, the resistance of the air being mk times the velocity. Show that the greatest height attained by the particle is $(V^2/g)\{\lambda - \log(1 + \lambda)\}$, where V is the terminal velocity of the particle and λV is the initialy velocity.* **[Lucknow 2006, 08, 09; Avadh 2006, 07; Meerut 2013, 2017; Nagpur 2010; Kanpur 2006; Purvanchal 2002]**

Sol. Refer Case II of Art. 12.3. Proceed upto equation (14) giving maximum height attained. In the given problem $U = \lambda V$. Hence (14) reduces to

$$h = (V^2/g)\log\{V/(V + \lambda V)\} + (V \cdot \lambda V)/g = V^2/g\,\{\log(1 + \lambda)^{-1} + \lambda\}$$

$$\text{or} \quad h = (V^2/g)\{\lambda - \log(1 + \lambda)\}, \text{ as desired.}$$

Ex. 2. *A particle of mass m, is falling under the influence of gravity through a medium whose resistance equals μ times the velocity. If the particle were released from rest, show that the distance fallen through in time t is $(gm^2/\mu^2)\{e^{-(\mu/m)t} - 1 + \mu t/m\}$* **[Bundelkhand 2008] [Purvanchel 2003, 06, 08]**

Sol. Refer figure of case I of Art. 12.3. Let the particle start from rest from a fixed point O. Let P be the position of the particle at any time t such that $OP = x$. Let v be its velocity at P. Here the force of resistnace is μv (given) which acts in vertical upward direction. The weight mg of the particle acts in vertically downward direction. Then the equation of motion of the particle at any time t is

$$m\ddot{x} = mg - \mu v \quad \text{or} \quad dv/dt = g - (\mu/m)v$$

or $$dt = \{1/(g - \mu v/m)\}\, dv$$

Integrating, $t = -(m/\mu) \log (g - \mu v/m) + A$, where A is a constant

Initially at O, when $t = 0$, $v = 0$ hence $A = (m/\mu) \log g$

$\therefore$ $t = -(m/\mu) \log (g - \mu v/m) + (m/\mu) \log g = -(m/\mu) \log (1 - \mu v/gm)$

or $\log (1 - \mu v/gm) = -\mu t/m$ or $1 - (\mu/gm)v = e^{-\mu t/m}$

or $v = dx/dt = (gm/\mu)(1 - e^{-\mu t/m})$ or $dx = (gm/\mu)(1 - e^{-\mu t/m})\, dt$

Integrating, $x = (gm/\mu)\{t + (m/\mu) e^{-\mu t/m}\} + B$, where B is a constant ...(1)

Initially at O, when $t = 0$, $x = 0$ so (1) gives $B = -gm^2/\mu^2$

Then (1) becomes $x = (gm/\mu)\{t + (m/\mu)e^{-\mu t/m}\} - gm^2/\mu^2$

or $$x = (gm^2/\mu^2)\{e^{-(\mu/m)t} - 1 + \mu t/m\}$$

Ex. 3. (*a*) *A particle of unit mass is projected upwards with velocity V in a medium for which the resistance is kv when the speed of the particle is v. Prove that the particle returns to the point of projection with speed V_1 such that* $V + V_1 = (g/k) \log \{(g + kV)/(g - kV_1)\}$ **[Meerut 1994]**

(*b*) ***A small stone of mass m is thrown vertically upwards with initial speed V. If the air resistance at speed v is mkv, where k is a positive constant, show that the stone returns to its starting point with velocity U given by the equation*** $g - kU = (g + kV)e^{-k(U+V)/g}$. **[Delhi Maths (H) 2004, 2007, 10]**

Sol. (*a*) Refer figure of case II of Art. 12.3. Let a particle of unit mass be projected with velocity V from the fixed point O'. Let Q be its position at any time t such that $O'Q = y$. Let v be its velocity at Q. Here kv is total resistance on the particle acting in the downward direction. The weight $1 \cdot g$ of the particle acts in vertically downward direction. So equation of motion of the particle is

$$1 \cdot \ddot{y} = -1 \cdot g - kv \quad \text{or} \quad v(dv/dy) = -(g + kv)$$

or $$k\, dy = -\frac{kv\, dv}{kv + g} \quad \text{or} \quad k\, dy = -\left(1 - \frac{g}{kv + g}\right) dv$$

Integrating, $ky = -v + (g/k) \log (kv + g) + A$, A is a constant.

Initially at O', when $y = 0$, $v = V$. So $A = V - (g/k) \log (kV + g)$

$\therefore$ $ky = -v + (g/k) \log (kv + g) + V - (g/k) \log (kV + g)$

or $$y = (V - v)/k + (g/k^2) \log \{(kv + g)/(kV + g)\} \qquad \text{...(1)}$$

Let O be the highest point reached by the particle such that $OO' = h$. Then, at O, when $y = h$, $v = 0$. So (1) gives

$$h = (V/k) + (g/k^2) \log \{g/(kV + g)\} \qquad \text{...(2)}$$

Now, refer figure of Case I of Art. 12.3. After coming to instantaneous rest at O, the particle falls vertically downwards. Let P be its position at any time t, the time being measured from the instant when the particle was at rest at O. Let $OP = x$ and let v be its velocity at P. Then the force of resistance will act in vertically upward direction while weight $1 \cdot g$ will act in vertically downward direction. So the equation motion during downward motion is

$$1 \cdot \ddot{x} = 1 \cdot g - kv \quad \text{or} \quad v(dv/dx) = g - kv \quad \text{or} \quad k\, dx = \{kv/(g - kv)\}\, dv$$

or $$k\, dx = \frac{g - (g - kv)}{g - kv} dv \quad \text{or} \quad k\, dx = \left(\frac{g}{g - kv} - 1\right) dv$$

Integrating, $kx = -(g/k) \log (g - kv) - v + B$, where B is a constant.

Initially at O, when $x = 0$, $v = 0$. So $B = (g/k) \log g$

$\therefore$ $kx = -(g/k) \log (g - kv) - v + (g/k) \log g$

or $$x = -(v/k) + (g/k^2) \log \{g/(g - kv)\} \qquad \text{...(3)}$$

By given problem the particle returns to O with velocity V_1 so that when $x = h$, $v = V_1$. Then (3) gives

$$h = -(V_1/k) + (g/k^2) \log \{g/(g - kV_1)\} \qquad \text{...(4)}$$

From (2) and (4), $\frac{V}{k}+\frac{g}{k^2}\log\frac{g}{kV+g}=-\frac{V_1}{k}+\frac{g}{k^2}\log\frac{g}{g-kV_1}$

or $V+V_1=\frac{g}{k}\left(\log\frac{g}{g-kV_1}-\log\frac{g}{g+kV}\right)=\frac{g}{k}\log\frac{g+kV}{g-kV_1}$

(*b*) Proceed as in part (*a*). Here equations of motion in upward and downward directions will be $m\ddot{y}=-mg-mkv$ and $m\ddot{x}=mg-mkv$. Here $V_1=U$ so we get

$$V+U=(g/k)\log\{(g+kV)/(g-kU)\}$$

or $\log\frac{g+kV}{g-kU}=\frac{k(V+U)}{g}$ or $\log\frac{g-kU}{g+kV}=-\frac{k(V+U)}{g}$

or $(g-kU)/(g+kV)=e^{-k(V+U)/g}$ or $g-kU=(g+kV)e^{-k(V+U)/g}$

EXERCISE

1. A heavy particle is projected in a resisting medium, the resistance varying as the velocity. If v_1 and v_2 are its velocities at any point in its upward and downward paths and t the interval between its passage through this point, prove that $v_1+v_2=gt$, $V-v_2=(V+v_1)\,e^{-gt/V}$, where V is the terminal velocity. **(Guwahati 2007)**

2. A particle is projected upwards in a medium whose resistance is gv/V where v is the velocity. If V be large compared to U, the velocity of projection, prove that the vertical height reached, the time of ascent and the time of descent are decreased respectively by the fractions $2U/3V$, $U/2V$, $U/6V$ of their values when there is no resistance and the particle returns to the point of projection with velocity $U(1-2U/3V)$.

12.4 Motion in a medium whose resistance varies as the square of the velocity

Case I. Descent. *A particle is falling under gravity in a medium whose resistance varies as the square of the velocity; to find the motion.* **[Kanpur 2018; Avadh 2011; Purvanchal 2008; Meerut 2013]**

For figure refer case I of Art. 12.3. Let a particle of mass m fall from rest under gravity from the fixed point O. Let P be the position of the particle at any time t such that $OP=x$. Let v be its velocity at P. Here kv^2 is the force of resistance per unit mass so that $F(=mkv^2)$ is resistance of the medium on the particle acting in the vertical upward direction. The weight mg of the particle acts in vertically downward direction. Then the equation of motion of the prticle is

$$m\ddot{x}=mg-mkv^2 \quad \text{or} \quad \ddot{x}=g(1-kv^2/g) \qquad \ldots(1)$$

Leg V be the terminal velocity so that when $v=V$, $\ddot{x}=0$.

Then, from (1), $0=g(1-kV^2/g)$ or $k=g/V^2$ ...(2)

Using (2), (1) becomes, $\ddot{x}=g(1-v^2/V^2)$ or $\ddot{x}=(g/V^2)(V^2-v^2)$ ...(3)

To find relation between v and x. Re-writing (3), we get

$$v\frac{dv}{dx}=\frac{g}{V^2}(V^2-v^2) \quad \text{or} \quad -\frac{2g}{V^2}dx=-\frac{2v\,dv}{V^2-v^2}$$

Integrating, $-(2g/V^2)x=\log(V^2-v^2)+A$, where A is a constant ...(3)′

Initially at O, when $x=0$, $v=0$ so $A=-\log V^2$

$\therefore$ $-(2g/V^2)x=\log(V^2-v^2)-\log V^2=\log\{(V^2-v^2)/V^2\}$, by (3)′ ...(4)′

or $(V^2-v^2)/V^2=e^{-2gx/V^2}$ or $v^2=V^2(1-e^{-2gx/V^2})$ ...(4)

Relation between v and t. Re-writing (3), we have

$$\frac{dv}{dt}=\frac{g}{V^2}(V^2-v^2) \quad \text{or} \quad \frac{g}{V^2}dt=\frac{dv}{V^2-v^2}$$

Integrating, $\frac{gt}{V^2}=\frac{1}{2V}\log\frac{V+v}{V-v}+B$, where B is a constant.

Initially at O, when t = 0, $v=0$. So $B=0$ and we get

$$\frac{gt}{V^2}=\frac{1}{2V}\log\frac{V+v}{V-v} \quad \text{or} \quad t=\frac{V}{g}\times\frac{1}{2}\log\frac{1+v/V}{1-v/V}=\frac{V}{g}\tanh^{-1}\frac{v}{V}$$

$[\because \tanh^{-1} z=(1/2)\times\log\{(1+z)/(1-z)\}]$ ...(5)

or $gt/V=\tanh^{-1}(v/V)$ or $v=V\tanh(gt/V)$

Relation between *x* and *t*. Re-writing (5), we have

$dx/dt=V\tanh(gt/V)$ or $dx=V\tanh(gt/V)\,dt$

Integrating, $x=(V^2/g)\log\cosh(gt/V)+C$, where C is a constant.

Initially at O, when $t=0$, $x=0$. So $C=0$ (as $\cosh 0=1$)

Thus. $x=(V^2/g)\log\cosh(gt/V)$ ...(6)

Case II. Ascent. *A particle is projected vertically upwards with a velocity U in a medium whose resistance varies as the square of the velocity : to find the motion.* **[Bundelkhand 2008 Garhwal 2002; Lucknow 2001; Purvanchel 2009, 11; Meerut 1998]**

For figure refer case II of Art. 12.3 Let a particle of mass m be projected vertically upwards with velocity U from the fixed point O'. Let Q be the position of the particle at any time t such that $O'Q=y$. Let v be its velocity at Q. Here kv^2 is the force of resistance per unit mass so that $F(=mkv^2)$ is resistance of the medium on the particle acting in vertically downward direction. The weight mg of the particle acts in vertical downward direction. Then the equation of motion of the particle is

$m\ddot{y}=-mg-mkv^2$ or $\ddot{y}=-g(1+kv^2/g)$ ...(7)

Let V be the terminal velocity of the particle during its downward motion, *i.e.*, the velocity when the acceleration of the particle during its downward motion is zero. Then $0=mg-mkV^2$ or $k=g/V^2$. So (7) becomes

$\ddot{y}=-g(1+v^2/V^2)$ or $\ddot{y}=-(g/V^2)(V^2+v^2)$ ...(8)

Relation between *v* and *y*. Re-writing (8), we have

$$v\frac{dv}{dy}=-\frac{g}{V^2}(V^2+v^2) \quad \text{or} \quad -\frac{2g}{V^2}dy=\frac{2v\,dv}{V^2+v^2}$$

Integrating, $-(2g/V^2)y=\log(V^2+v^2)+A$, where A is a constant. ...(9)

Initially at O', when $y=0$, $v=U$. So (9) gives $A=-\log(V^2+U^2)$

From (9), $-(2g/V^2)y=\log(V^2+v^2)-\log(V^2+U^2)$,

giving $y=(V^2/2g)\log\{(V^2+U^2)/(V^2+v^2)\}$ ...(10)

Suppose h is the maximum height attained at O so that $v=0$ when $y=h$. Then from (10), we have

$h=(V^2/2g)\log\{(V^2+U^2)/V^2\}$ ...(11)

Relation between *v* and *t*. Re-writing (3), we have

$$\frac{dv}{dt}=-\frac{g}{V^2}(V^2+v^2) \quad \text{or} \quad dt=-\frac{V^2}{g}\frac{dv}{V^2+v^2}$$

Integrating, $t=-(V/g)\tan^{-1}(v/V)+B$, where B is a constant. ...(12)

Initially at O', when $t=0$, $v=U$. So (12) gives $B=(V/g)\tan^{-1}(U/V)$

Then, by (12), $t=(V/g)\{\tan^{-1}(U/V)-\tan^{-1}(v/V)\}$ ...(13)

Let T be the time taken by the particle to reach its highest point O *i.e.*, let $t=T$ where $v=0$. Then (13) rduces to

$$T=(V/g)\tan^{-1}(U/V) \quad ...(14)$$

Corollary I. *A particle projected upwards with a velocity U with air-resistance which varies at the square of the velocity. Find the greatest height attained by the particle.* **[Kolkata 1993, 95]**

Proof. Refer Case II of Art. 12.4 and do upto relation (11).

Corollary II. *A particle is projected vertically upwards with velocity U, in a medium where resistance is kv^2 per unit mass for velocity v of the particle, show that the greatest height attained by the particle is $(1/2k)\times\{\log(g+kU^2)/g\}$.* **[Meerut 2003]**

Proof. Refer Case II of Art. 12.4 and do upto relation (11). But $k=g/V^2$ so that $V^2=g/k$. Thus, relation (11) becomes $h=(1/2k)\times\log\{(g+kU^2)/g\}$.

Corollary III. *If U and V be the velocity of projection and the terminal velocity respectively of a body rising against a resistance varying as the square of the velocity, prove that the time to reach the highest point is* $(V/g)\tan^{-1}(U/V)$. **[Lucknow 2000]**

Proof. Refer Case II of Art. 12.4. See relation between v and t upto (14).

Corollary IV. *If V be the terminal velocity of a particle of mass m falling against a resistance proportional to the square of the velocity, prove that the kinetic energy acquired in falling through a distance h from the position of rest is* $(mV^2/2)\times(1-e^{-2gh/V^2})$ **[Kanpur 2012]**

Proof : Refer Case I of Art. 12.4 upto equation (4). Putting $x = h$, velocity at depth h is given by

$$v^2 = V^2(1-e^{-2gh/V^2}) \qquad \text{...(1)}$$

$\therefore$ K.E. of particle $= (1/2)mv^2 = (mV^2/2)\times(1-e^{-2gh/V^2})$

TYPE 3 : SOLVED EXAMPLES

Ex. 1. *A particle falls from rest under gravity through a distance x in a medium whose resistance varies as the square of the velocity. If v be the velocity actually acquired by it, v_0 the velocity it would have acquired, had there been no resisting medium and V the terminal velocity, show that*

$$\frac{v^2}{v_0^2} = 1-\frac{1}{2}\frac{v_0^2}{V^2}+\frac{1}{2\cdot 3}\frac{v_0^4}{V^4}-\frac{1}{2\cdot 3\cdot 4}\frac{v_0^6}{V^6}+\ldots$$

Sol. Proceed as in Case I of Art. 12.4 upto equation (4).

Thus, $$v^2 = V^2(1-e^{-2gx/V^2}) \qquad \text{...}(i)$$

When there is no resisting medium, then the velocity acquired by the particle in falling freely through a distance x is given by $v_0^2 = 2gx$. Then (i) reduces to

$$v^2 = V^2(1-e^{-v_0^2/V^2}) = V^2\left\{1-\left(1-\frac{v_0^2}{V^2}+\frac{v_0^4}{V^4\,2!}-\frac{v_0^6}{V^6\,3!}+\ldots\right)\right\}$$

or $$v^2 = v_0^2-\frac{v_0^4}{2!V^2}+\frac{v_0^6}{3!V^4}-\frac{v_0^8}{4!V^6}+\ldots \quad \text{or} \quad \frac{v^2}{v_0^2} = 1-\frac{1}{2}\frac{v_0^2}{V^2}+\frac{1}{2\cdot 3}\frac{v_0^4}{V^4}-\frac{1}{2\cdot 3\cdot 4}\frac{v_0^6}{V^6}+\ldots$$

Ex. 2. (a) *A particle projected upwards with a velocity U, in a medium whose resistance varies as the square of the velocity, will return to the point of projection with velocity $v_1 = UV/(U^2+V^2)^{1/2}$ after a time $(V/g)\{\tan^{-1}(U/V)+\tanh^{-1}(v_1/V)\}$, where V is the terminal velocity.* **[Bundelkhand 2002, 11; Meerut 2014; Garhwal 2003; Meerut 2002]**

(b) *A particle is projected vertically upwards with a velocity U and the resistance kv^2 where v is the velocity. Show that the velocity v_1 with which the particle will return to the point of projection is given by $1/v_1^2 = 1/U^2+(k/g)$.* **[Meerut 2016, 18; Avadh 2009, 11, Bundelkhand 2011; Kanpur 2018]**

(c) *A body of mass m is projected vertically upward in a medium for which the resistance is mk^2v^2. If the initial velocity is v_0, show that the body returns to the point of projection with a velocity v_1 such that $v_1^2 = (gv_0^2)/(g+k^2v_0^2)$.*

Sol. (a) Proceed as in Case II of Art. 12.4 upto equation (14) for upward motion from O' to O. Now the particle comes to instantaneous rest at the highest point O and then falls down from rest under gravity in the given resisting medium. Now proceed as in Case I of Art. 12.4. upto relation (5). Let v_1 be the velocity of the particle when it reaches back the point of projection.

Thus at O', when $x = h$, $v = v_1$. So from (4)$'$, we get

$$-(2gh/V^2) = \log\{(V^2-v_1^2)/V^2\}$$

$\therefore$ By (11), $$-\frac{2g}{V^2}\times\frac{V^2}{2g}\log\frac{V^2+U^2}{V^2} = \log\frac{V^2-v_1^2}{V^2} \quad \text{or} \quad \log\frac{V^2}{V^2+U^2} = \log\frac{V^2-v_1^2}{V^2}$$

or $\quad V^2/(V^2+U^2) = (V^2-v_1^2)/V^2 \quad$ or $\quad V^4 = (V^2+U^2)(V^2-v_1^2)$

or $\quad v_1^2(V^2+U^2) = U^2V^2 \quad$ or $\quad v_1 = UV/(V^2+U^2)^{1/2} \qquad \text{...}(i)$

Let T' be time taken by the particle in moving from O to O'. Then at O', when $v = v_1$, $t = T'$. So (5) gives

$gT'/V = \tanh^{-1}(v_1/V)$ or $T' = (V/g)\tanh^{-1}(v_1/V)$ ...(ii)

Adding (14) and (*ii*), the required time from O' to O and back to O'

$= T + T' = (V/g)\{\tan^{-1}(U/V) + \tanh^{-1}(v_1/V)\}$ ...(iii)

(*b*) Proceed as in part (*a*) upto relation (*i*). Recall that $V^2 = g/k$. Putting this value of V^2 in (*i*), we get

$v_1^2(g/k + U^2) = U^2(g/k)$ or $v_1^2 g + v_1^2 U^2 k = U^2 g$ ...(iv)

Dividing by $U^2 v_1^2 g$, $1/U^2 + k/g = 1/v_1^2$,

(*c*) Proceed as in part (*b*) upto (*iv*). From (*iv*), we get

$v_1^2(g + U^2 k) = U^2 g$ or $v_1^2 = (U^2 g)/(g + U^2 k)$ ...(v)

For the present problem, we have v_0 in place of U and k^2 in plane of k. So (*v*) reduces to $v_1^2 = (v_0^2 g)/(g + v_0^2 k^2)$

Ex. 3. *A particle is projected upward with velocity U in a medium, resistance of which is $gU^{-2}\tan^2\alpha$ times the square of the velocity, α being a constant. Show that the particle will return to the point of projection with velocity u cos α after a time $ug^{-2}\cot\alpha\,[\alpha + \log\{\cos\alpha/(1 - \sin\alpha)\}]$*

[Meerut 2000; Kanpur 2000]

Sol. Proceed as in part (*a*) of Ex. 2 upto relations (*i*) and (*iii*), namely,

$$v_1 = UV/(V^2 + U^2)^{1/2} \quad \text{...(i)}$$

and total time $= (V/g)\{\tan^{-1}(U/V) + \tanh^{-1}(v_1/V)\}$ ...(iii)

In the present problem $k = gU^{-2}\tan^2\alpha$

$$\therefore \quad V^2 = \frac{g}{k} = \frac{g}{gU^{-2}\tan^2\alpha} = U^2\cot^2\alpha \quad \text{or} \quad V = U\cot\alpha \quad \text{...(iv)}$$

$$\therefore \text{ From (i), } \quad v_1 = \frac{U^2\cot\alpha}{(U^2 + U^2\cot^2\alpha)^{1/2}} = \frac{U^2\cot\alpha}{U\operatorname{cosec}\alpha} = U\cos\alpha \quad \text{...(v)}$$

Now, $\tan^{-1}(U/V) = \tan^{-1}(\tan\alpha) = \alpha$, by (*iv*) ...(vi)

$$\text{and} \quad \tanh^{-1}\frac{v_1}{V} = \tanh^{-1}\left(\frac{U\cos\alpha}{U\cot\alpha}\right) = \tanh^{-1}\sin\alpha = \frac{1}{2}\log\frac{1+\sin\alpha}{1-\sin\alpha}$$

$$\text{or} \quad \tanh^{-1}\frac{v_1}{V} = \frac{1}{2}\log\frac{(1+\sin\alpha)(1-\sin\alpha)}{(1-\sin\alpha)^2} = \frac{1}{2}\log\left(\frac{\cos\alpha}{1-\sin\alpha}\right)^2 = \log\frac{\cos\alpha}{1-\sin\alpha}$$

Thus, $\tanh^{-1}(v_1/V) = \log\{\cos\alpha/(1 - \sin\alpha)$ ...(vii)

Using (*iv*), (*vi*) and (*vii*), (*iii*) gives

the required time $= ug^{-1}\cot\alpha\,[\alpha + \log\{\cos\alpha/(1 - \sin\alpha)\}]$

Ex. 4. *A heavy particle is projected upward in a medium the resistance of which varies as the square of the velocity. It has a kinetic energy K in its upward path at a given point, when it passes the same point on the way down, show that its loss of kinetic energy is $K^2/(K + K')$, where K' is the limit to which the energy approaches in its downward course.* **[Meerut 1999, Kumaun 1993]**

Sol. Refer figures of Art. 12.3. Proceed as in Case II of Art. 12.4 upto equation (9). Let h be the maximum height attained at O during its upward motion so that when $y = O'O = h$, $v = 0$. So (9) gives $A = -\log V^2 - (2gh/V^2)$

$\therefore$ From (9), $-(2gy/V^2) = \log(V^2 + v^2) - \log V^2 - (2gh/V^2)$

so that $h - y = (V^2/2g)\log\{(V^2 + v^2)/V^2\}$ ...(i)

If v_1 is the velocity of the particle at the point R (between O' and O such that $OR = h'$), then at R, when $y = O'R = OO' - OR = h - h'$, $v = v_1$. So from (*i*),

$h - (h - h') = (V^2/2g)\log\{(V^2 + v_1^2)/V^2\}$ ...(ii)

After coming to instantaneous rest at O, the particle falls vertically downwards. Now proceed as in Case I of Art. 12.4 upto equation (4)′. If v_2 is the velocity of the particle at the point R, then from (4)′, putting $x = h'$, $v = v_2$, we get

$-(2g/V^2)h' = \log\{(V^2 - v_2^2)/V^2\}$ or $h' = (V^2/2g)\log\{V^2/(V^2 - v_2^2)\}$ ...(iii)

Equating the values of h' from (*ii*) and (*iii*), we have

$(V^2/2g)\log\{(V^2+v_1^2)/V^2\}=(V^2/2g)\log\{V^2/(V^2-v_2^2)\}$

$(V^2+v_1^2)/V^2=V^2/(V^2-v_2^2)$ or $v_2^2=(v_1^2V^2)/(V^2+v_1^2)$ ...(iv)

The required loss in K.E. $=(1/2)mv_1^2-(1/2)mv_2^2$

$$=\frac{1}{2}m\left(v_1^2-\frac{v_1^2V^2}{V^2+v_1^2}\right)=\frac{1}{2}mv_1^2\times\frac{v_1^2}{v_1^2+V^2}=\frac{1}{2}mv_1^2\,\frac{(1/2)mv_1^2}{(1/2)mv_1^2+(1/2)mV^2}$$

$=K\times\{K/(K+K')\}$, where $K=(1/2)mv_1^2$ and $K'=(1/2)mV^2$

$=K^2/(K+K')$, *where* K' is the limit to which the energy approaches in its downward course

Ex. 5. *A shell is fired vertically upward with speed v_0. The resistance is $mgcv^2$. Show that it attains its greatest height at time t, given by $\tan(gtc^{1/2})=v_0c^{1/2}$. Deduce that, no matter how large v_0 may be, t cannot exceed $(1/2)\pi g^{-1}c^{-1/2}$*

[Delhi B.A. (Prog.) III, 2014, Delhi B.Sc. (Hons.), 2003, 07]

Sol. Let a shell of mass m be projected vertically upward with velocity v_0 from the fixed point O'. Let Q be its position at any time t and let v be its velocity at Q. The shell is subjected to two forces mg and $mgcv^2$ both acting vertically downwards. Then the equation of motion of the particle is given by

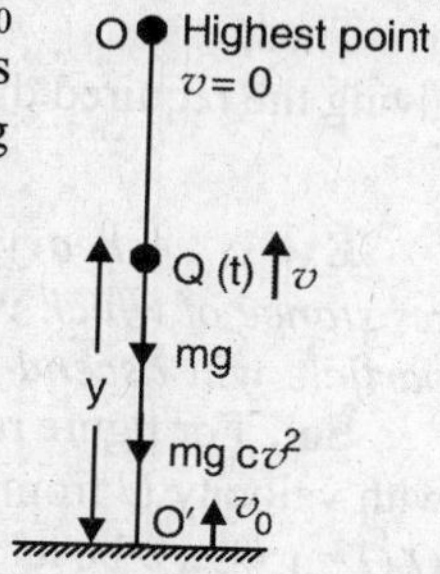

$m\ddot{y}=-mg-mgcv^2$ or $dv/dt=-g(1+cv^2)$

or $gdt=-\{1/(1+cv^2)\}\,dv$

Integrating, $gt=-(1/c^{1/2})\tan^{-1}(v\cdot c^{1/2})+A$, ...(1)

where A is a constant

Initially at O', when $t=0$, $v=v_0$. So (1) gives $A=(1/c^{1/2})\tan^{-1}(v_0\cdot c^{1/2})$

Then (1) gives $t=(1/gc^{1/2})\{\tan^{-1}(v_0c^{1/2})-\tan^{-1}(vc^{1/2})\}$ ...(2)

Let O be the highest point reached by the particle during its upward motion. Taking $v=0$ in (2), the required time to attain the greatest height $O'O$ is given by

$t=(1/gc^{1/2})\tan^{-1}(v_0c^{1/2})$ or $\tan(gt\,c^{1/2})=v_0c^{1/2}$ ...(3)

Deduction. Taking limit of both sides of (3) as $v_0\to\infty$, we get

$$t\to\lim_{v_0\to\infty}\frac{1}{gc^{1/2}}\tan^{-1}(v_0c^{1/2})\quad\text{or}\quad t\to\frac{1}{gc^{1/2}}\times\frac{\pi}{2}\quad\text{or } t\to\frac{1}{2}\times\pi\,g^{-1}c^{-1/2},$$

showing that no matter how large v_0 may be, t cannot exceed $(1/2)\pi\,g^{-1}\,c^{-1/2}$

Ex. 6. *A particle falls from rest in a medium in which the resistance is kv^2 per unit mass. Prove that the distance fallen in time t is $(1/k)\log\cosh\{t(gk)^{1/2}\}$.* **[Kumaun 2003]**

If the particle were ascending, show that at any instant its distance from the highest point of its path is $(1/k)\log\sec\{t(gk)^{1/2}\}$, where t now denotes the time it will take to reach its highest point. **[Meerut 2015]**

Sol. Ist part. Proceed as in Case I of Art. 12.4 upto (6). You need not discuss relation between v and x for this problem. Thus,

$$x=(V^2/g)\log\cosh(gt/V)$$

But $k=g/V^2$ or $V=(g/k)^{1/2}$ so $x=(1/k)\log\cosh\{t(gk)^{1/2}\}$,

Second part. For figure refer Case II of Art. 12.3. When the particle is ascending upwards, let y be its distance from the starting point O' after time T. Let v be its velocity at point Q where $O'Q=y$. The resistance mkv^2 of medium and weight mg of the particle both act in vertically downward direction. Then equation of the particle is

$m(d^2y/dT^2)=-mg-mkv^2$ or $dv/dT=-(g+kv^2)$

$$\text{or}\quad\frac{dv}{g+kv^2}=-dT\quad\text{or}\quad\frac{dv}{k\{(g/k)+v^2\}}=-dT$$

$$\text{Integrating,}\quad\frac{1}{k}\frac{1}{(g/k)^{1/2}}\tan^{-1}\frac{v}{(g/k)^{1/2}}=-T+A,\ \text{where } A \text{ is constant.}\ \ldots(1)$$

Let t' be the time from the point of projection O' to reach the highest point O. Then at O, when $T=t'$, $v=0$. So (1) gives $0=-t'+A$ or $A=t'$. So (1) gives

$$\frac{1}{(gk)^{1/2}}\tan^{-1}\frac{v}{(g/k)^{1/2}} = t' - T \quad \text{or} \quad \frac{v}{(g/k)^{1/2}} = \tan\{(t' - T)(gk)^{1/2}\}$$

or $$v = dy/dT = (g/k)^{1/2} \tan\{(t' - T)(gk))^{1/2}\} \quad ...(2)$$

Let O be the highest point reached by the particle such that $O'O = h$. Let x be the depth below O of the point Q distant y from O'. Then

$$x = h - y \quad ...(3)$$

Let t denote the time taken by the particle in moving from Q to O. Then

$$t = t' - T \quad ...(4)$$

From (3) and (4), $dx = -dy$ and $dt = -dT$. So $dx/dt = dy/dT$

$\therefore$ (2) becomes $\quad dx/dt = (g/k)^{1/2} \tan\{t(gk)^{1/2}\}$

Integrating, $\quad x = (g/k)^{1/2}.\{1/(gk)^{1/2}\} \log \sec\{t(gk)^{1/2}\} + B$, B is a const.

From (3) and (4), when $t = 0$, $x = 0$. So $B = 0$.

Thus, $\quad x = (1/k) \log \sec\{t(gk)^{1/2}\}$,

giving the required distance x of the particle from the highest point.

MISCELLANEOUS SOLVED EXAMPLES

Ex. 1. *A heavy particle is projected vertically upwards with velocity U in a medium, the resistance of which varies as the cube of the particle's velocity. Determine the height to which the particle will ascend.*

Sol. For figure refer case II of Art. 12.3. Let a particle of mass m be projected vertically upwards with velocity U from the fixed point O'. Let Q be the position of the particle at any time t such that $O'Q = y$. Let v be its velocity at Q. Here kv^3 is the force of resistance per unit mass so that $F(= mkv^3)$ is total resistance of the medium on the particle acting in vertically downward direction. The weight mg of the particle acts in vertically downward direction. Then the equation of motion of the particle is

$$m\ddot{y} = -mg - mkv^3 \quad \text{or} \quad \ddot{y} = -g(1 + kv^3/g) \quad ...(1)$$

Let V be the terminal velocity of the particle during its downward motion, *i.e.*, the velocity when the acceleration of the particle during its downward motion is zero. Then,

$$0 = mg - mkV^3 \quad ...(2)$$

Then, by (2), $\quad 0 = g - kV^3 \quad$ or $\quad k/g = 1/V^3 \quad ...(3)$

So (1) reduces to $\quad v(dv/dy) = -g(1 + v^3/V^3)$

or $$\{v/(v^3 + V^3)\}\, dv = -(g/V^3)\, dy \quad ...(4)$$

Let $$\frac{v}{v^3+V^3} = \frac{v}{(v+V)(v^2-vV+V^2)} = \frac{A}{v+V} + \frac{Bv+C}{v^2-vV+V^2} \quad ...(5)$$

or $$v = A(v^2 - vV + V^2) + (Bv + C)(v + V)$$

Equating the coefficients of v^2 and v and constant terms on both sides,

$$0 = A + B, \quad 1 = -AV + BV + C \quad \text{and} \quad 0 = AV^2 + CV.$$

Solving these, $\quad A = -(1/3V), \quad B = 1/3V, \quad C = 1/3$

$\therefore$ From (5), $$\frac{v}{v^3+V^3} = -\frac{1}{3V(v+V)} + \frac{v+V}{3V(v^2-Vv+V^2)} \quad ...(6)$$

Using (5) and (6), (4) can be re-written as

$$-\frac{g}{V^3}dy = \left[-\frac{1}{3V(v+V)} + \frac{v+V}{3V(v^2-Vv+V^2)}\right]dv$$

or $$\frac{3g}{V^2}dy = \left[\frac{1}{v+V} - \frac{1}{2}\frac{2v+2V}{v^2-vV+V^2}\right]dv = \frac{dv}{v+V} - \frac{1}{2}\frac{(2v-V)+3V}{v^2-vV+V^2}dv$$

or $$\frac{3g}{V^3}dy - \frac{dv}{v+V} + \frac{1}{2}\frac{2v-V}{v^2-vV+V^2}dv = -\frac{3V}{2}\frac{dv}{v^2-vV+V^2}$$

Integrating on both sides, we have

$$(3gy)/V^2 + D - \log(v + V) + (1/2)\log(v^2 - vV + V^2)$$

$$= -\frac{3V}{2}\int\frac{dv}{(v-V/2)+(V\sqrt{3}/2)^2}, \text{ where } D \text{ is a constant}$$

$$= -\frac{3V}{2}\times\frac{1}{(V\sqrt{3}/2)}\tan^{-1}\frac{v-V/2}{(V\sqrt{3}/2)} = -\sqrt{3}\tan^{-1}\left(\frac{2v-V}{V\sqrt{3}}\right)$$

$\therefore$ $(3gy)/V^2 + D = \log(v+V) - (1/2)\log(v^2 - vV + V^2) - \sqrt{3}\ \tan^{-1}\{(2v-V)/V\sqrt{3}\}$ ...(7)

Initially at O', when $y = 0$, $v = U$ (given). So (7) reduces to

$\therefore$ $D = \log(U+V) - (1/2)\log(U^2 - UV + V^2) - \sqrt{3}\ \tan^{-1}\{(2U-V)/V\sqrt{3}\}$ ...(8)

Subtracting (8) from (7), we have

$$\frac{3gy}{V^2} = \log\frac{v+V}{U+V} - \frac{1}{2}\log\frac{v^2-vV+V^2}{U^2-UV+V^2}$$
$$+\sqrt{3}[\tan^{-1}\{(2U-V)/V\sqrt{3}\} - \tan^{-1}\{(2v-V)/V\sqrt{3}\}] \quad ...(9)$$

Let h be the maximum height attained by the particle at O. Then at O, when $y = h$, $v = 0$. So (9) reduces to

$$\frac{3gh}{V^2} = \log\frac{V}{U+V} - \frac{1}{2}\log\frac{V^2}{U^2-UV+V^2} + \sqrt{3}\left(\tan^{-1}\frac{2U-V}{V\sqrt{3}} + \tan^{-1}\frac{V}{V\sqrt{3}}\right)$$

or
$$h = \frac{V^2}{3g}\left[\log\frac{(U^2-UV+V^2)^{1/2}}{U+V} + \sqrt{3}\left(\tan^{-1}\frac{2U-V}{V\sqrt{3}} + \frac{\pi}{6}\right)\right]$$

Ex. 2. *If the resistance vary as the 4th power of the velocity, the energy of mass at a depth x below the highest point when moving in a vertical line under gravity will be E tan (mgx/E) when rising and E tanh (mgx/E) when falling, where E is the terminal energy in the medium.*

[Bhopal 2004]

Sol. Refer figure of Case I of Art. 12.3. Let a particle of mass m be projected vertically upwards from a point O' and O be the highest point reached by the particle such that $OO' = h$. Let P be the point on the path of the particle at a depth x below O so that $O'P = h - x$.

The equation of motion of the particle when it is falling vertically downwards is

$$m\ddot{x} = mg - mkv^4 \quad \text{or} \quad \ddot{x} = g(1 - kv^4/g) \quad ...(1)$$

Let V be the terminal velocity. Then, $\ddot{x} = 0$ when $v = V$.

So (1) reduces to $\quad 0 = g(1 - kV^4/g) \quad$ or $\quad k = g/V^4$ ...(2)

Using (2), (1) gives $\quad v\dfrac{dv}{dx} = g\left(1 - \dfrac{v^4}{V^4}\right) \quad$ or $\quad \dfrac{2v\,dv}{V^4 - v^4} = \dfrac{2g\,dx}{V^4}$

or $\quad \dfrac{dz}{(V^2)^2 - z^2} = \dfrac{2g\,dx}{V^4}$, putting $v^2 = z$ so that $2v\,dv = dz$

Integrating, $\quad \dfrac{1}{V^2}\tanh^{-1}\dfrac{z}{V^2} = \dfrac{2gx}{V^4} + A \quad$ or $\quad \dfrac{1}{V^2}\tanh^{-1}\dfrac{v^2}{V^2} = \dfrac{2gx}{V^4} + A$,

where A is a constant. But at O, $x = 0$, $v = 0$. So $A = 0$

$\therefore$ $(1/V^2)\tanh^{-1}(v^2/V^2) = (2gx)V^2$ or $v^2 = V^2\tanh(2gx/V^2)$ ...(3)

Given, E = terminal energy in medium = $(1/2)mV^2$ ...(4)

Hence the kinetic energy of the particle at P when falling

$= (1/2)mv^2 = (1/2)mV^2\tanh(2gx/V^2)$, by (3)

$= E\tanh(mgx/E)$, since by (4), $V^2 = 2E/m$

We now consider motion of the particle from O' in vertically upward direction. Refer figure of case II of Art. 12.3. Let the particle be projected vertically upwards with velocity U from the fixed point O'. Let Q be the position of the particle at any time t such that $O'Q = y$. Let v be its velocity at Q. Then the equation of motion of the particle when it is moving vertically upwards is

$$m\ddot{y} = -mg - mkv^4 \quad \text{or} \quad \ddot{y} = -g(1 + kv^4/g)$$

or
$$\ddot{y} = -g(1 + v^4/V^4), \text{ using (2)}$$

or $v\dfrac{dv}{dy} = -\dfrac{g(V^4+v^4)}{V^4}$ or $\dfrac{2v\,dv}{V^4+v^4} = -\dfrac{2g\,dy}{V^4}$

or $\dfrac{dz}{(V^2)^2+z^2} = -\dfrac{2g\,dy}{V^4}$, putting $v^2 = z$ and $2v\,dv = dz$

Integrating, $\dfrac{1}{V^2}\tan^{-1}\dfrac{z}{V^2} = -\dfrac{2gy}{V^4} + B$, or $\dfrac{1}{V^2}\tan^{-1}\dfrac{v^2}{V^2} = -\dfrac{2gy}{V^4} + B$,

where B is a constant. But at O', $y = 0$, $v = U$. So $B = (1/V^2)\tan^{-1}(U^2/V^2)$

$$\therefore\ \frac{1}{V^2}\tan^{-1}\frac{v^2}{V^4} = -\frac{2gy}{V^4} + \frac{1}{V^2}\tan^{-1}\frac{U^2}{V^2} \text{ or } \frac{2gy}{V^2} = \tan^{-1}\frac{U^2}{V^2} - \tan^{-1}\frac{v^2}{V^2} \quad \ldots(5)$$

At the highest point O, $y = h$, $v = 0$. So (5) gives

$$(2gh)/V^2 = \tan^{-1}(U^2/V^2) \quad \ldots(6)$$

At P, $y = h - x$ and let $v = v_1$. Then, by (5), we get

$$\frac{2g(h-x)}{V^2} = \tan^{-1}\frac{U^2}{V^2} - \tan^{-1}\frac{v_1^2}{V^2} \quad \ldots(7)$$

Subtracting (7) from (6), $(2gx)/V^2 = \tan^{-1}(v_1^2/V^2)$

or $v_1^2/V^2 = \tan(2gx/V^2)$ or $v_1^2 = V^2\tan(2gx/V^2)$ ...(8)

Hence the kinetic energy of particle at P when rising

$= (1/2)mv_1^2 = (1/2)mV^2\tan(2gx/V^2)$, by (8)

$= E\tan(mgx/E)$, since by (4), $V^2 = 2E/m$

EXERCISE

1. A particle of mass m falls from rest at a distance a from the centre of the earth, the motion meeting a small resistance proportional to the square of the velocity v and the retardation being μ for unit velocity; show that the kinetic energy at a distance x from the centre is $mgr^2\,\{1/x - 1/a + 2\mu(1 - x/a) - 2\mu\log(a/x)\}$, the square of μ being neglected and r is the radius of the earth.

2. An attracting force varying as the distance acts on a particle initially at rest at a distance a. Show that if V be the velocity when the particle is at a distnace x and V' the velocity, when the resistance of air is taken into account, then

$$V' = V\left\{1 - \frac{k(2a+x)(a-x)}{3(a+x)}\right\}, \text{ nearly,}$$

the resistance being k times the square of velocity, k being very small.

3. A particle moves from rest at a distance a from a fixed point O under the action of a force equal to μ times the distance per unit mass. If the resistance of the medium in which it moves be k times the square of the velocity per unit mass then show that the square of its velocity when it is at a distance x from O is $(\mu x/k) - (\mu a/k)\,e^{2k(x-a)} + (\mu/2k^2)\,\{1 - e^{2k(x-a)}\}$. Show also that when it first comes to rest it will be at a distance b given by $(1 - 2bk)\,e^{2bk} = (1 + 2ak)\,e^{-2ak}$.

4. Two particles move in a medium whose resistance varies as square of the velocity. One is let fall from a height h and the other is projected upwards at the sme instant with unitial velocity sufficient to carry it to a height h. Show that the particles meet at a depth y below the highest point given by $\cosh\beta\cos(\alpha - \beta) = 1$ where $gv = V^2\log\cosh\beta$ and $gh = V^2\sec\alpha$, V being the terminal velocity.

12.5 Motion of a projectile in a resisting medium in which resistance varies as velocity. *A particle is projected under gravity and a resistance equal to mk (velocity), with a velocity u at an angle α to the horizon; to find the motion.* **[Bundelkhand 2007; Purvanchel 2003; Kumaun 1999; Garhwal 2001; Kanpur 2009]**

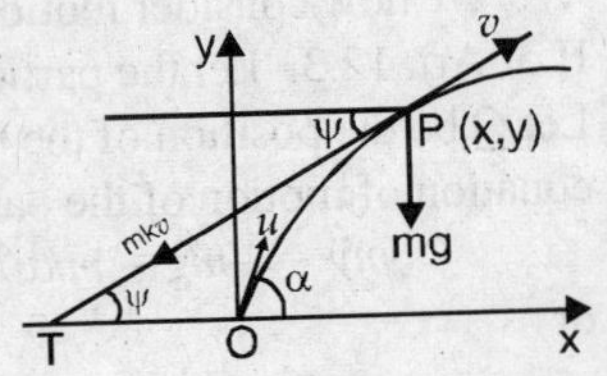

Let a particle be projected from a fixed point O with velocity u at an angle α to the horizon. Let O be taken as the origin and horizontal and vertical lines through O as x-axis and y-axis respectively. Let $P(x, y)$ be the position of the particle after time t and let tangent PT at P be inclined at an angle ψ to x-axis so that

$dy/dx = \tan\psi, \quad dy/ds = \sin\psi, \quad dx/ds = \cos\psi, \quad$ where $s =$ arc OP ...(1)

The external forces acting on the particle are its weight mg acting vertically downwards and the resistance mkv of the medium along PT in a direction opposite to the direction of motion. Then the equation of the motion along OX and OY respectively are given by

$m\ddot{x} = -mkv\cos\psi \quad$ or $\quad \ddot{x} = -k(ds/dt)\,(dx/ds) = -k\dot{x}$, by (1) ...(2)

and $m\ddot{y} = -mg - mkv\sin\psi$ or $\ddot{y} = -g - k(ds/dt) \times (dy/ds) = -g - k\dot{y}$, by (1) ...(3)

Re-writing (2), $\quad \ddot{x}/\dot{x} = -k$

Integrating, $\quad \log\dot{x} = -kt + A$, where A is a constant. ...(4)

Initially at O, when $t = 0$, $\dot{x} =$ horizontal component of initial velocity u of the particle $= u\cos\alpha$. So (4) gives $\quad A = \log(u\cos\alpha)$. Then (4) becomes

$\log\dot{x} = -kt + \log(u\cos\alpha) \quad$ or $\quad \log(\dot{x}/u\cos\alpha) = -kt$

or $\quad \dot{x} = u\cos\alpha\, e^{-kt} \quad$ or $\quad dx/dt = u\cos\alpha\, e^{-kt}$ ...(5)

Re-writing (3), $\quad k\ddot{y}/(g + k\dot{y}) = -k$

Integrating, $\quad \log(g + k\dot{y}) = -kt + B$, where B is a constant. ...(6)

Initially at O, when $t = 0$, $\dot{y} =$ vertical component of initial velocity u of the particle $= u\sin\alpha$. So (6) gives $B = \log(g + ku\sin\alpha)$. Then (6) becomes

$\log(g + k\dot{y}) = -kt + \log(g + ku\sin\alpha)$ or $\log\{(g + k\dot{y})/(g + ku\sin\alpha)\} = -kt$

so that $\quad g + k(dy/dt) = (g + ku\sin\alpha)\, e^{-kt}$ ...(7)

(5) and (7) give horizontal and vertical components of velocity at any time t.

Relation beteen x, *t* and *y*, *t*. Integrating (5), we get

$x = -(u/k)\cos\alpha\, e^{-kt} + C$, where C is a constant. ...(8)

Initially at O, wen $t = 0$, $x = 0$. So (8) gives $C = (u/k)\cos\alpha$

$\therefore$ (8) becomes $\quad x = (u/k)\cos\alpha\,(1 - e^{-kt})$ ...(9)

Integrating (7), $\quad gt + ky = -(1/k)(g + ku\sin\alpha)e^{-kt} + D$ ...(10)

Initially at O, when $t = 0$, $y = 0$. So (10) gives $\quad D = (1/k)(g + ku\sin\alpha)$

$\therefore$ (10) becomes $\quad gt + ky = (1/k)(g + ku\sin\alpha)(1 - e^{-kt})$ ...(11)

Equation (9) and (11) give the horizontal and vertical distances described by the particle at any time t. (9) and (11) together can be regarded as the parametric equation of the trajectory, t being parameter.

Equaton of trajectory in cartesian form. We now eliminate t from (9) and (11) as follows. From (9), we have

$$1 - \frac{kx}{u\cos\alpha} = e^{-kt} \quad \text{so that} \quad t = -\frac{1}{k}\log\left(1 - \frac{kx}{u\cos\alpha}\right) \qquad ...(12)$$

Substituting the values of $1 - e^{-kt}$ and t given by (12) in (11), we get

$$-\frac{g}{k}\log\left(1 - \frac{kx}{u\cos\alpha}\right) + ky = \frac{g + ku\sin\alpha}{k} \times \frac{kx}{u\cos\alpha}$$

or $$y = \frac{x}{ku\cos\alpha}(g + ku\sin\alpha) + \frac{g}{k^2}\log\left(1 - \frac{kx}{u\cos\alpha}\right) \qquad ...(13)$$

Remark 1. From (5) and (7), we find that when $t \to \infty$, then $dx/dt \to 0$ and $dy/dt \to -g/k = -V$, say. Then $V(= g/k)$ is known as the terminal velocity.

Remark 2. From (9) and (11), we find that when $t \to \infty$, then $x \to 0$ and $y \to -\infty$. Hence the path (13) has a vertical asymptote $x = (u/k)\cos\alpha$.

TYPE 4 : SOLVED EXAMPLES BASED ON ART. 12.5

Ex. 1. (*a*) *A particle of unit mass is projected with velocity V and inclination* α *in a medium whose resistance is k times velocity. Show that if k is small the equation of the path is approximately.*

$$y = x\tan\alpha - \frac{gx^2}{2V^2\cos^2\alpha} - \frac{kgx^3}{3V^3\cos^3\alpha}$$

[Delhi Maths (Hon.) 1999, 2002, 06]

Sol. Do as in Art. 12.5 upto equation (13) by taking $m = 1$ and $u = V$. Then

$$y = \frac{x}{kV \cos\alpha}(g + kV \sin\alpha) + \frac{g}{k^2}\log\left(1 - \frac{kx}{V\cos\alpha}\right) \quad \text{...(14)}$$

We know that $\log(1 - x) = -(x + x^2/2 + x^3/3 + ...)$, if $|x| < 1$ ...(15)

Given that k is small. Let $|(kx)/V\cos\alpha| < 1$. Then using (15), (14) gives

$$y = \frac{gx}{kV\cos\alpha} + x\tan\alpha - \frac{g}{k^2}\left(\frac{kx}{V\cos\alpha} + \frac{k^2x^2}{2V^2\cos^2\alpha} + \frac{k^3x^3}{3V^3\cos^3\alpha} + ...\right)$$

or $\quad y = x\tan\alpha - (gx^2/2V^2\cos^2\alpha) - (kgx^3/3V^3\cos^3\alpha)$, ...(16)

is the required equation of the path approximately by neglecting squares and higher powers of k.

Note : $k = 0$ gives ordinary equation to the trajectory for unrestricted motion.

Ex. 1. (b) *In example 1(a), if the particle is projected in the given resisted medium from a point on a plane inclined at an angle β to the horizontal, to find the range on the inclined plane (upto first order of approximations).*

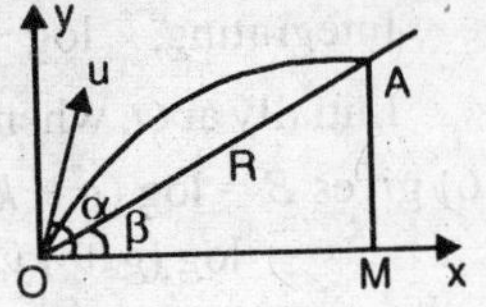

Sol. Let OA be the given inclined plane such that $\angle AOX = \beta$. Let the particle strike it at A, where $OA = R$. Then coordinates of A will be (OM, MA) *i.e.*, $(R\cos\beta, R\sin\beta)$. Now do as in Ex. 1(*a*) upto equation (16). Since A lies on (16), we must have

$$R\sin\beta = R\cos\beta\tan\alpha - \frac{gR^2\cos^2\beta}{2V^2\cos^2\alpha} - \frac{gkR^3\cos^3\beta}{3V^3\cos^3\alpha}$$

or $$\frac{gR\cos^2\beta}{2V^2\cos^2\alpha} + \frac{gkR^2\cos^3\beta}{3V^3\cos^3\alpha} = \frac{\sin\alpha\cos\beta - \cos\alpha\sin\beta}{\cos\alpha}$$

or $$\frac{gR\cos^2\beta}{2V^2\cos^2\alpha}\left(1 + \frac{2kR\cos\beta}{3V\cos\alpha}\right) = \frac{\sin(\alpha - \beta)}{\cos\alpha}$$

or $$R = \frac{2V^2\cos\alpha\sin(\alpha - \beta)}{g\cos^2\beta}\left(1 + \frac{2kR\cos\beta}{3V\cos\alpha}\right)^{-1}$$

or $$R = \frac{2V^2\cos\alpha\sin(\alpha-\beta)}{g\cos^2\beta}\left(1 - \frac{2kR\cos\beta}{3V\cos\alpha}\right),$$ to first order of approximation, k being small

or $$R\left\{1 + \frac{4kV\sin(\alpha-\beta)}{3g\cos\beta}\right\} = \frac{2V^2\cos\alpha\sin(\alpha-\beta)}{g\cos^2\beta}$$

or $$R = \frac{2V^2\cos\alpha\sin(\alpha-\beta)}{g\cos^2\beta}\left\{1 + \frac{4kV\sin(\alpha-\beta)}{3g\cos\beta}\right\}^{-1}$$

or $\quad R = \{2V^2\cos\alpha\sin(\alpha - \beta)/g\cos^2\beta\} \times [1 - \{4kV\sin(\alpha - \beta)/3g\cos\beta\}]$,

is the required range on the inclined plane (to the first order of aproximation, k being small).

Ex. 1. (c) *A particle of m is projected in a resisting medium in a direction making an angle α with the horizontal, the horizontal component of its velocity being u. If the resistance is mk times the velocity, where k is small, prove that the equation of the path can be written in the approximate form* $y = x\tan\alpha - (gx^2/2u^2)(1 + 2kx/3u)$ *where squares and higher powers of k are neglected.*

Sol. Proceed as Ex. 1(*a*) upto equation (16). Hence $u = V\cos\alpha$. Then (16) can be re-written as

$$y = x\tan\alpha - \frac{gx^2}{2u^2} - \frac{kgx^3}{3u^3} \quad \text{or} \quad y = x\tan\alpha - \frac{gx^2}{2u^2}\left(1 + \frac{2kx}{3u}\right)$$

Ex. 2. *If a particle be moving in a medium whose resistance varies as the velocity of the particle, show that the equation of the trajectory can, by a proper choice of axes, be put in the form* $y + ax = b\log x$. **[Garhwal 2004]**

Sol. Proceed as in Art. 12.5 upto equation (13). Now complete the solution as follows. Let

$$1 - (kx/u\cos\alpha) = X \quad \text{...(14)}$$

Then (13) gives $\quad y = (1/k^2)(1 - X)(g + ku \sin\alpha) + (g/k^2)\log X$

or $\quad y - (1/k^2)(g + ku \sin\alpha) = (g/k^2)\log X - (1/k^2)(g + ku \sin\alpha) X$

or $\quad Y = (g/k^2)\log X - (1/k^2)(g + ku \sin\alpha)X, \quad$ ...(15)

where $\quad Y = y - (1/k^2)(g + ku \sin\alpha) \quad$ ...(16)

Let $a = (1/k^2)(g + ku \sin\alpha)$ and $b = (g/k^2)$. Then (15) becomes

$Y + aX = b \log X$ which is of the form $y + ax = b \log x$.

Ex. 3. *A particle acted on by gravity is projected in a medium the resistance of which varies as the velocity. Show that its acceleration retains a fixed direction and diminishes without limit.*

Sol. First proceed upto equation (11) as in Art. 12.5. Now complete the solution as follows. Let θ be the direction of acceleration with the horizon. Then, using (2) and (3), we obtain

$$\tan\theta = \frac{\ddot{y}}{\ddot{x}} = \frac{-(g + k\dot{y})}{-k\dot{x}} = \frac{(g + ku \sin\alpha)e^{-kt}}{ku \cos\alpha\, e^{-kt}}, \text{ by (5) and (7)}$$

or $\quad \tan\theta = (g + ku \sin\alpha)/(ku \cos\alpha),$

which is independent of t and so it retains a fixed direction.

Let f be the magnitude of acceleration. Then, as before, we get

$$f = (\ddot{x}^2 + \ddot{y}^2)^{1/2} = \{(g + ku \sin\alpha)^2 e^{-2kt} + k^2u^2\cos^2\alpha\, e^{-2kt}\}^{1/2}$$

or $\quad f = (g^2 + k^2u^2 + 2gku \sin\alpha)^{1/2}/e^{kt}$

As t increases e^{kt} continually increases and consequently f decreases and as $t \to \infty, f \to 0$. Thus the acceleration diminishes without limit to zero.

Ex. 4. *A particle of unit mass is projected with velocity u at an inclination α above the horizon in a medium whose resistance is k times the velocity. Show that its direction will again make an angle α with the horizon after a time (1/k) log {1 + (2ku/g) sin α}*

Sol. First proceed upto equation (11) as in Art. 12.5. Now complete the solution as follows. Let θ be the direction of motion with the horizon after time t. Using (5) and (7), we get

$$\tan\theta = \frac{\dot{y}}{\dot{x}} = \frac{k\dot{y}}{k\dot{x}} = \frac{(g + ku \sin\alpha)e^{-kt} - g}{ku \cos\alpha\, e^{-kt}} \quad \text{...(12)}$$

By the given problem, the direction of motion makes an angle α with the horizon *i.e.*, $\pi - \alpha$ with horizon (measured anti-clockwise) *i.e.*, here $\theta = \pi - \alpha$. Putting $\theta = \pi - \alpha$ in (12), we have

$-\tan\alpha = \{(ku \sin\alpha + g)\, e^{-kt} - g\}/(ku \cos\alpha\, e^{-kt})$

or $\quad -ku \sin\alpha\, e^{-kt} = (ku \sin\alpha + g)\, e^{-kt} - g$

or $\quad ge^{kt} = 2ku \sin\alpha + g \quad$ or $\quad e^{kt} = 1 + (2ku/g)\sin\alpha$

or $\quad kt = \log\{1 + (2ku/g)\sin\alpha\} \quad$ or $\quad t = (1/k)\log\{1 + (2ku/g)\sin\alpha\}$

Ex. 5. *A particle is projected with a velocity whose horizontal and vertical components are U and V from a point in a medium whose resistance per unit of mass is k times the speed. Obtain the equation of the path and prove that if k is small, the horizontal range is approximately* $(2UV/g) - (8UV^2k/3g^2)$. **[I.A.S. 1992]**

Sol. First proceed as in Art. 12.5 upto equation (13). Here U and V are horizontal and vertical components of the velocity of projection u making an angle α with the horizon. So, here, we have

$u \cos\alpha = U \quad$ and $\quad u \sin\alpha = V \quad$ ...(14)

Equation of path. Using (14), (13) reduces to

$$y = (x/kU)(g + kV) + (g/k^2)\log(1 - kx/U),$$

which is the required equation of the path.

Approximate value of the horizontal range. Using (14), (9) and (11) reduce to

$x = (U/k)(1 - e^{-kt}) \quad$ ...(15)

and $\quad gt + ky = (1/k)(g + kV)(1 - e^{-kt}). \quad$ ...(16)

Clearly the horizontal range R (say) is the value of x obtained in time t during which $y = 0$. So range R is given by

$$R = (U/k)(1 - e^{-kt}) \quad ...(17)$$

where
$$gt = (1/k)(g + kV)(1 - e^{-kt}) \quad ...(18)$$

From (17), $1 - (Rk/U) = e^{-kt}$ or $-kt = \log(1 - Rk/U)$ ...(19)

Using (19), (18) gives $-\dfrac{g}{k}\log\left(1 - \dfrac{Rk}{U}\right) = \dfrac{g + kV}{k} \times \dfrac{Rk}{U}$

or
$$-\frac{g}{k}\left(-\frac{Rk}{U} - \frac{R^2k^2}{2U^2} - \frac{R^3k^3}{3U^3}\right) = \frac{R(g + kV)}{U},$$

[on expanding log $(1 - Rk/U)$ by result $\log(1 - x) = -x - (1/2)x^2 - (1/3)x^3 - ...$ and rejecting terms containing k^4 and higher powers, k being small]

or $Rg/U + (R^2kg)/2U^2 - (R^3k^2g)/3U^3 = Rg/U + (RkV)/U$

or $R = (2UV)/g - (2kR^2)/3U$ ...(20)

To the first order of approximation, $R = (2UV)/g$. Putting this value of R in R.H.S. of (20), we get

$$R = \frac{2UV}{g} - \frac{2k}{3U} \times \left(\frac{2UV}{g}\right)^2 \quad \text{or} \quad R = \frac{2UV}{g} - \frac{8UV^2k}{3g^2}$$

Ex. 6. *If the resistnace vary as the velocity and the range on the horizontal plane through the point of projection is maximum, show that the angle α which the direction of projection makes with the vertical is given by*

$\lambda(1 + \lambda\cos\alpha)/(\lambda + \cos\alpha) = \log(1 + \lambda\sec\alpha)$,

where λ is the ratio of the velocity of projection to terminal velocity. *(Kanpur 2000)*

Sol. Proceed as in Art. 12.5 upto equation (11) by taking θ in place of α. Then (9) and (11) respectively become

$$x = (u/k)\cos\theta(1 - e^{-kt}) \quad ...(12)$$

and
$$gt + ky = (1/k)(g + ku\sin\theta)(1 - e^{-kt}) \quad ...(13)$$

The range R is the value of x obtained in time t during which $y = 0$. So putting $x = R$ and $y = 0$ in (12) and (13), we get

$$R = (u/k)\cos\theta(1 - e^{-kt}), \quad ...(14)$$

where
$$gt = (1/k)(g + ku\sin\theta)(1 - e^{-kt}) \quad ...(15)$$

Now, range R is maximum when $dR/d\theta = 0$ ...(16)

Differntiating (14) and (15) w.r.t. θ and using (16), we get

$$0 = -(u/k)\sin\theta(1 - e^{-kt}) + (u/k)\cos\theta(ke^{-kt})(dt/d\theta) \quad ...(17)$$

and $g(dt/d\theta) = u\cos\theta(1 - e^{-kt}) + (1/k)(g + ku\sin\theta)(ke^{-kt})(dt/d\theta)$ ...(18)

From (18), $\{g - (g + ku\sin\theta)e^{-kt}\}(dt/d\theta) = u\cos\theta(1 - e^{-kt})$ ...(19)

From (17), $(u\cos\theta\, e^{-kt})(dt/d\theta) = (1/k)u\sin\theta(1 - e^{-kt})$ ...(20)

Dividing (19) by (20), $\dfrac{g - (g + ku\sin\theta)e^{-kt}}{u\cos\theta\, e^{-kt}} = \dfrac{k\cos\theta}{\sin\theta}$

or $g\sin\theta = e^{-kt}(uk\cos^2\theta + uk\sin^2\theta + g\sin\theta)$

or $\dfrac{g\sin\theta}{uk + g\sin\theta} = e^{-kt}$ or $t = -\dfrac{1}{k}\log\dfrac{g\sin\theta}{uk + g\sin\theta}$

Putting the above values of e^{-kt} and t in (15), we get

$$-\frac{g}{k}\log\frac{u\sin\theta}{uk + g\sin\theta} = \frac{g + ku\sin\theta}{k}\left(1 - \frac{g\sin\theta}{uk + g\sin\theta}\right) \quad ...(21)$$

If V is the terminal velocity, then $V = g/k$

Given $u/V = \lambda$ so $u = \lambda V = \lambda(g/k)$ or $uk = g\lambda$ ...(22)

Since by problem, the direction of motion makes an angle α with vertical, so

$$\theta = 90° - \alpha \quad ...(23)$$

Using (22) and (23), (21) reduces to

$$\frac{g}{k}\log\frac{g\lambda + g\cos\alpha}{g\cos\alpha} = \frac{g + g\lambda\cos\alpha}{k}\left(1 - \frac{g\cos\alpha}{g\lambda + g\cos\alpha}\right)$$

or $\quad \log(1 + \lambda \sec\alpha) = \{(1 + \lambda\cos\alpha)\lambda\}/(\lambda + \cos\alpha)$

12.6 Trajectory in a resisting medium whose resistance varies as square of velocity. *A particle is projected under gravity and a resistance equal to mk(velocity)²; to find the motion.*

(Rohilkhand 2008)

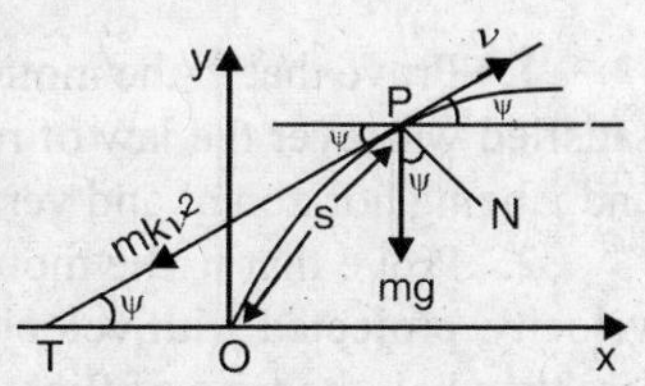

Let a particle be projected from a fixed point O. Let O be taken as the origin and horizontal and vertical lines through O as coordinate axes. Let $P(x, y)$ be the position of the particle after time t and let tangent PT at P be inclined at an angle ψ to x-axis. Let PN be inward drawn normal at P. Let v be the velocity of the particle at P along tangent TP and let u be the horizontal component of v so that $u = v\cos\psi$. The external forces acting on the particle are its weight mg acting vertically downwards and resistance mkv^2 of the medium along PT in a direction opposite the direction of motion. Here we shall write equations of motion parallel to x-axis and along the normal PN. These are respectively given by

$$m(du/dt) = -mkv^2\cos\psi \quad \text{or} \quad du/dt = -kv^2\cos\psi \qquad \ldots(1)$$

and $$m(v^2/\rho) = mg\cos\psi \quad \text{or} \quad v^2/\rho = g\cos\psi \qquad \ldots(2)$$

We now proceed to find the intrinsic equation (*i.e.*, relation between ρ and ψ) of the trajectory with help of (1) and (2). Since $v = ds/dt$ and $v\cos\psi = u$, (1) reduces to

$$du/dt = -k(ds/dt)\cdot u \quad \text{or} \quad (1/u)\,du = -k\,ds \qquad \ldots(3)$$

Integrating, $\quad u = Ae^{-ks}$, $\quad A$ being an arbitrary constant. $\qquad \ldots(4)$

Initially when $s = 0$, let $u = u_0$. So (4) gives $\quad A = u_0$.

Hence (4) reduces to $\quad u = u_0\,e^{-ks} \qquad \ldots(5)$

Since ψ is decreases as s increases, so $\quad \rho = -ds/d\psi \qquad \ldots(6)$

Using (6), (3) becomes $\quad g\cos\psi = v^2(-d\psi/ds)$

or $$g\cos\psi = -v\frac{ds}{dt}\frac{d\psi}{ds} = -v\frac{d\psi}{dt} = -v\frac{d\psi}{du}\frac{du}{dt}$$

or $\quad g\cos\psi = -v(d\psi/du)\,(-kv^2\cos\psi)$, using (1)

or $\quad g = kv^2\,(d\psi/du) \quad$ or $\quad g = k(u\sec\psi)^3\,(d\psi/du)$, as $v\cos\psi = u$

or $$\int u^{-3}du = \int (k/g)\sec^3\psi\,d\psi \qquad \ldots(7)$$

Using $\int \sec^n x\,dx = -\dfrac{\sec^{n-2} x\tan x}{n-1} + \dfrac{n-2}{n-1}\int \sec^{n-2} x\,dx$, for $n = 3$, (7) gives

$$-(1/2u^2) = (k/g)\,\{(1/2)\sec\psi\tan\psi + (1/2)\log(\sec\psi + \tan\psi)\} + B$$

or $\quad (1/u_0^2)\,e^{2ks} = -(k/g)\,\{\sec\psi\tan\psi + \log(\sec\psi + \tan\psi)\} + B$, by (5)

The above equation is intrinsic equation of the path. The value of the arbitrary constant B can be obtained by using the initial conditions of the problem.

12.7 Trajectory in a resisting medium whose resistance varies as *n*th power of velocity. *A particle is projected under gravity and a resistance equal to mk(velocity)ⁿ; to find the motion.*

Draw figure as in Art. 12.6. In that figure replace mkv^2 by mkv^n. Proceed exactly as before. Then (1) and (2) will reduce to

$$du/dt = -kv^n\cos\psi \qquad \ldots(1)$$

and $$v^2/\rho = g\cos\psi \qquad \ldots(2)$$

Since ψ decreases as s increases, $\quad \rho = -ds/d\psi \qquad \ldots(3)$

Also, u = horizontal component of velocity $= v\cos\psi \qquad \ldots(4)$

From (2) and (3), $\quad g\cos\psi = v^2(-d\psi/ds) = v(ds/dt)\,(-d\psi/ds)$

or $\quad g\cos\psi = -v(d\psi/dt) \quad$ or $\quad g\cos\psi = -v(d\psi/du)\,(du/dt)$

or $\quad g\cos\psi = -v(d\psi/du)\cdot(-kv^n\cos\psi)$, using (1)

or $g = kv^{n+1}\,(d\psi/du) \quad$ or $\quad g = k(u\sec\psi)^{n+1}\,(d\psi/du) \quad$ or $\quad u^{-n-1}\,du = (k/g)\sec^{n+1}\psi\,d\psi$.

Integrating, $\quad -\dfrac{1}{nu^n} = \dfrac{k}{g}\displaystyle\int \sec^{n+1}\psi\,d\psi + C$, where C is a constant or $$\frac{1}{(v \cos \psi)^n} = -\frac{nk}{g} \int \sec^{n+1} \psi \, d\psi + C,$$

which gives velocity in terms of ψ at any point.

EXERCISE

1. Prove that in the motion of a projectile in a resisting medium the equation $d^2y/dx^2 = -g/u^2$ is satisfied whatever the law of resistance, u being the horizontal component of velocity, the axes of x and y being horizontal and vertically upwards.

2. Prove that in the motion of a projectile in a resisting medium, resistance being k times the velocity, projected with velocity U at an angle α with the horizon, the time to the greatest height is less than half the time of flight.

3. A particle is projected undre gravity and a resistance $mk \times$ (velocity) with a velocity u at an angle α to the horizon. Find the greatest height attained by the particle. What will be the greatest height if $k = 0$? **[Calcutta 2003]**

12.8 Motion on a rough cycloid. *The base of a rough cycloidal arc is horizontal and vertex downwards. A particles slides down its arc; to discuss the motion.* **[Purvanchal 2006]**

Let $P(s, \psi)$ be the position of the particle at any time t. Let v be its velocity there and R the normal reaction along inward drawn normal PN of the curve on it. The other forces acting on it are its weight mg vertically downwards and friction μR along the tangent PT' in the direction opposite to the motion of the particle. Then the equation of motion along PT' and PN respectively are given by

$$mv(dv/ds) = \mu R - mg \sin \psi \qquad \ldots(1)$$

and $$m(v^2/\rho) = R - mg \cos \psi \qquad \ldots(2)$$

To eliminate R, putting the value of R given by (2) in (1), we get

$$mv(dv/ds) = \mu(mv^2/\rho + mg \cos \psi) - mg \sin \psi$$

or $$(1/2) \times (dv^2/ds) = \mu v^2/\rho + g(\mu \cos \psi - \sin \psi) \qquad \ldots(3)$$

Intrinsic equation of cycloid is $$s = 4a \sin \psi \qquad \ldots(4)$$

From (4), $$\rho = ds/d\psi = 4a \cos \psi \qquad \ldots(5)$$

Multiplying both sides (3) by 2ρ and using (5), we get

$$\frac{dv^2}{ds}\frac{ds}{d\psi} - 2\mu v^2 = 2g \cdot 4a \cos \psi (\mu \cos \psi - \sin \psi)$$

or $$dv^2/d\psi - 2\mu v^2 = 8ag \cos \psi \, (\mu \cos \psi - \sin \psi), \qquad \ldots(6)$$

which is linear equation of the first order in v^2.

Its Integrating factor $= e^{\int -(2\mu)\, d\psi} = e^{-2\mu\psi}$ and solution is

$$v^2 e^{-2\mu\psi} = C + 8ag \int e^{-2\mu\psi} \cos \psi (\mu \cos \psi - \sin \psi) \, d\psi \qquad \ldots(7)$$

$$= C + 8ag \int e^{-2\mu\psi} \{(\mu/2)(1 + \cos 2\psi) - (1/2) \sin 2\psi\} \, d\psi$$

$$= C + 4ag \left\{(-1/2) e^{-2\mu\psi} + \mu \int e^{-2\mu\psi} \cos 2\psi \, d\psi - \int e^{-2\mu\psi} \sin 2\psi \, d\psi\right\}$$

$$= C + 4ag \left\{-\frac{1}{2} e^{-2\mu\psi} + \frac{\mu e^{-2\mu\psi}}{4 + 4\mu^2}(-2\mu \cos 2\psi + 2 \sin 2\psi)\right.$$

$$\left. - \frac{e^{-2\mu\psi}}{4 + 4\mu^2}(-2\mu \sin 2\psi - 2 \cos 2\psi)\right\}, \text{ using the results}$$

$$\int e^{ax} \sin bx \, dx = \frac{e^{ax}(a \sin bx - b \cos bx)}{a^2 + b^2}, \quad \int e^{ax} \cos bx \, dx = \frac{e^{ax}(a \cos bx + b \sin bx)}{a^2 + b^2}$$

$$\therefore\ v^2 e^{-2\mu\psi} = C + 4ag\left[-\frac{e^{-2\mu\psi}}{2} + \frac{e^{-2\mu\psi}}{4+4\mu^2}\{4\mu \sin 2\psi + 2(1-\mu^2)\cos 2\psi\}\right] \quad ...(8)$$

The constant of integration in (8) can be obtained by using the given condition of the problem. Then (8) gives the velocity of the particle at any position. By putting the value of v^2 so obtained in (2), we can find reaction R.

TYPE 5 : SOLVED EXERCISES ON ART. 12.8

Ex. 1. *The base of a rough cycloidal arc is horizontal and vertex downwards; a bead slides along it, starting from rest at the cusp and coming to rest at the vertex. Show that* $\mu^2 e^{\mu\pi} = 1$.

[Gorakhpur 1997, Bhopal 2004]

Sol. Proceed exactly, as in Art. 12.8 and get result (8), *i.e.*,

$$v^2 e^{-2\mu\psi} = C + 4ag\left[-\frac{e^{-2\mu\psi}}{2} + \frac{e^{-2\mu\psi}}{2(1+\mu^2)}\{2\mu \sin 2\psi + (1-\mu^2)\cos 2\psi\}\right] \quad ...(8)$$

Initially at cusp A, when $\psi = \pi/2$, $v = 0$. So (8) gives

$$C = -4ag\{(-1/2)e^{-\mu\pi} - e^{-\mu\pi}(1-\mu^2)/2(1+\mu^2)\} \quad ...(9)$$

By problem, the particle comes to rest at vertex O so that when $\psi = 0$, $v = 0$. Then (8) gives

$$C = -4ag\{(-1/2) + (1-\mu^2)/2(1+\mu^2)\} \quad ...(10)$$

From (9) and (10), $$-4ag\left\{-\frac{e^{-\mu\pi}}{2} - \frac{e^{-\mu\pi}(1-\mu^2)}{2(1+\mu^2)}\right\} = -4ag\left\{-\frac{1}{2} + \frac{1-\mu^2}{2(1+\mu^2)}\right\}$$

or $$e^{-\mu\pi}\left(-\frac{1}{2} - \frac{1-\mu^2}{2(1+\mu^2)}\right) = \frac{1-\mu^2}{2(1+\mu^2)} - \frac{1}{2} \quad \text{or} \quad -\frac{e^{-\mu\pi}}{1+\mu^2} = -\frac{\mu^2}{1+\mu^2}$$

or $e^{-\mu\pi} = \mu^2$ or $\mu^2 e^{\mu\pi} = 1$

Ex. 2. *A particle slides in a vertical plane down a rough cycloidal arc whose axis is vertical and vertex downwards starting from a point where the tangent makes an angle* θ *with the horizon and coming to rest at the vertex; show that* $\mu e^{\mu\theta} = \sin\theta - \mu\cos\theta$. **(Purvanchel 2007)**

Sol. Proceed as in Art. 12.8 and get result (8), *i.e.*,

$$v^2 e^{-2\mu\psi} = C + 4ag\left[-\frac{e^{-2\mu\psi}}{2} + \frac{e^{-2\mu\psi}}{2(1+\mu^2)}\{2\mu \sin 2\psi + (1-\mu^2)\cos 2\psi\}\right] \quad ...(8)$$

Initially, given that, when $\psi = \theta$, $v = 0$. So (8) gives

$$C = -4ag\left[-\frac{e^{-2\mu\theta}}{2} + \frac{e^{-2\mu\theta}}{2(1+\mu^2)}\{2\mu \sin 2\theta + (1-\mu^2)\cos 2\theta\}\right] \quad ...(9)$$

Given, at vertex O, $v = 0$, $\psi = 0$. So (8) gives

$$C = -4ag\{(-1/2) + (1-\mu^2)/2(1+\mu^2)\} \quad ...(10)$$

Equating of two values of C from (9) and (10), we have

$$-\frac{e^{-2\mu\theta}}{2} + \frac{e^{-2\mu\theta}}{2(1+\mu^2)}\{2\mu\sin 2\theta + (1-\mu^2)\cos 2\theta\} = -\frac{1}{2} + \frac{1-\mu^2}{2(1+\mu^2)}$$

or $$\frac{e^{-2\mu\theta}}{2(1+\mu^2)}\{-1-\mu^2 + 2\mu\sin 2\theta + (1-\mu^2)\cos 2\theta\} = -\frac{2\mu^2}{2(1+\mu^2)}$$

or $-1 - \mu^2 + 4\mu\sin\theta\cos\theta + (1-\mu^2)(\cos^2\theta - \sin^2\theta) = -2\mu^2 e^{2\mu\theta}$

or $1 + \mu^2 - 4\mu\sin\theta\cos\theta - \cos^2\theta + \sin^2\theta + \mu^2\cos^2\theta - \mu^2\sin^2\theta = 2\mu^2 e^{2\mu\theta}$

or $(1 - \cos^2\theta) + \sin^2\theta + \mu^2(1 - \sin^2\theta) + \mu^2\cos^2\theta - 4\mu\sin\theta\cos\theta = 2\mu^2 e^{2\mu\theta}$

or $2\sin^2\theta + 2\mu^2\cos^2\theta - 4\mu\sin\theta\cos\theta = 2\mu^2 e^{2\mu\theta}$

or $(\sin\theta - \mu\cos\theta)^2 = \mu^2 e^{2\mu\theta}$ or $\sin\theta - \mu\cos\theta = \mu e^{\mu\theta}$.

Ex. 3. *A rough cycloid has its plane vertical and the line joining its cusps horizontal. A heavy particle slides down the curve from rest at a cusp and comes to rest at a point on the other side of the vertex where the tangent is inclined at 45° to the vertical. Show that the coefficient of friction satisfies the equation* $3\mu\pi + 4 \log (1 + \mu) = 2 \log 2$.

Hint : Proceed as in Art. 12.8 and get result (8). Then use conditions $v = 0$, $\psi = \pi/2$ and $v = 0$, $\psi = -(\pi/4)$ to get two values of C as in Ex. 1 and Ex. 2. Equate the two values of C and simplify to get the required result.

Ex. 4. *A particle starts from rest from cusp of a rough cycloid whose axis is vertical and vertex downwards. Show that its velocity at the vertex is to its velocity at the same point when the cycloid is smooth is* $(e^{-\mu\pi} - \mu^2)^{1/2} : (1 + \mu^2)^{1/2}$, *where* μ *is the coefficient of friction. Further show that the particle will certainly come to rest before reaching the vertex if* $\mu = 0.5$, *given that* $\log 2 = 0.69315$. **[Calcutta 2002]**

Sol. Proceed as in Art. 12.8 and get result (8), *i.e.*,

$$v^2 e^{-2\mu\psi} = C + 4ag\left[-\frac{e^{-2\mu\psi}}{2} + \frac{e^{-2\mu\psi}}{2(1+\mu^2)}\{2\mu \sin 2\psi + (1-\mu^2)\cos 2\psi\}\right] \quad ...(8)$$

Initially at cusp A, when $\psi = \pi/2$, $v = 0$. So (8) gives

$$0 = C + 4ag\{-(1/2)e^{-\mu\pi} - e^{-\mu\pi}(1 - \mu^2)/2(1 + \mu^2)\} \quad ...(9)$$

Subtracting (9) from (8), velocity at any point is given by

$$v^2 e^{-2\mu\psi} = 4ag\left[-\frac{e^{-2\mu\psi}}{2} + \frac{e^{-2\mu\psi}}{2(1+\mu^2)}\{2\mu \sin 2\psi + (1-\mu^2)\cos 2\psi\} + \frac{e^{-\mu\pi}}{2} + e^{-\mu\pi}(1-\mu^2)/2(1+\mu^2)\right] \quad ...(10)$$

If $v = v_1$ when $\psi = 0$ (*i.e.*, at vertex), then (10) gives

$$v_1^2 = 4ag\left[-\frac{1}{2} + \frac{1-\mu^2}{2(1+\mu^2)} + \frac{e^{-\mu\pi}}{2} + \frac{e^{-\mu\pi}(1-\mu^2)}{2(1+\mu^2)}\right]$$

or
$$v_1^2 = \{4ag/(1 + \mu^2)\}\,(e^{-\mu\pi} - \mu^2) \quad ...(11)$$

If v_2 is the velocity of the particle at the vertex when the cycloid is smooth, then putting $\mu = 0$ in (11), we get $v_2^2 = 4ag$. Then (11) gives

$$v_1^2 = \{v_2^2/(1 + \mu^2)\}\,(e^{-\mu\pi} - \mu^2) \quad \text{or} \quad v_1/v_2 = (e^{-\mu\pi} - \mu^2)^{1/2}/(1 + \mu^2)^{1/2}$$

Second part. If the particle comes to rest before reaching the vertex, then (11) suggests that

$$e^{-\mu\pi} - \mu^2 < 0 \quad \text{or} \quad \mu^2 > e^{\mu\pi} \quad \text{or} \quad \mu e^{\mu\pi/2} > 1 \quad ...(12)$$

or $\log \mu + (\mu\pi/2) \log e > \log 1$ or $\log (1/2) + (1/4) \times (22/7) > 1$

[$\because \mu = 0.5 = 1/2$ and $\pi = 22/7$)

or $(11/7) - \log 2 > 0$ or $0.785 - 0.69315 > 0$,

which is true. So if $\mu = 0.5$, the particle comes to rest before reaching the vertex.

Ex. 5. *A particle starts from rest from the cusp of a rough cycloid whose axis is vertical and vertex downwards. Show that it will come to rest before reaching the vertex if* $\mu e^{\mu\pi/2} > 1$.

Sol. Refer second part of Ex. 4 upto equation (12).

Ex. 6. *A particle slides on a rough wire in the form of the cycloid* $s = 4a \sin \psi$, *which is fixed in a vertical plane with the axis vertical and vertex downwards. It is projected from the vertex* ($\psi = 0$) *with speed* u, *so it comes to rest at the cusp* ($\psi = \pi/2$). *Show that* $e^{\mu\pi} = \mu^2 + (\mu^2 + 1)(u^2/4ag)$, μ *being the coefficient of friction.*

Sol. Draw figure as in Art. 12.8 in which indicate force μR along PT in place of PT' for the given problem. Proceed as in Art. 12.8. by replacing μ by $-\mu$ utpto relation (8). Thus, we shall get

$$v^2 e^{2\mu\psi} = C + 4ag\left[-\frac{e^{2\mu\psi}}{2} + \frac{e^{2\mu\psi}}{2(1+\mu^2)}\{-2\mu \sin 2\psi + (1-\mu^2)\cos 2\psi\}\right] \quad ...(8)$$

Initially at vertex, $v = u$ when $\psi = 0$. So (8) gives

$$u^2 = C + 4ag\{-(1/2) + (1-\mu^2)/2(1+\mu^2)\} \quad ...(9)$$

Given that at cusp $\psi = \pi/2$ and $v = 0$. So (9) gives

$$0 = C + 4ag\{(-1/2)\, e^{\mu\pi} - e^{\mu\pi}(1-\mu^2)/2(1+\mu^2)\} \quad ...(10)$$

Subtracting (10) from (9), we get

$$u^2 = 4ag\{-(1/2) + (1-\mu^2)/2(1+\mu^2) + (1/2)e^{\mu\pi} + e^{\mu\pi}(1-\mu^2)/2(1+\mu^2)\}$$

or $$u^2/4ag = -\mu^2/(1+\mu^2) + e^{\mu\pi}/(1+\mu^2) \quad \text{or} \quad e^{\mu\pi} = \mu^2 + (\mu^2+1)(u^2/4ag)$$

Ex. 7. *A particle slides down a rough cycloid whose axis is vertical and vertex downward. Prove that the time of reaching a certain point of the cycloid is independent of the starting point.*

Prove also that if λ is the angle of friction, and if the tangent at the starting point makes with the horizontal an angle greater than α, where α is the least positive angle, which satisfies the equation $\sin(\alpha - \lambda) = e^{(\alpha+\lambda)\tan\lambda} \sin 2\lambda$, then the particle will oscillate.

12.9 Motion of a particle on any rough plane curve

[Kanpur 2012, 2017]

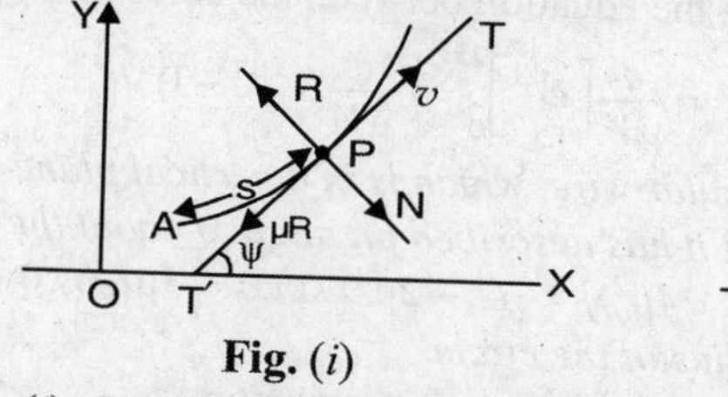

Fig. (i)

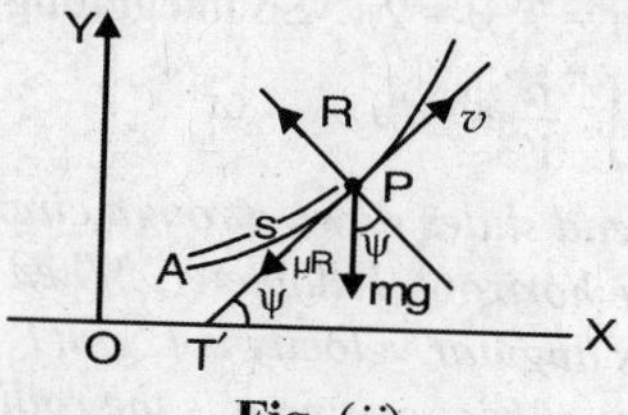

Fig. (ii)

Refer figure (*i*). Let s be the arc measured from a fixed point A on the curve, in the direction of motion, to the position of the particle P of mass m at any time t. Let R be the pressure measured inwards, μ the coefficient of friction and T, N components of the external force on the particle along the tangent and the outward normal. Then the equations of motion are

$$mv(dv/ds) = T - \mu R \quad \text{and} \quad m(v^2/\rho) = R - N \quad ...(1)$$

In case in which a heavy particle is sliding upwards on the curve in a vertical plane as shown in figure (*ii*), (1) become

$$mv(dv/ds) = -mg\sin\psi - \mu R \quad ...(2)$$

and $$m(v^2/\rho) = R - mg\cos\psi, \quad ...(3)$$

where ψ is the inclination of the tangent to the horizontal.

Substituting the value of R given by (3) in (2), we get

$$mv(dv/ds) = -mg\sin\psi - \psi\{m(v^2/\rho) + mg\cos\psi\}$$

or $$dv^2/d\psi + 2\mu v^2 = -2g\rho(\sin\psi + \mu\cos\psi) \quad ...(4)$$

From the intrinsic equation of the curve, say $s = f(\psi)$, we get $\rho = f'(\psi)$ and the linear equation (4) has an integrating factor $e^{2\mu\psi}$. Hence its solution is

$$v^2 e^{2\mu\psi} = -2g\int e^{2\mu\psi} f'(\psi)(\sin\psi + \mu\cos\psi)\,d\psi + C, \ C \text{ is a constants.} \quad ...(5)$$

But $v^2 = (ds/dt)^2 = (ds/d\psi)^2 (d\psi/dt)^2$. So we can get t in terms of ψ.

Remark. In applying the method of this article it should be noted carefully that the friction must always oppose the motion, and that in the case of a two-sided constraint (*e.g.* bead on a wire), if R changes sign during the motion, the motion will cease to be represented by (4), because it will become necessary to change the sign of R in (3) but not in (2).

TYPE 6 : SOLVED EXAMPLES BASED ON ART. 12.9

Ex. 1. *A particle is projected along the inner surface of a rough sphere and is acted on by no forces. Show that it will return to the point of projection at the end of time $(a/\mu V)(e^{2\mu\pi} - 1)$, where a is the radius of the sphere, V is the velocity of projection and μ is the coefficient of friction.*

Sol. Let A be the point of projection. Let P be the position of the particle at any time t such that $\angle AOP = \theta$ and are $AP = s$ so that $s = a\theta$ so that $ds/d\theta = a$. Since there are no forces, the particle will continue to move in the plane section *i.e.*, in a circle. Thus, ρ = radius of curvature of circle $= a$. The only external forces are R and μR as shown in the figure. So the equations of motion of the particle are given by

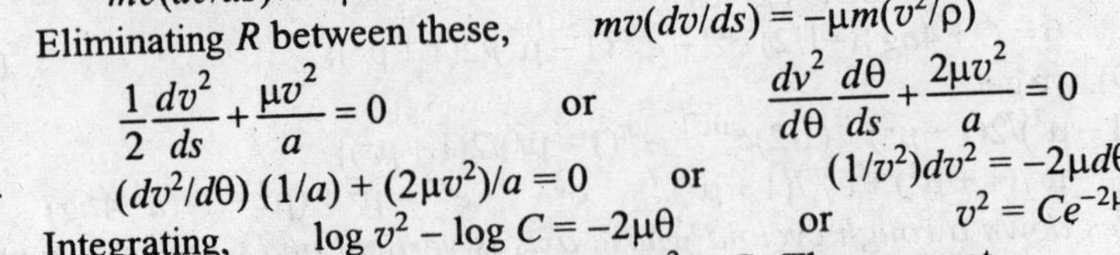

$$mv(dv/ds) = -\mu R \quad \text{and} \quad mv^2/\rho = R$$

Eliminating R between these, $\quad mv(dv/ds) = -\mu m(v^2/\rho)$

or $\quad \dfrac{1}{2}\dfrac{dv^2}{ds} + \dfrac{\mu v^2}{a} = 0 \quad$ or $\quad \dfrac{dv^2}{d\theta}\dfrac{d\theta}{ds} + \dfrac{2\mu v^2}{a} = 0$

or $\quad (dv^2/d\theta)(1/a) + (2\mu v^2)/a = 0 \quad$ or $\quad (1/v^2)dv^2 = -2\mu d\theta$

Integrating, $\quad \log v^2 - \log C = -2\mu\theta \quad$ or $\quad v^2 = Ce^{-2\mu\theta}$

Initially at A, when $\theta = 0$, $v = V$. So $V^2 = C$. Then, we get

$v^2 = V^2 e^{-2\mu\theta} \quad$ or $\quad v = V e^{-\mu\theta} \quad$ or $\quad (ds/dt)\, e^{\mu\theta} = V$

or $\quad \dfrac{ds}{d\theta}\dfrac{d\theta}{dt} e^{\mu\theta} = V \quad$ or $\quad a\dfrac{d\theta}{dt} e^{\mu\theta} = V \quad$ or $\quad dt = \dfrac{a}{V} e^{\mu\theta}\, d\theta$

Let T be the time after which the particle will return to the point of projection so that when $t = 0$, $\theta = 0$ and when $t = T$, $\theta = 2\pi$. So integrating the equation between the coresonding limits we get

$$\int_0^T dt = \int_0^{2\pi} \frac{a}{V} e^{\mu\theta}\, d\theta \quad \text{or} \quad T = \frac{a}{\mu V}\left[e^{\mu\theta}\right]_0^{2\pi} = \frac{a}{\mu V}(e^{2\mu\pi} - 1)$$

Ex. 2. *A bead slides down a rough circular wire, which is in a vertical plane starting from rest at the end of the horizontal diameter. When it has described an angle θ about the centre, show that the square of its angular velocity is $\{2g/a(1 + 4\mu^2)\} \times \{(1 - 2\mu^2)\sin\theta + 3\mu(\cos\theta - e^{-2\mu\theta})\}$, where μ is the coefficient of friction and a is the radius of the circle.*

Sol. Let AB be the horizontal diameter and the particle starts from A. Let P be its position at any time t such that $\angle AOP = \theta$ and arc $AP = s$ so that $s = a\theta$ and $ds/d\theta = a$. Also for circle of radius a, ρ = radius of curvature = a. The external forces acting on the particle are R, μR and mg as shown in the figure. Hence the usual equations of motion of the particle along tangent and normal are given by

$$mv(dv/ds) = mg\cos\theta - \mu R \quad \text{and} \quad mv^2/\rho = R - mg\sin\theta$$

Eliminating R between the above equations,

$$mv(dv/ds) = mg\cos\theta - \mu(mv^2/a + mg\sin\theta), \text{ as } \rho = a$$

or $\quad (1/2) \times (dv^2/ds) + \mu v^2/a = g(\cos\theta - \mu\sin\theta)$

or $\quad \dfrac{dv^2}{d\theta}\dfrac{d\theta}{ds} + \dfrac{2\mu v^2}{a} = 2g(\cos\theta - \mu\sin\theta)$

or $\quad (dv^2/d\theta) \times (1/a) + (2\mu v^2)/a = 2g(\cos\theta - \mu\sin\theta)$, as $ds/d\theta = a$

or $\quad dv^2/d\theta + 2\mu v^2 = 2ga(\cos\theta - \mu\sin\theta)$,

which is linear diff. eqn. whose I.F. $= e^{2\mu\theta}$ and so solution is

$$v^2 e^{2\mu\theta} = C + 2ga\int e^{2\mu\theta}(\cos\theta - \mu\sin\theta)\, d\theta$$

or $\quad v^2 e^{2\mu\theta} = C + 2ga\left[\int e^{2\mu\theta}\cos\theta\, d\theta - \mu\int e^{2\mu\theta}\sin\theta\, d\theta\right] \quad \ldots(1)$

To integrate the R.H.S. of (1), we use the following results :

$$\int e^{ax}\sin bx\, dx = e^{ax}(a\sin bx - b\cos bx)/(a^2 + b^2)$$

and $\quad \displaystyle\int e^{ax}\cos bx\, dx = e^{ax}(a\cos bx + b\sin bx)/(a^2 + b^2)$

$$\therefore\ v^2 e^{2\mu\theta} = C + 2ga\left[\frac{e^{2\mu\theta}(2\mu\cos\theta + \sin\theta)}{1 + 4\mu^2} - \frac{\mu e^{2\mu\theta}(2\mu\sin\theta - \cos\theta)}{1 + 4\mu^2}\right]$$

or $\quad v^2 e^{2\mu\theta} = C + 2ga\, e^{2\mu\theta}\{3\mu\cos\theta + (1 - 2\mu^2)\sin\theta\}/(1 + 4\mu^2) \quad \ldots(2)$

Initially at A, when $\theta = 0$, $v = 0$. So (2) gives $C = -(6ag\mu)/(1 + 4\mu^2)$.

$$\therefore \text{ (2) gives } \quad v^2 e^{2\mu\theta} = -\frac{6ag\mu}{1 + 4\mu^2} + \frac{2ga\, e^{2\mu\theta}}{1 + 4\mu^2}\{3\mu\cos\theta + (1 - 2\mu^2)\sin\theta\}$$

or $\quad v^2 = \{2ag/(1+4\mu^2)\} \times [-3\mu e^{-2\mu\theta} + 3\mu \cos\theta + (1-2\mu^2)\sin\theta\}$

or $\quad v^2 = \{2ag/(1+4\mu^2)\} \times \{(1-2\mu^2)\sin\theta + 3\mu(\cos\theta - e^{-2\mu\theta})\}$ ...(3)

But $s = a\theta$ so $v = ds/dt = a\, d\theta/dt$. So (3) gives

$$a^2(d\theta/dt)^2 = \{(2ag)/(1+4\mu^2)\} \times \{(1-2\mu^2)\sin\theta + 3\mu(\cos\theta - e^{-2\mu\theta})\}$$

or $$(d\theta/dt)^2 = \{2g/a(1+4\mu^2)\} \times \{(1-2\mu^2)\sin\theta + 3\mu(\cos\theta - e^{-2\mu\theta})\},$$

which gives the required square of the angular velocity.

Ex. 3. *A heavy bead slides on a rough circular wire of radius a in a vertical plane, the coefficient of friction being $1/\sqrt{2}$. The bead is projected from the lowest point; show that in order that it may just reach the highest point the velocity of projection must be $\{(2/3)\sqrt{6}\, ga\, e^{\sqrt{2}(\alpha+\pi/2)}\}^{1/2}$, where α is the acute angle $\tan^{-1}\sqrt{2}$.*

Sol. Let the particle be projected from the lowest point A of the circular wire with velocity u. Let P be its position at any time such that $\angle AOP = \theta$ and arc $AP = s$ so that $s = a\theta$ and $ds/d\theta = a$. The external forces acting on the particle are reaction R normally towards centre O, friction μR tangentially, opposite to the direction of motion PT and mg vertically downwards. Then the equation of motion, in the early stages of the motion along tangent and normal are given by

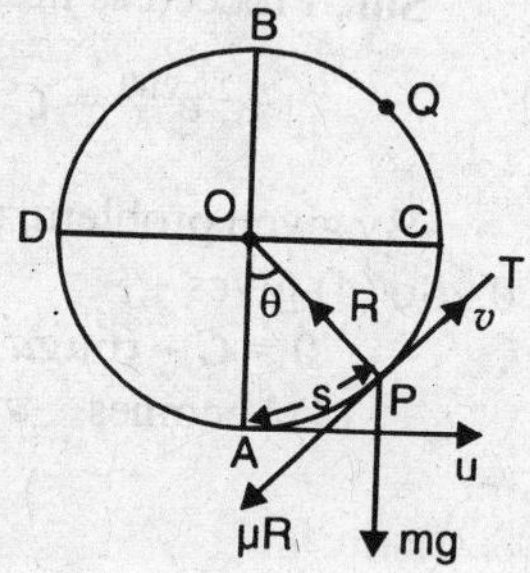

$$mv(dv/ds) = -mg\sin\theta - \mu R \quad ...(1)$$

and $\quad m(v^2/a) = R - mg\cos\theta$, as $\rho = a$ for circle ...(2)

Eliminating R between (1) and (2), we get

$$mv(dv/ds) = -mg\sin\theta - \mu(mg\cos\theta + mv^2/a)$$

or $\quad (1/2)\times(dv^2/ds) + mv^2/a = -g(\sin\theta + \mu\cos\theta)$

or $$\frac{dv^2}{d\theta}\frac{d\theta}{ds} + \frac{2\mu v^2}{a} = -2g(\sin\theta + \mu\cos\theta)$$

or $\quad dv^2/d\theta + 2\mu v^2 = -2ga(\sin\theta + \mu\cos\theta)$, ...(3)

[$\because$ $ds/d\theta = a \Rightarrow d\theta/ds = 1/a$]

It is linear differential equation whose I.F. $= e^{2\mu\theta}$ and solution is

$$v^2 e^{2\mu\theta} = C - 2ga\int e^{2\mu\theta}(\sin\theta + \mu\cos\theta)\, d\theta$$

$$= C - 2ag\left[\frac{e^{2\mu\theta}(2\mu\sin\theta - \cos\theta)}{1+4\mu^2} + \frac{\mu e^{2\mu\theta}(2\mu\cos\theta + \sin\theta)}{1+4\mu^2}\right],$$

$\therefore$ $$v^2 e^{2\mu\theta} = C - \frac{2ag\, e^{2\mu\theta}}{1+4\mu^2}\{3\mu\sin\theta - (1-2\mu^2)\cos\theta\} \quad ...(4)$$

or $$v^2 e^{\theta\sqrt{2}} = C - ga\sqrt{2}\, e^{\theta\sqrt{2}}\sin\theta, \text{ as } \mu = 1/\sqrt{2} \text{ (given)} \quad ...(4)'$$

Initially at A, when $\theta = 0$, $v = u$. Then (4)′ gives $C = u^2$.

So (4)′ becomes $$v^2 e^{\theta\sqrt{2}} = u^2 - ga\sqrt{2}\, e^{\theta\sqrt{2}}\sin\theta \quad ...(5)$$

According to the given data v decreases and vanishes at the highest point B where $\theta = \pi$. So there will be an angle θ between O and π for which $\cos\theta = -v^2/ag$ and from (2) for this value of θ, the pressure R vanishes and changes sign. For this point Q (say) onwards (3) ceases to represent the motion, because (1) still holds good, but instead of (2), we now have

$$mv^2/a = -R - mg\cos\theta \quad ...(6)$$

Eliminating R from (1) and (6) and doing as before, we get

$$dv^2/d\theta^2 - 2\mu v^2 = -2ga(\sin\theta - \mu\cos\theta) \quad ...(7)$$

which is linear diff. equation. Solving as before, we finally get

$$v^2 e^{-2\mu\theta} = A + \frac{2ga\, e^{-2\mu\theta}}{1+4\mu^2}\{3\mu\sin\theta + (1-2\mu^2)\cos\theta\}$$

or $$v^2 e^{-\theta\sqrt{2}} = A + ga\sqrt{2}\, e^{-\theta\sqrt{2}}\sin\theta, \text{ as } \mu = 1/\sqrt{2} \quad ...(8)$$

At the highest point B, when $\theta = \pi$, $v = 0$. So (8) gives $A = 0$.

$\therefore$ (8) becomes $v^2 e^{-\theta\sqrt{2}} = ga\sqrt{2}\, e^{-\theta\sqrt{2}} \sin\theta$ or $v^2 = ga\sqrt{2}\sin\theta$ ...(9)

Since the motion is continuous, the values of v^2 given by (5) and (9) must be the same at the point at which $R = 0$ *i.e.*, when v^2 is also $-ga\cos\theta$. Putting this value of v^2 in (9), we get

$$-\cos\theta = \sqrt{2}\sin\theta \quad \text{or} \quad (\cos\theta)/-\sqrt{2} = (\sin\theta)/1 = 1/\sqrt{3},$$

showing that $\theta = \alpha + \pi/2$, where α is the acute angle $\tan^{-1}\sqrt{2}$. Then from (5) and (9), we get

$$ga\sqrt{2}\, e^{\theta\sqrt{2}} \sin\theta = u^2 - ga\sqrt{2}\, e^{\theta\sqrt{2}} \sin\theta \quad \text{or} \quad u^2 = 2\sqrt{2}ga\, e^{\theta\sqrt{2}} \sin\theta$$

or $\quad u^2 = 2\sqrt{2}ga\, e^{\sqrt{2}(\alpha+\pi/2)} . (1/\sqrt{3})$ or $u = \{(2/3)\sqrt{6}ga\, e^{\sqrt{2}(\alpha+\pi/2)}\}^{1/2}$

Ex. 4. *A ring which can slide on a rough circular wire in a vertical plane is projected from the lowest point with such a velocity as will take it to the horizontal diameter. If the ring returns to the lowest point, show that its velocity on arrival is to its velocity of projection as* $(1 - 2\mu^2 - 3\mu\, e^{-\mu\pi})^{1/2} : (1 - 2\mu^2 + 3\mu\, e^{\mu\pi})^{1/2}$.

Sol. Proceed as in Ex. 3 (noting that here $\mu \neq 1/\sqrt{2}$) upto relation (4), *i.e.*,

$$v^2 e^{2\mu\theta} = C - \frac{2ag\, e^{2\mu\theta}}{1 + 4\mu^2}\{3\mu\sin\theta - (1 - 2\mu^2)\cos\theta\} \qquad ...(4)$$

By given problem, at the end C of the horizontal diameter of the circular wire, when $\theta = \pi/2$, $v = 0$. So (4) gives

$$0 = C - (6\mu ga\, e^{\mu\pi})/(1 + 4\mu^2) \quad \text{or} \quad C = (6\mu ga\, e^{\mu\pi})/(1 + 4\mu^2)$$

$\therefore$ (4) becomes $v^2 e^{2\mu\theta} = (6\mu ga\, e^{\mu\pi})/(1 + 4\mu^2)$

$$- \frac{2ag\, e^{2\mu\theta}}{1 + 4\mu^2}\{3\mu\sin\theta - (1 - 2\mu^2)\cos\theta)\} \qquad ...(5)$$

Initially at A, when $\theta = 0$, $v = u$ (say). Then (5) gives

$$u^2 = \{2ga(1 - 2\mu^2 + 3\mu\, e^{\mu\pi})\}/(1 + 4\mu^2) \qquad ...(6)$$

After reaching at C the ring retraces its circular path CA. Let at any time t particle be at P where $\angle AOP = \theta$. Then we shall have the same external forces except that the direction of the force of friction will be opposite to that it was while moving from A to C. Hence the velocity at any point can be obtained by replacing μ by $-\mu$ in (4). If u' be the velocity when the ring returns to A where $\theta = 0$, then writing $-\mu$ for μ in (6), we have

$$u'^2 = \{2ga(1 - 2\mu^2 - 3\mu\, e^{-\mu\pi}\}/(1 + 4\mu^2) \qquad ...(7)$$

From (6) and (7), $\quad u' : u = (1 - 2\mu^2 - 3\mu\, e^{-\mu\pi})^{1/2} : (1 - 2\mu^2 + 3\mu\, e^{\mu\pi})\}^{1/2}$

Ex. 5. *A bead moves along a rough curved wire which is such that its direction of motion is with constant angular velocity. Show that a possible form of wire is an equiangular spiral.*

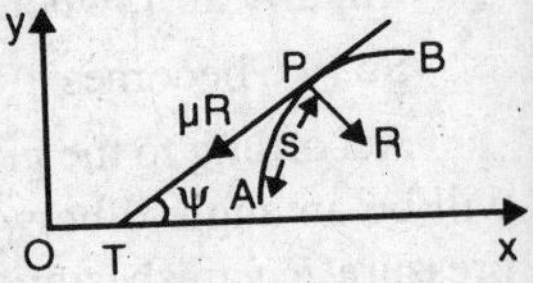

Sol. Let a bead move on rough wire AB and let at any time t its position be $P(s, \psi)$, where arc $AP = s$ and ψ is the angle which tangent PT makes with OX. Then equations of motion of the bead are

$$mv(dv/ds) = -\mu R \quad \text{and} \quad m(v^2/\rho) = R$$

Eliminating R beteween them, $\quad mv(dv/ds) = -\mu(mv^2/\rho)$

or $\quad \dfrac{dv}{ds}\rho = -\mu v \quad$ or $\quad \dfrac{dv}{ds}\dfrac{ds}{d\psi} = -\mu v \quad$ or $\quad \dfrac{dv}{v} = -\mu\, d\psi$

Integrating, $\quad \log v = -\mu\psi = \log C \quad$ or $\quad v = C\, e^{-\mu\psi}$

or $\quad \dfrac{ds}{dt} = C\, e^{-\mu\psi} \quad$ or $\quad \dfrac{ds}{d\psi}\dfrac{d\psi}{dt} = C\, e^{-\mu\psi} \qquad ...(1)$

Since the bead changes its direction of motion ψ with constant angular velocity, we have $d\psi/dt =$ constant $= k$, say.

$\therefore$ (1) becomes $\quad (ds/d\psi)\cdot k = C\, e^{-\mu\psi} \quad$ or $\quad k\, ds = C\, e^{-\mu\psi}\, d\psi$

Integrating, $\quad ks = -(C/\mu)\, e^{-\mu\psi} + d \quad$ or $\quad s = -(C/k\mu)\, e^{-\mu\psi} + (d/C)$

which is of the form $s = a + b\, e^{-\mu\psi}$ which represent the intrisic equation of equiangular spiral.

EXERCISE

1. A particle is projected horizontally with velocity V along the inside of a rough vertical circle from the lowest point. Prove that if it completes the circuit, it will return to the lowest point with a velocity v given by

$$v^2 = V^2 e^{-4\pi\mu} + \{2ag/(1+4\mu^2)\}(1-2\mu^2)(1-e^{-4\pi\mu})$$

2. A particle is projected horizontally from the lowest point of a rough sphere of radius a. After describing an arc less than a quadrant, it returns and comes to rest at the lowest point. Show that the initial velocity must be $\sin\alpha\,\{2ga(1+\mu^2)/(1-2\mu^2)\}^{1/2}$, where μ is the coefficient of friction and α is the arc through which the particle moves.

3. A small heavy bead is threaded on a rough circular wire which is clamped at the ends of a diameter, so that its plane is horizontal. The bead is projected with velocity V from one end of the clamps and just comes to rest at the other. Prove that the velocity of projection is $(ga \sinh 2\mu\pi)^{1/2}$, where a is the radius of the wire.

4. A particle falls from a position of limiting equilibrium near the top of a nearly smooth sphere. Show that it will leave the sphere at the point whose radius is inclined to the vertical at an angle $\alpha + \mu\{2 - (4\alpha/3 \sin\alpha)\}$, where $\cos\alpha = 2/3$ and μ is the small coefficient of friction.

5. A bead slides on a rough circular wire which is fixed in a vertical plane, the coefficient of friction between the bead and the wire being μ. If the bead starting from rest at the level of the centre of the circle comes to rest at the lowest point, prove that $3\mu = e^{\mu\pi}(1-2\mu^2)$.

12.10 Type 6 : Examples based on motion on a smooth curve under resistance

Ex. 1. (a) *A heavy particle of mass m slides on a smooth arc of a cycloid in a medium whose resistance is $mv^2/2c$, v being the velocity of the particle and c being the distance of the starting point from the vertex. If the axis is vertical and vertex upwards, find the velocity of the particle at the cusp.* **[I.A.S. 2002]**

(b) *A heavy bead of mass m slides on a smooth wire in the shape of a cycloid whose axis is vertical and vertex upward in a medium whose resistance is $mv^2/2c$ and the distance of the starting point from the vertex is c. Find the time of descent to the cusp.*

Sol. (a) Let $P(s, \psi)$ be the position of the particle at any time t. Let v be its velocity there and R the normal reaction of the curve on it. The other forces acting on it are its weight mg vertically downwards and resistance $mv^2/2c$ along the tangent PT in the direction opposite to the motion of the particle. Then the equation of motion along the tangent is

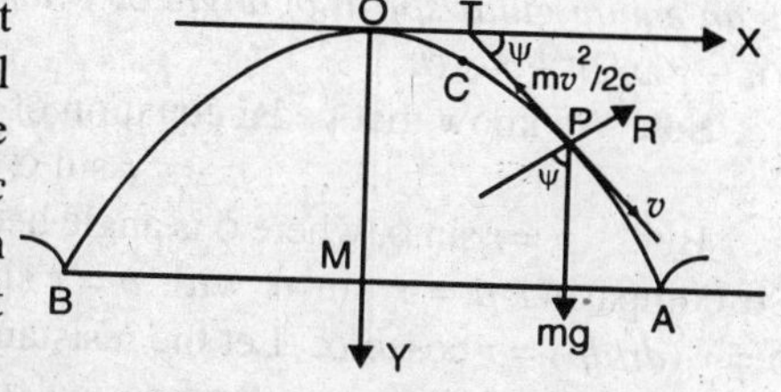

$$mv(dv/ds) = mg\sin\psi - (mv^2/2c) \quad \text{or} \quad dv^2/ds = 2g\sin\psi - (v^2/c) \qquad ...(1)$$

Intrinsic equation of the cycloid is $\qquad s = 4a\sin\psi. \qquad ...(2)$

Using (2), (1) gives $\qquad dv^2/ds + (v^2/c) = gs/2a, \qquad ...(3)$

which is linear equation whose I.F. $= e^{\int(1/c)\,ds} = e^{s/c}$ and solution is

$$v^2 e^{s/c} = A + \frac{g}{2a}\int s e^{s/c} ds = A + \frac{g}{2a}\left\{s(ce^{s/c}) - \int ce^{s/c} ds\right\}$$

or $\quad v^2 e^{s/c} = A + (g/2a)(sc\,e^{s/c} - c^2 e^{s/c})$, where A is a constant. ...(4)

Initially, when $s = c$ (given), $v = 0$. So (4) gives $A = 0$ and so by (4),

$v^2 e^{s/c} = (g/2a)\,c\,e^{s/c}(s-c) \quad$ or $\quad v = (gc/2a)^{1/2}(s-c)^{1/2}$ ...(5)

Let velocity at cusp A be v_1. Then, $v = v_1$ when $s = 4a$.

So (5) gives $\quad v_1 = (gc/2a)^{1/2}(4a-c)^{1/2}$, as required velocity.

(b) Proceed as in part (*a*) upto relation (5). Then proceed as follows. Re-writing (5), we get

$$\frac{ds}{dt} = \left(\frac{gc}{2a}\right)^{1/2}\cdot(s-c)^{1/2} \quad \text{or} \quad dt = \left(\frac{2a}{gc}\right)^{1/2}(s-c)^{-1/2}\,ds. \qquad ...(6)$$

Let T be the time taken by the particle to reach from initial position (where $s = c$, $t = 0$) to cusp (where $s = 4a$, $t = T$). So integrating (6) between the corresponding limits, we have

$$[t]_{t=0}^{T}=\left(\frac{2a}{gc}\right)^{1/2}\left[\frac{(s-c)^{1/2}}{(1/2)}\right]_{c}^{4a} \quad \text{or} \quad T=\left(\frac{2a}{gc}\right)^{1/2}\cdot 2(4a-c)^{1/2}$$

or $$T = \{8a(4a-c)/gc\}^{1/2}, \text{ as required.}$$

Ex. 2. *If the resistance of the air to a particle's motion be n times its weight and the particle be projected horizontally with velocity V, show that the velocity of the particle when it is moving at an inclination ϕ to the horizontal is $V(1-\sin\phi)^{(n-1)/2}(1+\sin\phi)^{-(n+1)/2}$.*

Sol. Let $P(s, \psi)$ be the position of given particle of mass m at any time t. Let PT and PN be tangent and normal to the given curve at P. Then equations of motion along the tangent and normal are

$$mv(dv/ds) = mg\sin\psi - mng \qquad \ldots(1)$$

or $$mv^2/\rho = mg\cos\psi \qquad \ldots(2)$$

Re-writing (1), $$v\,dv = g(\sin\psi - n)\,(ds/d\psi)\,d\psi$$

or $$v\,dv = g(\sin\psi-n)\left(\frac{v^2}{g\cos\psi}\right)d\psi \qquad \left[\because \frac{ds}{d\psi}=\rho=\frac{v^2}{g\cos\psi}, \text{ by (2)}\right]$$

or $$(1/v)\,dv = (\tan\psi - n\sec\psi)\,d\psi$$

Integrating, $$\log v = C + \log\sec\psi - n\log(\sec\psi+\tan\psi) \qquad \ldots(3)$$

Initially at O, $\psi = 0$, $v = V$. So (3) gives $C = \log V$

$\therefore$ from (3), $$\log v = \log V + \log\sec\psi - n\log\{(1+\sin\psi)/\cos\psi\}$$

or $$\frac{v}{V}=\sec\psi\cdot\left(\frac{\cos\psi}{1+\sin\psi}\right)^{n}=\frac{\cos^{n-1}\psi}{(1+\sin\psi)^{n}}=\frac{(1-\sin^2\psi)^{(n-1)/2}}{(1+\sin\psi)^{n}}$$

or $$v = V(1-\sin\psi)^{(n-1)/2}(1+\sin\psi)^{-(n+1)/2},$$

showing that the velocity when $\psi = \phi$ is given by

$$v = V(1-\sin\phi)^{(n-2)/2}(1+\sin\phi)^{-(n+1)/2}$$

Ex. 3. *A particle moving in a resisting medium is acted upon by a central force μ/r^n; if the path be an equiangular spiral of angle α, whose pole is at the centre of force, show that the resistance is $\{(n-3)\mu/2r^n\}\cos\alpha$.* **[Kanpur 2015]**

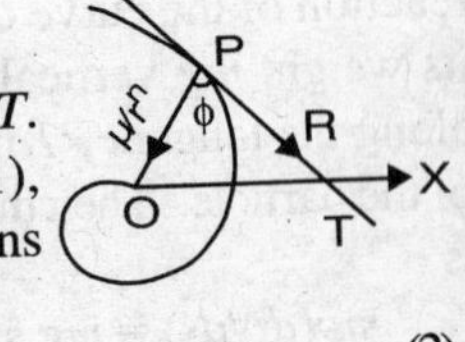

Sol. We know that pedal equation of an equiangular spiral of angle α is

$$p \;=\; r\sin\alpha \qquad \ldots(1)$$

But $p = r\sin\phi$, where ϕ is angle between radius vector OP and tangent PT. So comparing $p = r\sin\alpha$ with $p = r\sin\phi$, we find $\phi = \alpha$. Also, from (1), $\rho = r\cdot(dr/dp) = r\,\text{cosec}\,\alpha$. Let the resistance per unit mass be R. Then equations of motion along the tangent and normal are given by

$$v(dv/ds) = -(\mu/r^n)\cos\phi - R \quad \text{or} \quad v(dv/ds) = -(\mu/r^n)\cos\alpha - R \qquad \ldots(2)$$

and $$v^2/\rho = (\mu/r^n)\sin\phi \quad \text{or} \quad v^2/\rho = (\mu/r^n)\sin\alpha \qquad \ldots(3)$$

From (3), $$v^2 = (\mu/r^n)\sin\alpha\,.\,(r\,\text{cosec}\,\alpha) \quad \text{or} \quad v^2 = \mu/r^{n-1}$$

Differentiating both sides of the above equation w.r.t. 's', we get

$$2v\frac{dv}{ds}=-\frac{(n-1)\mu}{r^n}\frac{dr}{ds} \quad \text{or} \quad v\frac{dv}{ds}=-\frac{n-1}{2}\frac{\mu}{r^n}\cos\alpha \qquad \ldots(4)$$

$$[\because dr/ds = \cos\phi = \cos\alpha]$$

$\therefore$ from (2) and (4), $$-\frac{n-1}{2}\frac{\mu}{r^n}\cos\alpha = -\frac{\mu}{r^n}\cos\alpha - R$$

or $$R = -\frac{\mu}{r^n}\cos\alpha + \frac{n-1}{2}\frac{\mu}{r^n}\cos\alpha \quad \text{or} \quad R = \frac{n-3}{2}\times\frac{\mu\cos\alpha}{r^n}$$

Ex. 4. *A particle moves in a resisting medium with a given central acceleration P, the path of the particle being given, show that the resistance is $-(1/2p^2)\dfrac{d}{ds}\{p^3(dr/dp)P\}$.*

Sol. Let the particle be at P at any time t, PT and PN being tangent and normal there. OP is radius vector and PM is perpendicular to OP. Then $\angle OPT = \phi$ and so $\angle OPN = 90 - \phi$ and $\angle MPN = \phi$. Let v be the velocity of the particle at P. Its normal acceleration is v^2/ρ and transverse acceleration is $\frac{1}{r}\frac{d}{dt}\left(r^2 \frac{d\theta}{dt}\right)$. So equations of motion along normal PN and the transverse direction PM are given by

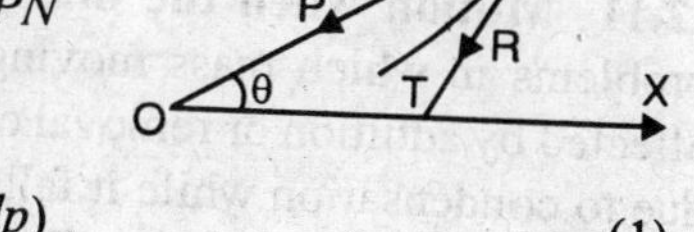

$$\frac{v^2}{\rho} = P \sin\phi \quad \text{or} \quad \frac{v^2}{r(dr/dp)} = \frac{Pp}{r} \quad \text{or} \quad v^2 = pP(dr/dp) \qquad \ldots(1)$$

and
$$\frac{1}{r}\frac{d}{dt}\left(r^2 \frac{d\theta}{dt}\right) = -R \sin\phi \qquad \ldots(2)$$

Let
$$r^2(d\theta/dt) = h = vp \qquad \ldots(3)$$

Then (2) becomes $(1/r)\,(dh/dt) = -(Rp/r)$ or $R = -(1/p)\,(dh/dt)$

or
$$R = -\frac{1}{p}\frac{dh}{ds}\frac{ds}{dt} = -\frac{v}{p}\frac{dh}{ds} = -\frac{h}{p^2}\frac{dh}{ds}, \text{ by (3)}$$

or
$$R = -\frac{1}{2p^2}\frac{dh^2}{ds} = -\frac{1}{2p^2}\frac{d}{ds}(v^2 p^2), \text{ using (3)}$$

or
$$R = -\frac{1}{2p^2}\frac{d}{ds}\left(p \cdot P \cdot \frac{dr}{dp} \cdot p^2\right) = -\frac{1}{2p^2}\frac{d}{ds}\left(p^3 \frac{dr}{dp} P\right), \text{ using (1)}$$

Ex. 5. (a) *If a body moves under a central force P in a medium which exerts a resistance equal to k times the velocity per unit of mass, prove that $u + (d^2u/d\theta^2) = (P/h^2u^2)\,e^{2kt}$, where h is the initial moment of momentum about the centre of force.*

(b) *A particle moves with central acceleration P in a medium of the resistance is k(velocity)^2. Show that the equation of its path is $d^2u/d\theta^2 + u = (P/h^2u^2)\,e^{2ks}$, where s the length of the arc described and h is the initial moment of momentum about the centre of force.*

Sol. (a) Draw a figure similar to Ex. 4. by replacing R by kv. Then proceed exactly as in Ex. 4 and get

$$v^2/\rho = P \sin\phi \qquad \ldots(1)$$

and
$$\frac{1}{r}\frac{d}{dt}\left(r^2 \frac{d\theta}{dt}\right) = -kv \sin\phi \qquad \ldots(2)$$

Let $r^2(d\theta/dt) = vp = H$ (say) = moment of momentum ...(3)

$\therefore$ (2) gives
$$\frac{1}{r}\frac{dH}{dt} = -k\frac{H}{p}\cdot\frac{p}{r}, \text{ using } p = r \sin\phi$$

or
$$dH/dt = -kH \quad \text{or} \quad (1/H)\,dH = -k\,dt$$

Integrating,
$$\log H = C - kt, \ C \text{ is a constant.} \qquad \ldots(4)$$

Initially, when $t = 0$, $H = h$ (given). So (4) gives $C = \log h$

$\therefore$ (4) becomes
$$\log H - \log h = -k \quad \text{or} \quad H = h\,e^{-kt} \qquad \ldots(5)$$

From (1),
$$P = \frac{v^2}{\rho \sin\phi} = \frac{H^2}{p^2} \times \frac{1}{r(dr/dp)} \times \frac{1}{(p/r)} \quad \text{or} \quad P = \frac{H^2}{p^3}\frac{dp}{dr} \qquad \ldots(6)$$

We know that $1/p^2 = u^2 + (du/d\theta)^2$, where $u = 1/r$

$$\therefore \ -\frac{2}{p^3}\frac{dp}{d\theta} = 2u\frac{du}{d\theta} + 2\frac{du}{d\theta}\frac{d^2u}{d\theta^2} \quad \text{or} \quad -\frac{2}{p^3}\frac{dp}{d\theta} = \frac{du}{d\theta}\left(u + \frac{d^2u}{d\theta^2}\right)$$

or
$$-\frac{1}{p^3}\frac{dp}{dr} = \frac{du}{dr}\left(u + \frac{d^2u}{d\theta^2}\right) \quad \text{or} \quad -\frac{1}{p^3}\frac{dp}{dr} = -\frac{1}{r^2}\left(u + \frac{d^2u}{dr}\right), \text{ as } u = \frac{1}{r}$$

or
$$(1/p^3)\,(dp/dr) = u^2(u + d^2u/d\theta^2) \qquad \ldots(7)$$

From (6) and (7), $P = H^2u^2\left(u + \frac{d^2u}{d\theta^2}\right) = h^2e^{-2kt}u^2\left(u + \frac{d^2u}{d\theta^2}\right)$, by (5)

or $(P/h^2u^2)\, e^{2kt} = u + (d^2u/d\theta^2)$

(*b*) Try yourself as in part (*a*).

12.11 Motion when the mass moving varies. In dynamics, we come across with number of problems in which mass moving varies with time. Accordingly, the motion of the body will be affected by addition or removal of the material. For example, a raindrop acquires an additional mass due to condensation while it falls through a cloud. Similarly, a rocket ejected loses its mass due to burning of fuel in the form of hot gases. In what follows, we propose to study such problems.

To derive the equation motion when mass varies **[Lucknow 1991]**

We know that the equation $P = mf$ holds only if the mass m remains constant throughout the motion. Let the moving mass does not remain constant during its motion. Let m and v be the mass and velocity of the moving mass at any time t. then, Newton's second law in its more fundamental form is given by ...(1)

$$P = d(mv)/dt$$

Suppose that a particle gains in time δt an incement δm of mass and that this increament δm was moving with a velocity u.

So in time δt the increment in the momentum of the particle

$$= m \cdot \delta v + \delta m\,(v + \delta v - u)$$

and the impulse of the force in this time is $P\,\delta t$.

Equating these, we get $\quad m\,\delta v + v\,\delta m + \delta m\,\delta v - u\,\delta m = P\,\delta t$

or $\quad m(\delta v/\delta t) + v(\delta m/\delta t) + \delta m(\delta v/\delta t) - u\,(\delta m/\delta t) = P$

Taking limit as $\delta t \to 0$ and noting that $\delta m(\delta v/\delta t) \to 0$, we get

$$m(dv/dt) + v(dm/dt) - u(dm/dt) = P$$

or $$d(mv)/dt = P + u(dm/dt) \qquad ...(2)$$

which is known as the *equation of motion when mass varies*

Particular case I : In derivation of (2), we assumed that the additional mass δm had velocity u in the same direction as that of velocity v of mass m. Now, if the additional mass δm moves in a direction just opposite to that of mass m, then new form of (2) will be

$$d(mv)/dt = P - u(dm/dt) \qquad ...(2A)$$

Particular case II : In derivation of (2), we assumed that the directions of v and P are the same. However, if they have opposite directions, then (2) will be of the form

$$d(mv)/dt = -P + u(dm/dt) \qquad ...(2B)$$

Remark. In some problems we shall use results (2), (2A), (2B) directly. Hence a proper choice of these results must be made in solving problems.

TYPE 7 : SOLVED EXAMPLES BASED ON ART. 12.11

Ex. 1. *A particle of mass M is at rest and begins to move under the action of a constant force F n a fixed direction. It encounters resistance of a stream of fine dust moving in the opposite direction with velocity V, which deposits on it at a constant rate ρ. Show that its mass will be m when it has travelled a distance $(k/\rho^2)\,[m - M\{1 + \log(m/M)\}]$; where $k = F - \rho V$.* **[I.A.S. 2001]**

Sol. When the mass of the particle is m, suppose that it gains in time δt an increment δm of mass and that this increment was moving with velocity V in the direction opposite to that of mass.

In time δt the increment in the momentum of the particle

$$= m\,\delta v + \delta m\{v + \delta v - (-V)\}$$

and the impulse of the force in this time is $F\,\delta t$.

Equating these, $\quad m\,\delta v + v\,\delta m + \delta m\,\delta v + V\,\delta m = F\,\delta t \qquad ...(1)$

Dividing throughout by δt and proceeding to the limit as $\delta t \to 0$, (1) gives

$$m(dv/dt) + (v + V)(dm/dt) = F, \quad \text{as} \quad \lim_{\delta t \to 0} \delta v(\delta m/\delta t) = 0$$

where $m = M + \rho t$, since the matter deposited in time t is ρt. Replacing m by $M + \rho t$ in the above equation, we get

$$(M + \rho t)\frac{dv}{dt} + (v + V)\frac{d}{dt}(M + \rho t) = F$$

or $$(M + \rho t)\frac{dv}{dt} + v\frac{d}{dt}(M + \rho t) + V\frac{d}{dt}(M + \rho t) = F$$

or $$\frac{d}{dt}\{v(M + \rho t)\} = F - V\frac{d}{dt}(M + \rho t) = F - V\rho, \text{ as } \frac{dM}{dt} = 0$$

or $$\frac{d}{dt}\{v(M + \rho t)\} = k, \text{ as } F - \rho V = k \text{ (given)}$$

Integrating w.r.t. 't', $v(M + \rho t) = C + kt$, where C is a constant. ...(2)

Initially when $t = 0$, $v = 0$. So (2) gives $C = 0$ and (2) becomes

$$v(M + \rho t) = kt \quad \text{or} \quad v = \frac{ds}{dt} = \frac{kt}{M + \rho t} \quad \text{or} \quad ds = \frac{kt\,dt}{M + \rho t}$$

Integrating, $s = A + \dfrac{k}{\rho}\displaystyle\int \frac{\rho t + M - M}{M + \rho t}\,dt$, where A is a constant.

or $$s = A + (k/\rho)\{t - (M/\rho)\log(M + \rho t)\} \quad ...(3)$$

Initially when $t = 0$, $s = 0$. So (3) gives $A = (Mk/\rho^2)\log M$

$\therefore$ (3) gives $s = (Mk/\rho^2)\log M + (k/\rho) \times \{t - (M/\rho)\log(M + \rho t)\}$

or $s = (k/\rho^2)(\rho t - M\log m) + (Mk/\rho^2)\log M$, as $M + \rho t = m$

$= (k/\rho^2)\{m - M - M\log(m/M)\}$ as $\rho t = m - M$

$= (k/\rho^2)[m - M\{1 + \log(m/M)\}]$, where $k = F - \rho V$

Ex. 2. *A rocket initially of total mass M, throws off every second a mass fM with constant velocity V relative to the rocket. Show that it cannot rise at once unless fV > g and that it cannot rise at all unless fV > λg, where λM is the mass of the case of the rocket.* **[Lucknow 1995, 1998]**

Sol. Let u, u' be the velocities communicated to the rocket and the mass thrown off in unit time by the explosion; then the relative velocity of the mass $= V = u' - (-u) = u' + u$. When the mass fM is thrown off, the remaining mass of the rocket $= M - fM = (1 - f)M$. At the start of explosion, we have

Momentum of the rocket = momentum of the mass thrown off

i.e., $(1 - f)Mu = fMu' = fM(V - u)$, as $V = u' + u$

Thus, $u = fV$ ...(1)

Since the gravity in unit time gives a velocity g, so the rocket will not rise at once unless

$g < u$ or $g < fV$, by (1)

Now, when all the powder is burnt (except the last element), the law of conservation of momentum gives

$\lambda Mu = fMu'$, where λM is the mass of the case

or $\lambda Mu = fM(V - u)$ or $u = fV/(\lambda + f)$ ...(2)

Now the velocity imparted by the gravity in unit time being g, rocket will not start at all unless

$g < u$ or $g < fV/(\lambda + f)$ or $fV > \lambda g + fg$

or $fV > \lambda g$, as required.

Ex. 3. (a) *A spherical raindrop, falling freely, receives in each instant an increase in volume equal to λ times its surface at that instant; find the velocity at the end of time t, and the distance fallen through in that time.* **[Guwahati 2007]**

Sol. When the raindrop has fallen through a distance x in time t, let its radius be r and its mass M. Then, we have

$$\frac{d}{dt}\left(M\frac{dx}{dt}\right) = Mg \quad ...(1)$$

Given that $$dM/dt = \rho \cdot 4\pi r^2 \lambda \quad ...(2)$$

Then $M = \frac{4}{3}\pi\rho r^3 \Rightarrow \frac{dM}{dt} = 4\pi\rho r^2 \frac{dr}{dt} \Rightarrow 4\pi\rho\lambda r^2 = 4\pi\rho r^2 \frac{dt}{dt}$, by (2)

$\therefore \quad dr/dt = \lambda$ and so $r = \lambda t + C$, where C is a constant. ...(3)

Initially, when $t = 0, r = a$, say. So (3) gives $C = a$.

Then (3) reduces to $r = a + \lambda t$ and so (1) becomes

$$\frac{d}{dt}\left[\frac{4}{3}\pi(a+\lambda t)^3 \rho \frac{dx}{dt}\right] = \frac{4}{3}\pi(a+\lambda t)^3 \rho g \text{ or } \frac{d}{dt}\left[(a+\lambda t)^3 \frac{dx}{dt}\right] = (a+\lambda t)^3 g \quad ...(4)$$

Integrating (4) w.r.t 't', we get

$(a + \lambda t)^3 \times (dx/dt) = \{(a + \lambda t)^4 g\}/4\lambda + A$, where A is a constant. ...(5)

Initially when $t = 0$, $dx/dt = 0$ (given). So (5) $\Rightarrow A = -a^4 g/4\lambda$

$$\therefore \text{(5) gives} \quad \frac{dx}{dt} = \frac{g}{4\lambda}\left[a + \lambda t - \frac{a^4}{(a+\lambda t)^3}\right], \quad ...(6)$$

which gives the velocity at the end of time t.

$$\text{Integrating (6),} \quad x = \frac{g}{4\lambda^2}\left[\frac{(a+\lambda t)^2}{2} + \frac{a^4}{2(a+\lambda t)^2}\right] + B, \ B \text{ is a const.} \quad ...(7)$$

Initially when $t = 0, x = 0$. So (7) gives $B = -ga^2/4\lambda^2$. So (7) gives

$$x = \frac{g}{8\lambda^2}\left[(a+\lambda t)^2 - 2a^2 + \frac{a^4}{(a+\lambda t)^2}\right] = \frac{g}{8\lambda^2}\left[a + \lambda t - \frac{a^2}{a+\lambda t}\right]^2$$

or $\quad x = (gt^2/8)\ \{(2a + \lambda t)/(a + \lambda t)\}^2$ ...(8)

which gives the distance fallen through in time t.

Ex. 3. (b) *A spherical rain drop, falling freely, receives in each instant an increase of volume equal to n times its surface at that instant. Find the velocity at the end of time t, and the distance fallen through in that time if n be the initial radius of the raindrop.* **[I.F.S. 2002]**

Sol. This problem is a particular case of Ex. 3. (*a*). Here $\lambda = n$ and $a = n$. With these values relations (6) and (8) give the required results as follows :

$$\text{Velocity} = \frac{dx}{dt} = \frac{g}{4n}\left\{n + nt - \frac{n^4}{(n+nt)^3}\right\} = \frac{g}{4}\left\{1 + t - \frac{1}{(1+t)^3}\right\}$$

and distance $= x = (gt^2/g)\ \{(2n + nt)/(n + nt)\}^2 = (gt^2/8)\ \{(2 + t)/(1 + t)\}^2$

Ex. 4. *A rocket of initial gross mass m_0 takes off vertically, exhausi gases having a constant relative speed v'. If m is the mass, and v the speed of the rocket at the instant t, show that $v = v' \log_e (m_0/m) - gt$.*

Sol. Taking the vertically upward direction as positive, we shall consider the change in the linear momentum, over the interval $[t, t + \delta t]$.

During this interval, suppose that a mass $(-\delta m)$ of the gases is exhausted and the velocity of the rocket becomes $v + \delta v$. Thus, we have

m + δm
Rocket
v+ δv
(– δm)
v– v′

Linear momentum at the instant $t = mv$

and linear momentum of the same system, at the instant $t + \delta t$

$= (m + \delta m)\ (v + \delta v) + (-\delta m)\ (v - v') = mv + m\ \delta v + \delta m\ \delta v - v'\ \delta m$

$\therefore$ Change in linear momentum $= (mv + m\ \delta v + \delta m\ \delta v - v'\ \delta m) - (mv)$

$= m\ \delta v + v'\ \delta m + \delta m\ \delta v$

Since the only external force acting is the weight mg, vertically donwards,

$\therefore$ Impulse of the external force over interval $[t, t + \delta t] = (-mg)\ \delta t$

Then, using the relation between linear momentum and impulse, we get

$$m\ \delta v + v'\ \delta m + \delta m\ \delta v = -mg\ \delta t$$

Dividing by $m\ \delta t$ and the taking limit as $\delta t \to 0$, we get

$$(dv/dt) + (v'/m) \times (dm/dt) = -g$$

Integrating w.r.t. 't', we get

$$v + v' \log m = -gt + C, \; C \text{ being a constant} \qquad ...(1)$$

Initially when $t = 0$, $m = m_0$ and $v = 0$. So (1) gives $\quad C = v' \log m_0$

Then, (1) gives $\quad v = v' \log (m_0/m) - gt$

Ex. 5. *A particle moving along a straight line has a mass m and speed v, at time t. The particle ejects matter backwards at a constant rate λm_0 with velocity u, relative to the particle itself. Show that the equation of motion of the particle is $m(dv/dt) - \lambda m_0 u = F$, where F is the external force acting on the particle.*

Sol. Proceed as in Ex. 4.

Ex. 6. *A spherical drop of liquid falling freely in a vapour acquires mass by condensation at a constant rate c. Show that the velocity after falling from rest in time t is $(gt/2) \times \{1 + M/(M + ct)\}$, where M is the initial mass of the drop.*

Sol. Let m be the mass of the drop at time t and v the velocity. Then, the equation of motion of the drop when its mass varies is given by

$$d(mv)/dt = P + u\,(dm/dt) \qquad ...(1)$$

Since the drop is falling freely under gravity, so $P = mg$. Again, since the additional mass is picked up from rest, so here $u = 0$. Hence, (1) reduces to

$$d(mv)/dt = mg \qquad ...(2)$$

Initial mass of the drop is M and the drop acquires additional mass by condensation at a constant rate c. Hence, we have

$$m = M + ct, \; M \text{ and } c \text{ being constants.} \qquad ...(3)$$

From (2) and (3), $\quad d\{v(M + ct)\}/dt = g(M + ct)$ or $d\{v(M + ct)\} = g(M + ct)dt$

Integraing, $\quad$ **$v(M + ct) = (g/2c) \times (M + ct)^2 + A$, A being a constant.** ...(4)

Initially, when $t = 0$, $v = 0$. So (4) gives $\quad A = -(gM^2)/2c$

Hence (4) becomes $\quad v(M + ct) = (g/2c) \times (M + ct)^2 - (gM^2)/2c$

or $\quad v = \dfrac{g(M + ct)^2}{2c} - \dfrac{gM^2}{2c\,(M + ct)} = \dfrac{g}{2c} \times \dfrac{(M + ct)^2 - M^2}{M + ct} = \dfrac{gt}{2} \times \dfrac{2M + ct}{M + ct}$

or $\quad v = \dfrac{gt}{2} \times \dfrac{(M + ct) + M}{M + ct} \quad$ or $\quad v = \dfrac{gt}{2}\left(1 + \dfrac{M}{M + ct}\right)$

Ex. 7. *A spherical raindrop of radius a cm falls from rest through a vertical height h, receiving throughout the motion an accumulation of condensed vapour at the rate of k grammes per square cm per second, no vertical force but gravity acting. Show that when it reaches the ground its radius will be $k(2h/g)^{1/2}\,\{(1 + (ga^2/2hk^2)^{1/2}\}$.* **[Bhopal 2004]**

Sol. Let m be the the mass, v the velocity and r the radius of the raindrop at time t. Then, we have

$$m = (4/3)\pi r^3 \rho \quad \text{so that} \quad dm/dt = 4\pi r^2 \rho (dr/dt) \qquad ...(1)$$

By given problem, rate of increase of mass $= dm/dt = k \times 4\pi r^2 \rho$ ...(2)

From (1) and (2), $4\pi r^2 \rho (dr/dt) = k \times 4\pi r^2 \rho \quad$ or $\quad dr = k\,dt$

Integrating, $\quad r = kt + A$, where A is a constant. ...(3)

Initially, when $t = 0$, let $r = a$. So (3) gives $A = a$.

Hence (3) becomes $\quad r = a + kt$ ...(4)

Now, the equation of motion of raindrop of varying mass is

$$d(mv)/dt = P + u(dm/dt) \qquad ...(5)$$

Since the drop is falling under gravity only, so $P = mg$. Again, since the additional mass is picked up from rest, so here $u = 0$. Hence (1) reduces to

$$d(mv)/dt = mg \qquad ...(6)$$

Using (4), (1) gives $\quad m = (4/3)\pi\rho(a + kt)^3$. Hence (6) gives

$$\frac{d}{dt}\{v \times (4/3)\pi\rho(a + kt)^3\} = g \times (4/3)\,\pi\rho(a + kt)^3$$

or $\quad d\{v(a + kt)^3\}/dt = g(a + kt)^3 \quad$ or $\quad d\{v(a + kt)^3\} = g(a + kt)^3 dt$

Integrating, $v(kt + a)^3 = (g/4k) \times (a + kt)^4 + B$, B being a constant. ...(7)

Initially at $t = 0$, $v = 0$. So (7) gives $B = -(ga^4)/4k$. Then, by (7),

$v(a + kt)^3 = (g/4k) \times \{(a + kt)^4 - a^4\}$ or $v = (g/4k) \times \{a + kt - a^4(a + kt)^{-3}\}$

If the drop has described a distance x in time t from rest, so that $v = dx/dt$, then the above equation becomes

$dx/dt = (g/4k) \times \{a + kt - a^4(a + kt)^{-3}\}$ or $dx = (g/4k)\{a + kt - a^4(a + kt)^{-3}\}dt$

Integrating, $x = (g/4k) \times \{(1/2k)(a + kt)^2 + (a^4/2k)(a + kt)^{-2}\} + C$.

or $x = (g/8k^2) \times \{(a + kt)^2 + a^4/(a + kt)^2\} + C$, C being a constant. ...(8)

Initially at $t = 0$, $x = 0$. So (8) gives $C = -(ga^2)/4k^2$. Then, by (8)

$$x = \frac{g}{8k^2}\left\{(a+kt)^2 + \frac{a^4}{(a+kt)^2} - 2a^2\right\} = \frac{g}{8k^2}\left\{(a+kt) - \frac{a^2}{a+kt}\right\}^2 = \frac{g}{8k^2} \times \left\{\frac{(a+kt)^2 - a^2}{a+kt}\right\}^2$$

or $x = (g/8k^2) \times (2akt + k^2t^2)^2 (a + kt)^{-2} = (gt^2/8)(2a + kt)^2 (a + kt)^{-2}$ ...(9)

Let the drop reaches the ground after time t_1 and let its radius be b on reaching the ground. Then, when $t = t_1$, $x = h$ and $r = b$.

$\therefore$ From (4), $b = a + kt_1$ or $kt_1 = b - a$ ...(10)

and from (9), $h = (gt_1^2/8)(2a + kt_1)^2 (a + kt_1)^{-2}$ or $8(h/g)(a + kt_1)^2 = t_1^2 (2a + kt_1)^2$

Taking square roots on both sides, $2(2h/g)^{1/2} (a + kt_1) = t_1(2a + kt_1)$

or $2(2h/g)^{1/2} b = (1/k)(b - a)(2a + b - a)$ or $2k(2h/g)^{1/2} b = b^2 - a^2$

or $b^2 - 2k(2h/g)^{1/2} b - a^2 = 0$, which is quadratic equation in b.

Solving it, $b = [2k(2h/g)^{1/2} \pm \{4k^2 \times (2h/g) + 4a^2\}^{1/2}]/2$

Thus, $b = k(2h/g)^{1/2} + k(2h/g)^{1/2} (1 + ga^2/2hk^2)^{1/2}$,

taking +ve value of b since radius of the drop cannot be –ve.

$\therefore$ The required radius of drop $= b = k(2h/g)^{1/2} \{1 + ga^2/2hk^2)^{1/2}\}$

Ex. 8. *A trailer full of sand is pulled by a constant force F. Sand leaks out at the rate of λ units of mass per second. Show that the velocity at the end of time t is $-(F/\lambda) \log (1 - \lambda t/M)$ and the distance moved during that time is $(FM/\lambda^2) \{(1 - \lambda t/M) \log (1 - \lambda t/M) + \lambda t/M\}$, where M is the initial mass of the trailer and its contents.*

Sol. Let m be the mass of trailer and its contents at time t and v the velocity. Then, the equation of motion of the trailer and its contents when its mass varies is given by

$$d(mv)/dt = P + u(dm/dt) \quad ...(2)$$

Here $P = F$(given) and $u = v$

Initial mass of the trailer and its contents is M and it is losing mass (due to leakage of sand) at the rate of λ units per second. Hence, we have

$$m = M - \lambda t, \ M \text{ and } \lambda \text{ being constants.} \quad ...(3)$$

From (1), (2) and (3), $d\{v(M - \lambda t)\}/dt = F + vd(M - \lambda t)/dt$

or $(M - \lambda t)(dv/dt) - v\lambda = F - v\lambda$ or $dv = \{F/(M - \lambda t)\}\, dt$

Integrating, $v = -(F/\lambda) \log (M - \lambda t) + A$, A being a constant. ...(4)

Initially, when $t = 0$, $v = 0$. So (4) gives $A = (F/\lambda) \log M$. So, by (4)

$v = -(F/\lambda)\{\log(M - \lambda t) - \log M\} = -(F/\lambda) \log (1 - \lambda t/M)$ ...(5)

If the trailer and its contents have described a distance x in time t from rest so that $v = dx/dt$, then (5) gives

$dx/dt = -(F/\lambda) \log (1 - \lambda t/M)$ or $dx = -(F/\lambda) \log (1 - \lambda t/M) dt$

Integrating, $x = -(F/\lambda)\{t \log (1 - \lambda t/M) - \int 1 \times \{1/(1 - \lambda t/M)\} \times (-\lambda/M)\}dt + B$

or $x = -(Ft/\lambda) \log (1 - \lambda t/M) - F\int\{t/(M - \lambda t)\}\, dt + B$

$$= -\frac{Ft}{\lambda} \log\left(1 + \frac{\lambda t}{M}\right) + \frac{F}{\lambda}\int \frac{(M - \lambda t) - M}{M - \lambda t} dt + B$$

$= -(Ft/\lambda)\log(1-\lambda t/M) + (F/\lambda)\int\{(1 - M/(M-\lambda t)\}\,dt + B$

$\therefore\quad x = -(Ft/\lambda)\log(1-\lambda t/M) + (F/\lambda)\{t + (M/\lambda)\log(M-\lambda t)\} + B$ ...(6)

Initially, when $t = 0$, $x = 0$. So (6) gives $B = -(FM/\lambda^2)\log M$. So, by (6)

$x = -(Ft/\lambda)\log(1-\lambda t/M) + (Ft/\lambda) + (FM/\lambda^2)\{\log(M-\lambda t) - \log M\}$

or $\quad x = (FM/\lambda^2)\log(1-\lambda t/M) - (Ft/\lambda)\log(1-\lambda t/M) + (Ft/\lambda)$

or $\quad x = (FM/t^2)\{(1-\lambda t/M)\log(1-\lambda t/M) + \lambda t/M\}$,

giving the required distance moved at the end of time t.

Ex. 9. *A mass in the form of a solid cylinder, of radius c, acted upon by no forces, moves parallel to its axis through a uniform cloud of fine dust, of volume density ρ, which is at rest. If the particles of dust which meet the mass adher to it, and if M and u be the mass and velocity at the beginning of the motion, prove that the distance x traversed in time t is given by the equation* $(M + \pi\rho c^2 x)^2 = M^2 + 2\pi\rho u c^2 tM$.

Sol. Let m be the mass of unit length of the cylinder at time t and v the velocity. Then equation of motion the given cylinder of varying mass is

$d(mv)/dt \;=\; P + u\,(dm/dt)$ ...(1)

Since no force is acting on the cylinder, so $P = 0$. Again since the dust is at rest, so velocity of the additional mass $= u = 0$. Hence (1) reduces to

$d(mv)/dt = 0 \quad$ so that $\quad mv = A$, A being a constant. ...(2)

Initially, when $m = M$, $v = u$ (given). So (2) gives $\quad A = Mu$.

Hence (2) becomes $\quad mv = Mu$, M and u being constants. ...(3)

Given rate of increase of mass $= dm/dt = \pi c^2\rho v$ ...(4)

From (3), $\quad v = (Mu)/m$. Hence (4) reduces to

$dm/dt = \pi c^2\rho \times (Mu/m) \quad$ or $\quad 2m\,dm = 2\pi c^2\rho Mu dt$

Integrating, $\quad m^2 = 2\pi c^2\rho\, Mut + B$, B being a constant. ...(5)

Initially, at $t = 0$, $m = M$. So (5) gives $B = M^2$. Then, by (5)

$m^2 = 2\pi c^2\rho u\, Mt + M^2 \quad$ or $\quad m = (2\pi c^2\rho u Mt + M^2)^{1/2}$ ...(6)

From (3), $m = (Mu)/v$. So (6) reduces to

$(Mu)/v = (2\pi c^2\rho u Mt + M^2)^{1/2} \quad$ or $\quad v = Mu(2\pi c^2\rho u Mt + M^2)^{-1/2}$ ...(7)

Now, if the hinder end of the cylinder has described a distance x from rest in time t, so that $v = dx/dt$. Then, by (7), we have

$dx/dt = Mu(2\pi c^2\rho u Mt + M^2)^{-1/2} \quad$ or $\quad dx = Mu(2\pi c^2\rho u Mt + M^2)^{-1/2}\,dt$

Integrating $\quad x = \{2Mu/(2\pi c^2\rho u M)\} \times (2\pi c^2\rho u Mt + M^2)^{1/2} + C$ ...(8)

Initially, when $t = 0$, $x = 0$. So (8) gives $C = -M/(\pi c^2\rho)$. So, by (8)

$x = (1/\pi c^2\rho)\{(2\pi c^2\rho u Mt + M^2)^{1/2} - M\}$ or $M + \pi c^2\rho x = (2\pi c^2\rho u Mt + M^2)^{1/2}$

Squaring both sides, we get $\quad (M + \pi\rho c^2 x)^2 = M^2 + 2\pi\rho u c^2 t\,M$.

Ex. 10. *A mass in the form of a solid cylinder, the area of whose cross-sectoin is A, moves parallel to its axis, being acted on by a constant force F, through a uniform cloud of fine dust of volume density ρ which is moving in a direction opposite that of the cylinder with constant velocity V. If all the dust that meets the cylinder clings to it, find the velocity and distance described in any time t, the cylinder being originally at rest and its initial mass m.*

Sol. Let M be mass of unit length of the cylinder at time t and v the velocity. Since the dust is moving with velocity in V in a direction opposite to that of the cylinder, hence the equation of motion of the cylinder of varying mass is given by (refer equation (2*A*) of Art. 12.11)

$d(Mv)/dt = F - V(dM/dt) \quad$ or $\quad d(Mv)/dt + d(VM)/dt = F$

or $\quad d(Mv + MV)/dt = F$, V and F being constants.

Integrating, $\quad Mv + MV = Ft + B$, B bieng a constant. ...(1)

Initially, at $t = 0$, $M = m$ and $v = 0$ (given). So (1) gives $B = mV$.

Hence (1) becomes $\quad M(v + V) = Ft + mV$. ...(2)

Again, rate of increase of mass $= dM/dt = A\rho\{v-(-V)\}$

$\therefore \quad dM/dt = A\rho(v+V) \quad \text{or} \quad v+V = (1/A\rho)\times(dM/dt)$ ...(3)

Substituting the value of $v+V$ from (3) in (2), we have

$(M/A\rho)\times(dM/dt) = Ft + mV \quad \text{or} \quad 2MdM = 2A\rho\,(Ft+mV)dt$ ...(4)

Integrating (4), $M^2 = 2A\rho\{(1/2)Ft^2 + mVt\} + C$, C being a constant. ...(5)

Initially, when $t=0$, $M=m$. So (5) gives $C=m^2$. Then, by (5)

$M^2 = A\rho(Ft^2 + 2mVt) + m^2 \quad \text{or} \quad M = (m^2 + 2m\rho AVt + A\rho Ft)^{1/2}$ ...(6)

Substituting the value of dM/dt from (3) in (4), we get

$(M/A\rho)\times A\rho(v+V) = Ft + mV \quad \text{or} \quad v+V = (Ft+mV)/M$

or $v = -V + \dfrac{Ft+mV}{M}$ or $v = -V + \dfrac{Ft+mV}{(m^2+2m\rho AVt + A\rho Ft^2)^{1/2}}$. ...(7)

[Putting value of M with help of (6)]

Relation (6) gives the required velocity of the cylinder at any time t. Again, if the hinder end of the cylinder has described distance x from rest in time t, so that $v = dx/dt$, then (6) reduces to

$dx/dt = -V + (m^2 + 2m\rho AVt + A\rho Ft^2)^{-1/2}\times(Ft+mV)$

or $dx = \{-V + (1/2A\rho)\times(m^2 + 2mrAVt + A\rho Ft^2)^{-1/2}\times(2m\rho AV + 2A\rho Ft)\}dt$

Integrating, $x = -Vt + (1/A\rho)\times(m^2 + 2m\rho AVt + A\rho Ft^2)^{1/2} + D$ ...(8)

Initially, when $t=0$, $x=0$. So (8) gives $D = -m/A\rho$. Then, by (8)

$x = -Vt + (1/A\rho)\times\{(m^2 + 2m\rho AVt + A\rho Ft^2)^{1/2} - m\}$,

which gives the required distance described in time t.

The acceleration $f\,(=dv/dt)$ of the cylinder can be obtained by differentiating both sides of (8) with respect of 't'. Thus, we have

$$f = dv/dt$$

$$= \frac{F\times(m^2+2m\rho AVt + A\rho Ft^2)^{1/2} - (Ft+mV)\times(1/2)\times(m^2+2m\rho AVt + A\rho Ft^2)^{-1/2}\times(2m\rho AV + 2A\rho Ft)}{(m^2+2m\rho AVt + A\rho Ft^2)}$$

$$= \frac{F(m^2+2m\rho AVt + A\rho Ft^2) - (Ft+mV)\times 2A\rho(mV+Ft)}{(m^2+2m\rho AVt + A\rho Ft^2)^{3/2}}$$

$$= \frac{F(m^2+2m\rho AVt + A\rho Ft^2) - 2A\rho(F^2t^2 + m^2V^2 + 2FtmV)}{(m^2+2m\rho AVt + A\rho Ft^2)^{3/2}}$$

Thus, $f = m^2(F - A\rho V^2)/(m^2 + 2m\rho AVt + A\rho Ft^2)^{3/2}$ ...(9)

Relation (9) shows that the motion is always in the direction of the force or opposite, according as

$f \lesseqgtr 0, \quad i.e. \quad (F - A\rho V^2) \lesseqgtr 0, \quad i.e. \quad F \lesseqgtr A\rho V^2.$

Ex. 11. *A rocket, originally of mass M, throws off every unit of time a mass eM with relative velocity V, and if M' be the mass of the case etc., show that it cannot rise at once unless eV > g, nor at all unless (eMV)/M' > g. If it just rises vertically at once, show that its greatest velocity is V log (M/M') (g/e) (1 – M'/M) and that the greatest height it reaches is*

$$(V^2/2g)\{\log(M/M')\}^2 + (V/e)\{1 - M'/M - \log(M/M')\}.$$

Sol. Let m be the mass of the rocket at any time t and v the velocity in vertically upward direction. Then the equation motion of the rocket of variable mass is given by

$$d(mv)/dt = -mg - u(dm/dt), \qquad ...(1)$$

since the rocket is moving vertically upwards and the force acting on is $P(=-mg)$ acting in vertically downward direction. Here u is the velocity of the ejected mass, which moves in vertically downward direction while the rocket is moving in vertically upward direction.

Since V is the relative velocity of the ejected mass with respect to the rocket, hence, by definition of relative velocity, we have

Relative velocity of ejected mass = Actual vel. of ejected mass – Actual vel. of rocket

$\therefore$ $$V = u-(-v) \Rightarrow u = V-v$$

$\therefore$ (1) gives, $$d(mv)/dt = -mg-(V-v)\,(dm/dt)$$

or $$m(dv/dt)+v\,(dm/dt) = -mg-V\,(dm/dt)+v\,(dm/dt)$$

$\therefore$ $$dv/dt=-g-(V/m)\times(dm/dt) \qquad ...(2)$$

M is the initial mass of the rocket and a mass is being ejected from the rocket at a constant rate eM per unit of time. Hence, we have

$$m = M-eMt = M(1-et),\ e \text{ and } M \text{ being constants} \qquad ...(3)$$

Substituting the value of m from (3) in (2), we have

$$\frac{dv}{dt}=-g-\frac{V}{M(1-et)}\frac{d}{dt}\{M(1-et)\} \text{ or } f=\frac{dv}{dt}=-g+\frac{Ve}{1-et}, \qquad ...(4)$$

which gives acceleration $f(=dv/dt)$ of the rocket at any time t. Let $f=f_o$, when $t=0$. Then from (4),

$$f_o=-g+Ve.$$

Clearly, the rocket cannot rise at once unless $f_o > o$ *i.e.* $-g+Ve>o$, *i.e.* $Ve>g$. In what follows, we assume that $Ve>g$ so that rocket lifts itself at $t=o$.

From (4), $$dv = \{-g+(Ve)/(1-et)\}\,dt.$$

Integrating, $$v=-gt-V\log(1-et)+A,\ A \text{ being a constant} \qquad ...(5)$$

Initially, when $t=0, v=0$. So (5) gives $A=0$. Then, by (5)

$$v=-gt-V\log(1-et). \qquad ...(6)$$

Let x be the distance moved by the rocket in vertically upward direction in time t so that $v=dx/dt$. So, (6) reduces to

$$dx/dt=-gt-V\log(1-et) \quad \text{or} \quad dx=\{-gt-V\log(1-et)\}\,dt.$$

Integrating, $$x=-(1/2)\,gt^2-V\int 1.\log(1-et)dt$$

or $$x=-(1/2)gt^2-V\left[t\log(1-et)-\int t\cdot\frac{(-e)}{1-et}dt\right]+B,\ B \text{ is a constant}$$

$$=-(1/2)gt^2-Vt\log(1-et)+V\int\frac{(1-et)-1}{1-et}dt+B$$

$$=-(1/2)gt^2-Vt\log(1-et)+V\int\{1-1/(1-et)\}\,dt+B$$

$\therefore$ $$x=(-1/2)gt^2-Vt\log(1-et)+V\{t+(1/e)\log(1-et)\}+B \qquad ...(7)$$

Initially, when $t=0$, let $x=0$. Then (7) gives $B=0$. so, by (7)

$$x=-(1/2)\,gt^2-Vt\log(1-et)+(V/e)\log(1-et)+Vt-V/e+V/e$$

$$=-(1/2)gt^2+(V/e-Vt)\log(1-et)-(V/e-Vt)+V/e$$

$$=-(1/2)gt^2+(V/e-Vt)\{\log(1-et)-1\}+V/e$$

$\therefore$ $$x=-(1/2)gt^2+(V/e)(1-et)\{\log(1-et)-1\}+V/e \qquad ...(8)$$

The mass of the case (or shell) of the rocket is given to be M'. Clearly, the rocket will be lightest when all fuel is burnt *i.e.*, when $m=M'$; Let $m=M'$ when $t=t'$. Then (3) reduces to

$$M'=M(1-et') \quad \text{so that} \quad et'=1-(M'/M) \qquad ...(9)$$

Let $x=h'$ when $t=t'$. Hence, from (8), we have

$$h'=-(1/2)\,gt'^2+(V/e)(1-et')\{\log(1-et')-1\}+V/e \qquad ...(10)$$

Substituting the value of t' from (9) in (10), we have

$$h'=-(g/2\,e^2)(1-M'/M)^2+(V/e)\times(M'/M)\{\log(M'/M)-1\}+V/e$$

or $h' = -(g/2e^2)(1 - M'/M)^2 - (V/e) \times (M'/M) \log (M/M') - (V/e) \times (M'/M) + V/e$ ...(11)

[Note that $\log (M'/M) = -\log (M/M')$]

When the rocket reaches at hight h' at $t = t'$, then the rocket will acquire maximum velocity v' (say). So putting $t = t'$ and $v = v'$ in (6), the maximum velocity of the rocket is given by

$$v' = -gt' - V \log (1 - et') \quad ...(12)$$

Substituting the value of t' from (9) in (12), we have

$$v' = -(g/e)(1 - M'/M) - V \log (M'/M) = V \log (M/M') - (g/e) \log (1 - M'/M) \quad ...(13)$$

Just after acquiring maximum velocity v' at a hight h', the rocket will now move freely vertically upwards against gravity and acquire a further maximum height H, given by

$$H = v'^2/2g = (1/2g)\{V \log (M/M') - (g/e)(1 - M'/M)\}^2$$

$$= (1/2g)[V^2 \{\log (M/M')^2 + (g^2/e^2)(1 - M'/M)^2 - (Vg/e) \log (M/M'). (1 - M'/M)]$$

$$\therefore \quad H = (V^2/2g)\{\log (M/M')\}^2 + (g/2e^2)(1 - M'/M)^2 - (V/e) \log (M/M') + (V/e)(M'/M) \log (M/M') \quad ...(14)$$

Hence the required greatest height ever reached by the rocket (using (13) and (14))

$$= h + H = -(V/e) \times (M'/M) + V/e + (V^2/2g)\{\log (M/M')\}^2 - (V/e) \log (M'/M)$$

$$= (V^2/2g)\{\log (M/M')\}^2 + (V/e)\{1 - M'/M - \log (M/M')\}$$

EXERCISE

1. A falling raindrop has its radius uniformly increased by access of moisture. If it have given to it a horizontal velocity, show that it then describes a hyperbola, one of whose asymptotes is vertical.

2. A ball, of mass m, is moving under gravity in a medium which deposits matter on the ball at a uniform rate μ. Show that the equation to the trajectory, referred to horizontal and vertical axes through a point on itself, may be written in the form $k^2uy = kx(g + kv) + gu(1 - e^{kx/u})$, where u, v are horizontal and vertical velocities at the origin and $mk = 2\mu$.

3. A chain, of length l, in coiled at the edge of a table. One end is fastened to a particle, whose mass is equal to that of the whole chain, and the other end is put over the edge. Show that, immediately after leaving the table, the particle is moving with velocity $(1/2) \times (5gl/6)^{1/2}$.

4. A uniform chain, of length l and mass ml, is coiled on the floor, and a mass mc is attached to one end and projected vertically upwards with velocity $\sqrt{2gh}$. Show that, according as the chain does or does not completely leave the floor, the velocity of the mass on finally reaching the floor again is the velocity due to a fall through a height $(1/3) \times \{2l - c + a^3/(l + c)^2\}$ or $a - c$, where $a^3 = c^2(c + 3h)$.

5. A uniform cord, of lenght l, hangs over a smooth pulley and a monkey whose weight is that of the lenght k of the cord, clings to one end and the system remains in equilibrium. If he starts suddenly, show that he will cease to ascend in space at the end of time

$$\{(l + k)/2g\}^{1/2} \text{ cash}^{-1} (1 + l/k).$$

6. Establish the formula $\boldsymbol{F}(t) = m(t) \times (d\boldsymbol{v}/dt) + \lambda \boldsymbol{u}$ for the motion of a particle of varying mass $m(t)$ with velocity v under a force $\boldsymbol{F}(t)$, matter being emitted at a constant rate λ with velocity $\boldsymbol{u}$ relative to the particle. **[Calcutta 2002]**

MISCELLANEOUS PROBLEMS ON CHAPTER 12

Ex. 1. *V is called terminal velocity when*

(a) neither upward nor downward acceleration vanishes (b) only upward acceleration vanishes (c) only downward acceleration vanishes (d) none of these. **[Kanpur 2001, 02]**

Sol. Ans. (c) Refer Art. 12.2

Ex. 2. *The equation of motion of a particle of mass m projected vertically upwards under gravity in a resisting medium whose resistance varies as the square of the velocity is*

(a) $mv(dv/dy) = -mg - mkv^2$ (b) $mv(dv/dy) = mg - mkv^2$

(c) $mv(dv/dy) = mg + mkv^2$ (d) $mv(dv/dy) = -mg + mkv^2$ **[Kanpur 2002, 03]**

Sol. Ans. (a). Refer case II of Art. 12.4

Ex. 3. *A heavy particle is projected in a resisting medium. If v be the velocity at any time and ϕ the inclination to the vertical of the direction of motion and f the retardation, prove that*

$$\frac{1}{v}\frac{dv}{d\phi} + \cot\phi + \frac{f}{g\sin\phi} = 0$$ **[Kanpur 2007]**

Sol. Let the given body move along the curve OPQ and let P be its position at any time. Let PT and PN be tangent and normal at P. By probelm, $\angle TPM = \phi$. Let $\angle PTX = \psi$.

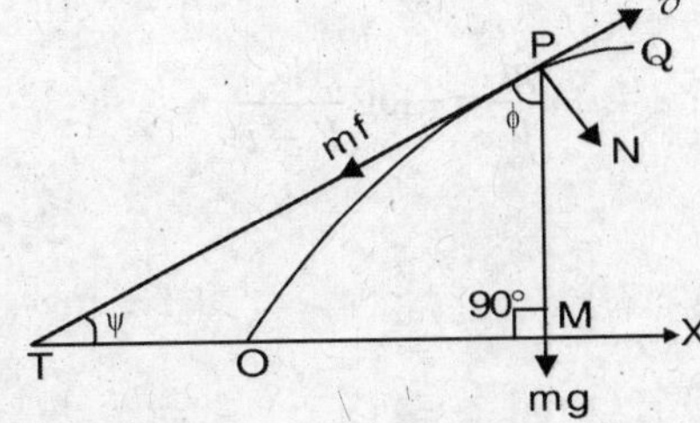

Equations of motion along TP and PN respectively are

$$mv/(dv/ds) = -mg\cos\phi - mf \quad \text{and} \quad m(v^2/\rho) = mg\sin\phi$$

Dividing the corresponding sides of the above equations, we get

$$v\frac{dv}{ds} \times \frac{\rho}{v^2} = -\frac{g\cos\phi + f}{g\sin\phi} \quad \text{or} \quad \frac{1}{v}\frac{dv}{ds}\left(-\frac{ds}{d\psi}\right) = -\cot\phi - \frac{f}{g\sin\phi}$$

[Since ψ decreases as s increases, we have $\rho = -(ds/d\psi)$]

$$\text{or} \quad -\frac{1}{v}\frac{dv}{d\psi} + \cot\phi + \frac{f}{g\sin\phi} = 0 \quad \text{or} \quad -\frac{1}{v}\frac{dv}{d\phi}\frac{d\phi}{d\psi} + \cot\phi + \frac{f}{g\sin\phi} = 0$$

$$\text{or} \quad \frac{1}{v}\frac{dv}{d\phi} + \cot\phi + \frac{f}{\sin\phi} = 0, \quad \text{as} \quad \psi + \phi = \pi/2 \quad \Rightarrow \quad \frac{d\phi}{d\psi} = -1$$

Ex. 4. *The equation of motion of a particle falling vertically downwards in a resisting medium is $m(d^2x/dt^2) = mg - mkv^2$. If V is the terminal velocity, then*

(a) $k = V^2/g$ (b) $k = g/V^2$ (c) $k = V/g$ (d) $k = g/v$

Sol. Ans. (b). Refer Art. 12.4, page 12.5 **(Bundelkhand 2009, Rohilkhand 2010)**

Ex. 5. *A particle falls under gravity (constant) from rest in a medium whose resistance varies as the velocity. Its equation of motion is given by*

(a) $m(d^2x/dt^2) = -mg - mkv$ (b) $m(d^2x/dt^2) = -mg + mkv$

(c) $m(d^2x/dt^2) = mg + mkv$ (d) $m(d^2x/dt^2) = mg - mkv$ **(Garhwal 2003)**

Sol. Ans. (d). Refer Art. 12.3 on page 12.2

Ex. 6. *The resisting force is*

(*a*) *Conservative* (*b*) *Non-conservative* (*c*) *Central* (*d*) *None of these*

Sol. Ans. (*b*). **(Avadh 2006, Garhwal 2004, Rohilkhand 2008, 09)**

Ex. 7. *By increasing the velocity of any body in any medium, the resistance of the medium*

(*a*) *Increases* (*b*) *Decreases* (*c*) *Remains the same* (*d*) *None of these*

(Garhwal 2002)

Sol. Ans. (*a*).

Ex. 8. *A particle falls under gravity in a resisting medium which varies as v, the velocity, then prove that the time t is given by*

$$t = \frac{V}{g}\log\frac{V}{V-v}$$ **[Agra 2014, 2018]**

Sol. Proceed as in Art. 12.3 upto equation (7), namely,

$$v = V(1 - e^{-(gt/V)})$$

giving $e^{-(gt/V)} = 1 - \frac{v}{V} = \frac{V-v}{V}$ or $-\frac{gt}{V} = \log\left(\frac{V-v}{V}\right) = \log\left(\frac{V}{V-v}\right)^{-1}$

or $-\frac{gt}{V} = -\log\frac{V}{V-v}$ giving $t = \frac{V}{g}\log\frac{V}{V-v}$

13

Motion in Three Dimensions

13.1 Cylindrical (polar) coordinates. Any point $P(x, y, z)$ whose projection on the xy-plane is $M(x, y)$ has the cylindrical coordinates (ρ, ϕ, z), where $\rho = OM$, $\phi = \angle XOM$ and $z = MP$. The limits on ρ, ϕ, z are : $0 \le \rho < \infty$, $0 \le \phi \le 2\pi$ and $-\infty < z < \infty$. From figure, we have

$$x = \rho \cos \phi, \quad y = \rho \sin \phi, \quad z = z. \quad \ldots(1)$$

With help of (1), expressing ρ, ϕ and z in terms of x, y, z, we get

$$\rho = (x^2 + y^2)^{1/2}, \quad \phi = \tan^{-1}(y/x), \quad z = z \quad \ldots(2)$$

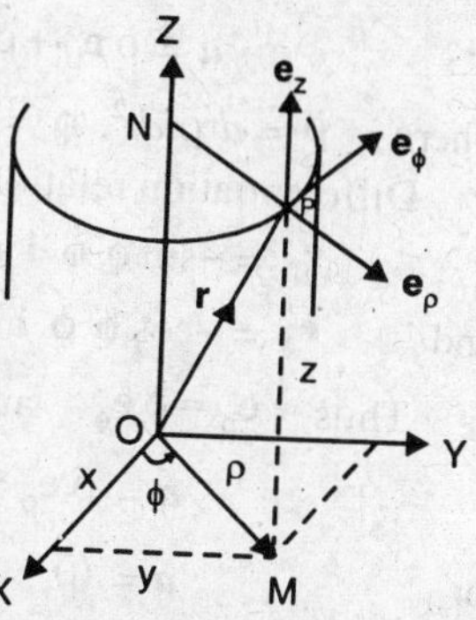

The coordinate surfaces in the cylindrical coordinate system are : (*i*) $\rho = C_1$ *i.e.*, $x^2 + y^2 = C_1^2$ *i.e.*, right circular cylinders having the z-axis as a common axis (*ii*) $\phi = C_2$ *i.e.*, $y = x \tan C_2$, *i.e.*, half planes through the z-axis (*iii*) $z = C_3$ *i.e.*, planes parallel to the xy-plane, as in the cartesian coordinate system. The point P is the point of intersection of these three coordinate surfaces. The coordinate curves for ρ, ϕ and z are respectively straight lines perpendicular to the z-axis, horizontal circles with centres on the z-axis and lines parallel to the z-axis.

The usual mutually perpendicular unit vectors $\mathbf{e}_\rho, \mathbf{e}_\phi, \mathbf{e}_z$ in cylindrical coordinate system are as follows. The unit vector $\mathbf{e}_\rho$ is normal to the cylindrical surface ρ = constant in the direction of increasing radius ρ. The unit vector $\mathbf{e}_\phi$ is tangential to the cylindrical surface, perpendicular to the half plane ϕ = constant and pointing in the direction of increasing azimuth angle ϕ. The third unit vector $\mathbf{e}_z$ is the usual cartesian unit vector $\mathbf{k}$ in the direction of increasing z. Then, we have*

$$\mathbf{e}_\rho = \cos\phi\,\mathbf{i} + \sin\phi\,\mathbf{j}, \qquad \mathbf{e}_\phi = -\sin\phi\,\mathbf{i} + \cos\phi\,\mathbf{j}, \qquad \mathbf{e}_z = \mathbf{k}, \quad \ldots(3)$$

where **i**, **j**, **k** are usual unit vectors in the cartesian system.

13.2 To find the components of velocity and acceleration of a moving point in terms of cylindrical coordinates

(Garakhpur 2008, 09)

Refer figure of Art. 13.1. Let (x, y, z) be the coordinates of any point P moving in space at any instant t. If (ρ, ϕ, z) be its cylindrical coordinates, then

$$x = \rho\cos\phi, \qquad y = \rho\sin\phi, \qquad z = z \quad \ldots(1)$$

Let **r** be the position vector of P. Then

$$\mathbf{r} = x\mathbf{i} + y\mathbf{j} + z\mathbf{k} \quad \text{or} \quad \mathbf{r} = \rho\cos\phi\,\mathbf{i} + \rho\sin\phi\,\mathbf{j} + z\mathbf{k} \quad \ldots(2)$$

To find the components of velocity at *P*. Now, the velocity vector **v** of P is the rate of change of displacement. Hence, using (2), we have

$$\mathbf{v} = d\mathbf{r}/dt = (\dot\rho\cos\phi - \rho\sin\phi\,\dot\phi)\mathbf{i} + (\dot\rho\sin\phi + \rho\cos\phi\,\dot\phi)\mathbf{j} + \dot z\,\mathbf{k},$$

where dot · denotes the derivative w.r.t. 't'. Re-writing it, we get

$$\mathbf{v} = \dot\rho(\cos\phi\,\mathbf{i} + \sin\phi\,\mathbf{j}) + \rho\dot\phi(-\sin\phi\,\mathbf{i} + \cos\phi\,\mathbf{j}) + \dot z\,\mathbf{k}$$

or

$$\mathbf{v} = \dot\rho\,\mathbf{e}_\rho + \rho\dot\phi\,\mathbf{e}_\phi + \dot z\,\mathbf{e}_z, \quad \ldots(3)$$

*Refer Art. 4.4B of Chapter 4 in Simplified Course in Vector Calculus by Dr. M.D. Raisinghania, published by S. Chand and Co. New Delhi where $\mathbf{e}_\rho$, $\mathbf{e}_\phi$, $\mathbf{e}_z$ (= **k**) are usual mutually perpendicular unit vectors in the cylindrical coordinates given by

$$\mathbf{e}_\rho = \cos\phi\,\mathbf{i} + \sin\phi\,\mathbf{j}, \qquad \mathbf{e}_\phi = -\sin\phi\,\mathbf{i} + \cos\phi\,\mathbf{j}, \qquad \mathbf{e}_z = \mathbf{k} \qquad \ldots(4)$$

If v_ρ, v_ϕ, v_z be the components of velocity of P in the directions of $\mathbf{e}_\rho, \mathbf{e}_\phi, \mathbf{e}_z$ respectively, then by (3) we have $v_\rho = d\rho/dt$, along NP, $v_\phi = \rho(d\phi/dt)$ perpendicular to the plane $ZOMP$ and dz/dt parallel to OZ. ...(5)

Note : **k** is constant unit vector $\Rightarrow d\mathbf{k}/dt = \mathbf{0}$. So $d\mathbf{e}_z/dt = d\mathbf{k}/dt = \mathbf{0}$.

To find the components of acceleration at *P*. Now, the acceleration vector $\boldsymbol{a}$ of P is the rate of change of velocity vector **v** of P. Hence, using (3),

$$\boldsymbol{a} = \frac{d\mathbf{v}}{dt} = \frac{d}{dt}(\dot\rho\,\mathbf{e}_\rho) + \frac{d}{dt}(\rho\dot\phi\,\mathbf{e}_\phi) + \frac{d}{dt}(\dot z\,\mathbf{e}_z)$$

or
$$\boldsymbol{a} = \ddot\rho\,\mathbf{e}_\rho + \dot\rho\,\dot{\mathbf{e}}_\rho + \dot\rho\dot\phi\,\mathbf{e}_\phi + \rho\ddot\phi\,\mathbf{e}_\phi + \rho\dot\phi\,\dot{\mathbf{e}}_\phi + \ddot z\,\mathbf{e}_z, \qquad \ldots(6)$$

where $\ddot\rho = d^2\rho/dt^2$, $\ddot\phi = d^2\phi/dt^2$ and $\ddot z = d^2z/dt^2$. Also $d\mathbf{e}_z/dt = d\mathbf{k}/dt = \mathbf{0}$

Differentiation relations (4) w.r.t. '*t*', we get

$$\dot{\mathbf{e}}_\rho = -\sin\phi\,\dot\phi\,\mathbf{i} + \cos\phi\,\dot\phi\,\mathbf{j} = \dot\phi\,(-\sin\phi\,\mathbf{i} + \cos\phi\,\mathbf{j}) = \dot\phi\,\mathbf{e}_\phi, \text{ by (4)}$$

and
$$\dot{\mathbf{e}}_\phi = -\cos\phi\,\dot\phi\,\mathbf{i} - \sin\phi\,\dot\phi\,\mathbf{j} = -\dot\phi\,(\cos\phi\,\mathbf{i} + \sin\phi\,\mathbf{j}) = -\dot\phi\,\mathbf{e}_\rho, \text{ by (4)}$$

Thus $\dot{\mathbf{e}}_\rho = \dot\phi\,\mathbf{e}_\phi$ and $\dot{\mathbf{e}}_\phi = -\dot\phi\,\mathbf{e}_\rho$. Hence (6) becomes

$$\boldsymbol{a} = \ddot\rho\,\mathbf{e}_\rho + \dot\rho\dot\phi\,\mathbf{e}_\phi + \dot\rho\dot\phi\,\mathbf{e}_\phi + \rho\ddot\phi\,\mathbf{e}_\phi - \rho\dot\phi^2\mathbf{e}_\rho + \ddot z\,\mathbf{e}_z$$

or
$$\boldsymbol{a} = (\ddot\rho - \rho\dot\phi^2)\mathbf{e}_\rho + (2\dot\rho\dot\phi + \rho\ddot\phi)\mathbf{e}_\phi + \ddot z\,\mathbf{e}_z \qquad \ldots(7)$$

If a_ρ, a_ϕ, a_z be the components of acceleration of P in the directions of $\mathbf{e}_\rho, \mathbf{e}_\phi, \mathbf{e}_z$ respectively, then by (7), we have

$$a_\rho = \ddot\rho - r\dot\phi^2, \text{ along } NP$$

$$a_\phi = 2\dot\rho\dot\phi + \rho\ddot\phi = \frac{1}{\rho}\frac{d}{dt}\left(\rho^2\frac{d\phi}{dt}\right), \text{ perpendicular to the plane } ZOMP$$

$$a_z = \ddot z = d^2z/dt^2, \text{ parallel to } OZ.$$

13.3 Spherical (polar) coordinates. Any point $P(x, y, z)$ whose projection on the xy-plane is $M(x, y)$ has the spherical coordinates (r, θ, ϕ) whre $r = OP$, $\theta = \angle ZOP$ and $\phi = \angle XOM$. The limits on r, θ, ϕ are : $0 \le r < \infty, 0 \le \theta \le \pi, 0 \le \phi \le 2\pi$. From figure, we have

$$x = r\sin\theta\cos\phi, \qquad y = r\sin\theta\sin\phi, \qquad z = r\cos\theta \qquad \ldots(1)$$

With help of (1), expressing r, θ and ϕ in terms of x, y, z, we have

$$r = (x^2 + y^2 + z^2)^{1/2}, \qquad \theta = \tan^{-1}\{(x^2 + y^2)^{1/2}/z\} \qquad \text{and} \qquad \phi = \tan^{-1}(y/x).$$

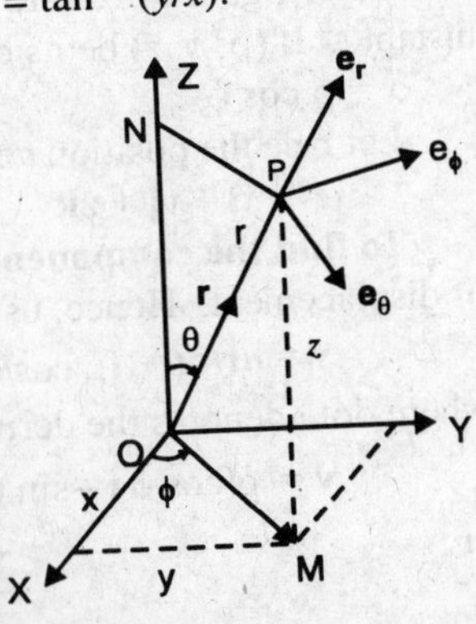

The coordinate surfaces in the spherical coordinate system are : (*i*) $r = C_1$ *i.e.*, $x^2 + y^2 + z^2 = C_1^2$ *i..e*, concentric spheres centred at the origin (*ii*) $\theta = C_2$ *i.e.*, $x^2 + y^2 = z^2(\tan^2 C_2)$ *i.e.*, right circular cones with axis as z-(polar) axis and vertices at the origin. (*iii*) $\phi = C_3$ *i.e.*, $y = x\tan C_3$ *i.e.*, half planes through the z-(polar) axis. The point P is the point of intersection of these three coordinate surfaces. The coordinate curves for, r, θ and ϕ are respectively straight lines passing through the origin, vertical circles with centre at the origin and the horizontal circles with centres on the z-axis.

Z
N
P
r
θ
z
O
Y
X
φ
y
M
$\mathbf{e}_r$
$\mathbf{e}_\phi$
$\mathbf{e}_\theta$
r

The usual mutually perpendicular unit vector $\mathbf{e}_r$, $\mathbf{e}_\theta$, $\mathbf{e}_\phi$ in spherical coordinate system are in the directions of the tangents to the r, θ and ϕ coordinate curves. These unit vectors are directed respectively in the direction of r-increasing, θ increasing and ϕ increasing respectively. Thus, we have*

$$\mathbf{e}_\rho = \sin\theta(\cos\phi\,\mathbf{i} + \sin\phi\,\mathbf{j}) + \cos\theta\,\mathbf{k},$$

$$\mathbf{e}_\theta = \cos\theta(\cos\phi\,\mathbf{i} + \sin\phi\,\mathbf{j}) - \sin\theta\,\mathbf{k}, \qquad \mathbf{e}_\phi = -\sin\phi\,\mathbf{i} + \cos\phi\,\mathbf{j}$$

*Refer Art. 4.4C of Chapter 4 in Simplified Course in Vector Calculus by Dr. M.D. Reisinghonia, published by S. Chand and Co. New Delhi

13.4 To find the components of velocity and acceleration of a moving point in terms of spherical coordinates **[Gorakhpur 2007, 09; Lucknow 2009, 11; Purvanchal 2009, 10]**

Refer figure of Art. 13.3. Let (x, y, z) be the coordinates of any point P moving in space at any instant t. If (r, θ, ϕ) be its spherical coordinates, then

$$x = r \sin\theta \cos\phi, \qquad y = r \sin\theta \sin\phi, \qquad z = r\cos\theta \qquad \ldots(1)$$

Let $\mathbf{r}$ be the position vector of P. Then

$$\mathbf{r} = x\mathbf{i} + y\mathbf{j} + z\mathbf{k}$$

or $$\mathbf{r} = r\sin\theta\cos\phi\,\mathbf{i} + r\sin\theta\sin\phi\,\mathbf{k} + r\cos\theta\,\mathbf{k}, \text{ by (1)} \qquad \ldots(2)$$

The usual mutually perpendicular unit vector $\mathbf{e}_r, \mathbf{e}_\theta, \mathbf{e}_\phi$ in spherical coordinates are given by

$$\mathbf{e}_r = \sin\theta\cos\phi\,\mathbf{i} + \sin\theta\sin\phi\,\mathbf{j} + \cos\theta\,\mathbf{k} \qquad \ldots(3)$$

$$\mathbf{e}_\phi = \cos\theta\cos\phi\,\mathbf{i} + \cos\theta\sin\phi\,\mathbf{j} - \sin\theta\,\mathbf{k} \qquad \ldots(4)$$

$$\mathbf{e}_\phi = -\sin\phi\,\mathbf{i} + \cos\phi\,\mathbf{j} \qquad \ldots(5)$$

To find the components of velocity at *P*. Now, the velocity vector $\mathbf{v}$ of P is the rate of change of displacement. But displacement $\mathbf{r} = r\,\mathbf{e}_r$, where $|\mathbf{r}| = r$.

$$\therefore \qquad \mathbf{v} = \frac{d\mathbf{r}}{dt} = \frac{d(r\,\mathbf{e}_r)}{dt} = \frac{dr}{dt}\mathbf{e}_r + r\frac{d\,\mathbf{e}_r}{dt} \qquad \ldots(6)$$

By (2), $\quad \mathbf{e}_r = \mathbf{r}/r = \sin\theta\cos\phi\,\mathbf{i} + \sin\theta\sin\phi\,\mathbf{j} + \cos\theta\,\mathbf{k}$

Differentiating both sides w.r.t. 't', we get

$$d\mathbf{e}_r/dt = (\cos\theta\cos\phi\;\dot{\theta} - \sin\theta\sin\phi\;\dot{\phi})\,\mathbf{i} + (\cos\theta\sin\phi\;\dot{\theta} + \sin\theta\cos\phi\;\dot{\phi})\mathbf{i} - \sin\theta\;\dot{\theta}\,\mathbf{k}$$

$$= \dot{\theta}\,(\cos\theta\cos\phi\,\mathbf{i} + \cos\theta\sin\phi\,\mathbf{j} - \sin\theta\,\mathbf{k}) + \sin\theta\;\dot{\phi}\,(-\sin\phi\,\mathbf{i} + \cos\phi\,\mathbf{j})$$

$$\therefore \qquad d\mathbf{e}_r/dt = \dot{\theta}\,\mathbf{e}_\theta + \sin\theta\,\dot{\phi}\,\mathbf{e}_\phi, \text{ using (4) and (5)} \qquad \ldots(7)$$

Using (7), (6) reduces to $\quad \mathbf{v} = \dot{r}\,\mathbf{e}_r + r(\dot{\theta}\mathbf{e}_\theta + \sin\theta\,\dot{\phi}\,\mathbf{e}_\phi) \qquad \ldots(8)$

If v_r, v_θ, v_ϕ be the components of velocity of P in the directions of $\mathbf{e}_r$, $\mathbf{e}_\theta$ and $\mathbf{e}_\phi$ respectively, then from (8), we have

$v_r = \dot{r} = dr/dt$, along OP

$v_\theta = r\dot{\theta} = r(d\theta/dt)$, perpendicular to OP in the plane $ZOMP$ in the direction of θ increasing

$v_\phi = r\sin\theta\;\dot{\phi} = r\sin\theta\,(d\phi/dt)$, in a direction perpendicular to the plane $ZOMP$ in the direction of ϕ increasing.

To find the components of acceleration at *P*. Now the acceleration vector $\boldsymbol{a}$ of P is the rate of change of velocity $\mathbf{v}$ of P. Using (8), we have

$$\boldsymbol{a} = \frac{d\mathbf{v}}{dt} = \frac{d}{dt}(\dot{r}\,\mathbf{e}_r) + \frac{d}{dt}(r\dot{\theta}\,\mathbf{e}_\theta) + \frac{d}{dt}(r\sin\theta\,\dot{\phi}\,\mathbf{e}_\phi)$$

or $$\boldsymbol{a} = \ddot{r}\,\mathbf{e}_r + \dot{r}\,\dot{\mathbf{e}}_r + \dot{r}\,\dot{\theta}\,\mathbf{e}_\theta + r\,\ddot{\theta}\,\mathbf{e}_\theta + r\,\dot{\theta}\,\dot{\mathbf{e}}_\theta + \dot{r}\sin\theta\,\dot{\phi}\,\mathbf{e}_\phi + r\cos\theta\,\dot{\theta}\dot{\phi}\,\mathbf{e}_\phi + r\sin\theta\,\ddot{\phi}\,\mathbf{e}_\phi + r\sin\theta\,\dot{\phi}\,\dot{\mathbf{e}}_\phi \qquad \ldots(9)$$

Differentiation relations (3), (4) and (5), w.r.t. 't', we get

$$\dot{\mathbf{e}}_r = d\mathbf{e}_r/dt = (\cos\theta\cos\phi\;\dot{\theta} - \sin\theta\sin\phi\;\dot{\phi})\mathbf{i} + (\cos\theta\sin\phi\;\dot{\theta} + \sin\theta\cos\phi\;\dot{\phi})\mathbf{j} - \sin\theta\;\dot{\theta}\;\mathbf{k}$$

$$= \dot{\theta}\,(\cos\theta\cos\phi\,\mathbf{i} + \cos\theta\sin\phi\,\mathbf{j} - \sin\theta\,\mathbf{k}) + \sin\theta\;\dot{\phi}\,(-\sin\phi\,\mathbf{i} + \cos\phi\,\mathbf{j})$$

Thus, $\quad \dot{\mathbf{e}}_r = \dot{\theta}\,\mathbf{e}_\theta + \sin\theta\,\dot{\phi}\,\mathbf{e}_\phi$, by (4) and (5) $\qquad \ldots(10)$

Now, $\quad \dot{\mathbf{e}}_\theta = d\mathbf{e}_\theta/dt = (-\sin\theta\cos\phi\;\dot{\theta} - \cos\theta\sin\phi\;\dot{\phi})\mathbf{i} + (-\sin\theta\sin\phi\;\dot{\phi} + \cos\theta\cos\phi\;\dot{\phi})\mathbf{j} - \cos\theta\;\dot{\theta}\;\mathbf{k}$

$= -\dot{\theta}\,(\sin\theta\cos\phi\,\mathbf{i} + \sin\theta\sin\phi\,\mathbf{j} + \cos\theta\,\mathbf{k}) + \cos\theta\,\dot{\phi}\,(-\sin\phi\,\mathbf{i} + \cos\phi\,\mathbf{j})$

Thus, $\dot{\mathbf{e}}_\theta = -\dot{\theta}\,\mathbf{e}_r + \cos\theta\,\dot{\phi}\,\mathbf{e}_\phi$, using (3) and (5) ...(11)

Finally, $\dot{\mathbf{e}}_\phi = d\mathbf{e}_\phi/dt = -\cos\phi\,\dot{\phi}\,\mathbf{i} - \sin\phi\,\dot{\phi}\,\mathbf{j}$

or $\dot{\mathbf{e}}_\phi = -\dot{\phi}\,(\cos\phi\,\mathbf{i} + \sin\phi\,\mathbf{j})$ ...(12)

Multiplying both sides of (3) and (4) by $\sin\theta$ and $\cos\theta$ respectively and then adding, we have $\mathbf{e}_r\sin\theta + \mathbf{e}_\theta\cos\theta = \cos\phi\,\mathbf{i} + \sin\phi\,\mathbf{j}$ ...(13)

Using (13), (12) reduces to

$$\dot{\mathbf{e}}_\phi = -\dot{\phi}\,(\mathbf{e}_r\sin\theta + \mathbf{e}_\theta\cos\theta) \qquad ...(14)$$

Using values of $\dot{\mathbf{e}}_r$, $\dot{\mathbf{e}}_\theta$ and $\dot{\mathbf{e}}_\phi$ given by (10), (11) and (14) in (9), we have

$$\boldsymbol{a} = \ddot{r}\,\mathbf{e}_r + \dot{r}(\dot{\theta}\,\mathbf{e}_\theta + \sin\theta\,\dot{\phi}\,\mathbf{e}_\phi) + \dot{r}\dot{\theta}\,\mathbf{e}_\theta + r\ddot{\theta}\,\mathbf{e}_\theta + r\dot{\theta}(-\dot{\theta}\,\mathbf{e}_r + \cos\theta\,\dot{\phi}\,\mathbf{e}_\phi)$$
$$+ \dot{r}\sin\theta\,\dot{\phi}\,\mathbf{e}_\phi + r\cos\theta\,\dot{\theta}\dot{\phi}\,\mathbf{e}_\phi + r\sin\theta\,\ddot{\phi}\,\mathbf{e}_\phi - r\sin\theta\,\dot{\phi}^2(\mathbf{e}_r\sin\theta + \mathbf{e}_\theta\cos\theta)$$

or

$$\boldsymbol{a} = \mathbf{e}_r(\ddot{r} - r\dot{\theta}^2 - r\dot{\phi}^2\sin^2\theta) + \mathbf{e}_\theta(2\dot{r}\dot{\theta} + r\ddot{\theta} - r\sin\theta\cos\theta\,\dot{\phi}^2)$$
$$+ \mathbf{e}_\phi(2\dot{r}\dot{\phi}\sin\theta + 2r\cos\theta\,\dot{\theta}\dot{\phi} + r\sin\theta\,\ddot{\phi}) \qquad ...(15)$$

If a_r, a_θ, a_ϕ be the components of acceleration in the direction of unit vectors $\mathbf{e}_r$, $\mathbf{e}_\theta$ and $\mathbf{e}_\phi$ respectively, *i.e.*, along *OP*, perpendicular to *OP* in the plane *ZOMP* in the direction of θ increasing, and perpendicular to the plane *ZOMP* in the direction of ϕ increasing, then we have

$$a_r = \ddot{r} - r\dot{\theta}^2 - r\dot{\phi}^2\sin^2\theta$$
$$a_\theta = 2\dot{r}\dot{\theta} + r\ddot{\theta} - r\sin\theta\cos\theta\,\dot{\phi}^2 = \frac{1}{r}\frac{d}{dt}(r^2\dot{\theta}) - r\sin\theta\cos\theta\,\dot{\phi}^2$$
$$a_\phi = 2\dot{r}\dot{\phi}\sin\theta + 2r\cos\theta\,\dot{\theta}\dot{\phi} + r\sin\theta\,\ddot{\phi} = \frac{1}{r\sin\theta}\frac{d}{dt}(r^2\sin^2\theta\,\dot{\phi})$$

13.5 Method of solving problems based on motion in three dimensions while using spherical (polar) coordinates. Recall the following results of Art. 13.4. The components v_r, v_θ, v_ϕ of velocity and the components a_r, a_θ, a_ϕ of acceleration of the moving particle are given by

$$v_r = \dot{r}, \qquad v_\theta = r\dot{\theta}, \qquad v_\phi = r\sin\theta\,\dot{\phi} \qquad ...1(a)$$

$$\left.\begin{aligned} a_r &= \ddot{r} - r\dot{\theta}^2 - r\dot{\phi}^2\sin^2\theta \\ a_\theta &= \frac{1}{r}\frac{d}{dt}(r^2\dot{\theta}) - r\sin\theta\cos\theta\,\dot{\phi}^2,\; a_\phi = \frac{1}{r\sin\theta}\frac{d}{dt}(r^2\sin^2\theta\,\dot{\phi}) \end{aligned}\right\} \qquad ...2(a)$$

The above results can be further simplified while solving problems related to a particular geometry of the prblem. We now discuss two important cases. Students are advised to remember these results for direct application in problems.

Particular Case I : Suppose the given particle moves in such a way that the position $P(r, \theta, \phi)$ of the particle at any time is such that r = constant = a, say. In that case coordinates of P become (a, θ, ϕ) and clearly $\dot{r} = 0 = \ddot{r}$. Hence results 1(a) and 2(a) take the following form

$$v_r = 0, \qquad v_\theta = a\dot{\theta}, \qquad v_\phi = a\sin\theta\,\dot{\phi} \qquad ...1(b)$$

$$a_r = -a\dot{\theta}^2 - a\dot{\phi}^2\sin^2\theta,\; a_\theta = a\ddot{\theta} - a\sin\theta\cos\theta\,\dot{\phi}^2,\; a_\phi = \frac{a}{\sin\theta}\frac{d}{dt}(\sin^2\theta\,\dot{\phi}) \qquad ...2(b)$$

Particular Case II : Suppose the given particle moves in such a way that the position $P(r, \theta, \phi)$ of the particle at any time is such that θ = constant = α, say. In that case coordinates of P become (r, α, ϕ) and clearly $\dot{\theta} = 0 = \ddot{\theta}$. Hence results 1($a$) and 2($a$) take the following form

$$v_r = \dot{r}, \qquad v_\theta = 0, \qquad v_\phi = r\sin\alpha\,\dot{\phi} \qquad ...1(c)$$

$$a_r = \ddot{r} - r\dot{\phi}^2\sin^2\alpha,\; a_\theta = -r\sin\alpha\cos\alpha\,\dot{\phi}^2,\; a_\phi = \frac{\sin\alpha}{r}\frac{d}{dt}(r^2\dot{\phi}) \qquad ...2(c)$$

13.6 Solvd examples based on particle Case I of Art. 13.5

Ex. 1. *A particle is attached to one end of a string, of length a, the other end of which is tied to a fixed point O. When the string is inclined at an acute angle α to the downward-drawn vertical the particle is projected horizontally and perpendicular to the string with a velocity V. Find the resulting motion.*

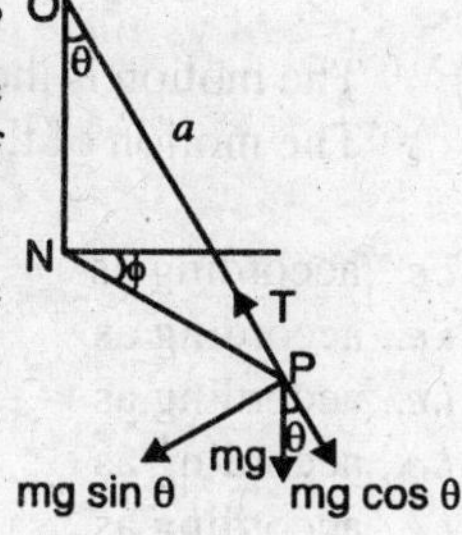

Sol. Let a particle of mass m be attached to one end P of the string OP whose other end is tied to a fixed point O. Let at any time t the string be inclined at an angle θ to the downward vertical and let spherical coordinates of P be (r, θ, ϕ) where $r = a$ = constant so that $\dot{r} = \ddot{r} = 0$. Hence the usual components of acceleration a_r, a_θ, a_ϕ of P in terms of spherical coordinates are given by

$$a_r = -a\dot{\theta}^2 - a\dot{\phi}^2 \sin^2\theta, \; a_\theta = a\ddot{\theta} - a \sin\theta \cos\theta\, \dot{\phi}^2 \text{ and}$$

$$a_\phi = \frac{a}{\sin\theta}\frac{d}{dt}(\sin^2\theta\dot{\phi})$$

Let T be tension in the string. Then equations of motion are given by

$$ma_r = -T + mg\cos\theta \; i.e., \; -ma(\dot{\theta}^2 + \sin^2\theta\,\dot{\phi}^2) = -T + mg\cos\theta \qquad \ldots(1)$$

$$ma_\theta = -mg\sin\theta \quad i.e., \quad a\ddot{\theta} - a\sin\theta\cos\theta\,\dot{\phi}^2 = -g\sin\theta \qquad \ldots(2)$$

and $\quad ma_\phi = 0 \qquad i.e., \qquad \dfrac{a}{\sin\theta}\dfrac{d}{dt}(\sin^2\theta\cdot\dot{\phi}) = 0 \qquad \ldots(3)$

Integrating (3), $\sin^2\theta\,\dot{\phi} = A$, where A is an arbitrary constant. ...(4)

Given that when the string is inclined at an acute angle α to the downward vertical the particle is projected horizontally and perpendicular to the string with a velocity V. Hence, initially, when $\theta = \alpha$, $v_\phi = V$. We know that $v_\phi = a\sin\theta\,\dot{\phi}$ and also when $\theta = \alpha$, $v_\phi = V$ and $\dot{\phi} = [\dot{\phi}]_0$, say. So $V = a\sin\alpha\,[\dot{\phi}]_0$ so that $[\dot{\phi}]_0 = V/a\sin\alpha$. Putting $\theta = \alpha$ and $\dot{\phi} = [\dot{\phi}]_0 = V/a\sin\alpha$ in (4), we get

$\sin^2\alpha\,\{V/a\sin\alpha\} = A \qquad$ or $\qquad A = (V/a)\sin\alpha$

∴ (4) reduces to $\sin^2\theta\,\dot{\phi} = (V/a)\sin\alpha$ or $\dot{\phi} = V\sin\alpha/a\sin^2\theta$ ...(5)

Using (5), (2) gives $\quad a\ddot{\theta} - a\sin\theta\cos\theta\,(V^2\sin^2\alpha/a^2\sin^4\theta) = -g\sin\theta$

or $\quad \dot{\theta}(d\dot{\theta}/d\theta) = (V^2\sin^2\alpha/a^2)(\sin\theta)^{-3}\cos\theta - (g/a)\sin\theta$, as $\ddot{\theta} = \dot{\theta}(d\dot{\theta}/d\theta)$

or $\quad 2\dot{\theta}\,d\dot{\theta} = \{(2V^2\sin^2\alpha/a^2)(\sin\theta)^{-3}\cos\theta - (2g/a)\sin\theta\}d\theta$

Integrating, $\quad \dot{\theta}^2 = -(V^2\sin^2\alpha/a^2\sin^2\theta) + (2g/a)\cos\theta + B$, ...(6)

where B is an arbitrary constant.

By problem, initially, when $\theta = \alpha$, $v_\theta = a\dot{\theta} = 0$ so that $\dot{\theta} = 0$

∴ (6) gives $\qquad 0 = -(V^2/a^2) + (2g/a)\cos\alpha + B \qquad$ ...(7)

Subtracting (7) from (6), we have

$$\dot{\theta}^2 = \frac{V^2\sin^2\alpha}{a^2}\left(\frac{1}{\sin^2\alpha} - \frac{1}{\sin^2\theta}\right) - \frac{2g}{a}(\cos\alpha - \cos\theta)$$

or $\quad \dot{\theta}^2 = (V^2/a^2\sin^2\theta)(\sin^2\theta - \sin^2\alpha) - (2g/a)(\cos\alpha - \cos\theta)$

or $\quad \dot{\theta}^2 = (V^2/a^2\sin^2\theta)(\cos^2\alpha - \cos^2\theta) - (2g/a)(\cos\alpha - \cos\theta)$ ...(8)

Let $\qquad V^2 = 4agn^2$ ...(9)

∴ (8) becomes $\quad \dot{\theta}^2 = (4gn^2/a\sin^2\theta)(\cos\alpha - \cos\theta)(\cos\alpha + \cos\theta) - (2g/a)(\cos\alpha - \cos\theta)$

or $\quad \dot{\theta}^2 = (2g/a)(\cos\alpha - \cos\theta)\{(2n^2/\sin^2\theta)(\cos\alpha + \cos\theta) - 1\}$ ...(10)

(10) shows that $\dot{\theta} = 0$ when $\theta = \alpha$ which corresponds to the initial position of the string. Hence by (10), $\dot{\theta}$ is again zero when

$(2n^2/\sin^2\theta)(\cos\alpha + \cos\theta) - 1 = 0$ or $2n^2(\cos\alpha + \cos\theta) - (1 - \cos^2\theta) = 0$

or $\cos^2\theta + 2n^2\cos\theta + (2n^2\cos\alpha - 1) = 0,$ ...(11)

giving $\cos\theta = -n^2 \pm (1 - 2n^2\cos\alpha + n^4)^{1/2}$

The lower sign gives an inadmissible value of θ. The only inclination at which $\dot{\theta}$ again vanishes is when $\theta = \beta$, where

$$\cos\beta = -n^2 + (1 - 2n^2\cos\alpha + n^4)^{1/2} \quad \text{...(12)}$$

The motion is therefore confined between values α and β, of θ.

The motion of the particle is always above or below the starting point,

according as $\beta \gtrless \alpha$, *i.e.*, according as $\cos\beta \lessgtr \cos\alpha$

i.e., according as $(1 - 2n^2\cos\alpha + n^4)^{1/2} \lessgtr n^2 + \cos\alpha$, using (12)

i.e., according as $1 - 2n^2\cos\alpha + n^4 \lessgtr n^4 + 2n^2\cos\alpha + \cos^2\alpha$

i.e., according as $1 - \cos^2\alpha \lessgtr 4n^2\cos\alpha$

i.e., according as $n^2 \gtrless (\sin^2\alpha/4\cos\alpha)$

i.e., according as $V^2/4ag \gtrless (\sin^2\alpha/4\cos\alpha)$, using (9)

i.e., according as $V^2 \gtrless ag\sin\alpha\tan\alpha$

To find tension of the string at any instant. In the foregoing it is assumed that T does not vanish during the motion. Let v the velocity of the particle at any instant at P.

$$\therefore v^2 = v_r^2 + v_\theta^2 + v_\phi^2 = 0 + (a\dot{\theta})^2 + (a\sin\theta\,\dot{\phi})^2 = a^2(\dot{\theta}^2 + \sin^2\theta\,\dot{\phi}^2) \quad \text{...(13)}$$

Using (13), (1) can be rewritten as

$$-ma(v^2/a^2) = -T + mg\cos\theta \quad \text{or} \quad T = m(g\cos\theta + v^2/a) \quad \text{...(14)}$$

Now, using the principle of energy, we have

Change in K.E. = Work done by external forces

i.e., $(1/2)mv^2 - (1/2)mV^2 = -mg(a\cos\alpha - a\cos\theta)$

so that $v^2 = V^2 - 2ga(\cos\alpha - \cos\theta)$ ...(15)

Using (15), (14) reduces to

$$T = m\{g\cos\theta + (1/a)(V^2 - 2ga\cos\alpha + 2ga\cos\theta)\}$$

or $$T = m\{V^2/a + g(3\cos\theta - 2\cos\alpha)\}$$

Ex. 2. *A heavy particle moves in a smooth sphere; show that, if the velocity be that due to the level of the centre, the reaction of the surface will vary as the depth below the centre.*

Sol. Let a heavy particle of mass m move inside a smooth sphere of radius a and centre O. Let AB be its horizontal diameter and OC be its vertical radius. At any time t, let P be the position of the particle such that $\angle POC = \theta$ and let spherical coordinates of P be (r, θ, ϕ) where $r = a$ = constant so that $\dot{r} = \ddot{r} = 0$. Hence the usual components of acceleration a_r, a_θ, a_ϕ of P in terms of spherical coordinates are given by

$$a_r = -a\dot{\theta}^2 - a\sin^2\theta\,\dot{\phi}^2,$$

$$a_\theta = a\ddot{\theta} - a\sin\theta\cos\theta\,\dot{\phi}^2, \quad a_\phi = \frac{a}{\sin\theta}\frac{d}{dt}(\sin^2\theta\,\dot{\phi})$$

Let R be the reaction of the sphere acting along PO. Then equations of motion are given by

$$ma_r = -R + mg\cos\theta \;\; i.e., \;\; -ma(\dot{\theta}^2 + \sin^2\theta\,\dot{\phi}^2) = -R + mg\cos\theta \quad \text{...(1)}$$

$$ma_\theta = -mg\sin\theta, \quad i.e., \quad a\ddot{\theta} - a\sin\theta\cos\theta\,\dot{\phi}^2 = -g\sin\theta \quad \text{...(2)}$$

and $$ma_\phi = 0 \quad i.e., \quad \frac{a}{\sin\theta}\frac{d}{dt}(\sin^2\theta\,\dot{\phi}) = 0 \quad \text{...(3)}$$

Let v be the velocity of the particle at any instant t. If v_r, v_θ, v_ϕ be the components of v in terms of spherical coordinates, then

$$v_r = \dot{r} = 0, \quad v_\theta = a\dot{\theta} \quad \text{and} \quad v_\phi = a\sin\theta\,\dot{\phi}.$$

Thus, $v^2 = v_r^2 + v_\theta^2 + v_\phi^2 = a^2\dot{\theta}^2 + a^2 \sin^2\theta\,\dot{\phi}^2 = a^2(\dot{\theta}^2 + \sin^2\theta\,\dot{\phi}^2)$

so that $\dot{\theta}^2 + \sin^2\theta\,\dot{\phi}^2 = v^2/a^2$ ...(4)

Using (4), (1) gives $-ma(v^2/a^2) = -R + mg\cos\theta$

or $R = mg\cos\theta + m(v^2/a)$ ...(5)

Given that velocity v at P is that due the level AB of the centre, *i.e.*, the velocity at P is that of a particle falling freely from AB upto P. Hence using $v^2 = u^2 + 2gh$, we get $v^2 = 0 + 2g \cdot ON = 2g \cdot a\cos\theta$. Then (5) reduces to

$$R = mg\cos\theta + (m/a)\times(2ga\cos\theta) = 3mg\cos\theta = (3mg/a)\times(a\cos\theta)$$

i.e., $R = (3mg/a)\times ON$, showing that the reaction R at P will vary as the depth ON below the centre O.

Ex. 3. *A particle is projected horizontally along the interior surface of a smooth hemisphere whose axis is vertical and whose vertex is downwards; the point of projection being at an angular distance* β *from the lowest point, show that the initial velocity so that the particle may just ascend to the rim of the hemisphere is* $(2ag\sec\beta)^{1/2}$*, a being the radius of the hemisphere.* **[Bhopal 2004]**

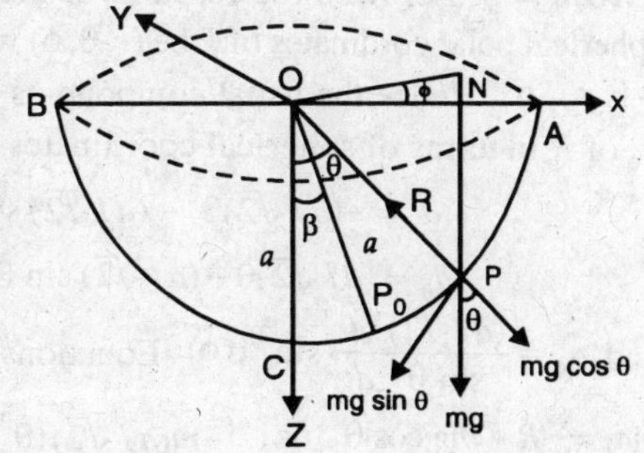

Sol. Let ABC be hemisphere of radius a whose axis OC is vertical and whose vertex C is downards. At any instant t, let P be the position of the given particle of mass m such that $\angle COP = \theta$. Let R be the reaction of the sphere along PO. Let spherical polar coordinates of P be (r, θ, ϕ) where $r = a$ so that $\dot{r} = \ddot{r} = 0$. Hence the usual components of acceleration a_r, a_θ, a_ϕ of P in terms of spherical coordinates are given by

$$a_r = -a\dot{\theta}^2 - a\sin^2\theta\,\dot{\theta}^2,$$

$$a_\theta = a\ddot{\theta} - a\sin\theta\cos\theta\,\dot{\phi}^2,\quad a_\phi = \frac{a}{\sin\theta}\frac{d}{dt}(\sin^2\theta\,\dot{\phi})$$

Equations of motion are given by

$ma_r = -R + mg\cos\theta$ *i.e.*, $-ma(\dot{\theta}^2 + \sin^2\theta\,\dot{\phi}^2) = -R + mg\cos\theta$ ...(1)

$ma_\theta = -mg\sin\theta$ *i.e.*, $a\ddot{\theta} - a\sin\theta\cos\theta\,\dot{\phi}^2 = -g\sin\theta$ ...(2)

and $ma_\phi = 0$ *i.e.*, $\dfrac{a}{\sin\theta}\dfrac{d}{dt}(\sin^2\theta\,\dot{\phi}) = 0$ ...(3)

Integrating (3), $\sin^2\theta\,\dot{\phi} = A$, where A is an arbitrary constant. ...(4)

Let initially the given particle be projected horizontally with velocity V along the interior surface the hemisphere from the point P_0 such that $\angle P_0OC = \beta$. Let v_r, v_θ, v_ϕ be the components of velocity of moving particle in spherical coordinates. Then, by the condition of the problem, $v_r = 0$, $v_\theta = 0$ and $v_\phi = V$. But $v_\phi = a\sin\theta\,\dot{\phi}$. So initially at P_0, when $\theta = \beta$, $v_\phi = V$, let $\dot{\phi} = [\dot{\phi}]_0$. Thus

$V = a\sin\beta\,[\dot{\phi}]_0$ so that $[\dot{\phi}]_0 = V/a\sin\beta$ ...(5)

Now, since $\dot{\phi} = [\dot{\phi}]_0$ when $\theta = \beta$, hence (4) gives

$A = \sin^2\beta\,[\dot{\phi}]_0$ or $A = \sin^2\beta\,(V/a\sin\beta) = (V/a)\sin\beta$, by (5) ...(6)

$\therefore$ (4) gives $\sin^2\theta\,\dot{\phi} = (V/a)\sin\beta$ or $\dot{\phi} = (V\sin\beta)/a\sin^2\theta$ ...(7)

Using (7), (2) gives $a\ddot{\theta} - a\sin\theta\cos\theta\,(V^2\sin^2\beta/a^2\sin^4\theta) = -g\sin\theta$

or $\dot{\theta}(d\dot{\theta}/d\theta) = (V^2\sin^2\beta/a^2)(\sin\theta)^{-3}\cos\theta - (g/a)\sin\theta$, as $\ddot{\theta} = \dot{\theta}(d\dot{\theta}/d\theta)$

or $2\dot{\theta}\,d\dot{\theta} = \{(2V^2\sin^2\beta/a^2)(\sin\theta)^{-3}\cos\theta - (2g/a)\sin\theta\}\,d\theta$

Integrating, $\dot{\theta}^2 = -(V^2\sin^2\beta/a^2\sin^2\theta) + (2g/a)\cos\theta + B$, ...(8)

where B is an arbitrary constant.

Initially at P_0, when $\theta = \beta$, $v_\theta = 0$ *i.e.*, $a\dot{\theta} = 0$ *i.e.*, $\dot{\theta} = 0$

$\therefore$ (8) gives $0 = -(V^2/a^2) + (2g/a)\cos\beta + B$ ...(9)

Subtracting (9) from (8), we have

$$\dot{\theta}^2 = (V^2/a^2)(1 - \sin^2\beta/\sin^2\theta) + (2g/a)(\cos\theta - \cos\beta)$$

By the problem, the particle just ascends to the rim A of the hemishpere. So, at A, when $\theta = \beta$, $\dot{\theta} = 0$. Hence the above relation reduces to

$0 = (V^2/a^2)(1 - \sin^2\beta) - (2g/a)\cos\beta$ or $(V^2\cos^2\beta)/a^2 = (2g/a)\cos\beta$

Thus, $V^2 = 2ga\sec\beta$ or $V = (2ga\sec\beta)^{1/2}$.

Ex. 4. *A heavy particle is projected horizontally along the inner surface of a smooth spherical shell of radius $a/\sqrt{2}$ with velocity $(7ag/3)^{1/2}$ at a depth $2a/3$ below the centre. Show that it will rise to a height $a/3$ above the centre, and that pressure on the sphere just vanishes at the highest point of the path.*

Sol. Let ABQ be gives spherical sheli of radius $a/\sqrt{2}$ so that $OP_0 = OP = a/\sqrt{2}$. Let AB be the vertical diameter. At any instant t, let P be the position of the given particle of mass m such that $\angle POB = \theta$. Let M be the projection of P on xy-plane and let $\angle XOM = \phi$. Let R be the reaction of the shell along PO. Let spherical polar cordinates of P be (r, θ, ϕ) whre $r = a/\sqrt{2}$ so that $\dot{r} = \ddot{r} = 0$. Hence the usual components of acceleration a_r, a_θ, a_ϕ of P in terms of spherical coordinates are given by

$$a_r = -(a/\sqrt{2})\dot{\theta}^2 - (a/\sqrt{2})\sin^2\theta\,\dot{\phi}^2,$$

$$a_\theta = (a/\sqrt{2})\ddot{\theta} - (a/\sqrt{2})\sin\theta\cos\theta\,\dot{\phi}^2$$

and $a_\phi = \dfrac{(a/\sqrt{2})}{\sin\theta}\dfrac{d}{dt}(\sin^2\theta\,\dot{\phi})$. Equations of motion are given by

$ma_r = -R + mg\cos\theta$ *i.e.,* $-m(a/\sqrt{2})(\dot{\theta}^2 + \sin^2\theta\,\dot{\phi}^2) = -R + mg\cos\theta$ ...(1)

$ma_\theta = -mg\sin\theta$ *i.e.,* $(a/\sqrt{2})\ddot{\theta} - (a/\sqrt{2})\sin\theta\cos\theta\,\dot{\phi}^2 = -g\sin\theta$ ...(2)

and $ma_\phi = 0$ *i.e.,* $\dfrac{a}{\sqrt{2}\sin\theta}\dfrac{d}{dt}(\sin^2\theta\,\dot{\phi}) = 0$ ...(3)

Integrating (3), $\sin^2\theta\,\dot{\phi} = A$, where A is an arbitrary constant. ...(4)

Let initially the given particle be projected horizontally with velocity $(7ag/3)^{1/2}$ along the inner surface of the shell from the point P_0 such that $\angle P_0OB = \alpha$ and $OL = 2a/3$, where P_0L is perpendicular to AB.

From ΔOP_0L, $\cos\alpha = \dfrac{OL}{OP_0} = \dfrac{2a/3}{a/\sqrt{2}} = \dfrac{2\sqrt{2}}{3}$. So $\sin\alpha = \dfrac{1}{3}$ ...(5)

Let v_r, v_θ, v_ϕ be the comonents of velocity of moving particle in spherical coordinates. Then, by the condition of the problem, $v_r = 0$, $v_\theta = 0$ and $v_\phi = (7ag/3)^{1/2}$. But $v_\phi = (a/\sqrt{2})\sin\theta\,\dot{\phi}$. So initially at P_0, when $\theta = \alpha$, $v_\phi = (7ag/3)^{1/2}$, let $\dot{\phi} = [\dot{\phi}]_0$. Thus, initially at P_0, we have

$(7ag/3)^{1/2} = (a/\sqrt{2})\sin\alpha\,[\dot{\phi}]_0$ or $(7ag/3)^{1/2} = (a/3\sqrt{2})\,[\dot{\phi}]_0$, by (5)

Thus, $[\dot{\phi}]_0 = 3(14g/3a)^{1/2}$ ...(6)

Now, at P_0, when $\theta = \alpha$, $\dot{\phi} = [\dot{\phi}]_0$. So (4) gives

$\sin^2\alpha\,[\dot{\phi}]_0 = A$ or $A = (1/9) \times 3 \times (14g/3a)^{1/2} = (1/3) \times (14g/3a)^{1/2}$

$\therefore$ (4) becomes $\sin^2\theta\,\dot{\phi} = (1/3) \times (14g/3a)^{1/2}$ ...(7)

or $\dot{\phi} = (1/3\sin^2\theta) \times (14g/3a)^{1/2}$

Then (2) gives $\dfrac{a}{\sqrt{2}}\ddot{\theta} = \dfrac{a\sin\theta\cos\theta}{\sqrt{2}} \times \dfrac{1}{9\sin^4\theta} \times \dfrac{14g}{3a} = -g\sin\theta$

or $\dot{\theta}\dfrac{d\dot{\theta}}{d\theta} = \dfrac{14g\cos\theta}{27a\sin^3\theta} - \dfrac{g\sqrt{2}\sin\theta}{a}$

or $$2\dot{\theta}\, d\dot{\theta} = \{(28g/27a)\,(\sin\theta)^{-3}\cos\theta - (2g\sqrt{2}/a)\sin\theta\}\, d\theta$$

Integrating, $$\dot{\theta}^2 = -(14g/27a\sin^2\theta) + (2\sqrt{2}g\cos\theta/a) + B, \quad \ldots(8)$$

where B is an arbitrary constant.

Initially at P_0, when $\theta=\alpha$, $v_\theta=0$ *i.e.*, $(a/\sqrt{2})\dot{\theta}=0$ *i.e.*, $\dot{\theta}=0$. So (8) gives

$$0 = -\frac{14g}{27a\sin^2\alpha} + \frac{2\sqrt{2}g\cos\alpha}{a} + B \quad \text{or} \quad B = \frac{14g\times 9}{27a} - \frac{2\sqrt{2}g}{a}\times\frac{2\sqrt{2}}{3}, \text{ by (5)}$$

$\therefore$ $$B = (14g/3a) - (8g/3a) \quad \text{or} \quad B = 2g/a$$

$\therefore$ (8) gives $$\dot{\theta}^2 = -(14g/27a\sin^2\theta) + (2\sqrt{2}g\cos\theta/a) + 2g/a \quad \ldots(9)$$

Suppose the particle rises upto Q where $\angle BOQ=\beta$. Then, at Q, $\theta=\beta$ and $\dot{\theta}=0$. Hence (9) gives

$$0 = -(14g/27a\sin^2\beta) + (2\sqrt{2}g\cos\beta/a) + (2g/a)$$

or $$27\sin^2\beta + 27\sqrt{2}\sin^2\beta\cos\beta - 7 = 0$$

or $$27(1-\cos^2\beta) + 27\sqrt{2}\cos\beta\,(1-\cos^2\beta) - 7 = 0$$

or $$27\sqrt{2}\cos^3\beta + 27\cos^2\beta - 27\sqrt{2}\cos\beta - 20 = 0$$

or $$9\sqrt{2}\cos^2\beta\,(3\cos\beta - 2\sqrt{2}) + 21\cos\beta\,(3\cos\beta - 2\sqrt{2}) + 5\sqrt{2}\,(3\cos\beta - 2\sqrt{2}) = 0$$

or $$(3\cos\beta - 2\sqrt{2})(9\sqrt{2}\cos^2\beta + 21\cos\beta + 5\sqrt{2}) = 0$$

or $$(3\cos\beta - 2\sqrt{2})(3\cos\beta + \sqrt{2})(3\sqrt{2}\cos\beta + 5) = 0$$

giving $\cos\beta = 2\sqrt{2}/3$ or $\cos\beta = -\sqrt{2}/3$ or $\cos\beta = -5/3\sqrt{2}$

Using (5), we see that $$\cos\beta = 2\sqrt{2}/3 \Rightarrow \cos\beta = \cos\alpha = \theta = \alpha,$$

which corresponds to the initial position P_0 of the particle where $\dot{\theta}=0$. Again $\cos\beta = -5/3\sqrt{2}$ is not particle because its numerical value is greater than one. So we take $\cos\beta = -\sqrt{2}/3$. Negative value shows that the particle will rise above centre upto Q, such that $\angle BOQ=\beta$. Hence $\angle AOQ=\pi-\beta$. From ΔOQN, we have

$$ON = OQ\cos(\pi-\beta) = -(a/\sqrt{2})\cos\beta = -(a/\sqrt{2})\times(-\sqrt{2}/3) = a/3,$$

which proves the first part of the problem.

Substituting the values of $\dot{\phi}$ and $\dot{\theta}$ given by (7) and (9) in (1), we get

$$R = mg\cos\theta + (ma/\sqrt{2})\dot{\theta}^2 + (ma/\sqrt{2})\sin^2\theta\,\dot{\phi}^2$$

$$= mg\cos\theta + (ma/\sqrt{2})\{-(14g/27a\sin^2\theta) + (2\sqrt{2}g\cos\theta/a) + (2g/a)\}$$
$$+ (ma/\sqrt{2})\sin^2\theta \times (1/9\sin^4\theta)\times(14g/3a)$$

$\therefore$ $$R = mg(3\cos\theta + \sqrt{2}), \text{ on simplification} \quad \ldots(10)$$

Pressure at the highest point Q can be obtained by putting $\theta=\beta$ in (10). Hence the pressure at Q $= mg(3\cos\beta + \sqrt{2}) = 0$, as $\cos\beta = -\sqrt{2}/3$. So pressure on the sphere just vanishes at the highest point Q of the path of the particle.

Ex. 5. *A particle moves on a smooth sphere under no forces except the pressure of the surface; show that its path is given by the equation cot θ = cot β cos φ where θ and φ are its angular coordinates.*

(Purvanchel 2007, 09; Kanpur 2012)

Sol. Let AB be the vertical diameter of given sphere of radius a. At any instant let P be the position of the particle such that $\angle POA=\theta$. Let R be the pressure of the sphere. Let spherical polar coordinates of P be (r, θ, ϕ) where $r = a$ so that $\dot{r} = \ddot{r} = 0$. The usual components of acceleration a_r, a_θ, a_ϕ of P in terms of spherical coordinates are

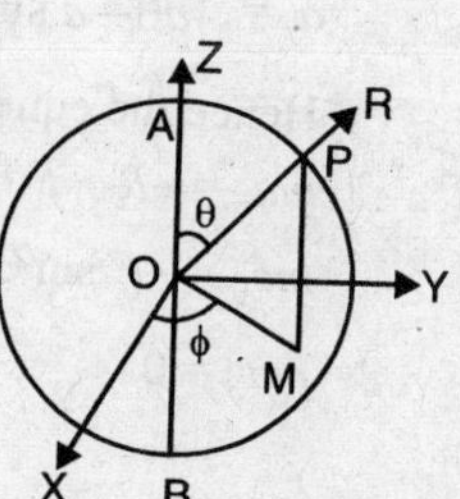

$$a_r = -a\dot{\theta}^2 - a\sin^2\theta\,\dot{\theta}^2, \quad a_\theta = a\ddot{\theta} - a\sin\theta\cos\theta\,\dot{\phi}^2,$$

and $a_\phi = \dfrac{a}{\sin\theta}\dfrac{d}{dt}(\sin^2\theta\,\dot\phi)$

Since the particle moves under no forces except the pressure R along OP, so there is no force on the particle in θ and ϕ directions. So we must have

$$a_\theta = 0 \quad \text{or} \quad a\ddot\theta - a\sin\theta\cos\theta\,\dot\phi^2 = 0 \qquad \text{...(1)}$$

and $$a_\phi = 0 \quad \text{or} \quad \frac{a}{\sin\theta}\frac{d}{dt}(\sin^2\theta\,\dot\phi) = 0 \qquad \text{...(2)}$$

Integrating (2), $\sin^2\theta\,\dot\phi = A$, A being a constant. ...(3)

$\therefore$ $d\phi/dt = A/\sin^2\theta$

Using (3), (1) gives $\ddot\theta - \sin\theta\cos\theta \times (A^2/\sin^4\theta) = 0$

or $\dot\theta\dfrac{d\dot\theta}{d\theta} = \dfrac{A^2\cos\theta}{\sin^3\theta}$ or $2\dot\theta\,d\dot\theta = 2A^2(\sin\theta)^{-3}\cos\theta\,d\theta$

Integrating, $\dot\theta^2 = -A^2/\sin^2\theta + B$, B is a constant. ...(4)

Initially, let when $\theta = \beta$, $\dot\theta = 0$. So (4) $\Rightarrow B = A^2/\sin^2\beta$

$\therefore$ (4) gives $\left(\dfrac{d\theta}{dt}\right)^2 = A^2\left(\dfrac{1}{\sin^2\beta} - \dfrac{1}{\sin^2\theta}\right) = A^2\dfrac{\sin^2\theta - \sin^2\beta}{\sin^2\beta\sin^2\theta}$

or $d\theta/dt = A(\sin^2\theta - \sin^2\beta)^{1/2}/\sin\beta\sin\theta$ ...(5)

Dividing (3) by (5), $\dfrac{d\phi}{d\theta} = \dfrac{A}{\sin^2\theta} \times \dfrac{\sin\beta\sin\theta}{(\sin^2\theta - \sin^2\beta)^{1/2}}$

or $d\phi = \dfrac{\operatorname{cosec}^2\theta\,d\theta}{(\operatorname{cosec}^2\beta - \operatorname{cosec}^2\theta)^{1/2}} = \dfrac{\operatorname{cosec}^2\theta\,d\theta}{(\cot^2\beta - \cot^2\theta)^{1/2}}$

Putting $\cot\theta = u$ so that $-\operatorname{cosec}^2\theta\,d\theta = du$

$\therefore$ $d\phi = -du/(\cot^2\beta - u^2)^{1/2}$

Integrating, $\phi = \cos^{-1}(u/\cot\beta) + C = \cos^{-1}(\cot\theta/\cot\beta) + C$ ...(6)

Initially, when $\theta = \beta$, let $\phi = 0$. So (6) gives $C = 0$

$\therefore$ (6) becomes $\phi = \cos^{-1}(\cot\theta/\cot\beta)$ or $\cot\theta = \cot\beta\cos\phi$,

which the required path of the particle.

Ex. 6. *A particle moves on the surface of a smooth sphere along a rhumb-line, being acted on by a force parallel to the axis of the rhumb-line. Show that the force varies inversely as the fourth power of the distance from the axis and directly as the distance from the meridian plane perpendicular to axis.*

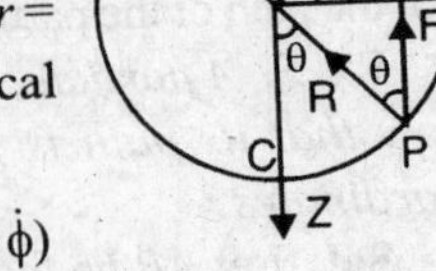

Sol. Let OC be vertical radius of given sphere of radius a and centre O. At any instant t, let P be the position of the given particle of mass m such that $\angle POC = \theta$. Let R be the reaction of the sphere and F the force parallel to the axis of the rhumb-line. Let spherical polar coordinates of P be (r, θ, ϕ) where $r = a$ so that $\dot r = \ddot r = 0$. So components of acceleration of P in terms of spherical coordinates are

$$a_r = -a\dot\theta^2 - a\sin^2\theta\,\dot\phi^2,\; a_\theta = a\ddot\theta - a\sin\theta\cos\theta\,\dot\phi^2,\; a_\phi = \frac{a}{\sin\theta}\frac{d}{dt}(\sin^2\theta\,\dot\phi)$$

Hence the equations of motion are given by

$$ma_r = -R + F\cos\theta \quad i.e., \quad -ma(\dot\theta^2 + \sin^2\theta\,\dot\phi^2) = -R + F\cos\theta \qquad \text{...(1)}$$

$$ma_\theta = -F\sin\theta \quad i.e., \quad m(a\ddot\theta - a\sin\theta\cos\theta\,\dot\phi^2) = -F\sin\theta \qquad \text{...(2)}$$

$$ma_\phi = 0 \quad i.e., \quad \frac{a}{\sin\theta}\frac{d}{dt}(\sin^2\theta\,\dot\phi) = 0 \qquad \text{...(3)}$$

A rhumb-line is defined as a curve on the sphere cutting all the meridians at a constant angle α. Its equation is given by

$(\dot{\phi} \sin \theta)/\dot{\theta} = \tan \alpha$ so that $\dot{\theta} = \dot{\phi} \sin \theta \cot \alpha$ ...(4)

Integrating (3), $\sin^2 \theta \, \dot{\phi} = A$, A being a constant. ...(5)

Substituting the value of $\dot{\phi}$ from (5) in (4), we get

$\dot{\theta} = (A/\sin^2 \theta) \cdot \sin \theta \cot \alpha$ or $\dot{\theta} = A \cot \alpha \operatorname{cosec} \theta$ ...(6)

Differentiating (6) w.r.t., 't', we get

$\ddot{\theta} = -A \cot \alpha \operatorname{cosec} \theta \cot \theta \, \dot{\theta} = -A^2 \cot^2 \alpha \operatorname{cosec}^2 \theta \cot \theta$, by (6) ...(7)

From (2), $F \sin \theta/am = \sin \theta \cos \theta \, \dot{\phi}^2 - \ddot{\theta}$ ...(8)

Substituting the values of $\dot{\phi}$ and $\ddot{\theta}$ from (5) and (7) in (8), we get

$$F \sin \theta/am = \sin \theta \cos \theta \times (A^2/\sin^4 \theta) + (A^2 \cot^2 \alpha \cos \theta)/\sin^2 \theta$$

or $$F \sin \theta/am = (A^2 \cos \theta/\sin^3 \theta) \times (1 + \cot^2 \alpha) = (A^2 \cos \theta \operatorname{cosec}^2 \alpha)/\sin^3 \theta$$

$$\therefore \quad F = a^4 m A^2 \operatorname{cosec}^2 \alpha \cdot (a \cos \theta) \cdot \{1/(a \sin \theta)^4\},$$

showing that $F \propto 1/(a \sin \theta)^4$ and also $F \propto a \cos \theta$ *i.e.*, force F varies as the fourth power of the distance from the axis and directly as the distance from the meridian plane perpendicular to the axis.

EXERCISE

1. A heavy particle is projected with velocity V from the end of a horizontal diameter of a sphere of radius a along the inner surface, the direction of projection making an angle β with the equator. If the particle never leaves the surface, prove that $3 \sin^2 \beta < 2 + (V^2/3ga)^2$.

2. A particle constrained to move on a smooth spherical surface is projected horizontally from a point at the level of the centre so that its angular velocity relative to the centre is ω. If $\omega^2 a$ be very great compared with g, show that its depth z below the level of the centre at time t is $(2g/\omega^2) \sin^2 (\omega t/2)$ approximately.

3. A particle describes a rhumb-line on a sphere in such a way that its longitude increases uniformly; show that the resultant acceleration varies as the cosine of the latitude and that its direction makes with the normal an angle equal to the latitude.

4. A particle moves on the inside of a smooth sphere, of radius a, under a force perpendicular to and acting from a given diameter, which equals $(\mu \sin \theta/\cos^4 \theta)$ when the particle is at an angular distance θ from that diameter; if, when the angular distance of the particle is γ, it is projected with velocity $(\mu a)^{1/2} \sec \gamma$ in a direction perpendicular to the plane through itself and the given diameter, show that its path is a small circle of the sphere, and find the reaction of the sphere.

5. A particle moves on the surface of a smooth sphere and is acted on by a force in the direction of the perpendicular from the particle on a diameter and equal to $\mu/(\text{distance})^3$. Show that it can be projected so that its path will cut the meridians at a constant angle.

6. A particle moves on the interior of a smooth sphere, of radius a, under a force producing an acceleration $\mu \omega^n$ along the perpendicular ω drawn to a fixed diameter. It is projected with velocity V along the great circle to which this diameter is perpendicular and is slightly disturbed from its path; show that the new path will cut the old one m times in a revolution, where $m^2 = 4(1 - \mu a^{n+1}/V^2)$.

13.7 Solved examples based on particular Case II of Art. 13.5

Ex. 1. *A particle moves on the inner surface of a smooth cone, of vertical angle 2α, being acted on by a force towards the vertex of the cone, and its direction of motion always cuts the generators at a constant angle β; find the motion and the law of force.*

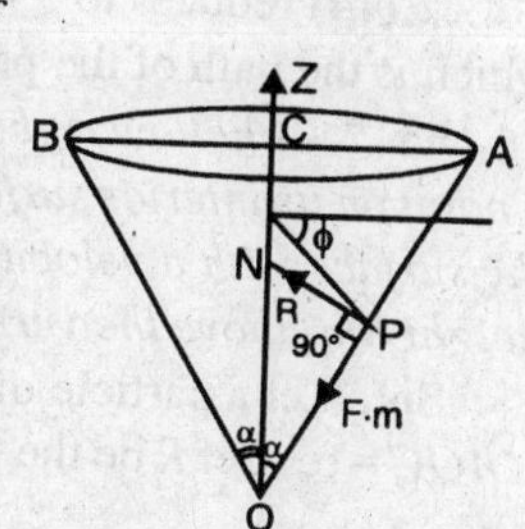

Sol. Let a particle of mass m move on the inner surface of the cone OAB whose axis is OC and $\angle AOC = \alpha$. Let $F \cdot m$ be the force and R the reaction of the cone acting along perpendicular PN to OA. At any time t, let P be the position of the particle such that $OP = r$. Let spherical polar coordinates of P be (r, θ, ϕ) where $\theta = \alpha$ so that $\dot{\theta} = 0$. Hence the usual components of velocity v_r, v_θ, v_ϕ and the components of acceleration a_r, a_θ, a_ϕ are

$$v_r = \dot{r}, \qquad v_\theta = 0, \qquad v_\phi = r \sin\alpha\, \dot{\phi} \qquad \ldots(1)$$

$$a_r = \ddot{r} - r\dot{\phi}^2 \sin^2\alpha, \quad a_\theta = -r \sin\alpha \cos\alpha\, \dot{\phi}^2, \quad a_\phi = \frac{\sin\alpha}{r}\frac{d}{dt}(r^2\dot{\phi}) \qquad \ldots(2)$$

Hence the equations of motion are

$$ma_r = -F\cdot m \qquad \text{or} \qquad \ddot{r} - r\sin^2\alpha\, \dot{\phi}^2 = -F \qquad \ldots(3)$$

$$ma_\theta = -R \qquad \text{or} \qquad mr \sin\alpha \cos\alpha\, \dot{\phi}^2 = R \qquad \ldots(4)$$

and $$ma_\phi = 0 \qquad \text{or} \qquad \frac{m \sin\alpha}{r}\frac{d}{dt}(r^2\dot{\phi}) = 0 \qquad \ldots(5)$$

Also, since the direction of motion always cuts OP at an angle β,

$\therefore$ $$\tan\beta = v_\phi / v_r = (r \sin\alpha\, \dot{\phi}) / \dot{r} \qquad \ldots(6)$$

Integrating (5), $\quad r^2\dot{\phi} = A$, A being a constant. $\qquad \ldots(7)$

Substituting the value of $\dot{\phi}$ from (7) in (6), we have

$$\dot{r} = r \sin\alpha \cot\beta \cdot (A/r^2) \qquad \text{or} \qquad \dot{r} = A(1/r)\sin\alpha\cot\beta \qquad \ldots(8)$$

Differentiating both sides of (8), we have

$$\ddot{r} = A(-1/r^2)\,\dot{r}\, \sin\alpha\cot\beta \qquad \text{or} \qquad \ddot{r} = -A^2(1/r^3)\sin^2\alpha\cot^2\beta, \text{ by (8)} \qquad \ldots(9)$$

Substituting the values of $\dot{\phi}$ and $\ddot{r}$ from (7) and (9) in (3), we get

$$-F = -A^2(1/r^3)\sin^2\alpha\cot^2\beta - r\sin^2\alpha\,(A^2/r^4) \quad \text{or} \quad F = (A^2/r^3)\sin^2\alpha\,(\cot^2\beta + 1)$$

or $$F = (A^2\sin^2\alpha\,\text{cosec}^2\beta)/r^3 \qquad \text{or} \qquad F = \mu/r^3, \qquad \ldots(10)$$

where $$\mu = A^2\sin^2\alpha\,\text{cosec}^2\beta = (A^2\sin^2\alpha)/\sin^2\beta, \text{ say} \qquad \ldots(11)$$

(10) shows that force varies inversely as cube of the distance from the vertex O. This gives the required law of force.

With help of (1), the velocity v at any time t at P is given by

$$v^2 = v_r^2 + v_\theta^2 + v_\phi^2 = \dot{r}^2 + (r\sin\alpha\,\dot{\phi})^2$$

or $$v^2 = A^2(1/r)^2\sin^2\alpha\cot^2\beta + r^2\sin^2\alpha\,(A/r^2)^2, \text{ using (7) and (8)}$$

or $$v^2 = (A^2/r^2)\sin^2\alpha\,(\cot^2\beta + 1) = (A^2\sin^2\alpha/\sin^2\beta)\cdot(1/r^2) = \mu/r^2, \text{ by (11)}$$

Thus, $\quad v = \sqrt{\mu}/r \quad$ *i.e.*, velocity varies inversely as OP.

Using (7), (4) reduces to

$$R/m = (r\sin\alpha\cos\alpha)\cdot(A/r^2)^2 = (A^2/r^3)\sin\alpha\cos\alpha \qquad \ldots(12)$$

$$= (A^2/r^3)\sin^2\alpha\,\text{cosec}^2\beta\cdot(\sin^2\beta\cos\alpha/\sin\alpha) = F\cdot(\sin^2\beta\cos\alpha/\sin\alpha), \text{ by (10)}$$

Thus, $$R/m = (F/\sin\alpha)\cdot\sin^2\beta\cos\alpha, \qquad \ldots(13)$$

which gives relation between R and F.

Re-writing (6), $\quad \dot{r}/\dot{\phi} = r\sin\alpha\cot\beta \quad$ or $\quad \dfrac{dr/dt}{d\phi/dt} = r\sin\alpha\cot\beta$

or $$dr/d\phi = r\sin\alpha\cot\beta \qquad \text{or} \qquad (1/r)\,dr = \sin\alpha\cot\beta\, d\phi \qquad \ldots(14)$$

Integrating, $\quad \log r = \phi\sin\alpha\cot\beta + B$, where B is a constant.

Let, initially when $r = r_0$ (say), $\phi = 0$. So (14) gives $\quad B = \log r_0$

$\therefore$ (14) reduces to $\quad \log r - \log r_0 = \phi\sin\alpha\cot\beta \quad$ or $\quad r = r_0\, e^{\phi\sin\alpha\cot\beta}$,

which is the path of the particle.

Ex. 2. *A hollow right circular cone is placed with its vertex downward and axis vertical, and at a point on its interior surface at a height h above the vertex a particle is projected horizontally along the surface with a velocity* $\{2gh/(n^2 + n)\}^{1/2}$. *Show that the lowest point of its path will bé at a height h/n above the vertex of the cone.*

Sol. Let a particle of mass m move on the inner surface of the cone AOB whose axis is OC and $\angle AOC = \alpha$. Let R be the reaction of the cone acting along perpendicular PN to OA. At any time t, let

P be the position of the particle such that $OP = r$. Let spherical coordinates of P be (r, θ, ϕ) where $\theta = \alpha$ so that $\dot{\theta} = 0$. Hence the usual components of acceleration in terms of spherical coordinates are given by

$$a_r = \ddot{r} - r\sin^2\alpha\,\dot{\phi}^2, \qquad a_\theta = -r\sin\alpha\cos\alpha\,\dot{\phi}^2, \qquad a_\phi = \frac{\sin\alpha}{r}\frac{d}{dt}(r^2\dot{\phi})$$

Hence the equations of motion are given by

$$ma_r = -mg\cos\alpha \qquad i.e., \qquad \ddot{r} - r\sin^2\alpha\,\dot{\phi}^2 = -g\cos\alpha \qquad \ldots(1)$$

$$ma_\theta = -R + mg\sin\alpha \quad i.e., \quad -mr\sin\alpha\cos\alpha\,\dot{\phi}^2 = -R + mg\sin\alpha \qquad \ldots(2)$$

and
$$ma_\phi = 0 \qquad i.e., \qquad \frac{\sin\alpha}{r}\frac{d}{dt}(r^2\dot{\phi}) = 0 \qquad \ldots(3)$$

Integrating (3), $\quad r^2\dot{\phi} = A$, A being a constant. $\qquad \ldots(4)$

Components of velocity at P are given by

$$v_r = \dot{r}, \qquad v_\theta = 0 \qquad \text{and} \qquad v_\phi = r\sin\alpha\,\dot{\phi} \qquad \ldots(5)$$

Given that initially the particle is projected horizontally with velocity $\{2gh/(n^2+n)\}^{1/2}$ from a point P_0 (say) such that OM_0 = height of P_0 above $O = h$. Hence initially, at P_0, we have $v_r = dr/dt = 0$, $v_\theta = 0$ and $v_\phi = \{2gh/(n^2+n)\}^{1/2}$. Now, at P_0, $r = OP_0 = h\sec\alpha$. Also, let at P_0, $\dot{\phi} = [\dot{\phi}]_0$. Hence, (5) gives

$$\{2gh/(n^2+n)\}^{1/2} = v_\phi = h\sec\alpha\sin\alpha\,[\dot{\phi}]_0,$$

giving
$$[\dot{\phi}]_0 = (\cos\alpha/h\sin\alpha)\cdot\{2gh/(n^2+n)\}^{1/2} \qquad \ldots(6)$$

Since at P_0, $\quad r = OP_0 = h\sec\alpha \quad$ and $\quad \dot{\phi} = [\dot{\phi}]_0$, (4) gives

$h^2\sec^2\alpha\,[\dot{\phi}]_0 = A \quad$ or $\quad A = (h/\sin\alpha\cos\alpha)\cdot\{2gh/(n^2+n)\}^{1/2}$, by (6)

$\therefore$ (4) gives $\quad \dot{\phi} = A/r^2 = (h/r^2\sin\alpha\cos\alpha)\cdot\{2gh/(n^2+n)\}^{1/2} \qquad \ldots(7)$

Substituting the value of $\dot{\phi}$ from (7) in (1), We have

$$\dot{r}\frac{d\dot{r}}{dr} - \frac{rh^2\sin^2\alpha}{r^4\sin^2\alpha\cos^2\alpha}\left(\frac{2gh}{n^2+n}\right) = -g\cos\alpha, \text{ as } \ddot{r} = \dot{r}\frac{d\dot{r}}{dr}$$

or
$$2\dot{r}\,d\dot{r} = \left\{\frac{4gh^3r^{-3}}{(n^2+n)\cos^2\alpha} - 2g\cos\alpha\right\}dr$$

Integrating,
$$\dot{r}^2 = -\frac{2gh^3}{r^2(n^2+n)\cos^2\alpha} - 2gr\cos\alpha + B, \qquad \ldots(8)$$

where B is an arbitrary constant.

But initially at P_0, when $r = OP_0 = h\sec\alpha$, $\dot{r} = 0$. So (8) gives
$0 = -(2gh)/(n^2+n) - 2gh + B \quad$ or $\quad B = 2gh(1+n+n^2)/(n^2+n)$

$$\therefore \text{(8) gives } \dot{r}^2 = -\frac{2gh^3}{r^2(n^2+n)\cos^2\alpha} - 2gr\cos\alpha + \frac{2gh(1+n+n^2)}{n^2+n} \qquad \ldots(9)$$

Now, from (9), $\dot{r}$ is again zero when

$$0 = -\frac{2gh^3}{(n^2+n)z^2} - 2gz + \frac{2gh(1+n+n^2)}{n^2+n}, \text{ where } z = r\cos\alpha$$

or
$$(n^2+n)z^3 - hz^2(1+n+n^2) + h^3 = 0$$

or
$$(n^2+n)z^3 - (n+n^2)hz^2 - z^2h + h^3 = 0$$

$$(n^2+n)z^2(z-h) - h(z-h)(z+h) = 0 \text{ or } (z-h)\{n(n+1)z^2 - hz - h^2\} = 0$$

or
$$(z-h)\{n(n+1)z^2 - (n+1)hz + nhz - h^2\} = 0$$

or
$$(z-h)\{(n+1)z(nz-h) + h(nz-h)\} = 0$$

Thus, $r \cos \alpha = h, h/n$ or $-h/(n+1)$, as $z = r \cos \alpha$

Now the negative value of $r \cos \alpha$ is inadmissible. Again $r \cos \alpha = h$ corresponds to the initial position P_0 of the particle. Hence the required lowest point of the path of the particle will be at height $= r \cos \alpha = h/n$ above the vertex O of the cone.

Ex. 3. *A particle moves on a smooth right circular cone under a force which is always in a direction perpendicular to the axis of the cone; if the particle describes on the cone a curve which cuts all the generators at a given constant angle, find the law of force and the initial velocity, and show that at any instant the reaction of the cone is proportional to the acting force.*

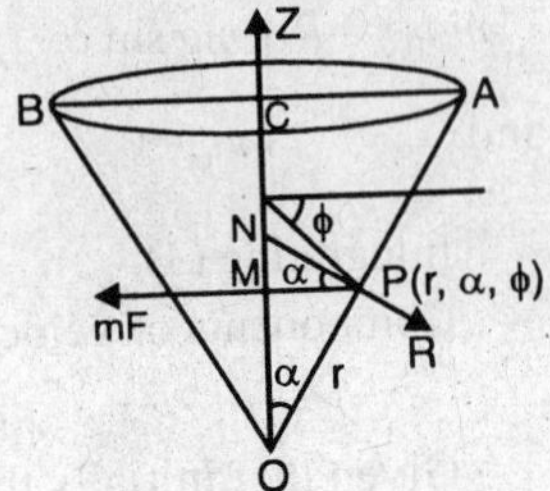

Sol. Let a particle of mass m move on the smooth right circular cone OAB whose axis OZ is vertical and $\angle AOZ = \alpha$. Let Fm be the force acting on the particle in the direction of the perpendicular PM to the axis OZ of the cone and let R be the reaction of the surface of the cone acting along the normal NP to the surface at P. Let $P(r, \alpha, \phi)$ be the spherical coordinates of P, where $OP = r$ and $\theta = \alpha$. Since $\theta = \alpha$, we have $\dot{\theta} = 0$ and hence the usual components of velocity v_r, v_θ, v_ϕ and components of acceleration a_r, a_θ, a_ϕ are given by

$$v_r = \dot{r}, \qquad v_\theta = 0, \qquad v_\phi = r \sin \alpha\, \dot{\phi} \quad \ldots(1)$$

$$\text{and } a_r = \ddot{r} - r \sin^2 \alpha\, \dot{\phi}^2, \quad a_\theta = -r \sin \alpha \cos \alpha\, \dot{\phi}^2, \quad a_\phi = \frac{\sin \alpha}{r}\frac{d}{dt}(r^2\dot{\phi}) \quad \ldots(2)$$

Hence the usual equations of motion are

$$ma_r = -mF \sin \alpha \quad i.e., \quad \ddot{r} - r \sin^2 \alpha\, \dot{\phi}^2 = -F \sin \alpha \quad \ldots(3)$$

$$ma_\theta = R - mF \cos \alpha \quad i.e., \quad -r \sin \alpha \cos \alpha\, \dot{\phi}^2 = R/m - F \cos \alpha \quad \ldots(4)$$

$$ma_\phi = 0, \quad i.e., \quad \frac{\sin \alpha}{r}\frac{d}{dt}(r^2\dot{\phi}) = 0 \quad \ldots(5)$$

Integrating (5), $\quad r^2\dot{\phi} = A$, A being a constant. $\quad \ldots(6)$

Let the particle describe on the cone a curve which cuts all the generators at a constant angle β, say. Then, we have

$$\tan \beta = v_\phi / v_r = (r \sin \alpha\, \dot{\phi})/\dot{r} \quad \ldots(7)$$

Let v be the velocity of the particle at P. Then, by (1)

$$v = (v_r^2 + v_\theta^2 + v_\phi^2)^{1/2} = (\dot{r}^2 + r^2 \sin^2 \alpha\, \dot{\phi}^2)^{1/2} \quad \ldots(8)$$

Substituting the value of $\dot{\phi}$ from (6) in (7), we get

$$\dot{r} = r \sin \alpha \cot \beta\, \dot{\phi} = r \sin \alpha \cot \beta \cdot (A/r^2) = (A/r) \sin \alpha \cot \beta \quad \ldots(9)$$

Differentiating both sides of (9) w.r.t. 't', we get

$$\ddot{r} = -(A/r^2)\,\dot{r} \sin \alpha \cot \beta = -(A^2/r^3) \sin^2 \alpha \cot^2 \beta, \text{ by (9)} \quad \ldots(10)$$

Substituting the values of $\dot{\phi}$ and $\ddot{r}$ from (6) and (10) in (3), we get

$$F \sin \alpha = r \sin^2 \alpha \cdot (A/r^2)^2 + (A^2/r^3) \sin^2 \alpha \cot^2 \beta = (A^2/r^3) \sin^2 \alpha \operatorname{cosec}^2 \beta$$

Thus, $$F = (A^2 \sin \alpha \operatorname{cosec}^2 \beta)/r^3 = \mu/r^3, \quad \ldots(11)$$

where $$\mu = A^2 \sin \alpha \operatorname{cosec}^2 \beta = \text{constant.} \quad \ldots(12)$$

Relation (11) gives the required law of force, *i.e.*, the force F acts towards the axis and varies inversely as the cube of the distance of the particle from the vertex O of the cube.

Substituting the values of $\dot{\phi}$ and $\dot{r}$ from (6) and (9) in (8), we get

$$v^2 = (A^2/r^2) \sin^2 \alpha \cot^2 \beta + r^2 \sin^2 \alpha \cdot (A/r^2)^2 = (A^2/r^2) \sin^2 \alpha \operatorname{cosec}^2 \beta$$

or $$v^2 = (\mu \sin \alpha)/r^2 \quad \text{or} \quad v = (\mu \sin \alpha)^{1/2}/r, \text{ using (12)}$$

Since $(\mu \sin \alpha)^{1/2}$ is a constant, it follows that the velocity at P varies inversely as the distance of P from the vertex O of the cone.

Substituting the value of $\dot{\phi}$ from (6) in (4), we get

$$R/m = F \cos \alpha - r \sin \alpha \cos \alpha \cdot (A/r^2)^2 = F \cos \alpha - (A^2 \sin \alpha / r^3) \cos \alpha$$

$$= F \cos \alpha - (F/\operatorname{cosec}^2 \beta) \cos \alpha, \text{ using (11)}$$

$$= F \cos \alpha\, (1 - \sin^2 \beta) = F \cos \alpha \cos^2 \beta$$

Thus $R = (m \cos\alpha \cos^2\beta) \times F$, showing the R is proportional to F, as required.

Ex. 4. *A thin straight hollow smooth tube is always inclined at an angle α to the upward drawn vertical, and revolves with uniform velocity ω about a vertical axis which intersects it. A heavy particle is projected from the stationary point of the tube with velocity (g/ω) cot α; show that in time t it has described a distance $(g \cos\alpha/\omega^2 \sin^2\alpha)(1 - e^{-\omega \sin\alpha \cdot t})$. Find also the reaction of the tube.*

Sol. Let OA be the given thin straight hollow smooth tube inclined at angle α to the upward vertical OZ and let it revolve with uniform velocity ω about OZ. Let, at any time t, coordinates of P be (r, θ, ϕ), where $\theta = \alpha$ so that $\dot{\theta} = 0$. Also, here $\dot{\phi} = \omega$ (given). Hence the usual components of acceleration of particle P are

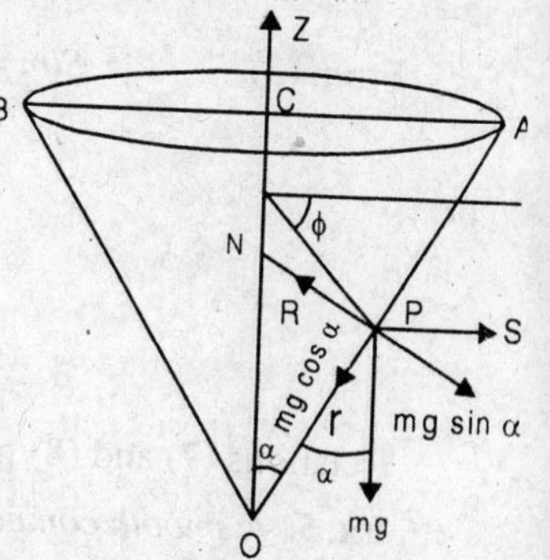

$$a_r = \ddot{r} - r\sin^2\alpha\,\omega^2,$$

$$a_\theta = -r\sin\alpha\cos\alpha\cdot\omega^2,\quad a_\phi = \frac{\sin\alpha}{r}\frac{d}{dt}(r^2\omega)$$

Let R and S be the reactions of the tube in and perpendicular to the vertical plane through the tube. Then the equations of motion are given by

$$ma_r = -mg\cos\alpha \quad \text{i.e.,} \quad \ddot{r} - r\omega^2\sin^2\alpha = -g\cos\alpha \qquad \ldots(1)$$

$$ma_\theta = -R + mg\cos\alpha \quad \text{i.e.,} \quad -r\sin\alpha\cos\alpha\cdot\omega^2 = -R/m + g\sin\alpha \qquad \ldots(2)$$

and $$ma_\phi = S \quad \text{i.e.,} \quad \frac{\sin\alpha}{r}\frac{d}{dt}(r^2\omega) = \frac{S}{m} \qquad \ldots(3)$$

Re-writing (2), $$\dot{r}(d\dot{r}/dr) = r\omega^2\sin^2\alpha - g\cos\alpha$$

or $$2\dot{r}\,d\dot{r} = (2r\omega^2\sin^2\alpha - 2g\cos\alpha)\,dr$$

Integrating, $\dot{r}^2 = r^2\omega^2\sin^2\alpha - 2gr\cos\alpha + A$, A being a constant. ...(4)

Initially at O, where $r = 0$, $\dot{r} = (g/\omega)\cot\alpha$ (given). So (4) gives $A = (g^2/\omega^2)\cot^2\alpha$. Hence (4) reduces to

$$\dot{r}^2 = r^2\omega^2\sin^2\alpha - 2gr\cos\alpha + (g^2/\omega^2)\cot^2\alpha$$

or $$\dot{r}^2 = \omega^2\sin^2\alpha\left\{r^2 - 2r\cdot\frac{g\cos\alpha}{\omega^2\sin^2\alpha} + \left(\frac{g\cos\alpha}{\omega^2\sin^2\alpha}\right)^2\right\} = \omega^2\sin^2\alpha\left(\frac{g\cos\alpha}{\omega^2\sin^2\alpha} - r\right)^2$$

$$\therefore \frac{dr}{dt} = \omega\sin\alpha\left(\frac{g\cos\alpha}{\omega^2\sin^2\alpha} - r\right) \quad \text{or} \quad \omega\sin\alpha\,dt = \frac{dr}{(g\cos\alpha/\omega^2\sin^2\alpha) - r}$$

Integrating, $\omega t\sin\alpha = -\log\{(g\cos\alpha/\omega^2\sin^2\alpha) - r\} + B$, ...(5)

Initially when $t = 0$, $r = 0$. So (5) gives $B = \log(g\cos\alpha/\omega^2\sin^2\alpha)$

$\therefore$ from (5), $\omega t\sin\alpha = -\log\{(g\cos\alpha/\omega^2\sin^2\alpha) - r\} + \log(g\cos\alpha/\omega^2\sin^2\alpha)$

or $-\omega t\sin\alpha = \log(g\cos\alpha/\omega^2\sin^2\alpha) - \log\{(g\cos\alpha/\omega^2\sin^2\alpha) - r\}$

or $$-\omega t\sin\alpha = \log\{1 - (r\omega^2\sin^2\alpha/g\cos\alpha)\}$$

or $e^{-\omega t \sin\alpha} = 1 - (r\omega^2 \sin^2 \alpha / g \cos\alpha)$ or $r = (g\cos\alpha/\omega^2 \sin^2\alpha)(1 - e^{-\omega t \sin\alpha})$ ...(6)

This gives the distance r described in time t.

Substituting the value of r from (6) in (2), we get

$$R/m = g\sin\alpha + \sin\alpha\cos\alpha\cdot\omega^2\cdot(g\cos\alpha/\omega^2\sin^2\alpha)(1 - e^{-\omega t}\sin\alpha)$$

$$R = (mg/\sin\alpha)(1 - \cos^2\alpha\cdot e^{-\omega t}\sin\alpha) \quad ...(7)$$

From (3), $S/m = \{(\sin\alpha)/r\}\cdot 2r\dot{r}\omega = 2\omega\sin\alpha\,(dr/dt)$

or $$S = 2\omega m\sin\alpha\frac{d}{dt}\left\{\frac{g\cos\alpha}{\omega^2\sin^2\alpha}(1 - e^{-\omega t\sin\alpha})\right\}, \text{ by (6)}$$

$$= 2\omega\, m\sin\alpha\cdot(g\cos\alpha/\omega^2\sin^2\alpha)\cdot(\omega\sin\alpha\, e^{-\omega t\sin\alpha})$$

$\therefore$ $$S = 2gm\cos\alpha\, e^{-\omega t\sin\alpha} \quad ...(8)$$

Relations (7) and (8) give reactions of the tube at any time t.

Ex. 5. *A smooth conical surface is fixed with its axis vertical and vertex downwards. A particle is in steady motion on its concave side in a horizontal circle and is slightly disturbed. Show that the time of a small oscillation about this state of steady motion is $2\pi(l/3g\cos\alpha)^{1/2}$, where α is the semi-vertical angle of the cone and l is the length of the generator of the circle of steady motion.*

Sol. Refer figure of Ex. 2. In that figure replace $OM_0 = h$ by $OP_0 = l$. Let a particle of mass m move on the inner surface of the cone AOB whose axis is OC and $\angle AOC = \alpha$. Let R be reaction of the cone acting along perpendicular PN to OA. At any t, let P be the position of the particle such that $OP = r$. Let spherical polar coordinates of P be (r, θ, ϕ) where $\theta = \alpha$ so that $\dot{\theta} = 0$. Hence the usual components of acceleration in terms of spherical coordinates are given by

$$a_r = \ddot{r} - r\sin^2\alpha\,\dot{\phi}^2, \qquad a_\theta = -r\sin\alpha\cos\alpha\,\dot{\phi}^2, \qquad a_\phi = \frac{\sin\alpha}{r}\frac{d}{dt}(r^2\dot{\phi})$$

Hence the equations of motion are given by

$$ma_r = -mg\cos\alpha \quad i.e., \quad \ddot{r} - r\sin^2\alpha\,\dot{\phi}^2 = -g\cos\alpha \quad ...(1)$$

$$ma_\theta = -R + mg\sin\alpha \quad i.e., \quad -mr\sin\alpha\cos\alpha\,\dot{\phi}^2 = -R + mg\sin\alpha \quad ...(2)$$

and $$ma_\phi = 0 \quad i.e., \quad \frac{\sin\alpha}{r}\frac{d}{dt}(r^2\dot{\phi}) = 0 \quad ...(3)$$

Integrating (3), $r^2\dot{\phi} = A$, A being a constant. ...(4)

Initially let particle be at P_0 such that $OP_0 = l$. In this position the particle is in steady motion in a horizontal circle moving with angular velocity ω, say. Thus, initially, when $r = l$, $\dot{\phi} = \omega$. Hence (4) gives $l^2\omega = A$. Then (4) becomes

$$r^2\dot{\phi} = l^2\omega \quad \text{or} \quad \dot{\phi} = l^2\omega/r^2 \quad ...(5)$$

Substituting the value of $\dot{\phi}$ from (5) in (1), we get

$$\ddot{r} = (l^4\omega^2 \sin\alpha \cos\alpha)/r^3 - g\cos\alpha \qquad \ldots(6)$$

But for steady motion, $\ddot{r} = 0$ when $r = l$. So (6) gives

$$0 = (l^4\omega^2 \sin\alpha\cos\alpha)/l^3 - g\cos\alpha \quad \text{or} \quad g\cos\alpha = l\omega^2 \sin\alpha\cos\alpha \qquad \ldots(7)$$

Putting $r = l + \rho$, where ρ is small, (6) gives

$$\frac{d^2\rho}{dt^2} = \frac{l^4\omega^2 \sin\alpha\cos\alpha}{(l+\rho)^3} - g\cos\alpha = l\omega^2 \sin\alpha\cos\alpha\left(1+\frac{\rho}{l}\right)^{-3} - g\cos\alpha$$

$= l\omega^2 \sin\alpha\cos\alpha\,(1 - 3\rho/l) - l\omega^2 \sin\alpha\cos\alpha$, using (7) and the fact that ρ is small.

or $\ddot{\rho} = -(3/l)g\cos\alpha\cdot\rho$, using (5)

which represents a S.H.M. of time period $2\pi(l/3g\cos\alpha)^{1/2}$.

EXERCISE

1. A smooth circular cone, of angle 2α, has its axis vertical and its vertex, which is pierced with a small hole, downwards. A mass M hangs at rest by a string which passes through the vertex, and a mass m attached to the upper end describes a horizontal circle on the inner surface of the cone. Find the time T of a complete revolution, and show that small oscillations about the steady motion take place in the time $T \operatorname{cosec}\alpha\{(M+m)/3m\}^{1/2}$.

2. A point moves with a constant velocity on a cone so that its direction of motion makes a constant angle with a plane perpendicular to the axis of the cone. Show that the resultant acceleration is perpendicular to the axis of the cone and varies inversely as the distance of the point from the axis.

3. At the vertex of a smooth cone of vertical cone 2α, fixed with its axis vertical and vertex downwards, is a centre of repulsive force $\mu/(\text{distance})^4$. A weightless particle is projected horizontally with velocity $(2\mu \sin^3\alpha/c^3)^{1/2}$ from a point, distant c from the axis, along the inside of the surface. Show that it will describe a curve on the cone whose projection on a horizontal plane is $1 - (c/r) = 3\tanh^2\{(\theta/2)\sin\alpha)\}$.

4. A particle moves on a smooth cone under the action of a force to the vertex varying as the square of the distance. If the cone be developed into a plane, show that the path becomes a conic section.

5. If a particle moves on the inner surface of a right circular cone under the action of a force from the vertex, the law of repulsion being $m\mu(a\cos^2\alpha/r^3 - 1/2r^3)$, where 2α is the vertical angle of the cone, and if it be projected from an apse at distance a with velocity $(\mu/a)^{1/2}\sin\alpha$, show that the path will be a parabola.

[show that the plane of the motion is parallel to a generator of the cone]

6. A particle is constrained to move on a smooth conical surface of vertical angle 2α and describes a plane curve under the action of an attraction to the vertex, the plane of the orbit cutting the axis of the cone at a distance a from the vertex. Show that the attractive force must vary as $(1/r^2) - (a/r^3)\cos\alpha$.

MISCELLANEOUS EXERCISE

1. A particle moves on a rough circular cylinder under the action of no external forces. Initially the particle has a velocity V in a direction making an angle α with the transverse plane of the cylinder, show that the space described in time t is $(a/\mu)\sec^2\alpha \log\{1 + (\mu Vt/a)\cos^2\alpha\}$.

2. Three masses m_1, m_2 and m_3 are fastened to a string which passes through a ring, and m_1, describes a horizontal circle as a conical pendulum while m_2 and m_3 hang vertically. If m_3 drops off, show that the instantaneous change of tension of the string is $(gm_1m_3)/(m_1 + m_2)$.

3. Investigate the motion of a conical pendulum when disturbed from its state of steady motion by a small vertical harmoic oscillation of the point of support. Can the steady motion be rendered unstable by such a disturbance ?

4. Find the acceleration of a particle in terms of spherical polar co-ordinates.

(Purvanchel 2003, 07)

5. A particle moving along a curve in three dimension, find its acceleration along the tangent to the curve.

(Purvanchel 2002, 06)

14

Moments and Products of Inertia

14.1 Moment of inertia (M.I.). Definition. **[Meerut 2005, 07, 09]**

The moment of inertia of a particle about a line is defined as $I = mr^2$, where m is the mass of the particle and r its perpendicular distance from the line. The moment of inertia of a system of particles is defined as the sum of the moments of inertia of the separate particles. Thus, we have

$$I = \sum_{i=1}^{n} m_i r_i^2, \qquad \ldots(1)$$

if the system consists of n particles of masses $m_1, m_2, \ldots, m_n$, situated at distances $r_1, r_2, \ldots, r_n$ from the line about which the moment of inertia is taken.

In the case of a rigid body (or a continuous distirbution of matter), the definition (1) takes the form

$$I = \int r^2 \, dm, \qquad \ldots(2)$$

whre the integral extends over all elements of the body.

Remark. The expression $\Sigma m_i r_i^2$ is also used to define the M.I. of a body about a point (or a plane) where r_i denotes distance from a point (or a plane).

14.2 Radius of gyration. Definition. **[Agra 2011]**

If the moment of inertia of a system of particles (or a body) about a line is expressed in the form $I = MK^2$, where M is the total mass of the system (or a body) and K a non-negative real number, then the number K so defined is known as the *radius of gyration* of the system (or a body) about the given line.

14.3A Perpendicular axis theorem. [Delhi B.Sc. (Hons) 2005, 10; Kanpur 2003, 06; Mumbai 2010]

If A, B are the moments of inertia of a plane lamina about two perpendicular lines on its plane, then the moment of inertia of the lamina, about an axis through their intersection and perpendicular to the plane is C given by C = A + B.

Proof. Consider a typical element δm at P whose coordinates referred to Ox and Oy as axes are (x, y). Then, clearly OZ is the line through the intersection of OX and OY and perpendicular to the plane. Then, by definition, we have

$$A = \Sigma y^2 \, \delta m \quad \text{and} \quad B = \Sigma x^2 \, \delta m \qquad \ldots(1)$$

Now, $C =$ of inertia about OZ

$$= \Sigma OP^2 \, \delta m = \Sigma(x^2 + y^2) \, \delta m = \Sigma y^2 \, \delta m + \Sigma x^2 \, \delta m = A + B, \text{ by (1)}$$

Note. The above theorem is true only for a plane lamina.

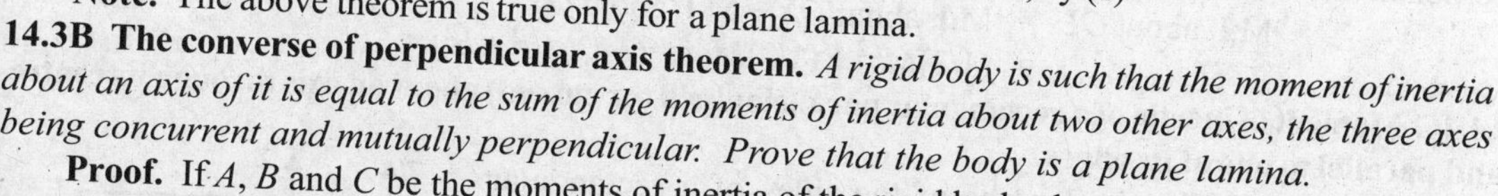

14.3B The converse of perpendicular axis theorem. *A rigid body is such that the moment of inertia about an axis of it is equal to the sum of the moments of inertia about two other axes, the three axes being concurrent and mutually perpendicular. Prove that the body is a plane lamina.*

Proof. If A, B and C be the moments of inertia of the rigid body about x-axis, y-axis and z-axis respectively, then given that

$$C = A + B \quad \text{or} \quad \int (x^2 + y^2) \, \delta m = \int (y^2 + z^2) \, \delta m + \int (z^2 + x^2) \, \delta m$$

or $\quad 2 \int z^2 \, \delta m = 0 \quad$ so that $\quad z^2 = 0 \quad$ and $\quad$ hence $z = 0$.

Since $z = 0$ for all points of the rigid body, it must be a plane lamina.

14.4 Moments of inertia : Some standard cases

14.4A Moment of inertia of a rod of length 2*a*

Case I. *About an axis through an end perpendicular to the rod.* **(Kanpur 2009; Agra 2010)**

14.2

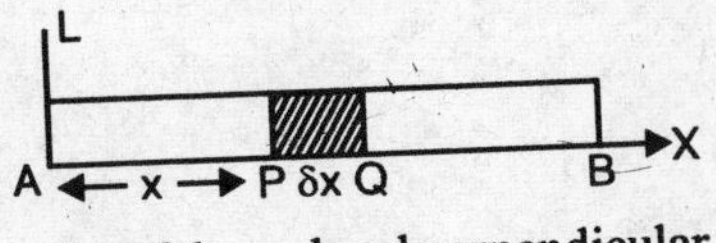

Let M be the mass of the rod AB so that its mass per unit length $= \rho = M/2a$. Consider an element PQ of breadth δx at a distance x from the end A. Then, δm = mass of the element $PQ = \rho\ \delta x = (M/2a)\ \delta x$.

$\therefore$ M.I. of this element PQ about the line AL passing through the end A of the rod and perpendicular to it $= x^2\ \delta m = (M/2a)x^2\ \delta x$

Hence M.I. of the rod AB about AL

$$= \int_0^{2a} (M/2a)x^2\, dx = (M/2a)[x^3/3]_0^{2a} = (4/3)Ma^2$$

(Meerut 2009)

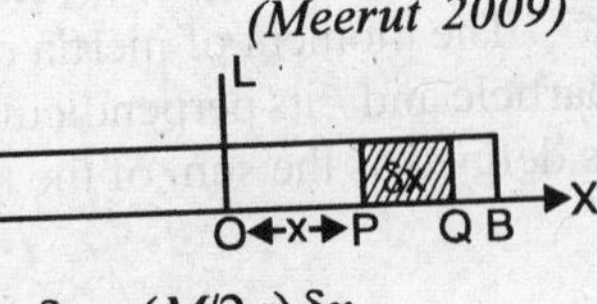

Case II. *About an axis through the middle point perpendicular to the rod.* Let OL be the line passing through the middle point O of the rod AB and perpendicular to it. Let M be the mass of the rod AB so that its mass per unit length $= \rho = M/2a$. Consider as element PQ of breadth δx at a distance x from O. Then, δm = mass of element $PQ = \rho\ \delta x = (M/2a)\ \delta x$.

$\therefore$ M.I. of the element PQ about $OL = x^2\ \delta m = (M/2a)x^2\ \delta x$

Hence M.I. of the rod AB above OL

$$= \int_{-a}^{a} (M/2a)x^2\, dx = (M/2a)[x^3/3]_{-a}^{a} = (1/3)Ma^2$$

14.4B Moment of inertia of a rectangular lamina

(Meerut 2005)

Case I. *About a line through centre parallel to a side.*

Let M be the mass of a rectangular lamina $ABCD$ such that $AB = 2a$ and $BC = 2b$. Hence, ρ = mass per unit area of the rectangle $= M/4ab$. Let OX and OY be straight lines parallel to AB and BC through the centre O of the rectangle. Consider an elementary area $PQRS$ at $P(x, y)$ of the rectangle. Then, we have

δm = mass of the element $PQRS$

$= \rho\ \delta x\ \delta y = (M/4ab)\ \delta x\ \delta y$

M.I. of this elementary mass δm about $OX = y^2\ \delta m = (M/4ab)y^2\ \delta x\ \delta y$

Hence M.I. of the rectangular lamina $ABCD$ about OX

$$= \int_{x=-a}^{a}\int_{y=-b}^{b} \frac{M}{4ab} y^2\, dx\, dy = \frac{M}{4ab}\int_{-a}^{a}\left[\frac{y^3}{3}\right]_{-b}^{b} dx = \frac{M}{4ab}\cdot\frac{2b^3}{3}\int_{-a}^{a} dx = \frac{Mb^2}{3}$$

Similarly, M.I. of the rectangle about $OY = (1/3)Mb^2$.

Case II. *Above a line through its centre and perpendicular to its plane.* (Agra 2009)

Let OL be a line through the centre and perpendicular to the rectangular lamina $ABCD$. Using perpendicular axis theorem (refer Art. 14.3A), we have

$$\text{M.I. about } OL = \text{M.I. about } OX + \text{M.I. about } OY$$
$$= (1/3)Mb^2 + (1/3)Ma^2 = (M/3)(a^2 + b^2)$$

14.4C Moment of inertia of a rectangular parallelopiped (or cuboid) about an axis through its centre and parallel to one of its edges.

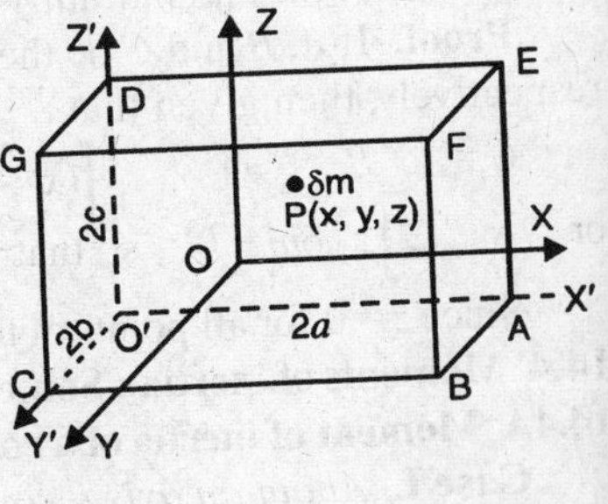

Let $2a$, $2b$, $2c$ be the lengths of the sides of a rectangular parallelopiped, and O its centre. Let OX, OY, OZ be the the axes through O and parallel to the edges of the rectangular parallelopiped whose mass is M (say). Then ρ = mass per unit volume $= M/8abc$. Consider an element $\delta x\ \delta y\ \delta z$ at the point $P(x, y, z)$. Its mass $= \delta m = \rho\ \delta x\ \delta y\ \delta z$ and distance of $P(x, y, z)$ from OX is $(y^2 + z^2)^{1/2}$.

$\therefore$ M.I. of δm at P about $OX = \rho(y^2 + z^2)\ \delta x\ \delta y\ \delta z$.

Hence M.I. of the rectangular parallelopiped about OX (which is parallel to the edge $2a$)

$$= \int_{x=-a}^{a}\int_{y=-b}^{b}\int_{z=-c}^{c} \rho(y^2+z^2)\,dx\,dy\,dz = \rho\int_{-a}^{a}\int_{-b}^{b}\left[y^2 z+\frac{1}{3}z^3\right]_{-c}^{c} dx\,dy$$

$$= \rho\int_{-a}^{a}\int_{-b}^{b} 2(y^2 c+\frac{1}{3}c^3)\,dx\,dy = 2\rho\int_{-a}^{a}\left[\frac{1}{3}y^3 c+\frac{1}{3}c^3 y\right]_{-b}^{b} dx$$

$$= \frac{2\rho}{3}\int_{-a}^{a} 2(b^3 c+c^3 b)\,dx = \frac{4\rho}{3}bc(b^2+c^2)[x]_{-a}^{a} = \frac{8\rho\,abc}{3}(b^2+c^2) = \frac{M(b^2+c^2)}{3}, \text{ as } \rho = \frac{M}{8abc}$$

Similarly, the M.I. of the rectangular parallelopiped about the straight lines OY and OZ through O and parallel to the edges $2b$ and $2c$ are $(1/3)M(c^2+a^2)$ and $(1/3)M(a^2+b^2)$ respectively.

Particular Case. For a cube of side, $2a$, we have $2a = 2b = 2c$. Hence M.I. of a cube about a line through its centre O and parallel to any one of its edges is $(2/3)Ma^2$.

14.4D Moment of inertia of a circular wire

[Agra 2010, 12; Meerut 2006, 08]

Case I. *About a diameter.* Let M be the mass of the circular wire of centre O and radius a. Then, ρ = mass per unit length of the wire $= M/2\pi a$. Consider an elementary arc $PQ (= a\,\delta\theta)$ of the wire. Then elementary mass $= \delta m = \rho a\,\delta\theta$ whose distance from the diameter $AB = PM = a \sin\theta$.

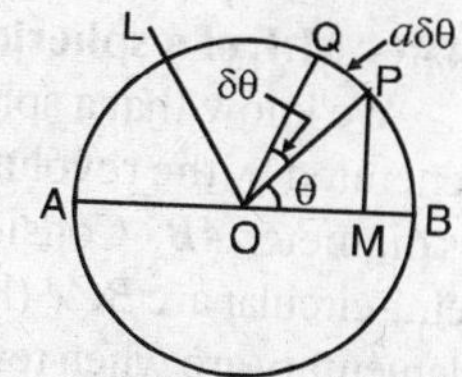

Now, M.I. of the element PQ about diameter AB

$= PM^2 \cdot \delta m = (a\sin\theta)^2 \cdot \rho a\,\delta\theta = \rho a^3 \sin^2\theta\,\delta\theta$

Hence, M.I. of the circular wire about the diameter AB

$$= \int_0^{2\pi} \rho a^3 \sin^2\theta\,d\theta = \frac{1}{2}\rho a^3\int_0^{2\pi}(1-\cos 2\theta)\,d\theta$$

$= (\rho a^3/2)\times[\theta-(1/2)\sin 2\theta]_0^{2\pi} = (1/2)\cdot\rho\cdot a^3\cdot 2\pi$

$= (1/2)\cdot(M/2\pi a)\cdot a^3\cdot 2\pi = (1/2)Ma^2$, as $\rho = M/2\pi a$

Case II. About a line through the centre and perpendicular to its plane. **[Mumbai 2010]**

Let OL be the line through the centre O and perpendicular to its plane. Then, M.I. of the elementary arc PQ about $ON = OP^2 \cdot \delta m = a^2 \cdot \rho a\,\delta\theta$.

Hence, M.I. of the circular wire about OL

$$= \int_0^{2\pi} \rho a^3 d\theta = \rho a^3 [\theta]_0^{2\pi} = \frac{M}{2\pi a}\cdot a^3\cdot 2\pi = Ma^2, \text{ as } \rho = \frac{M}{2\pi a}$$

14.4E Moment of inertia of a circular disc

[Kanpur 2003; Agra 2012]

Case I. *About a diameter.* Let M be the mass of the circular disc of centre O and radius a. Then, ρ = mass per unit area of the disc $= M/\pi a^2$. Consider an elementary area $r\,\delta\theta\,\delta r$ at $P(r,\theta)$ of the disc referred to the centre O as the pole and OX as initial line. Mass of this element $= \delta m = \rho\cdot r\,\delta\theta\,\delta r$ and distance of element from $OX = PM = r\sin\theta$.

Now M.I. of the element about the diameter $X'X$

$= PM^2 \cdot \delta m = (r\sin\theta)^2 \cdot \rho r\,\delta\theta\,\delta r = \rho\cdot r^3 \sin^2\theta\,\delta\theta\,\delta r$

Hence, M.I. of the circular disc about the diameter $X'X$ =

$$\int_{\theta=0}^{2\pi}\int_{r=0}^{a} \rho r^3 \sin^2\theta\,d\theta\,dr = \rho\int_0^{2\pi}\left[\frac{1}{4}r^4\right]_0^a \sin^2\theta\,d\theta = \frac{1}{4}\rho a^4\int_0^{2\pi}\frac{1}{2}(1-\cos 2\theta)\,d\theta$$

$= (1/8)\rho a^4[\theta-(1/2)\sin 2\theta]_0^{2\pi} = (1/8)(M/\pi a^2)a^4\cdot 2\pi = (1/4)Ma^2$

Case II. *About a line through the centre and perpendicular to its plane.* Let OL be the line through the centre O and perpendicular to its plane. Then,

M.I. of the mass of the elementary area $r\,\delta\theta\,\delta r$ about $OL = OP^2 \cdot \delta m = r^2 \cdot \rho r\,\delta\theta\,\delta r = \rho r^3\,\delta\theta\,\delta r$

Hence, M.I. of the circular disc about OL

$$= \int_0^{2\pi}\int_{r=0}^{a} \rho r^3\,d\theta\,dr = \rho\int_0^{2\pi}\left[\frac{1}{4}r^4\right]_0^a d\theta = \frac{1}{4}\rho a^4[\theta]_0^{2\pi}$$

$= (1/4)\cdot(M/\pi a^2)\cdot a^4\cdot 2\pi = (1/2)Ma^2$, as $\rho = M/\pi a^2$

14.4F Moment of inertia of a uniform triangular lamina about one side.

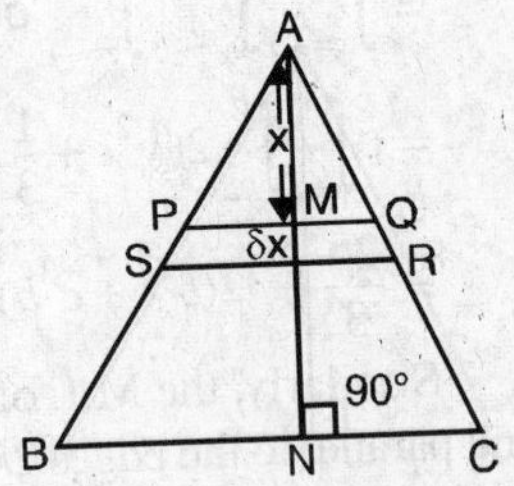

Let M be the mass and $h(=AN)$ be the height of a triangular lamina ABC. Let us divide the lamina ABC by strips parallel to BC. Let $PQRS$ be one such an elementary strip, of breadth δx and at a distance x from the vertex A of the triangle. Let $BC = a$. Since ABC and APQ are similar triangles, we have

$x/AN = PQ/BC$ so that $PQ = ax/h$.

Then, ρ = mass per unit area of $\Delta ABC = M/(ha/2) = 2M/ha$

Now, the mass of the elementary strip $= \delta m = \rho \cdot PQ \cdot \delta x = P(ax/h)\, \delta x$

and M.I. of this elementary strip about BC

$= \delta m \cdot MN^2 = (\rho ax/h)\, \delta x \cdot (h-x)^2$, as $MN = AN - AM = h - x$

Hence, M.I. of the triangle ABC about BC

$$= \int_0^h \frac{\rho a}{h}(h-x)^2 x\, dx = \frac{\rho a}{h}\int_0^h (h^2 x - 2hx^2 + x^3)\, dx = \frac{\rho a}{h}\left[\frac{1}{2}h^2x^2 - \frac{2}{3}hx^3 + \frac{1}{4}x^4\right]_0^h$$

$= (1/12)\rho a h^3 = (1/12)\cdot(2M/ha)\cdot ah^3 = (1/6)Mh^2$

(Kanpur 2002, 04)

14.4G M.I. of a spherical shell (or hollow sphere) about a diameter.

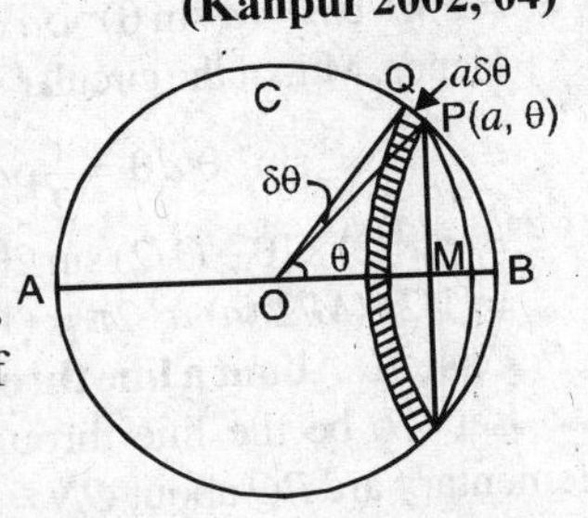

We know that a spherical shell (or hollow sphere) of radius a can be generated by the revolution of a semi-circular wire BCA of radius a about its diameter AB. Consider an elementary arc $PQ(= a\,\delta\theta)$ at $P(a, \theta)$ of the semi-circular arc BCA (Here O is taken as origin and OB as initial line). This elementary arc when revolved about AB, generates a circular ring of radius $PM(= a \sin\theta)$. Now, ρ = mass per unit surface area of the spherical shell (of mass M, say) $= M/4\pi a^2$. Let PM be perpendicular to AB.

Now, mass of the elementary ring $= \delta m = \rho \cdot 2\pi\, PM \cdot a\, \delta\theta$

or $\delta m = \rho \cdot 2\pi \cdot (a \sin\theta) \cdot a\, \delta\theta = 2\pi\rho a^2 \sin\theta\, \delta\theta$

Since the distance of every point of this elementary ring from the diameter AB is $a \sin\theta$, hence M.I. of the elementary ring about AB (a line through the centre M of the ring and perpendicular to its plane) $= \delta m \cdot PM^2$. [Refer Case II of Art. 14.4D]

$= (2\pi\rho a^2 \sin\theta\, \delta\theta) \cdot (a \sin\theta)^2 = 2\pi\rho a^4 \sin^3\theta\, \delta\theta$

Hence, M.I. of the shell about the diameter AB

$$= \int_0^{\pi} 2\pi\rho a^3 \sin^3\theta\, d\theta = 4\pi\rho a^4 \int_0^{\pi/2} \sin^3\theta\, d\theta = 4\pi\rho a^4 \cdot \frac{2}{3}$$, using Walli' formulas

$= (8\pi/3)\cdot(M/4\pi a^2)\cdot a^4 = (2/3)Ma^2$, as $\rho = M/4\pi a^2$

14.4H M.I. of a solid sphere about a diameter

[Agra 2011; Kanpur 2000, 04, 07; Meerut 2007; I.A.S. 1999; Gujrat 2004]

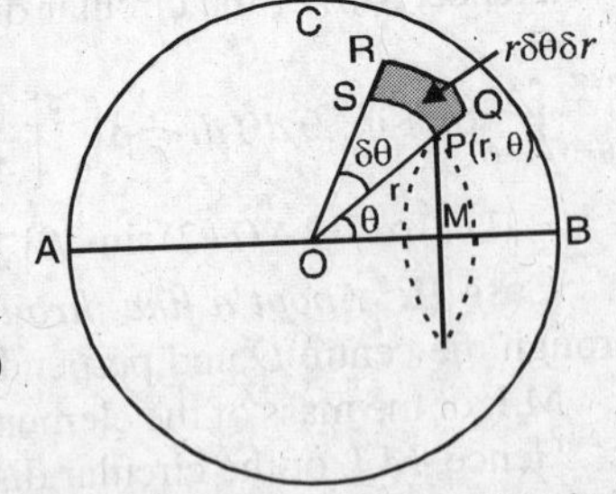

We know that a solid sphere of radius a can be generated by the revolution of a semi-circular area ABC of radius a about its diameter AB. Let M be mass of the solid sphere, then ρ = mass per unit volume $= M/(4\pi a^3/3) = 3M/4\pi a^3$. Let the centre O of semi-circular area be taken as origin and OB as initial line. Consider an elementary area $r\,\delta\theta\,\delta r$ at $P(r, \theta)$ of the semi-circular area. This element, when revolved about the diameter AB, generates a circular ring of radius $PM(= r \sin\theta)$ and cross-section $r\,\delta\theta\,\delta r$. Here PM is perpendicular to AB.

$\therefore$ δm = mass of the elementary ring $= \rho \cdot 2\pi r \sin\theta \cdot r\,\delta\theta\,\delta r$

M.I. of this elementary ring about AB (a line through the centre M of the ring and perpendicular to its plane) $= \delta m \cdot PM^2$ [Refer Case II of Art. 14.4D]

$= 2\pi\rho r^2 \sin\theta\,\delta\theta\,\delta r \cdot (r \sin\theta)^2 = 2\pi\rho r^4 \sin^3\theta\,\delta\theta\,\delta r$

Hence, M.I. of the sphere about the diameter AB

$$= \int_{\theta=0}^{\pi} \int_{r=a}^{a} 2\pi\rho r^4 \sin^3\theta \, d\theta \, dr = 2\pi\rho \int_0^{\pi} \left[\frac{1}{5} r^5\right]_0^a \sin^3\theta \, d\theta$$

$$= \frac{2\pi\rho a^5}{5} \int_0^{\pi} \sin^3\theta \, d\theta = \frac{4\pi\rho a^5}{5} \int_0^{\pi/2} \sin^3\theta \, d\theta = \frac{4\pi\rho a^5}{5} \cdot \frac{2}{3}$$ [Using Walli's formula]

$= (8/15)\pi a^5 \cdot (3M/4\pi a^3) = (2/5)Ma^2$, as $\rho = 3M/4\pi a^3$

14.4I M.I. of an elliptic disc

Case I. *About the major axis.* **(Meerut 2005, 10; Agra 2009)**

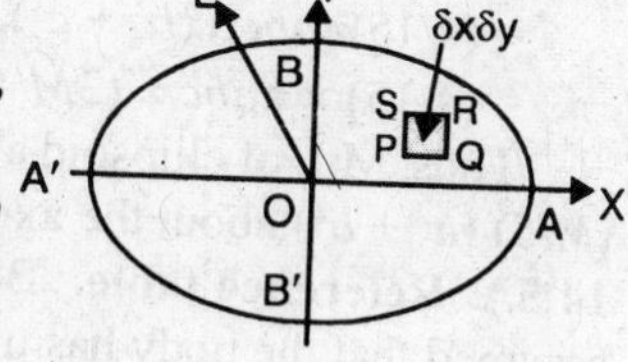

Let M be the mass of an elliptic disc of axes $2a$ and $2b$. Then, ρ = mass per unit area of the disc = $M/\pi ab$. Equation of the elipse is

$x^2/a^2 + y^2/b^2 = 1$ ...(1)

Consider an elementary area $PQRS$ (= $\delta x \, \delta y$) of the ellipse disc at $P(x, y)$. Then its mass $\rho = \delta m = \rho \, \delta x \, \delta y$.

$\therefore$ M.I. of the elementary mass δm about major axis AA' = = $y^2 \delta m = y^2 \rho \delta x \delta y$

$\therefore$ M.I. of the elliptic disc about major axis AA'

$= \int_{x=-a}^{a} \int_{y=-y_1}^{y_1} y^2 \rho \, dx \, dy$, where $y_1 = b(1 - x^2/a^2)^{1/2}$, by (1)

$$= \rho \int_{a}^{a} \left[\frac{1}{3} y^3\right]_{-y_1}^{y_1} dx = \frac{2}{3} \rho \int_{a}^{a} y_1^3 \, dx = \frac{2\rho}{3} \int_{a}^{a} b^3 \left(1 - \frac{x^2}{a^2}\right)^{3/2} dx$$

$$= \frac{2\rho ab^3}{3} \int_{-\pi/2}^{\pi/2} (1 - \sin^2\theta)^{3/2} \cdot a \cos\theta \, d\theta$$, putting $x = a \sin\theta$

$$= \frac{2\rho ab^3}{3} \int_{-\pi/2}^{\pi/2} \cos^4\theta \, d\theta = \frac{4\rho ab^3}{3} \int_0^{\pi/2} \cos^4\theta \, d\theta = \frac{4\rho ab^3}{3} \cdot \frac{3}{4} \cdot \frac{1}{2} \cdot \frac{\pi}{2}$$ [Using Walli's formula]

$= (1/4)\rho\pi b^3 a = (1/4) \times (M/\pi ab) \times (\pi b^3 a) = (1/4)Mb^2$

Case II. *About the minor axis.* **[Kanpur 2005]**

Proceed as in Case I. Then M.I. of the elliptic disc about minor axis $BB' = (1/4)Ma^2$

Case III. *About the line OL through the centre O and perpendicular to the plane of the lamina*

Using perpendicular theorem (See Art. 14.3 A), we get **[Meerut 2010]**

M.I. about OL = M.I. about AA' + M.I. about BB'

$= (1/4)Mb^2 + (1/4)Ma^2 = (1/4)\, M(a^2 + b^2)$

14.4J M.I. of an ellipsoid

[Meerut 2005; Purvanchel 2006]

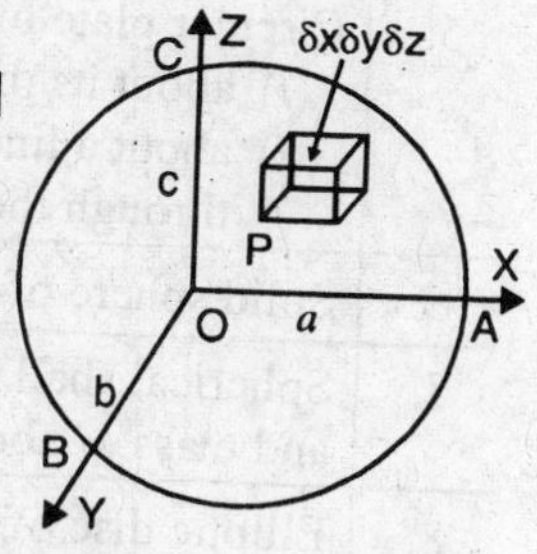

Let the equation of the ellipsoid be $x^2/a^2 + y^2/b^2 + z^2/c^2 = 1$ ...(1)

Let M be mass of this ellipsoid. Then

$$\rho = \text{mass per unit volume} = \frac{M}{(4/3)\pi abc} = \frac{3M}{4\pi abc}$$ ...(2)

Consider an elementary volume $\delta x \, \delta y \, \delta z$ at the point $P(x, y, z)$ in the positive octant. The mass of this element = $\delta m = \rho \, \delta x \, \delta y \, \delta z$. Its distance from $OX = (y^2 + z^2)^{1/2}$.

$\therefore$ Its M.I. about $OX = \rho \, \delta x \, \delta y \, \delta z \, (y^2 + z^2)$

Hence, M.I. of the ellipsoid about OX

$= 8 \iiint \rho(y^2 + z^2) \, dx \, dy \, dz$, where

$x^2/a^2 + y^2/b^2 + z^2/c^2 \le 1$ and the integration being extended over the positive octant of the ellipsoid.

Put $x^2/a^2 = u$, *i.e.*, $x = au^{1/2}$ so that $dx = (1/2)au^{-1/2}\, du$,

$y^2/b^2 = v$, *i.e.*, $y = bv^{1/2}$ so that $dy = (1/2)bv^{-1/2}\, dv$

and $z^2/c^2 = w$, *i.e.*, $z = cw^{1/2}$ so that $dz = (1/2)cw^{-1/2}\, dw$

Hence, the M.I. of the ellipsoid about OX

$= 8 \iiint \rho (1/8) abc \, (b^2 v + c^2 w) \, u^{-1/2} v^{-1/2} w^{-1/2} \, du \, dv \, dw$, where u, v, w are +ive, $u + v + w \le 1$

$$= abc\rho \iiint (b^2 u^{1/2-1} v^{3/2-1} w^{1/2-1} + c^2 u^{1/2-1} v^{1/2-1} w^{3/2-1})\, du\, dv\, dw$$

$$= abc\rho (b^2 + c^2) \frac{\Gamma(1/2)\Gamma(3/2)\Gamma(1/2)}{\Gamma(1/2+3/2+1/2+1)}, \text{ by Dirichlet's theorem*}$$

$$= abc\rho(b^2 + c^2)\frac{\sqrt{\pi}.\sqrt{\pi}\cdot\Gamma(3/2)}{\Gamma(7/2)} = \pi abc\rho(b^2+c^2)\cdot\frac{\Gamma(3/2)}{(5/2)\cdot(3/2)\Gamma(3/2)}$$

$= (4/15)\pi abc\rho\ (b^2 + c^2)$, as $\Gamma(1/2) = \sqrt{\pi}$ and $\Gamma(n+1) = n\Gamma(n)$

$= (4/15) \times \pi abc \times (3M/4\pi abc) \times (b^2 + c^2) = (1/5)M\ (b^2 + c^2)$, by (2)

Thus, M.I. of ellipsoid about the axis $2a$ is $(M/5)\ (b^2 + c^2)$ with similar results $(M/5)\ (a^2 + c^2)$ and $(M/5)\ (a^2 + b^2)$ about the axes $2b$ and $2c$ respectively.

14.5A Reference table. Below we give standard results already obtained. In each case, we have assumed that the body has uniform density. Remember them for direct applications later on.

S.No.	*Rigid body*	*M.I.*
1.	Uniform rod of length $2a$ and mass M (*i*) about a line ⊥ to the rod through an end (*ii*) about an axis ⊥ to the rod through the centre of rod	 $(4/3)Ma^2$ $(1/3)Ma^2$
2.	Rectangular lamina of sides $2a$, $2b$ and mass M (*i*) About a line through the centre of the lamina and parallel to the side $2a$ (*ii*) About an axis ⊥ to the lamina through the centre of the lamina	 $(1/3)Mb^2$ $(M/3)\ (b^2 + c^2)$
3.	Rectangular parallelopiped of edges $2a$, $2b$, $2c$ about a line parallel to the edge $2a$, through the centre of the paralellopiped	$(M/3)\ (b^2 + c^2)$
4.	Circular ring of radius a and mass M (*i*) About a diameter (*ii*) About a line ⊥ to the plane of the ring and through the centre of the ring	 $(M/2)a^2$ Ma^2
5.	Circular plate of radius a and mass M (*i*) about its diameter (*ii*) about a line ⊥ to the plane of the plate and through the centre of the plate	 $(M/4)a^2$ $(M/2)a^2$
6.	Solid sphere of radius a and mass M about a diameter	$(2/5)Ma^2$
7.	Spherical shall (or hollow sphere) of radius a and mass M about a diameter.	$(2/3)Ma^2$
8.	Elliptic disc of axes $2a$, $2b$ and mass M (*i*) about the axis $2a$ (*ii*) about a line ⊥ to the disc through its centre	 $(1/4)Mb^2$ $(M/4)\ (a^2 + b^2)$
9.	Ellipsoid of axes $2a$, $2b$ and $2c$ about $2a$	$(M/5)\ (b^2 + c^2)$
10.	Triangular lamina ABC about BC of mass M and height $h(=AN)$, N being the foot of ⊥ from A on BC.	$(1/6)Mh^2$

***Dirichlet's theorem.** If x, y, z are all +ve such that $x + y + z \le 1$, then

$$\iiint x^{l-1} y^{m-1} z^{n-1}\, dx\, dy\, dz = \frac{\Gamma(l)\Gamma(m)\Gamma(n)}{\Gamma(l+m+n+1)}$$

14.5B Routh's rule. **[Agra 2011; Meerut 2003]**

All the simple cases, considered above, except for the triangular lamina, possess axes of symmetry. The results of above reference table may be remembered with help of the following rule, due to Routh :

$$\text{Moment of inertia about an axis of symmetry} = \text{mass} \times \frac{\text{sum of squares of perpendicular semi-axes}}{3,\ 4,\ \text{or}\ 5},$$

where the denominator is 3, 4, or 5 according as the body is rectangular (including rod),elliptical (including circular) or ellipsoidal (including spherical).

14.6 Product of inertia (P.I.). Definition. Let P and Q be two planes, and let p, q denote the perpendicular distances from them of an elementary mass δm of the body. The distance is counted +ve or –ve according as δm lies on one side or the other side of the corresponding plane. Then the product of inertia of the body with respect to the planes P, Q is defined as the real number

$$\Sigma pq\,\delta m \qquad \text{or} \qquad \int pq\,\delta m, \qquad \ldots(1)$$

where the summation (or integral) extends over all elements of the body. The expression (1) is also used to define the product of inertia of a body about two lines P, Q, where p, q denote the distances of the element δm from the two lines.

Note. If the body is symmetrical about the lines P, Q then P.I. of the body vanishes.

14.7 Notations : Moments and product of inertia with respect three mutually perpendicular axes. Let (x, y, z) be the coordinates of the element δm of a body with respect to three mutually perpendicular axes OX, OY, OZ. Then the M.I. of the body about OX, OY, OZ denoted by A, B, C respectively and the product of inertia about the axes OY, OZ; OZ, OX and OX, OY denoted by F, H, F respectively are given by

$$A = \Sigma\delta m\,(y^2 + z^2), \qquad B = \Sigma\delta m\,(z^2 + x^2), \qquad C = \Sigma\delta m\,(x^2 + y^2),$$

$$F = \Sigma\delta m\,yz, \qquad G = \Sigma\delta m\,zx, \qquad H = \Sigma\delta m\,xy$$

14.8 Method of decomposition. We divide the body into a number of parts, for each of which the M.I. is known; by adding the moments of inertia of these parts, we obtain the required result. This is known as the *method of decomposition* and has been illustrated in the following example.

Example. *Show that the M.I. of a hollow sphere of mass M, and external and internal radii a, b respectively, about a diameter, is* $(2M/5) \times \{(a^5 - b^5)/(a^3 - b^3)\}$. **(Garhwal 93; Kumaun 96)**

Solution. The given hollow sphere can be treated as the difference of two solid spheres of radii a, b. If M_1, M_2 be their masses then, $M = M_1 - M_2$.

Now, the mass of a sphere being proportional to its volume and volume being proportional to cube of its radius, we get

$$\frac{M_1}{a^3} = \frac{M_2}{b^3} = \frac{M_1 - M_2}{a^3 - b^3} = \frac{M}{a^3 - b^3} \qquad \ldots(1)$$

Let L be the line of their common diameter, then M.I. of the hollow sphere about line l $= (2/5)M_1 a^2 - (2/5)M_2 b^2$, using result of Art. 14.4 H

$$= \frac{2}{5}\left[\frac{Ma^5}{a^3 - b^3} - \frac{Mb^5}{a^3 - b^3}\right] = \frac{2M}{5}\cdot\frac{a^5 - b^5}{a^3 - b^3}, \text{ using (1)}$$

TYPE 1 : SOLVED EXAMPLES BASED ON ART. 14.1 TO 14.8

Ex. 1. *Find the product of inertia of an elliptic quadrant about its axes.* **[Rohilkhand 1998]**

Sol. Let the equation of the ellipse be $\quad x^2/a^2 + y^2/b^2 = 1 \qquad \ldots(1)$

Let M be mass of the elliptic quadrant OAB. Then ρ = mass per unit area $= M/(\pi ab/4)$ so that $\rho = 4M/\pi ab$. Consider an elementary area $\delta x\,\delta y$ at the point $P(x, y)$. Then its mass $= \delta m = \rho\,\delta x\,\delta y$.

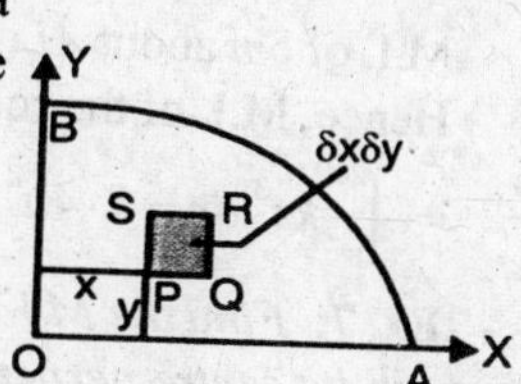

$\therefore$ Product of inertia of δm about OX, OY

$$= xy\,\delta m = \rho\,xy\,\delta x\,\delta y.$$

Hence, the required product of inertia about OX, OY

$= \int_{x=0}^{a} \int_{y=0}^{y_1} \rho xy \, dx \, dy$, where $y_1 = b(1 - x^2/a^2)^{1/2}$, by (1)

$$= \int_0^a \rho x \left[\frac{1}{2} y^2\right]_0^{y_1} dx = \frac{1}{2}\rho \int_0^a x y_1^2 \, dx = \frac{1}{2}\rho \int_0^a x b^2 (1 - x^2/a^2) \, dx$$

$$= \frac{\rho b^2}{2a^2} \int_0^a (xa^2 - x^3) \, dx = \frac{\rho b^2}{2a^2} \left[\frac{1}{2} a^2 x^2 - \frac{1}{4} x^4\right]_0^a = \frac{1}{8} \rho a^2 b^2$$

$= (1/8) \times (4M/\pi ab) a^2 b^2 = (Mab)/2\pi$, as $\rho = 4M/\pi ab$.

Ex. 2. *Find the product of inertia of a semi-circular wire about diameter and tangent at its extremity.* **[Agra 2011; Garhwal 2010; Kanpur 2009]**

Sol. Let APB be semi-circular wire whose AB is diameter and O is centre. Let AC be tangent at extremity A of the diameter AB. Then, $\angle CAB = 90°$. Consider an elementary arc $PQ (= a\,\delta\theta)$ at the point $P(a, \theta)$ of the semi-circular arc of mass M so that ρ = mass per unit length = $M/\pi a$. Again the mass of the element $PQ = \delta m = \rho a\, d\theta$. Its distance from AB is $PM (= a \sin\theta)$ and its distance from $AC = AM = AO + OM = a + a\cos\theta = a(1 + \cos\theta)$. Let PM be perpendicular to AB

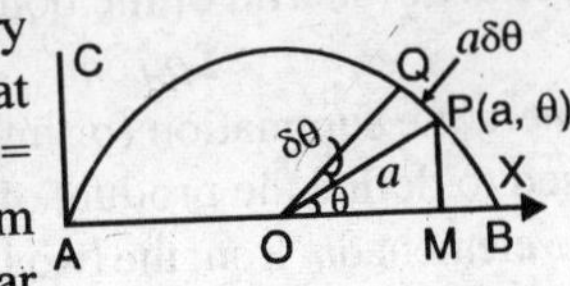

$\therefore$ Product of inertia of δm about AB and AC
$= \delta m \cdot PM \cdot AM = \rho a\, d\theta \cdot a \sin\theta \cdot a(1 + \cos\theta)$

Hence, product of inertia of the wire about AB, AC

$$= \int_0^{\pi} \rho a^3 \sin\theta (1 + \cos\theta) \, d\theta = \rho a^3 \int_0^{\pi} (\sin\theta + \sin\theta\cos\theta) \, d\theta$$

$= \rho a^3 [-\cos\theta + (1/2)\sin^2\theta]_0^{\pi} = 2\rho a^3 = 2(M/\pi a) a^3 = 2Ma^2/\pi$

Ex. 3. *Find the M.I. of the elliptic lamina $4x^2 + 9y^2 = 36$ about a straight line through its centre and perpendicular to the lamina, M being its mass.*

Sol. Comparing $4x^2 + 9y^2 = 36$ *i.e.*, $x^2/3^2 + y^2/2^2 = 1$ with $x^2/a^2 + y^2/b^2 = 1$, here $a = 3, b = 2$. Proceed as in Case III of Art. 14.4 I, the required M.I. $= (1/4)M(a^2 + b^2) = (1/4)M(3^2 + 2^2) = 13M/4$

Ex. 4. *ABC is a uniform equilateral triangular plate of mass M and side 2a. Show that its M.I. about its side is $(1/2)Ma^2$.*

Sol. Proceed as in Art. 14.4F. From figure, $h = AN = 2a \sin 60° = a\sqrt{3}$. Hence the required M.I. $= (1/6)Mh^2 = (1/6)M(a\sqrt{3})^2 = (1/2)Ma^2$.

Ex. 5. *Find the M.I. of the ellipsoid $x^2/9 + y^2/4 + z^2/16 = 1$, about the axis of x, M being its mass.*

Sol. Proceed as in Art. 14.4J. Here $a = 3, b = 2$ and $c = 4$. Hence the required M.I. $= (M/5) \times (b^2 + c^2) = (M/5) \times (2^2 + 4^2) = 4M$.

Ex. 6. *Find the M.I. of a thin rod of which the line density varies as the distance from one end about an axis passing through that end and at right angles to the rod.*

Sol. Refer figure of case I of Art. 14.4A. Let AB be a rod of length $2a$ and AL the line through the end A and perpendicular to the rod. Consider an element PQ of breadth δx at a distance x from the end A. Then, ρ = mass per unit length (*i.e.*, line density) $= k \times AP = kx$ (given), where k is constant of proportionality.

Now, $\quad \delta m$ = mass of the element $PQ = \rho\,\delta x = kx\,\delta x$

$$\therefore \text{M} = \text{mass of the rod} = \int_0^{2a} kx \, dx = \frac{k}{2}[x^2]_0^{2a} = 2a^2 k \Rightarrow k = \frac{M}{2a^2} \qquad \text{...(1)}$$

M.I. of δm about $AL = AP^2 \cdot \delta m = x^2 \cdot kx \, dx = kx^3 \, dx$

Hence, M.I. of the rod about AL

$$= \int_0^{2a} kx^3 \, dx = k \cdot 4a^4 = \frac{M}{2a^2} \cdot 4a^4 = 2Ma^2, \text{ using (1)}$$

Ex. 7. *Find the M.I. of a quadrant of the elliptic disc $x^2/a^2 + y^2 b^2 = 1$ of mass M about the line through its centre perpendicular to its plane, the density at any point is proportional to xy.*

Sol. For figure, refer Ex. 1. Consider an elementary area $\delta x\ \delta y$ at the point $P(x, y)$ in the quadrant OAB of the elliptic disc of mass M and axes of length $2a$ and $2b$. Given, ρ = mass per unit are at $P = kxy$, k being a constant

$\therefore\ \delta m$ = mass of the element $\delta x\ \delta y = \rho\ \delta x\ \delta y = kxy\ \delta x\ \delta y$

Then $\quad M = \int_{x=0}^{a} \int_{y=0}^{y_1} k\,xy\,dx\,dy$, where $y_1 = b(1 - x^2/a^2)^{1/2}$

or $\quad M = k \int_0^a x\left[\frac{1}{2}y^2\right]_0^{y_1} dx = kb^2 \int_0^a x\left(1 - \frac{x^2}{a^2}\right) dx = \frac{1}{8}ka^2b^2$

Hence $\quad k = 8M/a^2b^2 \qquad \ldots(1)$

Now, M.I. of the element $\delta x\ \delta y$ about the line OL (draw yourself in figure of Ex. 1) through the centre and perpendicular to the plane of the disc

$$= PO^2 \cdot \delta m = (x^2 + y^2)\cdot k\,xy\ \delta x\ \delta y$$

Hence, M.I. of the quadrant of the disc about OL

$$= \int_{x=0}^{a} \int_{y=0}^{y_1} k\,xy(x^2 + y^2)\,dx\,dy, \text{ where } y_1 = b(1 - x^2/a^2)^{1/2}$$

$$= k\int_0^a x\left[x^2\frac{y^2}{2} + \frac{y^4}{4}\right]_0^{y_1} dx = k\int_0^a \left[\frac{x^3b^2}{2}\left(1 - \frac{x^2}{a^2}\right) + \frac{xb^4}{4}\left(1 - \frac{x^2}{a^2}\right)^2\right] dx$$

$$= \frac{kb^2}{4}\int_0^a \left[b^2x + 2\left(1 - \frac{b^2}{a^2}\right)x^3 - \left(\frac{2}{a^2} - \frac{b^2}{a^4}\right)x^5\right] dx$$

$$= (kb^2/4)[(1/2)b^2x^2 + (1/2)x^4(1 - b^2/a^2) - (1/6)\times(2/a^2 - b^2/a^6)x^6]_0^a$$

$$= (1/24)\,ka^2b^2(a^2 + b^2) = (M/3)\,(a^2 + b^2), \text{ using (1)}$$

Ex. 8. (*a*) *Find the M.I. of a right solid cone of mass M, height h and radius of whose base is a, about its axis.*

(*b*) *Find the M.I. of a truncated cone about the axis, the radii being a and b.*

[Kanpur 2011]

Sol. (*a*) Let ODE be the given cone whose semi-vertical angle is α and mass per unit volume is ρ. Then $\quad M = (1/3)\pi\ (h \tan\alpha)^2 \cdot h\rho$

$\Rightarrow\quad \rho = 3M/\pi h^3 \tan^2\alpha \qquad \ldots(1)$

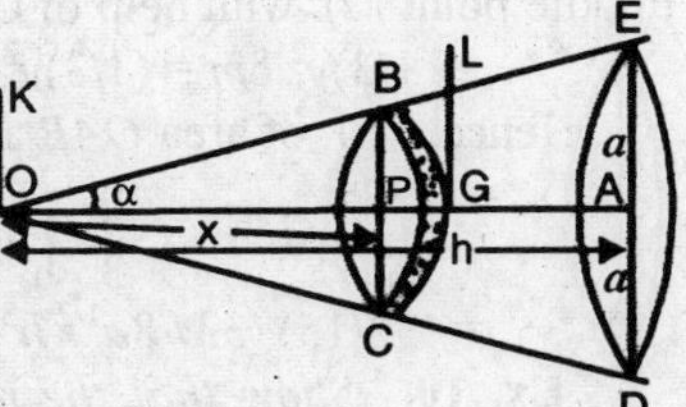

Consider an elementary disc BC of thickness δx, parallel to the base DE and at a distance x from the vertex O. Then, δm = mass of the elementary disc $BC = \pi(x \tan\alpha)^2 \cdot \delta x\ \rho$.

$\therefore$ M.I. of the elementary disc about the axis OA (which is a line through its centre P and perpendicular to its plane)

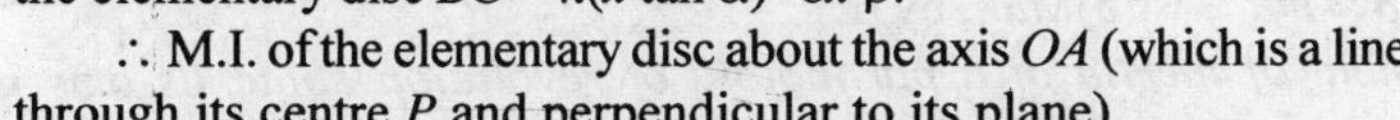

$$= (\rho\pi x^2 \tan^2\alpha\ \delta x)\cdot(PB^2/2), \text{ using Case II of Art. 14.4F}$$

$$= (\rho\pi x^2 \tan^2\alpha\ \delta x)\cdot(x\tan\alpha)^2\cdot(1/2) = (1/2)\rho\pi x^4 \tan^4\alpha\ \delta x$$

Hence, M.I. of the cone ODE about it axis OA

$$= \int_0^h \frac{1}{2}\rho\pi x^4 \tan^4\alpha\ dx = \frac{\rho\pi h^5 \tan^4\alpha}{10} = \frac{3M}{\pi h^3 \tan^2\alpha} \times \frac{\pi h^5 \tan^4\alpha}{10}, \text{ by (1)}$$

$$= (3/10)Mh^2 \tan^2\alpha = (3/10)Ma^2, \text{ as } a = h\tan\alpha$$

(*b*) The given truncated cone $DEE'D'$ can be treated as the difference of two solid cones ODE and $OD'E'$ of radii a, b respectively. If M_1 and M_2 be their masses, then $M = M_1 - M_2$. We shall proceed as in example of Art. 14.8. Now, let semi-vertical angle be α and ρ be uniform volume density of the matter. Then

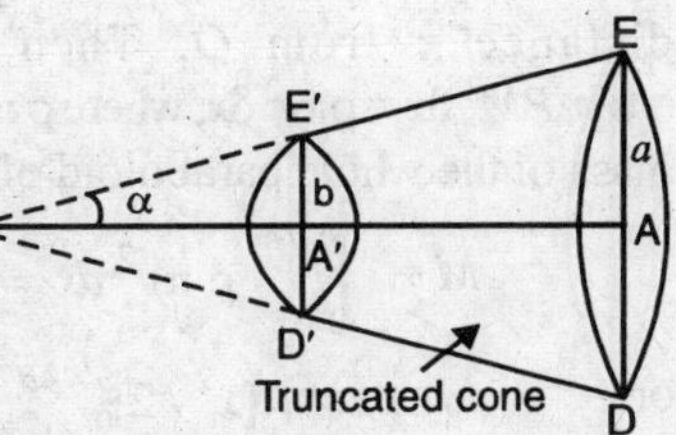

$$M_1 = \text{Mass of cone } ODE = \rho(1/3)\pi\ OA^2 \cdot AE$$

$$= (\rho/3)\pi\ (a\cot\alpha)^2 \cdot a$$

Similarly, $M_2 = (\rho/3)\pi\,(b\cot\alpha)^2 b$. Thus, $M_1/M_2 = a^3/b^3$

$$\therefore \quad \frac{M_1}{a^3} = \frac{M_2}{b^3} = \frac{M_1 - M_2}{a^3 - b^3} = \frac{M}{a^3 - b^3} \qquad \ldots(1)$$

Then M.I. of the truncated cone about its axis $A'A$

$= (3/10)M_1a^2 - (3/10)M_2b^2$, using result of part ($a$)

$$= \frac{3}{10}\left[\frac{Ma^5}{a^3 - b^3} - \frac{Mb^5}{a^3 - b^3}\right] = \frac{3M}{10}\frac{a^5 - b^5}{a^3 - b^3}, \text{ using (1)}$$

Ex. 9. *Show that the M.I. of a parabolic area (of latus rectum 4a) cut off by an ordinate at a distance h from the vertex is (3/7)Mh^2 about the tangent at the vertex and (4/5)Mah about the axis.* **[Meerut 1990]**

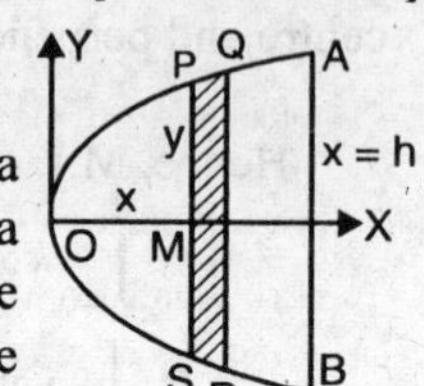

Sol. The equation of a parabola of latus rectum $4a$ can be writen as

$$y^2 = 4ax \qquad \ldots(1)$$

Let AOB be the portion of the parabola (1) cut off by an ordinate AB at a distance h from the vertex. Let $PQRS$ be an elementary strip parallel to OY at a distance x from O and of width δx and let coordinates of P be (x, y). Let ρ be the mass per unit area. Then, δm = mass of the elementary strip = $\rho \cdot 2y\,\delta x$. Let M be mass of the portion AOB of the parabola. Then

$$M = \int_0^h 2y\rho\,dx = 2\rho\int_0^h 2(ax)^{1/2}dx = 4\rho a^{1/2}\cdot\frac{2}{3}\cdot h^{3/2}$$

Hence, $\qquad \rho = 3M/8a^{1/2}h^{3/2} \qquad \ldots(2)$

Since the distance of each point of strip $PQRS$ from OY is x, we have

$\therefore$ M.I. of the strip $PQRS$ about $OY = x^2\,\delta m = x^2 \cdot 2\rho y\,\delta x$

Hence, M.I. of area AOB about the tangent OY at the vertex

$$= \int_0^h 2\rho y x^2\,dx = 2\rho\int_0^h x^2\cdot 2(ax)^{1/2}\,dx = 4\rho a^{1/2}\int_0^h x^{5/2}\,dx = \frac{8}{7}\rho a^{1/2}h^{7/2}$$

$= (8/7)\cdot(3M/8a^{1/2}h^{3/2})\cdot a^{1/2}h^{7/2} = (3/7)Mh^2$, by (2)

Next, M.I. of the strip $PQRS$ about OX (which is perpendicular to the strip and passes through its middle point M), with help of Case II, Art. 14.4A.

$= (1/3)y^2\,\delta m = (1/3)y^2 \cdot 2\rho y\,\delta x$

Hence, M.I. of area OAB about axis OX

$$= \int_0^h \frac{2}{3}\rho y^3\,dx = \frac{2\rho}{3}\int_0^h (4ax)^{3/2}dx = \frac{16\rho a^{3/2}}{3}\times\frac{2h^{5/2}}{5} = \frac{32}{15}\rho a^{3/2}h^{5/2}$$

$= (32/15)\cdot(3M/8a^{1/2}h^{3/2})\cdot a^{3/2}h^{5/2} = (4/5)Mah$, by (2)

Ex. 10. *Show that the moment of inertia of a paraboloid of revolution about its axis is (M/3) × the square of the radius of its base.* ***(Kanpur 2010)***

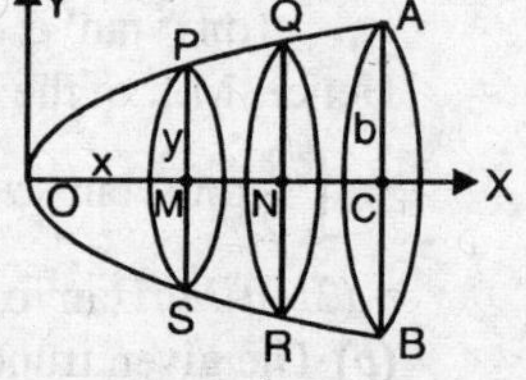

Sol. Let the paraboloid of revolution be generated by the revolution of the arc OPA of the parabola $\qquad y^2 = 4ax \qquad \ldots(1)$

about the x-axis OX. Let b be the radius of the circular base AB of the paraboloid of revolution. Then co-ordinates of A are (OC, b) which lie on (1) so that $b^2 = 4a\cdot OC$ so that $OC = b^2/4a$. Thus, x-coordinate of A is $b^2/4a$. Consider an elementary disc $PQRS$ of thickness δx, parallel to OY and at a distance x from O. Then, δm = mass of the elementary disc $= \rho\cdot\pi\cdot PM^2\cdot\delta x = \rho\pi y^2\,\delta x$, where ρ is mass per unit volume of the given paraboloid of revolution. Let M be the mass of the whole paraboloid of revolution AOB. Then

$$M = \int_0^{b^2/4a}\rho\,\pi y^2\,dx = 4a\pi\rho\int_0^{b^2/4a}x\,dx, \text{ as } y^2 = 4ax \text{ by (1)}$$

or $\qquad M = 4a\pi\rho\,[x^2/2]_0^{b^2/4a} = \rho\pi b^4/8a \;\Rightarrow\; \rho = 8aM/\pi b^4 \qquad \ldots(2)$

Now, M.I. of the elementary disc *PQRS* about *OX* (which is a line through its centre *M* and perpendicular to its plane)

$= \rho\pi y^2\, \delta x \cdot (y^2/2)$, using case II of Art. 14.4F

Hence M.I. of the paraboloid of revolution about the axis *OX*

$$= \int_0^{b^2/4a} \frac{\rho\pi}{2} y^4\, dx = \frac{16\pi\rho a^4}{2} \int_0^{b^2/4a} x^2\, dx, \text{ using (1)}$$

$= (\pi b^6/24a) \times \rho = (\pi b^6/24a) \times (8aM/\pi b^4) = (M/3) \times b^2$

$= (M/3) \times$ the square of the radius of the base.

Ex. 11. *Find the M.I. of the arc of a circle about (i) the diameter bisecting the arc (ii) an axis through the centre, perpendicular to its plane (iii) an axis through its middle point perpendicular to its plane.* **[Kanpur 2001; Agra 1996**

Sol. Let the given arc *AB* of a circle of radius *a* subtend an angle 2α at the centre *O*. Let *OM* be along the diameter bisecting the arc *AB* at *M*. Let *M* be mass of the arc *AB* and let ρ be its mass per unit length. Since arc $AB = 2\alpha a$, we have $\rho = M/2\alpha a$. Consider an elementary arc $PQ (= a\,\delta\theta)$ at point $P(a, \theta)$ of the arc *AB*. Then δm = mass of the elementary arc $= \rho a\, d\theta$.

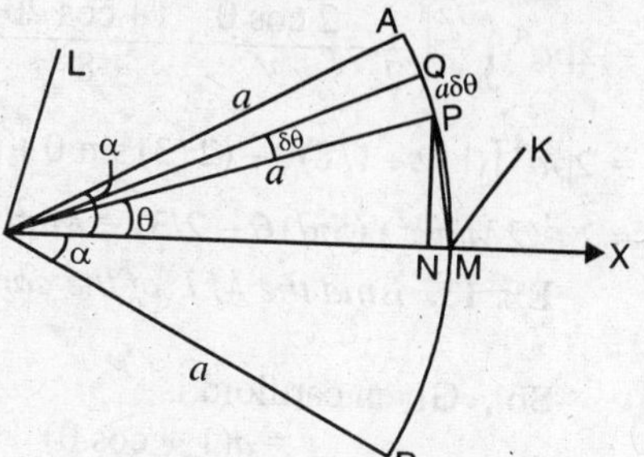

Part (*i*). Let *PN* be perpendicular to *OM*. Then,

$$PN = OP \sin\theta = a \sin\theta.$$

$\therefore$ M.I. of the elementary arc *PQ* about *OM*

$= PN^2 \cdot \delta m = (a \sin\theta)^2 \cdot \rho a\, \delta\theta = \rho a^3 \sin^2\theta\, \delta\theta$

Hence, M.I. of the arc *AB* about *OM*

$$= \int_{-\alpha}^{\alpha} \rho a^3 \sin^2\theta\, d\theta = \frac{\rho a^3}{2} \int_{-\alpha}^{\alpha} (1 - \cos 2\theta)\, d\theta = \frac{\rho a^3}{2} \left[\theta - \frac{1}{2}\sin 2\theta\right]_{-\alpha}^{\alpha}$$

$= (M/2\alpha a) \cdot (a^3/2) \cdot (2\alpha - \sin 2\alpha) = (Ma^2/2\alpha)\,(\alpha - \sin\alpha\cos\alpha)$

Part (*ii*). Let *OL* be a line through *O* perpendicular to the plane of the arc *AB*. Distance of *P* from $OL = OP = a$.

$\therefore$ M.I. of elementary mass δm at *P* about $OL = (\rho a\, \delta\theta) \cdot a^2$

Hence, M.I. of the arc *AB* about *OL*

$$= \int_{-\alpha}^{\alpha} \rho a^3\, d\theta = \rho a^3 \cdot 2\alpha = (M/2\alpha a) \cdot (2\alpha a^3) = Ma^2$$

Part (*iii*). Let *MK* be a line through the middle point *M* of the arc *AB* perpendicular to the plane of the arc. Now, from ΔPOM, by cosine formula of Trigonometry, we get

$PM^2 = OP^2 + OM^2 - 2OP \cdot OM \cos\theta = a^2 + a^2 - 2a^2\cos\theta = 2a^2(1 - \cos\theta)$

Thus, $PM^2 = 2a^2 \cdot 2\sin^2(\theta/2)$ so that $PM^2 = 4a^2\sin^2(\theta/2)$

Now, M.I. of elementary mass δm at *P* about *MK*

$= \delta m \cdot PM^2 = \rho a\, \delta\theta \cdot 4a^2 \sin^2(\theta/2) = 4\rho a^3 \sin^2(\theta/2)\, \delta\theta$

Hence, M.I. of the arc *AB* about *MK*

$$= \int_{-\alpha}^{\alpha} 4\rho a^3 \sin^2(\theta/2)\, d\theta = 8\rho a^3 \int_0^{\alpha} \sin^2(\theta/2)\, d\theta = 4\rho a^3 \int_0^{\alpha} (1 - \cos\theta)\, d\theta$$

$= 4\rho a^3[\alpha - \sin\alpha]_0^{\alpha} = 4(M/2\alpha a) \cdot a^3\,(\alpha - \sin\alpha) = (2Ma^2/\alpha) \times (\alpha - \sin\alpha)$

Ex. 12. *Show that the M.I. of a semi-circular lamina about a tangent parallel to the bounding diameter is $Ma^2(5/4 - 8/3\pi)$ where a is the radius and M is the mass of the lamina.* **[Kanpur 2008**

Sol. Let *AMB* be a semi-circular lamina of radius *a* and mass *M*. Let ρ be its mass per unit area. Then $\rho = M/(\pi a^2/2) = 2M/\pi a^2$.

Consider an elementary area $r\,\delta\theta\,\delta r$ at the point $P(r, \theta)$ of the lamina. Then δm = mass of this elementary area $= \rho r\,\delta\theta\,\delta r$.

Let CD be the tangent at middle M of arc AMB parallel to the bounding diameter AB of the given semi-circular lamina. Let PL and PN be perpendiculars from P on OM and CD respectively.

Now, PN = distance of the point P from $CD = LM = OM - OL = a - r\cos\theta$.

So M.I. of the elementary mass δm at P about CD

$$= PN^2 \cdot \delta m = (a - r\cos\theta)^2 \cdot \rho r\,\delta\theta\,\delta r$$

Hence, the M.I. of the lamina AMB about CD

$$= \int_{-\pi/2}^{\pi/2}\int_0^a (a - r\cos\theta)^2 \rho r\,d\theta\,dr = \rho\int_{-\pi/2}^{\pi/2}\int_0^a (a^2 r - 2ar^2\cos\theta + r^3\cos^2\theta)\,d\theta\,dr$$

$$= \rho\int_{-\pi/2}^{\pi/2}\left[\frac{a^2r^2}{2} - \frac{2ar^3\cos\theta}{3} + \frac{r^4\cos^2\theta}{4}\right]_0^a d\theta = \rho\int_{-\pi/2}^{\pi/2}\left[\frac{a^4}{2} - \frac{2a^4\cos\theta}{3} + \frac{a^4\cos^2\theta}{4}\right]d\theta$$

$$= 2\rho a^4\int_0^{\pi/2}\left(\frac{1}{2} - \frac{2\cos\theta}{3} + \frac{1+\cos 2\theta}{8}\right)d\theta$$

$$= 2\rho a^4\left[(1/2 + 1/8)\theta - (2/3)\sin\theta + (1/16)\sin 2\theta\right]_0^{\pi/2}$$

$$= 2a^4(2M/\pi a^2)\,(5\pi/16 - 2/3) = Ma^2(5/4 - 8/3\pi)$$

Ex. 13. *Find the M.I. of the cardioid $r = a(1 + \cos\theta)$ of density ρ, about the initial line.*

[Meerut 2005, 07]

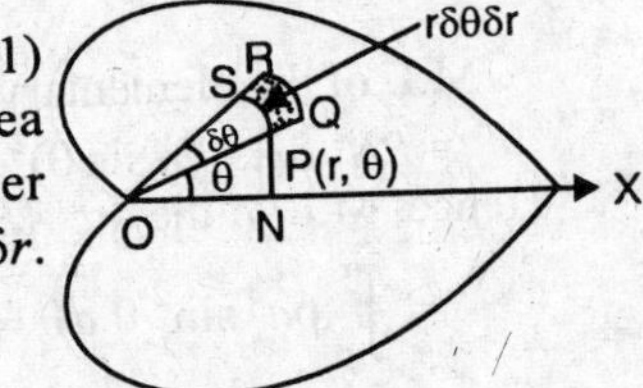

Sol. Given cardioid is

$$r = a(1 + \cos\theta) \qquad \ldots(1)$$

Consider an elementary area $PQRS$ $(= r\,\delta\theta\,\delta r)$ at $P(r, \theta)$ of the area bounded by (1). Let M be mass of the cardiod and ρ be its mass per unit area. Then δm = mass of the elementary area $PQRS = \rho r\,\delta\theta\,\delta r$.

$$\therefore\ M = 2\int_{\theta=0}^{\pi}\int_{r=0}^{a(1+\cos\theta)} \rho r\,d\theta\,dr = 2\rho\int_0^{\pi}\left[\frac{1}{2}r^2\right]_0^{a(1+\cos\theta)} d\theta$$

$$= \rho a^2\int_0^{\pi}(1+\cos\theta)^2\,d\theta = \rho a^2\int_0^{\pi}(1 + 2\cos\theta + \cos^2\theta)\,d\theta$$

$$= \rho a^2\int_0^{\pi}\{1 + 2\cos\theta + (1/2)(1+\cos 2\theta)\}\,d\theta = \rho a^2\left[(3/2)\theta + 2\sin\theta + (1/4)\sin 2\theta\right]_0^{\pi}$$

Thus, $\quad M = (3/2)\pi\rho a^2 \quad$ so that $\quad \rho = 2M/3\pi a^2 \qquad \ldots(2)$

Now, M.I. of the elementary mass δm at P about OX

$$= PN^2 \cdot \delta m = (r\sin\theta)^2 \cdot \rho r\,\delta\theta\,\delta r = \rho r^3\sin^2\theta\,\delta\theta\,\delta r$$

Hence, M.I. of the cardiod about OX

$$= \int_{\theta=-\pi}^{\pi}\int_{r=0}^{a(1+\cos\theta)} \rho r^3\sin^2\theta\,d\theta\,dr = \rho\int_{-\pi}^{\pi}\sin^2\theta\left[\frac{1}{4}r^4\right]_0^{a(1+\cos\theta)} d\theta$$

$$= \frac{1}{4}\rho a^4\int_{-\pi}^{\pi}(1+\cos\theta)^4\sin^2\theta\,d\theta = \frac{\rho a^4}{4}\int_{-\pi}^{\pi}\left(2\cos^2\frac{1}{2}\theta\right)^4 \times \left(2\sin\frac{1}{2}\theta\cos\frac{1}{2}\theta\right)^2 d\theta$$

$$= 32\rho a^4\int_{-\pi/2}^{\pi/2}\sin^2\phi\cos^{10}\phi\,d\phi, \qquad \text{putting } \frac{\theta}{2} = \phi \text{ and } d\theta = 2d\phi$$

$$= 64\rho a^4\int_0^{\pi/2}\sin^2\phi\cos^{10}\phi\,d\theta, \text{ as the integrand is an even function of } \phi$$

$$= 64\rho a^4 \cdot \frac{\Gamma(3/2)\Gamma(11/2)}{2\Gamma(7)} = \frac{21}{32}\rho\pi a^4 = \frac{21\pi a^4}{32} \times \frac{2M}{3\pi a^2} = \frac{7Ma^2}{16}, \text{ by (2)}$$

Ex. 14. *Find the M.I. of the area of lemniscate $r^2 = a^2\cos 2\theta$ (i) about its axis (ii) about a line through the origin in its plane and perpendicular to its axis (iii) about a line through the origin and perpendicular to its plane.*

[Agra 2012; Meerut 1995, 96; Purvanchal 2007]

Sol. Given curve is

$$r^2 = a^2 \cos 2\theta \qquad \ldots(1)$$

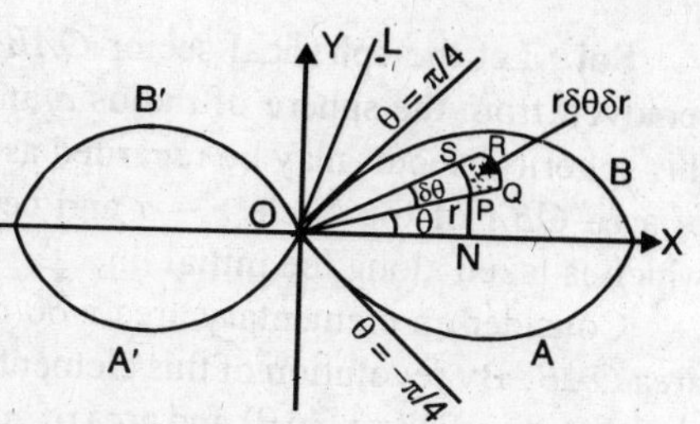

The curve (1) is as shown in the figure. It consists of two identical symmetrical loops. OAB and $OA'B'$. Loop OAB is formed between the lines $\theta = \pi/4$ and $\theta = -\pi/4$. Consider an elementary area $PQRS$ ($= r\,\delta\theta\,\delta r$) at $P(r, \theta)$ of the area of loop OAB. Let M be mass of two loops and ρ be mass per unit area.

Now, δm = mass of the elementary area $PQRS = \rho \cdot r\,\delta\theta\,\delta r$

$$\therefore \quad M = 2\int_{\theta=-\pi/4}^{\pi/4}\int_{r=0}^{a(\cos 2\theta)^{1/2}} \rho r\,d\theta\,dr = 2\rho\int_{-\pi/4}^{\pi/4}\left[\frac{1}{2}r^2\right]_0^{a(\cos 2\theta)^{1/2}} d\theta$$

$$= \rho a^2\int_{-\pi/4}^{\pi/4}\cos 2\theta\,d\theta = \rho a^2\,[(1/2)\sin 2\theta]_{-\pi/4}^{\pi/4} = \rho a^2$$

Thus $\quad M = \rho a^2 \quad$ so that $\quad \rho = M/a^2 \qquad \ldots(2)$

Part (*i*). M.I. of the elementary mass δm at P about the axis OX

$= PN^2 \cdot \delta m = (r\sin\theta)^2 \cdot \rho r\,\delta\theta\,\delta r = \rho r^3 \sin^2\theta\,\delta\theta\,\delta r$

Hence, M.I. of the lamniscate about OX

$$= 2\int_{\theta=-\pi/4}^{\pi/4}\int_{r=0}^{a(\cos 2\theta)^{1/2}} \rho r^3 \sin^2\theta\,d\theta\,dr = 2\rho\int_{-\pi/4}^{\pi/4}\left[\frac{r^4}{4}\right]_0^{(a\cos 2\theta)^{1/2}} \sin^2\theta\,d\theta$$

$$= \frac{\rho a^4}{2}\int_{-\pi/4}^{\pi/4}\cos^2 2\theta\sin^2\theta\,d\theta = 2\times\frac{\rho a^4}{2}\int_0^{\pi/4}\cos^2 2\theta\sin^2\theta\,d\theta$$

$$= \frac{\rho a^4}{2}\int_0^{\pi/4}\cos^2 2\theta(1-\cos 2\theta)\,d\theta = \frac{\rho a^4}{4}\int_0^{\pi/2}\cos^2\phi(1-\cos\phi)\,d\phi$$

[Putting $2\theta = \phi$ and $d\theta = (1/2)\,d\phi$]

$$= \frac{\rho a^4}{4}\left[\int_0^{\pi/2}\cos^2\phi\,d\phi - \int_0^{\pi/2}\cos^3\phi\,d\phi\right] = \frac{\rho a^4}{4}\left[\frac{\Gamma(3/2)\Gamma(1/2)}{2\Gamma(2)} - \frac{\Gamma(2)\Gamma(1/2)}{2\Gamma(5/2)}\right]$$

$= (M/a^2)\cdot(a^4/4)\,(\pi/4 - 2/3) = (Ma^2/4)\,(\pi/4 - 2/3)$, using (2)

Part (*ii*). M.I. of the elementary mass δm at P about OY (which is the line through the origin in the plane of (1) and perpendicular to its axis)

$= ON^2 \cdot \delta m = (r\cos\theta)^2 \cdot \rho r\,\delta\theta\,\delta r = \rho r^3 \cos^2\theta\,\delta\theta\,\delta r$

Hence, M.I. of the lamniscate about OY

$$= 2\int_{\theta=-\pi/4}^{\pi/4}\int_{r=0}^{a(\cos 2\theta)^{1/2}} \rho r^3\cos^2\theta\,d\theta\,dr = 2\rho\int_{-\pi/4}^{\pi/4}\left[\frac{r^4}{4}\right]_0^{(a\cos 2\theta)^{1/2}} \cos^2\theta\,d\theta$$

$= (Ma^2/4)\,(\pi/4 + 2/3)$, simplifying as a part (*i*)

Part (*iii*). Let OL be the line through the origin and perpendicular to the plane of (1). Then, we have

M.I. of the elementary mass δm at P about $OL = PO^2 \cdot \delta m = r^2 \cdot \rho r\,\delta\theta\,\delta r$

Hence, M.I. of the laminscate about OL

$$= 2\int_{\theta=-\pi/4}^{\pi/4}\int_{r=0}^{a(\cos 2\theta)^{1/2}} \rho r^3 d\theta\,dr = \frac{2\rho}{4}\int_{-\pi/4}^{\pi/4} a^4\cos^2 2\theta\,d\theta$$

$$= (\rho a^4/2)\times 2\int_0^{\pi/4}\cos^2 2\theta\,d\theta = \frac{\rho a^4}{2}\int_0^{\pi/4}(1+\cos 4\theta) = \frac{\rho a^4}{2}\left[\theta + \frac{1}{2}\sin 4\theta\right]_0^{\pi/4}$$

$= (M/a^2)\cdot(a^4/2)\,(\pi/4) = (\pi Ma^2)/8$, using (2)

Ex. 15. *From a uniform sphere of radius a, spherical sector of vertical angle 2α is removed. Show that the M.I. of the remainder of mass M about the axis of symmetry is* $(M/5)a^2\,(1 + \cos\alpha)\,(2 - \cos\alpha)$

[Garhwal 1992]

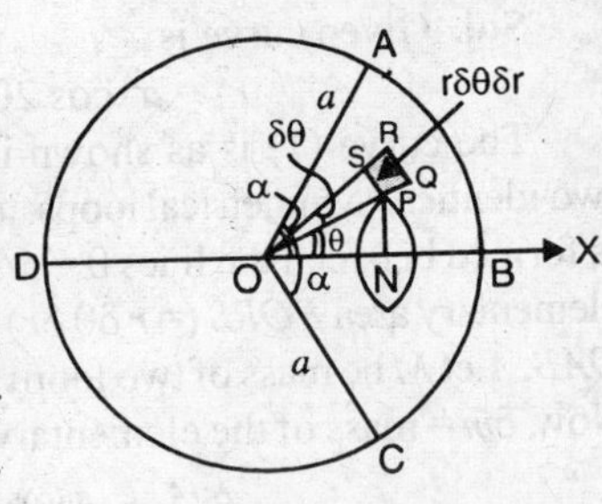

Sol. Let the spherical sector $OABCO$ of vertical angle 2α be removed from the sphere of radius a and cetnre O. The volume of this spherical sector may be regarded as generated by the revolution of area OBA of the circle $r = a$ and centre O about the radius OB which is taken along the initial line.

Consider an elementary area $r\,\delta\theta\,\delta r$ at the point $P(r, \theta)$ of the area OAB. By revolution of this elementary area about OB, a circular ring of radius $PN(= r \sin\theta)$ and area of cross-section $r\,\delta\theta\,\delta r$ is formed. Let ρ be mass per unit volume of the sphere. Then, δm = mass of the elementary ring $= \rho \cdot 2\pi\, r \sin\theta \cdot r\,\delta\theta\,\delta r$

Let M be mass of the remainder $OADCO$ of the sphere

$\therefore \quad M$ = mass of the sphere – mass of the spherical sector $OABCO$

$$= \frac{4}{3}\pi a^3\rho - \int_{\theta=0}^{\alpha}\int_{r=0}^{a} 2\pi\rho r^2 \sin\theta\, d\theta\, dr$$

$$= \frac{4}{3}\pi a^3\rho - 2\pi\rho\int_0^{\alpha}\left[\frac{1}{3}r^3\right]_a^0 \sin\theta\, d\theta = \frac{4}{3}\pi a^3\rho - \frac{2\pi a^3\rho}{3}[-\cos\theta]_0^{\alpha}$$

Thus, $\quad M = (4/3)\pi a^3\rho - (2/3)\pi a^3\rho\,(1-\cos\alpha) = (2\pi a^3\rho/3)\,(1+\cos\alpha)$ ...(1)

$\therefore \quad \rho = 3M/2\pi a^3(1+\cos\alpha)$

We shall now find M.I. of spherical sector $OADCO$ about OB.

M.I. of elementary ring about OB (a line through the centre N of the ring and perpendicular to its plane)

$= \delta m \cdot PN^2$, by Case II of Art. 14.4D

$= 2\pi\rho r^2 \sin\theta\,\delta\theta\,\delta r \cdot (r\sin\theta)^2 = 2\pi\rho r^4 \sin^3\theta\,\delta\theta\,\delta r$

Hence, M.I. of the spherical sector $OABCO$ about OB

$$= \int_{\theta=0}^{\alpha}\int_{r=0}^{\alpha} 2\pi\rho r^4 \sin^3\theta\, d\theta\, dr = \frac{2\pi\rho a^5}{5}\int_0^{\alpha}\sin^3\theta\, d\theta$$

$$= \frac{2\pi\rho a^5}{10}\int_0^{\alpha}\frac{3\sin\theta - \sin 3\theta}{4}\,d\theta, \text{ as } \sin 3\theta = 3\sin\theta - 4\sin^3\theta$$

$$= \frac{\pi\rho a^5}{10}\left[-3\cos\theta + \frac{1}{3}\cos 3\theta\right]_0^{\alpha} = \frac{\pi\rho a^5}{10}\left(-3\cos\alpha + \frac{1}{3}\cos 3\alpha + 3 - \frac{1}{3}\right) \quad ...(2)$$

Using the method of decomposition (refer Art. 14.8), M.I. of the remainder about OB, the axis of symmetry

M.I. of the sphere about OB – M.I. of the spherical sector $OABCO$ about OB

$$= \frac{2}{5}\left(\frac{4}{3}\pi a^3\rho\right)a^2 - \frac{\pi\rho a^5}{10}\left\{-3\cos\alpha + \frac{1}{3}(4\cos^3\alpha - 3\cos\alpha) + \frac{8}{3}\right\}$$

[Using result of Art. 14.4H and the above result (2)]

$= (8/15)\pi\rho a^5 - (1/30)\pi\rho a^5\,(-12\cos\alpha + 4\cos^3\alpha + 8)$

$= (\pi\rho a^5/30)\,(8 + 12\cos\alpha - 4\cos^3\alpha) = (\pi\rho a^5/15)\,(2 + 7\cos\alpha - \cos^3\alpha)$

$$= \frac{\pi a^5}{15}\cdot\frac{3M}{2\pi a^3(1+\cos\alpha)}(1+\cos\alpha)^2(2-\cos\alpha) = \frac{Ma^2}{5}(1+\cos\alpha)(2-\cos\alpha)$$

[on substituting the value of ρ from (1)]

EXERCISE

1. A parabolic arc is cut off by a double ordinate at a distance h from the vertex, show that for the M.I. about the tangent at the vertex, $K^2 = 3h^2/7$, where K is the radius of gyration.

2. Show that the M.I. about the x-axis of the portion of the parabola $y^2 = 4ax$ bounded by the x-axis and latus rectum, supposing the surface density at each point vary as the cube of the abscissa, is $(12/11)Ma^2$, where M is the mass of the portion.

3. Find the M.I. of a circular disc about an axis through its centre and perpendicular to its plane, assuming that the density at any point varies as the square of its distance from the centre.

4. Show that M.I. of one loop of the lemniscate $r^2 = a^2 \sin 2\theta$ about an axis perpendicular to its plane at the pole is $(1/8)\pi Ma^2$, M being mass of a loop.

5. Find the M.I. about the x-axis of the portion of the ellipsoid $x^2/a^2 + y^2/b^2 + z^2/c^2 = 1$, which lies in the positive octant, the law of volume denisty to be $\rho = \mu xyz$. **[Agra 1994]**

[Hint : Try like Art. 14.4I]

Ans. $(1/4)\ M(b^2 + c^2)$.

14.9 Theorem of parallel axis. **[Agra 2010; Kanpur 2002; Delhi 2004]**

If the moments and products of inertia about any ine through the centre of gravity of a body are known, to obtain the corresponding quantities for any parallel line.

Proof. Let GX, GY, GZ be any three axes through the centre of gravity G, and OX', OY', OC' parallel axes through any point O. Let the coordinates of any element m of the body situated at P be (x, y, z) and (x', y', z') referred to the two sets of axes. If (f, g, h) be the coordinates of G referred to OX', OY' and OZ', then

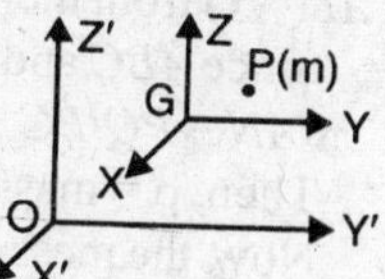

$$x' = x + f, \qquad y' = y + g \quad \text{and} \quad z' = z + h \qquad \ldots(1)$$

Hence the M.I. of the body about OX'

$= \Sigma m(y'^2 + z'^2) = \Sigma m\{(y + g)^2 + (z + h)^2\} = \Sigma m(y^2 + z^2 + g^2 + h^2 + 2yg + 2hz)$

$= \Sigma m(y^2 + z^2) + (f^2 + g^2)\Sigma m + 2g\ \Sigma my + 2h\ \Sigma mz \qquad \ldots(2)$

Now, $\Sigma my/\Sigma m$ = the y-coordinate of G referred to G as origin

Thus, $\Sigma my/\Sigma m = 0$ so that $\Sigma my = 0$. Similarly $\Sigma mz = 0$

$\therefore$ By (2), M.I. about $OX' = \Sigma m(y^2 + z^2) + (f^2 + g^2)\cdot M$, where $\Sigma m = M$

i.e., **M.I. of the body about OX' = M.I. of the body above GX + M.I. of a mass M placed at G about OX'** $\ldots(3)$

Again, the product of inertia about the axes OX' and OY'

$= \Sigma mx'y' = \Sigma m(x + f)(y + g) = \Sigma m(xy + fg + xg + fy)$

$= \Sigma mxy + fg\ \Sigma m + g\ \Sigma xm + f\ \Sigma ym$

$= \Sigma mxy + fg\ M$, as before $\Sigma m = M$, $\Sigma xm = 0$ and $\Sigma ym = 0$

Hence, **Product of inertia about OX', OY' = the product of inertia about GX, GY + the product of inertia of a mass M placed at G about the axes OX', OY'** $\ldots(4)$

From (3) and (4) we get the required results.

Corollary. From the first part of this article, it follows that of all axes drawn in a given direction the one through the centre of gravity is the one such that the moment of inertia about it is a minimum.

TYPE 2 : SOLVED EXAMPLES BASED ON ART. 14.9

Ex. 1. *Find the M.I. of a uniform circular disc of radius 2 cm and mass 10 gms about a tangent.*

Sol. By the theorem of parallel axis, M.I. of the given circular disc of mass $M(= 10$ gm) and radius $a(= 2$ cms) about the tnagent TT'

= M.I. of the disc about the parallel diameter AB

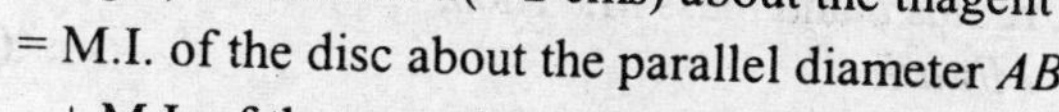

+ M.I. of the mass M placed at centre of gravity O of the disc about TT'

$= (1/4)Ma^2 + Ma^2 = (5/4)Ma^2 = (5/4)\cdot 10.2^2 = 50$ gm cm^2.

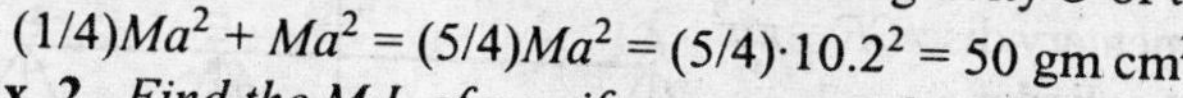

Ex. 2. *Find the M.I. of a uniform rectangular lamina of sides 2a and 2b about its sides.*

Sol. Refer figure of Art. 14.4B. Let M be mass of the lamina. Then by the theorem of parallel axis, we have

M.I. of lamina $ABCD$ about the side AB

= M.I. of $ABCD$ about the parallel axis OX through the centre of gravity O

+ M.I. of the total mass M placed at O about AB

$= (1/3)Mb^2 + Mb^2 = (4/3)Mb^2$

Ex. 3. *Find the M.I. of a rectangular parallelopiped about an edge.*

Sol. Refer figure of Art. 14.4C. Let $2a$, $2b$, $2c$ be the lengths of the edges of given parallelopiped of mass M, say. We shall determine the M.I. about an edge, say $O'A$ of length $2a$. In figure O is the centre of gravity of the parallelopiped. Then by the theorem of parallel axis, we have

M.I. of the rectangular paralelopiped about the edge $O'A$
= M.I. of the rectangular parallelopiped about a parallel axis OX
+ M.I. of the total mass M placed at centre of gravity O about $O'X'$
$= (1/3)M(b^2 + c^2) + M(b^2 + c^2) = (4/3)M(b^2 + c^2)$

Ex. 4. *Find the M.I. of the triangle ABC about a perpendicular to its plane through A.* **[Purvanchel 2005; Garhwal 1995; Kumaun 97; Meerut 2006]**

Sol. Let M be the mass and $h(= AN)$ be the height of a triangular lamina ABC. Let us divide the lamina ABC by strips parallel to BC. Let $PQRS$ be one such an elementary strip of breadth δx and at a distance at $AL(= x)$ from the vertex A of the triangle ABC. Let $BC = a$.

Since ABC and APQ are similar triangles, we have
$x/AN = PQ/BC$ or $x/h = PQ/a$ so that $PQ = ax/h$

Then, ρ = mass per unit area of $\Delta ABC = M/(ha/2) = 2M/ha$

Now, the mass of the elementary strip $= \delta m = \rho \cdot PQ \cdot \delta x = \rho(ax/h)\,\delta x$

Let M be middle point of PQ and AD be perpendicular to the plane of the triangle through A. Then, by the theorem of parallel axis, we have

M.I. of the elementary strip $PQRS$ about AD
= M.I. about a parallel line through the middle point M of PQ
+ (mass of strip)$\cdot AM^2$

$$= \frac{\rho a x\,\delta x}{h}\left\{\frac{1}{3}\left(\frac{ax}{2h}\right)^2 + AM^2\right\} = \frac{\rho a x\,\delta x}{h}\left\{\frac{1}{3}\left(\frac{ax}{2h}\right)^2 + \left(\frac{x \times AK}{h}\right)^2\right\}, \text{ by case II, Art. 14.4A}$$

[Since from similar triangles APQ and ABC, $AM/AK = x/h$]

Hence the required M.I. of the ΔABC about AD

$$= \frac{\rho a}{4h^3}\left(\frac{a^2}{3} + 4AK^2\right)\int_0^h x^3\,dx = \frac{\rho a h}{48}(a^2 + 12AK^2)$$

$= (2M/ha) \times (ah/48) \times (a^2 + 12AK^2) = (M/24)(a^2 + 12AK^2)$ ...(1)

Now, $AK^2 = AN^2 + NK^2 = AN^2 + (BK - BN)^2 = (AN^2 + BN^2) + BK^2 - 2BK \cdot BN$
$= AB^2 + (BC/2)^2 - 2 \cdot (BC/2) \cdot AB \cos B$

[As $PQ \parallel BC$ and M is mid point of $PQ \Rightarrow K$ is mid-point of BC]

$= c^2 + a^2/4 - ac \cdot \{(a^2 + c^2 - b^2)/2ac\} = (2b^2 + 2c^2 - a^2)/4$, where $AB = c$, $AC = b$

Hence from (1), the required M.I. about AD
$= (M/24)\{a^2 + 3(2b^2 + 2c^2 - a^2)\} = (M/12)(3b^2 + 3c^2 - a^2)$

Ex. 5. *Find the M.I. of a right circular cylinder about (i) its axis (ii) a straight line through its centre of graivty and perpendicular to its axis.* **[Garhwal 2010]**

Sol. Consider a cylinder of height $h (= HK)$, radius a and mass M. Then,

ρ = mass per unit volume $= M/\pi a^2 h$ ...(1)

Consider a circular disc EF of thickness δx at a distance $x(= OP)$ from O so that δm = mass of the elementary disc $EF = \pi a^2 \delta x \rho$. Here O is the centre of gravity of the cylinder.

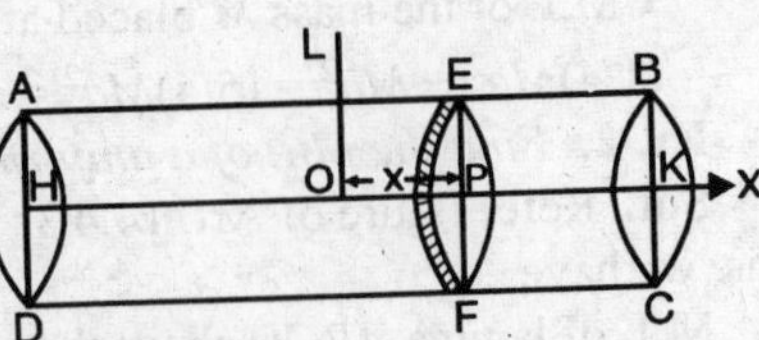

(*i*) Then, M.I. of the elementary disc EF about the axis HK (which is a line through its centre and perpendicular to its plane)
$= (\pi a^2 \delta x \rho) \cdot (a^2/2)$, using case II of Art. 14.4F.

Hence, M.I. of the cylinder about its axis HK

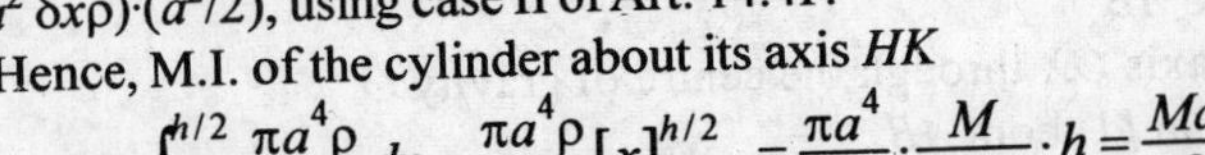

$$= \int_{-h/2}^{h/2} \frac{\pi a^4 \rho}{2}\,dx = \frac{\pi a^4 \rho}{2}[x]_{-h/2}^{h/2} = \frac{\pi a^4}{2} \cdot \frac{M}{\pi a^2 h} \cdot h = \frac{Ma^2}{2}, \text{ by (1)}$$

(*ii*) Let OL be the line through O and perpendicular to the axis HK. Then, by the theorem of parallel axis, M.I. of elementary disc EF about OL

= M.I. of the elementary disc about the parallel line EF through its centre of graivty P + M.I. of the

the total mass δm of the disc placed at P about OL

$= (1/4)a^2\,\delta m + x^2\,\delta m = (a^2/4 + x^2)\cdot\rho\pi a^2\,\delta x$

Hence the required M.I. of the cylinder about OL

$$= \int_{-h/2}^{h/2}\left(\frac{1}{4}a^2 + x^2\right)\rho\pi a^2\,dx = \rho\pi a^2\left[\frac{1}{4}a^2x + \frac{1}{3}x^3\right]_{-h/2}^{h/2} = \frac{\rho\pi a^2 h}{4}\left(a^2 + \frac{h^2}{3}\right)$$

$$= \frac{M}{\pi a^2 h}\cdot\frac{\pi a^2 h}{4}\left(a^2 + \frac{h^2}{3}\right) = \frac{1}{4}M\left(a^2 + \frac{1}{3}h^2\right)$$

Ex. 6. *Prove that the M.I. of a uniform right circular solid cone of mass M, height h and radius a, about a diameter of its base is (M/20) (3a² + 2h²).*

[Purvanchel 2005; Garhwal 1995; I.F.S. 2000; Kumaun 97; Meerut 93]

Sol. Refer figure of Ex. 8(*a*) on page 14.9. Let *ODE* be the given cone whose semi-vertical angle is α and mass per unit volume is ρ. Then

$M = (1/3)\pi\,(h\tan\alpha)^2\cdot h\rho \quad\Rightarrow\quad \rho = 3M/\pi h^3\tan^2\alpha$ ...(1)

Consider an elementary disc BC of thickness δx, parallel to the base DE and at a distance x from the vertex O. Then, δm = mass of the elementary disc $BC = \pi(x\tan\alpha)^2\,\delta x\rho$. Now, by the theorem of parallel axis, we have

M.I. of the disc BC about the diameter ED of the base of cone

= M.I. of the disc about the parallel diameter BC of the disc

+ M.I. of the total mass δm at the centre P about ED

$= (1/4)\,\delta m\cdot BP^2 + \delta m\cdot PA^2 = \rho\pi x^2\tan^2\alpha\,\{(1/4)x^2\tan^2\alpha + (h-x)^2\}$

Hence, the M.I. of the cone about the diameter of its base

$$= \int_0^h \rho\pi x^2\tan^2\alpha\left\{\frac{1}{4}x^2\tan^2\alpha + (h-x)^2\right\}dx$$

$$= \frac{\rho\pi\tan^2\alpha}{4}\int_0^h (x^4\tan^2\alpha + 4h^2x^2 - 8hx^3 + 4x^4)\,dx = \frac{\rho\pi h^5\tan^2\alpha}{60}(3\tan^2\alpha + 2)$$

$= (M/\pi a^2 h)\cdot(\pi h^5\tan^2\alpha/60)\cdot(3\tan^2\alpha + 2) = (1/20)Mh^2\,(3\tan^2\alpha + 2)$

$= (1/20)M\,(3a^2 + 2h^2)$, as $\tan\alpha = a/h$

Ex., 7. *A solid body, of density ρ, is in the shape of the solid formed by the revolution of the cardioid r = a(1 + cos θ) about the initial line; show that its moment of inertia about a straight line through the pole perpendicular to the initial line is (352/105)ρπa⁵.*

Sol. The given cardioid is given by curve *OCABO*. **Here** *OY* **is the** line through the pole O perpendicular to the initial line OX. Consider an elementary area $r\,\delta\theta\,\delta r$ at $P(r,\theta)$ of the area bounded by the cardioid. This element, when revolved about OA forms a circular ring PQ of radius $PL\ (= r\sin\theta)$ and cross section $r\,\delta\theta\,\delta r$.

Mass of this elementary ring $= \delta m = 2\pi r\sin\theta\cdot r\,\delta\theta\,\delta r$. Then by the theorem of parallel axis, M.I. of the elementary ring PQ about OY

= M.I. of the ring about the diameter PQ parallel to OY

+ M.I. of the total mass δm at the centre L about OY

$= \delta m\cdot(PL^2/2) + \delta m\cdot OL^2$, using result of Art. 14.4D

$= 2\pi r^2\rho\sin\theta\,\delta\theta\,\delta r\,\{(1/2)\,r^2\sin^2\theta + r^2\cos^2\theta\}$

Hence, M.I. of the given solid of revolution about OY

$$= 2\pi\rho\int_{\theta=0}^{\pi}\int_{r=0}^{a(1+\cos\theta)} r^4\sin\theta\left(\frac{1}{2}\sin^2\theta + \cos^2\theta\right)d\theta\,dr = \pi\rho\int_0^{\pi}\left[\frac{r^5}{5}\right]_0^{a(1+\cos\theta)}\sin\theta(1+\cos^2\theta)\,d\theta$$

$$= \frac{\pi\rho a^5}{5}\int_0^{\pi}(1+\cos\theta)^5(1+\cos^2\theta)\sin\theta\,d\theta = -\frac{\pi\rho a^5}{5}\int_2^0\{1+(u-1)^2\}^2\cdot u^5\,du$$

[Putting $1 + \cos\theta = u$ so that $-\sin\theta\,d\theta = du$]

$$= \frac{\pi\rho a^5}{5}\int_0^2 (2u^5 - 2u^6 + u^7)\,du = \frac{\pi\rho a^5}{5}\left[\frac{1}{3}u^6 - \frac{2}{7}u^7 + \frac{1}{8}u^8\right]_0^2$$

$$= (1/5) \times (\pi\rho a^5) \times (64/3 - 256/3 + 32) = (352/105)\,\pi\rho a^5$$

14.10 Moment of inertia of a body about a line. *Given the moments and products of inertia about three perpendicular axes; to find the moment of inertia about any line through their meeting point.* **[Mumbai 2010; Delhi Maths (H) 2008]**

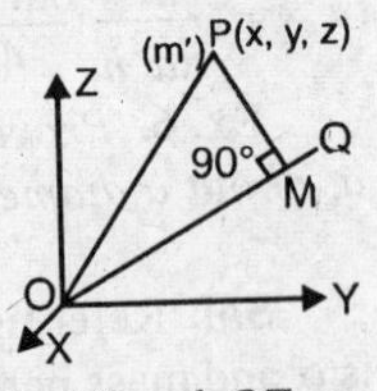

Let OX, OY, OZ be the three given axes, and let A, B and C be the moments of inertia with respect to them, and F, G and H the products of inertia with respect to the axes of y and z, of z and x and x and y respectively. Take any element m' of the body at $P(x, y, z)$. Then, we know that

$$\left.\begin{aligned} &A = \Sigma m'(y^2 + z^2), \quad B = \Sigma m'(z^2 + x^2), \quad C = \Sigma m'(x^2 + y^2), \\ &F = \Sigma m'yz, \qquad G = \Sigma m'zx \quad \text{and} \quad H = \Sigma m'xy \end{aligned}\right\} \quad \ldots(1)$$

Let the M.I. be required about OQ, whose direction cosines with respect to OX, OY and OZ are l, m and n. Draw PM perpendicular to the axis OQ. Then from coordinate geometry, we know that $OP^2 = x^2 + y^2 + z^2$ and $OM = lx + my + nz$. Now, from right angled triangle OPM, we have

$$PM^2 = OP^2 - OM^2 = (x^2 + y^2 + z^2) - (lx + my + nz)^2$$

$$= x^2(1 - l^2) + y^2(1 - m^2) + z^2(1 - n^2) - 2mnyz - 2nlzx - 2lmxy$$

$$= x^2(m^2 + n^2) + y^2(l^2 + n^2) + z^2(l^2 + m^2) - 2mnyz - 2nlzx - 2lmxy \qquad [\because l^2 + m^2 + n^2 = 1]$$

$$= l^2(y^2 + z^2) + m^2(z^2 + x^2) + n^2(x^2 + y^2) - 2mnyz - 2nlzx - 2lmxy$$

If I denotes the M.I. of the body about OQ, then

$$I = \Sigma m' \cdot PM^2 = l^2\,\Sigma m'(y^2 + z^2) + m^2\,\Sigma m'(z^2 + x^2) + n^2\,\Sigma m'(x^2 + y^2) - 2mn\,\Sigma m'yz - 2ln\,\Sigma m'zx - 2lm\,\Sigma m'xy$$

or

$$I = Al^2 + Bm^2 + Cn^2 - 2Fmn - 2Gln - 2Hlm, \text{ by (1)}$$

14.11 Moments and products of inertia of a plane lamina about a line. *If the moments and products of inertia of a plane lamina about two perpendicular axes in the plane are known, to find the M.I. about any other axis through their point of intersection and the product of inertia about this axis and a line perpendicular to it through the same point of inetersection.* **(Guwahati 2007)**

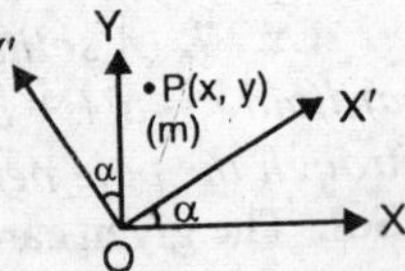

Consider an element m of the body at P whose coordinates referred to OX and OY as coordinate axes are (x, y). Let A, B be moments of inertia of the given plane lamina about OX and OY, and F product of inertia about of the lamina about OX and OY. Then, we have

$$A = \Sigma my^2, \qquad B = \Sigma mx^2 \qquad \text{and} \qquad H = \Sigma mxy \qquad \ldots(1)$$

Let (x', y') be the coordinates of the point P referred to new coordinates axes OX' and OY', where $\angle XOX' = \alpha$. Then, from coordinate geometry, we have

$$x' = x\cos\alpha + y\sin\alpha \qquad \text{and} \qquad y' = y\cos\alpha - x\sin\alpha \qquad \ldots(2)$$

Hence the moment of inertia of the lamina about $OX' = A' = \Sigma my'^2$.

$$= \Sigma m(y\cos\alpha - x\sin\alpha)^2 = \cos^2\alpha\,\Sigma my^2 - 2\sin\alpha\cos\alpha\,\Sigma mxy + \sin^2\alpha\,\Sigma mx^2$$

$$= A\cos^2\alpha - H\sin 2\alpha + B\sin^2\alpha, \text{ using (1)}$$

Similarly, $B' = \text{M.I. above } OY' = \Sigma mx'^2 = \Sigma m(x\cos\alpha + y\sin\alpha)^2$

$$= A\sin^2\alpha + H\sin 2\alpha + B\cos^2\alpha$$

Also, the product of inertia about OX' and OY'

$$= H = \Sigma mx'y' = \Sigma m(x\cos\alpha + y\sin\alpha)(y\cos\alpha - x\sin\alpha)$$

$$= \sin\alpha\cos\alpha\,(\Sigma my^2 - \Sigma mx^2) + (\cos^2\alpha - \sin^2\alpha)\,\Sigma mxy$$

$$= (1/2) \times (A - B)\sin 2\alpha + \text{H}\cos 2\alpha, \text{ using (1)}$$

14.12 Some useful propositions

Proposition I. *If A, B, C denote the moments of inertia and D, E, F the products of inertia about three mutually perpendicular axes, then the sum of any two of them is greater than the third.* **[Kanpur 2010]**

Proof. Since $A = \Sigma m(y^2 + z^2), \quad B = \Sigma m(z^2 + x^2), \quad C = \Sigma m(x^2 + y^2),$

$\therefore \quad A+B-C=\Sigma\{m(y^2+z^2)+m(z^2+x^2)-m(x^2+y^2)=2\Sigma mz^2$

Thus, $A+B-C>0$ showing that $A+B>C$. Hence proved.

Proposition II. *The sum of the moments of inertia about any three rectangular axes meeting at a given point is always constant and is equal to twice the M.I. about that point.*

Proof. Here $A+B+C=\Sigma m(y^2+z^2)+\Sigma m(z^2+x^2)+\Sigma m(x^2+y^2)$

or $\quad A+B+C=2\Sigma m(x^2+y^2+z^2)=2\Sigma mr^2$, where $r=(x^2+y^2+z^2)^{1/2}$,

showing that the sum $A+B+C$ is independent of the directions of axes and is equal to twice the M.I. about the given point.

Proposition III. *The sum of the moments of inertia of a body with reference to any plane through a given point and its normal at that point is constant and is equal to the M.I. of the body with respect to that point.*

Proof. Let the given point be taken as origin and plane as the plane of xy. Let C' be the M.I. of the body with respect to the xy-plane and C be the M.I. of the body about its normal at O (which is z-axis), then $C'=\Sigma mz^2$, $C=\Sigma m(x^2+y^2)$.

Hence $C+C'=\Sigma m(x^2+y^2+z^2)=\Sigma mr^2$, where $r=(x^2+y^2+z^2)^{1/2}$, showing that $C+C'$ is independent of the plane through O and is constant and equal to the M.I. of the body about O.

An important deduction of Proposition III. *If A', B', C' denote the moments of inertia of the body with respect to the planes YOZ, ZOX and XOY respectively, then $A'=(B+C-A)/2$, $B'=(C+A-B)/2$, $C'=(A+B-C)/2$*

Proof. From propositions II and III, we have

$$A+B+C=2\Sigma mr^2 \quad \text{and} \quad C+C'=\Sigma mr^2$$

$$\therefore \quad (A+B+C)-2(C+C')=0 \quad \text{so that} \quad C'=(A+B-C)/2$$

Similarly, we can show that $A'=(B+C-A)/2$, $B'=(C+A-B)/2$

Proposition IV. *To show that* $A>2F$, $B>2G$ *and* $C>2H$

Since A.M. > G.M. so $(y^2+z^2)/2=(y^2 \cdot z^2)^{1/2}$ or $y^2+z^2>2yz$

Hence $\Sigma m(y^2+z^2)>2\Sigma myz$ or $A>2F$

Similarly, we can show that $B>2G$ and $C>2H$.

TYPE 3 : SOLVED EXAMPLES BASED ON ART. 14.10 TO 14.12

Ex. 1. *Find the M.I. of a rectangle of mass M and sides 2a, 2b about a diagonal. Deduce that in the case of a square.* **[Agra 2006]**

Sol. Let $ABCD$ be the given rectangle with sides $AB=2a$ and $AD=2h$. Then,

$A=$ M.I. of the rectangle about $OX=(1/3)Mb^2$

$B=$ M.I. of the rectangle about $OY=(1/3)Ma^2$,

and $H=$ P.I. of the rectangle about OX and $OY=0$, as the rectangle is symmetrical about OX and OY.

Let the diagonal AC make an angle α with OX. Since $OC=(a^2+b^2)^{1/2}$, we have

$\sin\alpha=b/(a^2+b^2)^{1/2}$ and $\cos\alpha=a/(a^2+b^2)^{1/2}$.

Hence M.I. of the rectangle about the diagonal AC (See result of Art. 14.11)

$$=A\cos^2\alpha+B\sin^2\alpha-H\sin 2\alpha=\frac{Mb^2}{3}\cdot\frac{a^2}{a^2+b^2}+\frac{Ma^2}{3}\cdot\frac{b^2}{a^2+b^2}=\frac{2M}{3}\cdot\frac{a^2b^2}{a^2+b^2}$$

Deduction. For a square, $2b=2a$ so that $b=a$

$\therefore$ M.I. of the square about its diagonal $=(2M/3)\times(a^4/2a^2)=(1/3)Ma^2$.

Ex. 2. (a) *Find the M.I. of an elliptic area about a line, in its plane, passing through its centre and inclined at an angle θ to the major axis.*

(b) *Show that the M.I. of an elliptic area of mass M and semi-axes a and b about a diameter of length 2r is $Ma^2b^2/4r^2$.* **[Meerut 2007; Kanpur 2000, 07, 11]**

(c) *If k_1 and k_2 be the radii of gyration of an elliptic lamina about two conjugate diameters, then drove that $1/k_1^2+1/k_2^2=4(1/a^2+1/b^2)$.* **[Guwahati 2007]**

(*d*) *Show that M.I. of an ellipse of mass M and semi-axes a and b about a tangent is (5/4)Mp^2, where p is the perpendicular from the centre on the tangent.* **[Kanpur 2000]**

(*e*) *Show that the sum of the moments of inertia of an elliptic area about any two tangents at right angles is always the same.* **[Purvanchal 1996]**

Sol. (*a*) Let M be the mass and a, b be the semi-axes of the given elliptic area. Let PQ be a line passing through its centre O and making an angle θ with major axis OX. Then, we have

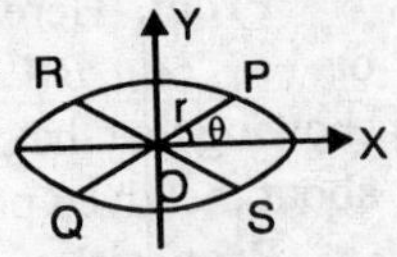

A = M.I. of the elliptic area about $OX = (1/4)Mb^2$

B = M.I. of the elliptic area about $OY = (1/4)Ma^2$

and H = P.I. of the elliptic area about OX, $OY = 0$, as the elliptic area is symmetrical about OX and OY.

Hence M.I. of the elliptic area about PQ (by Art. 14.11)

$= A \cos^2\theta + B \sin^2\theta - H \sin 2\theta = (1/4)Mb^2 \cos^2\theta + (1/4)Ma^2 \sin^2\theta$

$= (M/4)(b^2 \cos^2\theta + a^2 \sin^2\theta)$

(*b*) Refer figure of part (*a*). Here PQ is a diameter of length $2r$ of the ellipse

$$x^2/a^2 + y^2/b^2 = 1 \quad \text{...(1)}$$

Let PQ make an angle θ with OX. Then, as in part (*a*),

$$I = \text{M.I. of the elliptic area about } PQ = (M/4)(b^2 \cos^2\theta + a^2 \sin^2\theta) \quad \text{...(2)}$$

Now, $OP = r$ and so coordinates of P are $(r\cos\theta, r\sin\theta)$. Since P lies on (1), we have

$$(r^2/a^2)\cos^2\theta + (r^2/b^2)\sin^2\theta = 1 \text{ so that}$$

$$b^2 \cos^2\theta + a^2 \sin^2\theta = a^2b^2/r^2 \quad \text{...(3)}$$

Using (3), (2) gives $\quad I = (M/4)\cdot(a^2b^2/r^2) = Ma^2b^2/4r^2 \quad$ ...(4)

(*c*) Let PQ and RS be two conjugate diameters of an ellipse $x^2/a^2 + y^2/b^2 = 1$. Let $PQ = 2r_1$ and $RS = 2r_2$. Then, as in part (*b*), we have

$$Mk_1^2 = Ma^2b^2/4r_1^2 \quad \text{and} \quad Mk_2^2 = Ma^2b^2/4r_2^2$$

$$\therefore \quad \frac{1}{k^2} + \frac{1}{k_2^2} = \frac{4}{a^2b^2}(r_1^2 + r_1^2) = \frac{4}{a^2b^2}(a^2 + b^2) = 4\left(\frac{1}{a^2} + \frac{1}{b^2}\right)$$

[$\because r_1^2 + r_2^2 = a^2 + b^2$, by a property of conjugate diameters]

(*d*) Let the tangent AT be inclined at angle θ with OX. Then, its equation is given by

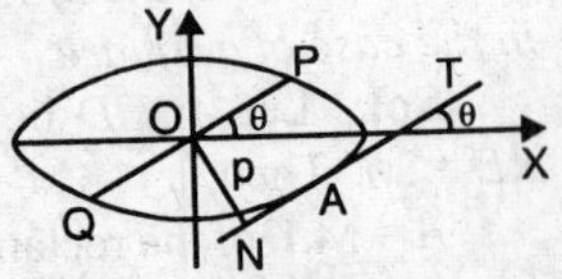

$$y = x \tan\theta + (a^2 \tan^2\theta + b^2) \quad \text{...(1)}$$

$\therefore \quad p = ON$ = perpendicular distance of (1) from centre $O(0, 0)$

$$= \frac{(a^2 \tan^2\theta + b^2)^{1/2}}{(1 + \tan^2\theta)^{1/2}} = (a^2 \sin^2\theta + b^2 \cos^2\theta)^{1/2} \quad \text{...(2)}$$

Let PQ be the diameter of the ellipse, parallel to (1). Then as in part (*a*),

I = M.I. of the elliptic area about $PQ = (M/4)(b^2 \cos^2\theta + a^2 \sin^2\theta)$

or $\quad I = (M/4)p^2$, using (2) $\quad$ ...(3)

Hence, by the theorem of parallel axis. (See Art. 14.9), the required M.I. about the tangent AT

= M.I. about PQ + M.I. of mass M placed at centre of gravity O of the elliptic area about the tangent AT

$= I + M \times (ON)^2 = (M/4)p^2 + Mp^2 = (5/4)Mp^2$, using (3)

(*e*) Proceed as in part (*d*). Then I_1 = M.I. about a tangent AT inclined at an angle θ to OX $= (5/4)Mp^2 = (5/4)M(a^2 \sin^2\theta + b^2 \cos^2\theta)$

Clearly, a tangent perpendicular to AT will make an angle $\pi/2 + \theta$ with OX. Hence, as before, I_2 = M.I. about this perpendicular tangent

$= (5/4)M\{a^2 \sin^2(\pi/2 + \theta) + b^2 \cos^2(\pi/2 + \theta)\} = (5/4)M(a^2 \cos^2\theta + b^2 \sin^2\theta)$

So the sum of moments of inertia about two perpendicular tangents $= I_1 + I_2 = (5/4)M(a^2 + b^2)$, which being independent of θ is always the same.

Ex. 3. *Show that the M.I. of right solid cone whose height is h and radius of whose base is a, is $3Ma^2(6h^2 + a^2)/20(h^2 + a^2)$ about a slant side, and $3M(h^2 + 4a^2)/80$ about a line through the centre of gravity of the cone perpendicular to its axis.* **[Agra 96; Garhwal 94; Kanpur 92; Meerut 96]**

Sol. Refer figure of Ex. 8(*a*) on page 14.9. Proceed as in Ex. 8(*a*) to show that

M.I. of the cone about its axis $OA = (3/10)Ma^2$ ...(1)

Using the theorem of parallel axis, M.I. of the cone about OK (which is a line through vertex O and perpendicular to OA)

$$= \int_0^h \pi x^2 \tan^2\alpha \cdot \rho\left(\frac{1}{4}x^2\tan^2\alpha + x^2\right)dx = \frac{\pi\tan^2\alpha\cdot\rho h^5}{5}\left(\frac{\tan^2\alpha}{4}+1\right)$$

$$= \frac{\pi h^5\tan^2\alpha}{5}\times\frac{3M}{\pi h^3\tan^2\alpha}\left(\frac{\tan^2\alpha}{4}+1\right) = \frac{3Mh^2}{5}\left(\frac{a^2}{4h^2}+1\right), \text{ as } \tan\alpha = \frac{a}{h}$$

$= (3M/20)\,(a^2 + 4h^2)$

Since the cone is symmetrical about OA, so the product of inertia about OA and OK is zero.

Using result of Art. 14.11, M.I. of the cone about slant height OE

$$= \frac{3Ma^2\cos^2\alpha}{10} + \frac{3M(a^2+4h^2)\sin^2\alpha}{20} = \frac{3Ma^2}{10}\times\frac{h^2}{a^2+h^2} + \frac{3M(a^2+4h^2)}{20}\times\frac{a^2}{a^2+h^2}$$

[Since $\tan\alpha = a/h$, so $\sin\alpha = a/(a^2+h^2)^{1/2}$, $\cos\alpha = h/(a^2+h^2)^{1/2}$]

$= 3Ma^2\,(6h^2 + a^2)/20(a^2 + h^2)$

Second part. Let GK be a line through the centre of gravity G of the cone and perpendicular to its axis OA. By the theorem of parallel axis, we get

M.I. about OK = M.I. about GL + M.I. of mass M placed at G about OK

$\Rightarrow$ M.I. about GL = M.I. about OK − M.I. of mass M at G about OK

$= (3M/20) \times (a^2 + 4h^2) - M(3h/4)^2$, as $OG = 3h/4$

$= (3M/80) \times (4a^2 + 16h^2 - 15h^2) = (3M/80) \times (h^2 + 4a^2)$

EXERCISE

1. Show that for a thin hemispherical shell of mass M and radius a, the M.I. about any line through the vertex is $(2/3)Ma^2$. **[Garhwal 1995]**

2. Show that the M.I. of an elliptic area of mass M and equation $ax^2 + 2hxy + by^2 + 2gx + 2fy + c = 0$ about a diameter parallel to the axis of x is $(-aM\Delta)/4(ab - h^2)^2$, where $\Delta = abc + 2fgh - af^2 - bg^2 - ch^2$. **[Rohilkhand 1991]**

3. A closed shell of total mass M, made of thin uniform sheet metal, is in the thin form of a righ circular cone, of slant height l and base radius r. Prove that M.I. of the shell about its axis of symmetry is $(1/2)Mr^2$, and that about a line through the vertex perpendicular to the axis is $(M/4)\,(2l^2 + 2rl - 3r^2)$.

4. Show that the M.I. of an ellipsoid of mass M and semi-axes a, b, c with regard to a diametral plane whose direction-cosines referred to principal planes are l, m, n is $(M/5)\,(a^2l^2 + b^2m^2 + c^2n^2)$. **[Rohilkhand 1995]**

14.13 Momental ellipsoid. Definition.

If A, B, C, F, G, H are usual moments and products of inertia about three mutually perpendicular axes through a point O, then the M.I. of the body about a line OQ, whose direction cosines are l, m, n is given by (refer Art. 14.10)

$$Al^2 + Bm^2 + Cn^2 - 2Fmn - 2Gln - 2Hlm$$

Let this line OQ turn in any manner about the given point O and be of such length that the M.I. is inversely proportional to the square of the length OQ, *i.e.*,

$$Al^2 + Bm^2 + Cn^2 - 2Fmn - 2Gln - 2Hlm = Mk^4/OQ^2, \quad ...(1)$$

where M is the mass of the body and k is a const. to keep the dimensions correct.

If $OQ = r$ and coordinates of Q be (x, y, z), refered to the axes OX, OY, OZ, then $x = lr, y = mr$ and $z = nr$. Putting $OQ = r$ and then multiplying both sides by r in (1), the cartesian equation of the locus of Q is given by

$$Ax^2 + By^2 + Cz^2 - 2Fyz - 2Gzx - 2Hxy = Mk^4, \quad ...(2)$$

which represents an ellipsoid since A, B, C are all essentially positive. (2) is known as the *momental ellipsoid* of the body at the point O.

Since the position of Q is obtained by a physical definition, which is independent of any particular axes of coordinates, we arrive at the same ellipsoid whatever be the axes OX, OY, OZ with which we start.

Remark. We know from coordinate geometry of three dimensions that for every ellipsoid there can be found three perpendicular diameters such that, if they be taken as axes of coordinates, the resulting equation of the ellipsoid has no terms involving yz, zx, or xy and the new axes of coordinates are then called the *principal axes* of the ellipsoid. Let the momental ellipsoid (1) when referred to its principal axes have as equation

$$A_1x^2 + B_1y^2 + C_1z^2 = Mk^4, \quad \text{...(3)}$$

The products of inertia about these new axes are clearly zero, because otherwise there would enter terms involving xy, yz or zx.

Hence we have the following important proposition : *For any body whatever there exists at each point O a set of three perpendicular axes (which are the three principal diameters of the momental ellipsoid at O) such that the products of inertia of the body about them, two at a time, all vanish.* **[Agra 1995, Kanpur 1996]**

These three new axes are called the *principal axes* of the body at the point O. Also, a plane through any two of these axes is called a *principal plane* of the body at O.

Corollary. *If the three principal moments at any point O are equal to each other, ellipsoid becomes a sphere.* **[Meerut 2004]**

We know from coordinate geometry of three dimensions that of the three principal axes of the ellipsoid, one is the maximum radius vector of the ellipsoid, and another is the minimum. Since the square of the radius vector of the momental ellipsoid is inversely proportional to the corresponding M.I. of the body, it follows that of three principal axes, one has the minimum M.I. and another has the maximum.

If the three principal moments of inertia at O are equal, the ellipsoid of inertia becomes a sphere, all radii of which are equal; in this case all moments of inertia about lines through O are equal.

As an example, the perpendicular from the centre of gravity of a cube on three coterminous faces are principal axes, because referred to them as axes, $\Sigma mxy = 0$, $\Sigma myz = 0$, $\Sigma mzx = 0$. Moreover, three moments of inertia about them are equal, each being $(2/3)Ma^2$, whre $2a$ is the side of the cube (refer particular case of Art. 14.4C)

Hence the M.I. about any line whose direction cosines are l, m, n through the centre of the cube is

$= (2/3)Ma^2l^2 + (2/3)Ma^2m^2 + (2/3)Ma^2n^2 = (2/3)Ma^2\,(l^2 + m^2 + n^2) = (2/3)Ma^2$, which is always the same, since it is independent of d.c.'s l, m, n of the line.

14.14 Momental ellipse. Definition. In the case of a plane lamina, if A, B are moments of inertia about the perpendicular axes OX and OY and H is the product of inertia about them, then the moment of inertia of the lamina about a line OQ making an angle θ with OX is given by (use result of Art. 14.11, by replacing α by θ and OX' by OQ)

$$A\cos^2\theta + B\sin^2\theta - 2H\sin\theta\cos\theta$$

Again if the point Q be such that the M.I. is inversely proportional to OQ^2 (*i.e.*, r^2 if $OQ = r$), then

$$A\cos^2\theta - 2H\sin\theta\cos\theta + B\sin^2\theta = Mk^4/r^2, \quad \text{...(1)}$$

where M is the mass of the body and k is some constant to keep the dimensions correct. Let (x, y) be the coordinates of Q referred to the axes OX and OY, then $x = r\cos\theta$, $y = r\sin\theta$ and so (1) reduces to

$$Ax^2 - 2Hxy + By^2 = Mk^4, \quad \text{...(2)}$$

which represents an ellipse (since A and B are essentially positive) and is known as *momental ellipse* at the point O.

Remark. Clearly the momental ellipse is the section of the momental ellipsoid at O by the plane of the lamina.

TYPE 4 : SOLVED EXAMPLES BASED ON ART. 14.13 AND 14.14

Ex. 1. *Show that the momental ellipsoid at the centre of an elliptic plane is $x^2/a^2 + y^2/b^2 + z^2(1/a^2 + 1/b^2) = const.$* **[Kanpur 2004. 07: Meerut 2007]**

Sol. Let M be the mass of the given elliptic plate of semi-axes a and b. Take the major axis and minor axis of the ellipse and a perpendicular line OZ as the axes of reference. Then,

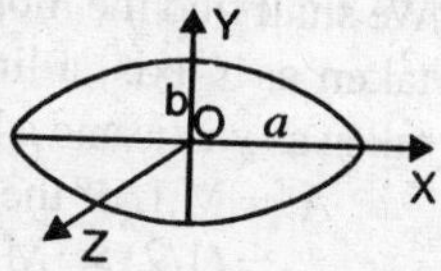

$$A = \text{M.I. about } OX = (1/4)Mb^2;$$
$$B = \text{M.I. about } OY = (1/4)Ma^2$$
and
$$C = \text{M.I. about } OZ = (1/4)M(a^2 + b^2)$$

If F, G and H are products of inertia about the axes y, z; z, x and x, y respectively, then due to symmetry of the ellipse about Ox and Oy, we have $F = G = H = 0$. Hence the equation of the momental ellipse at O, namely

$$Ax^2 + By^2 + Cz^2 - 2Fyz - 2Gzx - 2Hxy = \text{const.}$$

becomes
$$(1/4)Mb^2x^2 + (1/4)Ma^2y^2 + (1/4)M(a^2 + b^2)z^2 = \text{const.}$$

Dividing by $(1/4)Ma^2b^2$, the required equation of momental ellipsoid is

$$x^2/a^2 + y^2/b^2 + z^2(1/a^2 + 1/b^2) = \text{const.}$$

Ex. 2. *Show that the momental ellipsoid at the centre of an ellipsoid is $(b^2 + c^2)x^2 + (c^2 + a^2)y^2 + (a^2 + b^2)z^2 = const.$* **[Kanpur 2009; Purvanchel 2007; Meerut 2004]**

Sol. Referred to principal axes as the axes of reference, the equation of the given ellipsoid is $x^2/a^2 + y^2/b^2 + z^2/c^2 = 1$. Then, we have A = M.I. about $OX = (M/5)(b^2 + c^2)$, B = M.I. about $OY = (M/5)(c^2 + a^2)$ and C = M.I. about $OZ = (M/5)(a^2 + b^2)$. If F, G, H are the products of inertia about the axes y, z; z, x and x, y respectively, then due to symmetry of the given ellipsoid about all axes, $F = G = H = 0$. So the eqn. of the momental ellipsoid at the centre O, namely,

$$Ax^2 + By^2 + Cz^2 - 2Fyz - 2Gzx - 2Hxy = \text{const.}$$

becomes
$$(M/5)\{(b^2 + c^2)x^2 + (c^2 + a^2)y^2 + (a^2 + b^2)z^2\} = \text{const.}$$
or
$$(b^2 + c^2)x^2 + (c^2 + a^2)y^2 + (a^2 + b^2)z^2 = \text{const.}$$

Ex. 3. *Show that the equation of the momental ellipsoid at the corner of a cube of side $2a$ referred to its principal axes is $2x^2 + 11(y^2 + z^2) = c$, where c is a constant.* **[Kanpur 2005]**

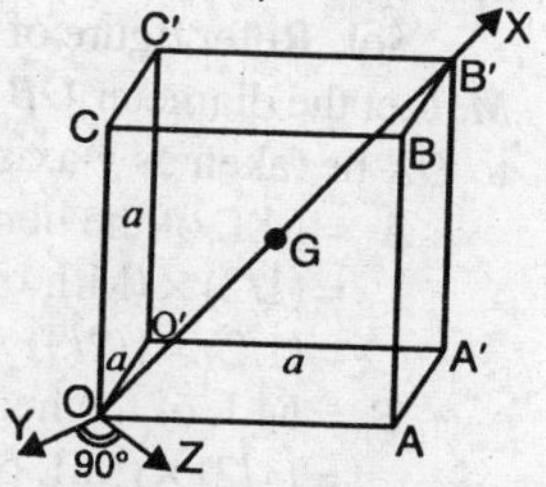

Sol. Let O be the corner of the cube of mass M, and G its centre. We are to find the equation of the momental ellipsoid at the centre O referred to its principal axes.

We know that (refer corollary of Art. 14.13) the M.I. of the given cube about any line through its centre of gravity is $(2/3)Ma^2$.

Take the line OG as the axis of x and the mutually perpendicular lines OY, OZ through O as the axes of y and z. Since the product of inertia of the cube about any two mutually perpendicular lines through G is zero and the coordinates of G referred to OX, OY, OZ as coordinate axes are $(a\sqrt{3}, 0, 0)$, the products of inertia about the axes OX, OY, OZ taken in pairs are also zero. Hence OX, OY, OZ are the principal axes of the momental ellipsoid at O. Now, we have

A = M.I. of the cube about $OX = (2/3)Ma^2$, since OX passes through the centre of gravity G of the cube

B = M.I. about OY = M.I. about a line through G parallel to OY + M.I. of mass M placed at G about OY

$$= (2/3)Ma^2 + M(a\sqrt{3})^2 = (11/3)Ma^2$$

Likewise, C = M.I. of the cube about $OZ = (11/3)Ma^2$

Since OX, OY, OZ are principle axes, so $F = G = H = 0$.

Hence, the equation of the momental ellipsoid at O is

$$Ax^2 + By^2 + Cz^2 - 2Fyz - 2Gzx - 2Hxy = \text{const.}$$

or $(2/3)Ma^2x^2 + (11/3)Ma^2y^2 + (11/3)Ma^2z^2 = \text{const.}$ or $2x^2 + 11(y^2 + z^2) = c$

Ex. 4. *Show that the momental ellipsoid at a point on the rim of a hemisphere is $2x^2 + 7(y^2 + z^2) - (15/4)xz$ = const.* **[Agra 2006, 2011; Kanpur 2004, 06, 10; Purvanchel 2005]**

Sol. Let O be a point on the rim of a hemisphere of radius a and mass M. We shall find the momental ellipsoid at O. Let the diameter OB through O be taken as x-axis, a line in the plane of the base and perpendicular to OX be taken as y-axis and a line perpendicular to the plane of the base as z-axis. Then,

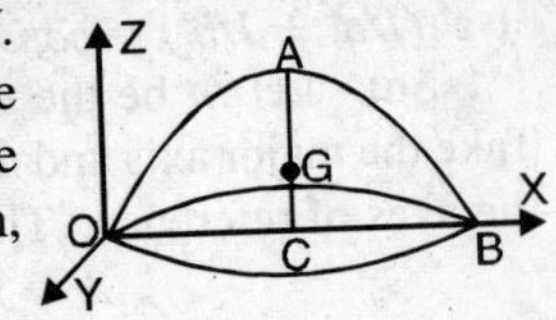

A = M.I. of the hemisphere about OX
$= (1/2) \times$ (M.I. of the whole sphere about OX), by symmetry
$= (1/2) \times \{(2/5) \cdot 2M \cdot a^2\} = (2/5)Ma^2$

B = M.I. of the hemisphere about OY
$= (1/2) \times$ (M.I. of the whole sphere about the line through the centre C of the sphere and parallel to OY, *i.e.*, a diameter of the sphere + M.I. of a mass $2M$ placed at C about OY)
$= (1/2) \times \{(2/5) \cdot 2M \cdot a^2 + 2Ma^2\} = (7/5)Ma^2$

Similarly, C = M.I. of the hemisphere about $OZ = (7/5)Ma^2$

Note that the coordinates of centre of gravity G of hemisphere are $(a, 0, 3a/8)$

$\therefore$ F = Product of inertia of the hemisphere about OY and OZ
= P.I. of the hemisphere about the lines through G, parallel to OY and OZ
+ P.I. of a mass M placed at G about OY and OZ
$= 0 + M \cdot 0 \cdot (3a/8) = 0$, using symmetry for the first term

Similarly, G = P.I. of the hemisphere about OZ and OX
$= 0 + M \cdot (3a/8) \cdot a = (3/8)Ma^2$

and H = P.I. of the hemisphere about OX and $OY = 0 + M \cdot a \cdot 0 = 0$

Hence the equation of the momental ellipsoid at O is

$$Ax^2 + By^2 + Cz^2 - 2Fyz - 2Gzx - 2Hxy = \text{const.}$$

i.e., $(2/5)Ma^2x^2 + (7/5)Ma^2y^2 + (7/5)Ma^2z^2 - 2 \cdot (3/8)a^2zx = \text{const.}$

or $2x^2 + 7(y^2 + z^2) - (15/4)xz = \text{const.}$, as required

Ex. 5. *Show that the momental ellipsoid at a point on the edge of a circular base of a thin hemispherical shell is $2x^2 + 5(y^2 + z^2) - 3zx$ = const.* **[Agra 2011; Kanpur 2011, 2014]**

Sol. Refer figure of Ex. 4. Let O be a point on the rim of a hemispherical shell of radius a and mass M. Let the diameter OB through O be taken as x-axis, a line in the plane of the base and perpendicular to OX be taken as y-axis and a line perpendicular to the plane of the base as z-axis. Then,

A = M.I. of the hemispherical shell about OX
$= (1/2) \times$ (M.I. of whole spherical shell about OX), by symmetry
$= (1/2) \times \{(2/3) \times 2M \times a^2\} = (2/3)Ma^2$.

B = M.I. of the hemispherical shell about OY
$= (1/2) \times$ (M.I. of whole spherical shell about OY), by symmetry
$= (1/2) \times$ (M.I. of whole spherical shell about the line through the centre C of the spherical shell and parallel to OY *i.e.* a diameter of the shell + M.I. of a mass $2M$ placed at C about OY)
$= (1/2) \times \{(2/3) \cdot 2M \cdot a^2 + 2M \cdot a^2\} = (5/3)Ma^2$

Similarly, C = M.I. of the hemispherical shell about $OZ = (5/3)Ma^2$

Note that the coordinates of centre of gravity of G hemispherical shell are $(a, 0, a/2)$. Hence, we have

F = P.I. of hemispherical shell about OY and OZ
= P.I. of the hemispherical shell about the lines parallel to OY, OZ through G + P.I. of mass M placed at G about OY, OZ
$= 0 + M \cdot 0 \cdot (a/2) = 0$, using symmetry for the first term

Similarly, G = P.I. of the hemispherical shell about OZ, OX
$$= 0 + M\cdot(a/2)\cdot a = (1/2)ma^2$$
and H = P.I. of the hemispherical shell about $OX, OY = 0 + M\cdot a\cdot 0 = 0$

Hence, the equation of the momental ellipsoid at O is
$$Ax^2 + By^2 + Cz^2 - 2Fyz - 2Gzx - 2Hxy = \text{const.}$$
i.e., $(2/3)Ma^2x^2 + (5/3)Ma^2y^2 + (5/3)Ma^2z^2 - 2\cdot(1/2)Ma^2zx = \text{const.}$

or $2x^2 + 5(y^2 + z^2) - 3zx = \text{const.}$, as required.

Ex. 6. *Find the length of a homogeneous solid circular cylinder of radius a, given that the momental ellipsoid at the mass centre is a sphere.* **[Delhi B.Sc. (Hons.) 1993, 96, 2005, 09]**

Sol. Let O be the mass centre (*i.e.*, centre of gravity) of a solid cylinder of mass M and length $h\,(= AB)$. Take the coordinate axis OX along the axis AB of the cylinder and two mutually perpendicular lines OY and OZ through O as the axes of y and z. Since the cylinder is symmetrical about these three axes, it follows that $F = G = H = 0$. Proceeding as in Ex. 5 on page 14.16, we have

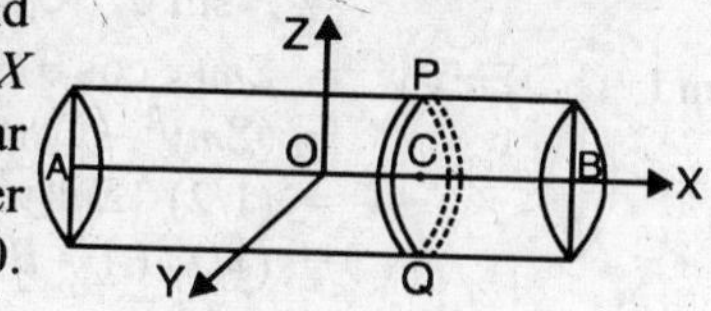

A = M.I. of the cylinder about $OX = (1/2)Ma^2$

B = M.I. of the cylinder about $OY = (1/4)M(a^2 + h^2/3)$

C = M.I. of the cylinder about $OZ = (1/4)M(a^2 + h^2/3)$

The equation of the ellipsoid at O is
$$Ax^2 + By^2 + Cz^2 = \text{const. as } F = G = H = 0$$
Since this momental ellipsoid is a sphere in our given problem, it follows that $A = B = C$. Now
$$A = B \Rightarrow (1/2)Ma^2 = (1/4)\,M(a^2 + h^4/3) \Rightarrow h = a\sqrt{3}.$$

EXERCISE

1. Show that the momental ellipsoid at any point O of a material straight rod is $y^2 + z^2 = \text{const.}$

2. The principal axes at the centre of gravity being the axes of reference, prove that the momental ellipsoid at the point (p, q, r) is $(A/M + q^2 + r^2)x^2 + (B/M + r^2 + p^2)y^2 + (C/M + p^2 + q^2)z^2 - 2qryz - 2rpzx - 2pqxy = \text{const.}$, when referred to its centre of gravity as origin. **[Garhwal 1994]**

3. If $S = Ax^2 + By^2 + Cz^2 - 2Fyz - 2Gzx - 2Hxy = \text{const.}$ be the equation of the momental ellipsoid at the centre of gravity O of a body referred to any rectangular axes through O, then prove that momental ellipsoid at the point (p, q, r) is $S + M[(qz - ry)^2 + (rx - pz)^2 + (py - qx)^2\} = \text{const.}$, where M is the mass of the body.

14.15 Principal axes and principal moments of inertia. Definition. [Agra 2010]

A set of rectangular axes $OXYZ$, are said to be *principal axes* at point O of the given body, if with respect to these axes $F = G = H = O$. Also, in that case, the moments of inertia about the axes are called *principal moments of inertia* at O.

14.16 Existence of principal axes.

To find whether a given straight line is, at any point of its length, a principal axis of a given material system, and if so, to find the other two principal axes. **[Kanpur 2010]**

Take the given straight line as the axis of z, with any point O on it as origin, and any two perpendicular lines OX, OY as the axes of x and y respectively. Assume that OZ is a principal axis at a point O' of its length and let OX', OY' be the other two principal axes where $O'X'$ is inclined at an angle θ to a line $O'X_1$ parallel to OX. Let $OO' = h$.

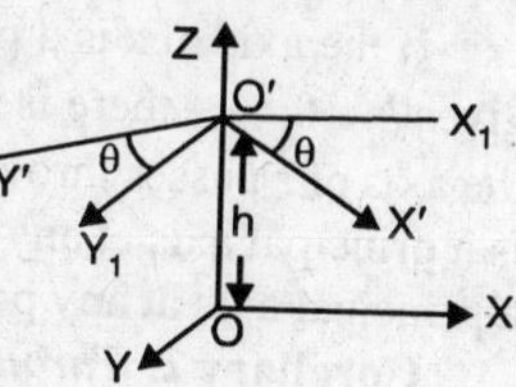

Let (x, y, z) be the coordinates of any particle m of the material system referred to OX, OY, OZ and (x', y', z') its coordinates referred to $O'X', O'Y', O'Z$. Then,
$$x' = x\cos\theta + y\sin\theta, \quad y' = -x\sin\theta + y\cos\theta \quad \text{and} \quad z' = z - h \qquad \ldots(1)$$
We know that the necessary and sufficient conditons, that $O'X'$, $O'Y'$, $O'Z$ may be the principal axes of the system, the products of inertia about them taken two at time must vanish, that is,
$$\Sigma my'z' = 0, \qquad \Sigma mz'x' = 0 \qquad \text{and} \qquad \Sigma mx'y' = 0 \qquad \ldots(2)$$

Let M be the mass of the whole system and $\overline{x}, \overline{y}$ be the x-ordinate and y-coordinate of the centre of gravity of the system, then we have

$$\overline{x} = (\Sigma mx)/\Sigma m = (\Sigma mx)/M \quad \text{and} \quad \overline{y} = (\Sigma my)/\Sigma m = (\Sigma my)/M \qquad \ldots(3)$$

$$\begin{aligned} \text{Now, } \Sigma my'z' &= \Sigma m(-x\sin\theta + y\cos\theta)(z-h), \text{ by (1)} \\ &= (\Sigma myz)\cos\theta - (\Sigma mzx)\sin\theta + h(\Sigma mx)\sin\theta - h(\Sigma my)\cos\theta \\ &= F\cos\theta - G\sin\theta + Mh(\overline{x}\sin\theta - \overline{y}\cos\theta), \text{ using (3)} \qquad \ldots(4) \end{aligned}$$

$$\begin{aligned} \Sigma mz'x' &= \Sigma m(z-h)(x\cos\theta + y\sin\theta), \text{ by (1)} \\ &= (\Sigma myz)\sin\theta + (\Sigma mzx)\cos\theta - h(\Sigma mx)\cos\theta - h(\Sigma my)\sin\theta \\ &= F\sin\theta + G\cos\theta - Mh(\overline{x}\cos\theta + \overline{y}\sin\theta), \text{ by (3)} \qquad \ldots(5) \end{aligned}$$

$$\begin{aligned} \text{and} \quad \Sigma mx'y' &= \Sigma m(x\cos\theta + y\sin\theta)(-x\sin\theta + y\cos\theta) \\ &= (\Sigma my^2 - \Sigma mx^2)\sin\theta\cos\theta + (\Sigma mxy)(\cos^2\theta - \sin^2\theta) \\ &= (1/2)\{\Sigma m(y^2+z^2) - \Sigma m(z^2+x^2)\}\sin 2\theta + (\Sigma mxy)\cos 2\theta \\ &= (1/2)(A-B)\sin 2\theta + F\cos 2\theta \qquad \ldots(6) \end{aligned}$$

If $\Sigma mx'y' = 0$ then (6) gives $(1/2)(A-B)\sin 2\theta + H\cos 2\theta = 0$

or $\tan 2\theta = 2H/(B-A)$ or $\theta = (1/2) \times \tan^{-1}\{2H/(B-A)\}$ ...(7)

Since $\Sigma my'z' = 0$, (4) $\Rightarrow Mh = (G\sin\theta - F\cos\theta)/(\overline{x}\sin\theta - \overline{y}\cos\theta)$

and as $\Sigma mz'x' = 0$, (5) $\Rightarrow Mh = (F\sin\theta + G\cos\theta)/(\overline{x}\cos\theta + \overline{y}\sin\theta)$

$$\therefore \quad Mh = \frac{G\sin\theta - F\cos\theta}{\overline{x}\sin\theta - \overline{y}\cos\theta} = \frac{F\sin\theta + G\cos\theta}{\overline{x}\cos\theta + \overline{y}\sin\theta} \qquad \ldots(8)$$

Using a property of equality of two ratios, each fraction of (8)

$$= \frac{(G\sin\theta - F\cos\theta)\sin\theta + (F\sin\theta + G\cos\theta)\cos\theta}{(\overline{x}\sin\theta - \overline{y}\cos\theta)\sin\theta + (\overline{x}\cos\theta + \overline{y}\sin\theta)\cos\theta} = \frac{G}{\overline{x}}$$

Also, each fraction of (8)

$$= \frac{(G\sin\theta - F\cos\theta)(-\cos\theta) + (F\sin\theta + G\cos\theta)\sin\theta}{(\overline{x}\sin\theta - \overline{y}\cos\theta)(-\cos\theta) + (\overline{x}\cos\theta + \overline{y}\sin\theta)\sin\theta} = \frac{F}{\overline{y}}$$

Equating the values of the above two fractions, $G/\overline{x} = F/\overline{y}$ ...(9)

Combining (8) with the above two results, we obtain

$$Mh = \frac{G}{\overline{x}} = \frac{F}{\overline{y}} \quad \text{so that} \quad h = \frac{G}{M\overline{x}} = \frac{F}{M\overline{y}} \qquad \ldots(10)$$

(9) is the condition that must hold so that OZ may be a principal axis at some point of its length, and then, if it be satisfied, (7) and (10) give the position of the point and the directions of the other principal axes at it.

Thus, we have proved that it is always possible to choose rectangular axes $O'X'$, $O'Y'$, $O'Z$ as principal axes. Then the moments of inertia about them are called *principal moments of inertia.* Thus, we have shown existence of principal axes and principal moments of inertia.

Corollary 1. *If an axis passes through the centre of gravity of a body, and is a principal axis at any point of its length, it is a principal axis at all points of its length.* **(Kanpur 2006)**

If the axis of z is a principal axis at O, then $F = G = H = 0$ and relation (7) of the above article gives $h = 0$, *i.e.*, there is no other such point as O', except when $\overline{x} = 0$ and $\overline{y} = 0$. In the latter case the axis of z passes through the centre of gravity and the value of h is indeterminate, *i.e.*, the axis of z is a principal axis at any point of its length. Hence if an axis passes through C.G. of the body and is a principal axis at any point of its length, it is a principal axis at all the points of its length.

Corollary 2. *Through each point in the plane of a lamina, there exist a pair of principal axes of the lamina.*

Let a line perpendicular to the lamina through any point O of it be taken as z-axis. Then for each point of the lamina. $z = 0$. Hence $F = G = O$. This shows that whatever be the point O in the plane of the lamina, condition (9) of article 14.16 is always satisfied. Again, $F = G = O$ and relation (10) give $h = 0$, showing that z-axis is a principal axis of the lamina at O, the other two principal axes will clearly be in the plane, whose directions are given by (refer relation (7)) $\tan 2\theta = 2H/(B - A)$.

14.17 General method of finding principal axes and principal moments of inertia. Let O be any point of a rigid body and $Oxyz$ be any axes at O. Let A, B, C, F, G, H be the moments and products of inertia of the rigid body with respect to $Oxyz$. By the existence theorem, there exists a principal frame $Ox'y'z'$ at O and let A', B', C' be the principal moments of inertia of the rigid body with respect to OX', OY' and OZ' respectively.

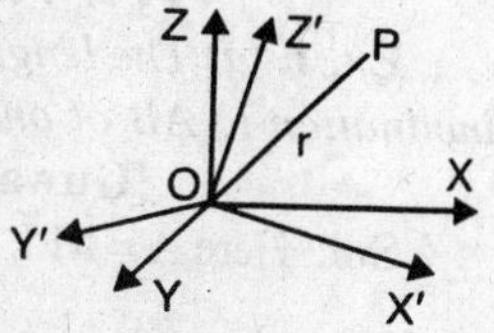

Let P be any point of the rigid body, where $OP = r$. Let (x, y, z) and (x', y', z') be cordinates of P referred to two sets of axes. Then, we have

$$r^2 = x^2 + y^2 + z^2 = x'^2 + y'^2 + z'^2 \qquad \ldots(1)$$

Since the direction cosines of OP are $x/r, y/r, z/r$ and $x'/r, y'/r, z'/r$ with respect to two sets of axes, hence the moment of inertia I of the rigid body about OP with respect to the two sets of axes is given by

$$I = A(x/r)^2 + B(y/r)^2 + C(z/r)^2 - 2F(yz/r^2) - 2G(zx/r^2) - 2H(xy/r^2)$$

and $$I = A'(x'/r)^2 + B'(y'/r)^2 + C'(z'/r)^2,$$

which can be re-written as

$$I = (Ax^2 + By^2 + Cz^2 - 2Fyz - 2Gzx - 2Hxy)/(x^2 + y^2 + z^2) \qquad \ldots(2)$$

and $$I = (A'x'^2 + B'y'^2 + C'z'^2)/(x'^2 + y'^2 + z'^2) \qquad \ldots(3)$$

So, no matter how the constant I is chosen, we have the identity

$$Ax^2 + By^2 + Cz^2 - 2Fyz - 2Gzx - Hxy - I(x^2 + y^2 + z^2)$$
$$= A'x'^2 + B'y'^2 + C'z'^2 - I(x'^2 + y'^2 + z'^2) \qquad \ldots(4)$$

Let us denote each side of this identity by ϕ so that

$$\phi(x, y, z) \equiv Ax^2 + By^2 + Cz^2 - 2Fyz - 2Gzx - 2Hxy - I(x^2 + y^2 + z^2)$$

and $$\phi(x', y', z') \equiv A'x'^2 + B'y'^2 + C'z'^2 - I(x'^2 + y'^2 + z'^2) \qquad \ldots(5)$$

Consider the equations $\quad \partial\phi/\partial x' = 0, \quad \partial\phi/\partial y' = 0, \quad \partial\phi/\partial z' = 0 \qquad \ldots(6)$

Using (6), (7) gives $\quad (A' - I)x' = 0, (B' - I)y' = 0, (C' - I)z' = 0 \qquad \ldots(7)$

Rejecting the trivial solution $x' = y' = z' = 0$, the non-trivial solution of equations (8) is given by $\ldots(8)$

$$I = A',\ x' = \text{arbitrary},\ y' = 0 = z'$$
$$I = B',\ y' = \text{arbitrary},\ x' = 0 = z'$$

and $$I = C',\ z' = \text{arbitrary},\ x' = 0 = y'$$

Thus the equations (7) have non-trivial solution provided I is equal to one of the principal moments of inertia; the corresponding values of x', y', z' give the principal axes.

Now, $$\frac{\partial\phi}{\partial x} = \frac{\partial\phi}{\partial x'}\frac{\partial x'}{\partial x} + \frac{\partial\phi}{\partial y'}\frac{\partial y'}{\partial x} + \frac{\partial\phi}{\partial z'}\frac{\partial z'}{\partial x} = 0, \text{ using (7)}$$

Similarly, $\quad \partial\phi/\partial y = 0 \quad$ and $\quad \partial\phi/\partial z = 0$.

Thus, (7) imply that $\quad \partial\phi/\partial x = 0, \quad \partial\phi/\partial y = 0, \quad \partial\phi/\partial z = 0 \qquad \ldots(9)$

If I, x', y', z' are chosen as above, (7) are satisfied and therefore (9) are satisfied. Using (5), (9) give

$$(A - I)x - Hy - Gz = 0$$
$$-Hx + (B - I)y - Fz = 0$$

and $$-Gx - Fy + (C - I)z = 0 \qquad \ldots(10)$$

Since these equations have a solution other than $x = y = z = 0$, it follows that

$$\begin{vmatrix} A - I & -H & -G \\ -H & B - I & -F \\ -G & -F & C - I \end{vmatrix} = 0 \qquad \ldots(11)$$

This is a cubic equation for I, and we have proved that its three roots are the three principal moments of inertia.

Working rule : Starting with genral axes $Oxyz$ with moments and products of inertia A, B, C, F, G, H, the three principal moments of inertia at O are the values of I satisfying (11); and the directions of the three principal axes are given by the ratios $x : y : z$ determined by (10) when the above values of I are substituted.

TYPE 4 : SOLVED EXAMPLES BASED ON ART. 14.15 TO 14.17

Ex. 1. **(a)** *The lengths AB and AD of the sides of a rectangle ABCD are 2a and 2b; show that the inclination to AB of one of the principal axes at A is $(1/2) \times \tan^{-1} \{3ab/2(a^2 - b^2)\}$.*

[Guwahati 2007; Purvanchel 2005; Delhi B.Sc. (Hons.), 2002, 2007; Kanpur 2013]

Sol. Here A = M.I. of the rectangle about AB

$= (1/3)Mb^2 + Mb^2 = (4/3)Mb^2$,

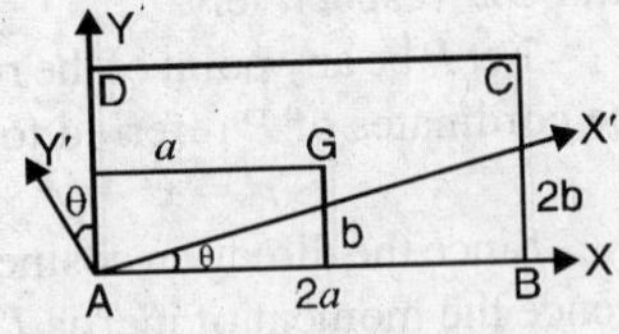

B = M.I. of the rectangle about AD

$= (1/3)Ma^2 + Ma^2 = (4/3)Ma^2$

and H = Product of inertia about $AB, AD = Mab$,

where M is the mass of the rectangle.

Hence, if θ is the inclination of a principle axis AX' to AB, then we have

$$\tan 2\theta = \frac{2H}{B - A} = \frac{2Mab}{(4/3)Ma^2 - (4/3)Mb^2} = \frac{3ab}{2(a^2 - b^2)} \Rightarrow \theta = \frac{1}{2}\tan^{-1}\frac{3ab}{2(a^2 - b^2)}$$

Ex. 1. **(b)** *Find the angle of inclination of the principal axis to the side of a square at its corner.*

Sol. Here $2b = 2a$. So $\theta = (1/2) \times \tan^{-1}(\infty) = (1/2) \times (\pi/2) = \pi/4$

Ex. 2. **(a)** *Show that at the centre of a quadrant of an ellipse, the principal axes in its plane are inclined at an angle $(1/2) \times \tan^{-1} \{4ab/\pi(a^2 - b^2)\}$ to the axes.*

[Kanpur 2002, 05, 07; Purvanchel 2007; Mumbai 2010; Meerut 2006]

Sol. Let AOB be the quadrant of ellipse

$$x^2/a^2 + y^2/b^2 = 1 \qquad \ldots(1)$$

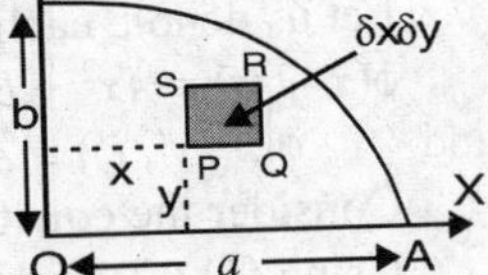

Consider an element $PQRS$ of area $\delta x\ \delta y$ at point $P(x, y)$. Let M be the mass and ρ the mass per unit area of the quadrant. Then

$$M = (1/4)\rho\pi ab \text{ so that } \rho = 4M/\pi ab \qquad \ldots(2)$$

$\therefore$ A = M.I. of the quadrant about OX

$$= \int_{x=0}^{a}\int_{y=0}^{(b/a)(a^2-x^2)^{1/2}} y^2\rho\, dx\, dy = \frac{\rho b^3}{4a^3}\int_0^a (a^2 - x^2)^{3/2}\, dx$$

$$= \frac{\rho b^3}{3a^3}\cdot a^4\int_0^{\pi/2}\cos^4\theta\, d\theta = \frac{\rho b^3 a}{3}\cdot\frac{3}{4}\cdot\frac{1}{2}\cdot\frac{\pi}{2}, \text{ by Walli's formula}$$

[Putting $x = a \sin\theta$ and $dx = a\cos\theta\, d\theta$]

$= (4M/\pi ab)\cdot(\pi b^3 a/16) = (1/4)Mb^2$,

B = M.I. of the quadrant about OY

$$= \int_{x=0}^{a}\int_{y=0}^{(b/a)(a^2-x^2)^{1/2}} x^2\rho\, dx\, dy = \frac{\rho b}{a}\int_0^a x^2(a^2 - x^2)^{1/2}\, dx$$

$$= \frac{\rho b}{a}\cdot a^4\int_0^{\pi/2}\sin^2\theta\cos^2\theta\, d\theta = \frac{\rho ba^3}{4}\int_0^{\pi/2}\sin^2 2\theta\, d\theta = \frac{\rho ba^3}{8}\int_0^{\pi/2}(1 - \cos 4\theta)\, d\theta$$

[Putting $x = a\sin\theta$ and $dx = a\cos\theta\, d\theta$]

$= (\rho ba^3/8)\,[\theta - (1/4)\sin 4\theta]_0^{\pi/2} = (4M/\pi ab) \times (ba^3/8) \times (\pi/2) = (1/4)Ma^2$

and H = Product of inertia of the quadrant about OX and OY

$$= \int_{x=0}^{a}\int_{y=0}^{(b/a)(a^2-x^2)^{1/2}} \rho\, xy\, dx\, dy = \frac{\rho b^2}{2a^2}\int_0^a x(a^2 - x^2)\, dx$$

$= (\rho b^2/2a^2) \times [(a^2/2)x^2 - (1/4)x^4]_0^a = (4M/\pi ab) \times (b^2/2a^2) \times (a^4/4) = Mab/2\pi$

So if one of the principal axes inclined at an angle θ to OX, then

$$\tan 2\theta = \frac{2H}{B - A} = \frac{(Mab)/\pi}{(1/4)Ma^2 - (1/4)Mb^2} = \frac{4ab}{\pi(a^2 - b^2)} \Rightarrow \theta = \frac{1}{2}\tan^{-1}\frac{4ab}{\pi(a^2 - b^2)}$$

Ex. 2. **(b)** *Find the principal axes of an elliptic area at any point of its bounding arc.*

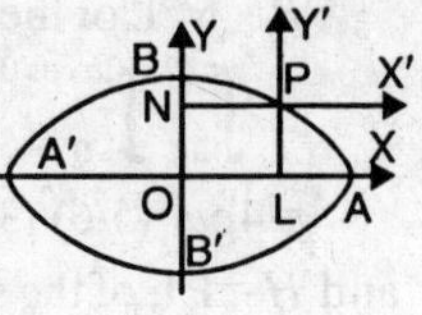

Sol. Let the equation of the ellipse be $x^2/a^2 + y^2/b^2 = 1$ and $P(a \cos \phi, b \sin \phi)$ be any point on it. Draw PL and PN perpendiculars to OX and OY so that $PL = b \sin \phi$ and $PN = a \cos \phi$. Draw axes PX' and PY' parallel to the axes OX, OY of the ellipse. Then,

$$A = \text{M.I. about } PX' = (1/4)Mb^2 + M\cdot(PL)^2 = (1/4)Mb^2 + Mb^2 \sin^2 \phi$$

$$B = \text{M.I. above } PY' = (1/4)Ma^2 + M\cdot PN^2 = (1/4)Ma^2 + Ma^2 \cos^2 \phi$$

and $\quad H = \text{P.I. about } PX \text{ and } PY = 0 + M \times PL \times PN = Mab \cos \phi \sin \phi,$

where M is the mass of the elliptic lamina.

If θ be the inclination of principal axis at P with PX', then

$$\tan 2\theta = \frac{2H}{B - A} = \frac{2Mab \cos \phi \sin \phi}{(M/4)(a^2 - b^2) + M(a^2 \cos^2 \phi - b^2 \sin^2 \phi)}$$

$$\therefore\ \theta = \frac{1}{2} \tan^{-1} \frac{8ab \tan \phi}{(a^2 - b^2)\sec^2 \theta + 4a^2 - 4b^2 \tan^2 \phi} = \frac{1}{2} \tan^{-1} \frac{8ab \tan \phi}{(5a^2 - b^2) + (a^2 - 5b^2)\tan^2 \phi}$$

Ex. 3. *Find the principal axes at a corner point of a solid cube.*

Sol. Proceed as in Ex. 3, page 14.23. Then we find that OX, OY, OZ are principal axes at corner O of the given solid cube, where OX passes through centre of gravity G of the cube.

Ex. 4. (*a*) *Show that one of the principal axes at a point on the circular rim of the solid hemisphere, is inclined at an angle* $\tan^{-1}$ *(1/3) to the radius through the point.*

(*b*) *Show that one of the principal axes at any point on the edge of the circular base of a thin hemispherical shell is inclined at an angle* $\pi/8$ *to the radius through the point.* **(Agra 1998)**

Sol. (*a*) Refer Ex. 4 on page 14.24. Let O be any point of the circular base of the solid hemisphere. Then,

$$A = (2/5)Ma^2,\ B = (7/5)Ma^2,\ C = (7/5)Ma^2,\ F = 0,\ G = (3/8)Ma^2 \text{ and } H = 0.$$

Since $F = H = 0$, so OY is a principal axis at O. The other two principal axes lie in xz-plane. If θ is the angle that any of them makes with OX, then

$$\tan 2\theta = \frac{2G}{C - A} = \frac{(3/4)Ma^2}{(7/5)Ma^2 - (2/5)Ma^2} = \frac{3}{4} \quad \text{or} \quad \frac{2\tan\theta}{1 - \tan^2\theta} = \frac{3}{4}$$

or $\quad 3\tan^2\theta + 8\tan\theta - 3 = 0 \quad$ so that $\quad \tan\theta = 1/3 \quad$ or $\quad -3$

But $\ \tan\theta = -3 \ \Rightarrow\ \theta > \pi/2$ which is not possible.

So we take $\tan\theta = 1/3$ and hence $\theta = \tan^{-1}(1/3)$

(*b*) Refer Ex. 5 on page 14.24. Let O be a point on the edge of the circular base of a thin hemisphere. Then, $A = (2/3)Ma^2$, $B = (5/3)Ma^2$, $C = (5/3)Ma^2$, $F = 0$, $G = (1/2)Ma^2$, $H = 0$.

Since $F = H = 0$, so OY is a principal axis at O. The other two principal axes lie in xz-plane. If θ is the angle that any of them makes with OX, then

$$\tan 2\theta = \frac{2G}{C - A} = \frac{Ma^2}{(5/3)Ma^2 - (2/3)Ma^2} = 1 \Rightarrow 2\theta = \frac{\pi}{4} \Rightarrow \theta = \frac{\pi}{8}.$$

Ex. 5. *Show that at an extremity of the bounding diameter of a semi-circular lamina the principal axis makes an angle (1/2) ×* $\tan^{-1}$ *(8/3π) to the diameter.* **[Kanpur 2001, 06; 11]**

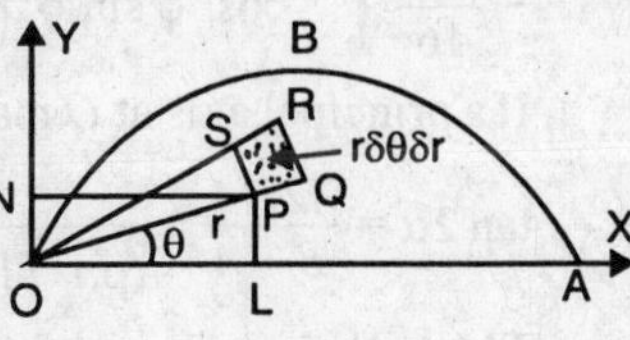

Sol. The polar equation of the semi-circle OAB of radius a referred to the diameter OA is initial line and O as pole is $r = 2a \cos\theta$. Let ρ be density the semi circular lamina and let $r\,\delta r\,\delta\theta$ be elementary area $PQRS$ at $P(r, \theta)$. Let OY be perpendicular to OA. Let PL and PN be perpendiculars from P on OX and OY respectively. Then $PL = r\sin\theta$ and $PN = r\cos\theta$. Now, we have

A = M.I. of the semi-circular lamina about OX

$$= \int_{\theta=0}^{\pi/2} \int_{r=0}^{2a\cos\theta} (r \sin\theta)^2 \cdot \rho r\, d\theta\, dr = \frac{1}{4}(2a)^4 \rho \int_0^{\pi/2} \sin^2\theta \cos^4\theta\, d\theta$$

$$= 4\rho a^4 \frac{\Gamma(3/2)\Gamma(5/2)}{2\Gamma(4)} = \frac{\pi\rho a^4}{8}$$

B = M.I. of the semi-circular lamina about OY

$$= \int_{\theta=0}^{\pi/2} \int_{r=0}^{2a\cos\theta} (r\cos\theta)^2 \cdot \rho r \, d\theta \, dr = \frac{1}{4}(2a)^4 \rho \int_0^{\pi/2} \cos^6\theta \, d\theta$$

$= 4\rho a^4 \cdot (5/6)\cdot(3/4)\cdot(1/2)\cdot(\pi/2) = (5/8)\pi\rho a^4$, by Walli's formula

and H = P.I. of the semi-circular lamina about OX, OY

$$= \int_{\theta=0}^{\pi/2} \int_{r=0}^{2a\cos\theta} (r\cos\theta)\cdot(r\sin\theta)\cdot \rho r \, d\theta \, dr = \frac{1}{4}\rho(2a)^4 \int_0^{\pi/2} \cos^5\theta \sin\theta \, d\theta$$

$= 4\rho a^4[-(1/6)\cos^6\theta]_0^{\pi/2} = (4\rho a^4)\cdot(1/6) = (2/3)\rho a^4$

If α is the angle that a principal axis at O makes with OX, then

$\tan 2\alpha = (2H)/(B-A) = 8/3\pi$ so that $\alpha = (1/2)\times\tan^{-1}(8/3\pi)$

Ex. 6. *Show that the principal axes at the node of a half loop of the lemniscate $r^2 = a^2\cos 2\theta$ are inclined to the initial line at angles $(1/2)\times\tan^{-1}(1/2)$ and $\pi/2 + (1/2)\times\tan^{-1}(1/2)$.*

[Garhwal 2010; Kanpur 2008; Meerut 2008; Purvanchal 2006]

Sol. In figure OBA represents half loop of the lemnhiscate $r^2 = a^2\cos 2\theta$ whose node is at O. Let ρ be the density of the lamina and let $r\,\delta\theta\,\delta r$ be elementary area $PQRS$ at $P(r, \theta)$. Let PL and PN be perpendiculars from P on OX and OY respectively. Then $PL = r\sin\theta$ and $PN = r\cos\theta$. Now, we have

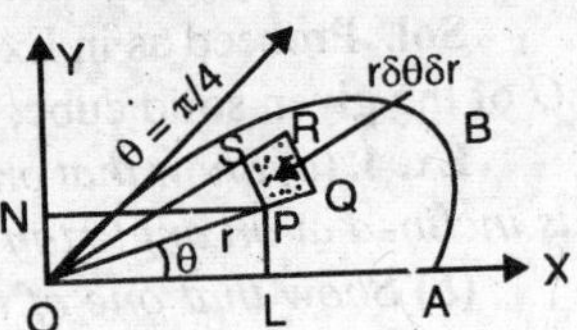

$\therefore$ A = M.I. of half the loop about OX

$$= \int_0^{\pi/4} \int_0^{a(\cos 2\theta)^{1/2}} (r\sin\theta)^2 \cdot \rho r \, d\theta \, dr = \frac{\rho a^4}{4} \int_0^{\pi/4} \cos^2 2\theta \sin^2\theta \, d\theta$$

$$= \frac{\pi a^4}{8} \int_0^{\pi/4} \cos^2 2\theta (1-\cos 2\theta) \, d\theta = \frac{\rho a^4}{16} \int_0^{\pi/2} (\cos^2\phi - \cos^3\phi) \, d\phi, \text{ if } 2\theta = \phi$$

$$= \frac{\rho a^4}{16}\left(\frac{1}{2}\cdot\frac{\pi}{2} - \frac{2}{3}\right) = \frac{\rho a^4}{192}(3\pi - 8), \text{ using Walli's formulas}$$

B = M.I. of the half loop about OY

$$= \int_0^{\pi/4} \int_0^{a(\cos 2\theta)^{1/2}} (r\cos\theta)^2 \cdot \rho r \, d\theta \, dr = \frac{\rho a^4}{4} \int_0^{\pi/4} \cos^2 2\theta \cos^2\theta \, d\theta$$

$$= \frac{\rho a^4}{8} \int_0^{\pi/4} \cos^2 2\theta (1+\cos 2\theta) \, d\theta = \frac{\rho a^4}{8} \int_0^{\pi/2} (\cos^2\phi + \cos^3\phi), \text{ if } 2\theta = \phi$$

$$= \frac{\rho a^4}{16}\left(\frac{\pi}{4} + \frac{2}{3}\right) = \frac{\rho a^4}{192}(3\pi + 8), \text{ using Walli's formulas}$$

and H = Product of inertia of the loop about OX, OY,

$$= \int_0^{\pi/4} \int_0^{a(\cos 2\theta)^{1/2}} (r\cos\theta)\cdot(r\sin\theta)\cdot \rho r \, d\theta \, dr = \frac{\rho a^4}{8} \int_0^{\pi/4} \cos^2 2\theta \sin 2\theta \, d\theta$$

$$= \frac{\rho a^4}{16} \int_0^{\pi/2} \cos^2\phi \sin\phi \, d\theta = \frac{\rho a^4}{48}, \text{ putting } 2\theta = \phi \text{ and } d\theta = (1/2)d\phi$$

If a principal axis at O makes an angle α with OX, then

$$\tan 2\alpha = \frac{2H}{B-A} = \frac{\rho a^4/24}{(\rho a^4/192)(3\pi + 8 - 3\pi + 8)} = \frac{1}{2} \Rightarrow \alpha = \frac{1}{2}\tan^{-1}\frac{1}{2}$$

The other principal axis being at right angles to the former makes an angle $\pi/2 + (1/2)\times\tan^{-1}(1/2)$ with OX.

Ex. 7. *A wire is in the form of a semi-circle of radius a. Show that at an end of its diameter the principal axes in its plane are inclined to the diameter at angles $(1/2)\times\tan^{-1}(4/\pi)$ and $(\pi/2) + (1/2)\times\tan^{-1}(4/\pi)$.*

Sol. Let C be the centre, M the mass, ρ the mass per unit length and OA the diameter of the given semi-circular wire OPA of radius a. Take OX and OY to be the axes of reference and O as an end of OA under consideration. Consider an elementary arc $PQ\ (= a\delta\phi)$ at P. Its mass is $\rho a\ \delta\phi$ where $\rho = M/\pi a$. If $PL(= a \sin\phi)$ and $PN(= a + a\cos\phi)$ be perpendiculars from P on OX and OY, then

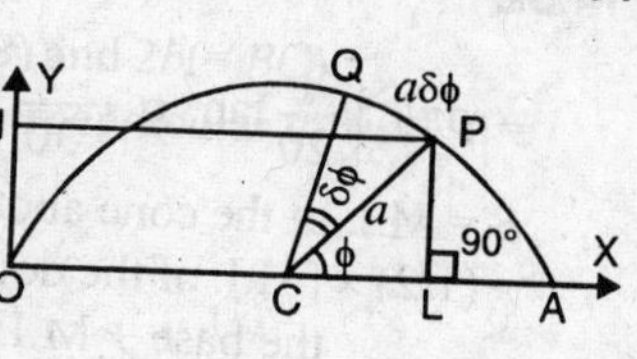

$$A = \text{M.I. of the wire about } OX = \int_0^{\pi} PL^2 \cdot \rho a\, d\phi = \int_0^{\pi} a^2 \sin^2\phi \cdot \rho a \cdot d\phi$$

$$= 2\rho a^3 \int_0^{\pi/2} \sin^2\phi\, d\phi = 2 \cdot \frac{M}{\pi a} \cdot a^3 \cdot \frac{1}{2} \cdot \frac{\pi}{2} = \frac{1}{2} Ma^2, \text{ by Walli's formula}$$

$$B = \text{M.I. of the wire about } OY = \int_0^{\pi} PN^2 \cdot \rho a\, d\phi = \int_0^{\pi} (a + a\cos\phi)^2 \cdot \rho a\, d\phi$$

$$= \rho a^3 \int_0^{\pi} (1 + 2\cos\phi + \cos^2\phi)\, d\phi = \frac{3}{2} Ma^2, \text{ on simplifying}$$

and H = Product of inertia about OX, OY = $\int_0^{\pi} PL \cdot PM \cdot \rho a\, d\phi$

$$= \int_0^{\pi} a\sin\phi \cdot (a + a\cos\phi) \cdot \rho a\, d\phi = \rho a^3 \left[-\cos\phi + \frac{1}{2}\sin^2\phi \right]_0^{\pi}$$

$= (M/\pi a) \cdot a^3 \cdot 2 = (2/\pi)Ma^2$

So, if a principal axis at O makes an angle θ with OX, then

$$\tan 2\theta = \frac{2H}{B - A} = \frac{(4/\pi)Ma^2}{(3/2)Ma^2 - (1/2)Ma^2} = \frac{4}{\pi} \Rightarrow \theta = \frac{1}{2}\tan^{-1}\frac{4}{\pi}$$

The other principal axis being perpendicular to the above axis makes, an angle $\pi/2 + (1/2) \times \tan^{-1}(4/\pi)$ with OX

Ex. 8. (a) *Prove that the equation of the momental ellipsoid at a point on the circular edge of a solid cone is* $(3a^2 + 2h^2)x^2 + (23a^2 + 2h^2)y^2 + 26a^2z^2 - 10ahxz = \text{const.}$, *where h is the height and a the radius of the base.*

(b) *Find the principal axes of a right circular cone at a point on the circumference of the base, and show that one of them will pass through its C.G. if the vertical angle of the cone is* $2\tan^{-1}(1/2)$.

[Agra 1992, Meerut 98; Rohilkhand 99]

Sol. (a) Let O be a point on the circular edge of the cone, at which we want to determine the momental ellipsoid. Let M be the mass, ρ mass per unit volume, h the height, a the radius and α the semi-vertical angle of the given solid cone. Then $M = (1/3)\pi\rho h^3 \tan^2\alpha \Rightarrow \rho = 3M(\rho h^3 \tan^2\alpha)$.

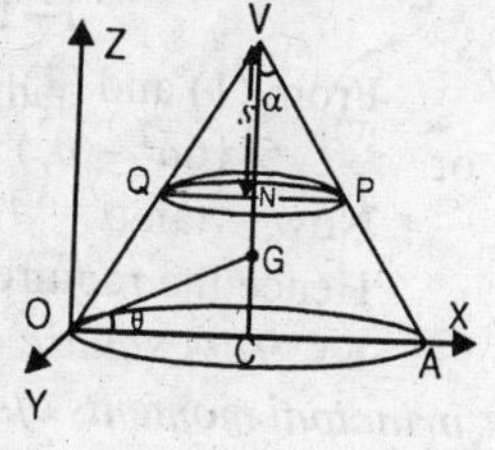

We choose the x-axis along the diameter OA of the base, axis OY perpendicular to OA in the plane of the base and axis OZ perpendicular to the base. Consider an elementary disc PQ parallel to the base at a distance s from the vertex V and thickness δs. Then,

δm = mass of the elementary disc = $\rho \cdot \pi (QN)^2 \cdot \delta s = \rho\pi (s\tan\alpha)^2\, \delta s$

M.I. of this elementary disc about OX

= Its M.I. about PQ + M.I. of mass δm placed at N about OX

$= (1/4)QN^2 \cdot \delta m + (CN)^2 \cdot \delta m = \{(1/4)s^2\tan^2\alpha + (h - s)^2\}\rho\pi s^2 \tan^2\alpha\ \delta s$

$\therefore A$ = M.I. of the cone about OX

$$= \int_0^h \left\{ \frac{1}{4}s^2\tan^2\alpha + (h - s)^2 \right\} \rho\pi s^2 \tan^2\alpha\, ds$$

$$= \rho\pi\tan^2\alpha \int_0^h \left(\frac{1}{4}s^4 \tan^2\alpha + h^2 s^2 - 2hs^3 + s^4 \right) ds$$

$$= \rho\pi\tan^2\alpha \Big[(1/20)s^5 \tan^2\alpha + h^2(s^2/3) - 2h(s^4/4) + s^5/5 \Big]_0^h$$

$$= \rho\pi h^5\left[\frac{1}{20}\tan^2\alpha + \frac{1}{30}\right] = \frac{3M}{\pi h^3 \tan^2\alpha}\cdot\pi h^5\left(\frac{1}{20}\tan^2\alpha + \frac{1}{30}\right) = \frac{M}{20}(3a^2 + 2h^2)$$

B = M.I. of the cone about OY = (1/2) × (M.I. of the double cone about OY)

= (1/2) × [M.I. of the double cone about an axis parallel to OY through C *i.e.*, about a diameter of the base + M.I. of mass $2M$ placed at C about OY]

= (1/2) × [2 × (1/20) $(3a^2 + 2h^2) + 2Ma^2\}$ = (1/20) × $M(23a^2 + 2h^2)$

Now, M.I. of the elementary disc PQ about OZ

= Its M.I. about VC + M.I. of mass δm at N about OZ

= (1/2) × $QN^2\cdot\delta m + OC^2\cdot\delta m = \{(1/2)\cdot s^2\tan^2\alpha + a^2\}\cdot\rho\pi s^2\tan^2\alpha\,\delta s$

∴ C = M.I. of the cone about OZ

$$= \rho\pi\tan^2\alpha\int_0^h\left(\frac{1}{2}s^4\tan^2\alpha + a^2s^2\right)ds = \rho\pi h^3\tan^2\alpha\left(\frac{1}{10}h^2\tan^2\alpha + \frac{1}{3}a^2\right)$$

$$= \frac{3M}{\pi h^3\tan^2\alpha}\cdot\pi h^3\cdot\tan^2\alpha\left(\frac{1}{10}h^2\cdot\frac{a^2}{h^2} + \frac{1}{3}a^2\right) = \frac{13}{10}Ma^2, \text{ as } \tan\alpha = \frac{a}{h}$$

There is symmetry about y-axis and the coordinates of centre of gravity G of the cone are $(a, 0, h/4)$. Hence $F = H = O$ and G = P.I. of the cone about OZ and OX $= 0 + M\cdot(\pi/4)\cdot a = (1/4)Mah$.

Hence the equation of the momental ellipsoid at O is

$Ax^2 + By^2 + Cz^2 - 2Fyz - 2Gzx - 2Hxy$ = const.

or $(M/20)(3a^2 + 2h^2)x^2 + (M/20)(23a^2 + 2h^2)y^2 + (13/10)Ma^2z^2 - 2\cdot(M/4)ahzx$ = const.

or $(3a^2 + 2h^2)x^2 + (23a^2 + 2h^2)y^2 + 26a^2z^2 - 10ahzx$ = const.

(b) Refer part (*a*). Let O be a point on the circular edge of the cone, at which we want to determine the principal axes of the cone. As in part (*a*), first prove that $A = (M/20)(3a^2 + 2h^2)$, $B = (M/20)(23a^2 + 2h^2)$, $C = (13/10)Ma^2$, $F = 0$, $G = (1/4)Mah$, $H = 0$. Since $F = 0$ and $H = 0$, the axis of y taken in the figure is the principal axis at O. The other two axes at O will be in the xz-plane. Let one of the principal axes make an angle θ with OX in xz-plane, then

$$\tan 2\theta = \frac{2G}{C - A} = \frac{(1/2)Mah}{(13/10)Ma^2 - (M/20)(3a^2 + 2h^2)} = \frac{10ah}{23a^2 - 2h^2} \quad \text{...(1)}$$

Now, if this line passes through the centre of gravity G of the cone, then $\tan\theta = GC/OC = h/4a$, as $GC = h/4$, $OC = a$

$$\therefore\quad \tan 2\theta = \frac{2\tan\theta}{1 - \tan^2\theta} = \frac{2\times(h/4a)}{1 - (h^2/16a^2)} = \frac{8ah}{16a^2 - h^2} \quad \text{...(2)}$$

From (1) and (2), $10ah/(23a^2 - 2h^2) = 8ah/(16a^2 - h^2)$

or $5(16a^2 - h^2) = 4(23a^2 - 2h^2)$ giving $h = 2a$

Now, $\tan\alpha = AC/VC = a/h = a/2a = 1/2 \Rightarrow \alpha = \tan^{-1}(1/2)$

Hence the required vertical angle = $2\alpha = 2\tan^{-1}(1/2)$

Ex. 9. *A square of side a has particles of masses m, 2m, 3m, 4m at its vertices. Show that the principal moments of inertia at the centre of the square are $2ma^2$, $3ma^2$, $5ma^2$, and find the directions of the principal axes.* **[Delhi BSc. (Hons.) 1994, 2000, 2001, 2004]**

Sol. Let Oxy be taken as rectangular axes at the centre of given square $ABCD$. Let particles of masses m, $2m$, $3m$, $4m$ be situated at A, B, C, D respectively.

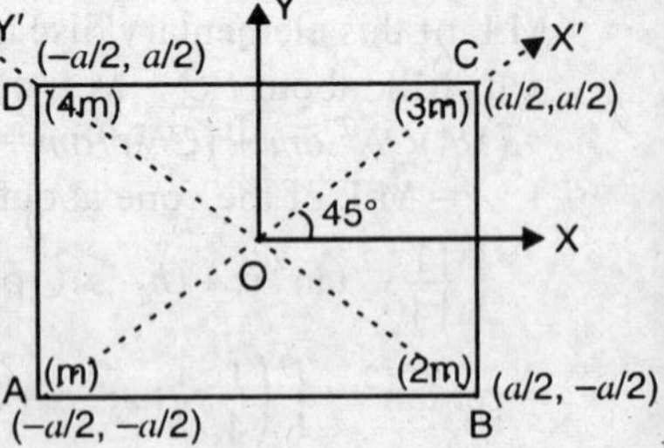

A = M.I. of the particles about OX

$= m(a^2/4) + 2m(a^2/4) + 3m(a^2/4) + 4m(a^2/4)$

Thus, $A = (5/2)ma^2$. Similarly, $B = (5/2)ma^2$

H = product of inertia about OX and OY

$= m(-a/2)(-a/2) + 2m(a/2)(-a/2) + 3m(a/2)(a/2)$
$\quad + 4m(-a/2)(a/2) = -(1/2)ma^2$

If one of the principal axes is inclined at an angle θ to OX, then

$\cot 2\theta = (B - A)/2H = 0 \;\Rightarrow\; 2\theta = 90^\circ \;\Rightarrow\; \theta = 45^\circ$,

showing that the principal axes at O are the diagonals AC and BD. Take OX' and OY' as the principal axes at O. Then principal moments of inertia are given by (refer Art. 14.11)

$A' = A\cos^2\theta + B\sin^2\theta - H\sin 2\theta$

$= (5/2)ma^2 \cos 45^\circ + (5/2)ma^2 \sin^2 45^\circ + (1/2)ma^2 \sin 90^\circ$

Thus, $A' = 3ma^2$, Similarly, $B' = A\sin^2\theta + H\sin 2\theta + B\cos^2\theta = 2ma^2$

Using perpendicular axis theorem (refer Art. 14.3), the third principal moment $= C' = A' + B' = 3ma^2 + 2ma^2 = 5ma^2$.

So $2ma^2$, $3ma^2$, $5ma^2$ are the principal moments of inertia.

Ex. 10. *The centre of mass of a plane lamina of mass M is G and GX, GY are the principal axes of inertia at G, Ox, Oy are axes in the plane of the lamina parallel to GX, GY respectively. Prove that the product of inertia of the lamina with respect to Ox, Oy is* $M\alpha\beta$, *where* α, β *are the co-ordinates of G with respect to Ox, Oy. If the principal moments of inertia of the lamina are* $Mn\alpha^2$ *about GX and* $Mm\beta^2$ *about GY, and one of the axes of inertia at O is inclined at an angle of 45° to Ox, prove* $(n-1)/(m-1) = \beta^2/\alpha^2$.

[Delhi B.Sc. (Hons.) 1999, 2002; 07]

Sol. Take any element δM of the lamina at P whose coordinates with respect to Ox, Oy and GX, GY are (x, y) and (X, Y) respectively. Since coordinates of G are (α, β), so

$$x = X + \alpha \quad \text{and} \quad y = Y + \beta.$$

$\therefore\ H =$ P.I. with respect to Ox, Oy

$= \Sigma\delta M\,xy = \Sigma\delta M(X+\alpha)(Y+\beta)$

or $H = \Sigma\delta M\,XY + \alpha\beta\,\Sigma\delta M + \alpha\,\Sigma y\,\delta M + \beta\,\Sigma X\,\delta M$ ...(1)

Since coordinates of G with respect to axes GX, GY are $(0, 0)$, so

$\Sigma Y\,\delta M/\Sigma\delta M = 0$ and $\Sigma X\,\delta M/\Sigma\delta M = 0$

Thus, $\Sigma Y\,\delta M = 0$ and $\Sigma X\,\delta M = 0$. Also $\Sigma\delta M = M$ ...(2)

Since GX, GY are principal axes at G, product of inertia of the plane lamina with respect to GX, GY is zero. Thus, $\Sigma XY\,\delta M = 0$. ...(3)

Using (2) and (3), (1) gives $H = M\alpha\beta$, as required.

Using theorem of parallel axis (refer Art. 14.9), we get

$A =$ M.I. of the lamina about $Ox = Mn\alpha^2 + M\beta^2$

and $B =$ M.I. of the lamina about $Oy = Mm\beta^2 + M\alpha^2$

Since $H \neq 0$, it follows that that Ox, Oy are not principal axes at O. Let Ox', Oy' be principal axes at O. Then, given that Ox' is inclined at 45° to Ox.

$\therefore\ \cot 2\theta = (B-A)/2H \Rightarrow \cot 90^\circ = (B-A)/2H \Rightarrow B = A$, as $\cot 90^\circ = 0$

Now, $B = A \Rightarrow Mm\beta^2 + M\alpha^2 = Mn\alpha^2 + M\beta^2 \Rightarrow (n-1)/(m-1) = \beta^2/\alpha^2$

Ex. 11. *Show that for a thin hemispherical solid of radius a and mass M, the principal moments of inertia at the centre of gravity are* $(83/320)Ma^2$, $(83/320)Ma^2$, $(2/5)Ma^2$.

Sol. Let G be the centre of gravity of the given hemispherical solid ABC. Let O be centre of its base and OB be its central radius. Then, $OG = 3a/8$. Take GX, GY as axes of x and y in plane parallel to the plane of base and GZ as z-axis along OB. Then, by theorem of parallel axis, we have

$A =$ M.I. about $GX =$ M.I. about $AC - M{\cdot}OG^2 = (2/5)Ma^2 - M(3a/8)^2$

Thus, $A = (83/320)Ma^2$. Similarly, $B =$ M.I. about $GY = (83/320)Ma^2$

and $C =$ M.I. about $GZ = (2/5)Ma^2$

Since coordinates of O are $(0, 0, -3a/8)$, we have

$F =$ P.I. about GY, GZ

$=$ P.I. about the parallel lines through $O -$ P.I. of mass M at O about GY, GZ

$= 0 - M{\cdot}0{\cdot}(-3a/8) = 0$

Similarly, $G = 0$, $H = 0$. Since $F = G = H = 0$, GX, GY, GZ are principal axes at G and so the principal moments are $(83/320)Ma^2$, $(83/320)Ma^2$, $(2/5)Ma^2$.

Ex. 12. *Show that for a thin hemispherical shell of radius a and mass M, principal moments of inertia at centre of gravity are $(5/12)Ma^2$, $(5/12)Ma^2$, $(2/3)Ma^2$.*

Hint. Refer figure of Ex. 11. As in Ex. 11, here $OG = a/2$.

A = M.I. about $GX = (2/3)Ma^2 - M.\ OG^2 = (2/3)Ma^2 - M(a/2)^2 = (5/12)Ma^2$

Similarly, $B = (5/12)Ma^2$ and $C = (2/3)Ma^2$. Also $F = G = H = 0$. So as before principal moments of inertia are $(5/12)Ma^2$, $(5/12)Ma^2$, $(2/3)Ma^2$.

Ex. 13. *If A andB be the moments of inertia of a uniform lamina about perpendicular axes OX and OY, lying in its plane, and H be the product of inertia of the lamina about these lines, show that the principal moments at O are equal to $[A + B \pm \{(A - B)^2 + 4H^2\}^{1/2}]/2$.*

Sol. Using given facts, the equation of the momental ellipse at O is

$$Ax^2 + By^2 - 2Hxy = 1 \qquad \ldots(1)$$

Let OX', OY' be the principal axes at O, and let A' and B' be the corresponding principal moments of inertia at O. Then, referred to the new axes, the equation of the momental ellipse at O is

$$A'x'^2 + B'y'^2 = 1 \qquad \ldots(2)$$

Let $\angle X'OX = \theta$. Then we find that axes $OX'Y'$ can be obtained by rotating the axes OXY through an angle θ. Since (1) and (2) represent the same ellipse, equating the invariants of (1) and (2), we have

$$A' + B' = A + B \qquad \text{and} \qquad A'B' = AB - H^2 \qquad \ldots(3)$$

From (3), we find that A', B' are roots of the quadratic equation

$$t^2 - (A' + B')t + A'B' = 0 \qquad \text{or} \qquad t^2 - (A + B)t + AB - H^2 = 0 \qquad \ldots(4)$$

Solving (4), $\quad t = [(A + B)] \pm \{(A + B)^2 - 4(AB - H^2)\}^{1/2}]/2$

$\therefore \quad A', B' = [A + B \pm \{(A - B)^2 + 4H^2\}^{1/2}]/2$, as required.

Ex. 14. *Find the moments of inertia of a solid homogeneous cube about an arbitrary line through its centre. What are the principal axes of inertia at a corner ? Prove that the equation of the momental ellipsoid at the corner is $2x^2 + 11(y^2 + z^2)$ = constant.* **[Delhi B.Sc. (Hons.) 1992, 92]**

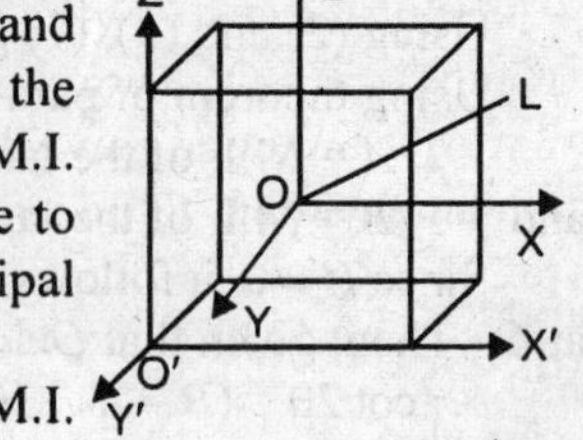

Sol. Let O' be the corner of the given cube of mass M and side of length $2a$. Let O be its centre. Let OX, OY, OZ be the axes through the centre and parallel to the edges of the cube. Let $O'X'$, $O'Y'$, $O'Z'$ be axes through the corner O' and parallel to the edges of the cube. Then, by Art. 14.4C, A_1 = M.I. of the cube about $OX = (2/3)Ma^2$. Similarly, $B_1 = C_1 = (2/3)Ma^2$. Due to symmetry, $F_1 = G_1 = H_1 = 0$. Hence, by definition OX, OY, OZ are principal axes at O.

Let l, m, n be direction cosines of any line OL through O. Then the M.I. of the cube about OL (refer Art. 14.10)

$= A_1l^2 + B_1m^2 + C_1n^2 - 2F_1mn - 2G_1ln - 2H_1lm$

$= (2/3)Ma^2(l^2 + m^2 + n^2) = (2/3)Ma^2$, as $l^2 + m^2 + n^2 = 1$

Now, coordinates of O are (a, a, a) with respect to $OX'Y'Z'$.

Using theorem of parallel axis, we have

A = M.I. about OX' = M.I. about $OX + M\cdot$(distance of O from OX')^2

$= A_1 + M\cdot(a^2 + a^2) = (2/3)Ma^2 + 2Ma^2 = (8/3)Ma^2$

Similarly, $\qquad B = C = (8/3)Ma^2$

Again, $\quad F$ = P.I. about OY', OZ'

= P.I. about OY, OZ + P.I. of mass M at O about OY', OZ'

$= 0 + m\cdot a\cdot a = ma^2$

Similarly, $\qquad G = H = ma^2$

Using result (11) of Art. 14.17, the required principal moments of inertia I at O are given by

$$\begin{vmatrix} A-I & -H & -G \\ -H & B-I & -F \\ -G & -F & C-I \end{vmatrix} = 0 \quad \text{or} \quad \begin{vmatrix} (8/3)Ma^2 - I & -ma^2 & -ma^2 \\ -ma^2 & (8/3)Ma^2 - I & -ma^2 \\ -ma^2 & -ma^2 & (8/3)Ma^2 - I \end{vmatrix} = 0$$

or $\begin{vmatrix} (2/3)Ma^2 - I & -ma^2 & -ma^2 \\ (2/3)Ma^2 - I & (8/3)Ma^2 - I & -ma^2 \\ (2/3)Ma^2 - I & -ma^2 & (8/3)Ma^2 - I \end{vmatrix} = 0,$ Applying $C_1 \to C_1 + C_2 + C_3$

or $\begin{vmatrix} (2/3)Ma^2 - I & -ma^2 & -ma^2 \\ 0 & (11/3)Ma^2 - I & 0 \\ 0 & 0 & (11/3)Ma^2 - I \end{vmatrix} = 0,$ Applying $R_2 \to R_2 - R_1$ and $R_3 \to R_3 - R_1$

or $\{(2/3)Ma^2 - I\}\,\{(11/3)Ma^2 - I\}^2 = 0$

so that $A' = (2/3)Ma^2$, $B' = (11/3)Ma^2 = C'$, are three required principal moments of inertia of the cube at the corner O' of the cube. Then, the required equation of the momental ellipsoid at O is given by $A'x^2 + B'y^2 + C'z^2 =$ const. or $(2/3)Ma^2x^2 + (11/3)Ma^2(y^2 + z^2) =$ const.

or $2x^2 + 11(y^2 + z^2) =$ const.

Using result (10) of Art. 14.17, the directions of three principal axes are given by the ratios $x : y : z$ determined by the following equations :

$$\left.\begin{aligned} (A - I)x - Hy - Gz &= 0 \\ -Hx + (B - I)y - Fz &= 0 \\ -Gx - Fy + (C - I)z &= 0 \end{aligned}\right\} \qquad \ldots(1)$$

Taking $I = A' = (2/3)Ma^2$ in (1), we get

$$2x - y - z = 0, \quad -x + 2y - z = 0, \quad -x - y + 2z = 0 \qquad \ldots(2)$$

On solving the first two equation of (2), we get

$x/3 = y/3 = z/3$ *i.e.*, $x = y = z$,

showing that the direction cosines of one principal axis at O' are $1/\sqrt{3}, 1/\sqrt{3}, 1/\sqrt{3}$. So it is equally inclined to the axes $O'X'$, $O'Y'$, $O'Z'$ and passes through the centre O of the cube.

Next, taking $I = B' = (11/3)Ma^2$, we find that each of the equation in (1) reduces to $x + y + z = 0$. It contains two free variables y and z. Therefore any two directions through O', perpendicular to each other and also perpendicular to first principal direction $x = y = z$, together are the required two other principal axes at O.

EXERCISE

1. A uniform rectangular plate whose sides are of lengths $2a$, $2b$ has a portion cut out in the form of a square whose centre is the centre of the rectangle and whose mass is half the mass of the plate. Show that the axes of greatest and least moment of inertia at a corner of the rectangle make angles θ, $\theta + \pi/2$ with a side, where $\tan 2\theta = 6ab/5(a^2 - b^2)$. **(Kanpur 2000)**

2. A uniform square lamina is bounded by the axes of x and y and the lines $x = 2c$, $y = 2c$, and a corner is cut off by the line $x/a + y/b = 1$. Show that the principal axes at the centre of the square are inclined to the axis of x at angles given by $\tan 2\theta = \{ab - 2(a + b)c + 3c^2\}/(a - b)(a + b - 2c)$.

3. If the vertical angle of cone is 90°, then show that the point at which a generator is a principal axis divides the generator in the ratio 3 : 7.

4. The length of the axis of a solid paraboloid of revolution is equal to the latus rectum of the generating parabola. Prove that one principal axis at a point in the circular rim meets the axis of revolution at an angle $(1/2)\tan^{-1}(2/3)$.

5. A uniform lamina is bounded by a parabolic arc of latus rectum $4a$ and a double ordinate at a distance b from the vertex. If $b = (1/3)a(7 + 4\sqrt{7})$, show that two of the principal axes at the end of a latus rectum are the tangent and normal there.

6. Three rods AB, BC, CD each of mass m and length $2a$ are such that each is perpendicular to the other two. Show that the principal moments of inertia at the centre of the mass are ma^2, $(11/3)ma^2$ and $4ma^2$. **(Purvanchel 2005, 06)**

7. A uniform solid circular cone of semi-vertical angle α and height h is cut in half by a plane through its axis. Show that the principal moments of inertia at the vertex for one of the halves are $(3/5)Mh^2\{1 + (1/4)\tan^2\alpha\}$, and $(3/10)Mh^2\{1 + (3/4)\tan^2\alpha\} \pm (3/10)Mh^2[\{1 - (1/4)\tan^2\alpha\}^2 + (64/9\pi^2)\tan^2\alpha]^{1/2}$

8. Two particles each of mass m are placed at the extremities of the minor axis of an elliptic area of mass M. Prove that principal axes at any point of the circumference of the ellpse will be the tangent and normal to the ellipse if $m/M = 5e^2/8(1 - e^2)$. **[Agra 1998; Garhwal 1994]**

9. Prove the existence of principal axes and principal moments of inertia at any point of a rigid body. **[Delhi B.Sc. (H) 2003]**

Hint. Refer Art. 14.16.

14.18 Equimomental or kinetically (or dynamically) equivalent bodies or system of bodies. Definition. **[Kanpur 2002; Calcutta 2003]**

If two bodies or system of bodies are such that they have equal moments of inertia about every line of space, then they are said to be *equimomental*.

14.19 Theorem. Necessary and sufficient conditions for the two systems to be equimomental. **(Kanpur 2004)**

If two systems are such that they have (i) the same total mass; (ii) the same centroid, (iii) the same principal axes and principal moments of inertia at the centroid, they are equimomental.

The conditions are necessary. let two systems S_1 and S_2 to be equimomental. Let G_1, G_2 be the centroids of the two systems, and M_1, M_2 their total masses. Since the systems are equimomental, they have equal moments of inertia I, say, about G_1G_2. Using theorem of parallel axis, the moments of inertia about any parallel line distant d from G_1G_2 are respectively $I + M_1d^2$ and $I + M_2d^2$. Since the systems are equimomental, we must have

$$I + M_1d^2 = I + M_2d^2$$

so that $M_1 = M_2 = M$, say. This proves condition (*i*).

We shall now prove condition (*ii*). Let J be the M.I. of either system about the line L_1 through G_1 perpendicular to G_1G_2. Using theorem of parallel axis, the M.I. of the system S_1 about a parallel line L_2 through G_2 is $J + Ma^2$, where $G_1G_2 = a$. Again, by theorem of parallel axis, we have

M.I. of the system S_2 about L_1 = M.I. of the system S_2 about $L_2 + M \cdot a^2$

$\Rightarrow$ M.I. of the system S_2 about $L_2 = J - Ma^2$.

Thus, the moments of inertia of the first and second systems about line L_2 are $J + Ma^2$ and $J - Ma^2$ respectively. Since the two systems are equimomental, we must have

$$J + Ma^2 = J - Ma^2 \quad \Rightarrow \quad a = 0,$$

showing that G_1 and G_2 coincide with common centre of gravity G, say and so condition (*ii*) is proved.

The systems S_1 and S_2 have equal moments of inertia about any line through their common centroid G. Thus the momental ellipsoids at G are the same and so are the principal axes and principal moments of inertia. This proves condition (*iii*).

The conditions are sufficient. Suppose that the stated conditions (*i*), (*ii*) and (*iii*) hold. Denote by M the total mass of each system, G the centroid, A, B, C the principal moments of inertia at G.

Then for both systems, the M.I. about the line L (say) through G having direction cosines l, m, n with respect to the principal axes is $Al^2 + Bm^2 + Cn^2$. Again, for the both systems by theorem of parallel axis, the M.I. about a parallel line L' distant h from the line L is given by $Al^2 + Bm^2 + Cn^2 + Mh^2$. Thus the two system have equal moments of inertia about any line of space and the systems are equimomental.

14.20 Theorem. *The moments and products of inertia of a uniform triangle about any lines are the same as the moments and products of inertia about the same lines, of three particles placed at the middle points of the sides, each equal to one-third of the mass of the triangle.*

[Purvanchel 2006; Delhi Maths (Hons.) 1995; Calcutta 1994]

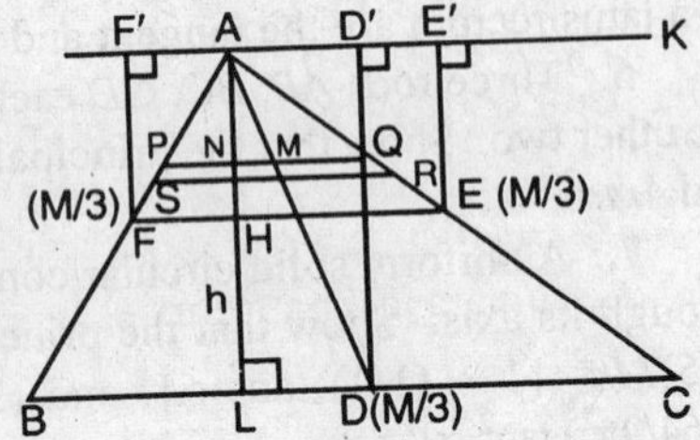

Proof. Let AL (= h) be height of the given triangle ABC and AK perpendicular to AL in the plane of the triangle. Let AD be the median of the triangle ABC of mass M and ρ the mass per unit area of the triangle. Then, if $BC = a$, then $M = (1\ 2)ah\rho$ so that $\rho = 2M/ah$.

Consider an elementary strip PQ parallel to BC of breadth δx and at a distance $x (= AN)$ from the vertex. From similar triangles APQ and ABC, we get $PQ/BC = AN/AL \Rightarrow PQ = xa/h$

$\therefore$ δm = mass of the strip $PQ = \rho \times PQ \times \delta x = (\rho a/h) \times x\, \delta x$

Now, M.I. of the strip PQ about AK = Its M.I. about PQ + M.I. of mass δm placed at C.G. M of PQ about AK

So, M.I. of the strip PQ about $AK = 0 + x^2\, \delta m = (\rho a/h)x^3\, \delta x$

$$\therefore \text{ M.I. of } \Delta ABC \text{ about } AK = \int_0^h \frac{\rho a}{h} x^3 dx = \frac{\rho a h^3}{4} = \frac{2M}{ah} \times \frac{ah^3}{4} = \frac{Mh^2}{2} \quad \ldots(1)$$

Also, M.I. of the strip PQ about AL = M.I. of the strip PQ about the parallel line through C.G. M of PQ + M.I. of mass δm placed at M about AL

So, M.I. of the strip PQ about AL

$$= \frac{1}{3}\left(\frac{1}{2}PQ\right)^2 \delta m + NM^2\, \delta m = \left\{\frac{1}{3}\left(\frac{ax}{2h}\right)^2 + NM^2\right\} \cdot \frac{\rho a}{h} x\, \delta x \quad \ldots(2)$$

From similar triangles ANM and ALD, we get

$$NM/LD = AN/AL = x/h \quad \Rightarrow \quad NM = (x/h)\, LD \quad \ldots(3)$$

Using (3), (2) may be re-written as

M.I. of the strip PQ about $AL = (\rho a/12h^3)\,(a^2 + 12LD^2)x^3\, \delta x$

$\therefore$ M.I. of the triangle ABC about AL

$$= \int_0^h \frac{\rho a}{12h^3}(a^2 + 12LD^2)x^3\, dx = \frac{\rho a(a^2 + 12LD^2)}{12h^3} \int_0^h x^3\, dx$$

$= (2M/ah) \cdot \{a(a^2 + 12LD^2)/12h^3\} \cdot (h^4/4) = (M/24)\,(a^2 + 12LD^2)$ ...(4)

Again, product of inertia about AK, AL

$$= \int_0^h (AN \cdot NM) \frac{\rho a}{h} x\, dx = \int_0^h x \cdot \frac{x \cdot LD}{h} \cdot \frac{\rho a x}{h} dx \text{, using (3)}$$

$= (1/4)\rho a h^2 \cdot LD = (1/4) \cdot (2M/ah) \cdot ah^2 \cdot LD = (1/2)Mh \cdot LD$ ...(5)

Now consider three particles each of mass $M/3$ placed at the mid-points D, E, F of the sides of ΔABC. We shall now find their moments and products of inertia about the some lines AK and AL.

M.I. of three particles each of mass $M/3$ placed at D, E, F about AK

$= (M/3) \cdot DD'^2 + (M/3) \cdot EE'^2 + (M/3) \cdot FF'^2 = (M/3)\,(h^2 + h^2/4 + h^2/4) = (1/2)Mh^2$ ...(6)

M.I. of three particles each of mass $M/3$ placed at D, E, F about AL

$= (M/3) \cdot LD^2 + (M/3) \cdot EH^2 + (M/3)FH^2 = (M/3) \cdot \{LD^2 + (CL/2)^2 + (BL/2)^2\}$

$= (M/3)\ \{LD^2 + (1/4)\,(CD + LD)^2 + (1/4)\,(BD - LD^2)\}$

$$= \frac{M}{3}\left[LD^2 + \frac{1}{4}\left\{\left(\frac{a}{2} + LD\right)^2 + \frac{1}{4}\left(\frac{a}{2} - LD\right)^2\right\}\right] \text{, as } CD = BD = \frac{BC}{2} = \frac{a}{2}$$

$= (M/3)\ \{LD^2 + a^2/8 + (1/2)LD^2\} = (M/24)\,(a^2 + 12LD^2)$ ...(7)

P.I. of three particles each of mass $M/3$ placed at D, E, F about AK, AL

$= (M/3) \cdot LD \cdot AL + (M/3) \cdot EH \cdot AH - (M/3) \cdot FH \cdot AH$

$= (M/3)\ \{LD \cdot h + (1/2)CL \cdot (h/2) - (1/2)BL \cdot (h/2)\}$

$= (1/12)Mh\,(4LD + CL - BL) = (1/12)Mh\ \{4LD + (CD + LD) - (BD - LD)\}$

$= (1/12)Mh\ \{4LD + (a/2 + LD) - (a/2 - LD)\} = (1/2) \cdot Mh \cdot DL$ ...(8)

Comparing results (1), (4) and (5) with the results (6), (7) and (8), we see that the moments and products of the three particles about AK, AL are thus the same as those of the triangle. Hence, by Art. 14.11, the moments of inertia about any line through A are the same; and also, by the same article, the products of inertia about any two perpendicular lines through A are the same.

Also, it is easily seen that the centre of gravity of the three particles coincides with the centre of gravity of the triangle.

Hence, by theorem of parallel axis (refer Art. 14.9), it follows that the moments and products of inertia about any lines through the common centre of gravity are the same for the two systems, and therefore also, by the same theorem, the moments and products about any two other perpendicular lines in the plane of the triangle are the same.

Finally, by the perpendicular theorem (refer Art. 14.3A), the M.I. about a perpendicular to the plane of the triangle through any point P is equal to the sum of moments of inertia about any two perpendicular lines through P in the plane of the triangle, and is thus the same for the two systems.

Note. The mechanical systems such as the triangle and the three particles, which are such that their moments of inertia about all lines are the same, are said to be *equi-momental* or *kinetically equivalent*.

TYPE 6 : SOLVED EXAMPLES BASED ON ART. 14.18 TO 14.20

Ex. 1. *Show that a rod of mass m, and length 2a, is equi-momental with a system of three particles of masses m/6 each at the end points and 2m/3 at the mid-point.*

Sol. Let O be the middle point of the rod. In either case, let $OXYZ$ be a set of rectangular axes with OX along AB. Then in either case, clearly,

$$A = 0, B = C = ma^2/3, F = G = H = 0,$$

showing that they have the same principal axes and same principal moments of inertia at centroid O. Again, O is the common C.G. and m is total mass in either case. Hence, a rod of mass m and length $2a$ is equimomental with three particles of masses $m/6$ each at the end points and $2m/3$ at the mid point.

Ex. 2. *ABCD is a uniform parallelogram of mass M. At the middle points of the four sides are placed particles each of mass M/6 and at the intersection of the diagonals a particle of mass M/3, show that these five particles and the parallelogram are equimomental systems.*

[Agra 1998; Meerut 1997]

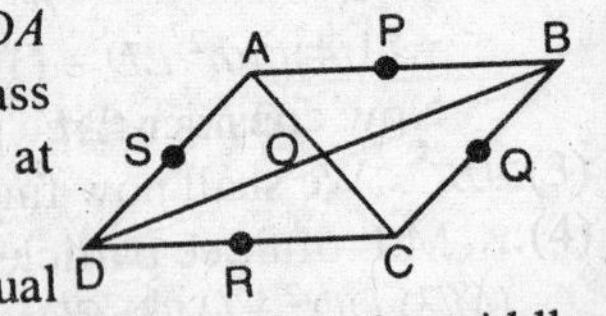

Sol. Let P, Q, R, S be the mid-points of the sides AB, BC, CD and DA respectively of the given parallelogram $ABCD$. Then, mass of ΔABD = mass of $\Delta BCD = M/2$. We know that diagonals AC and BD bisect each other at their point of intersection O.

Now, triangle ABD is equimomental to three particles each of mass equal to one third the mass of the triangle ABD *i.e.*, each of mass $(1/3) \times (M/2)$ $(= M/6)$ placed at the middle points P, O, S of its sides. Similarly, the triangle BCD is equimomental to three particles each of mass $M/6$ placed at the middle points Q, R and O of its sides.

Hence the parallelogram $ABCD$ (which is sum of two triangles ABD and BCD) of mass M is equimomental to the particles each of mass $M/6$ placed at the middle points P, Q, R, S of its sides and a particle of mass $(M/6 + M/6)$ *i.e.* $M/3$ placed at O. This proves the required result.

Ex. 3. *Proved that a hoop of mass m and radius $a/\sqrt{2}$ is equimomental with a circular plate of mass m and radius a.* **[Delhi Maths. (Hons.) 1998]**

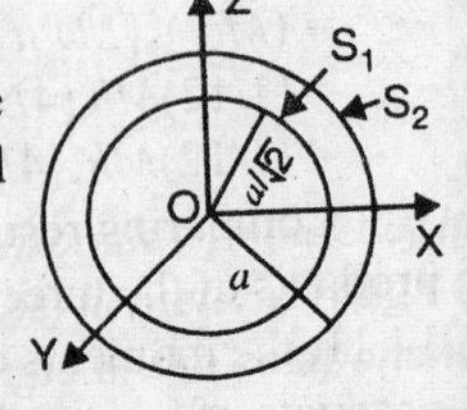

Sol. Let S_1 denote a circular hoop of radius $a/\sqrt{2}$ and S_2 denote a circular plate of radius a, both having the same mass m and common centre of gravity O. Let $OXYZ$ be a set of rectangular axes for both systems. By symmetry, we see that OX, OY, OZ are principal axes for both the systems at O. The principal moments of inertia A_1, B_1, C_1 for system S_1 are given by

$$A_1 = B_1 = (1/2)m\,(a/\sqrt{2}\,)^2 = (1/4)ma^2,\ C_1 = A_1 + B_1 = (1/2)ma^2$$

Again, for the system S_2, the principal moments of inertia A_2, B_2, C_2 are

$$A_2 = B_2 = (1/4)ma^2 \text{ and } C_2 = A_2 + B_2 = (1/2)ma^2$$

Since the two system S_1 and S_2 have the same total mass, the same centroid, the same principal axes and principal moments of inertia at the centroid O, it follows that S_1 and S_2 are equimomental, as required.

Ex. 4. *Obtain the M.I. of triangular lamina ABC about a straight line through A (or any vertex) in the plane of the triangle.* **(Kanpur 2008)**

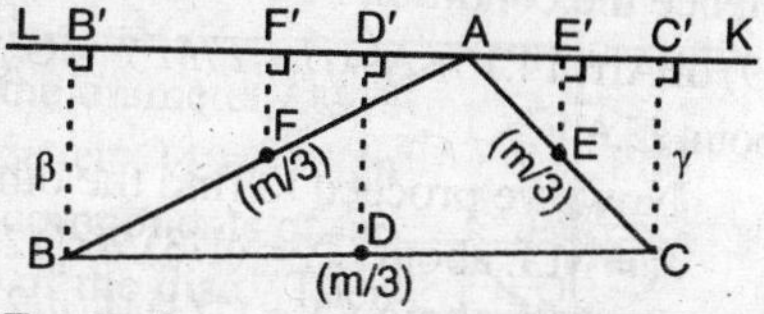

Sol. let β and γ be the distances of two vertices *B* and *C* of the triangle *ABC* from the straight line *LK* (passing through *A*) about which M.I. is to found.

Let *m* be mass of the triangle. Then the triangle is equimomental to three particles each of mass *m*/3 placed at the mid-points *D, E, F* of the sides. The perpendicular distances of *D, E, F* from *LK* are given by $DD' = (\beta + \gamma)/2$, $EE' = CC'/2 = \gamma/2$ and $FF' = BB'/2 = \beta/2$. Then, we have

M.I. of the triangle *ABC* about *LK*

= sum of the moments of inertia of masses *m*/3 each at *D, E, F* about *LK*

$= (m/3)\cdot DD'^2 + (m/3)\cdot EE'^2 + (m/3)\cdot FF'^2$

$= (m/3) \times \{(1/4) \times (\beta + \gamma)^2 + \gamma^2/4 + \beta^2/4\} = (m/6) \times (\beta^2 + \gamma^2 + \beta\gamma)$

Ex. 5. (a) *If α, β, γ be the distances of the vertices of a triangle of mass m from any straight line in its plane, show that the M.I. of the triangle about this line is* $(m/6) \times (\alpha^2 + \beta^2 + \gamma^2 + \beta\gamma + \gamma\alpha + \alpha\beta)$. **[Kanpur 2009]**

(b) *If α, β, γ and h be the distances of the vertices and the centre of inertia of a uniform triangular lamina of mass m from any line, prove that the M.I. about that line is* $(m/12) \times (\alpha^2 + \beta^2 + \gamma^2 + 9h^2)$.

(c) *Show that a uniform triangular lamina of mass m is equimomental with three particles, each of mass m/12 placed at the angular points and a particle of mass 3m/4 placed at the centre of inertia of the triangle.*

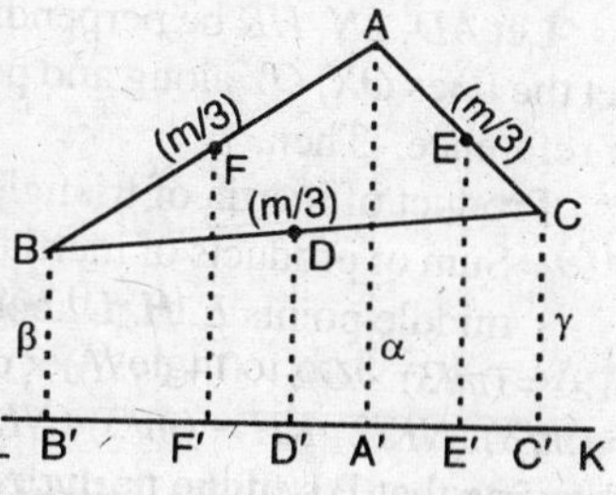

Sol. (*a*) Let *ABC* be given triangle and let *AA', BB', CC'* be perpendiculars from *A, B, C* on the given line *LK*. Then, here $AA' = \alpha$, $BB' = \beta$, $CC' = \gamma$. (given)

Let *DD', EE', FF'* be perpendiculars from the middle points *D, E, F* of the sides of the triangle on *LK*. Then $DD' = (BB' + CC')/2 = (\beta + \gamma)/2$.

Similarly, $EE' = (\alpha + \gamma)/2$, $FF' = (\alpha + \beta)/2$

Since the triangle *ABC* of mass *m* is equimomental to three particles each of mass *m*/3 placed at the mid-points *D, E, F*, hence the M.I. of triangle *ABC* about *LK*

= the sum of the moments of inertia of masses *m*/3 each of *D, E, F* about *LK*

$= (m/3)\cdot DD'^2 + (m/3)\cdot EE'^2 + (m/3)\cdot FF'^2$

$= (m/3)\ \{(\beta + \gamma)^2/4 + (\alpha + \gamma)^2/4 + (\alpha + \beta)^2/4\} = (m/6)\ (\alpha^2 + \beta^2 + \gamma^2 + \beta\gamma + \gamma\alpha + \alpha\beta)$

(*b*) Refer part (*a*). Also, here $h = (\alpha + \beta + \gamma)/3$ ...(1)

As in part (*a*), M.I. about line *LK*

$= (m/6)\ (\alpha^2 + \beta^2 + \gamma^2 + \beta\gamma + \gamma\alpha + \alpha\beta) = (m/12)\ (2\alpha^2 + 2\beta^2 + 2\gamma^2 + 2\beta\gamma + 2\gamma\alpha + 2\alpha\beta)$

$= (m/12)\ \{\alpha^2 + \beta^2 + \gamma^2 + (\alpha + \beta + \gamma)^2\} = (m/12)\ (\alpha^2 + \beta^2 + \gamma^2 + 9h^2)$, by (1)

(*c*) Here, total mass of four particles $= 3m/12 + 3m/4 = m =$ mass of the triangle and C.G. of the four particles coincides with that of the triangular lamina.

Now, let α, β, γ and *h* be the distances of the vertices and the C.G. of the triangular lamina from the line *LK* [refer figure of part (*a*)] Then, M.I. of the four particles about *LK*

$= (m/12)\alpha^2 + (m/12)\beta^2 + (m/12)\gamma^2 + (3m/4)\cdot h^2 = (m/12)\ (\alpha^2 + \beta^2 + \gamma^2 + 9h^2)$

= M.I. of the triangle about *LK*, refer part (*b*)

Ex. 6. *At the vertex C of a triangle ABC, which is right angled at C, show that the principal axes are a perpendicular to the plane and two others inclined to the sides at an angle* $(1/2)\tan^{-1}\{ab/(a^2 - b^2)\}$. **[Agra 2009; Garhwal 2010]**

Sol. Let *M* be the mass of the triangle *ABC*, then the triangle is equimomental to three particles each of mass *M*/3 placed at the middle points *D*, *E* and *F* of the sides of the triangle.

Take *C* as origin and *CX, CY* be taken as axes of reference. Here we are to show that a line through *C* perpendicular to the plane is a principal axis at *C*. Let this axis be regarded as axis of *z*.

Clearly, z-coordinate of C.G. of the triangle is zero and so $F = 0$ and $G = 0$. Hence the condition $F/\bar{y} = G/\bar{x}$ is satisfied for the axis CZ. (Refer equation (9) of Art. 14.16). Also $h = F/M\bar{y} = G/M\bar{x} = 0$, *i.e.*, axis of z is principal axis at the point C.

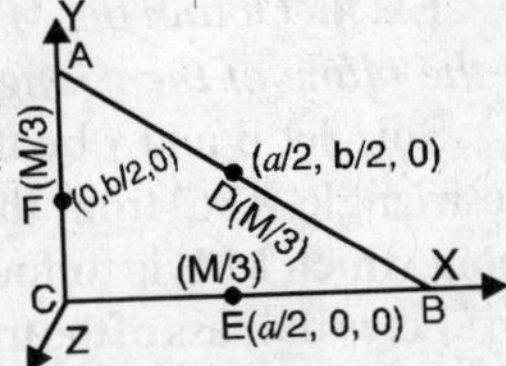

Now, we proceed to find the other two principal axes through C.

$A =$ M.I. about $CX = (1/3)M(b/2)^2 + (1/3)M(b/2)^2 = (1/6)Mb^2$

$B =$ M.I. about $CY = (1/3)M(a/2)^2 + (1/3)M(a/2)^2 = (16)Ma^2$,

and $H =$ product of inertia about CX, CY

$= (M/3)\times(a/2)\times(0) + (M/3)\times 0\times(b/2) + (M/3)\times(a/2)\times(b/2) = (1/12)Mab$

If θ is the angle that a principal axis makes with CX, then

$$\tan 2\theta = \frac{2H}{B-A} = \frac{(m/6)ab}{(1/6)Ma^2 - (1/6)Mb^2} \quad \text{or} \quad \theta = \frac{1}{2}\tan^{-1}\frac{ab}{a^2-b^2}$$

Ex. 7. *ABC is a triangular area and AD is perpendicular to BC and AE is a median. O is the middle point of DE; show that BC is a principal axis of the triangle at O.* **[Kanpur 2014]**

Sol. Let M be the mass of the triangle ABC. Then the triangle is equi-momental to three particles each of mass $M/3$ placed at the middle points E, H, L of the sides of the triangle.

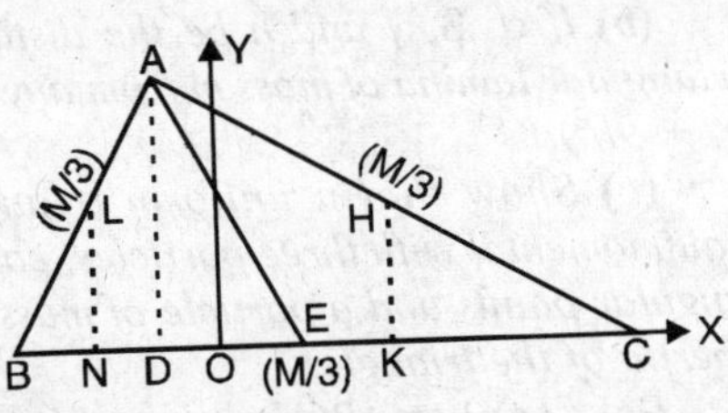

Let AD, LN, HK be perpendicular from A, L and H on BC. Let the lines OX, OY along and perpendicular to BC be the axes of reference. Then,

Product of inertia of triangle ABC about OX, OY

= Sum of products of inertia of masses each equal to $m/3$ at middle points E, H, L about OX, OY

$= (m/3)\times OE\times 0 + (m/3)\times OK\times KH + (m/3)\times(-ON)\times NL$

$= (m/3)\cdot(NK/2)\cdot KH - (m/3)\cdot(NK/2)\cdot KH = 0$, as $OK = ON = NK/2$ and $LN = KH$ showing that P.I. of the triangle ABC vanishes about BC and perpendicular to BC at O. Hence BC is a principal axis of the triangle ABC at O.

EXERCISE

1. Particles each equal to one-quarter of the mass of an elliptic area are placed at the middle points of the chords joining the extremities of any pair of conjugate diameters. Prove that these four particles are equimomental to the elliptic area.

2. Show that the M.I. of a regular polygon of n sides and mass M about any straight line through its centre is $Mc^2(2 + \cos 2\pi/n)/24(1 - \cos 2\pi/n)$, where c is the length of each side.

3. Show that any lamina is dynamically equivalent to the three particles, each one-third of the mass of the lamina, placed at the corners of a maximum triangle inscribed in an ellipse, whose equation referred to the principal axes at the centre of inertia is $x^2/B + y^2/A = 2$, where mA and mB are the principal moments of inertia about OX and OY and m is the mass.

4. Show that there is a momental ellipse at an angular point of triangular area which touches the opposite side at its middle point and bisects adjacent sides.

5. Show that there is a momental ellipse at the centre of inertia of a uniform triangle which touches the sides of the triangle at the middle points.

OBJECTIVES QUESTIONS

1. The M.I. of a uniform elliptic disc of mass M and of major and minor axes of lengths $2a$ and $2b$ about a line through the centre and perpendicular to its plane is (*a*) $(M/4)(a^2 + b^2)$
(*b*) $(M/2)(a^2 + b^2)$ (*c*) $Ma^2/2$ (*d*) $Mb^2/2$ **(Agra 2006; Kanpur 2003, 04)** **Ans.** (*a*)

2. The radius of gyration of a uniform circular disc of radius a about a line through its centre perpendicular to its plane is **Ans.** (*c*)

(*a*) $a/2$ (*b*) a (*c*) $a/\sqrt{2}$ (*d*) $a/\sqrt{3}$.

3. Moment of inertia of a circular disc of radius a and mass M about its diameter is
(*a*) Ma^2 (*b*) $(Ma^2/2)$ (*c*) $(Ma^2)/3$ (*d*) $(Ma^2)/4$. **[Kanpur 2001]**

Sol. Ans. (d). Refer case I of Art. 14.4E.

4. Moment of inertia of a hollow sphere of radius a and mass M about a diameter is
(*a*) $(3/2) \times Ma^2$ (*b*) $(2/3) \times Ma^2$ (*c*) $(2/5) \times Ma^2$ (*d*) None of these. **[Kanpur 2004]**

Sol. Ans. (*a*). Refer Art. 14.4 G.

5. Moment of inertial of a solid sphere of radius a and mass M about a diameter is
(*a*) Ma^2 (*b*) $(Ma^2)/5$ (*c*) $(2Ma^2)/5$ (*d*) $(3Ma^2)/5$. **[Kanpur 2002]**

Sol. Ans (c). Refer Art. 14.4H

6. A pair of axes about which the product of inertia is zero is called
(*a*) principal axes (*b*) parallel axes (*c*) perpendicular axes (*d*) none of these **[Kanpur 2001, 02]**

Sol. Ans (c). Refer Art. 14.15

7. If referred to three mutually perpendicular lines OX, OY and OZ as coordinate axes, the equation of the momental ellipsoid of a body at the point O is $3x^2 + 4y^2 + 5z^2 = 1$, then the product of inertia of the body about OX, OY is (*a*) 3 (*b*) 4 (*c*) 5 (*d*) 0. **[Agra 2006]**

Sol. Ans. (*d*) Refer remark of Art. 14.13

8. At vertex C of triangle ABC, which is right angled at C, the principal axes are inclined to sides at an angle
(*a*) $\tan^{-1} \{ab/(a^2 - b^2)\}$ (*b*) $\tan^{-1} \{2ab/(a^2 - b^2)\}$
(*c*) $(1/2) \times \tan^{-1} \{ab/(a^2 - b^2)\}$ (*d*) None of these **[Garhwal 2010]**

Hint. Ans. (c). Refer solved Ex.6, page 14.39.

MISCELLANEOUS PROBLEMS

1. Find the product of inertia of a solid disc in the form of quadrant of a circle about its bounding radii. **[Kanpur 2006]**

Hint : Proceed like Ex. 1 page 14.7 with $b = a$. **Ans.** $(1/2\pi) \times Ma^2$

2. Show that the product of inertia of a right angled triangle of sides $2a$, $2b$ containing the right angle is $(1/3) \times M\,ab$. **[Kanpur 2000]**

3. Show that the moment of inertia of a uniform cube about any axis through its centre is the same. **[Kanpur 2004]**

Hint: Refer corollary of Art. 14.13

4. Prove that $\tan 2\theta = 2H/(B - A)$, where the relation determines the inclination of the principal axes to the axes of coordinates. **[Kanpur 2001, 02, 03, 04]**

Hint: Refer equation (7) of Art. 14.16

5. A rectangular lamina $ABCD$ is such that $AB = 2a$ and $AD = 2b$. Find the directions of the principal axes at A. **[Delhi Maths (H) 2007]**

Hint : Refer Ex. 1(*a*), page 14.28. The required directions are given by θ and $\theta + \pi/2$, where $\theta = (1/2) \times \tan^{-1}\{3ab/2(a^2 - b^2\}$

6. Show that at the vertex C of a triangle ABC which is right angled at C the principal axis in the plane is inclined to the side at an angle $(1/2) \times \tan^{-1} \{ab/(a^2 - b^2)\}$. **[Kanpur 2003]**

7. Show that a uniform triangle of mass M is kinematically equivalent to three particles each of mass $M/3$ placed at the middle points of the sides **[Kanpur 2005]**

Hint : Refer Art. 14.18 and Art. 14.20

8. *Three rods AB, BC, CD each of mass m and length 2a are such that each is perpendicular to the other two. Show that the principal moments of inertia at the centre of mass are* ma^2, $(11/3) \times ma^2$ *and* $4ma^2$. **[Purvanchal 2005, 06]**

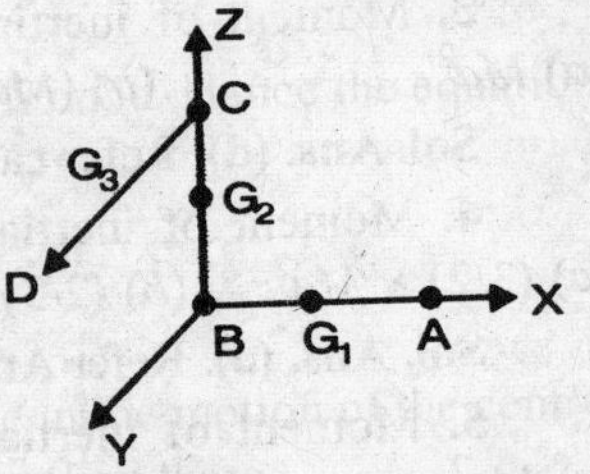

Sol. Draw BY parallel to CD. Let G_1, G_2, G_3 be the middle points of AB, BC and CD respectively. Let BA, BY and BC be taken along OX, OY and OZ respectively. Then, co-ordinates of G_1, G_2 and G_3 are $(a, 0, 0)$, $(0, 0, a)$ and $(0, a, 2a)$.

Let $G(\bar{x}, \bar{y}, \bar{z})$ be the centre of mass of the given rods each of mass m. Then.

$$\bar{x} = \frac{m \times a + m \times 0 + m \times 0}{m + m + m} = \frac{a}{3}, \quad \bar{y} = \frac{m \times 0 + m \times 0 + m \times a}{m + m + m} = \frac{a}{3}, \quad \bar{z} = \frac{m \times 0 + m \times a + m \times 2a}{m + m + m} = a$$

Thus, coordinates of G are $(a/3, a/3, a)$

Let GX', GY', GZ' be axes parallel to BA, BY and BC respectively and passing through the centre of mass G.

Then, with reference to new axes GX', GY', GZ',

coordinates of G_1 are $(a - a/3, 0 - a/3, 0 - a)$, *i.e.*, $(2a/3, -a/3 - a)$

coordinates of G_2 are $(0 - a/3, 0 - a/3, a - a)$, *i.e.*, $(-a/3, - a/3, 0)$

and coordinates of G_3 are $(0 - a/3, a - a/3, 2a - a)$, *i.e.*, $(-a/3, 2a/3, a)$

Now, using the theorem of parallel axis, we have

A_1 = M.I. of the three rods about GX'

= M.I. of rod AB about GX' + M.I. of rod BC about GX' + M.I. of rod CD about GX'

$$= [0 + m\{(-a/3)^2 + (-a)^2\}] + [ma^2/3 + m\{(-a/3)^2 + 0^2\}] + [ma^2/3 + m\{(2a/3)^2 + a^2\}]$$

$= (10/3) \times ma^2$

B_1 = M.I. of three rods about GY'

$= [ma^2/3 + m\{(-a)^2 + (2a/3)^2\}] + [ma^2/3 + m\{0^2 + (-a/3)^2\}] + m\{a^2 + (-a/3)^2\}$

$= (10/3) \times ma^2$

C_1 = M.I. of the three rods about GZ'

$$= [1/3ma^2 + m\{(2a/3)^2 + (-a/3)^2\}] + [m\{(-a/3)^2 + (-a/3)^2\}]$$
$$+ [ma^2/3 + m\{(-a/3)^2 + (2a/3)^2\}] = 2ma^2$$

D_1 = P.I. of the three rods about GY', GZ'

$= m \times (-a/3) \times (-a) + m \times (-a/3) \times 0 + m \times (2a/3) \times a = ma^2$

E_1 = P.I. of the three rods about GZ', GX'

$= m \times (-a) \times (2a/3) + m \times 0 \times (-a/3) + m \times a \times (-a/3) = -ma^2$

F_1 = P.I. of the three rods about GX', GY'

$= m \times (2a/3) \times (-a/3) + m \times (-a/3) \times (-a/3) + m \times (-a/3) \times (2a/3) = -(1/3) \times ma^2$

Hence, the momental ellipsoid at mass centre G is given by,

$$A_1 x^2 + B_1 y^2 + C_1 z^2 - 2D_1\, yz - 2E_1\, zx - 2F_1\, xy = 3\, mk^4$$

or $\quad (10/3) \times ma^2 x^2 + (10/3) \times ma^2 y^2 + 2ma^2 z^2 - 2ma^2 yz + ma^2 zx + (2/3) \times ma^2 xy = 3mk^4$

or $\quad (1/3) \times ma^2 (10x^2 + 10y^2 + 6z^2 - 6yz + 6zx + 2xy) = 3mk^4 \quad \ldots (1)$

Reducing $10x^2 + 10y^2 + 6z^2 - 6yz + 6zx + 2xy$ by means of the discriminant cubic, namely,

$$\lambda^3-(a+b+c)\lambda^2+(ab+bc+ca-f^2-g^2-h^2)\lambda-(abc+2fgh-af^2-bg^2-ch^2)=0,$$

we obtain $\lambda^3-26\lambda^2+201\lambda-396=0$ or $(\lambda-3)(\lambda-11)(\lambda-12)=0$. Thus, $\lambda=3,11,12$.

Hence referred to the principal axes through the centre of mass G, (1) reduce to

$(1/3)\times ma^2(3x^2+11y^2+12z^2)=3mk^4$ or $ma^2x^2+(11/3)\times ma^2y^2+4\,ma^2z^2=3mk^4$

The coefficients of x^2, y^2, z^2 should be the pricipal moments of inertia. Therefore, the required principal moments of inertia at the centre of mass are ma^2, $(11/3)\times ma^2$ and $4ma^2$.

Ex. 9. Obtain an expression for the moment of inertia of a rigid body about a line with direction cosine α, β, γ. **[Delhi Maths. (H) 2008]**

Ex. 10. Fill up the gaps:

(*i*) M.I of sphere of radius *a* about its diameter is **........** **[Agra 2012]**

(*ii*) A uniform triangular lamina of mass M is equimomental to three particles of man M/3 placed at the .... of the sides of the triangle **[Agra 2012]**

Sol. Part (*i*). Ans. $(2/5)\times Ma^2$. Refer Art. 14.14H, Page 14.4

Part (*ii*). Ans. Middle points. Refer Art. 14.20, page 14.36

Resultant [illegible] $= 0$. Thus [illegible]

[illegible]

The vector [illegible] perpendicular [illegible]

[illegible]

[illegible]

(iii) Ans. [illegible] points [illegible]

15

D'Alembert's principle The general equations of motion

15.1 Introduction. If (x, y, z) be the co-ordinates of a moving particle of mass m, at any time t and X, Y, Z be the components of the external forces acting on it parallel to the axes, then by Newton's second law of motion the equations of motion of the particle are

$$m(d^2x/dt^2) = X, \qquad m(d^2y/dt^2) = Y, \qquad m(d^2z/dt^2) = Z \qquad \dots(1)$$

Now a rigid body is formed by material particles whose mutual distance do not change on account of the effect of forces. For a rigid body, we assume that (*i*) the mutual actions between two particles act along the straight line joining them (*ii*) the action and reaction between two particles are equal and opposite.

While considering the motion of a rigid body, we have to write down the equations of motion of all the particles of the rigid body according to equations given by (1). But here the external forces acting on a particle of the body include, together with external forces (such as its weight), the unknown inner forces acting due to the action of the rest of the body on it.

D'Alembert gave a technique by which all the necessary equations may be obtained without writing down the equations of motion of all particles of a rigid body and without considering the unknown forces. It is known as D'Alembert's principle and is based on the following law. *The internal actions and reactions of any system of rigid bodies in motion are in equilibrium amongst themselves.*

For future reference, we define the following two important terms :

Impressed forces. **[Meerut 2012; Agra 2012]**

The external forces acting on a body are called impressed forces. For example, the weight of the body is the impressed force on the body. Again, if a body is tied to a string then the tension in the string is an impressed force on the body.

Effective force. **[Agra 2010, 12; Meerut 2005, 2012]**

The effective force on a particle in any direction is defined as the product of its mass and its acceleration in that direction. Thus, if a particle is at point (x, y, z) at any time t, then the effective force on this particle at time t are $m(d^2x/dt^2)$, $m(d^2y/dt^2)$, $m(d^2z/dt^2)$ parallel to the co-ordinate axes.

15.2 D'Alembert's principle. *The reversed effective forces acting on the body and the external forces of the system are in equilibrium.* **[Agra 2011, 12, 13; Nagpur 2010; Kanpur 2004; 11, 12, 14; Guwahati 2007; Purvanchal 2005; Calcutta 2003; Meerut 2008, 10, 11; Agra 2010, 12]**

Proof. Let a rigid body be in motion. At any time t, let $\mathbf{r}$ be the position vector of a particle of mass m and $\mathbf{F}$ and $\mathbf{R}$ the external and internal forces respectively acting on it, then by Newton's second law of motion

$$m(d^2\mathbf{r}/dt^2) = \mathbf{F} + \mathbf{R} \qquad \text{or} \qquad \mathbf{F} + \mathbf{R} - m(d^2\mathbf{r}/dt^2) = \mathbf{0},$$

showing that the forces $\mathbf{F}$, $\mathbf{R}$ and $-m(d^2\mathbf{r}/dt^2)$ are in equilibrium.

Applying the same reasoning to every particle of the rigid body, the forces $\Sigma\mathbf{F}$, $\Sigma\mathbf{R}$ and $\Sigma\{-(m)\ (d^2\mathbf{r}/dt^2)\}$ are in equilibrium, where the summation extends to all the particles of the rigid body.

But the internal forces (*i.e.,* mutual actions and reactions between the particles of the rigid body) form pairs of equal and opposite forces and so these balance among themselves, that is, their vector sum must vanish so that $\Sigma\mathbf{R} = \mathbf{0}$.

Hence the forces $\Sigma \mathbf{F} + \Sigma\{-(m)(d^2\mathbf{r}/dt^2)\}$ are in equilibrium,

that is, $\Sigma \mathbf{F} + \Sigma\{-(m)(d^2\mathbf{r}/dt^2)\} = \mathbf{0}$,

showing that the reversed effective forces acting on the body and the impressed (or external) forces on the system are in equilibrium.

15.3 Working rule to solve problems by using D'Alembert's principle

We reduce the given problem of dynamics to the problem of statics. Accordingly, we mark the external forces acting on the rigid body and the effective forces acting in opposite directions on the rigid body. Now we solve the resulting problem by methods of statics. For example, if all the forces (external and reversed effective forces) are coplanar, then we equate to zero the resolved parts of all these forces in two mutually perpendicular directions and take moments about a suitable point in the plane of the forces.

15.4 General equations of motion of a rigid body. *To deduce the general equations of motion of a rigid body from D'Alembert's principle.*

[Kanpur 2005; Meerut 2001; Purvanchel 2007; Agra 2011]

Let a rigid body be in motion under the action of given external forces. If $\mathbf{r}$ be the position vector at any time t of a particle of mass m of the body referred to some fixed origin and $\mathbf{F}$ be the external force acting on it, then by D'Alembert's principle, we have

$$\Sigma \mathbf{F} + \Sigma\{-(m)(d^2\mathbf{r}/dt^2)\} = \mathbf{0} \quad \text{or} \quad \Sigma m(d^2\mathbf{r}/dt^2) = \Sigma \mathbf{F} \quad \ldots(1)$$

Multiplying vectorially by $\mathbf{r}$, (1) reduces to

$$\Sigma m\mathbf{r} \times (d^2\mathbf{r}/dt^2) = \Sigma \mathbf{r} \times \mathbf{F} \quad \ldots(2)$$

(1) and (2) are the general vector equations of motion of a rigid body.

Deduction of general equations of motion in scalar (cartesian) form

Let (x, y, z) be the cartesian coordinates of the particle m at any t and X, Y, Z be the components of force $\mathbf{F}$ parallel to the coordinate axes. Then $\mathbf{r} = x\mathbf{i} + y\mathbf{j} + \mathbf{j}\mathbf{k}$ and $\mathbf{F} = X\mathbf{i} + Y\mathbf{j} + Z\mathbf{k}$. Also, $d^2\mathbf{r}/dt^2 = \ddot{x}\mathbf{i} + \ddot{y}\mathbf{j} + \ddot{z}\mathbf{k}$, where $\ddot{x} = d^2x/dt^2$, $\ddot{y} = d^2y/dt^2$ and $\ddot{z} = d^2z/dt^2$. Substituting the above values of $\mathbf{r}$, $\mathbf{F}$ and $d^2\mathbf{r}/dt^2$ in (1) and (2), we get

$$\Sigma m(\ddot{x}\mathbf{i} + \ddot{y}\mathbf{j} + \ddot{z}\mathbf{k}) = \Sigma(X\mathbf{i} + Y\mathbf{j} + Z\mathbf{k}) \quad \ldots(3)$$

and $\Sigma m(x\mathbf{i} + y\mathbf{j} + z\mathbf{k}) \times (\ddot{x}\mathbf{i} + \ddot{y}\mathbf{j} + \ddot{z}\mathbf{k}) = \Sigma(x\mathbf{i} + y\mathbf{j} + z\mathbf{k}) \times (X\mathbf{i} + Y\mathbf{j} + Z\mathbf{k})$

i.e., $\Sigma m\{(y\ddot{z} - z\ddot{y})\mathbf{i} + (z\ddot{x} - x\ddot{z})\mathbf{j} + (x\ddot{y} - y\ddot{x})\mathbf{k}\} = \Sigma\{(yZ - zY)\mathbf{i} + (zX - xZ)\mathbf{j} + (xY - yX)\mathbf{k}\}$ ...(4)

Equating the coefficients of $\mathbf{i}, \mathbf{j}, \mathbf{k}$, (3) and (4) give

$$\Sigma m\ddot{x} = \Sigma X, \quad m\ddot{y} = \Sigma Y, \quad m\ddot{z} = \Sigma Z \quad \ldots(5)$$

$$\Sigma m(y\ddot{z} - z\ddot{y}) = \Sigma(yZ - zY), \quad \Sigma m(z\ddot{x} - x\ddot{z}) = \Sigma(zX - xZ) \quad \Sigma m(x\ddot{y} - y\ddot{x}) = \Sigma(xY - yX) \quad \ldots(6)$$

Equation (5) and (6) are six general equations of motion.

Equations in (5) show that the sums of the components of the effective forces parallel to the axes of co-ordinates are respectively equal to the sums of the components of the external (impressed) forces parallel to the same axes.

Equations in (6) show that the sums of the moments of the effective forces about co-ordinate axes are respectively equal to the sums of the moments of the impressed forces about the same axes.

15.5 Motion of the centre of inertia. *To show that the centre of inertia of a body moves as if all the mass of the body were collected at it, and as if all the external forces acting on the body were acting on it in directions parallel to those in which they act.*

[Agra 2006; Garhwal 1999; Calcutta 2003; Meerut 2001, 2004]

Proof. At time t let $\overline{\mathbf{r}}$ be the position vector of the centre of inertia and $\mathbf{r}$ the position vector of a particle of mass m of a rigid body of mass M. Then,

$$\Sigma m = M \quad \text{and} \quad \overline{\mathbf{r}} = (\Sigma m\mathbf{r})/\Sigma m = (\Sigma m\mathbf{r})/M \quad \ldots(1)$$

From (1), $\quad d^2\overline{\mathbf{r}}/dt^2 = (1/M)\,\Sigma m(d^2\mathbf{r}/dt^2)$ ...(2)

Vector equation for the motion of a rigid body is

$$\Sigma m(d^2\mathbf{r}/dt^2) = \Sigma \mathbf{F}, \quad \ldots(3)$$

where **F** is the external force acting on particle of mass m at time t.

From (2) and (3), $d^2\bar{\mathbf{r}}/dt^2 = (1/M)\,\Sigma\mathbf{F}$ or $M(d^2\bar{\mathbf{r}}/dt^2) = \Sigma\mathbf{F}$, ...(4)

which is the vector from of the equation of motion of a particle of mass M placed at the centre of inertia of the body and acted upon by the forces $\Sigma\mathbf{F}$.

Deduction of the equations of motion of the centre of inertia in scalar (cartesian) form. Let $(\bar{x}, \bar{y}, \bar{z})$ be the cartesian coordinates of the centre of inertia and X, Y, Z be the components of force **F** parallel to the coordinate axes. Then $\bar{\mathbf{r}} = \bar{x}\,\mathbf{i} + \bar{y}\,\mathbf{j} + \bar{z}\,\mathbf{k}$ and $\mathbf{F} = X\mathbf{i} + Y\mathbf{j} + Z\mathbf{k}$. Also, $d^2\bar{\mathbf{r}}/dt^2 = (d^2\bar{x}/dt^2)\mathbf{i} + (d^2\bar{y}/dt^2)\mathbf{j} + (d^2\bar{z}/dt^2)\mathbf{k}$. Putting these values in (4), we get

$$M\{(d^2\bar{x}/dt^2)\mathbf{i} + (d^2\bar{y}/dt^2)\mathbf{j} + (d^2\bar{z}/dt^2)\mathbf{k}\} = \Sigma(X\mathbf{i} + Y\mathbf{j} + 2\mathbf{k})$$

Equating coefficients of **i**, **j**, **k** on both sides, we get

$$M(d^2\bar{x}/dt^2) = \Sigma X,\ \ M(d^2\bar{y}/dt^2) = \Sigma Y,\ \ M(d^2\bar{z}/dt^2) = \Sigma Z, \qquad \text{...(5)}$$

But (5) are the equations of motion of a particle of mass M, placed at the centre of inertia of the body, and acted by the forces $\Sigma X, \Sigma Y, \Sigma Z$ parallel and equal to the external forces acting on the different points of the body.

Remark. The proposition of the above article is known as the *principle of conservation of motion of translation.* From this we see that the motion of centre of gravity is independent of rotation.

15.6 Motion relative to the centre of inertia. *To show that the motion of body about its centre of inertia is the same as it would be if the centre of inertia were fixed and the same forces acted on the body.*

[Guwahati 2007; Kanpur 2008, 09; Calcutta 2003; Meerut 2002; Purvanchal 2006; Agra 2012]

Proof. Consider a rigid body in motion. At time t, let **r** be the position vector of the centre of inertia of a rigid body of mass M referred to O as fixed origin. Let m be the mass of a particle P whose position vector referred to a fixed origin O is **r** and whose position vector with regard to the centre of inertia **G** is **r**′. Then in figure,

$\mathbf{r} = \overrightarrow{OP}$, $\quad \bar{\mathbf{r}} = \overrightarrow{OG}$, $\quad \mathbf{r}' = \overrightarrow{GP}$ $\quad$ so that we have

$\mathbf{r} = \bar{\mathbf{r}} + \mathbf{r}'$ $\quad$ and so $\quad d^2\mathbf{r}/dt^2 = d^2\bar{\mathbf{r}}/dt^2 + d^2\mathbf{r}'/dt^2$ ...(1)

The moment vector equation of motion for a rigid body is

$$\Sigma m\mathbf{r} \times (d^2\mathbf{r}/dt^2) = \Sigma\mathbf{r} \times \mathbf{F},$$

where **F** is the external force acting on the particle m at any time t. ...(2)

Using (1), (2) reduces to $\Sigma\{m(\bar{\mathbf{r}} + \mathbf{r}') \times (d^2\bar{\mathbf{r}}/dt^2 + d^2\mathbf{r}'/dt^2)\} = \Sigma\{(\bar{\mathbf{r}} + \mathbf{r}') \times \mathbf{F}\}$

or $\quad \bar{\mathbf{r}} \times (d^2\bar{\mathbf{r}}/dt^2)(\Sigma m) + \Sigma m\mathbf{r}' \times (d^2\mathbf{r}'/dt^2) + \bar{r} \times \Sigma m(d^2\mathbf{r}'/dt^2)$

$\quad + (\Sigma m\mathbf{r}') \times (d^2\bar{\mathbf{r}}/dt^2) = \bar{\mathbf{r}} \times \Sigma\mathbf{F} + \Sigma\mathbf{r}' \times \mathbf{F}$ ...(3)

Since the position vector of G referred to G as origin is **0**,

hence $\quad (\Sigma m\mathbf{r}')/\Sigma m = \mathbf{0}$, $\quad$ *i.e.*, $\quad \Sigma m\mathbf{r}' = \mathbf{0}$ $\quad$ and so $\quad \Sigma m(d^2\mathbf{r}'/dt^2) = \mathbf{0}$ ...(4)

Again, the equation of motion of the centre of inertia gives

$M(d^2\mathbf{r}/dt^2) = \Sigma\mathbf{F}$ $\quad$ or $\quad \Sigma m(d^2\mathbf{r}/dt^2) = \Sigma\mathbf{F}$ ...(5)

Using (4) and (5), (3) reduces to

$$\bar{\mathbf{r}} \times \Sigma\mathbf{F} + \Sigma m\mathbf{r}' \times (d^2\mathbf{r}'/dt^2) = \bar{\mathbf{r}} \times \Sigma\mathbf{F} + \Sigma\mathbf{r}' \times \mathbf{F}$$

or $\quad \Sigma m\mathbf{r}' \times (d^2\mathbf{r}'/dt^2) = \Sigma\mathbf{r}' \times \mathbf{F},$ ...(6)

which is the vector equation of a rigid body when the centre of inertia is regarded as fixed point.

Scalar (cartesian) form of the above equation (6). If (x, y, z) and (x', y', z') are the coordinates of the particle of mass m at P at any time t referred to the rectangular axes OX, OY, OZ through the fixed origin O and the parallel axes GX', GY', GZ' through the centre of inertia G respectively, then $\mathbf{r} = x\mathbf{i} + y\mathbf{z} + z\mathbf{k}$ and $\mathbf{r}' = x'\mathbf{i} + y'\mathbf{j} + z'\mathbf{k}$. Let $(\bar{x}, \bar{y}, \bar{z})$ be the coordinates of G referred to the axes OX, OY, OZ through O so that $\bar{r} = \bar{x}\,\mathbf{i} + \bar{y}\,\mathbf{j} + \bar{z}\,\mathbf{k}$. Let X, Y, Z be the components of the external

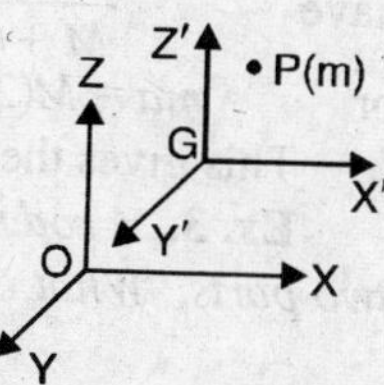

force **F** parallel to the axes so that $\mathbf{F} = X\mathbf{i} + Y\mathbf{j} + Z\mathbf{k}$. Substituting the values of $\mathbf{r}'$ and $\mathbf{F}$ in (6), we get

$$\Sigma m\{(x'\mathbf{i}+y'\mathbf{j}+z'\mathbf{k}) \times (\ddot{x}'\mathbf{i}+\ddot{y}'\mathbf{j}+\ddot{z}'\mathbf{k})\} = \Sigma\{(x'\mathbf{i}+y'\mathbf{j}+z'\mathbf{k}) \times (X\mathbf{i}+Y\mathbf{j}+Z\mathbf{k})\}$$

or $$\Sigma m\{(y'\ddot{z}'-z'\ddot{y}')\mathbf{i}+(z'\ddot{x}'-x'\ddot{z}')\mathbf{j}+(x'\ddot{y}'-y'\ddot{x}')\mathbf{k}\} = \Sigma\{(y'Z-z'Y)\mathbf{i}+(z'X-x'Z)\mathbf{j}+(x'Y-y'X)\mathbf{k}\}$$

Equating coefficients of **i**, **j**, **k** on both sides, we get

$$\Sigma m(y'\ddot{z}'-z'\ddot{y}') = \Sigma(y'Z-z'Y), \qquad \Sigma m(z'\ddot{x}'-x'\ddot{z}') = \Sigma(z'X-x'Z)$$

and $$\Sigma m(x'\ddot{y}'-y'\ddot{x}') = \Sigma(x'Y-y'X)$$

But the above three equations are of the same form as three equations in (6) of Art. 16.4, and are thus the same equations as we should have obtained if we had regarded the centre of inertia as a fixed point. Hence the required proposition is proved.

Remark 1. The proposition of the above article is known as the *principle of conservation of motion of rotation.* From this we see that the motion round the centre of inertia is independent of its motion of translation.

Remark 2. In articles 15.5 and 15.6 we have proved that the motion of translation and the motion of rotation can be considered independently.

SOLVED EXAMPLES

Ex. 1. *A circular board is placed on a smooth horizontal plane and a boy runs round the edge of it at a uniform rate. What is the motion of the centre of the board?*

[Kanpur 2002; Meerut 2005, 07, 10]

Sol. Let O be the centre, a the radius and M the mass of the circular board. Let initially, the boy of mass, say m, be situated at B on the edge of the board so that $OB = a$. Let G be the centre of gravity of the system. Then

$$OG = (M{\cdot}0 + m{\cdot}a)/(M+m) = ma/(M+m)$$

Since there is no external horizontal force acting on the system, by D'Alembert's principle the centre of gravity G of the system will remain at rest when the boy runs round the edge of the board. Hence as the boy runs round the edge of the board at a uniform rate, the centre O of the board describes a circle round G as centre and OG as radius.

Ex. 2. *A rough uniform board of mass m and length 2a, rests on a smooth horizontal plane and a man of mass M walks on it from one end to the other. Find the distance through the board moves in this time.*

Kanpur 2000; Agra 2012; Garhwal 91; Meerut 2004, 06, 08]

Sol. Here the board and the man constitute a material system. Since there is no external horizontal force acting on the system, by D'Alembert's principle the centre of gravity of the system will remain at rest. In fact, as the man moves forward, the board slips backwards, thus keeping the position of the centre of gravity unchanged.

Let AB be the position of the board (with C as middle point) when the man is at A. In this position, the distance of the C.G. of the system from A

$$= (M{\cdot}0 + m{\cdot}a)/(M+m) = ma/(M+a), \text{ as } AC = BC = a$$

Let $A'B'$ be the position of the board when the man reaches the other end of the board. If the board slips through a distance x in this time (*i.e.*, $AA' = x$) then in this position, the distance of centre of gravity of the system from A

$$= \frac{M \cdot AB' + m \cdot AC'}{M+m} = \frac{M(2a-x)+m(a-x)}{M+m}, \; C' \text{ being mid point of } A'B'$$

But the distance of C.G. is the same from A in initial and final positions of the man, so we must have

$$\frac{ma}{M+m} = \frac{M(2a-x)+m(a-x)}{M+m}$$

or $$ma = M(2a-x) + m(a-x) \quad \text{or} \quad x = 2Ma/(M+m)$$

This gives the required distance moved by the board.

Ex. 3. *A rod revolving on a smooth horizontal plane about one end, which is fixed, breaks into two parts. What is the subsequent motion of the two parts.* **[Kanpur 2001; Meerut 1995, 98]**

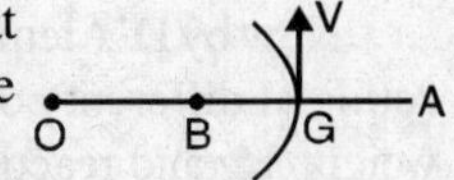

Sol. Let the given rod OA revolve about the fixed end O. If it breaks, say at B, then we are to discuss the motion of two parts, namely OB and AB. Clearly the part OB will continue to rotate about O with the same angular velocity as before.

The part AB of the rod also has the same angular velocity about O and its centre of gravity G has a linear velocity (V, say) along the tangent at G to the circle with centre O and OG as radius. Therefore the part AB will fly off along the tangent line at G to the circle with centre at O and OG as radius. Moreover, since the motion of a body about its centre of inertia is the same as if centre of inertia were fixed and all the forces were acting on it (refer Art. 15.6), the part AB will continue to rotate about G with the same angular velocity.

Ex. 4. *A plank of mass M is initially at rest along a line of greatest slope of a smooth plane inclined at an angle α to the horizon and man of mass M′, starting from the upper end, walks down the plank so that it does not move; show that he gets to other end in time* $\{2M'a/(M + M')g \sin\alpha\}^{1/2}$, *where a is the length of the rod.*

[Agra 2006; Calcutta 2003; Kanpur 2010; Meerut 2000, 05; Purvanchel 2007; Guwahati 2007]

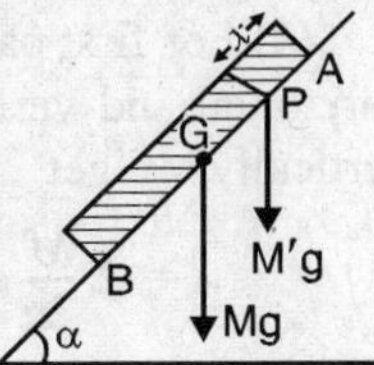

Sol. The plank and the man constitute a material system. The figure shows the given plank AB resting along the line of greatest slope of a given inclined plane. A man of mass M' starts moving down the plank from its upper end A. Let the man move down the plank through a distance $x(=AP)$ in time t. Since the plank does not move, hence if $\bar{x}$ is the distance of the C.G. of the system from A in this position, we have

$$\bar{x} = \frac{M \cdot AG + M' \cdot AP}{M + M'} = \frac{M \times (a/2) + M'x}{M + M'}$$

Differentiating the above above relation twice w.r.t. 'x', we get

$$\frac{d^2\bar{x}}{dt^2} = \frac{M'}{M + M'} \frac{d^2x}{dt^2} \qquad \ldots(1)$$

Now for the motion of translation of the system, we should consider only the motion of its C.G., supposing as if all the external forces act on it. Now the total weight $(M + M')g$ will act vertically downward at the C.G. of the system. Hence the equation of motion of the C.G. of the system along the plane is $\quad (M + M')(d^2\bar{x}/dt^2) = (M + M')g \sin\alpha$

or $\quad M'(d^2x/dt^2) = (M + M')g \sin\alpha$, using (1) $\qquad \ldots(2)$

Integrating (2), $\quad M'(dx/dt) = (M + M')g\, t \sin\alpha + C \qquad \ldots(3)$

But initially, when $t = 0$, $dx/dt = 0$. So (3) gives $C = 0$.

Then, (3) becomes $\quad M'(dx/dt) = (M + M')gt \sin\alpha \qquad \ldots(4)$

Integrating (4), $\quad M'x = (1/2) \times (M + M')g\, t^2 \sin\alpha + C' \qquad \ldots(5)$

But initially, when $t = 0$, $x = 0$. So (5) gives $C' = 0$

∴ (5) becomes $\quad M'x = (1/2) \times (M + M')gt^2 \sin\alpha$

so that $\quad t = \{2M'x/(M + M')g \sin\alpha\}^{1/2} \qquad \ldots(6)$

Putting $x = a$ in (6), the required time $= \{2M'a/(M + M')g \sin\alpha\}^{1/2}$

Ex. 5. (a) *A uniform rod OA, of length 2a, free to turn about its end O, revolves with uniform angular velocity ω about the vertical OZ through O, and is inclined at a constant angle α to OZ, show that the value of α is either zero or* $\cos^{-1}(3g/4a\omega^2)$.

[Agra 2009, 11; Guwahati 2007; Kanpur 2011, 12, 13]

(b) *A rod, of length 2a, revolves with uniform angular velocity ω about a vertical axis through a smooth joint at one extremity of the rod so that it describes a cone of semi-vertical angle α, show that* $\omega^2 = 3g/(4a \cos\alpha)$. *Prove also that the direction of reaction at the hinge makes with the vertical an angle* $\tan^{-1}\{(3/4)\tan\alpha\}$. **[Meerut 2000, 2011; Kanpur 2011]**

Sol. (a) Take an element $PQ(=\delta x)$ at a distance x from O, such that $OP = x$. The mass of the element PQ is $(M/2a)\,\delta x$. Draw PL perpendicular to OZ. Then element PQ will describe a circle of radius $PL(= x \sin\alpha)$ about L. Hence the effective force on this element PQ is $(M/2a)\,\delta x \cdot PL\omega^2$ along PL. So the reversed effective force on the element PQ is $(M/2a)\,\delta x \cdot x \sin\alpha \cdot \omega^2$ along LP as shown in the figure.

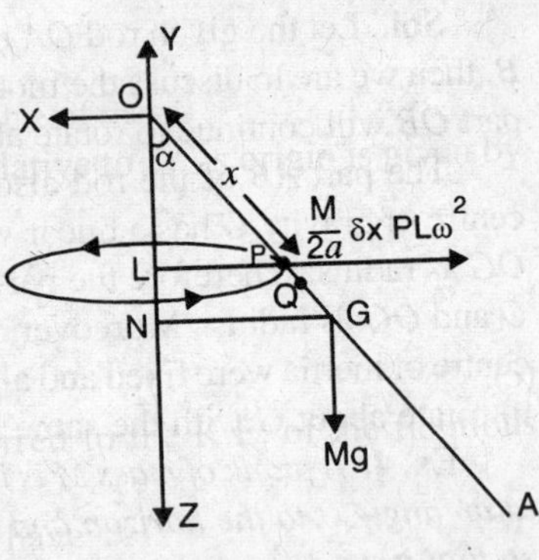

Now by D'Alembert's principle all the reversed effective forces acting at different points of the rod, and the external forces, namely weight Mg and reaction at O are in equilibrium. To avoid reaction at O, taking moments about O, we have

$$Mg \cdot a \sin\alpha - \{\Sigma(M/2a)\,\delta x \cdot \omega^2 x \sin\alpha\} \cdot x \cos\alpha = 0$$

or
$$Mg\, a \sin\alpha - \frac{M\omega^2 \sin\alpha \cos\alpha}{2a} \int_0^{2a} x^2\, dx = 0$$

or
$$Mg\, a \sin\alpha - (M/2a)\omega^2 \sin\alpha \cos\alpha\, \{(2a)^3/3\} = 0$$

or
$$Mg\, a \sin\alpha\, \{1 - (4a/3g)\omega^2 \cos\alpha\} = 0$$

giving either $\sin\alpha = 0$ *i.e.,* $\alpha = 0$ or $\cos\alpha = 3g/4a\omega^2$ ...(1)

Hence, the rod is inclined at an angle zero or $\cos^{-1}(3g/4a\,\omega^2)$

Remark. If $\omega^2 < 3g/4a$, then $\cos\alpha > 1$ and so in this case second value of α is not possible and hence $\alpha = 0$ is the only possible value.

(*b*) For first part refer part (*a*). To find the direction of reaction at the hinge O, let X, Y be the horizontal and vertical components of reaction at O. Then resolving the forces horizontally and vertically, we get

$$X = \Sigma \frac{M}{2a}\,\delta x\, \omega^2 x \sin\alpha = \int_0^{2a} \frac{M\omega^2 \sin\alpha}{2a}\, x\, dx = Ma\omega^2 \sin\alpha$$

and
$$Y = Mg$$

Let the reaction at O make an angle θ with vertical. Then

$$\tan\theta = \frac{X}{Y} = \frac{Maw^2 \sin\alpha}{Mg} = \frac{a \sin\alpha}{g} \cdot \frac{3g}{4\cos\alpha}, \text{ using (1)}$$

or $\tan\theta = (3/4)\tan\alpha$ so that $\theta = \tan^{-1}\{(3/4)\tan\alpha\}$

Ex. 6. *A rod of length 2a, is suspended by a string of length l, attached to one end; if the string and rod revolve about the vertical with uniform angular velocity, and their inclination to the vertical be θ and φ respectively, show that* $3l/a = (4\tan\theta - 3\tan\phi)\sin\phi/(\tan\phi - \tan\theta)\sin\theta$.

[Kanpur 2011, I.F.S. 2003; Meerut 2002, 07; Purvanchel 2005

Sol. Let AB be the given rod of length $2a$ and mass M. Suppose the rod is suspended by a string OA of length l. Let the string and the rod revolve about the vertical line OZ with uniform angular velocity ω and their inclination to the vertical be θ and ϕ respectively as in the figure.

Take an element $PQ(= \delta x)$ at a distance x from A. Then mass of this element is $(M/2a)\delta x$. Drop perpendicular PL to the vertical line OZ through O. Then $PL = PK + KL = PK + AH = x \sin\phi + l \sin\theta$

As the rod revolves about OZ, the element δx will describe a circle of radius PL in the horizontal plane. As shown in the figure, the reversed effective force on element δx along LP will be $(M/2a)\,\delta x\,\omega^2\, PL$ *i.e.,* $(M/2a)\,\delta x\,\omega^2\,(x\sin\phi + l\sin\theta)$. Again the external forces acting on the rod are (*i*) tension T at A along AO and (*ii*) its weight Mg acting vertically downwards at its middle point G.

For equilibrium of the rod resolving horizontally and vertically the forces acting on the rod, we get

$$T\sin\theta = \Sigma \frac{m}{2a}\,\delta x\,\omega^2 (x\sin\phi + l\sin\theta) = \int_0^{2a} \frac{ma}{2a}\,\omega^2 (x\sin\phi + l\sin\theta)\, dx$$

or $T\sin\theta = (M/2a)\omega^2[(l/2)x^2 \sin\phi + lx\sin\theta]_0^{2a} = M\omega^2(a\sin\phi + l\sin\theta)$ ...(1)

and $T\cos\theta = Mg$ ...(2)

Finally, taking moments about A of all forces acting on the rod, we get

$$-Mg \cdot NG + \Sigma(M/2a)\,\delta x\,\omega^2\,(x\sin\phi + l\sin\theta) \cdot AK = 0$$

or
$$Mg\, a\sin\phi = \int_0^{2a} \frac{M\omega^2}{2a}(x\sin\phi + l\sin\theta) \cdot x\cos\phi\, dx$$

or $Mg\,a \sin \phi = (M/2a)\,\omega^3 \cos \phi\,[(1/3)x^3 \sin \phi + (1/2)l\,x^2 \sin \theta]_0^{2a}$

or $Mg\,a \sin \phi = a\,M\omega^2 \cos \phi \cdot \{l \sin \theta + (4/3)\,a \sin \phi\}$

or $g \tan \phi = (1/3)\omega^2\,(3l \sin \theta + 4a \sin \phi)$ ...(3)

Dividing (1) by (2), $\tan \theta = (\omega^2/g)\,(l \sin \theta + a \sin \theta)$ ...(4)

To eliminate ω^2, dividing (3) by (4), we get

$$\frac{g \tan \phi}{\tan \theta} = \frac{g}{3} \cdot \frac{3l \sin \theta + 4 \sin \phi}{l \sin \theta + a \sin \phi}$$

or $3 \tan \phi\,(l \sin \theta + a \sin \phi) = \tan \theta\,(3l \sin \theta + 4a \sin \phi)$

or $3l \sin \theta\,(\tan \phi - \tan \theta) = a \sin \phi\,(4 \tan \theta - 3 \tan \phi)$

or $3l/a = (4 \tan \theta - 3 \tan \phi) \sin \phi/(\tan \phi - \tan \theta) \sin \theta$

Ex. 7. *Two uniform spheres, each of mass M and radius a, are firmly fixed to the ends of two uniform rods, each of mass m and length l, and the other ends of the rods are freely hinged to a point O. The whole system revolves as in the Governor of a Steam Engine, about a vertical line through O with the angular velocity ω. Show that when the motion is steady, the rods are inclined to the vertical at an angle θ, given by the equation*

$$\cos \theta = g\{M(l + a) + (1/2)ml)\}/\omega^2\{M(l + a^2) + (1/3)ml^2\}$$

Sol. Let OA, OA' be two rods, each of length l and mass m hinged freely at a point O. Let two spheres each of mass M and radius a (with centres at C and C') be attached to the other ends of the rods. Given that the whole system revolves about the vertical line, OZ say; with angular velocity ω. When motion is steady, θ is the inclination of each rod to OZ.

In what follows we shall consider motion of rod OA and sphere with centre at C only. Consider an element $PQ(= \delta x)$ of the rod at a distnace x from O. Mass of this element is $(m/l)\,\delta x$. Here the reversed effective force on this element is $(m/l)\,\delta x\,\omega^2 \cdot PN$ *i.e.*, $(m/l)\,\delta x\,\omega^2 \cdot x \sin \theta$ along NP. Again the reversed effective force on the sphere with centre C is $M\omega^2 \cdot CL$ *i.e.*, $M\omega^2\,(a + l) \sin \theta$ along LC. Figure shows the above mentioned effective forces. The external forces on the rod OA and the sphere with centre C are the weights mg acting at centre of gravity G of the rod and Mg acting at centre C of the sphere. Both these external forces are acting in vertically downwards direction.

Consider equilibrium of the system consisting of rod OA and sphere attached to it. To avoid reaction at O, taking moments about O, we get

$$\Sigma(m/l)\,\delta x\,\omega^2\,x \sin \theta \cdot ON + Mw^2(a + l) \sin \theta \cdot OL - mg \cdot DG - Mg \cdot LC = 0$$

$$\int_0^l \frac{m}{l}\omega^2 x \sin \theta\,(x \cos \theta)\,dx + M\omega^2 (a + l) \sin \theta \cdot (a + l) \cos \theta$$

$$- mg \cdot (l/2) \sin \theta - Mg \cdot (a + l) \sin \theta = 0$$

$$[\omega^2\{(1/3)ml^2 + M(a + l)^2 \cos \theta - g\{(1/2)ml + M(a + l)] \sin \theta = 0$$

Hence either $\sin \theta = 0$ *i.e.*, $\theta = 0$ which is not possible

or $\omega^2\{(1/3)ml^2 + M(a + l)^2\} \cos \theta - g\{(1/2)ml + M(a + l)\} = 0$

so that $\cos \theta = g\{M(a + l) + (1/2)ml\}/\omega^2\{M(a + 1)^2 + (1/3)ml^2\}$

Ex. 8. *A thin circular disc of mass M and radius a, can turn freely about a thin axis OA, which is perpendicular to its plane and passes through a point O of its circumference. The axis OA is compelled to move in a horizontal plane with angular velocity ω about its end A. Show that the inclination θ to the vertical of the radius of the disc through O is* $\cos^{-1}(g/a\omega^2)$ *unless* $\omega^2 < g/a$, *and then θ is zero.*

Sol. Let C be the centre of the given thin circular disc of mass M and radius a. Here the plane of the disc is perpendicular to the axis of rotation OA. Now, when the disc rotates round OA, the plane of the disc will be slightly tilted. Suppose the radius OC of the disc in this tilted position makes an angle θ with the vertical.

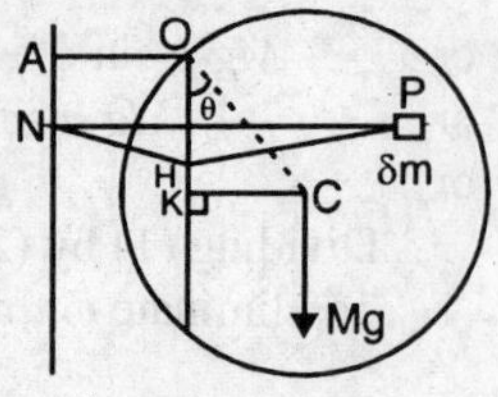

Take an element mass δm at P. Let PN and PH be perpendiculars to the verticals through A and O respectively. Now, the element δm will describe a circle of radius PN with constant angular velocity ω about N and so the reversed effective force acting on the element at P is $\delta m\ \omega^2\ NP$.

Again force $\delta m \cdot \omega^2 \cdot NP$ along NP is equivalent to two component forces $\delta m \cdot \omega^2\ NH$ along NH and $\delta m \cdot \omega^2 \cdot HP$ along HP.

Now moment of force $\delta m \cdot \omega^2 \cdot NH$ about OA is zero because the force $\delta m \cdot \omega^2 \cdot NH$ is parallel to OA. Hence, taking moments about OA, we obtain

$\Sigma \delta m \cdot \omega \cdot HP \cdot OH - Mg \cdot CK = 0$ or $\omega^2\, \Sigma \delta m \cdot HP \cdot OH = Mg \cdot a \cdot \sin\theta$ ...(1)

Now, $\Sigma \delta m \cdot HP \cdot OH$ = product of inertia of disc about OH and horizontal line through O

= Product of inertia of disc about parallel lines through C $+ M \cdot (a \sin\theta)\ (a\cos\theta)$ (Using theorem of parallel axis)

$= 0 + Ma^2 \sin\theta\cos\theta$

Hence (1) reduces to $\omega^2 \cdot Ma^2 \sin\theta\cos\theta = Mg\, a \sin\theta$

so that $\sin\theta\,(a\omega^2\cos\theta - g) = 0$

giving $\sin\theta = 0$ or $\cos\theta = g/a\omega^2 \Rightarrow \theta = 0$ or $\theta = \cos^{-1}(g/a\omega^2)$

If $\omega^2 < g/a$ *i.e.,* $g/a\omega^2 > 1$, then the second value is not possible. Hence when $\omega^2 < g/a$, $\theta = 0$ is the only possible value.

EXERCISE

1. Find the motion of the rod OAB, with two masses m and m' attached to it at A and B respectively, when it moves round the vertical as a conical pendulum with uniform angular velocity, the angle θ which the rod makes with the vertical being constant.

Ans. For a possible motion, the angular velocity ω is given by

$$\omega^2 = (ma + m'b)g/(ma^2 + m'b^2)\cos\theta$$

2. A thin heavy disc can turn freely about an axis in its own plane, and this axis revolves horizontally with a uniform angular vleocity ω about a fixed point on itself. Show that the inclination θ of the plane of the disc to the vertical is $\cos^{-1}(gh/k^2\omega^2)$, where h is the distance of the centre of inertia of the disc from the axis and k is the radius of gyration of the disc about the axis. If $\omega^2 < gh/k^2$, then the plane of the disc is vertical.

15.7. Impulse of an implusive force. Definition. When the forces acting on a body are very great and act for a very short time, we measure their effects by their impulses. If the short time during which an impulsive force X acts be T, then its impulse is $\int_0^T X\,dt$.

5.8 General equations of motion under impulsive forces. ***To find the general equation of motion of rigid body acted on by a number of impulsive forces.*** **(Guwahati 2007; Purvanchel 2005)**

In the case of impulsive forces, the following general equations of motion of a rigid body (refer equations (5) and (6) in Art. 15.4)

$$\Sigma m(d^2x/dt^2) = \Sigma X \quad ...(1)$$

$$\Sigma m(d^2y/dt^2) = \Sigma Y \quad ...(2)$$

$$\Sigma m(d^2z/dt^2) = \Sigma Z \quad ...(3)$$

$$\Sigma m(y(d^2z/dt^2) - z(d^2y/dt^2)) = \Sigma(yZ - zY) \quad ...(4)$$

$$\Sigma m\{z(d^2x/dt^2) - x(d^2z/dt^2)\} = \Sigma(zX - xZ) \quad ...(5)$$

and $$\Sigma m(x(d^2y/dt^2) - y(d^2x/dt^2)) = \Sigma(xY - yX) \quad ...(6)$$

take a new form as discussed below.

Let u, v, w and u', v', w' be the velocities parallel to the axes respectively before and after the action of impulsive forces on the particle of mass m situated at point $P(x, y, z)$ of the rigid body. Let X', Y', Z' be the resolved parts of the impulse of force on particle of mass m parallel to the axes. Let the impulsive force acts for a time T.

Integrating (1) w.r.t. 't' between the limits 0 to T, we get

$$\left[\Sigma m \frac{dx}{dt}\right]_0^T = \int_0^T (\Sigma X) dt = \Sigma \int_0^T X\, dt$$

or $\qquad \Sigma m(u' - u) = \Sigma X' \qquad$ or $\qquad \Sigma mu' - \Sigma mu = \Sigma X' \qquad$ ...(7)

Similarly, (2) and (3) lead to

$$\Sigma mv' - \Sigma mv = \Sigma Y' \qquad \text{...(8)}$$

and

$$\Sigma mw' - \Sigma mw = \Sigma Z' \qquad \text{...(9)}$$

Equations (7), (8) and (9) state that the total change of momentum parllel to the axes is equal to the sum of all external impulses in these directions.

Now, integrating (4) w.r.t. 't' between the limits 0 to T, we get

$$\left[\Sigma m \{y(dz/dt) - z(dy/dt)\}\right]_0^T = \Sigma \left\{ y \int_0^T Z\, dt - x \int_0^T Y\, dt \right\},$$

where we have used the fact that the time interval T is so small that the body has not moved during this interval and hence x, y, z have been treated as constants. The above equation further simplifies to

$$\Sigma m\{y(w' - w) - z(v' - v)\} = \Sigma(yZ' - zY')$$

or

$$\Sigma m(yw' - zv') - \Sigma m(yw - zv) = \Sigma(yZ' - zY') \qquad \text{...(10)}$$

Similarly, (5) and (6) lead to

$$\Sigma m(zu' - xw') - \Sigma m(zu - xw) = \Sigma(zX' - xZ') \qquad \text{...(11)}$$

and

$$\Sigma m(xv' - yu') - \Sigma m(xv - yu) = \Sigma(xY' - yX') \qquad \text{...(12)}$$

Equations (10), (11) and (12) state that the total change in the moment of momentum about the axes is equal to the sum of the moments about these axes of all external impulses. **[Meerut 2001]**

Miscellaneous Problems

1. Fill up the gap: The reversed effective forces together with the impressed forces keep the system in equilibrium is known as .... principle. (Kanpur 2002, 03)

[**Ans.** D'Alembert's. Refer Art 15.2]

2. If $\vec{r}$ is the position vector of a particle of mass m at any time, then the effective force on the particle is $d^2\vec{r}/dt^2$. (T/F) **(Agra 2006, 2009)**

[**Ans. False.** Refer Art 15.1]

3. Suppose a rigid body is in motion. At any time t, let $\vec{r}$ be the position vector of mass m of the body and $\vec{F}$ and $\vec{R}$ the external and internal forces respectively acting on it, then

(*a*) $\Sigma \vec{F} + \Sigma m(d^2\vec{r}/dt^2) = \vec{0}$ $\qquad$ (*b*) $\Sigma \vec{F} + \Sigma(-m(d^2\vec{r}/dt^2) = \vec{0}$

(*c*) $\Sigma \vec{R} + \Sigma m(d^2\vec{r}/dt^2) = \vec{0}$ $\qquad$ (*d*) $\Sigma \vec{R} + \Sigma\{-m(d^2\vec{r}/dt^2)\} = \vec{0}$ $\qquad$ **(Agra 2006)**

[**Ans.** (*b*) Refer Art. 15.2]

Ex. 4. Fill up the gaps:

(i) The effective force on a particle of mass m along x-axis is .... **(Agra 2011)**

(ii) The effective force on a particle is defined as the .... of the mass m and its acceleration **(Agra 2012)**

Sol. (*i*). **Ans.** $m(d^2x/dt^2)$. Refer Art. 15.1, page 15.1

(*ii*). **Ans.** Product. Refer page 15.1

Ex. 5. True of False: *If $\vec{r}$ be the position vector of a particle of mass m at any time t, then the reversed effective force on the particle is $-m\,(d\,\vec{r}/dt)$* **(Agra 2012)**

Sol. Ans. False. Correct value of the reversed effective force is $-(d^2\,\vec{r}/dt^2)$ as discussed in Art. 15.2, page 15.1.

$= (x^2 + y^2 + z^2)(\omega_1 \mathbf{i} + \omega_2 \mathbf{j} + \omega_3 \mathbf{k}) - \ldots$

If $\mathbf{h} = h_1\mathbf{i} + h_2\mathbf{j} + h_3\mathbf{k}$, then relation (2) reduces to

$$h_1\mathbf{i} + h_2\mathbf{j} + h_3\mathbf{k} = \Sigma\delta m[\{(y^2+z^2)\omega_1 - xy\omega_2 - xz\omega_3\}\mathbf{i} + \ldots]$$

or $h_1\mathbf{i} + h_2\mathbf{j} + h_3\mathbf{k} = (A\omega_1 - H\omega_2 - G\omega_3)\mathbf{i} + (B\omega_2 - F\omega_3 - H\omega_1)\mathbf{j} + (C\omega_3 - G\omega_1 - F\omega_2)\mathbf{k}$

where A, B, C, F, G, H are the moments and products of inertia of the rigid body with respect to coordinate axes Ox, Oy, Oz.

Hence the angular momenta h_1, h_2, h_3 of the rigid body about the x-, y- and z-axes are respectively given by

$$h_1 = A\omega_1 - H\omega_2 - G\omega_3$$
$$h_2 = -H\omega_1 + B\omega_2 - F\omega_3$$
$$h_3 = -G\omega_1 - F\omega_2 + C\omega_3 \qquad \ldots(3)$$

Re-writing the equations (3) in matrix form, we get

$$\begin{bmatrix} h_1 \\ h_2 \\ h_3 \end{bmatrix} = \begin{bmatrix} A & -H & -G \\ -H & B & -F \\ -G & -F & C \end{bmatrix} \begin{bmatrix} \omega_1 \\ \omega_2 \\ \omega_3 \end{bmatrix}$$

The symmetrical 3×3 matrix on the R.H.S. is also known as *inertia matrix*.

Corollary 1. If Ox, Oy, Oz are principal axes, then $F = G = H = 0$ and formulae (3) are greatly simplified. They become

$$h_1 = A\omega_1, \qquad h_2 = B\omega_2, \qquad h_3 = C\omega_3 \qquad \ldots(4)$$

Hence the angular momentum $\mathbf{h}$ of the rigid body is given by

$$\mathbf{h} = A\omega_1\mathbf{i} + B\omega_2\mathbf{j} + C\omega_3\mathbf{k} \qquad \ldots(5)$$

where A, B, C are principle moments of inertia at O.

Corollary 2. *If a rigid body is constrained to rotate about a fixed axis, then the angular momentum vector is not along the axis of rotation unless the latter is a principal axis of inertia.*

Let the z-axis be the axis of rotation so that $\omega_1 = \omega_2 = 0$ and $\omega_3 = \omega$. Then equations (3) give

$$h_1 = -G\omega, \qquad h_2 = -F\omega, \qquad h_3 = C\omega$$

so that $\mathbf{h} = (-G\mathbf{i} - F\mathbf{j} + C\mathbf{k})\omega$, which is not along z-axis. $\qquad \ldots(6)$

Clearly, if $F = G = 0$, *i.e.*, z-axis is a principal axis of inertia, then $\mathbf{h} = C\omega\mathbf{k}$ giving

$$h = C\omega \text{ where } h = |\mathbf{h}| \qquad \ldots(7)$$

Relation (7) shows that the angular momentum is equal to the product of M.I. about the axis of rotation and the angular velocity.

Corollary 3. *To find an expression of the kinetic energy of the rigid body in terms of its angular momentum.*

If A, B, C be the principal moments of inertia at O, then angular momentum $\mathbf{h}$ is given by

$$\mathbf{h} = A\omega_1\mathbf{i} + B\omega_2\mathbf{j} + C\omega_3\mathbf{k}$$

Also,

$$\boldsymbol{\omega} = \omega_1\mathbf{i} + \omega_2\mathbf{j} + \omega_3\mathbf{k}$$

$$\therefore \quad \boldsymbol{\omega}\cdot\mathbf{h} = A\omega_1^2 + B\omega_2^2 + C\omega_3^2 = 2T \qquad \text{or} \qquad T = (\boldsymbol{\omega}\cdot\mathbf{h})/2,$$

where T is the K.E. of the rigid body with a fixed point rotating with constant angular velocity $\boldsymbol{\omega}$.

17.6 Angular momentum of a rigid body in general.

The angular momentum of a rigid body (or any moving system) about a point O is the sum of the angular momentum about O of a particle moving with the mass centre and …

16

Motion about a fixed axis

16.1 Moment of the effective forces about the axis of rotation

[Agra 2010; Garhwal 1994, 95; Meerut 1990]

Let the fixed axis of rotation of a rigid body be a perpendicular OZ at O to the plane of the paper. Let a plane ZOA through OZ and fixed in space, cutting the plane of the paper along OA, be taken as the plane of reference. Let the plane ZOG through OZ and fixed in the body cut the plane of paper along OG.

Consider an element of mass m at the point P of the rotating body and let the plane ZOQ through OZ and P cut the plane of the paper long OQ. Let the angle between ZOQ and ZOG be α. Clearly, as the body rotates about OZ the angle α remains always the same. At time t, let the planes ZOG and ZOQ make angles θ and ϕ respectively with the plane ZOA, *i.e.*, $\angle AOG = \theta$ and $\angle AOQ = \phi$.

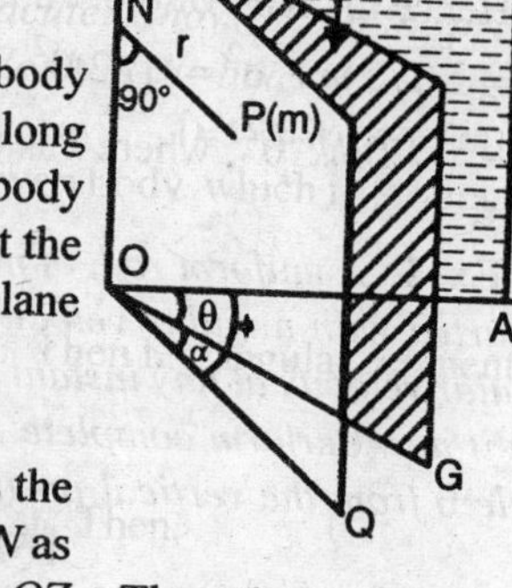

Now, $\quad \theta + \alpha = \phi \quad$ so that $\quad \dot{\theta} = \dot{\phi} \quad$ and $\quad \ddot{\theta} = \ddot{\phi} \quad$ ...(1)

If $PN(= r)$ be the perpendicular distance of P from OZ, then as the body rotates about OZ, the point P describes a circle of radius r and N as centre. Clearly the plane of this circle will be perpendicular to OZ. Then the components of acceleration of mass m at P are $r\dot{\phi}^2$ along PN and $r\ddot{\phi}$ perpendicular to PN. Hence the effective forces on the particle of m at P are $mr\dot{\phi}^2$ along PN and $mr\ddot{\phi}$ perpendicular to PN. Using (1), now the effective forces on the particle of m are $mr\dot{\theta}^2$ along PN and $mr\ddot{\theta}$ perpendicular to PN.

Since the effective force $mr\dot{\theta}^2$ cuts the axis OZ at N, hence its moment about OZ will be zero. So moment of the effective forces on mass m about OZ

$$= (mr\ddot{\theta}) \cdot NP = (mr\ddot{\theta}) \cdot r = mr^2\ddot{\theta}$$

Hence the moment of the effective forces of the whole body about OZ

$= \Sigma mr^2\ddot{\theta} = \ddot{\theta}\,\Sigma mr^2$, as $\ddot{\theta}$ is the same for all particles of the body

$= Mk^2\ddot{\theta}$, where k is the radius of gyration of the body about OZ

16.2 Equation of motion of the body about the axis of rotation.

[Agra 2006; Meerut 1993, 95, 2004; Garhwal 98]

While rotation of a rigid body about a fixed axis (OZ, say), the impressed forces include, besides the external forces, the reaction on the axis of rotation OZ. But the moment of the reaction on axis OZ about OZ is zero. Also by D'Alembert's principle the reversed effective forces and the impressed forces are always in equilibrium. Hence, the moment of the effective forces about the axis OZ is equal to the moment of the impressed forces about OZ. Thus if L be the moment of the impressed forces about the axis of rotation OZ, in the sense which would cause θ to increase, we have

$$Mk^2\ddot{\theta} = L, \qquad ..(1)$$

which is known as the *equation of motion* of the body about the axis of rotation.

Remark. When (1) is integrated twice, it will give θ and $d\theta/dt$ in terms of the time t. The arbitrary constants which appear in the integration will be known if we are given the position of the plane ZOG and its angular velocity at any time.

16.3 Moment of momentum about the axis of rotation. Refer figure of Art. 16.1. The velocity of the particle of mass m at P is $r\dot{\phi}$ perpendicular to NP.

Hence the moment of momentum of the particle m at P about OZ

$= (mr\dot{\phi}) \cdot NP = (mr\dot{\phi}) \cdot r = mr^2\dot{\theta}$, as $\dot{\phi} = \dot{\theta}$

Thus, moment of momentum of the whole body about OZ

$= \Sigma mr^2\dot{\theta} = \dot{\theta}\Sigma mr^2$, as $\dot{\theta}$ is the same for all particles of the body

$= mk^2\dot{\theta}$, where k is the radius of gyration of the body about OZ

16.4 Kinetic energy (K.E.) of a body rotating about a fixed axis **(Agra 2010, 13)**

Refer figure of Art. 16.1. The velocity of the particle of mass m at P is $r\dot{\phi}$ perpendicular to NP. So

K.E. of the particle of mass m at P $= (1/2)m(r\dot{\phi})^2 = (1/2)mr^2\dot{\theta}^2$, as $\dot{\phi} = \dot{\theta}$

Hence the K.E. of the whole body

$= \Sigma(1/2)mr^2\dot{\theta}^2 = (1/2)\dot{\theta}^2\,\Sigma mr^2$, as $\dot{\theta}$ is the same for all particles of the body

$= (1/2)mk^2\dot{\theta}^2$, where k is the radius of gyration of the body about OZ

TYPE 1 : SOLVED EXAMPLES

Ex. 1. *A uniform rod, of mass M and length 2a, can turn freely about one end which is fixed. It is started with angular velocity ω from the position in which it hangs vertically. Then (i) find its angular velocity at any instant (ii) find the least angular velocity with which it must begin to move so that it may perform complete revolution in a vertical plane and (iii) find the time of describing an angle θ from the vertical position in case (ii).*

[Agra 2009, 12; Kanpur 2010, 12, 14, Meerut 2007; Purvanchel 2007]

Sol. (*i*) Let OA be the given rod which turns freely about one end O which is fixed. The vertical line through O is the line fixed in space and OA is the line fixed in the body. Here the axis of rotation is the horizontal line through O, perpendicular to the plane in which the rod rotates. At time t let OA make an angle θ with the vertical.

Taking moments about O, the equation motion is

$$Mk^2\ddot{\theta} = -Mg \cdot GN, \qquad \ldots(1)$$

where negative sign is taken on R.H.S. as the moment of Mg is in the direction of θ decreasing. Also, here $GN = a \sin\theta$, $k^2 = (4/3)a^2$.

$\therefore$ (1) becomes $M(4/3)a^2\,\dot{\theta}(d\dot{\theta}/d\theta) = -Mga\sin\theta$, as $\ddot{\theta} = \dot{\theta}(d\dot{\theta}/d\theta)$

or $$2\dot{\theta}\,d\dot{\theta} = -(3g/2a)\sin\theta\,d\theta$$

Integrating, $$\dot{\theta}^2 = (3g/2a)\cos\theta + C, \text{ } C \text{ is a constant.} \qquad \ldots(2)$$

But initialy, when $\theta = 0$, $\dot{\theta} = \omega$. So (2) gives $C = \omega^2 - (3g/2a)$

$\therefore$ (2) becomes $$\dot{\theta}^2 = \omega^2 - (3g/2a)(1 - \cos\theta), \qquad \ldots(3)$$

(3) gives the angular velocity of rod at any instant.

(*ii*) The required least angular velocity ω with which the rod should begin to move in order to make complete revolutions in vertical plane is given by (3) by using the facts that $\dot{\theta} = 0$ when $\theta = \pi$. So, from (3), $0 = \omega^2 - (3g/a)$ or $\omega = (3g/a)^{1/2}$ which is the required angular velocity.

(*iii*) When $\omega = (3g/a)^{1/2}$, (3) reduces to

$$\left(\frac{d\theta}{dt}\right)^2 = \frac{3g}{a} - \frac{3g}{2a}(1 - \cos\theta) = \frac{3g}{2a}(1 + \cos\theta) = \frac{3g}{a}\cos^2\frac{\theta}{2} \qquad \ldots(4)$$

so that $$dt = (3g/a)\cos\theta/2\,d\theta,$$

where +ve sign is taken on R.H.S. as θ increases with increase in t.

If t is the time of describing an angle θ, we get

$$t = \left(\frac{a}{3g}\right)^{1/2} \int_0^\theta \sec\frac{\theta}{2}\, d\theta = \left(\frac{a}{3g}\right)^{1/2} \left[2 \log \tan\left(\frac{\theta}{4} + \frac{\pi}{4}\right)\right]_0^\theta$$

or
$$t = 2(a/3g)^{1/2} \log \tan(\theta/4 + \pi/4)$$

Ex. 2. *A uniform rod AB is freely movable on a rough inclined plane whose inclination to the horizon is i and whose coefficient of friction is μ, about a smooth pin fixed through the end A; the bar is held in the horizontal position in the plane and allowed to fall from this position. If θ be the angle through which it falls from rest, show that (sin θ)/θ = μ cot i.* **(Kanpur 2006, 2008)**

Sol. Let at any time t, the rod AB fall through an angle θ from its original horizontal position AO. If R is the reaction of the plane, then $R = Mg \cos i$. (see Fig. (i)). Now refer fig. (ii). The external forces acting on rod are (i) some unknown reaction at A (ii) the resolved part $Mg \sin i$ of its weight Mg acting downwards along the line of the greatest slope of the inclined plane (iii) the frictional force μR, *i.e.*, $\mu Mg \cos i$ acting through the C.G. G of rod AB in a direction at right angles to AB.

Fig. (i)

Fig. (ii)

In order to avoid reaction at A, taking moments about A, we get

$$Mk^2\ddot{\theta} = Mg \sin i \cdot AC - \mu Mg \cos i \cdot AG$$
$$= Mg \sin i \cdot a \cos\theta - \mu Mg \cos i \cdot a$$

or
$$2k^2(\dot{\theta}\, d\dot{\theta}/d\theta) = ga(\sin i \cos\theta - \mu \cos i), \text{ as } \ddot{\theta} = \dot{\theta}(d\dot{\theta}/d\theta)$$

or
$$2k^2\dot{\theta}\, d\dot{\theta} = 2ga(\sin i \cos\theta - \mu \cos i)\, d\theta$$

Integrating, $\qquad k^2\dot{\theta}^2 = 2ga(\sin i \sin\theta - \mu\theta \cos i) + C$ ...(1)

Initially, when $\theta = 0$, $\dot{\theta} = 0$ so (1) gives $C = 0$

So (1) becomes $\qquad k^2\dot{\theta}^2 = 2ga(\sin i \sin\theta - \mu\theta \cos i)$ ...(2)

Clearly, the rod will come to rest when $\dot{\theta} = 0$, *i.e.*,

$$0 = 2ga(\sin i \sin\theta - \mu\theta \cos i) \quad \text{or} \quad (\sin\theta)/\theta = \mu \cot i$$

Ex. 3. *A fine string has two masses M and M′ tied to its ends and passes over a rough pulley of mass m, whose centre is fixed. If the string does not slip over the pulley, show that M will descend with acceleration $(M - M')g/\{M + M' + (mk^2/a^2)\}$, where a is the radius and k is the radius of* ***through which it falls from rest, show that (sin θ)/θ = μ cot i.*** **(Kanpur 2006, 2008)**

If the pulley be not sufficiently rough to prevent sliding and M be the descending mass, show that its acceleration is $(M - M'e^{\mu\pi})g/(M + M'e^{\mu\pi})$ and that the pulley will now spin with angular acceleration equal to $2MM'ga(e^{\mu\pi} - 1)/mk^2(M + M'e^{\mu\pi})$

Sol. Case I. When string does not slip. Let in time t, the mass M move through a distance x downwards and the mass M' move through a distance x upwards. Also let θ be the angle turned by the pulley during the same time, then we have

$$x = a\theta \quad \text{so that} \quad \dot{x} = a\dot{\theta} \quad \text{and} \quad \ddot{x} = a\ddot{\theta} \quad \text{...(1)}$$

Let T and T' be the tensions in the two parts of the string. Then the equations of motion of M and M' are

$$M\ddot{x} = Mg - T \quad \text{...(2)}$$
$$M'\ddot{x} = T' - M'g \quad \text{...(3)}$$

Again, by taking moments about the fixed horizontal axis passing through the centre O of the pulley, the equation of motion of the pulley is

$$mk^2\ddot{\theta} = Ta - T'a \quad \text{or} \quad m(k^2/a)\ddot{\theta} = T - T' \quad \text{...(4)}$$

or
$$m(k^2/a^2)\ddot{x} = T - T', \text{ as by (1), } \ddot{\theta} = \ddot{x}/a \quad \text{...(5)}$$

Adding (2), (3) and (5), we get

$\{M + M' + (mk^2/a^2)\}\ \ddot{x} = (M - M')g$ or $\ddot{x} = (M - M')g/\{M + M' + (mk^2/a^2)\}$

This gives the required acceleration of mass M.

Case II. When the pulley is not rough enough to prevent sliding

In this case the equation (1) will not hold good. If μ is the coefficient of friction, then in this case it is known from statics that

$$T = T' e^{\mu\pi} \quad \text{...(6)}$$

Solving (2), (3) and (6) for T, T' and $\ddot{x}$, we get

$$T = (2MM'g\, e^{\mu\pi})/(M + M' e^{\mu\pi}), \qquad T' = (2MM'g)/(M + M' e^{\mu\pi}) \quad \text{...(7)}$$

and $$\ddot{x} = (M - M' e^{\mu\pi})g/(M + M' e^{\mu\pi}) \quad \text{...(8)}$$

Equation (8) gives the required acceleration of mass M.

Substituting the values of T and T' from (7) in (4), we get

$$\ddot{\theta} = \{2MM'ga\,(e^{\mu\pi} - 1)\}/mk^2(M + M'e^{\mu\pi}),$$

which is required angular acceleration of the pulley.

Ex. 4. *A vertical circular plate of radius a is capable of revolving about a smooth horizontal axis through its centre, a rough perfectly flexible chain whose mass is equal to that of the plate and whose length is equal to its circumference hangs over its rim in equilibrium, if one end be slightly displaced, show that the velocity of the chain when the other end reaches the plate is* $(\pi ag/6)^{1/2}$.

Sol. At any time t, let x be distance described by the chain and θ be the angle turned by the pulley in this time. If v is the velocity of chain, then

$$x = a\theta \quad \Rightarrow \quad v = \dot{x} = a\dot{\theta} \quad \text{...(1)}$$

Let M be the mass of the plate or that of the string.

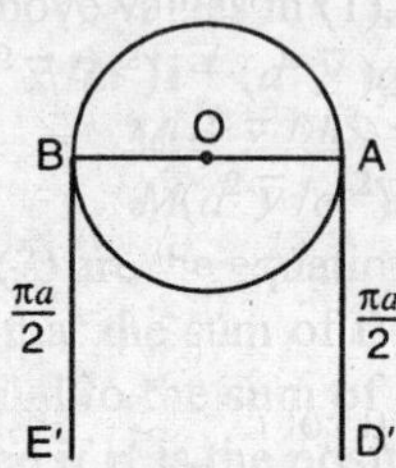

(Initial position of the Chain)

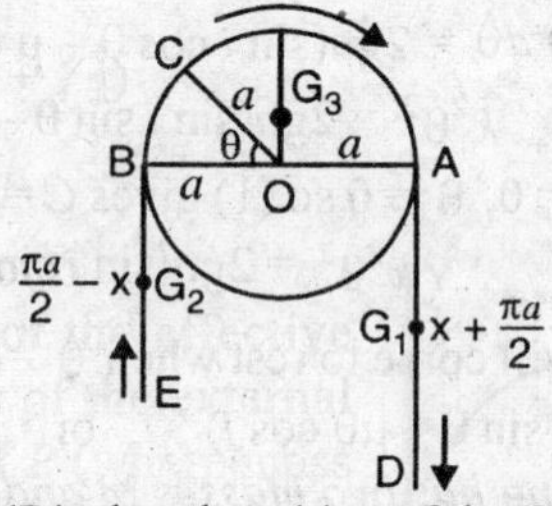

(Displaced position of the Chain)

Now, the total K.E. of the system at time t

= K.E. of the chain + K.E. of the pulley

$= (1/2)Mv^2 + (1/2)\,Mk^2\dot{\theta}^2 = (1/2)Mv^2 + (1/2)M\,(a^2/2) \times (v/a)^2 = (3/4)Mv^2$

[$\because$ For circular plate, $k^2 = a^2/2$. Also, by (1), $\dot{\theta} = v/a$]

Let w be the weight per unit length of the chain. Then the weight of the portion AD of the chain is $(x + \pi a/2)w$ and the depth of its C.G. G_1 below AB is $(1/2)\,(x + \pi a/2)$, the weight of the portion BE of the chain is $(\pi a/2 - x)w$ and the depth of its C.G. G_2 below AB is $(1/2)\,(\pi a/2 - x)$ and the weight of the semi-circular chain ACB is πaw and the height of its C.G. G_3 above AB is $2a/\pi$ (which is a standard result of C.G. of semi-circular wire).

If x_1 is the depth of C.G. of whole chain below AB at time t. Then

$$x_1 = \frac{(x + \pi a/2)w \cdot (1/2)(x + \pi a/2) + (\pi a/2 - x)w \cdot (1/2)(\pi a/2 - x) + \pi aw \cdot (-2a/\pi)}{(x + \pi a/2)w + (\pi a/2 - x)w + \pi aw}$$

or $$x_1 = (a^2\pi^2 + 4x^2 - 8a^2)/8\pi a \quad \text{...(2)}$$

Putting $x = 0$ in (2), the depth x_2 of C.G. of chain in the initial position (*i.e.*, when $t = 0$) is given by

$$x_2 = (a^2\pi^2 - 8a^2)/8\pi a \quad \text{...(3)}$$

Hence the displacement of the C.G. of the chain in the vertical downward direction in time $t = x_1 - x_2 = 4x^2/8\pi a = x^2/2\pi a$

Hence the work done by the system in time t

= Work done by the chain + Work done by the plate = $(Mg)\cdot(x^2/2\pi a)$ ...(4)

$\therefore$ By the energy equation *i.e.*, Change in K.E. = The work done by the forces,

or $(3/4)Mv^2 = (Mgx)^2/2\pi a$ or $v^2 = (2gx^2)/3\pi a$, ...(5)

which gives the velocity of the chain in terms of distance x. When the end E reaches the pulley, then $x = \pi a/2$. Putting $x = \pi a/2$ in (5), the required velocity v is given by

$$v^2 = \left(\frac{2g}{3\pi a}\right)\times\left(\frac{\pi a}{2}\right)^2 = \frac{\pi a g}{6} \quad \text{so that} \quad v = \left(\frac{\pi a g}{6}\right)^{1/2}$$

Ex. 5. *A uniform string of length 20 feet and mass 40 lbs, hangs in equal length over circular pulley of mass 10 lbs and small radius, the axis of pulley being horizontal. Masses of 40 lbs and 35 lbs are attached to the ends of the string and motion takes place. Show that the time taken by the small mass to reach the pulley is (1/4)* $\sqrt{15}$ *log* $(9 + 4\sqrt{5})$ *seconds.*

Sol. At any time t, let x be the distance described by the string and θ be the angle turned by the pulley in this time. If v is the velocity of string, then

$$x = a\theta \quad \Rightarrow \quad v = \dot{x} = a\dot{\theta} \qquad ...(1)$$

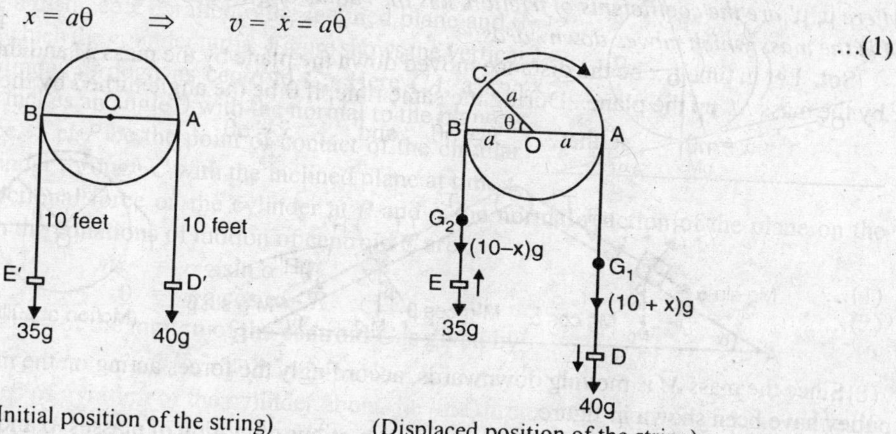

(Initial position of the string) (Displaced position of the string)

Now, the total K.E. of the system at time t

= K.E. of string + K.E. of weights at D and E + K.E. of the pulley

$= (1/2)\cdot 40v^2 + (1/2)\cdot 35v^2 + (1/2)\cdot 40v^2 + (1/2)\cdot 10\cdot k^2\dot{\theta}^2$

$= (115/2)v^2 + 5(a^2/2)\cdot(v/a)^2$, as $k^2 = a^2/2$ and $\dot{\theta} = v/a$, by (1)

$= (115/2)v^2 + (5/2)v^2 = 60v^2$

Let w be the weight per unit length of the string. Then, the weight of the portion AD of the chain is $(10 + x)g$ and the depth of its C.G. G_1 below AB is $(10 + x)/2$ and the weight of the portion BE of the string is $(10 - x)g$ and and the depth of its C.G. G_2 below AB is $(10 - x)/2$. If x_1 is the depth of C.G. of whole string below AB, then

$$x_1 = \frac{(10+x)w\cdot(1/2)(10+x)+(10-x)w\cdot(1/2)(10-x)}{(10+x)w+(10-x)w} = \frac{100+x^2}{20} \qquad ...(2)$$

Putting $x = 0$ in (2), the depth x_2 of C.G. of string in the initial position (*i.e.*, when $t = 0$) is given by $x_2 = 50$.

Hence the displacement of the C.G. of the string in the vertical downward direction in time $t = x_1 - x_2 = x^2/20$. Also, note that distance moved by weight 40 lbs is x in vertically downward direction whereas the distance moved by the weight 35 lbs is x in vertically upward direction.

Hence the work done by the system in time t

= Work done by weights at D and E + Work done by weight of string

$= 40gx - 35gx + 40g\cdot(x^2/20) = gx(5 + 2x)$

$\therefore$ By the energy equation, *i.e.*, change in K.E. = the work done by the forces

$$60v^2 = gx(5 + 2x) = 64(x^2 + 5x/2), \text{ as } g = 32 \text{ feet/sec}^2$$

or $v = \frac{dx}{dt} = \frac{4}{\sqrt{15}}\left(x^2 + \frac{5x}{2}\right)^{1/2}$ or $dt = \frac{\sqrt{15}}{4}\frac{dx}{(x^2 + 5x/2)^{1/2}}$ ...(3)

When the end E reaches the pulley, we have $x = 10$ feet. Hence integrating (3) between the limits $x = 0$ to $x = 10$, the required time t is given by

$$t = \frac{\sqrt{15}}{4}\int_0^{10} \frac{dx}{\{(x+5/4)^2 - (5/4)^2\}^{1/2}} = \frac{\sqrt{15}}{4}\left[\log\left[x + \frac{5}{4} + \left\{\left(x+\frac{5}{4}\right)^2 - \frac{25}{16}\right\}^{1/2}\right]\right]_0^{10}$$

or $t = (1/4)\sqrt{15}\log(9 + 4\sqrt{5})$ seconds, as desired.

Ex. 6. *Two unequal masses, M and M′, rest on two rough planes inclined at angles α and β to the horizon; they are connected over a small pulley, of mass m and radius a, which is placed at the common vertex of the two planes; show that the acceleration of either mass is*

g{M(sin α – μ cos α) – M′(sin β + μ′ cos β}/(M + M′ + mk²/a²),

where μ, μ′ are the coefficients of friction, k is the radius of gyration of the pulley about its axis, and M is the mass which moves downwards. **[Kanpur 2009]**

Sol. Let in time t, x be the distance moved down the plane by the mass M and the same distance x by the mass M' up the plane. During the same time, if θ be the angle turned by the pulley, than

$$x = a\theta \quad \text{so that} \quad \dot{x} = a\dot{\theta} \quad \text{and} \quad \ddot{x} = a\ddot{\theta} \qquad \text{...(1)}$$

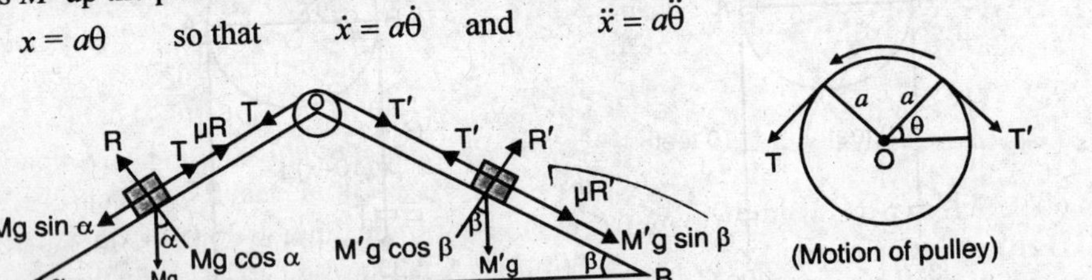

(Motion of pulley)

Since the mass M is moving downwards, accordingly the forces acting on the masses M, M' and pulley have been shown in figure.

Using Newton's second law of motion, the equations of motion of masses M and M' respectively, are given by

$$M\ddot{x} = Mg\sin\alpha - \mu R - T \qquad \text{...(2)}$$

and

$$M'\ddot{x} = T' - M'g\sin\beta - \mu' R' \qquad \text{...(3)}$$

Finally, taking moments about the horizontal axis of rotation of the pulley passing through its centre O, the equation of the pulley is given by

$$mk^2\ddot{\theta} = T\cdot a - T'a \quad \text{or} \quad m(k^2/a^2)\ddot{x} = T - T', \text{ by (1)} \qquad \text{...(4)}$$

Adding (2), (3) and (4), we get

$$(M + M' + mk^2/a^2)\,\ddot{x} = Mg\sin\alpha - \mu Mg\cos\alpha - M'g\sin\beta - \mu' M'g\cos\beta$$

$$(\because R = Mg\cos\alpha \text{ and } R' = M'g\cos\beta)$$

or $\ddot{x} = g\{M(\sin\alpha - \mu\cos\alpha) - M'(\sin\beta + \mu'\cos\beta)\}/(M + M' + mk^2/a^2)$

giving the required acceleration of either mass.

[Mumbai 2010; Meerut 2004]

16.5 Compound pendulum. Definition.

A system consisting of a rigid body free to oscillate, under gravity, about a fixed horizontal axis, is known as a *compound pendulum*.

Time of a complete oscillation of a compound pendulum. *To show that* $2\pi(k^2/hg)^{1/2}$ *is the time of a complete small oscillation of a compound pendulum, where k is the radius of gyration of the body about the fixed axis and h is the distance of centre of inertia of the body from the fixed axis.*

[Guwahati 2003, 07; Kanpur 2003, 04; Meerut 1993, 96]

Proof. Let the body be free to turn about a horizontal axis NN'. Let the plane through the centre of inertia G and perpendicular to the axis NN' of rotation be the plane of the paper.

The axis of rotation passes through O. Let θ be the angle between the vertical line OZ and the line OG, so that θ is the angle the plane fixed in the body makes with a plane fixed in space.

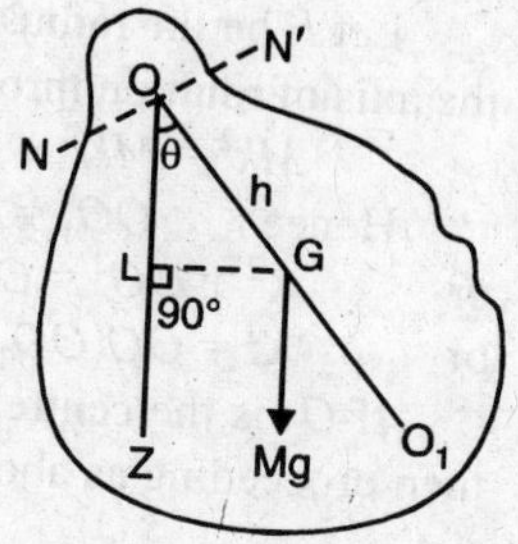

The impressed forces on the body are (*i*) its weight Mg acting vertically downwards at G, and (*ii*) the reaction of the body about the fixed axis.

To avoid the reaction of the fixed axis NN', taking moments about the NN', we have

$$Mk^2\ddot{\theta} = -Mg \cdot GL = -Mg \cdot h \sin\theta, \qquad \text{...(1)}$$

where –ve sign is taken as the moment of Mg is in the direction of θ decreasing.

If θ be so small that its cubes and higher powers may be neglected, *i.e.*, $\sin\theta = \theta$ approximately, then (1) becomes

$$k^2\ddot{\theta} = -gh\theta \quad \text{or} \quad \ddot{\theta} = -(gh/k^2)\theta,$$

showing that the motion of the body about the fixed axis is S.H.M. Hence the required time of a complete oscillation of the compound pendulum is given by ...(2)

$$T = \frac{2\pi}{(gh/k^2)^{1/2}} \quad \text{or} \quad T = 2\pi\left(\frac{k^2}{gh}\right)^{1/2} \qquad \text{...(3)}$$

16.6 Simple equivalent pendulum. Definition. **[Meerut 2005; Mumbai 2010]**

A simple pendulum having the same periodic time as the compound pendulum is known as a simple equivalent pendulum.

If l be the length of a simple pendulum, then period of this pendulum $= T_1 = 2\pi\,(l/g)^{1/2}$. Again, the time of one oscillation of a comound pendulum $= T_2 = 2\pi\,(k^2/gh)^{1/2}$. If the simple pendulum of length l is the simple equivalent pendulum, then we must have

$$T_1 = T_2 \quad \text{or} \quad 2\pi(l/g)^{1/2} = 2\pi(k^2/gh)^{1/2} \quad \text{so that} \quad l = k^2/h \qquad \text{...(1)}$$

In the case of a compound pendulum, the quantity k^2/h is known as the *length of simple equivalent pendulum.*

Remark. From the above discussion, we conclude that corresponding to any compound pendulum, it is always possible to construct a simple pendulum of length given by the above formula (1), such that the two pendulums *oscillate in unison.*

16.7 Centre of suspension and centre of oscillation **[Kanpur 2010; Agra 2006; Meerut 2004]**

Centre of suspension. Definition. The point O where the axis of rotation cuts the plane perpendicular to it through centre of gravity G of the body, is called the *centre of suspension.* Refer figure of Art. 16.5.

Centre of Oscillation. Definition. If O is the centre of suspension and G the centre of gravity of the body, then the point O_1 on OG produced such that $OO_1 = k^2/h$, is called the *centre of oscillation* of the body. See fig. of Art. 16.5.

Remark. From the above definition of centre of oscillation, we find that if whole mass of the body were collected at the centre of oscillation and hanged from the centre of suspension O by a string of length k^2/h, the time of oscillation of tnis simple pendulum will be the same as that of a body swinging about O.

It is to be noted that the time of oscillation $T = 2\pi(l/g)^{1/2}$ of a simple pendulum depends upon the length l of the string and not upon the quantity of mass attached at the other end of the string. Hence in the case of a simple equivalent pendulum, we have an important result : *If besides the mass of the body an additional mass is attached at the centre of oscillation; the period of oscillation will be the same as prior to attaching the additional mass.*

16.8 The centre of suspension and the centre of oscillation of a compound pendulum are convertible. (i.e. reversible) **[Agra 2009, 11; Kerala 2002: Meerut 2008, 10, 12]**

Let O be the centre of suspension and O_1 the centre of oscillation of a compound pendulum. Then $OO_1 = k^2/h$, where k is the radius of gyration of the body about the axis of rotation through O and $h = OG$, where G is the centre of gravity of the body.

We want to show that if we suspend the body from O_1 instead of from O, then the body will swing with the point O as its centre of oscillation.

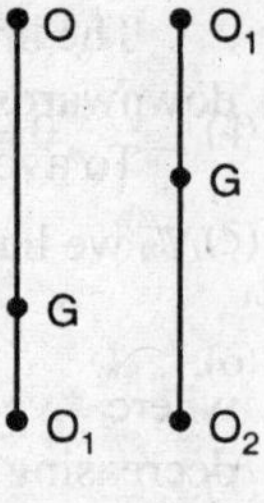

Let K be the radius of gyration of the body about an axis through G and parallel to the axis of rotation through O. Then, by the theorem of parallel axis,

$$Mk^2 = MK^2 + M \cdot OG^2 \quad \text{or} \quad k^2 = K^2 + OG^2$$

Hence $\quad OO_1 = K^2/h = (K^2 + OG^2)/OG$, as $h = OG$

or $\quad K^2 + OG^2 = OG \cdot OO_1 \quad$ or $\quad K^2 = OG \cdot OO_1 - OG^2$

or $\quad K^2 = OG(OO_1 - OG) \quad$ or $\quad K^2 = OG \cdot O_1G \quad$ ...(1)

If O_2 is the centre of oscillation of the body when the centre of suspension is O_1, then proceeding as above, we can show that

$$K^2 = O_1G \cdot O_2G \quad \text{...(2)}$$

Comparing (1) and (2), we conclude that O_2 and O are the same points, *i.e.*, if O_1 is the centre of suspension, O is the centre of oscillation. Hence the centre of suspension and centre of oscillation are convertible *i.e.*, reversible.

16.9 Minimum time of oscillation of a compound pendulum. **[Agra 2009**

Let k be the radius of gyration of the body about the axis of rotation and let K be its radius of gyratior about a line through centre of gravity, parallel to the axis of rotation. Refer fig. of Art. 16.5. Then, by the theorem of parallel axis, we have

$$Mk^2 = MK^2 + Mh^2 \quad \text{so that} \quad k^2 = K^2 + h^2$$

Let l be the length of simple equivalent pendulum. Then

$$l = k^2/h = (K^2 + h^2)/h = h + (K^2/h) \quad \text{...(1)}$$

Clearly the time of oscillation of a compound pendulum will be least when the length l of the simple equivlent pendulum is minimum.

From (1), $\quad dl/dh = 1 - K^2/h^2 \quad$ ...(2)

From maximum or minimum value of l, $\quad dl/dh = 0$,

i.e., $\quad 1 - K^2/h^2 = 0 \quad$ so that $\quad h = K$

From (2), $\quad d^2l/dh^2 = 2K^2/h^3$, which is +ve when $h = K$.

Hence l is minimum when $h = K$. Putting $h = K$ in (1), the minimum value of $l = K + (K^2/K) = 2K$ and the minimum time of oscillation of a compound is given by

$$T = 2\pi(l/g)^{1/2} = 2\pi(2K/g)^{1/2}$$

Remark. If $h = 0$ or infinity, *i.e.*, if the axis of suspension either passes through the centre of gravity or be at infinity, then (1) shows that the corresponding equivalent pendulum is of infinite length and hence the time of oscillation is infinite.

TYPE 2 : SOLVED EXAMPLES

Ex. 1. *Find the length of the simple equivalent pendulum, when a circular disc of radius a oscillates about a fixed horizontal axis which is tangent to it.*

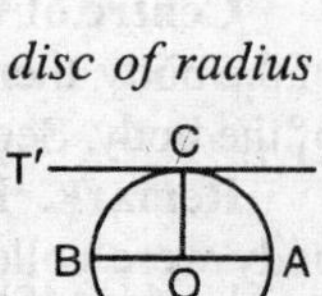

Sol. Let TT' be a fixed horizontal axis which is tangent to the disc at C. Let AB be the diameter of the disc such that AB and TT' are parallel. Let k be the radius of gyration of the disc about TT' and let M be mass of the disc. Then, by theorem of parallel axis, we have

M.I. of disc about TT' = M.I. about AB + M.I. of mass M at O about TT'

or $\quad Mk^2 = M(a^2/4) + Ma^2 \quad$ so that $\quad k^2 = 5a^2/4$

Also, $\quad h$ = distance of C.G. of the disc from axis $TT' = a$

Hence the length of the equivalent simple pendulum is given by

$$l = k^2/h = 5a^2/4a = 5a/4$$

Ex. 2. (a) *A rectangular plate swings in a vertical plane about one of its corners. If its period is one second, find the length of the diagonal.* **(I.A.S. 2005)**

(b) *Show that a square plate of side 2b suspended from one of the corners oscillates in unison with simple pendulum of length $(4b\sqrt{2})/3$.* **[Delhi B.Sc. (Hons.) 1997]**

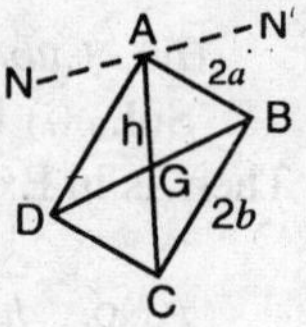

Sol. (*a*) Let the rectangular plate *ABCD* of sides of lengths $2a(= AB)$ and $2b(= BC)$ swing in its vertical plane about a horizontal line NN' through *A* and perpendicular to its plane as axis of rotation. Let *G* be C.G. and *M* the mass of the plate.

Now, $h =$ depth of C.G of the plate below *A*

$= AG = AC/2 = (1/2)(AB^2 + BC^2)^{1/2} = (1/2)(4a^2 + 4b^2)^{1/2} = (a^2 + b^2)^{1/2}$ ...(1)

Let *k* be the radius of gyration of the plate about the axis of rotation NN'. Then by the theorem of parallel axis, we have

M.I. of the plate about NN' = M.I. of the plate about a line through *G* and parallel to NN' + M.I. of Mass *M* at *G* about NN'

$\therefore \quad Mk^2 = (M/3)(a^2 + b^2) + M\cdot h^2 = (M/3)(a^2 + b^2) + M(a^2 + b^2)$, by (1)

Hence, $\quad k^2 = (4/3)(a^2 + b^2) \quad$ or $\quad k^2 = (4/3)h^2$ by (1) ...(2)

Hence, period of oscillation $= T = 2\pi(k^2/gh)^{1/2} = 2\pi(4h/3g)^{1/2}$, by (2)

By given problem, period of oscillation $= T = 1$ second

So $\quad 1 = 2\pi(4h/3g)^{1/2} \quad$ or $\quad 16\pi^2 h = 3g \quad$ or $\quad h = 3g/16\pi^2$

Hence the length of the diagonal $= AC = 2 \times AG = 2h = 3g/8\pi^2$

(*b*) Here *ABCD* is square so that $2a = 2b$ *i.e.*, $b = a$. Then as in part (*a*),

$$h = (b^2 + b^2)^{1/2} = b\sqrt{2},\ k^2 = (4/3)h^2$$

Hence the required length of the equivalent simple pendulum

$= l = k^2/h = 4h^3/3h = (4/3)h = (4b\sqrt{2})/3$

Ex. 3. (*a*) *Find the time of oscillation of a comound pendulum, consisting of a rod of mass m and length a carrying at one end a sphere of mass m_1 and diameter 2b, the other end of the rod being fixed.*

(b) *A compound pendulum consists of a rod of mass m and length a carries at one end a sphere of mass m_1 and diameter 2b, the other end of the rod being fixed. Prove that the length of an equivalent pendulum is*

$[(1/3)\,ma^2 + m_1\{(2/5)b^2 + (a + b)^2\}]/\{ma/2 + m_1(a + b)\}$ **[Meerut 1997]**

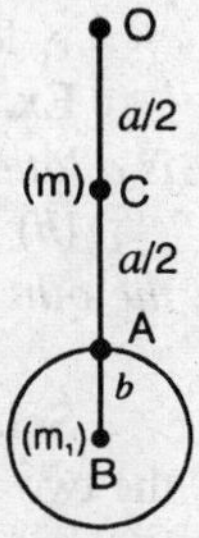

Sol. (*a*) Let *OA* be the rod of mass *m* and length *a* attached at *A* to a sphere of radius *b* and mass m_1. Let *C* be the mid-point of *OA*. Let *h* be the depth of C.G. of the system below *O*, then

$$h = \frac{m \cdot OC + m_1 \cdot OB}{m + m_1} = \frac{m \cdot (a/2) + m_1(a+b)}{m + m_1}. \qquad ...(1)$$

where *C* is the middle point of rod.

Let *k* be the radius of gyration of the system about the axis through *O*, then

$$(m + m_1)k^2 = (4/3)m\,(a/2)^2 + \{(2/5)m_1 b^2 + m_1 \cdot OB^2\}$$

or
$$k^2 = \frac{(1/3)ma^2 + m_1\{(2/5)b^2 + (a+b)^2\}}{m + m_1} \qquad ...(2)$$

The required time of oscillation of the compound pendulum.

$= 2\pi(k^2/gh)^{1/2}$, where k^2 and *h* are given by (1) and (2).

(*b*) Procced as in part (*a*) to find *h* and k^2. Then, the required length of the equivalent simple pendulum

$$= \frac{k^2}{h} = \frac{(1/3)\,ma^2 + m_1\,\{(2/5)b^2 + (a+b)^2\}}{m(a/2) + m_1(a+b)}, \text{ by (1) and (2)}$$

Ex. 4. *Find the length of the simple equivalent pendulum in the following cases, the axis being horizontal*

(*i*) *A circular wire of radius a, free to oscillate about a tangent to it.*

(*ii*) *A circular wire of radius a, free to oscillate about an axis, perpendicular to the plane of the wire, at any point of its arc.*

(*iii*) *Cube of side 2a, free to oscillate about one of its edge as an axis.*

(*iv*) *Cube of side 2a, free to oscillate about a diagonal of one of its faces.*

Sol. (*i*) Refer figure of Ex. 1. wherein *ABC* is given circular wire in place of a circular disc. Then as in Ex. 1, we have

$$Mk^2 = M(a^2/2) + Ma^2 \qquad \text{so that} \qquad k^2 = 3a^2/2$$

Also, h = distance of C.G. of the wire from TT' = a

Hence, the required length of the equivalent simple pendulum

$$= k^2/h = 3a^2/2a = 3a/2$$

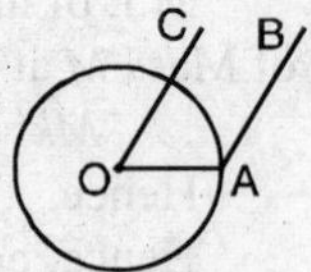

(*ii*) Let *AB* be the given axis where *AB* is perpendicular to the plane of the wire at any point *A*. Let *OC* be parallel to *AB* and perpendicular to the plane of the wire. Let *k* be the radius of gyration of the wire about *AB* and let *M* mass of the wire. Then by, theorem of parallel axis, we have

$$Mk^2 = Ma^2 + Ma^2 \qquad \text{so that} \qquad k^2 = 2a^2$$

Also, h = distance of C.G. *O* of the wire from *AB* = a

∴ length of the simple equivalent pendulum = $k^2/a = 2a$.

(*iii*) Refer solved Ex. 3 of page 14.15 of in Chapter 14. Here for cube, $a = b = c$. Let *k* be the radius of gyration of the cube about edge *OA* and let *M* be mass of the cube. Then, we have

$Mk^2 = (4/3)\, M(a^2 + a^2)$ so that $k^2 = (8/3)a^2$

h = distance of C.G. of the cube from $OA = (a^2 + a^2)^{1/2} = a\sqrt{2}$

∴ length of the simple equivalent pendulum = $k^2/h = 4a\sqrt{2}/3$

(*iv*) Refer figure of Art. 14.4*C* on page 14.2 of Chapter 14. Let $2a$ be the size of an edge of given cube (*ABFE, CO′DG*) and let *O* be its centre of gravity. Let the axis of rotation be the diagonal *AC* of face *ABCO′*. Let *k* be the radius of gyration of the cube about *AC* and let *M* be mass of the cube. We know that M.I. of a cube about any line through *O* is $2Ma^2/3$. Hence by theorem of parallel axis, we have

$$Mk^2 = 2Ma^2/3 + Ma^2 \qquad \text{so that} \qquad k^2 = 5a^2/3$$

Also, h = distance of C.G. *O* of the cube from *AC* = a

∴ length of the simple equivalent pendulum = $k^2/h = 5a/3$

Ex. 5. (***a***) *Find the length of the simple equivalent pendulum for an elliptic lamina when the axis is a latus rectum.* **[Kanpur 2014]**

(***b***) *An elliptic lamina is such that when it swings about one latus rectum as a horizontal axis, the other latus rectum passes through the centre of oscillation, prove that the eccentricity is 1/2.*

[Agra 2011; Meerut 2010; Kanpur 2009; Guwahati 2007]

Sol. (***a***) Let the latus rectum L_1L_2 be the axis. Let *C* be the centre and *O*, *O′* be the two foci of the elliptic lamina. Let *k* be the radius of gyration about L_1L_2 and *M* be mass of the elliptic lamina. Then

Mk^2 = M.I. about $L_1L_2 = M(a^2/4) + M{\cdot}OC^2$

or $Mk^2 = M(a^2/4) + M(ae)^2$ so that $k^2 = a^2(e^2 + 1/4)$,

where *e* is the eccentricity of the elliptic lamina.

Also, h = Distance of C.G. *C* of the lamina from $L_1L_2 = ae$

∴ l = length of the simple equivalent pendulum

$$= k^2/h = a(e + 1/4e)$$

(***b***) Refer figure of part (*a*). Let L_1L_2 and $L_1'L_2'$ be two latera recta of the elliptic lamina and L_1L_2 be the horizontal axis of rotation. Given that the other latus rectum $L_1'L_2'$ passes through the centre of oscillation *O′*.

Hence l = length of simple equivalent pendulum = $OO' = 2ae$

Now, proceed as in part (*a*) to show that $l = a(e + 1/4e)$

∴ $2ae = a(e + 1/4e)$ or $2e^2 = e^2 + 1/4$ or $e = 1/2$

Ex. 6. *A uniform elliptic board swings about a horizontal axis at right angles to the plane of the board and passing through one focus. If the centre of oscillation be other focus, prove that its eccentricity is* $(2/5)^{1/2}$.

Hint. Proceed as in part (*b*) of Ex. 5.

Ex. 7. *Three uniform rods AB, BC, CD each of length a, are freely jointed at B and C and suspended from the points A and D which are in the same horizontal line and at a distance a apart. Prove that when the rods move in a vertical plane, the length of equivalent pendulum is 5a/6.*

Sol. Here the system forms a compound pendulum swinging about the horizontal AD with rods AB, CD vertical and BC horizontal. Let m be mass of each rod. Let G_1, G_2, G_3 be centres of gravity of rods AB, BC and CD respectively. Given $AB = BC = CD = a$. Hence $AG_1 = DG_3 = a/2$ and $EG_2 = a$. Let k be the radius of the gyration of the system about AD. Then

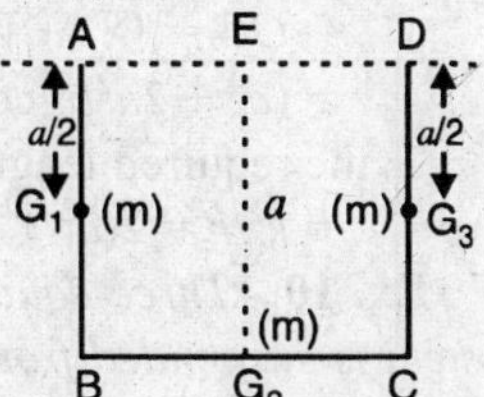

M.I. of the system about AD = sum of the moments of inertia of three rods about AD

or $$3mk^2 = (4/3)\, m(a/2)^2 + ma^2 + (4/3)m\,(a/2)^2 \quad \Rightarrow \quad k^2 = 5a^2/9$$

Let h be the depth of the C.G. of the system below AD.

Then $$h = \frac{m \cdot AG_1 + m \cdot EG_2 + m \cdot DG_3}{m + m + m} = \frac{m(a/2) + ma + m(a/2)}{3m} = \frac{2a}{3}$$

$\therefore$ length of simple equivalent pendulum $= k^2/h = 5a/6$

Ex. 8. *A uniform triangular lamina can oscillate in its own plane about the angle A. Prove that the length of the simple equivalent pendulum is $\{3(b^2 + c^2) - a^2\}/4\{2(b^2 + c^2) - a^2\}^{1/2}$, the axis through A being horizontal.* **(Purvanchel 2005)**

Sol. Here the given lamina ABC oscillates in its own plane about a line AH, which is perpendicular to the plane of the lamina. Let M be the mass of the lamina. Let k be the radius of gyration of ΔABC about AH.

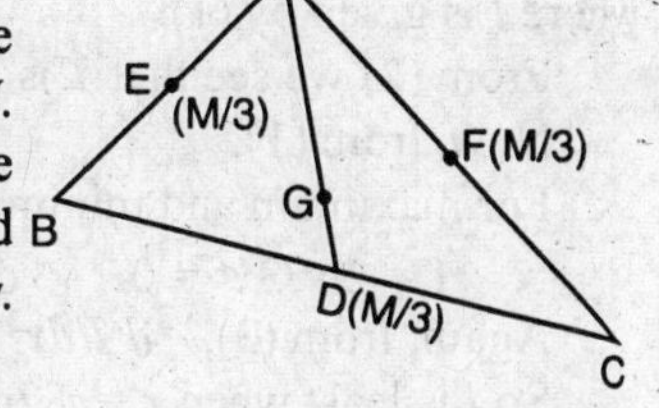

We know that M.I. of the triangle ABC about axis AH is equal to the sum of the moments of inertia of three particles each of mass $M/3$ placed at the middle points E, D and F of the sides AB, BC and CA respectively.

$$\therefore \quad Mk^2 = (M/3)AE^2 + (M/3)AD^2 + (M/3)AF^2 \quad \ldots(1)$$

Now, $AE = (1/2)AB = c/2$, $\quad AF = (1/2)AC = b/2$

Also, by geometry, $AB^2 + AC^2 = 2(AD^2 + BD^2)$ and $BD = a/2$

$\therefore \quad c^2 + b^2 = 2(AD^2 + a^2/4)$ so that $AD^2 = (2b^2 + 2c^2 - a^2)/4$ ...(2)

So (1) gives $Mk^2 = (M/3) \times (c^2/4) + (M/3) \times (1/4) \times (2b^2 + 2c^2 - a^2) + (M/3) \times (b^2/4)$

or $k^2 = (1/12) \times (c^2 + 2b^2 + 2c^2 - a^2 + b^2) = (1/12) \times \{3(b^2 + c^2) - a^2\}$

Also, h = Distance of C.G. G of the lamina below AH

$= AG = (2/3)AD = (1/3)\,(2b^2 + 2c^2 - a^2)^{1/2}$, using (2)

$\therefore$ length of the simple equivalent pendulum

$= k^2/h = \{3(b^2 + c^2) - a^2\}/4\{2(b^2 + c^2) - a^2\}^{1/2}$

Ex. 9. ***A bent lever, whose arms are of lengths a and b, the angle between them being α, makes a small oscillations in its own plane about the fulcrum, show that the length of the correspoinding simple pendulum is $2(a^3 + b^3)/3(a^4 + 2a^2b^2 \cos\alpha + b^4)^{1/2}$.*** **[Kanpur 2008; Meerut 2008]**

Sol. Let the bent lever with arms $OA(= a)$ and $OB(= b)$ and $\angle AOB = \alpha$, oscillate in its own plane about the fulcrum O. Let G_1 and G_2 be centres of gravity of OA and OB respectively. Take OA and a perpendicular line OY as the axes of x and y. Let $G_2N \perp OX$. Then, $ON = (b/2)\cos\alpha$, $NG_2 = (b/2)\sin\alpha$. Then the coordinates of G_1 and G_2 are $(a/2, 0)$ and $((b/2)\cos\alpha, (b/2)\sin\alpha)$ respectively. Let w be the weight per unit length of each arm so that weights of OA and OB are aw and bw respectively.

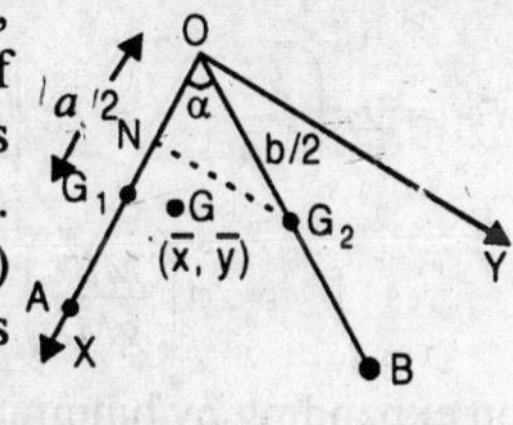

Let k be the radius of gyration about the axis of rotation through O.

Then, $$(aw + bw)k^2 = (4/3)aw\cdot(a/2)^2 + (4/3)bw\cdot(b/2)^2$$

Thus, $$k^2 = (a^3 + b^3)/3(a + b)$$

Let $(\bar{x}, \bar{y})$ be the coordinates of C.G. 'G' of the lever from O.

Then, $\bar{x} = \dfrac{aw\cdot(a/2)+bw\cdot(b/2)\cos\alpha}{aw+bw}$, $\bar{y} = \dfrac{aw\cdot 0+bw\cdot(b/2)\sin\alpha}{aw+bw}$

Thus, $\bar{x} = (a^2 + b^2 \cos\alpha)/2(a+b)$, $\bar{y} = b^2 \sin\alpha/2(a+b)$

$\therefore$ h = perpendicular distance of G from the axis of rotation

$= OG = (\bar{x}^2 + \bar{y}^2)^{1/2} = \{(a^2 + b^2\cos\alpha)^2 + (b^2\sin\alpha)^2\}^{1/2}/2(a+b)$

$= (a^4 + 2a^2b^2\cos\alpha + b^4)^{1/2}/2(a+b)$

$\therefore$ the required length of the simple equivalent pendulum

$= k^2/h = 2(a^3 + b^3)/3(a^4 + 2a^2b^2\cos\alpha + b^4)^{1/2}$

Ex. 10. *Three equal particles are attached to a weighless rod at equal distance apart. The system is suspended from, and is free to turn about, a point of the rod distance x from the middle particle. Find the time of a small oscillation and show that it is least when x = .82a nearly.*

Sol. Let there be a weightless rod AC on which are attached three equal particles of mass m each at A, B and C so that $AB = BC = a$. Let the system rotate about the axis OD, such that $OB = x$. Let k be the radius of gyration of the system about OD. Then

$$3mk^2 = mx^2 + m(a+x)^2 + m(a-x)^2 \quad \text{or} \quad k^2 = (3x^2 + 2a^2)/3$$

A(m)
a – x
O — D
x
B(m)
a
C(m)

Also, h = distance of C.G. of the system from $O = OB = x$

Now, l = the length of the equivalent simple pendulum

or $l = k^2/h = (3x^2 + 2a^2)/3x = x + 2a^2/3x$ ...(1)

Hence T = the required time of small oscillation $= 2\pi(l/g)^{1/2}$, ...(2)

where l is given by (1).

From (2) we see that T is least if l is least.

Now, from (1), $dl/dx = 1 - 2a^2/3x^2$ ...(3)

For maximum and minimum value of l, $dl/dx = 0$

$\therefore$ $1 - 2a^2/3x^2 = 0$ or $x = a\sqrt{6}/3 = (2.499/3)a = .82a$ nearly

Again, from (3), $d^2l/dx^2 = 4a^2/3x^2$, which is positive

So l is least when $x = .82a$ and T given by (2) is least when $x = .82a$ nearly.

Ex. 11. *A simple circular pendulum is formed of a mass M suspended from a fixed point by a weightless wire of length l. If a mass m, very small compared with M, be knotted on to the wire at a distance a from the point of suspension, show that the time of small vibration of the pendulum is approximately diminished by (ma/2Ml) × (1 – a/l) of itself.* **[Kanpur 1996]**

Sol. Let T_1 be the time of oscillation before knotting the mass m at point B(where $OB = a$) of the given wire $OA(= l)$. Then, we have

$$T_1 = 2\pi(l/g)^{1/2} \qquad ...(1)$$

O
a
B (m)
A (M)

After knotting the mass m at B, let k be the radius of gyration of the system about the axis of rotation through O and h be the distance of C.G. of the system from O. Then, we have

$(m+M)k^2 = M\cdot l^2 + ma^2$ so that $k^2 = (Ml^2 + ma^2)/(m+M)$

and $h = (Ml + ma)/(M+m)$

Let T_2 be the time of oscillation of the pendulum in second case. Then

$$T_2 = 2\pi\left(\frac{k^2}{gh}\right)^{1/2} = 2\pi\left\{\frac{Ml^2 + ma^2}{g(Ml + ma)}\right\}^{1/2}$$

$$= 2\pi\left(\frac{Ml^2}{gMl}\right)^{1/2}\left(1 + \frac{ma^2}{Ml^2}\right)^{1/2} \times \left(1 + \frac{ma}{Ml}\right)^{-1/2} = 2\pi\left(\frac{l}{g}\right)^{1/2}\left(1 + \frac{ma^2}{2Ml^2}\right)\left(1 - \frac{ma}{2Ml}\right),$$

on expanding by binomial theorem and neglecting squares and higher powers of small quantity m/M.

$$\therefore \quad T_2 = T_1\left(1 - \frac{ma}{2Ml} + \frac{ma^2}{2Ml^2}\right) = T_1 - \left(\frac{ma}{2Ml} - \frac{ma^2}{2Ml^2}\right)T_1$$

Hence $T_1 - T_2 = (ma/2Ml) \times (1 - a/l)T_1$,

showing that the period T_1 is diminished by $(ma/2Ml) \times (1 - a/l)$ of itself.

Ex. 12. **(a)** *A solid homogeneous cone of height h and semi-vertical angle α, oscillate about a diameter of its base. Show that the length of the simple equivalent pendulum is $(h/5) \times (2 + 3 \tan^2 \alpha)$.*
[Purvanchel 2006; Agra 2006: Kanpur 2005. 07, 11; Meerut 2006, 2007, 09

(b) *A solid homogeneous cone of height h and vertical angle 2α oscillates about a horizontal axis through its vertex. Show that the length of the simple equivalent pendulum is $(h/5) \times (4 + \tan^2 \alpha)$.*
[Purvanchel 2005; Kanpur 2006; Garhwal 1994; Meerut 2004]

Sol. **(a)** Let O be the vertex of the given cone of OAB which oscillates about a diameter AB of its base. Let M be mass and ρ mass per unit volume of the cone. Then

$$M = (1/3)\pi h^3 \tan^2 \alpha \cdot \rho. \quad ...(1)$$

Consider an elementary disc FE of thickness δx, parallel to AB and at a distance x from O. Then, we have

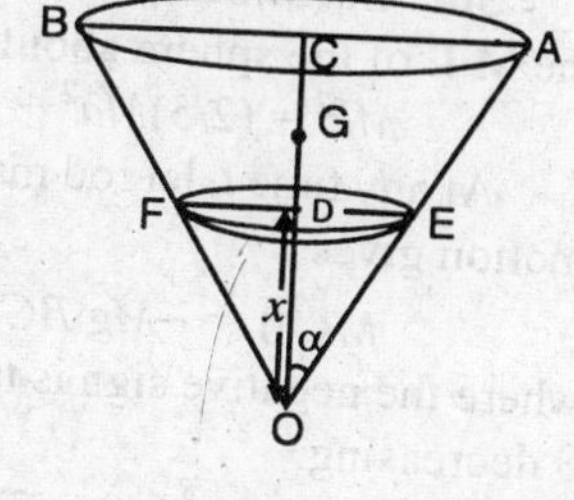

δm = mass of the disc $= \rho \pi x^2 \tan^2 \alpha \; \delta x$

∴ M.I. of the disc about $AB = (1/4)\, \delta m \cdot DE^2 + \delta m \cdot DC^2$
$= \rho \pi x^2 \tan^2 \alpha \, \{(1/4) x^2 \tan^2 \alpha + (h - x)^2\}$

Let k be the radius of gyration of the cone about AB

∴ Mk^2 = M.I. of the cone about AB

$$= \int_0^h \rho \pi \tan^2 \alpha \{(1/4) x^4 \tan^2 \alpha + (h - x)^2 x^2\}\, dx$$

$$= \frac{1}{4} \rho \pi \tan^2 \alpha \int_0^h (x^4 \tan^2 \alpha + 4h^2 x^2 - 8hx^3 + 4x^4)\, dx$$

$$= (1/4) \rho \pi \tan^2 \alpha \, \{(1/5) h^5 \tan^2 \alpha + (4/3) h^5 - 2h^5 + (4/5) h^5\}$$

$$= (1/60) \rho \pi \tan^2 \alpha \cdot h^5 (3 \tan^2 \alpha + 2) = (M/20) h^2 (3 \tan^2 \alpha + 2), \text{ by (1)}$$

Thus, $\quad k^2 = (h^2/20) \times (3 \tan^2 \alpha + 2)$

Also, $\quad h'$ = the distance of C.G. 'G' of the cone below $AB = h/4$

∴ length of simple equivalent pendulum $= k^2/h' = (h/5) \times (2 + 3 \tan^2 \alpha)$

(b) Let O be the vertex of the given cone OAB which oscillates about the horizontal axis OL through O. Let M be mass and ρ mass per unit volume of the cone. Then

$$M = (1/3)\pi h^3 \tan^2 \alpha \cdot \rho. \quad ...(1)$$

Consider an elementary disc EF of thickness δx, parallel to OL and at a distance x from O. Then, we have

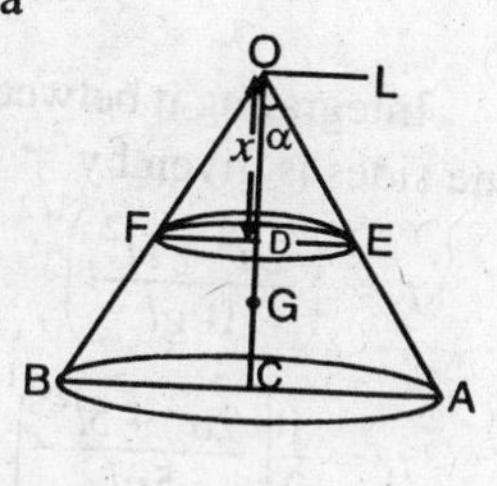

δm = mass of the disc $= \rho \pi x^2 \tan^2 \alpha \; \delta x$

∴ M.I. of the disc about $OL = (1/4)\, \delta x \cdot DE^2 + \delta m \cdot OD^2$
$= \rho \pi x^2 \tan^2 \alpha \, \{(1/4) x^2 \tan^2 \alpha + x^2\}$

Let k be the radius of gyration of the cone about OL

∴ Mk^2 = M.I. of the cone about OL

$$= \int_0^h \rho \pi \tan^2 \alpha \left(\frac{1}{4} \tan^2 \alpha + 1\right) x^4 dx = \frac{1}{5} \rho \pi h^5 \tan^2 \alpha \left(\frac{1}{4} \tan^2 \alpha + 1\right)$$

Then, by (1), $\quad Mk^2 = (3/20)\, M (\tan^2 \alpha + 4) h^2$ or $k^2 = (3h^2/20) \times (\tan^2 \alpha + 4)$

Also, $\quad h'$ = the distance of C.G. 'G' of the cone below $OL = 3h/4$

∴ length of simple equivalent pendulum $= k^2/h' = (h/5) \times (4 + \tan^2 \alpha)$

Ex. 13. *A weightless straight rod ABC, of length 2a, is movable about the end A which is fixed and carries two particles of the same mass, one fastened to the middle point C and the other to the end B of the rod. If the rod be held in a horizontal position and then let go, show that its angular velocity when vertical is $(6g/5a)^{1/2}$ and that 5a/3 is the length of the simple equivalent pendulum.*

Sol. Here the given rod ACB starts from horizontal position AX. Let it be inclined at an angle θ from the vertical position at any time t. Given $AC = CB = a$. Let two particles of mass m each be placed at C and B. Let k be the radius of gyration of the system about axis of rotation. Then

A - - - - - - - X
θ a
C (m)
a
D B(m)

$2mk^2 = ma^2 + m \times (2a)^2 \qquad$ so that $\qquad k^2 = 5a^2/2$

Also, h = distance of C.G. of the system from $A = (ma + m \cdot 2a)/(m + m) = 3a/2$

$\therefore$ l = length of the equivalent simple pendulum = $k^2/h = 5a/3$

Now, by energy equation, we have

Change in K.E. = work done or $(1/2)\cdot(2mk^2\dot{\theta}^2) = 2mg\cdot(3/2)a\cos\theta$

or $(5/2)a^2\dot{\theta}^2 = 3ga\cos\theta$ or $\dot{\theta}^2 = (6g/5a)\cos\theta$, ...(1)

which gives the angular velocity of rod in any position. Putting $\theta = 0$, the angular velocity when the rod is vertical is given by $\dot{\theta}^2 = 6g/5a$ or $\dot{\theta} = (6g/5a)^{1/2}$.

Ex. 14. *A sphere of radius a, is suspended by a fine wire from a fixed point at a distance l from its centre, show that the time of small oscillation is given by* $2\pi\left(\frac{5l^2+2a^2}{5lg}\right)^{1/2}\left\{1+\frac{1}{4}\sin^2\left(\frac{1}{2}\alpha\right)\right\}$, *where α represents the amplitude of the vibration.* **[Kanpur 2000, 09; Meerut 2012; Rohilkhand 94]**

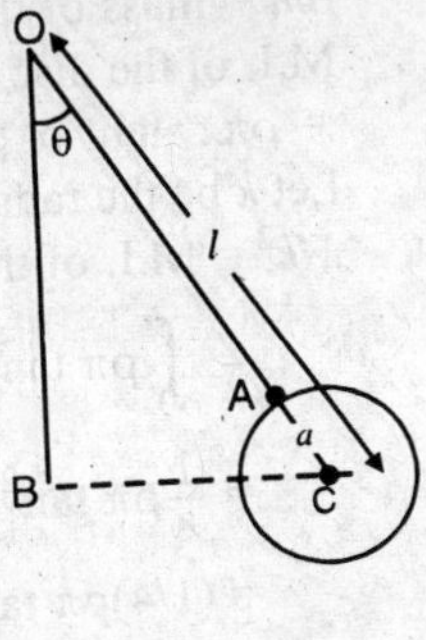

Sol. Let M be the mass of the sphere of centre C such that $OC = l$. If Mk^2 is the M.I. of the sphere about the axis of rotation through O, then

$$Mk^2 = (2/5)Ma^2 + Ml^2 = M(l^2 + 2a^2/5)$$

At any time t, let rod make an angle θ with the vertical. Then the equation of motion gives

$$Mk^2\ddot{\theta} = -Mg\cdot BC \quad \text{or} \quad M(l^2 + 2a^2/5)\,\ddot{\theta} = -Mg\cdot l\sin\theta,$$

where the negative sign is taken because the moment of Mg is in the direction of θ decreasing.

Since $\ddot{\theta} = \dot{\theta}(d\dot{\theta}/d\theta)$, we get

$$\dot{\theta}\frac{d\dot{\theta}}{d\theta} = -\frac{5gl}{2a^2+5l^2}\sin\theta \quad \text{or} \quad 2\dot{\theta}\,d\dot{\theta} = -\frac{10gl\sin\theta}{2a^2+5l^2}\,d\theta$$

Integrating, $\dot{\theta}^2 = (10\,gl\cos\theta)/(2a^2 + 5l^2) + C$ ...(1)

But when $\theta = \alpha$, $\dot{\theta} = 0$. So (1) gives $C = -(10\,gl\cos\alpha)/(2a^2 + 5l^2)$

$\therefore$ (1) becomes $\dot{\theta}^2 = 10\,gl\,(\cos\theta - \cos\alpha)/(2a^2 + 5l^2)$

so that $\dot{\theta} = d\theta/dt = \{10\,gl\,(\cos\theta - \cos\alpha)/(2a^2 + 5l^2)\}^{1/2}$

or $$dt = \left(\frac{2a^2+5l^2}{10gl}\right)^{1/2}\frac{d\theta}{(\cos\theta - \cos\alpha)^{1/2}}$$

Integrating it between $\theta = 0$ to $\theta = \alpha$, time t_1 from the vertical position to the extreme position on one sides is given by

$$t_1 = \left(\frac{2a^2+5l^2}{10gl}\right)^{1/2}\int_0^{\alpha}\frac{d\theta}{\{1-2\sin^2\theta/2-(1-2\sin^2\alpha/2)\}^{1/2}}$$

$$t_1 = \frac{1}{2}\left(\frac{2a^2+5l^2}{5gl}\right)^{1/2}\int_0^{\alpha}\frac{d\theta}{(\sin^2\alpha/2-\sin^2\theta/2)^{1/2}} \quad ...(2)$$

Let $\sin\theta/2 = \sin\alpha/2\sin\phi$ so that $(1/2)\cos\theta/2\,d\theta = \sin\alpha/2\cos\phi\,d\phi$ ...(3)

and $\{(2a^2 + 5l^2)/5gl\}^{1/2} = A$ ...(4)

$$\therefore \text{(2) becomes} \quad t_1 = A\int_0^{\pi/2}\frac{\cos\phi\sin\alpha/2\,d\phi}{\cos\theta/2\cdot\cos\phi\sin\alpha/2} = A\int_0^{\pi/2}\frac{d\phi}{\cos\theta/2}$$

or $$t_1 = A\int_0^{\pi/2}\frac{d\phi}{(1-\sin^2\theta/2)^{1/2}} = A\int_0^{\pi/2}\left(1-\sin^2\frac{\theta}{2}\right)^{-1/2}d\phi$$

$$= A\int_0^{\pi/2}\left(1-\sin^2\frac{\alpha}{2}\sin^2\phi\right)^{-1/2}d\phi\text{, using (3)}$$

$$= A\int_0^{\pi/2}\left(1+\frac{1}{2}\sin^2\frac{\alpha}{2}\sin^2\phi\right)d\phi\text{,}$$ [expanding by binomial theorem and neglecting higher

powers of small quantity $\sin \alpha/2$ as α is very small]

Thus, $t_1 = A[\pi/2 + (1/2) \sin^2 \alpha/2 \cdot (\pi/4)]$,

$\therefore$ Time of one small complete oscillation

$$= 4t_1 = 4\cdot(\pi/2)A\ \{1 + (1/4)\sin^2 \alpha/2\}$$

$$= 2\pi\{(5l^2 + a^2)/5gl\}^{1/2}\ \{1 + (1/4)\sin^2 \alpha/2\}, \text{ by (4)}$$

EXERCISE

1. A pendulum is supported at O, and P is the centre of oscillation. Show that, if an additional weight is rigidly attached at P, the period of oscillation is unaltered.

Hint : Refer remark of Art. 16.7.

2. A flat circular disc of radius a has a hole in it of radius b whose centre is at a distance c from the centre of the disc ($c < a - b$). The disc is free to oscillate in a vertical plane about a smooth horizontal circular rod of radius b passing through the hole. Show that the length of the equivalent pendulum is $c + (a^4 - b^4)/2a^2c$.

3. A uniform rod of mass m and length $2a$ can oscillate about a horizontal axis through one end. A circular disc of mass $24m$ and radius $a/3$ can have its centre clamped in any point of the rod and its plane contains the axis of rotation. Show that, for oscillation under gravity the length of the simple equivalent pendulum lies between $a/2$ and $2a$.

4. A uniform wire, in the form of an arc of a circle of given radius; is swinging about a horizontal axis through the middle point of the arc perpendicular to the plane of the arc. Show that the time of small oscillation is independent of the length of the arc, and the length of an equivalent simple pendulum is equal to the diameter of the circle.

5. Two bodies can move freely and independently under the action of gravity about the same horizontal axis, their masses are m, m' and the distances of their centres of gravity from the axis are h and h'. If the lengths of their equivalent simple pendulum be l, l', prove that when fastened together the length of the equivalent simple pendulum will be $(mhl + m'h'l')/(mh + m'h')$.

16.10 Reaction of the axis of rotation. *To find the reaction of the axis of rotation when a rigid body is rotating about it, given that the forces and the body are symmetrical with respect to the plane through the centre of gravity perpendicular to the fixed axis and gravity is the only external force.*

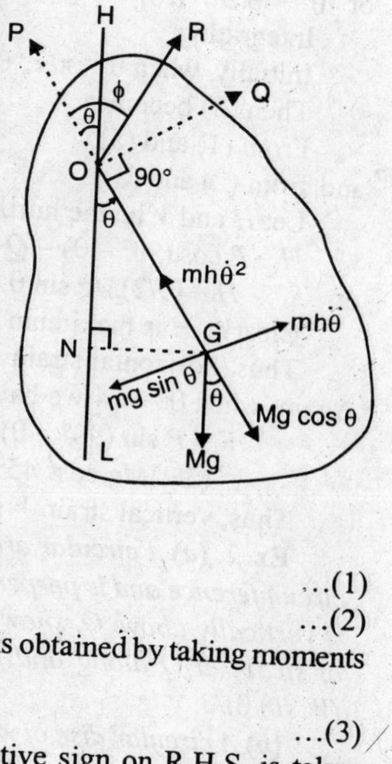

Here both the forces and the body are symmetrical with respect to the plane through the centre of gravity 'G' perpendicular to the fixed axis *i.e.,* with respect to the plane of the paper and let O be the point where the axis of rotation meets this plane.

By symmetry, the actions of the axis on the body must reduce to a single force R acting at O in the plane of the paper; let the components of this single force be P and Q, along and perpendicular to GO produced. Let the direction of R make an angle ϕ with GO produced.

Since G describes a circle around O of radius $h(= OG)$, hence the components of its acceleration along and perpendicular to GO will be $h\dot{\theta}^2$ and $h\ddot{\theta}$ respectively. We know that (Refer Art. 15.5, Chapter 15) the motion of the centre of gravity G is the same as it would be if it were a particle of mass M acted on by all the external forces applied to it parallel to their original directions. Hence equations of motion of the C.G. 'G' are given by

$$Mh(d\theta/dt)^2 = P - Mg\cos\theta \quad \text{...(1)}$$

and

$$Mh(d^2\theta/dt^2) = Q - Mg\sin\theta \quad \text{...(2)}$$

Again, the equation of motion of the body about the axis of rotation is obtained by taking moments about the axis of rotation through O. Thus,

$$Mk^2(d^2\theta/dt^2) = -Mg\cdot GN \quad \text{or} \quad Mk^2(d^2\theta/dt^2) = -Mg\cdot h\sin\theta, \quad \text{...(3)}$$

where k is the radius of gyration about the axis of rotation and negative sign on R.H.S. is taken because the moment of Mg is in the direction of θ decreasing.

From (1), (2) and (3) we get P, Q, R and ϕ as follows.

Eliminating $d^2\theta/dt^2$ between (2) and (3), we get Q.

If (3) be integrated and the resulting constant determined from the initial condition, then, we get P by using (1).

$\therefore \quad R = \text{resultant reaction at } O = (P^2 + Q^2)^{1/2}$

and $\quad \tan\phi = Q/P \quad$ so that $\quad \phi = \tan^{-1}(Q/P)$

Remark. On resolving P and Q horizontally and vertically, the components of reaction in horizontal and vertical directions are given by $P \sin\theta - Q\cos\theta$ and $P\cos\theta + Q\sin\theta$ respectively.

TYPE 3 : SOLVED EXAMPLES

Ex. 1. *A thin uniform rod has one end attached to a smooth hinge and is allowed to fall from the horizontal position. Show that the horizontal strain on the hinge is greatest when the rod is inclined at an angle of 45° to the vertical, and that the vertical strain is then 11/8 times the weight of the rod.* **[Meerut 2005, 09, 10]**

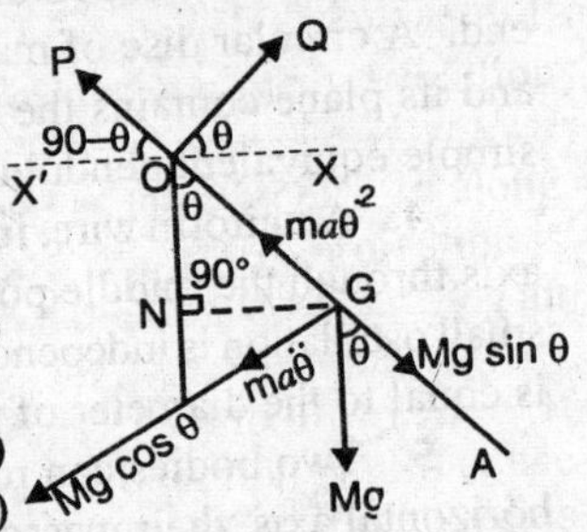

Sol. Let $2a$ be the length and M the mass of the rod OA, attached to a smooth hinge at O and held initially in the horizontal position XOX'. Let the rod make an angle θ with vertical at any time t. The centre of gravity 'G' of the rod describes a circle of radius a about O. Let P, Q be the components of the single force of reaction along and perpendicular to GO. Then the equations of motion of G along and perpendicular to GO are given by

$$Ma(d\theta/dt)^2 = P - Mg\cos\theta \quad \ldots(1)$$

and $$Ma(d^2\theta/dt^2) = Q - Mg\sin\theta \quad \ldots(2)$$

Let k be the radius of gyration of the rod about a fixed axis through O and perpendicular to the rod. Then

$$Mk^2 = (4/3)Ma^2 \quad \text{so that} \quad k^2 = (4/3)a^2$$

Taking moments about O, the equation of motion of the rod is

$$Mk^2(d^2\theta/dt^2) = -Mg{\cdot}GN \quad \text{or} \quad M(4a^2/3)(d^2\theta/dt^2) = -Mg{\cdot}a\sin\theta \quad \ldots(3)$$

or $$\ddot{\theta} = \dot{\theta}(d\dot{\theta}/d\theta) = -(3g/4a)\sin\theta \quad \text{or} \quad 2\dot{\theta}\,d\dot{\theta} = -(3g/2a)\sin\theta\,d\theta \quad \ldots(4)$$

Integrating, $$\dot{\theta}^2 = (3g/2a)\cos\theta + C$$

Initially, when $\theta = \pi/2$, $\dot{\theta} = 0$. So (3) gives $C = 0$

Then (3) becomes $$(d\theta/dt)^2 = (3g/2a)\cos\theta \quad \ldots(5)$$

From (1) and (5), $$P = (5/2)Mg\cos\theta \quad \ldots(6)$$

and from (2) and (3), $$Q = (1/4)Mg\sin\theta \quad \ldots(7)$$

Let H and V be the horizontal and vertical strains on the hinge. Then,

$$H = P\cos(90^\circ - \theta) - Q\cos\theta = (5/2)Mg\cos\theta\sin\theta - (1/4)Mg\sin\theta\cos\theta$$

or $$H = (9/2)Mg\sin\theta\cos\theta \quad \text{or} \quad H = (9/8)Mg\sin 2\theta \quad \ldots(8)$$

Clearly, H is maximum when $\sin 2\theta = 1$ or $2\theta = \pi/2$ or $\theta = \pi/4$

Thus, horizontal strain H is maximum when the rod is inclined at angle of 45° to the vertical.

Again, when $\theta = \pi/4$ we have

$$V = P\sin(90^\circ - \theta) + Q\sin\theta = (5/2)Mg\cos^2\theta + (1/4)Mg\sin^2\theta$$
$$= (5/2)Mg\cos^2 45^\circ + (1/4)Mg\sin^2 45^\circ = (11/8)Mg, \text{ as } \theta = 45^\circ$$

Thus, vertical strain V is 11/8 times the weight of the rod.

Ex. 2. (a) *A circular area can turn about a horizontal axis which passes through a point O of its circumference and is perpendicular to its plane. If motion commences when the diameter through O is vertically above O, show that when the diameter has turned through an angle θ, the components of strain at O along and perpendicular to this diameter are respectively W(7 cos θ − 4)/3 and (W sin θ)/3.* **[Meerut 1993]**

(b) *A circular disc of weight W can turn freely about a horizontal axis perpendicular to its plane which passes through a point O on its circumference. If it starts from rest with the diameter vertically* *above O, show that the resultant pressure on the axis when that diameter is horizontal and vertically below O are respectively (W/3) $\sqrt{17}$ and 11 W/3. Further prove that the axis must be able to bear at least 11/3 times the weight of the disc.* **[Kanpur 2009]**

Sol. **(*a*)** Let a be the radius of the disc of mass M. Initially the diameter OA is vertically above O along OL. See fig. (*i*). Let at any time t diameter OA make an angle θ with the vertical as shown in fig. (*ii*). Let P and Q be the components of the strain at O along and perpendicular to OG, where G is the centre of the disc. Then the equations of motion of G along and perpendicular to GO are given by

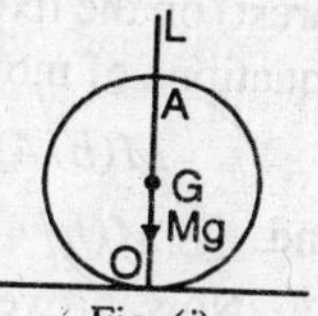

Fig. (*i*)

$$Ma(d\theta/dt)^2 = Mg\cos\theta - P \qquad ...(1)$$

and
$$Ma(d^2\theta/dt^2) = Mg\sin\theta - Q \qquad ...(2)$$

Let k be the radius of gyration of the disc about an axis through O perpendicular to the disc. Then,

$$Mk^2 = M(a^2/2) + Ma^2 \quad \text{or} \quad k^2 = (3/2)a^2$$

Hence, taking moments about the axis of rotation, we get

$$Mk^2(d^2\theta/dt^2) = Mg{\cdot}ON$$

or
$$M \times (3a^2/2) \times (d^2\theta/dt^2) = Mg{\cdot}a\sin\theta$$

so that
$$d^2\theta/dt^2 = (2g/3a)\sin\theta \qquad ...(3)$$

Fig. (*ii*)

Re-writing (3), $\dot\theta(d\dot\theta/d\theta) = (2g/3a)\sin\theta$

or
$$2\dot\theta\, d\dot\theta = (4g/3a)\sin\theta\, d\theta$$

Integrating it,
$$\dot\theta^2 = -(4g/3a)\cos\theta + C \qquad ...(4)$$

But initially, when $\theta = 0$, then $\dot\theta = 0$. So (4) gives $C = 4g/3a$

So (4) becomes
$$(d\theta/dt)^2 = (4g/3a)(1 - \cos\theta) \qquad ...(5)$$

From (1) and (5), $P = Mg\cos\theta - (4Mg/3)(1 - \cos\theta) = (W/3)(7\cos\theta - 4)$

[$\because$ Weight of disc $= W = Mg$]

From (2) and (3), $\quad Q = Mg\sin\theta - (2Mg/3)\sin\theta = (W/3)\sin\theta$

(*b*) Proceed first as in part (*a*) to show that the components of the reaction on the axis of rotation, along and perpendicular to the diameter OA are given by

$$P = (W/3)(7\cos\theta - 4) \quad \text{and} \quad Q = (W/3)\sin\theta \qquad ...(6)$$

To find the resultant pressure on the axis when the diameter is horizontal. In this situation of the rod, $\theta = \pi/2$. So (6) gives $P = -(4/3)W$ and $Q = W/3$. Hence the required resultant pressure

$$= (P^2 + Q^2)^{1/2} = \{(16/9 + 1/9)W^2\}^{1/2} = (W/3)\sqrt{17}.$$

To find the resultant pressure on the axis when the diameter OA is vertically below. In this situation $\theta = \pi$. So (6) gives $P = -(11W/3)$ and $Q = 0$. Hence the resultant pressure $= (P^2 + Q^2)^{1/2} = P = (11/3)W$, numerically.

Last part of the problem. The resultant pressure R on the axis of rotation at any time t is given by

$$R = (P^2 + Q^2)^{1/2} = [(W^2/9)\{(7\cos\theta - 4)^2 + \sin^2\theta\}]^{1/2}, \text{ by (6)}$$
$$= (W/3)(48\cos^2\theta - 56\cos\theta + 17)^{1/2},$$

which is clearly maximum when $\cos\theta = -1$ *i.e*, when $\theta = \pi$.

$\therefore$ Maximum pressure of the axis $= (W/3)(48\cos^2\pi - 56\cos\pi + 17) = 11W/3$

So the axis must be able to bear at least 11/3 times weight of the disc.

Ex. 3. *A right cone of vertical angle 2α can turn freely about an axis passing through the centre of its base and perpendicular to the axis; if the cone starts from rest with its axis horizontal, show tha when, axis is vertical, the thrust on the fixed axis to the weight of the cone is*

$$\{1 + (1/2)\cos^2\alpha\} : \{1 - (1/3)\cos^2\alpha\}.$$

Sol. Let the given cone be of mass M and let it turn freely round a horizontal diameter AB of the base. Let horizontal line OX be the initial position of the axis of the cone. Now, let afte time t, the axis OC of the cone make an angle θ with OX. Let h be the height of the cone and G its centre of gravity. Then $OG = h/4$. So G will describe a circle of radius $h/4$ round O. Let P and Q be the components of the reaction (*i.e.*, the thrust) on the fixed axis AOB, along and perpendicular to GO. Then equations of motion are given by

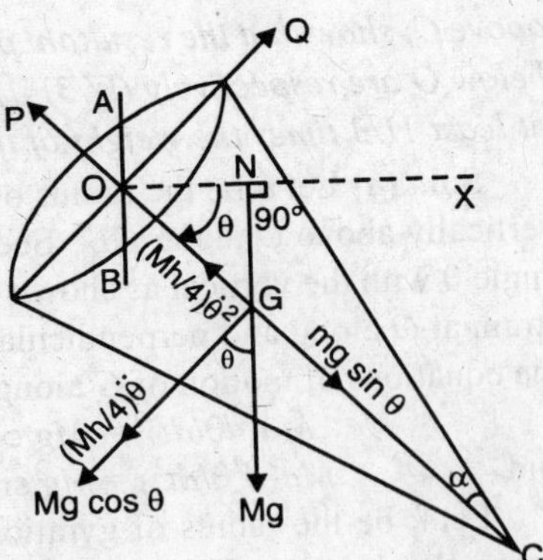

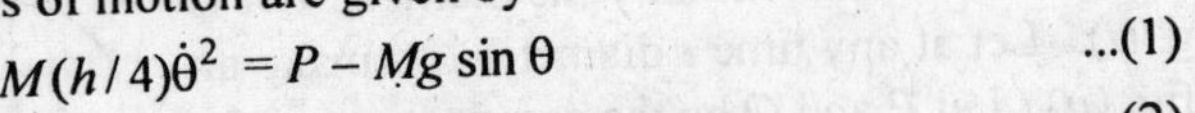

$$M(h/4)\dot{\theta}^2 = P - Mg\sin\theta \quad \text{...(1)}$$

and $$M(h/4)\ddot{\theta} = Mg\cos\theta - Q \quad \text{...(2)}$$

Now, prove as in Ex. 12(*a*), page 16.13 that

$$k^2 = h^2(2 + 3\tan^2\alpha)/20$$

Hence moment equation about O is given by

$$Mk^2\ddot{\theta} = Mg{\cdot}ON$$

or $$(M/20)h^2(2 + 3\tan^2\alpha)\ddot{\theta} = Mg{\cdot}(h/4)\cos\theta$$

or $$h\ddot{\theta} = \{5g/(2 + 3\tan^2\alpha)\}\cos\theta \quad \text{...(3)}$$

From (3), $$h\dot{\theta}\frac{d\dot{\theta}}{d\theta} = \frac{5g\cos\theta}{2 + 3\tan^2\theta} \quad \text{or} \quad 2h\dot{\theta}\,d\dot{\theta} = \frac{10g\cos\theta\,d\theta}{2 + 3\tan^2\alpha}$$

Integrating it, $$h\dot{\theta}^2 = (10g\sin\theta)/(2 + 3\tan^2\alpha) + C \quad \text{...(4)}$$

Initially when OC is horizontal, $\theta = 0$ and $\dot{\theta} = 0$. So (4) gives $C = 0$

$\therefore$ (4) $$h\dot{\theta}^2 = (10g\sin\theta)/(2 + 3\tan^2\alpha) \quad \text{...(5)}$$

Substituting the value of $h\dot{\theta}^2$ from (5) in (1), we get

$$P = \frac{M}{4}\cdot\frac{10g\sin\theta}{2 + 3\tan^2\alpha} + Mg\sin\theta = \frac{9 + 6\tan^2\alpha}{2(2 + 3\tan^2\alpha)}\cdot Mg\sin\theta$$

Next, substituting the value of $h\ddot{\theta}$ from (3) in (2), we get

$$Q = Mg\cos\theta - \frac{M}{4}\cdot\frac{5g\cos\theta}{2 + 3\tan^2\alpha} = \frac{3 + 12\tan^2\alpha}{4(2 + 3\tan^2\alpha)}\cdot Mg\cos\theta$$

When the axis of the cone is vertical *i.e.*, when $\theta = \pi/2$, then

$$P = (9 + 6\tan^2\alpha)Mg/2(2 + 3\tan^2\alpha) \quad \text{and} \quad Q = 0$$

Hence the required resultant thrust R on the axis in this position is given by

$$R = (P^2 + Q^2)^{1/2} = P = \frac{9 + 6\tan^2\alpha}{2(2 + 3 + \tan^2\alpha)}\cdot Mg = \frac{9\cos^2\alpha + 6\sin^2\alpha}{2(2\cos^2\alpha + 3\sin^2\alpha)}\cdot Mg$$

$$\therefore \frac{R}{Mg} = \frac{6(\cos^2\alpha + \sin^2\alpha) + 3\cos^2\alpha}{2[3(\cos^2\alpha + \sin^2\alpha) - \cos^2\alpha]} = \frac{6 + 3\cos^2\alpha}{2(3 - \cos^2\alpha)} = \frac{1 + (1/2)\cos^2\alpha}{1 - (1/3)\cos^2\alpha}$$

Ex. 4. *A uniform semi-circular arc of mass m and radius a, is free at its ends to two points in the same vertical line and is rotating with constant angular velocity ω. Show that the horizontal thrust on the upper end is* $m(g + \omega^2 a)/\pi$. **[Agra 2009, 12; Kanpur 2007; Meerut 2007, 2008, 12]**

Sol. Here the given semi-circular arc ACB, of mass m, centre O and radius a is rotating with constant angular velocity ω about the vertical line AB, A, B being the ends of the arc. We know that $OG = 2a/\pi$, where G is the centre of gravity of the circular arc ACB. Now, G will describe about O a circle of radius $2a/\pi$ with constant angular velocity ω. Since ω is constant, therefore the only effective force on G is $m(2a/\pi)\omega^2$ along GO.

Let the horizontal and vertical components of reaction at the upper end A be P and Q and those at the lower end be P' and Q' respectively.

Taking moment about B, we have

Moment of the effective forces = Moment of the external forces

$$\therefore \quad m(2a/\pi)\omega^2{\cdot}BO = P{\cdot}BA - mg{\cdot}OG$$

or $$m(2a/\pi)\omega^2 a = P{\cdot}2a - mg{\cdot}(2a/\pi) \quad \text{or} \quad P = m(g + \omega^2 a)/\pi$$

EXERCISE

1. A heavy homogeneous cube, of weight W, can swing about an edge which is horizontal; it starts from rest beings displaced from its unstable position of equilibrium; when the perpendicular from the centre of gravity upon the edge has turned through an angle θ, show that the components of the action at the hinge along and right angles to this perpendicular are $(W/2)(3 - 5\cos\theta)$ and $(W/4)\sin\theta$. **[Meerut 2008]**

2. A uniform rod of length $2a$ and weight W is turning about its end O and starts from the position in which it was vertically above O. When it has turned through an angle θ, show that the horizontal and vertical reactions are $(3W/4)(2 - 3\cos\theta)$ and $(W/4)(1 - 3\cos\theta)^2$.

3. A regular tetrahedon, of mass M, swings about one edge which is horizontal. In the initial position the perpendicular from the centre of mass upon the edge is horizontal. Show that, when this line makes an angle θ with the vertical, the vertical component of thrust is

$$(Mg/7)(2\sin^2\theta + 17\cos^2\theta).$$

16.11 Motion about a fixed axis. Impulsive forces

By Art. 15.8 of Chapter 15, we know that the change in the moment of momentum about the fixed axis is equal to the moment L of the impulsive forces about this axis. Again, by Art. 16.3, the moment of momentum of the body about the axis is $mk^2\dot{\theta}$, where $\dot{\theta}$ is the angular velocity and mk^2 the M.I. about the axis. Hence, if ω and ω' be the angular velocities about the axis just before and just after the action of the impulsive forces, this change is $mk^2(\omega' - \omega)$ and hence

$$mk^2(\omega' - \omega) = L.$$

TYPE 4 : SOLVED EXAMPLES

Ex. 1. *A uniform beam AB can turn about its end A and is in equilibrium; find the point of its length where a blow must be applied to it so that impulse at A may be in each case (1/n)th of that of the blow.*

Sol. Let AB be the beam of mass M and length $2a$, which is free to turn about A. Let a blow (impulse) P be applied at the point C at a distance x from A. Then by hypothesis it will produce an impulsive action P/n at A. Let ω be the angular velocity of the beam just after the blow. Let k be the radius of gyration of the rod about an axis through A and perpendicular to AB.

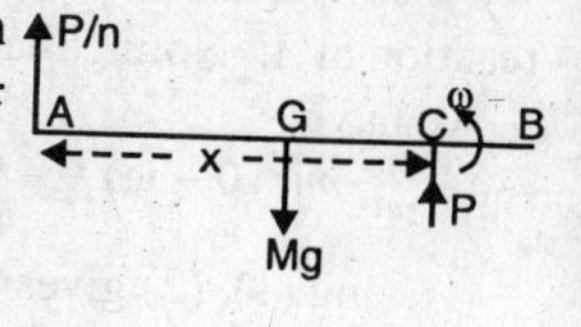

Then, $\quad Mk^2 = Ma^2/3 + Ma^2 \quad$ or $\quad k^2 = 4a^2/3$

Equations of motion are given by

$$Mk^2(\omega - 0) = Px \quad \text{or} \quad M(4a^2/3)\omega = Px \quad \ldots(1)$$

and $$Ma(\omega - 0) = P + P/n \quad \text{or} \quad Ma\omega = P(1 + 1/n) \quad \ldots(2)$$

Dividing (1) by (2), $\quad \dfrac{4a}{3} = \dfrac{x}{(1+1/n)} \quad$ or $\quad x = \dfrac{4a}{3}\left(1+\dfrac{1}{n}\right)$

Remark. If the direction of the impulsive action P/n is opposite that as shown in the above figure, then $x = (4a/3)(1 - 1/n)$.

Ex. 2. *A uniform rod OA, of mass M and length 2a, rests on a smooth table and is free to turn about a smooth pivot at its end O; in contact with it at a distance b from O is an inelastic particle of mass m; a horizontal blow, of impulse P, is given to the rod at a distance x from O in a direction perpendicular to the rod; find the resultant instantaneous angular velocity of the rod and the impulsive actions at O and on the particle.*

Sol. Let an inelastic particle of mass m be in contact with rod OA at B such that $OB = b$. When the horizontal blow of impulse P is given at a point C such that $OC = x$, then suppose S be the impulse of the action between the rod and the particle and ω the angular velocity of the rod just after the blow. Then taking moments of the system about O, we get

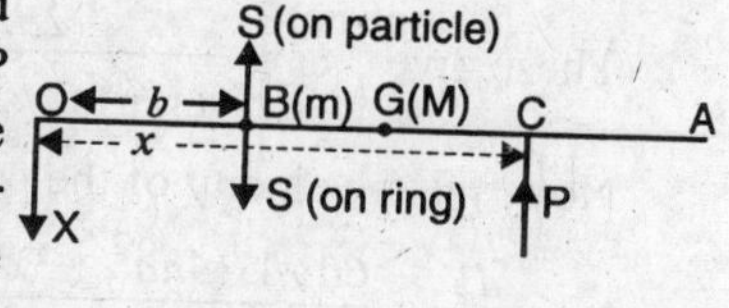

$$(m + M)k^2(\omega - 0) = P \cdot x, \quad \ldots(1)$$

where $(m + M)k^2 = (4/3)Ma^2 + mb^2$ ...(2)

Again a velocity $b\omega$ will be generated in the mass by the impulse S on it. Hence the equation of motion of the particle is given by

$$mb\omega = S \quad ...(3)$$

Dividing (1) by (2), $\omega = Px/\{(4/3)Ma^2 + mb^2\}$ ...(4)

Using (4), (3) gives $S = mbPx/\{(4/3)Ma^2 + mb^2\}$ ...(5)

Again, let X be the action at O on the rod. Then, since the change in the motion of the centre of gravity 'G' of the rod is the same as it all the impulsive forces were applied there.

$\therefore$ $Ma(\omega - 0) = P - S - X$

$\therefore$ $X = P - mb\omega - Ma\omega = P - (Ma + mb)\omega$, by (3)

or $X = P - Px(Ma + mb)/\{(4/3)Ma^2 + mb^2\}$, by (4)

Ex. 3. *A rod, of mass m and length 2a, which is capable of free motion about one end A falls from a vertical position and when it is horizontal strikes a fixed inelastic obstacle at a distance b from the end A. Show that the impulse of the blow is $m(2a/b)(2ga/3)^{1/2}$ and that the impulse of the reaction at A is $m(3ga/2)^{1/2}(1 - 4a/3b)$ vertically upwards.*

Sol. Let given rod capable of motion about the fixed end A falls from its vertical position AB' and strike a fixed inelastic obstacle at O at a distance $AO = b$, from A, after coming to its horizontal position AB. Let ω be the angular velocity of the rod just before striking the obstacle. Then by energy equation, we get

Change in K.E. = Work done

or $(1/2)mk^2\omega^2 - 0 = mg{\cdot}a$ or $(1/2)m{\cdot}(4a^2/3){\cdot}\omega^2 = mga$

which gives $\omega = (3g/2a)^{1/2}$ ...(1)

Let P be the impulse when the rod strikes the inelastic obstacle at O and let X be the impulsive reaction at A. Noting that the rod comes to rest after striking the obstacle and taking moments about A, we have

$mk^2(0 - \omega) = -P{\cdot}AO$ or $m(4a^2/3)\omega = Pb$ ...(2)

Using (1), (2) gives $P = \dfrac{4}{3}\dfrac{ma^2}{b}\cdot\left(\dfrac{3g}{2a}\right)^{1/2} = m\cdot\dfrac{2a}{b}\left(\dfrac{2ga}{3}\right)^{1/2}$ ...(3)

Now, for the motion of the centre of gravity G of the rod, we get

$m(0 - a\omega) = -P - X$ or $X = ma\omega - P$

or $X = ma\left(\dfrac{3g}{2a}\right)^{1/2} - m\cdot\dfrac{2a}{b}\left(\dfrac{2ga}{3}\right)^{1/2} = m\left(\dfrac{3ga}{2}\right)^{1/2}\left(1 - \dfrac{4a}{3b}\right)$, by (1) and (3)

Ex. 4. *A rod of mass nM, is lying on a horizontal table, and has one end fixed; a particle of mass M is in contact with it. The rod receives a horizontal blow at its free end; find the position of the particle so that it may start moving with the maximum velocity. In this case show that the kinetic energies communicated to the rod and mass are equal.* **[Rajasthan 1998]**

Sol. Let AB be the rod of length $2a$ and let the given particle of mass M be placed at C such that $AC = x$. Suppose an impulse P is applied to the rod at the free end B, so as to generate an angular velocity ω. Taking moments of the system about O, we have

$$(nM + M)k^2(\omega - 0) = P{\cdot}2a \quad \text{where} \quad (nM + M)k^2 = nM(4a^2/3) + Mx^2$$

These give $\omega = \dfrac{2aP}{nM(4a^2/3) + Mx^2} = \dfrac{6aP}{M(4na^2 + 3x^2)}$ ...(1)

Now, V = velocity of the particle = $x\omega = (6aPx)/M(4na^2 + 3x^2)$

$$\therefore \quad \frac{dV}{dx} = \frac{6ap}{M}\frac{1\cdot(4na^2 + 3x^2) - x\cdot 6x}{(4na^2 + 3x^2)^2} = \frac{6aP}{M}\cdot\frac{4na^2 - 3x^2}{(4na^2 + 3x^2)^2} \quad ...(2)$$

For a maximum or a minimum value of V, we have

$$dV/dx = 0 \quad \text{or} \quad 4na^2 - 3x^2 = 0 \quad \text{or} \quad x = 2a(n/3)^{1/2} \qquad \ldots(3)$$

From (2), $$\frac{d^2V}{dx^2} = \frac{6ap}{M} \cdot \frac{-6x(4na^2+3x^2)^2 - (4na^2-3x^2)\cdot 2(4na^2+3x^2)\cdot 6x}{(4na^2+3x^2)^4}$$

$$= -108aPx\,(4na^2 - x^2)/M(4na^2+3x^2)^3,$$

which is negative when $x = 2a(n/3)^{1/2}$. Hence V is maximum when $x = 2a(n/3)^{1/2}$. Thus the given particle will move with maximum velocity if it is kept at C, where $AC = x = 2a(n/3)^{1/2}$. Again, in this case, we have

K.E. of the rod = $(1/2)nMk^2\omega^2 = (1/2)nM\,(4a^2/3)\omega^2 = (2/3)nMa^2\omega^2$

and K.E. of the particle = $(1/2)M\,(x\omega)^2 = (1/2)M\,(4a^2n/3)\,\omega^2 = (2/3)nMa^2\omega^2$

This shows that when $x = 2a(n/3)^{1/2}$, K.E. of rod = K.E. of particle.

EXERCISE

1. A uniform thin rod AB, of length $2a$, lies at rest on a smoth horizontal table and is freely hinged at A, a fixed point. It is set in motion by a horizontal blow P applied to it in a direction perpendicular to its length at a distance $5a/4$ from A and magnitude of impulse at A is Q. The rod is then brought to rest by a horizontal blow R applied at B perpendicular to AB. Prove that $P/16 = Q/1 = R/10$.

2. A straight rod, of mass m and length $2l$, swinging about one end as a compound pendulum, starts from rest in a horizontal position and when vertical is struck by a blow at its middle point which reverses and halves its angular velocity. Prove that the impulse of the blow is $m(6gl)^{1/2}$.

3. Two wheels spindles in fixed bearing suddenly engage so that their angular velocities become inversely proportional to radii and in opposite directions. One wheel of radius a and M.I. I_1 has angular velocity ω initially; the other, of radius b and M.I. I_2 initially at rest. Show that their new angular velocities are $b^2I_1\omega/(I_1b^2 + I_2a^2)$ and $I_1ab\omega/(I_1b^2 + I_2a^2)$.

4. The door of a railway carriage stands open at right angles to the length of the train when the latter starts to move with an acceleration f, the door being supposed to be smoothly hinged to the carriage and to be uniform and of breadth $2a$, show that its angular velocity when it has turned through an angle θ is $(3f \sin \theta/2a)^{1/2}$.

5. A uniform inelastic beam can resolve about its centre of gravity in a vertical plane and is at rest inclined at an angle α to the vertical. A particle of given mass is let fall from a given height above the centre and hits the beam in a given point P; find the position of P so that the resulting angular velocity may be a maximum.

6. A uniform beam, of mass m and length $2l$, is horizontal and can turn freely about its centre which is fixed. A ball, of mass m' and moving with velocity u, hits beam at one end. If coefficient of restitution for the impact be e, show that the angular velocity of the beam after impact is $3m'(1+e)u/(m+3m')l$ and that the velocity of the ball is then $u(em - 2m')/(m + 3m')$.

16.12 Centre of percussion. [Kanpur 2006, 2014; Meerut 2006; Purvanchal 2005]

If a body, rotating about a given axis, is so struck that there is no imulsive pressure on the axis, then any point on the line of action of the force is called a *centre of* percussion.

Note. The line of action of the force is called the *line of percussion*. Again when the line of action of the blow is given, the axis about which the body begins to turn is called the *axis of spontaneous rotation*. This clearly coincides with the position of the fixed axis is the first case.

16.13 Centre of percussion of a rod. [Kanpur 2006, 2014; Meerut 2008; Agra 2011, 13]

Let a rod OA, of length $2a$ and mass M, suspended freely from one end O be struck by a horizontal blow P at a point C such that $OC = x$.

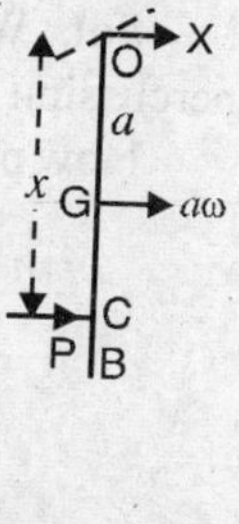

Let ω be the instantaneous angular velocity communicated to the rod and X the impulsive action at O. Then the centre of gravity G of the rod will move with velocity $a\omega$.

Consider the motion of the centre of gravity G. Now, equating the change in the momentum in the horizontal direction to the total external impulse in that direction (refer Art. 15.8, Chapter 15), we get

$$M(a\omega - 0) = P + X \qquad \ldots(1)$$

Let k be the radius of gyration of the rod about O. Then Mk^2 = M.I. of the rod about O = $M(a^2/3) + Ma^2$ so that $k^2 = 4a^2/3$. Taking moments about the axis through O, we get

$$Mk^2(\omega - 0) = P \cdot x \quad \text{...(2)}$$

If the blow has been given through the centre of percussion, there should be no impulse on the axis, *i.e.*, $X = 0$. Then (1) reduces to $P = Ma\omega$. Putting this value of P in (2), we get $Mk^2\omega = Ma\omega x$ or $x = k^2/a$. Using Art. 16.6, we have

$OC = x = k^2/a$ = length of the equivalent simple pendulum.

In this case, C, the required point, coincides with the centre of oscillation, *i.e.*, the centre of percussion with regard to the fixed axis coincides with the centre of oscillation with regard to the same axis.

16.14 Working rule to find the centre of percussion of a body for a fixed axis. General Case

Step I. Find the depth h of the centre of gravity of the body below the fixed axis of rotation.

Step II. Find the point where the fixed axis is principal axis of the body. The results of Art. 14.15 and 14.16 of Chapter 14 may be used.

Step III. Take a distance k^2/h (*i.e.*, a length equal to equivalent simple pendulum) below it and draw an axis perpendicular to the plane containing the fixed axis and centre of gravity.

Step IV. Any point on this line is a centre of precussion of the body for the given fixed axis.

Remark. Since the direction of the blow passes through centre of oscillation, it follows that for a lamina suspended freely and capable of rotation about a perpendicular axis, the centre of oscillation coincides with the centre of percurssion.

TYPE 5 : SOLVED EXAMPLES

Ex. 1. *Find the position of the centre of percussion of a uniform rod with one end fixed.*

Sol. Let a rod OB of mass M and length $2a$ be suspended freely from fixed point O. Then

h = depth of its C.G. from $O = OG = a$

Let k be the radius of gyration of the rod about the axis of rotation through O and perpendicular to it. Then

Mk^2 = M.I. of the rod about the axis of rotation

or $Mk^2 = M(a^2/3) + Ma^2$ so that $k^2 = 4a^2/3$.

$\therefore$ The required distance of the centre of percussion from O

$= k^2/h = 4a^2/3a = 4a/3$

axis
O
G
C
B

Ex. 2. *Find the position of the centre of percussion of a uniform circular plate with axis a horizontal tangent.*

Sol. Let M be the mass, G the centre and TOT' the horizontal tangent of a the given circular plate of radius a (say). Then, h = depth of its centre of gravity below 0 = $OG = a$. Let k be the radius of gyration of the plate about TT'. Then

Mk^2 = M.I. of the plate about $TT' = M(a^2/4) + Ma^2$

Thus, $Mk^2 = (5Ma^2)/4$ so that $k^2 = 5a^2/4$

$\therefore$ The required distance of the centre of percussion below O

$= k^2/h = 5a^2/4a = 5a/4$

T' O T
a
G
A

Ex. 3. *A pendulum is constructed of a solid sphere, of mass m_1 and radius b, which is attached to the end of a rod, of mass m and length a. Show that there will be no strain on the axis if the pendulum be struck at a distance $[m_1\{(2/5)b^2 + (a + b)^2\} + (1/3)ma^2]/\{m_1(a + b) + (1/2)ma\}$ from the axis.*

Sol. We know that there will be no strain on axis of the pendulum if it be struck at its centre of percussion which is known to be at a depth k^2/h below O.

Now proceed as in Ex. 3, page 16.9 to prove that

$$h = \frac{m_1(a+b) + (1/2)ma}{m + m_1} \quad \text{and } k^2 = \frac{m_1\{(2/5)b^2 + (a+b)^2\} + (1/3)ma^2}{m + m_1}$$

$\therefore$ depth of the centre of percussion below O

$= k^2/h = [m_1\{(2/5)b^2 + (a+b)^2\} + (1/3)ma^2]/\{m_1(a+b) + (1/2)ma\}$

Ex. 4. (*a*) *Find the centre of percussion of a triangle ABC which is free to move about its side BC.* **[Purvanchal 2007; Agra 2012; Meerut 2007, 09, 10]**

(*b*) *Find how can equilateral triangular lamina must be struck so that it may commence to rotate about a side.* **[Meerut 2005, 06]**

Sol. (*a*) We shall use working rule 16.14 for finding the required centre of percussion as follows :

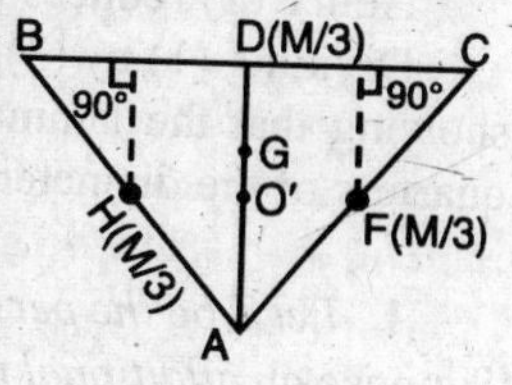

Draw AD perpendicular to BC and let E, F and H be middle points of sides BC, AC and AB respectively. Let HL and FN be perpendiculars to BC. Let G be the centre of gravity of triangle ABC. If $AD = p$, then h = distance of G below the axis $BC = GK = p/3$.

Let O be the middle point of DE. Then proceed as in Ex. 7 on page 14.40 of Chapter 14 to show that BC is the principal axis of the triangle at O. Let k^2 be the radius of gyration of the triangle about BC. Let M be the mass of the triangle. Then, we have

Mk^2 = M.I. of the triangle about BC = sum of M.I. of three particles each of mass $M/3$ placed E, F, H

or $\quad Mk^2 = (M/3)\cdot 0^2 + (M/3)\cdot HL^2 + (M/3)\cdot FN^2 = (M/3)\cdot(p/2)^2 + (M/3)\cdot(p/2)^2$

or $\quad Mk^2 = 2 \times (M/3) \times (p^2/4) \quad$ so that $\quad k^2 = p^2/6$

Hence $\quad k^2/h = (p^2/6) \times (3/p) = p/2$

In the triangle draw OO' perpendicular to BC to meet AE in O', so that $OO' = (1/2)\, AD = p/2$. Hence O' is the required centre of percussion. Thus, the middle point O' of the median AE is the centre of percussion.

(*b*) Let the equilateral triangle ABC rotate about side BC. Then the blow should be given at the centre of percussion when BC is the axis of rotation of the lamina.

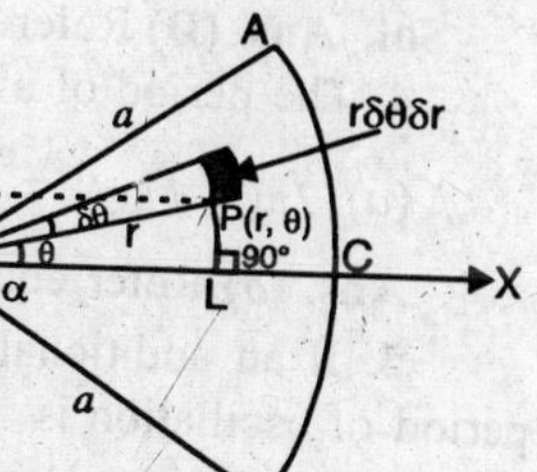

This problem is a particular case of part (*a*) where $AB = BC = CA$. Note that here points D, E, O and K of part (*a*) conicide as shown in the new figure for given equilateral triangle. Let $AD = p$. Then as in part (*a*), prove that $h = DG = p/3$, $k^2 = p^2/6$ and BC is the principal axis of the triangle at middle point D of triangle ABC.

Hence if O' is the centre of percussion, we have

DO' = depth of centre of percussion below BC along the bisecting median AD of $BC = k^2/h = p/2$.

So O' is at middle point of DA. Thus blow should be given at the middle point of the median bisecting the side about which triangular lamina rotates.

Ex. 5. *Find the position of the centre of percusion of a sector of a circle, axis in the plane of the sector, perpendicular to its symmetrical radius, and passing through the centre of the circle.* **[Meerut 2005]**

Sol. Let OAB be the sector of a circle of radius a such that $\angle AOB = 2\alpha$. Let x-axis be taken along its symmetrical radius OC and y-axis perpendicular to OX and in the plane of the sector. Consider an elementary area $r\,\delta\theta\,\delta r$ at $P(r, \theta)$ and ρ be mass per unit area of the sector. Then δm = mass of the elementary area $= \rho r\,\delta\theta\,\delta r$ and so M.I. of the elementary mass about $OY = \delta m.\ PN^2 = \delta m \cdot OL^2 = \rho r\,\delta\theta\delta r \cdot (r\cos\theta)^2 = \rho r^3 \cos^2\theta\,\delta\theta\,\delta r$.

Let k^2 be the radius of gyration of the sector about OY,

$\therefore \quad Mk^2$ = M.I. of the sector about OY

$$= \int_{\theta=-\alpha}^{\alpha}\int_{r=0}^{a} \rho r^3 \cos^2\theta\, d\theta\, dr = \rho \int_{-\alpha}^{\alpha} \cos^2\theta \left[\frac{1}{4}r^4\right]_0^a d\theta$$

$$= \frac{a^4\rho}{8}\int_{-\alpha}^{\alpha}(1+\cos 2\theta)\,d\theta = \frac{a^4\rho}{8}\left[\theta+\frac{1}{2}\sin 2\theta\right]_{-\alpha}^{\alpha} = \frac{2a^4\rho}{8}\left(\alpha+\frac{1}{2}\sin 2\alpha\right)$$

But M = mass of the sector $= a^2\alpha\rho$

$\therefore$ $a^2\alpha\rho k^2 = (a^2\rho/4)\,(\alpha + \sin\alpha\cos\alpha)$ or $k^2 = (a^2/4\alpha)\,(\alpha + \sin\alpha\cos\alpha)$

Also, h = distance of C.G. from $OY = (2a\sin\alpha)/3\alpha$ (Remember)

Hence the distance of centre of percussion on OX from O

$= k^2/h = 3a(\alpha + \sin\alpha\cos\alpha)/8\sin\alpha$

Ex. 6. *A uniform circular lamina rests on a smooth horizontal plane. Show that it will commence to turn about a point O on its circumference if it be struck by a horizontal blow whose line of action is perpendicular to diameter through O and at a distance from O equal to three quarters of the diameter of lamina.*

Sol. Let M be the mass, a the radius and C the centre of given uniform circular lamina which rests on a smooth horizontal plane. Let P be the horizontal impulse through B perpendicular to the diameter OA such that $OB = x$. Let the impulse P produce an impulsive reaction X at O and an angular velocity ω about a line perpendicular to the disc through O. Taking moments about O, we have

$Mk^2\omega = P\cdot OB$, where $k^2 = (1/2)a^2 + a^2 = 3a^2/2$

Thus, $M(3a^2/2)\omega = P\cdot x$ ...(1)

Now velocity generated in the centre after impulse will be $a\omega$.

But, the change in the linear momentum = The resultant of impulsive forces

$\therefore$ $M(a\omega - 0) = P + X$ ...(2)

If there is no impulsive action at O, then $X = 0$.

Hence (2) reduces to $Ma\omega = P$ ...(3)

Dividing (1) by (3), $x = 3a/2 = (3/4)\times(2a)$,

showing that the lamina will commence to turn about O if be struck at a distance equal to three quarters of the diameter from O at right angles to it.

OBJECTIVE QUESTIONS

1. *Let T be the period of oscillation of a compound pendulum and O be its centre of oscillation. Suppose an additional mass m is attached at O. Then the period of oscillation of this new compound pendulum is*

(A) T/m *(B) Tm* *(C) $T/m^{1/2}$* *(D) None of the above*

Sol. Ans. (D) Period is T. Refer remark of Art. 16.6.

2. *If the time of complete oscillation of a compound pendulum is $2\pi(k^2/gh)^{1/2}$, then the length of simple equivalent pendulum is* **[Agra 2006]**

(A) k^2/g *(B) k^2/h* *(C) k^2/gh* *(D) k/h*

Sol. Ans. (B) Refer Art. 16.6.

3. *The depth of the centre of percussion from O of a uniform rod, of length 2a, with one end fixed at O is* *(A) a* *(B) a/2* *(C) 2a/3* *(D) 4a/3*

Sol. Ans. (D) Refer Ex. 1 on page 16.22.

4. The period of a compound pendulum is

(*a*) $2\pi(h^2/kg)^{1/2}$ (*b*) $2\pi(k^2/hg)^{1/2}$ (*c*) $2\pi(hg/k^2)$ (*d*) None of these **[Kanpur 2004]**

Ans. (*b*) Refer equation (3) of Art. 16.5

5. If an additional weight is attached to the centre of oscillation of a pendulum, then the period of oscillation is

(*a*) dependent on the weight attached
(*b*) neither alters nor depends on the weight (*c*) remains unaltered (*d*) none of these. **[Kanpur 2002, 03]**

Ans. (c). Refer remark of Art. 16.7

MISCELLANEOUS PROBLEMS

1. Define (*i*) Centre of suspension

   (*ii*) Centre of oscillation **[Agra 2006]**

   (*iii*) Centre of percussion **[Agra 2006]**

   [Purvanchal 2005, Kanpur 2006, Meerut 2005]

2. Find the centre of percussion of a rod. **[Purvanchal 2005, Kanpur 2005, 06]**

3. A plank of mass M, radius of gyration k and length $2b$ can swing like a sea-saw across a perfectly fixed cylinder of radius a. At its ends hang two particles, each of mass m by string of length l. Show that, as the system swings, the lengths of its simple equivalent pendulum are I and $(MK^2 + 2mb^2)/\{a(M + 2m)$. **[Meerut 2007]**

Ex. 4. *If the time of complete oscillation of a compound pendulum is* $2\pi \times (k^2/gh)^{1/2}$, *then the length of simple equivalent pendulum is*

(*a*) k^2/g (*b*) k^2/h (*c*) $k^2/(gh)$ (*d*) k/h **(Agra 2012)**

Sol. Ans. (*b*). Refer Art. 16.6, page 16.7.

Ex. 5. True of false?

(*i*) *Length of simple equivalent pendulum is* k^2/h. **(Agra 2011)**

(*ii*) *The centre of suspension and the centre of oscillation are convertible.* **(Agra 2011)**

(*iii*) *The centre of percussion of a rod never coincides with centre of oscillation.* **(Agra 2012)**

Sol. (*i*) **Ans.** True. Refer Art. 16.6, page 16.7

(*ii*) **Ans.** True. Refer Art. 16.8, page 16.7

(*iii*) **Ans.** False, Refer Remark on page 16.22

17

Some aspects of rigid body dynamics

17.1 Kinetic energy and angular momentum. We shall compute these quantities when the system is a rigid body moving parallel to a plane. Let us first suppose that a rigid body is rotating about a fixed axis. Let ω be the instantaneous value of the angular velocity. Then the kinetic energy of a particle of mass m_i situated at a distance r_i from the axis is $(1/2)mr_i^2\omega^2$. Assume that the rigid body consists of n particles. Then the kinetic energy T of the rigid body is given by

$$T=\sum_{i=1}^{n}\frac{1}{2}mr_i^2\omega^2=\frac{1}{2}\omega^2\sum_{i=1}^{n}m_ir_i^2=\frac{1}{2}I\omega^2,$$

where I is the M.I. of the rigid body about the axis.

Again, the angular momentum of a particle about the axis of rotation is $m_ir_i^2\omega$, and hence the angular momentum h of the rigid body is given by

$$h=\Sigma m_i\omega r_i^2=\omega\ \Sigma m_ir_i^2=I\omega$$

Let us now assume that the rigid body no longer rotates about a fixed axis but moves in a general manner parallel to a fixed fundamental plane. Then we can easily show that *the kinetic energy and angular momentum, both relative to the mass centre, of a rigid body moving parallel to a plane are $T=(1/2)I\omega^2$ and $h=I\omega$, where ω is the angular velocity of the body and I its M.I. about an axis through the mass centre perpendicular to the plane of motion.*

In what follows we shall state and prove König theorem in general three-dimensional form. This theorem helps us to complete the calculation of the kinetic energy of a rigid body moving parallel to a plane.

Theorem of König. *The kinetic energy of a moving system is equal to the sum of (i) the kinetic energy of a fictitious particle moving with the mass centre and having a mass equal to the total mass of the system and (ii) the kinetic energy of the motion relative to the mass centre.*

[Delhi B.Sc. (H) 1997

Proof. Let O be a fixed origin and O' the centre of mass of the given system of n particles. Let $\overrightarrow{OO'}=\mathbf{r}_0$. Let P_i be position of the particle of mass m_i. Let the position vectors of P_i relative to O and O' be $\mathbf{r}_i$ and $\mathbf{r}_i'$ respectively. Then, by triangle law of vector addition, we get

$$\mathbf{r}_i=\mathbf{r}_0+\mathbf{r}_i' \qquad \ldots(1)$$

Relative to O, the kinetic energy T of the system is given by

$$T=\sum_{i=1}^{n}\frac{1}{2}m_i\left(\frac{d\mathbf{r}_i}{dt}\right)^2=\sum_{i=1}^{n}\frac{1}{2}m_i\left(\frac{d\mathbf{r}_0}{dt}+\frac{d\mathbf{r}_i'}{dt}\right)^2, \text{ using (1)}$$

or $$T=\sum_{i=1}^{n}\frac{1}{2}m_i\left\{\left(\frac{d\mathbf{r}_0}{dt}\right)^2+\left(\frac{d\mathbf{r}_i'}{dt}\right)^2+2\left(\frac{d\mathbf{r}_0}{dt}\cdot\frac{d\mathbf{r}_i'}{dt}\right)\right\}$$

or $$T=\frac{1}{2}\left(\sum_{i=1}^{n}m_i\right)\left(\frac{d\mathbf{r}_0}{dt}\right)^2+\frac{1}{2}\sum_{i=1}^{n}m_i\left(\frac{d\mathbf{r}_i'}{dt}\right)^2+\frac{d\mathbf{r}_0}{dt}\cdot\left(\sum m_i\frac{d\mathbf{r}_i'}{dt}\right) \qquad \ldots(2)$$

Let $\sum_{i=1}^{n}m_i=m=$ mass of the system of n particles

v_0 = speed of the centre of mass = $|d\mathbf{r}_0/dt|$

$T_0 = (1/2)m(d\mathbf{r}_0/dt)^2 = (1/2)mv_0^2$ = K.E. of the mass centre

$T' = (1/2)\Sigma m_i(d\mathbf{r}'_i/dt)^2$ = K.E. of the system relative to the centre of mass.

Now, by definition, the position vector of the mass centre O' relative to O' as origin is given by

$$\mathbf{0} = \left(\sum_{i=1}^{n} m_i\mathbf{r}'_i\right) \Big/ \left(\sum_{i=1}^{n} m_i\right) \quad \text{so that} \quad \sum_{i=1}^{n} m_i\left(\frac{d\mathbf{r}'_i}{dt}\right) = \mathbf{0}$$

Using the above facts, (2) may be re-written as

$$T = T_0 + T', \qquad \ldots(3)$$

which proves the theorem. In the above discussion we have referred to the K.E. of the fictitious particle as the kinetic energy of the mass centre.

Remark. For a rigid body moving parallel to a fixed plane, we have

$$T = (1/2)mv_0^2 + (1/2)I\omega^2,$$

where m = mass of the body, v_0 = speed of the mass centre, I = M.I. about the mass centre and ω = angular velocity.

17.2 The kinetic energy of a rigid body with a fixed point. *To determine an expression for the K.E. of a rigid body with a fixed point rotating with constant angular velocity.*

[Meerut 2007; Kanpur 2006, 07, 10; Delhi Maths (H) 2000, 2002, 2005, 2011]

Let a rigid body rotate above axis through a fixed point O with angular velocity $\boldsymbol{\omega}$. Let δm be the mass of a particle P of the body and $\mathbf{r}$ be the position vector of P relative to O as origin. Let $\mathbf{v}$ be the velocity of the particle P. Then the K.E. of the body is given by

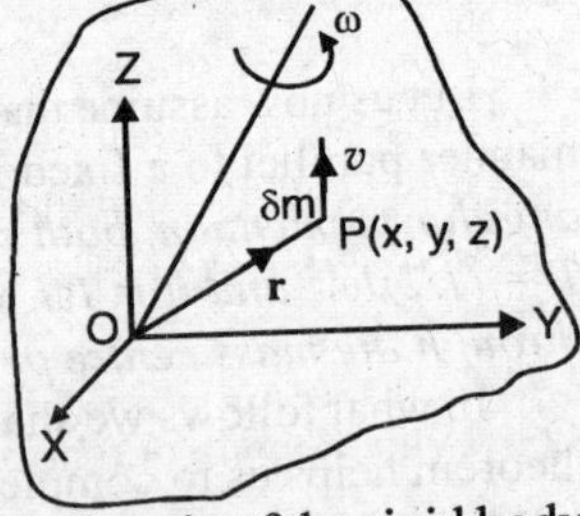

$$T = \frac{1}{2}\Sigma\delta m v^2 = \frac{1}{2}\Sigma\delta m\, \mathbf{v}^2 \quad (\text{if } v = |\mathbf{v}|)$$

where the summation extends over all particles of the body. Since

$$\mathbf{v} = \boldsymbol{\omega}\times\mathbf{r}, \text{ we have}$$

$$T = \frac{1}{2}\Sigma\delta m\,(\boldsymbol{\omega}\times\mathbf{r})^2 \qquad \ldots(1)$$

We now propose to re-write (1) in terms of the moments and products of inertia of the rigid body at O. Let $Oxyz$ be any rectangular axes at O, and $\mathbf{i}, \mathbf{j}, \mathbf{k}$ unit vectors along them. Resolving vectors in the directions of these axes,

$$\mathbf{r} = x\mathbf{i} + y\mathbf{j} + z\mathbf{k} \quad \text{and} \quad \boldsymbol{\omega} = \omega_1\mathbf{i} + \omega_2\mathbf{j} + \omega_3\mathbf{k} \qquad \ldots(2)$$

$$\therefore \quad \boldsymbol{\omega}\times\mathbf{r} = \begin{vmatrix} \mathbf{i} & \mathbf{j} & \mathbf{k} \\ \omega_1 & \omega_2 & \omega_3 \\ x & y & z \end{vmatrix}$$

or $$\boldsymbol{\omega}\times\mathbf{r} = (\omega_2 z - \omega_3 y)\mathbf{i} + (\omega_3 x - \omega_1 z)\mathbf{j} + (\omega_1 y - \omega_2 x)\mathbf{k}$$

or $$(\boldsymbol{\omega}\times\mathbf{r})^2 = |\boldsymbol{\omega}\times\mathbf{r}|^2 = (\omega_2 z - \omega_3 y)^2 + (\omega_3 x - \omega_1 z)^2 + (\omega_1 y - \omega_2 x)^2$$

or $$(\boldsymbol{\omega}\times\mathbf{r})^2 = (y^2+z^2)\omega_1^2 + (z^2+x^2)\omega_2^2 + (x^2+y^2)\omega_3^2 - 2yz\omega_2\omega_3 - 2zx\omega_3\omega_1 - 2xy\omega_1\omega_2 \qquad \ldots(3)$$

Hence, by substituting from (3) in (1), we have

$$2T = \omega_1^2\Sigma\delta m(y^2+z^2) + \omega_2^2\Sigma\delta m(z^2+x^2) + \omega_3^2\Sigma\delta m(x^2+y^2)$$
$$- 2\omega_2\omega_3\Sigma\delta m\, yz - 2\omega_3\omega_1\Sigma\delta m\, zx - 2\omega_1\omega_2\Sigma\delta m\, xy$$

or $$T = (1/2)\times(A\omega_1^2 + B\omega_2^2 + C\omega_3^2 - 2F\omega_2\omega_3 - 2G\omega_3\omega_1 - 2H\omega_1\omega_2), \qquad \ldots(4)$$

where A, B, C, F, G, H are the moments and products of inertia for axes $Oxyz$.

If the axes $Oxyz$ are principal axes of inertia, then $F = G = H = 0$ and so by (4), we obtain

$$T = (A\omega_1^2 + B\omega_2^2 + C\omega_3^2)/2, \qquad \ldots(5)$$

where A, B, C are the principal moments of inertia with respect to $Oxyz$ as principal axes of inertia at O.

Corollary 1. *If a rigid body rotates about a fixed line L through O, then K.E. of the body is given by $T = (1/2)\times C\omega^2$, where C is the M.I. of the body about the line L.*

Proof. Taking Oz along the given line L, we see that $\omega_1 = \omega_2 = 0$ and $\omega_3 = \omega$ and so (4) reduces to

$$T = (1/2)\times C\omega^2.$$

Corollary 2. If $|\omega| = \omega$, then the direction cosines α, β, γ of the line of axis of rotation through O are given by

$$\alpha = w_1/w, \qquad \beta = w_2/w, \qquad \gamma = w_3/w \qquad \ldots(6)$$

Using (6), (4) can be re-written as

$$T = (1/2) \times (A\alpha^2 + B\beta^2 + C\gamma^2 - 2F\beta\gamma - 2G\gamma\alpha - 2H\alpha\beta)\omega^2$$

or
$$T = (1/2) \times \omega^2 I,$$

where
$$I = A\alpha^2 + B\beta^2 + C\gamma^2 - 2F\beta\gamma - 2G\gamma\alpha - 2H\alpha\beta \qquad \ldots(7)$$

is the M.I. of the rigid body about the axis of rotation through the fixed point O having direction cosines α, β, γ.

17.3 The kinetic energy of a rigid body in general

We propose to determine the K.E. T of a rigid body moving quite generally in space. Applying the theorem of König (refer Art. 17.1), we have

$$T = (1/2) \times mv_0^2 + T', \qquad \ldots(1)$$

where m = mass of body, v_0 = speed of mass centre

and T' = K.E. of motion relative to the centre

Let us treat the mass centre as a base point in the body. Hence the motion relative to mass centre is that of a rigid body turning about a fixed point. Then T' is given by (Using result (5) of Art. 17.2 with a proper interpretation of the symbols)

$$T = (1/2) \times mv_0^2 + (1/2) \times (A\omega_1^2 + B\omega_2^2 + C\omega_3^2),$$

where $\omega_1, \omega_2, \omega_3$ = Components of the angular velocity $\boldsymbol{\omega}$ in the directions of principal axes of inertia at the mass centre

and A, B, C = principal M.I. at the centre of mass.

ILLUSTRATIVE SOLVED EXAMPLES

Ex. 1. (a) *Find the kinetic energy of a disc of mass m and radius a, rolling along the ground with speed v.* **[Delhi Maths (H) 1997]**

(b) *What is the kinetic energy of a homogeneous circular cylinder, of mass m and radius a, rolling on a plane with linear velocity v ?*

Sol. (a) Let O be the centre of the given disc which is rolling along the ground in the vertical plane (say, xy-plane). Let the centre O move with velocity v along x-axis. Let OZ be perpendicular to the plane of the disc. Then, due to symmetry, $Oxyz$ must be a principal frame at centre of mass O.

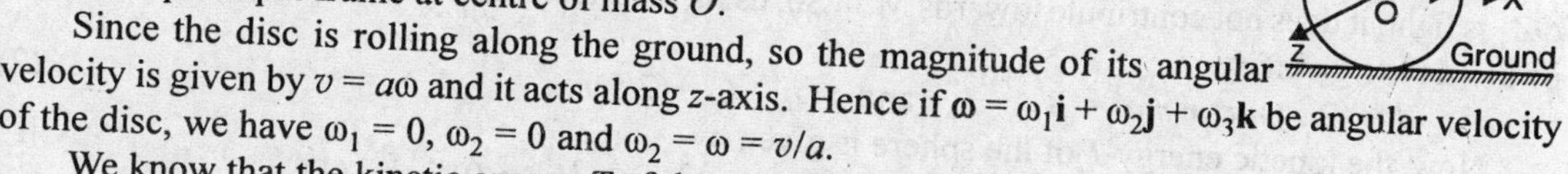

Since the disc is rolling along the ground, so the magnitude of its angular velocity is given by $v = a\omega$ and it acts along z-axis. Hence if $\boldsymbol{\omega} = \omega_1\mathbf{i} + \omega_2\mathbf{j} + \omega_3\mathbf{k}$ be angular velocity of the disc, we have $\omega_1 = 0$, $\omega_2 = 0$ and $\omega_2 = \omega = v/a$.

We know that the kinetic energy T of the circular disc (which is moving system for the present problem) is given by

$$T = (1/2) \times mv^2 + (1/2) \times (A\omega_1^2 + B\omega_2^2 + C\omega_3^2)$$

or
$$T = (1/2) \times mv^2 + (1/2) \times C(\omega^2/a^2) \text{ as } A = B = 0 \qquad \ldots(2)$$

But C = M.I. of the circular disc about z-axis $= (1/2) \times ma^2$

$\therefore$ from (2), $T = (1/2) \times mv^2 + (1/2) \times (ma^2/2) \times (v^2/a^2) = (3/4) \times mv^2$

(b) Consider the vertical section of the circular cylinder through its centre mass O, perpendicular to the axis of cylinder as shown in the figure of part (*a*). Here the vertical section is in xy-plane and z-axis is along the axis of the cylinder. Then proceed exactly as in part (*a*) upto equation (2).

Thus, $T = (1/2) \times mv^2 + (1/2) \times C \times (v^2/a^2)$, as $A = B = O$

But C = M.I. of cylinder about its axis $OZ = (1/2) \times ma^2$

$\therefore$ $T = (1/2) \times mv^2 + (1/2) \times (ma^2/2) \times (v^2/a^2) = (3/4) \times mv^2$

Ex. 2. *A solid sphere of mass m and radius a rolls without sliding with a uniform linear velocity v along a straight line on a smooth horizontal table. Prove that its total kinetic energy is (7/10) × mv².*

Sol. Consider the vertical section of the given solid sphere through its centre of mass O, as shown in the figure of Ex. 1. (*a*). Here the vertical section is in *xy*-plane and *z*-axis is along the diameter of the sphere.

Proceed as in part (*a*) of Ex. 1. upto equation (2).

$\therefore \qquad T = (1/2) \times mv^2 + (1/2) \times C \times (v^2/a^2)$, as $A = B = O$.

But $\qquad C =$ M.I. of the sphere about diameter $OZ = (2/5)ma^2$

$\therefore \qquad t = (1/2) \times mv^2 + (1/2) \times (2ma^2/5) \times (v^2/a^2) = (7/10) \times mv^2$

Ex. 3. *A governor consists of two equal spheres, of mass m and radius a. They are fixed to the ends of equal light rods, each of length c – a, which are hinged to a collar on a vertical axle. By means of a light linkage and sliding collar, the equality of the inclinations to the vertical of the two rods is ensured. If this angle of inclination is θ and the angular velocity of governer about its vertical axis is ω, show that the kinetic energy is* $A(\dot{\theta}^2 + \omega^2 \sin^2 \theta) + c\omega^2 \cos^2 \theta$ *where* $A = m(c^2 + 2a^2/5)$ *and* $C = 2ma^2/5$.

[Delhi Maths (Hons.) 1999, 2004, 2009]

Sol. Let G_1 and G_2 be centres of the two given spheres and OA_1, OA_2 be light rods, each of length $c - a$. Due to symmetry, the kinetic energy of both spheres will be the same. So we proceed to find kinetic energy of the first sphere with centre G_1 as follows.

Consider a fixed point *O* on the sliding collar of the vertical axel. We select rectangular axes *Oxyz* in such a way that the angular velocity vector $\boldsymbol{\omega}$ (whose magnitude is ω) about the vertical axel always lies in the *xz*-plane and *y*-axis is perpendicular to *xz*-plane at *O*.

If $\boldsymbol{\omega}$ is the angular velocity vector, then its components along *Oz* and *Ox* are $\omega \cos \theta$ and $\omega \sin \theta$ respectively. Since the rod is attached to the sliding collar, hence it moves up and down. Therefore θ changes with time and so $\dot{\theta}$ is the component of $\boldsymbol{\omega}$ along *OY*.

Clearly, the angular velocity vector of the system is given by

$$\boldsymbol{\omega} = \omega_1 \mathbf{i} + \omega_2 \mathbf{j} + \omega_3 \mathbf{k} = (\omega \sin \theta)\mathbf{i} + \dot{\theta}\,\mathbf{j} + (\omega \cos \theta)\mathbf{k} \qquad \ldots(1)$$

so that $\qquad \omega_1 = \omega \sin \theta, \qquad \omega_2 = \dot{\theta} \qquad$ and $\qquad \omega_3 = \omega \cos \theta$

Due to symmetry, *Oxyz* is a principal frame at *O* of the system for the first sphere only. Since rod OA_1 is light, it does not contribute towards M.I. So, using parallel axis theorem, the principal moments of inertia are given by

$$A = (2/5)ma^2 + mc^2 = B \qquad \text{and} \qquad C = (2/5)ma^2 \qquad \ldots(2)$$

Now, the kinetic energy *T* of the sphere is given by

$$T = (A\omega_1^2 + B\omega_2^2 + C\omega_3^2)/2 = (A\omega^2 \sin^2 \theta + A\dot{\theta}^2 + C\omega^2 \cos^2 \theta)/2, \text{ by (1)}$$

$\therefore$ Required total K.E. $= 2T = A(\dot{\theta}^2 + \omega^2 \sin^2 \theta) + C\omega^2 \cos^2 \theta$, where *A* and *C* are given by (2).

Ex. 4. *A car is turning a corner, the middle point of the back axle describing a circle of radius r. If the length of the axle is 2a and the wheels are regarded as uniform discs, each of radius b, prove that the ratio of the kinetic energies of the back wheels is* $\{6(r + a)^2 + b^2\}/\{6(r - a)^2 + b^2\}$.

[Delhi Maths (H) 1992, 95, 98, 2003, 06, 08]

Sol. Let the corner *O* be taken as the fixed point. Let *A* and *B* be the centres of the two back wheels, each of radius *b*. The length of the axel *AB* is 2*a*. Let *G* be the middle point of *AB*. Then, by problem *G* describes a circle of radius *r* about *O*. Clearly, $AG = BG = a$ and so $OA = r - a$ and $OB = r + a$.

Let us first find the kinetic energy of wheel with centre *A*.

Due to symmetry, the axes *Oxyz* is a principal frame at *O*. The system consisting of two wheels is rotating about the *y*-axis. If θ is the angle through which *A* rotates about *y*-axis in a circle of radius $(r - a)$ with centre *O*, then $\dot{\theta}$ is the constant angular velocity which acts along *y*-axis.

Furthermore, simultaneously the wheel rotates about *x*-axis (which passes through *A*). If ϕ is th angle through which any point of the wheel rotates, then $\dot{\phi}$ is the angular velocity of any point P of this wheel which acts along negative direction of x-axis. The angular velocity $\boldsymbol{\omega}$ of the first wheel is given by

$$\boldsymbol{\omega} = -\dot{\phi}\mathbf{i} + \dot{\theta}\mathbf{j} + 0\mathbf{k} \qquad \text{...(1)}$$

By the condition for pure rolling, we have

arc AA' = arc PP' so that $(r-a)\theta = b\phi$

$\therefore\ (r-a)\dot{\theta} = b\dot{\phi}$ or $\dot{\phi} = \{(r-a)/b\}\,\dot{\theta}$ ...(2)

So by (1) $\boldsymbol{\omega} = \omega_1\mathbf{i} + \omega_2\mathbf{j} + \omega_3\mathbf{k} = -\{(r-a)/b\}\,\dot{\theta}\mathbf{i} + \dot{\theta}\mathbf{j} + 0\mathbf{k}$

so that $\omega_1 = -\{(r-a)/b\}\,\dot{\theta}$, $\omega_2 = \dot{\theta}$ and $\omega_3 = 0$ ...(3)

Using parallel axis theorem, the principal moments of inertia of the first wheel about $Oxyz$ are given by

$$A = (1/2)mb^2,\ B = C = (1/4)mb^2 + m(r-a)^2,$$

where m is mass of the wheel of radius b. ...(4)

Using (3) and (4), the K.E. of the first wheel is given by

$$T_1 = (A\omega_1^2 + B\omega_2^2 + C\omega_3^2)/2$$

$$= \frac{1}{2}\left[\frac{1}{2}mb^2 \times \frac{(r-a)^2\dot{\theta}^2}{b^2} + \left\{\frac{1}{4}mb^2 + m(r-a)^2\right\}\dot{\theta}^2 + 0\right]$$

or $$T_1 = (1/8)\,m\dot{\theta}^2\ \{6(r-a^2) + b^2\} \qquad \text{...(5)}$$

Finally, we now find K.E. of the second wheel with centre B which is moving in a circle of radius $(r+a)$ with centre O. As before results (3) will be true for the second wheel also. However, we have new values of A, B, C for the second wheel as follows :

$$A = (1/2)mb^2, \qquad B = C = (1/4)mb^2 + m(r+a)^2 \qquad \text{...(6)}$$

Using (3) and (6), the K.E. of the second wheel is

$$T_2 = (1/2)\,(A\omega_1^2 + B\omega_2^2 + C\omega_3^2) = (1/8)\,m\dot{\theta}^2\ \{6(r+a)^2 + b^2\}$$

From (5) and (7), $T_2 : T_1 = \{6(r+a)^2 + b^2\}/\{6(r-a)^2 + b^2\}$ ...(7)

17.4 Angular momentum of a particle and of a system of particles. Definitions. Consider a particle of mass m, moving with velocity $\mathbf{v}$ relative to some frame of reference. The linear momentum is $m\mathbf{v}$. Then the *angular momentum* $\mathbf{h}$, about any point O, is defined as the moment of $m\mathbf{v}$ about O, *i.e.*,

$$\mathbf{h} = \mathbf{r} \times m\mathbf{v},$$

where $\mathbf{r}$ is the position vector of the particle relative to O. ...(1)

For a system of n particles, the angular momentum relative to a point O is defined as

$$\mathbf{h} = \sum_{i=1}^{n} (\mathbf{r}_i \times m\mathbf{v}_i), \qquad \text{...(2)}$$

where m_i, $\mathbf{r}_i$, $\mathbf{v}_i$ denote the mass, position vector (relative to a point O), and velocity of the particle, respectively.

17.5 To find an expression for angular momentum of a rigid body turning about a fixed point and fixed axes.

[Delhi Maths (Hons.) 1996]

Let O be a fixed point of the rigid body, which is rotating with angular velocity $\boldsymbol{\omega}$ about an axis through O. Consider a rectangular frame of axes at O. Let δm be the mass of a typical particle at $P(x, y, z)$, $\mathbf{r}$ its position vector and $\mathbf{v}$, its velocity. Then, let

$$\mathbf{r} = x\mathbf{i} + y\mathbf{j} + z\mathbf{k} \quad \text{and} \quad \boldsymbol{\omega} = \omega_1\mathbf{i} + \omega_2\mathbf{j} + \omega_3\mathbf{k} \qquad \text{...(1)}$$

Then, the angular momentum $\mathbf{h}$ of the rigid body about O is given by

$$\mathbf{h} = \Sigma(\mathbf{r} \times \delta m\ \mathbf{v}) = \Sigma\delta m\ (\mathbf{r} \times \mathbf{v}),$$

where the summation extends over all particles in the body. ...(2)

Since $\mathbf{v} = \boldsymbol{\omega} \times \mathbf{r}$, we have

$$\mathbf{r} \times \mathbf{v} = \mathbf{r} \times (\boldsymbol{\omega} \times \mathbf{r}) = (\mathbf{r}\cdot\mathbf{r})\boldsymbol{\omega} - (\mathbf{r}\cdot\boldsymbol{\omega})\mathbf{r}$$

$$= (x^2+y^2+z^2)(\omega_1\mathbf{i}+\omega_2\mathbf{j}+\omega_3\mathbf{k}) - (x\omega_1+y\omega_2+z\omega_3)(x\mathbf{i}+y\mathbf{j}+z\mathbf{k}), \text{ using (1)}$$

If $h = h_1\mathbf{i} + h_2\mathbf{j} + h_3\mathbf{k}$, then relation (2) reduces to

$$h_1\mathbf{i}+h_2\mathbf{j}+h_3\mathbf{k} = \Sigma\delta m[\{(y^2+z^2)\omega_1 - xy\omega_2 - xz\omega_3\}\mathbf{i} + \{(z^2+x^2)\omega_2 - yz\omega_3 - yx\omega_1\}\mathbf{j} + \{(x^2+y^2)\omega_3 - zx\omega_1 - zy\omega_2\}\mathbf{k}]$$

or $$h_1\mathbf{i}+h_2\mathbf{j}+h_3\mathbf{k} = (A\omega_1 - H\omega_2 - G\omega_3)\mathbf{i} + (B\omega_2 - F\omega_3 - H\omega_1)\mathbf{j} + (C\omega_3 - G\omega_1 - F\omega_2)\mathbf{k},$$

where A, B, C, F, G, H are the moments and products of inertia of the rigid body with respect to coordinate axes ox, oy, oz.

Hence the angular momenta h_1, h_2, h_3 of the rigid body about the x-, y- and z-axes are respectively are given by

$$\begin{aligned} h_1 &= A\omega_1 - H\omega_2 - G\omega_3 \\ h_2 &= -H\omega_1 + B\omega_2 - F\omega_3 \\ h_3 &= -G\omega_1 - F\omega_2 + C\omega_3 \end{aligned} \qquad \ldots(3)$$

Re-writing the equations (3) in matrix form, we get

$$\begin{bmatrix} h_1 \\ h_2 \\ h_3 \end{bmatrix} = \begin{bmatrix} A & -H & -G \\ -H & B & -F \\ -G & -F & C \end{bmatrix} \begin{bmatrix} \omega_1 \\ \omega_2 \\ \omega_3 \end{bmatrix} \qquad \ldots(4)$$

The symmetrical 3×3 matrix on the R.H.S. is also known as *inertia matrix.*

Corollary 1. If Ox, Oy, Oz are principal axes, then $F = G = H = O$ and formulae (3) are greatly simplified. They become ...(5)

$$h_1 = A\omega_1, \qquad h_2 = B\omega_2, \qquad h_3 = C\omega_3$$

Then, the angular momentum $\mathbf{h}$ of the rigid body O is given by ...(6)

$$\mathbf{h} = A\omega_1\mathbf{i} + B\omega_2\mathbf{j} + C\omega_3\mathbf{k},$$

where A, B, C are principle moments of inertia at O.

Corollary 2. *If a rigid body is constrained to rotate about a fixed axis, then the angular momentum vector does not lie along the axis of rotation, unless the latter is a principal axis of inertia.*

Proof. Let z-axis be the axis of rotation so that $\omega_1 = \omega_2 = 0$ and $\omega_3 = \omega$. Then equations (3) give

$$h_1 = -G\,\omega, \qquad h_2 = -F\,\omega, \qquad h_3 = C\,\omega$$

Then $\mathbf{h} = -\omega(G\mathbf{i} + F\mathbf{j} - C\mathbf{k})$, which is not along z-axis.

Clearly, if $F = G = 0$, *i.e.*, z-axis is a principal axis of inertia, then $\mathbf{h} = C\omega\mathbf{k}$ giving $h = C\omega$, where $h = |\mathbf{h}|$ ...(7)

Relation (7) shows that the angular momentum is equal to the product of M.I. about the axis of rotation and the angular velocity.

Corollary 3. *To find an expression of the kinetic energy of the rigid body in terms of its angular momentum.*

If A, B, C be the principal moments of inertia at O, then angular momentum $\mathbf{h}$ is given by

$$\mathbf{h} = A\omega_1\mathbf{i} + B\omega_2\mathbf{j} + C\omega_3\mathbf{k}$$

Also, $$\boldsymbol{\omega} = \omega_1\mathbf{i} + \omega_2\mathbf{j} + \omega_3\mathbf{k}$$

$$\therefore \quad \boldsymbol{\omega}\cdot\mathbf{h} = A\omega_1^2 + B\omega_2^2 + C\omega_3^2 = 2T \quad \text{or} \quad T = (\boldsymbol{\omega}\cdot\mathbf{h})/2,$$

where T is the K.E. of the rigid body with a fixed point rotating with constant angular velocity $\boldsymbol{\omega}$.

17.6 Angular momentum of a rigid body in general.

To show that the angular momentum of a moving system about a point O is the sum of the following parts : (i) the angular momentum about O of a particle moving with the mass centre and having a mass equal to the total mass of the system (ii) the angular momentum of the system about the mass centre. **[Delhi Maths (H) 1994]**

Proof. For figure, refer Art. 17.1. Let O be a fixed origin and O' the centre of mass of the given of n particles. Let $\overrightarrow{OO'} = \mathbf{r}_0$. Let P_i be the position of particle of mass m_i. Let the position vectors of P_i relative O and O' be $\mathbf{r}_i$ and $\mathbf{r}'_i$ respectively, Then,

$$\mathbf{r}_i = \mathbf{r}_0 + \mathbf{r}'_i \qquad \ldots(1)$$

From (1), $\quad d\mathbf{r}_i/dt = (d\mathbf{r}_0/dt) + (d\mathbf{r}'_i/dt) \quad$ or $\quad \mathbf{v}_i = \mathbf{v}_0 + \mathbf{v}'_i$ ...(2)

where $\mathbf{v}_i = d\mathbf{r}_i/dt$, $\mathbf{v}'_i = d\mathbf{r}'_i/dt$ and $\mathbf{v}_0 = d\mathbf{r}_0/dt$ = velocity of centre of mass.

The angular momentum **h** of the system about O is given by

$$\mathbf{h} = \Sigma(\mathbf{r}_i \times m_i\mathbf{v}_i) = \Sigma m_i(\mathbf{r}_i \times \mathbf{v}_i)$$
$$= \Sigma m_i(\mathbf{r}_0 + \mathbf{r}_i') \times (\mathbf{v}_0 + \mathbf{v}_i') \text{ using (1) and (2)}$$
$$= \Sigma m_i(\mathbf{r}_0 \times \mathbf{v}_0 + \mathbf{r}_0 \times \mathbf{v}_i' + \mathbf{r}_i' \times \mathbf{v}_0 + \mathbf{r}_i' \times \mathbf{v}_i')$$

$$\therefore\ \mathbf{h} = \mathbf{r}_0 \times (\Sigma m_i)\mathbf{v}_0 + \mathbf{r}_0 \times (\Sigma m_i\mathbf{v}_i') + (\Sigma m_i\mathbf{r}_i') \times \mathbf{v}_0 + \Sigma(\mathbf{r}_i' \times m_i\,\mathbf{v}_i') \qquad ...(3)$$

Now, by definition, the position vector of the mass centre O' relative to O' as origin is given by

$$\mathbf{0} = (\Sigma m_i\mathbf{r}_i')/\Sigma m_i \quad \text{so that} \quad \Sigma m_i\mathbf{r}_i' = \mathbf{0} \qquad ...(4)$$

Diff. w.r.t. 't', (4) gives $\qquad \Sigma m_i\mathbf{v}_i' = \mathbf{0}$ ...(5)

Now, $\Sigma m_i = m$ = mass of whole system of n particles. ...(6)

Also, $\qquad \mathbf{h}' = \Sigma(\mathbf{r}_i' \times m_i\mathbf{v}_i')$, ...(7)

where $\mathbf{h}'$ is the angular momentum of the system about O'.

Using (4), (5), (6) and (7), (3) reduces to

$\mathbf{h} = \mathbf{r}_0 \times m\mathbf{v}_0 + \mathbf{h}'$, which proves the required result.

ILLUSTRATIVE SOLVED EXAMPLES

Ex. 1. *A body turns about a fixed point. Prove that the angle between its angular velocity vector and its angular momentum vector (about the fixed point) is always acute. Show that, if the principal moments of inertia A, B, C are all different, then the angle vanishes only if the body is turning about a principal axis.* **[Delhi Maths (H) 1995]**

Sol. Let Ox, Oy, Oz be principal axes at a fixed point O of the given body, which is turning about O with angular velocity $\boldsymbol{\omega}$ given by

$$\boldsymbol{\omega} = \omega_1\mathbf{i} + \omega_2\mathbf{j} + \omega_3\mathbf{k} \qquad ...(1)$$

Let A, B, C be the principal moments of inertia of the body at O. Then the angular momentum **h** of the body about O is given by

$$\mathbf{h} = A\omega_1\mathbf{i} + B\omega_2\mathbf{j} + C\omega_3\mathbf{k} \qquad ...(2)$$

Let θ be the angle between the vectors $\boldsymbol{\omega}$ and **h** given by (1) and (2). Then,

$$\cos\theta = \frac{\mathbf{h}\cdot\boldsymbol{\omega}}{|\mathbf{h}||\boldsymbol{\omega}|} = \frac{A\omega_1^2 + B\omega_2^2 + C\omega_3^2}{(\omega_1^2+\omega_2^2+\omega_3^2)^{1/2}\cdot(A^2\omega_1^2 + B^2\omega_2^2 + C^2\omega_3^2)^{1/2}} \qquad ...(3)$$

Since A, B, C are always positive, (3) $\Rightarrow$ $\cos\theta > 0$, showing that θ is always an acute angle. This proves the first part of the problem.

Second part. Let the body be turning about a principal axis, say z-axis. Then $\omega_1 = \omega_2 = 0$ and $\omega_3 = \omega$ (say) and so (3) reduces to

$$\cos\theta = (C\omega^2)/(C\omega)\cdot(\omega) = 1, \text{ as } C \neq 0,$$

showing that $\theta = 0$ *i.e.*, angle vanishes when body rotates about a principle axis.

Ex. 2. *A rod, of length 2a and mass m, turns about one end O, describing a cone with semi-vertical angle α. It compltes a revolution in time T. Find the magnitude and direction of the angular momentum of the rod about O.* **[Delhi Math (Hons.) 1998, 2002, 2005, 2008, 2010]**

Sol. Let the rod OD turn about O, describing a cone ODE with semi-vertical angle α. Then the end D of the rod moves in a circle and completes a revolution in time T. Let ω be the constant angular velocity of the rod, then

$$2\pi = \omega T \quad \text{so that} \quad \omega = 2\pi/T \qquad ...(1)$$

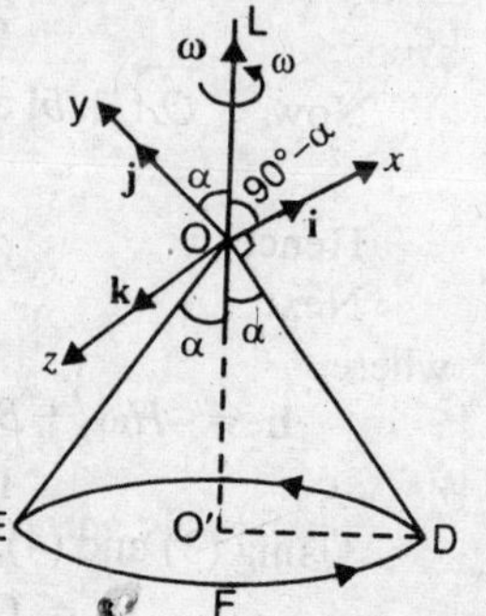

We select y-axis along the rod DO, x-axis through O and perpendicular to DO. Let x-axis and y-axis be so chosen that the unit vectors **i**, **j** and angular velocity vector $\boldsymbol{\omega}$ are always coplanar. Let z-axis be chosen through O and perpendicular to plane of **i**, **j** and $\boldsymbol{\omega}$. Then $Oxyz$ forms a right handed orthgonal triod of axes. By symmetry, $Oxyz$ will be a principal frame for the rod at O. Then the principal moments of inertia A, B, C are given by

$$A = C = (4/3)ma^2 \quad \text{and} \quad B = 0 \qquad ...(2)$$

The angular velocity $\boldsymbol{\omega}$ of the rod is given by

$$\boldsymbol{\omega} = \omega\cos(90^\circ - \alpha)\mathbf{i} + (\omega\cos\alpha)\mathbf{j} + (\omega\cos 90^\circ)\mathbf{k} = \omega_1\mathbf{i} + \omega_2\mathbf{j} + \omega_3\mathbf{k}$$

Thus, $\omega_1 = \omega \sin \alpha, \qquad \omega_2 = \omega \cos \alpha, \qquad \omega_3 = 0$ ...(3)

Hence the angular momentum $\mathbf{h}$ of the rod about O is given by

$$\mathbf{h} = A\omega_1\mathbf{i} + B\omega_2\mathbf{j} + C\omega_3\mathbf{k} = (4/3)ma^2\omega \sin \alpha\, \mathbf{i}, \text{ by (2) and (3)}$$

or
$$\mathbf{h} = (4/3)ma^2 \times (2\pi/T) \sin \alpha\, \mathbf{i}, \text{ by (1)}$$

showing that the angular momentum vector lies in a vertical plane containing the rod. It acts in direction perpendicular to the rod and its magnitude is

$$(8\pi ma^2 \sin \alpha)/3T.$$

Ex. 3. *OA is a light rod of length b which turns with angular velocity Ω about an axis OB perpendicular to it. A is the middle point of a rod CD of mass m and length 2a, hinged to OA at A in such a way that CD is always coplanar with OB. If θ denotes the angle OAC, prove that the componenets of the angular momentum of CD about O in the directions OA, OB and a direction perpendicular to them are, respectively,* $-(1/3) \times ma^2\Omega \sin\theta \cos\theta$, $m\Omega\{b^2 + (a^2/3) \times \cos^2\theta\}$ *and* $(1/3) \times ma^2\dot{\theta}$. **[Delhi Math (H) 1995]**

Sol. Let x-axis and y-axis be taken along OA and OB respectively. Let z-axis be taken perpendicular to both of them.

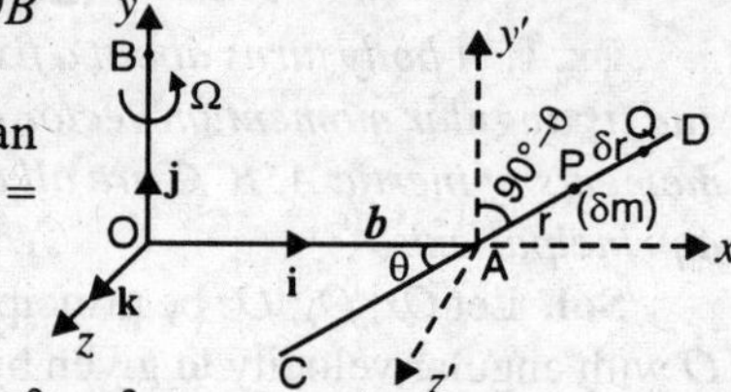

Now mass per unit length of the rod CD is $m/2a$. Consider an element PQ of breadth δr at a distance r from A. Then, δm = mass of the element $PQ = (m/2a)\,\delta r$

$\therefore\quad A$ = M.I. of the rod CD about AX

$$= \int_{-a}^{a}\left(\frac{m}{2a}\right)(r\sin\theta)^2\,dr = \frac{m\sin^2\theta}{2a}\int_{-a}^{a} r^2\,dr = \frac{1}{3}ma^2\sin^2\theta$$

B = M.I. of the rod CD about Ay' (parallel to OB)

$$= \int_{-a}^{a}\left(\frac{m}{2a}\right)(r\cos\theta)^2\,dr = \frac{m\cos^2\theta}{2a}\int_{-a}^{a} r^2\,dr = \frac{1}{3}ma^2\cos^2\theta$$

C = M.I. about Az', which is $\perp$ to Ax and Ay'

$= A + B = (1/3)ma^2(\sin^2\theta + \cos^2\theta) = (1/3)ma^2$

H = Product of inertia of the rod CD about Ax and Ay'

$$= \int_{-a}^{a}\left(\frac{m}{2a}\right)\cdot(r\sin\theta)(r\cos\theta)\,dr = \frac{1}{3}ma^2\sin\theta\cos\theta$$

$F = G = O$, as $z = 0$, motion being in xy-plane

Now, the angular velocity $\boldsymbol{\omega}$ of the system is given by

$\boldsymbol{\omega}$ = Angular velocity of the centre of mass A with respect to O
+ Angular velocity of the rod CD about A

or
$$\boldsymbol{\omega} = \Omega\mathbf{j} + \dot{\theta}\,\mathbf{k} = \omega_1\mathbf{i} + \omega_2\mathbf{j} + \omega_3\mathbf{k}, \text{ say}$$

$\therefore\quad \omega_1 = 0, \qquad \omega_2 = \Omega \qquad \text{and} \qquad \omega_3 = \dot{\theta}$ ...(1)

Let $\mathbf{h}$ be the angular momentum of rod CD about O. Then

$\mathbf{h}$ = Angular momentum $\mathbf{h}_0$ of the centre of mass A about O
+ Angular momentum $\mathbf{h}_A$ of the rod CD relative to mass centre A

i.e.,
$$\mathbf{h} = \mathbf{h}_0 + \mathbf{h}_A$$

Now, $\overrightarrow{OA} = b\mathbf{i}$ and $\mathbf{v}_A$ = velocity of the centre of mass about O

$$= \Omega\mathbf{j} \times b\mathbf{i} = -\Omega b\mathbf{k}, \text{ as } \mathbf{j} \times \mathbf{i} = -\mathbf{k}$$

Hence,
$$\mathbf{h} = b\mathbf{i} \times m\mathbf{v}_A = b\mathbf{i} \times (-m\Omega b\mathbf{k}) = \Omega mb^2\mathbf{j}, \quad \text{...(3)}$$

Next,
$$\mathbf{h}_A = h_1\mathbf{i} + h_2\mathbf{j} + h_3\mathbf{k}, \text{ (refer Art. 17.5)}$$

where,
$$\mathbf{h}_1 = A\omega_1 - H\omega_2 - G\omega_3 = -H\Omega,$$

$$h_2 = -H\omega_1 + B\omega_2 - F\omega_3 = B\Omega \quad \text{and} \quad h_3 = -G\omega_1 - F\omega_2 + C\omega_3 = C\dot{\theta}$$

$\therefore$
$$\mathbf{h}_A = -H\Omega\mathbf{i} + B\Omega\mathbf{j} + C\dot{\theta}\,\mathbf{k} \quad \text{...(4)}$$

Using (3) and (4), (2) reduces to

$$\mathbf{h} = \Omega mb^2\mathbf{j} - H\Omega\mathbf{i} + B\Omega\mathbf{j} + C\dot{\theta}\,\mathbf{k}$$

or $\qquad \mathbf{h} = -(1/3)ma^2\Omega \sin\theta \cos\theta\, \mathbf{i} + m\Omega\, \{b^2 + (a^2/3)\cos^2\theta\}\mathbf{j}$
$\qquad + (1/3)ma^2\dot{\theta}\, \mathbf{k}$, using values of A, B, C, F, G, H.

From the above expression of **h**, we see that the componenets of angular momentum of CD about O in the directions OA, OB and a direction perpendicular to them are respectively,

$$-(1/3) \times ma^2\Omega \sin\theta\cos\theta,\; m\Omega\,[b^2 + (a^2/3)\times\cos^2\theta\} \text{ and } (1/3)\times ma^2\dot{\theta}.$$

Ex. 4. *A uniform circular disc of radius a and mass m is rigidly mounted on one end of a thin light shaft CD, of length b. The shaft is normal to the disc at its centre C. The disc rolls on a rough horizontal plane, D being fixed in this plane by a smooth universal joint. If the centre of disc rotates about the vertical through D with angular velocity n, find the angular velocity, kinetic energy, and angular momentum of the disc about D.* **[Delhi Maths (Hons.) 2000, 2003, 10]**

Sol. Let EF be the given circular disc of radius a and mass m which is mounted on one end C of a thin light shaft CD. CD is perpendicular to the plane of the disc at is centre C. Let the disc roll on rough horizontal plane DE. Let E be the point of contact of the disc with the plane and $\angle CDE = \alpha$. From ΔCDE, $DE = (a^2 + b^2)^{1/2}$ and so

$$\sin\alpha = a/(a^2 + b^2)^{1/2} \quad \text{and} \quad \cos\alpha = b/(a^2 + b^2)^{1/2} \qquad \ldots(1)$$

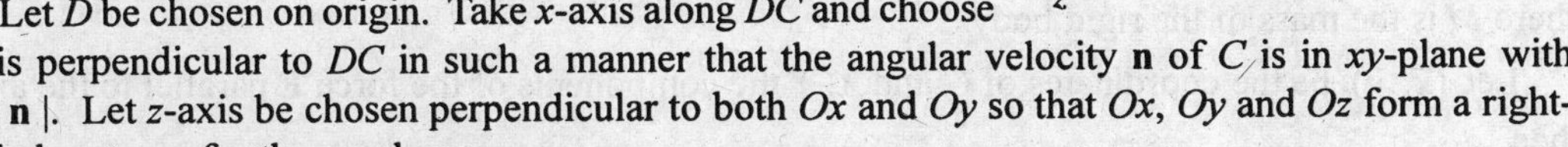

Let D be chosen on origin. Take x-axis along DC and choose y-axis perpendicular to DC in such a manner that the angular velocity **n** of C is in xy-plane with $n = |\,\mathbf{n}\,|$. Let z-axis be chosen perpendicular to both Ox and Oy so that Ox, Oy and Oz form a right-handed system of orthogonal axes.

The angular velocity **n** of C about DN is given by

$$\mathbf{n} = n\cos(90° - \alpha)\mathbf{i} + n\cos\alpha\,\mathbf{j} = (n\sin\alpha)\mathbf{i} + (n\cos\alpha)\mathbf{j} \qquad \ldots(2)$$

Due to symmetry, $Oxyz$ is a principal frame of axes. Hence, the principal moments of inertia A, B, C of the rigid body (which is circular disc EF here because mass of rod CD being zero will not contribute towards M.I.) about the axes Dx, Dy, Dz at D are given by

$$A = (1/2)ma^2,\; B = (1/4)ma^2 + mb^2 = (1/4)m\,(a^2 + 4b^2) = C \qquad \ldots(3)$$

As shown in the figure, when the disc rolls on the plane, the point E will describe a circle of radius a about C. Simultaneously C will also describe a circle of radius $DE(= (a^2 + b^2)^{1/2})$ about D.

Let θ and ϕ be the respective angles described by E about C and D. Using the condition for pure rolling, we obtain

$$a\theta = (a^2 + b^2)^{1/2}\phi \quad \text{so that} \quad \dot{\theta} = \frac{(a^2+b^2)^{1/2}}{a}\dot{\phi} = \frac{(a^2+b^2)^{1/2}}{a}n, \qquad \ldots(4)$$

[$\because$ E rotates with constant angular velocity about $D \Rightarrow \dot{\phi} = n$]

Let $\boldsymbol{\omega}$ be the angular velocity of the disc about D. Then

$\boldsymbol{\omega}$ = Angular velocity of C about D + Angular velocity of any point E of disc about C

or $\qquad \boldsymbol{\omega} = \mathbf{n} + \dot{\theta}\,(-\mathbf{i}) = (n\sin\alpha)\mathbf{i} + (n\cos\alpha)\mathbf{j} - \dot{\theta}\,\mathbf{i}$, using (2)

or $\qquad \boldsymbol{\omega} = \dfrac{na}{(a^2+b^2)^{1/2}}\mathbf{i} + \dfrac{nb}{(a^2+b^2)^{1/2}}\mathbf{j} - \dfrac{(a^2+b^2)^{1/2}}{a}n\mathbf{i}$, using (1) and (4)

or $\qquad \boldsymbol{\omega} = -\{nb^2/a(a^2 + b^2)^{1/2}\}\mathbf{i} + \{nb/(a^2 + b^2)^{1/2}\}\mathbf{j} + 0\mathbf{k}, \qquad \ldots(5)$

which is required angular velocity of the disc about D.

Let $\boldsymbol{\omega} = \omega_1\mathbf{i} + \omega_2\mathbf{j} + \omega_3\mathbf{k}$. Then (5) gives

$$\omega_1 = -nb^2/a(a^2 + b^2)^{1/2}, \qquad \omega_2 = nb/(a^2 + b^2)^{1/2}, \qquad \omega_3 = 0 \qquad \ldots(6)$$

Let **h** be the angular momentum of the disc about D. Then

$$\mathbf{h} = \omega_1 A\,\mathbf{i} + \omega_2 B\,\mathbf{j} + \omega_3 C\,\mathbf{k}$$

or $\quad \mathbf{h} = -\dfrac{nb^2}{a(a^2+b^2)^{1/2}} \times \dfrac{ma^2}{2}\mathbf{i} + \dfrac{nb}{(a^2+b^2)^{1/2}} \times \dfrac{m(a^2+4b^2)}{4}\mathbf{j}$, using (3) and (6)

or $\qquad \mathbf{h} = -\{mnab^2/2(a^2 + b^2)^{1/2}\}\mathbf{i} + \{mnb(a^2 + 4b^2)/4(a^2 + b^2)^{1/2}\}\mathbf{j},$

which is the required angular momentum of the disc about D.

Let T be the kinetic energy of the disc about D. Then

$$T = (A\omega_1^2 + B\omega_2^2 + C\omega_3^2)/2 \quad \text{or} \quad 2T = A\omega_1^2 + B\omega_2^2 + C\omega_3^2$$

or $$2T = \frac{ma^2}{2} \times \frac{n^2b^4}{a^2(a^2+b^2)} + \frac{m(a^2+4b^2)}{4} \times \frac{n^2b^2}{(a^2+b^2)}, \text{ using (3) and (6)}$$

or $$T = \frac{mn^2b^2}{8(a^2+b^2)}(2b^2+a^2+4b^2) \quad \text{or} \quad T = \frac{mn^2b^2(a^2+6b^2)}{8(a^2+b^2)}$$

17.7 Equation of motion of a rigid body in two dimensions. *To find the equations of motion in two dimensions when the forces acting are finite.*

We know that motion of a rigid body is a combination of the following two independent motions : (*i*) the motion of a centre of gravity and (*ii*) the motion about the centre of gravity. Now (*i*) states that the motion of the centre of gravity G is such that the total mass M of the rigid body is concentrated at the C.G. 'G' of the body and all the external forces are transferred parallel to themselves to act at G. Therefore, if $\bar{\mathbf{r}}$ be the position vector of the C.G. and $\mathbf{F}$ the external force acting at any point m of the body, then

$$M(d^2\bar{\mathbf{r}}/dt^2) = \Sigma\mathbf{F}, \qquad \text{...(1)}$$

where M is the mass of the rigid body.

Let $(\bar{x}, \bar{y})$ be the coordinates of G and X, Y the components of the force $\mathbf{F}$ parallel to the axes. Then

$$\bar{\mathbf{r}} = \bar{x}\,\mathbf{i} + \bar{y}\,\mathbf{j} \quad \text{and} \quad \mathbf{F} = X\mathbf{i} + Y\mathbf{j}$$

Substituting the above values in (1), we get

$$M\{(d^2\bar{x}/dt^2)\mathbf{i} + (d^2\bar{y})dt^2)\mathbf{j}\} = \Sigma(X\mathbf{i} + Y\mathbf{j})$$

so that $$M(d^2\bar{x}/dt^2) = \Sigma X \qquad \text{...(2)}$$

and $$M(d^2\bar{y}/dt^2) = \Sigma Y \qquad \text{...(3)}$$

Equation (2) and (3) are the equations of motion of G.

Again, (*ii*) states that the sum of the momenta, of the effective forces about G, is equal to the sum of the moments of the external forces about G. Hence if $\mathbf{r}'$ is the position vector of P (where mass m of the rigid body is placed) relative to G and $\mathbf{F}$ the external force acting on it, then

$$\Sigma\mathbf{r}' \times m\frac{d^2\mathbf{r}'}{dt^2} = \Sigma\mathbf{r}' \times \mathbf{F} \quad \text{or} \quad \frac{d}{dt}\Sigma m\mathbf{r}' \times \frac{d\mathbf{r}'}{dt} = \Sigma\mathbf{r}' \times \mathbf{F} \qquad \text{...(4)}$$

Let ϕ be the angle that the line joining P to G makes with a line GB fixed in space, and θ the angle that a line GA fixed in the body, makes with GB. Let $\angle AGP = \alpha$ so that $\phi = \theta + \alpha$, where α is constant.

Now, $$\phi = \theta + \alpha \Rightarrow \dot{\phi} = \dot{\theta}$$

Let $GP = r' = |\mathbf{r}'|$. Then the velocity of P relative to G.

$= r'\dot{\phi} = r'\dot{\theta}$ in a direction perpendicular to $\mathbf{r}'$ in the plane AGP.

If $\mathbf{e}_1$, $\mathbf{e}_2$ be the unit vectors along and perpendicular to $\mathbf{r}'$ in the plane AGP, then we have

$$\mathbf{r}' = r'\,\mathbf{e}_1 \quad \text{and} \quad d\mathbf{r}'/dt = r'\dot{\theta}\,\mathbf{e}_2 \qquad \text{...(5)}$$

Let $\mathbf{n}$ be the unit vector normal to the plane AGP

$\therefore\ \Sigma m\mathbf{r}' \times (d\mathbf{r}'/dt)$

$= \Sigma m(r'\mathbf{e}_1) \times (r'\dot{\theta}\mathbf{e}_2)$, using (5)

$= \dot{\theta}\,(\Sigma mr'^2)\mathbf{n}$, where $\mathbf{e}_1 \times \mathbf{e}_2 = \mathbf{n}$ ...(6)

$= \dot{\theta}\,(Mk^2)\mathbf{n}$, where k is the radius of gyration of the body about G ...(7)

Also, $\Sigma\mathbf{r}' \times \mathbf{F} = \Sigma p'F\mathbf{n}$, by definition of momentum, where p' is in the length of the perpendicular from G upon the direction of the force **F** and $F = |\mathbf{F}|$. Using (6) and (7), (4) may be re-written as

$$\frac{d}{dt}(Mk^2\dot{\theta})\mathbf{n} = (\Sigma p'F)\mathbf{n} \qquad \text{so that} \qquad Mk^2\ddot{\theta} = \Sigma p'F \qquad \ldots(8)$$

Let (x', y') the co-ordinates of P relative to G. Then, $p'F = x'Y - y'X$. So (8) reduces to

$$Mk^2\ddot{\theta} = \Sigma(x'Y - y'X) \qquad \ldots(9)$$

Equations (2), (3) and (9) are the required equations of motion of a rigid body moving in two dimensions, under finite forces.

ILLUSTRATIVE SOLVED EXAMPLES

Ex. 1. (a) *Uniform solid circular cylinder rolls down an inclined rough plane. Discuss the motion.* **[Kanpur 2005; Guwahati 2007; Delhi Maths (H) 2007]**

Sol. Let the cylinder be initially at rest with its point A in contact with O on the plane. At any time t, let the centroid C of the cylinder describe a distance x parallel to the inclined plane and θ the angle through which the cylinder turns. Figure shows the vertical section of the cylinder through its centroid C. Here CA, a line fixed in the body makes an angle θ with the normal to the plane, a line fixed in space. Let P be the point of contact of the circular section of the cylinder through C with the inclined plane at time t.

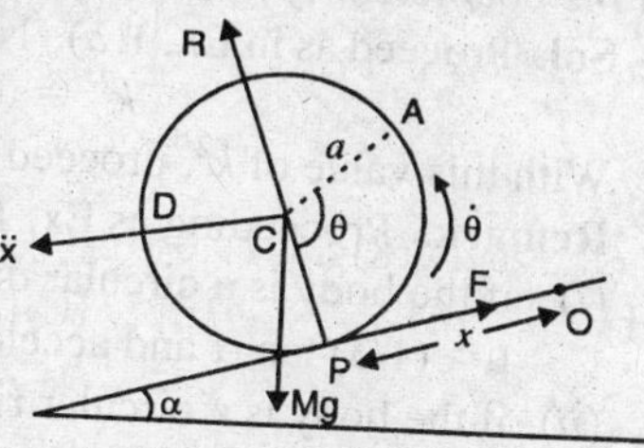

If F is the frictional force on the cylinder at P and R the normal reaction of the plane on the cylinder at P, then the equations of motion of centroid C are

$$m\ddot{x} = mg \sin\alpha - F \qquad \ldots(1)$$

$$0 = mg\cos\alpha - R \qquad \ldots(2)$$

Also, the equation of the motion of the centroid C is given by

$$mk^2\ddot{\theta} = F\cdot a, \qquad \ldots(3)$$

where k is the radius of gyration of the cylinder about the line through C perpendicular the plane of rotation Thus, $\qquad k^2 = a^2/2 \qquad \ldots(4)$

In order that there may be no sliding of the cylinder, *i.e.*, rolling may occur without slipping, we have arc PA = distance OP or $a\theta = x$ or $a\ddot{\theta} = \ddot{x}$ $\qquad \ldots(5)$

From (3) and (5), $\qquad mk^2(\ddot{x}/a) = F\cdot a \qquad \ldots(6)$

or $\qquad mk^2(\ddot{x}/a) = a(mg\sin\alpha - m\ddot{x})$, using (1)

or $\qquad \ddot{x}(k^2/a + a) = ga\sin\alpha$ or $\ddot{x} = (a^2 g\sin\alpha)/(k^2 + a^2) \qquad \ldots(7)$

Integrating (7), $\qquad \dot{x} = (a^2 g\, t\sin\alpha)/(k^2 + a^2) + C \qquad \ldots(8)$

where C is an arbitrary constant.

Initially at O, when $t = 0$, $\dot{x} = 0$. So $C = 0$ by (8)

Thus, (8) gives $\qquad dx/dt = \{(a^2 g\sin\alpha)/(k^2 + a^2)\}t$

Integrating, $\qquad x = (a^2 g t^2 \sin\alpha)/2(a^2 + k^2) + D, \qquad \ldots(9)$

where D is an arbitrary constant.

Initially at O, when $t = 0, x = 0$. So (9) gives $D = 0$

Thus, by (9), $\qquad x = (a^2 g t^2 \sin\alpha)/2(a^2 + k^2) \qquad \ldots(10)$

Using (4), (6) reduces to $\quad F = (m\ddot{x})/2 \quad$ or $\quad m\ddot{x} = 2F$

Then (1) becomes $\quad 2F = mg\sin\alpha - F \quad$ or $\quad F = (1/3)mg\sin\alpha \qquad \ldots(11)$

Also, from (2) $\qquad R = mg\cos\alpha \qquad \ldots(12)$

If μ be the coefficient of friction, then rolling may occur without slipping if

$F < \mu R$ or $\mu > F/R$ or $\mu > (1/3)\tan\alpha$, by (11) and (12)

Ex. 1. (b) *A uniform solid cylinder is placed with its axis horizontal on a plane, with inclination α to the horizontal. Show that the least coefficient of friction between it and plane, so that it may roll and may not slide is (1/3) tan α.* **[Guwahati 2007; Kanpur 2005]**

Sol. Proceed as in the last part of solution of Ex. 1. (*a*). We have proved $\mu > (1/3) \tan \alpha$ and so the required least value of μ is $(1/3) \tan \alpha$.

Ex. 1. (c) *A uniform hollow cylinder is placed with its axis horizontal on a plane, with inclination α to the horizontal. Show that the least coefficient of friction between it and plane, so that it may roll and may not slide is (1/2) tan α.*

Sol. Proceed as in Ex. 1(*a*). Note that for hollow circular cylinder equation (4) becomes

$$k^2 = a^2 \qquad \text{...(4)}'$$

With this value of *k*, proceed yourself to show that $\mu > (1/2) \tan \alpha$. Hence the least value of μ is $(1/2) \tan \alpha$.

Ex. 2. *A uniform solid sphere rolls rolls down a line of greatest slope of a plane, rough enough to prevent sliding at angle α to the horizontal. Find the motion and show that $\mu > (2/7) \tan \alpha$, where μ is the coefficient of friction between the sphere and the plane.* **[Kanpur 2007]**

Sol. Proceed as in Ex. 1(*a*). Note that for solid sphere equation (4) becomes

$$k^2 = (2/5)a^2 \qquad \text{...(4)}'$$

With this value of k^2, proceed to prove that $\mu > (2/7) \tan \alpha$.

Remark. Proceeding as Ex. 1(*a*) we can show that

(*i*) if the body is a circular disc so that $k^2 = a^2/2$, then
$\mu > (1/3) \tan \alpha$ and acceleration is $\ddot{x} = (2/3)g \sin \alpha$

(*ii*) if the body is a circular ring so that $k^2 = a^2$, then
$\mu > (1/2) \tan \alpha$ and acceleration is $\ddot{x} = (1/2)g \sin \alpha$

(*iii*) if the body is a hollow sphere so that $k^2 = (2/3)a^2$, then
$\mu > (2/5) \tan \alpha$ and acceleration $= \ddot{x} = (1/2)g \sin \alpha$

Ex. 3. *Discuss the motion of uniform solid circular cylinder rolling down a rough inclined plane and show that it rolls down with constant acceleration.* **(Delhi Maths (H) 2007)**

[**Hint.** Refer Ex. 1(a). Here (7) $\Rightarrow$ *acceleration is constant*]

Miscellaneous Examples

1. *A rigid body turns about some fixed point O. If ω_1, ω_2, ω_3* be angular velocities and A, B, C are the moments of inertia about the principal axes OA, OB, OC. Find the kinetic energy of the body.

[**Hint.** Refer equation (5) of Art. 17.2] **[Kanpur 2006, 07; Meerut 2007]**

2. Prove that the kinetic energy of a rigid body moving in two dimensions can be given as $1/2) \times mv^2 + (1/2) \times mk^2 \dot{\theta}^2$ **[Purvanchal 2006]**

MISCELLANEOUS PROBLEMS AND RESULTS ON THE ENTIRE BOOK

Ex.1. *Two cars start off to race with velocities v_1 and v_2 and travel in a straight line with uniform accelerations f_1 and f_2. If the race ends in a dead heat, prove that the length of the course is $\{2(v_1 - v_2)(v_1 f_2 - v_2 f_1)\}/(f_1 - f_2)^2$.* **[Delhi B.A. (Prog) III 2010]**

Sol. Let the length of the course be d. Since the race ends in a dead heat, it follows that the cars reach the destination at the same time. Suppose that each car takes time T to describe the distance d. Then, using formula $s = ut + (1/2) \times ft^2$,

for the first car, we have $\quad d = v_1 T + (1/2) \times f_1 T^2 \quad \ldots(1)$

and for the second car, we have $\quad d = v_2 T + (1/2) \times f_2 T^2 \quad \ldots(2)$

Substracting (2) from (1), $\quad 0 = (v_1 - v_2)\, T + (1/2) \times (f_1 - f_2) T^2$

or $\quad T\{v_1 - v_2 + (1/2) \times (f_1 - f_2)\, T\} = 0 \quad$ so that $\quad v_1 - v_2 + (1/2) \times (f_1 - f_2)\, \mathrm{T} = 0, \quad \ldots(3)$

since $T = 0$ corresponds to the initial position. Now, from (3), we have

$(1/2) \times (f_1 - f_2)\, T = -(v_1 - v_2) \quad$ so that $\quad T = -\{2(v_1 - v_2)/(f_1 - f_2)\} \quad \ldots(4)$

Substituting the value of T given by (4) in (1), we have

$$d = -\frac{2v_1(v_1 - v_2)}{f_1 - f_2} + \frac{1}{2} \times \frac{4f_1(v_1 - v_2)^2}{(f_1 - f_2)^2} = \frac{2(v_1 - v_2)}{f_1 - f_2}\left\{-v_1 + \frac{f_1(v_1 - v_2)}{f_1 - f_2}\right\}$$

or $$d = \frac{2(v_1 - v_2)}{f_1 - f_2} \times \frac{v_1 f_2 - v_2 f_1}{f_1 - f_2} \quad \text{or} \quad d = \frac{2(v_1 - v_2)(v_1 f_2 - v_2 f_1)}{(f_1 - f_2)^2}$$

Ex.2. *A uniformly accelerated automobile passes two telegraph poles with velocities 10 and 20 m.p.h respectively. Calculate its velocity when it is half way between the poles.*

[Delhi B.Sc. (Prog) III 2009]

Sol. Let the two given poles be situated at points A and B such that $AB = d$ miles. Let M be the middle point of AB so that $AB = d/2$ miles. Let $v_A (= 10 \text{ m.p.h})$, v_m and $v_B (= 20 \text{ m.p.h.})$ be the velocities of the automobile at A, M and B respectively. Let f be the uniform accleration of the automobile. Then using the formula $v^2 = u^2 + 2fs$,

for motion from A to B, we have $\quad v_B^2 = v_A^2 + 2fd \quad \ldots(1)$

and for motion from A to M, we have $\quad v_M^2 = v_A^2 + 2f \times (d/2) \quad$ or $\quad 2v_M^2 = 2v_A^2 + 2fd \quad \ldots(2)$

Subtracting (1) from (2), $\quad 2v_M^2 - v_B^2 = v_A^2 \quad$ so that $\quad v_M^2 = (v_B^2 + v_A^2)/2$

Then, $\quad v_M^2 = \{(20)^2 + (10)^2\}/2 = 500 \quad$ so that $\quad v_m = 5\sqrt{10}$ m.p.h.

Ex.3. *A particle moves in a straight line with uniform acceleration and its distances from O on the line (not necessarily the position t = 0) at t_1, t_2, t_3 are d_1, d_2, d_3 respectively. Prove that if t_1, t_2, t_3 form an A.P., whose common difference is d and d_1, d_2, d_3 are in G.P., then acceleration is $(\sqrt{d_1} - \sqrt{d_3})^2/d^2$.* **[Delhi B.A. (Prog.) III, 2014]**

Sol. Let the given particle start from a point A, with velocity u, where the distance of A from the origin is a. Let f be uniform acceleration of the particle. Then, we have

$$d_1 = a + ut_1 + (1/2) \times f t_1^2 \quad \ldots(1)$$

$$d_2 = a + ut_2 + (1/2) \times f t_2^2 \quad \ldots(2)$$

$$d_3 = a + ut_3 + (1/2) \times f t_3^2 \quad \ldots(3)$$

Since t_1, t_2, t_3 form an A.P. with common difference d, we have

$t_2 - t_1 = t_3 - t_2 = d$ so that $t_1 + t_3 - 2t_2 = 0$ ...(4)

Since d_1, d_2, d_3 form an G..P. we have $d_2 = \sqrt{d_1 d_3}$ ...(5)

Adding (1) and (3), $d_1 + d_3 = 2a + u\,(t_1 + t_3) + (1/2) \times f(t_1^2 + t_3^2)$ ...(6)

Again, from (2), $2d_2 = 2a + 2ut_2 + (1/2) \times f \times (2t_2^2)$ ...(7)

Subtracting (7) from (6), $d_1 + d_3 - 2d_2 = u\,(t_1 + t_3 - 2t_2) + (f/2) \times (t_1^2 + t_3^2 - t_2^2)$

or $d_1 + d_3 - 2d_2 = (u \times 0) + (f/2) \times (t_1^2 + t_3^2 - t_2^2)$, using (4)

Thus, $f = 2\,(d_1 + d_3 - 2d_2) / (t_1^2 + t_3^2 - 2t_2^2)$ ...(8)

Now, using (5), $d_1 + d_3 - 2d_2 = d_1 + d_3 - 2\sqrt{d_1 d_3} = (\sqrt{d_1} - \sqrt{d_3})^2$ ...(9)

Again, $t_1^2 + t_3^2 - 2t_2^2 = (t_3^2 - t_2^2) - (t_2^2 - t_1^2) = (t_3 - t_2)(t_3 + t_2) - (t_2 - t_1)(t_2 + t_1)$

$= d\,(t_3 + t_2) - d\,(t_2 + t_1)$, using (4)

$= d\,(t_3 + t_2 - t_2 - t_1) = d\,\{(t_3 - t_2) + (t_2 - t_1)\} = d \times 2d$, by (4)

Then, $t_1^2 + t_3^2 - 2t_2^2 = 2d^2$ ...(10)

Using (9) and (10), (8) yields $f = (\sqrt{d_1} - \sqrt{d_3})^2 / d^2$

Ex.4. *A stone is dropped into an empty pit of depth h is heard to strike the bottom after t seconds. Prove that $2h\,\{1 + (gt/v)\} = gt^2$ when v, the velocity of sound is so large in magnitude as compared with h that $(h/v)^2$ can be neglected* **(Pune 2010)**

Sol. Let t_1 be the time taken by the stone to reach the bottom of the pit. Then, using $h = ut + (1/2) \times gt^2$, we have

$h = 0 + (1/2) \times gt_1^2$ so that $t_1 = (2\,h/g)^{1/2}$ ...(1)

Let t_2 be the time taken by the sound to reach the top of the pit, Then, we have

$h = vt_2$ so that $t_2 = h/v$ ...(2)

Given that $t_1 + t_2 = t$, Hence adding (1) and (2), we have

$(2\,h/g)^{1/2} + h/v = t$ or $(2h/g)^{1/2} = t - (h/v)$ or $2h/g = t^2 - (2ht/v) + h^2/v^2$

giving $2\,h/g = t^2 - (2ht/v)$, since by hypothesis $(h/v)^2$ can be neglected

Thus, $2h = gt^2 - (2htg)/v$ so that $2h\,\{1 + (gt/v)\} = gt^2$

Ex.5 *A body sliding down a smooth inclined plane is observed to cover equal distances, each equal to l, in consecutive intervals of time t_1 and t_2. Show that the inclination of the plane is $\sin^{-1}\{2l\,(t_1 - t_2) / gt_1t_2\,(t_1 + t_2)\}$* **[G.N.D.U. Amritsar 2010]**

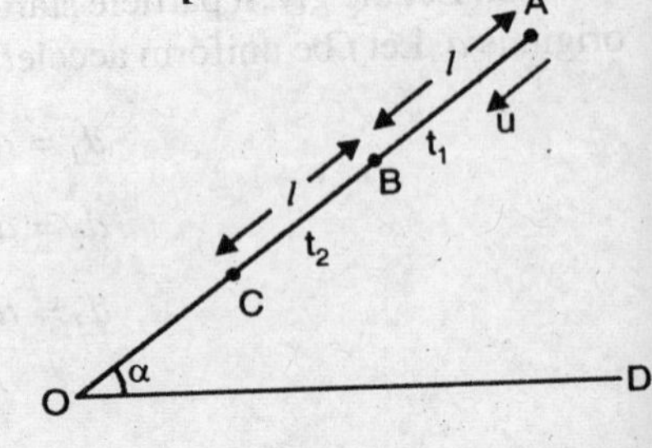

Sol. Let α be the inclination of the plane to the horizon. Let u be the velocity of the body at A. Let $AB = BC = l$ so that time from A to B is t_1 and time from B to C is t_2. Hence, the time taken by the body to move from A to C is $t_1 + t_2$. Then using formula $s = ut + (1/2) \times ft^2$, for motion from A to B, we have $\qquad l = ut_1 + (1/2) \times g \sin\alpha \times t_1^2 \qquad \ldots(1)$

and for motion from A to C, we have $\qquad 2l = u(t_1 + t_2) + (1/2) \times g \sin\alpha \times (t_1 + t_2)^2 \ldots(2)$

From (1) and (2), $\quad \frac{l}{t_1} = u + \frac{1}{2} \times g \sin\alpha \times t_1 \quad$ and $\quad \frac{2l}{t_1 + t_2} = u + \frac{1}{2} \times g \sin\alpha \times (t_1 + t_2)$

$\Rightarrow \quad \frac{2l}{t_1 + t_2} - \frac{l}{t_1} = \frac{1}{2} \times (g \sin\alpha) \times t_2 \qquad \Rightarrow \quad \frac{l(2t_1 - t_1 - t_2)}{t_1(t_1 + t_2)} = \frac{1}{2} \times (g \sin\alpha) \times t_2$

$\Rightarrow \quad \sin\alpha = \frac{2l(t_1 - t_2)}{gt_1 t_2 (t_1 + t_2)} \qquad \Rightarrow \quad \alpha = \sin^{-1}\left\{\frac{2l(t_1 - t_2)}{gt_1 t_2 (t_1 + t_2)}\right\}$

Ex. 6. *In Atwood's machine there are two masses $m_1 = 10$ kg and $m_2 = 20$ kg, which are hanging over a smooth pulley by an elastic, light string, then find the acceleration produced.*

(Pune 2010)

Sol. Atwood's a machine. In this machine two equal weights are attached to the ends of a string and suspended over a light pulley running with a little friction as possible. Over one of the weights resting on the platform an additional weight called the rider, is placed. The platform is dropped instantaneously and the system begins to move.

As shown in the adjoining figure, the mass m_1 moves up and mass m_2 moves down with the same acceleration f, say. Since the pulley is smooth, the tension T of the string is same throughout the string.

Equations of motion of the masses m_1 and m_2 are given by

$T - m_1 g = m_1 f \qquad$ and $\qquad m_2 g - T = m_2 f$

Adding these, $(m_2 - m_1) g = (m_1 + m_2) f$ so that $f = \{(m_2 - m_1) g\} / (m_1 + m_2)$

Here, $m_1 = 10$ kg and $m_2 = 20$kg. Hence, $f = (10\,g)/30 = (g/3)$ m/sec^2

Ex.7 *State Newton's second law of motion and deduce the fundamental equation of motion.*

(G.N.D.U. Amritsar 2010)

Sol. Statement of Newton's second law of motion. *The rate of change of momentum is directly proportional to the impressed forces and takes place in the direction of the force*

Deduction of the fundamental equation of motion, namely, $P = mf$

Let P be the impressed force acting upon a body of mass m which is moving with velocity v in the direction of the force P. Then, by Newton's second law of motion, we have

$P \,\alpha\, \frac{d}{dt}(mv) \qquad$ so that $\qquad P = k\frac{d}{dt}(mv)$, k being a constant scalar $\qquad \ldots(1)$

If we assume the mass to be independent of time, then (1) yields

$$P = kmf, \text{ where } f(= dv/dt) \text{ is acceleration} \qquad \ldots(2)$$

If the unit of force be so chosen that it produces unit acceleration while acting on a unit mass, i.e., if $P = 1$, when m = 1 and f = 1, then (2) reduces to $k = 1$. Hence (1) yields

$P = mf$, which is known as the fundamental equation of motion.

Ex. 8. *Let velocity of point P be* $\mathbf{u} = 10\mathbf{i} + 2\mathbf{j}$ *and let velocity of Q be* $\mathbf{v} = 2\mathbf{i} + 2\mathbf{j}$. *Then, find the magnitude of velocity of P with respect to Q.*

(Pune 2010)

Sol. Let **w** be the velocity of P with respect to Q.

Then, $\qquad \mathbf{w} = \mathbf{u} - \mathbf{v} = (10\,\mathbf{i} + 2\,\mathbf{j}) - (2\mathbf{i} + 2\,\mathbf{j}) = 8\,\mathbf{i}$

Hence, the required magnitude of velocity of P with respect to $Q = |\mathbf{w}| = |8\,i| = 8$.

Ex.9. *Prove that the work down in elastic string AB of natural length 'l' and modulus 'λ' from tension T_1 to T_2 is given by $(T_2^2 - T_1^2)/2\lambda$.*

(Pune 2010)

Sol. Let $AB\,(=1)$ be the natural length of an elastic string. Let T_1 and T_2 be the tensions in the string when its extended lengths are $AP_1\,(=1+x_1)$ and $AP_2\,(=1+x_2)$ respectively. Then, extension $= AP_2 - AP_1 = x_2 - x_1$. Again, using formula $T = (\lambda x)/l$ for tension in an elastic string, we have

A B T_1 P_1 T_2 P_2 ; 1 ; x_1 ; x_2

$$T_1 = \lambda x_1 \quad \text{and} \quad T_2 = \lambda x_2, \quad \text{as} \quad l = AB = 1.$$

Thus, we have $\quad x_1 = T_1/\lambda \quad$ and $\quad x_2 = T_2/\lambda \quad$ ...(1)

Now, the required work done = mean of tension × extension

[Refer result of solved example on page 2.1]

Thus, the required workdone

$$= (1/2)\times(T_1+T_2)\times(x_2-x_1) = (1/2)\times(T_1+T_2)(T_2/\lambda - T_1/\lambda) = (T_2^2 - T_1^2)/2\lambda$$

Ex.10. *Define conservative forces, kinetic energy and potential energy. Derive expression for kinetic energy and deduce that the rate of change of kinetic energy with respect to space is equal to the magnitude of the impressed forces* **[G.N.D.U. Amritsar 2010]**

Sol. Refer Art. 2.8, Art 2.10 and Art 2.11 for the first part.

Deduction. The rate of change of kinetic energy w.r.t. space

$$= \frac{d}{ds}\left(\frac{1}{2}mv^2\right) = \frac{1}{2}m\times 2v\frac{dv}{ds} = mf, \quad \text{as} \quad f = v\frac{dv}{ds}$$

= The magnitude of the impressed forces.

Ex. 11. *Define force potential. If force* $\mathbf{F} = x\mathbf{i} + y\mathbf{j} + z\mathbf{k}$ *is acting on a particle, then find force potential.* **(Pune 2010)**

Sol. In what follows we shall require the statement of the following theorem:

Theorem. *A necessary and sufficient condition that a field of force* $\mathbf{F} = X\mathbf{i} + Y\mathbf{j} + Z\mathbf{k}$ *be conservative is that there exists a differentiable scalar function* $V(x, y, z)$ *such that*

$$X = -(\partial V/\partial x), \quad Y = -(\partial V/\partial y) \quad \text{and} \quad Z = -(\partial V/\partial z).$$

The function $V(x, y, z)$ so defined is known as the *potential function* (or *force potential*) of the consecutive field.

In the given problem, $\quad \mathbf{F} = x\mathbf{i} + y\mathbf{j} + z\mathbf{k} \quad$ ...(1)

Let $V(x, y, z)$ be the required force potential. Then, we have

$$x = -(\partial V/\partial y), \quad y = -(\partial V/\partial y) \quad \text{and} \quad z = -(\partial V/\partial z) \quad ...(2)$$

Now, $\quad dV = (\partial V/\partial x)\,dx + (\partial V/\partial y)\,dy + (\partial V/\partial z)\,dz = -(x\,dx + y\,dy + z\,dz)$, using (2)

Integrating, $\quad V = -(x^2 + y^2 + z^2)/2$

Ex. 12. *Show that* $\mathbf{F} = yz\mathbf{i} + zx\mathbf{j} + xy\mathbf{k}$ *is a conservative field and find the potential function with zero value at (1, 1, 1).*

Sol. Clearly there exists a scalar function $\quad V = C - xyz \quad$ ...(1)

which is single valued and is such that

$$-\partial V/\partial x = yz = x - \text{component of the force}$$

$$-\partial V/\partial y = zx = y - \text{component of the force}$$

and $$-\partial V/\partial z = xy = z - \text{component of the force}$$

Hence, by definition, the given field in conservative and the function defined by (1) is the associated potential function. Since $V = 0$ when $x = y = z = 1$ and hence by (1), C = 1. Thus, the required potential function is $V = 1 - xyz$.

Ex.13. *A particle of mass m moves on a straight line under the influence of a force directed towards the origin O on the line and proportional to the distance from O; the force at unit distance is of magnitude mk^2. The particle passes O with velocity u, If x is its coordinate at time t and v its velocity at that instant, show that $v^2 + k^2 x^2 = u^2$.* **[Delhi B.Sc. (Prog) III 2008]**

Sol. By hypothesis, the equation of motion of the given particle is

$m\ddot{x} = -mk^2x$ or $v\,(dv/dx) = -k^2x$ or $2v\,dv = -2k^2x\,dx$

Integrating, $v^2 = C - k^2x^2$, C being an arbitrary constant ...(1)

Given that when $x = 0$, $v = u$. Hence, (1) gives $C = u^2$. therefore, (1) yields

$v^2 = u^2 - k^2x^2$ or $v^2 + k^2x^2 = u^2$

Ex. 14. *If a particle moves along a straight line according to the law $v^2 = ax - bx^3$, prove that $27\,bv^4 = 4\,(a - 2f)\,(a + f)^2$, f being acceleration* **[Delhi B.A. (Prog) III 2010, 16 17]**

Sol. Given $v^2 = ax - bx^3$ ...(1)

Differentiating (1) w.r.t. 'x', $2v\,(dv/dx) = a - 3bx^2$

Hence $2f = a - 3bx^2$, as $f = v\,(dv/dx)$ ...(2)

Now, R.H.S. of the required result

$= 4\,(a - 2f)\,(a + f)^2 = 4\,\{a - (a - 3bx^2)\}\,\{a + (a - 3bx^2)/2\}^2$, using (2)

$= 4 \times (3b\,x^2) \times \{\,(3a - 3bx^2)^2/4\,\} = 27\,bx^2\,(a - bx^2)^2 = 27b\,(ax - bx^3)^2$

$= (27b) \times (v^2)^2 = 27\,bv^4$, using (1)

$=$ L.H.S. of the required result.

Thus, $27bv^4 = 4\,(a - 2f)\,(a + f)^2$

Ex. 15. *If a particle is moving according to the law $v^2 = 2\,(x \sin x + \cos x)$ where x is the distance described, find its acceleration.* **(Delhi B.A. (Prog) III 2011)**

Sol. Given $v^2 = 2x \sin x + 2 \cos x$ ... (1)

Differentiating (1) w.r.t. 'x', $2v\,(dv/dx) = 2 \sin x + 2 \cos x - 2 \sin x$

Thus $f = \cos x$, as $f = v\,(dv/dx) =$ acceleration

Hence, The required acceleration is $\cos x$.

Ex. 16. *Show that the particle from infinity will reach the surface of the earth with a velocity 11.4 km per second* **(Kanpur 2009, 10, 13)**

Hint: Proceed as in Ex. 6 (*b*), page 3.4 of chapter 3

Ex. 17. *A particle is shot upwards from the earth's surface with a velocity 1.6 km/sec. Considering variations in gravity, find roughly in kilometers greatest height attained. Take the radius of earth as 6380 km and g = 9.8 m/sec²* **(Meerut 2010)**

Sol. Let r be the radius of the earth. Suppose the particle is projected vertically upwards from the surface of the earth with velocity u and let the greatest height attained be H above the surface of the earth. Let P be the position of the particle at any time t and x the distance of P from the centre of the earth. Since P is outside the surface of the earth, the equation of motion of P is given by

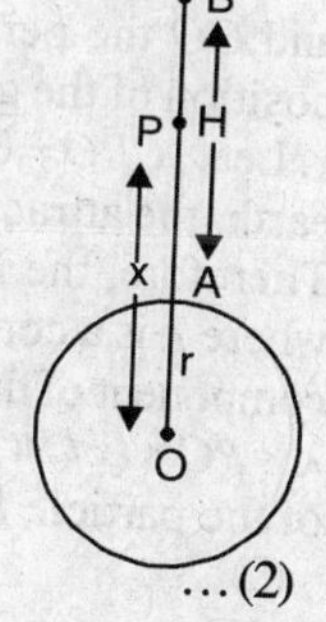

$d^2x/dt^2 = -(\mu/x^2)$, μ being a constant ...(1)

But on the surface of earth, $x = r$ and $d^2x/dt^2 = -g$. So (1) $\Rightarrow \mu = r^2$.

Hence (1) reduces to $d^2x/dt^2 = -(gr^2/x^2)$ or $v\,(dv/dx) = -(gr^2/x^2)$

Thus, $2v\,dv = -\{(2gr^2)/x^2\}\,dx$

Integrating, $v^2 = (2gr^2)/x + C$, C being an arbitrary constant ...(2)

But, when $x = r$, $v = u$ (given). Hence, (2) gives $C = u^2 - 2gr$

Hence, (2) reduce to $\qquad v^2 = (2gr^2)/x + u^2 - 2gr \qquad \ldots(3)$

Again, let $v = 0$, when $x = OB = OA + AB = r + H$. Hence, (3) gives

$0 = (2gr^2)/(r+H) + u^2 - 2gr \quad$ or $\quad 0 = 2gr^2 + u^2(r+H) - 2gr(r+H)$

or $\quad H(2gr - u^2) = u^2 r \quad$ or $\quad H = (u^2 r)/(2gr - u^2) \qquad \ldots(4)$

In our given problem, $r = 6380$ km, $u = 1.6$ km/sec and $g = 9.8$ m/sec^2 = $10^{-3} \times 9.8$ km/sec^2. Hence, (4) yields

$$H = \frac{1.6 \times 1.6 \times 6380}{2 \times 10^{-3} \times 9.8 \times 6380 - (1.6 \times 1.6)} \text{ km} = 133 \text{ km (approximately)}$$

Ex. 18. *In S.H.M., let u, v, w be the velocities at distances a, b, c from a fixed point on a line which is not centre of the force. Let T be the period of S.H.M. Find the relation between T, a, b, c,* **(Kanpur 2009)** *u^2, v^2 and w^2.*

Sol. Proceed as in Ex. 16 (a), page 4.8. The required is given by

$$\frac{4\pi^2}{T^2}(a-b)(b-c)(c-a) = \begin{vmatrix} u^2 & v^2 & w^2 \\ a & b & c \\ 1 & 1 & 1 \end{vmatrix}$$

Ex.19. *If we assume that the earth is homogeneous sphere of radius a, the attraction on a particle inside the earth produces an acceleration which varies directly the distance from the particle would take centre of the earth and the acceleration on the surface is g, then find the time a to move down a smooth straight tube from the surface of the earth to centre?* **(I.A.S. 2009)**

Sol. Let AA' be the straight tube passing through the centre of the earth. Let the particle be at P at any time t such that $OP = x$. Then, according to the given problem, we have

$$\ddot{x} = -\lambda x, \text{ where } \lambda \text{ is a constant of proportionality} \qquad \ldots(1)$$

Since (1) holds at A, where $\ddot{x} = -g$ (given) and $x = a$, we have $-g = -\lambda a$ so that $\lambda = g/a$. Hence, (1) reduces to

$$\ddot{x} = -(g/a) \times x,$$

which is the standard equation of a S.H.M. of period T gives by

$$T = 2\pi/(g/a)^{1/2} = 2\pi \times (a/g)^{1/2}$$

Therefore, the required time for motion from A to $O = T/4 = (\pi/2) \times (a/g)^{1/2}$.

Ex. 20. *Show that, if a smooth tunnel be cut between any two points on the earth's surface, it would be traversed by a particle starting from rest under the attraction of gravity in about 42.5 minutes nearly (by taking radius of earth = a = 4000 × 1760 × 3 feet and g = 32.2 feet /sec^2*

Sol. Let C be the centre of the earth, AA' the tunnel and CO the perpendicular from C on AA'. Let P be position of the given particle at any time t, where $OP = x$. Let $\angle CPO = \theta$, We know that inside the surface of the earth, the attraction varies as the distance from centre. Therefore, the force acting on P is $\lambda \times PC$ along PC, where λ is a constant of proportionality. Therefore, the component of this force along PO is $\lambda \times PC \times \cos\theta$, i.e., $\lambda \times PC \times (x/PC)$, i.e., λx. Hence, the equation of motion of the particle P is given by

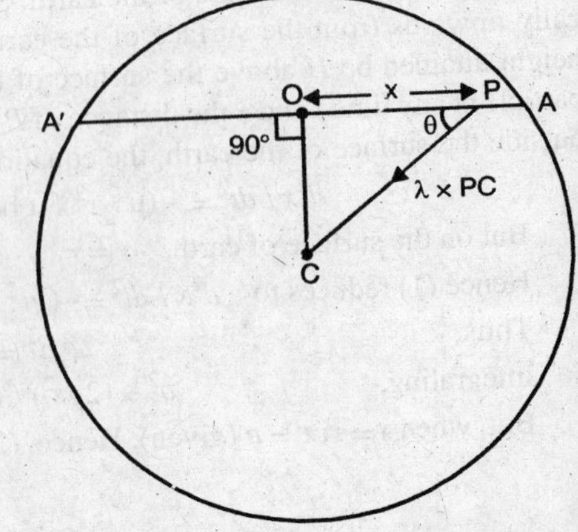

$$\ddot{x} = -\lambda x \qquad \ldots(1)$$

Now, on the surface of the earth, $\ddot{x} = -g$ and $x = a$. Hence, (1) gives $\lambda = g/a$

Therefore, (1) reduces to
$$\ddot{x} = -(g/a) \times x, \qquad \ldots(2)$$
which is the standard equation of S.H.M. of period $= T = 2\pi/(g/a)^{1/2}$. Therefore,

$$\text{the required time} = \frac{T}{2} = \pi\left(\frac{a}{g}\right)^{1/2} = 3.1416 \times \left(\frac{4000 \times 1760 \times 3}{32.2}\right)^{1/2} = 42.5 \text{ minutes}, \quad \text{(nearly)}$$

Ex. 21. *A vertical string extends a distance l when a given weight is attached to its lower point. The weight is pulled down further a distance a (< l) and let go. Find the period of simple harmonic motion that ensues and show that the maximum velocity of the weight is a $(g/l)^{1/2}$.*

(Delhi B.A. (Prog.) III, 2013; Mumbai 2010)

Sol. Let l_0 (= OA) be the natural length of the string as shown in figure (i). As shown in figure (ii), let B be the position of equilibrium such that extension $= AB = l$

Then, the tension T_0 in the string will balance the weight mg of the particle. Thus, we have

$$mg = T_0 \Rightarrow mg = \lambda(l/l_0) \Rightarrow \lambda = (mgl_0)/l \quad \ldots(1)$$

As shown in figure (*iii*), let the weight be pulled down further upto position C where $BC = a$ (< l) and then let go. Let P be the position of the weight at any subsequent time t such that $BP = x$.

Fig. (*i*)

Fig. (*ii*)

Fig. (*iii*)

Then, equation of motion of the mass is given by

$$m\ddot{x} = mg - T = mg - \lambda\frac{l + x}{l_0} = mg - \frac{\lambda l}{l_0} - \frac{\lambda x}{l_0} = -\frac{\lambda x}{l_0}, \text{ using (1)}$$

or
$$m\ddot{x} = -(mgl_0/l) \times (x/l_0) \quad \text{or} \quad \ddot{x} = -(g/l) \times x, \text{ using (1)}$$

which is standard equation of S.H.M. of the form $\ddot{x} = -\mu x$ where $\mu = g/l$. Hence the motion is S.H.M. about B where amplitude $= BC = a$, periodic time $= 2\pi/\sqrt{\mu} = 2\pi \times (l/g)^{1/2}$ and maximum velocity $= a\sqrt{\mu} = a(g/l)^{1/2}$. (Refer results (*i*), (*iii*) and (*iv*) of Art. 4.1)

Ex. 22. *Prove that for a particle moving along a curve, the tangential and normal components of velocity are* $(\dot{s}, 0)$ and of acceleration are $(dv/dt, v^2/\rho)$, *where v is the speed and ρ the radius of curvature at the point.*

[(Delhi B.Sc. (Prog) III 2008]

Sol. Refer Art. 5.7, page 5.21. From equation (2) of Art. 5.21, the tangential and normal components of velocity vector v are $(\dot{s}, 0)$. Again, from equation (3) of Art. 5.21, The tangential and normal components of acceleration vector f are $(dv/dt, v^2/\rho)$. For complete solution, entire Art. 5.7 must be proved.

Ex. 23. *Prove that for a particle moving in a plane and occupying the position P (r, θ) at any instant t, the radial and transverse components of velocity are* $(\dot{r}, r\dot{\theta})$ *and of acceleration are*
$$\left(\ddot{r} - r\dot{\theta}^2, \frac{1}{r}\frac{d}{dt}(r^2\dot{\theta})\right)$$

[Delhi B.Sc. (Prog). III 2009]

Sol. First give complete proof of Art 5.6, page 5.10. The radial and transverse components are (v_r, v_θ), i.e., $(\dot{r}, r\dot{\theta})$. Again, the radial and transverse components of a acceleration are (f_r, f_θ), i.e., $\left(\ddot{r} - r\dot{\theta}^2, \frac{1}{r}\frac{d}{dt}(r^2\dot{\theta})\right)$.

Ex. 24 *A point is describing a circle of radius a with velocity v. If ω and ω′ are its angular velocities about two points which are inverse points with respects to the circle, prove that ω + ω′ = v/a* **(Kanpur 2007)**

Sol. Let A and A' be two inverse points with respect to a circle with O as its centre. Then, by definition of inverse point, we have

$OA \times OA' = a^2$ so that $OA = a^2 / OA'$ ...(1)

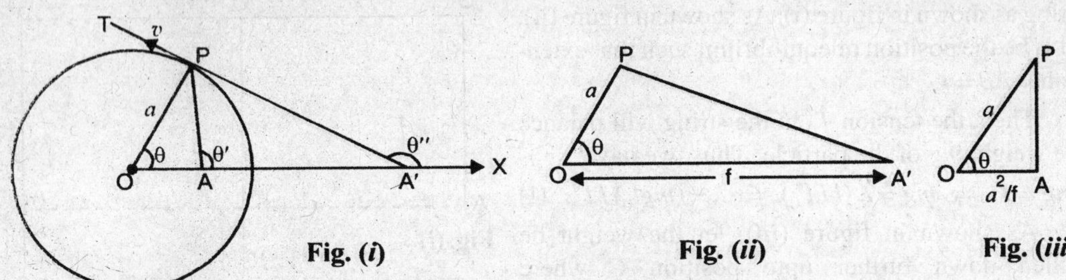

Fig. (*i*) Fig. (*ii*) Fig. (*iii*)

Let $OA' = f$. Then, from (1), $OA = a^2 / f$. From triangles OAP and $OA'P$, we have

$OA / OP = OP / OA'$ and including $\angle ACP = \angle A'CP$ ...(2)

From (2) and plane geometry, it follow that

$\angle OAP = \angle OPA'$ and $\angle OPA = \angle OA'P$ ...(3)

Let $\angle POX = \theta$, $\angle PAX = \theta'$ and $\angle PA'X = \theta''$ ... (4)

Then, ω = angular velocity of the point P about $A = d\theta' / dt$ ...(5)

and ω' = angular velocity of the point P about $A' = d\theta'' / dt$ ...(6)

Thus, $\omega + \omega' = d\theta' / dt + d\theta'' / dt = d(\theta' + \theta'') / dt$ ...(7)

Now, from figure (*i*), $\angle OA'P = \pi - \theta''$ and so (3) $\Rightarrow$ $\angle OPA = \pi - \theta''$

Again, from OAP, θ' being an external angle, we have

$\theta' = \angle POX + \angle OPA = \theta + \pi - \theta''$ so that $\theta' + \theta'' = \theta + \pi$ ...(8)

Since the point P describes a circle of radius a with velocity v, we have (refer Art. 5.3, page 5.4)

$d\theta / dt = pv / r^2$ or $d\theta/dt = a v / a^2$, since $p = r = a$ for point P

Thus, $d\theta/dt = v / a$ ...(10)

Using (10), (9) reduces to $\omega + \omega' = v/a$

Ex. 25. *A particle moves along the curve $x = t^2 + 1, y = t^2, z = 2t + 5$, where t is the time. Find the components of velocity at t = 1 in the direction* **i** + **j** + 3**k**.

Sol. Velocity vector $= v = (dx / dt)\,\mathbf{i} + (dy / dt)\,\mathbf{j} + (dz / dt)\,\mathbf{k} = 2t\,\mathbf{i} + 2t\,\mathbf{j} + 2\mathbf{k}$

[Using the given values of x, y and z]

Hence, the required component of velocity in the direction $\mathbf{i} + \mathbf{j} + 3\mathbf{k}$

$$= (2t\,\mathbf{i} + 2t\,\mathbf{j} + 2\mathbf{k}) \cdot (\mathbf{i} + \mathbf{j} + 3\mathbf{k}) = 2t + 2t + 6$$

Thus, when $t = 1$, the required component of velocity $= 2 + 2 + 6 = 10$ units

Ex. 26. *Derive tangential and normal components of velocity and acceleration of a particle moving in a plane having intrisic coordinates (s, ψ).*

Hint: Refer Art. 5.7, page 5.21 **[Mumbai 2010]**

Ex. 27. *If a particle is moving with uniform speed 20 m/sec, then find its tangential and normal components of velocity.* **(Pune 2010)**

Hint. Use Art 5.7, page 5.21. Hence, tangential and normal components of velocity are 20 m/sec and 0 respectively.

Ex. 28. *A point moving in a plane has position vector* **r** *with respect to a fixed point O. If OP makes angle θ with fixed line through O, show that the acceleration vector* **f** *of the particle is given by*

$$\mathbf{f} = (\ddot{r} - r\dot{\theta}^2)\hat{\mathbf{r}} + \frac{1}{r}\frac{d}{dt}(r^2\dot{\theta})\hat{\boldsymbol{\theta}}.$$ **(Mumbai 2010)**

Hint. Proceeding as in Art. 5.6, page 5.10, we arrive at

$$\mathbf{f} = (\ddot{r} - r\dot{\theta}^2)\mathbf{e}_r + (2\dot{r}\dot{\theta} + r\ddot{\theta})\mathbf{e}_\theta \qquad \ldots(1)$$

Writing $\hat{\mathbf{r}}$ and $\hat{\boldsymbol{\theta}}$ in place of $\mathbf{e}_r$ and $\mathbf{e}_\theta$ respectively and noting that

$$2\dot{r}\dot{\theta} + r\ddot{\theta} = \frac{1}{r}\frac{d}{dt}(2r\dot{r}\dot{\theta} + r^2\ddot{\theta}) = \frac{1}{r}\frac{d}{dt}(r^2\dot{\theta}),$$

we see that (1) reduces to the required form of **f**.

Ex. 29. *Find the greatest distance that a stone can be thrown inside a horizontal tunnel 10 feet high with a velocity of projection of 80 ft/sec. Find the corresponding time of flight.*

[Delhi B.Sc. (Prog) III 2010]

Sol. Let α be the angle of projection. Here, we take $g = 32\ \text{ft}/\text{sec}^2$.

Now, maximum height $= 10 = \dfrac{u^2 \sin^2\alpha}{2g}$ so that $\sin^2\alpha = \dfrac{2\times 10\times 32}{80\times 80} = \dfrac{1}{10}$

Thus, $\sin\alpha = 1/\sqrt{10}$ and hence $\cos\alpha = (1-\sin^2\alpha)^{1/2} = (1-1/10)^{1/2} = 3/\sqrt{10}$.

Hence, R = the required greatest distance $= (u^2 \sin 2\alpha)/g$

or $$R = \frac{2u^2 \sin\alpha\cos\alpha}{g} = \frac{2\times 80\times 80\times(1/\sqrt{10})\times(3/\sqrt{10})}{32} = 120 \text{ feet}$$

Again, T = The required corresponding time $= \dfrac{2u\sin\alpha}{g} = \dfrac{2\times 80\times(1/\sqrt{10})}{32} = \dfrac{\sqrt{10}}{2}$ sec

Ex. 30. (*i*) *If the maximum horizontal range of a particle is R, show that the greatest height attained is R/4.*

(ii) A boy can throw a ball 60 meters. How long is the ball in the air and what height does it attain. **[Delhi B.Sc. (Prog) III 2010]**

Sol. Part (i) : Refer solved Ex.2(a), page 6.5

Part (ii): Since the boy can throw a ball 60 metres, so here R = 60 meters. Hence, by part (*i*), the maximum height attained = *R*/4 = 15 meters. But $R = u^2/g$ and so $u^2 = Rg = 60\,g$ and $\alpha = 45°$, which corresponds to the maximum horizontal range *R*. Let T be the corresponding time of flight. Thus, we have

$$T = \frac{2u\sin\alpha}{g} = \frac{2\times(60g)^{1/2}\times\sin 45°}{g} = \left(\frac{120}{g}\right)^{1/2} = \left(\frac{120}{9.8}\right)^{1/2} = 3.5 \text{ sec (approximately).}$$

Ex. 31. *A particle is projected at an angle of 75° and after 5 seconds it appears to have an elevation of 45° from the point of projection. What is the velocity of projection.* **(I.A.S. 2009)**

Sol. Let u be the velocity of projection. Given that the given particle is projected at an angle of α (= 75°) and after t seconds it appears to have elevation of θ (= 45°) from the point of projection. Then, we know that (refer result (*) of Art. 6.5, page 6.5)

$$\tan\theta = \frac{u\sin\alpha - gt}{u\cos\alpha} \quad \text{so that} \quad \tan 45° = \frac{u\sin 75° - 5g}{u\cos 75°}$$

or $u(\sin 75° - \cos 75°) = 5g$ or $u(\sin 75° - \sin 15°) = 5g$

or $u \times 2\cos 45° \sin 30° = 5g$ or $u/\sqrt{2} = 5g$ or $u = 5\sqrt{2}\,g.$

Ex.32. *Prove that if the time of flight of a bullet over a horizontal range R is T seconds, then inclination of the direction of projectile to the horizontal is $\tan^{-1}(gT^2/2R)$*

(Delhi B.A. (Prog) III 2011)

Sol. Let u be the velocity of projection and α be the angle of projection of the given bullet. Then, we know that

$$T = \frac{2u\sin\alpha}{g} \quad \text{so that} \quad gT^2 = \frac{4u^2\sin^2\alpha}{g} \qquad \text{...(1)}$$

and $$R = \frac{2u^2\sin\alpha\cos\alpha}{g} \quad \text{so that} \quad 2R = \frac{4u^2\sin\alpha\cos\alpha}{g} \qquad \text{... (2)}$$

From (1) and (2), $$\frac{gT^2}{2R} = \frac{4u^2\sin^2\alpha}{g} \times \frac{g}{4u^2\sin\alpha\cos\alpha} = \tan\alpha.$$

Hence, we have $$\alpha = \tan^{-1}(gT^2/2R).$$

Ex. 33. *The velocity of a particle at the maximum height is $\sqrt{2/5}$ times the velocity it had at half the greatest height. What is the angle of projection?* **(I.A.S. 2009)**

Sol. Let u and α be the velocity of projection and angle of projection respectively of the given particle. Let H be the maximum height attained by the particle. Then, we have

$$H = (u^2\sin^2\alpha)/2g \qquad \text{...(1)}$$

Let v_1 and v_2 be the velocities of the particle at the maximum height H and at height $H/2$ respectively. Then, we have

$$v_1 = u\cos\alpha \qquad \text{...(2)}$$

and $$v_2^2 = u^2 - 2g \times (H/2) = u^2 - (u^2\sin^2\alpha)/2, \text{ using (1)} \qquad \text{...(3)}$$

Given that $v_1 = \sqrt{2/5}\,v_2$ so that $5v_1^2 = 2v_2^2$

or $$5u^2\cos^2\alpha = 2\{u^2 - (u^2\sin^2\alpha)/2\}, \text{ using (2) and (3)}$$

or $5\cos^2\alpha = 2 - (1 - \cos^2\alpha)$ or $4\cos^2\alpha = 1$ giving $\cos\alpha = 1/2$ so that $\alpha = 60°$

Ex. 34. *If a particle is projected with velocity 30m/sec from origin, Then find farthest point on horizontal that particle can reach.* **(Pune 2010)**

Sol. Using result (iii) of Art. 6.4 and noting that $u = 30$ m/sec,

the required maximum horizontal range $= u^2/g = (30)^2/9.8 = 91.8$ meters

Ex.35. *Show that there are two values of time for which projectile is at same height from horizontal.* **(Pune 2010)**

Sol. Let O be the point of projection and let u be the velocity of projection. Let P and Q be at the same height h so that $PA = QB = h$. Let t be time taken by the particle to reach a height h. Then, we have

$$h = (u \sin \alpha) \times t - (1/2) \times gt^2 \quad \text{or} \quad gt^2 - (2u \sin \alpha)\, t + 2h = 0, \quad \ldots(1)$$

which is quadratic in t. Hence, there are two values of time for which projectile is at the same height. If t_1 and t_2 $(t_2 > t_1)$ be roots of (1), then the times taken by the projectile in moving from O to P and O to Q will be t_1 and t_2 respectively.

Ex.36. *A particle is projected with speed $(4gh/3)^{1/2}$ from a fixed point O on a table of height h standing on a horizontal floor. The particle reaches the floor at a point whose horizontal distance is 2h from O. Find the two possible angles of projection.* **[Mumbai 2010]**

Sol. Take O as origin and let horizontal and vertical lines passing through O be taken as co-ordinate axes as shown in the adjoining figure. Let M be the foot of the table such that $OM = h$. Let P be point where the particle strikes the floor MP such that $MP = 2h$. Here $u\ (= (4gh/3)^{1/2})$ is the velocity of projection, Let α be the required angle of projection.

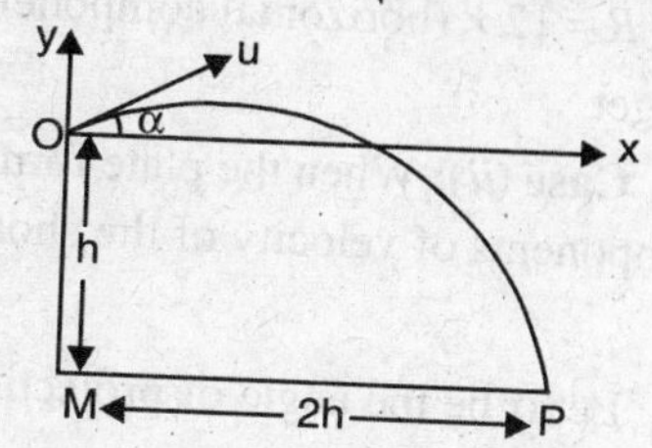

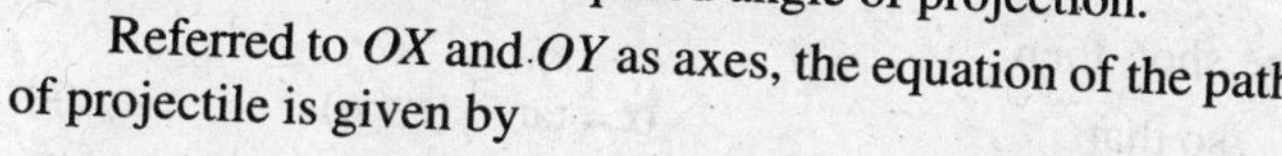

Referred to OX and OY as axes, the equation of the path of projectile is given by

$$y = x \tan \alpha - (gx^2) / (2u^2 \cos^2 \alpha) \quad \ldots(1)$$

Since $P\,(2h, -h)$ lies on (1), we have

$$-h = 2h \tan \alpha - \frac{g \times 4h^2}{2 \times (4gh/3) \times \cos^2 \alpha} \quad \text{or} \quad -1 = 2 \tan \alpha - \frac{3}{2}(1 + \tan^2 \alpha)$$

or $\quad 3 \tan^2 \alpha - 4 \tan \alpha + 1 = 0 \quad$ or $\quad (\tan \alpha - 1)(3 \tan \alpha - 1) = 0$

or $\quad \tan \alpha = 1 \quad$ or $\quad 1/3 \quad$ so that $\quad \alpha = 45° \quad$ or $\quad \alpha = \tan^{-1}(1/3)$

Ex. 37. *A particle just goes over a wall of height 'h' meters and a distance 'd' meters apart from the point of projection, and later it hits a mark at a height 'h' meters and distance '2d' meters. Show that the velocity of projection is given by $4v^2/g = (4d^2 + 9h^2)/h$*

(Delhi B.A. (Prog) III 2009, 16, 17)

Sol. Let v and α be the velocity of projection and angle of projection respectively of the particle. Then, the equation of the path of the particle is given by

$$y = x \tan \alpha - (gx^2) / (2v^2 \cos^2 x) \quad \ldots(1)$$

According to the given problem, the given particle just goes over the walls AN and BM as shown in the figure. Hence, the points $A\ (d, h)$ and $B\ (2d, h)$ must satisfy (1). Hence, using (1), we have

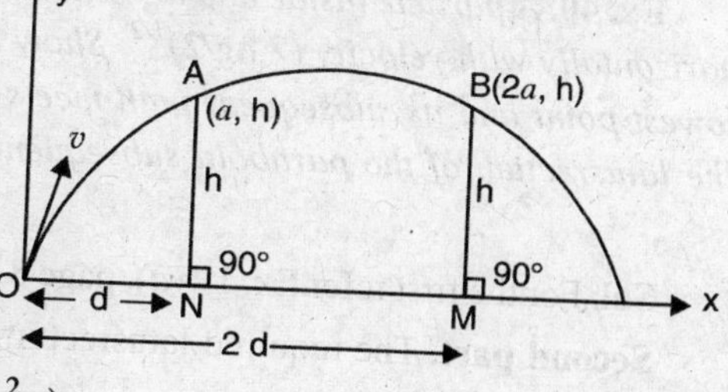

$$h = d \tan \alpha - (gd^2) / (2v^2 \cos^2 \alpha) \quad \ldots(2)$$

$$h = 2d \tan \alpha - (4gd^2) / (2v^2 \cos^2 \alpha) \quad \ldots(3)$$

Subtracting (2) from (3), $0 = d \tan \alpha - (3gd^2)/(2v^2 \cos^2 \alpha) \quad \Rightarrow \quad v^2 = (3gd)/(2 \sin \alpha \cos \alpha) \quad \ldots(4)$

Multiplying both sides of (2) by 4, we get $\quad 4h = 4d \tan \alpha - (4gd^2)/(2v^2 \cos^2 \alpha) \quad \ldots(5)$

Subtracting (3) from (5), $\quad 3h = 2d \tan \alpha \quad$ so that $\quad \tan \alpha = 3h/2d \quad \ldots(6)$

From (6), $\quad \sin \alpha = 3h/(4d^2 + 9h^2)^{1/2} \quad$ and $\quad \cos \alpha = 2d/(4d^2 + 9h^2)^{1/2}$

Substituting the above values of $\sin \alpha$ and $\cos \alpha$ in (4), we have

$$v^2 = \frac{3gd(4d^2+9h^2)}{2\times 3h\times 2d} \quad \text{or} \quad \frac{4v^2}{g} = \frac{4d^2+9h^2}{h}$$

Ex. 38. *A gun is fired from a moving platform, and the ranges of the shot are observed to be R and S, when the platform is moving backward and forward respectively, with velocity V. Prove that the elevation of the gun is* $\tan^{-1}\{g(R-S)^2/4V^2(R+S)\}$ **(Delhi B.A. (Prog) III 2010, 15, 16, 17)**

Sol. Let u and v be the horizontal and vertical components of velocity of the shot when the platform is at rest.

Case (i) : When the platform is moving forward with velocity V, the horizontal and vertical components of velocity of the shot are $u + V$ and v respectively. Then, using the following formula.

$R = \{2 \times$ (horizontal component velocity) $\times$ (vertical component of velocity)$\}/g$, ...(1)

we get $$R = \{2(u+V)v\}/g \qquad ...(2)$$

Case (*ii*): When the plate form is moving backward with velocity V, the horizontal and vertical components of velocity of the shot are $u - V$ and v respectively. Then, using formula (1), we have

$$S = \{2(u-V)v\}/g \qquad ...(3)$$

Let α be the angle of projection of the shot, then

$\tan\alpha = v/u$ so that $\alpha = \tan^{-1}(v/u)$ ...(4)

From (1) and (2), $R + S = (4uv)/g$ and $R - S = (4Vv)/g$ ... (5)

From (5), $\dfrac{(R-S)^2}{R+S} = \dfrac{16V^2v^2}{g^2}\times\dfrac{g}{4uv}$ so that $\dfrac{v}{u} = \dfrac{g(R-S)^2}{4V^2(R+S)}$ ... (6)

From (4) and (6), $$\alpha = \tan^{-1}\{g(R-S)^2/4V^2(R+S)\}$$

Ex. 39. *A particle is placed on the outside of a smooth vertical circle. If particle starts from a point whose angular distance is α from the highest point of the circle. Show that it will fly off the curve when* $\cos\theta = (2/3)\times\cos\alpha$ **(Agra 2010, Meerut 2008)**

Hint. Refer Ex. 6 (*a*), page 8.3

Ex. 40. *A particle inside and at the lowest point of a fixed hollow sphere of radius a is projected horizontally with velocity $(7ag/2)^{1/2}$. Show that it will leave the sphere at a height 3a/2 above the lowest point and its subsequent path meets the sphere again at the point of projection. Find also the latusrectum of the parabola subsequently described by the particle.*

(Delhi B.A. (Prog) III 2010)

Sol. For the first refer Ex. 19 (*a*), page 8.15.

Second part. The required latusrectum of the parabola subsequently described.

$$= \frac{2V^2\cos^2(\pi-\alpha)}{g} = \frac{2V^2\cos^2\alpha}{g} = \frac{2\times(ag/2)\times\cos^2 120^\circ}{g} = \frac{ag\times(1/4)}{g} = \frac{a}{4}$$

Ex. 41. *If the bob of a simple pendulum is projected from the position of stable equilibrium with velocity equal to that due to falling from the highest point of the circle, show that the time of describing any angle θ is* $(l/g)^{1/2}\times\log\{\sec(\theta/2)+\tan(\theta/2)\}$. **(Delhi B.A. (Prog) III 2007, 2012)**

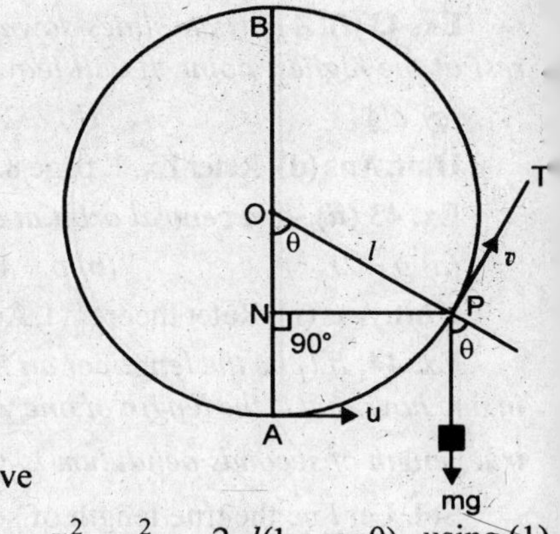

Sol. Let A, O and B be the lowest point, centre and the highest point of the circular path PAB described by the *bob* of the given pendulum. Let $AO = OP = l$, where P is the position of the bob at any time t and $\angle AOP = \theta$. Let v be the velocity of the bob at P along the tangent PT. Let PN be the perpendicular from P on AB. Then, from ΔOPN, we have

$$AN = OA - ON = l - l\cos\theta = l(1-\cos\theta) \quad \text{... (1)}$$

Let u be the velocity of the projection of the bob at A. Consider the motion of the bob from A to P.

Then, by the principle of conservation of energy, we have

$$(1/2)\times mv^2 - (1/2)\times mu^2 = (-mg)\times AN \quad \text{or} \quad v^2 - u^2 = -2gl(1-\cos\theta), \text{ using (1)}$$

Then, $\quad v^2 = u^2 - 2gl + 2gl\cos\theta \quad$ ... (2)

According to the given problem, velocity u at A is equal to that due to falling from the highest point B. Hence, considering the motion of the bob falling under gravity from B to A, we have

$$u^2 = 0 + 2g \times BA \quad \text{so that} \quad u^2 = 4gl \quad \text{... (3)}$$

From (2) and (3), $\quad v^2 = 2gl(1+\cos\theta) \quad$ or $\quad v^2 = 4gl\cos^2(\theta/2) \quad$ ... (4)

Let arc $AP = s$ so that $s = l\theta$. Hence, $\quad v = ds/dt = l(d\theta/dt) \quad$ ... (5)

From (4) and (5), $l^2(d\theta/dt)^2 = 4gl\cos^2(\theta/2) \quad$ or $\quad d\theta/dt = 2(g/l)^{1/2}\cos(\theta/2) \quad$ ... (6)

where we have taken positive sign before the radical sign because θ increases as t increases.

Separating variables, (6) yields $\quad dt = (1/2)\times(l/g)^{1/2}\sec(\theta/2)\,d\theta \quad$ ... (7)

Integrating both sides of (7) between the corresponding limits $t = 0$ to $t = t$ and $\theta = 0$ to $\theta = \theta$, we have

$$\int_0^t dt = (1/2)\times(l/g)^{1/2}\int_0^\theta \sec(\theta/2)\,d\theta \quad \text{or} \quad t = (l/g)^{1/2}\left[\log\{\sec(\theta/2)+\tan(\theta/2)\}\right]_0^\theta$$

Thus, $\quad t = \text{the required time} = (l/g)^{1/2}\log\{\sec(\theta/2)+\tan(\theta/2)\}$

Ex. 42 (*a*). *Define a conical pendulum. Show that the least angular velocity required by a particle to move in conical pendulum is* $\sqrt{g/l}$. **(G.N.D.U. Amritsar 2010)**

Sol. Proceed as in Art. 8.10 upto equation (6) and thus get $\omega \geq (g/l)^{1/2}$. From this result, it follows that the least angular velocity $= (g/l)^{1/2}$.

Ex. 42 (*b*). *A heavy particle is fixed to one end of an inextensible string in order to complete the circle, then the tension at the lowest point should be*

(a) $\geq 3\,mg$ *(b)* $\geq 4\,mg$ *(c)* $\geq 6\,mg$ *(d)* $\geq 5\,mg$ **(Garhwal 2010)**

Sol. Ans (c). Refer Art. 8.5, page 8.8. The particle will make complete circular revolution if $u^2 \geq 5ga$. Again, the tension T is given by (on replacing R by T in equation (7) of Art. 8.3 for the given problem)

$$T = (m/a)\times(u^2 + ga), \quad \text{... (1)}$$

Now, $u^2 \geq 5ga \quad \Rightarrow \quad u^2 + ga \geq 6ga \quad \Rightarrow \quad (m/a)\times(u^2+ga) \geq (m/a)\times 6ga$

Hence, $\quad$ (1) $\quad \Rightarrow \quad T \geq 6mg$

Ex. 43 (*i*). *A particle slides down the outside of a smooth vertical circle of radius a starting from rest at the highest point. It will leave the circle after describing vertical distance equal to*

(a) a/4 *(b) a/2* *(c) 2a/3* *(d) a/3* **(Agra 2010)**

Hint. Ans (d). Refer Ex. 1, page 8.2

Ex. 43 (*ii*). *In a central orbit at an apse*

(a) $p = 2r$ *(b)* $p = 1/r$ *(c)* $p = r$ *(d)* $p = r^2$ **(Agra 2010)**

Hint. Ans (c). Refer theorem I, Art. 9.1, page 9.12

Ex. 44. *If* l_1 *be the length of an imperfectly adjusted seconds pendulum which gains n seconds in one hour and* l_2 *the length of one which loses n seconds in one hour, at the same place, show that true length of seconds pendulum is* $(4l_1l_2)/(l_1 + l_2 + 2\sqrt{l_1l_2})$. **(Delhi B.A. (Prog) III 2011)**

Sol. Let l be the true length of seconds pendulum.

Time for one beat = $\pi(\text{length}/g)^{1/2}$ sec and 1 hour = 3600 seconds

By given data, $$3600 = (3600 + n)\,\pi\,(l_1/g)^{1/2} \quad \text{... (1)}$$

$$3600 = (3600 - n)\,\pi\,(l_2/g)^{1/2} \quad \text{... (2)}$$

and $$3600 = 3600\,\pi\,(l/g)^{1/2} \quad \text{... (3)}$$

From (1) and (2), $(3600 + n)\sqrt{l_1} = (3600 - n)\sqrt{l_2}$ or $n(\sqrt{l_1} + \sqrt{l_2}) = 3600\,(\sqrt{l_2} - \sqrt{l_1})$

Thus, $$n = 3600\,(\sqrt{l_2} - \sqrt{l_1})/(\sqrt{l_1} + \sqrt{l_2}) \quad \text{... (4)}$$

From (2) and (3) $3600\sqrt{l} = (3600 - n)\sqrt{l_2}$ or $\sqrt{l} = (1 - n/3600)\sqrt{l_2}$

or $$\sqrt{l} = \left(1 - \frac{\sqrt{l_2} - \sqrt{l_1}}{\sqrt{l_2} + \sqrt{l_1}}\right)\sqrt{l_2} = \frac{2\sqrt{l_1}\sqrt{l_2}}{\sqrt{l_2} + \sqrt{l_1}}, \text{ using (4)} \quad \text{... (5)}$$

Squaring, both sities of (5), we obtain

$$l = (4l_1l_2)/(l_1 + l_2 + 2\sqrt{l_1l_2}).$$

Ex. 45. *Derive the equation* $d^2u/d\theta^2 + u = P/h^2u^2$ *for motion of a particle describing central orbit under an attractive force 'P' per unit mass where u = 1/r.* **(Dehli B.A. (Prog) III 2009)**

Sol. Let F be the acceleration produced by the attractive force 'P' acting on a unit mass. Thus, $P = F \times 1 = F$. Now, proceed as in Art. 9.3 A and get $d^2u/d\theta^2 + u = F/h^2u^2$.

Ex. 46. *Find the law of force if particle describes conic* $l/r = 1 + e\cos\theta$ *with focus as pole.* **(Pune 2010)**

Hint. Proceed as in Art. 9.9 upto equation (3). Thus, $F = \mu/r^2$, showing that the force varies inversely as the square of the distance of the particle from pole.

Ex. 47. *Find the law of force for a particle under central force describing the orbit* $r = a\,(1 - \cos\theta)$ **(Mumbai 2010)**

Sol. Given $r = a(1 - \cos\theta)$ or $1 = 2au\sin^2(\theta/2)$, where $u = 1/r$ ... (1)

Taking logarithm of both sides of (1), $0 = \log 2a + \log u + 2\log\sin(\theta/2)$ ... (2)

Differentiating (2) w.r.t. 'θ', $$0 = \frac{1}{u}\frac{du}{d\theta} + \frac{2}{\sin(\theta/2)} \times \frac{1}{2}\cos\frac{\theta}{2} \Rightarrow \frac{du}{d\theta} = -u\cot\frac{\theta}{2} \quad \text{... (3)}$$

From (3), $$\frac{d^2u}{d\theta^2} = -\left[\frac{du}{d\theta}\cot\frac{\theta}{2} + u\times\left(-\frac{1}{2}\times\operatorname{cosec}^2\frac{\theta}{2}\right)\right] = u\cot^2\frac{\theta}{2} + \frac{u}{2}\operatorname{cosec}^2\frac{\theta}{2}, \text{ using (3)}$$

Substituting the above value of $d^2u/d\theta^2$ in the equation of central orbit $F/h^2u^2 = u + (d^2u/d\theta^2)$, we have

$$\frac{F}{h^2u^2} = u + u\cot^2\frac{\theta}{2} + \frac{u}{2}\operatorname{cosec}^2\frac{\theta}{2} = u\left(\operatorname{cosec}^2\frac{\theta}{2} + \frac{1}{2}\operatorname{cosec}^2\frac{\theta}{2}\right) = \frac{3}{2}u\operatorname{cosec}^2\frac{\theta}{2}$$

Thus, $$F = h^2u^3 \times (3/2)\times \operatorname{cosec}^2(\theta/2) = h^2u^3\times(3/2)\times 2au, \text{ using (1)}$$

Hence, $F = (3ah^2)/r^4$, as $u = 1/r$. Therefore the force varies inversely as the fourth power of the distance from the pole.

Ex. 48. *A particle describes an orbit with a central acceleration $\mu u^3 - \lambda u^5$, being projected from an apse at distance a with velocity equal to that from infinity, show that the path is $r = a\cosh(\theta/n)$, where $n^2 + 1 = 2\mu a^2/\lambda$* **(Delhi B.A. (Part III) 2011)**

Sol. Here, the central acceleration $= F = \mu u^3 - \lambda u^5$, where $u = 1/r$ ... (1)

If V is the velocity at a distance 'a' acquired in falling from infinity under the same acceleration, then we know that

$$V^2 = -2\int_\infty^a F\,dr = -2\int_\infty^a\left(\frac{\mu}{r^3} - \frac{\lambda}{r^5}\right)dr = -2\left[-\frac{\mu}{2r^2} + \frac{\lambda}{4r^4}\right]_\infty^a = \frac{\mu}{a^2} - \frac{\lambda}{2a^4} \quad \text{... (2)}$$

The differential equation of the path is given by

$$h^2(du^2/d\theta^2 + u) = F/u^2 \quad \text{or} \quad h^2(d^2u/d\theta^2 + u) = \mu u - \lambda u^3, \text{ using (1)} \quad \text{... (3)}$$

Multiplying both sides of (3) by $2(du/d\theta)$ and integrating, we get

$$v^2 = h^2\{(du/d\theta)^2 + u^2\} = \mu u^2 - (\lambda/2)\times u^4 + A, \ A \text{ being a constant} \quad \text{... (4)}$$

Given that $$n^2 + 1 = (2\mu a^2)/\lambda \quad \text{or} \quad 2\mu a^2 - \lambda = \lambda n^2 \quad \text{... (5)}$$

Now, applying (4) at initial point, namely, apse, where $r = a$ *i.e.* $u = 1/a$, $du/d\theta = 0$, and $v = V = (\mu/a^2 - \lambda/2a^4)^{1/2}$, we obtain

$$(\mu/a^2) - (\lambda/2a^4) = h^2/a^2 = (\mu/a^2) - (\lambda/2a^4) + A \quad \Rightarrow \quad A = 0, \quad h^2 = \mu - (\lambda/2a^2)$$

Re-writing, the above value of h^2, $\quad h^2 = (2\mu a^2 - \lambda)/2a^2 = \lambda n^2/2a^2$, using (5).

Setting $A = 0$ and $h^2 = \lambda n^2/2a^2$ in (4), we have

$$(\lambda n^2/2a^2)\times\{(du/d\theta)^2 + u^2\} = \mu u^2 - (\lambda u^4/2)$$

or $$\frac{\lambda n^2}{2a^2}\left(\frac{du}{d\theta}\right)^2 = \left(\mu - \frac{\lambda n^2}{2a^2}\right)u^2 - \frac{\lambda u^4}{2} \quad \text{or} \quad \frac{\lambda n^2}{2a^2}\left(\frac{du}{d\theta}\right)^2 = \frac{2\mu a^2 - \lambda n^2}{2a^2}u^2 - \frac{\lambda u^4}{2}$$

or $$(\lambda n^2/2a^2)\times(du/d\theta)^2 = (\lambda u^2/2a^2) - (\lambda u^4/2), \quad \text{since} \quad (5) \Rightarrow 2\mu a^2 - \lambda n^2 = \lambda$$

or $$n^2\left(\frac{du}{d\theta}\right)^2 = u^2(1 - a^2u^2) \quad \text{or} \quad \frac{n\,du}{u\sqrt{1 - a^2u^2}} = d\theta$$

Integrating, $$n\operatorname{sech}^{-1}(au) = \theta + B, \ B \text{ being an arbitrary constant} \quad \text{...(6)}$$

Initially, when $u = 1/a$, $\theta = 0$. Hence, (6) gives $B = 0$. Then, (6) reduces to

$$n\operatorname{sech}^{-1}(au) = \theta \quad \text{or} \quad \operatorname{sech}^{-1}(a/r) = \theta/n \quad \text{or} \quad a/r = \operatorname{sech}(\theta/n)$$

Thus, $r = a\cosh(\theta/n)$, which is required equation of the path.

Ex. 49. *A particle is projected with a law of force μ/r^3 from an apse at a distance 'a' with velocity $(2/a) \times (\mu/3)^{1/2}$, show that the orbit is $r \cos(\theta/2) = a$* **(Agra 2000, 02, 09)**

Sol. Here, the central force per unit mass $= F = \mu/r^3 = \mu u^3$, where $u = 1/r$

Hence, the differential equation of the central orbit, namely,

$h^2\{(d^2u/d\theta^2)+u\} = F/u^2$ reduces to $h^2\{(d^2u/d\theta^2)+u\} = \mu u$ ... (1)

Multiplying both sides of (1) by $2(du/d\theta)$ and then integrating, we get

$$v^2 = h^2\{(du/d\theta)^2 + u^2\} = \mu u^2 + A, \; A \text{ being an arbitrary constant} \quad \text{... (2)}$$

Initially at an apse, $r = a$, $u = 1/a$, $du/d\theta = 0$ and $v = (2/a) \times (\mu/3)^{1/3}$. Hence, (2) yields

$4\mu/3a = h^2/a^2 = \mu/a^2 + A$ giving $h^2 = 4\mu/3$ and $A = \mu/2a^2$ ... (3)

Using (3), (2) reduces to $(4\mu/3) \times \{(du/d\theta)^2 + u^2\} = \mu u^2 + \mu/3a^2$

or $$\left(\frac{du}{d\theta}\right)^2 + u^2 = \frac{3u^2}{4} + \frac{1}{4a^2} \quad \text{or} \quad \left(\frac{du}{d\theta}\right)^2 = \frac{1}{4a^2} - \frac{u^2}{4} = \frac{1-u^2a^2}{4a^2} \quad \text{... (4)}$$

But $u = 1/r$ so that $du/dr = -(1/r^2)\times(dr/d\theta)$. Then, (4) reduces to

$$\frac{1}{r^4}\left(\frac{dr}{d\theta}\right)^2 = \frac{r^2-a^2}{4a^2r^2} \quad \text{or} \quad \frac{dr}{d\theta} = \frac{r\sqrt{r^2-a^2}}{2a} \quad \text{or} \quad \frac{dr}{r\sqrt{r^2-a^2}} = \frac{1}{2}d\theta$$

Integrating, $\sec^{-1}(r/a) = \theta/2 + B$, B being an arbitrary constant ... (5)

If θ be measured from the initial radius vector, then $\theta = 0$ when $r = a$. Hence, (5) yields $B = 0$.

Therefore, (5) reduces to

$\sec^{-1}(r/a) = \theta/2$ or $r/a = \sec(\theta/2)$ or $r\cos(\theta/2) = a$.

Ex. 50. *Derive the energy equation $(du/d\theta)^2 + u^2 = 2(E-V)/h^2$ for a particle moving under a central conservative force of attraction.* **(Delhi Maths (Hons.) III 2009)**

Sol. Suppose that a particle of mass m is moving under an attractive central force F per unit mass. Let the centre of force be taken as pole. Let $P(r, \theta)$ be the position of the particle at any time t. Let $\dot{r}$ and $r\dot{\theta}$ be radial and transverse components of velocity of the particle at any time t.

Since the force is central, therefore its momentum about the origin is zero. Hence, using the principle of angular momentum, it follows that the angular momentum h of the particle about the origin must remain constant and thus we obtain

$$r^2\dot{\theta} = h, \text{ where } h \text{ is a constant} \quad \text{...(1)}$$

Now, the kinetic energy T of the particle (of unit mass) is given by

$T = (1/2) \times 1 \times \{(\dot{r})^2 + (r\dot{\theta})^2\}$ so that $2T = \dot{r}^2 + r^2\dot{\theta}^2$ ...(2)

Since $u = 1/r$, hence (1) reduces to $\dot{\theta} = hu^2$ ...(3)

Now, $$\dot{r} = \frac{dr}{dt} = \frac{du^{-1}}{dt} = -u^{-2}\frac{du}{dt} = -u^{-2}\frac{du}{d\theta}\frac{d\theta}{dt} = -h\frac{du}{d\theta}, \text{ using (3)} \quad \text{...(4)}$$

Using (3) and (4), (2) reduces to

$2T = h^2\{(du/d\theta)^2 + u^2\}$ or $T = (h^2/2) \times \{(du/d\theta)^2 + u^2\}$ ...(5)

Let the attractive constant force F be a function of r alone and hence the system is conservative. Let V be the potential energy per unit mass at $P(r, \theta)$. Then, we have

$dV/dr = F$ so that $V = \int_{r_0}^{r} F dr, r_0$ being a constant ...(6)

Using the principle of conservation of energy, we obtain

$$T + V = E, \quad \text{where } E \text{ is a constant.}$$

or $(h^2/2) \times \{(du/d\theta)^2 + u^2\} = E - V$ or $(du/d\theta)^2 + u^2 = 2(E - V)/h^2$, which is the desired *energy equation.*

Ex. 51. *A particle is moving under a central force $5/r^2$ per unit msass acting towards the origin. Initially, $r = 5$, $v = 1$ and the particle is projected in a direction making an angle $\sin^{-1}(0.8)$ to the radius vector. Find the apsidal distances and the speed at the apses.* **[Mumbai 2010]**

Sol. Here, the central acceleration $= F = 5/r^2 = 5u^2$, where $u = 1/r$.

The differential equation of the central orbit is $h^2 \{(d^2u/d\theta^2) + u\} = F/u^2$

or $$h^2 \{(d^2u/d\theta^2) + u\} = (5u^2)/u^2 = 5 \quad \ldots(1)$$

Multiplying both sides of (1) by $2\,(du/d\theta)$ and then integrating, we get

$$v^2 = h^2 \{(du/d\theta)^2 + u^2\} = 10u + A, \; A \text{ being an arbitrary constant} \quad \ldots(2)$$

According to the given problem, initially when $r = 5$, $\phi = \sin^{-1}(0.8)$ so that $\sin\phi = 0.8 = 8/10 = 4/5$. Therefore, using formula $p = r\sin\phi$, we get $p = 5 \times (4/5) = 4$ so that $1/p^2 = 1/16$.

Hence, $(du/d\theta)^2 + u^2 = 1/p^2$ reduces to $(du/d\theta)^2 + u^2 = 1/16$ ...(3)

Now, using the facts that when $u = 1/r = 1/5$, $v = 1$, $(du/d\theta)^2 + u^2 = 1/16$, (2) reduces to

$1 = h^2 \times (1/16) = 10 \times (1/5) + A$ so that $h^2 = 16$ and $A = -1$.

Then, from (2), $$v^2 = 16\{(du/d\theta)^2 + u^2\} = 10u - 1 \quad \ldots(4)$$

At an apre, $du/d\theta = 0$. Hence, the required apsidal distances can be obtained by setting $du/d\theta = 0$ in (4). Thus, we have

$16u^2 = 10u - 1$ or $16u^2 - 10u + 1 = 0$ giving $u = 1/8, 1/2$; so that $r = 8, 2$

Now, when $u = 1/8$, from (4), $v^2 = 16(0 + 1/64) = 1/4 \Rightarrow v = 1/2$

and when $u = 1/2$, from (4), $v^2 = 16(0 + 1/4) = 4 \Rightarrow v = 2$

Thus, the required apsidal distances are **8** and **2** and the speeds at the corresponding apses are 1/2 and 2 respectively.

Ex. 52 (*i*). *At an apse*

(a) $d^3r/d\theta^3 = 0$ *(b)* $du/d\theta = 0$ *(c)* $d^2r/d\theta^2 = 0$ *(d)* $du/d\theta = 1$ **(Garhwal 2010)**

Sol. Ans. (*b*). Refer Art. 9.11, page 9.12.

Ex. 52 (*ii*). *Prove that in a central field, the areal velocity is conserved* **(Nagpur 2010)**

Ex. 52 (*iii*). *Show that angular velocity and areal velocity of particle moving in a central orbit are constant.* **(Pune 2010)**

Hint. For Ex. 52 (*ii*) and Ex. 52 (*iii*); refer Art. 9.4 and Art 9.5

Ex. 52 (*iv*). *If the central orbit be an ellipse, the focus being the centre of force, find the law of force, the velocity on the path and period. Prove with usual notations, $v^2 = \mu(2/r - 1/a)$ and $h^2 = \mu a(1 - e^2)$* **(Delhi B.A. (Prog) III 2010)**

Hint. Refer Art. 9.9 on page 9.4

Ex. 52 (*v*). *For a particle describing a central orbit, prove in usual notations.*

(i) $h = pv = |\mathbf{r} \times v|$ *(ii)* $v^2 = h^2\{u^2 + (du/d\theta)^2\}$ **(Mumbai 2010)**

Hint. (*i*). From figure of Art 9.5, page 9.2, $|\mathbf{r} \times v| = rv\sin\phi$. But from $\Delta\, ONP$, $p = r\sin\phi$ and hence $|\mathbf{r} \times v| = pv$. Again, from equation (2) of Art. 9.5, $h = pv$. Therefore, $h = pv = |\mathbf{r} \times v|$

Part. (*ii*). Refer equation (3) of Art. 9.5.

Ex. 52 (*vi*) *According to kepler's third law of motion, which identify is true?*

(a) $T^2 \propto a^3$ *(b)* $T^2 \propto 1/a^3$ *(c)* $a^2 \propto T^2$ *(d)* $a^2 \propto 1/T^2$ **(Agra 2009)**

Hint. Ans. (*a*). Refer deduction 3, page 10.2

Ex. 53. *A particle describes an elliptic orbit under a central force towards one focus S. If v_1 is the speed at the end B of the minor axis and v_2, v_3 the speeds at the ends A, A′ of the major axis, show that $v_1^2 = v_2 v_3$.* **(Delhi B.A. (Prog) III 2010)**

Sol. Let $ABA'B'$ be the elliptic orbit described by the particle moving under the given central force generating central acceleration μ/(distance)². Let B be an end of minor axis and A, A' be ends of major axis of the elliptic orbit. Let v_1, v_2 and v_3 be velocities at B, A and A' respectively. Let the equation of the elliptic orbit be $x^2/a^2 + y^2/b^2 = 1$. Then, from the adjoing figure, we have

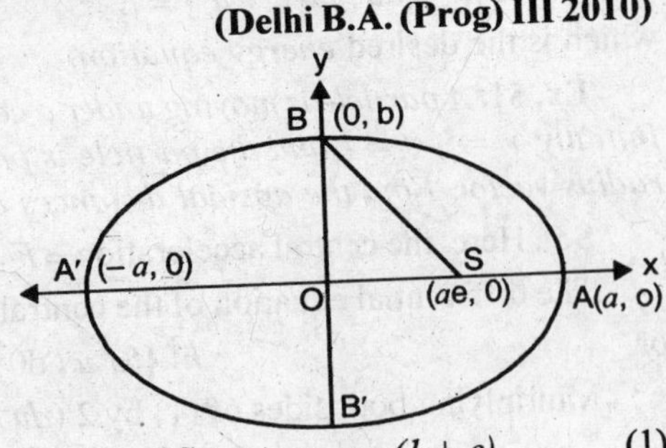

$$SA = OA - OS = a - ae = a(1-e) \quad \text{and} \quad SB = OA' + OS = a + ae = a(1+e) \quad \text{... (1)}$$

Now, from Δ OSA,
$$SB = (OS^2 + OB^2)^{1/2} = (a^2e^2 + b^2)^{1/2} = a \quad \text{... (2)}$$

[∵ In an ellipse, $b^2 = a^2(1-e^2)$ so that $a^2e^2 + b^2 = a^2$]

We know that velocity v of the particle at any distance r from the focus S is given by (refer equation (3) of Art. 10.2, page 10.1)

$$v^2 = \mu\,(2/r - 1/a) \quad \text{... (3)}$$

Applying formula (3) at points B, A and A' and using (1) and (2), we get

$$v_1^2 = \mu(2/SB - 1/a) = \mu(2/a - 1/a) = \mu/a \quad \text{...(4)}$$

$$v_2^2 = \mu\left(\frac{2}{SA} - \frac{1}{a}\right) = \mu\left\{\frac{2}{a(1-e)} - \frac{1}{a}\right\} = \frac{\mu}{a} \times \frac{1+e}{1-e} \quad \text{... (5)}$$

and
$$v_3^2 = \mu\left(\frac{2}{SA'} - \frac{1}{a}\right) = \mu\left\{\frac{2}{a(1+e)} - \frac{1}{a}\right\} = \frac{\mu}{a} \times \frac{1-e}{1+e} \quad \text{... (6)}$$

Multiplying the corresponding sides of (5) and (6), we have

$$v_2^2 v_3^2 = \mu^2 / a^2 = v_1^2, \text{ using (4).}$$

Ex. 54. *A particle sides down the curve of a smooth cycloid whose axis is vertical and vertex upwards. Discuss the motion.* **(Kanpur 2007/08)**

Hint. Refer Ex. 13, page 11.7

Ex. 55. *Find the periodic time in cycloidal motion.* **(Agra 2009)**

Hint. Refer equation (5) of Art. 11.4, page 1.2

Ex. 56. *A particle slides down smooth cycloid from rest at the cusp, the axis being vertical and vertex downwards, prove that*

(i) The magnitude of the acceleration is equal to g at every point of the path

(ii) Pressure when the particle arrives at vertex is equal to twice the weight of the particle

(iii) The time of descent to the vertex is $\pi \times (a/g)^{1/2}$ **(Meerut 2010)**

Hint. For parts (*i*) and (*ii*), refer Ex. 5(*d*), page 11.4

Part (*iii*) Refer Art. 11.4, page 11.4. The required time t' is given by

$$t' = T/4 = (1/4) \times 4\pi \times (a/g)^{1/2} = \pi \times (a/g)^{1/2}$$

Ex. 57. *Show that the curve which is such that a particle will slide down each of its tangent to the horizontal axis in a given time is a cycloid whose axis vertical.*

Sol. Let OX be the given horizontal axis, APB be the given curve and $P(x, y)$ be any point on the curve APB. Let the tangent PT make an angle ψ with OX. Then, the component of acceleration g of the particle along the tangent PT is $g \sin \psi$.

Let t be the time taken by the particle to move from P to T.

Considering motion along PT and using formula $s = ut + (1/2) \times ft^2$, we have

$PT = (1/2) \times g \sin \psi \times t^2$ or $y \operatorname{cosec} \psi = (1/2) \times gt^2 \sin \psi$

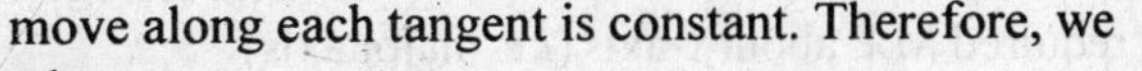

Thus, $$y = (1/2) \times gt^2 \sin^2 \psi \qquad \text{... (1)}$$

Given that the time taken by the particle to move along each tangent is constant. Therefore, we take $(1/2) \times gt^2 = \text{constant} = 2a$, say. Then, (1) reduces to

$$y = 2a \sin^2 \psi \qquad \text{... (2)}$$

Differentiating both sides of (2), w.r.t. 's', we have $dy/ds = (4a \sin \psi \cos \psi) \times (d\psi / ds)$

or $\sin \psi = (4\, a \sin \psi \cos \psi) \times (d\psi / ds)$, as $dy / ds = \sin \psi$

Thus, $$ds = 4a \cos \psi \, d\psi \qquad \text{... (3)}$$

Integrating (3), $s = 4a \sin \psi + C$, C being an arbitrary constant ... (4)

Assume that $\psi = 0$ when $s = 0$. Hence, (4) yields $C = 0$. Hence, (4) reduces to $s = 4a \sin \psi$, which is a cycloid whose axis is vertical.

Ex. 58. *A particle is projected upwards with velocity U in a resisting medium whose resistance varies as the velocity. Find the relation between the time and velocity.* **(Kanpur 2009)**

Hint. For solution refer Art. 12.3, page 12.2

Ex. 59. *Find the product of inertia of a solid disc in the form of quadrant of a circle about its bounding radii.* **(Kanpur 2010)**

Sol. Let the given solid be the quadrant of a circle given by

$$x^2 + y^2 = a^2 \qquad \text{... (1)}$$

Let ρ be the mass per unit area of the given solid disc of mass, M (say). Then, we have

$\rho = M/(\pi a^2/4)$ so that $\rho = 4M/\pi a^2$ ... (2)

Consider an elementary area $\delta x\, \delta y$ of mass δm at the point $P(x, y)$. Then, $\delta m = \delta x\, \delta y\, \rho$. Now, we have

product of inertia of mass δm about OX and $OY = xy\, \delta m = \rho xy\, \delta x\, \delta y$

Hence, the required product of inertia of the given solid about OX and OY

$$= \int_{x=0}^{a} \int_{y=0}^{y_1} \rho xy \, dx\, dy, \text{ where } y_1 = (a^2 - x^2)^{1/2}, \text{ using (1)}$$

$$= \rho \int_0^a x \left[y^2/2 \right]_0^{y_1} dx = \frac{1}{2} \rho \int_0^a x y_1^2 \, dx = \frac{1}{2} \rho \int_0^a x(a^2 - x^2)\, dx, \text{ as } y_1 = (a^2 - x^2)^{1/2}$$

$$= \frac{1}{2} \rho \left[\frac{a^2 x^2}{2} - \frac{x^4}{4} \right]_0^a = \frac{\rho a^4}{8} = \frac{a^4}{8} \times \frac{4M}{\pi a^2} = \frac{Ma^2}{2\pi}, \text{ using (2)}$$

Ex. 60. *State the theorem of parallel axis for moment of inertia* **(Agra 2010)**

Hint. Refer Art. 14.9.

Ex. 61. *Write T for true and F for false statements:*

(i) Any line perpendicular to the plane lamina at any point O of it is a principle axis of the lamina at O. **(Agra 2009)**

(ii) The moment of inertia of a uniform circular disc of mass M and radius a about any diameter of the disc is $(Ma^2)/2$. **(Agra 2009)**

Sol. (*i*). **Ans.** T. Refer corollary 2 of Art. 14.16, page 14.25.

(*ii*). **Ans.** F. Refer case I, Art. 14.4 E, page 14.3

Ex. 62. *Find the moment of inertia of an ellipse $x^2/9 + y^2/4 = 1$ about x-axis.*

Hint. Proceed as in case I of Art. 14.4 I, page 14.5. Note that here $a = 3$ and $b = 2$. Hence, the required M.I. about x-axis (*i.e.* major axis)

$$= (1/4) \times Mb^2 = (1/4) \times 4M = M, \text{ where } M \text{ is mass of the elliptic disc.}$$

Ex. 63. *A uniform lamina is bounded by a parabolic arc, of latusrectum 4a, and a double ordinate at a distance b from the vertex. If $b = a(7 + 4\sqrt{7})/3$, show that two of the principle axes at the end of a latus rectum are the tangent and normal there.*

Sol. Let the equation of given parabolic arc COD be given by

$$y^2 = 4ax \qquad \text{... (1)}$$

Let the given lamina COD be bounded by double ordinate CD such that $OE = b$. Let AB be the latusrectum of the parabola (1) so that $OS = a$ and $AS = BS = 2a$. Hence, the coordinates of the end A of the latus rectum are $(a, 2a)$. Let coordinates of any point P of the lamina be (x, y) and let $PQQ'P'$ be an elementary area $\delta x \delta y$ of the lamina COD.

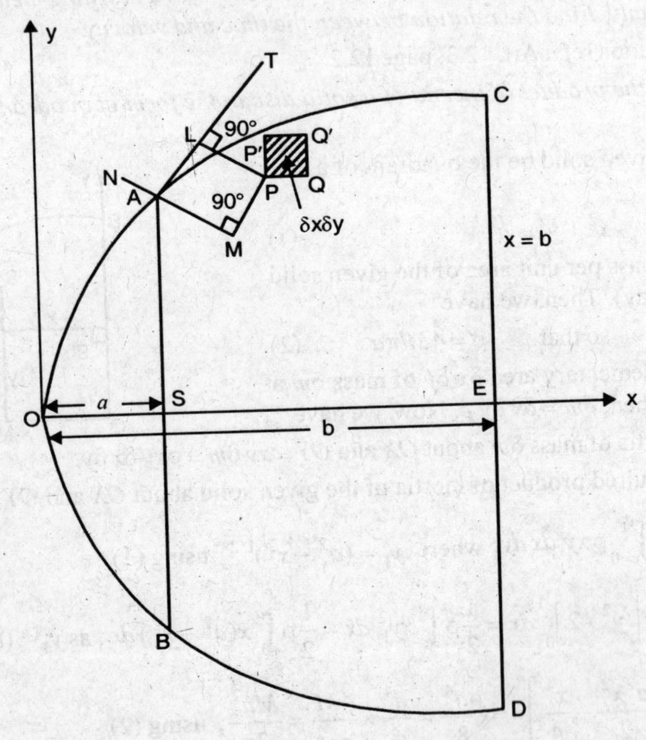

Let AT and AN be tangent and normal respectively at A. Let PM and PL be perpendiculars from P on AN and AT respectively.

From (1), $2(dy/dx) = 4a$ so that $dy/dx = 2a/y$

Hence, the gradient of tangent AT at $A(a, 2a) = (dy/dx)_{(a,\, 2a)} = 2a/2a = 1$.

Therefore, the equation of the tangent AT is given by

$$y - 2a = 1 \times (x - a) \quad \text{or} \quad y - x - a = 0 \qquad \text{... (2)}$$

We know that gradient of tangent $AT \times$ gradient of normal $AN = -1$.

Hence, $1 \times$ gradient of normal $AN = -1 \Rightarrow$ gradient of $AN = -1$.

Therefore, the equation of the normal AN is given by

$$y - 2a = (-1) \times (x - a) \quad \text{or} \quad y + x - 3a = 0 \qquad \text{... (3)}$$

Let ρ be the mass per unit area. Then, $\delta m =$ mass of $PQQ'P' = \rho\, \delta x\, \delta y$

Now, we have $PL = (y - x - a)/\sqrt{2}$ and $PM = (y + x - 3a)/\sqrt{2}$

Hence, the product of inertia of δm about AT and AN

$$= PL \times PM \times \delta m = \{(y - x - a)(y + x - 3a)/2\} \times \rho\, \delta x\, \delta y$$

By definition of the principle axes, AT and AN will be the principle axes of the given lamina COD if the product of inertia about AT and AN vanish, that is, if

$$\int_{x=0}^{b} \int_{y=-2\sqrt{ax}}^{2\sqrt{ax}} \frac{\rho(y - x - a)(y + x - 3a)}{2}\, dx\, dy = 0$$

or $$\int_0^b \int_{-2\sqrt{ax}}^{2\sqrt{ax}} (y^2 - 4ay + 3a^2 + 2ax - x^2)\, dx\, dy = 0$$

or $$\int_0^b \left[\frac{1}{3} y^3 - 2ay^2 + y(3a^2 + 2ax - x^2)\right]_{-2\sqrt{ax}}^{2\sqrt{ax}} dx = 0$$

or $$\int_0^b \left\{(1/3) \times 2 \times (2\sqrt{ax})^3 + 2\sqrt{ax}\,(3a^2 + 2ax - x^2)\right\} dx = 0$$

or $$\int_0^b \{(4/3) \times a^{3/2} x^{3/2} + 3a^{5/2} x^{1/2} + 2a^{3/2} x^{3/2} - a^{1/2} x^{5/2}\}\, dx = 0$$

or $$\left[(8/15) \times a^{3/2} x^{5/2} + 2a^{5/2} x^{3/2} + (4/5) \times a^{3/2} x^{5/2} - (2/7) \times a^{1/2} x^{7/2}\right]_0^b = 0$$

or $$(4/3) \times a^{3/2} b^{5/2} + 2a^{5/2} b^{3/2} - (2/7) \times a^{1/2} b^{7/2} = 0 \quad \text{or} \quad 3b^2 - 14\,ab - 21\,a^2 = 0$$

giving $$b = \{14a \pm (196a^2 + 4 \times 3 \times 21a^2)^{1/2}\}/6 = (14a \pm 8a\sqrt{7})/6$$

Since $b > 0$, we have $$b = a(7 + 4\sqrt{7})/3$$

Ex. 64. *Prove that the moment of inertia I of a rigid body about an axis making angles α, β, γ with coordinate axes is given by*

$$I = A\cos^2\alpha + B\cos^2\beta + C\cos^2\gamma - 2F\cos\beta\cos\gamma - 2G\cos\alpha\cos\gamma - 2H\cos\alpha\cos\beta$$

(Mumbai 2010)

Hint. Use Art. 14.10 with $l = \cos\alpha$, $m = \cos\beta$, $n = \cos\gamma$

Ex. 65. *Prove the existence of principle axes and principle moment of inertia at any point of the rigid body* **(Dehli 2009)**

Hint. Refer Art. 14.15 and Art. 14.16, page 14.25

Ex. 66. *Find the expressions for moment of momentum, kinetic energy and equations of motion for a rigid body.* **(Agra 2010)**

Ex. 67. *State and prove D' Alembert's principle and deduce the general equations for a rigid body.* **(Agra 2010)**

Hint. Refer Art. 15.2 and Art. 15.3.

Ex. 68. *Fill in the blanks: If* $\vec{r}$ *is the position vector of a particle of mass m at any time t, then the reversed effective force on the particle is ....* **(Agra 2009, 10)**

Sol. Ans. $(-m)(d^2\vec{r}/dt^2)$. Refer Art. 15.2

Ex. 69. *Moment of momentum of a rigid body about the fixed axis is*

(a) $mk^2(d\theta/dt)$ *(b)* $(1/3)\times mk^2(d^2\theta/dt^2)$

(c) $(1/2)\times mk^2(d\theta/dt)^2$ *(d)* $(1/3)\times mk^2(d\theta/dt)^2$ **(Agra 2009, 10)**

Hint. Ans. (*a*). Refer Art. 16.3, page 16.2

Ex. 70. Write *'T'* for true and *'F'* for false statements

(i) The minimum time of oscillation of a compound pendulum is $2\pi\times(2k/g)^{1/2}$ *where k is the radius of gyration of the rigid body.* **(Agra 2009)**

(ii) The centre of suspension and the centre of oscillation of a compound pendulum are convertible. **(Agra 2009)**

Sol. (*i*). **Ans.** '*F*'. Refer Art. 6.9 page 16.8 (*ii*). **Ans.** '*T*'. Refer Art. 16.8, page 16.7.

Ex. 71. *A uniform rod of length 2a is held horizontally. Its one end is fixed and the other just released. Show that when the movable end comes at the lowest position, the angular velocity of the rod is* $(3g/2a)^{1/2}$. **(Mumbai 2010)**

Hint. Proceed as in Ex. 1, page 16.16 upto equation (5). Then, the required angular velocity is given by equation (5) on replacing θ by 0. Therefore the required angular velocity is given by

$$(d\theta/dt)^2 = 3g/2a \quad \text{so that} \quad d\theta/dt = (3g/2a)^{1/2}$$

Ex. 72. *Prove that the kinetic energy of a rigid body about a point O is equal to the kinetic energy of the body rotating about the centre of mass G plus kinetic energy of a single mass M placed at G about O.* **(Mumbai 2010)**

Ex. 73. *A uniform cylinder is placed with its axis horizontal on a plane, whose inclination to the horizon is α. Show that the least coefficient of friction between it and the plane, so that it way roll and not slide is (tan α)/3.* **(Mumbai 2005)**

Hint. Refer Ex. 1(a), page 17.11.

Ex.74. *A heavy particle of mass m is projected from the lowest point of a small vertical circle with velocity u and moves along the inner side of the circle of radius 'a'. Discuss the motion when* $u^2 \le 2ag$. **[Delhi B.SC. (Hons.) III 2010]**

Sol. Consider the motion of a heavy particle of mass m placed on the inside of a smooth vertical circle (with centre O and radius *a*) and projected with velocity *u* from its lowest point *A*. Let *P* be the position of the particle at any time *t*, such that $\angle POA = \theta$, Let v be velocity of the particle along *PT*. Let *PN* be perpendicular on OA. Then, we have

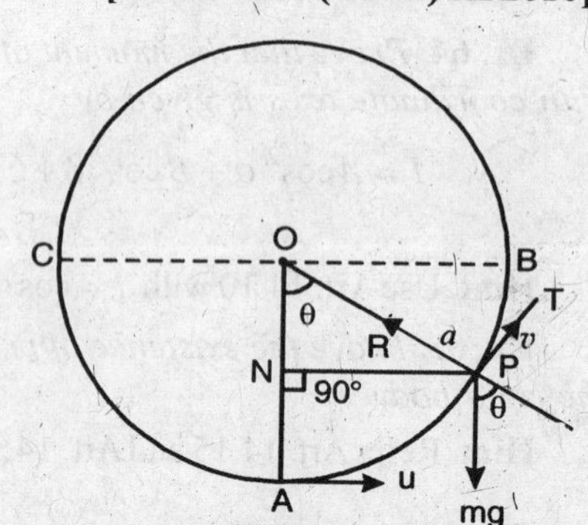

$$h = AN = AO - ON = a - a\cos\theta = a(1-\cos\theta)$$

so that $\cos\theta = (a-h)/a$ ...(1)

For motion from A to P, by the principle of conservation of energy, we have

$(1/2) \times mv^2 - (1/2) \times mu^2 = (-mg) \times h$ or $v^2 = u^2 - 2ga(1 - \cos\theta)$ ...(2)

The external forces acting on the particle are its weight mg acting vertically downwards and the reaction R on the circle, directed along PO. If arc $AP = s$, then the components of acceleration along PT and PO are d^2s/dt^2 and v^2/a respectively. The equation of motion in the inward normal sense is given by

$(mv^2)/a = R - mg\cos\theta$ or $R = mg\cos\theta + (m/a) \times v^2$

or $R = mg\cos\theta + (m/a) \times \{u^2 - 2ga(1 - \cos\theta)\}$, using (2)

or $R = (m/a) \times (u^2 - 2ga + 3ga\cos\theta)$ ...(3)

From (1) and (2), we have $v^2 = u^2 - 2gh$ ...(4)

Substituting the value of cos θ from (1) in (3), we have

$R = (m/a) \times \{u^2 - 2ga + 3g(a - h)\}$ or $R = (m/a) \times \{u^2 + ga - 3gh\}$ ...(5)

Let $h = h_1$ when $v = 0$. Hence, (4) yields $h_1 = u^2/2g$ ...(6)

Let $h = h_2$ when $R = 0$. Hence (5) yields $h_2 = (u^2 + ga)/3g$ ...(7)

Let $\theta = \theta_1$ when $v = 0$, Hence, (2) yields $\cos\theta_1 = (2ag - u^2)/2ag$ ...(8)

Let $\theta = \theta_2$ when $R = 0$, Hence, (3) yields $\cos\theta_2 = (2ag - u^2)/3ag$ ...(9)

Discussion of the motion when $u^2 \le 2ag$

Case (i) Let $u^2 = 2ag$ Then, (8) and (9) $\Rightarrow$ $\cos\theta_1 = \cos\theta_2 = 0$ $\Rightarrow$ $\theta_1 = \theta_2 = \pi/2$

Hence, if the particle is projected with velocity $u = \sqrt{2ag}$, then it will rise upto the level of the horizontal diameter BC through O and will oscillate about A in the semi-circular are BAC.

Case (ii) Let $u^2 < 2ag$. Then, (9) and (10) $\Rightarrow$ $h_1 < h_2$,

showing that the velocity vanishes before reaction has vanished. Against from (1), it follows that when $u^2 < 2ag$, then $\cos\theta_1$ is positive and hence θ_1 is an acute angle. Hence if a particle is projected with velocity $u < \sqrt{2ag}$, then it will oscillate about A and will not rise upto the horizontal diameter BC passing through O.

Ex. 75. *Show that the orbit described under a central attractive force varying directly as the distance is an ellipse having its centre at the centre of force.* **(Delhi B.Sc (Hons) III 2011)**

Sol. Suppose a particle moves in an orbit AB under a central attractive force λr (λ being a constant) which is always directed towards a fixed point O in the plane. Take the centre of force O as the pole and horizontal line OX in the plane of the orbit as initial line. Let OY be perpendicular to OX. Let $P(r, \theta)$ be the position of the particle at any time t and let cartesian coordinates of P be (x, y).

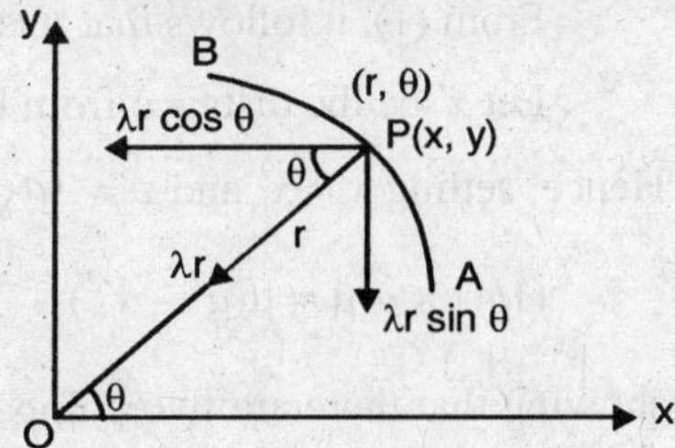

Resolved parts of acceleration of the particle at P are d^2x/dt^2 and d^2y/dt^2 along OX and OY respectively. Again, the resolved parts of force λr acting at P along OX and OY are $-\lambda r\cos\theta$ and $-\lambda r\sin\theta$ (i.e. $-\lambda x$ and $-\lambda y$, as $x = r\cos\theta$ and $y = r\sin\theta$) along OX and OY respectively. Hence, the equations of motion along OX and OY respectively are given by

$m(d^2x/dt^2) = -\lambda x$ and $m(d^2y/dt^2) = -\lambda y$

or $(D^2 + k^2)x = 0$ and $(D^2 + k^2)y = 0$, where $D \equiv d/dx$ and $k^2 = \lambda/m$ ...(1)

The general solutions of the differential equations given by (1) are

$x = a_1 \cos kt + b_1 \sin kt$, a_1 and b_1 being arbitrary constants ...(2)

$y = a_2 \cos kt + b_2 \sin kt$, a_2 and b_2 being arbitrary constants ...(3)

Multiplying (1) by a_2 and (2) by a_1 and then subtracting, we have

$$a_2 x - a_1 y = (a_2 b_1 - a_1 b_2) \sin kt \qquad ...(4)$$

Again, multiplying (1) by b_2 and (2) by b_1 and then subtracting, we have

$$b_2 x - b_1 y = (a_1 b_2 - a_2 b_1) \cos kt \qquad ...(5)$$

Squaring and adding (4) and (5), the equation the required orbit described by the particle is given by

$$(a_2 x - a_1 y)^2 + (b_2 x - b_1 y)^2 = (a_1 b_2 - a_2 b_1)^2$$

or $$(a_2^2 + b_2^2) x^2 + (a_1^2 + b_1^2) y^2 - 2(a_1 a_2 + b_1 b_2) xy - (a_1 b_2 - a_2 b_1)^2 = 0 \qquad ...(6)$$

Standard equation of a conic is given by $ax^2 + 2hxy + by^2 + 2gx + 2fy + c = 0$ ...(7)

Comparing (6) and (7), we obtain

$$a = a_2^2 + b_2^2, \quad h = -(a_1 a_2 + b_1 b_2), \quad b = a_1^2 + b_1^2, \quad g = f = 0, \quad c = -(a_1 b_2 - a_2 b_1)^2 \qquad ...(8)$$

Here $$h^2 - ab = (a_1 a_2 + b_1 b_2)^2 - (a_2^2 + b_2^2)(a_1^2 + b_1^2) = -(a_1 b_2 - a_2 b_1)^2$$

which is negative. Hence (6) represents an ellipse.

Now, we know that the co-ordinates of centre of (7) are $\left(\dfrac{hf - bg}{ab - h^2}, \dfrac{gh - af}{ab - h^2}\right)$ ...(9)

Using (8), we find that the coordinates of the centre of ellipse (7) are (0 ,0) which is the centre of force. Thus, the orbit (6) described by given force is an ellipse having its centre at the centre of force.

Ex. 76. *At what distance from the centre the velocity in a simple harmonic motion will be half of the maximum?* **(G.N.D.U. Amritsan 2010; Rohilkhand 2008)**

Sol. Let the centre of the motion be at origin. If a is the amplitude of the motion, then the velocity v of the particle at a distance x from the centre is given by

$$v^2 = \mu(a^2 - x^2), \ \mu \text{ being a constant} \qquad ...(1)$$

From (1), it follows that v is maximum when $x = 0$. Hence, maximum velocity $= a\sqrt{\mu}$

Let x' be the distance from the centre of the point where the velocity is half of the maximum.

Hence, setting $x = x'$ and $v = (a\sqrt{\mu})/2$ in (1), we get

$$(1/4) \times a^2 \mu = \mu(a^2 - x'^2) \quad \text{or} \quad x'^2 = (3a^2/4) \quad \text{or} \quad x' = \pm\{a\sqrt{3}/2\},$$

showing that there are two points, each at a distance $(a\sqrt{3})/2$ from the centre, where the velocity is half the maximum.

Ex. 77. *A light elastic string AB of length l is fixed at A and is such that if a weight w be attached to B, the string will be stretched to a length 2l. If weight w/4 be attached to B and let fall from the level of A, prove that (i) the magnitude of the S.H.M. that executes is 3l/4 (ii) the distance through which it falls is 2l (iii) the period of oscillation is*

$$(l/4g)^{1/2} \ \{4\sqrt{2} + \pi + 2\sin^{-1}(1/3)\}.$$

(Meerut 2000, 2013)

Sol. Here AB (= l) is the natural length of an elastic string whose one end is fixed at A. Let λ be the modulus of elasticity of the string. If a weight w be attached to B, it extends the string to a length $2l$ while hanging in equilibrium. Hence, we have

$$w = \lambda\{(2l - l)/l\} \quad \text{so that} \quad \lambda = w \qquad ...(1)$$

Again, when a weight $w/4$ is attached at B, then let the weight be in equilibrium at C where BC = d, (say). Then, considering the equilibrium of weight at C, we have

$$w/4 = \lambda \times (d/l) \quad \text{or} \quad d = l/4, \quad \text{as} \quad \lambda = w \qquad ...(2)$$

vel = 0, l, v', d, x, w/4

Now, the weight $w/4$ is dropped from A and so it falls the distance AB (= l) freely under gravity. If v' be the velocity acquired by the weight at B, we have $v' = \sqrt{2gl}$ in the downward direction. When this weight falls below B, the string begins to extend beyond its natural length and the tension begins to operate. The velocity of the weight continues increasing upto C, after which it starts decreasing. Suppose the weight comes to instantaneous rest at D, where CD = a. During the motion of the weight below B, let P be its position at any time t, where CP = x. Note that, we have taken C as origin and CP is the direction of x increasing.

At P, the forces acting on the weight (of mass $w/4g$) are (i) its weight $w/4$ acting vertically downwards and (i) the tension $(\lambda/l) \times (d + x)$ acting vertically upwards. Then, the equation of motion of the weight $w/4$ is given by

$$\frac{w}{4g}\frac{d^2x}{dt^2} = \frac{w}{4} - \frac{\lambda}{l}(d + x) \quad \text{or} \quad \frac{w}{4g}\frac{d^2x}{dt^2} = \frac{w}{4} - \frac{w}{l}(d + x) = \frac{w}{4} - \frac{w}{4} - \frac{wx}{l}, \text{ using (1) and (2)}$$

or
$$d^2x/dt^2 = -(4g/l) \times x, \qquad ...(3)$$

which is the equation of a S.H.M., with centre at C and amplitude CD (= a). Note that the equation (2) holds good so long as the string is stretched, *i.e.,* for the motion of the weight from B to D, Let $v = dx/dt$. Then $d^2x/dt^2 = v(dv/dx)$ and hence (3) reduces to

$$v(dv/dx) = -(4g/l) \times x \quad \text{or} \quad 2v\,dv = -(8g/l) \times x\,dx \qquad ...(4)$$

Integrating (4), $\quad v^2 = -(4g/l) \times x^2 + E$, E being an arbitrary constant $\qquad ...(5)$

But at B, $x = -l/4$ and $v = \sqrt{2gl}$ and hence (5) reduces to

$2gl = -(4g/l) \times (l^2/16) + E \quad$ giving $\quad E = (9gl)/4$. Hence, (4) reduces to

$$v^2 = -\frac{4g}{l}x^2 + \frac{9gl}{4} = \frac{4g}{l}\left(\frac{9}{16}l^2 - x^2\right), \qquad ...(6)$$

which gives velocity at any point between B and D.

At D, x = CD = a and $v = 0$ and so (6) yields $a = 3l/4$, showing that the required amplitude of the S.H.M. is $3l/4$. Again, the total distance through which the weight falls

$$= AB + BC + CD = l + (l/4) + (3l/4) = 2l.$$

Let t_1 be the time taken by the weight to fall freely under gravity from A to B. Then, using formula $v = u + gt$, we have $\sqrt{2gl} = 0 + gt_1$ so that $\quad t_1 = (2l/g)^{1/2}$.

Again, let t_2 be the time taken by the weight to fall from B to D while moving in S.H.M. From (6), we have

$$v = dx/dt = (4g/l)^{1/2}\sqrt{(9l^2/16) - x^2}, \qquad ...(7)$$

where we have taken positive sign because as t increases, x also increases. Re-writting (7), we get

$$dt = \left(\frac{l}{4g}\right)^{1/2} \frac{dx}{\{(3l/4)^2 - x^2\}^{1/2}} \qquad ...(8)$$

Integrating (8) between the limits $t = 0$ to $t = t_2$ and the corresponding limits $x = -l/4$ to $x = 3l/4$, we have

$$\int_0^{t_2} dt = \left(\frac{l}{4g}\right)^{1/2} \int_{-l/4}^{3l/4} \frac{dx}{\{(3l/4)^2 - x^2\}^{1/2}} = \left(\frac{l}{4g}\right)^{1/2} \left[\sin^{-1}\frac{x}{(3l/4)}\right]_{-l/4}^{3l/4}$$

Thus, $\quad t_2 = (l/4g)^{1/2}\{\sin^{-1}1 - \sin^{-1}(-1/3)\} = (l/4g)^{1/2}\{\pi/2 + \sin^{-1}(1/3)\}$

Hence the total time taken to fall from A to D.

$$= t_1 + t_2 = (2l/g)^{1/2} + (l/4g)^{1/2}\{\pi/2 + \sin^{-1}(1/3\} = (l/4g)^{1/2}\{\pi/2 + 2\sqrt{2} + \sin^{-1}(1/3\}$$

Note that after instantaneous rest at D, the weight begins to move upwards. From D to B it moves in S.H.M. as discussed above. At B the string becomes slack and S.H.M. ceases. The velocity of the weight at B is $\sqrt{2gl}$ in the vertically upward direction. Above B the weight rises freely under gravity and comes to instantaneous rest at A. Thus, it oscillates again and again from A to D.

Hence, the required period of one complete oscillation

$$= 2 \times (\text{time from A to D}) = 2(t_1 + t_2) = (l/4g)^{1/2}\{\pi + 4\sqrt{2} + 2\sin^{-1}(1/3)\}.$$

Ex. 78. *A point P is moving along a fixed line AB with uniform velocity v. Show that the angular velocity about a point O is inversely proportional to OP².* **(Bundelkhand 2007)**

Sol. Let P be the position at any time t of a point moving along a fixed line AB with uniform velocity v. Let OP $= r$ and let OM be perpendicular to AB so that OM $= p$, where p is a constant because the straight line AB and the point O are both fixed.

B
v
P
r
90°
O
M
p
A

Hence, the required angular velocity ω of the point P about O is given by $\omega = vp/r^2 = vp/\text{OP}^2$. Therefore, ω is proportional to OP² because vp is a constant.

Ex. 79. *The velocities of a particle along and perpendicular to the radius vector are λr and μθ; find the path and show that the accelerations along and perpendicular to the radius vector are $\lambda^2 r - (\mu^2\theta^2/r)$ and $\mu\theta(\lambda + \mu/r)$.* **(Gujrat University 2004; Guwahati 2007; Lucknow 2010; Purvanchal 2011; Rohilkand 2006, 09)**

Sol. Given that $\quad$ radial velocity $= dr/dt = \lambda r$ $\quad$ ...(1)

and $\quad$ transverse velocity $= r(d\theta/dt) = \mu\theta$ $\quad$ ...(2)

Dividing (1) by (2), we have $\quad \dfrac{dr}{r\,d\theta} = \dfrac{\lambda r}{\mu\theta} \quad$ or $\quad \dfrac{\mu}{\lambda}\dfrac{dr}{r^2} = \dfrac{d\theta}{\theta}$

Integrating, $\quad -(\mu/\lambda r) = \log\theta - \log C$, where C in an arbitrary constant

or $\quad \theta/C = e^{-(\mu/\lambda r)} \quad$ or $\quad \theta = Ce^{-(\mu/\lambda r)}$, $\quad$ ...(3)

which in the required equation of the path.

Now, radial acceleration

$$= \frac{d^2r}{dt^2} - r\left(\frac{d\theta}{dt}\right)^2 = \frac{d}{dt}\left(\frac{dr}{dt}\right) - \frac{1}{r} \times \left(r\frac{d\theta}{dt}\right)^2 = \frac{d}{dt}(\lambda r) - \frac{1}{r} \times (\mu\theta)^2, \text{ using (1) and (2)}$$

$$= \lambda(dr/dt) - (\mu^2\theta^2/r) = \lambda^2 r - (\mu^2\theta^2/r), \text{ using (1)}$$

Again, transverse acceleration

$$= \frac{1}{r}\frac{d}{dt}\left(r^2\frac{d\theta}{dt}\right) = \frac{1}{r}\frac{d}{dt}\left(r \times r\frac{d\theta}{dt}\right) = \frac{1}{r}\frac{d}{dt}(r \times \mu\theta), \text{ using (2)}$$

$$= \frac{\mu}{r}\left(\frac{dr}{dt}\theta + r\frac{d\theta}{dt}\right) = \frac{\mu}{r}(\lambda r\theta + \mu\theta) = \mu\theta\left(\lambda + \frac{\mu}{r}\right), \text{ using (1) and (2)}$$

Ex. 80. *A particle at the end of a string of length a, the upper end of which is fixed, is projected horizontally with a velocity which is k times due to a fall a. If the particle leaves the circular path, show that it does so at a height $(1 + 2k^2)/3$ above the lowest point. Also find the greatest height that the particle attains during the entire motion.*

Sol. Let OA be the equilibrium position of the string with end O fixed. If a particle falls freely under gravity through a distance a, then the velocity acquired is $\sqrt{2ga}$. Hence, by the given problem, initially the particle is projected horizontally from A with a velocity u given by

$$u = k\sqrt{2ga}.$$

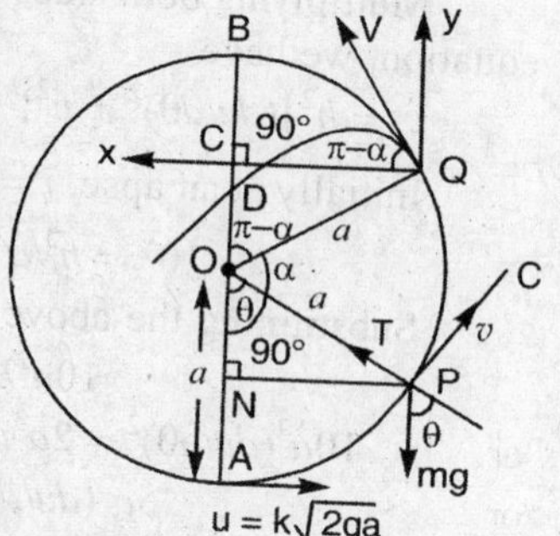

Let P be the position of the particle at any time t, such that $\angle AOP = \theta$ and let v be velocity at P along tangent PC. Let T be the tension in the string along PO and let mg be the weight of the particle acting vertically downwards.

For motion from A to P, by the principle of conservation of energy, we have

$$(1/2) \times mv^2 - (1/2) \times m \times (2gak^2) = (-mg) \times AN, \text{ where PN is perpendicular to OA}$$

or $$v^2 = 2gak^2 - 2g(OA - OM) = 2gak^2 - 2g(a - a\cos\theta).$$

Thus, $$v^2 = 2gak^2 - 2ga + 2ga\cos\theta \qquad ...(1)$$

Equation of motion of the particle in the inward drawn normal PO is given by

$$(mv^2)/a = T - mg\cos\theta \quad \text{so that} \quad T = mg\cos\theta + (mv^2)/a$$

or $$T = mg\cos\theta + (m/a) \times (2gak^2 - 2ga + 2ga\cos\theta), \text{ using (1)}$$

Thus, $$T = (m/a) \times (2gak^2 - 2ga + 3ga\cos\theta) \qquad ...(2)$$

Let the particle leave the circular path at Q where $\angle AOQ = \alpha$. Let V be the velocity of the particle at Q. Setting T = 0 and $\theta = \alpha$ in (2), we obtain

$$0 = 2gak^2 - 2ga + 3ga\cos\alpha \quad \text{so that} \quad \cos\alpha = (2/3) \times (1 - k^2) \qquad ...(3)$$

Let QC be perpendicular to vertical diameter AB. Then, the required height AC of Q above A is given by

$$AC = AO + OC = a + a\cos(\pi - \alpha) = a - a\cos\alpha = a - a \times (2/3) \times (1 - k^2), \text{ using (3)}$$

Thus, $$AC = a\{1 - (2/3) \times (1 - k^2)\} = (a/3) \times (1 + 2k^2).$$

Next, at Q, v = V and $\theta = \alpha$ and hence from (1) we have

$$V^2 = 2gak^2 - 2ga + 2ga\cos\alpha = 2gak^2 - 2ga + (2ga) \times (2/3) \times (1 - k^2), \text{ by (3)}$$

or $$V^2 = (2ga/3) \times (3k^2 - 3 + 2 - 2k^2) \quad \text{or} \quad V^2 = (2ga/3) \times (k^2 - 1) \qquad ...(4)$$

After crossing Q, the particle leaves the circular path and traces out a parabolic path with velocity V and $(\pi - \alpha)$ as angle of projection. Then, the maximum height H attained by the particle above horizontal line QC is given by

$$H = \frac{V^2\sin^2(\pi-\alpha)}{2g} = \frac{V^2\sin^2\alpha}{2g} = \frac{V^2}{2g}(1-\cos^2\alpha) = \frac{2ga(k^2-1)}{3\times 2g}\left\{1 - \frac{4(1-k^2)^2}{9}\right\},$$

[using (3) and (4)]

$$= (1/27) \times a(k^2 - 1)\{9 - 4(1 - 2k^2 + k^4)\} = (a/27) \times (k^2 - 1)(5 + 8k^2 - 4k^4)$$

Hence, the required height attained by the particle above A

$$= AC + H = (a/3) \times (1 + 2k^2) + (a/27) \times (k^2 - 1)(5 + 8k^2 - 4k^4)$$

$$= (a/27) \times (4 + 15k^2 + 12k^4 - 4k^6) = (a/27) \times (1 + 2k^2)^2 (4 - k^2)$$

Ex. 81. *A particle moves under a central force $m\lambda(3a^3u^4 + 8au^2)$. It is projected from an apse at a distance a from the centre of force with velocity $\sqrt{10\lambda}$. Show that the second apsidal* *distance is half the first, and that the equation to the path is* $2r = a\{1 + \operatorname{sech}(\theta/\sqrt{5})\}$.

Sol. Here, the central force per unit mass = $F = \lambda(3a^3u^4 + 8au^2)$, where $u = 1/r$.

Hence, the differential equation of the central orbit is given by

$$h^2(d^2u/d\theta^2 + u) = F/u^2 = \lambda(3a^3u^2 + 8a)$$

Multiplying both sides of the above equation by $2(du/d\theta)$ and then integrating the resulting equation, we have

$$v^2 = h^2\{(du/d\theta)^2 + u^2\} = 2\lambda(3a^3u^3 + 8au) + A, \text{ A being an arbitrary constant} \quad ...(1)$$

Initially at an apse, $r = a$, $u = 1/a$, $du/d\theta = 0$ and $v = \sqrt{10\lambda}$. Hence, (1) gives

$$10\lambda = h^2/a^2 = \lambda(2 + 16) + A \quad \text{giving} \quad h^2 = 10a^2\lambda \quad \text{and} \quad A = -8\lambda$$

Substituting the above values of h and A in (1), we obtain

$$10a^2\lambda\{u^2 + (du/d\theta)^2\} = \lambda(2a^3u^3 + 16au) - 8\lambda$$

or $$10a^2(du/d\theta)^2 = 2a^3u^3 - 10a^2u^2 + 16au - 8 = 2\{a^2u^2(au-1) - 4au(au-1) + 4(au-1)\}$$

or $$5a^2(du/d\theta)^2 = (au-1)(a^2u^2 - 4au + 4) = (au-1)(au-2)^2 \quad ...(2)$$

Determination of the second apsidal distance. At an apse $du/d\theta = 0$.

Hence, (2) $\Rightarrow$ $0 = (au-1)(au-2)^2$ $\Rightarrow$ $u = 1/a$ and $u = 2/a$ $\Rightarrow$ $r = a$ and $r = a/2$

Now $r = a$ is the first apsidal distance. Hence, the second apsidal distance is $a/2$ which is half of the first.

Determination of the equation of the path. From (2), we have

$$a\sqrt{5}\,\frac{du}{d\theta} = -(au-2)\sqrt{au-1} \quad \text{or} \quad \frac{d\theta}{\sqrt{5}} = -\frac{a\,du}{(au-2)\sqrt{au-1}}$$

Substituting $au - 1 = z^2$ so that $a\,du = 2z\,dz$, the above equation yields

$$\frac{d\theta}{\sqrt{5}} = -\frac{2z\,dz}{(z^2-1)} \quad \text{or} \quad \frac{d\theta}{2\sqrt{5}} = \frac{dz}{1-z^2}$$

Integrating, $\theta/2\sqrt{5} + B = \tanh^{-1}z$, where B is an arbitrary constant

or $$(\theta/2\sqrt{5}) + B = \tanh^{-1}\sqrt{au-1}, \text{ since } z = \sqrt{au-1} \quad ...(3)$$

But initially, when $u = 1/a$, $\theta = 0$. Hence, (3) yields $B = 0$. Hence, (3) becomes

$$\theta/2\sqrt{5} = \tanh^{-1}\sqrt{au-1} \quad \text{or} \quad \tanh(\theta/2\sqrt{5}) = \sqrt{au-1} \quad ...(4)$$

Using well known formula cash $2A = (1 + \tanh^2 A)/(1 - \tanh^2 A)$, (4) gives

$$\cosh\frac{\theta}{\sqrt{5}} = \frac{1 + \tanh^2(\theta/2\sqrt{5})}{1 - \tanh^2(\theta/2\sqrt{5})} = \frac{1 + au - 1}{1 - (au-1)} = \frac{au}{2 - au}$$

or $\operatorname{sech}(\theta/\sqrt{5}) = (2 - au)/au = (2/au) - 1$ or $1 + \operatorname{sech}(\theta/\sqrt{5}) = 2r/a$, as $u = 1/r$

or $2r = a\{1 + \operatorname{sech}(\theta/\sqrt{5})\}$, which is the required equation of the path.

Ex. 82. *A particle moves with central acceleration* $\mu(r + 2a^3/r^2)$ *being projected from an apse at a distance a with twice the velocity for a circle at that distance, find the other apsidal distance and show that equation to the path is*

$$\frac{\theta}{2} = \tan^{-1}(t\sqrt{3}) - \frac{1}{\sqrt{5}}\tan^{-1}(t\sqrt{5/3}), \quad \textit{where} \quad t^2 = \frac{r-a}{3a-r}.$$

Sol. Here, the central acceleration = $F = \mu(r + 2a^3/r^2) = \mu(1/u + 2a^3u^2)$, where $r = 1/u$

If V is the velocity for a circle at a distance a under the same acceleration, we know that (refer Art. 9.7)

$V^2 = a[F]_{r=a}$ giving $V^2 = a \times \mu(a + 2a^3/a^2)$ so that $V = a\sqrt{3\mu}$

Hence, at apse, the velocity of projection of the particle $= 2V = 2a\sqrt{3\mu}$

The differential equation of the central orbit is $h^2(u^2 + d^2u/d\theta^2) = F/u^2$

or $$h^2(u + d^2u/d\theta^2) = (\mu/u^2) \times (1/u + 2a^3u^2) = \mu(1/u^3 + 2a^3)$$

Multiplying both sides of the above equation by $2(du/d\theta)$ and integrating, we get

$$h^2\{u^2 + (du/d\theta)^2\} = 2\mu\{-(1/2u^2) + 2a^3u\} + A, \text{ where A is a constant}$$

or $$v^2 = h^2\{u^2 + (du/d\theta)^2\} = \mu(4a^3u - 1/u^2) + A \qquad ...(1)$$

But initially, when at apse, $r = a$, $u = 1/a$, $du/d\theta = 0$ and $v = 2a\sqrt{3\mu}$.

Hence, (1) $\Rightarrow$ $12\mu a^2 = h^2/a^2 = \mu(4a^2 - a^2) + A \Rightarrow h^2 = 12\mu a^2$ and $A = 9\mu a^2$

Substituting the above values of h^2 and A in (1), we have

$12\mu a^4\{u^2 + (du/d\theta)^2\} = \mu(4a^3u - 1/u^2) + 9\mu a^2$ or $12a^4(du/d\theta)^2 = -12a^4u^2 - 1/u^2 + 4a^3u + 9a^2$

Putting $u = 1/r$ so that $du/d\theta = -(1/r^2) \times (dr/d\theta)$, the above equation yields

$$\frac{12a^4}{r^4} \times \left(\frac{dr}{d\theta}\right)^2 = -\frac{12a^4}{r^2} - r^2 + \frac{4a^3}{r} + 9a^2$$

or $$12a^4(dr/d\theta)^2 = -12a^4r^2 - r^6 + 4a^3r^2 + 9a^2r^4 = r^2(-r^4 + 9a^2r^2 + 4a^3r - 12a^4)$$
$$= r^2\{-r^3(r-a) - r^2a(r-a) + 8a^2r(r-a) + 12a^3(r-a)\}$$
$$= r^2(r-a)(-r^3 - r^2a + 8a^2r + 12a^3)$$
$$= r^2(r-a)\{-r^2(r-3a) - 4ar(r-3a) - 4a^2(r-3a)\}$$

or $$12a^4(dr/d\theta)^2 = r^2(r-a)(3a-r)(r^2 + 4ar + 4a^2) = r^2(r-a)(3a-r)(r+2a)^2$$

Hence, $$2\sqrt{3}a^2(dr/d\theta) = r(r+2a)\{(r-a)(3a-r)\}^{1/2} \qquad ...(2)$$

At an apse, we have $du/d\theta = 0$ or $dr/d\theta = 0$. Putting $dr/d\theta = 0$ in (2), we obtain

$0 = r(r+2a)\{(r-a)(3a-r)\}^{1/2}$ giving $r = a$ and $r = 3a$,

noting that only positive values of r are permissible for an apse. But $r = a$ is the first apsidal distance and hence the second required apsidal distance is $r = 3a$.

Determination of the equation of the path of the particle

From (2), we have $$\frac{1}{2}d\theta = \frac{a^2\sqrt{3}\,dr}{r(r+2a)\{(r-a)(3a-r)\}^{1/2}} \qquad ...(3)$$

Given that $(r-a) = (3a-r)t^2$ or $r = \dfrac{a(3t^2+1)}{1+t^2}$ so that $dr = \dfrac{4atdt}{(1+t^2)^2}$. So (3) yields

$$\frac{1}{2}d\theta = \frac{\{(a^2\sqrt{3} \times 4at)/(1+t^2)^2\}dt}{\dfrac{a(3t^2+1)}{1+t^2}\left\{\dfrac{a(3t^2+1)}{1+t^2} + 2a\right\}\left\{3a - \dfrac{a(3t^2+1)}{1+t^2}\right\}t}$$

or $$\frac{1}{2}d\theta = \frac{2\sqrt{3}(t^2+1)}{(3t^2+1)(5t^2+3)}dt \quad \text{or} \quad \frac{1}{2}d\theta = \sqrt{3}\left(\frac{1}{3t^2+1} - \frac{1}{5t^2+3}\right)dt$$

[On resolving into partial fractions]

Integrating, $$\frac{\theta}{2} + B = \tan^{-1}(t\sqrt{3}) - \left(\sqrt{3} \times \frac{1}{\sqrt{3}\sqrt{5}}\right)\tan^{-1}\left(\frac{t\sqrt{5}}{\sqrt{3}}\right),$$ where B is a constant or $$\theta/2 + B = \tan^{-1}(t\sqrt{3}) - (1/\sqrt{5})\tan^{-1}(t\sqrt{5/3}) \quad ...(4)$$

But Initially, when $r = a$, let $\theta = 0$. Also, from the given relation $t^2 = (r-a)/(3a-r)$, initially $t = \{(a-a)/(3a-a)\}^{1/2} = 0$. Hence, (4) yields $B = 0$. Therefore, (4) reduces to

$$\theta/2 = \tan^{-1}(t\sqrt{3}) - (1/\sqrt{5})\tan^{-1}\{(t\sqrt{5/3})\}, \text{ where } t^2 = (r-a)/(3a-r).$$

This is the desired equation of the path.

Ex. 83. *Show that only law for a central attraction for which the velocity in a circle at a distance is equal to the velocity acquired in falling from infinity to the distance is that oj inverse cube.*

Sol. Let F be the central attraction, and let $F = f'(r)$. [Note carefully]

For the particle falling from infinity under the assumed law, when it is at a distance r from the centre of force, the equation of motion will be

$$v(dv/dr) = -f'(r) \quad \text{or} \quad 2v\,dv = -2f'(r)dr$$

Integrating it, $v^2 = -2f(r) + C$, C being an arbitrary constant ...(1)

Again, let v be the velocity of the particle moving in a circle at a distance r. Then in the inward normal direction, the equation of motion is given by

$v^2/\rho = f'(r)$ or $v^2/a = f'(r)$, as $\rho = a$ for a circle. Thus, we have

$$v^2 = r\,f(r) \quad ...(2)$$

Now, according to the given problem, the velocities given by (1) and (2) are equal. Therefore,

$$r f'(r) = -2f(r) + C \quad \text{or} \quad r^2 f'(r) + 2r\,f(r) = Cr$$

Integrating, $r^2 f(r) = (1/2) \times Cr^2 + A$, A being an arbitrary constant

or $f(r) = (1/2) \times C + A/r^2$ so that $f'(r) = -(2A/r^3)$

Hence, by hypothesis $F = f'(r) = -(2A/r^3)$ ...(3)

Since 2A is *a* constant, (3) shown that central attraction F varies inversely as cube of the distance.

Ex. 84. *A particle acted on by a central attractive force μu^3 is projected with a velocity $\sqrt{\mu}/a$ at an angle $\pi/4$ with its initial distance a from the centre of force. Show that its orbit is the equiangular spiral $r = ae^{-\theta}$.*

Sol. Here, the central attaction $= F = \mu u^3$.

The differential equation of the central orbit is

$$h^2(u^2 + d^2u/d\theta^2) = F/u^2 = \mu u \quad ...(1)$$

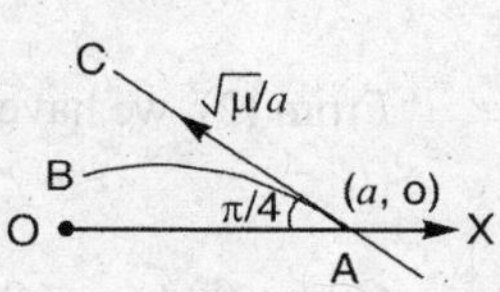

Multiplying both sides of (1) by $2(du/d\theta)$ and then integrating the resulting equation, we get

$$v^2 = h^2\{u^2 + (du/d\theta)^2\} = \mu u^2 + A, \text{ where A is a constant} \quad ...(2)$$

Initially, when $r = a$, $u = 1/a$, $v = \sqrt{\mu}/a$ and $\phi = \pi/4$ ...(3)

Now, $p = r\sin\phi \Rightarrow p = a\sin(\pi/4) = a/\sqrt{2}$ and hence

$$1/p^2 = u^2 + (du/d\theta)^2 = 2/a^2 \quad ...(4)$$

Using (3) and (4), (2) reduces to $\mu/a^2 = 2h^2/a^2 = \mu/a^2 + A$

Hence $h^2 = \mu/2$ and $A = 0$. With these values, (2) reduces to

$$(\mu/2) \times \{u^2 + (du/d\theta)^2\} = \mu u^2 \quad \text{or} \quad (du/d\theta)^2 = u^2 \quad ...(5)$$

Putting $u = 1/r$ and $du/d\theta = -(1/r^2)(dr/d\theta)$ in (5), we get

$$(1/r^4) \times (dr/d\theta)^2 = 1/r^2 \quad \text{or} \quad (dr/d\theta)^2 = r^2$$

giving $$dr/d\theta = -r, \quad ...(6)$$

where we have taken negative sign because r decreases while θ increases as the particle starts moving along AB from the point A.

Integrating (6) $\quad \log r = -\theta + B$, where B is a constant ...(7)

But, initially, when $r = a$. let $\theta = 0$. Hence, (7) yields $\quad B = \log a$

Therefore, (7) becomes $\quad \log r - \log a = -\theta \quad$ or $\quad r = ae^{-\theta}$,

which in the required equation of the path and it is an equiangular spiral.

Ex. 85. *A particle moves with a central acceleration $\mu/(\text{distance})^5$ and projected from the apse at a distance a with a velocity equal to n times that which would be acquired in falling from infinity; show that the other apsidal distance is $a/(n^2-1)^{1/2}$.* **(Purvanchal 2011)**

If n = 1 and particle be projected in any direction, show that the path is a circle passing through the centre of force. **(Rohilkhand 2011)**

Sol. Here, $\quad$ the central acceleration $= F = \mu/r^5 = \mu u^5$, where $u = 1/r$

Let V be the velocity from infinity to a distance a from the centre under the same acceleration. Then, from Art. 9.7, we have

$$V^2 = -2\int_\infty^a F\,dr = -2\int_\infty^a \frac{\mu}{r^5}\,dr = -2\left[-\frac{\mu}{4r^4}\right]_\infty^a = \frac{\mu}{2a^4} \quad \text{so that} \quad V = \sqrt{\frac{\mu}{2a^4}}$$

The differential equation of the central orbit is

$$h^2(u + d^2u/d\theta^2) = F/u^2 = \mu u^5/u^2 = \mu u^3$$

Multiplying both sides by $2(du/d\theta)$ and integrating, we get

$$h^2\{u^2 + (du/d\theta)^2\} = (2\mu u^4)/4 + A, \text{ where A is a constant}$$

or
$$v^2 = h^2\{u^2 + (du/d\theta)^2\} = (\mu u^4)/2 + A \qquad ...(1)$$

But initially at an apse, when $r = a$, *i.e.*, $u = 1/a$, $du/d\theta = 0$ and $v = nV = n\sqrt{(\mu/2a^4)}$

Hence, (1) yields

$(n^2\mu)/2a^4 = h^2/a^2 = \mu/2a^4 + A \quad$ so that $\quad h^2 = (n^2\mu)/2a^2 \quad$ and $\quad A = \{(n^2-1)\mu\}/2a^4$

Substituting the above values of h^2 and A in (1), we obtain

$$\frac{n^2\mu}{2a^2}\left\{u^2 + \left(\frac{du}{d\theta}\right)^2\right\} = \frac{\mu u^4}{2} + \frac{(n^2-1)\mu}{2a^4} \quad \text{or} \quad \left(\frac{du}{d\theta}\right)^2 = \frac{a^4u^4 - a^2n^2u^2 + n^2 - 1}{n^2a^2} \qquad ...(2)$$

At an apse, we have $du/d\theta = 0$. So from (2), the apsidal distances are given by

$a^4u^4 - a^2n^2u^2 + n^2 - 1 = 0 \quad$ or $\quad (a^4/r^4) - (a^2n^2/r^2) + n^2 - 1 = 0$

or $\quad (n^2-1)r^4 - a^2n^2r^2 + a^4 = 0$, $\quad$ which is a quadratic equation in r^2.

Let $r_1^{\,2}$ and $r_2^{\,2}$ be its roots. Then, by theory of equation, we have

$r_1^{\,2}\, r_2^{\,2} = a^4/(n^2-1) \quad$ or $\quad r_1r_2 = a^2/(n^2-1)^{1/2} \qquad ...(3)$

Now, given that the first apsidal distance, $r_1 = a$. Then, from (3), we have

$ar_2 = a^2/(n^2-1)^{1/2} \quad$ so that $\quad r_2 = a/(n^2-1)^{1/2}$

Thus, the second apsidal distance $= r_2 = a/(n^2-1)^{1/2}$

Second Part. When $n = 1$ and the particle is projected in any direction, say at an angle α to the radius vector, then at the point of projection, we have $\phi = \alpha$, $\quad p = r\sin\phi = a\sin\alpha$

Therefore, $\quad 1/p^2 = u^2 + (du/d\theta)^2 = 1/(a\sin\alpha)^2$

Thus, in the present second part, initially when $r = a$, *i.e.*, $u = 1/a$, we have

$v = V = \sqrt{\mu/2a^4} \quad$ and $\quad u^2 + (du/d\theta)^2 = 1/(a^2\sin^2\alpha)$

Hence, for this second part, (1) takes the following new form

$\mu/2a^4 = h^2(a^2\sin^2\alpha) = \mu/2a^2 + A$ so that $h^2 = (\mu \sin^2\alpha)/2a^2$ and $A = 0$

Substituting the above new values of h^2 and A in (1), we obtain

$$\frac{\mu \sin^2\alpha}{2a^2}\left[u^2 + \left(\frac{du}{d\theta}\right)^2\right] = \frac{\mu u^4}{2} \quad \text{or} \quad u^2 + \left(\frac{du}{d\theta}\right)^2 = \frac{a^2u^4}{\sin^2\alpha} \quad \text{or} \quad \left(\frac{du}{d\theta}\right)^2 = \frac{a^2u^4}{\sin^2\alpha} - u^2$$

Putting $u = 1/r$ so that $du/d\theta = -(1/r^2) \times (dr/d\theta)$, the above equation reduces to

$$\left(-\frac{1}{r^2}\frac{dr}{d\theta}\right)^2 = \frac{a^2}{r^4\sin^2\alpha} - \frac{1}{r^2} \quad \text{or} \quad \left(\frac{dr}{d\theta}\right)^2 = a^2\operatorname{cosec}^2\alpha - r^2$$

or $$\frac{dr}{d\theta} = (a^2\operatorname{cosec}^2\alpha - r^2)^{1/2} \quad \text{or} \quad d\theta = \frac{dr}{(a^2\operatorname{cosec}^2\alpha - r^2)^{1/2}}$$

Integrating, $\theta + B = \sin^{-1}\{r/(a \operatorname{cosec}\alpha)\}$, B being a constant ...(4)

Initially, when $r = a$, let $\theta = 0$. Hence, (4) yields $B = \sin^{-1}(\sin\alpha) = \alpha$.

Hence, (4) reduces to $\theta + \alpha = \sin^{-1}\{r/(a \operatorname{cosec}\alpha)\}$

or $r = (a \operatorname{cosec}\alpha)\sin(\theta + \alpha)$ or $r = (a \operatorname{cosec}\alpha)\cos\{\pi/2 - (\theta + \alpha)\}$

or $r = (a \operatorname{cosec}\alpha)\cos\{(\theta + \alpha) - \pi/2\}$ or $r = (a \operatorname{cosec}\alpha)\cos\{\theta - (\pi/2 - \alpha)\}$

or $r = (a \operatorname{cosec}\alpha)\cos(\theta - \beta)$, where $\beta = \pi/2 - \alpha$

which represents a circle of diameter a cosecα and pole on its circumference. Therefore, the path of the particle is a circle through the centre of force.

Ex. 86. *A particle is acted on by a central repulsive force which varies as the nth power of the distance. If the velocity at any point be equal to that which would be acquired in falling from the centre to the point, show that the equation to the path is of the form*

$$r^{(n+3)/2}\cos\{(n+3) \times (\theta/2)\} = \text{constant}.$$

Sol. Since the particle is acted on by a central repulsive force which varies as the nth power of the distance, hence the central acceleration F is given by

$$F = -\mu(\text{distance})^n = -\mu r^n = -(\mu/u^n)$$

While falling in a straight line from rest from the centre of force if v is the velocity of the particle at a distance x from the centre, then

$$v(dv/dx) = \mu x^n \quad \text{or} \quad v\,dv = \mu x^n\,dx \qquad ...(1)$$

Let V be the velocity of the particle acquired in falling from the centre to a distance r. Then, from (1), we have

$$\int_0^V v\,dv = \int_0^r \mu x^n\,dx \quad \text{or} \quad \frac{V^2}{2} = \mu\left[\frac{x^{n+1}}{n+1}\right]_0^r \quad \text{or} \quad V^2 = \left(\frac{2\mu}{n+1}\right)r^{n+1}$$

The differential equation of the central orbit is given by

$$h^2(u + d^2u/d\theta^2) = F/u^2 = -\frac{\mu/u^n}{u^2} = -\mu u^{-n-2} \qquad ...(2)$$

Multiplying both sides of (2) by $2(du/d\theta)$ and then integrating, we get

$$v^2 = h^2\{u^2 + (du/d\theta)^2\} = -\frac{2\mu u^{-n-1}}{(-n-1)} + A = \frac{2\mu}{(n+1)u^{n+1}} + A \qquad ...(3)$$

Given that $v = V$ so that $v^2 = (2\mu)/\{(n+1)u^{n+1}\}$. Hence, (3) yields

$$\frac{2\mu}{(n+1)u^{n+1}} = \frac{2\mu}{(n+1)u^{n+1}} + A \quad \text{so that} \quad A = 0.$$ Therefore, (3) reduces to

$$h^2\left\{u^2+\left(\frac{du}{d\theta}\right)^2\right\}=\frac{2\mu}{(n+1)\,u^{n+1}} \quad \text{or} \quad u^2+\left(\frac{du}{d\theta}\right)^2=\frac{2\mu}{(n+1)h^2\,u^{n+1}}=\frac{k^2}{u^{n+1}} \quad ...(4)$$

where $k^2 = 2\mu/\{(n + 1)h^2\}$ = constant. Re-writing (4), we have

$$\left(\frac{du}{d\theta}\right)^2=\frac{k^2}{u^{n+1}}-u^2=\frac{k^2-u^{n+3}}{u^{n+1}} \quad \text{or} \quad \frac{du}{d\theta}=-\frac{(k^2-u^{n+3})^{1/2}}{u^{(n+1)/2}}$$

Separating variables, $$d\theta=-\frac{u^{(n+1)/2}}{(k^2-u^{n+3})}\,du \quad ...(5)$$

Substituting $u^{(n+3)/2} = y$ so that $\{(n + 3)/2\}\, u^{(n+1)/2}\, du = dy$, (5) becomes

$$d\theta=-\frac{2dy}{(n+3)\,(k^2-y^2)} \quad \text{or} \quad \frac{n+3}{2}\,d\theta=-\frac{dy}{(k^2-y^2)}$$

Integrating, $\{(n + 3)\theta\}/2 + B = \cos^{-1}(y/k) = \cos^{-1}\{u^{(n+3)/2}/k\}$ ...(5)

Let us select k such that when $u = 1/a$, $\theta = 0$, $(1/k) \times (1/a)^{(n+3)/2} = 1$.

Therefore, (6) yields $0 + B = \cos^{-1}1$ so that $B = 0$. Hence, (6) reduces to

$$\frac{(n+3)\theta}{2}=\cos^{-1}\left\{\frac{u^{(n+3)/2}}{k}\right\} \quad \text{or} \quad u^{(n+3)/2}=k\cos\left\{\frac{(n+3)\theta}{2}\right\}, \quad \text{where} \quad u=\frac{1}{r}$$

or $r^{(n+3)/2}\cos\{(n + 3) \times (\theta/2)\} = 1/k$ = constant, which is the required equation to the path.

Ex. 87. *A particle moves with a central acceleration $\mu(3u^3 + a^2u^5)$ being projected from a distance a at an angle 45° with a velocity equal to that in a circle at the same distance. Prove that the time to centre of force is* $(a^2/\sqrt{2\mu}) \times (2 - \pi/2)$.

Sol. Here, the central acceleration $= F = \mu(3u^3 + a^2u^5)$ ...(1)

Let V be the velocity of projection. Then, by From I of Art 9.7, we have

$$V^2 = a[F]_{r=a} = a[F]_{u=1/a} = a\mu(3/a^3 + a^2/a^5) = 4\mu/a^2 \quad ...(2)$$

We know that the differential equation central orbit is

$$h^2(d^2u/d\theta^2 + u) = F/u^2 \quad \text{or} \quad h^2(d^2u/d\theta^2 + u) = \mu(3u + a^2u^3), \text{ using (1)}$$

Multiplying both sides by $2(du/d\theta)$ and then integrating, we have

$$v^2 = h^2\{(du/d\theta)^2 + u^2\} = 2\mu\{(3u^2/2) + (a^2u^4/4)\} + A \quad ...(3)$$

According to the given problem, initially $\phi = 45°$. Also, initially $r = a$. Hence, we get

$$p = r\sin\phi \Rightarrow p = a\sin 45° \Rightarrow p = a/\sqrt{2} \Rightarrow 1/p^2 = 2/a^2 \Rightarrow (du/d\theta)^2 + u^2 = 2/a^2$$

Thus, when $u = 1/r = 1/a$, $(du/d\theta)^2 + u^2 = 2/a^2$ and $v^2 = V^2 = 4\mu/a^2$, by (2)

Hence, (2) yields $4\mu/a^2 = 2h^2/a^2 = (3\mu/a^2) + (\mu/2a^2) + A \Rightarrow h^2 = 2\mu$ and $A = \mu/2a^2$

Therefore, (3) yields $2\mu\{(du/d\theta)^2 + u^2\} = 2\mu\{(3u^2/2) + (a^2u^4/4)\} + \mu/2a^2$

or $$2\{(du/d\theta)^2 + u^2\} = 3u^2 + (a^2u^4/2) + (1/2a^2) \quad ...(4)$$

Putting $u = 1/r$ and $du/d\theta = -(1/r^2)(dr/d\theta)$ in (4), we obtain

$$2\{(1/r^4) \times (dr/d\theta)^2 + (1/r^2)\} = (3/r^2) + (a^2/2r^4) + (1/2a^2)$$

or $(1/r^4) \times (dr/d\theta)^2 = (3/2r^2) + (a^2/4r^4) + (1/4a^2) - (1/r^2)$

or $(dr/d\theta)^2 = (3r^2/2) + (a^2/4) + (r^4/4a^2) - r^2 = (r^2/2) + (a^2/4) + (r^4/4a^2)$

or $$(dr/d\theta)^2 = (2a^2r^2 + a^4 + r^4)/(4a^2) = \{(a^2 + r^2)/2a\}^2,$$

giving $$dr/d\theta = -\{(a^2 + r^2)/2a\} \quad ...(5)$$

Also, we know that $r^2(d\theta/dt) = h$ or $h\,dt = r^2d\theta$

or $$h\,dt = -r^2\{2a/(a^2 + r^2)\}dr, \text{ using (5)} \quad ...(6)$$

Let T be the required time. Integrating (6) between the limits $t = 0$ to $t = T$ and the corresponding limits $r = a$ to $r = 0$, we heve

$$h\int_0^T dt = -2a\int_a^0 \frac{r^2}{a^2+r^2}\,dr$$

or $$hT = 2a\int_0^a \left(1 - \frac{a^2}{a^2+r^2}\right)dr = 2a[r - a\tan^{-1}(r/a)]_0^a = 2a\{a - a\times(\pi/2)\}$$

or $$T = (a^2/h)\times(2-\pi/2) = (a^2/\sqrt{2\mu})\times(2-\pi/2), \text{ as } h = \sqrt{2\mu}$$

Ex. 88. *A particle is projected with velocity $(2\mu/3c^3)^{1/2}$ from a point P in a field of attractive force μ/r^4 to a point O distance c from P, where r denotes the distance from O. If the direction of projection makes an angle 45° with PO, prove that the orbit is a cardioid and the particle will arrive at O after a time $(3c^5/\mu)^{1/2} \times \{(3\pi/4) - 2\}$.*

Sol. Here, the central acceleration $= F = \mu/r^4 = \mu u^4$.

The differential equation of the central orbit is $h^2(d^2u/d\theta^2 + u) = F/u^2 = \mu u^2$

Multiplying both sides by $2(du/d\theta)$ and then integrating it, we get

$$v^2 = h^2\{(du/d\theta)^2 + u^2\} = (2\mu u^3/3) + A, \quad \text{where A is a constant} \qquad ...(1)$$

Now, Initially at P, when $r = c$, $u = 1/c$, $v = (2\mu/2c^3)^{1/2}$, $\phi = 45°$. Now, we have

$$p = r\sin\phi \Rightarrow p = c\sin 45° = c/\sqrt{2} \Rightarrow 1/p^2 = 2/c^2 \quad \text{or} \quad d^2u/d\theta^2 + u = 2/c^2$$

Using the above values in (1), we have

$$(2\mu/3c^3) = (2h^2/c^2) = (2\mu/3c^3) + A \quad \text{so that} \quad h^2 = (\mu/3c) \quad \text{and} \quad A = 0. \qquad ...(2)$$

Substituting the above values of h^2 and A in (1), we obtain

$$(\mu/3c)\times\{(du/d\theta)^2 + u^2\} = (2\mu u^3/3) \quad \text{or} \quad (du/d\theta)^2 = 3cu^3 - u^2 \qquad ...(3)$$

Putting $u = 1/r$ so that $du/d\theta = -(1/r^2)\}\times(dr/d\theta)$, (3) reduces to

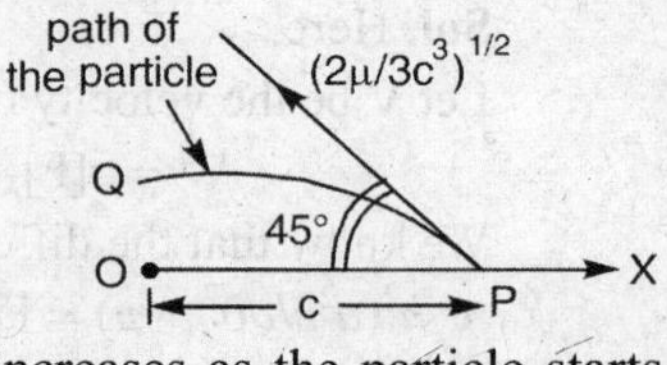

$$(1/r^4)\times(dr/d\theta)^2 = (2c/r^3) - (1/r^2)$$

or $$(dr/d\theta)^2 = 2cr - r^2 = c^2 - (r-c)^2$$

or $$dr/d\theta = -\{c^2 - (r-c)^2\}^{1/2}, \qquad ...(4)$$

path of the particle $(2\mu/3c^3)^{1/2}$ Q 45° O c P X

where we have taken negative sign because r decreases while θ increases as the particle starts moving from the starting point P. Separating variables, (4) reduces to

$$d\theta = -\frac{dr}{\{c^2-(r-c)^2\}^{1/2}} \quad \text{so that} \quad \theta + B = \cos^{-1}\{(r-c)/c\} \qquad ...(5)$$

But initially at P, when $r = c$, let $\theta = 0$. Then, (5) yields $B = \cos^{-1}0 = \pi/2$

Hence, (5) reduces to $\theta + \pi/2 = \cos^{-1}\{(r-c)/c\}$ or $(r-c)/c = \cos(\pi/2 + \theta)$

or $$r = c\{1 + \cos(\pi/2+\theta)\} \quad \text{or} \quad r = c(1-\sin\theta), \quad \text{which is a cardioid} \qquad ...(6)$$

Determination of the required time. We have $h^2 = r^2(d\theta/dt)$

or $$dt = \frac{r^2}{h}d\theta = \frac{c^2(1-\sin\theta)^2}{(\mu/3c)^{1/2}}d\theta, \quad \text{using (2) and (6)}$$

or $$dt = (3c^5/\mu)^{1/2}(1-\sin\theta)^2\,d\theta \qquad ...(7)$$

At the starting point P, $\theta = 0$. Also, at the point O, $r = 0$.

Putting $r = 0$ in the equation of the path $r = c(1 - \sin\theta)$, we get

$1 - \sin\theta = 0$ so that $\sin\theta = 1$ giving $\theta = \pi/2$.

Thus, at the point O, $\theta = \pi/2$. Let T be the time taken by the particle to move from P to O. Then, integrating (3) between the limits $t = 0$ to $t = T$ and the corresponding limits $\theta = 0$ tc

$\theta = \pi/2$, we have

$$\int_0^T dt = \left(\frac{3c^5}{\mu}\right)^{1/2} \int_0^{\pi/2} (1-\sin\theta)^2 d\theta = \left(\frac{3c^5}{\mu}\right)^{1/2} \int_0^{\pi/2} (1 - 2\sin\theta + \sin^2\theta)\, d\theta$$

or
$$T = \left(\frac{3c^5}{\mu}\right)^{1/2} \times \left[\int_0^{\pi/2} d\theta - 2\int_0^{\pi/2} \sin\theta\, d\theta + \int_0^{\pi/2} \frac{1-\cos 2\theta}{2} d\theta\right]$$

or
$$T = \left(\frac{3c^5}{\mu}\right)^{1/2} \times \left\{[\theta]_0^{\pi/2} - 2[-\cos\theta]_0^{\pi/2} + \frac{1}{2}\left[\theta - \frac{\sin 2\theta}{2}\right]_0^{\pi/2}\right\}$$

Thus, $T = (3c^5/\mu)^{1/2} (\pi/2 - 2 + \pi/4) = (3c^5/\mu)^{1/2} \times \{(3\pi/4) - 2\}$

Ex. 89. *A particle moves in a curve under a central attraction so that its velocity at any point is equal to that in a circle at the same distance and under the same attraction. Show that the law of force is that of inverse cube, and the path is an equiangular spiral.* **(Purvanchal 2009)**

Sol. Let the central acceleration $= F = u^2 f'(u)$m where $u = 1/r$ [Note carefully] ...(1)

The differential equation of the central orbit is given by

$$h^2(d^2u/d\theta^2 + u) = F/u^2 = f'(u)$$

Multiplying both sides by $2(du/d\theta)$ and then integrating it, we obtain

$$v^2 = h^2\{(du/d\theta)^2 + u^2\} = 2f(u) + A, \text{ where A is a constant} \quad ...(2)$$

Given that the velocity of the particle at any point in the orbit is equal to that in a circle at the same distance under the same acceleration. Therefore, we have

$v^2/r = F$ so that $v^2 = rF = ru^2 f'(u) = uf'(u)$, using (1)

Substituting the above value of v^2 in (2), we obtain

$$uf'(u) = 2f(u) + A \quad \text{or} \quad \frac{1}{u^2}f'(u) - \frac{2}{u^3}f(u) = \frac{A}{u^3} \quad \text{or} \quad \frac{d}{du}\left\{\frac{1}{u^2}f(u)\right\} = \frac{A}{u^3}$$

Integrating, $(1/u^2) \times f(u) = -(A/2u^2) + B$ giving $f(u) = Bu^2 - (A/2)$ ...(3)

Differentiating both sides of (3) w.r.t. 'u', we have

$f'(u) = 2Bu$ and so (1) yields $F = u^2 \times 2Bu = 2Bu^3 = 2B/r^3$, ...(4)

showing that the acceleration varies inversely as the cube of the distance from the centre.

To determine the required equation of the path

The differential equation central orbit in pedal form is $(h^2/p^3) \times (dp.dr) = F$

or
$$\frac{h^2}{p^3}\frac{dp}{dr} = \frac{2B}{r^3} \quad \text{or} \quad -\frac{2h^2}{p^3}dp = -\frac{4B}{r^3}dr, \quad \text{using (4)}$$

Integrating, $h^2/p^2 = 2B/r^2 + C$, where C is a constant. ...(5)

Assuming that $p \to \infty$ when $r \to \infty$, (5) yields $C = 0$. Hence, (5) becomes

$h^2/p^2 = 2B/r^2$ or $p = ar$, where $a\ (= h/\sqrt{2B})$ is a constant.

This is the pedal equation of an equiangular spiral and hence the required path is an equiangular spiral.

Ex. 90. *a particle moves with central acceleration $(\mu u^2 + \lambda u^3)$ and the velocity of projection at distance R is V; show that the particle will ultimately go off to infinity if $V^2 > (2\mu/R) + (\lambda/R^2)$.*

Sol. Here, the central acceleration $= F = \mu u^2 + \lambda u^3$.

The differential equation of the central orbit is given by

$$h^2(d^2u/d\theta^2 + u) = F/u^2 = (\mu u^2 + \lambda u^3)/u^2 = \mu + \lambda u$$

Multiplying both sides by $2(du/d\theta)$ and then integrating, we get

$$v^2 = h^2\{(du/d\theta)^2 + u^2\} = 2\mu u + \lambda u^2 + A, \text{ where A is constant} \quad ...(1)$$

But initially, when $r = R$, *i.e.*, $u = 1/R$ $v = V$. Hence, (1) yields

$$V^2 = 2\mu/R + \lambda/R^2 + A \quad \text{so that} \quad A = V^2 - (2\mu/R) - (\lambda/R^2) \quad ...(2)$$

From (1), we have $\quad h^2\{(du/d\theta)^2 + u^2\} = 2\mu u + \lambda u^2 + A$, where A is given by (2)

Thus, $\quad h^2(du/d\theta)^2 = (\lambda - h^2)u^2 + 2\mu u + A$

or $$h^2\left(\frac{du}{d\theta}\right)^2 = (\lambda - h^2)\left\{u^2 + \frac{2\mu}{\lambda - h^2}u + \frac{\mu^2}{(\lambda - h^2)^2} + \frac{A}{\lambda - h^2} - \frac{\mu^2}{(\lambda - h^2)^2}\right\}. \quad \textbf{[Note Carefully]}$$

or $$h\frac{du}{d\theta} = (\lambda - h^2)^{1/2}\left\{\left(u + \frac{\mu}{\lambda - h^2}\right)^2 + k\right\}^{1/2}, \text{ where } \quad k = \frac{A}{\lambda - h^2} - \frac{\mu^2}{(\lambda - h^2)^2} \quad ...(3)$$

Separating variables, $$d\theta = \frac{h}{(\lambda - h^2)^{1/2}} \frac{du}{\left\{\left(u + \frac{\mu}{\lambda - h^2}\right)^2 + k\right\}^{1/2}}$$

Integrating, $$\theta + B = \frac{h}{(\lambda - h^2)^{1/2}} \log\left[\left(u + \frac{\mu}{\lambda - h^2}\right) + \left\{\left(u + \frac{\mu}{\lambda - h^2}\right)^2 + k\right\}^{1/2}\right] \quad ...(4)$$

Now, the particle will ultimately go off to infinity if θ is real when $r \to \infty$, *i.e*, when $u = 0$.
Putting $u = 0$ in (4) and substituting the value of k given by (3), we obtain

$$\theta + B = \{h/(\lambda - h^2)\} \log\left[\{\mu/(\lambda - h^2)\} + \{A/(\lambda - h^2)\}^{1/2}\right] \quad ...(5)$$

Assuming $\lambda > h^2$, it follows that (5) will give real values of θ provided $A > 0$. Thus, in order that the particle will ultimately go off to infinity if

$$A > 0 \quad \text{or} \quad V^2 - (2\mu/R) - (\lambda/R^2) > 0, \quad \text{using (3)}$$

Thus, the required condition is given by $\quad V^2 > (2\mu/R) + (\lambda/R^2)$

Ex. 91. *A particle slides down the arc of a smooth cycloid whose axis is vertical and vertex lowest, starting at rest from the cusp. Prove that the time occupied in falling from the first half of the vertical height is equal to the time falling down the second half.* **(Meerut 2007)**

Sol. Let O be the vertex of a smooth cycloid AOA′ and OM its axis. Suppose a particle of mass m slides down the arc of the cycloid starting at rest from cusp A. Let P be the position of the particle at any time t where arc OP = s and let ψ be the angle which the tangent PT at P to the cycloid makes with the tangent OX at the vertex O of the cycloid.

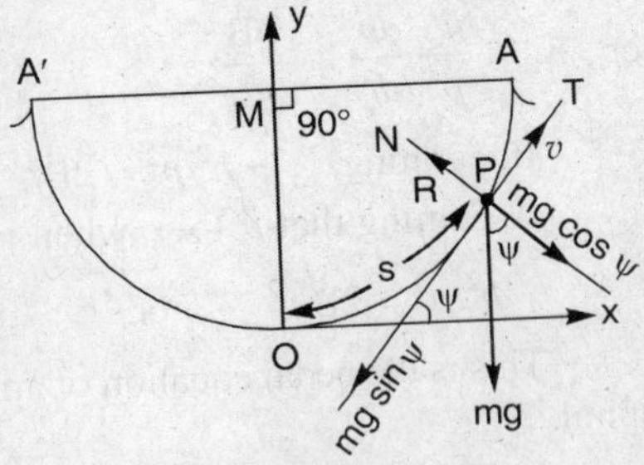

The forces acting on the particle at P are (*i*) the weight mg acting vertically downwards and (*ii*) the normal reaction R acting along the inward drawn normal PN at P. Resolving these forces along the tangent at P, the tangential equation of motion of the particle is given by

$$m(d^2s/dt^2) = -mg \sin\psi, \quad ...(1)$$

Now, the intrisic equation of the cycloid is $\quad s = 4a \sin\psi \quad ...(2)$

Using (2), (1) reduces to $$d^2s/dt^2 = -(g/4a) \times s \quad ...(3)$$

Multiplying both sides of (3) by $2(ds/dt)$ and then integrating, we get

$$(ds/dt)^2 = A - (gs^2/4a), \text{ A being an arbitrary constant} \quad ...(4)$$

Initially at the cusp A, $s = 4a$ and ds/dt = velocity = 0.

Therefore, from (4), we have $A = (g/4a) \times (4a)^2 = 4ag$

Thus, (5) reduces to
$$\left(\frac{ds}{dt}\right)^2 = 4ag - \frac{gs^2}{4a} = \frac{g}{4a}(16a^2 - s^2)$$

giving
$$ds/dt = -(1/2) \times (g/a)^{1/2} \times (16a^2 - s^2)^{1/2} \quad ...(5)$$

where we have taken negative sign because the particle is moving in direction of s decreasing.

Separating variables, (5) gives
$$dt = -2\sqrt{\frac{a}{g}}\frac{ds}{(16a^2 - s^2)^{1/2}} \quad ...(6)$$

The vertical height of the cycloid is $2a$. At the point where the particle has fallen down the first half of the vertical height, we have $y = a$.

Putting $y = a$ in the well known result $s^2 = 8ay$, we get $s^2 = 8a^2$ or $s = 2a\sqrt{2}$

Therefore, integrating (6) from $s = 4a$ to $s = 2a\sqrt{2}$, the time t_1 taken by the particle in falling down the first half of the vertical height of the cycloid is given by

$$t_1 = -2\sqrt{\frac{a}{g}}\int_{s=4a}^{2a\sqrt{2}}\frac{ds}{(16a^2 - s^2)^{1/2}} = 2\sqrt{\frac{a}{g}}\left[\cos^{-1}\frac{s}{4a}\right]_{4a}^{2a\sqrt{2}} = 2\sqrt{\frac{a}{g}}\left(\cos^{-1}\frac{1}{\sqrt{2}} - \cos^{-1}1\right)$$

or
$$t_1 = 2(a/g)^{1/2}(\pi/4 - 0) = (\pi/2) \times (a/g)^{1/2} \quad ...(7)$$

Next, integrating (6) from $s = 2a\sqrt{2}$ to $s = 0$, the time t_2 taken by the particle in falling down the second half of the vertical height of the cycloid is given by

$$t_2 = -2\sqrt{\frac{a}{g}}\int_{s=2a\sqrt{2}}^{0}\frac{ds}{(16a^2 - s^2)^{1/2}} = -\sqrt{\frac{a}{g}}\left[\cos^{-1}\frac{s}{4a}\right]_{2a\sqrt{2}}^{0} = 2\sqrt{\frac{a}{g}}\left(\cos^{-1}0 - \cos^{-1}\frac{1}{\sqrt{2}}\right)$$

or
$$t_2 = 2(a/g)^{1/2} \times (\pi/2 - \pi/4) = (\pi/2) \times (a/g)^{1/2} \quad ...(8)$$

From (7) and (8), $t_1 = t_2$. Therefore, the time occupied in falling down the first half of the vertical height is equal to the time of falling down the second half.

Ex. 92. *If a particle starts from the vertex of a cycloid whose axis is vertical and vertex upwards, prove that the velocity at any point varies as the distance of that point from the vertex measured along the arc.*

Sol. Let a particle of mass m, starting from rest at O slide down the arc of a smooth cycloid whose axis OM is vertical and vertex O is upwards. Let P be the position of the particle at time t such that arc OP = s and the tangent at P to the cycloid makes an angle ψ with the tangent OX at the vertex O. The forces acting on the particle at P are: (*i*) weight mg acting vertically downwards and (*ii*) the reaction R acting along the outward drawn normal PN. Hence the equation of motion of the particle along the tangent PT is given by

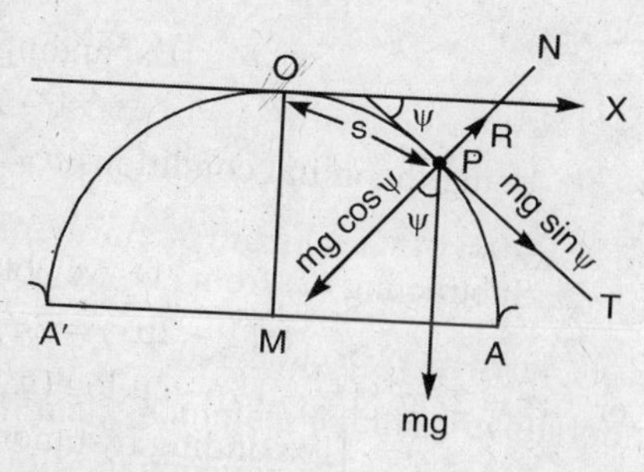

$$m(d^2s/dt^2) = mg \sin\psi \quad \text{or} \quad d^2s/dt^2 = (gs/4a), \quad \text{as} \quad s = 4a \sin\psi$$

Multiplying both sides by $2(ds/dt)$ and then integrating, we get

$$(ds/dt)^2 = v^2 = (g/4a) \times s^2 + C, \quad \text{C being an arbitrary constant} \quad ...(1)$$

But intially at cusp O, when $s = 0$, $v = 0$. Hence, (1) gives $C = 0$.

Therefore, (1) reduces to $v^2 = (g/4a) \times s^2$ or $v = (g/4a)^{1/2} \times s$ ...(2)

Since $(g/4a)^{1/2}$ is a constant, (2) shows that the velocity v at any point varies as the distance s of that point from the vertex O. This is what we wished to prove.

Ex. 93. *A particle of mass m falls from rest at a distance a from the centre of the earth, the motion meeting a small resistance proportional to the square of the velocity v and the retardation being μ for unit velocity. Show that the kinetic energy at a distance x from the centre is $mgr^2\{1/x - 1/a + 2(1 - x/a) - 2\mu\log(a/x)\}$, the square of μ being neglected and r is the radius of the earth.*

Sol. Let a particle of mass m fall from rest at a distance a from the centre O of the earth. Let v be the velocity of the particle at time t at the point P, where $OP = x$. Then, the two acceleration, *i.e.*, the forces acting on the unit mass of the particle are as given below:

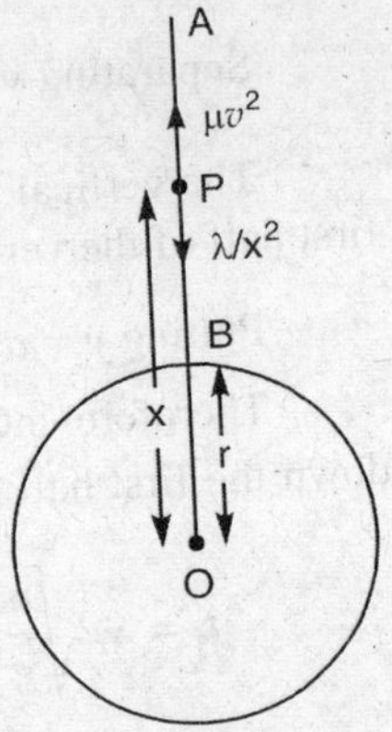

A, μv^2, P, λ/x^2, B, x, r, O

(*i*) The attraction of the earth towards its centre is λ/x^2, where λ is a constant. Since the attraction (acceleration) on the surface of the earth is g, we have $g = \lambda/r^2$, because on the surface of earth, $x = r$. Thus, $\lambda = r^2 g$.

Therefore, the attraction of the earth towards the centre (*i.e.* in the direction of x decreasing) is given by $(r^2 g)/x^2$.

(*ii*) The resistance of the medium on the particle is kv^2, acting against the direction of motion. But, we are given that for $v = 1$, the retardation due to the resistance is μ. Thus, $\mu = k \times 1^2$ so that $k = \mu$.

Therefore, the retardation on the particle due to the resistance of the medium is given by μv^2 acting in the direction of x increasing.

Now, the equation of motion of the particle is given by

$$\frac{d^2 x}{dt^2} = -\frac{r^2 g}{x^2} + \mu v^2 \quad \text{or} \quad v\frac{dv}{dx} = -\frac{r^2 g}{x^2} + \mu v^2 \quad \text{or} \quad -\frac{1}{2}\frac{dv^2}{dx} = -\frac{r^2 g}{x^2} + \mu v^2$$

or $$(dv^2/dx) - (2\mu) \times v^2 = -(2r^2 g/x^2),$$

which is a linear differential equation whose I.F. $= e^{\int(-2\mu)\,dx} = e^{-2\mu x}$ and its solution is

$$v^2 e^{-2\mu x} = C - \int \frac{2r^2 g}{x^2} e^{-2\mu x}\,dx, \text{ C being an arbitrary constant}$$

or $$v^2(1 - 2\mu x) = C - 2r^2 g \int \frac{1 - 2\mu x}{x^2}\,dx$$

[Expanding $e^{-2\mu x}$ and neglecting the squares and higher powers of μ]

or $$v^2(1 - 2\mu x) = C + 2r^2 g\,(1/x + 2\mu \log x) \quad ...(1)$$

Using the initial conditions $v = 0$ when $x = a$, (1) yields

$$0 = C + 2r^2 g\,(1/a + 2\mu \log a) \quad ...(2)$$

Subtracting (2) from (1), we obtain

$$v^2(1 - 2\mu x) = 2r^2 g\,(1/x - 1/a + 2\mu \log x - 2\mu \log a)$$

or $$v^2 = 2r^2 g\,\{1/x - 1/a - 2\mu \log (a/x)\}\,(1 - 2\mu x)^{-1} = 2r^2 g\,\{1/x - 1/a - 2\mu \log (a/x)\} \times (1 + 2\mu x)$$

[Expanding by binomial theorem and neglecting squares and higher powers of μ]

or $$v^2 = 2r^2 g\,\{1/x - 1/a + 2\mu x(1/x - 1/a) - 2\mu \log (a/x)\}, \quad \text{neglecting } \mu^2$$

or $$v^2 = 2r^2 g\,\{1/x - 1/a + 2\mu(1 - x/a) - 2\mu \log (a/x)\} \quad ...(3)$$

Hence, the required kinetic energy of the particle at a distance x from the centre.

$= (1/2) \times mv^2 = mgr^2\{1/x - 1/a + 2\mu(1 - x/a) - 2\mu\log(a/x)\}$, using (3)

Ex. 94. *A particle moves from rest at a distance a from a fixed point O under the action of a force to O equal to μ times the distance per unit of mass. If the resistance of the medium in which it moves be k times the square of the velocity per unit mass, then show that the square of the velocity when it is at a distance x from O is*

$$(\mu x/k) - (\mu a/k) \times e^{2k(x-a)} + (\mu/2k^2) \times \{1 - e^{2k(x-a)}\}$$

Show also that when it first comes to rest it will be at a distance b given by

$$(1 - 2bk)e^{2bk} = (1 + 2ak)e^{-2ak}$$

Sol. Let a particle of mass m start from rest at a distance a from the fixed point O. Let v be the velocity of the particle at time t at the point P where OP = x. Now, the particle at P is subject to the following forces:

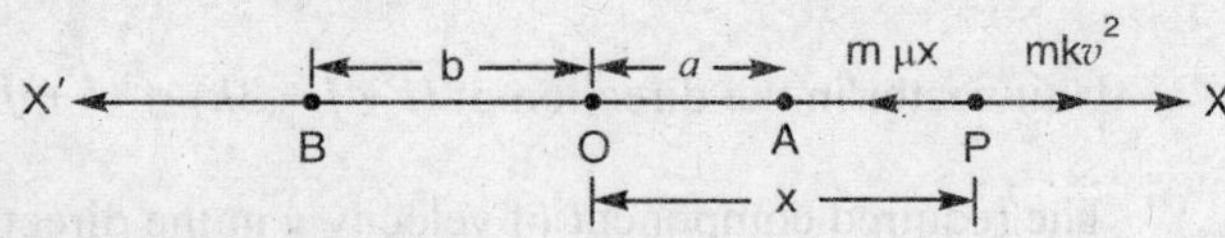

(*i*) the force $m\mu x$ towards O, *i.e.*, in the direction of x increasing and (*ii*) the resistance of the medium = mkv^2 acting opposite to the direction of motion, *i.e.*, in the direction of x increasing.

Hence, the equation of motion of the particle is given by

$$m(d^2x/dt^2) = -m\mu x + mkv^2 \qquad \text{or} \qquad v(dv/dx) = -\mu x + kv^2$$

$$\text{or} \qquad \frac{1}{2}\frac{dv^2}{dx} = -\mu x + kv^2 \qquad \text{or} \qquad \frac{dv^2}{dx} - 2kv^2 = -2\mu x,$$

which is a linear differential equation in variables v^2 and x whose I.F. = $e^{\int(-2k)\,dx} = e^{-2kx}$ and its solution is given by

$$v^2 e^{-2kx} = C - \int(2\mu x)e^{-2kx}\,dx, \text{ where C is an arbitrary constant}$$

$$\text{or} \qquad v^2e^{-2kx} = C - 2\mu\left[x \times \frac{e^{-2kx}}{(-2k)} - \int\left\{1 \times \frac{e^{-2kx}}{(-2k)}\right\}dx\right] = C + \frac{\mu}{k}\left(x\,e^{-2kx} + \frac{1}{2k}e^{-2kx}\right) \qquad \text{... (1)}$$

Now, initially, when $x = a$, $v = 0$. So, (1) $\Rightarrow 0 = C + (\mu/k) \times \{ae^{-2ka} + (1/2k) \times e^{-2ka}\}$ ...(2)

Subtracting (2) from (1), we have

$$v^2e^{-2kx} = \frac{\mu}{k}\left(xe^{-2kx} + \frac{e^{-2kx}}{2k} - ae^{-2ka} - \frac{e^{-2ka}}{2k}\right) = \frac{\mu x}{k} + \frac{\mu}{2k^2} - \frac{\mu a}{k}e^{2k(x-a)} - \frac{\mu}{2k^2}e^{2k(x-a)}$$

$$\text{or} \qquad v^2 = (\mu x/k) - (\mu a/k) \times e^{2k(x-a)} + (\mu/2k^2) \times \{1 - e^{2k(x-a)}\}, \qquad \text{...(3)}$$

which proves the first part of the given problem.

Second Par: Let v = V when x = 0. Then, from (4) we have

$$V^2 = -(\mu a/k) \times e^{-2ka} + (\mu/2k^2) \times (1 - e^{-2ka}) \neq 0$$

Since V ≠ 0, it follows that the particle does not come to rest at the centre of force O. Therefore, the particle moves to the left of O with velocity V. As the particle moves to the left of O, the force of attraction and resistance of the medium will act towards O and hence the velocity of the particle will go on decreasing.

Let the particle come to an instantaneous rest at a distance b from O on its left. Putting $v = 0$ and $x = -b$ in (3), we obtain

$$0 = -(\mu b/k) - (\mu a/k) \times e^{2k(-b-a)} + (\mu/2k^2) \times \{1 - e^{2k(-b-a)}\}$$

$$\text{or} \qquad (1/2k^2) - (b/k) = \{(a/k) + (1/2k^2)\}e^{-2kb-2ka}$$

$$\text{or} \qquad (1 - 2bk) = (2ak + 1)\,e^{-2ak}\,e^{-2bk} \qquad \text{or} \qquad (1 - 2bk)\,e^{2bk} = (1 + 2ak)\,e^{-2ak}$$

Ex. 95. *A particle moves along the curve $x = t^3 + 1$, $y = t^2$, $z = 2t + 5$, where t is the time. Find the components of velocity and acceleration at $t = 1$ in the direction* $\mathbf{i} + \mathbf{j} + 3\mathbf{k}$.

(Agra 2010)

Sol. Given that $\mathbf{r} = x\mathbf{i} + y\mathbf{j} + z\mathbf{k} = (t^3 + 1)\mathbf{i} + t^2\mathbf{j} + (2t + 5)\mathbf{k}$. ...(1)

Velocity vector $= \mathbf{v} = d\mathbf{r}/dt = 3t^2\mathbf{i} + 2t\mathbf{j} + 2\mathbf{k}$ ...(2)

and acceleration vector $= \mathbf{a} = d^2\mathbf{r}/dt^2 = d\mathbf{v}/dt = 6t\mathbf{i} + 2\mathbf{j}$ ...(3)

At $t = 1$, $\mathbf{v} = 3\mathbf{i} + 2\mathbf{j} + 2\mathbf{k}$ and $\mathbf{a} = 6\mathbf{i} + 2\mathbf{j}$

Unit vector in the direction of $(\mathbf{i} + \mathbf{j} + 3\mathbf{k}) = (\mathbf{i} + \mathbf{j} + 3\mathbf{k})/\sqrt{11} = \hat{\mathbf{b}}$, say

The required component of velocity $\mathbf{v}$ in the direction of $\hat{\mathbf{b}}$

$$= \mathbf{v} \cdot \hat{\mathbf{b}} = (3\mathbf{i} + 2\mathbf{j} + 2\mathbf{k}).\{(\mathbf{i} + \mathbf{j} + 3\mathbf{k})/\sqrt{11}\} = (3 + 2 + 6)/\sqrt{11} = \sqrt{11}$$

Similarly, the required component of acceleration $\mathbf{a}$ in the direction of $\hat{\mathbf{b}}$

$$= \mathbf{a} \cdot \hat{\mathbf{b}} = (6\mathbf{i} + 2\mathbf{j} + 0\mathbf{k}).\{(\mathbf{i} + \mathbf{j} + 3\mathbf{k})/\sqrt{11}\} = (6 + 2 + 0)/\sqrt{11} = 8/\sqrt{11}$$

Ex. 96. *A particle moves along a curve whose parametric equation are $x = e^{-t}$, $y = a \cos 3t$, $z = b \sin 3t$, where t is the time*

(*i*) *Determine the velocity and acceleration at time t* **(Rohilkhand 2009)**

(*ii*) *Find the magnitude of the velocity and acceleration at $t = 0$.* **(Meerut 2004)**

Sol. Here $\mathbf{r} = x\mathbf{i} + y\mathbf{j} + z\mathbf{k} = e^{-t}\mathbf{i} + (a \text{ cost } 3t)\mathbf{j} + (b \sin 3t)\,\mathbf{k}$ ...(1)

(*i*) From (1), velocity $\mathbf{v} = d\mathbf{r}/dt = -e^{-t}\mathbf{i} - (3a \sin 3t)\,\mathbf{j} + (3b \cos 3t)\,\mathbf{k}$

and acceleration $\mathbf{a} = d\mathbf{v}/dt = e^{-t}\mathbf{i} - (9a \text{ cost} 3t)\mathbf{j} - (9b \sin 3t)\,\mathbf{k}$

(*ii*) At $t = 0$, $\mathbf{v} = -\mathbf{i} + 3b\mathbf{k}$ and $\mathbf{a} = \mathbf{i} - 9a\mathbf{j}$

Hence, magnitude of $\mathbf{v} = |\mathbf{v}| = (1 + 9b^2)^{1/2}$ and magnitude of acceleration $= |\mathbf{a}| = (1 + 81a^2)^{1/2}$.

Ex. 97. *Calculate in miles per second the least velocity which will carry the particle from earth's surface to infinity.* **(Agra 2012)**

Sol. Let a be the radius of the earth. Suppose the particle is projected vertically upwards from the surface of the earth. Let P be the position of the particle at any time t where OP $= x$. Since P is outside the surface of the earth, the equation of motion of the particle is given by

$$d^2x/dt^2 = -(\mu/x^2),\ \mu \text{ being a constant} \quad ...(1)$$

But on the surface of earth, $x = a$ and $d^2x/dt^2 = -g$.

So, (1) $\Rightarrow$ $g = \mu/a^2$ so that $\mu = ga^2$

Hence, (1) reduces to $d^2x/dt^2 = -(ga^2/x^2)$ or $v(dv/dx) = -(ga^2/x^2)$

Separating variables, we have $2v dv = -\{(2ga^2/x^2\}\ dx$

Integrating, $v^2 = (2ga^2)/x + C$. ...(2)

Let $v = 0$ when $x \to \infty$. Then, (2) gives C = 0. Hence, (2) reudces to

$$v^2 = (2ga^2)/x \quad ...(3)$$

Putting $x = a$ and $v = u$ in (3), $u = (2ga)^{1/2}$, ...(4)

which gives the least velocity u at the earth's surface which will carry the particle to infinity.

Now, we know that $a = 4000$ miles $= 4000 \times 3 \times 1760$ ft and $g = 32$ ft/sec^2.

Hence, from (4), $u = \{2 \times 4000 \times 3 \times 1760 \times 32\}^{1/2}$ ft/sec

or $$u = 8 \times 200 \times 4\sqrt{33} \text{ ft/sec} = \frac{8 \times 200 \times 4\sqrt{33}}{3 \times 1760} \text{ miles/sec} = 7 \text{ miles/sec, nearly.}$$

Ex. 98. *A particle moves on an equiangular spiral $r = ae^{m\theta}$ with uniform speed v. Find its radial and transverse velocities and accelerations*

(Meerut 2014; Kanpur 2014, Lucknow 2008; Rohilkhand 2006)

Sol. The given equation of the path of the particle is $\qquad r = e^{m\theta}$ ...(1)

From (1), $\qquad dr/dt = a\,e^{m\theta} \times m \times (d\theta/dt) = mr(d\theta/dt)$, using (1) ...(2)

Since the particle moves with uniform speed v, we have

$$v^2 = \left(\frac{dr}{dt}\right)^2 + \left(r\frac{d\theta}{dt}\right)^2 = \left(mr\frac{dr}{dt}\right)^2 + \left(r\frac{d\theta}{dt}\right)^2, \text{ using (2)}$$

or $\qquad v^2 = r^2(1 + m^2) \times (d\theta/dt)^2 \qquad$ so that $\qquad r(d\theta/dt) = v/(1 + m^2)^{1/2}$

Hence, $\qquad$ the transverse velocity $= r(d\theta/dt) = v/(1 + m^2)^{1/2}$ ...(3)

and $\qquad$ the radial velocity $= \dfrac{dr}{dt} = mr\dfrac{d\theta}{dt} = \dfrac{mv}{(1+m^2)^{1/2}}$, using (3) ...(4)

Now, radial acceleration of the particle

$$= \frac{d^2r}{dt^2} - r\left(\frac{d\theta}{dt}\right)^2 = 0 - \frac{1}{r}\left(r\frac{d\theta}{dt}\right)^2, \text{ since } \quad (4) \Rightarrow \quad \frac{dr}{dt} = \text{constant} \quad \Rightarrow \quad \frac{d^2r}{dt^2} = 0$$

$= -[v^2/\{r(1 + m^2)\}]$, using (3)

Finally, transverse acceleration of the particle

$$= \frac{1}{r}\frac{t}{dt}\left(r^2\frac{d\theta}{dt}\right) = \frac{1}{r}\frac{d}{dt}\left(r \times r\frac{d\theta}{dt}\right) = \frac{1}{r}\frac{d}{dt}\left(r \times \frac{v}{(1+m^2)^{1/2}}\right), \text{ using (3)}$$

$$= \frac{1}{r} \times \frac{v}{(1+m^2)^{1/2}}\frac{dr}{dt} = \frac{1}{r} \times \frac{v}{(1+m^2)^{1/2}} \times \frac{mv}{(1+m^2)^{1/2}} = \frac{mv^2}{r(1+m^2)}, \text{ using (4)}$$

Ex. 99. *A particle moves in a plane under a constant acceleration μa parallel to OX and an acceleration $-2\mu y$ parallel to OY, where OX and OY are rectangular axes. If the particle starts from rest at a point (0, a), find the path.*

(Meerut 2004)

Sol. Here the equations of motion of the particle are given by

$$d^2x/dt^2 = \mu a \qquad \text{...(1)}$$

and

$$d^2y/dt^2 = -2\mu y \qquad \text{...(2)}$$

Multiplying both sides of (1) by $2(dx/dt)$ and then integrating, we get

$(dx/dt)^2 = 2\mu ax + C$, C being an arbitrary constant ...(3)

Initially, when $x = 0$, we have $dx/dt = 0$. Therefore, (3) yields $\qquad C = 0.$

Hence, (3) reduces to $\qquad (dx/dt)^2 = 2\mu ax \qquad$ or $\qquad dx/dt = (2\mu ax)^{1/2}$ ...(4)

where we have taken positive sign because x increases as t increases

Now, multiplying both sides of (2) by $2(dy/dt)$ and then integrating, we get

$(dy/dt)^2 = -2\mu y^2 + D$, D being an arbitrary constant ...(5)

Initially, when $y = a$, $dy/dt = 0$. Therefore, (5) yields $\qquad D = 2\mu a^2$

Hence, (5) reduces to $\qquad (dy/dt)^2 = 2\mu(a^2 - y^2) \quad$ or $\quad dy/dt = -\{2\mu(a^2 - y^2)\}^{1/2}$, ...(6)

where we taken negative sign because y decreases as t increases.

Dividing (6) by (4), $\qquad \dfrac{dy}{dx} = -\dfrac{(a^2 - y^2)^{1/2}}{\sqrt{ax}} \qquad$ or $\qquad -\dfrac{dy}{(a^2 - y^2)^{1/2}} = \dfrac{1}{\sqrt{a}}x^{-1/2}\,dx$

Integrating, $\qquad \cos^{-1}(y/a) = (2/\sqrt{a}) \times x^{1/2} + \text{E}$, where E is a constant ...(7)

Initially, when $x = 0, y = a$. Hence, (7) yields E = 0. Therefore, (7) reduces to

$\cos^{-1}(y/a) = 2(x/a)^{1/2}$ or $y = a\cos\{2(x/a)^{1/2}\}$, which is the required equation of the path.

Ex. 100. *What is the value (in miles/sec.) of the least velocity which will carry a body from the earth's surface to infinity?* **(Agra 2012)**

(*a*) *10 miles/sec.* (*b*) *9 miles/sec* (*c*) *8 miles/sec* (*d*) *7 miles/sec.*

Sol. Ans. (*d*). Refer solved Ex. 97 on page M.40.

Ex. 101. *A particle describes an ellipse under a force $\mu/(\text{distance})^2$ towards the focus. If it was projected with velocity V from a point distant r from the centre of force, show that its periodic time is* $(2\pi/\sqrt{\mu}) \times (1/r - V^2/\mu)^{-3/2}$. **(Purvanchal 2011, Rohilkhand 2007)**

Sol. Proceed as in Art. 9.9, page 9.4 upto equation (8). Thus, we have

$$v^2 = \mu(2/r - 1/a) \quad \text{...}(i)$$

and

$$T = (2\pi/\sqrt{\mu}) \times a^{3/2} \quad \text{...}(ii)$$

For the given problem $v = V$ and hence (*i*) yields $V^2 = \mu(2/r - 1/a)$

giving $1/a = 2/r - V^2/\mu$ so that $a = (2/r - V^2/\mu)^{-1}$ ...(*iii*)

Form (*ii*) and (*iii*), $T = (2\pi/\sqrt{\mu}) \times (2/r - V^2/\mu)^{-3/2}$, which is the required periodic time.

Ex. 102. *Time period T of planetary motion is:* (*a*) $T = (2\pi/\mu) \times a^{3/2}$ (*b*) $T = (2\pi/\sqrt{\mu}) \times a^{3/2}$

(*c*) $T = 2\pi \times (a/\mu)$ (*d*) $T = 2\pi \times (a/\mu)^{3/2}$ **(Agra 2011)**

Sol. Ans. (*b*). Refer Deduction 3. of Art 10.4, page 10.2.

Ex. 103. *A particle moves along a straight line so that after t seconds, its distance from a fixed point O on the line is given by $s = (t-1)^2(t-2)$. Find its distance from O, when its velocity is zero.* **(Delhi B.Sc. (Prog). III. 2012)**

Sol. Given $s = (t-1)^2 . (s-2)$ ...(1)

From (1), $v = \text{velocity} = ds/dt = (t-1)^2 + 2(t-1)(t-2) = (t-1)(3t-5)$ ... (2)

From (2), $v = 0 \Rightarrow (t-1)(3t-5) = 0 \Rightarrow t = 1$ or $t = 5/3$.

When $t = 1$, from (1), $s = 0$. Therefore, the given particle returns to the fixed point O at $t = 1$.

When $t = 3/5$, from (1), $s = (3/5 - 1)^2 (3/5 - 2) = -(28/125)$. ...(3)

The negative sign on R.H.S. of (3) shows that when $t = 3/5$, the given particle is situated on the negative side of the fixed point O.

Ex. 104. *What is the time to describe the ellipse?* **(Agra 2012)**

(*a*) $T^2 \alpha a^3$ (*b*) $T \alpha a^2$ (*c*) $T \alpha a^3$ (*d*) $T^2 \alpha a^2$

Sol. Ans. (*a*). Refer Art. 10.4, page 10.2.

Ex. 105. *A particle is projected vertically upwards. Prove that if the resistance of air were constant and equal to (1/n)th of its weight, the time of ascent and descent would be $(n-1)^{1/2} : (n+1)^{1/2}$.*

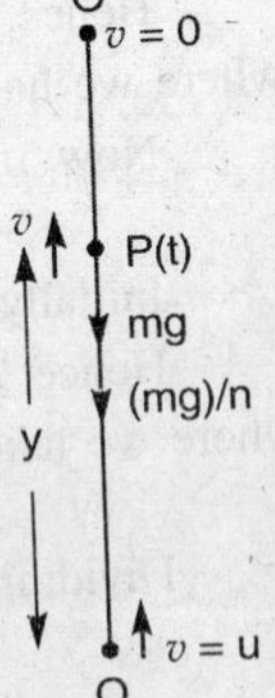

Sol. Let a particle of mass m be projected vertically upwards with velocity u from the fixed point O. Let P be the position of the particle at any time t such that OP = y. Let v be its velocity at P. The particle at P is subject to the following two forces (*i*) weight mg acting in vertically downward direction (*ii*) the force of resistance $(mg)/n$ acting in vertically downward direction. Then, the equation of motion of the particle at any time t is given by

$$m(d^2y/dt^2) = -mg - (mg/n) \quad \text{or} \quad d^2y/dt^2 = -\{(n+1)/n\}g \quad \text{...(1)}$$

Re-writing (1), $$\frac{dv}{dt} = -\frac{n+1}{n}g \quad \text{or} \quad dt = -\left(\frac{n}{n+1}\right)\frac{1}{g}\,dv \quad \text{...(2)}$$

Let O′ be the point of maximum height, *i.e.*, the velocity of the particle becomes zero at O′. Let OO′ = h and let t_1 be the time of ascent from O to O′.

Integrating (2) between the limits $t = 0$ to $t = t_1$ and the corresponding limits $v = u$ to $v = 0$, we have

$$\int_0^{t_1} dt = -\left(\frac{n}{n+1}\right)\frac{1}{g}\int_u^0 du \quad \text{giving} \quad t_1 = \left(\frac{n}{n+1}\right)\times\frac{u}{g} \qquad ...(3)$$

Again, re-writing (1), $v\dfrac{dv}{dy} = -\left(\dfrac{n+1}{n}\right)g$ or $dy = -\left(\dfrac{n+1}{n}\right)\dfrac{v}{g}\,dv$ ...(4)

Integrating (4) between the limits $y = 0$ to $y = h$ and the corresponding limits $v = u$ to $v = 0$, we have

$$\int_0^h dy = -\left(\frac{n}{n+1}\right)\frac{1}{g}\int_u^0 v\,dv \quad \text{giving} \quad h = \left(\frac{n}{n+1}\right)\times\frac{u^2}{2g} \qquad ...(5)$$

Motion from O′ to O: After reaching the highest point O′, the particle of mass m will fall from rest under gravity from O′. Let Q be the position of the particle at any time t such that O′Q = x. Let v be its velocity at Q. The particle at Q is subject to the following two forces: (*i*) weight mg acting in vertically downward direction (*ii*) the force of resistance $(mg)/n$ acting in vertically upward direction. Then the equation of motion of the particle at any time is given by

O′ $v = 0$ x (mg)/n Q(t) v mg $v = u'$ O

$$m\frac{d^2x}{dt^2} = mg - \frac{mg}{n} \quad \text{or} \quad \frac{d^2x}{dt^2} = \left(\frac{n-1}{n}\right)g \qquad ...(6)$$

Re-writing (6), $v\dfrac{dv}{dx} = \dfrac{n-1}{n}g$ or $dx = \left(\dfrac{n-1}{n}\right)\dfrac{v}{g}\,dv$ ...(7)

Suppose the particle reaches back at O with velocity u'. Integrating (7) between the limits $x = 0$ to $x = h$ and the corresponding limits $v = 0$ to $v = u'$, we have

$$\int_0^h dx = \left(\frac{n}{n-1}\right)\frac{1}{g}\int_0^{u'} v\,dv \quad \text{or} \quad \left(\frac{n}{n+1}\right)\frac{u^2}{2g} = \left(\frac{n}{n-1}\right)\frac{u'^2}{2g}, \text{ using (5)}$$

or $u'^2 = \{(n-1)/(n+1)\}\,u^2$ so that $u' = \{(n-1)/(n+1)\}^{1/2}u$ ...(8)

Again, (6) can be re-written as

$$\frac{dv}{dt} = \left(\frac{n-1}{n}\right)g \quad \text{or} \quad dt = \left(\frac{n}{n-1}\right)\frac{1}{g}\,dv \qquad ...(9)$$

Let t_2 be the time of descent from O′ to O. Then, integrating (9) between the limits $t = 0$ to $t = t_2$ and the corresponding limits $v = 0$ to $v = u'$, we have

$$\int_0^{t_2} dt = \left(\frac{n}{n-1}\right)\frac{1}{g}\int_0^{u'} dv \quad \text{or} \quad t_2 = \left(\frac{n}{n-1}\right)\frac{u'}{g} = \left(\frac{n}{n-1}\right)\frac{1}{g}\times\left(\frac{n-1}{n+1}\right)^{1/2} u, \text{ using (8)}$$

Thus, $$t_2 = \frac{u}{g}\times\frac{n}{(n-1)^{1/2}\,(n+1)^{1/2}} \qquad ...(10)$$

From (3) and (7), $$\frac{t_1}{t_2} = \frac{n}{n+1}\times\frac{u}{g}\times\frac{g(n-1)^{1/2}\,(n+1)^{1/2}}{nu} = \frac{(n-1)^{1/2}}{(n+1)^{1/2}}$$

Therefore, $$t_1 : t_2 = (n-1)^{1/2} : (n+1)^{1/2}.$$

Ex. 106. *In an oscillating pendulum the tension of the string when the bob is in its lowest position is n times the weight of the bob. Prove that the angle of swing on each side of the vertical is* $\cos^{-1}\{(3-n)/2\}$. **(Agra 2002, 2012)**

Sol. Let O be the fixed point, l the length of the string and M the lowest position of the bob of mass m. Let P be the position of the bob at any time t such that $\angle MOP = \theta$. Let arc MP $= s$ so that $s = l\theta$. Hence $\dot{s} = l\dot{\theta}$ and $\ddot{s} = l\ddot{\theta}$. Let v be velocity at P. But $v = \dot{s}$. Hence, $v = l\dot{\theta}$.

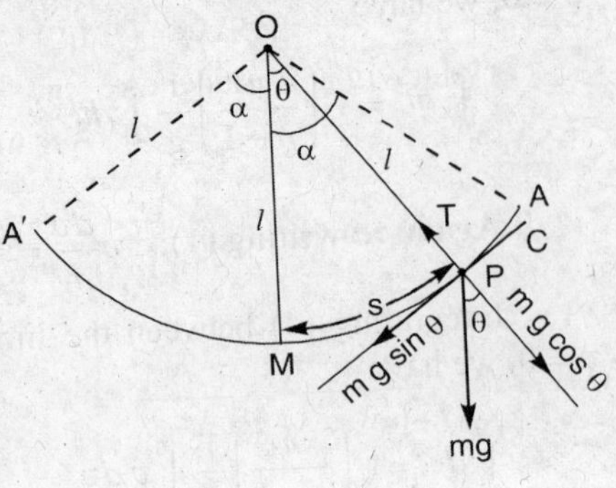

The forces acting on the bob at P are: (*i*) its weight mg acting in vertically downward direction (*ii*) tension T of the string acting along PO. Let PC be the tangent at P. Then, equations of motion along PC and PO are given by

$$m\ddot{s} = -mg\sin\theta \quad \text{or} \quad l\ddot{\theta} = -g\sin\theta \qquad \text{...(1)}$$

and $$m(v^2/l) = T - mg\cos\theta \quad \text{or} \quad ml\dot{\theta}^2 = T - mg\cos\theta \qquad \text{...(2)}$$

But $$\ddot{\theta} = \frac{d^2\theta}{dt^2} = \frac{d}{dt}\left(\frac{d\theta}{dt}\right) = \frac{d\dot{\theta}}{dt} = \frac{d\dot{\theta}}{d\theta}\frac{d\theta}{dt} = \dot{\theta}\frac{d\dot{\theta}}{d\theta}. \qquad \text{...(3)}$$

Using (3), (1) reduces to $l\dot{\theta}(d\dot{\theta}/d\theta) = -g\sin\theta$ or $2l\dot{\theta}\,d\dot{\theta} = -2g\sin\theta\,d\theta$

Integrating $$l\dot{\theta}^2 = 2g\cos\theta + C, \quad \text{C being an arbitrary constant} \qquad \text{...(4)}$$

Suppose the bob swings from position A to A′ such that $\angle MOA = \angle MOA' = \alpha$. Clearly velocity at A is zero and so $v = l\dot{\theta} = 0 \Rightarrow \dot{\theta} = 0$ when $\theta = \alpha$. Using this condition at point A, (4) reduces to $0 = 2g\cos\alpha + C$ so that $C = -2g\cos\alpha$. Hence, (4) reduces to

$$l\dot{\theta}^2 = 2g(\cos\theta - \cos\alpha)$$

Substituting the above value of $l\dot{\theta}^2$ in (2), we have

$$2mg(\cos\theta - \cos\alpha) = T - mg\cos\theta \qquad \text{...(5)}$$

Given that at lowest point M, when $\theta = 0$, $T = n \times mg$. Hence, (5) becomes

$$2mg(1-\cos\alpha) = (n \times mg) - mg \quad \text{or} \quad 2 - 2\cos\alpha = n - 1$$

or $$\cos\alpha = (3-n)/2 \quad \text{so that} \quad \alpha = \cos^{-1}\{(3-n)/2\},$$

which gives the required angle of swing on each side of the vertical line OM.

Ex. 107. *A particle slides outside a smooth fixed vertical circle starting from rest at the top, and another is projected from the lowest point, inside the circle with just sufficient velocity to carry it to the top, prove that both leave circle at the same point and proceed to describe the parts of the same parabola.*

Sol. Let A be the top position of the circle with centre O and radius a and let AOB be its vertical diameter. As shown in Fig. (*i*), let P be the position of the first particle at any time t, having velocity v along tangent PT where $\angle AOP = \theta$. Let PN be perpendicular to AB. Then,

$$AN = a - a\cos\theta = a(1-\cos\theta).$$

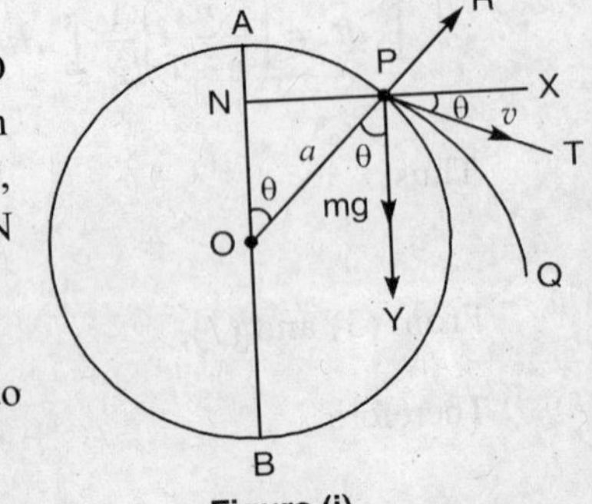

Figure (i)

We now apply the principle of conservation of energy to the motion from A to P and obtain

$$(1/2) \times mv^2 - (1/2) \times m \times 0^2 = mga(1 - \cos\theta)$$

Thus, $$v^2 = 2ga - 2ga\cos\theta \quad ...(1)$$

The external forces acting on the particle are: (*i*) the normal reaction R along OP and (*ii*) weight mg acting in vertically downward direction. Hence, the equation of motion in the inward normal sense PO is given by

$$m(v^2/a) = mg\cos\theta - R \quad \text{or} \quad R = mg\cos\theta - (m/a) \times (2ga - 2ga\cos\theta), \text{ using (1)}$$

or $$R = mg(\cos\theta - 2 + 2\cos\theta) \quad \text{or} \quad R = mg(3\cos\theta - 2) \quad ...(2)$$

When the particle leaves the circle, R = 0 and so from (2), we have

$$3\cos\theta - 2 = 0 \quad \text{giving} \quad \cos\theta = 2/3 \quad ...(3)$$

If v_1 be the velocity of the particle at the time of leaving the circle, then putting $\cos\theta = 2/3$ in (1), we have

$$v_1^2 = 2ga - (2ga) \times (2/3) = (2/3) \times ga \quad ...(4)$$

After leaving the circular path at P, the particle will describe parabolic path PQ with velocity v_1 and angle of projection θ given by (4) and (3) respectively.

As shown in Fig. (*ii*), let us now consider motion of the second particle of mass m placed on the inside of the same circle and projected with velocity u from B. Given that velocity u is just sufficient to carry the particle to the top A. Clearly, velocity at A is zero. Using the principle of conservation of energy to the motion from B to A, we obtain

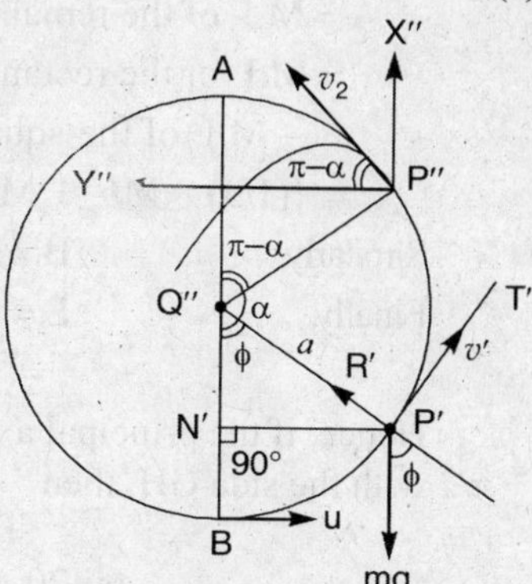

Figure (ii)

$$(1/2) \times m \times 0^2 - (1/2) \times m \times u^2 = (-mg) \times 2a \quad \text{or} \quad u = \sqrt{4ga} \quad ...(5)$$

Let P′ be the position of the second particle at any time t, such that $\angle BOP' = \phi$. Let v' be the velocity of the particle along tangent P′T′. Let P′N′ be perpendicular to AB. Then,

$$BN' = OB - ON' = a(1 - \cos\phi).$$

For motion from B to P′, by the principle of conservation of energy, we get

$$(1/2) \times mv'^2 - (1/2) \times mu^2 = (-mg) \times BN' \quad \text{or} \quad v'^2 = 4ga - 2ga(1 - \cos\phi)$$

Thus, $$v'^2 = 2ga(1 + \cos\phi) \quad ...(6)$$

The external forces acting on the particle are: (*i*) the normal reaction R′ acting along P′O (*ii*) weight mg of the particle acting in vertically downward direction. Hence, the equation of motion in the inward normal sense P′O is given by

$$m(v'^2/a) = R' - mg\cos\phi \quad \text{or} \quad R' = mg\cos\phi + (m/a) \times 2ga(1 + \cos\phi), \text{ by (6)}$$

or $$R' = mg(2 + 3\cos\phi)$$

Let the particle leave the circle at P″ where $\phi = \alpha$ and R′ = 0. Hence,

$$2 + 3\cos\alpha = 0 \quad \text{giving} \quad \cos\alpha = -2/3 \quad ...(7)$$

Re-writing (7), we have $$\cos(\pi - \alpha) = 2/3 \quad ...(8)$$

Putting $\phi = \alpha$ in (6), the velocity v_2 of the second particle is given by

$$v_2^2 = 2ga(1 + \cos\alpha) = 2ga(1 - 2/3) = (2/3) \times ga, \text{ using (7)} \quad ...(9)$$

Since (3) and (8) $\Rightarrow \theta = \pi - \alpha$, it follows that points P and P″ coincide. Hence the two particles will leave the circle at the same point. Again, $v_1 = v_2$ and $\theta = \pi - \alpha$ imply that soon after leaving the circular path, while describing parabolic paths, both the particles will possess the same velocity of projection and the same angle of projection (in magnitude). Hence, the given two particles must proceed to describe the parts of the same parabola.

Ex. 108. *A uniform rectangular plate whose sides are 2a, 2b has a portion cut out in the form of a square whose centre is the centre of the rectangle and whose mass is half the mass of the* *plate. Show that the axis of greatest and least moment of inertia at a corner of the rectangle make angles* θ, $\theta + \pi/2$ *with a side where* $\tan 2\theta = (6ab)/\{5(a^2 - b^2)\}$ **(Meerut 2000)**

Sol. Let M be the mass of the given rectangular plate GHIJ, where GH = $2a$ and GJ = $2b$ and let $2c$ be the side of the square PQRS cut out from it such that O is centre of GHIJ as well as that of PQRS. Also, given that mass of the square plate is M/2.

Now, in usual symbols, using result of Art 14.8, we have

A = M.I. of the remaining portion about GH

= M.I. of the rectangle about AB

– M.I. of the square about AB

$= \{(1/3) \times Mb^2 + Mb^2\} - \{(1/3) \times (M/2) \times c^2 + (M/2) \times b^2\} = (M/6) \times (5b^2 - c^2)$

Similarly, B = M.I. of the remaining portion about GJ = $(M/6) \times (5a^2 - c^2)$

Finally, F = product of inertia of the remaining portion about GH and GJ

$= (0 + Mab) - \{0 + (Mab)/2\} = (1/2) \times Mab$

Hence, if the principal axes in the plane of the rectangle at the corner G make angles θ and $\theta + \pi/2$ with the side GH, then

$$\tan 2\theta = \frac{2F}{B - A} = \frac{Mab}{(M/6) \times (5a^2 - 5b^2)} = \frac{6ab}{5(a^2 - b^2)}.$$

Ex. 109. *If the vertical angle of the cone is 90°, prove that the point at which a generator is a principal axis divides the generator in the ratio 3 : 7.*

Sol. Let UVW be given cone of vertical angle 90° and let $h(=$ VK) be its height. Let the generator UV be the principal axis of the cone at the point O. Let us consider the section of the cone passing through UV and VK. Choose OX and OY, the x-axis and y-axis, perpendicular to VK and parallel to VK respectively in the section under consideration and let OZ be the z-axis perpendicular to the section being considered.

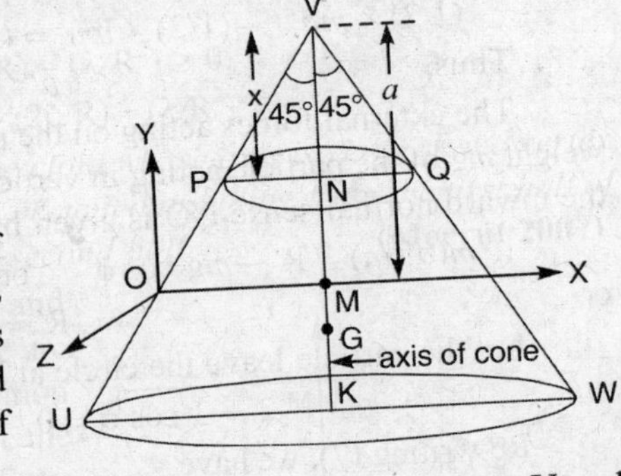

In what follows, we shall use results of Art. 14.15, page 14.25.

Since the z-coordinate of the centre of gravity G of the cone is zero, it follows that D = E = O. Therefore, OZ is a principal axis at O. Hence, the remaining principal axes at O are the generator UV and the line through O and perpendicular to UV in the above mentioned section of the cone.

Consider an elementary circular disc PQ of width δx at a distance x from the vertex V and perpendicular to VK. Here VN = x. The radius of the disc PQ is $x \tan 45°$, *i.e.*, x and hence the mass of the elementary disc PQ = $\delta m = \rho\, \pi x^2 \delta x$

M.I. of the elementary disc PQ about OX.

$= (1/4) \times PN^2 \times \delta m + MN^2\, \delta m = \{x^2/4 + (VM - x)^2\}\, \rho\pi x^2\, \delta x$

$= \{x^2/4 + (a - x)^2\} \times \rho\pi x^2\, \delta x$, where VM = a, say

Hence, A = M.I. of the cone about OX = $\int_0^h \{x^2/4 + (a - x)^2\}\, \rho\pi x^2\, dx$

Thus, $A = \pi\rho \int_0^h \{(5/4) \times x^4 - 2ax^3 + a^2x^2\}\, dx = (1/12) \times \pi\rho h^3(3h^2 - 6ah + 4a^2)$ ... (1)

Again, B = M.I. of the cone about OY $= \int_0^h (x^2/2 + a^2)\pi\rho x^2\, dx$

$$= \pi\rho \int_0^h (x^4/2 + x^2a^2)\, dx = (1/30) \times \pi\rho h^3(3h^2 + 10a^2) \quad \text{... (2)}$$

Again, here θ = the angle between the principal axis UV and OX = 45° and hence $2\theta = 90°$. Therefore, result $\tan 2\theta = (2F)/(B - A)$ reduces to

$$\tan 90° = (2F)/(B - A) \;\Rightarrow\; B - A = 0 \;\Rightarrow\; A = B$$

Hence, $(1/12) \times \pi\rho h^3(3h^2 - 6ha + 4a^2) = (1/30) \times \pi\rho h^3(3h^2 + 10a^2)$, using (1) and (2)

or $9h^2 = 30ah \;\Rightarrow\; a = 3h/10 \;\Rightarrow\;$ VM $= 3h/10$, as a = VM

From, similar triangles VOM and VUK, we obtain

$$\frac{VO}{VU} = \frac{VM}{VK} = \frac{VM}{h} = \frac{3}{10}, \quad \text{as} \quad VM = \frac{3h}{10}$$

$$\Rightarrow \quad VO = \frac{3}{10}UV \quad \text{and} \quad OU = UV - VO = UV - \frac{3}{10}UV = \frac{7}{10}UV$$

Therefore, $\qquad$ VO : OU = 3 : 7.

Ex. 110. *Two particles each of mass m are placed at the extremities of the minor axis of an elliptic area of mass M. Prove that principal axis at any point of the circumference of the ellipse will be the tangent and normal to the ellipse if $m/M = (5e^2)/\{8(1 - 2e^2\}$.*

Sol. Let PQRS be an elliptic area of mass M. Let $O'(x, y)$ be a point on the circumference of the ellipse

$$x^2/a^2 + y^2/b^2 = 1$$

referred to the axes through the centre O. ...(1)

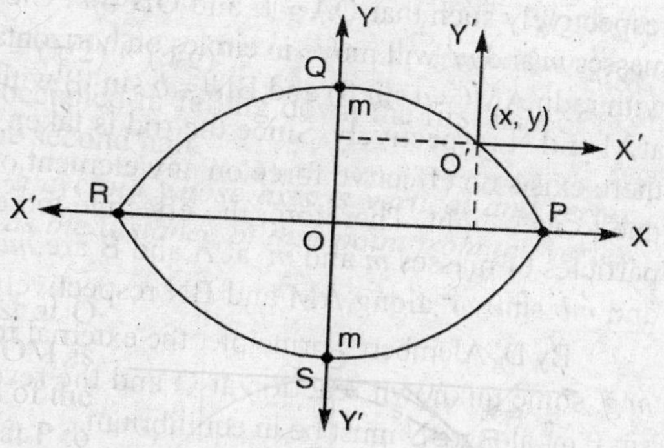

Let O′X′ and O′Y′ be axes parallel to OX and OY respectively. Referred to the axes O′X′, O′Y′ the coordinates of the centre O are $(-x, -y)$ and those of the ends Q and S of minor axis are $(-x, b - y)$ and $(-x, -b - y)$ respectively.

According to our problem, two particles each of mass m are placed at Q and S. Then, in usual symbols, we have

A = moment of inertia of the elliptic lamina and the two particles about O′X′

$= M\{(b^2/4) + y^2\} + m\{(b - y)^2 + (b + y)^2\}$

$= (M/4) \times (b^2 + 4y^2) + 2m(b^2 + y^2)$

B = moment of inertia of the elliptic lamina and the two particles about O′Y′

$= M\{(a^2/4) + x^2\} + m(x^2 + x^2) = (M/4) \times (a^2 + 4x^2) + 2mx^2$

F = product of inertia of the elliptic lamina and the two particles about O′X′, O′Y′

$= M(-x)(-y) + m\{(-x)(b - y) + (-x)(-b - y)\} = (M + 2m)xy$

If θ be the inclination with O′X′ of a principal axis at P, then

$$\tan 2\theta = \frac{2F}{B - A} = \frac{2(M + 2m)xy}{\{(M/4) \times (a^2 + 4x^2) + 2mx^2\} - \{(M/4) \times (b^2 + 4y^2) + 2m(b^2 + y^2)\}}$$

or
$$\tan 2\theta = \frac{8(M+2m)xy}{M(4x^2-4y^2+a^2-b^2)+8m(x^2-y^2-b^2)} \quad ...(2)$$

The equation of the tangent at P(x, y) to the ellipse (1) is given by
$$(xX/a^2) + (yY/b^2) = 1, \quad ...(3)$$
where (X, Y) are current coordinates. Then, the slope of the tangent at (x, y) is $-(b^2x/a^2y)$. Hence, if the tangent (3) at P is the principal axis, then
$$\tan\theta = -(b^2x/a^2y). \quad ...(4)$$

Hence, $$\tan 2\theta = \frac{2\tan\theta}{1-\tan^2\theta} = -\frac{2(b^2x/a^2y)}{1-(b^4x^2/a^4y^2)} = -\frac{2xya^2b^2}{a^4y^2-b^4x^2}, \text{ using (4)} \quad ...(5)$$

Now, (2) and (5) $\Rightarrow$ $$\frac{8(M+2m)xy}{M(4x^2-4y^2+a^2-b^2)+8m(x^2-y^2-b^2)} = -\frac{2xya^2b^2}{a^4y^2-b^4x^2}$$

or $$M\{4(x^2/a^2+y^2/b^2)+1\}(a^2-b^2) = 8m\{2b^2-a^2(x^2/a^2+y^2/b^2)\}$$

or $$5M(a^2-b^2) = 8m(2b^2-a^2), \quad \text{using (1)}$$

or $$\frac{m}{M} = \frac{5}{8}\times\frac{a^2-b^2}{2b^2-a^2} = \frac{5}{8}\times\frac{a^2e^2}{a^2-2a^2e^2} = \frac{5}{8}\times\frac{e^2}{1-2e^2}, \text{ as } b^2 = a^2(1-e^2)$$

Ex. 111. *Find the motion of the rod OAB, with two masses m and m′ attached to it at A and B respectively, when it moves round the vertical as a conical pendulum with uniform angular velocity, the angle θ which the rod makes with the vertical being constant.*

Sol. Let the given light rod OAB revolve round the vertical line OZ, (say) with uniform angular velocity ω, making a constant angle θ with OZ. Given that two masses m and m' are attached to the rod AB at A and B respectively such that OA = a and OB = b. Clearly, the masses m and m' will move in circles on horizontal planes with radii AM(= $a\sin\theta$) and BN(= $b\sin\theta$) with centres at M and N respectively. Since the rod is taken as light, there exists no effective force on any element of the rod due to its weight. Therefore, the effective forces on the particles of masses m and m' at A and B are $ma\sin\theta\,\omega^2$ and $mb\sin\theta\,\omega^2$ along AM and BN respectively.

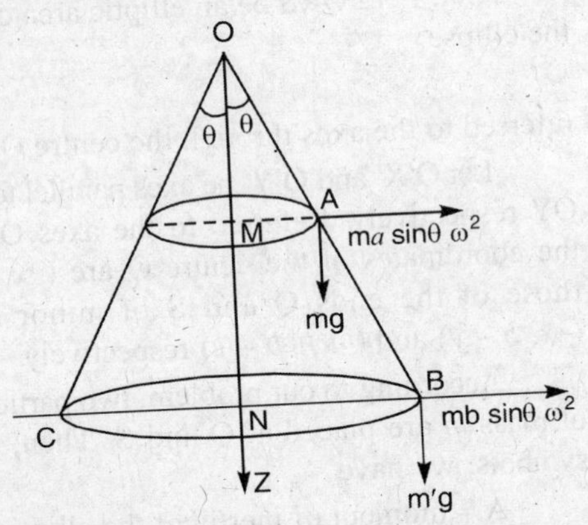

By D' Alembert's principle, the external forces mg, $m'g$, some unknown reaction at O and the reversed effective forces $ma\sin\theta\,\omega^2$ along AM and $mb\sin\theta\,\omega^2$ along BN must be in equilibrium.

In order to avoid unknown reaction at O, we take moments about O and obtain
$$(ma\sin\theta\,\omega^2)\times OM + (m'b\sin\theta\,\omega^2)\times ON - mg\times AM - m'g\times BN = 0$$
or $$(ma\sin\theta\times a\cos\theta + m'b\sin\theta\times b\cos\theta)\omega^2 = g(ma\sin\theta + m'b\sin\theta) \quad ...(1)$$
Since $\sin\theta \neq 0$, (1) reduces to $$\omega^2 = \{(ma+m'b)g\}/\{(ma^2+m'b^2)\cos\theta\},$$
which determines the motion of rod.

Ex. 112. *A flat circular disc of radius a has a hole in it of radius b whose centre is at a distance c from the centre of the disc (c < a – b). The disc is free to oscillate in a vertical plane about a smooth horizontal circular rod of the radius b passing through the hole. Show that the length of the equivalent pendulum is* $c + \{(a^4-b^4)/2a^2c\}$.

Sol. Let O be the centre of given circular disc of radius a and mass m_1. Again, let O′ be the centre of circular hole of radius b and mass m_2. Given that OO′ = c.

Since the given disc is free to oscillate about a smooth horizontal circular rod of radius b passing through the hole, it follows that O′ is the centre of suspension and the horizontal line through O′ and perpendicular to the given disc is the axis of rotation. Taking ρ as the density of the disc, we have $m_1 = \pi a^2 \rho$ and $m_2 = \pi b^2 \rho$. Again, we have

h = depth of the C.G. of the disc with hole below O′

$$= \frac{m_1 \times \text{O}'\text{O} - m_2 \times 0}{m_1 - m_2} = \frac{\pi a^2 \rho c}{\pi a^2 \rho - \pi b^2 \rho}$$

Thus, we have $\qquad h = (a^2 c)/(a^2 - b^2)$. ... (1)

Let k be the radius of gyration of the disc with hole about the axis of rotation. Then,

$(m_1 - m_2)k^2$ = M.I. of the disc with hole about the axis of rotation

= (M.I. of the disc without hole – M.I. of the hole), about the axis of rotation

or $\quad (\pi a^2 \rho - \pi b^2 \rho)k^2 = \{m_1 \times (a^2/2) + m_1 \times \text{O}'\text{O}^2\} - m_2 \times (b^2/2) = m_1(a^2/2 + c^2) - m_2 \times (b^2/2)$

or $\quad \pi \rho k^2(a^2 - b^2) = \pi a^2 \rho\,(a^2/2 + c^2) - mb^2 \rho \times (b^2/2) \;\Rightarrow\; k^2 = (a^4 + 2a^2c^2 - b^4)/2(a^2 - b^2)$ ... (2)

Hence, the required length of the simple equivalent pendulum

$$= \frac{k^2}{h} = \frac{a^4 + 2a^2c^2 - b^4}{2a^2 c} = c + \frac{1}{2} \times \left(\frac{a^4 - b^4}{a^2 c}\right), \text{ using (1) and (2)}$$

Ex. 113. *A uniform thin rod AB, of length 2a, lies at rest on a smooth horizontal table and is freely hinged at A, a fixed point. It is set in motion by a horizontal blow P applied to it in a direction perpendicular to its length at a distance 5a/4 from A and the magnitude of the impulse at A is Q. The rod is then brought to rest by a horizontal blow R applied at B perpendicular to AB. Prove that P/16 = Q/1 = R/10.*

Sol. Given that the rod AB of length $2a$ (having mass M, say) lies on a smooth horizontal table and it is hinged at A. The rod is set in motion by a horizontal blow P applied at C such that AC = $5a/4$. Given that the magnitude of the impulse at A is Q. Let ω be the angular velocity of the rod just after the impulse. Then, the equations of motion of the rod AB are given by

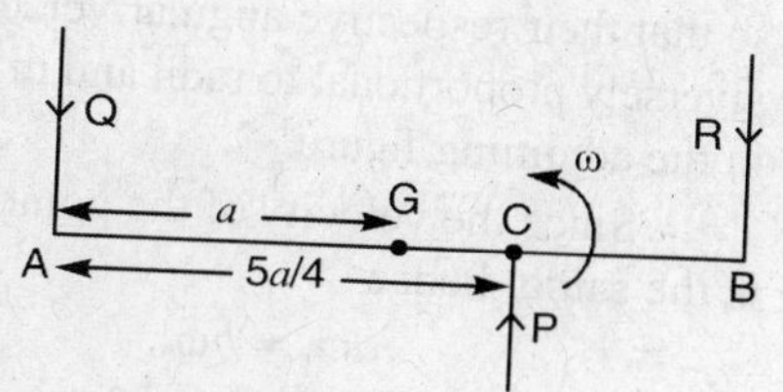

$\text{M}k^2(\omega - 0) = \text{P} \times \text{AC}$ or $\text{M} \times (4/3) \times a^2\omega = \text{P} \times (5a/4)$ or $\text{M}aw = (15/16) \times \text{P}$ ...(1)

and $\quad \text{M}a(\omega - 0) = \text{P} - \text{Q}$ $\quad$ or $\quad \text{M}a\omega = \text{P} - \text{Q}$ ...(2)

Then, (1) and (2) $\Rightarrow$ $\text{P} - \text{Q} = (15/16) \times \text{P}$ giving $\text{P}/16 = \text{Q}/1$ ...(3)

Since the rod is brought to rest by a horizontal blow R applied at B perpendicular to the rod, we have

$\text{M}k^2(0 - \omega) = -(\text{R} \times \text{AB})$ $\quad$ or $\quad \text{M} \times (4/3) \times a^2\omega = \text{R} \times 2a$ $\quad$ or $\quad \text{R} = (2/3) \times \text{M}a\omega$

Substituting the value of $\text{M}a\omega$ given by (1) in the above equation, we obtain

$\text{R} = (2/3) \times (15/16) \times \text{P}$ $\quad$ so that $\quad \text{P}/16 = \text{R}/10$ ...(4)

From (3) and (4), we obtain $\quad \text{P}/16 = \text{Q}/1 = \text{R}/10.$

Ex. 114. *A straight rod, of mass m and length 2l, swinging about one end as a compound pendulum, starts from rest in a horizontal position and when vertical is struck by a blow at *the middle point which reverses and halves its angular velocity. Prove that the impulse of the blow is* $m\sqrt{(6gl)}$.

Sol. Let OA be the given rod of mass m and length $2l$, swinging about one end O as a compound pendulum. Let horizontal position OX be its initial position of rest. Let ω be the angular velocity of the rod when it is in the vertical position. Then, by energy equation, we get

$$(1/2) \times mk^2\omega^2 - 0 = mgl \quad \text{or} \quad (m/2) \times (4/3) \times l^2\omega^2 = mgl$$

giving $$\omega = (3g/2l)^{1/2} \qquad ...(1)$$

According to the given problem, when the rod comes in the vertical position it is struck by a blow at its middle point G (where OG = GA = l). Let P be the impulse of the blow. Then, according to the given problem $\omega/2$ is the angular velocity of rod in the direction opposite to ω after impulse.

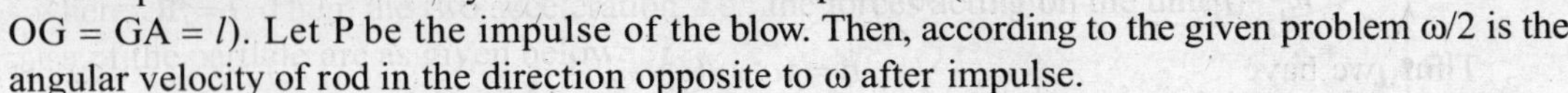

Since the change in moment of momentum about the fixed axis = moment of the impulses about it,

we have, $(4/3) \times ml^2\{\omega/2 - (-\omega)\} = \text{P} \times l$ giving $\text{P} = 2ml\omega \times (3g/2l)^{1/2}$, by (1)

Thus, the required impulse of the blow = $m\sqrt{(6gl)}$.

Ex. 115. *Two wheels spindles in fixed bearing suddenly engage so that their angular velocities become inversely proportional to radii and in opposite directions. One wheel of radius a and moment of inertia* I_1, *has angular velocity* ω *initially; the other, of radius b and moment of inertia* I_2 *initially is at rest. Show that their angular velocities are*

$$(b^2 I_1\omega)/(I_1 b^2 + I_2 a^2) \quad \text{and} \quad (I_1 ab\omega)/(I_1 b^2 + I_2 a^2).$$

Sol. Let A_1, be the given wheel of radius a and moment of inertial I_1 and A_2 the wheel of radius b and moment of inertial I_2. Given that initially the wheel A_1 is rotating with angular velocity ω while the wheel A_2 is at rest. According to the given problem, the wheels A_1 and A_2 spindles in fixed bearing suddenly engage so that their respective angular velocities ω_1 and ω_2 (say) become inversely proportional to radii and in opposite directions as shown in the adjoining figure.

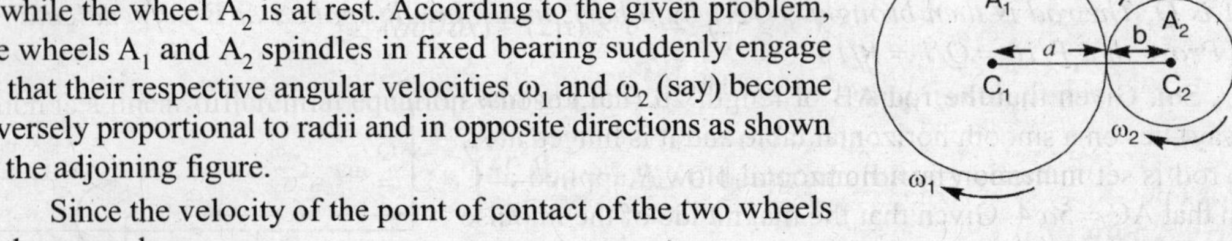

Since the velocity of the point of contact of the two wheels is the same, hence

$$a\omega_1 = b\omega_2 \quad \text{giving} \quad \omega_2 = (a/b) \times \omega_1 \qquad ...(1)$$

Let R be the impulsive force, then

the equation of motion for the wheel A_1 is $\quad I_1(\omega - \omega_1) = R \times a \qquad ...(2)$

and the equation of motion for the wheel A_2 is $\quad I_2(\omega_2 - 0) = R \times b \qquad ...(3)$

Dividing (2) by (3) yields $\{I_1(\omega - \omega_1)\}/(I_2\omega_2) = a/b$ or $\omega_2 a I_2 = b(\omega - \omega_1)I_1$

Substituting the value of ω_2 given by (1) in the above equation, we get

$(a\omega_1/b) \times aI_2 = b(\omega - \omega_1)I_1$ or $(I_1 b^2 + I_2 a^2)\omega_1 = I_1 b^2\omega$

and hence $\omega_1 = \dfrac{I_1 b^2 \omega}{I_1 b^2 + I_2 a^2}$ and hence $(1) \Rightarrow \omega_2 = \dfrac{a\omega_1}{b} = \dfrac{I_1 ab}{I_1 b^2 + I_2 a^2}$

Ex. 116. *A particle of mass m is constrained to execute a S.H.M. under a force towards O of magnitude* $m\omega^2 x$, *x being the particle's displacement from O. When passing through O, its velocity is V. Find the distance travelled and the time that elapses when its velocity has become V/2 in the same direction.* **[Delhi B.Sc. (Prog.) III, 2012]**

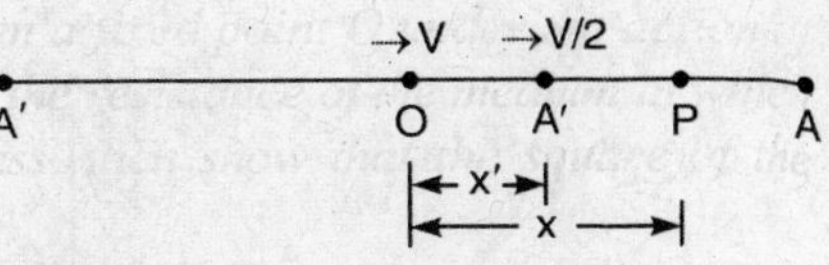

Sol. Let O be the fixed point in the line A′OA and let P denote the position of particle at any time t, executing S.H.M. with centre at O. Let the particle start from O with velocity V and let A′ be the position of the particle at time $t = t'$ such that OA′ $= x'$ and let velocity of the particle be V/2 at A′ as shown in the adjoining figure. The force acting on the particle at P is $m\omega^2 x$ acting along PO. Hence, the equation of motion of the particle of mass m is given by

$$m(d^2x/dt^2) = -m\omega^2 x \quad \text{or} \quad (D^2 + \omega^2)x = 0, \quad \text{where } D \equiv d/dt \qquad \dots (1)$$

The general solution of (1) is given by

$$x = C_1 \cos \omega t + C_2 \sin \omega t, \; C_1 \text{ and } C_2 \text{ being arbitrary constants} \qquad \dots (2)$$

Differentiating both sides of (2) w.r.t. 't', we have

$$dx/dt = v = -C_1 \omega \sin \omega t + C_2 \omega \cos \omega t \qquad \dots (3)$$

Initially at O, when $t = 0$, $v = V$ and $x = 0$. Hence, (2) and (3) yield

$$O = C_1 \quad \text{and} \quad V = C_2\omega \quad \text{so that} \quad C_1 = 0 \quad \text{and} \quad C_2 = V/\omega$$

Substituting the above values of C_1 and C_2 in (2) and (3), wt get

$$x = (V/\omega) \times \sin \omega t \qquad \dots (4)$$

and

$$v = V \cos \omega t \qquad \dots (5)$$

When particle is at A′, $v = V/2$ and $t = t'$. Hence (5) reduces to

$$V/2 = V \cos \omega t' \quad \text{so that} \quad \omega t' = \pi/3 \quad \text{and} \quad \text{hence} \quad t' = \pi/3\omega \qquad \dots (6)$$

Again, at A′, when $t = t'$, $x = x'$. Hence, (4) reduces to

$$x' = (V/\omega) \times \sin \omega t' = (V/\omega) \times \sin (\pi/3) = (V/\omega) \times (\sqrt{3}/2) = (V\sqrt{3})/2\omega$$

Therefore, the required distance is $(V\sqrt{3})/2\omega$ and the corresponding time is $\pi/3\omega$.

INDEX

NOTES

NOTES

NOTES